Conversion to and from the SI System

A. Conversion to SI

Name of Unit	Symbol	Multiply by	SI Unit	Symbol
inches	in	25.4	millimeters	mm
feet	ft	0.305	meters	m
yards	yd	0.914	meters	m
miles	mi	1.609	kilometers	km
square yards	yd^2	0.836	square meters	m^2
acres		0.405	hectares	ha
cubic yards	yd^3	0.765	cubic meters	m^3
quarts (liq)	qt	0.946	liters	l
ounces (avdp)	oz	28.350	grams	g
pounds	lb	0.454	kilograms	kg
Fahrenheit	°F	$\frac{5}{9}$ (after subtracting 32)	Celsius	°C

B. Conversion from SI

SI Unit	Symbol	Multiply by	To Find Unit	Symbol
millimeters	mm	0.039	inches	in
meters	m	3.281	feet	ft
meters	m	1.094	yards	yd
kilometers	km	0.621	miles	mi
square meters	m^2	1.196	square yards	yd^2
hectares	ha	2.471	acres	
cubic meters	m^3	1.308	cubic yards	yd^3
liters	l	1.057	quarts (liq)	qt
grams	g	0.035	ounces (avdp)	oz
kilograms	kg	2.205	pounds	lb
Celsius	°C	$\frac{9}{5}$ (then add 32)	Fahrenheit	°F

Introductory College Physics

Introductory College Physics

ATAM P. ARYA
West Virginia University

Macmillan Publishing Co., Inc.
New York

Collier Macmillan Publishers
London

MACMILLAN PUBLISHING CO., INC.
866 Third Avenue, New York, New York 10022

COLLIER MACMILLAN CANADA, LTD.

Library of Congress Cataloging in Publication Data

Arya, Atam Parkash.
Introductory college physics.

Includes index.
1. Physics. I. Title
QC21.2.A78 530 78-8587
ISBN 0-02-304000-9

Printing: 1 2 3 4 5 6 7 8 Year: 9 0 1 2 3 4 5

To Pam and Kam

Preface

THIS book is written for freshman and sophomore students who require a basic course in general physics, without the use of calculus. It includes students in forestry, agriculture, life sciences, premedicine, medical technology, pharmacy, home economics, and other related areas. No previous background of physics is essential. The level of mathematical preparation assumed includes simple algebra and plane trigonometry. (The most commonly used trigonometry and mathematical relations are summarized in Appendices B and C, respectively.) The length of the text is suitable for two semesters or three quarters courses. There is enough flexibility in length and topic selection so as to make the book suitable for the variety of students mentioned above. Each chapter opens with an introduction which briefly gives the highlights of the chapter, and is followed by the main body, a summary, and a large number of questions and problems. Throughout each chapter the important equations have been boxed for convenience.

The major aim I had in writing this textbook was to expose students both to the basic fundamentals of physics and to many diverse topics, classical as well as modern, that will allow them to gain insight into how physics has evolved, its present status, and its future direction. This has been accomplished by first treating the standard material thoroughly and then incorporating topics of recent interest from various branches of physics. Thus, the fundamental topics such as rotational motion, vibrational motion, kinetic theory of gases, entropy and second law of thermodynamics, Doppler effect, electromagnetic waves, and wave phenomenon in optics have been given comprehensive treatment.

Examples of the distinguishing topics discussed include the following: different *types of forces* (Chapter 4); *space physics, satellites* and *rocket propulsion* (Chapter 11); theory of *relativity* (Chapter 8); *semiconductor devices, laser, holography, superconductivity* (Chapter 33); and *nuclear medicine* (Chapter 34). Topics such as the *ear, eye,* and *eye defects* including *glaucoma,* and *resolving power* have been discussed in some length. Topics of present-day interest, such as *thermal pollution, noise pollution, nuclear fission, fusion,* and *nuclear reactors,* have been included. Throughout I have tried to keep the level uniform while blending the above topics with standard classical material.

In developing physical principles, I have tried to follow closely the exposition used by most lecturers in elementary physics courses designed for majors in areas mentioned in the beginning. Usually I have started each concept from basic principles, and then followed a logical development in a thorough way (blending the physical idea with mathematical derivations) leading to the final result. The mathematical results so derived have been elaborated by discussing their application to various situations.

I have included a variety of topics in the text to meet the demands and tastes of different instructors and students. I have tried to find an appropriate level of instruction that strikes a balance between general discussion, technical information, mathematics, and everyday applications of physics. Instead of completely deleting topics of marginal interest, they have been presented with less emphasis. Also, I have incorporated historical developments into the text in a natural and unobstructive manner, whenever possible, without interfering with the smooth flow of the text.

It has been said that solving problems at the general physics level is not only essential, but one of the best ways to learn and appreciate physics. To help students get started on problem solving I have included a generous number of solved examples, each followed by an exercise closely modeled after the example. The answers to the exercises are provided on the spot for easy checking. Most students will find it useful to read the example, solve the exercise to build confidence, and then proceed to do the problems.

I have included a large sampling of problems of varying difficulty. I have tried to achieve a solid balance of problems in each chapter ranging from straightforward plug-in to the two or three step challengers. The problem selection blends with the topic coverage and provides adequate practice with the concepts and formulas.

The SI units are used throughout the text with occasional use of the CGS and British units. The starred sections may be read out of sequence and, together with Chapters 31 to 34, also help in adjusting the length of the course. Diagrams are designed to help students grasp complicated abstract concepts. Sets of discussion questions and summaries serve as a reviewing device at the end of each chapter.

STUDENT STUDY GUIDE

A separate *Study Guide* is available to supplement the text. The purpose of the *Study Guide* is to identify for the student the objectives and the skills needed to solve problems similar to those presented at the end of each chapter, and to provide simple examples to enable the student to practice the skills.

The author welcomes, from both instructors and students, any new ideas and thoughts that will improve the book. Such suggestions will be taken into consideration in publication of future editions.

ACKNOWLEDGMENTS

Completing a work of such magnitude involves more than just the author. The author is indebted to Professor W. H. Kelly of Michigan State University, Professor Jack Prince of Bronx Community College of the City of New York, Professor Michael M. Shurman of the University of Wisconsin–Milwaukee, Professor Simon George of California State University at Long Beach, and Professor Dale Thomas of Hillsburg Community College at Tampa, Florida for reading the manuscript and suggesting changes for improvement. Professor Oleg D. Jefimenko was kind enough to discuss various topics while the manuscript was in developmental stages and provided many attractive photographs. Invaluable help came from Professor David K. Walker of Waynesburg College, Waynesburg, Pennsylvania. His thorough reading of the manuscript in the final stages of publication is highly appreciated. My thanks also to Professors Fred M. Goldberg, Milton E. McDonnell, John E. Littleton, and W. E. Vehse, all of

West Virginia University, for helping in various phases of the manuscript. My special thanks to Mr. Edward F. Novak for help in preparing the solutions manual, and to Mr. Jason M. Cook and Mr. Robert L. West, Jr. for making several photographs.

My thanks to the Editorial and Production departments of Macmillan, and especially to Mr. John Beck and Mr. Leo Malek whose keen interest and promptness saw it to completion.

Finally, my appreciation goes to my wife, Pauline, who not only typed the manuscript and proofread it several times, but also suggested and made changes which improved the manuscript.

A.P.A.

West Virginia University
Morgantown, WV
August 1978

Contents

CHAPTER
10 Vibrational Motion 188

CHAPTER
11 Gravitation and Space Physics 215

CHAPTER
12 Mechanics of Fluids 239

PART III Heat and Thermodynamics 267

Part I
Preliminaries

1
Scope of Physics and Terminology

To start with we shall briefly mention the scope of physics, its division into classical and modern physics, and its interrelation to other sciences. We shall review various units for measurement in different systems, with special emphasis on the modernized metric system called System International. Since we shall be using scalar as well as vector quantities throughout the text, it is also appropriate that we summarize the mathematical techniques of vector algebra.

1.1 Growth and Scope of Physics

As LATE as the middle of the nineteenth century, *physics* was called *natural philosophy*. It was concerned with all scientific investigations of natural phenomena which lead to the formulation of the basic laws of nature. Because of an enormous increase in the volume of scientific knowledge over the last three centuries, it has been found expedient to divide natural science into several disciplines—physics, astronomy, chemistry, geology, biology, and so on. Before the nineteenth century, it was possible for one scientist to learn and pursue many aspects of the science of his time. This was no longer true by the close of the century. It is remarkable to note that the amount of scientific knowledge that has been added to the field of physics alone in the present century far exceeds the total amount of knowledge (both in quality and quantity) that had been accumulated up to the close of the previous century. However, it should not come as a great surprise to find that physicists of today are much less confident about the true nature of the laws governing physical phenomena than were the physicists of the 1890s.

Now we ask ourselves a question: What is physics? According to one definition, physics deals with the *study of matter and energy or matter and motion*. Such investigations lead to the mathematical formulation of equations to represent different physical phenomena or events. Thus, the study of physics involves investigating such things as the laws of motion, the structure of space and time, the nature and type of forces that hold different materials together, the interaction between different particles, the interaction of electromagnetic radiation (such as light, x-rays, gamma rays, etc.) with matter, and so on.

The knowledge of physics that had accumulated previous to about the year 1890 is referred to as *classical physics*. Classical physics was further subdivided into such branches as mechanics, thermodynamics, sound, optics, electricity, magnetism, and electromagnetic fields.

By 1890, the laws of classical physics in their present form were well formulated. It was believed that there were no more new laws to be discovered. The only task remaining for the physicists was to improve upon the accuracy of the experimentally measured quantities.

About the beginning of the present century many new experimental facts revealed that the laws as formulated by previous investigators needed modifications under two different circumstances. First, it was found that even though the laws of classical physics adequately explained macroscopic systems

(systems visible to the naked eye), they were inadequate in explaining the behavior of microscopic systems (systems of the size of an atom or smaller, i.e., $\sim 10^{-10}$ m). This led to the development of *quantum mechanics*, in 1925, by E. Schrödinger, W. Heisenberg, M. Born, and P. A. M. Dirac. This development is the direct result of the quantum hypothesis proposed by Max Planck in 1900. Second, classical theory also failed to give the correct answers when applied to fast-moving particles. This led to the development of Einstein's *special theory of relativity*, in 1905, for particles that are moving with speeds approaching the speed of light c ($c = 3 \times 10^8$ m/s $\approx$ 186,000 mi/s).

At present, physics is divided into categories, as described below.

Classical Physics

Classical physics includes the study of mechanics, thermodynamics, sound, optics, electricity, magnetism, and electromagnetic fields. It is the study of macroscopic objects moving with speeds much smaller than the speed of light.

Modern Physics

Modern physics is further divided into the following subclasses.

Basic Concepts of Modern Physics This is essentially the development of ideas and experiments starting about 1900 and continuing up to the present day. These ideas led to the eventual development of new physical theories, such as relativity, quantum mechanics, field theories, and particle physics.

Relativistic Mechanics This deals with objects moving with speeds that approach the speed of light.

Quantum Theory or Wave Mechanics This is the mathematical structure that predicts and describes the behavior of particles of microscopic size, usually taken to be in the range $\sim 10^{-10}$ to 10^{-14} m.

Relativistic Quantum Mechanics This is a combination of relativistic mechanics and wave mechanics; that is, it is a physical theory (described as a mathematical structure) that predicts and describes the behavior of microscopic particles moving with speeds approaching that of light.

Physical Theories of the Future There are already some indications that present theories will have to be modified and new mathematical structures developed to describe the behavior of extremely small particles, submicroscopic particles, moving with the speed of light. There may be particles moving with speeds greater than the speed of light (such particles have been named *tachyons*). If these particles are found, our "modern" theories may become obsolete.

In this text we shall be concerned mostly with classical physics, but some aspects of modern physics, including relativity and quantum mechanics, will be discussed on a very elementary level.

Before closing this section, we must briefly state the scope of physical science, life science, medical science, and their relation to physics. Physics is essential to all sciences, providing them with basic principles and fundamental laws of nature. Technological developments in engineering are glamorous and flourishing on their own, yet all are built on basic laws that have been discovered in physics. We may now ask a second question: What has physics to do with health science? As Sir George Pickering has pointed out, medicine is neither an art nor a science, but a technology. Practicing a specialized technology in a particular field lasts for a limited period. Having basic knowledge, one can benefit from or contribute to new developments. Examples of such developments are the diagnostic techniques and treatments (together with the use of computers) such as radiation therapy, nuclear medicine, thermography, electric shock treatments, pacemakers, artificial hearts, and so on. The interaction between physics and medicine is not of recent origin. Some of the great names in physics, such as Bernoulli, Gilbert, Von Helmholtz, and Young, were physicians. Galileo, Foucault, and others were students of medicine.

Finally, let us briefly acquaint ourselves with such fields as astronomy, geology, chemistry, and biology.

Astronomy Astronomy deals with the laws governing motion, structure, and origin of the solar-system, structure of stars and galaxies that are parts of our universe. We may say that astronomy is that branch of physics which deals with very large objects and very large distances.

Geology Geology is the science that deals with the macroscopic structure, origin, development, and behavior patterns of earth matter. The earth is about 4.5 billion years old, and most of the ideas and theories in geology have resulted from direct observations. Even though geologists must do some experiments in the laboratory using a knowledge of chemistry, physics, and biology, it may be said that the whole earth is one big laboratory which provides geologists with a variety of objects large and small, young and old.

Chemistry Chemistry deals with molecular structure, synthesis of compounds, and chemical reaction rates under various conditions. Methods of physics are used quite frequently in physical chemistry.

Biology Biology deals with living material and life processes which are quite complex. Biochemistry and biophysics are the results of overlapping of the different fields. As an example of the dependence of a biologist on techniques of physics and chemistry, let us consider the DNA molecule, which carries and transmits the hereditary characteristics of living matter from generation to generation. The structure of this molecule was originally investigated by using the techniques of x-ray crystallography. To understand the basic cell structure, the biologist must understand the laws of physics and of chemical reaction rates as applied to molecular structure.

1.2. The Modernized Metric System

Scientific investigation may be called a two-step process: (1) the development of generalizations that arise from repeated observations of physical phenomena and (2) the explanation of these generalizations using logical reasoning. Such investigations eventually lead to the formulation of hypotheses, theories, and laws. The task of experimental physicists is not only to make measurements of physical phenomena but also to test the theories and laws of physics, and to improve on previous experimental work. Measurements involve such quantities as velocity, force, energy, temperature, electric current, magnetic field, and hundreds of others. The most surprising aspect is that all these quantities can be expressed in terms of a few basic quantities, such as length, mass, and time. These quantities are called *fundamental* or *basic quantities* (*base units*); all others that are expressed in terms of these are called *derived quantities*.

Three Basic Standards—Length, Mass, and Time

In order to communicate, scientists must agree on the definition of fundamental quantities. As we shall see shortly, there are three different sets of units being used. The most prevalent is that in which length is measured in *meters*, mass in *kilograms*, and time in *seconds*—hence the name *MKS system* (or *metric system*). So our first task is to define these three fundamental quantities, or *standards*.

STANDARD OF LENGTH—METER The *standard meter*, which has been in use for many years, is defined as the distance between the two marks on the ends of a platinum–iridium alloy metal bar kept in a temperature-controlled vault near Paris.

In 1960, by international agreement, the General Conference on Weights and Measures changed the standard of length to an atomic constant by the following procedure. A glass tube is filled with krypton gas in which an electrical discharge is maintained. The individual atoms of krypton of mass number 86 emit orange-red light. All waves of this orange-red light have the same wavelengths. The aim is to find how many such wavelengths will fit the length of the standard meter (see Figure 1.1). Thus the *standard meter* is defined to be equal to exactly 1,650,763.73 wavelengths of orange-red light emitted in a vacuum from krypton-86 atoms.

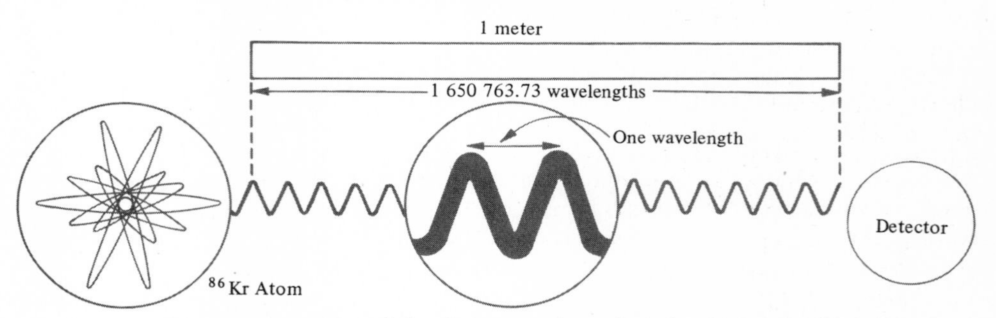

FIGURE 1.1. *Meter defined in terms of wavelengths of orange-red light from krypton 86 atoms.* (Courtesy of the National Bureau of Standards.)

This latest standard of length has many advantages over the old one: (1) it cannot be destroyed and is independent of small temperature variations; and (2) it can be reproduced anywhere without much trouble, and is much more accurate.

STANDARD OF TIME—SECOND In the past the spinning motion of the earth about its own axis, as well as its orbital motion about the sun, have been used to define a second. Thus, a second is defined to be 1/86,400 ($= 1/24 \times 60 \times 60$) of a mean solar day (1 mean solar day = successive appearances of the sun overhead averaged over 1 year).

The orbital motion of the earth around the sun can be calculated much more reliably than its spinning motion. In 1956, the International Conference of Weights and Measures redefined the second as follows: 1 second is defined to be exactly equal to the fraction 1/31,556,925.9747 of the tropical year 1900.

The determinations by astronomical observations were never clear-cut and satisfactory. In October 1967, the time standard was redefined in terms of an *atomic clock*, which makes use of the periodic atomic vibrations of certain

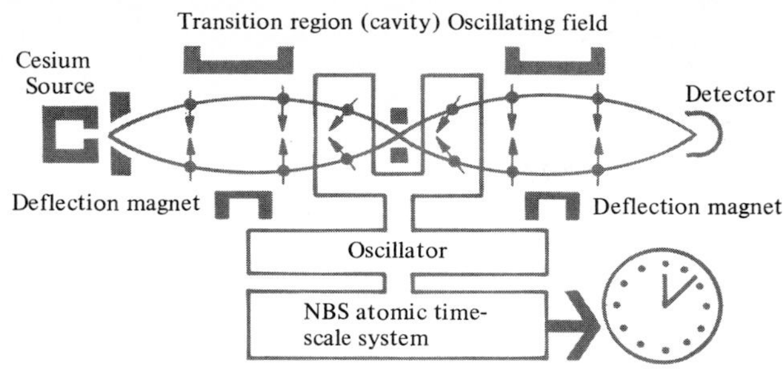

Figure 1.2. *Schematic diagram of an atomic beam spectrometer or "clock". When 9,192,631,770 oscillations (defined in a particular way) have occurred, the clock indicates that 1 second has passed.* (Courtesy of the National Bureau of Standards.)

atoms. Note that no matter how the standard is defined, the method always involves some sort of repetitive or periodic motion. Figure 1.2 shows an atomic clock which uses cesium atoms of mass number 133 and is in operation at the U.S. Bureau of Standards. According to the cesium clock, a *second* is defined to be exactly equal to the time interval of 9,192,631,770 vibrations of radiation from cesium atoms of mass number 133.

This method has an accuracy of 1 part in 10^{11}. It is possible that two cesium clocks running over a period of 5000 years will differ by only 1 second. The definition of a cesium second may be temporary, because it is thought that improved clocks may eventually come into existence.

Standard of Mass—Kilogram Originally, 1 kilogram mass was defined to be equal to the mass of 1 liter ($=10^{-3}$ m^3) of water at 4°C (water has its maximum density at 4°C). This unit is no longer used; another arbitrarily chosen object, as defined below, is used as a standard kilogram.

A platinum–iridium cylinder is carefully stored in a repository at the International Bureau of Weights and Measures in Sèvres (near Paris), France. The mass of the cylinder is defined to be exactly equal to a *kilogram*. This is the only base unit still defined by an artifact.

Figure 1.3 shows a platinum–iridium cylinder kept at the National Bureau of Standards in the United States. This was compared with the international standard in about 1885 and again in 1948. The mass has an accuracy of 1 part in 10^8. The exact replica of the standard kilogram is obtained by using the equal-arm balance.

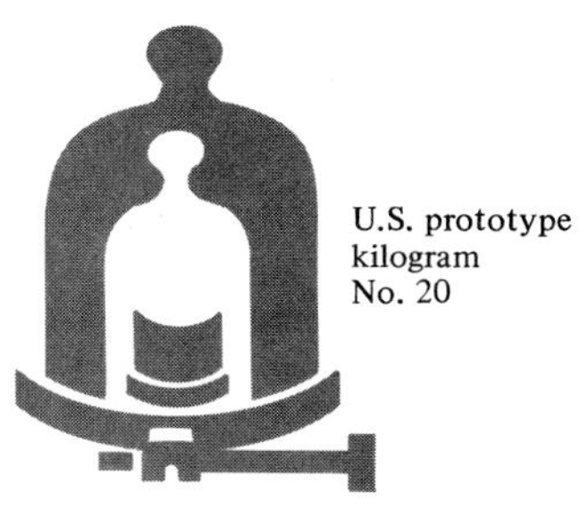

Figure 1.3. *Duplicate of the standard kilogram in the custody of the National Bureau of Standards serves as the mass standard for the United States.* (Courtesy of the National Bureau of Standards.)

Different Systems of Units

Système International (Modernized MKS System) As explained earlier, the meter, kilogram, and second constitute the MKS system. At the 1960 Conference on Weights and Measures, it was agreed to include the unit of electric current, called an *ampere* (A), as an independent unit. Thus, the MKS system becomes the MKSA, where A refers to an ampere.

The International System of Units, abbreviated SI after the French *Système International,* is a modernized version of the metric system established by international agreement. For convenience it uses seven base units:

1. Length, in *meters* (m).
2. Mass in *kilograms* (kg).
3. Time, in *seconds* (s).
4. Electric current, in *amperes* (A).
5. Temperature, in *Kelvin* (K).
6. Amount of substance, in *moles* (mol).
7. Luminous intensity, in *candelas* (cd).

The SI also uses two supplementary units:

1. Plane angle, in *radians* (rad).
2. Solid angle, in *steradians* (sr).

It may be pointed out that the MKS system is commonly used in scientific work, and it is becoming more and more accepted in industry and commerce throughout the world. Because of its simplicity and ease in manipulation, it is expected to be universally accepted sometime in the near future. We, on our part, will try to emphasize this system more than any other throughout this text.

The CGS or Gaussian System In this system the unit of length is the *centimeter* ($= 10^{-2}$ m), the unit of mass is the *gram* ($= 10^{-3}$ kg), and the unit of time is the *second*. Even though it appears that the CGS and MKS systems are very similar, some electric and magnetic equations take considerably different forms in the two systems.

The British System In this system the unit of length is the *foot* and the unit of time is the *second*. This system does not use mass as a basic unit; instead, *force* is used, the unit of which is the *pound* (lb). The unit of mass derived from the pound is called a *slug* (= 32.17 lb mass = 14.59 kg), as we shall see in Chapter 4. The unit of temperature in the British system is *degree Fahrenheit*, as will be defined in Chapter 13.

Some useful relations for conversion between the SI and British systems, and some of their multiples and submultiples, are given in Appendix A. Example 1.1 illustrates this.

Example 1.1 A car is moving with a speed of 50 mi/h. Convert this speed into km/h and m/s.

Since 1 mi = 1.609 km,

$$50\ \text{mi/h} = \left(\frac{50\ \cancel{\text{mi}}}{\text{h}}\right)\left(\frac{1.609\ \text{km}}{1\ \cancel{\text{mi}}}\right) = 80.45\ \text{km/h}$$

Also, 1 km = 1000 m and 1 h = 3600 s:

$$80.45\ \text{km/h} = \left(\frac{80.45\ \cancel{\text{km}}}{\cancel{\text{h}}}\right)\left(\frac{1000\ \text{m}}{1\ \cancel{\text{km}}}\right)\left(\frac{1\ \cancel{\text{h}}}{3600\ \text{s}}\right) = 22.35\ \text{m/s}$$

Exercise 1.1 A jet airplane is flying at a speed of 700 mi/h. Convert this into km/h, km/s, and m/s. [*Ans.*: 1130 km/h, 0.310 km/s, 310 m/s.]

1.3. Dimensional Analysis

Most physical quantities may be expressed in terms of length L, mass M, and time T. The quantities L, M, and T are dimensions. Thus, a quantity expressed as $L^aM^bT^c$ means that its length dimension is raised to the power of a, mass dimension is raised to the power of b, and time dimension is raised to the power of c. Thus, the dimension of area, which is length $\times$ width, is L^2, that of volume, L^3. Speed, which has units of length/time, will have dimensions L/T or LT^{-1}.

In order to add or subtract two quantities in physics, they must have the same dimensions. Similarly, no matter which system of units is used, all mathematical relations and equations must be dimensionally correct. That is, the quantities on both sides of an equation must have the same dimensions. We illustrate this point in Example 1.2.

Since equations must be dimensionally correct, dimensional analysis may be used to (1) check the correctness of the form of the equation, and (2) to check an answer computed from an equation for plausibility in a given situation. Of course, it is not possible to determine the multiplicative constants by dimensional analysis.

EXAMPLE 1.2 Show that the following equation is dimensionally correct.

$$x = v_0t + \tfrac{1}{2}at^2 \qquad \text{(i)}$$

where x is distance, v_0 is velocity (which has units of length/time), and a is acceleration (which has units of length/time^2). Thus, x has dimensions of $[L]$, v_0 that of $[L]/[T]$, a that of $[L]/[T]^2$, and t that of $[T]$. Equation i in dimensional form becomes

$$[L] = \frac{[L]}{[T]}[T] + \frac{1}{2}\frac{[L]}{[T]^2}[T]^2 \qquad \text{(ii)}$$

or

$$[L] = [L] + \tfrac{1}{2}[L] = \tfrac{3}{2}[L]$$

Except for the constant factor $\frac{3}{2}$, both sides have the same dimensions, that is, L.

EXERCISE 1.2 Show that the following equation is dimensionally correct.

$$v^2 = v_0^2 + 2ax$$

where v and v_0 are velocities and have units of length/time, acceleration a has units of length/time^2, and distance x has units of length. [*Ans.*: L^2T^{-2}.]

1.4. Scalar and Vector Quantities

Most of the physical quantities that we encounter in physics are either *scalar quantities* or *vector quantities.* A *scalar* quantity has *magnitude only; no direction is associated with it.* Examples are pure numbers and physical

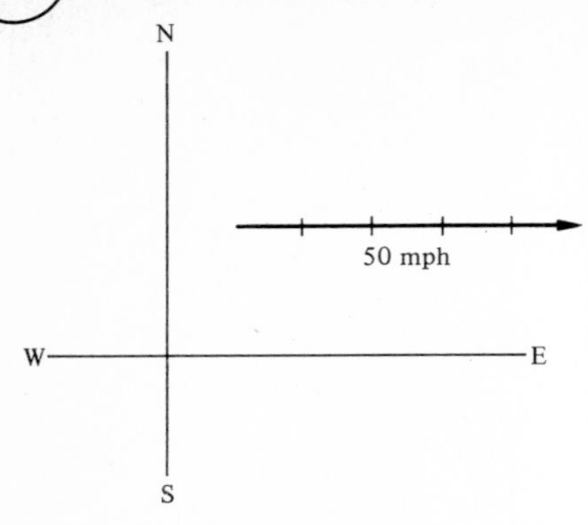

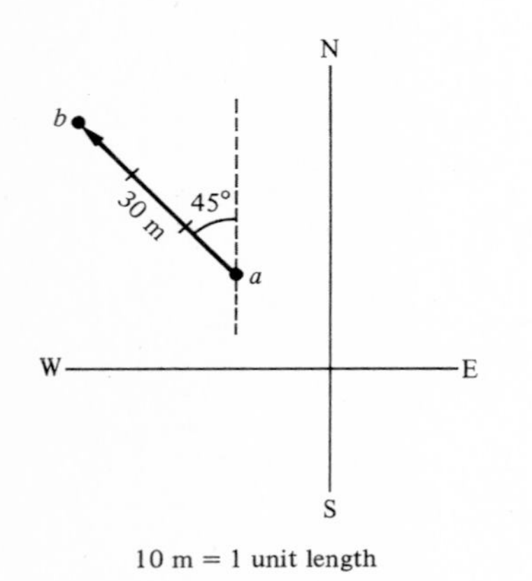

FIGURE 1.4. *Graphical representation of vector quantities.*

quantities such as mass, time, volume, distance, and speed. To completely specify a scalar quantity, one has to state the magnitude and the corresponding unit, if any. For example, 10 books, 5 chairs, 3 tables, 9 spoons, 50 kg, 1000 s, and 75 cm^3 are all scalar quantities. These quantities may be added, subtracted, multiplied, or divided by the simple arithmetic rules.

A *vector* quantity has *both magnitude and direction*, and it follows the rules of vector algebra to be discussed shortly. Examples include displacement, velocity, acceleration, force, and electric field strength. To make a complete statement about a vector quantity, one has to state both magnitude and direction. For example, if we say that a car is moving at 50 miles per hour, we are talking only about its speed or its magnitude, which is a scalar quantity. It does not tell us in which direction the car is moving. On the other hand, if we say that the car is moving at 50 miles per hour due east, we are talking about the velocity of the car, which has a magnitude of 50 miles per hour and is directed east, as shown in Figure 1.4(a); hence, velocity is a vector quantity. Similarly, if we say that an object is displaced (changes its position with respect to its surrounding) by 30 m due northwest, it means that the object was moved from its initial position a distance of 30 m along the line which makes an angle of 45° with the north position, as shown in Figure 1.4(b).

Thus, in order to show vectors in a diagram, we draw a line using some convenient scale (10 miles per hour to 1 inch or 10 miles per hour to 1 cm) and then draw an arrowhead at one end of this line. The length of this line represents the magnitude of the vector quantity, while the direction of the arrow represents the direction. Figure 1.4 demonstrates these points: instead of using north, south, east, and west as reference directions, we shall use the *XY* coordinate axes as a reference.

The next question is how to distinguish letters representing vector quantities from those representing scalar quantities. In printed material such as a textbook, a vector quantity is often represented by a boldface letter. For example, force is represented by **F**, velocity by **v**, acceleration by **a**. Without reference to any particular quantity, we will use letters **R**, **A**, **B**, and **C** to denote different vectors. Since it is hard to use boldface letters in handwriting, an alternative is to denote vectors by putting arrows above the letters, that is, $\vec{R}$, $\vec{A}$, $\vec{B}$, $\vec{C}$, . . . If we want to consider only the magnitude of the vectors, then *R*, *A*, *B*, *C*, . . . or $|\mathbf{R}|$, $|\mathbf{A}|$, $|\mathbf{B}|$, $|\mathbf{C}|$, . . . represent the magnitudes of the corresponding vector quantities.

1.5. Vector Algebra

To understand the real significance and use of vectors, we must know how to add, subtract, multiply, and divide vector quantities. We investigate these properties in some detail.

Vector Equality

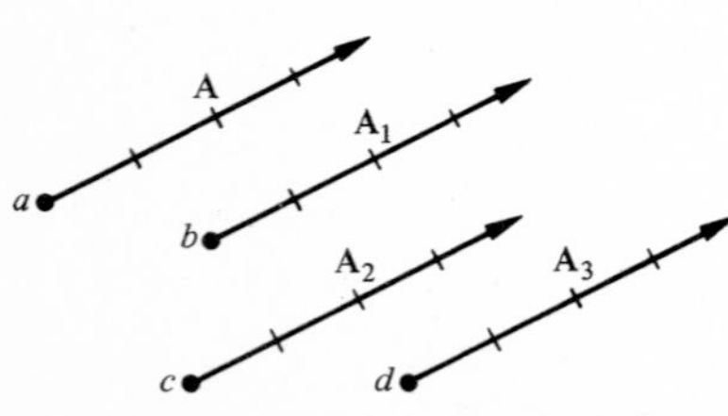

FIGURE 1.5. *Four equivalent vectors.*

Consider the four vectors $\mathbf{A}$, $\mathbf{A}_1$, $\mathbf{A}_2$, and $\mathbf{A}_3$ shown in Figure 1.5. All these vectors have equal magnitudes (equal to four arbitrary units in this case) and are parallel to each other. Such vectors are said to be *equal*, and this fact is written

$$\mathbf{A} = \mathbf{A}_1 = \mathbf{A}_2 = \mathbf{A}_3 \tag{1.1}$$

The only difference in these vectors is that they are located at different points, such as *a, b, c,* and *d,* respectively. The property stated by Equation 1.1 means that any vector can be replaced by any other vector which is parallel to the first vector and is equal in magnitude. If, in addition, equal vectors produce the same effect, they are said to be *equivalent vectors*.

Vector Addition by Geometrical Method

Let us see how we can add two vectors **A** and **B**, such as shown in Figure 1.6.

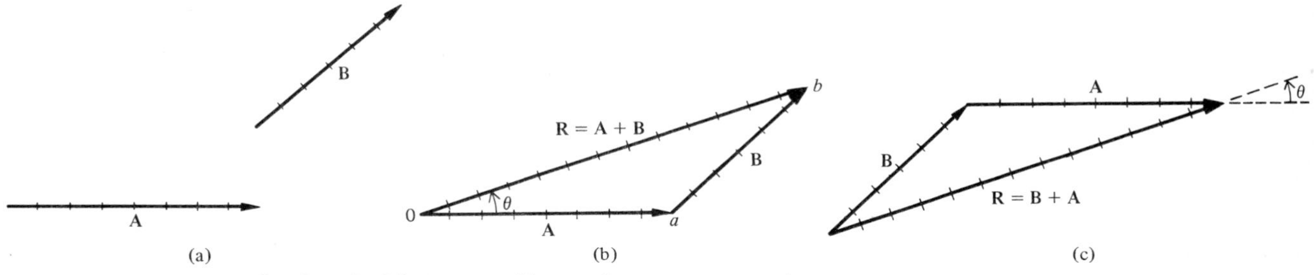

Figure 1.6. *Graphical method for vector addition of two vectors* **A** *and* **B**.

At point *O* draw a vector equal to vector **A**. At the head of this vector, which is at point *a*, draw another vector equal to vector **B**. Then the line joining the tail of **A** to the head of **B**, that is, line *Ob*, represents the sum of the two vectors **A** and **B** and is called the *resultant vector* or simply the *resultant*, **R**, which is written

$$\boxed{\mathbf{R} = \mathbf{A} + \mathbf{B}} \tag{1.2}$$

Note that the sum of the magnitudes of **A** and **B** is not equal to the magnitude of **R**, $|\mathbf{R}| \neq |\mathbf{A}| + |\mathbf{B}|$. This is the basic difference between vector quantities and scalar quantities. The magnitude of **R** can be found by using the scale used for drawing **A** and **B**. The direction of **R** is denoted by an angle θ which it makes with vector **A**.

Furthermore, it makes no difference whether we add **B** to **A** or **A** to **B** as shown in Figure 1.6(c); that is, the order of addition is of no consequence. This property of vector addition is known as the *commutative law* and is mathematically written

$$\mathbf{A} + \mathbf{B} = \mathbf{B} + \mathbf{A} \tag{1.3}$$

We can extend the method above to the addition of more than two vectors. To find the sum of three vectors **A**, **B**, and **C** shown in Figure 1.7, we use the following procedure. Draw a vector *Oa* parallel and equal to **A**. At the head of this vector, that is, at point *a*, draw a vector *ab* parallel and equal to **B**. Repeat this for vector **C** at point *b*. Then the line *Oc* is the resultant of these three vectors. From Figure 1.7(b) it is clear that

$$\mathbf{R} = \mathbf{R}_1 + \mathbf{C} = (\mathbf{A} + \mathbf{B}) + \mathbf{C} \tag{1.4}$$

Thus, the procedure is to find $\mathbf{R}_1$, which is the resultant of vectors **A** and **B**, and

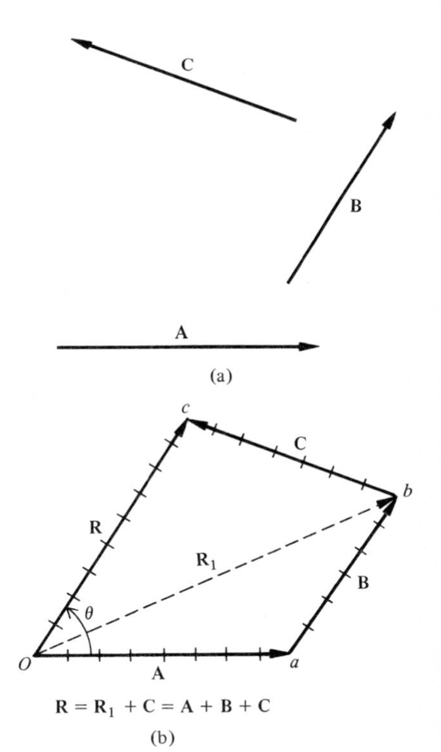

Figure 1.7. *Graphical method for vector addition of more than two vectors.*

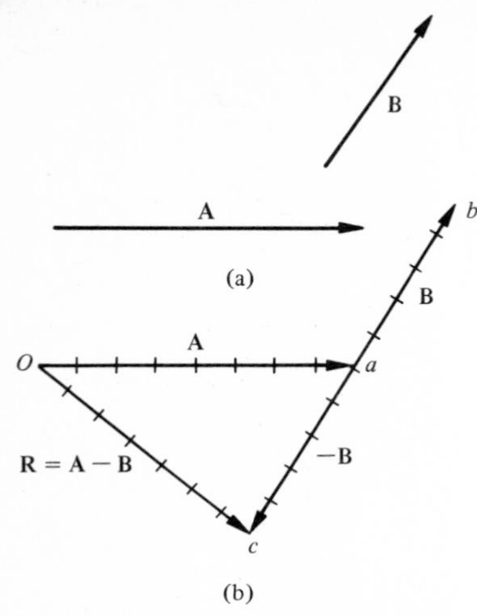

FIGURE 1.8. *Graphical method for vector subtraction.*

then combine $\mathbf{R}_1$ with **C** to find the resultant **R**. As a matter of fact, one does not have to draw the resultant $\mathbf{R}_1$. We can simply draw the tail of **B** with the head of **A**, the tail of **C** with the head of **B**; then the resultant is given by the line joining the tail of **A** with the head of **C**. This procedure may be extended to as many vectors as we want. In all cases an arbitrary scale is established and the magnitude of the resultant is measured by using the same scale and measuring the angle that the resultant makes with the chosen reference direction.

Once again, the order in which these vectors are added does not make any difference. Thus,

$$(\mathbf{A} + \mathbf{B}) + \mathbf{C} = \mathbf{A} + (\mathbf{B} + \mathbf{C}) \tag{1.5}$$

That is, the sum of $(\mathbf{A} + \mathbf{B})$ added to **C** gives the same resultant as the sum of **A** added to $(\mathbf{B} + \mathbf{C})$. This property of vector addition is called the *associative law*.

Vector Subtraction

Consider vectors **A** and **B** shown in Figure 1.8. We are interested in subtracting vector **B** from **A**; that is, we want to find $\mathbf{A} - \mathbf{B}$. This can be done geometrically if we write this in a slightly different form:

$$\mathbf{A} - \mathbf{B} = \mathbf{A} + (-\mathbf{B})$$

Thus, to find $\mathbf{A} - \mathbf{B}$, we add to **A** a vector that is equal in magnitude and opposite in direction to **B**. Note that if **B** is a vector from a to b, then $-\mathbf{B}$ is a vector from b to a or a to c as shown in Figure 1.8. Thus, the resultant **R** as shown in Figure 1.8 is

$$\boxed{\mathbf{R} = \mathbf{A} - \mathbf{B} = \mathbf{A} + (-\mathbf{B})} \tag{1.6}$$

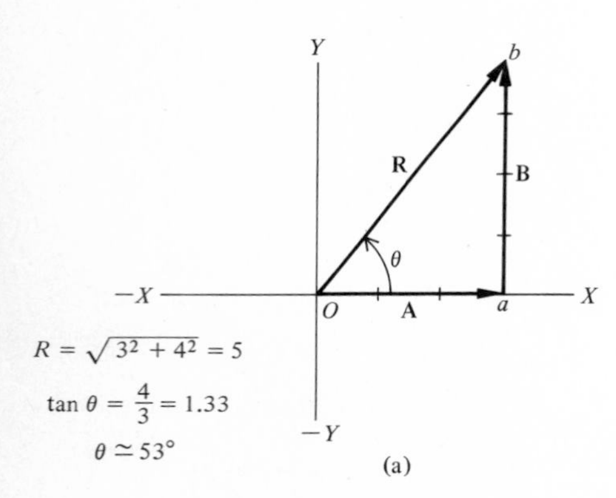

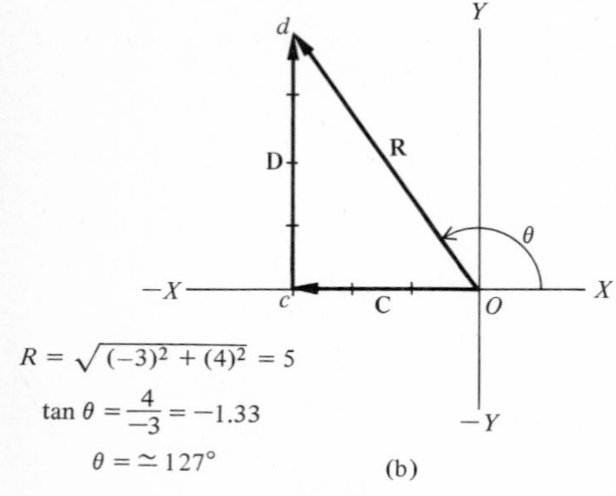

FIGURE 1.9. *Two simple examples of vector addition by the analytical method.*

Vector Addition by Analytical Method

The geometrical method of obtaining the resultant of several vectors discussed above is very cumbersome and not very accurate. The accuracy is a function of the scale used in making the diagram and the tools (pencils, rulers, protractor, etc.). A simple and more accurate analytical method is thus desirable. Before we go through the formal presentation, we will discuss two simple examples.

In Figure 1.9(a) a man walks from point O to point a. His displacement is given by vector **A**, which has a magnitude of 3 units and is in the direction of $+X$-axis. This is followed by another displacement of 4 units along ab, which is represented by a vector **B** and is parallel to the $+Y$-axis. We are interested in finding the net displacement. The vector **R** gives the magnitude and the direction of the resultant displacement; that is,

$$|\mathbf{R}| = \sqrt{A^2 + B^2} = \sqrt{3^2 + 4^2} = \sqrt{25} = 5$$

while (see Appendix B for trigonometrical relations)

$$\tan\theta = \frac{B}{A} = \frac{4}{3} = 1.33 \qquad \text{or} \qquad \theta = 53.1°$$

Thus, the resultant displacement is 5 units and it makes an angle of 53.1° with the $+X$-axis. Another example for calculating the resultant of two vectors **C** and **D** is shown in Figure 1.9(b).

In a more general situation, let us consider a vector **R** shown in Figure 1.10. By drawing perpendiculars from point P (the head of the vector) to the X and Y axes, we obtain $\mathbf{R}_x$ and $\mathbf{R}_y$, two vectors along the X and Y axes, respectively. These two vectors are called the *rectangular components* of the vector **R**, and the magnitudes of these are given by

$$R_x = R\cos\theta \tag{1.7}$$

$$R_y = R\sin\theta \tag{1.8}$$

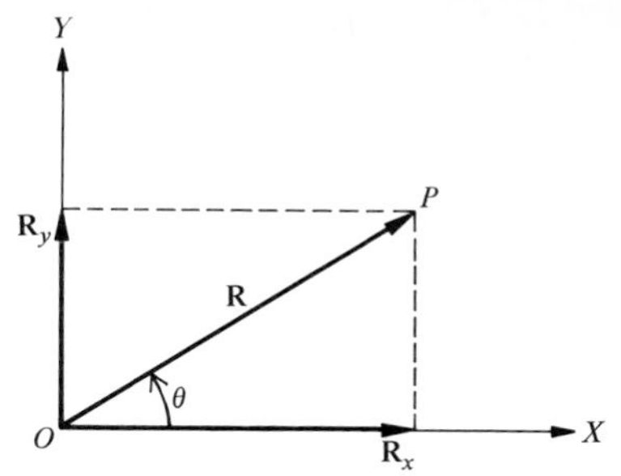

FIGURE 1.10. *Components* $\mathbf{R}_x$ *and* $\mathbf{R}_y$ *of a vector* **R**.

If R and θ are given, we can find R_x and R_y. On the other hand, if R_x and R_y are given, we can find R and θ by the following relation:

$$R = \sqrt{R_x^2 + R_y^2} \tag{1.9}$$

$$\tan\theta = \frac{R_y}{R_x} \tag{1.10}$$

This means in order to locate point P, we should be given either R and θ or R_x and R_y. We shall use this method to find a resultant of several vectors. Note that the advantage lies in the fact that all the R_x components in the same direction can be added like scalars, and, similarly, the R_y components.

Let us consider three vectors $\mathbf{R}_1$, $\mathbf{R}_2$, and $\mathbf{R}_3$ acting on an object O as shown in Figure 1.11(a). These vectors may be forces or some other vector quantities. We are interested in finding the resultant. Draw the XY coordinates as in Figure 1.11(b) and at point O draw $\mathbf{R}_1$, $\mathbf{R}_2$, and $\mathbf{R}_3$ parallel to the vectors in (a). Let us resolve each vector into its rectangular components as shown in Figure 1.11(b). That is:

Vector	Angle	X-Component	Y-Component
$\mathbf{R}_1$	θ_1	$R_{1x} = R_1\cos\theta_1$	$R_{1y} = R_1\sin\theta_1$
$\mathbf{R}_2$	θ_2	$R_{2x} = R_2\cos\theta_2$	$R_{2y} = R_2\sin\theta_2$
$\mathbf{R}_3$	θ_3	$R_{3x} = R_3\cos\theta_3$	$R_{3y} = R_3\sin\theta_3$

Let the sum of the X and Y components be denoted by R_X and R_Y. This means that

$$R_X = R_{1x} + R_{2x} + R_{3x} = \sum_{n=1}^{3} R_{nx} \tag{1.11}$$

and

$$R_Y = R_{1y} + R_{2y} + R_{3y} = \sum_{n=1}^{3} R_{ny} \tag{1.12}$$

where $\sum_{n=1}^{3}$ denotes the sum of all components from $n = 1$ to $n = 3$. Thus, the resultant of the three vectors is given by

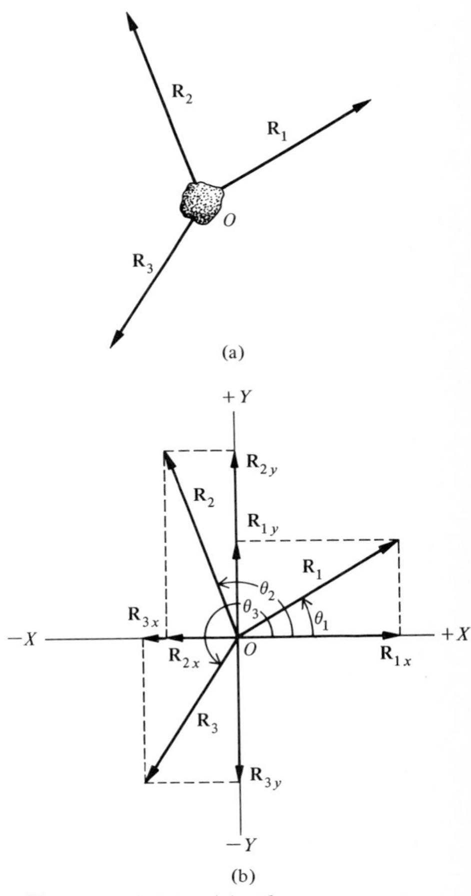

FIGURE 1.11. *(a) Three vector quantities acting at a point O. (b) The resultant of three vectors by the analytical method.*

$$R = \sqrt{R_X^2 + R_Y^2}$$
$$\tan\theta = \frac{R_Y}{R_X} \tag{1.13}$$

The preceding method is simple, quick, and accurate and may be extended to any number of vectors.

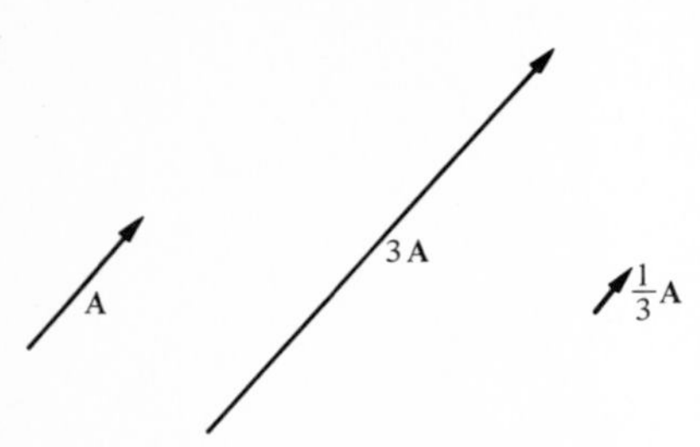

FIGURE 1.12. *Multiplication of a vector* **A** *by a scalar* $s = 3$ *and* $s = \frac{1}{3}$.

Multiplication of a Vector by a Scalar

Multiplication of a vector **A** by s, a scalar number, results in another vector $s\mathbf{A}$, whose magnitude is s times the magnitude of **A**; that is, sA or $s|\mathbf{A}|$. If $s = 3$, $s\mathbf{A} = 3\mathbf{A}$ and if $s = \frac{1}{3}$, $s\mathbf{A} = \frac{1}{3}\mathbf{A}$, as shown in Figure 1.12.

Multiplication of Two Vectors

Multiplication of vectors is quite complex, but it is useful in many physical situations of interest. We shall use two different types of multiplication: (1) scalar or dot product and (2) vector or cross product. We shall introduce these later when they are needed.

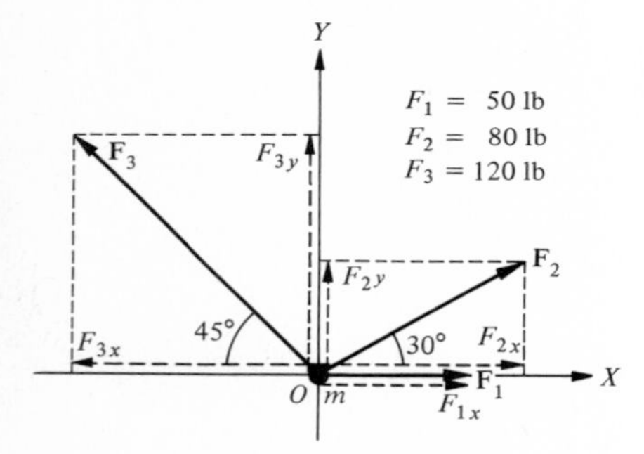

Ex. 1.3

EXAMPLE 1.3 Three forces are acting on mass m as shown in the accompanying figure. Calculate the resultant force. (Note that the mass will move in the direction of the resultant force.)

The magnitudes of the three forces as shown in the figure are $F_1 = 50$ lb, $F_2 = 80$ lb, and $F_3 = 120$ lb. F_1 makes an angle of $\theta_1 = 0°$ with the X-axis, and resolved into two components (dashed lines) gives

$$F_{1x} = F_1 \cos\theta_1 = (50 \text{ lb}) \cos 0° = 50 \text{ lb} \times 1 = 50 \text{ lb}$$
$$F_{1y} = F_1 \sin\theta_1 = (50 \text{ lb}) \sin 0° = 50 \text{ lb} \times 0 = 0$$

As shown in the figure, F_{1y} is zero.

Force F_2 makes an angle of $\theta_2 = 30°$ with the X-axis, and resolved into two components (dashed lines) gives

$$F_{2x} = F_2 \cos\theta_2 = (80 \text{ lb}) \cos 30° = 80 \text{ lb} \times 0.866 = 69 \text{ lb}$$
$$F_{2y} = F_2 \sin\theta_2 = (80 \text{ lb}) \sin 30° = 80 \text{ lb} \times 0.50 = 40 \text{ lb}$$

Force F_3 makes an angle of $\theta_3 = 135°$ with the $+X$-axis or $45°$ with the $-X$-axis, and resolved into two components (dashed lines) gives

$$F_{3x} = F_3 \cos\theta_3 = (120 \text{ lb}) \cos 135° = (120 \text{ lb})(-\cos 45°) = (120 \text{ lb})(-0.707)$$
$$= -85 \text{ lb}$$
$$F_{3y} = F_3 \sin\theta_3 = (120 \text{ lb}) \sin 135° = (120 \text{ lb}) \sin 45° = (120 \text{ lb}) \times 0.707$$
$$= 85 \text{ lb}$$

Note that instead of using $\theta_3 = 135°$ and resolving F_3 along the $+X$-axis and $+Y$-axis, we could have used $\theta_3 = 45°$ and resolved F_3 along the $-X$-axis and $+Y$-axis so that $F_{3x} = -120 \cos 45°$ and $F_{3y} = 120 \sin 45°$ obtaining the same results.

Let the resultant force **F** have the components F_x and F_y so that

$$F_x = F_{1x} + F_{2x} + F_{3x} = (50 + 69 - 85)\text{ lb} = 34\text{ lb}$$
$$F_y = F_{1y} + F_{2y} + F_{3y} = (0 + 40 + 85)\text{ lb} = 125\text{ lb}$$

Hence, the resultant force **F** is

$$F = \sqrt{F_x^2 + F_y^2} = \sqrt{(34)^2 + (125)^2}\text{ lb} = 130\text{ lb}$$

Suppose that the resultant **F** makes an angle θ with the $+X$-axis so that $F_x = F\cos\theta$ and $F_y = F\sin\theta$. From the definition of $\tan\theta$, dividing the second by the first,

$$\tan\theta = \frac{F_y}{F_x} = \frac{125}{34} = 3.68, \qquad \theta = 74.8°$$

Exercise 1.3 A man and a boy are pulling a box located at the origin. The man applies a force of 100 lb along the X-axis while the boy applies a force of 50 lb making an angle of 120° with the $+X$-axis. What is the resultant force and in what direction will the box move? [*Ans*.: $F = 86.6$ lb; $\theta = 30°$.]

SUMMARY

Physics is classified into two groups, *classical physics* and *modern physics*.

Most systems of units use three basic quantities—length, mass, and time. In the *MKS system* the standard of length is the meter, the standard of mass is the kilogram, and the standard of time is the second. The *SI system* uses seven basic quantities—m, kg, s, A, K, mol, and cd. The *CGS system* uses the cm, g, and s as base units. The *British system* uses the foot, the pound, and the second as the units of length, force, and time; and the slug is the derived unit of mass.

The dimensions of length, mass, and time are denoted by L, M, and T, respectively. All equations must have the same dimensions on both sides.

Scalar quantities have magnitudes only, while *vector quantities* have both magnitude and direction. The components of a vector **R** making an angle θ with the X-axis are $R_x = R\cos\theta$ and $R_y = R\sin\theta$. If R_X and R_Y are the sum of the X and Y components of vectors $\mathbf{R}_1, \mathbf{R}_2, \mathbf{R}_3, \ldots$, the resultant vector is given by $R = \sqrt{R_X^2 + R_Y^2}$ and $\tan\theta = R_Y/R_X$.

QUESTIONS

1. Can you name some other branches of natural science besides the ones listed in Section 1.1?
2. What were the shortcomings of classical physics that led to the development of quantum mechanics and the special theory of relativity in the beginning of the present century?
3. What do you think are the advantages of the SI system over the MKS system? Over the British system?
4. Compare and contrast units and dimensions.
5. Can we use the dimensional-analysis method to check the accuracy of an equation with regards to its multiplicative and additive constants?
6. Suppose that overnight all lengths, widths, and heights are reduced to one-half. How would you be able to detect this happening?

7. Name some scalar and vector quantities other than those listed in Section 1.4.
8. For the addition of vectors, which method is more accurate—graphical or analytical—and why?
9. Can we treat components of vector quantities as scalar quantities?
10. If two vectors have different magnitudes, can their resultant be zero?
11. Does a vector quantity of zero magnitude have any meaning?
12. If $|\mathbf{A} + \mathbf{B}| = |\mathbf{A} - \mathbf{B}|$, what is the relation between the two vectors **A** and **B**?

PROBLEMS

1. Convert the following quantities into metric units: (a) the height of Mount Everest, which is 29,030 ft; (b) a distance of 100 miles; (c) the distance between New York and London, which is 3473 miles; and (d) the height of the Washington Monument, which is 555 ft $5\frac{1}{8}$ in.
2. (a) The radius of the earth is 3960 miles. Express this in kilometers. (b) The circumference of the earth is 40,067.32 km. Express this in miles.
3. A piece of land is 100 ft $\times$ 80 ft. Express both lengths and the area in metric units.
4. Show that 1 U.S. gallon is equal to 3.8 liters and 1 liter is equal to 0.263 gallon. (1 U.S. gal = 231 in^3.)
5. A 5 lb-mass bag of sugar is how many kilograms?
6. The mass of the earth is 5.98×10^{24} kg. Express this in British units of lb-mass.
7. (a) The density of water is 1 g/cm^3. Express this in kg/m^3. (b) The density of air at standard temperature and pressure is 1.293 kg/m^3. Express this in lb-mass/ft^3.
8. Suppose that the age of the earth is 4 billion years. Express this in seconds using powers of 10.
9. Some radioactive atoms live as long as 1.39×10^{10} yr, while others live as briefly as 0.000 000 000 01 s. Express the first in terms of seconds and the latter in powers of 10.
10. Express the following speeds in km/h and m/s: (a) 25 mi/h; (b) 40 mi/h; (c) 55 mi/h; and (d) 65 mi/h.
11. A spaceship can move with a speed ranging from 7 to 14 mi/s. Express this range in (a) m/s; (b) km/s; and (c) km/h.
12. Show that the following equation is dimensionally correct.

$$x = \frac{v + v_0}{2} t$$

v and v_0 are velocities and have units of length/time, x that of length, and t that of time.

13. What are the dimensions of (a) the circumference of a circle; (b) density (mass/volume); and (c) momentum (mass $\times$ velocity)?
14. Acceleration has units of length/time^2. What are the dimensions of (a) force (mass $\times$ acceleration); and (b) pressure (force/area)?
15. Consider a displacement of 16 km ($\sim$10 mi) due east and a second displacement of 32 km ($\sim$20 mi) making an angle of 30° north of east. Find

the resultant of these two vectors graphically by using a scale of 4 km = 1 cm.

16. A boy moves 10 blocks west, followed by 8 blocks northeast. Find the resultant of these by the graphical method.

17. A vector of 100 units makes an angle of 60° with the X-axis. Find the X and Y components of this vector.

18. An automobile moves 30 km east and then 40 km north. Find the magnitude and direction of the resultant of these two vectors.

19. Consider a displacement of 50 km west and 40 km south. Calculate the resultant.

20. A force of 80 lb making an angle of 30° with the horizontal is applied to a lawn mower. What net force acts on the lawn mower in the horizontal direction?

21. A boy pulls a sled with a force of 60 lb making an angle of 37° with the horizontal. Calculate the vertical and horizontal components.

22. A man and boy are pulling a box with forces of 80 and 40 lb, respectively, making an angle of 135° with each other. Find the resultant and the direction in which the box will move. Assume that the man acts along the X-axis.

23. The following three forces are acting at the origin: a force of 50 lb along the X-axis, a second force of 80 lb making an angle of 45°, and a third force of 20 lb making an angle of 135° with the X-axis. Find the resultant of these three forces by the component method.

24. Find by (a) the graphical method; and (b) the component method the resultant of the three forces shown.

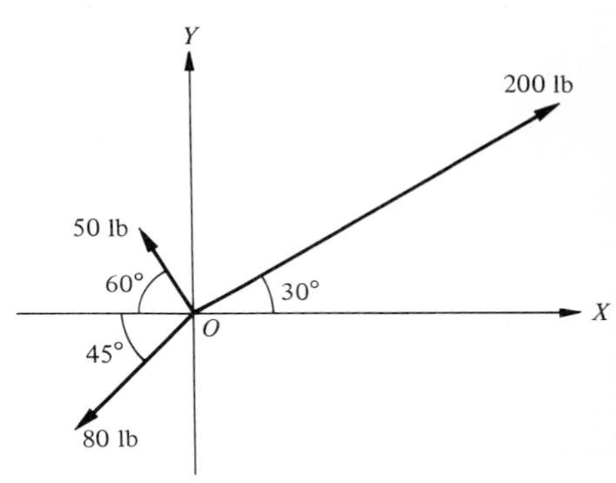

P. 1.24

25. A man is pulling on a box with a force of 80 lb along the $-Y$-axis while a boy is pulling with a force of 40 lb along a direction making an angle of 45° with and above the $-X$-axis. Find the resultant of these two forces and the direction in which the box will move.

26. In Problem 25, in what direction should the man apply the force so that the box will move along the $-X$-axis?

27. A man pulls on a box with a force of 100 lb along the $+X$-axis, while a boy pulls with a force of 80 lb making an angle of 150° with the $+X$-axis. What direction should a second boy apply a force of 60 lb so that the box will move along the $+Y$-axis?

28. If $|\mathbf{A}| = 20$ units and $|\mathbf{B}| = 12$ units, graphically find $\mathbf{A} + \mathbf{B}$ and $\mathbf{A} - \mathbf{B}$.

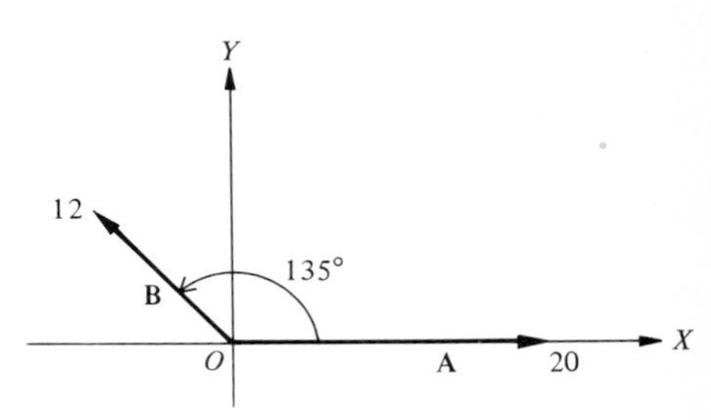

P. 1.28

2 Matter—From Atoms to Galaxies

The development of modern physics—atomic, molecular, solid, nuclear, and particle physics—has taken place mostly through the investigations of atomic and nuclear models, radioactivity, and nuclear reactions. In order to have a better appreciation of the interrelation and interdependence between classical and modern physics, we should have some exposure to these fields.

2.1. Physics, Man, and Matter

ALL KNOWLEDGE and investigations of physics started with man's sense experiences with his immediate surroundings. From there, physics departed and expanded in both directions in size, mass, and time scales—from very large to very small objects, from very massive objects to objects of almost zero mass, and from everlasting to very short lived objects. We shall illustrate these points.

Cosmology involves the study of very large objects such as galaxies and the universe with sizes $\sim 10^{20}$ m and $\sim 10^{26}$ m, respectively. The opposite extreme is the study of particle physics, which deals with particles as small as and smaller than 10^{-15} m. Thus, there is a range of sizes from 10^{26} m to 10^{-15} m, that is, a factor of 10^{41} between the two extremes. For comparison purposes the structure of the world in terms of size, mass, and time scales is illustrated in Tables 2.1 and 2.2.

Let us now compare the mass of a person, which is typically ~ 80 kg, to other objects. The mass of a typical planet may be as high as 10^{27} kg, while, on the other extreme, the mass of a typical elementary (or subatomic) particle may be as small as $\sim 10^{-30}$ kg. Once again, there is a factor of $\sim 10^{57}$ or more between the two extremes.

Let us now consider the time scale. The age of the earth is supposed to be $\sim 1.3 \times 10^{17}$ s ($\approx 4.1 \times 10^{9}$ yr), while the lifetime of a typical elementary particle may be as small as $\sim 10^{-10}$ s ($\approx 3.2 \times 10^{-18}$ yr). Thus, a span of the time scale between the two extremes is $\sim 10^{27}$.

Still another interesting feature is the temperature scale. The highest temperature is the temperature of the interior of the stars, $\sim 10^{7}$ K (degrees Kelvin) while the lowest temperatures that have been achieved in the laboratory are $\sim 10^{-5}$ K. There is a range of 10^{12} between the two extremes.

We might be tempted to think that since all knowledge started from man, the structure of man must have been studied in much detail and must be well known. This is not true. Living things are among the most complex of all and least understood, even though much more time and effort have been spent in their study as compared to the study of astronomy and cosmology, on the one hand, and particle physics, on the other.

Substances on this earth and the universe take one of three forms of matter—solids, liquids, and gases. There have always been two foremost priorities of physics in its vast investigations: (1) to find the basic building blocks of all matter; and (2) to investigate the nature of forces that hold different constitu-

TABLE 2.1
Extreme Range of Size and Mass

Object	Size	Mass	Field
Size of universe	$\sim 10^{26}$ m	$\sim 10^{51}$ kg	Cosmology
Galaxy	$\sim 10^{20}$ m	$\sim 10^{41}$ kg	Astronomy
Planets and stars	$\sim 10^{7}$–10^{12} m	$\sim 10^{27}$–10^{32} kg	Astrophysics
Plants and animals	$\sim 10^{-7}$–10^{2} m	~ 0.01–10^{3} kg	Biology
Man	~ 1.8 m	~ 80 kg	
Giant molecules	$\sim 10^{-7}$ m	$< 10^{-5}$ kg	Biochemistry
Molecules	$\sim 10^{-9}$ m	$\sim 10^{-25}$ kg	Chemistry
Atom	$\sim 10^{-10}$ m	$\sim 10^{-26}$ kg	Atomic physics
Nucleus	$\sim 10^{-14}$ m	$\sim 10^{-26}$ kg	Nuclear physics
Elementary particles	$\leqslant 10^{-15}$	$\leqslant 10^{-27}$ kg	Particle physics

ents of matter together. Before starting a systematic study of physics, we shall devote this chapter to some special topics of modern physics. The discussion in this chapter will help us to incorporate parts of modern physics with our study of classical physics and foresee the connection between them. Terms such as mass, force, charge, and so on, will be formally defined later, but for the time being we shall make use of them, assuming that the reader has already heard of them and has some understanding of such terms.

One quantity that describes both mass and size of an object is density. *Density*, ρ, is defined as mass per unit volume. Thus, if an object of mass M has volume V, its density is given by

$$\boxed{\rho = \frac{M}{V}} \tag{2.1}$$

The units of density are kilogram/meter^3. As with mass, length and time, density also has extreme variation. Typical densities may vary from 1 kg/m^3 ($= 10^{-3}$ g/cm^3) for gases, $\sim 10^3$ kg/m^3 ($= 1$ g/cm^3) for liquids, ~ 2 to

TABLE 2.2
Extreme Range of Time Scale

System	Seconds	Years[a]
Age of earth	$\sim 1.3 \times 10^{17}$	$\sim 4 \times 10^{9}$
Time man has been on earth	$\sim 1.6 \times 10^{13}$	$\sim 10^{5}$
Typical mean life of long-lived radioactive atom	$\sim 5 \times 10^{10}$	$\sim 1.6 \times 10^{3}$
Average life of man	$\sim 2.2 \times 10^{9}$	~ 60
Time period of earth around sun (1 yr)	3.16×10^{7}	1
Period of heart beat	~ 1	$\sim 3 \times 10^{-8}$
Typical period of sound oscillations	$\sim 5 \times 10^{-5}$	$\sim 1.6 \times 10^{-12}$
Typical time of rotation of molecule	$\sim 10^{-12}$	$\sim 3 \times 10^{-20}$
Typical mean life of an elementary particle	$\sim 10^{-16}$	$\sim 3 \times 10^{-24}$

[a] 1 year = 3.16×10^{7} s.

$20 \times 20^3\ \text{kg/m}^3$ ($= 2$ to $20\ \text{g/cm}^3$) for solids, and $2 \times 10^{17}\ \text{kg/m}^3$ ($= 2 \times 10^{14}\ \text{g/cm}^3$) for nuclear material.

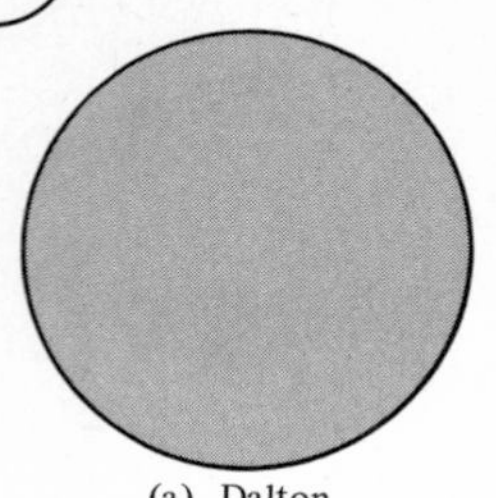

(a) Dalton

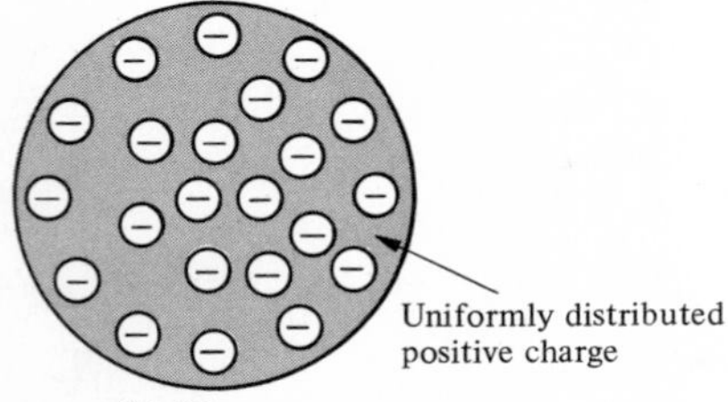

(b) Thomson

FIGURE 2.1. *(a) Dalton's atomic model and (b) the plum-pudding model of the atom as suggested by J. J. Thomson.*

2.2. Inner Mysteries of Atoms and Nuclei

Dalton's Atomic Theory

The basic desire of scientists to investigate the structure of matter has led to the formulation of many different models of the atom, as shown in Figures 2.1 and 2.3–2.5. According to Dalton's *atomic theory of matter* (1808), the atom was assumed to be the smallest indivisible constituent of all matter. The word *atom* is derived from the Greek word *atomos,* meaning indivisible. It was assumed that the atom had no structure. It was known at that time that there were some 80 different kinds of atoms, each corresponding to different elements. Furthermore, they all had different masses. It was also known that atoms combined together to form molecules, which in turn were the building blocks of all matter. Nobody seriously questioned Dalton's theory until about the beginning of the twentieth century.

The Plum-Pudding Model of the Atom

As late as 1897 the atom was still considered to be an indivisible unit. The discovery of the electron by J. J. Thomson in that year led to the speculation that the atom might be made up of positive and negative charges. By using cathode-ray apparatus, he showed that electrons were constituents of all atoms and that their mass was only a small fraction of the mass of any atom. Actually, the mass of an electron was found to be only 1/1837 of the mass of the hydrogen atom. Furthermore, it was found that all electrons carried a negative charge of the same magnitude, $-e$.

Because most matter is electrically neutral, and all atoms contain electrons, they must also contain a requisite number of positively charged particles, called *protons,* to make the atoms neutral. The charge on the proton is one positive unit, $+e$. The mass of a proton was found to be very slightly smaller than the mass of a hydrogen atom and is 1836 times the mass of an electron. Since the mass of electrons is almost negligible, protons must provide most of the mass of an atom. It was also known that the radius of an atom was $\sim 10^{-10}$ m, while the radius of an electron (from theoretical calculations) was $\sim 10^{-15}$ m.

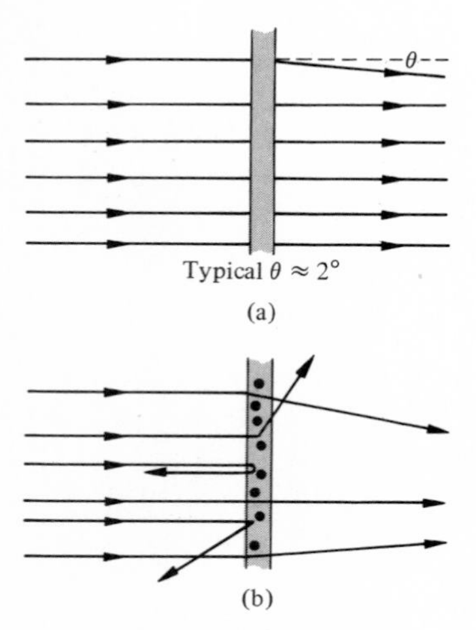

FIGURE 2.2. *Scattering of α-particles by thin metal foils (a) as predicted by the plum-pudding model and (b) as observed experimentally.*

In 1907, J. J. Thomson proposed a model of the structure of the atom. This model is called the *plum-pudding model* and is shown in Figure 2.1(b). According to this model, the positive charge of an atom, and hence the mass, is uniformly distributed over a sphere of radius $\sim 10^{-10}$ m with the electrons embedded in it as shown.

This plum-pudding model failed a crucial test in 1909, when H. Geiger and E. Marsden reported the results of their experiments on scattering (deviation from an incident path) of alpha particles (heavy charged particles, as we shall explain shortly) from a thin foil of metal such as silver and gold. According to the plum-pudding model, alpha particles should scatter by no more than a couple of degrees, as shown in Figure 2.2(a), while experimentally some are observed to scatter by 90° or more, as shown in Figure 2.2(b). Thus, the plum-pudding model of the atom was rejected in favor of a nuclear model of the atom suggested by Rutherford.

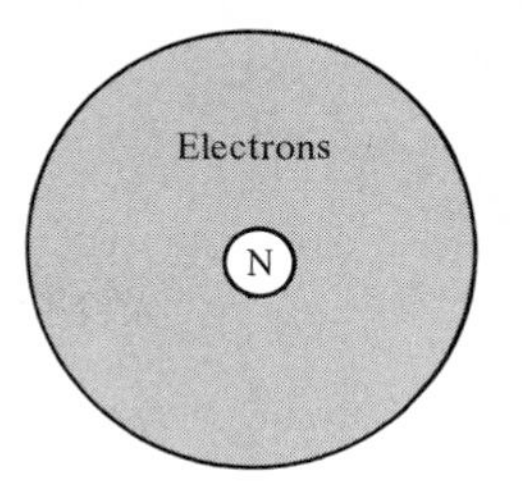

Figure 2.3. *Rutherford nuclear model of the atom.*

Rutherford Nuclear Model of the Atom

In 1911, Ernest Rutherford suggested a simple model of the atom that satisfactorily explained the experimental results of alpha-particle scattering. According to this model, as shown in Figure 2.3: *All positive charges of an atom and hence almost all the mass is assumed to be concentrated in a small volume at the center of the atom, called the nucleus.*

The size of the nucleus is $\sim 10^{-14}$ m, while the electrons are assumed to be distributed around the nucleus in such a way as to account for the radius of an atom being $\sim 10^{-10}$ m. Because electrons are almost point particles of negligible mass, the space around the nucleus in an atom is almost empty, while the mass of an atom is mostly concentrated in the nucleus.

Bohr Model of the Atom

Rutherford's nuclear model of the atom does not say anything about the arrangement of the electrons in an atom. An answer to this problem was given by Niels Bohr in 1913. Bohr suggested the planetary model of the atom. Just as the planets go around the sun, the electrons in an atom go around the nucleus in certain definite circular orbits, as shown in Figure 2.4. No electron can exist

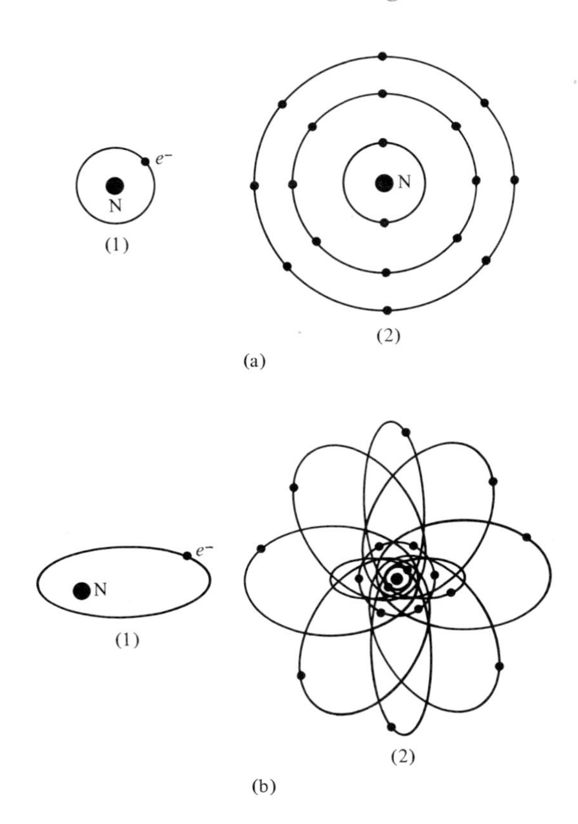

Figure 2.4. *(a) Bohr model and (b) Bohr-Sommerfeld model for the structure of (1) a hydrogen atom and (2) an argon atom.*

between any two orbits. As we shall see later, the planets are held in their orbits by the gravitational force between the sun and the planets, while the electrons are held in their orbits due to the electrical force between the electrons and the nucleus.

This Bohr model of the arrangement of the electrons in atoms [as shown in Figure 2.4(a)] was improved by A. Sommerfeld [as shown in Figure 2.4(b)], who assumed that the electrons could also move in elliptical orbits. These models were quite successful in explaining many experimentally observed facts, such as observed spectrum lines, ionization energies, and chemical behavior of different elements, as will be discussed in later chapters. However, the Bohr model had to yield to still another improved model—the quantum-mechanical model.

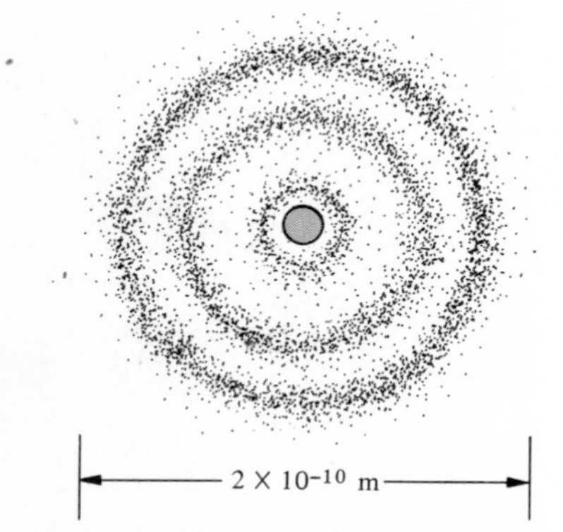

Figure 2.5. *Quantum-mechanical model of an atom.*

Quantum-Mechanical Model

According to the *quantum-mechanical model*, the electrons in an atom do not move around the nucleus in definite orbits, but are free to move anywhere in the atom. In quantum mechanics we talk in terms of probabilities of things to happen or probabilities of finding them at certain locations at given times. Thus, we would ask: What is the probability of finding an electron at different distances from a nucleus? Where the probability is large, the electron is most likely to be found; and where the probability is small, it is less likely to be found. Figure 2.5 shows the results of quantum-mechanical calculations for the probability of finding the electron around the nucleus. According to this figure, where the density of the points is large, the probability of finding the electron is large; where the density is small, the probability is also small.

One essential difference between the quantum-mechanical model and the Bohr model is clear. Even though the probability of finding the electron is high near the Bohr orbit radii in the quantum-mechanical model, the probability is not zero between these orbits. As a matter of fact, we may say that the Bohr model is a special case of the quantum-mechanical model. The Bohr model is much easier to visualize and hence is still frequently used in many situations. For much more accurate and correct predictions, the quantum-mechanical model must be used until we find a still better model.

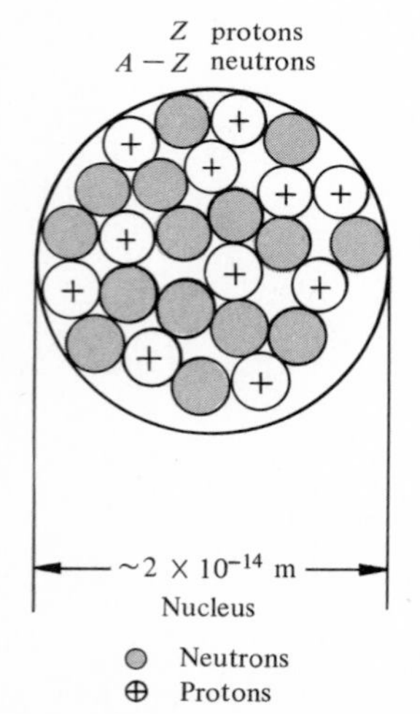

Figure 2.6. *Neutron–proton model of the nucleus.*

Neutron and Neutron–Proton Model of the Nucleus

The neutron was discovered by James Chadwick (1891–1974) in 1932. It is a neutral particle with zero net charge, and its mass is slightly larger than the mass of the proton. The discovery of the neutron led to the establishment of the neutron–proton model of the nucleus. Accordingly, when we say that an atom has mass number A and atomic number Z, we mean the following.

The nucleus of an atom consists of Z protons and N neutrons, so that $N + Z = A$, and the nucleus is surrounded by Z electrons. Thus, the *mass number* is the number of protons plus neutrons in the nucleus, while the *atomic number* Z is the number of protons in the nucleus or the number of electrons in the neutral atom (see Figure 2.6).

An atom of any element is denoted by ${}^{A}_{Z}X$, where X is the symbol of the element. Thus, ${}^{235}_{92}U$ means that a uranium nucleus of mass number 235 has 92 protons and (235 − 92 =) 143 neutrons, while there are 92 electrons surrounding the nucleus. Similar meaning holds for ${}^{16}_{8}O$, ${}^{32}_{16}S$, ${}^{12}_{6}C$, ${}^{56}_{26}Fe$, and so on. Elements are distinguished by having different Z values for their atoms. Atoms of an element which have the same Z, but different N are called *isotopes*. For example, oxygen has 3 isotopes ($^{16}_{8}O$, $^{17}_{8}O$ and $^{18}_{8}O$), while zinc has 5 isotopes ($^{64}_{30}Zn$, $^{66}_{30}Zn$, $^{67}_{30}Zn$, $^{68}_{30}Zn$ and $^{70}_{30}Zn$). There are about 100 different known elements and ~1500 different known isotopes; that is, there are about 1500 different types of nuclei that are known.

Unlike the atomic radius, the radius of a nucleus varies with A and is given by an expression

$$\boxed{R = (1.35 \times 10^{-15})A^{1/3}\ \mathrm{m}} \tag{2.2}$$

where A is the mass number. Thus the size of the nucleus depends upon A and for a typical case of $A = 150$ is $\sim 10^{-14}$ m. On the other hand, the size of the atom is almost constant $\sim(1 - 3) \times 10^{-10}$ m.

Atomic Masses

The masses of the electrons, protons, neutrons, atoms, and nuclei are so small that it is not convenient to express them in kg or g. In 1960 the General Conference on Weights and Measures introduced the ***unified mass unit*** (u), in terms of which the masses of different particles are given. Accordingly *one unified mass unit u is equal to one-twelfth of the mass of a neutral carbon 12* ($^{12}_{6}C$) *atom.*

By knowing the definition of a mole and Avogadro's number, we can find the relation between 1 u and a kg: 1 *mole* (1 kmole) of a substance is that amount which has a mass in grams (kilograms) equal to its atomic weight [or whose mass in grams (kilograms) equals the molecular weight]. One mole of any substance contains the same number of atoms (or molecules). This number is called *Avogadro's number*, N_A, and is equal to 6.02252×10^{23}/mole (6.02252×10^{26}/kmole). Thus, 1 mole (1 kmole) of carbon means that 12 g (12 kg) of atomic carbon contains Avogadro's number N_A of atoms. Therefore, the mass of 1 carbon atom is 12 g/N_A, and hence 1 u is

$$1\ \mathrm{u} = \frac{1}{12}\frac{12\ \mathrm{g}}{N_A} = \frac{1}{12}\frac{12\ \mathrm{g}}{6.02252 \times 10^{23}} = 1.66043 \times 10^{-24}\ \mathrm{g}$$

or

$$\boxed{1\ \mathrm{u} = 1.66 \times 10^{-27}\ \mathrm{kg}} \tag{2.3}$$

The mass of the electron, proton, and neutron, together with their charges, are given in Table 2.3.

TABLE 2.3
Properties of Protons, Neutrons, and Electrons

Name	Symbol	Charge	Mass
Proton	p	$+1e$	1.00759 u[a]
Neutron	n	$0e$	1.00898 u
Electron	e	$-1e$	0.00055 u

[a] $1\ \mathrm{u} = 1.66043 \times 10^{-27}$ kg.

These atoms and nuclei are so small that it has not been possible to observe them directly. All their dimensions are deduced from indirect experimental observations. The first person who came close to photographing individual atoms was E. W. Müller. With a field-ion microscope, Müller obtained images on a florescent screen in which individual atoms can be distinguished. The photograph in Figure 2.7 shows images of tungsten atoms. Recently, A. V. Crewe of the University of Chicago has also been successful in obtaining images of individual atoms.

Atoms combine together to form *molecules*. If the atoms and/or molecules arrange themselves in a symmetrical repeated pattern, it results in the formation of *crystals*. The crystal structures of various elements and compounds differ in two respects: (1) the arrangement of the atoms, and (2) the spacing between the atoms. Figure 2.8 shows a few examples of crystal structures. It is quite clear that differrent crystals will have different densities, defined as mass per unit volume. It is also clear that the building blocks of all matter are

FIGURE 2.7. *With a field-ion microscope E. W. Mueller obtained an image on a fluorescent screen of a single atom of tungsten metal magnified more than 2 million times.* (Courtesy of Erwin W. Mueller, Evan-Pugh Professor Emeritus, The Pennsylvania State University.)

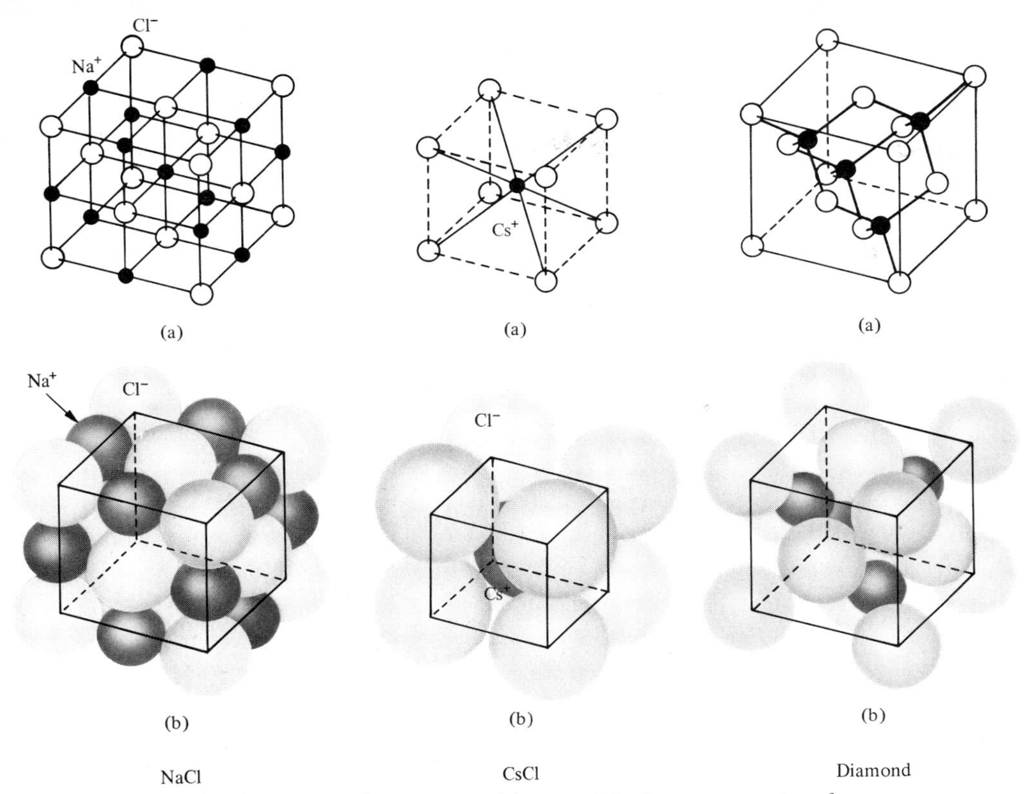

FIGURE 2.8. *Some crystal structures: (a) geometrical arrangement and (b) scale model.*

atoms, molecules, and crystals or, in other words, electrons, protons, and neutrons.

The density of nuclear matter is

$$\rho = 2 \times 10^{17}\ \mathrm{kg/m^3} = 2 \times 10^{14}\ \mathrm{g/cm^3} \simeq 4 \times 10^{9}\ \mathrm{tons/in^3} \qquad (2.4)$$

which is tremendously high as compared to the density between 1 and 20 $\mathrm{g/cm^3}$ for most solids. Of course, such densities are not found on earth except in nuclei. In outer space, certain stars, such as white dwarfs, are known to have densities as high as $\sim 1\ \mathrm{ton/in^3}$ ($5 \times 10^4\ \mathrm{g/cm^3} = 5 \times 10^7\ \mathrm{kg/m^3}$). These high densities have resulted from the collapse of the matter in these stars. Can you imagine what would happen if most of the space in atoms were not empty? The earth would reduce to a much smaller size. If the nuclei and electrons of all the atoms on this earth are packed together without leaving any empty space in atoms, it would result in a sphere of radius ~ 200 m!

EXAMPLE 2.1 The mass of a proton is $\sim 1\ \mathrm{u} = 1.66 \times 10^{-27}$ kg, while it is considered to be a sphere of radius $\sim 1.35 \times 10^{-15}$ m. Calculate the density of nuclear matter.

The density ρ of proton matter, or general nuclear matter, is ρ = mass/volume = M/V, where M is the mass of the proton and its volume $V = \frac{4}{3}\pi R^3$, R being the radius of the proton. Therefore,

$$\rho = \frac{M}{\frac{4}{3}\pi R^3} = \frac{1.66 \times 10^{-27}\ \text{kg}}{\frac{4}{3}\pi(1.35 \times 10^{-15}\ \text{m})^3} \simeq 2 \times 10^{17}\ \text{kg/m}^3\ (\simeq 4 \times 10^9\ \text{tons/in}^3)$$

Exercise 2.1 The mass of the earth is 5.98×10^{24} kg and its radius is 6.37×10^6 m. Calculate its density. What would be the density of earth matter if the earth's radius was only 100 m? [*Ans.*: 5.52×10^3 kg/m^3; 1.43×10^{18} kg/m^3.]

2.3. The Ever-Growing List of Elementary Particles

Thus far we have encountered three subatomic particles: the electron, proton, and neutron. These three and another particle, called a *photon* (discussed below), were the only *four particles* known in the early 1930s. The particles were called the *fundamental* or *elementary particles*. In subsequent years, the list of elementary particles grew at a much faster rate, leading to 14 by 1947, and 32 by 1957. For a while everyone thought that these 32 particles were the ultimate building blocks of all matter. But surprisingly, this list kept on growing, to about 64 by 1965, and does not seem to be stopping yet. The list has grown manyfold, and runs into hundreds.

Hundreds of experiments performed over the last 75 years led to the discovery of the numerous elementary particles. At this point we shall briefly discuss a photon and some other elementary particles.

Electron (e^-), Proton (p), and Neutron (n) These three particles have already been discussed in detail in the last section.

Photon ($h\nu$) In 1905, Einstein suggested that electromagnetic radiation, such as x-rays, gamma rays, radio waves, microwaves, and light waves, must also behave as corpuscles or particles. That is, they carry energy in packets or bundles or quanta, and are usually called photons. Each quantum or photon has an energy $h\nu$, where h is a constant known as Planck's constant and ν is its frequency (or repetition rate of the wave). Frequency ν may have a value of 10^8 repetitions per second for radio waves and 10^{22} for gamma rays. All photons travel at the speed of light in a vacuum (3×10^8 m/s).

Positron (e^+) The positron has the same mass as the electron, but has a charge equal and opposite to the electron, that is, e^+. Actually, the electron and the positron are said to form a particle–antiparticle pair, the electron being the particle and the positron being its antiparticle. (A positron is also called an *antielectron* and is denoted by $\overline{e^-}$, where the bar denotes an antiparticle.)

Neutrinos (ν) These particles have almost zero mass when at rest, no charge, and move with speeds almost equal to the speed of light. Their counterparts are called *antineutrinos* ($\overline{\nu}$).

The names of some of the other particles are the antiproton, antineutron, sigma, lambda, and xi. A group of particles called *mesons* include the mu mesons, pi mesons, and K mesons.

Not all the atoms occurring in nature are stable. As we shall discuss shortly many of them are unstable and frequently give out one or two particles. Also, new stable and unstable atoms (with Z even greater than 100) and the elementary particles can be produced in the laboratories by means of nuclear reactions.

*2.4. Radioactivity

Like many other discoveries, radioactivity was discovered accidentally by the French physicist Henri Becquerel in 1896. He was studying a compound of uranium that emits radiation when activated by sunlight. Once, after carefully wrapping the uranium compound in thick paper, he left it near photographic plates in a dark room. When the plates were developed, they showed images of uranium crystals. He repeated this experiment and came to the conclusion that some sort of radiation that affected the photograph plate (even after passing through the wrapping) was coming out of the uranium compound. Further investigations by Becquerel, Marie Curie, Frederich Soddy, and many others revealed that many other isotopes with $A > 210$ emitted such radiation. This phenomenon is called *radioactivity*, while the process itself is called *radioactive decay*. It was found that it is the nucleus that emits such radiation, and such elements are called *radioactive elements* or *radioactive nuclides*. The original radioactive nucleus is called the *parent nucleus* and the residual nucleus is called the *daughter nucleus*.

Further investigation of radioactivity revealed that the radiation emitted consisted of three different components or types: (1) alpha rays (α-rays), (2) beta rays (β-rays), and (3) gamma rays (γ-rays). One of the distinguishing factors is their penetrating power. For example, α-rays can be stopped by a thin sheet of paper; β-rays can penetrate a thin sheet of aluminum, but a few millimeters of aluminum will stop them; while low-energy γ-rays can be stopped only by thin sheets of lead, say $\frac{1}{10}$ in thick. That is, gamma rays are $\sim$100 times more penetrating than β-rays, while beta rays are $\sim$100 times more penetrating than α-rays. It takes a lead sheet several inches thick to stop γ-rays of high energies.

Alpha Rays (α) In the process of α-decay, the parent nucleus, say ${}^{A}_{Z}X$, which is unstable, decays suddenly and spontaneously into a daughter nucleus, ${}^{A-4}_{Z-2}Y$, and an α-particle (${}^{4}_{2}He$), that is,

$$\boxed{{}^{A}_{Z}X \longrightarrow {}^{A-4}_{Z-2}Y + {}^{4}_{2}He \quad (\alpha\text{-ray})} \tag{2.5}$$

where ${}^{4}_{2}He$ is a helium nucleus consisting of two protons and two neutrons. An atom that has lost its two electrons is said to be doubly ionized. Thus, we may say that α-particles are fast-moving, doubly ionized helium atoms. Note that both sides of the equation have the same total value of Z; that is, the charge is unchanged or is conserved during the radioactive decay process. Also, the mass number remains unchanged during this process. If we add the masses of the daughter nucleus and the α-particle, it is found to be slightly less than the mass of the parent. This difference in the mass appears in the form of energy of the daughter and the α-particle. Most α-particles are emitted with velocities between $\sim 1.5 \times 10^7$ m/s and $\sim 2.2 \times 10^7$ m/s. Two typical examples of alpha decay are

$$^{238}_{92}U \longrightarrow {}^{234}_{90}Th + {}^{4}_{2}He \tag{2.6}$$

$$^{222}_{86}Rn \longrightarrow {}^{218}_{84}Po + {}^{4}_{2}He \tag{2.7}$$

where U, Th, Rn, and Po are uranium, thorium, radon, and polonium, respectively.

Beta Rays (β^-) It has been shown that beta particles (represented by β^- or $_{-1}^{0}e$) are fast-moving electrons. Thus, when an unstable parent nucleus, say $^{A}_{Z}X$, decays by beta emission, the daughter nucleus is $_{Z+1}^{\ A}Y$, that is,

$$\boxed{^{A}_{Z}X \longrightarrow {}_{Z+1}^{\ A}Y + {}_{-1}^{0}e(\beta^-) + \bar{\nu}} \tag{2.8}$$

To conserve charge, one neutron is converted into a proton with simultaneous emission of a fast-moving electron and an antineutrino. Note that β^- and $\bar{\nu}$ do not exist inside the nucleus; they are created at the time of emission. Once again the mass of the daughter, beta, and antineutrino together is less than that of the parent. This difference in mass appears in the form of energy. The velocities of the β-particles emitted are up to 2.97×10^8 m/s. Typical examples of beta decay are

$$^{60}_{27}Co \longrightarrow {}^{60}_{28}Ni + {}_{-1}^{0}e + \bar{\nu} \tag{2.9}$$

$$^{64}_{29}Cu \longrightarrow {}^{64}_{30}Zn + {}_{-1}^{0}e + \bar{\nu} \tag{2.10}$$

Gamma Rays (γ) Gamma rays are quanta or packets of electromagnetic radiation energy called *photons*, as explained earlier. They do not have any charge or rest mass. Hence, in gamma decay the parent and the daughter have the same charge and the same mass number. The actual mass of the daughter is always less than the parent. This mass difference appears as the energy of the photon. The unstable parent nucleus, which gives γ-rays, is denoted by an asterisk on the upper right-hand corner and is called the *excited nucleus*. Thus,

$$\boxed{^{A}_{Z}X^* \longrightarrow {}^{A}_{Z}X + \gamma} \tag{2.11}$$

As mentioned earlier, the photons move with the speed of light (3×10^8 m/s). A few typical examples of γ-decay are the following:

$$\begin{aligned} ^{60}_{28}Ni^* &\longrightarrow {}^{60}_{28}Ni + \gamma \\ ^{137}_{56}Ba^* &\longrightarrow {}^{137}_{56}Ba + \gamma \\ ^{222}_{86}Rn^* &\longrightarrow {}^{222}_{86}Rn + \gamma \end{aligned} \tag{2.12}$$

*2.5. Nuclear Reactions

Suppose that a beam of fast-moving particles is incident on a target of some material. Depending upon the type of particle (say, a proton, neutron, electron, etc.) and its energy, it can come close to the target nucleus. The interaction between the particle and the nucleus can result in many different things. The incident particle may (1) be captured by the nucleus, (2) knock out another particle from the nucleus, (3) simply change its direction, or (4) lose some energy. In many cases γ-rays may also be emitted. The residual nuclei are usually different from the target nuclei. The nuclei may change in atomic number or mass number or both. Such a process of interaction between the incident particle and the target nucleus is called a *nuclear reaction*, and the change in the target nucleus is called a *transmutation*.

Suppose that a particle x is incident on a target nucleus X, which after interaction results in a recoil nucleus Y and an outgoing particle y. Such a nuclear reaction may be represented by a nuclear equation of the form

$$x + X \longrightarrow Y + y \tag{2.13}$$

In short, we may write a nuclear reaction also as $X(x, y)Y$. A correct nuclear reaction equation must have the same number of neutrons and protons on the left side as on the right side; that is, it must be balanced.

The first nuclear reaction was investigated by Ernest Rutherford in 1919 and is represented by

$${}^{4}_{2}\text{He} + {}^{14}_{7}\text{N} \longrightarrow {}^{17}_{8}\text{O} + {}^{1}_{1}\text{H} \tag{2.14}$$

where ${}^{4}_{2}$He are the α-particles obtained from a radioactive source, while ${}^{1}_{1}$H is the outgoing proton. The first reaction using protons which were accelerated in the laboratory was performed in 1930 by John Cockcroft and Ernest Walton and is represented by

$$p + {}^{7}_{3}\text{Li} \longrightarrow {}^{4}_{2}\text{He} + \alpha \tag{2.15}$$

The two equations above may also be written as ${}^{14}_{7}\text{N}(\alpha, p){}^{17}_{8}\text{O}$ and ${}^{7}_{3}\text{Li}(p, \alpha){}^{4}_{2}\text{He}$. The recoil nucleus Y in most cases is unstable, hence radioactive. Thus, we have a method of producing artificially radioactive nuclei even though they may not exist naturally. Some examples of reactions are ${}^{10}_{5}\text{B}(\alpha, p){}^{13}_{6}\text{C}$, ${}^{20}_{10}\text{Ne}(n, \alpha){}^{17}_{8}\text{O}$, ${}^{28}_{14}\text{Si}(\gamma, n){}^{27}_{14}\text{Si}$, ${}^{14}_{7}\text{N}(n, p){}^{14}_{6}\text{C}$, and ${}^{35}_{17}\text{Cl}(n, \gamma){}^{36}_{17}\text{Cl}$.

SUMMARY

The extreme size, mass, and time scales observed are $\sim 10^{-15}$ to $\sim 10^{26}$ m, $\sim 10^{-27}$ to $\sim 10^{32}$ kg, and $\sim 10^{-16}$ to $\sim 10^{17}$ s. The *density* of any material is defined as $\rho = M/V$.

The presently accepted model of the atom is the *nuclear model* of the atom. The nucleus located at the center of an atom occupies a very small fraction of the total atomic volume. The atomic size is constant $\sim(1 - 3) \times 10^{-10}$ m, while the size of the nucleus is given by $R = 1.35 \times 10^{-15} A^{1/3}$ m. The nucleus consists of Z protons and N neutrons, so that $Z + N = A$ while Z electrons surrounding the nucleus make up the atom.

One *mole* of any substance contains N_A $(= 6.02 \times 10^{23})$ atoms or molecules. One *unified mass unit*, u, is defined to be one-twelfth the mass of a neutral carbon atom and is equal to 1.66×10^{-27} kg.

A *photon* is a quantum or packet of energy moving with the speed of light.

* The phenomenon of emission of radiation from unstable nuclei is called *radioactivity*. The most common types of radiation emitted are α-, β-, and γ-rays.

* Whenever a fast-moving incident particle interacts with a target nucleus, the nuclear reaction represented by $x + X \longrightarrow Y + y$ or $X(x, y)Y$ takes place.

QUESTIONS

1. Compare the mass, height, and average life of a man with that of the two extremes discussed in the text.
2. List the year and the names of the scientists who proposed various classical models of the atom.
3. According to the planetary model of the solar system, planets go around the sun in elliptical orbits. Which model of the atom is similar to this?
4. Compare the volume of the nucleus with that of the atom. What do you conclude about the density of the material in the atom surrounding the nucleus (excluding the nucleus)?
5. What is the difference between the Bohr model and the quantum mechanical model of the atom?
6. Consider an atom of mass number A. What is the ratio of its nuclear radius to its atomic radius?
7. What is the difference between $^{235}_{92}U$ and $^{238}_{92}U$?
8. From comparing the properties of the electron and positron, what can you say about the antiproton, which is the antiparticle of the proton?

*9. How can you tell if a given radioactive source is emitting α-, β-, or γ-rays? You are given an instrument which can detect all types of radiation indiscriminately.

*10. Is it possible to produce gold by starting from a cheaper metal and using a nuclear reaction?

*11. List the names of the scientists who initially contributed to the following fields: (a) atomic models; (b) radioactivity; and (c) nuclear reactions.

PROBLEMS

1. How many protons should be placed side by side to make a cube that is 1 mm on each side? What will be the mass of this cube? The density of nuclear matter is 2×10^{17} kg/m^3.
2. How many hydrogen atoms (mass $= 1.67 \times 10^{-27}$ kg and radius $= 10^{-10}$ m) should be placed side by side to form a cube 1 mm on each side? What will be the mass and density of this cube?
3. Repeat Problem 2 assuming that the hydrogen atoms are placed 3×10^{-10} m apart.
4. What would be the radius of the earth if its size were reduced so that its density is the same as that of nuclear matter?
5. Pulsars are very dense stars. Suppose that the mass of a particular pulsar is equal to the mass of the sun ($= 2 \times 10^{30}$ kg). If the density of pulsar matter is 2×10^{6} kg/m^3, what is the volume and the radius of this star?
6. How many protons, neutrons, and electrons are there in an atom of $^{107}_{47}Ag$? Calculate the ratio of the mass of all the electrons to the total mass of the protons and neutrons. Can you say that for all practical purposes the mass of an atom is almost equal to the mass of the nucleus?
7. The nuclear radius is given by $R = 1.35 \times 10^{-15} A^{1/3}$ m. Calculate the radii and volumes for nuclei with $A = 27$, 64, 125, and 216.
8. All atoms have nearly constant volume and have a diameter of $\sim 10^{-10}$ m.

What fraction of the total volume is occupied by nuclei with $A = 27, 64, 125$, and 216?

9. The atomic mass of $^{16}_{8}O$ is 15.994915 u. Calculate the difference between the atomic mass of oxygen and its constituents, which are 8 protons, 8 neutrons, and 8 electrons. Calculate this mass difference in u, kg, and g.

10. Calculate the difference in the atomic mass of $^{238}_{92}U$ and its constituents: protons, neutrons, and electrons. The atomic mass of ^{238}U is 238.048608 u.

***11.** The following isotopes decay by emitting α-particles: $^{238}_{92}U$, $^{240}_{94}Pu$, and $^{212}_{84}Po$. (a) Write equations representing these decays; and (b) identify the daughter nuclei produced.

***12.** The end product of a certain nucleus decaying by α-emission is the daughter nucleus radium of mass number 224. Identify the parent.

***13.** The following nuclei decay by β^--emission: ^{32}P, ^{137}Cs, and ^{212}Bi. (a) Write equations describing these decays and (b) identify the daughter nuclei.

***14.** Iodine-131 decays by β^--emission. The resulting daughter nucleus decays by γ-emission. Write equations describing these decays and identify the daughters in each case.

***15.** Cobalt-60 decays by β^--emission. The resulting daughter nucleus decays by γ-emissions. Write equations describing these decays and identify the daughters in each case.

***16.** The daughter nucleus resulting from the α-decay of a parent isotope is $^{208}_{82}Pb$. Name the parent isotope.

***17.** The daughter nucleus resulting from the β^--decay of a parent isotope is $^{32}_{16}S$. Name the parent isotope.

***18.** When $^{59}_{27}Co$ is bombarded with neutrons, it absorbs the neutrons and changes into Ni with simultaneous emission of γ-rays. Write an equation describing this nuclear reaction.

***19.** Complete and write reaction equations for the following reactions, where d is a deuteron, a heavier isotope of hydrogen whose nucleus has 1 p and 1 n (that is, $^{2}_{1}H$): $^{10}_{5}B(\alpha, p)$, $^{27}_{13}Al(\alpha, d)$, $^{9}_{4}Be(d, n)$, and $^{15}_{7}N(d, n)$.

***20.** An α-particle is absorbed by a $^{60}_{28}Ni$ target. The nuclear reaction results in the emission of a neutron and a proton. Write the nuclear reaction equation.

***21.** A proton is absorbed by a $^{9}_{4}Be$ target. The nuclear reaction emits a neutron. Write the nuclear reaction equation.

Part II
Mechanics

3 Translational Motion

Quantities such as displacement, velocity, and acceleration are defined for objects that move with translational motion. The relation between these quantities and time is derived from basic principles. Special cases of freely falling bodies and projectile motion under the influence of acceleration due to gravity are investigated in detail. In closing we shall briefly describe the meaning of relative velocity.

3.1. Mechanics and Types of Motion

MECHANICS is one of the oldest and most familiar branches of physics. It deals with bodies at rest or in motion, and the conditions of rest or motion when bodies are under the influence of internal or external forces. The laws of mechanics apply to a wide range of objects—microscopic to macroscopic—such as the motion of electrons in atoms and that of planets in space.

The study of mechanics may be divided into kinematics and dynamics. *Kinematics* is concerned with a description of the motion (or trajectories) of objects, disregarding the forces producing the motion. *Dynamics* is concerned with the forces that produce changes in motion or changes in other properties, such as size and shape of objects.

There are three common types of motions that a body may exhibit: (1) *translational motion,* (2) *rotational motion,* and (3) *vibrational motion.* Examples of translational motion include a car moving in a straight line, an object falling vertically or moving on a curved path under the influence of the force of gravity, and an electron moving in a straight path in a television tube. Examples of rotational motion include the earth and other planets moving around the sun, a rotating wheel, a moving merry-go-round, and an electron moving in a circle around a nucleus in an atom. Examples of vibrational motion include a clock pendulum moving back and forth, the motion of a swing, a string in a musical instrument, and some motions of atoms in molecules. These examples are those of pure translational, pure rotational, and pure vibrational motions. There are other situations in which objects have motions that are combinations of two or three of these pure motions.

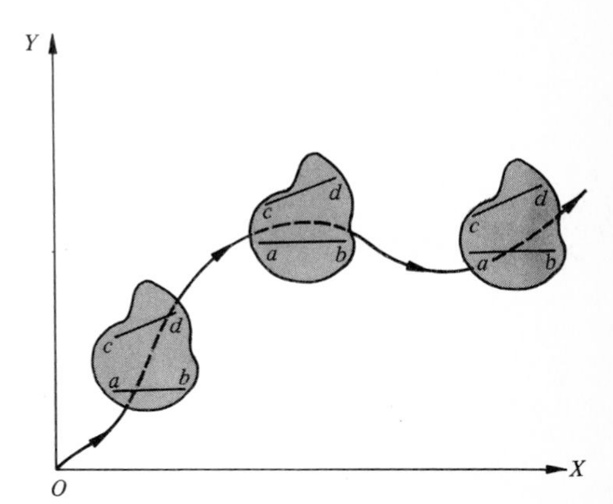

FIGURE 3.1. *Translational motion of an object. Note that even though the x- and y-coordinates change with time, the lines ab and cd on the object remain parallel to their initial position.*

Presently, we shall concentrate on the study of translational motion in some detail. A body is said to have only *translational motion* if any line drawn on the body remains parallel to itself at all times throughout the motion of the body, such as the lines *ab* and *cd* drawn on a body shown in Figure 3.1.

The problem of analyzing motion can be simplified if we consider a mathematical *point object* which is defined as an object that has zero size. Even though all real objects have finite size, they may be treated as point objects under certain conditions. For example, the size of the earth and size of the sun are each very small compared to the distance between them, and hence they may be treated as point objects when considering their relative motion. Also, an object with no rotational motion can be treated as a point object.

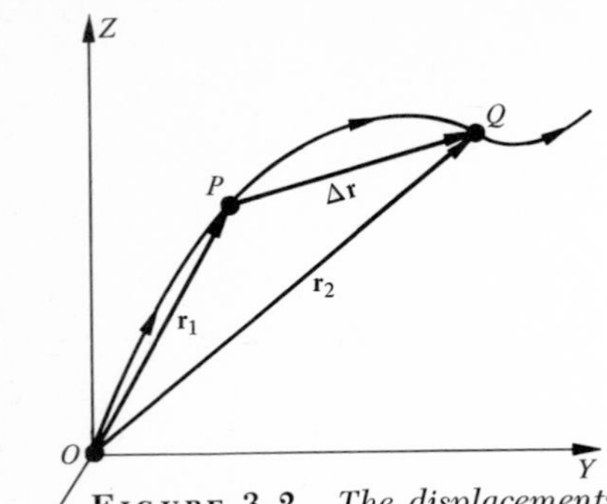

FIGURE 3.2. *The displacements of a particle when it is at P and Q are* $\mathbf{r}_1$ *and* $\mathbf{r}_2$ *from O, while Q is at* $\Delta\mathbf{r}$ *from P.*

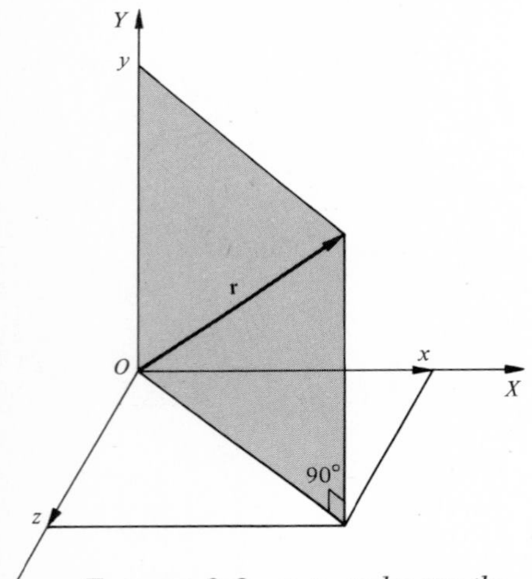

FIGURE 3.3. *x, y, and z are the three components of a displacement vector* **r**.

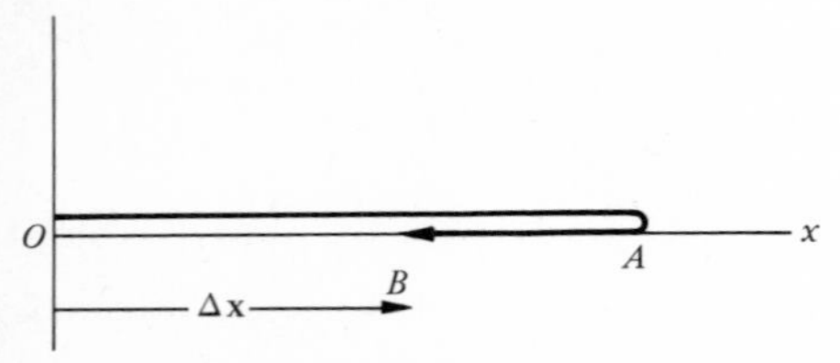

FIGURE 3.4. *The net displacement of an object when it moves from O to A and then to B is* $|\Delta\mathbf{x}| = OB$*, while the total distance traveled is* $\Delta s = OA + AB$.

3.2. Velocity and Acceleration

Velocity

A body is said to be in motion if it changes its position with respect to its surroundings. Consider an object that has been set into translational motion from point O (the origin of the coordinate system XYZ) at time $t = 0$ and follows a curved path shown in Figure 3.2. Let the object be at point P at time t_1 and at point Q at time t_2. The object when at P is said to have a displacement $\mathbf{r}_1$ with respect to the origin O, and when at Q it is said to have a displacement of $\Delta\mathbf{r}$ with respect to P. The displacement is a vector quantity since it has magnitude as well as direction. Any displacement vector **r** can be resolved into three rectangular components x, y, and z, as shown in Figure 3.3. To make matters simple, for the time being, let us restrict our discussion to one-dimensional motion; that is, all displacements are along the X-axis only.

The time rate of change of position of a body is called its *velocity*. Suppose that a body changes its position from $\mathbf{x}_1$ to $\mathbf{x}_2$, resulting in a displacement of $\Delta\mathbf{x} = \mathbf{x}_2 - \mathbf{x}_1$ in time Δt. The average velocity, $\bar{\mathbf{v}}$ (or $\mathbf{v}_{av}$), is defined as

$$\bar{\mathbf{v}} = \frac{\Delta\mathbf{x}}{\Delta t} \tag{3.1}$$

The direction of the average velocity vector is the same as that of the displacement vector. The magnitude of the displacement vector, denoted by $|\Delta\mathbf{x}|$ or Δx may be different from the actual distance Δs which a body may travel. For example, in Figure 3.4, a body travels along the X-axis from O to A and then back to B. The displacement $\Delta\mathbf{x}$ is equal to OB, but the distance Δs traveled by the body is equal to $OA + AB$.

The *average speed* of an object is defined as the ratio of the total distance traveled, Δs, to the time interval, Δt; that is,

$$\text{average speed} = \frac{\Delta s}{\Delta t} \tag{3.2}$$

Note that the speed is a scalar quantity (it has magnitude only), while the average velocity is a vector quantity. For example, in Figure 3.4, if the body returned to its original position, its average velocity will be zero (because $\Delta\mathbf{x} = 0$), while its average speed is definitely not equal to zero (because the total distance traveled is not equal to zero, $\Delta s = OA + AO$). If the body has equal displacements in equal intervals of time, it is said to have *uniform velocity*.

Of more interest is the velocity of a particle at a particular point or at a particular instant of time along its path, called the instantaneous velocity of the particle. We define *instantaneous velocity*, **v**, at a point to be the ratio of $\Delta\mathbf{x}$ to Δt provided that Δt is taken to be so small that it approaches zero, but the interval $\Delta\mathbf{x}$ still includes the point under consideration. We say that in the limit as $\Delta t \to 0$, the instantaneous velocity **v** is written as

$$\mathbf{v} = \lim_{\Delta t \to 0} \frac{\Delta\mathbf{x}}{\Delta t} \tag{3.3}$$

The meaning of this will become clear shortly when we discuss the graphical representation.

The absolute magnitude of the instantaneous velocity without reference to its direction is called the *speed* of the object at that point. For example, an automobile traveling due north at 60 mi/h and another automobile traveling 60 mi/h due south both have the same speeds (= 60 mi/h) but have different velocities. Also, if a body has uniform velocity, both the average velocity and the instantaneous velocity are equal.

When considering only the magnitude of the velocity, we may write $\bar{v} = \Delta x/\Delta t$. The units of these quantities—average velocity, average speed, uniform velocity, instantaneous velocity, and speed—are the same, that is, length/time = L/T, given by m/s, ft/s, mi/h, and so on, in different units.

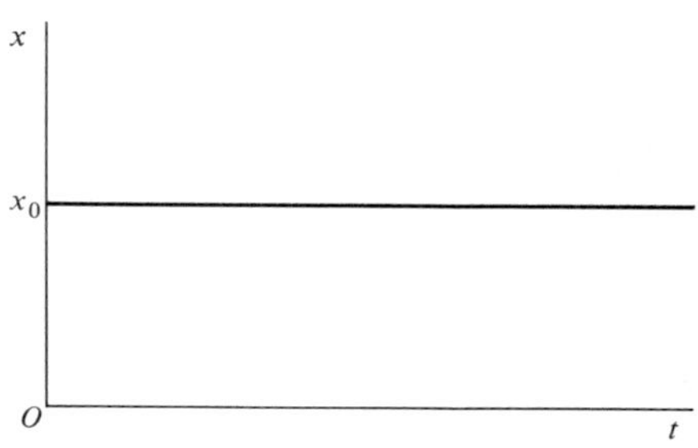

Figure 3.5. *Plot of x versus t for an object at rest at x_0.*

Acceleration

Another quantity commonly used in motion is acceleration—average acceleration and instantaneous acceleration. The *average acceleration*, $\bar{\mathbf{a}}$, is defined as the time rate of change of velocity. That is, if a body has velocity $\mathbf{v}_1$ at time t_1, and has velocity $\mathbf{v}_2$ at time t_2, the change in velocity $\Delta\mathbf{v} = \mathbf{v}_2 - \mathbf{v}_1$ takes place in time $\Delta t = t_2 - t_1$. Thus,

$$\bar{\mathbf{a}} = \frac{\Delta \mathbf{v}}{\Delta t} \tag{3.4}$$

and the direction of the average acceleration vector $\bar{\mathbf{a}}$ is the same as that of $\Delta\mathbf{v}$. The *instantaneous acceleration*, $\mathbf{a}$, at a point may be defined as the ratio of $\Delta\mathbf{v}$ to Δt when in the limit $\Delta t \to$ zero, but $\Delta\mathbf{v}$ still includes the point under consideration. That is,

$$\mathbf{a} = \lim_{\Delta t \to 0} \frac{\Delta \mathbf{v}}{\Delta t} \tag{3.5}$$

The graphical representation of this will be discussed shortly. Like $\bar{\mathbf{a}}$, $\mathbf{a}$ is a vector quantity. When considering only the magnitude, $\bar{a} = \Delta v/\Delta t$. Since the units of velocity are length/time and acceleration is the rate of change of velocity, the units of acceleration will be (1/time) (length/time) = L/T^2, that is, m/s^2, ft/s^2, mi/h^2, and so on. A body is said to have *deceleration* (negative acceleration) if the magnitude of its velocity decreases with time, that is, if the body is slowing down.

Once again, if the body has a uniform acceleration, the instantaneous and average acceleration are equal to the uniform acceleration. In what follows, we shall be concerned only with (1) uniform acceleration, and (2) motion in one dimension.

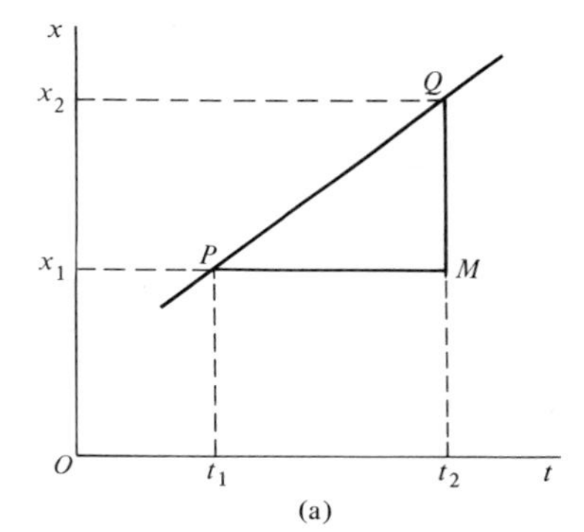

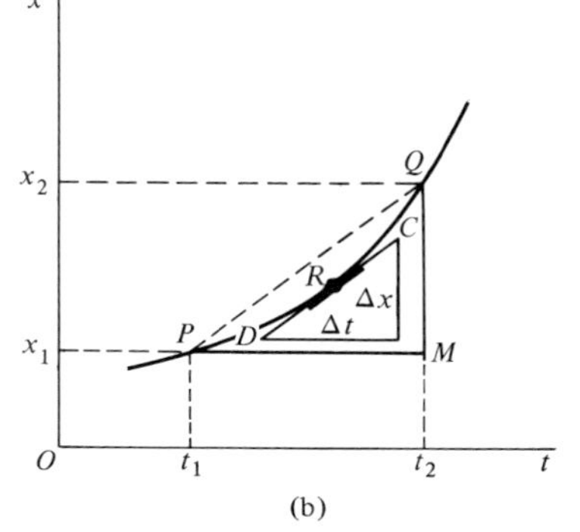

Figure 3.6. *Plot of x versus t for an object moving with (a) uniform velocity (straight line PQ) and (b) variable velocity (curved line PQ).*

Graphical Representation of Velocity and Acceleration

Suppose that an object is located at $x = x_0$ and is not moving at all. This means it has the same coordinate x_0 as time t changes. This is represented graphically by drawing a straight line parallel to the time axis as shown in Figure 3.5. On the other hand, if the body is moving with uniform velocity along the X-axis, the plot of x versus t will be a straight line PQ as shown in Figure 3.6(a). The average velocity $\bar{v}$ can be found by calculating the slope of this line; that is,

$$\bar{v} = \frac{QM}{PM} = \frac{x_2 - x_1}{t_2 - t_1} = \text{slope of } PQ$$

$\bar{v}$ has the same value no matter which two points are taken on the straight line to calculate the slope. In this case the average velocity is also equal to the instantaneous velocity at any point between x_1 and x_2.

Let us now consider the case in which the body has a nonuniform velocity. In this case, the plot of x versus t will be a nonlinear plot such as PQ shown in Figure 3.6(b). The average velocity $\bar{v}$ between any two points, say P and Q, is given by the slope of the chord PQ (dashed line PQ) that is,

$$\bar{v} = \frac{QM}{PM} = \frac{x_2 - x_1}{t_2 - t_1}$$

The instantaneous velocity at any point, say at R, as defined by Equation 3.3, may be calculated by the following method. In the neighborhood of R the path may be treated as a straight line (for a very small time interval), as shown by the boldface portion (obtained by joining two nearby points). This portion, when extended, results in a straight line CD, called the tangent to the path PQ at point R. The instantaneous velocity v at R is equal to the slope of this tangent. That is,

$$v_{(\text{at } R)} = \text{slope of the tangent to the curve at } R = \frac{\Delta x}{\Delta t} \tag{3.6}$$

This value of $\Delta x/\Delta t$ does not change if $\Delta t \rightarrow 0$. If the same procedure is followed at P and Q, the instantaneous velocity at Q is much larger than at P. This is clear because the curve is steeper at Q than at P and hence the slope is much larger at Q than at P.

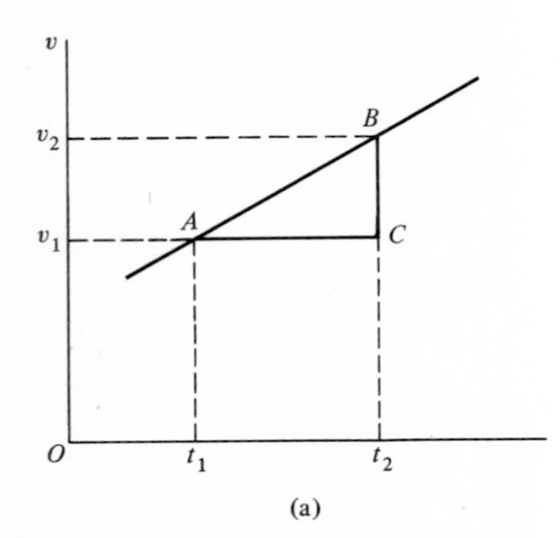

(a)

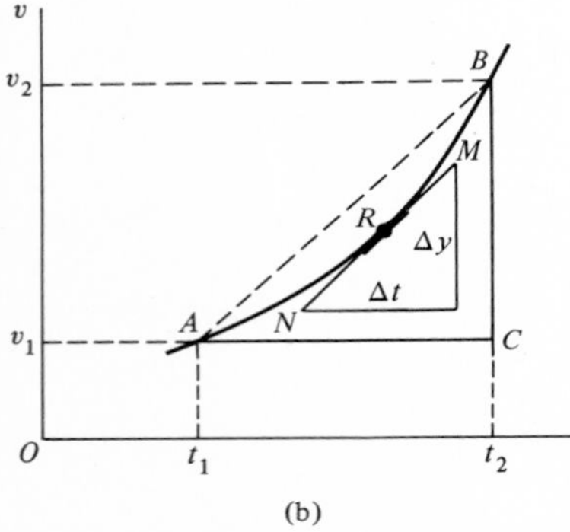

(b)

FIGURE 3.7. *Plot of v versus t for an object moving with (a) uniform acceleration (straight line AB) and (b) variable acceleration (curved line AB).*

We can extend this graphical procedure to the case of acceleration, as shown in Figure 3.7. Figure 3.7(a) shows the plot of v versus t when the body is moving with uniform acceleration. As before, the instantaneous and the average accelerations are the same in this case and are given by the slope of the line AB. That is,

$$\bar{a} = \frac{BC}{AC} = \frac{v_2 - v_1}{t_2 - t_2} = \text{slope of } AB$$

If the body is moving with variable acceleration, the plot of v versus t will be a curved path, as shown in Figure 3.7(b). The instantaneous acceleration at any point, say R, is found by drawing the tangent at that point and finding the slope of the tangent. Thus,

$$a_{(\text{at } R)} = \text{slope of the tangent to the curve at } R = \frac{\Delta v}{\Delta t} \tag{3.7}$$

This value of $\Delta v/\Delta t$ does not change if $\Delta t \rightarrow 0$. Once again a is much larger at B than at A because the curve is much steeper at B than at A.

EXAMPLE 3.1 The displacement of a particle is described by an equation $x = A + Bt^2$, where $A = 10$ m and $B = 12$ m/s^2 while x is in meters and t is in seconds. Calculate (a) the distance traveled between the time interval $t = 2$ s and $t = 4$ s; and (b) the average velocity between this time interval.

(a) Beginning with $x = A + Bt^2 = 10\text{ m} + (12\text{ m/s}^2)t^2$,

for $t_1 = 2$ s, $x = x_1 = 10\text{ m} + 12\text{ m/s}^2(2\text{ s})^2 = 58\text{ m}$
for $t_2 = 4$ s, $x = x_2 = 10\text{ m} + 12\text{ m/s}^2(4\text{ s})^2 = 202\text{ m}$
Therefore, the distance traveled or displacement, Δx, is

$$\Delta x = x_2 - x_1 = (202 - 58)\text{ m} = 144\text{ m}$$

(b) Since $\Delta x = 144$ m and $\Delta t = 4\text{ s} - 2\text{ s} = 2$ s, the average velocity $\bar{v}$ is

$$\bar{v} = \frac{\Delta x}{\Delta t} = \frac{144\text{ m}}{2\text{ s}} = 72\text{ m/s}$$

EXERCISE 3.1 The displacement of a certain particle is described by the equation $x = Ct + Dt^2$, where $C = 2$ m/s and $D = 0.1$ m/s^2, while x is in meters and t is in seconds. Calculate (a) the displacement between the time interval $t = 2$ s and $t = 5$ s; and (b) the average velocity in this time interval. [*Ans.*: (a) $\Delta x = 8.1$ m; (b) $\bar{v} = 2.7$ m/s.]

EXAMPLE 3.2 The velocity of a particle is described by the equation $v = Ct + Dt^2$, where $C = 0.1$ m/s^2 and $D = 0.02$ m/s^3. Calculate (a) the change in velocity during the time interval $t = 3$ s and $t = 6$ s; and (b) the average acceleration during this interval.
(a) Beginning with $v = Ct + Dt^2 = (0.1\text{ m/s}^2)t + (0.02\text{ m/s}^3)t^2$,

for $t_1 = 3$ s, $v = v_1 = (0.1\text{ m/s}^2)(3\text{ s}) + (0.02\text{ m/s}^3)(3\text{ s})^2 = 0.48\text{ m/s}$
for $t_2 = 6$ s, $v = v_2 = (0.1\text{ m/s}^2)(6\text{ s}) + (0.02\text{ m/s}^3)(6\text{ s})^2 = 1.32\text{ m/s}$

Therefore, change in velocity Δv is

$$\Delta v = v_2 - v_1 = (1.32 - 0.48)\text{ m/s} = 0.84\text{ m/s}$$

(b) Since $\Delta v = 0.84$ m/s and $\Delta t = 6\text{ s} - 3\text{ s} = 3$ s, the average acceleration $\bar{a}$ is

$$\bar{a} = \frac{\Delta v}{\Delta t} = \frac{0.84\text{ m/s}}{3\text{ s}} = 0.28\text{ m/s}^2$$

EXERCISE 3.2 Repeat Example 3.2 for the case in which the velocity of the particle is given by $v = 2Ct + 3Dt^2$. [*Ans.*: $\Delta v = 2.22$ m/s, $\bar{a} = 0.74$ m/s^2.]

3.3. Equations of Rectilinear Motion

Making use of the definitions of velocity and acceleration, we can derive some useful relations among x, v, a, and t for the motion of a body in a straight line. We shall assume that the acceleration a is constant throughout the motion, and hence the instantaneous acceleration will be equal to the average acceleration at every point. Starting with the definition of average velocity given by Equation 3.1, we may write, for its magnitude,

$$\bar{v} = \frac{\Delta x}{\Delta t} = \frac{x_2 - x_1}{t_2 - t_1}$$

Let us say that $x_1 = 0$ at $t_1 = 0$, that is, we are starting from the origin, and that $t_2 = t$ when $x_2 = x$. We get

$$x = \bar{v}t \tag{3.8}$$

That is, the distance x traveled in time t is equal to the product of $\bar{v}$ and t.

Let a body with initial velocity v_0 at $t = 0$ be moving with constant acceleration a and have a velocity v at time t. From the definition of acceleration,

$$a = \frac{\Delta v}{\Delta t} = \frac{v - v_0}{t - 0}$$

or

$$v = v_0 + at \tag{3.9}$$

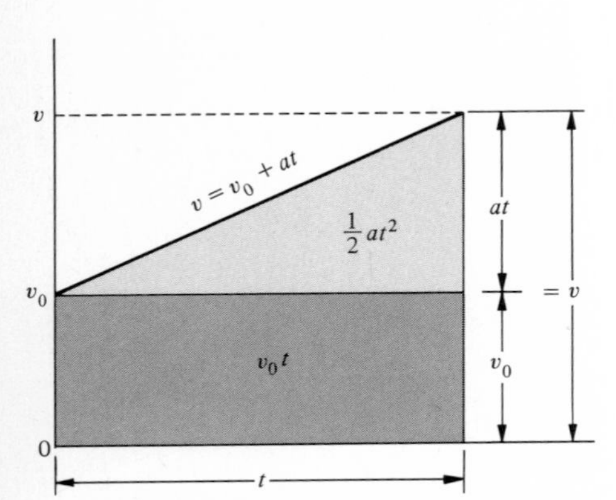

FIGURE 3.8. *The velocity-versus-time graph of a particle moving with a uniform acceleration. Note that the area under the velocity–time graph equals the displacement of the particle.*

which, when compared with $y = c + mx$, shows that this is the equation of a straight line as long as a is a constant. This is shown from the plot of v versus t in Figure 3.8. The velocity v at any time is equal to the sum of two segments along the v-axis—the initial velocity v_0 and the change in velocity in time t equal to at.

The next step is to find the displacement of a particle in time t that has an initial velocity v_0 at $t = 0$, final velocity v at time t, and is moving with constant acceleration a. As long as the acceleration is constant, the average velocity $\bar{v}$ in any time interval equals one-half the sum of the velocities at the beginning and the end of the interval. [This may be deduced from Figure 3.7(a).] Thus,

$$\bar{v} = \frac{v_0 + v}{2} \tag{3.10}$$

Hence, the displacement in time t is

$$x = \bar{v}t = \frac{v_0 + v}{2}t \tag{3.11}$$

Substituting for v from Equation 3.9, we get

$$x = \frac{v_0 + (v_0 + at)}{2}t$$

or

$$x = v_0t + \tfrac{1}{2}at^2 \tag{3.12}$$

It is of interest to note that the first term on the right, v_0t, is the area of the rectangle in Figure 3.8, while the second term, $\frac{1}{2}at^2$, is the area of the triangle in Figure 3.8, that is, the sum of these two areas is equal to the total displacement x. Thus, it may be stated that the displacement x of a body is equal to the area under the curve of a velocity–time graph.

Another relation for x may be derived by eliminating t between Equations 3.9 and 3.11. From Equation 3.9 $t = (v - v_0)/a$, hence

$$x = \bar{v}t = \left(\frac{v_0 + v}{2}\right)\left(\frac{v - v_0}{a}\right) = \frac{v^2 - v_0^2}{2a}$$

or

$$v^2 = v_0^2 + 2ax \tag{3.13}$$

Thus far we have derived the following relations for rectilinear motion with constant acceleration, which are not all independent. (Also, $x = \bar{v}t$ holds for all motions.)

$$\boxed{\begin{aligned} v &= v_0 + at \\ x &= \bar{v}t = \frac{v_0 + v}{2}t \\ x &= v_0t + \tfrac{1}{2}at^2 \\ v^2 &= v_0^2 + 2ax \end{aligned}} \tag{3.14}$$

Example 3.3 A jumbo jet starts from rest with an acceleration of 3 m/s² and makes a run for 35 s before taking off. What is the minimum length of the runway and what is the velocity of the jet at takeoff? Convert these quantities into British units.

Since $v_0 = 0$, $a = 3\ \text{m/s}^2$, and $t = 35$ s, we can calculate x and v from Equation 3.14:

$$x = v_0t + \tfrac{1}{2}at^2 = (0)(35\ \text{s}) + \tfrac{1}{2}(3\ \text{m/s}^2)(35\ \text{s})^2 = 1838\ \text{m} = 1.838\ \text{km}$$

Since 1 km = 0.62 mi,

$$x = (1.838)(0.62\ \text{mi}) = 1.14\ \text{mi}$$

Also, from Equation 3.14,

$$v = v_0 + at = 0 + (3\ \text{m/s}^2)(35\ \text{s}) = 105\ \text{m/s}$$

Since $1\ \text{m} = 10^{-3}$ km and $1\ \text{s} = \frac{1}{3600}$ h,

$$\begin{aligned} v &= (105)(10^{-3}\ \text{km}) \times 3600/\text{h} = 378\ \text{km/h} \\ &= (378)(0.62\ \text{mi})/\text{h} \simeq 234\ \text{mi/h} \end{aligned}$$

Note that the same results could be obtained by using the relations

$$x = \bar{v}t = \frac{v_0 + v}{2}t \qquad \text{and} \qquad v^2 = v_0^2 + 2ax$$

Exercise 3.3 A Boeing 727 jet lands with a speed of 160 m/s and runs for 32 s before coming to a stop. What is the acceleration and the minimum length of the runway? [*Ans.*: $a = -5$ m/s^2; $x = 2560$ m.]

3.4. Freely Falling Bodies

A common example of translational motion with almost constant acceleration which we experience in everyday life is that of a body falling toward the earth. It was first shown by Galileo Galilei (1564–1642) that if we ignore the effect of air resistance, all bodies fall with a nearly constant magnitude of acceleration toward the center of the earth. A truly freely falling body is one that moves through a vacuum. But for compact and dense bodies moving at low speeds, the air resistance is negligible and their motion closely approximates the motion of free fall. For example, if a sheet of paper and a book are let go from the surface of a table, the sheet of paper will take much longer to reach the floor. Now, if we crush the paper into a small compact ball (air resistance will be less for the ball than for a sheet of paper) and let the book and this ball be released simultaneously from the top of the table, both of them will arrive at the floor at about the same time. This demonstrates that they have the same acceleration, and that this acceleration is independent of the mass of the object. [This constant acceleration (as we shall see in Chapter 11) is due to the attractive force between the earth and the body under consideration.]

The acceleration of a freely falling body is denoted by the symbol **g** and is called the acceleration due to gravity (it is incorrect to call **g** "gravity" or the force of gravity). The value of **g** is found to be

$$g = 9.8 \text{ m/s}^2 = 32.2 \text{ ft/s}^2$$

and it is a vector quantity directed toward the center of the earth.

Strictly speaking, g is not a constant, but varies slightly from place to place on the surface of the earth and at small heights above the surface of the earth. The value of g differs no more than 0.4 percent at different places on earth, and even for altitudes up to 80 km (~50 mi), the value of g decreases only by about 2 percent from its value at the earth's surface. For all practical purposes (in this chapter) we will neglect air resistance as well as small variations in g, and will use the value of g given above.

The equations derived for rectilinear motion along the X-axis can be used for the case of freely falling bodies as well. Before we can use these equations, we must decide on the sign convention for direction of the motion. We shall take all distances and velocities upward as positive and downward as negative from the reference point. Of course, there is only one acceleration here, and it will always be negative according to our sign convention; that is, $a = a_y = -g = -9.8$ m/s^2. Thus, Equations 3.14 take the form (for vertical motion in the Y-direction), after replacing x by y and a by $-g$,

$$\boxed{\begin{aligned} v &= v_0 - gt \\ y &= \bar{v}t = \frac{v_0 + v}{2}t \\ y &= v_0 t - \tfrac{1}{2}gt^2 \\ v^2 &= v_0^2 - 2gy \end{aligned}} \qquad (3.15)$$

Note that under the effect of acceleration due to gravity, downward-falling bodies increase in velocity, while those projected upward are slowed down, come to rest, and then start falling downward with increasing velocity, as will be demonstrated in the following examples.

Example 3.4 A ball is thrown upward with a velocity of 29.4 m/s from a tower that is 98 m high. Calculate (a) the time it takes to reach its highest point; (b) the highest distance reached; (c) the distance and velocity after $t = 2$, 4, 6 s; and (d) the time it takes to reach the ground.

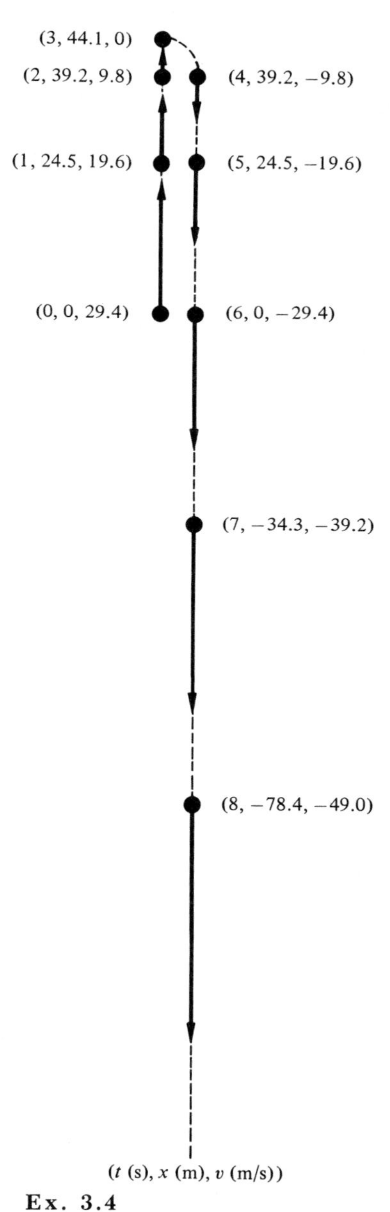

Ex. 3.4

Let us assume that the origin of our coordinate system is at the starting point on the tower, and that the upward direction is positive and the downward is negative. The quantities given are $v_0 = 29.4$ m/s, $y_0 = -98$ m, and $a = -g = -9.8$ m/s^2.

(a) At the highest point $v = 0$ from Equation 3.15, $v = v_0 - gt$; therefore,

$$t = \frac{v - v_0}{-g} = \frac{0 - (29.4 \text{ m/s})}{-9.8 \text{ m/s}^2} = 3 \text{ s}$$

(b) If $y = y_{max}$ from Equation 3.15, $y = v_0 t - \frac{1}{2}gt^2$:

$$y_{max} = (29.4 \text{ m/s})(3 \text{ s}) - \tfrac{1}{2}(9.8 \text{ m/s}^2)(3 \text{ s})^2 = 44.1 \text{ m}$$

(c) After $t = 2$ s, $v = v_2$,

$$v_2 = v_0 - gt = (29.4 \text{ m/s}) - (9.8 \text{ m/s}^2)(2 \text{ s}) = 9.8 \text{ m/s}$$

while after $t = 2$ s, $y = y_2$,

$$y_2 = v_0 t - \tfrac{1}{2}gt^2 = (29.4 \text{ m/s})(2 \text{ s}) - \tfrac{1}{2}(9.8 \text{ m/s}^2)(2 \text{ s})^2 = 39.2 \text{ m}$$

Similarly, using the preceding two equations, we can get v_4, v_6, and y_4, y_6. Thus, $v_0 = 29.4$ m/s, $v_2 = 9.8$ m/s, $v_4 = -9.8$ m/s, $v_6 = -29.4$ m/s; and $y_0 = 0$, $y_2 = 39.2$ m, $y_4 = 39.2$ m, $y_6 = 0$. These quantities are plotted in the accompanying figure. Note that except for the sign, the magnitudes of the quantities are the same going up as coming down. Also, it takes the same amount of time to go up as it does to come down.

(d) When $y = y_0 = -98$ m, $v_0 = 29.4$ m/s, and we want to find the time for the ball to go up, come to a stop, and then turn back to reach the ground. From Equation 3.15, $y = v_0 t - \frac{1}{2}gt^2$, we get

$$-98 \text{ m} = (29.4 \text{ m/s}^2)t - \tfrac{1}{2}(9.8 \text{ m/s}^2)t^2$$

or

$$t^2 - 6t - 20 = 0$$

Remembering that if $ax^2 + bx + c = 0$,

$$x = \frac{-b \pm \sqrt{b^2 - 4ac}}{2a}$$

Therefore,

$$t = \frac{-(-6) \pm \sqrt{(-6)^2 - 4(1)(-20)}}{2} = \frac{6 \pm \sqrt{116}}{2} = \frac{6 + 10.8}{2} = 8.4 \text{ s}$$

Note that we have ignored the negative square root, which gives t negative.

Exercise 3.4 A ball is dropped from a height of 100 m with a zero initial velocity. Calculate (a) the velocity at $t = 2$ s and 4 s; (b) the distance traveled after $t = 2$ s and 4 s; (c) the time it takes before it hits the ground; and (d) the velocity with which it hits the ground. [*Ans.*: (a) -19.6 m/s, -39.2 m/s; (b) 19.6 m, 78.4 m; (c) 4.52 s; (d) -44.3 m/s.]

3.5. Projectile Motion

Any object launched in an arbitrary direction in space with an initial velocity, which then follows a path determined by the acceleration due to gravity, is called a *projectile*, and such a motion is called *projectile motion*. The path of the projectile is called its *trajectory*. Examples of projectile motion include that of a golf ball in air, a bullet shot from a gun, and an object released from an airplane.

It is clear that projectile motion is both in the horizontal direction (along the X-axis) and in the vertical direction (along the Y-axis); that is, the motion is in the XY-plane. Let us first consider motion in a plane in general. The motion of any particle in a plane can be split into two rectilinear motions and treated somewhat independently of each other. Suppose that a particle is moving along a curved path PQ at a distance **r** from the origin, and that the position vector makes an angle ϕ with the X-axis, as shown in Figure 3.9. From Section 1.5, the relations between r and ϕ and x and y are known to be

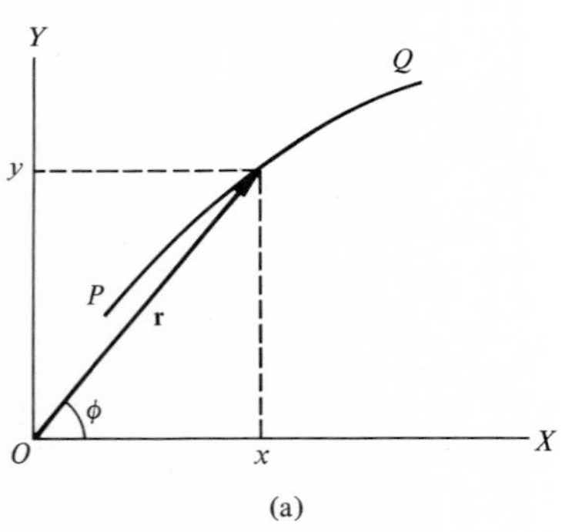

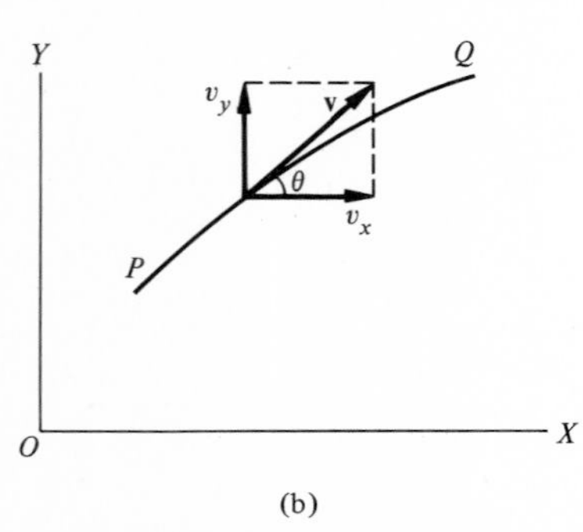

Figure 3.9. *X- and Y-components of (a) a displacement vector* **r** *and (b) a velocity vector* **v** *of a particle moving in the XY-plane.*

$$x = r\cos\phi \qquad y = r\sin\phi \tag{3.16}$$

and

$$r = \sqrt{x^2 + y^2} \qquad \tan\phi = \frac{y}{x} \tag{3.17}$$

Now suppose that this particle at **r** has velocity **v** which makes an angle θ with the X-axis, as in Figure 3.9. From Section 1.5 we also get

$$v_x = v\cos\theta \qquad v_y = v\sin\theta \tag{3.18}$$

and

$$|\mathbf{v}| = v = \sqrt{v_x^2 + v_y^2} \qquad \tan\theta = \frac{v_y}{v_x} \tag{3.19}$$

Similarly, acceleration **a** can be split into two components, a_x and a_y.

Once we have split **r**, **v**, and **a** into their components, the motion along the two axes can be treated independently of each other by using the following two sets of equations (note that v_{0x} and v_{0y} are components of v_0, which is the velocity at $t = 0$):

$$\text{(3.20)} \quad \begin{aligned} v_x &= v_{0x} + a_x t && \text{(a)} \\ x &= \bar{v}_x t = \frac{v_{0x} + v_x}{2} t && \text{(b)} \\ x &= v_{0x} t + \tfrac{1}{2} a_x t^2 && \text{(c)} \\ v_x^2 &= v_{0x}^2 + 2a_x x && \text{(d)} \end{aligned}$$

$$\begin{aligned} v_y &= v_{0y} + a_y t && \text{(a)} \\ y &= \bar{v}_y t = \frac{v_{0y} + v_y}{2} t && \text{(b)} \\ y &= v_{0y} t + \tfrac{1}{2} a_y t^2 && \text{(c)} \\ v_y^2 &= v_{0y}^2 + 2a_y y && \text{(d)} \end{aligned} \quad \text{(3.21)}$$

Let us apply these equations to projectile motion in which we assume that g is constant and that there is no acceleration along the X-axis; that is, $a_x = 0$ and $a_y = -g$. These conditions reduce the equations above to

$$\text{(3.22)} \quad \begin{aligned} v_x &= v_{0x} \\ x &= v_{0x} t \end{aligned}$$

$$\begin{aligned} v_y &= v_{0y} - gt \\ y &= \bar{v}_y t = [(v_{0y} + v_y)/2]t \\ y &= v_{0y} t - \tfrac{1}{2} g t^2 \\ v_y^2 &= v_{0y}^2 - 2gy \end{aligned} \quad \text{(3.23)}$$

To be more specific, consider the motion of the projectile shown in Figure 3.10.

Suppose that an object is thrown from point O with an initial velocity $\mathbf{v}_0$, making an angle θ_0 with the horizontal, as shown in Figure 3.10. The velocity $\mathbf{v}_0$ can be resolved into two components v_{0x} $(= v_0 \cos \theta_0)$ and v_{0y} $(= v_0 \sin \theta_0)$. At any time t the velocity of the projectile is $\mathbf{v}$, with the following two components from Equations 3.22 and 3.23:

$$v_x = v_{0x} = v_0 \cos \theta_0 \tag{3.24}$$

$$v_y = v_{0y} - gt = v_0 \sin \theta_0 - gt \tag{3.25}$$

We are taking upward direction as positive and downward as negative. The angle θ_0 is called the *angle of departure*. From Equations 3.24 and 3.25, we conclude that the *projectile motion consists of two independent motions, the horizontal motion with constant velocity and the vertical motion with constant acceleration*. This is also clear from Figure 3.10, where it is shown that all through the path of the projectile, the horizontal velocity is $v_{0x} = v_0 \cos \theta_0$ and is constant; while the vertical component of velocity v_y has initial value $v_{0y} = v_0 \sin \theta_0$, becomes zero as the projectile reaches its maximum height H, and starts increasing again (although in the negative direction) as the projectile continues to fall. The magnitude and the direction of the velocity of the projectile at any instant is given by Equations 3.19.

The coordinates (x, y) of the projectile at any time t are given by the following relations, from Equations 3.22 and 3.23:

$$x = v_{0x} t = (v_0 \cos \theta_0) t \tag{3.26}$$

$$y = v_{0y} t - \tfrac{1}{2} g t^2 = (v_0 \sin \theta_0) t - \tfrac{1}{2} g t^2 \tag{3.27}$$

If we make a plot of y versus x for different values of t, the resulting curve is

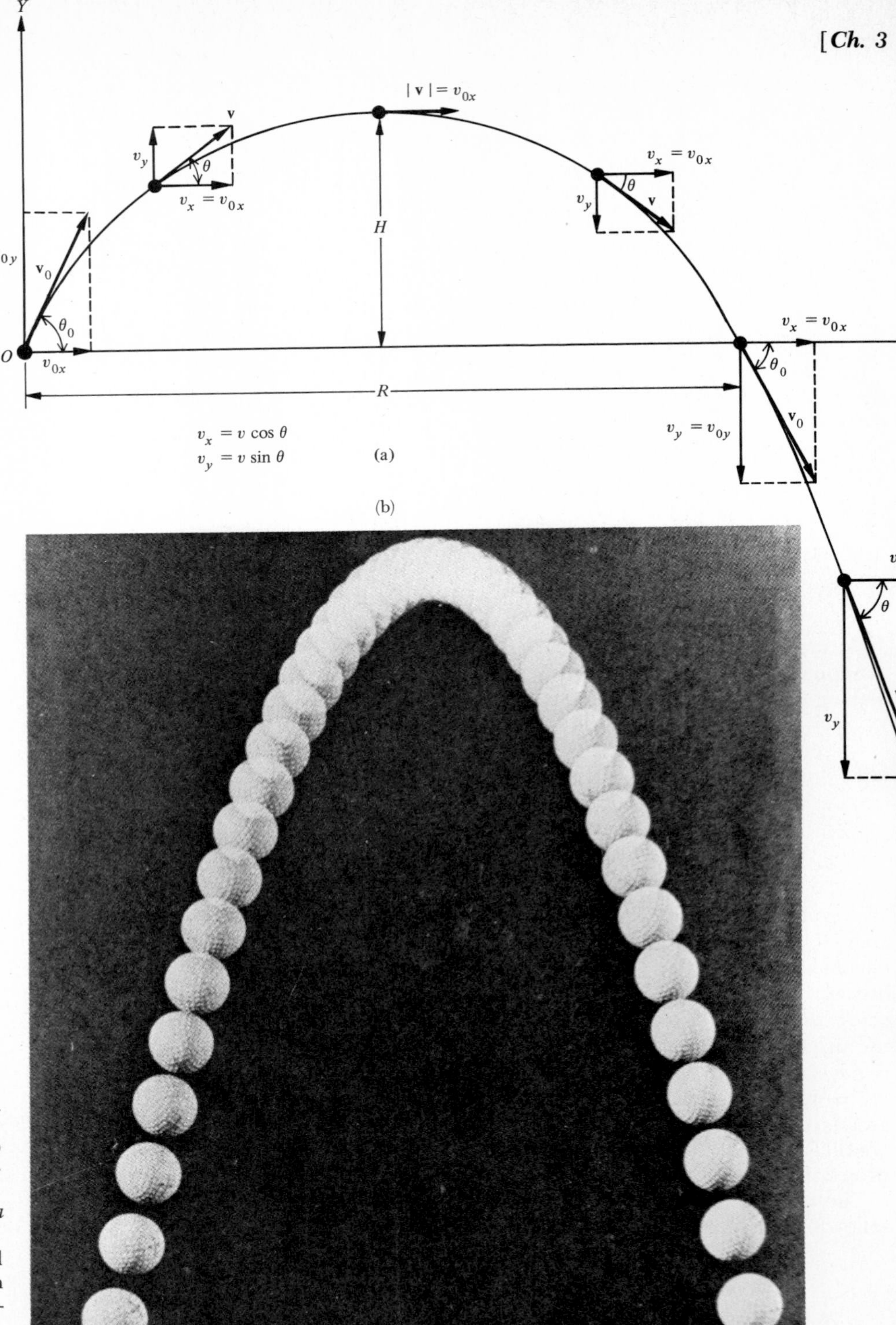

FIGURE 3.10. *(a) Trajectory of an object starting with a velocity* $\mathbf{v}_0$ *at an angle* θ_0 *with the horizontal. (b) Stroboscopic photograph (multi-flash photograph) of a projectile—a ball thrown in the air at an angle.* [Part (b) from *PSSC Physics*, 3rd ed., 1971, copyright by Education Development Center, Inc. Reprinted by permission of D. C. Heath & Company.]

a parabola. That is, the trajectory of a projectile is a parabola, as shown in Figure 3.10.

Figure 3.11 shows an experimental set up in which two metallic balls are

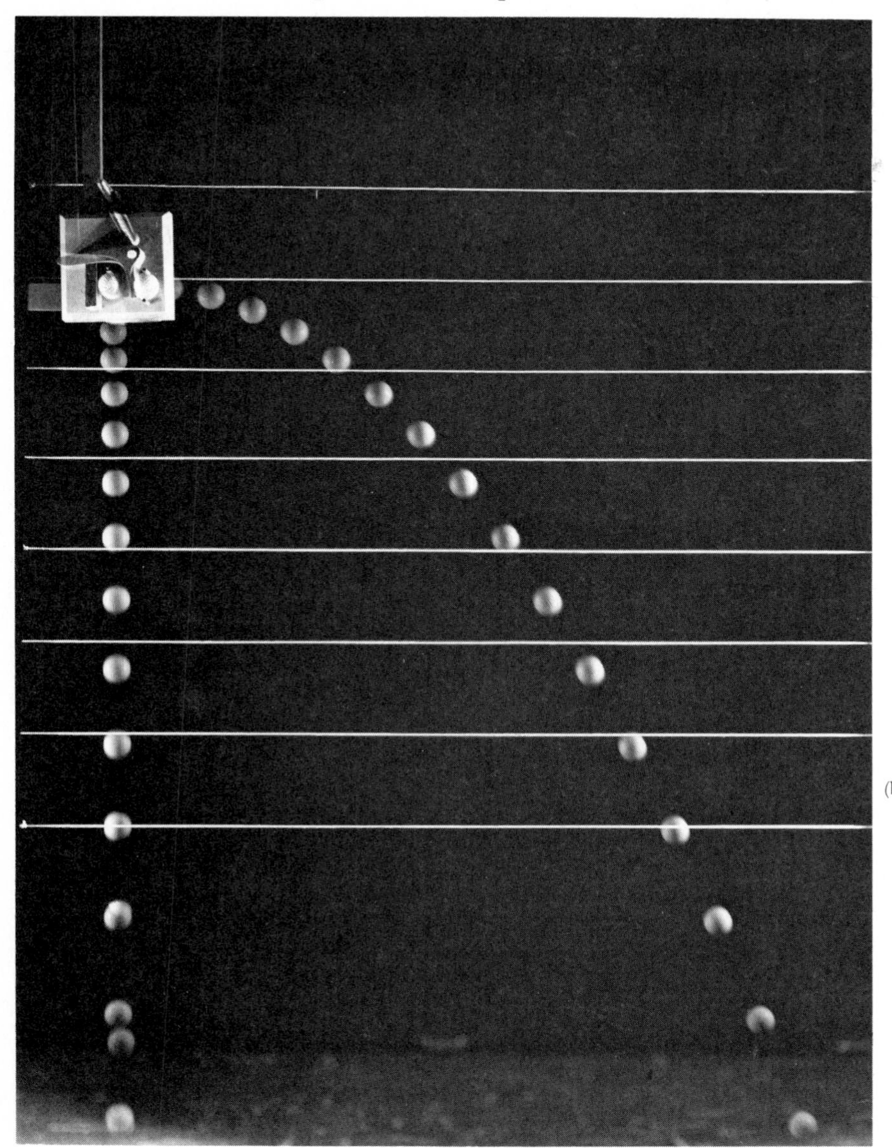

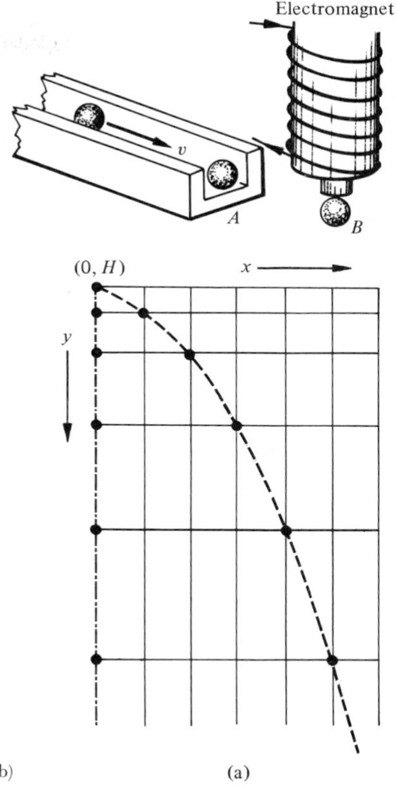

(b) (a)

Figure 3.11. *(a) Two objects, A and B, when released simultaneously reach the ground simultaneously, demonstrating that the vertical velocities are independent of horizontal velocities. (b) Stroboscopic photograph (multiflash photograph) of two balls released simultaneously—one simply dropped vertically and the other given an initial horizontal velocity* [Part (b) from *PSSC Physics*, 3rd ed., 1971, copyright by Education Development Center, Inc. Reprinted by permission of D. C. Heath & Company.]

thrown from a certain height H, that is, from the position (O, H). Ball A is given a horizontal velocity v_{0x}. As soon as it passes the point (O, H), another ball, B, which was suspended by being attached to an electromagnet, is released and falls vertically with an initial velocity of zero. We can show that the horizontal motion of ball A is unaffected while the vertical motions of both balls have the same vertical accelerations, $-g$. It is clear from the graphical representation that both balls travel the same vertical distances in the same time interval while ball A covers equal distances in equal intervals of time in the horizontal direction. Both balls will hit the ground at the same time.

The range R of the projectile as shown in Figure 3.10 can be calculated by making use of the fact that R is equal to that value of x for which $y = 0$. Thus, from Equation 3.27, if $y = 0$, $t = 2v_0 \sin\theta_0/g$. Substituting this value of t in Equation 3.26 and setting $x = R$, we get

$$R = v_{0x}t = v_0 \cos\theta_0 \frac{2v_0 \sin\theta_0}{g} = \frac{2v_0^2 \sin\theta_0 \cos\theta_0}{g}$$

$$\boxed{R = \frac{v_0^2 \sin 2\theta_0}{g}} \tag{3.28}$$

In Equation 3.28 g is constant and if v_0 is fixed, the value of R will depend upon the value of θ_0. By varying θ_0, R can be changed. It is clear that R will be maximum ($= R_{max}$) if $\sin 2\theta_0 = 1$, that is, if $2\theta_0 = 90°$ or $\theta_0 = 45°$. Thus, for $\theta_0 = 45°$, the maximum range is

$$R_{max} = \frac{v_0^2}{g} \tag{3.29}$$

The maximum height y_{max} ($= H$) is reached when the vertical component of velocity v_y becomes zero. According to the last equation in (3.23), v_y is given by

$$v_y^2 = v_{0y}^2 - 2gy = (v_0 \sin\theta_0)^2 - 2gy \tag{3.30}$$

Substituting $y = y_{max} = H$ and $v_y = 0$ yields $v_0^2 \sin^2\theta_0 = 2gH$ or

$$\boxed{H = \frac{v_0^2 \sin^2\theta_0}{2g}} \tag{3.31}$$

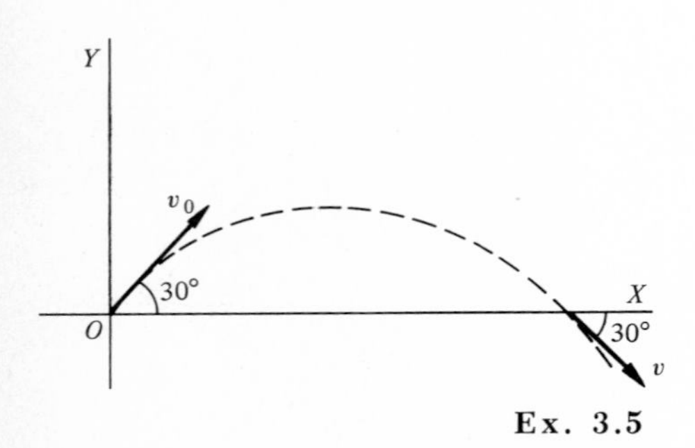

Ex. 3.5

Example 3.5 A stone is thrown with a velocity of 12 m/s, making an angle of 30° above the horizontal. Calculate (a) the time before it hits the ground; (b) the range; (c) the velocity just before impact; and (d) the maximum height that it reaches during its flight.

The quantities given are $v_0 = 12$ m/s, $a_x = 0$, $a_y = -g = -9.8$ m/s^2, $\theta_0 = 30°$, $\cos 30° = 0.87$, and $\sin 30° = 0.5$. Thus,

$$v_{0x} = v_0 \cos 30° = (12 \text{ m/s})(0.87) = 10.44 \text{ m/s}$$

$$v_{0y} = v_0 \sin 30° = (12 \text{ m/s})(0.5) = 6.0 \text{ m/s}$$

(a) From Equation 3.23, $y = v_{0y}t - \frac{1}{2}gt^2$, where $y = 0$:

$$0 = (6 \text{ m/s})t - \tfrac{1}{2}(9.8 \text{ m/s}^2)t^2$$

$$t = 0 \qquad \text{or} \qquad t = \frac{2(6 \text{ m/s})}{9.8 \text{ m/s}^2} = 1.22 \text{ s}$$

$t = 0$ corresponds to the starting point, where $x = 0$ and $y = 0$, while $t = 1.22$ s still corresponds to $y = 0$ but $x \neq 0$.

(b) $R = x = v_{0x}t + \frac{1}{2}a_x t^2 = v_{0x}t = (10.44\ \text{m/s})(1.22\ \text{s}) = 12.74\ \text{m}$

(c) $v_x = v_{0x} = 10.44\ \text{m/s} =$ the horizontal velocity, which remains constant

$$v_y = v_{0y} - gt = 6\ \text{m/s} - (9.8\ \text{m/s}^2)(1.22\ \text{s}) = -6\ \text{m/s} = -v_{0y}$$

As shown in the accompanying figure, the final velocity is the same in magnitude as the initial except that it is downward, making the same angle with the horizontal as before.

(d) At maximum height $v_y = 0$ from Equation 3.23, $v_y^2 = v_{0y}^2 - 2gy$:

$$0 = (6\ \text{m/s})^2 - 2(9.8\ \text{m/s}^2)y$$

or

$$y = \frac{(6\ \text{m/s})^2}{2(9.8\ \text{m/s}^2)} = 1.84\ \text{m}$$

Exercise 3.5 A stone is thrown with a horizontal velocity of 12 m/s from a building that is 20 m high, as shown in the accompanying figure. Calculate (a) the time when it hits the ground; (b) the horizontal and vertical components of the velocity just before it hits the ground; (c) the angle that it makes with the horizontal before hitting the ground (see the figure); and (d) the horizontal distance from the building when it hits the ground. [*Ans.*: (a) 2.02 s; (b) 12 m/s, −19.8 m/s; (c) 58.8°; (d) 24.24 m.]

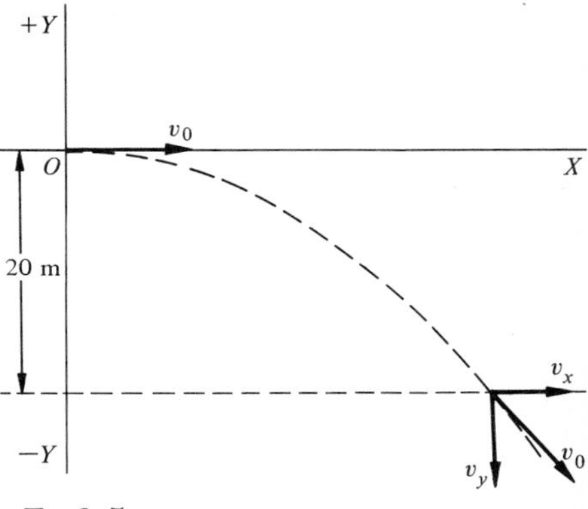

E. 3.5

*3.6. Relative Motion

Suppose that two cars, A and B, A being ahead of B, are moving along the $+X$-axis as shown in Figure 3.12(a). The velocity of A relative to an observer O on the roadside is v_{AO} and that of B relative to O is v_{BO}. We want to find the relative velocity v_{AB} defined as the velocity of A as observed by an observer in B. The answer is quite simple:

$$v_{AB} = v_{AO} - v_{BO} \tag{3.32}$$

Thus, if v_{AO} is 60 mi/h and v_{BO} is 50 mi/h, the velocity of A relative to B will be

$$v_{AB} = 60\ \text{mi/h} - 50\ \text{mi/h} = 10\ \text{mi/h}$$

Nothing is lost by writing Equation 3.32 in a more general form if we remember that $v_{BO} = -v_{OB}$; that is,

$$v_{AB} = v_{AO} - (-v_{OB})$$

or

$$v_{AB} = v_{AO} + v_{OB} \tag{3.33}$$

$$\begin{bmatrix}\text{velocity of } A\\ \text{relative to } B\end{bmatrix} = \begin{bmatrix}\text{velocity of } A\\ \text{relative to } O\end{bmatrix} + \begin{bmatrix}\text{velocity of } O\\ \text{relative to } B\end{bmatrix}$$

This form has an advantage since it can be extended to the case when motion is

(a)

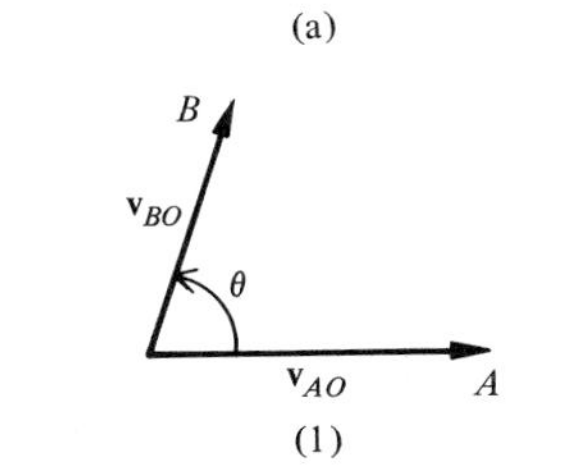

(1)

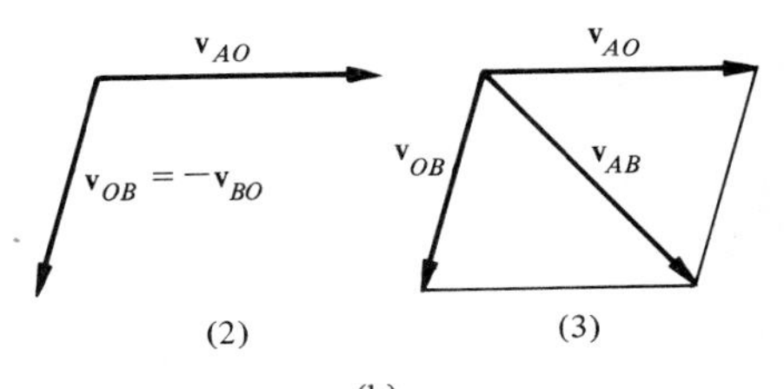

(2) (3)

(b)

Figure 3.12. *Method for obtaining relative velocity of A with respect to B, (a) when moving along the X-axis and (b) when making an angle θ with each other.*

not in a straight line. Thus, if A were moving along the X-axis while B were moving in a direction making an angle θ with the X-axis, the velocity of A relative to B would be

$$\mathbf{v}_{AB} = \mathbf{v}_{AO} + \mathbf{v}_{OB} \tag{3.34}$$

Note that $\mathbf{v}_{OB} = -\mathbf{v}_{BO}$. The addition above can be carried out geometrically or analytically, as explained in Chapter 1. For example, suppose that $\mathbf{v}_{AO}$ is along the X-axis and $\mathbf{v}_{BO}$ along a line making an angle θ with the X-axis, as shown in Figure 3.12 (b1). To get $\mathbf{v}_{AB}$, we find the resultant of $\mathbf{v}_{AO} + \mathbf{v}_{OB}$. But $\mathbf{v}_{OB} = -\mathbf{v}_{BO}$. Thus, we can find $\mathbf{v}_{AO} - \mathbf{v}_{BO}$; that is, we add to $\mathbf{v}_{AO}$ a vector equal in magnitude and opposite in direction to $\mathbf{v}_{BO}$, as shown in Figure 3.12(b2). The addition of these two vectors gives $\mathbf{v}_{AB}$, as shown in Figure 3.12(b3).

Using the analytical method, we solve Equation 3.34 by first writing it in component form, that is,

$$v_{ABx} = v_{AOx} + v_{OBx} \tag{3.35a}$$

$$v_{ABy} = v_{AOy} + v_{OBy} \tag{3.35b}$$

Therefore the resultant will be given by

$$v_{AB} = \sqrt{v_{ABx}^2 + v_{ABy}^2} \quad \text{and} \quad \tan\theta = \frac{v_{ABy}}{v_{ABx}} \tag{3.36}$$

In general, if A, B, C, and D are in motion, the velocity of A relative to D is

$$\mathbf{v}_{AD} = \mathbf{v}_{AB} + \mathbf{v}_{BC} + \mathbf{v}_{CD} \tag{3.37}$$

Example 3.6 An aircraft has a velocity of 320 km/h due north, and the wind velocity is 100 km/h due east, both with respect to the earth. Calculate the relative velocity of the aircraft with respect to the wind.

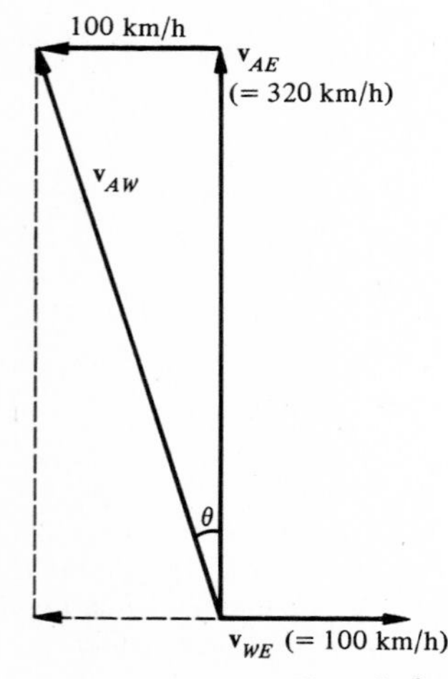

Ex. 3.6

Let A denote the aircraft, W the wind, and E the earth. The quantities given are $v_{AE} = 320$ km/h due north and $v_{WE} = 100$ km/h due east. We want to find the relative velocity of the aircraft with respect to the wind v_{AW}, that is, $\mathbf{v}_{AW} = \mathbf{v}_{AE} + \mathbf{v}_{EW}$. Since $\mathbf{v}_{EW} = -\mathbf{v}_{WE}$, $\mathbf{v}_{AW} = \mathbf{v}_{AE} + (-\mathbf{v}_{WE})$. To find $\mathbf{v}_{AW}$, we add to the velocity of the aircraft a velocity equal in magnitude and opposite in direction to that of the wind, as shown. Thus, the airspeed of the aircraft is

$$v_{AW} = \sqrt{(320 \text{ km/h})^2 + (100 \text{ km/h})^2} = 335 \text{ km/h}$$

$$\tan\theta = \frac{100}{320} = 0.312 \qquad \text{or} \qquad \theta = 17.3° \text{ W of N}$$

[To do the same problem analytically, we write in component form,

$$v_{AWx} = v_{AEx} + v_{EWx} \tag{i}$$

$$v_{AWy} = v_{AEy} + v_{EWy} \tag{ii}$$

where $v_{AEx} = 0$, $v_{AEy} = 320$ km/h, $v_{WEx} = 100$ km/h, and $v_{WEy} = 0$. Also

$v_{EWx} = -v_{WEx} = -100$ km/h. Substituting these in Equation i and ii

$$v_{AWx} = 0 + (-100 \text{ km/h}) \quad \text{and} \quad v_{AWy} = 320 \text{ km/h}$$

Thus, $$v_{AW} = \sqrt{(-100 \text{ km/h})^2 + (320 \text{ km/h})^2} = 335 \text{ km/h}$$

and $$\tan\theta = v_{AWx}/v_{AWy} = 100/320 = 0.312 \quad \text{or} \quad \theta = 17.3°]$$

Thus, if the aircraft heads northwest at $\theta = 17.3°$ with an airspeed of 335 km/h with respect to the air, it will travel northward at 320 km/h with respect to the earth. If we wanted to find the relative velocity of the wind with respect to the aircraft, we would add to the velocity of the wind a velocity equal in magnitude and opposite in direction to that of the aircraft.

EXERCISE 3.6 An aircraft is headed north with an airspeed of 240 km/h (150 mi/h) while the wind is blowing east with a velocity of 80 km/h with respect to the earth. What is the relative velocity of the aircraft with respect to the earth? In what direction should the aircraft head in order to travel north? [*Ans.*: $v_{AE} = 226$ km/h; $\theta = 19.5°$W of N.]

SUMMARY

A body is said to be in motion if it changes its position with respect to its surroundings. The time rate of change of position of a body is *average velocity* $\bar{\mathbf{v}} = \Delta\mathbf{x}/\Delta t$. The *instantaneous velocity* $\mathbf{v}$ at a point is the ratio of $\Delta\mathbf{x}/\Delta t$ provided that Δt is very small but the interval $\Delta\mathbf{x}$ still includes the point under consideration. The *average speed* is the ratio of the total distance traveled, Δs, to the time interval Δt.

The *average acceleration* is defined as the rate of change of velocity $\bar{\mathbf{a}} = \Delta\mathbf{v}/\Delta t$. The *instantaneous acceleration* $\mathbf{a}$ is the ratio of $\Delta\mathbf{v}/\Delta t$ provided that Δt is very small. A negative acceleration is called *deceleration*.

For uniformly accelerated motion of objects, the following relations hold: $v = v_0 + at$, $x = \bar{v}t = [(v_0 + v)/2]t$, $x = v_0 t + \frac{1}{2}at^2$ and $v^2 = v_0^2 + 2ax$. Similar relations hold for motion along the Y-axis. For vertical motion or projectile motion, $a = a_y = -g = -32 \text{ ft/s}^2 = -9.8 \text{ m/s}^2$.

The path of a projectile acted upon by a constant acceleration is a parabola. The components of velocity are $v_x = v_0\cos\theta_0$ and $v_y = v_0\sin\theta_0 - gt$ or $v = \sqrt{v_x^2 + v_y^2}$ and $\tan\theta_0 = v_y/v_x$. The range of the projectile is $R = v_0^2\sin 2\theta_0/g$.

* If A, B, C, and D are in motion, the relative velocity of A with respect to D is $\mathbf{v}_{AD} = \mathbf{v}_{AB} + \mathbf{v}_{BC} + \mathbf{v}_{CD}$.

QUESTIONS

1. Give at least two examples each (besides those given in the text) of (a) translational motion; (b) rotational motion; and (c) vibrational motion.
2. The radius of an atom is $\sim 10^{-10}$ m, while its nucleus has a radius of $\sim 10^{-14}$ m. Under what circumstances can you treat a nucleus as a point object?
3. What does the speedometer on an automobile register—average speed, speed, or velocity?

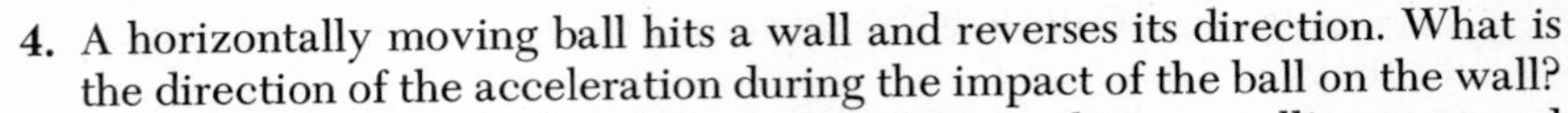

4. A horizontally moving ball hits a wall and reverses its direction. What is the direction of the acceleration during the impact of the ball on the wall?
5. X-rays are produced when fast-moving electrons hit a metallic target and are slowed down. What is the direction of the acceleration during this process?
6. By drawing tangents at P and Q in Figure 3.6(b), show (approximately) that the velocity at P is smaller than at Q.
7. By drawing tangents at A and B in Figure 3.7(b), show (approximately) that acceleration at B is greater than at A.
8. Make a plot of x versus t for an object that moves with (a) a uniform negative velocity; and (b) a variable negative velocity.
9. Make a plot of v versus t for an object that moves with (a) a uniform negative acceleration; and (b) a variable negative acceleration.
10. If the velocity-versus-time graph shown in Figure 3.8 was not a straight line but a curve, could you still say that the area under this curve is equal to the displacement?
11. Can you suggest an angle which a boy should make with the horizontal to have a maximum broad jump?
12. The plot of x versus t for a particle is shown in the figure. What can you say about the instantaneous velocity of the object at points 1 through 7?

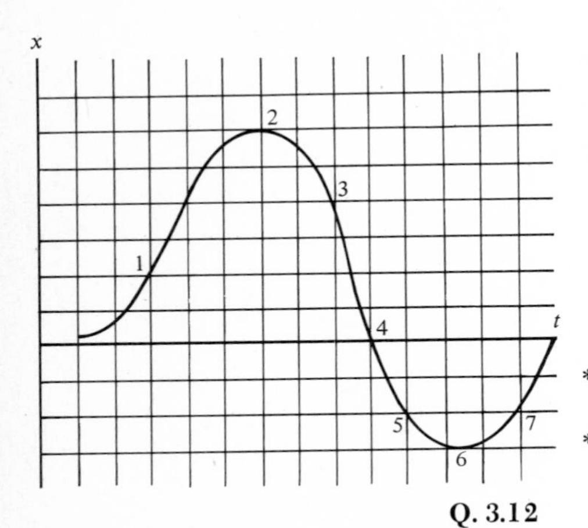

Q. 3.12

*13. Rain is falling vertically while you are walking east. In what direction should you hold your umbrella?
*14. An object is thrown vertically from a train moving east. What is the path of this object when observed by (a) a woman in the train; and (b) a woman standing on the road.

PROBLEMS

1. The displacement of a certain particle is described by the equation $x = A + Bt^2$, where $A = 10.0$ m and $B = -0.05$ m/s^2. Calculate (a) the displacement during the interval $t = 2$ s to $t = 4$ s; and (b) the average velocity during this interval. (c) Make a plot of x versus t.
2. The motion of a particle is given by $y = At + Bt^2$, where $A = 0.5$ m/s and $B = 0.2$ m/s^3. Calculate (a) the displacement between $t = 2$ s and $t = 3$ s; and (b) the average velocity in this interval.
3. The velocity of a particle is described by the equation $v = C + Dt$, where $C = 0.1$ m/s and $D = 0.05$ m/s^2. Calculate (a) the change in velocity during the time interval $t = 2$ s to $t = 4$ s; and (b) the average acceleration during this time interval. (c) Make a plot of v versus t.
4. The velocity of a particle is described by the equation $v = Ct + Dt^2$, where $C = 0.2$ m/s^2 and $D = 0.01$ m/s^3. Calculate (a) the change in velocity during the time interval $t = 2$ s to $t = 4$ s; and (b) the average acceleration during this time interval.
5. A certain car is equipped with an engine which, starting from rest, can reach a speed of 80 km/h (22.2 m/s $\approx$ 49.7 mi/h) in 12 s. What is the average acceleration and the distance it travels during this time?
6. A sports car traveling at 9 m/s ($\sim$20 mi/h) accelerates and attains a speed of 27 m/s ($\sim$60 mi/h) in 5 s. Calculate the average acceleration and the distance covered in 5 s.
7. The driver of a car traveling at a speed of 60 mi/h slams on his brakes and

stops within 4 s. Calculate the deceleration and the distance the car travels in these 4 s.

8. An automobile is traveling at a speed of 25 m/s when a driver observes an obstacle ahead and applies the brakes. The deceleration is 5 m/s^2 and the reaction time of the driver is 0.75 s. How far does the car travel before it comes to a stop? (1 m/s = 3.6 km/h = 2.237 mi/h.)

9. A body starting from rest has an acceleration of 2 m/s^2. How far does it travel in 8 s and what will be its velocity? What will be its velocity when it has traveled 50 m?

10. An electron traveling at a speed of 8×10^7 m/s passes through a sheet of aluminum 1 mm thick and emerges with a speed of 2×10^5 m/s. Calculate (a) the deceleration through the sheet assuming constant deceleration; and (b) the time taken to get through the sheet.

11. The take-off speed of an airplane is 40 m/s (144 km/h = 90 mi/h). What should be the acceleration if the total length of the runway is 800 m? Calculate the take-off time.

12. A car traveling at 72 km/h (20 m/s = $\simeq$45 mi/h) can be stopped (a) on a dry road within a distance of 60 m; (b) on a wet road within a distance of 90 m; and (c) on an icy road within a distance of 240 m. Calculate the deceleration and the stopping time in each case. If the reaction time of the driver is 0.75 s, how far will the car travel in each case before it comes to a stop?

13. A truck starts from rest with an acceleration of 1.5 m/s^2 while a car 150 m behind starts from rest with an acceleration of 2 m/s^2. (a) How long will it take before both the truck and car are side by side; and (b) how much distance is traveled by each?

14. A car is 30 m behind a truck when both are traveling at a constant speed of 90 km/h. The driver of the car decides to pass the truck and accelerates at 2 m/s^2. How long will it take before the car is 30 m ahead of the truck? What will be its speed? How much distance has it traveled in this time?

15. A ball is released from a height of 50 m. How long will it take to reach the ground and what will be its speed just before it hits the ground?

16. A ball is thrown upward from the ground with a speed of 16 m/s. (a) How far does it rise? (b) How long does it take to reach the highest point?

17. An object is released from a point 300 ft high. What is the position and speed of the object after 1 s, 2 s, 3 s, and 4 s? What is the average speed between (a) 1 s and 2 s; and (b) 3 s and 4 s?

18. An object is thrown upward with a speed of 20 m/s from a 10-story building that is 30 m high. Where will the object be after 2 s and 4 s? What will be its speed when it hits the ground?

19. A ball is thrown upward from the top of a 64-ft-high building with a speed of 16 ft/s. How long will it take and what will be its speed when it hits the ground?

20. A stone is dropped into a 50-m-deep well. How long will it take before the stone hits the water? If sound travels at a speed of 340 m/s, how long after the stone is released will we hear the splash?

21. A stone is dropped from a tower 98 m high. With what speed should a second stone be thrown 1 s later so that they both hit the ground at the same time?

22. From a building 25 m high a ball is dropped with zero initial velocity. One second later another ball is thrown downward with a velocity of 19.8 m/s. When and where will the second ball catch up with the first one?

23. A boy drops a ball from the top of a 48-ft-high building and at the same time another boy throws a ball upward from the ground with a speed of 32 ft/s. (a) How much time has elapsed when they pass each other? (b) Where do these two balls meet? (c) What are their speeds?

24. If the value of g were changed from 9.8 m/s² to 4.9 m/s², (a) how long will it take for a ball to fall through 98 m in each case; and (b) what will be its velocity when it reaches the ground?

25. A shell is fired with a velocity $v_x = 200$ m/s and $v_y = 49$ m/s. Calculate (a) the initial velocity and departure angle; (b) the range; and (c) the speed and the angle with which it hits the ground.

26. A baseball is hit with a velocity of 25 m/s (~80 ft/s), making an angle of 45° with the horizontal. Calculate (a) the time in the air; (b) the horizontal range; (c) the maximum height; and (d) the velocity with which it hits the ground.

27. A ball is thrown with a velocity of 10 m/s, making an angle θ with the horizontal. Calculate the range and the maximum height reached for $\theta = 30°$, 45°, and 60°.

28. A ball is thrown horizontally with a velocity of 20 m/s from the top of a building 50 m high. Calculate (a) the time of flight; (b) the horizontal range; and (c) the velocity with which it hits the ground.

29. A stone is thrown horizontally with a velocity of 15 m/s from the top of a bridge that is 30 m above the water. Calculate (a) the time the stone is in the air; (b) the horizontal range; and (c) the speed and the angle with which it hits the water.

30. A boy throws a ball horizontally, with a velocity of 8 m/s, from the top of a building. The ball falls to the ground 12 m away from the building. What is the height of the building?

31. A ball is thrown making an angle of 30° above the horizontal with a velocity of 20 m/s from the top of a building 50 m high. Calculate (a) the time of flight; (b) the horizontal range; and (c) the velocity with which it hits the ground.

32. A football player has to kick a football a total horizontal distance of 35 yd for it to cross a goalpost that is 12 ft high. What should be the minimum velocity if the departure angle is 45°?

33. A ball moves on a tabletop that is 1.5 m high with a velocity of 2 m/s for a distance of 0.75 m before leaving the table. Calculate (a) the total time the ball moves before hitting the floor; (b) the horizontal distance traveled; and (c) the velocity with which it hits the floor.

***34.** Car *A* is traveling with a speed of 72 km/h relative to the earth, while car *B* is traveling behind it with a velocity 88 km/h with respect to the earth. What is the relative velocity of *B* with respect to *A* (a) before *B* passes *A*; and (b) after *B* passes *A*? If these cars were moving in opposite directions, what would be the relative velocities while (c) approaching; and (d) receding from each other?

***35.** A ship is moving 30 m/s with respect to the earth, while a man on the ship is moving with velocity 2 m/s with respect to the ship. What is the relative velocity of the man with respect to the earth?

***36.** A train is moving at 40 m/s toward the east with respect to the earth. A boy on the train is running at 4 m/s and kicks a ball with a speed of 10 m/s due east with respect to him. What is the relative velocity of the ball with respect to an observer outside the train?

***37.** An aircraft has a speed of 320 km/h in still air. The wind is blowing eastward at 100 km/h. If the pilot wishes to fly north, in what direction should he head? What will be the airspeed?

***38.** A helicopter has an airspeed of 50 mi/h due south while the wind is blowing at 30 mi/h due east. What is the relative velocity of the helicopter with respect to the earth? How far will it travel in 2 h?

***39.** A train is moving at 20 m/s while rain is falling vertically with a speed of 10 m/s with respect to the ground. What is the relative velocity of the raindrops with respect to a passenger in the train?

***40.** A motorboat can run at 12 m/s in still water. Suppose that the water in a river flows at 2 m/s with respect to the boat. If the boat heads straight across the river, where will it end up on the opposite bank? (The river is 0.8 km wide.) How long will it take to cross the river? In what direction should the boat head so that it reaches a point directly across the river from the starting point?

4
Forces—Laws of Motion

We start with the study of dynamics—the branch of physics which deals with forces that produce a change in motion, size, and shape of an object. We shall familiarize ourselves with the concepts of force and linear momentum. After briefly explaining Newton's laws of motion, we shall investigate the relationship between the applied forces and the resulting motion. At the end we shall deal with the ever-present frictional force.

4.1. Forces That Hold Everything Together

THE PURPOSE of investigating the structure of nuclei and atoms has been to find the building blocks of all matter. The next thing that comes to our mind is: What holds these building blocks of matter together? For example, what makes the planets stay in their orbits? What holds the neutrons and protons together in the nucleus? What keeps the electrons from leaving the atom? The answer to all these questions lies in a commonly spoken and familiar word "force."

From our daily observations we find that force has different meaning to different people. For example, how hard a professional baseball batter hits a ball depends upon the magnitude of the force he is capable of delivering. An individual may apply enough force to move a car by pushing it. These are examples of "muscle force." The motion of an electron around the nucleus in an atom is due to what is called the *electrical force*. Matter is held together by electrical forces. The motion of the planets around the sun is due to the gravitational force. These are just a few of the innumerable examples of forces between objects. All these motions are the result of *interactions* between different objects. *Force* is a mathematical concept describing an interaction between two or more objects.

It does not always follow that an application of force will result in a motion or change in motion. For example, we may push a wall; that is, there is an interaction between us and the wall, and hence there is a force, but the wall may not move at all. At the other extreme, an interaction between two objects may change the motion of both the objects.

Force may be described as *a push or pull—resulting from the interaction between objects—which produces or tends to produce motion, stops or tends to stop motion.*

A natural question to ask is: How many types of elementary forces exist? Fortunately, the answer is simple. No matter what the situation, all known forces or interactions may be classified into one of four categories.

1. Gravitational forces.
2. Electromagnetic forces (electric, magnetic, elastic, chemical bonding).
3. Nuclear forces (or strong interactions).
4. Weak forces (or weak interactions).

Gravitational Forces

Gravitational force is the result of the interaction between the masses of different objects. The force is directly proportional to the masses and inversely proportional to the square of the distance between them (see Chapter 11). This force is the weakest of all four forces. If the masses are large enough, this force becomes appreciable. It is the gravitational force between different planets and the sun, which holds the planets in their orbits. Even though the distances between the planets are very large, the masses are so enormous that the gravitational forces are quite appreciable. There are gravitational forces between atoms, electrons, protons, plants, animals, and anything else that has mass. These masses are so small, however, that gravitational forces are negligible.

Electromagnetic Forces

Electromagnetic forces, also called *Coulomb forces,* are the results of an interaction between electrical charges. Electrostatic forces are the result of interactions between charges at rest. (It is well known that unlike charges attract each other; while like charges repel each other; see Chapter 19.) These forces are much stronger than gravitational forces. Electrical forces between bigger masses, such as the earth, moon, and planets, are zero because these objects are electrically neutral. The absence (or cancellation) of electrical forces permits the relatively much weaker gravitational force to manifest itself.

When charges are in motion, they produce magnetic forces. The elastic forces and the chemical or bonding force responsible for chemical reactions are a manifestation of electromagnetic forces.

Nuclear Forces

Nuclear forces, which are the strongest of all the forces, are always attractive [except that when neutrons and protons are very close ($\sim 10^{-16}$ m), they repel] and extend over a very short range, say up to the size of the nucleus, about 1.4×10^{-15} m. Nuclear forces are responsible for holding the constituents of the nuclei together. In a nucleus there are neutrons and protons. Owing to their masses, they have a gravitational force of attraction, but it is of extremely small magnitude. The positive charges on the protons which produce electrical repulsion, should disrupt the nucleus. In general, this does not happen because nuclear forces between the neutron–proton and proton–proton pairs are attractive and much stronger than the repulsive electrical forces of the protons. Also, nuclear forces do not extend much beyond the surface of the nucleus, and hence do not interfere or participate in the molecular or crystalline structures.

Weak Forces

Weak forces are a special kind of force, responsible for such processes as β-decay in radioactivity. They are weaker than nuclear and electromagnetic forces but stronger than gravitational. The range of this interaction is almost zero; that is, it does not extend much beyond the point of its origin.

To get an idea of the relative strength of these four forces, if we assign nuclear forces a strength of 1 unit, the relative strengths of the other forces are as follows:

nuclear: 1 electromagnetic: $\frac{1}{137}$ ($\sim 10^{-2}$)

weak: 10^{-13} gravitational: 10^{-39}

At present these four are the only forces known to be responsible for holding everything together in the universe.

4.2. Linear Momentum, Impulse, and Force

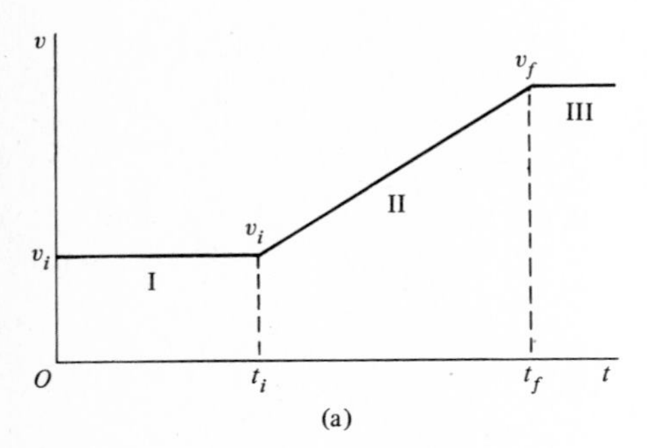

One of the most important quantities that we encounter very frequently in physics is *linear momentum*, or simply *momentum*, **p**. The linear momentum of an object is defined as the product of the mass m of an object and its velocity **v**. That is,

$$\boxed{\mathbf{p} = m\mathbf{v}} \tag{4.1}$$

As indicated, momentum is a vector quantity and its direction is the same as that of the velocity. The magnitude of **p** increases with increasing m and **v**, and will be zero if **v** is zero. Written in component form, Equation 4.1 becomes

$$p_x = mv_x, \qquad p_y = mv_y, \qquad p_z = mv_z \tag{4.2}$$

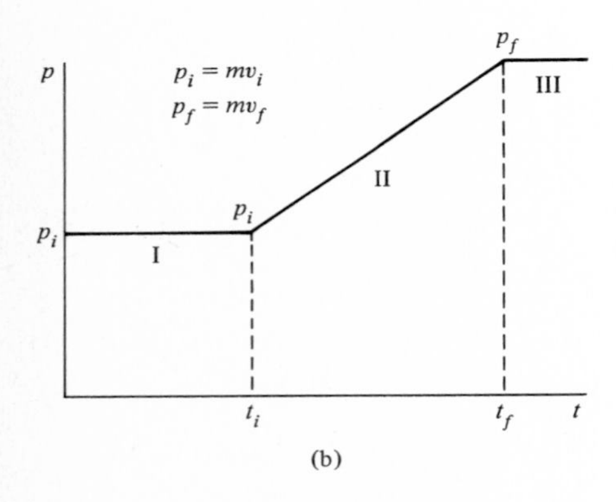

We shall limit our discussion to only one dimension, writing $p = mv$ without the subscripts. In the SI system, momentum has units of kg-m/s. Figure 4.1(a) shows a plot of velocity versus time for a particle of mass m moving along the X-axis. The particle is moving with a uniform velocity v_i between the time interval $t = 0$ and $t = t_i$, as indicated by the horizontal line I. Between $t = t_i$ and $t = t_f$ the particle has a uniform acceleration (line II) and reaches a final velocity v_f at $t = t_f$. After this time the velocity remains constant ($= v_f$), once again as shown by line III. If we multiply the ordinates of this plot by m, it results in a new plot of linear momentum, p versus t, as shown in Figure 4.1(b). It is clear from this plot that the momentum of the object remains constant before t_i and after t_f. During the time interval $t_f - t_i$, there must have been some interaction which causes a change in the momentum—a uniform increase of p with t in this case.

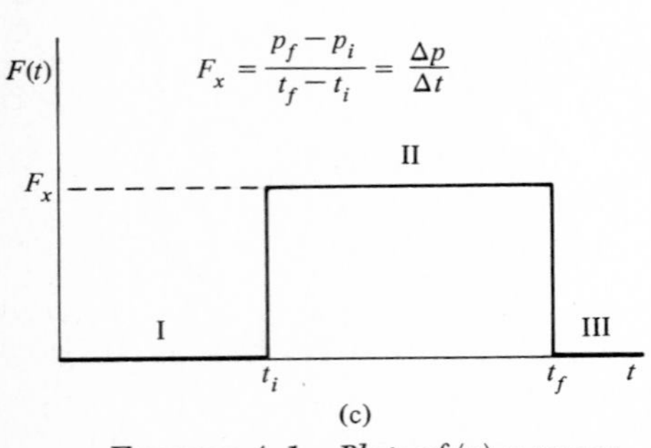

Figure 4.1. *Plots of (a) v versus t, (b) p versus t, and (c) F versus t for a particle moving along the X-axis.*

Let us calculate the rate of change of momentum, $\Delta p/\Delta t$, at different points along the curve in Figure 4.1(b). Before time t_i and after time t_f, the change in momentum is zero, and hence $\Delta p/\Delta t$ is zero. After time t_i and before time t_f, the increase in momentum is uniform, and hence $\Delta p/\Delta t$ is constant. The plot of $\Delta p/\Delta t$ versus t is shown in Figure 4.1(c). The quantity $\Delta p/\Delta t$ is denoted by F and is called *force*. That is,

$$F = \frac{\Delta p}{\Delta t} \tag{4.3}$$

Equation 4.3 may also be written as

$$\Delta p = F\,\Delta t \tag{4.4}$$

That is, the change in momentum $\Delta p = p_f - p_i$ is equal to the product of the force F and the time interval $\Delta t = t_f - t_i$. It is important to note that the change in momentum is equal to the area under the $F(t)$ versus t curve as shown in Figure 4.1(c), where $F(t)$ means that F is a function of time. The product of a force F and the time interval Δt is called the *impulse*. The impulse is also equal to the area under the force-versus-time curve. This leads to the *impulse–momentum theorem,* which states that *the change in momentum is equal to the total impulse.* The units of impulse are the same as that of momentum, that is, kg-m/s.

It is clear from Equation 4.4 that the same change in momentum will result as long as the area under the $F(t)$ versus t curve remains the same as shown in Figure 4.2. The area under curve A is equal to the area under curve B. The average force $\overline{F}_A$ applied in curve A is much larger (applied over interval Δt_A) while in curve B the average applied Force $\overline{F}_B$ is much smaller (applied over a longer interval of time). That is,

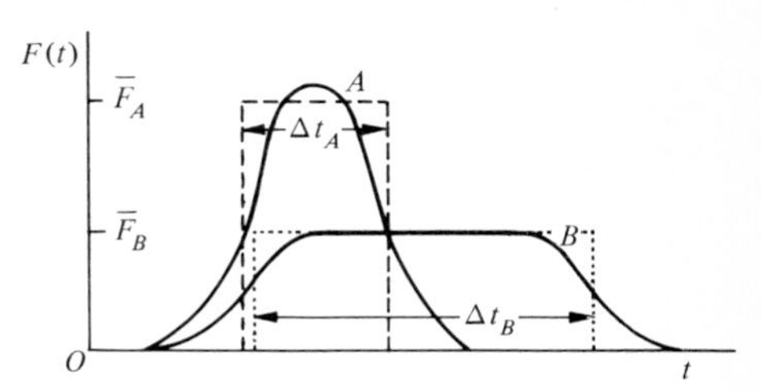

Figure 4.2. *Plot of $F(t)$ versus t for two different cases, both having the same impulse. Dashed and dotted lines are approximations to the solid curves.*

$$\overline{F}_A\,\Delta t_A = \overline{F}_B\,\Delta t_B \tag{4.5}$$

where $\overline{F}_A > \overline{F}_B$ and hence $\Delta t_A < \Delta t_B$. A common example of such a situation is that of an outfielder catching a ball that has been hit hard by a batter. The outfielder has a choice to stop the ball in a short time interval if he applies a large force, or stop it over a longer time interval by applying a smaller force. As one commonly observes, a fielder stretches or lets his hand go back while catching a ball so that he has to apply a smaller force, to avoid injury to his hand.

There are many situations in which it is not possible to know $F(t)$, but the total impulse can be found by measuring the change in momentum and using Equation 4.4.

Example 4.1 An electron of mass 9.1×10^{-31} kg with speed 4×10^3 m/s passes through an electric field for 2×10^{-7} s and emerges with a speed of 5×10^5 m/s. Calculate (a) the initial momentum; (b) the final momentum; (c) the impulse; and (d) the rate of change of momentum or average force.

(a) $p_i = mv_i = (9.1 \times 10^{-31}\text{ kg})(4 \times 10^3\text{ m/s}) = 3.64 \times 10^{-27}$ kg-m/s

(b) $p_f = mv_f = (9.1 \times 10^{-31}\text{ kg})(5 \times 10^5\text{ m/s}) = 4.55 \times 10^{-25}$ kg-m/s

(c) Impulse $= \Delta p = p_f - p_i = (4.55 \times 10^{-25} - 3.64 \times 10^{-27})$ kg-m/s
$= 4.51 \times 10^{-25}$ kg-m/s

(d) $\Delta p/\Delta t = (4.51 \times 10^{-25}\text{ kg-m/s})/(2 \times 10^{-7}\text{ s}) = 2.26 \times 10^{-18}\text{ kg-m/s}^2$

Note that this is an average force which the electric field exerts on the electron. The unit kg-m/s^2 is given the name newton, as we shall see shortly.

Exercise 4.1 A pitcher throws a baseball of mass 150 g with a speed of 25 m/s. The batter hits the ball with a speed of 40 m/s in a direction straight toward the pitcher. The duration of contact between the bat and the ball is 0.015 s. Calculate (a) the initial momentum; (b) the final momentum; (c) the impulse; and (d) the rate of change of momentum or the average force. [*Ans.*: (a) 3.75 kg-m/s; (b) -6 kg-m/s; (c) -9.75 kg-m/s; (d) -650 kg-m/s^2.]

4.3. Newton's First Law

Aristotle (384–322 B.C.) and most other scholars before Galileo Galilei (1564–1642) were of the opinion that a constant external force must be applied continuously to an object in order to keep it moving with uniform velocity. Through experimenting with highly polished bodies and surfaces, Galileo showed that no force was necessary to keep an object moving with constant velocity. It is the presence of frictional force that tends to stop moving objects—the smaller the frictional force between the object and the surface on which it is moving, the larger the distance it will travel before stopping. We may say that if the frictional force were reduced to zero (and in the absence of any other forces), the object will travel forever with constant velocity. These ideas were put together by Sir Isaac Newton (1642–1727) in the form of the first of three laws of motion.

> **Newton's first law:** ***Every object continues in its state of rest or uniform motion in a straight line unless a net external force acts on it to change that state.***

What the first law seems to imply is that the state of uniform motion of a body is as natural as its state of rest. We can go a step further and say that according to Newton's first law, there is no fundamental difference between an object at rest and an object moving with constant velocity because in both cases there is no net external force or interaction acting on the body. A body or a particle that satisfies Newton's first law of motion is called a *free particle,* a particle that is not subject to any net external force or interaction. A body is usually under the action of more than one force. Thus, when we say that no net external force acts on a body, we mean that the resultant of all the external forces acting on the body is zero.

There are two different experimental pieces of equipment available in laboratories for demonstrating frictionless motion and Newton's laws of motion. The first of these is a two-dimensional air table, shown in Figure 4.3(a). The tabletop has several thousand minute holes with soft jets of air or CO_2 gas issuing through them. When a plastic puck is placed on such a surface, it floats on a cushion of air. The friction between the puck and the surface of the air layer is so low that if the puck is set in motion, there is hardly any reduction in its speed over very large distances on such a surface.

The second device is the linear air track, shown in Figure 4.3(b), which consists of two parts—an inverted V-shaped surface having hundreds of fine holes with air streaming out, forming a layer on which are placed inverted V-shaped sliders. Once a V-shaped slider is set in motion, it rides on a "nearly frictionless" layer of air with constant velocity. It moves back and forth several times on the air track before coming to a stop.

4.4. Newton's Second Law

Newton's second law tells us what happens to the state of motion of a body if the net external force acting on the body is not zero. That is, when a body is not isolated, it interacts with surrounding objects.

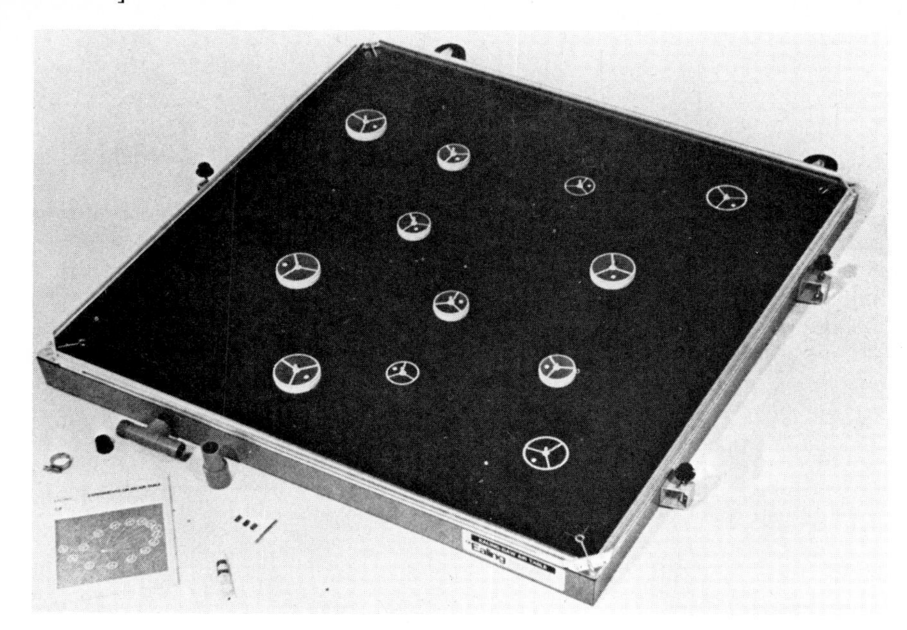

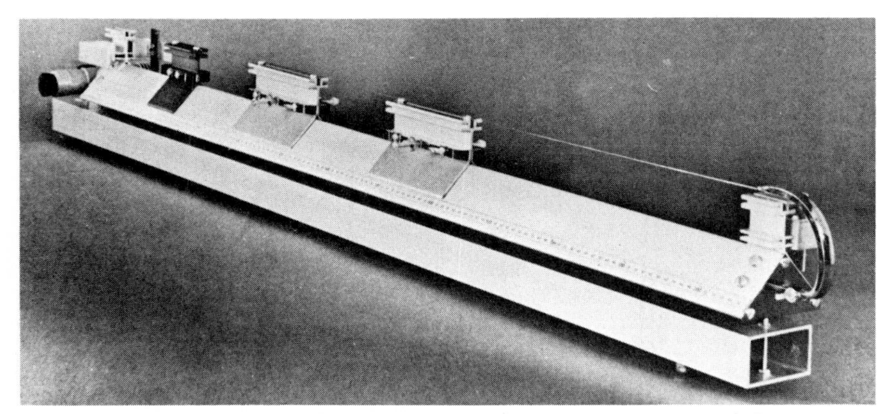

Figure 4.3. *(a) Frictionless two-dimensional air table. (b) A linear air track with inverted V-shaped sliders riding on a layer of air streaming through fine holes.* (Courtesy of The Ealing Corporation, South Natick, Mass.)

Newton's second law: ***The rate of change of momentum of an object is directly proportional to the force applied and takes place in the direction of the force.***

Choosing the units to be such that the constant of proportionality is unity, the statement of the law may be written mathematically as

$$\boxed{\mathbf{F} = \frac{\Delta \mathbf{p}}{\Delta t}} \tag{4.6}$$

where $\mathbf{p} = m\mathbf{v}$ is the momentum of an object of mass m which is moving with a velocity $\mathbf{v}$. This is the relation already given in Equation 4.3.

Equation 4.6 takes a simpler form if the moving mass m of the object remains constant at all speeds. As we shall see in Chapter 8, the mass of an object increases with its speed. For speeds that are low compared to the speed of light c ($= 3 \times 10^8$ m/s or 186,000 mi/s), the variation in mass is negligible. Hence, for the time being we assume this to be the case, thereby setting $m = m_0 =$ constant, where m_0 is the mass of the object when it is at rest. Since $\mathbf{p} = m\mathbf{v}$, Equation 4.6 takes the form

$$\mathbf{F} = \frac{\Delta \mathbf{p}}{\Delta t} = \frac{\Delta(m_0 \mathbf{v})}{\Delta t} = m_0 \frac{\Delta \mathbf{v}}{\Delta t} = m_0 \mathbf{a} \tag{4.7}$$

Dropping the subscript zero on m, Newton's second law takes the form

$$\boxed{\mathbf{F} = m\mathbf{a}} \qquad \text{if } m \text{ is constant} \tag{4.8}$$

That is, *force is equal to mass times acceleration provided m is constant.*

Note that $\mathbf{F}$ stands for net external force and hence may be a resultant vector sum of many forces acting on a body and may be written as $\Sigma\mathbf{F}$. Equation 4.8 may be written in component form as

$$\sum F_x = ma_x, \qquad \sum F_y = ma_y, \qquad \sum F_z = ma_z \tag{4.9}$$

Inertia, Mass, and Force

It is clear from Equation 4.8 that if the mass is large, a large force is required for motion to begin; and if the mass is small, a small force is required. This mass property of an object (or matter in general) that shows a reluctance to change its state of motion when a force is applied is called *inertia*. The mass m may be referred to as the *inertial mass* m_i. (Newton's first law is also referred to as the *law of inertia*.)

We know that objects are designated masses by comparing them with the replica of the standard mass by means of an equal-arm balance. Newton's second law provides still another means of comparing masses. Suppose an unknown mass m and standard mass m_s are both acted upon by equal forces so that

$$F = ma \qquad \text{and} \qquad F = m_s a_s$$

that is,

$$ma = m_s a_s$$

or

$$m = \left(\frac{a_s}{a}\right) m_s \tag{4.10}$$

where a and a_s are the accelerations produced in masses m and m_s, respectively.

This can be accomplished as illustrated in Figure 4.4. When the spring attached to the mass m_s has its natural unstretched length l, no force acts on the mass, as shown in Figure 4.4(a). Suppose now that we stretch the spring by an amount Δl. A force F_0 produced by the stretching will produce an acceleration of a_s in the mass m_s, as shown in Figure 4.4(b). Similarly, if the same spring attached to the mass m is stretched by the same amount Δl, it produces an acceleration a in the mass m. Knowing a, a_s, and m_s, we can calculate m.

The arrangement of Figure 4.4 can be used to define a unit of force. Suppose, for example, that in Figure 4.4(b), m_s is 1 unit of standard mass. If the spring is stretched so that it results in a_s to be 1 unit of acceleration, then the force produced by the spring is 1 unit. Similarly, if a_s has 2 units of acceleration, the force will have 2 units. This method is used for calibrating a spring to measure forces by noting the extension of the spring.

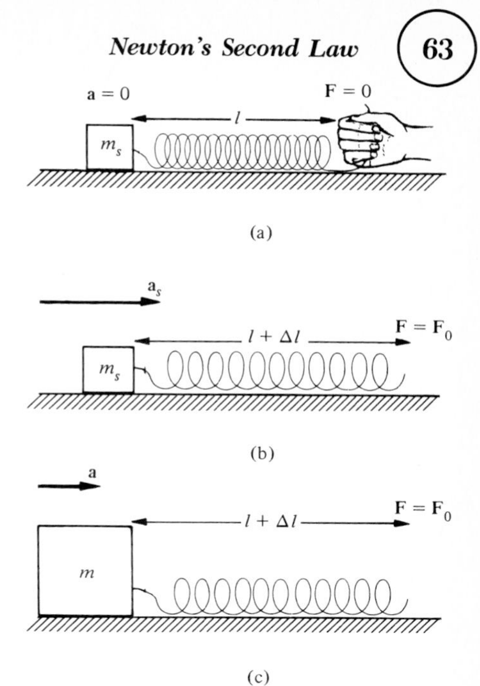

FIGURE 4.4. *Comparing unknown mass m with standard mass m_s by applying equal forces to the same spring and measuring accelerations.*

Weight of an Object

We discussed in Chapter 3 that freely falling objects are under the action (or force) of gravitational pull, which produces an acceleration g (= 9.8 m/s² or 32.2 ft/s²). Thus, we may say that an object of mass m is acted upon by a force $\mathbf{F} = m\mathbf{g}$. This force acting on an object due to the gravitational pull is called the *weight* of the object and is denoted by **W**; that is,

$$\boxed{\mathbf{W} = m\mathbf{g}} \qquad (4.11)$$

Thus, if the weight of an object is given, its mass is $m = \mathbf{W}/\mathbf{g}$.

Units of Force

From the units of mass and acceleration, we can derive the units of force. Corresponding to the three basic units of mass, length, and time, there are three units of force.

In the SI system the unit of force is the *newton* (N) and is defined as the force that will produce in a mass of 1 kg an acceleration of 1 m/s²; that is, 1 N = 1 kg-m/s².

In the CGS system the unit of force is the *dyne*, defined as the force that produces in a mass of 1 g an acceleration of 1 cm/s². Since 1 kg = 10^3 g and 1 m = 10^2 cm,

$$1 \text{ N} = 10^5 \text{ dyn}$$

Unlike the preceding systems, in which the units of mass are basic units while the units of force are derived, in the British engineering system the force is the basic unit and the unit of mass is the derived unit. Thus, in the engineering system, the basic unit of force is the pound and that of acceleration is ft/s², while the unit of mass (the derived unit) is called the slug. The *slug* is defined as that mass which when acted upon by a force of 1 lb will have an acceleration of 1 ft/s². Very often the pound is used mistakenly as a unit of mass. To avoid this, pound is sometimes called lb-force. The relation between the three force units is

$$1 \text{ N} = 10^5 \text{ dyn} = 0.225 \text{ lb}$$

TABLE 4.1
Force Units and Weights

System	Mass	Acceleration	Force	g	Weight
SI (MKS)	1 kg	1 m/s^2	1 N	9.8 m/s^2	9.8 N
CGS	1 g	1 cm/s^2	1 dyn	980 cm/s^2	980 dyn
British	1 slug	1 ft/s^2	1 lb	32.2 ft/s^2	32.2 lb

Also referring to Equation 4.11, we may say that a mass of 1 kg weighs 9.8 N, while a mass of 1 slug weighs 32.2 lb. This discussion is summarized in Table 4.1.

EXAMPLE 4.2 A car of mass 1200 kg moving at 22 m/s is brought to rest over a distance of 50 m. Calculate (a) the deceleration of the car; (b) the retarding force; and (c) the time it takes to stop.

(a) To calculate the deceleration a, let us use the equation $v^2 - v_0^2 = 2ax$, where $v_0 = 22$ m/s, $v = 0$, and $x = 50$ m. Thus,

$$a = \frac{v^2 - v_0^2}{2x} = \frac{0 - (22\ \text{m/s})^2}{2(50\ \text{m})} = -4.84\ \text{m/s}^2$$

The negative sign means that a is in the opposite direction to x.
(b) The retarding force F is given by

$$F = ma = (1200\ \text{kg})(-4.84\ \text{m/s}^2) = -5808\ \text{N}$$

F is in the opposite direction to x.
(c) From $v = v_0 + at$, we can calculate the time t the car takes to stop:

$$t = \frac{v - v_0}{a} = \frac{0 - (22\ \text{m/s})}{(-4.84\ \text{m/s}^2)} = 4.55\ \text{s}$$

EXERCISE 4.2 A 3000-lb automobile is traveling at 88 ft/s. How far will it travel before coming to a stop if the maximum retarding force applied by the brakes is 1000 lb? How long does it take before it stops? (Note that $m = W/g$.) [*Ans.*: 361 ft; 8.20 s.]

4.5. Newton's Third Law

The first law of motion concerns the state of motion of an object when the net external force acting on it is zero; the second law deals with the motion of the object when the net external force acting on it is not zero. The third law states the relation between the forces resulting from the interaction between two or more objects. For simplicity, let us consider two objects, A and B, that are completely isolated from their surroundings; and that A interacts with B and B interacts with A. This means that there is a force $\mathbf{F}_A$ acting on A due to the body B, and there is a force $\mathbf{F}_B$ on B due to the body A. According to Newton's third law, $\mathbf{F}_A$ is equal in magnitude and opposite in direction to $\mathbf{F}_B$;

that is,

$$\mathbf{F}_A = -\mathbf{F}_B \tag{4.12}$$

One of these forces, say $\mathbf{F}_A$, may be called *action* while the other force, $\mathbf{F}_B$ in this case, may be called *reaction*, or vice versa. This implies that we cannot say which is the cause or which is the effect. In words:

> **Newton's third law: To every action there is always an equal and opposite reaction;** ***that is, whenever one body exerts a certain force on a second body, the second body exerts an equal and opposite force on the first.***

This definition implies that the forces always exist in pairs. We may be misled into thinking that since the forces come in action–reaction pairs, the net force will be always equal to zero and hence can never produce a net acceleration. But this is not true. The reason is that the action and the reaction do not act on the same body—they always act on different bodies. To understand this, let us consider the following examples.

In Figure 4.5 is shown a mass m which is tied to one end of a string and hangs vertically. The other end of the string is tied to the ceiling. The earth pulls the mass m with a force $\mathbf{W}$ (equal to its weight) while the mass pulls the earth with an equal and opposite force $\mathbf{W}'$; thus $\mathbf{W}$ and $\mathbf{W}'$ form an action–reaction pair. Note that they act on different bodies. Next the string pulls the mass with a

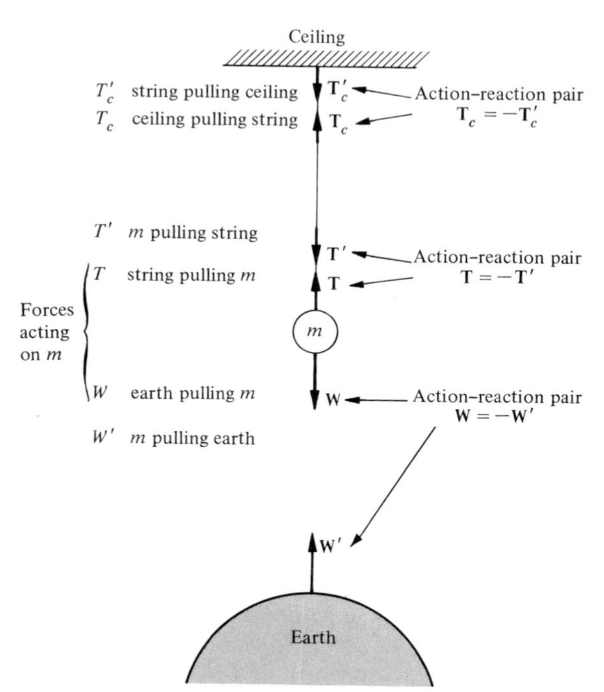

FIGURE 4.5. *Action–reaction pairs acting on mass m tied with a string to a ceiling.*

force **T**, called the *tension* in the string, while the mass pulls the string with an equal and opposite force **T′**. Thus, **T** and **T′** form another action–reaction pair. Furthermore, the ceiling pulls the string with a force $\mathbf{T}_c$, while the string pulls on the ceiling with force $\mathbf{T}_c'$. $\mathbf{T}_c$ and $\mathbf{T}_c'$ form another action–reaction pair.

The forces acting on the mass m are its weight **W** and tension **T** due to the string. Therefore, using Newton's second law,

$$\mathbf{T} + \mathbf{W} = m\mathbf{a}$$

But the mass is at rest, that is, $\mathbf{a} = 0$, so

$$\mathbf{T} + \mathbf{W} = 0 \qquad \text{or} \qquad \mathbf{T} = -\mathbf{W} \tag{4.13}$$

Thus, the tension in the string is a measure of the weight of the mass m and is the force with which the string pulls the mass. Note that even if **T** and **W** are equal and opposite, they do not form an action–reaction pair (they are acting on the same body).

As a second example, Figure 4.6 shows action–reaction pairs acting on a system with zero acceleration, $\mathbf{a} = 0$. Note that every part of this system is pulled by the earth with a force **W**, and the earth is pulled by this system with an equal and opposite force **W′**. This action–reaction pair is not shown in the figure.

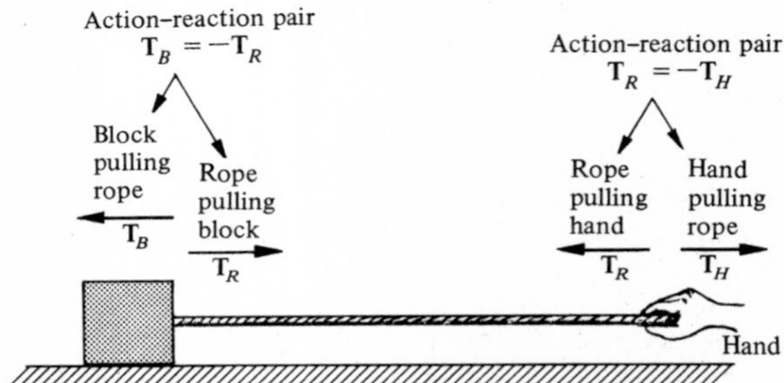

FIGURE 4.6. *Action–reaction pairs acting on a block pulled by a hand with* $\mathbf{a} = 0$.

EXAMPLE 4.3 An inextensible string connecting objects of masses m_1 and m_2 passes over a frictionless pulley as shown in the accompanying figure. Calculate (a) the acceleration of the system; and (b) the tension in each string. Calculate these quantities for $m_1 = 10$ kg and $m_2 = 5$ kg.

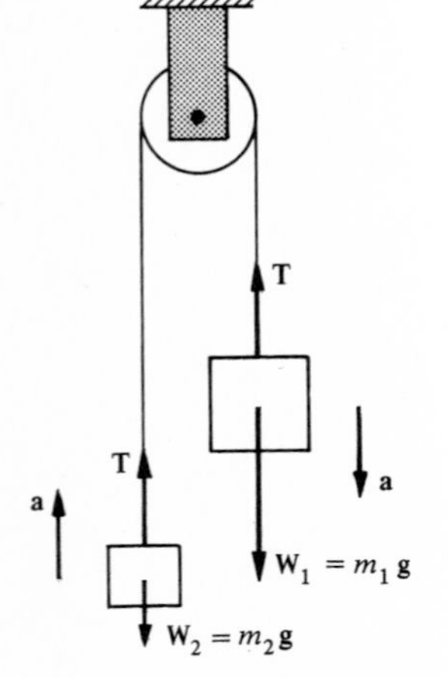

Ex. 4.3

Since the string is inextensible, both masses have the same acceleration a. Also, the pulley is frictionless, hence the tension T at both ends of the string is the same.

For mass m_1 moving downward with acceleration a downward as shown, we write $\Sigma F_y = m_1 a$ as

$$m_1 g - T = m_1 a \tag{i}$$

and for mass m_2 moving upward with acceleration a upward,

$$T - m_2 g = m_2 a \tag{ii}$$

Adding Equations i and ii gives $(m_1 - m_2)g = (m_1 + m_2)a$,

or

$$a = \frac{m_1 - m_2}{m_1 + m_2} g \tag{iii}$$

Substituting for a in either Equation i or ii and solving for T gives

$$T = \frac{2m_1 m_2}{m_1 + m_2} g \tag{iv}$$

Thus, for $m_1 = 10$ kg and $m_2 = 5$ kg, we get

$$a = \frac{(10 - 5)\text{kg}}{(10 + 5)\text{kg}}(9.8 \text{ m/s}^2) = 3.3 \text{ m/s}^2$$

and

$$T = \left(\frac{2(10)(5) \text{ kg}^2}{(10 + 5) \text{ kg}}\right)(9.8 \text{ m/s}^2) = 65.3 \text{ N}$$

The directions of a and T are as indicated in the figure.

Exercise 4.3 An inextensible string connecting masses m_1 and m_2 passes over a frictionless pulley as shown in the accompanying figure. Mass m_1 moves on a frictionless surface while m_2 moves vertically. Calculate (a) the acceleration of the system; and (b) the tension in each string. What are these values if $m_1 = 10$ kg and $m_2 = 5$ kg? [*Ans.*: (a) 3.27 m/s²; (b) 32.7 N.]

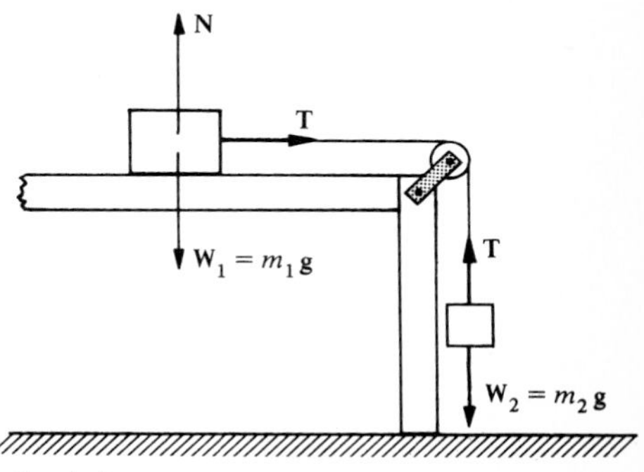

E. 4.3

4.6. Mass, Weight, and Weightlessness

An object will undergo acceleration **a** when a force **F** acts on it, that is, according to Newton's second law,

$$\mathbf{F} = m_i \mathbf{a} \tag{4.14}$$

where m_i is the *inertial mass* of the object. If this same object were suspended, the force of gravitational attraction would be equal to the weight **W**; that is,

$$\mathbf{W} = m_g \mathbf{g} \tag{4.15}$$

where **g** is the acceleration due to gravity, and m_g is called the *gravitational mass* of the object. We may say that the inertial mass of an object is a measure of its reluctance to change its state of motion when a force is applied, while the gravitational mass is a measure of the pull on the object exerted by the earth. All experimental evidence indicates that these two masses are completely equal and identical. This has led to an important principle, called the *principle of equivalence,* which is the foundation of the "general theory of relativity." In all our discussions where gravitational forces are also involved, we shall assume that

$$m_i = m_g = m \tag{4.16}$$

(This relation is true according to the theory of relativity.)

Thus, if the gravitational mass is equal to the inertial mass, it should be possible to add the accelerations **a** and **g**, thereby changing the effective (or apparent) value of **g**. If we could create a situation such that $\mathbf{a} + \mathbf{g} = 0$ (or $a - g = 0$), the weight of an object will be zero; that is, the object will be weightless. That such is the case can be understood by considering the following example.

Consider a mass m tied to a spring balance that is attached to the ceiling of an elevator as shown in Figure 4.7. The reading of the spring balance indicates the tension in the spring, which in turn is related to the weight **W** of the mass m. The reading of the spring balance is denoted by $\mathbf{W}'$ and is called the *apparent weight* of the body. Its value depends upon the acceleration of the elevator.

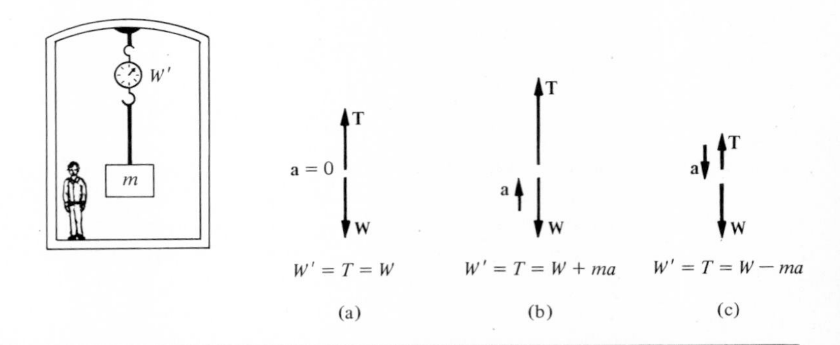

FIGURE 4.7. *Mass m observed by an observer in an elevator: (a) elevator at rest, (b) elevator moving upward with acceleration* **a**, *and (c) the elevator moving downward with acceleration* **a**.

ELEVATOR AT REST OR MOVING WITH UNIFORM VELOCITY. In either case $\mathbf{a} = \Delta\mathbf{v}/\Delta t = 0$ and, as shown in Figure 4.7(a), $T = W$. Since the apparent weight W' indicated by the spring balance is equal to the tension T, $W' = T = W$. That is, the apparent weight is equal to the gravitational force on the object according to an observer inside the elevator.

ELEVATOR WITH UNIFORM ACCELERATION MOVING UPWARD. Let the elevator move upward with acceleration **a**. Taking the upward direction as positive, the situation is as shown in Figure 4.7(b), and we may write

$$T - W = ma \qquad \text{or} \qquad T = W + ma$$

The apparent weight $= W' = T = W + ma = mg + ma$:

$$W' = T = m(g + a) \qquad a \text{ upward} \tag{4.17}$$

and the rope must supply support not only for the gravitational pull but for the whole apparent weight. This means that if $a = 3g$, $W' = m(g + 3g) = m(4g)$, that is, the apparent value of g has been changed from $1g$ to $4g$. This is the situation experienced by the astronauts during the takeoff process in rockets.

Elevator with Uniform Acceleration Moving Downward. The effective value of g can be decreased if the elevator moves downward with uniform acceleration a as shown in Figure 4.7(c). In this situation we may write

$$T - W = -ma \qquad \text{or} \qquad T = W - ma$$

The apparent weight $= W' = T = W - ma = mg - ma$:

$$W' = T = m(g - a) \qquad a \text{ downward} \tag{4.18}$$

It is clear that the apparent weight W' is less than the gravitational force on the object.

An interesting situation arises when $a = g$ and Equation 4.18 takes the form

$$W' = T = m(g - g) = 0 \tag{4.19}$$

That is, according to the observer in the elevator, the object appears to be *weightless*. Of course, this should not be surprising because when $a = g$, the motion corresponds to a free fall, in which case the rope must not exert any force. Hence, the spring balance will read zero; that is, the apparent weight is zero, or the object seems to be weightless.

We shall see in Chapter 11 that weightlessness can also be achieved by objects moving in circular orbits, as experienced by astronauts in satellites going around the earth in orbital paths.

Example 4.4 An object of mass 50 kg is hanging by a rope attached to the ceiling of an elevator. Calculate the tension in this rope if the elevator has an acceleration of (a) 4 m/s² upward; (b) 9.8 m/s² upward; (c) 4 m/s² downward; (d) 9.8 m/s² downward. (e) Calculate the tension if the elevator is moving with uniform velocity of 5 m/s upward or downward.

If acceleration a is upward, $T - W = ma$ or $T = m(g + a)$.

(a) For $a = 4\ \text{m/s}^2$, $T = 50\ \text{kg}\,(9.8\ \text{m/s}^2 + 4\ \text{m/s}^2) = 690\ \text{N}$

(b) For $a = 9.8\ \text{m/s}^2$, $T = 50\ \text{kg}\,(9.8\ \text{m/s}^2 + 9.8\ \text{m/s}^2) = 980\ \text{N}$

If acceleration a is downward, $T - W = -ma$ or $T = m(g - a)$.

(c) For $a = 4\ \text{m/s}^2$, $T = 50\ \text{kg}\,(9.8\ \text{m/s}^2 - 4\ \text{m/s}^2) = 290\ \text{N}$

(d) For $a = 9.8\ \text{m/s}^2$, $T = 50\ \text{kg}\,(9.8\ \text{m/s}^2 - 9.8\ \text{m/s}^2) = 0$

(e) Moving with uniform velocity means that $a = 0$. That is, $T - W = 0$ and hence, whether moving up or down,

$$T = W = mg = 50\ \text{kg}\,(9.8\ \text{m/s}^2) = 490\ \text{N}$$

Exercise 4.4 An elevator is moving upward with an acceleration of 3 m/s² when a spring balance attached to the ceiling and with an object hanging from it reads 100 N. (a) What is the magnitude of the mass? (b) What will the spring balance read if the elevator were moving downward with the same acceleration? (c) Under what conditions will the spring balance read the true weight of the object? [*Ans.*: (a) 7.8 kg; (b) 53 N; (c) if elevator is moving with constant speed.]

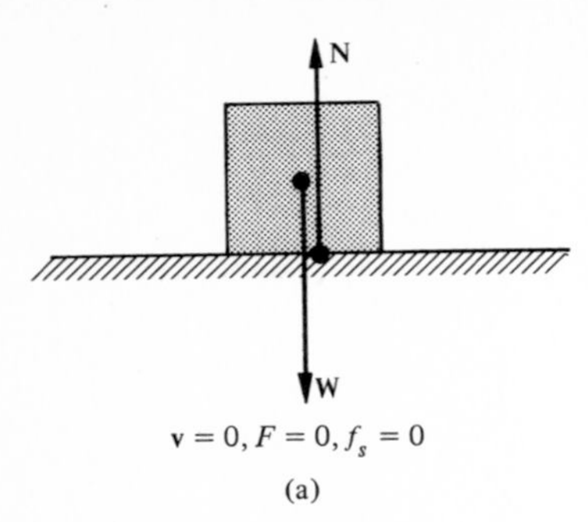

$v = 0, F = 0, f_s = 0$

(a)

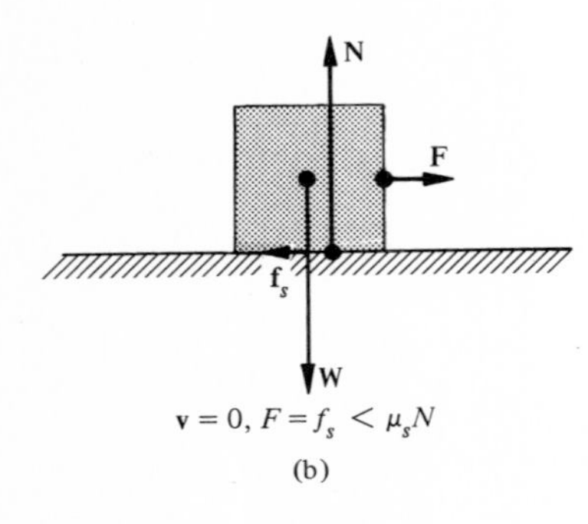

$v = 0, F = f_s < \mu_s N$

(b)

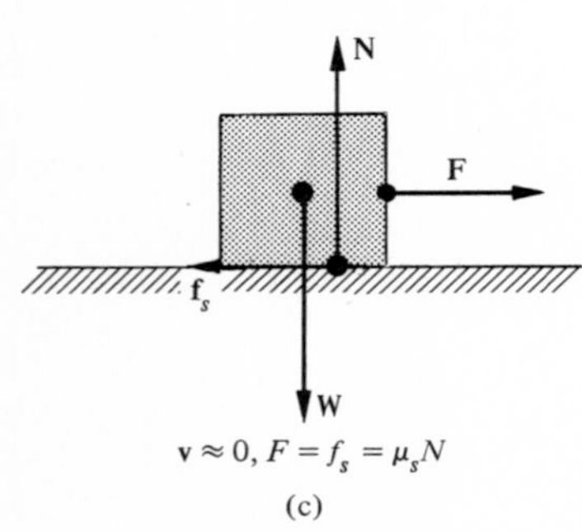

$v \approx 0, F = f_s = \mu_s N$

(c)

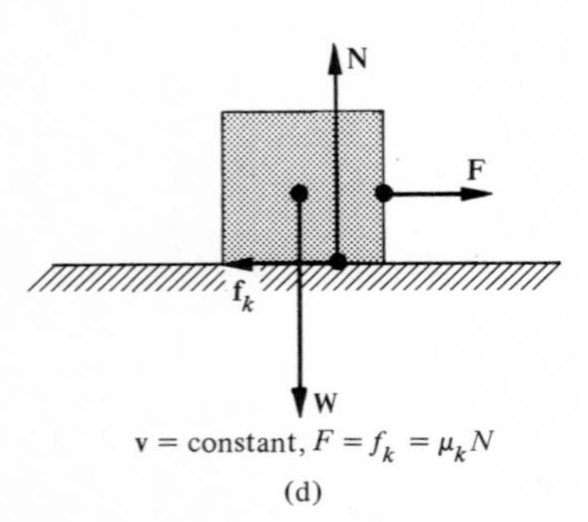

$v = \text{constant}, F = f_k = \mu_k N$

(d)

FIGURE 4.8. *Block of weight* **W** *under the action of forces* **F**, **N**, *and* $\mathbf{f}_s$ *(or* $\mathbf{f}_k$*).*

4.7. Frictional Forces

Another force of common occurrence is the frictional force that results when the surfaces of two bodies are in contact. Consider a block resting on a table. According to the second law of motion, if a force, no matter how small, is applied to this block, it should start moving. But this does not happen. A certain minimum force is needed to set the block in motion. This means that before the motion starts, there must be an equal and opposite force which prevents the block from moving. This opposition or resistance to the motion is called *friction*, while the resisting force is called *frictional force*. The frictional force is an obstruction to any type of motion and thereby causes wear and tear on the moving parts of all types of machines. Engineers and manufacturers of such machines are always busy trying to find ways to reduce friction. On the other hand, without friction it would be impossible to walk, push, pull, roll, operate automobiles, and so on.

Consider a block of mass M, and hence of weight $\mathbf{W} = M\mathbf{g}$, resting on a horizontal surface of a tabletop, as shown in Figure 4.8(a). The block is in equilibrium under the action of its weight **W** and an upward force **N** exerted on the block by the surface; that is, $\mathbf{N} - \mathbf{W} = 0$. The lines of action of **W** and **N** are shown slightly displaced for clarity.

Let us apply a small, horizontal force **F** as shown in Figure 4.8(b), so that the block does not move. Thus, **F** must be balanced by an equal and opposite force $\mathbf{f}_s$, as shown. There are four forces, **W**, **N**, **F**, and $\mathbf{f}_s$ acting on the block. The force **N** is called the *normal reaction* and is the force exerted by the surface on the block and is always normal to the surface. The horizontal force $\mathbf{f}_s$ is called the force of *static friction*. We may point out that the friction force $\mathbf{f}_s$ is self-adjusting; that is, when **F** is zero, $\mathbf{f}_s$ will be zero, and as **F** increases, so does $\mathbf{f}_s$, until **F** reaches such a value that $\mathbf{f}_s$ cannot increase any more, and hence the motion starts. (Note that the point of application of **N** depends upon **F** but is always normal to the surface.) In Figure 4.8(c), the applied force **F** has reached the maximum frictional force and the motion is just about to start. It is found experimentally that if the normal reaction **N** is increased by using a heavier block or by adding weights at the top of the block, the force **F** needed to just start the motion will have to be increased. This means that the frictional force is directly proportional to the normal reaction; that is,

$$f_s \propto N$$

or

$$\boxed{f_s \leq \mu_s N} \quad \text{for } v = 0 \tag{4.20}$$

where the constant of proportionality μ_s is called the *coefficient of static friction*. f_s reaches a maximum value of $\mu_s N$ when $F = f_s = \mu_s N$.

Further increase in **F** results in the accelerated motion of the block. Once the motion starts the frictional force $\mathbf{f}_k$, called the *kinetic frictional force*, comes into play and is less than $\mathbf{f}_s$. Thus, for the block to move with uniform velocity, the force **F** may be decreased as shown in Figure 4.8(d), so that

$$\boxed{F = f_k = \mu_k N} \quad \text{for constant } v \tag{4.21}$$

where μ_k is the *coefficient of kinetic friction* and is always less than μ_s. That the force of sliding or kinetic friction $\mathbf{f}_k$ is less than the force of static friction $\mathbf{f}_s$ may be observed in everyday experience from the fact that less force is needed to keep a body in motion than is needed to start the body from rest. Table 4.2 shows a few values of μ_s and μ_k.

We may note that if the direction of the applied force is reversed, the direction of $\mathbf{f}_s$ also reverses, without changing the direction of **N**.

Let us consider the following commonly encountered situation. A block is placed on an inclined plane as shown in Figure 4.9. In such cases it is convenient to take the X-axis parallel to the surface of the inclined plane and the Y-axis perpendicular to it, as shown. The weight **W** may be resolved into two components, $W \cos\theta$ and $W \sin\theta$. Suppose that the angle θ of the inclined plane was such that $W \sin\theta$ is equal to f_s and that the block just barely starts to move. Thus,

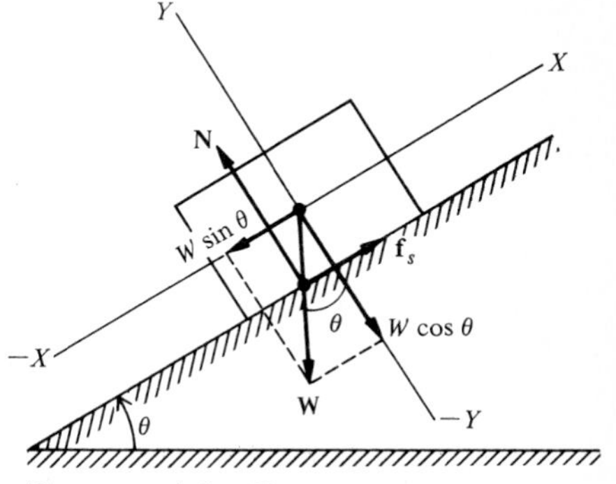

FIGURE 4.9. *Forces acting on a block of mass M on an inclined plane.*

$$\sum F_x = W \sin\theta_s - f_s = 0 \tag{4.22}$$

$$\sum F_y = N - W \cos\theta_s = 0 \tag{4.23}$$

$$f_s = \mu_s N \tag{4.24}$$

On rearranging,

$$W \sin\theta_s = f_s \qquad \text{and} \qquad W \cos\theta_s = N$$

Dividing one by the other gives

$$\frac{\sin\theta_s}{\cos\theta_s} = \tan\theta_s = \frac{f_s}{N} \tag{4.25}$$

TABLE 4.2
Typical Values of Average Coefficients of Friction

Material	μ_s	μ_k
Steel on steel (mild)	0.74	0.57
Steel on steel (hard)	0.78	0.42
Aluminum on steel	0.61	0.47
Copper on steel	0.53	0.36
Cast iron on cast iron	1.10	0.15
Nickel on nickel	1.10	0.53
Glass on glass	0.94	0.4
Copper on glass	0.68	0.53
Teflon on Teflon	0.04	0.04
Teflon on steel	0.04	0.04

Comparing with Equation 4.24, we may say that

$$\boxed{\mu_s = \tan\theta_s} \tag{4.26}$$

Thus, the coefficient of static friction may be found by slowly increasing the angle of the inclined plane and measuring the angle θ_s at which the block just starts sliding with slight tapping. The tangent of this angle is μ_s.

So far we have considered sliding friction, that is, one flat solid surface moving over another flat surface. Even though a flat surface may look very smooth and to be of negligible friction, on examining it under magnification we find that the surface is not smooth at all, as is clear from Figure 4.10. Of course, all solids are combinations of molecules, and hence unevenness results from the arrangement of these molecules. When two surfaces are placed together, there occurs some sort of welding between these uneven portions, owing to the interaction between the molecules. The interaction between the molecules is referred to as *cohesion* or *adhesion*, depending upon whether the two surfaces are alike (cohesion-like molecules) or different (adhesion-unlike molecules). Thus when one body is pulled over another, the frictional force is associated with the rupturing and re-forming of these tiny welds. As the motion continues, the welds between the surfaces keep on welding and rupturing continuously.

The friction can be reduced by rolling one body over the other, and such friction is called *rolling friction*. Also, friction may be reduced by using lubricants. So far we have been talking about the motion of a solid over another

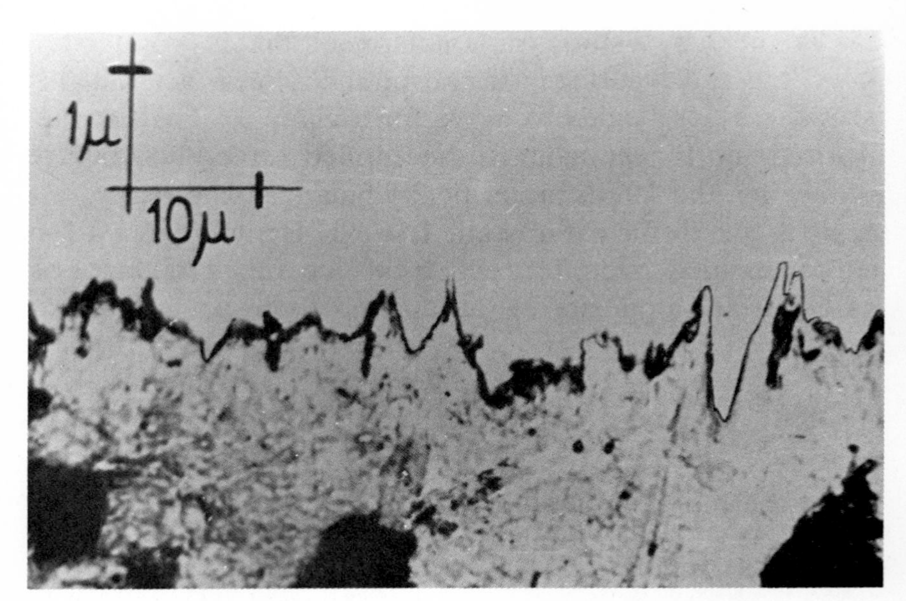

FIGURE 4.10. *Highly magnified view of a polished steel surface. The vertical scale is magnified 10 times more than the horizontal scale.* (From F. P. Bowden and D. Tabor, *Friction and Lubrication of Solids,* Oxford University Press, Oxford, 1950.)

solid. When we consider fluids (liquids or gases), the term *viscosity* is used instead of friction.

Example 4.5 An automobile of mass 1500 kg has to be given an acceleration of 2.0 m/s^2 on a rough surface where the coefficient of rolling friction is 0.15. Calculate (a) the force needed to accelerate; and (b) the force needed to keep the car moving at a constant velocity. Suppose that the engine was shut off when the automobile was moving at 22 m/s. (c) What will be the deceleration; and (d) how far will it travel before stopping?

(a) To accelerate the car, the applied force F must exceed the frictional force f. That is,

$$F - f = ma \tag{i}$$

where $f = \mu N = \mu mg$, or

$$\begin{aligned} F &= ma + \mu mg \\ &= (1500\text{ kg})(2.0\text{ m/s}^2) + (0.15)(1500\text{ kg})(9.8\text{ m/s}^2) \\ &= 3000\text{ N} + 2205\text{ N} = 5205\text{ N} \end{aligned}$$

(b) When moving at constant velocity, $a = 0$. From Equation i,

$$F = f = \mu N = \mu mg = 2205\text{ N}$$

(c) Once the engine is shut off, the only force is the retarding force of friction f. From Equation i, if $F = 0$,

$$a = -\frac{f}{m} = -\frac{\mu mg}{m} = -\mu g = -(0.15)(9.8\text{ m/s}^2) = -1.47\text{ m/s}^2$$

(d) From the relation $v^2 - v_0^2 = 2ax$, where $v_0 = 22$ m/s, $v = 0$, and $a = -1.47$ m/s^2, the value of x is

$$x = \frac{v^2 - v_0^2}{2a} = \frac{0 - (22\text{ m/s})^2}{2(-1.47\text{ m/s}^2)} = 164.6\text{ m}$$

Exercise 4.5 An automobile of mass 1500 kg is moving at 10 m/s. The engine is supplying a force of 2500 N while the frictional force is 1000 N. Calculate the acceleration. How long and how far will the car travel before its speed is doubled? [*Ans.*: $a = 1$ m/s^2; $x = 150$ m, $t = 10$ s.]

Example 4.6 A block of mass 40 kg is being pulled up an inclined plane with $\theta = 37°$ by a force $T = 400$ N which is parallel to the surface of the inclined plane, as shown in the accompanying figure. Calculate (a) the force exerted by the surface on the block; (b) the acceleration of the block; and (c) the time it will take to pull the block 10 m. The coefficient of friction is 0.3.

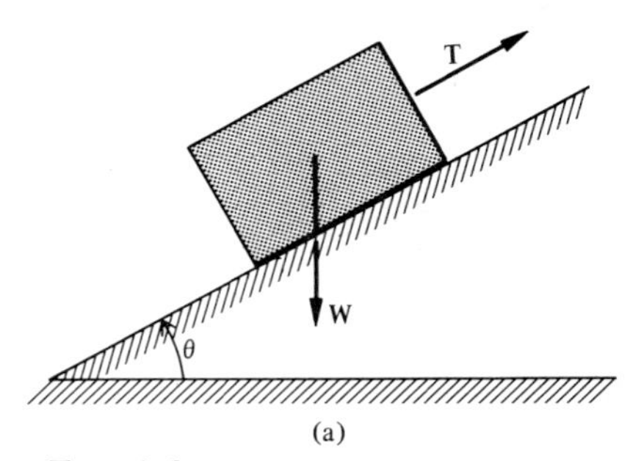

(a)

Ex. 4.6

(a) Let the X-axis be parallel to the surface of the inclined plane and the Y-axis be perpendicular to the plane. **T** is parallel to the $+X$-axis; **W** can be resolved into $W \sin\theta$ and $W \cos\theta$; the friction force **f** acts along the $-X$-axis; while the normal reaction **N** exerted by the plane on the block acts along the $+Y$-axis as shown in the accompanying figure. Equations $\Sigma F_x = ma_x$ and $\Sigma F_y = ma_y$ will take the form

$$\sum F_x = T - W\sin\theta - f = ma_x \qquad \text{(i)}$$

$$\sum F_y = N - W\cos\theta = 0 \qquad \text{(ii)}$$

$$f = \mu N \qquad \text{(iii)}$$

(b)

Since $W = mg = (40\text{ kg})(9.8\text{ m/s}^2) = 392\text{ N}$, and as given $\theta = 37°$, from Equation ii,

$$N = W\cos\theta = mg\cos 37° = 392\text{ N}(0.8) = 313.6\text{ N}$$

(b) From Equations i and iii,

$$\begin{aligned} ma_x &= T - W\sin\theta - \mu N = 400\text{ N} - 392\text{ N}\sin 37° - 0.3(313.6\text{ N}) \\ &= 400\text{ N} - 392\text{ N}(0.602) - 0.3(313.6\text{ N}) = 69.4\text{ N} \end{aligned}$$

$$a_x = \frac{69.4\text{ N}}{40\text{ kg}} = 1.74\text{ m/s}^2$$

(c) $v_0 = 0$, $x = 10$ m, $a = 1.74\text{ m/s}^2$, find t. From the relation

$$\begin{aligned} x &= v_0 t + \tfrac{1}{2}at^2 \\ 10\text{ m} &= \tfrac{1}{2}(1.74\text{ m/s}^2)t^2 \\ t^2 &= 11.5\text{ s}^2 \\ t &= 3.4\text{ s} \end{aligned}$$

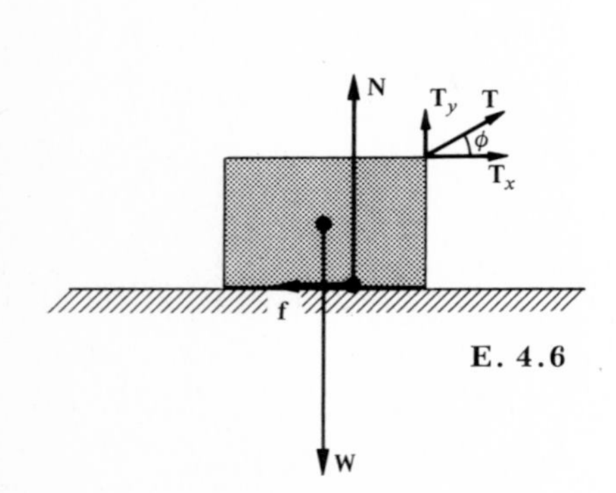

E. 4.6

EXERCISE 4.6 Repeat Example 4.6 if $\theta = 0°$ and T makes an angle of $\phi = 37°$ with the horizontal as shown in the accompanying figure. [*Ans.*: (a) $N = 151.2$ N; (b) $a = 6.86\text{ m/s}^2$; (c) $t = 1.71$ s.]

SUMMARY

Force is defined as a push or pull that produces, or tends to produce motion, stops or tends to stop motion. There are only four known types of forces: nuclear, electromagnetic, weak, and gravitational, with relative strengths of 1, $\frac{1}{137}$, 10^{-13}, and 10^{-39}, respectively.

Momentum $p = mv$, force $F = \Delta p/\Delta t$, and the impulse is $F\,\Delta t$. The *impulse-momentum theorem* states that the change in momentum is equal to the total impulse.

Newton's three laws of motion are:

1. Every object continues in its state of rest or uniform motion in a straight line unless a net external force acts on it to change that state.
2. The time rate of change of momentum is directly proportional to the force

applied and takes place in the direction of the force. That is, $\mathbf{F} = \Delta\mathbf{p}/\Delta t$ or $\mathbf{F} = m\mathbf{a}$.

3. To every action there is always an equal and opposite reaction.

The unit of force is the *newton*, defined as that force which produces for a mass of 1 kg an acceleration of 1 m/s^2.

The *weight* of an object of mass m is $W = mg$. If an elevator is accelerated upward, the apparent weight is $W' = T = m(g + a)$, and if the acceleration is downward, $W' = T = m(g - a)$.

The *force of friction* is directly proportional to the normal reaction, $f_s \leq \mu_s N$ for $v = 0$ and $f_k = \mu_k N$ for constant v.

QUESTIONS

1. Give some examples of situations in which gravitational, electromagnetic, or nuclear forces act.
2. Explain the roles that the various forces play in a stable atom.
3. Suppose that the plot of v versus t is shown in the figure. Make approximate plots of p versus t and F versus t.
4. To hit a ball hard, the batter always swings the bat before just hitting the ball. Explain why?
5. Explain why you tend to fall backward when an airplane is taking off and forward when it is landing (before coming to a stop)?
6. Is it possible to have two bodies that would have the same masses but different apparent weights?
7. What are the dimensions of g, N, and μ?
8. In terms of force, length, and time, what are the dimensions of mass?
9. Name the action–reaction pairs in situations that are different from those listed in the text.
10. In a tug-of-war, the harder you push against the ground, the better chance you have of winning. Explain.
11. Why is the friction between two pieces of wood reduced when they are smooth and polished?
12. Without friction it would be impossible to produce many types of motion—walking, pushing, pulling, and rolling. Explain why.
13. Is there any limit to polishing a surface to reduce friction? (After a certain amount of polishing, friction increases. Why?)
14. Is it possible to have μ greater than 1?
15. A block of mass m on an inclined plane slides downward even if there is an external force acting upward parallel to the plane. What is the direction of the frictional force?

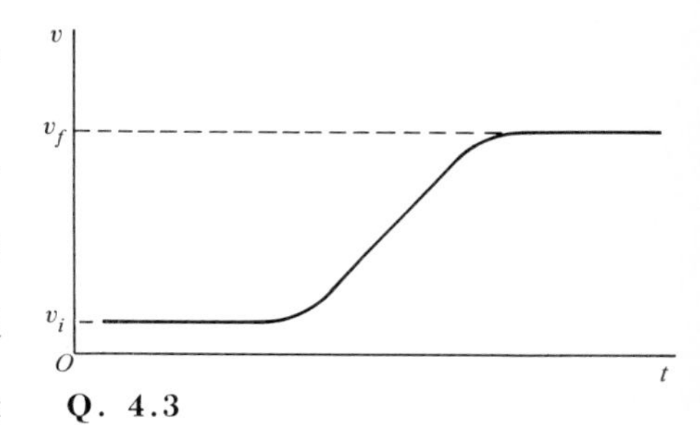

Q. 4.3

PROBLEMS

1. Calculate the linear momentum in the following situations: (a) a golf ball of mass 70 g moving with a velocity of 8 m/s; (b) a bullet of mass 10 g moving with a speed of 2000 m/s; and (c) an electron of mass 9.1×10^{-31} kg moving with a speed of 3×10^7 m/s.
2. A golf ball of mass 70 g when struck by a golf club acquires a speed of 75 m/s. The contact between the club and the ball lasts 10^{-4} s. Calculate

(a) the initial momentum; (b) the final momentum; (c) the impulse; and (d) the rate of change of momentum or average force.

3. A tennis ball of mass 100 g hits a wall with a speed of 20 m/s and rebounds with the same speed. Calculate the impulse and the average force imparted to the wall. The contact time is 10^{-3} s.
4. An electron traveling at 2.6×10^7 m/s (mass = 9.1×10^{-31} kg) hits the television screen and comes to rest within 1 mm. Calculate the following: (a) the initial momentum; (b) the final momentum; (c) the impulse; and (d) the rate of change of momentum or average retarding force.
5. A gas molecule of mass 6.68×10^{-26} kg moving with a speed of 1200 m/s hits head on the surface of a piston of an engine and rebounds with the same speed. The contact time is only 0.001 s. Calculate (a) the initial momentum; (b) the final momentum; (c) the impulse imparted to the piston; and (d) the average force exerted on the piston by the molecule.
6. Each of the four engines of a Boeing 747 jet can develop a thrust of 2×10^5 N. If its mass is 3×10^5 kg, what acceleration will the aircraft develop before taking off?
7. An airplane has a weight of 24,000 lb and it can accelerate to 8 ft/s^2. If it has four engines, what thrust should be supplied by each?
8. A car of mass 1500 kg starting from rest can reach a speed of 80 km/h within 10 s. Calculate the accelerating force of the car engine.
9. A sled with a boy sitting on it is moving with a speed of 5 m/s and comes to a stop within 50 m. Calculate the deceleration and the retarding force. The mass of the sled and the boy is 50 kg.
10. A car weighing 4000 lb and traveling at 50 mi/h is brought to rest within 160 ft after the brakes are applied. Calculate the deceleration and the retarding force.
11. Can a car of mass 1200 kg traveling at 15 m/s stop within a distance of 40 m if a braking force of 3600 N is applied?
12. A bullet of mass 2 g moving at 400 m/s strikes a wooden target and penetrates a distance of 5 cm before coming to rest. Calculate the deceleration of the bullet, and the retarding force applied by the wood.
13. A bullet of mass 2 g moving at 300 m/s penetrates 10 cm into a sandbag. What is the average retarding force?
14. A body starting from rest moves a distance of 60 m in 5 s over a smooth surface when a constant horizontal force of 50 N is applied. What is the mass of the body?
15. A force of 100 N acts on a body of mass 10 kg for a period of 5 s. Calculate (a) the acceleration; (b) the final velocity; and (c) the distance it travels in this time.
16. An electron (mass = 9.1×10^{-31} kg) starting from rest travels a distance of 2 cm and achieves a final velocity of 10^7 m/s. Calculate (a) the acceleration; (b) the accelerating force; and (c) the time it takes to travel 2 cm.
17. What is the weight of a body at the surface of the moon when its weight on the surface of the earth is 1 N? What is its mass? (g on the moon is 1.62 m/s^2.)
18. What should be the force of the air resistance opposing a descending parachute if we want its acceleration reduced to 2 m/s^2? The mass of the parachute and the aviator is 75 kg.

19. What should be the tension in a rope pulling vertically upward a mass of 100 kg so that starting from rest it acquires a velocity of 4 m/s in 1.5 s?

20. A force of 20 N making an angle of 30° with the horizontal is applied to a block of 100 kg. Calculate (a) the acceleration of the block; and (b) the force with which the ground pushes the block.

21. A mass of 100 kg is dropped through a height of 50 m. What is the acceleration of the earth just before the mass hits the earth? Is this acceleration negligible as compared to the acceleration of the mass? The mass of the earth is 5.98×10^{24} kg.

22. A mass of 100 kg tied with a cable is being raised upward. What will be the tension in the cable if the mass is moving with an acceleration of 5 m/s^2?

23. When an elevator is not moving, a man standing on a spring scale reads 170 lb. What will the scale read if the elevator were accelerated upward at 8 ft/s^2?

24. A 70-kg woman is standing in an elevator. What will be the force exerted by the floor of the elevator on the woman when the elevator is moving (a) with an acceleration of 2 m/s^2 (1) upward and (2) downward; and (b) with a uniform velocity of 2m/s (1) upward and (2) downward?

25. The total mass of an elevator with a 80-kg man in it is 1000 kg. This elevator moving upward with a speed of 8 m/s, is brought to rest over a distance of 20 m. Calculate (a) the tension T in the cables supporting the elevator; and (b) the force exerted on the man by the elevator floor.

26. There is a tension of 50,000 N in the cable of an elevator that is moving upward and has a weight of 40,000 N. Calculate (a) the acceleration; and (b) the distance it travels in 3 s.

27. A man of mass 70 kg is sliding down a rope that can sustain a tension of 600 N. Calculate the smallest acceleration possible without breaking the rope.

28. Calculate the horizontal force that must be applied to a 5-kg block to give it a velocity of 5 m/s in 2 s provided that (a) the block is on ice with no frictional force present; and (b) on a rough surface with a coefficient of friction of 0.12.

29. An object of mass 2000 kg is being pulled by a tractor with a constant velocity of 5 m/s on a rough surface. If the coefficient of friction is 0.8, what is the magnitude and direction of (a) frictional forces; and (b) the applied force?

30. A pickup truck of mass 3000 kg going at 25 m/s ($\approx$55 mi/h) has its brakes locked and starts skidding with the engine shut off. If the coefficient of friction is 0.20, how far will the truck travel before stopping?

31. The coefficient of friction between a block of wood of mass 20 kg and the table surface is 0.25. (a) What force is needed to give it an acceleration of 0.2 m/s^2? (b) What forces are acting when it is moving with a constant speed of 2 m/s? (c) If no external force is applied, how far will the block travel before coming to rest?

32. A block of mass m is sliding down a frictionless plane inclined at an angle θ with the horizontal. What are the normal reaction and acceleration for θ equal to 30°, 60°, or 90°?

33. A block of mass m is sliding down an inclined plane making an angle θ with the horizontal. Find the acceleration of the block and the normal reaction if $\theta = 30°$, $\mu = 0.5$, and $m = 10$ kg.

34. If $\mu_s = 0.72$, what will be the angle of the inclined surface when a mass on the surface will start sliding?

35. What is the coefficient of static friction if a block sitting on an inclined plane starts sliding when the surface of the incline makes an angle of 40° with the horizontal?

36. A block of mass 1 kg is projected up a 30° inclined plane with a velocity of 10 m/s. How far and for how long will the block move before coming to a stop? The coefficient of friction is 0.42.

37. In the pulley problem of Example 4.3, if $m_1 = 20$ kg, $m_2 = 10$ kg, what are a and T? How long will it take for the 20-kg block to fall through a distance of 2 m?

38. A block of mass 30 kg resting on a horizontal table is connected by a string over a frictionless pulley to a vertically hanging mass of 10 kg. The coefficient of friction between the block and the tabletop is 0.3. Calculate the acceleration of each block and the tension in the string.

5 Torque, Equilibrium, and Center of Mass

In order for a rigid body to be in translational as well as rotational equilibrium, it must satisfy the two conditions of equilibrium, which we investigate now. In addition to the role played by the force, we must understand the effect of torque as applied to a rigid body. Finally, we describe the motion of the center of mass of a system of particles under the influence of external forces.

5.1. Bodies in Equilibrium

WHEN A body at rest is acted upon by one or more forces simultaneously, several things may happen. The body may change its shape or size, or it may change its state from one of rest to one of translational or rotational motion. If the body is already in motion, its linear velocity or rotational velocity may increase or decrease. In this chapter we shall assume that we are dealing with rigid bodies. A *rigid body* is one in which the mean distances between different atoms and molecules remain constant. This means that the forces acting on the rigid bodies will not produce any changes in their shape and size. Thus, under the action of external forces, a rigid body may remain at rest, change its translational state of motion, change its rotational state of motion, or both states of motion.

A body is said to be in *equilibrium* if the external forces act in such a way that there is no change in its translational and rotational states. A body that remains at rest or moves with a uniform speed in a straight line is said to be in *translational equilibrium*. On the other hand, if under the action of external forces the body does not rotate at all or rotates at constant angular speed, it is said to be in *rotational equilibrium*.

Before investigating conditions that lead to translational and rotational equilibrium, it is necessary to understand how these motions are produced. In Chapter 4 we saw how the application of force produces translational motion. In the next section we shall study the causes of rotational motion.

5.2. Torque—Moment of Force

The effect of an external force acting on a rigid body allowed to rotate about an axis is to produce rotational acceleration about this axis. A quantity that is a measure of this rotational effect produced by the force is called the *torque* τ or moment of force, as defined below. Consider a rod mounted in such a way that it is capable of rotating in a plane perpendicular to an axis passing through O, as shown in Figure 5.1. To produce rotation, let us apply a force $\mathbf{F}$ acting in the same plane as the rod and at a perpendicular distance d from the axis of rotation. The rotational effects depend upon two factors: (1) the magnitude of the force F, and (2) the perpendicular distance d from the axis of rotation to the line of action of the force called the *lever arm* or *moment arm*. The product of

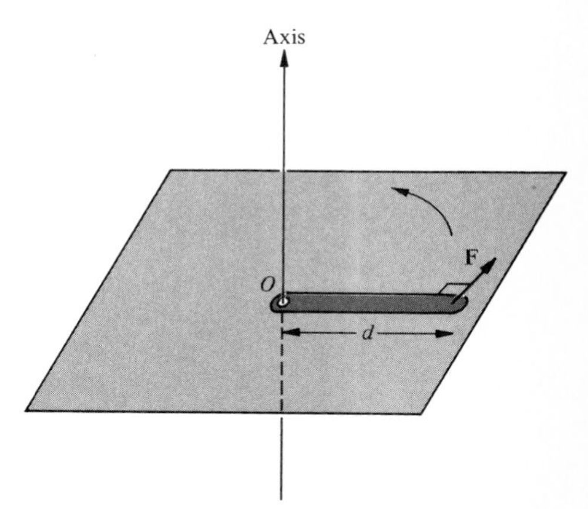

FIGURE 5.1. *Force* **F**, *acting in the plane at a distance d from an axis through O, produces rotational acceleration of the rod about the axis.*

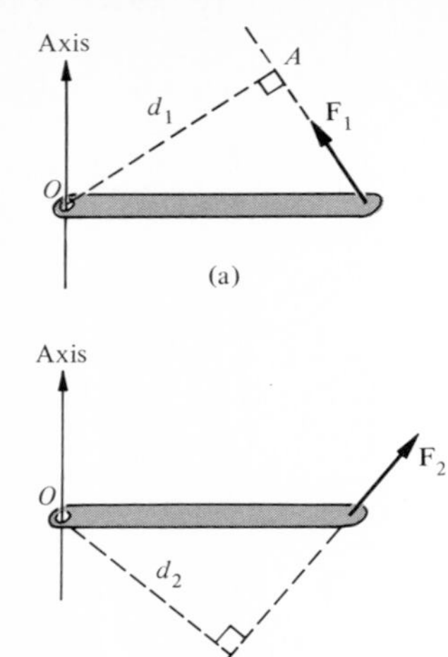

Figure 5.2. *The magnitude of the torque is given by (a) $\tau = F_1 d_1$ and (b) $\tau = F_2 d_2$.*

the applied force F and the moment arm d is called the *torque* or *moment of force* τ.

$$\text{torque} = \text{force} \times \text{moment arm}$$

$$\tau = Fd \tag{5.1}$$

Suppose that the force **F** is not perpendicular to the rod, as shown in Figure 5.2(a) and (b). In order to find the magnitude of the torque, we must find the moment arm by drawing perpendiculars OA and OB, as marked by d_1 and d_2, respectively, in these two cases. Thus, $\tau = F_1 d_1$ is shown in Figure 5.2(a) and $\tau = F_2 d_2$ in Figure 5.2(b).

To arrive at the definition of the torque in a more general way, consider a rigid body capable of rotating about an axis passing through O and acted upon by a force **F** at point P, as shown in Figure 5.3. Let us resolve **F** into two

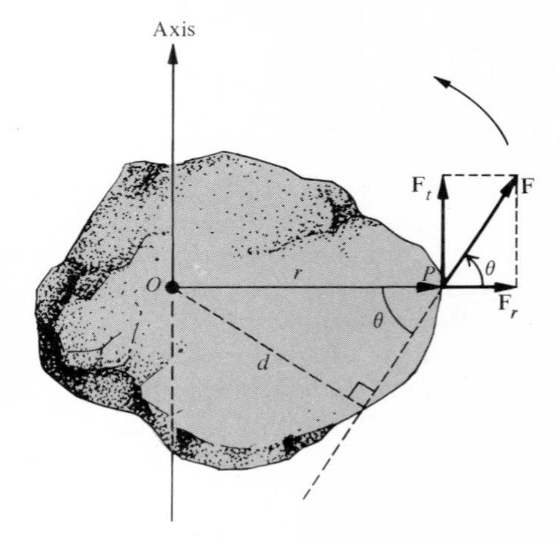

Figure 5.3. *Torque produced by the force* **F** *is equal to $\tau = F_t r = Fd \sin \theta$.*

components—the radial component $\mathbf{F}_r$ and the tangential component $\mathbf{F}_t$. The line of action of the radial component F_r passes through the axis, and hence the moment arm of F_r is zero, which means that the torque due to F_r is zero. Hence, F_r has no tendency to produce rotational motion. The tangential component $F_t = F \sin \theta$, for which r is the moment arm, produces rotational motion and the torque is given by

$$\tau = F_t r = Fr \sin \theta$$

But $r \sin \theta = d$, where d is the perpendicular distance from O to the line of action of the force **F**, and we may write the torque as

$$\boxed{\tau = F_t r = Fr \sin \theta = Fd} \tag{5.2}$$

Note that $\tau = Fd$ agrees with the definition of torque as stated above. Thus, we may define torque as

$$\text{torque} = \text{force} \times \text{perpendicular distance}$$

or

$$\text{torque} = \text{tangential component of force} \times \text{distance}$$

The units of torque are those of force multiplied by distance. Thus, the SI units of torque are N-m, the CGS units are dyn-cm, and the British units are lb-ft.

Next let us briefly discuss the sign convention used for the quantity torque. In Figure 5.4(a) the applied force F has a tendency to rotate the rod counterclockwise, while in Figure 5.4(b) the direction of the applied force with respect to the rotation axis is such that the rod has a tendency to rotate clockwise. *A torque producing a counterclockwise rotation is positive, while a torque producing a clockwise rotation is negative.*

As we shall see later, torque is a vector quantity and hence has direction as well as magnitude. For the time being it is enough to use the sign convention above to designate the direction of a torque.

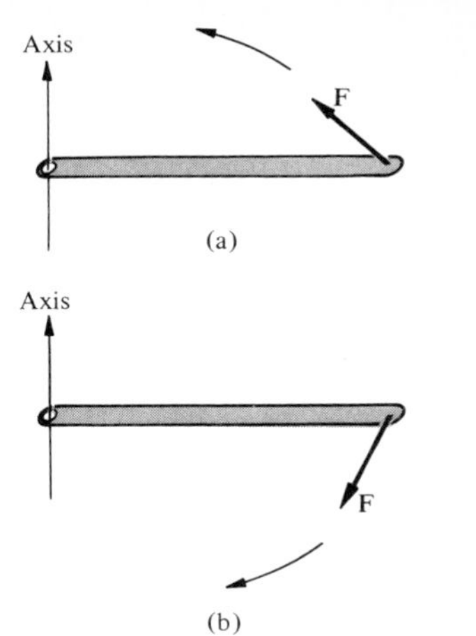

Figure 5.4. *(a) Force* **F** *produces a counterclockwise rotation (τ positive), while in (b) force* **F** *produces a clockwise rotation (τ negative).*

Example 5.1 A rod 1 m long is pivoted about O. Forces $F_1 = 4$ N and $F_2 = 4$ N are applied at points A and B as shown in (a) of the accompanying figure. Calculate the net torque about an axis through O. If the force F_1 makes an angle of 30° as shown in (b), what is the net torque?

Let τ_1 and τ_2 be the two torques produced by the forces F_1 and F_2, respectively. $F_1 = 4$ N, $d_1 = 1$ m, $F_2 = 4$ N, $d_2 = 0.5$ m. Thus,

$$\tau_1 = +F_1 d_1 = +(4 \text{ N})(1 \text{ m}) = +4 \text{ N-m} \qquad \text{(counterclockwise)}$$
$$\tau_2 = -F_2 d_2 = -(4 \text{ N})(0.5 \text{ m}) = -2 \text{ N-m} \qquad \text{(clockwise)}$$

The net torque

$$\tau = \tau_1 + \tau_2 = 4 \text{ N-m} + (-2 \text{ N-m}) = +2 \text{ N-m (counterclockwise)}$$

If F_1 acts at an angle of 30°, the perpendicular distance is calculated as follows:

$$\frac{OC}{OA} = \sin\theta$$

$$d = OC = OA \sin 30° = (1 \text{ m})(0.5) = 0.5 \text{ m}$$

Thus,

$$\tau_1 = F_1 d = (4 \text{ N})(0.5 \text{ m}) = 2 \text{ N-m}$$

while τ_2 is still the same $= -2$ N-m; thus,

$$\tau = \tau_1 + \tau_2 = 2 \text{ N-m} - 2 \text{ N-m} = 0$$

Exercise 5.1 In Example 5.1, if $F_1 = 12$ N, $F_2 = 8$ N, and $\theta = 60°$, calculate τ_1, τ_2, and the net torque τ. [*Ans.*: +12 N-m, −4 N-m, −6.4 N-m]

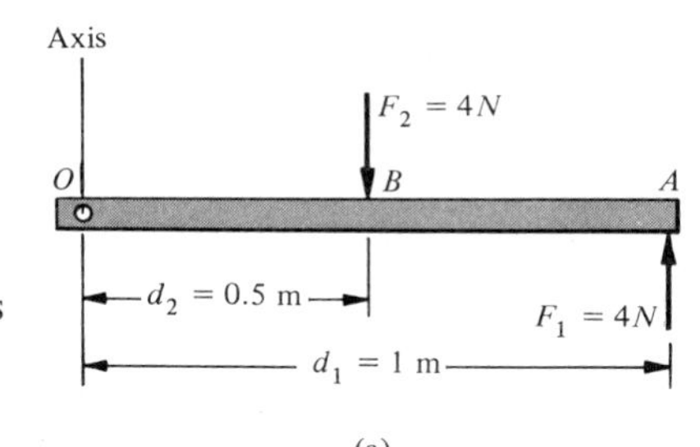

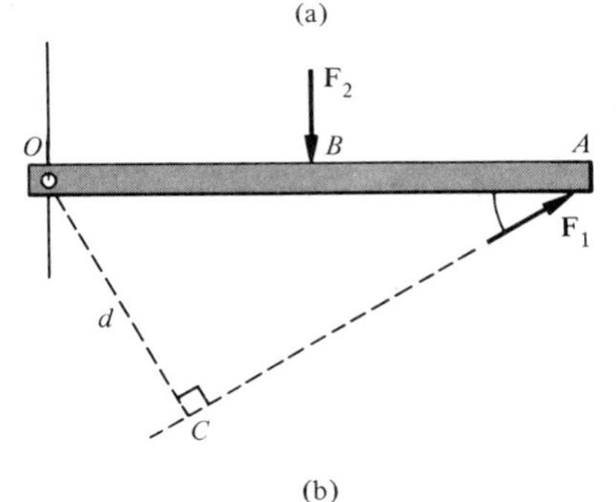

Ex. 5.1

5.3. Center of Gravity

Any object is a collection of a large number of particles. Suppose that their masses are $m_1, m_2, m_3, \ldots$, and their weights due to gravitational pull are $\mathbf{w}_1$, $\mathbf{w}_2$, $\mathbf{w}_3, \ldots$, respectively. The directions of these weights are toward the center of the earth. Since the radius of the earth is very large as compared to the size of the object, these weights may be considered to be parallel to each

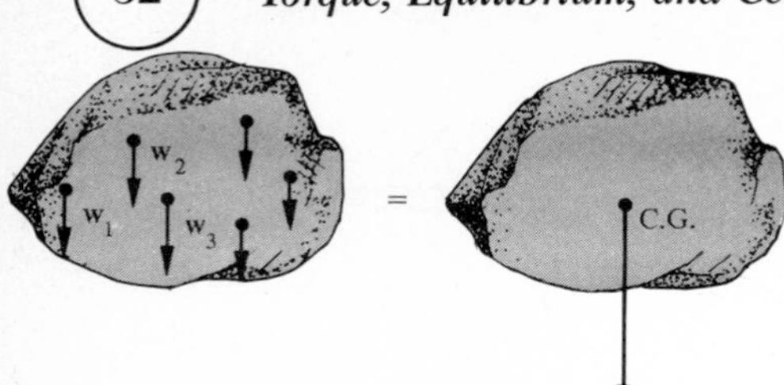

FIGURE 5.5. *Center of gravity of an irregularly shaped object.*

other, as shown in Figure 5.5. In considering equilibrium problems, if we had to consider the forces acting on each of these masses and had to calculate the torques due to them, it would be impossible to make any progress in solving even the simplest problems. But the difficulty can be overcome by introducing the center of gravity.

Since the individual weights are parallel to each other, their resultant total weight **W** is given by

$$\mathbf{W} = \mathbf{w}_1 + \mathbf{w}_2 + \mathbf{w}_3 + \cdots \tag{5.3}$$

and is parallel to the individual weights. We define the *center of gravity* (C.G.) as a point in the body where the weight **W** of the body may be supposed to act (without changing the state of the body). This is shown in Figure 5.5 and the weight **W** given by Equation 5.3 is directed downward and acts through the C.G. For a regular-shaped body, the center of gravity is the geometrical center. For example, the center of gravity of a square plate of uniform thickness is at its center (where the diagonals cross), and the center of gravity of a uniform meter stick is (halfway) at the center.

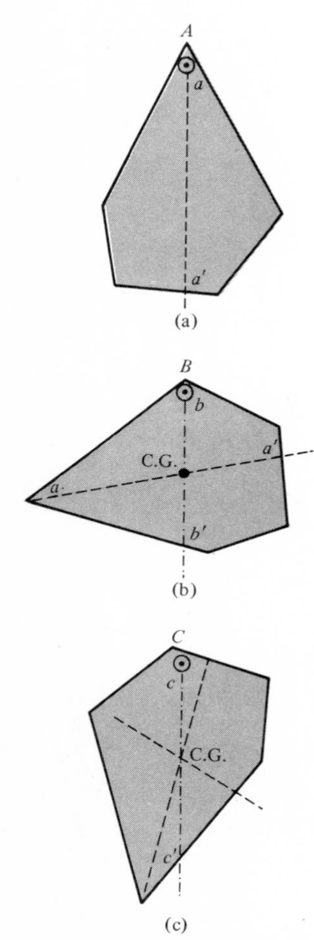

FIGURE 5.6. *Finding the center of gravity of an irregularly shaped object.*

For irregularly shaped bodies, the center of gravity may be located as shown in Figure 5.6. The irregular-shaped body is pivoted through point A. The vertical dashed line aa' is drawn passing through A, as shown in Figure 5.6(a). Now the body is pivoted through another point, such as B shown in Figure 5.6(b), and a vertical bb' is drawn passing through B. The point where aa' and bb' intersect is the center of gravity, C.G. This can be verified by pivoting the body through still another point C, as shown in Figure 5.6(c). The vertical line cc' passing through C also passes through C.G.

In the following sections, we shall be dealing with objects of finite sizes and weights. For all practical purposes we can replace an object by its weight acting with its line of action passing through the center of gravity.

5.4. First Condition of Equilibrium

A body that remains at rest or moves with uniform velocity is said to be in *translational equilibrium*. The condition under which a body will be in translational equilibrium is referred to as the first condition of equilibrium. From Newton's first law we know that a body will remain at rest or in uniform motion unless an external force is applied. Thus, in the absence of any external force, the body will be in translational equilibrium. We can arrive at this requirement from Newton's second law, according to which

$$\mathbf{F}_{\text{ext}} = m\mathbf{a}$$

where $\mathbf{F}_{\text{ext}}$ is the sum of the external forces equal to $\mathbf{F}_1 + \mathbf{F}_2 + \mathbf{F}_3 + \cdots$ acting on a body of mass m resulting in an acceleration **a**. If the body is in translational equilibrium, **a** must be equal to zero, which means that

$$\mathbf{F}_{\text{ext}} = 0 \qquad \text{or} \qquad \mathbf{F}_1 + \mathbf{F}_2 + \mathbf{F}_3 + \cdots = 0 \tag{5.4}$$

Thus, the first condition of equilibrium requires that a body will be in translational equilibrium only if the vector sum of all the external forces acting on the body (at rest or in translational motion) *is zero*. Equation 5.4 may be written in the component form as

$$F_{1x} + F_{2x} + F_{3x} + \cdots = \sum F_{nx} = 0$$

$$F_{1y} + F_{2y} + F_{3y} + \cdots = \sum F_{ny} = 0 \tag{5.5}$$

$$F_{1z} + F_{2z} + F_{3z} + \cdots = \sum F_{nz} = 0$$

Most of our discussion will be limited to coplanar forces acting on a rigid body, in which case the first condition of equilibrium given by Equation 5.5 reduces to

$$\boxed{\sum F_{nx} = 0, \qquad \sum F_{ny} = 0} \tag{5.6}$$

Let us consider a rigid body, shown in Figure 5.7. In Figure 5.7(a) only a single force $\mathbf{F}_1$ acts on it. If the body is at rest, the effect of this force will be to produce translational or rotational motion; or if the body is already in motion, the effect of the force will be to produce acceleration. In any case, the body will not be in equilibrium under the action of a single external force $\mathbf{F}_1$. Suppose that the body is under the action of two forces $\mathbf{F}_1$ and $\mathbf{F}_2$, as shown in Figure 5.7(b) or (c). The body will be in translational equilibrium only if these two forces are equal in magnitude and opposite in direction. Furthermore, if the line of action of these two forces is the same as shown in Figure 5.7(b), the body will be in rotational equilibrium as well. If the line of action is not the same as in Figure 5.7(c), the body will still have translational equilibrium, but may not be in rotational equilibrium, depending upon the position of the axis (as we shall discuss in the next section).

Let us now consider the case in which three forces $\mathbf{F}_1$, $\mathbf{F}_2$, and $\mathbf{F}_3$ are acting on the body as shown in Figure 5.8(a). For equilibrium the resultant of these three forces must be zero. That is,

$$\mathbf{F}_1 + \mathbf{F}_2 + \mathbf{F}_3 = 0$$

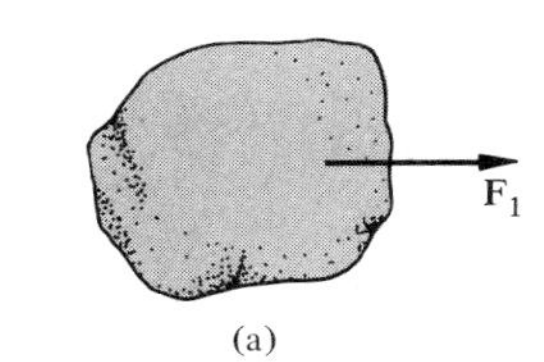

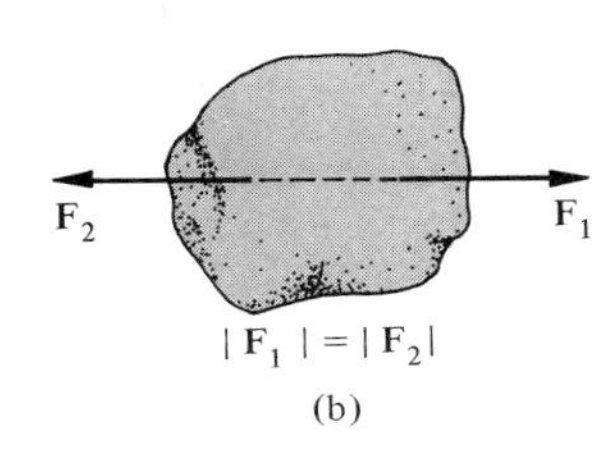

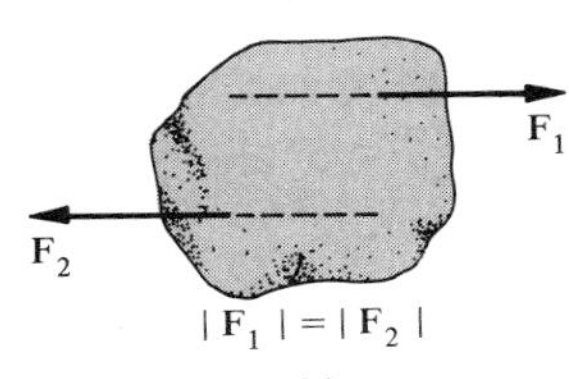

FIGURE 5.7. *(a) A rigid body acted upon by a single force $\mathbf{F}_1$ will not be in equilibrium. (b) Two equal and opposite forces acting along the same line of action will result in translational as well as rotational equilibrium. (c) A body under the action of two forces may not be in rotational equilibrium, but is in translational equilibrium.*

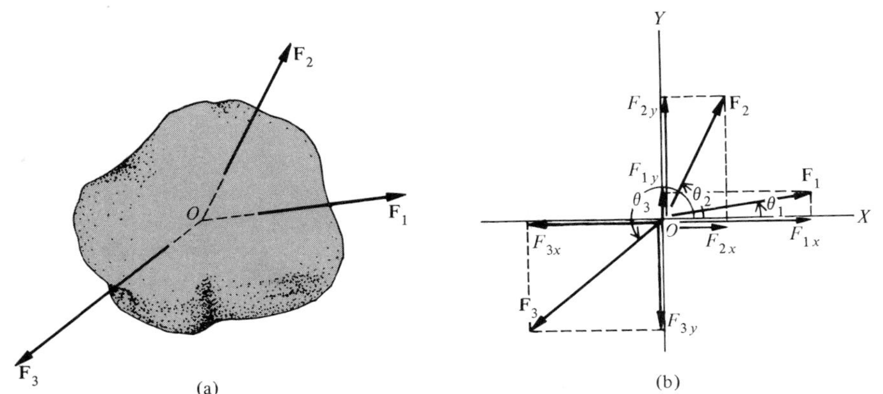

FIGURE 5.8. *(a) Three concurrent forces acting on a body will not produce rotational motion. (b) Shown are the components of the forces acting on a body in part (a).*

Since the forces are in one plane, the equation above, in component form as shown in Figure 5.8(b), becomes

$$\sum F_{nx} = F_{1x} + F_{2x} + F_{3x} = F_1 \cos\theta_1 + F_2 \cos\theta_2 + F_3 \cos\theta_3 = 0$$

$$\sum F_{ny} = F_{1y} + F_{2y} + F_{3y} = F_1 \sin\theta_1 + F_2 \sin\theta_2 + F_3 \sin\theta_3 = 0$$

Of course, if these conditions are satisfied, the body will be in translational equilibrium, but not necessarily in rotational equilibrium. For the body to be in rotational equilibrium, the second condition of equilibrium to be discussed in the next section must be satisfied. As we can see, if the lines of action of all three forces intersect at a point, that is, if the forces are concurrent, they do not produce any torque about the axis through this point, and hence the body will automatically be in rotational equilibrium.

To solve problems involving translational equilibrium, it will be helpful to use the following procedure:

1. Draw a sketch of the system under consideration, showing dimensions, angles, and forces (weights, tensions, etc.).
2. Choose a point in the system (where the lines of action for each force intersect) or the center of the object as a particle in equilibrium. Draw a separate diagram for this point or particle, showing all the external forces acting on it. This is called the *free-body diagram*. Draw as many free-body diagrams for different particles as may be necessary.
3. Resolve the forces into X- and Y-components. Equate the sum of all the X-components to zero and the sum of all the Y-components to zero. Repeat this for each free-body diagram. Solve these equations for the unknowns. The preceding points will be illustrated in the following example.

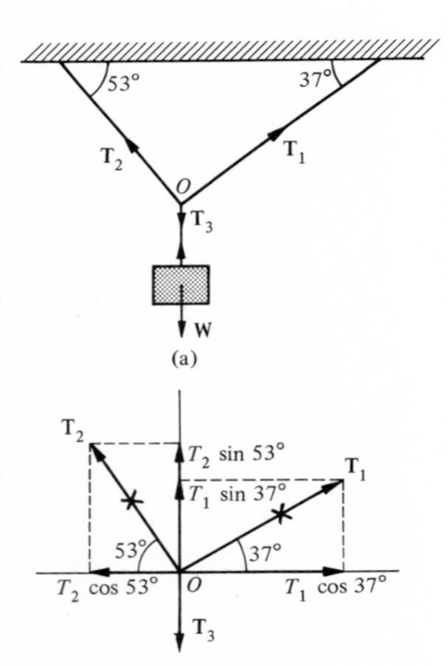

Ex. 5.2

Example 5.2 If the weight W in the accompanying figure is 980 N, calculate the tension T_1, T_2, and T_3 in the three ropes.

1. Part (a) of the figure shows the system under consideration and the angles, and forces.
2. We choose point O and weight W as two equilibrium points and draw free-body diagrams as shown in parts (b) and (c).
3. Resolve forces into X- and Y-components for each equilibrium point separately as discussed below, and solve for the unknowns.

Considering the equilibrium of point O [see part (b)],

$$\sum F_x = 0 \qquad T_1 \cos 37° - T_2 \cos 53° = 0 \tag{i}$$

$$\sum F_y = 0 \qquad T_1 \sin 37° + T_2 \sin 53° - T_3 = 0 \tag{ii}$$

Considering the equilibrium of weight W [see part (c)],

$$\sum F_y = 0 \qquad T_3 - W = 0 \tag{iii}$$

Thus,

$$T_3 = W = 980 \text{ N}$$

From Equations i and ii,

$$T_1(0.8) - T_2(0.6) = 0 \qquad \text{(iv)}$$

$$T_1(0.6) + T_2(0.8) - 980 \text{ N} = 0 \qquad \text{(v)}$$

From Equation iv,

$$T_1 = \frac{0.6}{0.8} T_2 = 0.75 T_2 \qquad \text{(vi)}$$

Substituting in Equation v,

$$(0.75)T_2(0.6) + T_2(0.8) - 980 \text{ N} = 0$$

$$T_2 = 784 \text{ N}$$

From Equation vi,

$$T_1 = (0.75)(784 \text{ N}) = 588 \text{ N}$$

Thus, $T_1 = 588$ N, $T_2 = 784$ N, and $T_3 = 980$ N.

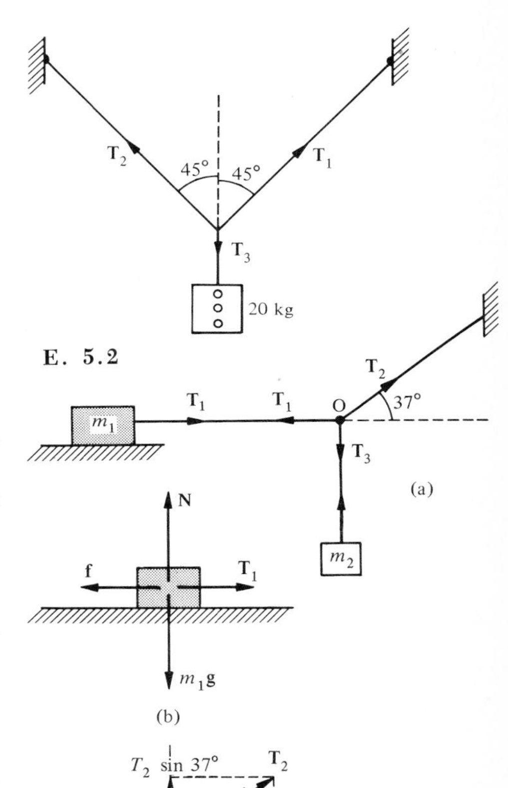

E. 5.2

(a)

(b)

EXERCISE 5.2 A 20-kg traffic light is suspended as shown. Calculate the tension in each of the three cables. [*Ans.*: $T_1 = T_2 = 139$ N; $T_3 = 196$ N.]

EXAMPLE 5.3 In the accompanying figure, $m_1 = 20$ kg, and the coefficient of friction between m_1 and the horizontal surface is 0.3. What is the value of m_2 before the mass m_1 starts sliding? Also, calculate the tensions T_1, T_2, and T_3.

1. The figure (a) shows the system under consideration and the dimensions, angles, and forces.
2. We choose mass m_1, point O, and mass m_2 as three equilibrium points and draw free-body diagrams as shown in parts (b), (c), and (d).
3. Resolve forces into X- and Y-components for each equilibrium point separately as discussed below and solve for the unknowns.

For mass m_1 to be in equilibrium as shown in part (b),

$$\sum F_x = 0 \qquad T_1 - f = 0 \qquad \text{(i)}$$

$$\sum F_y = 0 \qquad N - m_1 g = 0 \qquad \text{(ii)}$$

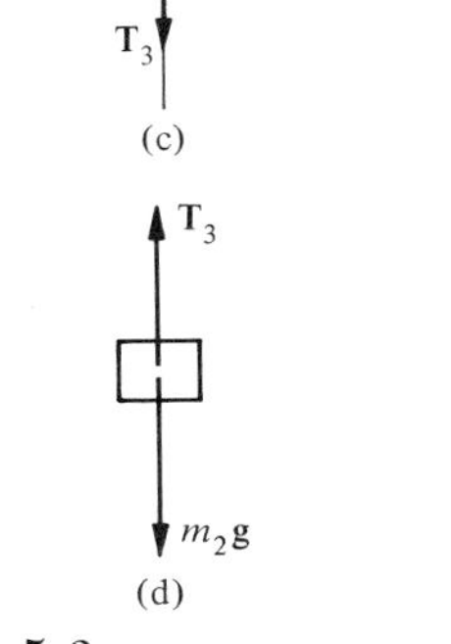

(c)

(d)

Ex. 5.3

For point O to be in equilibrium as shown in part (c),

$$\sum F_x = 0 \qquad T_2 \cos 37° - T_1 = 0 \qquad \text{(iii)}$$

$$\sum F_y = 0 \qquad T_2 \sin 37° - T_3 = 0 \qquad \text{(iv)}$$

For mass m_2 to be in equilibrium as shown in part (d),

$$\sum F_y = 0 \qquad T_3 - m_2 g = 0 \qquad \text{(v)}$$

From Equations i and ii,

$$T_1 = f = \mu N = \mu m_1 g = (0.3)(20\text{ kg})(9.8\text{ m/s}^2) = 58.8\text{ N}$$

From Equation iii,

$$T_2 = T_1/\cos 37° = 58.8\text{ N}/0.80 = 73.5\text{ N}$$

From Equation iv,

$$T_3 = T_2 \sin 37° = (73.5\text{ N})(0.60) = 44.2\text{ N}$$

From Equation v,

$$m_2 = T_3/g = \frac{44.2\text{ N}}{9.8\text{ m/s}^2} = 4.5\text{ kg}$$

EXERCISE 5.3 In Example 5.3, if $m_1 = 100$ kg and $m_2 = 20$ kg, what is the minimum value of μ so that the system will remain in equilibrium? [*Ans.*: $\mu = 0.267$.]

5.5. Second Condition of Equilibrium

If a body does not rotate at all or rotates at *constant* angular speed, it is said to be in rotational equilibrium. The second condition of equilibrium is the statement of conditions under which the body will be in *rotational equilibrium*. We have seen in Section 5.2 that rotational motion is produced by the application of a torque.

Thus, the second condition of equilibrium requires that a body will be in rotational equilibrium only if the sum of all the external torques acting on the body about any arbitrary axis is zero. That is, if $\boldsymbol{\tau}_1, \boldsymbol{\tau}_2, \boldsymbol{\tau}_3, \ldots$ are the external torques acting on the body, for equilibrium we may write

$$\boldsymbol{\tau}_1 + \boldsymbol{\tau}_2 + \boldsymbol{\tau}_3 + \cdots = 0$$

$$\sum \boldsymbol{\tau}_{\text{ext}} = 0 \tag{5.7}$$

Since the clockwise torques oppose counterclockwise torques, we may state the second condition of equilibrium in a slightly different manner. *The body will be in rotational equilibrium provided that the sum of the counterclockwise torques is equal to the sum of the clockwise torques.* That is,

$$\left(\sum \tau_{\text{ext}}\right)_{\text{counterclockwise}} = \left(\sum \tau_{\text{ext}}\right)_{\text{clockwise}} \tag{5.8}$$

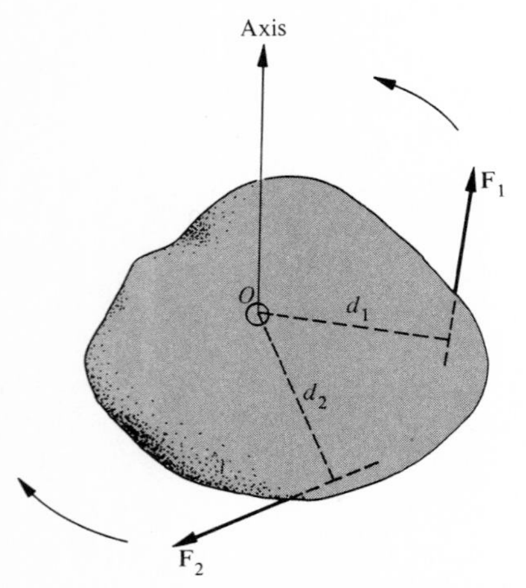

FIGURE 5.9. *The body will be in rotational equilibrium only if* $\Sigma\, \tau_{\text{ext}} = F_1 d_1 - F_2 d_2 = 0$.

For example, say that a body in Figure 5.9 is acted upon by two forces. The force $\mathbf{F}_1$ produces a counterclockwise torque $\tau_1 = F_1 d_1$, and force $\mathbf{F}_2$ produces a clockwise torque $\tau_2 = -F_2 d_2$, about an axis through O along the Z-axis. For this body to be in rotational equilibrium, from Equation 5.7,

$$\sum \tau_{\text{ext}} = \tau_1 + \tau_2 = F_1 d_1 - F_2 d_2 = 0 \qquad \text{or} \qquad F_1 d_1 = F_2 d_2$$

which is Equation 5.8.

It is necessary to point out (without proving) at this stage that if the sum of all the torques about an axis is zero, the sum of the torques will be zero about any other parallel axis provided that the sum of all the forces is zero.

Thus, we may now summarize both conditions of equilibrium. Since we shall be limiting our discussion to the motion of a body in the XY-plane only, the forces will be in the XY-plane, while the axis of rotation will be parallel to the Z-axis. In such circumstances the conditions of translational and rotational equilibrium take the form

$$\sum F_{nx} = 0, \qquad \sum F_{ny} = 0, \qquad \sum \tau_{nz} = 0 \tag{5.9}$$

respectively.

In solving problems we follow the same three steps outlined in the previous section in addition to equating to zero the sum of the torques about an axis, say about an axis through A by $\Sigma_A\, \tau_{nz} = 0$. This helps in solving for the unknowns, as will be demonstrated in the following examples.

Example 5.4 One end of a uniform rod of mass 2 kg and length 1 m is placed at a pivot point A, as shown in the accompanying figure. A mass of 4 kg is hung from a point 30 cm from A, while at the end B a force F is applied to hold the system in equilibrium as shown. Calculate the force F and the reaction R of the pivot.

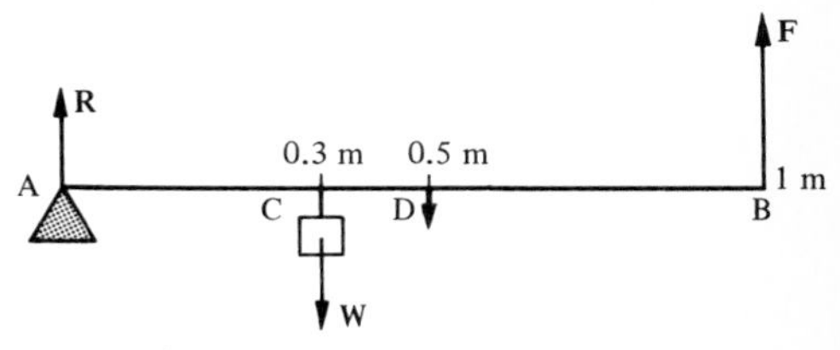

Ex. 5.4

The weight of the rod $w = (2\text{ kg})(9.8\text{ m/s}^2) = 19.6\text{ N}$ acts downward at the center $(x = 0.5\text{ m})$ of the rod. The weight W of mass 4 kg is $W = (4\text{ kg})(9.8\text{ m/s}^2) = 39.2\text{ N}$ and acts downward at $x = 30\text{ cm} = 0.3\text{ m}$. From the first condition of equilibrium,

$$\sum F_y = 0 \qquad F + R - W - w = 0$$

or

$$F + R - 39.2\text{ N} - 19.6\text{ N} = 0 \tag{i}$$

We now apply the second condition of equilibrium by taking torques about any point. Let us consider point A (note that F produces a counterclockwise torque; W and w produce clockwise torques; while R, passing through A, produces no torque):

$$\sum_A \tau_{\text{counterclockwise}} = \sum_A \tau_{\text{clockwise}}$$

$$F \times AB = W \times AC + W \times AD \tag{ii}$$

Thus, from Equation ii,

$$F = \frac{(39.2\text{ N})(0.3\text{ m}) + (19.6\text{ N})(0.5\text{ m})}{1.0\text{ m}} = 21.6\text{ N}$$

Combining this with Equation i,

$$21.6\text{ N} + R - 39.2\text{ N} - 19.6\text{ N} = 0$$
$$R = 37.2\text{ N}$$

We could have obtained the same results by taking torques about any other point, say D. Then w will produce no torque, W and F will produce counterclockwise torques, and R will produce a clockwise torque. Thus,

$$F \times DB + W \times CD = R \times DA$$
$$F(0.5\text{ m}) + W(0.2\text{ m}) = R(0.5\text{ m})$$
$$R - F = \frac{(39.2\text{ N})(0.2\text{ m})}{0.5\text{ m}} = 15.7\text{ N}$$

Combining with Equation i,

$$R + F = 58.8\text{ N}$$

we get $R = 37.2$ N, $F = 21.6$ N.

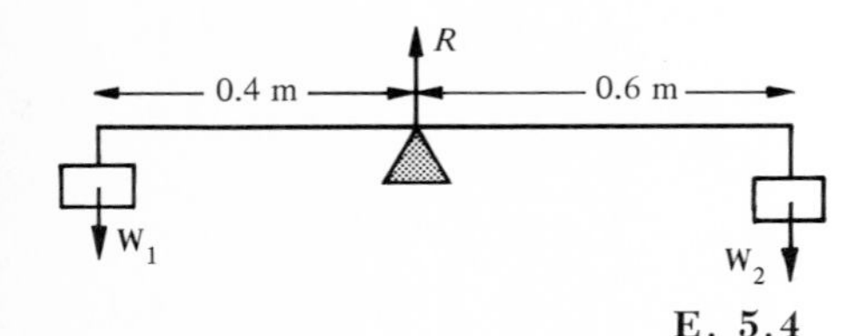

E. 5.4

Exercise 5.4 For a weightless rod 1 m long shown in the accompanying figure, if the mass of weight W_1 is 10 kg, calculate the mass of the weight W_2 and the reaction R of the pivot on the rod. [*Ans.*: $m_2 = 6.67$ kg, $R = 163.4$ N.]

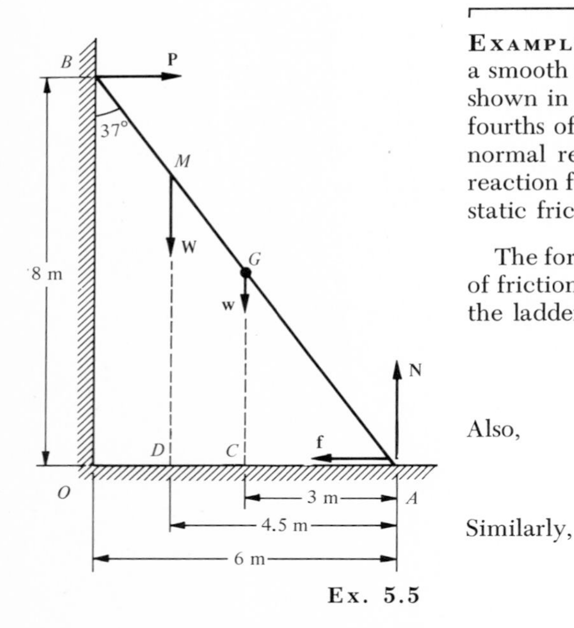

Ex. 5.5

Example 5.5 A uniform ladder of mass 20 kg and length 10 m leans against a smooth (frictionless) vertical wall making an angle of 37° with the wall as shown in the accompanying figure. Suppose that a 70-kg man climbs three-fourths of the way up the ladder. (a) Calculate the frictional force f and the normal reaction N exerted by the ground on the ladder. (b) Calculate the reaction force on the ladder exerted by the wall. (c) What is the coefficient of static friction if the ladder is just about to slip in this situation?

The forces acting on the ladder are shown in the figure, where f is the force of friction, N the normal reaction, P the reaction of the wall, w the weight of the ladder, and W the weight of the man.

$$w = mg = (20\text{ kg})(9.8\text{ m/s}^2) = 196\text{ N}$$
$$W = Mg = (70\text{ kg})(9.8\text{ m/s}^2) = 686\text{ N}$$

Also,

$$OA = AB \sin 37° = (10\text{ m})(0.6) = 6\text{ m}$$
$$OB = AB \cos 37° = (10\text{ m})(0.8) = 8\text{ m}$$

Similarly,

$$AC = AG \sin 37° = (5\text{ m})(0.6) = 3\text{ m}$$
$$AD = AM \sin 37° = (7.5\text{ m})(0.6) = 4.5\text{ m}$$

Applying the first condition of equilibrium

$$\sum F_x = 0 \qquad P - f = 0 \tag{i}$$

$$\sum F_y = 0 \qquad N - W - w = 0 \tag{ii}$$

In order to apply the second condition of equilibrium, we take torques about A. Note that the moment arms for f and N are zero.

$$\sum \tau_A = 0 \qquad W(AD) + w(AC) - P(OB) = 0 \tag{iii}$$

From ii,

$$N = W + w = 686 \text{ N} + 196 \text{ N} = 882 \text{ N}$$

From iii,

$$(686 \text{ N})(4.5 \text{ m}) + (196 \text{ N})(3 \text{ m}) - P(8 \text{ m}) = 0$$

$$P = \frac{(3087 + 588)}{8 \text{ m}} \text{N-m} = 459.4 \text{ N}$$

Hence from i,

$$f = P = 459.4 \text{ N}$$

Since the ladder is just about to slip, f is the maximum force of friction, hence $f = \mu_s N$. Therefore, $\mu_s = f/N = 459.4 \text{ N}/882 \text{ N} = 0.52$.

Exercise 5.5 A uniform ladder of mass 30 kg and length 8 m leans against a vertical frictionless wall, making an angle of 30° with the vertical. Calculate the force exerted by the floor (f_s and N) and the reaction of the wall when (a) no one is on the ladder; and (b) a 40-kg boy is three-fourths of the way up the ladder. [*Ans.*: (a) $f_s = P = 84.9$ N, $N = 294$ N; (b) $f_s = P = 254.6$ N, $N = 686$ N.]

Example 5.6 A man of mass 70 kg is standing vertically as shown in the accompanying figure. With what force F_B should a boy pull his arm so that the man will just about topple over? Assume no slipping.

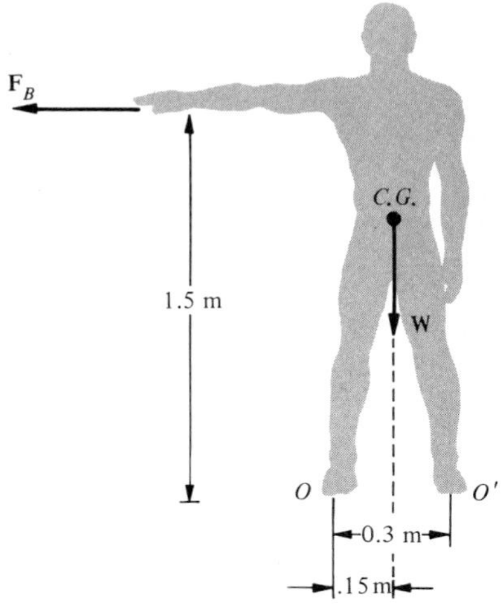

Ex. 5.6

The forces acting on the man are his weight $W(= Mg)$ and the force F_B applied by the boy as shown. The typical height of a man from foot to shoulder is 1.5 m ($\approx$ 5 ft) while the distance OO' between his feet is 0.3 m (= 30 cm $\approx$ 1 ft), as shown in figure. Let point O be the point about which the body rotates. The body will topple over if the counterclockwise torque τ_B produced by the boy about point O is equal to the clockwise torque τ_W produced by the weight $W = Mg$. The moment arm of F_B is 1.5 m, while that of W is 0.3/2 m = 0.15 m. Thus, (for reaction of the floor at O, $\tau = 0$)

$$\tau_B = \tau_W \qquad \text{or} \qquad F_B(1.5 \text{ m}) = (Mg)(0.15 \text{ m})$$

or

$$F_B = \frac{(70 \text{ kg})(9.8 \text{ m/s}^2)(0.15 \text{ m})}{1.5 \text{ m}} = 68.6 \text{ N} \qquad \text{and} \qquad \tau_B = \tau_W = 102.9 \text{ N-m}$$

Exercise 5.6 A 70-kg man is walking on a narrow wall. If he bends more than 15° sideways, he will fall. Assuming that his C.G. is 1 m high, calculate the torque that acts before the man falls. [*Ans.*: 177.7 N-m]

EXAMPLE 5.7 A man is holding a mass m of 10 kg in his hand as shown in the accompanying figure. The forces acting on his arm are (1) the weight $w = mg$ downward, (2) the force F_m exerted by the biceps muscles making an angle θ_m with the $+X$-axis, and (3) the reaction force F_r at the elbow making an angle of θ_r with the $+X$-axis. As shown, the value of $\theta_m = (180° - 120°) = 60°$. Calculate F_m, F_r, and θ_r. (Neglect the weight of the arm)

Applying the first condition of equilibrium, from figure (b)

$$\sum F_x = 0 \qquad F_m \cos\theta_m - F_r \cos\theta_r = 0 \tag{i}$$

$$\sum F_y = 0 \qquad F_m \sin\theta_m - F_r \sin\theta_r - w = 0 \tag{ii}$$

In order to apply the second condition of equilibrium, let us take the torques about O. Since F_r passes through this point, it produces no torque. The vertical component of F_m is $F_m \sin\theta_m$, and the moment arm is 5 cm, while the weight W has a moment arm of 35 cm. Thus, from the condition,

$$\tau_{\text{clockwise}} = \tau_{\text{counterclockwise}}$$

$$F_m \sin\theta_m(0.05\text{ m}) = w(0.35\text{ m}) \tag{iii}$$

We are given $\theta_m = 60°$, so $\cos 60° = 0.5$, $\sin 60° = 0.87$ and $w = mg = (10\text{ kg})(9.8\text{ m/s}^2) = 98$ N. Substituting these in i, ii, and iii,

$$F_m(0.5) - F_r \cos\theta_r = 0 \tag{iv}$$

$$F_m(0.87) - F_r \sin\theta_r - 98\text{ N} = 0 \tag{v}$$

$$F_m(0.87)(0.05\text{ m}) = (98\text{ N})(0.35\text{ m}) \tag{vi}$$

From Equation vi,

$$F_m = \frac{(98\text{ N})(0.35\text{ m})}{(0.87)(0.05\text{ m})} = 788\text{ N}$$

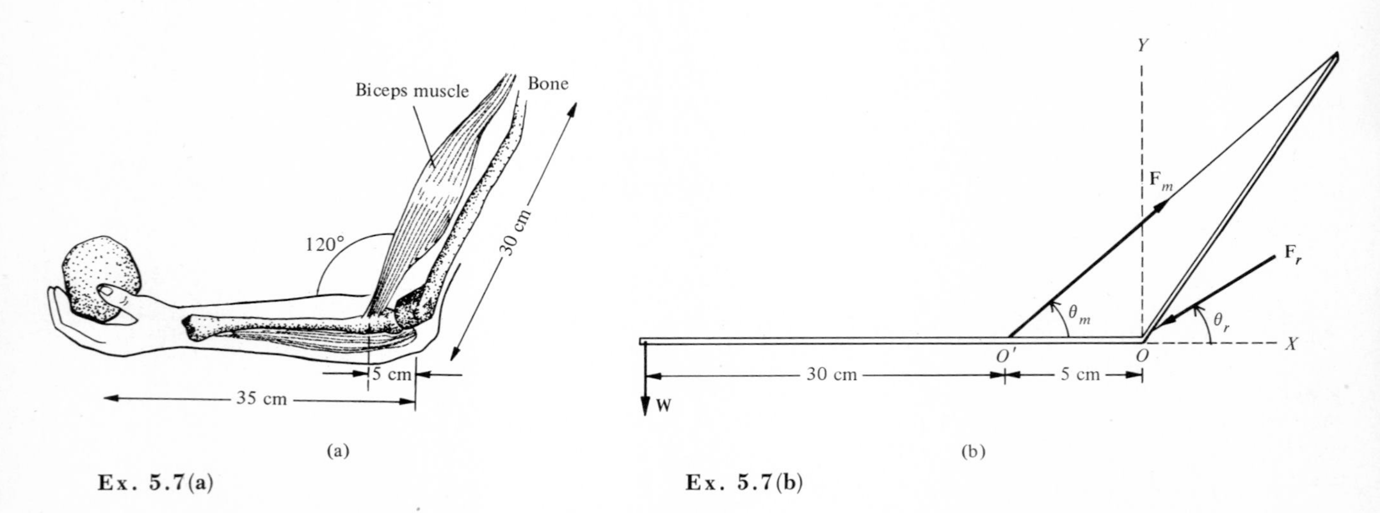

Ex. 5.7(a)

Ex. 5.7(b)

Substituting this in Equations iv and v and rearranging,

$$F_r \cos\theta_r = (788 \text{ N})(0.5) = 394 \text{ N} \qquad \text{(vii)}$$

$$F_r \sin\theta_r = (788 \text{ N})(0.87) - 98 \text{ N} = 588 \text{ N} \qquad \text{(viii)}$$

Squaring and adding these equations,

$$F_r^2(\cos^2\theta_r + \sin^2\theta_r) = (394 \text{ N})^2 + (588 \text{ N})^2 = 501{,}000 \text{ N}^2$$

Since $\cos^2\theta_r + \sin^2\theta_r = 1$,

$$F_r = \sqrt{501{,}000 \text{ N}^2} = 708 \text{ N}$$

Dividing Equation viii by vii,

$$\tan\theta_r = \frac{588 \text{ N}}{394 \text{ N}} = 1.49 \qquad \theta_r = 56^\circ$$

Hence,

$$F_m = 788 \text{ N}, \qquad F_r = 708 \text{ N} \qquad \text{and} \qquad \theta_r = 56^\circ$$

Exercise 5.7 Consider the situation shown in the accompanying figure, where $w = mg$, m being the mass of the arm, which is 2 kg; that is the man is holding his forearm in horizontal position. If $\theta_m = 60^\circ$, what are the values of F_m, F_r, and θ_r? [*Ans.*: 67.6 N, 51.8 N, and 49.2°.]

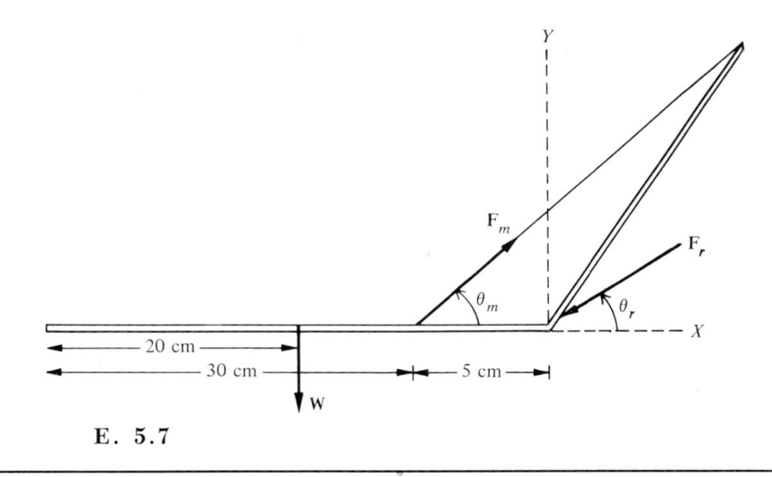

E. 5.7

*5.6. Center of Mass and Its Motion

We have been treating objects as particles having mass but no size; that is, we treat them as point masses. Thus, when we apply force on such an object, we have to consider only the translational motion; the rotational motion may be ignored. Suppose that we have an object of finite size or have many point masses at different locations. In either case we are dealing with a system of particles. Let us apply an external force to such a system. It is possible to find one point in the system which will move in the same way that a single particle (of mass equal to the sum of the masses of all the particles in the system) would move when subjected to the same external forces. This point is called the *center of mass* of the system. In many cases it is convenient to deal with the motion, momentum, and energy in terms of the center of mass of the system rather than of the individual particles.

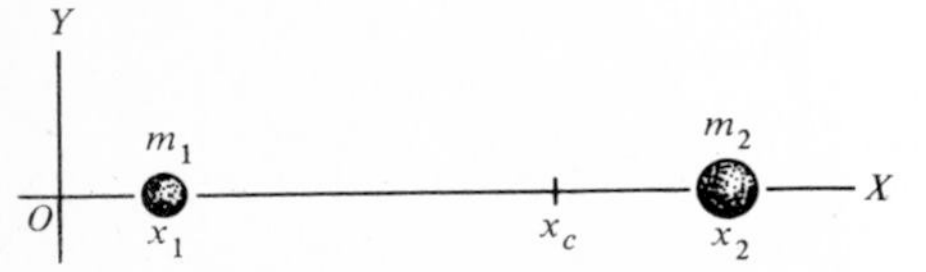

Figure 5.10. *The center of mass of two masses m_1 and m_2 lies between the masses on the line joining them.*

Center of Mass

As a simple example, consider two masses m_1 and m_2 at points x_1 and x_2 on the X-axis. The center of mass x_c will lie between the two masses and will be closer to the heavier mass, as shown in Figure 5.10. We define the center of mass, x_c by the condition

$$(m_1 + m_2)x_c = m_1x_1 + m_2x_2 \tag{5.10}$$

That is, we can replace m_1 and m_2 by a single mass $M = m_1 + m_2$ if it is placed at x_c. We may write Equation 5.10 as

$$x_c = \frac{m_1x_1 + m_2x_2}{m_1 + m_2} \tag{5.11}$$

Suppose that we have two masses m_1 located at (x_1, y_1) and m_2 at (x_2, y_2). The center of mass (x_c, y_c) of this system is

$$(m_1 + m_2)x_c = m_1x_1 + m_2x_2 \qquad \text{and} \qquad (m_1 + m_2)y_c = m_1y_1 + m_2y_2$$

or

$$x_c = \frac{m_1x_1 + m_2x_2}{m_1 + m_2} \qquad \text{and} \qquad y_c = \frac{m_1y_1 + m_2y_2}{m_1 + m_2} \tag{5.12}$$

The center of mass (x_c, y_c) still lies on the line joining the two masses. But this will not be the case if we are dealing with more than two masses which are not located in a straight line.

In a more general case, where the masses $m_1, m_2, m_3, \ldots$ are located at points (x_1, y_1, z_1), (x_2, y_2, z_2), (x_3, y_3, z_3), ..., respectively, the coordinates (x_c, y_c, z_c) of the center of mass are given by

$$x_c = \frac{m_1x_1 + m_2x_2 + m_3x_3 + \cdots}{m_1 + m_2 + m_3 + \cdots}, \qquad y_c = \frac{m_1y_1 + m_2y_2 + m_3y_3 + \cdots}{m_1 + m_2 + m_3 + \cdots},$$

$$z_c = \frac{m_1z_1 + m_2z_2 + m_3z_3 + \cdots}{m_1 + m_2 + m_3 + \cdots} \tag{5.13}$$

We may point out that the center of mass always coincides with the center of gravity if the value of g is the same over the system of particles under consideration.

Example 5.8 Three pieces of wood *A*, *B*, and *C* of uniform thickness and mass *m*, *m*, and 2*m*, respectively, are joined together as shown in the accompanying figure. Calculate the center of mass.

Since the pieces are of uniform thickness, their centers of mass must coincide with the geometrical center. As shown in part (b), the coordinates of the three

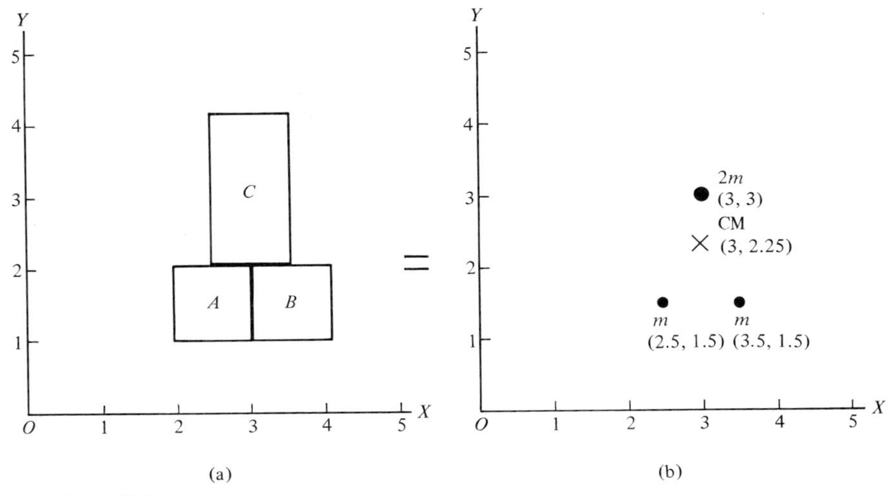

Ex. 5.8

points are (2.5, 1.5), (3.5, 1.5), and (3, 3) for masses m, m, and $2m$, respectively. Thus, according to Equation 5.13, the center of mass of the system is

$$x_c = \frac{m_1x_1 + m_2x_2 + m_3x_3}{m_1 + m_2 + m_3} = \frac{m(2.5) + m(3.5) + 2m(3)}{m + m + 2m} = \frac{12m}{4m} = 3$$

$$y_c = \frac{m_1y_1 + m_2y_2 + m_3y_3}{m_1 + m_2 + m_3} = \frac{m(1.5) + m(1.5) + 2m(3)}{m + m + 2m} = \frac{9m}{4m} = 2.25$$

$$(x_c, y_c) = (3, 2.25) \qquad \text{as shown in part (b)}$$

Exercise 5.8 In Example 5.8, if the three pieces have equal masses and the dimensions are still the same, calculate the center of mass. [*Ans.*: (3, 2).]

Motion of the Center of Mass

Once again consider a system containing masses $m_1, m_2, \ldots, m_n$ located at $\mathbf{r}_1(x_1, y_1, z_1)$, $\mathbf{r}_2(x_2, y_2, z_2)$, . . . , $\mathbf{r}_n(x_n, y_n, z_n)$, respectively. In vector notation the center of mass $\mathbf{r}_c$ is given by

$$(m_1 + m_2 + \cdots + m_n)\mathbf{r}_c = m_1\mathbf{r}_1 + m_2\mathbf{r}_2 + \cdots + m_n\mathbf{r}_n$$

or

$$\boxed{M\mathbf{r}_c = m_1\mathbf{r}_1 + m_2\mathbf{r}_2 + \cdots + m_n\mathbf{r}_n} \tag{5.14}$$

Note that Equation 5.14 is actually three equations, one for each x, y, and z coordinate, and $M = (m_1 + m_2 + m_3 + \cdots + m_n)$.

Suppose that mass m_1 has velocity $\mathbf{v}_1$, m_2 has velocity $\mathbf{v}_2$, . . . , and mass m_n has velocity $\mathbf{v}_n$. The velocity $\mathbf{v}_c$ of the center of mass is given by

$$M\mathbf{v}_c = m_1\mathbf{v}_1 + m_2\mathbf{v}_2 + \cdots + m_n\mathbf{v}_n \tag{5.15}$$

which is again a set of three equations. Equation 5.15 states that mass M, which is equal to the sum of the individual masses, when moving with the velocity equal to the velocity of the center of mass, has momentum $\mathbf{P}_c$, which is equal to the sum of the momenta $\mathbf{p}_i$ of all the individual particles. That is,

$$\boxed{\mathbf{P}_c = \mathbf{p}_1 + \mathbf{p}_2 + \cdots + \mathbf{p}_n} \tag{5.16}$$

where

$$\mathbf{P}_c = M\mathbf{v}_c \tag{5.17}$$

Let us assume further that mass m_1 has acceleration $\mathbf{a}_1$, m_2 has $\mathbf{a}_2$, . . . , m_n has $\mathbf{a}_n$. The acceleration $\mathbf{a}_c$ of the center of mass is given by

$$M\mathbf{a}_c = m_1\mathbf{a}_1 + m_2\mathbf{a}_2 + \cdots + m_n\mathbf{a}_n \tag{5.18}$$

By using Newton's second law, $\mathbf{F} = m\mathbf{a}$, we may write Equation 5.18 as

$$M\mathbf{a}_c = \mathbf{F}_1 + \mathbf{F}_2 + \cdots + \mathbf{F}_n \tag{5.19}$$

where $\mathbf{F}_1$ is the external force acting on m_1, $\mathbf{F}_2$ is the external force acting on m_2, and so on. If we denote the sum of the external forces by $\mathbf{F}_{\text{ext}}$, we may write

$$\boxed{M\mathbf{a}_c = \mathbf{F}_{\text{ext}}} \tag{5.20}$$

which states that *the center of mass of the system of particles moves as if all the mass of the system were located at the position of the center of mass and all the external forces were applied at the center of mass.*

At this point, we may ask the following question: What happens to the internal forces exerted by the particles on each other? According to Newton's third law of action and reaction, for each internal force on any one mass their is an equal and opposite force acting on some other mass. Thus, when the system is considered as a whole, the sum of the action forces is equal and opposite that of the sum of the reaction forces. Consequently, in any given system, *the sum of the internal forces is always zero.* Thus, in Equation 5.20, $\mathbf{F}_{\text{ext}}$ represents the sum of the external forces only.

EXAMPLE 5.9 Consider three masses, 6 kg, 2 kg, and 4 kg, located at (8, 6) (−2, 2), and (3, −1), respectively, as shown (where distances are in m). The forces acting on these masses are 15 N, 10 N, and 5 N, as shown. Find (a) the center of mass of the system; and (b) the acceleration of the center of mass.

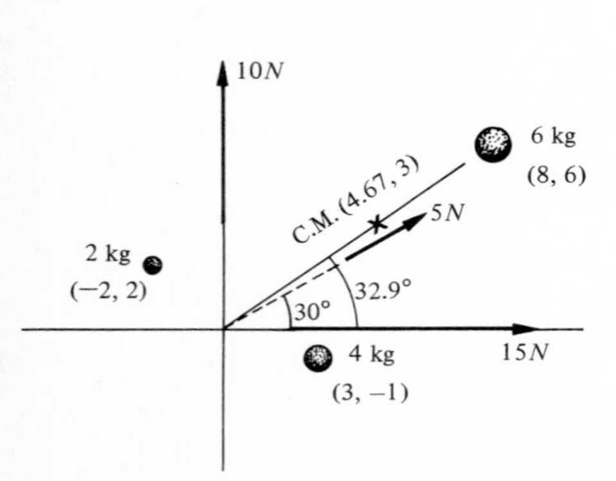

Ex. 5.9

Using Equation 5.13, the center of mass (x_c, y_c) is given by

$$x_c = \frac{6(8) + 2(-2) + 4(3)}{6 + 2 + 4}\,\text{m} = 4.67\ \text{m}$$

$$y_c = \frac{6(6) + 2(2) + 4(-1)}{6 + 2 + 4}\,\text{m} = 3\ \text{m}$$

The center of mass is located at C (4.67, 3), as shown.

In order to find the acceleration of the center of mass, we must find the resultant of the three forces.

$$F_x = (15 + 5\cos 30°)\text{ N} = 19.33\text{ N}$$
$$F_y = (10 + 5\sin 30°)\text{ N} = 12.5\text{ N}$$

Hence, the magnitude of the resultant external force is

$$F_{\text{ext}} = \sqrt{F_x^2 + F_y^2} = \sqrt{(19.33)^2 + (12.5)^2}\text{ N} = 23\text{ N}$$

while its direction is

$$\tan\theta = \frac{F_y}{F_x} = \frac{12.5\text{ N}}{19.33\text{ N}} = 0.65 \qquad \text{or} \qquad \theta = 32.9°$$

From Equation 5.20, the acceleration of the center of mass is, where $M = (6 + 2 + 4)$ kg $= 12$ kg,

$$a_c = \frac{F_{\text{ext}}}{M} = \frac{23\text{ N}}{12\text{ kg}} = 1.92\text{ m/s}^2$$

and it makes an angle of 32.9°, with the X-axis as shown.

EXERCISE 5.9 Consider two masses, 5 kg located at (2, 0) m and 10 kg at (0, 4) m. The 5-kg mass is acted upon by a 10-N force toward the $+X$-axis, and the 10 kg is acted upon by a 20-N force toward the $+Y$-axis. Find (a) the center of mass; and (b) the acceleration of the center of mass. [*Ans*.: (a) $x_c = 0.67$ m, $y_c = 2.67$ m; (b) $a_c = 1.49\text{ m/s}^2$, $\theta = 63°$ with $+X$-axis.]

SUMMARY

A *rigid body* is one in which the mean distances between different atoms and molecules remain constant. A body is said to be in *translational equilibrium* if it remains at rest or moves with a uniform speed in a straight line. A body is said to be in *rotational equilibrium* if it does not rotate at all or rotates at constant angular speed.

Torque is defined as the product of force and moment arm, $\tau = Fr\sin\theta = Fd$. The torques producing counterclockwise rotations are positive, while those producing clockwise rotations are negative.

The *first condition of equilibrium* requires that a body will be in translational equilibrium only if the vector sum of all the external forces acting on the body is zero, $\Sigma\,\mathbf{F}_{\text{ext}} = 0$. The *second condition of equilibrium* requires that a body will be in rotational equilibrium only if the sum of all the external torques acting on the body is zero, $\Sigma\tau_{\text{ext}} = 0$. For the special case of the motion of a body in the XY-plane, the conditions of equilibrium are

$$\sum F_{nx} = 0, \qquad \sum F_{ny} = 0, \qquad \sum \tau_{nz} = 0$$

* The *center of mass* of a system of particles is

$$x_c = (m_1x_1 + m_2x_2 + \cdots)/(m_1 + m_2 + \cdots)$$
$$y_c = (m_1y_1 + m_2y_2 + \cdots)/(m_1 + m_2 + \cdots)$$
$$z_c = (m_1z_1 + m_2z_2 + \cdots)/(m_1 + m_2 + \cdots)$$

The motion of the center of mass is described by the following equations $\mathbf{P}_c = \mathbf{p}_1 + \mathbf{p}_2 + \cdots$ and $M\mathbf{a}_c = \mathbf{F}_{ext} = \mathbf{F}_1 + \mathbf{F}_2 + \cdots$.

QUESTIONS

1. Give two examples of (a) rigid bodies; and (b) nonrigid bodies.
2. Give one example of the following: (a) a body in translational equilibrium; (b) a body in rotational equilibrium; (c) a body both in translational as well as in rotational equilibrium; and (d) a body neither in translational nor in rotational equilibrium.
3. Show that a given force, depending upon its line of action on a rigid body, can produce clockwise or counterclockwise torque.
4. Consider a rigid body acted upon by three forces of equal magnitudes. What should be the lines of action of these forces so that the body will be in (a) translational equilibrium; and (b) rotational equilibrium?
5. For the case shown in the figure in Problem 5.16 (page 98), is it possible to achieve equilibrium if $m_2 > m_1$ provided that μ for m_1 and the surface is less than unity?
6. A ladder is resting against a vertical wall. (a) Could a man climb the ladder if there were no friction between the ladder and the floor? (b) Is a ladder more likely to slip when the man is on the lower half of the ladder or the upper half? (c) Does friction between the ladder and the vertical wall help or hinder climbing?
7. Why will a car starting from rest accelerate faster if its tires do not spin?
8. Suppose the internal forces between a pair of particles did not act along the same line. How will this affect the conditions of equilibrium?
9. Why do you lean forward when pushing a heavy mass?
10. A ball thrown upward comes to rest when it reaches the highest point. Is the ball in equilibrium at this point?

*11. Where is the center of mass of (a) a square with a mass m at each of its corners; and (b) a square plate of uniform thickness?

*12. What is the location of the center-of-mass of (a) a solid circular plate of uniform thickness; and (b) a solid circular plate with a hole at the center?

PROBLEMS

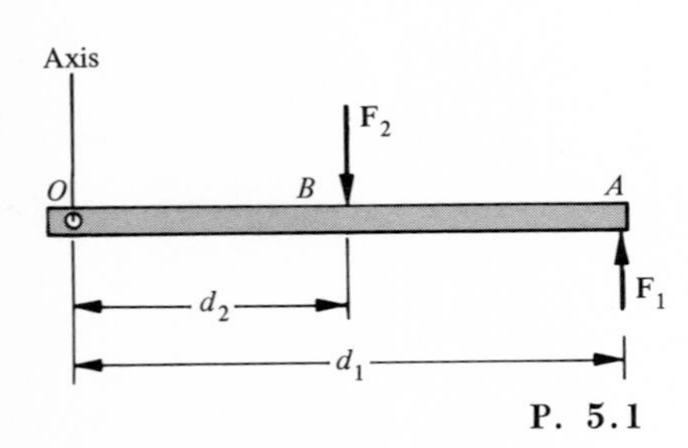

P. 5.1

1. In Figure P. 5.1, if $F_1 = 10$ N, $F_2 = 20$ N, $d_1 = 1$ m, and $d_2 = 0.4$ m, calculate the values of τ_1, τ_2, and the net torque about an axis through O.
2. In the figure in Problem 5.1, if $F_1 = 10$ N acting downward and $F_2 = 20$ N acting upward ($d_1 = 1$ m and $d_2 = 0.4$ m), calculate τ_1, τ_2, and the net torque.
3. In Figure P. 5.3, if $F_1 = 10$ N and $F_2 = 5$ N, calculate the net torque for (a) $\theta = 0°$; (b) $\theta = 45°$; (c) $\theta = 90°$; (d) $\theta = 135°$; and (e) $\theta = 180°$.
4. In the figure in Problem 5.3, if $F_1 = 12$ N, $F_2 = 12$ N, $d_1 = 1.0$ m, and $d_2 = 0.5$ m, what angle should F_1 make with the rod so that the net torque is zero?
5. In Figure P. 5.5, calculate τ_1, τ_2, and the net torque due to forces F_1 and F_2.
6. For the situation shown, calculate the tension T_1, T_2, and T_3.
7. A 100-kg mass is hung from the center of a wire stretched between two points. Calculate the tension in the wires in terms of angle θ. (*Hint:* Note

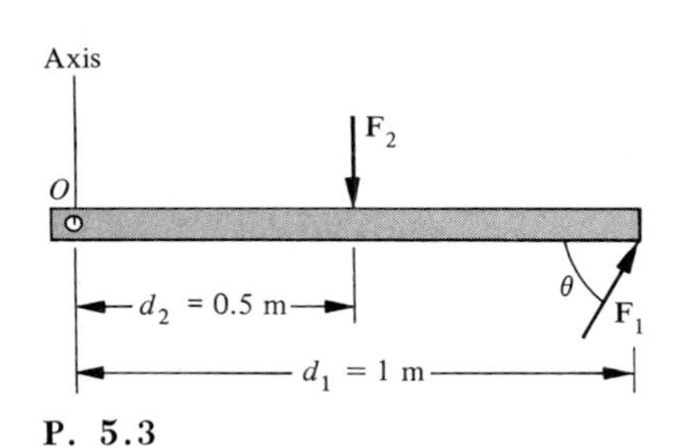

P. 5.3

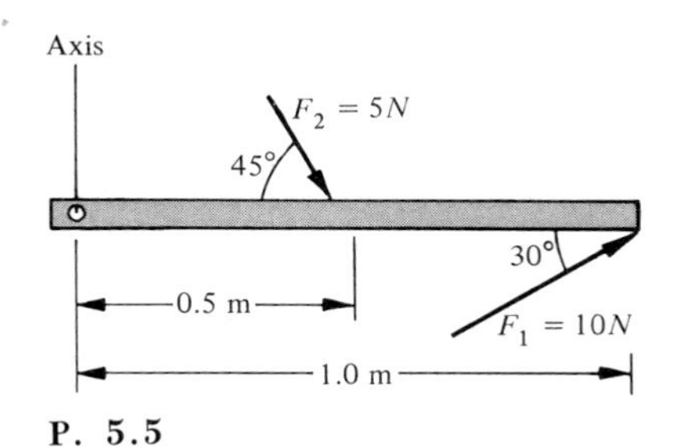

P. 5.5

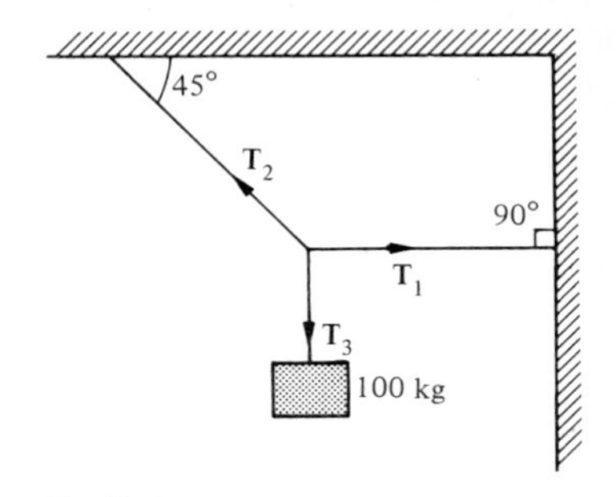

P. 5.6

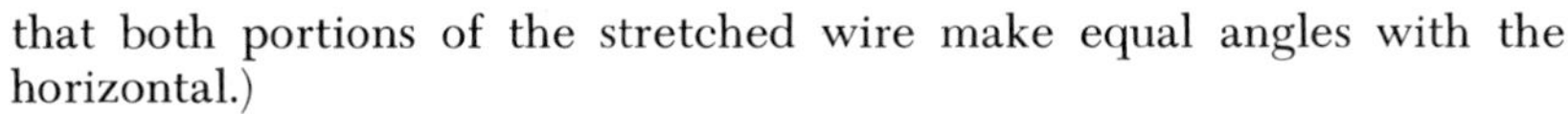
that both portions of the stretched wire make equal angles with the horizontal.)

8. For the situation shown, if the tension in one of the cable is double the other, calculate T_1 and T_2.
9. A 40-kg boy sitting in a swing is pulled to one side with a horizontal force F until the rope of the swing makes an angle of 30° with the vertical, as shown. Calculate F and the tension T in the rope when held in this position.
10. For the case shown, calculate the tension T in the cable and the reaction R exerted on the strut by the pivot. The mass m is 100 kg.
11. Repeat Problem 10 for the situation shown in Figure P. 5.11.

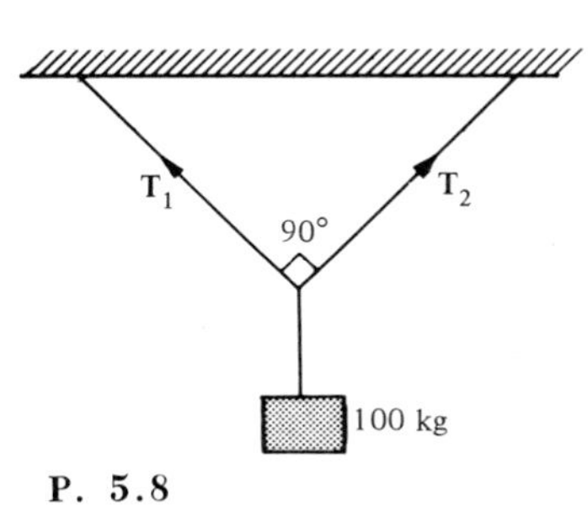

P. 5.8

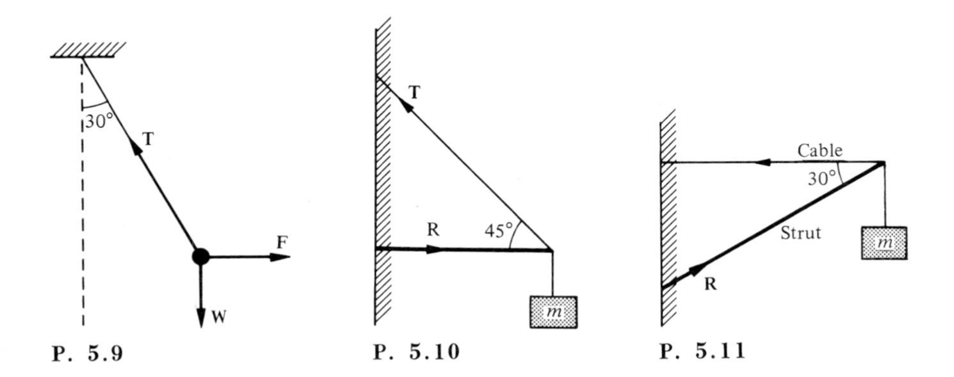

P. 5.9   P. 5.10   P. 5.11

12. A boy and a sled together have a mass of 40 kg. How much force is needed to pull the sled (a) starting from rest; and (b) once in motion to keep in motion? The coefficients of static and kinetic friction are 0.2 and 0.1, respectively.
13. A block of mass 4 kg is to be pulled by a force F applied parallel to the horizontal surface. F is found to be 20 N to start the motion; once in motion, only 12 N is needed to maintain the motion. Calculate the static and kinetic coefficients of friction.
14. Calculate the force T needed to pull at constant speed a block of weight 150 N when the force is applied at an angle of 30° above the horizontal. The coefficient of friction is 0.3.

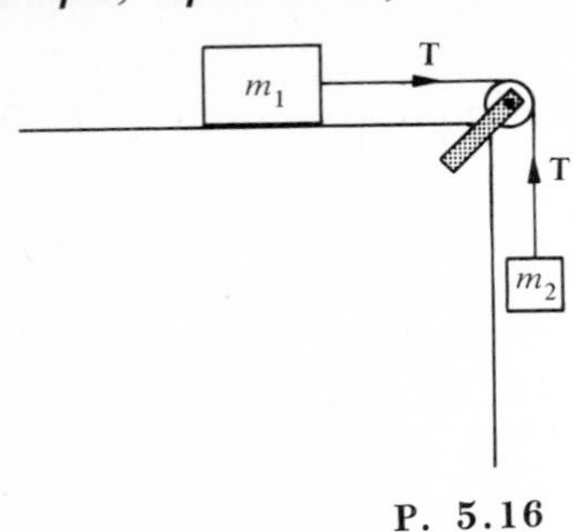

P. 5.16

15. A block of mass 100 kg is being moved steadily by pulling on a rope attached to the block making an angle of 30° with the horizontal. If the tension in the rope is 120 N, what are the force of friction and the coefficient of friction?
16. In the situation shown, if the coefficient of friction is 0.3, $m_1 = 20$ kg, what is the maximum value of m_2 for equilibrium? What is the value of the tension T?
17. In the situation shown, a block of mass $m_1 = 5$ kg is being pulled up steadily by a weight of mass $m_2 = 2$ kg. If the surface is frictionless, calculate (a) the angle θ of the incline; (b) the normal force exerted on the mass m_1 by the plane surface; and (c) the tension in the string.
18. Calculate the minimum value of m shown for which the system will remain in equilibrium. The coefficient of friction between the 80-kg block and the horizontal surface is 0.3.
19. Calculate the magnitudes and directions (clockwise or counterclockwise) of the torques produced by the forces F, N, and f as shown, if the torques are taken about an axis passing perpendicular to the plane through (a) A; (b) B; (c) C; and (d) D.

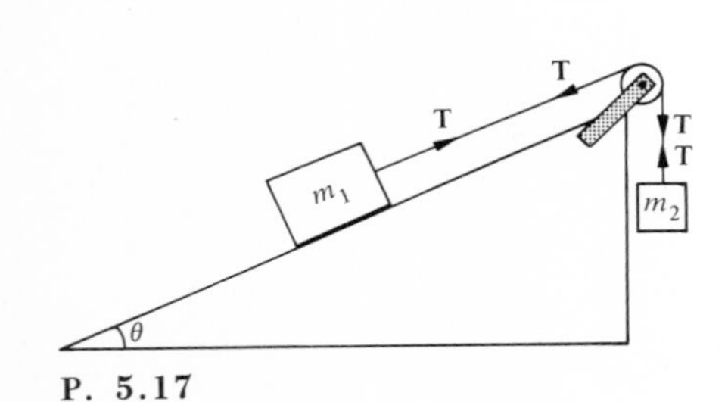

P. 5.17

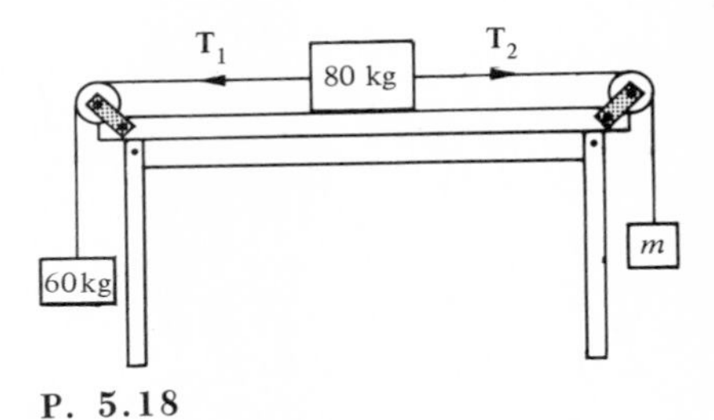

P. 5.18

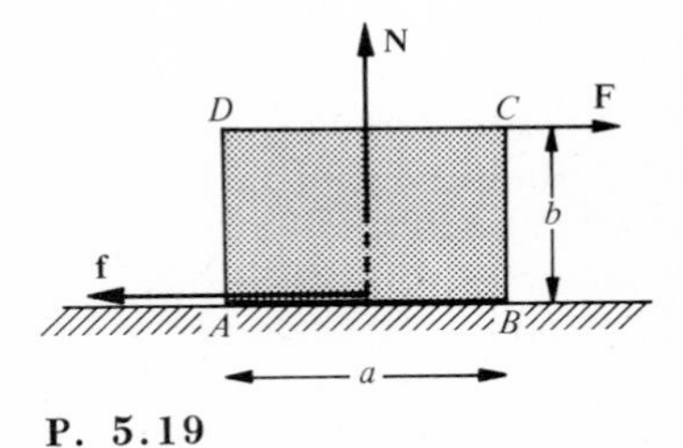

P. 5.19

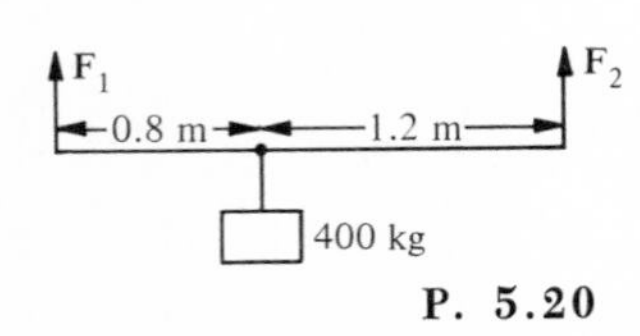

P. 5.20

20. Forces F_1 and F_2 are needed to carry a mass of 400 kg as shown. If the mass of the rod is negligible, calculate F_1 and F_2.
21. In Problem 20, calculate F_1 and F_2 if the mass of the rod is 20 kg and the rod is uniform.
22. A 40-kg boy sits at one end of a 6-m seesaw. At what point from the other end should a 50-kg boy sit so that the seesaw will be in equilibrium?
23. Three forces act at the corners of a triangle as shown. What is the resultant torque about the axis passing through O and perpendicular to the plane?
24. A rectangular block 0.5 m wide and 1.0 m high has a mass of 50 kg, as shown. At $y = 0.75$ m, what should be the value of F so that the block will tilt over?
25. The weight W of a uniform thin piece of wood is 200 N and the piece is resting on the 1.0 m side as shown. Calculate the value of a force T so that the wood piece will start to rotate about the pivot point P.
26. A uniform ladder weighing 400 N and 12 m long leans against a smooth vertical wall with its lower end 4 m away from the wall. A man weighing 600 N climbs halfway up the ladder. Calculate (a) the horizontal and vertical forces exerted by the floor on the ladder (that is, f and N); (b) the reaction P on the ladder by the wall; and (c) the coefficient of friction in this situation.

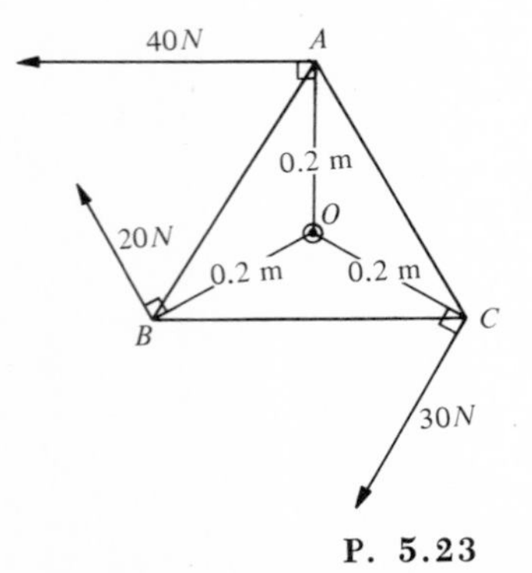

P. 5.23

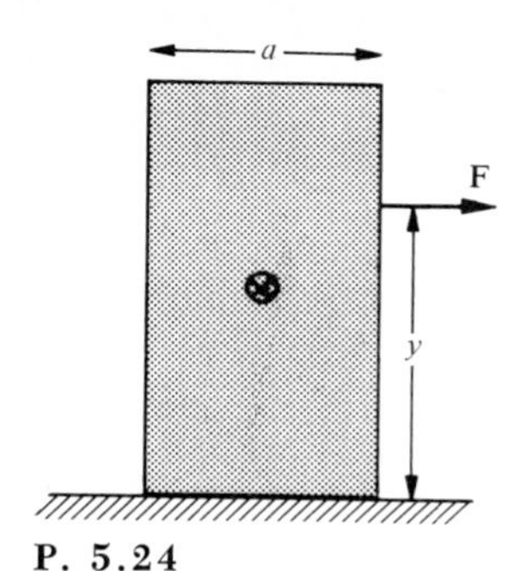

P. 5.24

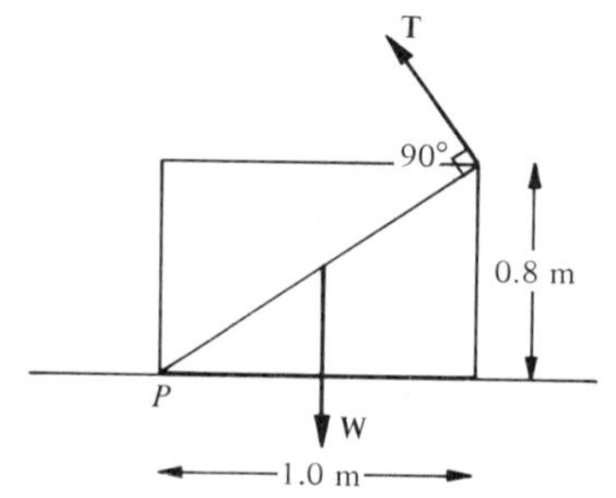

P. 5.25

27. In Problem 26, can the man climb all the way up the ladder if $\mu = 0.70$?

28. A ladder of mass 20 kg, length 6 m, and having a center of gravity located 2 m from the lower end, leans against a smooth vertical wall making an angle of 30° with the vertical. A 70-kg man can climb the ladder three-fourths of its length before it starts to slip. Under these conditions calculate f_s and N due to the ground, and the normal reaction P of the wall. Also calculate the resultant of N and f_s.

29. A 25-ft ladder weighing 50 lb has its center of gravity 10 ft from the bottom end and leans against a smooth vertical wall with its bottom end 12 ft from the wall. If the maximum horizontal frictional force available is 70 lb, how far up the ladder can a 160-lb man climb before the ladder starts slipping?

30. A uniform rod 6 m long has a mass of 4 kg. A weight of 20 N is tied at one end and a weight of 80 N at the other. Find the magnitude, direction, and point of application of a single force that will balance the other forces.

***31.** Find the position of the center of mass of the system of four masses shown.

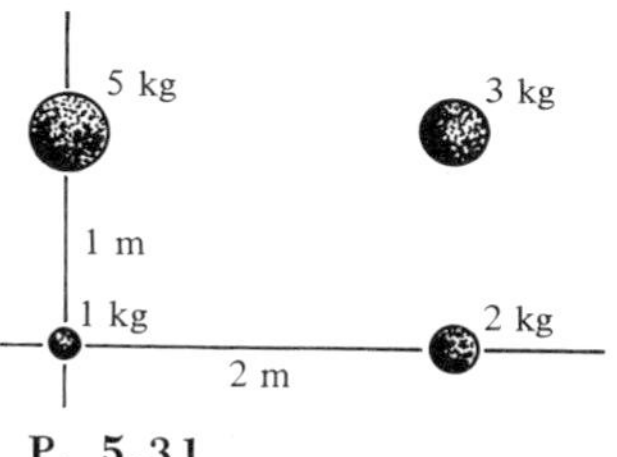

P. 5.31

***32.** A meter rod has a mass of 2 kg attached to one end, 4 kg to the other, and 6 kg at the center. Find the center of mass of this system. Where is the center of gravity?

***33.** Consider a disk of radius R and a small disk of radius $r = R/2$ placed as shown. Find the center of mass.

***34.** Find the center of mass of a steel plate cut in the form shown. Assume that the mass of a plate of area a^2 is m_0.

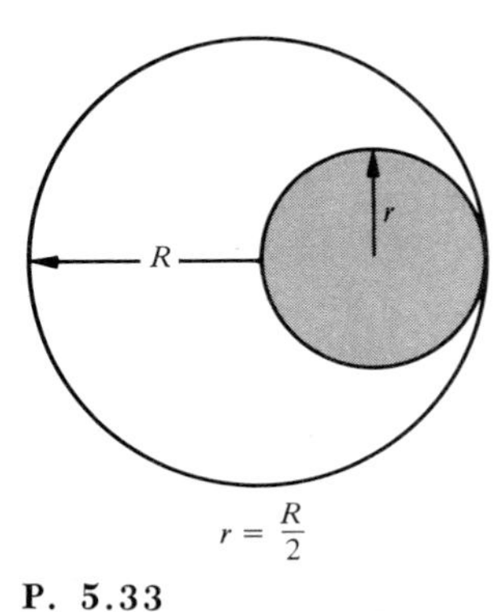

P. 5.33

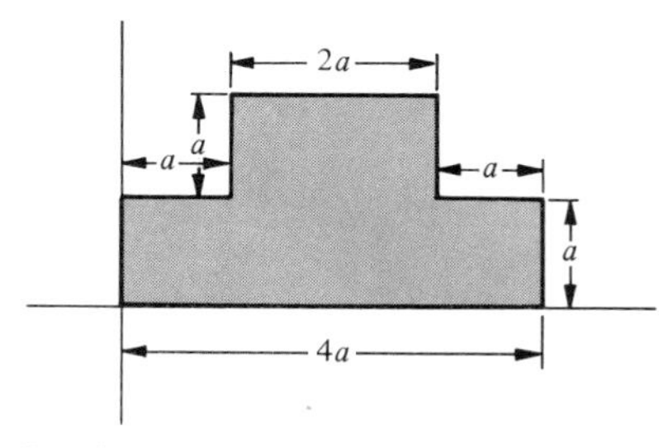

P. 5.34

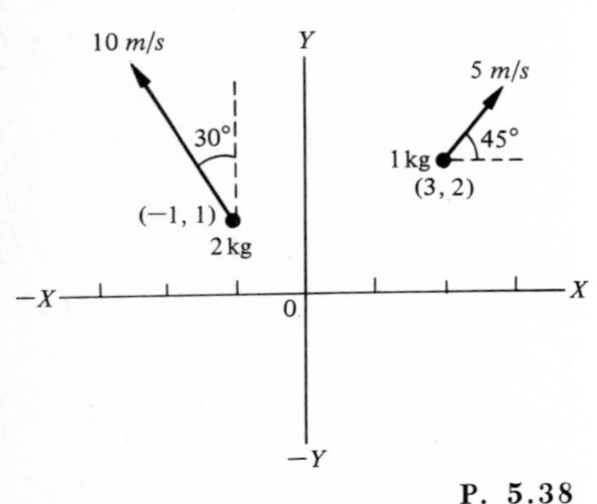

P. 5.38

*35. The separation between the two atoms of a HCl molecule is 1.27×10^{-10} m. Find the position of the center of mass with respect to either atom. $m_H = 1.008$ u, $m_{Cl} = 35.453$ u.

*36. The electron and the proton in the hydrogen atom are separated by a distance of 0.53×10^{-10} m. Find the position of the center of mass with respect to the proton. The mass of the proton is 1836 times the mass of the electron. Can you say that the center of mass coincides with the position of the proton?

*37. A projectile is fired with a speed of 300 m/s, making an angle of 37° with the horizontal. In its parabolic path it explodes into two pieces of mass $2M/5$ and $3M/5$, with the smaller piece landing at a distance of 5000 m. Where will the other piece land?

*38. Three particles, subject only to their internal interaction, are moving in the XY-plane. The velocities of two of these particles are as shown. What should be the velocity of the third particle (of mass 5 kg) so that the center of mass is at rest?

*39. Three masses, of 1 kg, 2 kg, and 3 kg, are located at (3, 0), (0, 4), and (−3, −3), respectively. The forces acting on these masses are 10 N, 5 N, and 8 N, respectively, as shown. Find (a) the center of mass of the system; and (b) the acceleration of the center of mass. (x, y) are in meters.

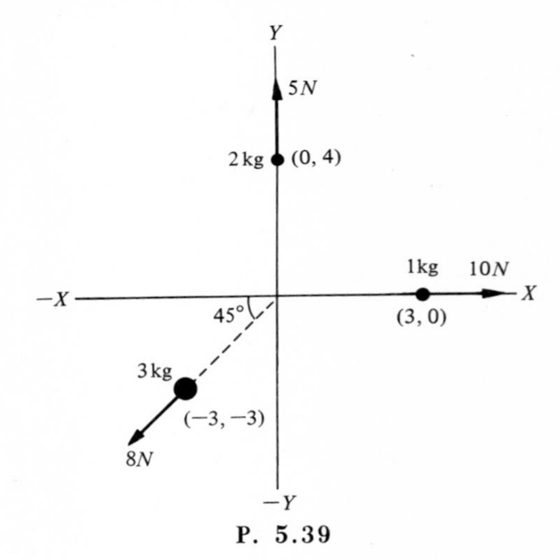

P. 5.39

6 Work and Energy

We shall probe into the scientific meaning of the commonly used terms of work and energy, and their relation to force. The term "energy" has a much wider scope than will be implied in this chapter. We shall limit ourselves to the discussion of mechanical energy and its subdivision into kinetic energy and potential energy—gravitational potential energy and elastic potential energy. We shall close the chapter with a discussion of the principle of conservation of mechanical energy.

6.1. Work

The intuitive meaning of work is quite different from the scientific definition of work. In everyday activity the term "work" is used equally for mental work and for physical work involving muscular force. You may read a book, or exert yourself mentally in thinking about a simple or difficult problem, you might be holding a weight without moving, or carrying a load and moving with uniform horizontal velocity—in all these cases, according to the scientific definition, you are not doing any work at all.

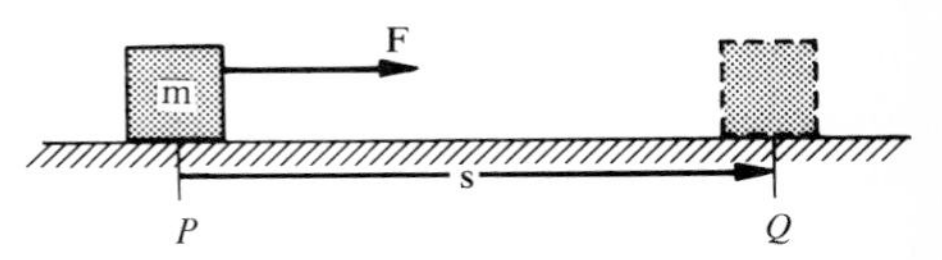

Figure 6.1. *Force* **F** *moves the block from P to Q.*

According to physics, work is said to be done only if the applied force or its component acts in the direction of a displacement. Consider a block of mass m placed at point P and acted upon by a force **F** as shown in Figure 6.1. The block is displaced through a distance s to a new position Q. *The work W done by the force* **F** *is equal to the product of the magnitudes of the force and the displacement, provided that the force is parallel to the displacement;* that is,

$$W = Fs \tag{6.1}$$

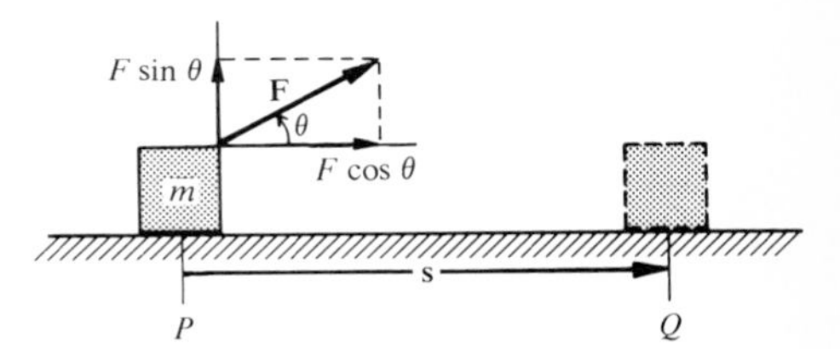

Figure 6.2. *Force* **F** *moves the block from P to Q.*

A more common situation is as shown in Figure 6.2, where the force **F** applied is not parallel to the displacement. In such cases the force is resolved into two components—$F\cos\theta$ parallel to **s** and $F\sin\theta$ perpendicular to **s**. *The work W done by the force* **F** *is defined as the product of the component of the force parallel to the displacement and the displacement;* that is,

$$W = (F\cos\theta)s \tag{6.2}$$

According to this general definition the amount of work done depends not only upon the magnitudes but also on the angle between **F** and **s**. Three extreme cases are shown in Figure 6.3. In (a), **F** and **s** are parallel and θ is $0°$, $\cos 0° = 1$. Thus, the work done is positive, $W = Fs$. In (b), **F** and **s** are perpendicular and hence θ is $90°$, $\cos 90° = 0$. Thus, the work done is zero. In (c), **F** is applied opposite to **s** and hence $\theta = 180°$, $\cos 180° = -1$. The work done is negative, $W = -Fs$. This happens when an object is moving in one direction while the force is in the opposite direction, retarding the motion. A typical example is the work done by the frictional forces, which always act in a direction opposite that of the displacement and hence is negative.

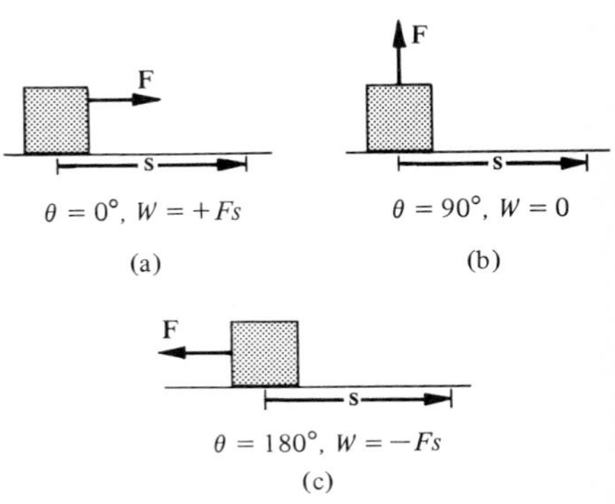

Figure 6.3. *Work done by a force* **F** *in three special cases.*

Referring to Equation 6.2, the work done is zero not only when θ is $90°$, that is, when **F** acts perpendicular to **s**, but also when **s** is zero. Thus, one may be holding a heavy weight but not doing any work because the displacement is zero. Similarly, one may be carrying a weight and moving parallel to the surface of the earth, but not doing any work because the force of gravitational pull is vertical and is perpendicular to the horizontal displacements; that is, θ is 90°.

According to the definition, the unit of work should be those of the product of force and displacement. In the MKS or SI system, the unit of force is the newton and that of displacement is the meter; hence, the unit of work will be the newton-meter, called the *joule*. *One **joule** of work is said to be done when a force of 1 newton acts through a displacement of 1 meter in the direction of the force.* Similarly, in the CGS system, the unit of work is the *erg* (force of 1 dyn applied through 1 cm), while in the engineering system the unit of work is the *foot-pound* (force of 1 lb applied through 1 ft). These units are summarized in Table 6.1. Note the following relation between the joule and the erg:

Table 6.1
Units of Work

System	Units of Force × Distance	Name	Symbol
SI	newton-meter	joule	J
CGS	dyne-centimeter	erg	erg
British engineering	pound-foot	foot-pound	ft-lb

$$1 \text{ joule} = 1 \text{ N-m} = \frac{1 \text{ kg-m-m}}{\text{s}^2}$$

$$= \frac{10^3 \text{ g } 10^2 \text{ cm } 10^2 \text{ cm}}{\text{s}^2} = \frac{10^7 \text{ g-cm-cm}}{\text{s}^2} = 10^7 \text{ dyn-cm}$$

$$\boxed{1 \text{ joule} = 10^7 \text{ ergs}} \tag{6.3}$$

In atomic and nuclear physics the most commonly used unit of work (or energy) is the *electron volt* (eV), which is related to the joule by

$$\boxed{1 \text{ eV} = 1.6021 \times 10^{-19} \text{ J}} \tag{6.4}$$

while 1 million electron volts (1 MeV) is equal to 10^6 eV. This unit will be explained in detail in Chapter 20.

Let us go back to Equation 6.2. In the notation of vector multiplication, Equation 6.2 may be written as (this is one of the two vector multiplications that we shall introduce)

$$\boxed{W = \mathbf{F} \cdot \mathbf{s} = Fs \cos \theta} \tag{6.5}$$

where $\mathbf{F} \cdot \mathbf{s}$ is the *dot product* or *scalar product* of two vectors and is equal to a scalar quantity W. θ is the angle between the two vectors.

According to Equation 6.2, if **F** and **s** are parallel and the body is being pushed along the X-axis starting from an initial position x_1 to the final position x_2 so that $s = x_2 - x_1$, and $\cos\theta = \cos 0° = 1$, the work done in this case is given by

$$W = Fs = F(x_2 - x_1) \tag{6.6}$$

where F is a constant force acting along the X-axis. The plot of F versus x will be as shown in Figure 6.4. From this plot and Equation 6.6, it is clear that the magnitude of the work done is equal to the area of a rectangle shown shaded in Figure 6.4. We may conclude that

$$\boxed{\text{work done} = \text{area under the } F\text{-versus-}x \text{ curve}} \tag{6.7}$$

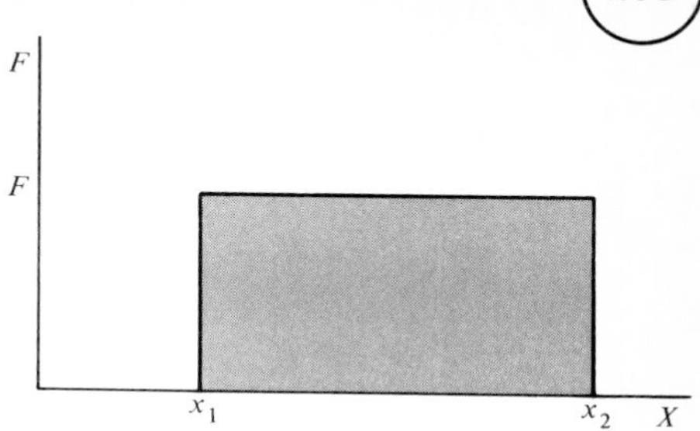

FIGURE 6.4. *The work done by a constant force is equal to the area of the shaded rectangle.*

Let us calculate the work done in the case when the applied force is not constant. Let F_x be the variable force acting on an object that is being pushed along the X-axis from $x = x_i$ to $x = x_f$. The variation in the force can arise from many causes. Suppose that an object has to be pushed along a rough surface at constant speed. This means that a variable force will have to be applied to make up for the uneven resistance of the surface to the motion of the object. The plot of F_x versus x in such cases will be as shown in Figure 6.5. In order to calculate the amount of work done by this force, let us divide the total displacement between x_i and x_f into very narrow strips of widths Δx_1, Δx_2, Δx_3, . . . , as shown in Figure 6.5. Even though the force F_x is variable, it may be considered to be almost constant during any one of these small displacements. For example, say for displacement Δx_2 the force is F_{x_2}, and hence the work done for Δx_2 displacement will be

$$\Delta W_2 = F_{x_2}\,\Delta x_2$$

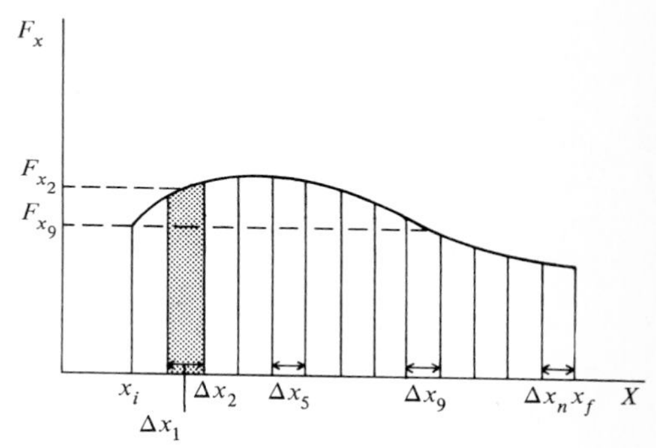

FIGURE 6.5. *Variable force moves an object from x_i to x_f.*

With similar expressions for all other displacements Δx_1, Δx_2, Δx_3, . . . , the forces applied are F_{x_1}, F_{x_2}, F_{x_3}, . . . , hence the total work done from the initial position x_1 to the final position x_f will be

$$W_{i\to f} = F_{x_1}\,\Delta x_1 + F_{x_2}\,\Delta x_2 + F_{x_3}\,\Delta x_3 + \cdots + F_{x_n}\,\Delta x_n$$

$$W_{i\to f} = \sum_{j=1}^{n} F_{x_j}\,\Delta x_j = \text{area under the curve} \tag{6.8}$$

EXAMPLE 6.1 A force **F** making an angle θ with the horizontal is pulling a block of mass M to the right as shown in the accompanying figure. The other forces acting are the reaction force **N**, the force of gravity **W** on the mass M, and the frictional force **f**; the coefficient of friction is μ. If the block moves through a distance x, calculate the work done by the different forces. Calculate the work done if $M = 10$ kg, $\theta = 30°$, $F = 100$ N, g $= 9.8$ m/s^2, $x = 2$ m, and $\mu = 0.2$.

The force **F** can be resolved into two components—$F\cos\theta$ parallel to the surface and $F\sin\theta$ perpendicular to the surface. To start with, if the body is in motion along the X-axis,

$$\sum F_x = F\cos\theta - f \geqslant 0 \tag{i}$$

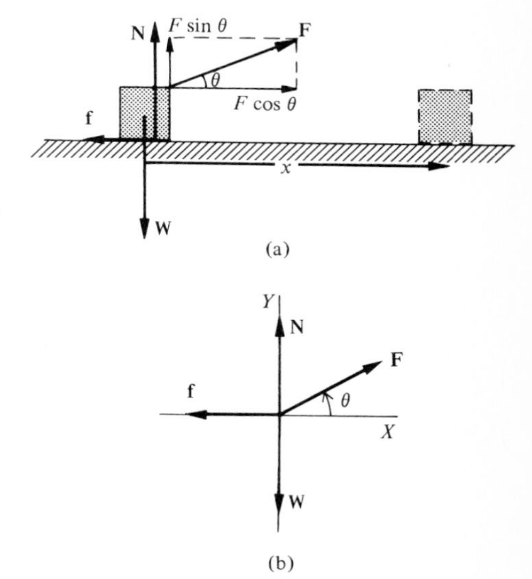

Ex. 6.1

$$\sum F_y = F\sin\theta + N - W = 0 \tag{ii}$$

Also,

$$f = \mu N. \tag{iii}$$

From Equation ii,

$$N = W - F\sin\theta$$

From Equation iii,

$$f = \mu N = \mu(W - F\sin\theta)$$

Let us first calculate the work done W_F by the force **F**. The component of **F** parallel to x is $F\cos\theta$. Therefore,

$$W_F = (F\cos\theta)x = Fx\cos\theta = (100\text{ N})(2\text{ m})\cos 30° = 173.2\text{ J}$$

The component $F\sin\theta$, being perpendicular to x, does no work. Remembering that f is opposite to x, the work done against friction W_f is given by

$$\begin{aligned} W_f &= -fx = -\mu(W - F\sin\theta)x \\ &= -0.2[(10\text{ kg})(9.8\text{ m/s}^2) - (100\text{ N})(\sin 30°)](2\text{ m}) \\ &= -0.2(98\text{ N} - 50\text{ N})(2\text{ m}) = -19.2\text{ J} \end{aligned}$$

No work is done by the normal reaction force **N** and the weight **W** because these are perpendicular to the displacement. Thus, the total work done is

$$\begin{aligned} W &= W_F + W_f + W_N + W_W = (Fx\cos\theta) - fx + 0 + 0 = 173.2\text{ J} - 19.2\text{ J} \\ &= 154\text{ J} \end{aligned}$$

Exercise 6.1 A block of mass 5 kg is pulled on a horizontal surface a distance of 3 m by a force of 50 N, which makes an angle of 45° with the horizontal. Calculate the total work done if (a) the surface is frictionless and (b) the coefficient of friction for the surface is 0.3. [*Ans.*: (a) $W = 106$ J; (b) $W = 93.8$ J.]

6.2. Power

Besides knowing the total amount of work done, it is important to know the time taken to do this work, whether it takes a year, a month, a day, or an hour. *Power* is the rate at which work is done by a working agent. Thus, if an amount of work ΔW is done over a time interval of Δt, the average power $\overline{P}$ is defined as

$$\overline{P} = \frac{\text{work done}}{\text{time interval}} = \frac{\Delta W}{\Delta t} \tag{6.9}$$

while the instantaneous power P is defined as that value of $\overline{P}$ for which Δt is very small, so that Δt approaches zero; that is,

$$\boxed{P = \lim_{\Delta t \to 0} \frac{\Delta W}{\Delta t}} \tag{6.10}$$

In SI units, if an agent does work at a rate of 1 J/s, it is said to have a power of 1 *watt*. This and other units of power are given in Table 6.2. A watt is a very small unit of power; hence the kilowatt (= 1000 watts) is commonly used.

Table 6.2
Units of Power

System of Units	Unit	Name
SI	joules/second (J/s)	watts (W)
CGS	ergs/second (ergs/s)	ergs/s
British engineering	foot-pounds/second (ft-lb/s)	ft-lb/s
	550 ft-lb/s	1 hp

Still another unit of power, which does not belong to any one of the three common units but is used more frequently, is horsepower. A working agent develops 1 *horsepower* (hp) if it is capable of doing 550 ft-lb of work per second.

$$1 \text{ hp} = 550 \text{ ft-lb/s} = 33{,}000 \text{ ft-lb/min} = 0.746 \text{ kW} \simeq 0.75 \text{ kW}$$

The origin of the unit "horsepower" goes back to the times of James Watt (1736–1819). According to Watt, a powerful horse could lift 550 lb a distance of 1 ft in 1 s and work steadily at this rate during a working day.

According to Equation 6.9 or 6.10, work may be defined as

$$\text{work} = \text{power} \times \text{time}$$

We can use the units of power and time to define a new unit of work. Thus, if power is stated in units of kilowatt and time in hours, the unit of work will be the kilowatt-hour. We may define 1 *kilowatt-hour* to be equal to the amount of work done in 1 hour by an agent working at a constant rate of 1 kilowatt. Thus,

$$1 \text{ kWh} = \left(1000 \frac{\text{J}}{\text{s}} h\right)\left(3600 \frac{\text{s}}{\text{h}}\right) = 3.6 \times 10^6 \text{ J} = 3.6 \text{ MJ}$$

where M stands for mega (= 10^6).

Let us go back to Equation 6.9 and express power in terms of force and velocity. From the definition of work, $\Delta W = F\,\Delta s$ provided that F is in the direction of the displacement Δs. Therefore,

$$\overline{P} = \frac{\Delta W}{\Delta t} = F\frac{\Delta s}{\Delta t} = F\overline{v}$$

while for instantaneous velocity v, the instantaneous power P is

$$\boxed{P = Fv} \qquad (6.11)$$

Note that P, like W, is also a scalar quantity.

Example 6.2 An electric motor is used to pump water from a well 20 m deep at the rate of 120 ℓ/min. Calculate the minimum power needed to do this if the motor is only 50 percent efficient.

Knowing that 1 ℓ = 1000 cm³ and 1 cm³ of water has a mass of 1 g, we can calculate the amount of water to be pumped.

$$\left(120\ \frac{\ell}{\text{min}}\right)\left(\frac{1\text{ g}}{\text{cm}^3}\right) = \left(\frac{120{,}000\text{ cm}^3}{60\text{ s}}\right)\left(\frac{1\text{ g}}{\text{cm}^3} \times \frac{1\text{ kg}}{1000\text{ g}}\right) = 2\text{ kg/s}$$

The force acting on 2 kg of water is

$$F = mg = (2\text{ kg})(9.8\text{ m/s}^2) = 19.6\text{ N} \qquad \text{downward}$$

The motor should supply a force equal and opposite to this and move it an average distance of 20 m in 1 s; that is, $\bar{v} = 20$ m/s. Thus, the average power needed is

$$\bar{P}' = F\bar{v} = (19.6\text{ N})(20\text{ m/s}) = 392\text{ watts}$$

Since the motor is only 50% efficient, the motor should have a power of

$$\bar{P} = (2)(392\text{ W}) = 784\text{ W} = 0.784\text{ kW} = 1.05\text{ hp}$$

Exercise 6.2 A mass of 10 kg is lifted vertically upward through a distance of 20 m in 4 s. Calculate (a) the work done; and (b) the rate at which the work is being done. [*Ans.*: (a) 1960 J; (b) 490 W.]

6.3. Kinetic Energy

The effects of applying a force on an object are many. If the applied force acting on an object exceeds the frictional force, it may change the velocity of the object. On the other hand, if the applied force is equal to or greater than the weight of an object, the object may be moved to a higher position. In both cases the force has done work that has not been wasted, but has been stored in the object. In the first case, this work is in the form of motion of the object, whereas in the second case, the work is in the form of the position of the object. In both cases the work stored may be used to do other work. We say that the objects possess energy.

Energy is the ability to do work; or, energy is *stored work*. We shall limit ourselves to the discussion of mechanical energy, which may be in the form of kinetic energy (energy due to motion) and/or potential energy (energy due to position). In this section we discuss kinetic energy.

A moving object is capable of doing work because of its motion; hence, we say that the object has kinetic energy. "Kinētikos" in Greek means "to move"; hence, *kinetic energy means energy due to motion.* The energy is stored in the object when work is done to change its velocity from a lower value to a higher value, or from rest to a certain velocity.

Let a mass m moving with an initial velocity v_i be acted upon by a force F over a distance s. Suppose that the velocity of the mass changes from v_i to v_f. From the relation $v_f^2 = v_i^2 + 2as$,

$$s = \left(\frac{1}{2a}\right)(v_f^2 - v_i^2)$$

Also, if m is constant, $F = ma$. Multiplying the preceding two equations, we get

$$Fs = \tfrac{1}{2}mv_f^2 - \tfrac{1}{2}mv_i^2 \tag{6.12}$$

But the left side of this equation is equal to the work W done by the net force F during the displacement s. That is, $W = Fs$, hence,

$$W = \tfrac{1}{2}mv_f^2 - \tfrac{1}{2}mv_i^2 \tag{6.13}$$

This equation states that the work done on an object is equal to the change in the quantity $\tfrac{1}{2}mv^2$ of the object. We call $\tfrac{1}{2}mv^2$ the *kinetic energy K*,

$$\boxed{K = \tfrac{1}{2}mv^2} \tag{6.14}$$

Thus, Equation 6.13 takes the form

$$\boxed{W = K_f - K_i = \tfrac{1}{2}mv_f^2 - \tfrac{1}{2}mv_i^2} \tag{6.15}$$

which is called the *work–energy theorem*. This equation implies that the work done by the applied force is equal to the change in kinetic energy. Note that the units of energy are the same as that of work as given in Table 6.1.

EXAMPLE 6.3 A 10-g bullet traveling at 200 m/s strikes a wooden target and penetrates 10 cm before coming to rest. Calculate the average retarding force.

The kinetic energy K of the bullet before hitting the wooden target is

$$K = \tfrac{1}{2}mv^2 = \tfrac{1}{2}(0.01 \text{ kg})(200 \text{ m/s})^2 = 200 \text{ J}$$

This energy is used in doing work in penetrating the wood. If F is the average retarding force and x is the distance through which it acts, that is, the distance the bullet travels, the work done $W = -Fx$. Therefore,

$$W = -Fx = K_f - K_i = 0 - K = -K$$

or

$$F = \frac{K}{x} = \frac{200 \text{ J}}{0.10 \text{ m}} = 2000 \text{ N}$$

EXERCISE 6.3 The brakes in a 4-ton truck fail when the truck is traveling at 50 mi/h. The driver shuts off the motor, drives onto a gravel road, and comes to a stop within 200 ft. Calculate the average retarding force. [*Ans.*: 3363 lb.]

6.4. Gravitational Potential Energy

We know that every object on the surface of the earth is attracted by the earth. If an object has mass m, it will experience a force that is called the weight $w = mg$ of the object. For all practical purposes, we assume that the value of g remains constant near the surface of the earth, so that the weight of the object remains constant. Also the space in which the mass is acted upon by the gravitational force is called the *gravitational force field* or *gravitational field*. (We shall discuss this in detail in Chapters 11 and 19.)

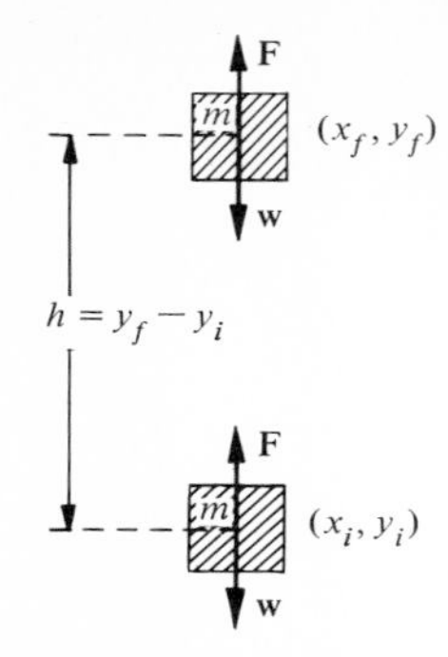

FIGURE 6.6. *Mass m is displaced through a height h in a gravitational field.*

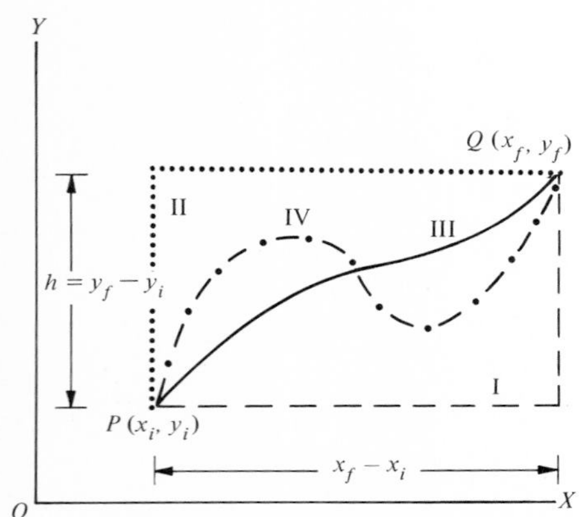

FIGURE 6.7. *Change in potential energy when a mass m is moved from P to Q is independent of the path.*

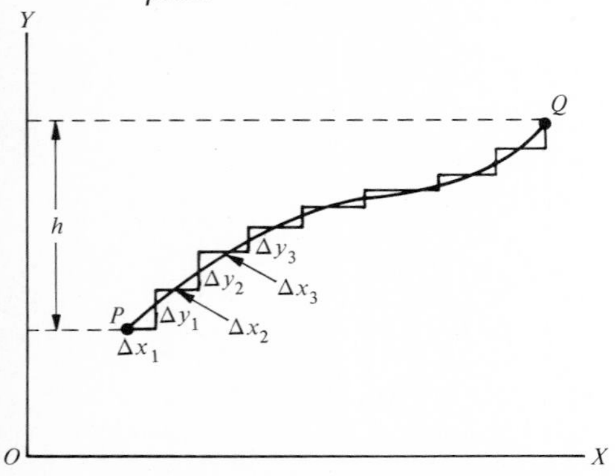

FIGURE 6.8. *No work is done along horizontal displacements, while the vertical displacements add up to h.*

Let an object of mass m in a gravitational field be at position (x_i, y_i) and suppose that it is being pulled with a force $w = mg$ acting downward. To move this object vertically upward, we must apply a force F upward which is equal to the weight w, as shown in Figure 6.6. Suppose further that the object is moved to a final position, (x_f, y_f). Thus, the vertical displacement in the direction of the applied force F is h. The work done by the force F is

$$W = Fh$$

But $F = w = mg$ and $h = y_f - y_i$. Therefore,

$$W = mgy_f - mgy_i \tag{6.16}$$

This equation states that the work done on an object which changes its position is characterized by the quantity mgy, called the *potential energy* V:

$$\boxed{V = mgy} \tag{6.17}$$

This is also called the *gravitational potential energy*. Thus, Equation 6.16 takes the form

$$\boxed{W = V_f - V_i = mgy_f - mgy_i = mgh} \tag{6.18}$$

This is also called the *work–energy theorem*. This equation implies that the work done by the applied force is converted into potential energy. It may be pointed out that Equation 6.18 is true only if the velocity v is constant when work is being done so that there is no change in kinetic energy.

Note that a change in potential energy depends on the initial and final value of y, that is, on $h\ (= y_f - y_i)$. Suppose that a body of mass m is moved from point P to point Q. It is immaterial whether we follow path I, path II, path III, path IV, or any other path (Figure 6.7); the same amount of work will have to be done. This is quite clear from Figure 6.8. Say the mass is moved from P to Q along path III of Figure 6.7. This path can be split into horizontal and vertical steps Δx_1, Δy_1, Δx_2, Δy_2, . . . as shown in Figure 6.8. When moving along Δx_1, Δx_2, . . . , the force is at right angles to these displacements ($\theta = 90°$, $\Delta W = F\,\Delta x \cos 90° = 0$) and no work is done. The vertical displacements $\Delta y_1 + \Delta y_2 + \cdots$ all add up to $h = y_f - y_i$. Thus, the total work done is mgh.

From this we conclude that the work done by or against the gravitational force field is independent of the path an object follows in changing its position. Any force field in which the work done between two points by and against the field is independent of the path and depends only on the initial and final positions of the object is called a *conservative force field*. The force associated with this field is called a *conservative force*. Thus, a gravitational force field is a conservative force field, while the gravitational force $w(= mg)$ is a conservative force.

A typical example of a nonconservative force is the frictional force. The work done in moving an object from one point to another will certainly depend upon the path followed if the frictional forces are present.

6.5. Conservation of Mechanical Energy

Let us consider two points $A(x_i, y_i)$ and $B(x_f, y_f)$ in a gravitational field. Let the velocity of an object of mass m be v_i at point A. Suppose that a force F is applied to move this object to point B, where its velocity is v_f. Let the work done by the force in this process be W. This work is used in changing the potential energy as well as kinetic energy; that is,

$$W = \text{final mechanical energy} - \text{initial mechanical energy}$$

or

$$\begin{aligned} W &= (K_f + V_f) - (K_i + V_i) \\ &= (\tfrac{1}{2}mv_f^2 + mgy_f) - (\tfrac{1}{2}mv_i^2 + mgy_i) \end{aligned} \tag{6.19}$$

Suppose no external force is acting on the mass m when it is moved from point A to B. The only force present is the gravitational force, $w = mg$. Since $W = 0$ Equation 6.19 takes the form

$$\tfrac{1}{2}mv_f^2 + mgy_f = \tfrac{1}{2}mv_i^2 + mgy_i = \text{constant}$$

$$K_f + V_f = K_i + V_i \qquad \text{if } \mathbf{F}_{\text{ext}} = 0 \tag{6.20}$$

which says that, *in a conservative force field, the sum of the kinetic and potential energy remains constant.* That is, the total mechanical energy is constant in a conservative force field. Note that Equation 6.20 may be written

$$(K_f - K_i) = -(V_f - V_i) \tag{6.21}$$

It is important to note that in such situations, when a mass m moves from a point of high gravitational potential energy to a point of low gravitational potential energy, the kinetic energy of m will increase, and vice versa. (When moving from a point of high potential energy to a point of low potential energy, the gravitational force does the work; while in the opposite case, the work is done against the gravitational force.)

EXAMPLE 6.4 The bob of a 2-m-long pendulum has a mass of 0.5 kg. The bob is pulled to a side until it makes an angle of 30° with the vertical, as shown. Calculate (a) the change in potential energy; and (b) the work done in moving the bob. What will be the speed of the bob when it passes the lowest point after being released?

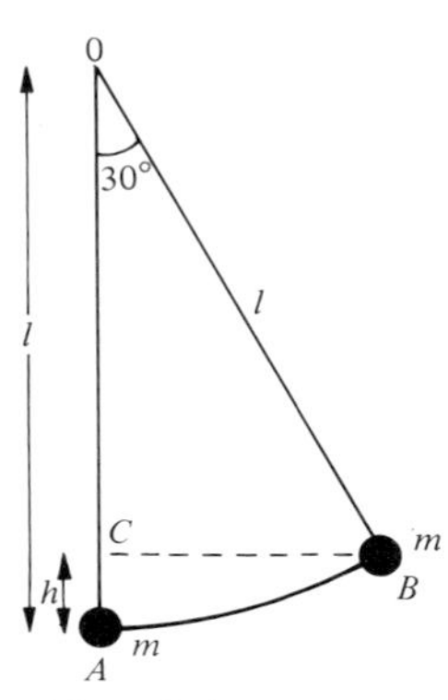

Ex. 6.4

(a) As the mass m is moved from A to B, it changes height by h, where $h = AC = OA - OC$. But $OA = OB = l$ and $OC/OB = \cos\theta$ or $OC = OB\cos\theta = l\cos\theta$:

$$h = OA - OC = l - l\cos\theta = l(1 - \cos\theta) \tag{i}$$

Thus, the gain in the potential energy V of mass m will be for $l = 2$ m and $\theta = 30°$:

$$\begin{aligned} V &= mgh = mgl(1 - \cos\theta) \qquad \text{(ii)} \\ &= (0.5 \text{ kg})(9.8 \text{ m/s}^2)(2 \text{ m})(1 - \cos 30°) \\ &= (0.5)(9.8)(2)(1 - 0.87) \text{ J} = 1.3 \text{ J} \end{aligned}$$

(b) All the work done is stored as potential energy. Therefore,

$$W = V = mgh = 1.3 \text{ J} \tag{iii}$$

If the mass m is now released from point B, when it reaches the lowest point A, it has lost all its potential energy, which appears as kinetic energy at point A. Also, $W = 0$ because tension is perpendicular to the displacement. Thus, Equation 6.20 takes the form

$$K_B + V_B = K_A + V_A$$

where $K_B = 0$ and $V_A = 0$. Thus,

$$V_B = K_A \quad \text{or} \quad mgh = \tfrac{1}{2}mv^2$$

or

$$v = \sqrt{2gh} \tag{iv}$$

From Equation i,

$$h = l(1 - \cos\theta) = (2 \text{ m})(1 - \cos 30°) = (2 \text{ m})(1 - 0.87) = 0.26 \text{ m}$$

Therefore, from Equation iv,

$$v = \sqrt{2(9.8 \text{ m/s}^2)(0.26 \text{ m})} = 2.3 \text{ m/s}$$

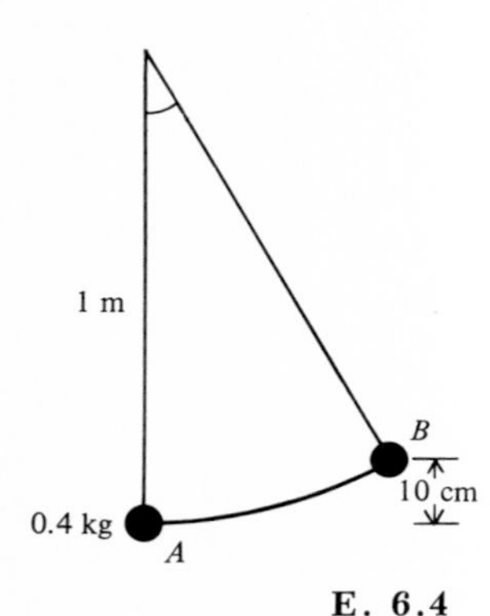

E. 6.4

Exercise 6.4 The bob of a 1-m-long pendulum has a mass of 0.4 kg. The bob is pulled from position A to position B until the center of gravity is raised by 10 cm. Calculate (a) the change in potential energy; (b) the angle which the bob makes with the vertical; and (c) the speed as it passes through A after being released from B. [*Ans.*: (a) 0.39 J; (b) 25.8°; (c) 1.4 m/s.]

Example 6.5 A mass of 100 kg at A is moving with 10 m/s as shown in the accompanying figure. Suppose that the track from A to D is frictionless, while after D the coefficient of friction is 0.6. The mass comes to a stop at E. Calculate (a) the potential energy; (b) the kinetic energy; and (c) the total energy at point A, B, C, and D. (d) Calculate the distance DE.

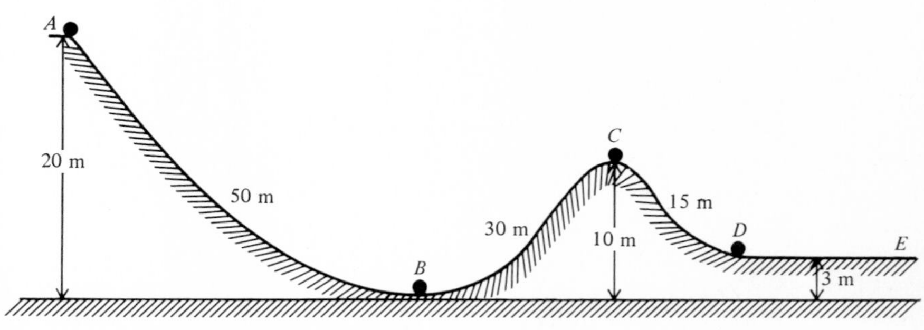

Ex. 6.5 and E. 6.5

(a) The potential energy at various points is

$$V_A = mgh_A = (100 \text{ kg})(9.8 \text{ m/s}^2)(20 \text{ m}) = 19{,}600 \text{ J}$$
$$V_B = mgh_B = (100 \text{ kg})(9.8 \text{ m/s}^2)(0) = 0$$
$$V_C = mgh_C = (100 \text{ kg})(9.8 \text{ m/s}^2)(10 \text{ m}) = 9800 \text{ J}$$
$$V_D = mgh_D = (100 \text{ kg})(9.8 \text{ m/s}^2)(3 \text{ m}) = 2940 \text{ J}$$

(b) and (c) Let us calculate the kinetic energy and total energy at different points:

$$K_A = \tfrac{1}{2}mv_A^2 = \tfrac{1}{2}(100 \text{ kg})(10 \text{ m/s})^2 = 5000 \text{ J}$$

Thus, the total energy at A is

$$E_A = K_A + V_A = 5000 \text{ J} + 19{,}600 \text{ J} = 24{,}600 \text{ J}$$

Applying the conservation principle to points A, B, C, and D, we get

$$E_A = E_B = E_C = E_D = 24{,}600 \text{ J}$$

Also, from $E_A = E_B$, $K_A + V_A = K_B + V_B$,

$$K_B = K_A + V_A - V_B = 5000 \text{ J} + 19{,}600 \text{ J} - 0 = 24{,}600 \text{ J}$$

From, $K_B = \tfrac{1}{2}mv_B^2 = 24{,}600$ J:

$$v_B = \sqrt{\frac{(2)(24{,}600 \text{ J})}{100 \text{ kg}}} = 22.2 \text{ m/s}$$

Similarly,

$$K_A + V_A = K_C + V_C$$
$$K_C = E_A - V_C = 24{,}600 \text{ J} - 9800 \text{ J} = 14{,}800 \text{ J}$$
$$v_C = \sqrt{\frac{(2)(14{,}800 \text{ J})}{100 \text{ kg}}} = 17.2 \text{ m/s}$$

Also,

$$K_A + V_A = K_D + V_D$$
$$K_D = E_A - V_D = 24{,}600 \text{ J} - 2940 \text{ J} = 21{,}660 \text{ J}$$
$$v_D = \sqrt{\frac{2(21{,}660 \text{ J})}{100 \text{ kg}}} = 20.8 \text{ m/s}$$

(d) The total energy at D is $E_D = 24{,}600$ J, of which the potential energy is $V_D = 2940$ J while the kinetic energy is $K_D = 21{,}660$ J. Since between D and E the potential energy remains constant, the kinetic energy must be used up doing work against friction. Let f be the force of friction and $DE = x$. Therefore,

$$K_D = W_f = fx = \mu Nx = \mu mgx$$

or

$$x = \frac{K_D}{\mu mg} = \frac{21{,}660 \text{ J}}{(0.6)(100 \text{ kg})(9.8 \text{ m/s}^2)} = 36.84 \text{ m}$$

EXERCISE 6.5 A mass of 10 kg at A in the above example is released with zero speed. Suppose that the track from A to E has an average frictional force of 10 N. Calculate (a) the work done against the frictional force between AB, BC, and CD; (b) the speed of the mass at points B, C, and D; and (c) the distance DE the mass will travel before coming to a stop. [*Ans.*: (a) 500 J, 300 J, 150 J; (b) 17.1 m/s, 6 m/s, 12 m/s; (c) 71.6 m.]

6.6. Elastic Potential Energy

We shall now discuss another form of mechanical potential energy, called *elastic potential energy,* which is the result of the distortion of a system caused by an applied force. When a force is applied to a rod, a straight wire, or any other system, there results a change in length, volume, area, shape, or some other form of distortion.

> **Hooke's law—*the distortion of a system is directly proportional to the applied distorting force and is in the direction of the force.***

If the distortion is the displacement, Hooke's law may be written

$$\mathbf{F}_{ext} = +k\mathbf{x} \tag{6.22}$$

where $\mathbf{F}_{ext}$ is the external applied force producing the displacement $\mathbf{x}$ and k is the proportionality constant, called the *force constant,* the force needed to produce a unit displacement, that is, force per unit displacement. The units of k are newtons per meter, pounds per foot, or pounds per inch. The relation of Equation 6.22 is true only if the magnitude of the displacement x does not exceed the elastic limit, that is, if the system returns to its original position after the applied force is removed.

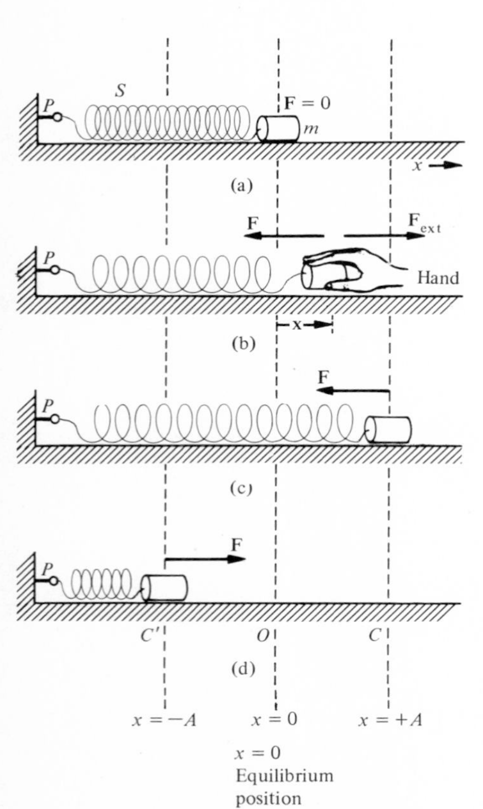

FIGURE 6.9. *Spring under the action of external force* $\mathbf{F}_{ext}$ *and restoring force* **F**.

Let us consider a spring S which is tied to a pin P at one end and to a mass m at the other, as shown in Figure 6.9(a). The mass m is lying on a frictionless table surface. The spring in this position is unstretched and is said to be in an equilibrium position, denoted by $x = 0$. Let us stretch this spring by applying an external force $\mathbf{F}_{ext}$ as shown in Figure 6.9(b). The displacement x of the spring is directly proportional to this applied force and is in the direction of this force. The larger the force we apply, the larger will be the displacement x, and hence the system obeys Hooke's law. As soon as the external force is applied, there comes into play another force **F**, called the *restoring force,* which is equal and opposite to the applied force $\mathbf{F}_{ext}$, as shown in Figure 6.9(b). The existence of the restoring force is evident from the fact that when the external force is applied, it is balanced by the restoring force, and hence the system does not accelerate. Also, it is the restoring force that brings the system back to its original position when the applied external force is removed (Figure 6.9(c)). Since $\mathbf{F} = -\mathbf{F}_{ext}$, Equation 6.22 may be written

$$\boxed{\mathbf{F} = -k\mathbf{x}} \tag{6.23}$$

That is, the restoring force is directly proportional to the displacement and is in the opposite direction to the displacement.

Equation 6.23 also applies to the case when the spring is compressed, as shown in Figure 6.9(d). The compressing force is to the left and the restoring force is to the right.

Let us now calculate the work done in stretching (or compressing) a spring. Let us apply a force F and stretch the spring through a distance x. Ignoring the signs, we may write $F = kx$, and the plot of F versus x is a straight line as shown in Figure 6.10. The work done in stretching the spring through a distance x is equal to the area under the F-versus-x curve. This is shown shaded in Figure 6.10 and is equal to $\frac{1}{2}(F)x = \frac{1}{2}(kx)x = \frac{1}{2}kx^2$. But this work done is stored as energy, called the *elastic potential*, U. That is,

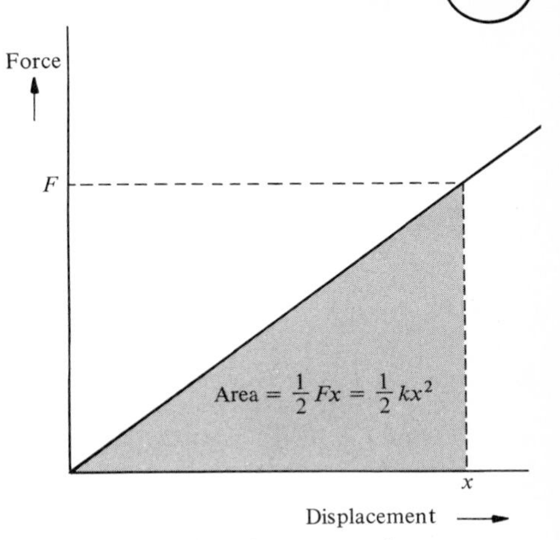

FIGURE 6.10. *F versus x for a spring. The work done is represented by the shaded area.*

$$\boxed{U = \tfrac{1}{2}kx^2} \tag{6.24}$$

Suppose that we stretch a spring through $x = A$; its elastic potential energy will be $\frac{1}{2}kA^2 = U_A$. If we now release the spring, for any value of x between $x = 0$ and $x = A$ the potential energy will be $\frac{1}{2}kx^2$, while the kinetic energy K of the mass m will be $\frac{1}{2}mv^2$, where v is the velocity of the mass. The sum of these two energies must remain constant (because only conservative forces are acting). Thus, for any two points i and f, the conservation of energy requires

$$K_i + U_i = K_f + U_f = \text{constant} = U_A \tag{6.25a}$$

or

$$\tfrac{1}{2}mv_i^2 + \tfrac{1}{2}kx_i^2 = \tfrac{1}{2}mv_f^2 + \tfrac{1}{2}kx_f^2 \tag{6.25b}$$

In general, the principle of conservation of mechanical energy requires (including kinetic, potential and elastic energies) that,

$$\boxed{K_i + V_i + U_i = K_f + V_f + U_f} \tag{6.26}$$

EXAMPLE 6.6 A spring stretches 10 cm when a force of 600 N is applied. Calculate the potential energy of the spring when (a) it stretches 12 cm; (b) a force of 800 N is applied; and (c) a mass of 20 kg is suspended from it vertically downward.

From Equation 6.22, $F = kx$, where $F = 600$ N and $x = 10$ cm $= 0.10$ m. Thus, the force constant k is

$$k = \frac{F}{x} = \frac{600 \text{ N}}{0.10 \text{ m}} = 6000 \text{ N/m}$$

(a) If $x = 12$ cm $= 0.12$ m from Equation 6.24, the potential energy of the spring is

$$U = \tfrac{1}{2}kx^2 = \tfrac{1}{2}(6000 \text{ N/m})(0.12 \text{ m})^2 = 43.2 \text{ J}$$

(b) If $F = 800$ N, the spring will stretch, so

$$x = \frac{F}{k} = \frac{800 \text{ N}}{6000 \text{ N/m}} = 0.133 \text{ m}$$

Thus, the potential energy of the spring will be

$$U = \tfrac{1}{2}kx^2 = \tfrac{1}{2}(6000\ \text{N/m})(0.133\ \text{m})^2 = 53.1\ \text{J}$$

(c) If $m = 20$ kg, the force is equal to the weight of this mass,

$$F = mg = (20\ \text{kg})(9.8\ \text{m/s}^2) = 196\ \text{N}$$

This force will stretch the spring by the amount

$$x = \frac{F}{k} = \frac{196\ \text{N}}{6000\ \text{N/m}} = 0.033\ \text{m}$$

Thus, the potential energy of the spring will be

$$U = \tfrac{1}{2}kx^2 = \tfrac{1}{2}(6000\ \text{N/m})(0.033\ \text{m})^2 = 3.27\ \text{J}$$

We could directly substitute the value of $x = F/k$ to calculate U without calculating x.

$$U = \tfrac{1}{2}kx^2 = \tfrac{1}{2}k(F/k)^2 = \tfrac{1}{2}F^2/k = \frac{1}{2}\frac{(196\ \text{N})^2}{6000\ \text{N/m}} = 3.27\ \text{J}$$

EXERCISE 6.6 A mass of 1.2 kg attached to a vertical spring stretched the spring by 8 cm. Calculate the spring constant and the potential energy of the spring. How much additional mass should be added to stretch the spring an additional 16 cm? [*Ans.*: 147 N/m, 0.47 J, 2.4 kg.]

SUMMARY

The *work* W done by the force F is defined as the product of the component of the force parallel to the displacement and the displacement; $W = \mathbf{F} \cdot \mathbf{s} = Fs \cos\theta$ = the area under the F-versus-x curve. Various units of work are the joule, the erg, and the ft-lb.

The rate at which work is done is called *power*, $P = \Delta W/\Delta t = Fv$. The units of power are the watt, the erg/s, and the ft-lb/s. Another unit of work is the kWh.

Energy is the ability to do work; energy is stored work. *Kinetic energy*, K, means energy due to motion; $K = \tfrac{1}{2}mv^2$ and $W = K_f - K_i$.

The gravitational potential energy V is the energy due to the position; $V = mgh$ and $W = V_f - V_i$. Any force field in which the work done between two points is independent of the path is called a *conservative force field*. The conservation of mechanical energy requires that $K_f + V_f = K_i + V_i$.

The *elastic potential energy* U is also due to position. For a system that obeys Hooke's law, $F = -kx$, the potential energy is $\tfrac{1}{2}kx^2$. Also, the conservation of energy requires that $K_f + V_f + U_f = K_i + V_i + U_i$.

QUESTIONS

1. Name four types of energy other than mechanical energy.
2. Name at least two systems, each of which operates on the following types of energy: (a) kinetic; (b) gravitational potential; and (c) elastic potential.
3. In which case is more work done: (1) moving from the first floor to the third floor using a winding staircase, or (2) using a straight staircase?

4. A boy carries groceries from a checkout counter to a car on a level path. Is he doing any work?
5. Car *A* has double the power of car *B*. Operating at full power both travel a distance of 100 miles. Which car does more work? Why?
6. If in a given system the mechanical energy is not conserved, what happens to the energy lost?
7. A swimmer is trying to swim against the water current in a river. From the bank he looks to be stationary. Is he doing any work?
8. Consider two springs *A* and *B*, spring *A* being stiffer than spring *B*; that is, $k_A > k_B$. On which spring is more work done (a) if equal force is applied on both; and (b) if both springs are stretched through the same distance?
9. Suppose that a spring is compressed and held at a height of h. What forms of energy will it have when (a) at height h; (b) just after being released; (c) at height h_1 ($< h$); and (d) just before hitting the ground? What happens to the energy when it hits the ground?
10. Give an example in which all three forms of mechanical energy are conserved.

PROBLEMS

1. Calculate the amount of work done in moving a 50-kg block through a distance of 10 m by applying a force of 100 N.
2. Calculate the amount of work done in moving a 50-kg block through a distance of 10 m by applying a force of 100 N which makes an angle of 60° with the horizontal.
3. A block is being pulled by a 10-N force while the frictional force is 1.2 N. If the block moves through a distance of 10 m, calculate the work done (a) by the pulling force; (b) by the frictional force; and (c) by the net force.
4. Calculate the work done when the body moves (a) from *A* to *B* along path I; (b) from *C* to *D* along path II; and (c) from *E* to *F* along path III.
5. A mass of 10 kg is lifted a distance of 5 m vertically upward at constant speed. Calculate the work done by an external force. What is the work done by the gravitational force?
6. Calculate the work done in pushing a mass m along the incline through length L and height h if (a) there is no frictional force; and (b) there is an average frictional force f.
7. A sled of mass 10 kg with a boy of mass 30 kg sitting on it slides a distance of 50 m down a hill that makes an angle of 30° with the horizontal. Calculate the work done by the force of gravity, assuming no friction.
8. In Problem 7, if the coefficient of friction is 0.3, calculate (a) the work done by the force of gravity; (b) the work done by the force of friction; and (c) the total work done.
9. Calculate the power supplied by a motor that lifts a mass of 100 kg vertically upward through a distance of 20 m in 1 min. What is the force supplied by the motor?
10. When a jet airplane flies at a constant speed of 300 m/s (~670 mph), its engines develop a thrust of 16,000 N. Calculate the total power in watts, kilowatts, and horsepower.
11. An electric motor lifts a 500-lb object at a speed of 3 m/s. Calculate the power of this motor in horsepower, watts, and kilowatts.

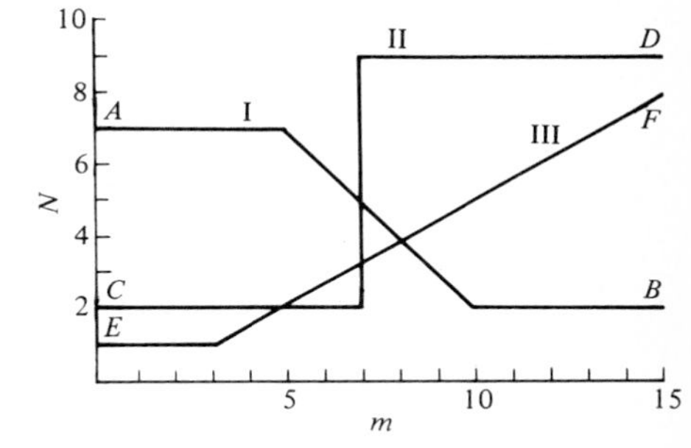

P. 6.4

12. Using the definition of horsepower, calculate the speed with which a motor of 1 hp can pull a mass of 60 kg upward.
13. In how many seconds should a 80-kg man climb 18 m to reach the fourth floor so that he will be working at the rate of 2 kW?
14. A 12-hp motor has to be operated 8 h/day. How much will it cost at the rate of 4 cents/kWh?
15. A ski tow is 400 m long and makes an angle of 37° with the horizontal. Estimate the power needed to move 50 riders at a time, each having a mass of 70 kg with the rope moving at 10 km/h.
16. A 6000-kg truck moves on a surface at a steady rate of 25 m/s (~55 mi/h) along a horizontal surface for which the coefficient of friction is 0.60. What is the power of the engine and how much work is done by the engine in 1 h?
17. An elevator of mass 1000 kg is moving upward with a uniform velocity of 3 m/s. Calculate the power supplied by the motor. Suppose that it took a distance of 2 m to reach this speed. Calculate the work done in changing the kinetic and potential energy.
18. A car of mass 1600 kg is moving with a uniform speed of 88 km/h (~55 mi/h). Calculate its kinetic energy in J and in kWh.
19. Calculate the kinetic energy in the following cases: (a) an electron of mass 9.1×10^{-31} kg traveling at 10^7 m/s; (b) a bullet of mass 2 g traveling at 400 m/s; and (c) an automobile of mass 1500 kg traveling at 20 m/s (~45 mi/h).
20. Suppose that the engine of an automobile is shut off when it is moving with a speed v. The automobile comes to a stop in a distance x as a result of frictional forces. Using the work–energy principle, show that $x = v^2/2\mu g$.
21. In Problem 20, calculate x if $\mu = 0.07$ and (a) $v = 20$ m/s; and (b) $v = 30$ m/s.
22. Calculate the change in potential energy of an object of mass 2 kg that is lifted through a height of 10 m above the ground level. If this 2-kg mass is allowed to fall, what will be its kinetic energy before hitting the ground?
23. When a 40-kg boy climbs a hill, his potential energy increases by 9.8×10^4 J. What is the height of the hill?
24. A 50-kg woman climbs the stairs to the fifth floor of a building that is 20 m from the ground. (a) How much work is done by the woman against gravity? (b) Calculate the change in her potential energy. (c) If she climbs the distance in 1 min, calculate her power in kW.
25. A block of mass m starting from rest is allowed to slide down a hill of height h and length L. If the average frictional force is f, show that the speed of the mass when it reaches the bottom is $v = \sqrt{(2/m)(mgh - fL)}$.
26. A tennis ball dropped from a height of 2 m rebounds only 1.5 m after hitting the ground. What is the change in its potential energy? What is the velocity just before and after the ball hits the ground?
27. A 1-kg mass falls from a height of 10 m into a sandbox. What is the speed of the mass just before hitting the sandbox? If it travels a distance of 2 cm into the sand before coming to rest, what is the average retarding force?
28. A stone of mass 1 kg is thrown at 10 m/s upward making an angle of 37° with the horizontal from a building that is 20 m high. By using the law of conservation of energy, calculate the speed when the stone hits the ground.

29. The driver of a 1500-kg car shuts off its motor when at the top of a hill 35 m high. The car travels 200 m and reaches the bottom of the hill. (a) What is its final speed if there is no friction? (b) What will be the final speed if there were an average retarding force of 80 N?

30. Can a car traveling at 25 m/s ($\sim$56 mi/h or $\sim$90 km/h) on a level road climb over a hill of 20 m height with its motor turned off providing there is no friction? What is the minimum speed to achieve this?

31. Consider a curved frictionless path that is a quarter of a circle of radius R, as shown. A body of mass m slides down this path from point A to B. What is its velocity v at B? Calculate v for $m = 2$ kg and $R = 2$ m.

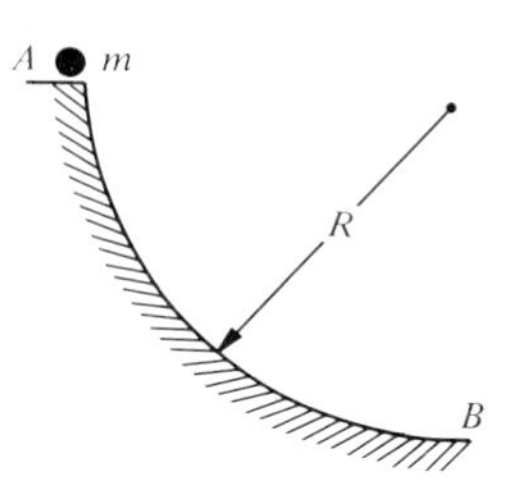

P. 6.31

32. In Problem 31, suppose that the track is not frictionless. When the mass of 2 kg reaches point B, its speed is 4 m/s. What is the work done against friction? Calculate the force of friction.

33. A block of mass 8 kg is being pulled a distance of 10 m up a 37° inclined plane by a force $F = 100$ N parallel to the inclined surface. The coefficient of friction is 0.30. Calculate the work done (a) by the force F; (b) against the force of friction; (c) the change in kinetic energy; and (d) the change in potential energy.

34. Consider the situation shown. If the 20-kg mass is released, calculate the following: (a) the change in potential energy of each mass and (b) the kinetic energy of the 20-kg mass just before it hits the floor, (c) the work done by the gravitational force in moving the 20-kg mass through 5 m; and (d) the work done in moving the 5-kg mass. (e) What happens to the difference in (c) and (d)?

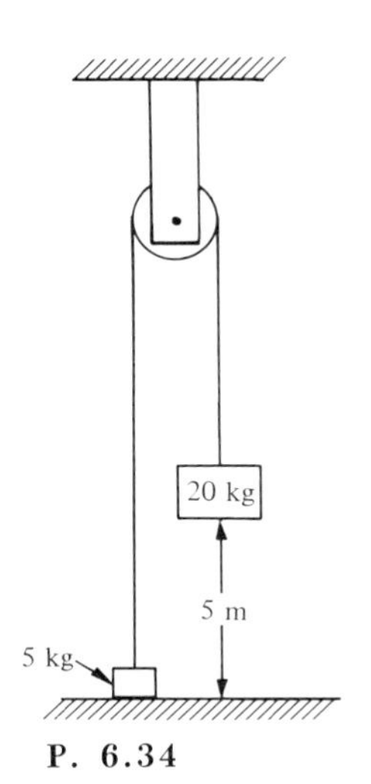

P. 6.34

35. The force F (in newtons) needed to stretch a spring a distance x (in meters) from the equilibrium position is given by $F = (50 \text{ N/m})x$. (a) Calculate the force needed to hold the spring at $x = 0.01$ m, 0.1 m, and 0.5 m. (b) Calculate the work that must be done to stretch the spring through $x = 0.01$ m, 0.1 m, and 0.5 m.

36. Calculate the work done in stretching a spring 12 cm from its equilibrium position. The spring constant is 8 N/cm.

37. What will be the spring constant if it stretches 10 cm when it has a potential energy of 5600 J?

38. A force of 800 N is needed to stretch a spring through 4 cm. If a force of 1200 N is applied to this spring: (a) How much will the spring stretch? (b) How much work is done in stretching the spring? (c) What will be the potential energy stored in the spring?

39. A man of mass m falls through a height h and lands on a spring having a spring constant k. Show that the spring will compress through a distance $x = (2mgh/k)^{1/2}$. (Note that $x << h$.) If $m = 10$ kg, $h = 1$ m, and $k = 98{,}000$ N/m, what will x be?

P. 6.39

40. Suppose that a 1-kg mass moving with velocity v strikes a spring of spring constant 4000 N/m and sticks to it. The spring compresses 10 cm. If half of the initial energy is used in inelastic processes, what is the initial velocity v?

P. 6.40

41. A spring gun has a spring constant of 80 N/cm. The spring is compressed 12 cm by a ball of mass 15 g. How much is the potential energy of the spring? If the trigger is pulled, what will the velocity of the ball be?

7 Conservation Laws and Symmetry Principles

Elastic and inelastic collisions are investigated by the application of the laws of conservation of linear momentum and kinetic energy. The principle of recoil is discussed with its application to alpha decay. A brief discussion of the relation between the conservation laws and symmetry principles is given.

7.1. Conservation Laws and Collisions

When two or more objects come close enough so that there is some sort of interaction between them, with or without the presence of external forces, we say that a *collision* has taken place between the objects. After the collision, the velocities and the directions of the colliding objects may or may not be the same as before the collision. The subject of collisions is most important in modern experimental physics—atomic physics, solid-state physics, nuclear physics, high-energy physics, and other branches. Very often one is interested in describing the nature of the interaction (or the type of force) between microscopic particles. If the particles are incident on a target, the paths and the energies of the interacting particles will be changed. By measuring the energies and the angular distribution of these scattered particles, we can gain information about their structure and the nature of the forces involved.

Applicable to collision problems are the conservation laws, some of which are (1) conservation of mass, (2) conservation of charge, (3) conservation of linear momentum, (4) conservation of angular momentum (see Chapter 9), and (5) conservation of energy. Some of the less familiar conservation laws, applicable on the microscopic scale in nuclear physics, will not be discussed here. In the following sections we shall limit our discussion to the analysis of those collisions in which the conservation laws of linear momentum and kinetic energy are obeyed.

Let us apply Newton's second law to a system of particles. Let particles of mass $m_1, m_2, \ldots$ be moving with velocity $v_{1x}, v_{2x}, \ldots$ along the X-axis so that their initial linear momenta are $(p_{1x} = m_1 v_{1x})_i$, $(p_{2x} = m_2 v_{2x})_i, \ldots$, respectively. Thus, the total initial linear momentum P_{xi} along the X-axis is

$$P_{xi} = (p_{1x} + p_{2x} + p_{3x} + \cdots)_i \tag{7.1}$$

Suppose that these particles interact with each other, resulting in a change in their velocity to $u_{1x}, u_{2x}, \ldots$ all along X-axis, and their final linear momentum to be $(p_{1x} = m_1 u_{1x})_f$, $(p_{2x} = m_2 u_{2x})_f, \ldots$, respectively. Thus, the total final linear momentum P_{xf} along the X-axis is

$$P_{xf} = (p_{1x} + p_{2x} + p_{3x} + \cdots)_f \tag{7.2}$$

The change in the momentum is

$$\Delta P_x = P_{xf} - P_{xi} \tag{7.3}$$

According to Newton's second law, $F_x = \Delta P_x/\Delta t$. If the external force $F_x = 0$, we get

$$\frac{\Delta P_x}{\Delta t} = 0 \quad \text{or} \quad \Delta P_x = P_{xf} - P_{xi} = 0$$

That is,

$$\boxed{P_{xi} = P_{xf} = \text{constant}} \quad \text{if } F_x = 0 \tag{7.4}$$

If we extend this to interactions of particles in three dimensions, Equation 7.4 takes the form

$$\boxed{\mathbf{P}_i = \mathbf{P}_f = \text{constant}} \quad \text{if } \mathbf{F} = 0 \tag{7.5}$$

or in component form,

$$(p_{1x} + p_{2x} + \cdots)_i = (p_{1x} + p_{2x} + \cdots)_f$$

with similar expressions for y and z components. Equation 7.5 states the

Law of conservation of linear momentum: ***If there are no external forces acting on a system, the total linear momentum of the system remains constant, in magnitude and direction.***

That is, the linear momentum of the system does not change. No matter what the nature of the collision may be, the total linear momentum of the system before the collision is always equal to the total linear momentum of the system after the collision. (Note that the total net internal forces are equal to zero because of Newton's third law.)

Let us now consider the kinetic energy of the system, Let $K_{1i}, K_{2i}, K_{3i}, \ldots$ be the kinetic energies of the individual particles before the collision; and let $K_{1f}, K_{2f}, K_{3f}, \ldots$ be the kinetic energies after the collision. The total initial kinetic energy K_i of the system before collision is

$$K_i = K_{1i} + K_{2i} + K_{3i} + \cdots = \tfrac{1}{2}m_1v_1^2 + \tfrac{1}{2}m_2v_2^2 + \tfrac{1}{2}m_3v_3^2 + \cdots \tag{7.6}$$

while the total final kinetic energy K_f of the system after the collision is

$$K_f = K_{1f} + K_{2f} + K_{3f} + \cdots = \tfrac{1}{2}m_1u_1^2 + \tfrac{1}{2}m_2u_2^2 + \tfrac{1}{2}m_3u_3^2 + \cdots \tag{7.7}$$

Unlike the linear momentum, the total kinetic energy is not always conserved.

If $K_i = K_f$; it means the *total energy is conserved*. Of course, the momentum is always conserved in any collision. *Elastic collisions* are collisions in which both linear momentum and kinetic energy are conserved.

In all other collisions, $K_f \neq K_i$, that is, the kinetic energy may be lost or gained in the collision. *Inelastic collisions* are collisions in which kinetic energy is not conserved, but linear momentum is conserved.

Still another quantity which is commonly used in collision problems is the *disintegration energy* Q, defined as the difference in the final kinetic energy K_f and the initial kinetic energy K_i; that is,

$$Q = K_f - K_i = (\tfrac{1}{2}m_1u_1^2 + \tfrac{1}{2}m_2u_2^2 + \cdots)_f - (\tfrac{1}{2}m_1v_1^2 + \tfrac{1}{2}m_2v_2^2 + \cdots)_i \tag{7.8}$$

If $K_f = K_i$, $Q = 0$ and such collisions are elastic collisions. If $K_f \neq K_i$, $Q \neq 0$ and such collisions are inelastic collisions. We explain these in detail shortly.

7.2. Elastic Collisions

As defined earlier, elastic collisions are those in which both linear momentum and kinetic energy are conserved. We shall discuss these collisions first in one dimension and then in two dimensions. The most common situation is one in which only two particles are involved, and we shall limit our discussion to such cases.

Elastic Collisions in One Dimension

This is the case of head-on collisions in which all motions occur along a straight line. Let us consider a simple situation shown in Figure 7.1(a), in which a particle of mass m_1 moving with velocity v_1, called the *incident particle*, approaches a second particle of mass m_2 moving with velocity v_2, called the *target*. After the collision, let the velocities of the two particles be u_1 and u_2 along the X-axis, as shown in Figure 7.1(b). According to the conservation of linear momentum Equation 7.5 and kinetic energy Equation 7.8, where $Q = 0$ for a perfectly elastic collision, we may write (assuming that momentum is positive to the right)

$$m_1v_1 + m_2v_2 = m_1u_1 + m_2u_2 \tag{7.9}$$

and

$$\tfrac{1}{2}m_1v_1^2 + \tfrac{1}{2}m_2v_2^2 = \tfrac{1}{2}m_1u_1^2 + \tfrac{1}{2}m_2u_2^2 \tag{7.10}$$

After rearranging, Equation 7.9 may be written as

$$m_1(v_1 - u_1) = m_2(u_2 - v_2) \tag{7.11}$$

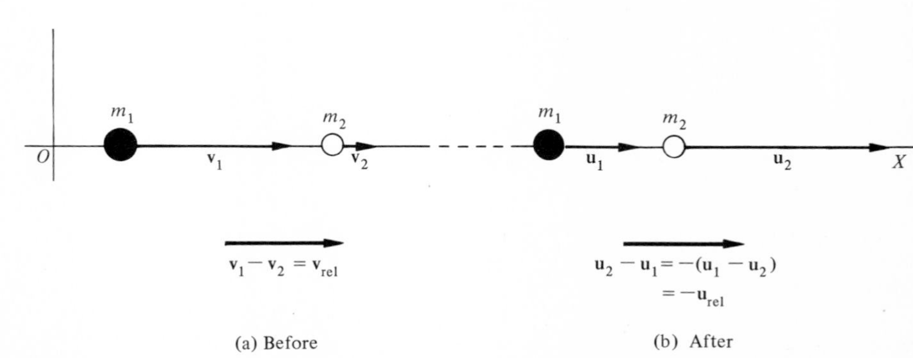

FIGURE 7.1. *Elastic collision in one dimension between two masses, m_1 and m_2. Note that $\mathbf{v}_{rel} = -\mathbf{u}_{rel}$.*

and Equation 7.10 as

$$m_1(v_1^2 - u_1^2) = m_2(u_2^2 - v_2^2)$$

or

$$m_1(v_1 - u_1)(v_1 + u_1) = m_2(u_2 - v_2)(u_2 + v_2) \tag{7.12}$$

Dividing Equation 7.12 by Equation 7.11, we get

$$v_1 + u_1 = v_2 + u_2$$

or

$$\boxed{\begin{array}{c} v_1 - v_2 = -(u_1 - u_2) \\ \text{speed of approach} = \text{speed of recession} \end{array}} \tag{7.13}$$

That is, in perfectly elastic collisions, the relative speed of approach of two objects is always equal to the relative speed of recession. In terms of relative velocities, Equation 7.13 may be written as $v_{rel} = -u_{rel}$.

Let us consider two special cases: (a) $m_1 = m_2$ and (b) $v_2 = 0$.

Case (a). When $m_1 = m_2$, Equation 7.9 takes the form

$$v_1 + v_2 = u_1 + u_2 \tag{7.14}$$

Solving Equations 7.13 and 7.14, we get

$$u_1 = v_2 \qquad \text{and} \qquad u_2 = v_1 \tag{7.15}$$

which implies that when two particles of equal mass collide elastically, they simply exchange velocities, as shown in Figure 7.2(a).

Case (b). When $v_2 = 0$, that is, when the target of mass m_2 is at rest, Equations 7.9 and 7.13 take the forms

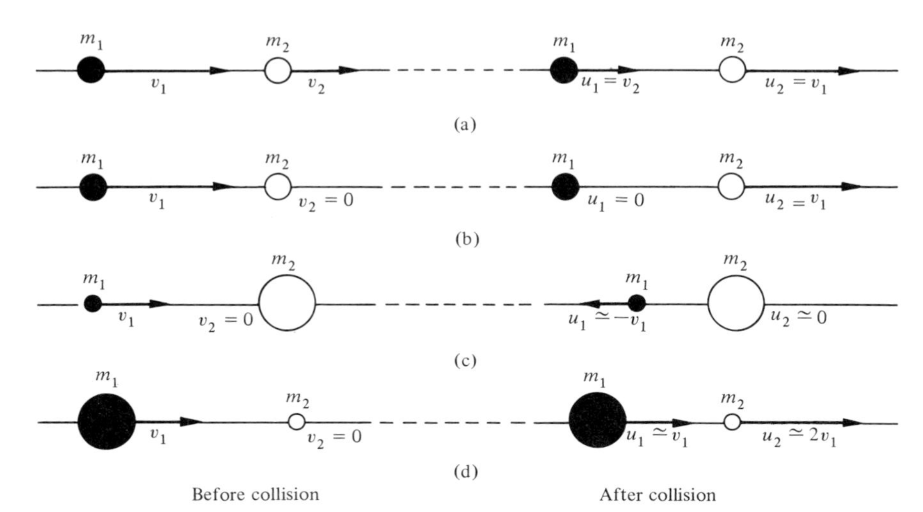

Figure 7.2. *Elastic collisions in one dimension between two masses, m_1 and m_2. (a) $m_1 = m_2$; (b) $v_2 = 0$, $m_1 = m_2$; (c) $v_2 = 0$, $m_2 >> m_1$; and (d) $v_2 = 0$, $m_2 << m_1$.*

$$m_1v_1 = m_1u_1 + m_2u_2 \tag{7.16a}$$

and

$$v_1 = u_2 - u_1 \tag{7.16b}$$

Solving these equations for u_1 and u_2 in terms of v_1, we obtain

$$u_1 = \frac{m_1 - m_2}{m_1 + m_2}v_1 \tag{7.17a}$$

and

$$u_2 = \frac{2m_1}{m_1 + m_2}v_1 \tag{7.17b}$$

These two equations have a few interesting consequences for different values of m_1 and m_2.

1. Suppose that $m_1 = m_2$. In this case Equation 7.17 yields

$$u_1 = 0 \qquad \text{and} \qquad u_2 = v_1 \tag{7.18}$$

This implies that the incident particle, which was moving with velocity v_1, comes to rest while the target particle that was at rest starts moving with a velocity v_1, as shown in Figure 7.2(b). Thus, the incident particle transfers all its kinetic energy to the target particle, provided that the masses of the incident and the target particles are equal.

A practical example is that of the slowing down of fast neutrons. In a nuclear reactor or nuclear power plant, a large number of fast-moving neutrons are produced. In order to slow down these neutrons, the nuclear fuel is surrounded by water. Water consists in part of protons which have mass almost equal to that of neutrons. Thus, when the fast neutrons strike the protons, the neutrons practically come to rest while the protons start moving with speeds equal to that of the incident neutrons. There are many reasons for slowing down fast neutrons. First, we do not want the neutrons to leave the nuclear fuel, because these neutrons are needed to be absorbed by the nuclear fuel to produce fission (and also more neutrons) and hence energy. Second, if these neutrons escape the nuclear reactor, they are dangerous to the personnel working around the plant (radiation damage, Chapter 34). Note, very fast moving protons are stopped over a very short distance, whereas neutrons can travel a very large distance before stopping. The materials used for slowing down neutrons are called moderators; examples include water, deuterium, and carbon.

2. Suppose that $m_2 >> m_1$. From Equation 7.17 we see that

$$u_1 \simeq -v_1 \qquad \text{while } u_2 \simeq 0 \tag{7.19}$$

This means that the small incident object merely bounces off in the opposite direction while the heavy target remains almost motionless, as shown in Figure 7.2(c).

3. Suppose that $m_2 << m_1$. In this case Equation 7.17 yields

$$u_1 \simeq v_1 \qquad \text{and} \qquad u_2 \simeq 2v_1 \tag{7.20}$$

That is, the incident particle keeps on moving without losing much energy, while the target particle moves with a velocity double that of the incident particle, as shown in Figure 7.2(d).

An example of such a collision is when fast-moving protons or other heavy particles collide head-on with electrons at rest. The electrons then start to move with double the incident particle velocity. Such fast-moving electrons have been named *delta rays*.

EXAMPLE 7.1 Show that the fractional loss in kinetic energy of an incident neutron in a head-on collision with an atomic nucleus initially at rest is $\approx 4A/(A+1)^2$, where A is the atomic mass number of the nucleus.

Let m_1, v_1, and K_{1i} be the mass, velocity, and kinetic energy of the incident neutron. Let m_2 be the mass of the atomic nucleus at rest. After the collision, let u_1 and u_2 be the velocities of the neutron and the nucleus, respectively. Thus, the initial and final kinetic energies K_{1i} and K_{1f} of the neutron are given by

$$K_{1i} = \tfrac{1}{2}m_1v_1^2 \qquad \text{and} \qquad K_{1f} = \tfrac{1}{2}m_1u_1^2$$

Therefore, substituting for u_1 from Equation 7.17,

$$\frac{K_{1f}}{K_{1i}} = \frac{u_1^2}{v_1^2} = \left(\frac{m_1 - m_2}{m_1 + m_2}\right)^2$$

and the fractional energy loss (frac) is

$$\text{frac} = \frac{K_{1i} - K_{1f}}{K_{1i}} = \frac{v_1^2 - u_1^2}{v_1^2} = 1 - \left(\frac{m_1 - m_2}{m_1 + m_2}\right)^2 = \frac{4m_1m_2}{(m_1 + m_2)^2}$$

Since the masses are approximately equal to the mass numbers, $m_1 \simeq 1$, and $m_2 \simeq A$, and the equations above take the form

$$\frac{K_{1f}}{K_{1i}} = \left(\frac{A-1}{A+1}\right)^2 \qquad \text{and} \qquad \text{frac} = \frac{4A}{(A+1)^2}$$

As an illustration, let us calculate the energy loss by a neutron when it collides with (1) uranium ($A = 235$), (2) sulfur ($A = 32$), (3) carbon ($A = 12$), (4) deuterium (a deuterium nucleus consists of a proton and a neutron; hence $A = 2$), and (5) a proton ($A = 1$).

$$\text{frac}_U = (4 \times 235)/(235 + 1)^2 = 0.017 = 1.7\%$$
$$\text{frac}_S = (4 \times 32)/(32 + 1)^2 = 0.118 = 11.8\%$$
$$\text{frac}_C = (4 \times 12)/(12 + 1)^2 = 0.284 = 28.4\%$$
$$\text{frac}_D = (4 \times 2)/(2 + 1)^2 = 0.889 = 88.9\%$$
$$\text{frac}_H = (4 \times 1)/(1 + 1)^2 = 1.000 = 100\%$$

As is clear, the neutron loses more energy in a single collision with a proton than in one involving any other atom. It is for this reason that water (containing hydrogen and hence protons) is used for slowing down neutrons in most nuclear reactors.

EXERCISE 7.1 Suppose that a neutron of velocity v_1 collides head-on three different times with nuclei of mass number A. After three collisions, what is the ratio of the final kinetic energy of the neutron to its initial kinetic energy? Do this for a specific case of (a) carbon; and (b) deuterium. [*Ans.*: (a) $K_f/K_i = 0.367$; (b) $K_f/K_i = 0.00137$.]

*Elastic Collisions in Two Dimensions

Let us consider the most common situation shown in Figure 7.3 in which a particle of mass m_1 (the incident particle) moving with velocity $\mathbf{v}$ along the X-axis collides with a particle of mass m_2 (the target), which is at rest. Let $\mathbf{u}_1$ and $\mathbf{u}_2$ be the velocities of the particles of masses m_1 and m_2, respectively, after the collision. Because the collision takes place in the XY-plane, we may write Equation 7.5, the conservation of linear momentum, in the component form as

$$(p_x)_i = (p_x)_f \qquad (7.21)$$
$$(p_y)_i = (p_y)_f$$

The direction of the incident particle is along the X-axis while after the collision the particle m_1 moves with velocity $\mathbf{u}_1$ making an angle θ with the X-axis, and m_2 moves with velocity $\mathbf{u}_2$ making an angle ϕ with the X-axis as shown in Figure 7.3. Thus, Equation 7.21 takes the form

$$m_1 v = m_1 u_1 \cos\theta + m_2 u_2 \cos\phi \qquad (7.22)$$
$$0 = m_1 u_1 \sin\theta - m_2 u_2 \sin\phi \qquad (7.23)$$

Because we are dealing with elastic collisions, the total initial kinetic energy K_i before the collision is equal to the total final kinetic energy K_f after the

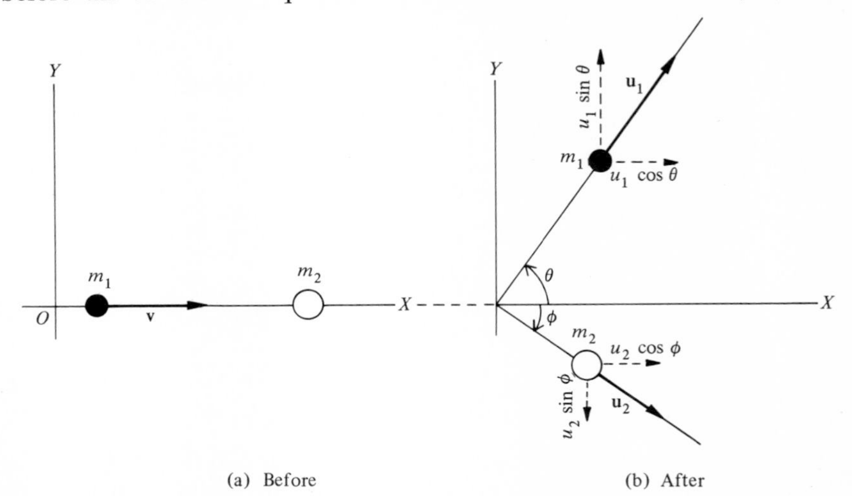

FIGURE 7.3. *Elastic collision between an incident particle of mass m_1 and a target particle of mass m_2: (a) before collision and (b) after collision. (c) Collision between two billiard balls. (d) Alpha particle moving from the left collides with a particle at rest and then they separate.* [*Part (c) courtesy of the Education Development Center, Inc., Newton, Mass.*]

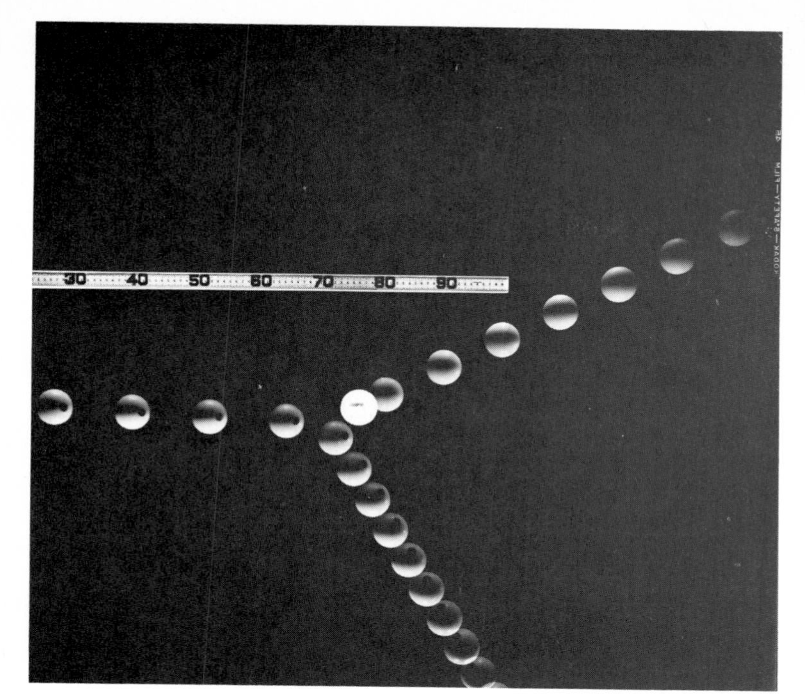

(c)

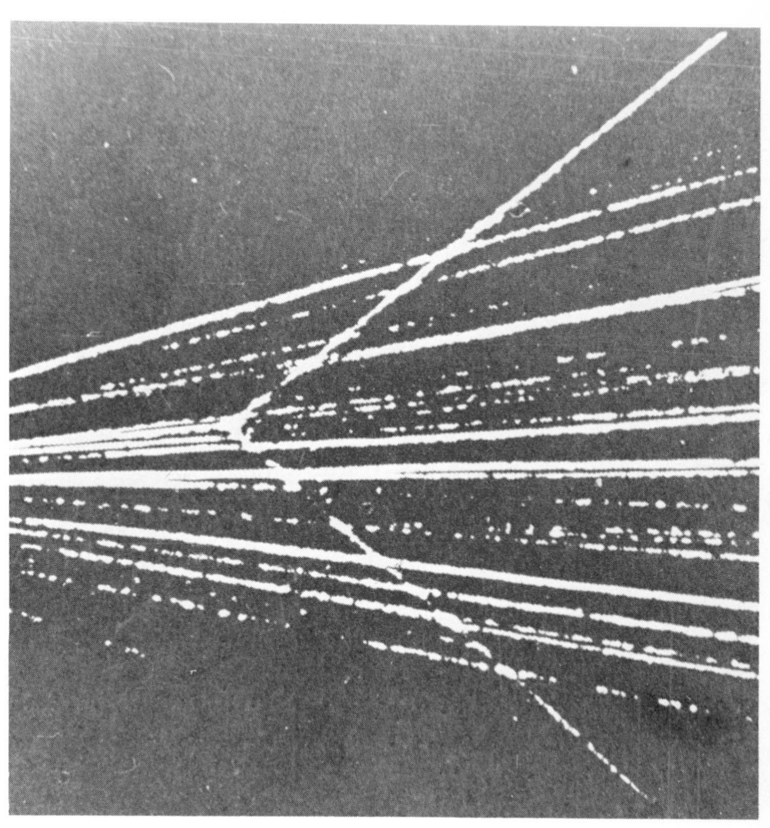

(d)

collision, that is, $K_i = K_f$ or $K_{1i} + K_{2i} = K_{1f} + K_{2f}$, where in this case $K_{2i} = 0$. Let $K = K_{1i} = \frac{1}{2}m_1v^2$, $K_{1f} = K_1 = \frac{1}{2}m_1u_1^2$, and $K_{2f} = K_2 = \frac{1}{2}m_2u_2^2$. Therefore, $K = K_1 + K_2$ and

$$\boxed{\tfrac{1}{2}m_1v^2 = \tfrac{1}{2}m_1u_1^2 + \tfrac{1}{2}m_2u_2^2} \tag{7.24}$$

The three equations 7.22, 7.23, and 7.24, which contain three known quantities, m_1, m_2, and v, are not sufficient to solve for the four unknown quantities u_1, u_2, θ, and ϕ. Thus, the problem is undetermined and there are many possible outcomes of a collision with a given set of values for m_1, m_2, and v. If, however, one of the four quantities u_1, u_2, θ, or ϕ is fixed, the remaining three are also fixed and hence the problem can be solved.

Let us consider two special cases.

Glancing Collisions. These are the collisions for which θ is very small while ϕ is almost 90°. For $\theta = 0$ and $\phi = 90°$, Equations 7.22 and 7.23 yield

$$v = u_1 \qquad \text{and} \qquad u_2 = 0 \tag{7.25}$$

$u_2 = 0$ means that $K_2 = 0$, and we may conclude that the incident particle in a glancing collision does not lose any energy and is deflected almost undeviated from its initial path.

Head-on Collisions. These are collisions for which $\phi = 0$; that is, the target particle moves in the direction of the incident particle. Equations 7.22 and 7.23 take the form

$$\begin{aligned} m_1v - m_2u_2 &= m_1u_1 \cos\theta \\ 0 &= m_1u_1 \sin\theta \end{aligned} \tag{7.26}$$

while the equation for the kinetic energy $K = K_1 + K_2$ remains the same.

7.3. Inelastic Collisions

As defined earlier, inelastic collisions are those in which only linear momentum is conserved while the total final kinetic energy is not equal to the total initial kinetic energy. Let us first consider a special case of a perfectly inelastic collision as shown in Figure 7.4. When the colliding bodies stick together after the collision and move as a single mass, the collision is a *perfectly* (or *completely*) *inelastic collision*. Referring to Figure 7.4, a particle of mass m_1 moving with velocity **v** collides with a particle of mass m_2 at rest. After the collision the two particles "stick" together and move with a velocity **u** along

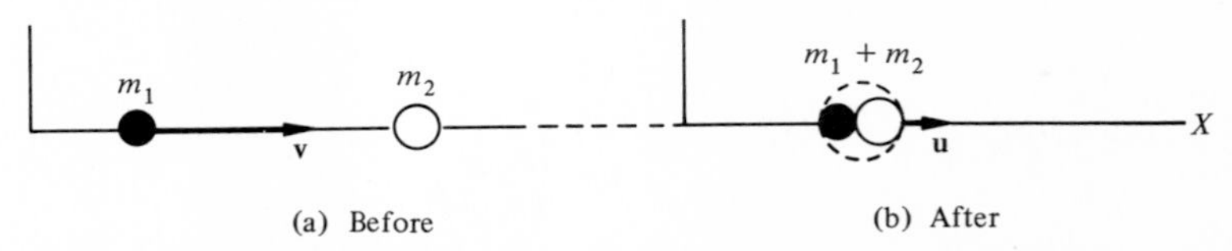

Figure 7.4. *Completely inelastic collision between an incident particle of mass m_1 and a target particle of mass m_2.*

the direction of the incident particle, that is, along the X-axis. According to the conservation of linear momentum,

$$m_1 v = (m_1 + m_2)u \tag{7.27}$$

The kinetic energy before the collision is

$$K_i = K_{1i} + K_{2i} = \tfrac{1}{2}m_1 v^2 \tag{7.28}$$

and the final kinetic energy K_f after the collision is

$$K_f = \tfrac{1}{2}(m_1 + m_2)u^2 \tag{7.29}$$

Dividing Equation 7.29 by Equation 7.28 and substituting for (u/v) from Equation 7.27, we get

$$\frac{K_f}{K_i} = \frac{m_1 + m_2}{m_1}\left(\frac{u}{v}\right)^2 = \frac{m_1 + m_2}{m_1}\left(\frac{m_1}{m_1 + m_2}\right)^2$$

or

$$\boxed{\frac{K_f}{K_i} = \frac{m_1}{m_1 + m_2}} \tag{7.30}$$

It is clear from this equation that in a perfectly inelastic collision the final kinetic energy is always less than the initial kinetic energy. The loss in the kinetic energy may appear as heat energy.

A more general case of an inelastic collision is one in which a particle of mass m_1 moving with velocity $\mathbf{v}_1$ and kinetic energy $K_1 = \frac{1}{2}m_1 v_1^2$ collides with a particle of mass m_2 with velocity $\mathbf{v}_2$ and kinetic energy $K_2 = \frac{1}{2}m_2 v_2^2$. After the collision two particles of masses m_3 and m_4 with velocities $\mathbf{u}_3$ and $\mathbf{u}_4$ and kinetic energies $K_3 = \frac{1}{2}m_3 u_3^2$ and $K_4 = \frac{1}{2}m_4 u_4^2$ are given out. In such collisions or reactions the final kinetic energy $K_f = K_3 + K_4$. The difference in the final and the initial kinetic energies is stated in terms of the *disintegration energy* Q or the *Q-value* of the collision (or the reaction) given by (Equation 7.8)

$$\boxed{Q = K_f - K_i = (K_3 + K_4) - (K_1 + K_2)} \tag{7.31}$$

If the final kinetic energy is less than the initial kinetic energy, Q will be negative and the collision or reaction is an *endoergic reaction*. On the other hand, if the final kinetic energy is greater than the initial kinetic energy, Q will be positive and the collision is an *exoergic reaction*. In the case of endoergic reactions the decrease in kinetic energy appears in one of the following processes: (1) in changing the internal energies of the particles; (2) in increasing the final masses so that $(m_3 + m_4) > (m_1 + m_2)$ (see Chapter 8); and (3) in producing heat (as in the case of a ballistic pendulum discussed in Example 7.2) or the like.

In exoergic reactions the final mass decreases, so $(m_3 + m_4) < (m_1 + m_2)$, while the kinetic energies of the final masses are greater than the kinetic energies of the initial masses.

We may point out once again that if $K_f = K_i$, $Q = 0$; hence, the collision or the reaction is elastic. If $Q \neq 0$, the collision is inelastic.

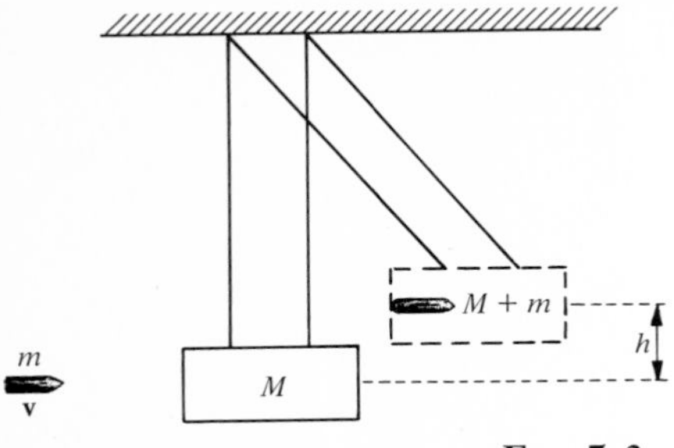

Ex. 7.2

Example 7.2 *The Ballistic Pendulum:* This is a device for measuring the speed of bullets. Suppose that a bullet of mass m moving with a velocity v strikes a block of mass M hanging vertically by two strings as shown in the accompanying figure. After striking, the bullet becomes embedded in the block and the two rise to a height h. Calculate v in terms of m, M, and h.

Before the bullet strikes the block, its momentum is mv. Let the velocity of the block–bullet combination after the collision be V, so that the momentum will be $(m + M)V$. From conservation of linear momentum,

$$mv = (m + M)V \tag{i}$$

After the collision the total kinetic energy of the system is $\frac{1}{2}(m + M)V^2$. All this kinetic energy is converted into potential energy $(m + M)gh$ when the block–bullet combination rises to a height h, as shown. Thus, from the conservation of energy,

$$\tfrac{1}{2}(m + M)V^2 = (m + M)gh \tag{ii}$$

Note that $\frac{1}{2}mv^2 \neq \frac{1}{2}(m + M)V^2$ because the collision between the bullet and the block is inelastic; hence, the kinetic energy is not conserved. From Equation ii,

$$V = \sqrt{2gh} \tag{iii}$$

Substituting this in Equation i and solving for v,

$$mv = (m + M)\sqrt{2gh} \qquad \text{or} \qquad v = \frac{m + M}{m}\sqrt{2gh} \tag{iv}$$

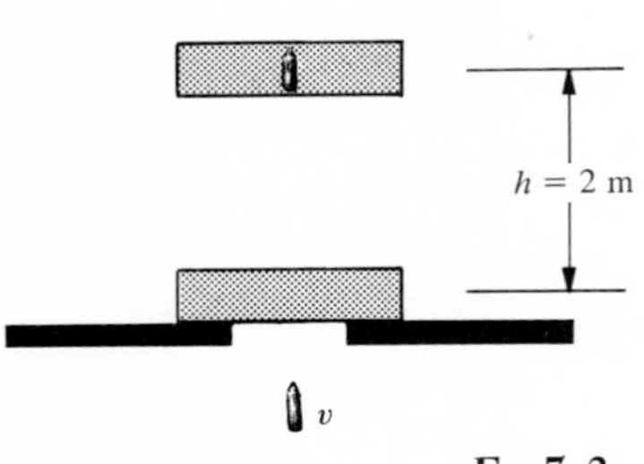

E. 7.2

Exercise 7.2 A 20-g bullet traveling with a velocity v becomes embedded into a 4-kg block that rests on a table top as shown in the accompanying figure. If the block–bullet combination rises to a height of 2 m, what is the initial velocity of the bullet? [*Ans.*: 1258 m/s.]

Example 7.3 Suppose that a nucleus of mass m_N moving with a velocity v_N along the $+Y$-axis, as shown in the accompanying figure, absorbs a slow-moving neutron of mass m_n and moving with velocity v_n along the $+X$-axis. After the neutron has been captured, what is the velocity of the system? (This is a common situation in nuclear reactions caused by slow neutrons.)

The conservation of linear momentum in the two dimensional case $\mathbf{p}_i = \mathbf{p}_f$ takes the form

$$(p_{1x} + p_{2x})_i = (p_{1x} + p_{2x})_f$$
$$(p_{1y} + p_{2y})_i = (p_{1y} + p_{2y})_f$$

For the situation shown, these equations take the form

$$m_n v_n = (m_N + m_n)v_{xf}$$
$$m_N v_N = (m_N + m_n)v_{yf}$$

or

$$v_{xf} = \frac{m_n v_n}{(m_N + m_n)} \qquad \text{and} \qquad v_{yf} = \frac{m_N v_N}{(m_N + m_n)} \tag{i}$$

Thus,

$$v_f = \sqrt{v_{xf}^2 + v_{yf}^2} \qquad \text{and} \qquad \tan\theta = \frac{v_{yf}}{v_{xf}} \tag{ii}$$

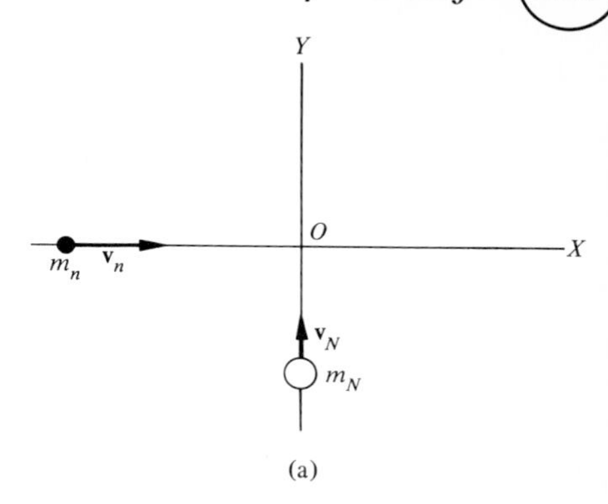

(a)

Let us consider slow neutrons which have a typical velocity of 2800 m/s incident on argon atoms for which $A = 40$. The argon atoms are moving with a velocity of 310 m/s. For $m_n = 1$ u, $m_N = 40$ u, $v_n = 2800$ m/s, and $v_N = 310$ m/s, from Equations i and ii we obtain

$$v_{xf} = \frac{(1\text{ u})(2800\text{ m/s})}{(40 + 1)\text{ u}} = 68\text{ m/s} \qquad \text{and} \qquad v_{yf} = \frac{(40\text{ u})(310\text{ m/s})}{(40 + 1)\text{ u}} = 302\text{ m/s}$$

$$v_f = \sqrt{(68)^2 + (302)^2}\text{ m/s} = 310\text{ m/s}$$

$$\tan\theta = \frac{302}{68} = 4.44 \qquad \theta \simeq 77°$$

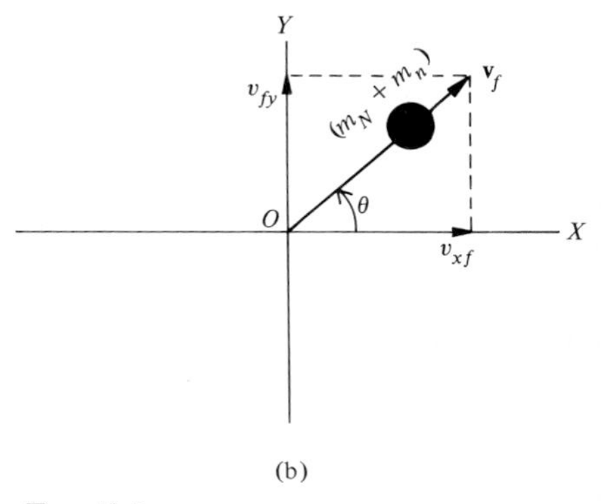

(b)

Ex. 7.3

Exercise 7.3 Suppose that a proton moving with a velocity of 2500 m/s along the $+Y$-axis absorbs a neutron moving with a velocity of 2500 m/s along the $+X$-axis, thus forming a deuteron. Calculate the final velocity of the deuteron (mass of neutron = 1.00898 u, mass of proton = 1.00759 u, and mass of deuteron = 2.01410 u). [*Ans.*: 1770 m/s, 45°.]

7.4. Recoil and Alpha Decay

Consider two masses m_1 and m_2 held together at rest by a compressed spring, as shown in Figure 7.5(a). If the system is released, the spring will exert equal and opposite forces on the two masses until the spring reaches its natural length. At this point we assume that the spring detaches itself from the masses. The masses start to move or recoil in the opposite direction with velocities u_1 and u_2, respectively, along a straight line as shown in Figure 7.5(b). Since the initial momentum is zero, conservation of momentum requires that the total final momentum must also be zero; that is, $p_i = p_f$ becomes

$$0 = p_{1f} + p_{2f} = m_1 u_1 + m_2 u_2$$

$$\boxed{p_{1f} = -p_{2f} \qquad \text{or} \qquad \frac{u_1}{u_2} = -\frac{m_2}{m_1}} \tag{7.32}$$

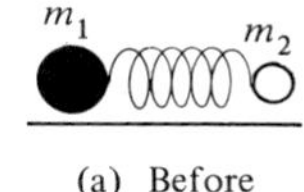

(a) Before

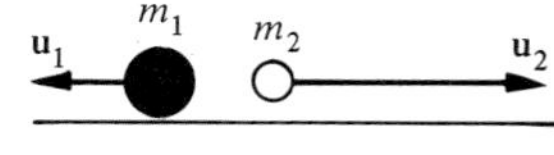

(b) After

Figure 7.5. *Compressed spring holding masses m_1 and m_2 as shown in (a) when released makes the two masses recoil in opposite directions with equal and opposite momentum, as shown in (b).*

Thus, the final velocities are inversely proportional to their masses while their momenta are equal and opposite.

The original kinetic energy is zero, but the spring being compressed has potential energy called the elastic potential energy E_s, which must be equal to the final kinetic energy. Thus,

$$E_s = K_f = \tfrac{1}{2}m_1u_1^2 + \tfrac{1}{2}m_2u_2^2 \tag{7.33}$$

The ratio of the kinetic energies K_1 and K_2 of masses m_1 and m_2 while recoiling is, using Equation 7.32,

$$\frac{K_1}{K_2} = \frac{\tfrac{1}{2}m_1u_1^2}{\tfrac{1}{2}m_2u_2^2} = \frac{m_1}{m_2}\left(-\frac{m_2}{m_1}\right)^2 \quad \text{or} \quad \boxed{\frac{K_1}{K_2} = \frac{m_2}{m_1}} \tag{7.34}$$

That is, in recoil the kinetic energies are inversely proportional to their masses.

A practical example is that of a radioactive nucleus which emits α-particles. Alpha (α)-particles are fast-moving doubly ionized helium (${}_2^4\mathrm{He}^{2+}$) atoms. The situation is as follows. The nucleus, called the *parent*, has too much internal energy and is unstable. In order to release this extra energy, the parent nucleus emits an α-particle, and what is left of the parent nucleus is called the *daughter nucleus*. The daughter nucleus and the α-particle recoil in a manner similar to that described above. The process is depicted in Figure 7.6 and is written

$${}_Z^AX \longrightarrow {}_{Z-2}^{A-4}Y + \alpha(= {}_2^4\mathrm{He}^{2+}) \tag{7.35}$$

where X is the parent nucleus with mass number A and atomic number Z; Y is the daughter nucleus with mass number $A - 4$ and atomic number $Z - 2$; and α is the alpha particle. The parent nucleus is called an *alpha-radioactive nucleus* while the process of emission is called *alpha decay* or *alpha disintegration.* From the conservation of linear momentum and referring to Figure

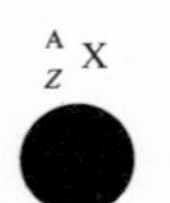

(a) Before decay

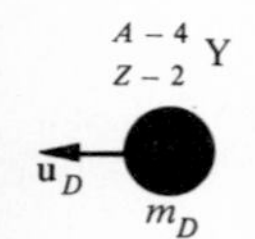

(b) After decay

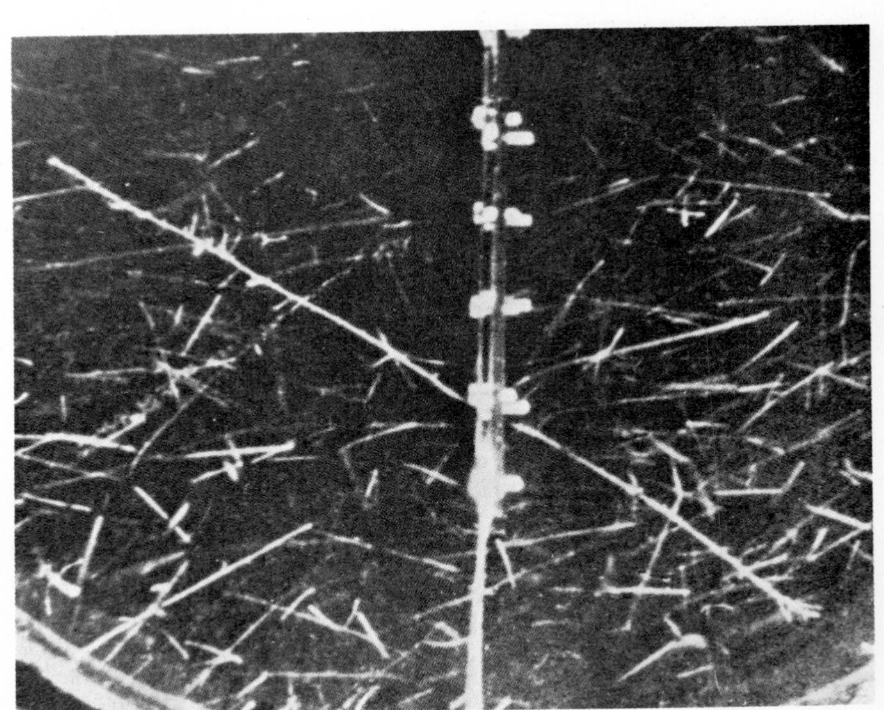

(c)

FIGURE 7.6. *(a) Parent radioactive nucleus before decay. (b) Daughter nucleus and the α-particle recoil in opposite directions with equal and opposite momentum. (c) Recoil of two fission fragments of almost equal masses from the fission (division) of unstable uranium nucleus. [Part (c) from Bøggild, Brostrøm, and Lauritsen, Phys. Rev.* **59,** *275 (1941).]*

7.6, we may write (subscripts D and α refer to the daughter nucleus and the α-particle, respectively):

$$m_D u_D + m_\alpha u_\alpha = 0$$

or

$$\frac{u_\alpha}{u_D} = -\frac{m_D}{m_\alpha} \tag{7.36}$$

while the ratio of the kinetic energies are

$$\frac{K_\alpha}{K_D} = \frac{\frac{1}{2}m_\alpha u_\alpha^2}{\frac{1}{2}m_D u_D^2} = \frac{m_\alpha}{m_D}\left(\frac{u_\alpha}{u_D}\right)^2 = \frac{m_\alpha}{m_D}\left(-\frac{m_D}{m_\alpha}\right)^2 = \frac{m_D}{m_\alpha} \tag{7.37}$$

Since the mass of the α-particle is $\simeq 4$ atomic mass units while the mass of the daughter is usually greater than 200 atomic mass units, the α-particle takes away most of the available energy and the recoiling daughter takes a very small amount, as illustrated in Example 7.4. Note that the Q-value of such a reaction is

$$Q = K_f - K_i = K_f - 0 = K_D + K_\alpha = \tfrac{1}{2}m_D u_D^2 + \tfrac{1}{2}m_\alpha u_\alpha^2 \tag{7.38}$$

EXAMPLE 7.4 A uranium 238 nucleus at rest decays (or disintegrates) by emitting an α-particle of velocity 1.4×10^7 m/s, leaving behind the daughter nucleus, thorium 234. Find (a) the recoil speed of the daughter; and (b) the Q-value or disintegration energy of the reaction.

(a) According to Equation 7.36,

$$u_D = -\left(\frac{m_\alpha}{m_D}\right)u_\alpha = -\left(\frac{4\text{ u}}{234\text{ u}}\right)(1.4 \times 10^7\text{ m/s}) = -2.4 \times 10^5\text{ m/s}$$

That is, the daughter recoils with a speed of 2.4×10^5 m/s in a direction opposite to that of the α-particle.

(b) The kinetic energy of the α-particle is given by

$$K_\alpha = \tfrac{1}{2}m_\alpha u_\alpha^2 = \tfrac{1}{2}(4)(1.67 \times 10^{-27}\text{ kg})(1.4 \times 10^7\text{ m/s})^2$$
$$= (6.55 \times 10^{-13}\text{ J})\left(\frac{1}{1.6 \times 10^{-19}\text{ J/eV}}\right) = 4.09 \times 10^6\text{ eV} = 4.09\text{ MeV}$$

From Equation 7.37,

$$K_D = \left(\frac{m_\alpha}{m_D}\right)K_\alpha = \left(\frac{4\text{ u}}{234\text{ u}}\right)(4.09\text{ MeV}) = 0.07\text{ MeV}$$

Hence, from Equation 7.38, the Q-value is

$$Q = K_\alpha + K_D = (4.09 + 0.07)\text{ MeV} = 4.16\text{ MeV}$$

EXERCISE 7.4 A radium nucleus of mass 222 u decays by emitting an α-particle of velocity 1.5×10^7 m/s. (a) What is the daughter nucleus and its speed? (b) What are the kinetic energies of the daughter nucleus and the α-particle? (c) What is the disintegration energy or Q-value? [*Ans.*: (a) $^{218}_{86}\text{Rn}$, 2.70×10^5 m/s; (b) 0.08 MeV, 4.70 MeV; (c) 4.78 MeV.

* 7.5. Symmetry Principles

In physics we deal with many laws, such as Newton's laws, the gravitational law, and Coulomb's law, and interrelated to these laws are the *conservation laws,* such as conservation of momentum and conservation of energy. Furthermore, we stated that there are four different types of forces—gravitational, electromagnetic, weak, and nuclear—which are responsible for the structure of all systems, microscopic or macroscopic. A natural question that comes to mind is: What laws and forces or other quantities are basic to the understanding of physics? Before we answer this question, we must understand what we mean by symmetry and symmetry principles. As we shall see, there is a close relationship between conservation laws and symmetry principles.

In general, a system is said to have *symmetry* when some characteristic in the system remains unchanged even though the system is changed in a certain respect. For example, we could talk about *translational symmetry,* as shown in Figure 7.7. In this case, if the system is given a linear displacement, the system remains invariant under linear displacement or translation. Thus, in Figure 7.7(a), the infinite railroad track is unchanged if it is shifted by any integral number of ties. Similarly, in Figure 7.7(b) a linear one-dimensional array of atoms shows translational symmetry. We can also talk about *rotational symmetry* when a system remains unchanged (or invariant) when rotated. For

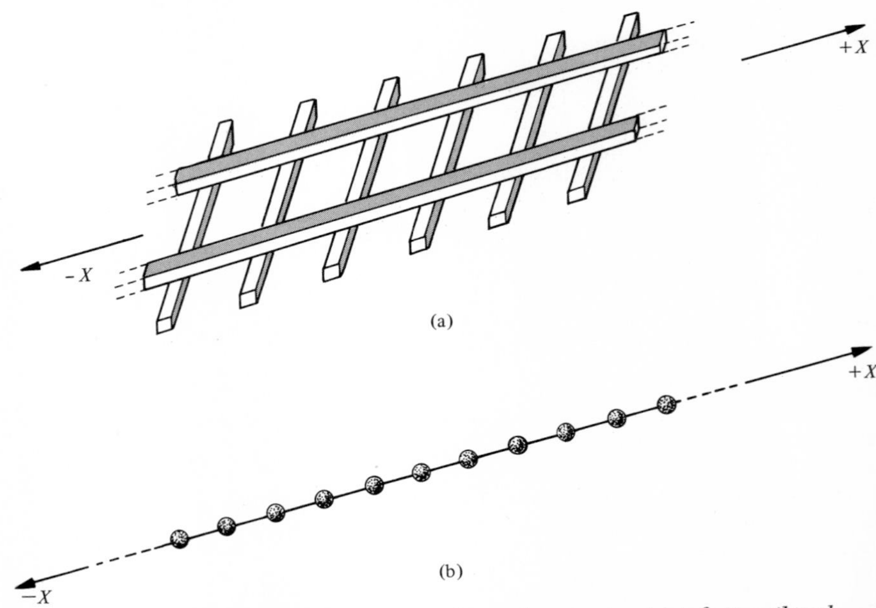

FIGURE 7.7. *Systems showing translational symmetry: (a) infinite railroad tracks and (b) linear array of atoms.*

example, a disk when rotated about an axis as shown in Figure 7.8(a) shows a rotational symmetry. Similarly, in Figure 7.8(b) a cube when rotated through intervals of 90° shows rotational symmetry. We can also talk about *reflection* (or *inversion*) *symmetry*, according to which systems remain unchanged under reflection. That is, when the coordinates (x, y, z) are changed to $(-x, -y, -z)$, the system remains invariant. These spatial symmetry principles may be extended to *symmetry in time*, according to which the laws of nature remain unchanged as time passes.

Let us now see the connection between the symmetry principles and the conservation laws. Any *conservation law* is a statement that some physical quantity of a system remains unchanged during all possible physical processes. Each conservation law is a part of some more general principle called an invariance principle. An *invariance principle* is a statement that all laws of nature remain unchanged for some particular change of physical condition. Of course, each of these invariance principles is due to some *basic symmetry* in nature, as we explain next.

Suppose that a scientist at MIT in the United States, one at Geneva in Switzerland, and one at Serpukhov in the USSR perform the same experiment and get the same answer (within statistical error); that is, they get the same results even though the experiment is performed at different places. This is because of one of the symmetries of nature—the *homogeneity of space*. If the same experiment is repeated later at any time, we find the same results. This is not strange because it is another one of the symmetries of nature—the *homogeneity of time*—a symmetry that we have come to accept unconsciously. These two together result in what is called the *space–time invariance principle*, according to which the laws of nature are the same at all points in space and for all times. It is through this invariance principle that we can derive the laws of conservation of momentum and energy. To be more specific, the conservation of linear momentum is due to the *homogeneity of space* (i.e., uniformity of space). This means that the system remains invariant under displacement, as shown in Figure 7.7. The conservation of angular momentum is due to the *isotropy of space* (uniformity of direction). That is, the system remains invariant under rotation, as shown in Figure 7.8. The conservation of energy is the result of *homogeneity of time*. In short, we may say that the three most important conservation laws—linear momentum, angular momentum, and energy—can be explained in terms of symmetry of space–time. Can you imagine what will happen if there were no space–time symmetry or that space were not uniform? It would mean that one could not predict the results of any experiment at different places and at different times.

We have mentioned only briefly symmetry and invariance principles. To make the microscopic systems in high-energy physics understandable, other symmetries and invariance principles have been introduced. Basically, there are two approaches to the problems of understanding physics: (1) the classical approach and (2) the modern approach. According to the classical approach, we consider laws that describe the structures and motions of the systems as basic, while conservation principles are derived from these basic laws. According to the modern approach, uniformity of space is considered basic. This symmetry leads to the invariance principle, from which the conservation laws may be deduced. Theoreticians prefer the latter approach.

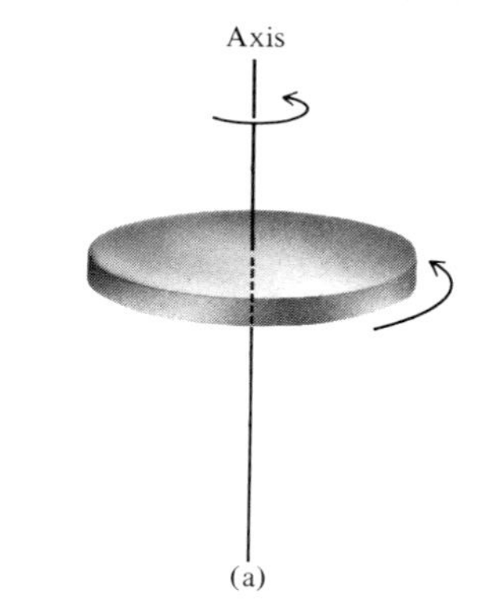

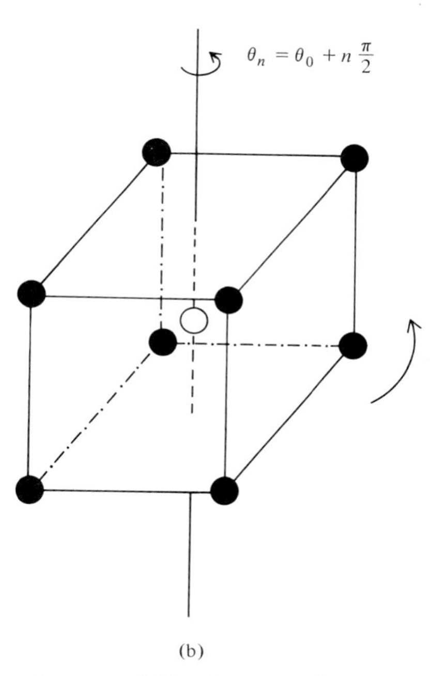

FIGURE 7.8. *Systems showing rotational symmetry: (a) solid disk and (b) simple crystal.*

SUMMARY

Interaction between two or more particles is a *collision*. In any collision, momentum is always conserved, $\mathbf{P}_i = \mathbf{P}_f$ = constant if $\mathbf{F} = 0$. If $K_i = K_f$, the total kinetic energy is conserved; if $K_i \neq K_f$, the kinetic energy is not conserved. In *elastic collisions* both momentum and kinetic energy are conserved, while in *inelastic collision* only momentum is conserved and kinetic energy is not. That is, for elastic collisions the disintegration energy is $Q = 0$ while for an inelastic collision, $Q \neq 0$.

In elastic collisions, in one dimension, $v_1 - v_2 = -(u_1 - u_2)$; speed of approach = speed of recession.*

*In two-dimensional elastic collisions $m_1 v = m_1 u_1 \cos\theta + m_2 u_2 \cos\phi$, $0 = m_1 u_1 \sin\theta - m_2 u_2 \sin\phi$, and $\frac{1}{2}m_1 v^2 = \frac{1}{2}m_1 u_1^2 + \frac{1}{2}m_2 u_2^2$.

In perfectly inelastic collisions $m_1 v = (m_1 + m_2)u$ or $K_f/K_i = m_1/(m_1 + m_2)$. In recoiling systems $m_1 u_1 + m_2 u_2 = 0$ and $K_1/K_2 = m_2/m_1$. Specifically, in α-decay, $K_\alpha/K_D = m_D/m$, while the disintegration energy is given by $Q = K_f - K_i = K_D + K_\alpha$.

*The *conservation of linear momentum* is the result of homogeneity of space, while *conservation of energy* is the result of homogeneity of time.

QUESTIONS

1. Give two examples of (a) elastic collisions; and (b) inelastic collisions.
2. How do you account for the kinetic energy lost in inelastic collisions? Give examples.
3. Consider an isolated system of three particles. Using Newton's third law, how do you show that the total momentum of the system is conserved?
4. If a shell explodes in the air, is its momentum conserved? Is its kinetic energy conserved?
5. Why is only light material such as water, deuterium, and carbon used in slowing down fast neutrons, and not heavy material?
6. Give one example of each of the collisions shown in Figure 7.2.
7. A ball strikes a smooth wall and rebounds with the same speed. Is linear momentum conserved? Explain.
8. Is it possible for an object (a) to have momentum without having energy; and (b) to have energy without having momentum?
9. When a neutron and proton combine together to form a deuteron, the final kinetic energy is greater than the initial kinetic energy. From where does this kinetic energy come from? (*Hint:* Calculate the initial and final masses.)
10. Give some practical examples of glancing and head-on collisions.
11. In the ballastic pendulum experiment, the kinetic energy just after the collision is less than the initial kinetic energy of the bullet. Where does the energy loss go?
12. When a soda bottle is opened after it has been shaken a little bit, why does its top fly away while the bottle itself remains almost stationary? Is linear momentum conserved?

PROBLEMS

1. Calculate the linear momenta and kinetic energies in the following cases: (a) an automobile of mass 1500 kg moving at 80 km/h; (b) a golf ball of mass 60 g moving at 30 m/s; (c) a bullet of mass 10 g moving at 300 m/s; and (d) an electron of mass 9.1×10^{-31} kg moving at 10^6 m/s.
2. A boy and a girl, each of mass 40 kg, are riding a 30-kg cart on a frictionless surface with a uniform velocity of 2 m/s. Suddenly, the boy jumps off the cart in the opposite direction. What is the final velocity of the cart and the girl in the cart?
3. Using Equations 7.9 and 7.13, show that

$$u_1 = \frac{(m_1 - m_2)v_1 + 2m_2v_2}{m_1 + m_2} \quad \text{and} \quad u_2 = \frac{(m_2 - m_1)v_2 + 2m_1v_1}{m_1 + m_2}$$

4. Using Equations 7.17, find the ratio of u_1/u_2 and K_1/K_2 if (a) $m_1 = 2m_2$ and (b) $m_1 = \frac{1}{2}m_2$.
5. A proton with a velocity of 2×10^6 m/s collides head-on and causes an elastic collision with the following (all of which are at rest): (a) an oxygen atom; (b) a neutron; and (c) an electron. Calculate the velocities u_1 and u_2 in each case. (The mass of a proton is 1 u, the mass of an electron is 1/1836 u, the mass of a neutron is 1 u, and the oxygen atom is 16 u.)
6. A hydrogen atom (mass = 1 u, where 1 u = 1.67×10^{-27} kg), moving with a velocity 1.5×10^7 m/s, has a head-on elastic collision with a sodium atom (mass = 23 u) at rest. Calculate the velocity and the kinetic energy of the hydrogen atom after the collision. What is the amount of energy lost by the hydrogen atom in this collision?
7. A block of mass 100 g, moving with a velocity 20 cm/s, has a perfectly elastic collision with mass m at rest. After the collision, the 100-g mass is moving with 5 cm/s in the initial direction. Find m and its velocity after the collision.
8. A neutron of kinetic energy 1 MeV is moving through space filled with (a) deuterium; and (b) carbon. Calculate the final energy of the neutrons after 8 collisions in each case.
9. Calculate the final energy of a neutron of 1 MeV after 5 collisions with (a) U-235 and (b) C-12.

* 10. Using Equations 7.22, 7.23, and 7.24, show that $K_2 = [(4m_1m_2)/(m_1 + m_2)^2]\ K$ and $K_1 = K - K_2 = [(m_1 - m_2)/(m_1 + m_2)]^2K$. (*Hint:* Square and add the two equations and use $\sin^2\theta + \cos^2\theta = 1$.)

* 11. A neutron (mass = 1 u) moving with 10^4 m/s after making an elastic collision with a carbon atom (mass = 12 u) is deflected through 90°. Calculate the recoil angle for the carbon atom. Calculate the velocities and the kinetic energies of the neutron and the carbon atom after the collision.

* 12. A hydrogen atom (mass = 1 u) with velocity 10^4 m/s after making an elastic collision with a helium atom (mass = 4 u) is deflected through an angle of 90°. Calculate the kinetic energies of the hydrogen and helium after the collision.

* 13. A particle of mass 3.34×10^{-27} kg ($\simeq$2 u) makes an elastic collision with another particle of mass 3.34×10^{-25} kg ($\simeq$200 u) at rest and is deflected

through an angle of 90°. Calculate the recoil angle and the recoil energy of the target particle after recoil if the velocity of the incident particle is 10^8 cm/s.

14. Show that if a tennis ball makes a perfect elastic collision with the ground, it rebounds back to the same height from which it was thrown.
15. In an air-track experiment a rider of mass 100 g moving with 10 cm/s collides with another of mass 500 g at rest. What are the velocities of the riders after the collision if (a) the collision is perfectly elastic; and (b) the collision is perfectly inelastic?
16. A 2-kg rifle fires a bullet of 20 g with a velocity of 350 m/s. Calculate the recoil velocity of the rifle. If the mass of the man holding the rifle is 70 kg, what will be the combined recoil velocity of the man and the rifle?
17. By some unfortunate accident an astronaut of mass 80 kg with 5 kg of equipment is separated from his spaceship and is about 10 m away from it. The only way he can catch up with the spaceship is to throw his equipment in a direction opposite to that of the spaceship with a velocity of 2 m/s. How long will it take for him to catch up with the spaceship?
18. An automobile of mass 1000 kg traveling at 20 m/s collides with another automobile of mass 2000 kg at rest. If the two stick together, what is their velocity after the collision?
19. A 20-g bullet becomes embedded in a 10-kg block suspended from the ceiling by a rope. After the collision the block and the bullet swing to a height of 25 cm above the equilibrium position. Calculate the original speed of the bullet.
20. A 2-kg mass moving along the $+X$-axis with a speed of 40 cm/s strikes a 4-kg mass moving in the opposite direction. After the collision the two blocks stick together and are at rest. Find the speed of the second mass before the collision.
21. A 10-g bullet moving horizontally with a speed of 100 m/s strikes and is embedded in a 25-kg block sitting on a horizontal surface. The block slides 3 m before coming to rest. Find the frictional force retarding the motion of the block on the horizontal surface.
22. A particle of mass m_1 and kinetic energy K makes a perfectly inelastic collision with a particle of mass m_2 at rest. Calculate the kinetic energy of the combined masses after the collision if (a) $m_1 = m_2$; (b) $m_1 = 0.1\, m_2$; (c) $m_1 = 10m_2$; and (d) $m_1 = 100m_2$. Is there any significant conclusion one may draw from the ratios?
23. A bullet of mass 10 g moving with velocity v becomes embedded in a ballistic pendulum of mass 2 kg. The pendulum rises a height of 10 cm. Find (a) the velocity of the bullet; and (b) the energy lost in the collision.
24. A 20-g bullet moving with a velocity of 200 m/s strikes a block of mass 2 kg at rest. After the bullet becomes embedded, how far will the block travel if the coefficient of friction is 0.20? How much energy is lost in the collision between the bullet and the block?
25. A 20-g bullet moving with velocity v becomes embedded in a block of mass 980 g as shown. The block is tied to a spring that compresses 10 cm. Find (a) the final velocity of the block-bullet combination; (b) the initial velocity of the bullet; and (c) the kinetic energy lost in the collision. The spring constant is 1000 N/m.

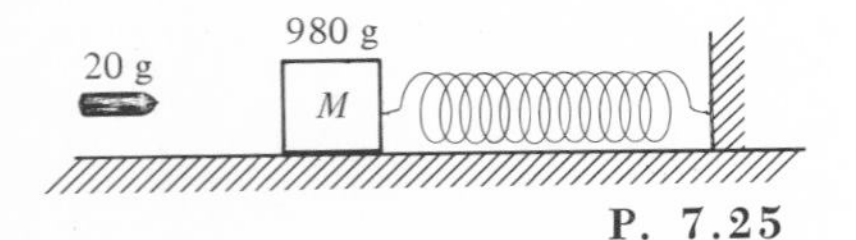

P. 7.25

26. A neutron (mass $= 1.67 \times 10^{-27}$ kg), moving with a velocity of 10^5 m/s, makes a head-on collision with a boron nucleus (mass $= 18.37 \times 10^{-27}$ kg) originally at rest. What is the ratio of the final kinetic energy of the system to the initial energy provided that the collision is perfectly inelastic?

27. A thorium nucleus ($A = 232$) emits an α-particle ($A = 4$) of energy 4 MeV. (a) Find the recoil velocity and energy of the daughter nucleus; and (b) the Q-value of the reaction.

28. An unstable nucleus at rest ($A = 150$) decays by emitting a fast-moving electron (called a β-particle) and another particle called a neutrino. These particles when emitted at right angles to each other, as shown, have momenta 2.1×10^{-22} kg-m/s and 9×10^{-23} kg-m/s, respectively. Find the recoil velocity of the nucleus. (This process in which an unstable nucleus decays by emitting an electron and a neutrino is called beta decay.)

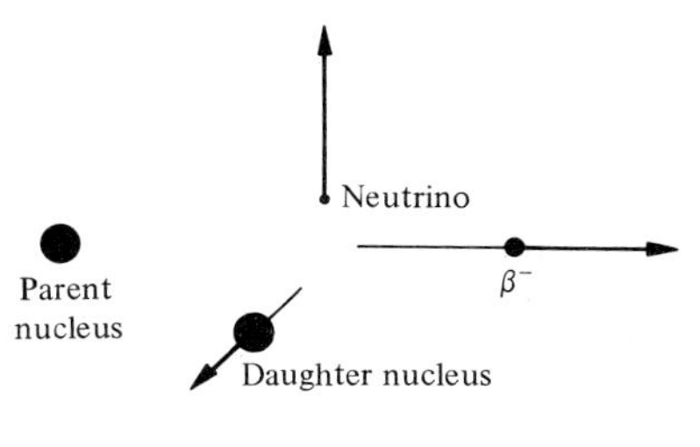

P. 7.28

8 Special Relativity— Mechanics at High Speeds

Failure of classical theory to explain many new experimental facts, such as the Michelson–Morley experiment, led to the development of the special theory of relativity by Albert Einstein in 1905. We shall show how this theory changed the concepts of length, mass, and time which led to the very important relation $E = mc^2$.

8.1. Need for a New Theory

A CLOSE look at the study of mechanics reveals that the entire subject of mechanics is based on a few simple laws such as Newton's laws of motion and gravitation. Basically, the subject of mechanics was well understood before the close of the nineteenth century. Perhaps the physicists at that time believed that no further improvements or modifications were needed. But in the beginning of the twentieth century, new experimental evidence brought into light certain difficulties. Newton's second law of motion, which was thought to be universally applicable at all speeds, was found to give correct results only for objects moving at low speeds but failed when applied to objects moving at high speeds—speeds approaching the speed of light, $c = 3 \times 10^8$ m/s. For example, a particle (say a proton) of mass m and moving with velocity v_i will have kinetic energy $K_i = \frac{1}{2}mv_i^2$ according to classical theory. Suppose that $v_i = \frac{9}{10}c$, and let us increase the velocity to $v_f = \frac{99}{100}c$. According to classical theory, $K_f/K_i = v_f^2/v_i^2 = (99/100c)^2/(9/10c)^2 = (11/10)^2 = 1.21$. That is, $K_f = 1.21K_i$, while experimentally it is found that $K_f = 3.1K_i$. This means that the expression for kinetic energy must be different at very high speeds. But this is not the only case in which Newtonian mechanics fails. The magnitudes of length, time interval, and mass are the same according to different observers when they are moving at low relative velocities. As we shall see, this is not true if the relative velocities approach that of the speed of light.

In 1905, Albert Einstein formulated the *special theory of relativity*. This theory, dealing with objects moving with uniform relative velocities, removes the above-mentioned difficulties. The *general theory of relativity*, introduced by Einstein in 1915, deals with accelerated systems and is still under development. We shall limit ourselves to the special theory of relativity.

8.2. Galilean Transformations and Newtonian Relativity

A *reference system* or *reference frame* is a set of coordinates (such as the three mutually perpendicular X, Y, and Z axes) attached to a system relative to which positions in space can be measured. We define *inertial systems* (*inertial frame of reference* or *simply inertial frames*) as those systems in which the law of inertia or Newton's law holds good. (Newton's law—a system in a state of rest or moving with uniform velocity, remains in that state unless acted upon by a net external force.) Relative to an inertial frame, any object on which no net external force acts will keep moving at constant speed in a straight line. As a matter of fact, any system that is moving with a uniform velocity relative to any inertial system is itself an inertial system. A physical happening or phenomenon in an inertial system is called an *event*. We are interested in transferring the space and time coordinates of an event from one inertial system to another moving with uniform relative velocity. In classical mechanics this can be accomplished by using a Galilean transformation described below.

Let us imagine two observers in two different inertial systems, S and S′, respectively. Let XYZ and $X'Y'Z'$ be the two sets of coordinate axes located in S and S′, respectively. The inertial frame S′ is moving with velocity v with respect to the system S to the right along the XX'-axes, as shown in Figure 8.1. Let us consider an event taking place at some point P and described by coordinates (x, y, z, t) and (x', y', z', t') corresponding to two observers in systems S and S′, respectively. Thus, using Figure 8.1, we obtain *the Galilean coordinate transformation equations*

$$\begin{aligned} x' &= x - vt \\ y' &= y \\ z' &= z \\ t' &= t \end{aligned} \qquad \text{or} \qquad \begin{aligned} x &= x' + vt \\ y &= y' \\ z &= z' \\ t &= t' \end{aligned} \tag{8.1}$$

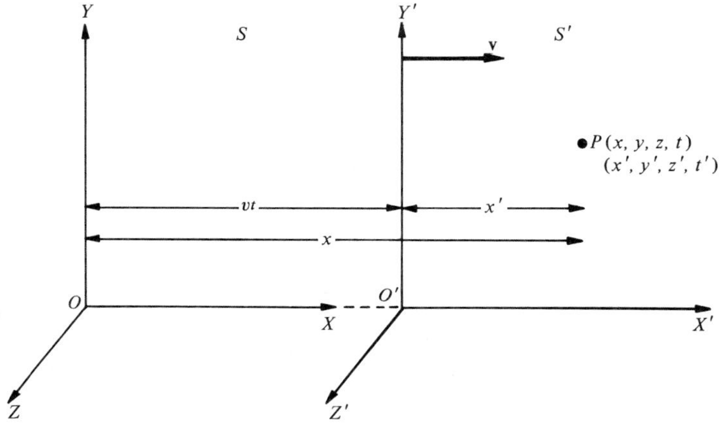

FIGURE 8.1. *Space–time coordinates of an event happening at P and measured in two different inertial systems in relative motion.*

It is important to observe that according to Newton's concept of the fourth coordinate time, we have assumed that the time coordinates use the same universal clock and do not depend on the space coordinates, and hence $t' = t$. Also, the rate of flow of time is the same for different observers in relative motion; that is, $\Delta t' = \Delta t$.

For a small displacement of point P along the XX' axes, we may write using x' and t' of Equation 8.1 $\Delta x' = \Delta x - v\,\Delta t$, $\Delta t' = \Delta t$. Dividing one by the other, we get

$$\frac{\Delta x'}{\Delta t'} = \frac{\Delta x}{\Delta t} - v$$

or

$$\boxed{u'_x = u_x - v} \tag{8.2}$$

which is the velocity transformation equation, u_x being the velocity of P with respect to the S-frame and u'_x the velocity of P respect to the S'-frame. Note that since the motion is only along the XX'-axes, $u'_y = u_y$ and $u'_z = u_z$ and we will not carry these any further.

If the velocity of point P changes by a small amount, and since v = constant, $\Delta v = 0$; also, since $\Delta t = \Delta t'$, we may write from Equation 8.2,

$$\frac{\Delta u'_x}{\Delta t'} = \frac{\Delta u_x}{\Delta t}$$

or

$$\boxed{a'_x = a_x} \tag{8.3}$$

or in general

$$\mathbf{a}' = \mathbf{a} \tag{8.4}$$

This equation implies that the acceleration does not vary from one inertial system to another; that is, the acceleration is *invariant* under Galilean transformation. In other words, we may state that the form of the equation of motion describing an event is the same in all inertial systems; that is, the form remains invariant (unchanged). For example, if a particle of mass m at point P is acted upon by a force $\mathbf{F}$ in an inertial system S, we may write

$$\mathbf{F} = m\mathbf{a} \tag{8.5}$$

Using Equation 8.4 and also assuming that the mass m is an invariant quantity, the observer in system S' will conclude that

$$\mathbf{F}' = m\mathbf{a}' \tag{8.6}$$

Equations 8.5 and 8.6 reveal that the form of *Newton's law remains unchanged or invariant under Galilean transformation.*

The discussion above may be summarized by formally stating the

> **Principle of relativity for Newtonian mechanics (or the principle of Newtonian relativity):** ***The laws of mechanics are the same in all inertial frames of reference.***

This means that there is no single preferred frame of reference. Nature does not seem to distinguish between a reference frame which is at rest from one that is moving with uniform relative velocity. The two are indistinguishable. Thus, the results of an experiment performed at the railroad platform, or in a train moving with uniform velocity, or in an airplane moving with uniform velocity, will all be the same. For example, the laws of conservation of linear momentum and kinetic energy will have the same form in all inertial systems.

Before leaving this section we must answer one more question. Where do we find a truly inertial system? We cannot say that a set of coordinate axes drawn on the earth represents an inertial system, because the earth itself has a small acceleration (owing to its rotational and orbital motions) and hence is not a true inertial system. A better approximation is an inertial system that is located with its origin at the center of the sun. But the sun itself is a part of a spiral galaxy that is rotating resulting in an acceleration of 10^{-10} m/s^2. The next place to look for a true inertial system is in distant stars that have almost zero acceleration. Thus, an ideal inertial system is a coordinate frame of reference fixed in space with respect to the "fixed stars." But for all practical purposes, a set of coordinate axes attached to the earth may be regarded as an inertial system, provided that we neglect the small acceleration.

EXAMPLE 8.1 Motion of a particle in a system S is described by the equations $u_x = u_{0x} + at$ and $u_x^2 = u_{0x}^2 + 2ax$, where a is the acceleration. Describe the motion of this particle in the S′-system moving with velocity v with respect to the S-system along the XX'-axis.

From Equations 8.1, 8.2, and 8.3,

$$t = t', \qquad u_x = u'_x + v, \qquad u_{0x} = u'_{0x} + v, \qquad a = a' \tag{i}$$

Substituting these in $u_x = u_{0x} + at$, we get

$$(u'_x + v) = (u'_{0x} + v) + a't'$$

or

$$u'_x = u'_{0x} + a't' \tag{ii}$$

Also, from Equation 8.1,

$$x = x' + vt' \tag{iii}$$

Substituting Equations i and iii in $u_x^2 = u_{0x}^2 + 2ax$,

$$(u'_x + v)^2 = (u'_{0x} + v)^2 + 2a'(x' + vt')$$

$$u'^2_x + \cancel{v^2} + 2u'_x v = u'^2_{0x} + \cancel{v^2} + 2u'_{0x}v + 2a'x' + 2a'vt'$$

Substituting for u'_x from Equation ii,

$$u'^2_x = 2v(u'_{0x} + a't') = u'^2_{0x} + 2u'_{0x}v + 2a'x' + 2a'vt'$$

or

$$u'^2_x = u'^2_{0x} + 2a'x' \tag{iv}$$

Equations ii and iv prove that the form of the equations remains invariant.

EXERCISE 8.1 Show that if the motion of a particle in system S is described by $x = u_{0x}t + \frac{1}{2}at^2$, its motion in S′ is described by $x' = u'_{0x}t' + \frac{1}{2}a't'^2$.

8.3. Conflict Between Galilean Transformations and Electromagnetism

Before discussing what the conflict is, let us review the status of electricity and magnetism by the close of the nineteenth century. In 1860, James C. Maxwell presented his theory of electromagnetism, which summarized all the laws of electricity and magnetism into four equations, now called the *Maxwell's equations*. His theory also predicted the existence of electromagnetic waves, which propagated through space with a speed of 3×10^8 m/s. Since this is the same speed as the measured speed of light, scientists thought that light might be an electromagnetic wave. When in 1888, Heinrich Hertz produced electromagnetic waves in the laboratory, this achievement, coupled with additional experimental and theoretical calculations, seemed to establish beyond doubt that light consists of electromagnetic waves.

Maxwell's theory did not require that there be a medium for the propagation of electromagnetic waves. However, because of other mechanical wave phenomena that do require a medium (e.g., air is needed to support sound vibrations, while water is needed to support water waves), physicists thought it necessary to assign a medium to propagate light and other electromagnetic waves. Nineteenth-century physicists named this medium *luminiferous ether* or just *ether*. They assumed that ether fills all space as well as vacuum. And they had to assign some strange properties to ether. (It was transparent to light and had negligible density but was as rigid as steel.) It was assumed that Maxwell's equations were valid in any reference frame that was at rest with respect to the ether. Such a frame was called an *ether frame* or *rest frame* or *absolute frame*. The velocity of the electromagnetic waves or light waves in this absolute frame, which we may call the inertial frame S at rest, was found to be always the same, $c = 2.997925 \times 10^8$ m/s.

With these ideas in mind, scientists devoted their efforts to the investigation of the following two points. First, can we prove the existence of the absolute reference frame? Second, are the Galilean transformations valid for electromagnetism? There were many experiments performed in order to prove or disprove these ideas. One final experiment that gave negative results and hence a final blow to these ideas was the Michelson–Morley experiment, reported by A. A. Michelson and E. M. Morley in 1887. Let us discuss these points in some detail.

Search for Absolute Reference Frame— The Michelson–Morley Experiment

The purpose was to confirm the existence of an absolute frame of reference (stationary ether frame) by measuring the absolute velocity of the earth with respect to the stationary ether. They assumed that the ether was at rest and was named the inertial system S, while the earth moving with velocity v, say along the X-axis, was the inertial system S'.

Suppose to start with that at $t = t' = 0$, the origins of the two inertial systems S and S' coincide, and at this time two light signals, one traveling along the XX' axes and the other along the YY' axes, are emitted. These signals each travel a distance L and are then reflected back from mirrors M_1 and M_2 as shown in Figure 8.2, where they are received by the inertial system S' of the

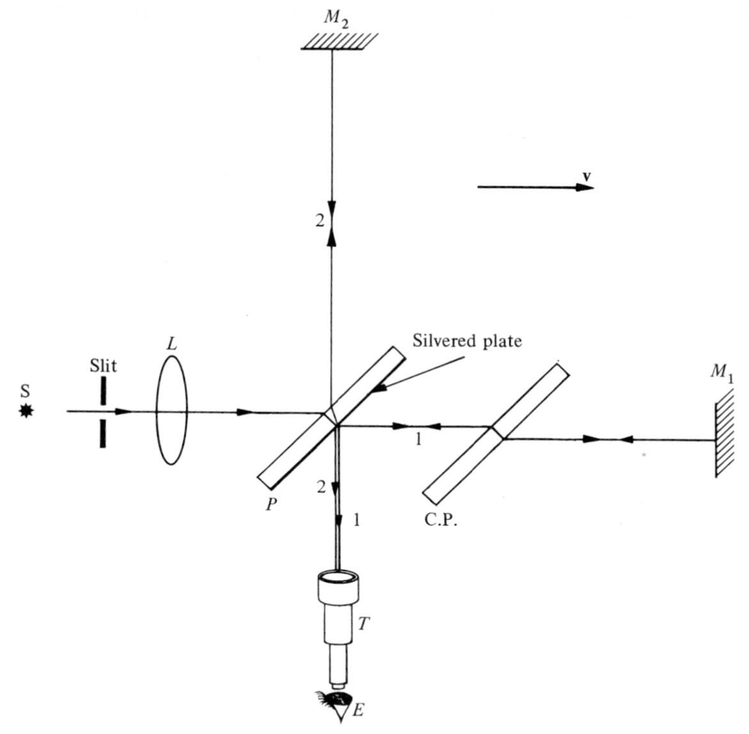

FIGURE 8.2. *Schematic diagram of the Michelson–Morley experiment.*

earth. (The experimental set up makes use of the Michelson interferometer to be discussed in Chapter 29. For the time being, we will ignore the technical details.) If we use Galilean transformations, the velocity of the light signal traveling from P to M_1 is $c - v$, while from M_1 to P (after reflection from M_1), its velocity is $c + v$. Thus, the time t_1 for this round trip is (Figure 8.3)

$$t_1 = \frac{L}{c - v} + \frac{L}{c + v} = \frac{2Lc}{c^2 - v^2} \tag{8.7}$$

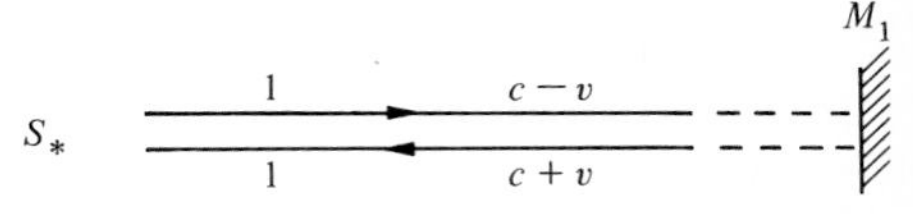

FIGURE 8.3. *Path of beam 1 in its round trip.*

The second light signal follows the path shown in Figure 8.4. The component of velocity perpendicular to the direction of motion of the signal generating source fixed on earth is obtained by vector addition and is $\sqrt{c^2 - v^2}$ as shown in Figure 8.4. Thus, the time t_2 which signal 2 takes to make the round trip PM_2P is

$$t_2 = \frac{2L}{\sqrt{c^2 - v^2}} \tag{8.8}$$

The difference between the time taken by these two light signals to make round trips is

$$\Delta t = t_1 - t_2 = \frac{2Lc}{c^2 - v^2} - \frac{2L}{\sqrt{c^2 - v^2}} \tag{8.9}$$

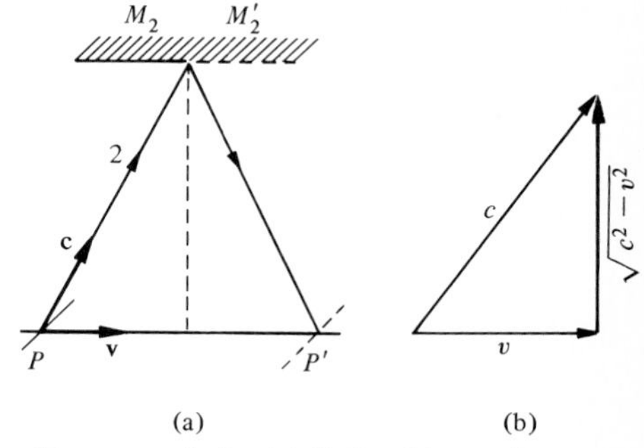

FIGURE 8.4. *(a) Path of beam 2 to make a round trip and (b) the resultant of velocities* **c** *and* **v**.

Thus, knowing L, c, and measuring Δt, we should be able to find v, the velocity of the earth with respect to the ether.

Michelson and Morley performed the above experiments at different places, different times of the year, and different heights, but they always found $\Delta t = 0$ (even though the equipment was able to detect extremely small values of Δt) and hence $v = 0$. That is, they could not detect the relative velocity of the earth with respect to the ether.

The negative result of the Michelson–Morley experiment was a big disappointment because this meant that it was not possible to measure the absolute velocity of the earth with respect to the rest frame of the ether; that is, we could not locate the absolute rest frame.

However, we can explain the negative result if we give up (1) the Galilean transformations, and (2) assume that the speed of light is invariant; that is, the velocity of light is the same in all inertial frames. Note that the fact that the velocity of light is constant has been experimentally verified. According to these assumptions, the relative velocities $c - v$, $c + v$, and $\sqrt{c^2 - v^2}$ have no meaning and are all equal to c. Thus, from Equations 8.7, 8.8, and 8.9, the time difference will be

$$\Delta t = t_1 - t_2 = \frac{2L}{c} - \frac{2L}{c} = 0 \tag{8.10}$$

Thus, the Galilean transformations hold good for classical mechanics, but not for electromagnetism.

8.4 Einstein's Postulates and Lorentz Transformations

Einstein's Postulates

In 1905 Albert Einstein developed the special theory of relativity, as reported in his paper "On the Electrodynamics of Moving Bodies." This theory deals with the physics of objects moving with high speeds and answers some of the difficulties we have been encountering so far. The theory is based on two basic postulates:

Postulate I. The Principle of Equivalence (or Relativity): ***The laws of physics are the same in all inertial frames.***

Postulate II. The Principle of the Constancy of the Speed of Light: ***The speed of light in free space (vacuum) is always a constant equal to c, and is independent of the relative motion of the inertial systems, the source, and the observer.***

Postulate I is the generalization of the fact that all physical laws, mechanics as well as electromagnetism, are invariant under all transformations; while postulate II states an experimental fact. As we shall see, these simple postulates have far-reaching consequences. These postulates also deny the existence of any absolute inertial frame (such as the ether frame).

In the following we derive the transformation equations, which are applicable to inertial reference frames. Such transformations were suggested by

Einstein and are called *Lorentz transformations,* because they were originated by H. A. Lorentz in 1890 in his electromagnetic theory of matter.

Lorentz Coordinate Transformations

Consider two inertial systems S and S′. We regard S to be at rest while S′ is moving with a velocity v with respect to S along the XX'-axes, as shown in Figure 8.5. To start with, the coordinates in the two inertial systems coincide at

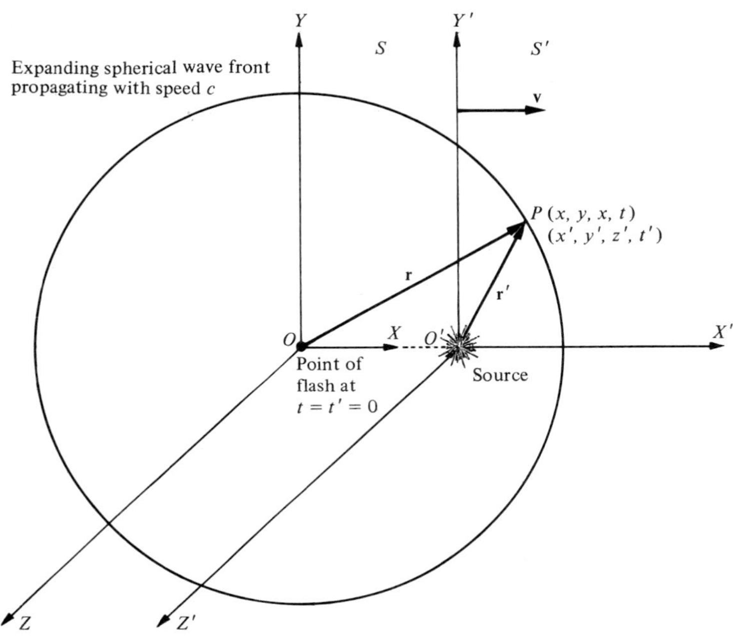

FIGURE 8.5. *Coordinates of point P on the expanding spherical wave-front as measured by observers in two inertial systems in relative motion.*

$t = t' = 0$, and at this moment a flash of light is emitted from the common origin. The flash travels as a spherical electromagnetic wave just like an expanding bubble. Later, a point such as P on the spherical surface is at a distance r from the origin of the system S and at a distance r' from the origin of system S′. Since according to postulate II of the special theory of relativity, the speed of light c is the same in both the S and S′ systems, in order for r and r' to be different, the times t and t' taken by the two signals must be different, so that

$$r = ct \qquad \text{and} \qquad r' = ct' \tag{8.11}$$

Note that in Newtonian mechanics, the speed of light will be different in different systems, but $t = t'$. It is through the assumption that $t \neq t'$ in relativity that transformations different than Galilean are obtained. Using the relation $r^2 = x^2 + y^2 + z^2$, Equation 8.11 may be written as

$$x^2 + y^2 + z^2 = c^2t^2 \qquad \text{and} \qquad x'^2 + y'^2 + z'^2 = c^2t'^2 \tag{8.12}$$

which represent spherical waves propagating with speed c, shown in Figure 8.5.

Since the relative motion of the two systems is along only the XX'-axes, the directions perpendicular to the motion are unaffected. That is,

$$y = y' \qquad \text{and} \qquad z = z' \tag{8.13}$$

Assuming that P is located along the XX' axes, so that $y = y' = 0$, $z = z' = 0$, we may write Equations 8.12 as

$$x^2 - c^2t^2 = 0 \qquad \text{and} \qquad x'^2 - c^2t'^2 = 0 \tag{8.14}$$

For this to be true, we must have

$$x^2 - c^2t^2 = x'^2 - c^2t'^2 \tag{8.15}$$

We are interested in finding x' and t' in terms of x and t. Without going into the details, we state that the relation of Equation 8.15 holds only if we assume that

$$x' = \frac{x - vt}{\sqrt{1 - (v^2/c^2)}} \tag{8.16}$$

and

$$t' = \frac{t - (v/c^2)x}{\sqrt{1 - (v^2/c^2)}} \tag{8.17}$$

That these relations are true may be verified by substituting for x' and t' in the right side of Equation 8.15, which results in the left-hand side. Thus, the Lorentz coordinate transformation equations (from system S to S′) are

$$\boxed{\begin{aligned} x' &= \frac{x - vt}{\sqrt{1 - (v^2/c^2)}} \\ y' &= y \\ z' &= z \\ t' &= \frac{t - (v/c^2)x}{\sqrt{1 - (v^2/c^2)}} \end{aligned}} \qquad \text{or} \qquad \begin{aligned} x' &= \gamma(x - vt) \\ y' &= y \\ z' &= z \\ t' &= \gamma\left(t - \frac{vx}{c^2}\right) \end{aligned} \tag{8.18}$$

where

$$\gamma = 1/\sqrt{1 - (v^2/c^2)} \tag{8.19}$$

The coordinates (x, y, z, t) in the S-system can be obtained from the coordinates (x', y', z', t') in the S′-system by replacing v by $-v$ and interchanging the primed and unprimed coordinates in Equations 8.18. That is, the resulting inverse transformations are

$$\boxed{\begin{aligned} x &= \frac{x' + vt'}{\sqrt{1 - (v^2/c^2)}} \\ y &= y' \\ z &= z' \\ t &= \frac{t' + (v/c^2)x'}{\sqrt{1 - (v^2/c^2)}} \end{aligned}} \qquad \text{or} \qquad \begin{aligned} x &= \gamma(x' + vt') \\ y &= y' \\ z &= z' \\ t &= \gamma\left(t' + \frac{vx'}{c^2}\right) \end{aligned} \tag{8.20}$$

It is clear from Equations 8.18 and 8.20 that unlike Newtonian mechanics, in relativistic mechanics t depends upon the values t' and x', and t' upon that of t and x.

When dealing with low velocities, these transformations must reduce to Galilean transformations of Newtonian or classical mechanics. Thus, as $v \to 0$, $v/c \ll 1$, $v^2/c^2 \lll 1$, and also $v/c^2 \lll 1$. These conditions reduce Equations 8.18 to

$$x' = x - vt, \qquad y' = y, \qquad z' = z, \qquad t' = t \tag{8.21}$$

These are the Galilean transformations of Equation 8.1. Thus, as v approaches zero, the Lorentz transformations reduce to Galilean transformations,

$$v \to 0 \text{ (Lorentz transformation)} = \text{(Galilean transformation)} \tag{8.22}$$

On the other extreme, if v is greater than c, the quantity $\sqrt{1 - (v^2/c^2)}$ will be imaginary; hence, the space and time coordinates will be imaginary. This is physically not possible, and we may conclude: *In vacuum no object can move with velocity greater than the velocity of light.*

Lorentz Velocity Transformations

We are now in a position to derive the Lorentz velocity transformation equations which are the relativistic counterpart of the classical Galilean velocity transformations given by Equation 8.2. Let us say that the inertial system S′ is moving with velocity v along the XX' axes with respect to the inertial system S. Suppose that a particle at point P is moving in space and has velocity u_x as measured by an observer in the S-system, and has velocity u'_x as measured by an observer in the S′-system. We want to find the relation between u_x $(= \Delta x/\Delta t)$ and u'_x $(= \Delta x'/\Delta t')$. From Equation 8.18, we get

$$\Delta x' = \gamma(\Delta x - v\,\Delta t) \qquad \text{and} \qquad \Delta t' = \gamma(\Delta t - v\,\Delta x/c^2)$$

and dividing one by the other,

$$u'_x = \frac{\Delta x'}{\Delta t'} = \frac{\Delta x - v\,\Delta t}{\Delta t - \dfrac{v\,\Delta x}{c^2}} = \frac{\dfrac{\Delta x}{\Delta t} - v}{1 - \dfrac{v}{c^2}\dfrac{\Delta x}{\Delta t}}$$

or

$$\boxed{u'_x = \frac{u_x - v}{1 - \dfrac{vu_x}{c^2}}} \tag{8.23}$$

which is the *Lorentz velocity transformation* equation. We can obtain the reverse transformation by replacing v by $-v$, and interchanging the primed and unprimed quantities. That is,

$$\boxed{u_x = \frac{u'_x + v}{1 + \dfrac{vu'_x}{c^2}}} \tag{8.24}$$

As $v \to 0$, $vu_x/c^2 \to 0$ and $vu'_x/c^2 \to 0$, and Equations 8.23 and 8.24 reduce to the Galilean equations $u'_x = u_x - v$ and $u_x = u'_x + v$.

Making use of the results above, we shall demonstrate in the following example that, in vacuum, no object moves with a speed greater than the speed of light.

EXAMPLE 8.2 An observer standing on the earth watches two rockets, A and B, approaching each other with a speed of $0.95c$. Show that the relative velocity of B with respect to A is less than c.

In order to calculate the relative velocity of B with respect to A, consider the earth to be system S and the rocket A to be system S' moving with a velocity $v = +0.95c$ along the XX' axes. Rocket B can be considered to be an object moving in space with a velocity $u_x = -0.95c$, as shown in the accompanying figure. In order to find the relative velocity u'_x of B with respect to A, we use Equation 8.23, that is,

$$u'_x = \frac{u_x - v}{1 - (vu_x/c^2)} = \frac{-0.95c - 0.95c}{1 - (0.95c)(-0.95c)/c^2} = -0.9987c$$

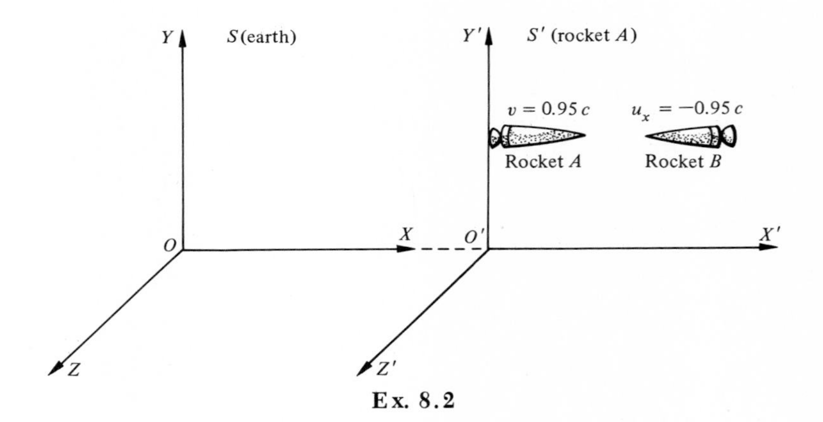

Ex. 8.2

which is less than c. This is the velocity with which the rocket B is moving along the $-X$-axis as observed by the rocket A. If we had used the Galilean transformations, the relative velocity obtained would have been $u'_x = u_x - v = -0.95c - 0.95c = -1.90c$, which is greater than c and is not possible according to special relativity.

EXERCISE 8.2 A proton is moving with a velocity $0.9c$ along the $+X$-axis and an electron with a velocity $0.6c$ along the $-X$-axis, according to the observer in the laboratory. What is the relative velocity of the electron with respect to the proton? What will be the relative velocity if either of the two were moving with a velocity c? [*Ans.*: $0.974c$, c, c.]

8.5. Length, Time, and Simultaneity in Relativity

The use of Lorentz transformations instead of Galilean transformations has many great consequences. Most important, the meanings of length, time, and simultaneity are altogether different in relativistic mechanics from those in classical mechanics. In relativity, there is no absolute length, time, and simultaneity. These quantities depend on the relative motion of the observers (except for the speed of light in a vacuum, which is the same in all inertial systems). Let us demonstrate these points.

Length Contraction

Consider two observers in the inertial systems S and S′ that are at rest at some time, say $t = t' = 0$, and their coordinate systems coincide. The observer in the S′ system has a rod of length L_0 lying parallel to the X'-axis, where $L_0 = x_2' - x_1'$, as shown in Figure 8.6. For the observer in the S-system the

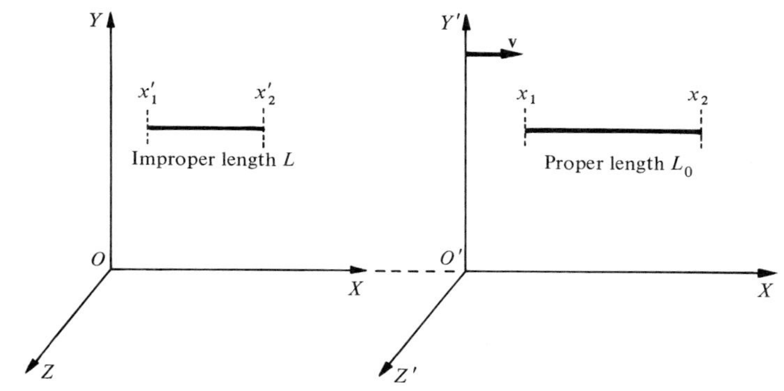

FIGURE 8.6. *Length of a rod measured in two different inertial systems in relative motion.*

length of the rod is also L_0, when there is no relative motion between the two observer. Now suppose that the S′-system starts moving with velocity **v** along the XX' axes. Once again the two observers make measurements of the length of the rod that is lying in the S′-system. For the observer in S′, the length of the rod is still the same, that is, $L_0 = x_2' - x_1'$, while for the observer in S the length L of the rod is $L = x_2 - x_1$. In order to express L in terms of L_0, we must express x_2 and x_1 in terms of x_2' and x_1' by using Lorentz transformations, Equation 8.18. That is,

$$x_2' = \gamma(x_2 - vt_2) \qquad \text{and} \qquad x_1' = \gamma(x_1 - vt_1)$$

On subtracting the second equation from the first

$$x_2' - x_1' = \gamma[(x_2 - x_1) - v(t_2 - t_1)] \tag{8.25}$$

or

$$L_0 = \gamma[L - v(t_2 - t_1)]$$

But $t_1 = t_2$ because the observer in the S-system measures the two ends of the rod simultaneously. Thus, the length of the rod lying in the S′-system and measured by the observer in the S-system is

$$L = \frac{L_0}{\gamma} = L_0\sqrt{1 - \frac{v^2}{c^2}} \tag{8.26}$$

Since $\gamma = 1/\sqrt{1 - (v^2/c^2)}$ is always greater than unity, the length L is always less than L_0. Thus, the rod lying in the S'-system always looks contracted when measured by an observer in the S-system.

The observed contraction effect is reciprocal. That is, if S has a rod of length L_0 at rest, to an observer moving with the S'-system the rod will appear to be contracted to length $L_0\sqrt{1 - (v^2/c^2)}$.

From the discussion above we conclude: *The length of an object is maximum when measured by an observer who is at rest relative to the object, and the length will appear contracted by a factor $\sqrt{1 - (v^2/c^2)}$ to an observer who is in motion relative to the object.*

The maximum length L_0 is called the *proper length* and the frame in which it is at rest is called the *proper reference frame*. The length L is called *improper length*, and the reference frame in which it is measured is called the *improper reference frame*.

Example 8.3 According to an observer at rest in the system S', a rod 1 m long is at rest along the X'-axis. What is the length of this rod according to an observer in the system S? The system S' is moving with a velocity $v = (\sqrt{3}/2)c$ with respect to S along the XX'-axes.

The situation is as shown in Figure 8.6. It is given that $L_0 = 1$ m and $v = (\sqrt{3}/2)c$. Hence, the length of the rod as observed from system S is L, given by

$$L = \frac{L_0}{\gamma} = \sqrt{1 - \frac{v^2}{c^2}}L_0 = \sqrt{1 - \frac{3}{4}}L_0 = L_0/2 = 1\text{ m}/2 = 0.5\text{ m}$$

Note that if $v = c$, $L = 0$ and the rod seems to be reduced to a point.

Exercise 8.3 A rod is at rest along the X-axis in the system S. According to an observer in system S' who is moving with velocity $0.5c$ along the XX' axes, the length of the rod is 0.75 m. What is the length of the rod according to an observer at rest in the system S? [*Ans.*: 0.866 m.]

Dilation of Time (Slowing Down of Clocks)

Just like the length, the magnitude of the time interval will be different in different reference frames; that is, the time interval is not absolute. Consider systems S and S', with S' moving with a velocity **v** along the XX'-axes. Both systems are equipped with clocks to register time intervals between two different events happening in space. But both the clocks are being observed by an observer in system S. We want to find the relation between the time interval $t_2 - t_1$ in S and $t_2' - t_1'$ in S'. From the fourth equation in 8.20,

$$t_1 = \gamma\left(t'_1 + \frac{vx'_1}{c^2}\right) \quad \text{and} \quad t_2 = \gamma\left(t'_2 + \frac{vx'_2}{c^2}\right)$$

subtracting the first from the second,

$$t_2 - t_1 = \gamma\left[(t'_2 - t'_1) + \frac{v}{c^2}(x'_2 - x'_1)\right]$$

According to this equation, the observer in S looks at the clock of S′ and compares it with his own. Note that the clock of S′ must be at rest during the time S is making the observation; that is, the position of the clock should remain the same, $x'_1 = x'_2$. Thus, the equation above reduces to

$$t_2 - t_1 = \gamma(t'_2 - t'_1) \quad \text{or} \quad \Delta t = \gamma \Delta t' \tag{8.27}$$

Let us define $t'_2 - t'_1 = \Delta t' = T_0$ as the time interval between the two events as recorded by an observer in S′ who is at rest with respect to the clock, and $t_2 - t_1 = \Delta t = T$ as the time interval as recorded by an observer in S who is in motion (with velocity $-\mathbf{v}$ with respect to S′ and its clock), we obtain the following relation from the equation above:

$$\boxed{T = \gamma T_0 = \frac{T_0}{\sqrt{1 - (v^2/c^2)}}} \tag{8.28}$$

Since γ or $1/\sqrt{1 - (v^2/c^2)}$ is always greater than unity, the interval T is longer than T_0. To the observer in S, the clock of S′ seems to be running slow as compared to his own clock. The time dilation effect is reciprocal. That is, to an observer in the S′-system who is looking at the clock in the S-system, the clock appears to run slow as compared to his own clock.

From the discussion above we may conclude: *A clock runs at its fastest when it is at rest relative to the observer and appears to be slowed down by a factor* $\gamma(= 1/\sqrt{1 - (v^2/c^2)})$ *when it is in motion relative to the observer.*

Once again, as in the case of length, T_0 is called *proper time* while T is called *improper time.*

Example 8.4 Mu mesons (or muons) are unstable particles that have the same charge as electrons, but their masses are 207 times the rest mass of the electron. Fast-moving muons form a part of cosmic radiation and can be produced in the laboratory as well. The lifetime of muons when almost at rest in the laboratory is found to be 2.2×10^{-6} s. If a beam of cosmic-ray muons has a speed of $0.98\,c$ relative to the earth, find (a) the apparent lifetime of the muons; and (b) the distance traveled by the beam during this time.

Let S be the laboratory reference frame in which the mesons are at rest so that $T_0 = 2.2 \times 10^{-6}$ s. Let S′ be the frame of reference of the mesons in cosmic rays which have a velocity of $v = 0.98\,c$. Thus, the apparent time T of the mesons in S′ as observed by an observer in S will be

$$T = \gamma T_0 = \frac{T_0}{\sqrt{1 - (v^2/c^2)}} = \frac{2.2 \times 10^{-6}\ \text{s}}{\sqrt{1 - (0.98)^2}} = \frac{2.2 \times 10^{-6}}{0.199} = 11.06 \times 10^{-6}\ \text{s}$$

The apparent distance traveled by the mesons in this time according to an observer in the S-frame is

$$x = vt = (0.98)(3 \times 10^8 \text{ m/s})(11.06 \times 10^{-6} \text{ s}) = 3252 \text{ m}$$

while to the observer in the S'-frame, the mesons travel a distance of

$$x = (0.98)(3 \times 10^8 \text{ m/s})(2.2 \times 10^{-6} \text{ s}) = 646.8 \text{ m}$$

These values have been checked experimentally, and hence verify the time dilation.

Exercise 8.4 The time period of a pendulum is 2 s. What is the time period of this pendulum when measured by an observer moving with a speed of 0.8 c with respect to the inertial system of the pendulum? [*Ans.:* 3.33 s.]

Simultaneity

According to the Galilean transformations, Equation 8.1, $t' = t$ or $\Delta t' = \Delta t$. This implies that, according to classical or Newtonian mechanics, the time interval has the same magnitude in different inertial frames. Thus, if two events P_1 and P_2 are simultaneous (occur at the same time) in one inertial frame, they are simultaneous in all inertial frames, that is, the *simultaneity* is *absolute*. This would have been true if the information could be conveyed with an infinite speed. Since we know that nothing can move with speed greater than the speed of light, there is no such thing as absolute simultaneity, just as shown above that there is no absolute length or time. That is, length, time, and simultaneity are all relative.

Thus, it is possible that events which may be simultaneous to one observer may not be simultaneous to another observer in a different inertial frame. This is because the time coordinates depend upon the space coordinates also, as shown by the Lorentz transformation equations. Similarly, two events happening at a given location at different times according to one observer may, according to another observer in another inertial frame, appear to be happening at different locations as well.

8.6. Relativistic Mechanics

In order for Newtonian mechanics to be invariant under the Lorentz transformations, the expression for mass, momentum, Newton's laws, and energy must be modified as we explain below.

Mass and Momentum

According to classical or Newtonian mechanics, the inertial mass of the body is constant, independent of velocity and is denoted by m_0, called the *rest mass*. Accordingly, the linear momentum is given by $p = m_0 v$. In relativistic mechanics, the observed mass is a function of velocity. Although at small velocities the mass of an object does not change appreciably from the rest mass m_0, at higher velocity it is not so. Using the law of conservation of linear momentum together with the Lorentz transformations, the following expression for the variation in mass with velocity is obtained:

$$m = \gamma m_0 = \frac{m_0}{\sqrt{1-(v^2/c^2)}} \tag{8.29}$$

where m_0 is the rest mass and m is the measured mass of an object when it is moving with respect to an observer and is called the *relativistic mass*. That is, m_0 and not m is an invariant quantity. For the values

$$v/c = 0.01 \quad 0.05 \quad 0.10 \quad 0.50 \quad 0.90 \quad 0.93 \quad 0.99 \quad 1.00$$

the corresponding values for m/m_0:

$$m/m_0 = 1.0005 \quad 1.0013 \quad 1.005 \quad 1.15 \quad 2.30 \quad 5.00 \quad 7.10$$

It is clear that at low velocities $m/m_0 \simeq 1$; that is, the relativistic mass m does not differ appreciably from the classical rest mass m_0. If $v \to c$, $m \to \infty$, meaning that when a material particle travels at the speed of light, would we observe its mass to be infinite? Of course, that would be impossible, and hence this implies that c is a limiting velocity impossible for a material particle to achieve.

The expression for the variation in mass with velocity as given by Equation 8.29 has been experimentally verified over many years and the agreement between theory and experiment as shown in Figure 8.7 is excellent.

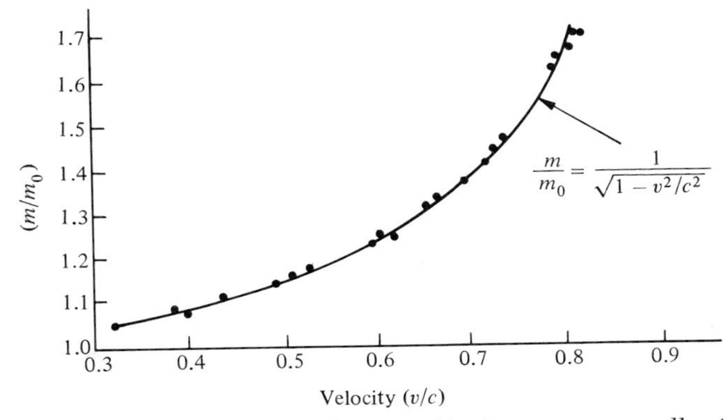

FIGURE 8.7. *The experimental points (dots) agree very well with the theoretical values (solid curve) of the variation of mass with velocity.*

In view of the discussion above, we redefine linear momentum. According to relativity theory a particle of rest mass m_0 moving with velocity v has momentum p given by

$$p = mv = \frac{m_0}{\sqrt{1-(v^2/c^2)}}v = \gamma m_0 v \tag{8.30}$$

Kinetic and Total Energy

In view of the changes in the definition of mass and momentum as given by Equations 8.29 and 8.30, it becomes essential to examine what changes must be incorporated in the classical definition of Newton's second law, kinetic energy, and total energy.

Thus, according to relativistic mechanics, Newton's second law is stated as: The force F acting on a body is equal to the time rate of change of relativistic momentum. That is,

$$F = \frac{\Delta p}{\Delta t} = \frac{\Delta(mv)}{\Delta t} = \frac{\Delta(\gamma m_0 v)}{\Delta t} = m_0 \frac{\Delta(\gamma v)}{\Delta t} \tag{8.31}$$

Thus, m_0 and not m is a constant. Using this definition of force and the theory of relativity, we obtain the following expression for kinetic energy:

$$\boxed{K = mc^2 - m_0c^2 = \gamma m_0 c^2 - m_0 c^2 = (\gamma - 1)m_0 c^2} \tag{8.32}$$

This expression for kinetic energy is quite different from the classical expression, $K = \frac{1}{2}mv^2$.

We must now show that for $v/c << 1$, Equation 8.32 reduces to the classical expression for kinetic energy. To show this expand γ by the binomial theorem and keep only the first two terms, because v/c is small in the classical limit. Thus,

$$\gamma = \frac{1}{\sqrt{1 - (v^2/c^2)}} = \left(1 - \frac{v^2}{c^2}\right)^{-\frac{1}{2}} = \left(1 + \frac{1}{2}\frac{v^2}{c^2} - \cdots\right) \simeq \left(1 + \frac{1}{2}\frac{v^2}{c^2}\right)$$

Substituting for γ in Equation 8.32, for $v/c << 1$,

$$K = m_0c^2(\gamma - 1) \simeq m_0c^2\left[\left(1 + \frac{1}{2}\frac{v^2}{c^2}\right) - 1\right] \simeq \frac{1}{2}mv^2$$

If we define $\Delta m = m - m_0 =$ increase in the mass of the particle, we may write Equation 8.32 as

$$K = (m - m_0)c^2 = \Delta mc^2 \tag{8.33}$$

That is, the kinetic energy increases with increasing relativistic mass. Thus, it is possible to assign energies to masses. It is usual to call the quantity $E_0 = m_0c^2$ as the *rest-mass energy,* while $E = mc^2 = K + m_0c^2$ is the *relativistic mass energy* or *total energy* and is equal to the sum of the kinetic energy and rest-mass energy. That is,

$$\boxed{E = K + E_0 = mc^2 = \gamma m_0 c^2 = \gamma E_0} \tag{8.34}$$

where

$$\boxed{E = mc^2 \qquad \text{and} \qquad E_0 = m_0c^2} \tag{8.35}$$

The relation $E = mc^2$ is *Einstein's mass–energy relation.*

EXAMPLE 8.5 At what speed will the mass of a proton become double its rest mass? What will be the kinetic energy of the proton at this speed?

From Equation 8.29, the relation between the rest mass m_0 and moving mass m is $m = m_0/\sqrt{1 - (v^2/c^2)}$. If $m = 2m_0$, we get $1/\sqrt{1 - (v^2/c^2)} = 2$. Squaring and rearranging, $v^2/c^2 = \frac{3}{4}$, or

$$v = \sqrt{3}/2\, c = 0.866\, c = (0.867)(3 \times 10^8 \text{ m/s}) = 2.598 \times 10^8 \text{ m/s}$$

Note that this is true for a particle of any mass: electron, proton, neutron, and so on.

From Equation 8.33 the kinetic energy is given by $K = (m - m_0)c^2$, but $m = 2m_0$. Therefore,

$$K = (2m_0 - m_0)c^2 = m_0c^2$$

For a proton, $m_0 = 1.67252 \times 10^{-27}$ kg while $c = 2.997925 \times 10^8$ m/s.

$$K = m_0c^2 = (1.67252 \times 10^{-27} \text{ kg})(2.997925 \times 10^8 \text{ m/s})^2$$
$$= 15.03186 \times 10^{-11} \text{ kg-m}^2/\text{s}^2$$

Since kg-m^2/s^2 = kg-m-m/s^2 = N-m = J,

$$K = 15.03186 \times 10^{-11} \text{ J}$$

Another common unit of energy is the electron volt (eV), and it is related to the joule by 1eV $= 1.6021 \times 10^{-19}$ J. Therefore,

$$K = \frac{15.03186 \times 10^{-11} \text{ J}}{1.6021 \times 10^{-19} \text{ J/eV}} = 9.3826 \times 10^8 \text{ eV} = 938.26 \text{ MeV}$$

Thus, when a proton doubles its rest mass, it has a velocity of 2.598×10^8 m/s and kinetic energy of 938.26 MeV. Note that since $E_0 = m_0c^2$, for a proton the energy equivalent to the rest-mass energy is 938.26 MeV. The rest mass of a proton in atomic mass units u is

$$m_p = 1.0072766 \text{ u}$$

The energy equivalent of 1 u is

$$1 \text{ u} = 931.478 \text{ MeV}/c^2$$

Exercise 8.5 Calculate the (a) moving mass; (b) kinetic energy; and (c) momentum of (1) a neutron and (2) an electron that is moving with a speed of $0.9\,c$ ($m_n = 1.67482 \times 10^{-27}$ kg). [*Ans.*: (1) (a) 3.84×10^{-27} kg; (b) 1.22 GeV; (c) 1.04×10^{-18} kg-m/s; (2) (a) 2.09×10^{-30} kg; (b) 0.663 MeV; (c) 5.64×10^{-22} kg-m/s.]

*8.7 Some Puzzles Yet Unsolved

Let us consider two interesting cases, the twin paradox and tachyons.

The Twin Paradox

The *twin paradox*, or the *clock paradox*, in relativity has been investigated in some depth in the last two decades and may be stated as follows:

Let us consider two clocks, R and M, which are synchronized and are at rest in a given inertial frame. Suppose that the clock M starts moving with a uniform velocity v with respect to clock R. After making a round trip to a nearby planet, the clock M returns to clock R. According to the theory of time dilation, the time interval T_M on the moving clock M is related to the time interval T_R on the rest clock R by the equation

$$T_R = \frac{T_M}{\sqrt{1-(v^2/c^2)}} \quad \text{that is,} \quad T_M < T_R \tag{8.36}$$

The moving clock M loses time compared to the stationary clock R, which is the same thing as saying that the moving clock has been running slow.

In everyday language the problem can be stated more simply. Let R and M represent twins Robert and Mike. Robert stays at home on earth while Mike visits a nearby planet in a spaceship. After T_M years when Mike comes back home, he finds his brother Robert has grown much older; that is, $T_R = \gamma T_M$, where γ is always greater than 1.

The moving twin Mike ages much less than his twin Robert, who is at rest. But is it really true? According to special relativity, the motion is relative, Mike is moving with velocity v with respect to Robert, while Robert is moving with velocity $-v$ with respect to Mike. Thus, the aging effect should be reciprocal. That is, the paradox is that Robert expects Mike to be younger while Mike expects Robert to be younger. But the situation cannot be readily treated according to the special theory of relativity. The spaceship in which Mike travels must accelerate and decelerate while starting, stopping, and turning around; that is, Mike does not remain in the same inertial frame all through his journey. Therefore, reference frames are not truly reciprocal.

Aging corresponds to a biological clock, while we are used to physical clocks. Actually, there is no difference between the two; for example, we could take heartbeats to be a clock, and slowing down the heartbeat will correspond to slowing down a clock. The difficulty in testing these lies in achieving high-enough speeds.

Tachyons

One of the implications of the theory of relativity has been that nothing, not even a small particle, can be made to travel faster than the speed of light. On the other hand, there is no reason why a particle cannot exist which is already traveling faster than light (so that we need not supply infinite energy to speed it up). Such particles are called *tachyons*, from the Greek word *tachys*, meaning "swift."

Scientists assume that the rest mass of a tachyon is imaginary while momentum and energy are real. They have to assume this because the rest mass is not directly observable while momentum and energy are. Another interesting property of a tachyon is that, as it loses energy, it speeds up, until it is traveling infinitely fast. When it gets to this point, it has no energy at all.

Physicists from Princeton University and elsewhere have tried to produce tachyons in the laboratory, but without any success to date.

SUMMARY

In mechanics where $v << c$, the *Galilean transformation* can be applied. This is not true for electromagnetism, the Galilean transformations together with the stationary ether hypothesis, cannot explain the negative results of the Michelson–Morley experiment.

The *special theory of relativity* developed by Albert Einstein in 1905 is based on two basic postulates: the principle of equivalence and the principle of the constancy of the speed of light. The relativity theory replaces the Galilean transformations by the following *Lorentz transformations*.

$$x' = \gamma(x - vt)$$
$$y' = y$$
$$z' = z \qquad \text{where } \gamma = \frac{1}{\sqrt{1 - (v^2/c^2)}}$$
$$t' = \gamma\left(1 - \frac{vx}{c^2}\right)$$

and

$$u'_x = \frac{u_x - v}{1 - (vu_x/c^2)}$$

The most important relations of classical and relativistic mechanics are summarized in the following.

Quantity	Classical Expression	Relativistic Expression
Length	$L = L_0$	$L = L_0\sqrt{1 - (v^2/c^2)}$
Time	$T = T_0$	$T = T_0 / \sqrt{1 - (v^2/c^2)}$
Mass	$m = m_0$	$m = m_0/\sqrt{1 - (v^2/c^2)}$
Momentum	$p = m_0 v$	$p = mv = \gamma m_0 v$
Kinetic energy	$K = \frac{1}{2}m_0 v^2$	$K = mc^2 - m_0c^2$ $= (\gamma - 1)m_0c^2$
Rest energy	$E_0 = 0$	$E_0 = m_0c^2$
Total energy	$E = K$	$E = mc^2 = E_0 + K$

QUESTIONS

1. If the principle of equivalence were not true, what additional information would be needed for scientists to communicate with each other?
2. Can you think of an alternative explanation for the negative results of the Michelson–Morley experiment?
3. Is it possible to find another set of equations that will accomplish the same thing as the Lorentz transformation equations?
4. What do you expect to be the speed of light in a medium other than a vacuum? Is it greater or smaller than c?
5. If $v > c$, the space and time coordinates will be imaginary. What will be the consequences of such results? Explain.
6. How will the density of a cube change if its length contracts only on one side?

7. Considering simultaneity, what is the relation of the past and future with the present in terms of space and time?
8. What will be the mass of an object that is moving with speed c? Is this possible?
*9. If tachyons were discovered, would it help in communicating with the rest of the unreachable parts of the universe and how?

PROBLEMS

1. A ship (reference frame S′) is sailing at sea with a speed of 13.4 m/s with respect to the shore (reference frame S). A man is walking on the deck of the ship with a speed of +1.34 m/s with respect to the ship. Using Galilean transformations, fill in the following table.

	Frame S′ $x'\ y'\ z'\ t'\ u'$	Frame S $x\ y\ z\ t\ u$
Coordinates of the ship	0 0 0 t	
Coordinates of the man	0 0 t	

2. The velocity of a particle in a system S is given by $u_x = a + bx + ct^2$, where a, b, and c are constants. Using Galilean transformations, describe this motion in a system S′ that is moving with a velocity of v along the XX'-axes.
3. Consider in one dimension the collision of two particles in the system S. The conservation of linear momentum requires

$$m_1u_1 + m_2u_2 = m_1U_1 + m_2U_2$$

where u_1, u_2 and U_1, U_2 are velocities before and after the collision along the X-axis. Show that in the S′-system, this equation takes the form

$$m_1u_1' + m_2u_2' = m_1U_1' + m_2U_2'$$

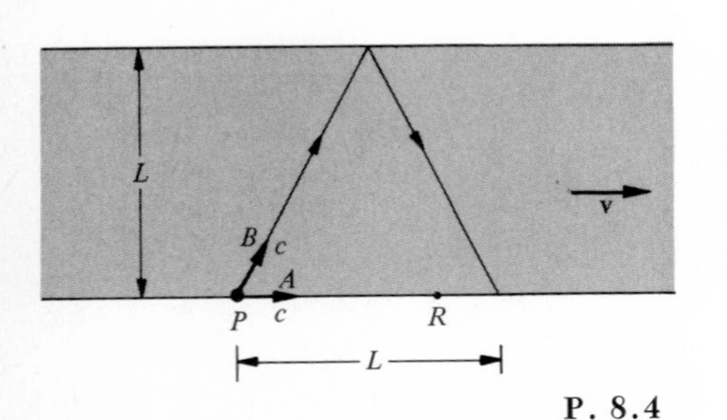

P. 8.4

4. Water in a river flows with a velocity v with respect to the bank as shown. The width of the river is L. The speed of boats A and B in still water is c with respect to the bank. Boat A starts from P and makes a round trip PRP in time t_1, while boat B goes across the river and returns in time t_2 (not necessarily to the same point). Calculate t_1 and t_2 in terms of v, c, and L. If we know t_1 and t_2, how do we calculate v? What happens if v is greater than c?
5. A river 3 miles wide has a current of 4 mi/h. A motorboat A goes up stream 3 mi and returns, while a motorboat B goes across the river and returns. The speed of the boats is 10 mi/h in still water. Calculate the times taken by the two boats to make round trips. What happens if the water current were more than 10 mi/h?
6. A rod 1 m in length is lying along the X-axis in the inertial system S. This length is being measured by an observer in a system S′ which is moving along the XX'-axes with a velocity v. What should be the value of v so that to an observer in S′ the length of the rod measures to be (a) 0.99 m; (b) 0.9 m; (c) 0.5 m; and (d) 0.1 m?

7. A cube is moving with a velocity v in the direction parallel to one of its edges. Show that the volume V of the cube and its density ρ, viewed by a stationary observer, are given by

$$V = V_0 \sqrt{1 - v^2/c^2} \qquad \text{and} \qquad \rho = \rho_0 \frac{1}{1 - (v^2/c^2)}$$

where m_0 and V_0 are the rest mass and rest volume of the cube, respectively, and $\rho_0 = m_0/V_0$.
8. As observed from a spaceship moving toward the earth with a speed of $0.75c$, a man runs a race for 10 s. How long does the runner think the race lasted? (*Hint:* $\Delta t'$ is given, calculate Δt.)
9. According to an observer on earth, an athlete runs a distance of 100 m in 10 s. According to an observer in a superjet moving with $0.95c$ parallel to athlete, calculate the distance the athlete ran and the time it took him.
10. A laser beam starting from earth reaches the moon in 1.21 s as measured by an observer on earth. How long will it take if the observation was made by an observer moving with a velocity of $0.5c$ parallel to the beam?
11. An observer in a spaceship is approaching a flashing light signal with a speed of $c/2$. Calculate the relative velocity of the light signal with respect to the spaceship.
12. Two particles are approaching each other. The velocities of these particles as measured by an observer on earth are $0.9c$ and $-0.9c$. Calculate the relative velocity of one particle with respect to the other according to (a) Galilean transformations; (b) Lorentz transformations.
13. An electron moving with a speed of $0.9c$ passes a proton moving with speed of $0.6c$. What is the speed of the electron with respect to the proton, and the speed of the proton with respect to the electron?
14. Find the speed at which (a) an electron; and (b) a proton has a mass 10 times its rest mass.
15. What should be the speed of an electron so that its mass will be (a) $1.1m_0$; (b) $10m_0$; and (c) $100m_0$?
16. Calculate the energy increase if the speed of an electron changes from (a) $0.1c$ to $0.5c$; (b) $0.5c$ to $0.9c$; and (c) $0.9c$ to $0.99c$.
17. A proton has a rest mass of 1.67×10^{-27} kg ($= 938$ MeV/c^2), which is given a kinetic energy of 10 GeV. Find the speed, mass, and momentum of the proton.
18. An electron is moving at a speed of $0.98c$. Find its kinetic energy, total energy, mass, and momentum.
19. A proton is moving at a speed of $0.98c$. Find its kinetic energy, total energy, mass, and momentum.
20. Show that the speed of a particle of rest mass m_0 and the total energy E is given by $v = c\sqrt{1 - (E_0/E)^2}$. [Start with $m = m_0/\sqrt{1 - (v^2/c^2)}$.]
21. Calculate the mass and momentum of a 1 GeV (a) electron and (b) proton.
22. The energy received by the earth from the sun is 1.33×10^3 W/m^2. Using Einstein's mass-energy relation, calculate the fractional rate at which the sun is losing mass. The mass of the sun is 2×10^{30} kg, and its distance from the earth is 1.55×10^{11} m.

9

Rotational and Circular Motion

Next to translational motion, the rotational motion of rigid bodies about fixed axes is the most common motion. Many of the features of rotational motion may be derived from analogy with translational motion, such as the equations of motion. Circular motion with a special emphasis on the concept of centripetal force will be discussed in detail. The relations between such quantities as torque, moment of inertia, rotational kinetic energy, work, and angular momentum will be introduced. We shall conclude with a brief discussion of rotation about a moving axis.

9.1 Rotational Motion and Angular Displacement

IN ORDER to understand pure rotational motion consider a point P in a three-dimensional rigid body shown in Figure 9.1. A rigid body is said to have pure *rotational motion* if every particle of the body (such as point P) moves in a circle, the centers of which lie on a straight line called the axis of rotation (such as the Z-axis in Figure 9.1). The shaded area in Figure 9.1 is the horizontal XY-plane containing P and perpendicular to the Z-axis.

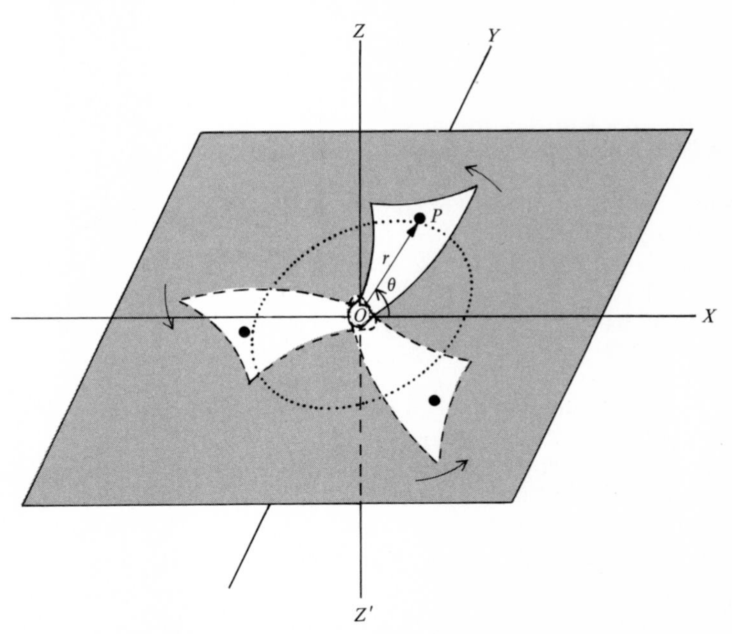

FIGURE 9.1. *Rotation of a rigid body about the Z-axis.*

When a particle moves in a circular path, such as point P, it is convenient to describe its position in terms of a radius vector $\mathbf{r}$ and its angular position θ. In Figure 9.1 the angle θ is between the line OP and the X-axis. The angle may be measured in degrees, revolutions, or radians. As shown in Figure 9.2(a), the arc length s on the circumference of a circle of radius r subtends an angle of θ radians which is given by

$$\boxed{\theta = \frac{s}{r}} \tag{9.1}$$

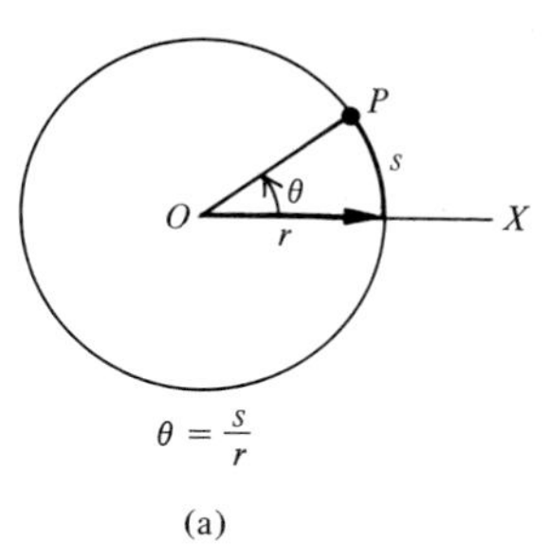

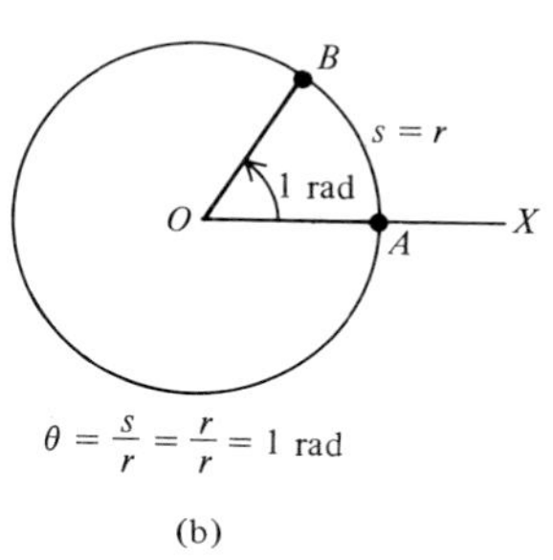

FIGURE 9.2. *(a) Angle θ subtended by an arc length s. (b) Definition of radian.*

Naturally if $s = r$, $\theta = 1$ radian $= 1$ rad. Thus, by definition, *1 radian is the angle subtended by an arc length equal to the radius of the circle.* (Note that s/r has no units, even though θ is given a unit of radian.) Since the circumference of a circle is $2\pi r$, the circle will contain $2\pi r/r = 2\pi$ radians. Thus,

$$360° = 2\pi \text{ rad} = 6.28 \text{ rad} = 1 \text{ revolution}$$

$$1 \text{ rad} = \frac{360}{2\pi} = 57.3 \text{ degrees} \tag{9.2}$$

9.2 Equations of Rotational Motion

Let a point P in a rigid body rotating about the Z-axis make an angle θ_1 with the X-axis at time t_1 and angle θ_2 at time t_2. The average angular velocity $\bar{\omega}$ is defined as

$$\boxed{\bar{\omega} = \frac{\theta_2 - \theta_1}{t_2 - t_1} = \frac{\Delta\theta}{\Delta t}} \tag{9.3}$$

The *instantaneous angular velocity* ω is defined as the limit approached by the ratio $\Delta\theta/\Delta t$ as Δt approaches zero, that is,

$$\omega = \lim_{\Delta t \to 0} \frac{\Delta\theta}{\Delta t} \tag{9.4}$$

The units of angular velocity are rad/s, deg/s, or rev/s. Different particles in the rigid body have different linear velocities, but they all have the same angular velocity. Hence, ω is a characteristic of the whole rigid body.

If a rigid body has angular velocity ω_1 at time t_1 and ω_2 at time t_2, the *average angular acceleration* $\bar{\alpha}$ is defined as

$$\boxed{\bar{\alpha} = \frac{\omega_2 - \omega_1}{t_2 - t_1} = \frac{\Delta\omega}{\Delta t}} \tag{9.5}$$

while the *instantaneous acceleration* α is defined as the limit approached by the ratio $\Delta\omega/\Delta t$ as Δt approaches zero, that is,

$$\alpha = \lim_{\Delta t \to 0} \frac{\Delta\omega}{\Delta t} \tag{9.6}$$

The units of α are rad/s^2, deg/s^2, or rev/s^2. Like ω, α is also a characteristic of the whole body.

We may say that the rotational quantities θ, ω, and α are analogous to the translational quantities s, v, and a. Further analogies exist between the equations of motion. Table 9.1 shows the corresponding equations with constant accelerations for the translational and rotational motions, respectively. We shall illustrate the use of these equations in Example 9.1.

TABLE 9.1
Equations for Motion With Constant Acceleration

Linear Motion (along X-axis)	Angular Motion (about Z-axis)
$v = v_0 + at$	$\omega = \omega_0 + \alpha t$
$x = \left(\frac{v + v_0}{2}\right)t$	$\theta = \left(\frac{\omega + \omega_0}{2}\right)t$
$x = v_0 t + \frac{1}{2}at^2$	$\theta = \omega_0 t + \frac{1}{2}\alpha t^2$
$v^2 = v_0^2 + 2ax$	$\omega^2 = \omega_0^2 + 2\alpha\theta$

EXAMPLE 9.1 A wheel starting with an angular velocity of 10 rad/s rotates for 5 s with a uniform acceleration of 2 rad/s². Calculate (a) the final angular velocity; (b) the average angular velocity; and (c) the angular displacement in this time interval.

At $t = 0$, $\omega_0 = 10$ rad/s, $\alpha = 2$ rad/s², and at $t = 5$ s, from Table 9.1,

$$\omega = \omega_0 + \alpha t = 10 \text{ rad/s} + (2 \text{ rad/s}^2)(5 \text{ s}) = 20 \text{ rad/s}$$

$$\bar{\omega} = \frac{\omega_0 + \omega}{2} = \frac{10 + 20}{2} \text{ rad/s} = 15 \text{ rad/s}$$

$$\theta = \omega_0 t + \tfrac{1}{2}\alpha t^2 = (10 \text{ rad/s})(5 \text{ s}) + \tfrac{1}{2}(2 \text{ rad/s}^2)(5 \text{ s})^2 = 75 \text{ rad}$$

EXERCISE 9.1 A body with a uniform angular velocity of 1 rev/s at $t = 0$ has an angular acceleration of 0.2 rev/s². After 10 s, what will be (a) the final angular velocity; (b) the average angular velocity; and (c) the angular displacement. Calculate these quantities in radians. [*Ans.*: (a) $\omega = 6\pi$ rad/s; (b) $\bar{\omega} = 4\pi$ rad/s; (c) $\theta = 40\pi$ rad.]

9.3. Tangential and Radial Accelerations

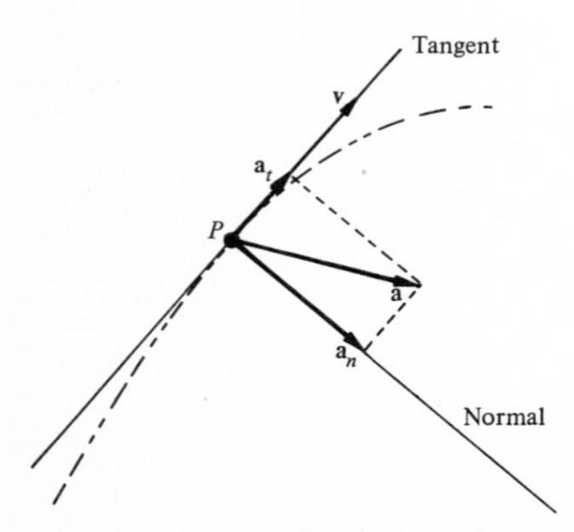

FIGURE 9.3. *Resolution of acceleration* **a** *into normal and tangential components* $\mathbf{a}_n$ *and* $\mathbf{a}_t$*, respectively.*

Suppose that a particle is moving on a curved path in an *XY*-plane and its instantaneous velocity **v** at any point *P* is along the tangent to the path at the point, while the acceleration is on the concave side of the path, as shown in Figure 9.3. Usually, we resolve **a** into rectangular components $\mathbf{a}_x$ and $\mathbf{a}_y$ along the fixed set of axes *X* and *Y*. In many situations it is convenient and is preferred to resolve **a** into *radial* (*or normal*) *and tangential rectangular components* $\mathbf{a}_n$ and $\mathbf{a}_t$ along the normal and the tangent, respectively, at the point in question as shown in Figure 9.3. Note that unlike $\mathbf{a}_x$ and $\mathbf{a}_y$, $\mathbf{a}_n$, and $\mathbf{a}_t$ components *do not have fixed directions* in space, but these two components have physical significance.

The normal component $\mathbf{a}_n$ is due to the change in the direction of the velocity **v** while the tangential component $\mathbf{a}_t$ is due to the change in the magnitude of the velocity **v**. This point may be made clear by reference to Figure 9.4. Let $\mathbf{v}_1$ and $\mathbf{v}_2$ represent the velocities of a particle at two nearby points *P* and *Q* on a curved path, as shown in Figure 9.4(a). The vector $\Delta\mathbf{v}$ is the change in velocity as shown in Figure 9.4(b). $\Delta\mathbf{v}$ may now be resolved into normal component $\Delta\mathbf{v}_n$ and tangential component $\Delta\mathbf{v}_t$. Hence, we may define the two accelerations as

$$\mathbf{a}_n = \frac{\Delta\mathbf{v}_n}{\Delta t} \tag{9.7}$$

and

$$\mathbf{a}_t = \frac{\Delta\mathbf{v}_t}{\Delta t} \tag{9.8}$$

As is clear from Figure 9.4, if $\mathbf{v}_1$ and $\mathbf{v}_2$ were in the same direction, $\Delta\mathbf{v}_n$ will be zero; hence, $\mathbf{a}_n$ will be zero. Thus, $\mathbf{a}_n$ arises from the change in the direction of the velocity. On the other hand, if the magnitude of $\mathbf{v}_1$ were equal to the magnitude of $\mathbf{v}_2$, $\Delta\mathbf{v}_t$ will be zero; hence, $\mathbf{a}_t$ will be zero. Thus, $\mathbf{a}_t$ arises from the change in magnitude of the velocity vector $\mathbf{v}$.

Of course, there must be some force $\mathbf{F}$ that acts on the particle; otherwise, no acceleration will be possible. Let this force be resolved into two components, $\mathbf{F}_n$ the normal component and $\mathbf{F}_t$ the tangential component. The component $\mathbf{F}_n$ is the transverse force which deviates the particle from its straight-line path and makes it move in a curved path. From Newton's second law,

$$\mathbf{F}_n = m\mathbf{a}_n \tag{9.9}$$

while the tangential component of the force $\mathbf{F}_t$ causes a change in the magnitude of the velocity of the particle. Once again, from Newton's second law we may write

$$\mathbf{F}_t = m\mathbf{a}_t \tag{9.10}$$

Before leaving this discussion, let us first find an expression for the tangential acceleration a_t. Consider a particle of mass m moving in a circle. If the angle θ is measured in radians, the arc length s and radius r in Figure 9.5 are given by the relation

$$\boxed{s = r\theta} \tag{9.11}$$

The linear velocity v_t, which is always tangent to the circle, is found from the equation above by writing $\Delta s = r\,\Delta\theta$ and dividing by Δt; that is,

$$\frac{\Delta s}{\Delta t} = r\frac{\Delta\theta}{\Delta t}$$

or

$$\boxed{v_t = r\omega} \tag{9.12}$$

v_t and ω are shown in Figure 9.5. Similarly, the tangential acceleration, a_t of the particle is given by writing the equation above as $\Delta v_t = r\,\Delta\omega$ and dividing by Δt; that is,

$$\frac{\Delta v_t}{\Delta t} = r\frac{\Delta\omega}{\Delta t}$$

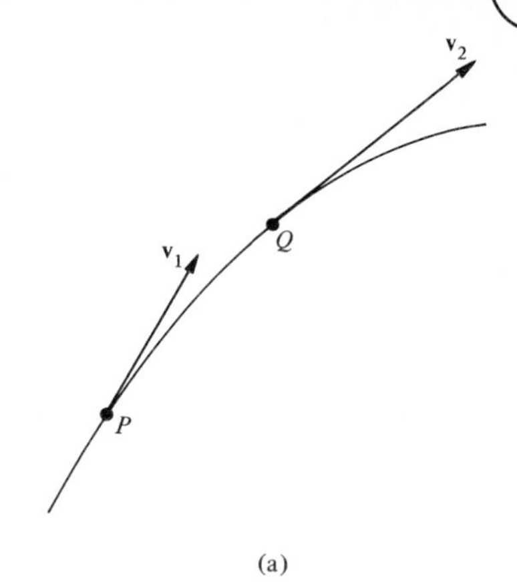

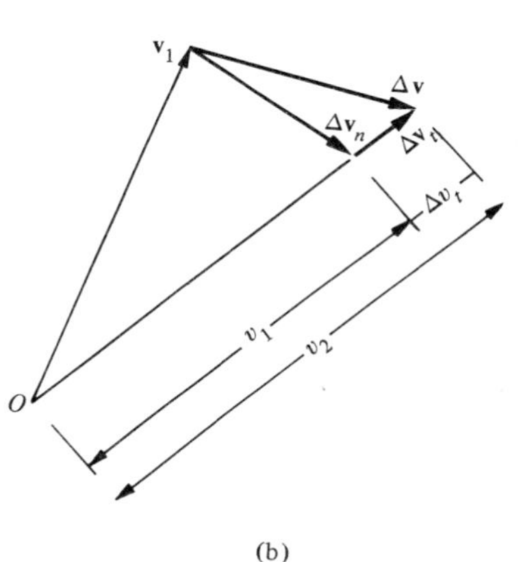

Figure 9.4. *(a) Velocity of a particle at P and Q on the curved path. (b) The change in velocity* $\Delta\mathbf{v}$ *is resolved into normal and tangential components.*

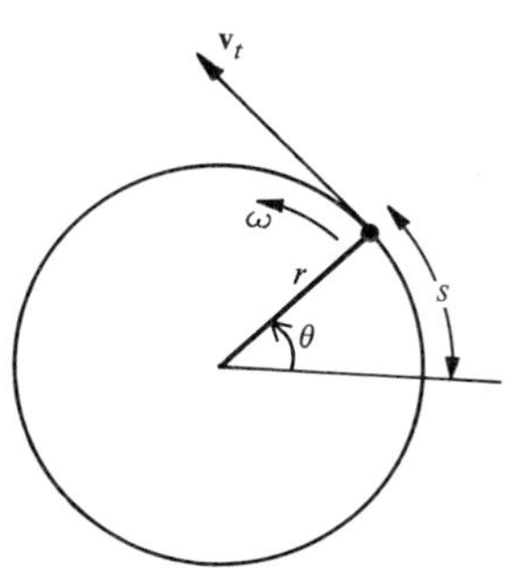

Figure 9.5. *Tangential velocity and angular velocity of a particle moving in a circle.*

or

$$a_t = r\alpha \tag{9.13}$$

We shall see the usefulness of these relations in many problems. An expression for the radial acceleration will be derived in the next section.

9.4. Centripetal Force

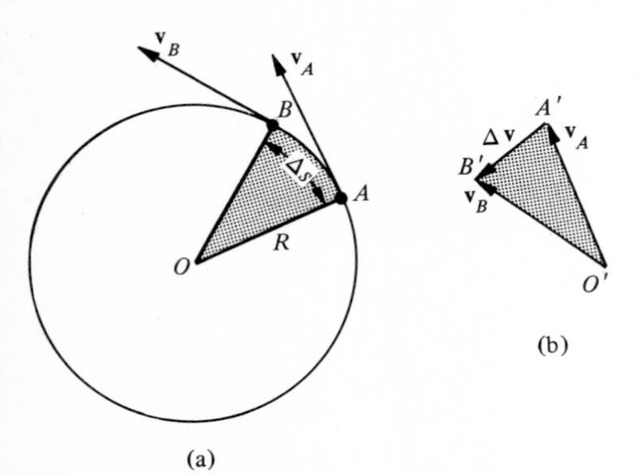

FIGURE 9.6. *Calculation of radial acceleration for a particle moving in a circular motion with constant speed.*

Let us consider a special case of a particle of mass m moving in a circle with a uniform speed as shown in Figure 9.6(a). There is only a normal acceleration, which points radially inward from the particle to the center of the circle; hence, it is called the *radial acceleration* $\mathbf{a}_r$ (instead of $\mathbf{a}_n$). (The tangential acceleration is zero because there is no change in the magnitude of $\mathbf{v}$.)

The magnitude of $\mathbf{a}_r$ may be calculated by reference to Figure 9.6. At point O', draw $O'A'$ equal and parallel to $\mathbf{v}_A$ and $O'B'$ equal and parallel to $\mathbf{v}_B$. Since $|\mathbf{v}_A| = |\mathbf{v}_B|$ and $O'A' = O'B'$, the triangles OAB and $O'A'B'$ in Figure 9.6(a) and (b), respectively, are similar triangles (for small angle $\Delta s \approx$ AB, a straight line). Thus, since $\Delta\mathbf{v}_n = \Delta\mathbf{v}$ for uniform circular motion,

$$\frac{A'B'}{O'A'} = \frac{AB}{OA} \qquad \text{or} \qquad \frac{\Delta v}{v_A} = \frac{\Delta s}{R}$$

But $\Delta s = v_A \Delta t$; hence,

$$\frac{\Delta v}{v_A} = \frac{v_A \Delta t}{R} \qquad \text{or} \qquad \frac{\Delta v}{\Delta t} = \frac{v_A^2}{R}$$

or in the limit $\Delta t \to 0$, $|\mathbf{v}_A| = |\mathbf{v}_B| = |\mathbf{v}| = v$, the radial acceleration a_r is given by

$$a_r = \frac{v^2}{R}$$

The direction of this radial acceleration is toward the center along the radius. Hence, this is called a *radial*, *central*, or *centripetal* (meaning seeking a center) *acceleration*. Also since $v/R = \omega$, $v^2/R = Rv^2/R^2 = R\omega^2$. Thus,

$$a_r = \frac{v^2}{R} = R\omega^2 = v\omega \tag{9.14}$$

where v is the tangential velocity.

Of course, this acceleration must be produced by some unbalanced force. This force is needed to change the direction of the particle moving in a circular path, although its speed remains the same. This force is called *centripetal force* F_r (also denoted as F_c) and is directed toward the center along the radius. Thus, from Newton's law $F_r = ma_r$ or

$$F_r = m\frac{v^2}{R} = mR\omega^2 \tag{9.15}$$

Without this force the particle will move in a straight line instead of in a circular path.

EXAMPLE 9.2 A particle of mass 0.5 kg moves in a circle of radius 2 m with acceleration **a** as shown in the accompanying figure. Calculate a_r, a_t, v, F_r, and F_t for both cases, if $a = 10\ \text{m/s}^2$.

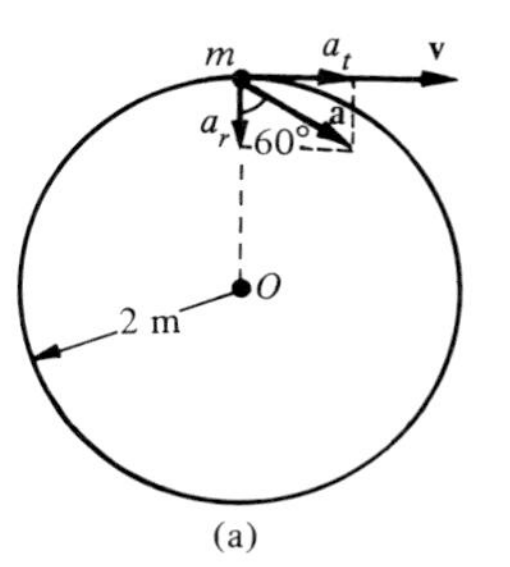

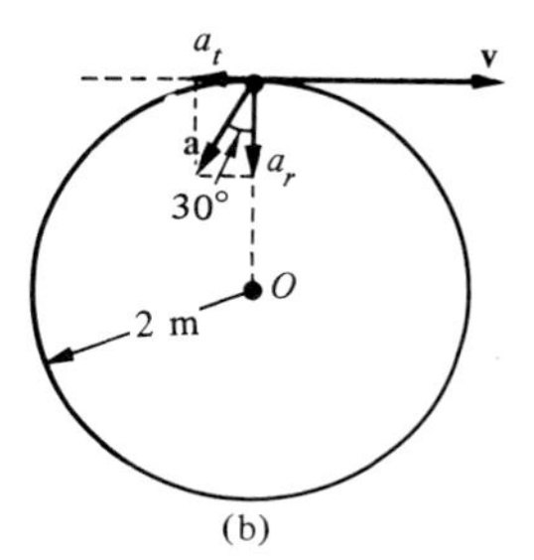

Ex. 9.2

(a)

$$a_r = a \cos 60° = (10\ \text{m/s}^2)(0.5) = 5\ \text{m/s}^2$$
$$a_t = a \sin 60° = (10\ \text{m/s}^2)(0.87) = 8.7\ \text{m/s}^2$$

Since $a_r = v^2/r$, we get $v = \sqrt{a_r r} = \sqrt{(5\ \text{m/s}^2)(2\ \text{m})} = 3.16\ \text{m/s}$

$$F_r = ma_r = (0.5\ \text{kg})(5\ \text{m/s}^2) = 2.5\ \text{N}$$
$$F_t = ma_t = (0.5\ \text{kg})(8.7\ \text{m/s}^2) = 4.35\ \text{N}$$

(b)

$$a_r = a \cos 30° = (10\ \text{m/s}^2)(0.87) = 8.7\ \text{m/s}^2$$
$$a_t = -a \sin 30° = -(10\ \text{m/s}^2)(0.5) = -5\ \text{m/s}^2$$

The negative sign of a_t means that a_t is in the opposite direction to **v** as shown in part (b). Again, $a_r = v^2/r$, so that

$$v = \sqrt{a_r r} = \sqrt{(8.7\ \text{m/s}^2)(2\ \text{m})} = 4.17\ \text{m/s}$$
$$F_r = ma_r = (0.5\ \text{kg})(8.7\ \text{m/s}^2) = 4.35\ \text{N}$$
$$F_t = ma_t = (0.5\ \text{kg})(-5\ \text{m/s}^2) = -2.5\ \text{N}$$

EXERCISE 9.2 Repeat Example 9.2 in case (a) $\theta = 0$; and (b) $\theta = 45°$. [*Ans.*: (a) $a_r = 10\ \text{m/s}^2$, $a_t = 0$, $v = 4.47\ \text{m/s}$, $F_r = 5\ \text{N}$, $F_t = 0$; (b) $a_r = 7.07\ \text{m/s}^2$, $a_t = -7.07\ \text{m/s}^2$, $v = 3.76\ \text{m/s}$, $F_r = 3.54\ \text{N}$, $F_t = -3.54\ \text{N}$.]

We shall apply the preceding ideas to the following situations: the banked road, motion in a vertical circle, and the centrifugal force.

The Banked Road

Figure 9.7 shows an object, say an automobile, moving with velocity **v** and rounding a curve of radius R on a level road. Since the object is moving in a circular path, there must be a centripetal force $\mathbf{F}_c$ present. The other forces acting are the normal reaction **N** and the weight of the object, $\mathbf{w} = m\mathbf{g}$. The force **N** is equal and opposite to $m\mathbf{g}$. Hence, the centripetal force $\mathbf{F}_c$ must be provided by the frictional force f between the automobile and the ground. In the case of a railway car rounding a curve, the centripetal force is the force exerted by the rails against the flanges on the wheels of the car. In both the cases above, if the force of friction or the force against the flanges is not enough, the object will not stay on the curved path but will fly away tangentially. Since $F_c = mv^2/R$, the value of v must be low, so that F_c does not exceed f.

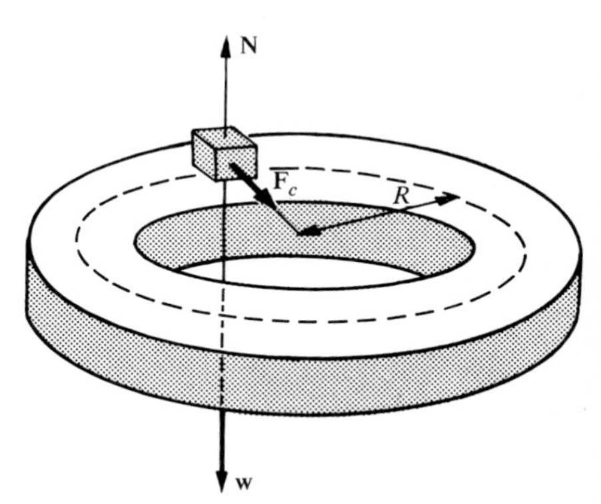

FIGURE 9.7. *Forces acting on an object moving on a level curved road.*

In order not to rely on friction (or in the case of rail cars to reduce wear on the flanges), the roads are usually banked, as shown in Figure 9.8. The banking angle θ is achieved by raising the level as we move from the inside to the

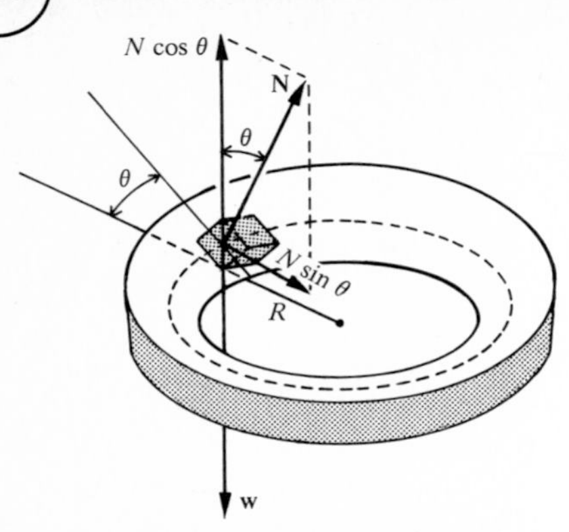

FIGURE 9.8. *Forces acting on an object moving on a banked curved road.*

outside of the curved road, as shown in Figure 9.8. Assume the surface is frictionless. The forces acting are still only two, the normal force **N** and the weight $\mathbf{w} = m\mathbf{g}$. But **N** can be resolved into two components, $N\cos\theta$, which equals mg, and $N\sin\theta$, which provides the centripetal force $F_c = mv^2/R$. Thus,

$$N\cos\theta = mg \qquad \text{and} \qquad N\sin\theta = \frac{mv^2}{R}$$

Dividing the second equation by the first,

$$\boxed{\tan\theta = \frac{v^2}{Rg}} \tag{9.16}$$

which indicates that the banking angle depends on v and R and is independent of mass. The roads are banked for some average speed v for a curved path of radius R. The object will fly off tangentially if v^2 exceeds $Rg\tan\theta$.

Motion in a Vertical Circle

A ball of mass m is tied to a string of length R and whirled in a vertical circle of radius R about the center O as shown in Figure 9.9. The motion is circular but is not uniform since the mass speeds up while coming down and slows down while going up.

Once again at any instant the forces acting are the tension **T** in the string and weight $\mathbf{w} = m\mathbf{g}$. As shown in Figure 9.9, $m\mathbf{g}$ may be resolved into tangential and normal components $mg\sin\theta$ and $mg\cos\theta$, respectively. Thus, the tangential and normal forces F_t and F_n, respectively, are

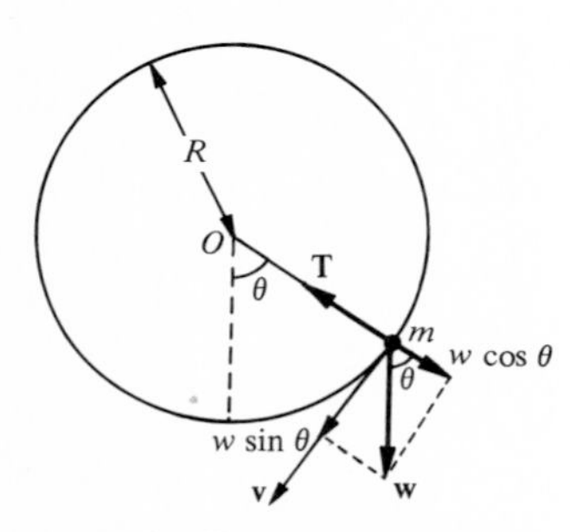

FIGURE 9.9. *Forces acting on an object of mass m moving in a vertical circle.*

$$F_t = mg\sin\theta$$

$$F_n = T - mg\cos\theta$$

But the normal force must provide the necessary centripetal force. That is,

$$F_n = T - mg\cos\theta = \frac{mv^2}{R}$$

or

$$\boxed{T = m\left(\frac{v^2}{R} + g\cos\theta\right)} \tag{9.17}$$

Let us consider two special cases of this equation. As shown in Figure 9.10 at point A, $\theta = 0°$, $T = T_A$, and $v = v_A$. For $\cos 0° = 1$ Equation 9.17 takes the form

$$T_A = m\left(\frac{v_A^2}{R} + g\right) \tag{9.18}$$

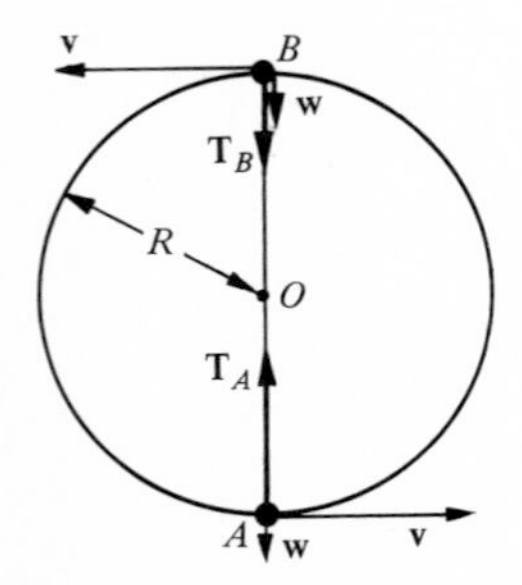

FIGURE 9.10. *Tension in the string when the mass m moving in a vertical circle is at the lowest and the highest positions.*

which means that the tension T should be large enough not only to overcome the weight mg but also to provide the centripetal force mv_A^2/R. On the other hand, at point B, $\theta = 180°$, $T = T_B$, and $v = v_B$. For $\cos 180° = -1$ Equation 9.17 takes the form

$$T_B = m\left(\frac{v_B^2}{R} - g\right) \tag{9.19}$$

In this case T_B is less than T_A because mg provides a part of the required centripetal force, as shown in Figure 9.10.

To be at the highest point B the mass m must have a certain minimum critical velocity v_0 at B; otherwise, the string will become slack. The value of v_0 can be found from Equation 9.19 by substituting $T_B = 0$; that is,

$$0 = m\left(\frac{v_0^2}{R} - g\right)$$

or

$$v_0 = \sqrt{Rg} \tag{9.20}$$

If the velocity is less than this, the ball does not make a complete circle. It leaves the circular path, follows a parabolic path for a short while, and then joins the circular path as shown in Figure 9.11. An example is the case of a ball looping-the-loop in a vertical circle. Also, if a bucket of water is moved fast enough in a vertical circle, the water will not spill.

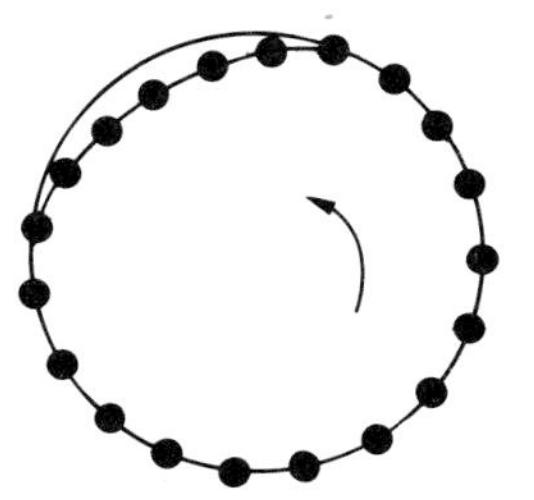

FIGURE 9.11. *Path of a ball moving in a vertical circle with insufficient velocity to reach the highest point.*

Centrifugal Force

Let us apply Newton's third law to the circular motion we have been considering so far. When a particle is moving in a circle, a centripetal force acts on it, directed toward the center. According to Newton's third law, there must be an equal and opposite force. This force is exerted by the rotating body on whatever is causing the body to rotate and is directed away from the center and is called the *centrifugal force* (meaning fleeing the center). A common example is that of a mass tied to a string and made to whirl in a horizontal circle. The hand applies a tension T which is the centripetal force while the mass applies a force on the hand, which is felt by a pull on the hand and is the centrifugal force. Thus, centripetal force and centrifugal force form an action–reaction pair. One of the many applications of centrifugal force is the separation of heavy and light liquids such as milk from cream—in a cream separator the cream collects at the center and milk at the rim when the separator is running at a high angular velocity. This is because cream is lighter and needs less centrifugal force and according to $F = m\omega^2 R$, collects at small R. Similarly, the corpuscles can be separated from blood.

EXAMPLE 9.3 A maximum speed at which a car can negotiate a curve of radius 100 m is 48 km/h ($\approx$ 30 mi/h). What is the coefficient of friction? What should be the banking angle if we do not want to depend upon friction?

If m is the mass of the car, from Figure 9.7 we get

$$N = mg \qquad \text{and} \qquad f = F_c = \frac{mv^2}{R}$$

But $f = \mu N = \mu mg = mv^2/R$. Substituting v = 48 km/h = 13.33 m/s, we get

$$\mu = \frac{v^2}{Rg} = \frac{(13.33 \text{ m/s})^2}{(100 \text{ m})(9.8 \text{ m/s}^2)} = 0.18$$

If there is no friction, the road must be banked as shown in Figure 9.8; hence, from Equation 9.16,

$$\tan \theta = v^2/Rg = 0.18 \qquad \text{or} \qquad \theta = 10.2°$$

EXERCISE 9.3 What is the safe speed of an automobile on a curved path of radius 110 m if (a) the road has no banking but the coefficient of friction is 0.4; and (b) the coefficient of friction is zero but it is banked at 20°? [*Ans.*: (a) 20.8 m/s; (b) 19.8 m/s.]

9.5 Torque and Moment of Inertia

We now consider rotational dynamics, that is, the relation between the forces and the resulting angular acceleration. Consider a particle of mass m at a distance r from the origin O. This particle is acted upon by a force F resulting in a motion in the XY-plane, as shown in Figure 9.12. As explained in Section 5.2 and demonstrated in Figure 5.3 as well as in Figure 9.12, the moment of force or torque τ according to Equation 5.2 is given by

$$\boxed{\tau = rF_t = rF \sin \theta = dF} \qquad (9.21)$$

That is, τ is equal to the product of r and the tangential component $F_t (= F \sin \theta)$ of the force F, as shown in Figure 9.12(a) or the product of F and the moment arm $d\ (= r \sin \theta)$, as demonstrated in Figure 9.12(b).

In Equation 9.21, both $\mathbf{r}$ and $\mathbf{F}$ are vectors. $\boldsymbol{\tau}$ is also a vector whose magnitude is equal to $rF \sin \theta$ and whose direction is along the $+Z$-axis for counterclockwise torques and along the $-Z$-axis for clockwise torques; that is, $\boldsymbol{\tau}$ is perpendicular to the plane containing $\mathbf{r}$ and $\mathbf{F}$, which is the XY plane in this case.

These characteristics of $\boldsymbol{\tau}$ are summarized by saying that $\boldsymbol{\tau}$ is a vector resulting from the *cross product* or *vector product* of the vectors $\mathbf{r}$ and $\mathbf{F}$, that is

$$\boxed{\boldsymbol{\tau} = \mathbf{r} \times \mathbf{F}} \qquad (9.22)$$

with the magnitude of $\boldsymbol{\tau}$ being equal to $|\mathbf{r}|\ |\mathbf{F}| \sin \theta = rF \sin \theta$, where θ is the angle between $\mathbf{r}$ and $\mathbf{F}$. The direction of $\boldsymbol{\tau}$ is found by the right-hand rule. Put the fingers of your right hand along the vector $\mathbf{r}$. Circle your fingers toward $\mathbf{F}$ through the smallest possible angle θ. The direction of the vector $\boldsymbol{\tau}$ is in the direction of the outstretched thumb as illustrated in Figure 9.13. Note that the order of multiplication is important because $\mathbf{F} \times \mathbf{r} = -\mathbf{r} \times \mathbf{F} = -\boldsymbol{\tau}$, as shown in Figure 9.13(b). (Compare this cross-product with the dot product $W = \mathbf{F} \cdot \mathbf{r} = Fr \cos \theta$.)

Let us go back to Equation 9.21, $\tau = rF_t$. From Equation 9.13, since $a_t = r\alpha$, we may write

$$\tau = rF_t = rma_t = rm(r\alpha)$$

that is,

$$\tau = mr^2\alpha \qquad (9.23)$$

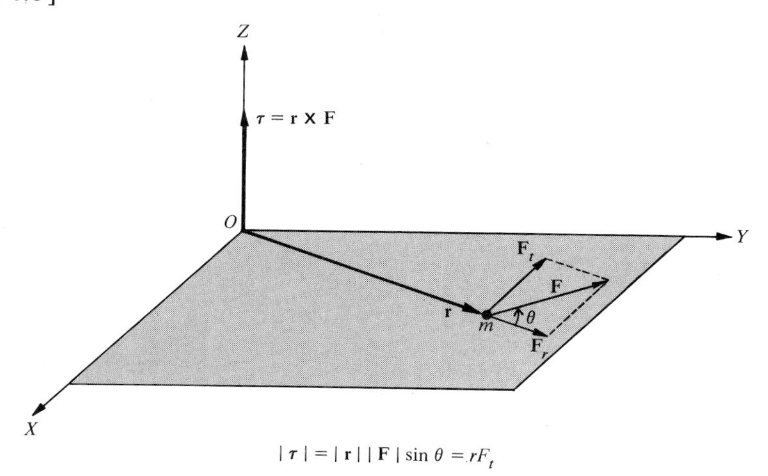

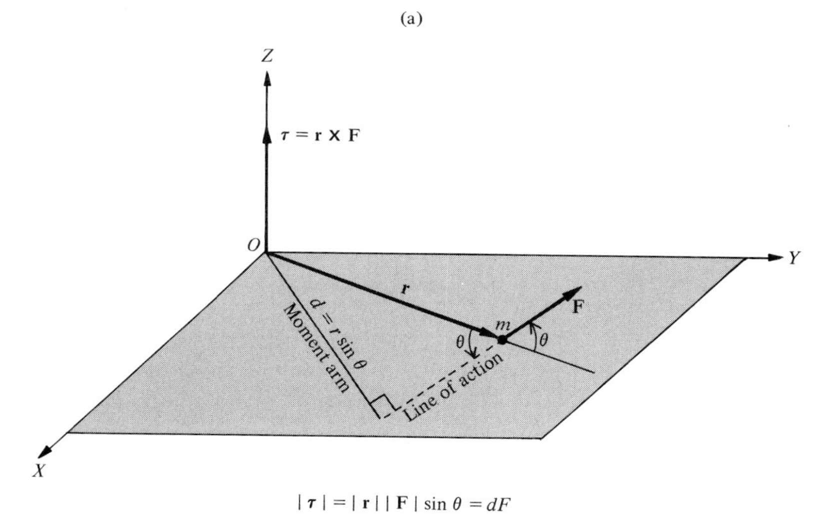

FIGURE 9.12. *(a) A particle of mass m moving with velocity* **v** *in the XY-plane under the action of a force* **F** *has a torque* **τ** *acting on it. (b) The relation among force, torque, and moment arm.*

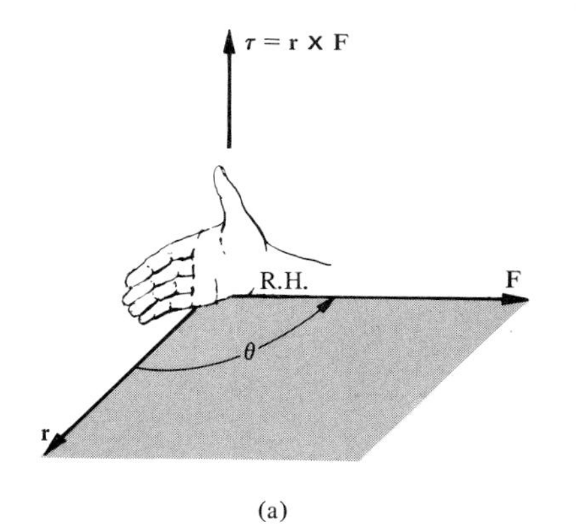

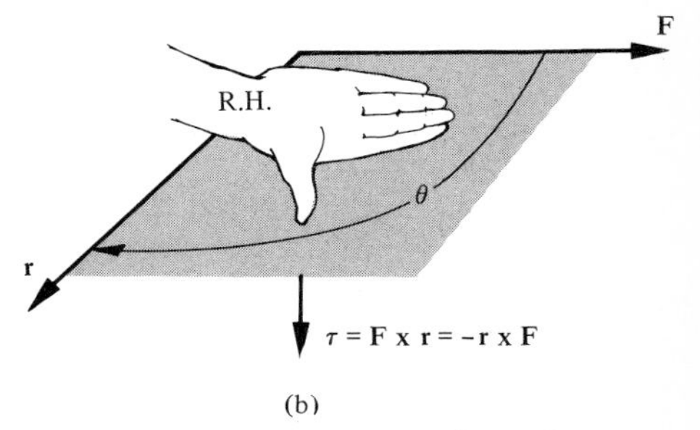

FIGURE 9.13. *Cross-product of two vectors. Note that* **r × F** = **−F × r**.

or in vector notation,

$$\boldsymbol{\tau} = mr^2\boldsymbol{\alpha} \tag{9.24}$$

It is clear from Figures 9.12 and 9.13 that $\boldsymbol{\tau}$ and $\boldsymbol{\alpha}$ are in the same direction along the Z-axis. The quantity mr^2 is defined as the moment of inertia I of point mass m about the axis through O, that is,

$$I = mr^2 \tag{9.25}$$

Also, $\boldsymbol{\alpha} = \Delta\boldsymbol{\omega}/\Delta t$, where the angular velocity is regarded as a vector quantity $\boldsymbol{\omega}$ with its direction along the axis about which it is rotating. Thus, Equation 9.24 takes the form

$$\boldsymbol{\tau} = I\boldsymbol{\alpha} = I\frac{\Delta\boldsymbol{\omega}}{\Delta t} \tag{9.26}$$

This is analogous to the equation for the translational motion,

$$\mathbf{F} = m\mathbf{a} = m\frac{\Delta\mathbf{v}}{\Delta t} \tag{9.27}$$

Note that the torque $\boldsymbol{\tau}$ in rotational motion plays an analogous role to the role the force $\mathbf{F}$ plays in translational motion. Also, comparing the equations above shows that I is analogous to m, where I is the moment of inertia for the rotational motion and m for the translational motion although the units of m and I are different.

The results obtained in Equation 9.26 can be extended to the case of a rigid body instead of a point mass. Thus, we may write

$$\boldsymbol{\tau}_{\text{ext}} = I\boldsymbol{\alpha} \tag{9.28}$$

where $\boldsymbol{\tau}_{\text{ext}}$ is the vector sum of all the torques acting on the system and I is the moment of inertia of the rigid body about the axis of rotation through O, given by

$$I = m_1r_1^2 + m_2r_2^2 + \cdots$$

where $m_1, m_2, \ldots$ are the masses of the rigid body at distances $r_1, r_2, \ldots,$ respectively, from O. Thus,

$$I = \sum_{i=1}^{n} m_i r_i^2 \tag{9.29}$$

Thus, Equation 9.28 is the desired equation for the rotational motion of the rigid body. The moment of inertia I may be calculated by using Equation 9.29. Some values of I for selected geometries are listed in Figure 9.14. Since $I = \tau/\alpha$, the *moment of inertia* I is also equal to the torque, which will produce a unit acceleration in the rigid body.

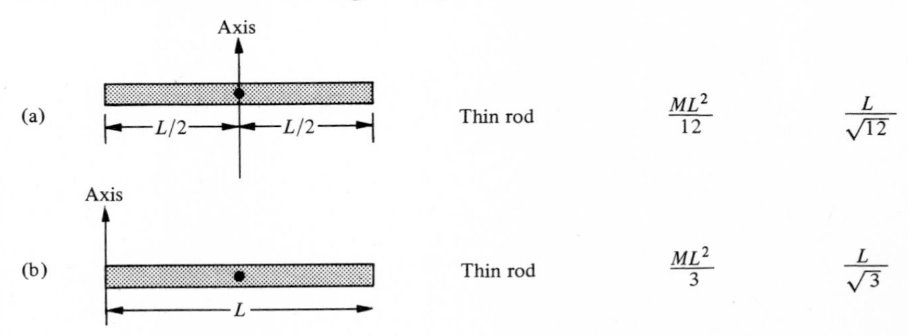

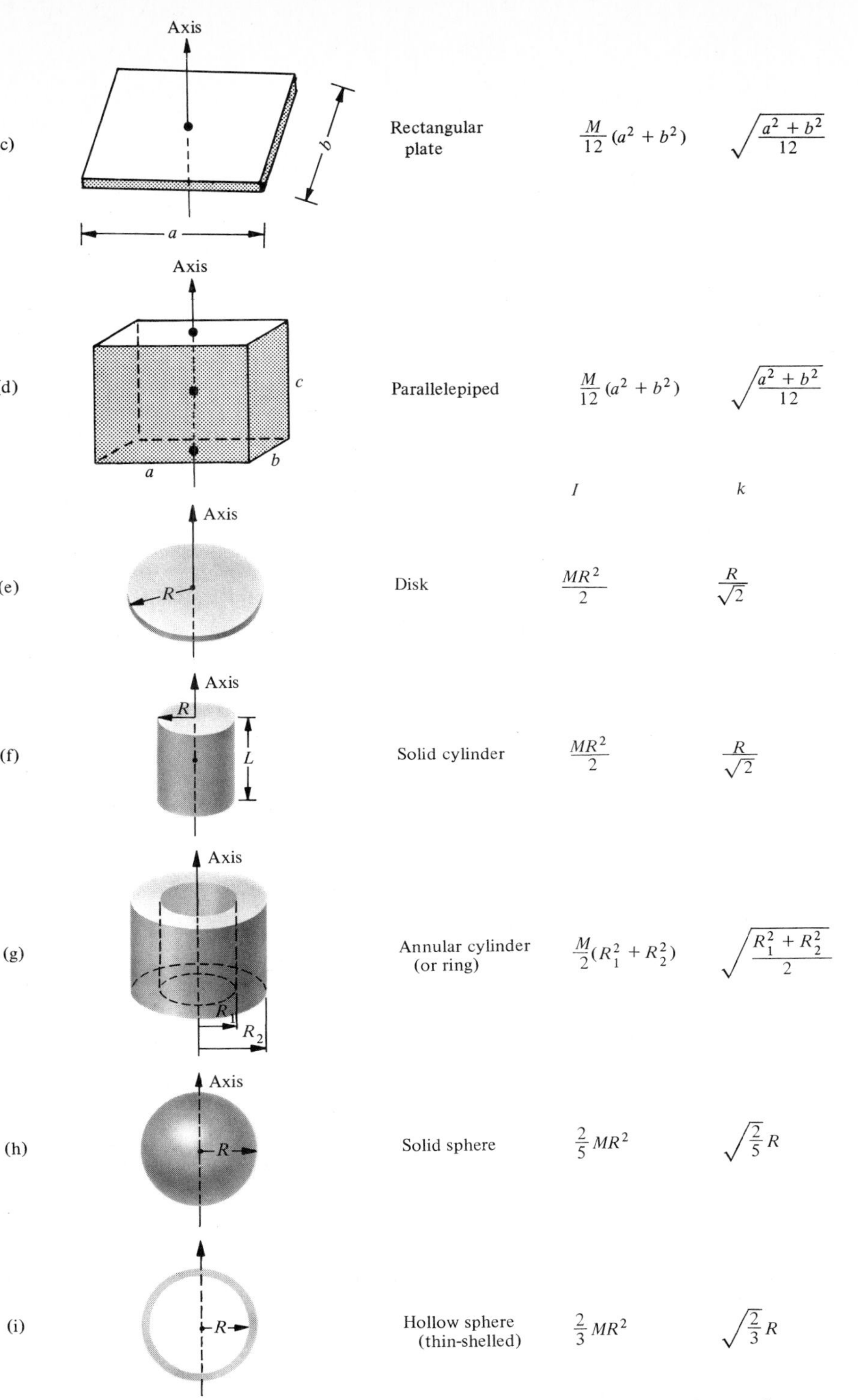

FIGURE 9.14. *Moment of inertia I, and radius of gyration k of different objects.*

A parallel-axis theorem which is very useful in many situations is stated (without proof) as follows. Suppose that a body of uniform mass density has mass M and its moment of inertia about an axis through its center of mass is I_{CM}. The moment of inertia I_A of this body about another axis A which is parallel to the axis through CM and is at a given distance d from it is given by

$$I_A = I_{CM} + Md^2 \tag{9.30}$$

In general, the moment of inertia of a body about any given axis may be written

$$I = Mk^2 \tag{9.31}$$

where M is the mass of the body and k is the *radius of gyration*. k may be found by comparing Equation 9.31 with Equation 9.29. In a way k is the effective value of r so that the whole body may be replaced by a point mass m located at $r = k$ without changing the moment of inertia of the body.

Unlike the case of a single particle, in a rigid body there are internal forces due to interactions between different particles. But internal forces come in equal and opposite pairs which are collinear; hence, the moments and torques due to these cancel each other, and need not be taken into consideration.

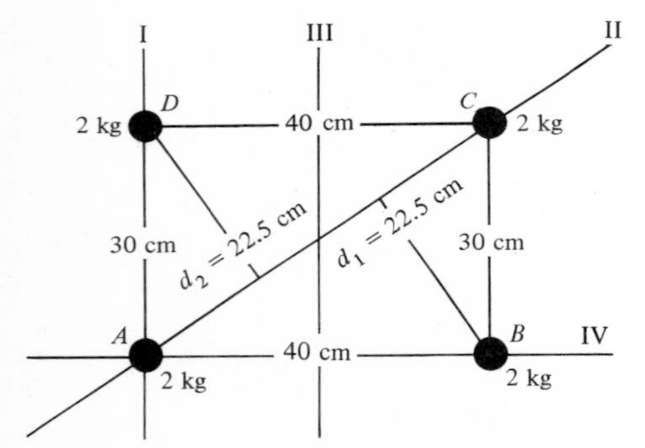

Ex. 9.4

Example 9.4 For the system of masses shown in the accompanying figure, calculate the moment of inertia about the axes I and II. Also, calculate the moment of inertia about an axis through A perpendicular to the plane of the figure.

The masses A and D about axis I make no contribution because the axis passes through them; hence, $r = 0$. Thus, from Equation 9.29,

$$\begin{aligned} I &= \sum m_i r_i^2 = m_B r_B^2 + m_C r_C^2 \\ &= (2\text{ kg})(0.4\text{ m})^2 + (2\text{ kg})(0.4\text{ m})^2 = 0.64\text{ kg-m}^2 \end{aligned}$$

The masses C and A about axis II make no contribution; hence,

$$I = m_B d_1^2 + m_D d_2^2 = (2\text{ kg})(0.225\text{ m})^2 + (2\text{·kg})(0.225\text{ m})^2 = 0.20\text{ kg-m}^2$$

The mass A about an axis through A makes no contribution. The mass C is at a distance of $AC = \sqrt{(0.4\text{ m})^2 + (0.3\text{ m})^2} = 0.5$ m. Thus,

$$\begin{aligned} I_A &= m_B r_B^2 + m_C r_C^2 + m_D r_D^2 \\ &= (2\text{ kg})(0.4\text{ m})^2 + (2\text{ kg})(0.5\text{ m})^2 + (2\text{ kg})(0.3\text{ m})^2 = 1.0\text{ kg-m}^2 \end{aligned}$$

Exercise 9.4 In the figure shown in Example 9.4, calculate the moment of inertia (a) about axis III; (b) about axis IV; and (c) about an axis through the center of the system and perpendicular to the plane of the figure. [*Ans.*: (a) 0.32 kg-m^2, (b) 0.36 kg-m^2, (c) 0.5 kg-m^2.]

EXAMPLE 9.5 Consider a system of two masses and a pulley as shown in the accompanying figure. If $m_1 = 12$ kg, $m_2 = 8$ kg, the mass of the pulley $m = 10$ kg and its radius $r = 10$ cm, calculate the tensions T_1, T_2 ($< T_1$) and acceleration a (assume the pulley to be a solid disk).

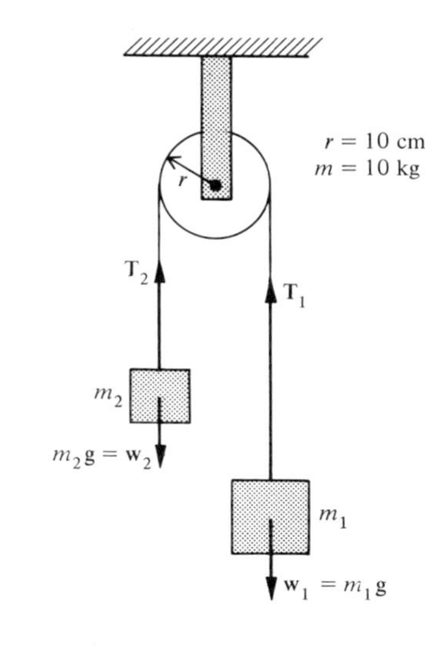

Ex. 9.5

Since the pulley has a finite mass, the two tensions T_1 and T_2 are not equal. This results in a torque $(T_1 - T_2)r$, which produces a clockwise rotation of the pulley. If I is the moment of inertia of the pulley and α is its acceleration, we may write for the motion of the pulley

$$\tau = I\alpha$$

or

$$(T_1 - T_2)r = I\alpha \qquad \text{(i)}$$

where I (for disk) $= \frac{1}{2}mr^2 = \frac{1}{2}(10 \text{ kg})(0.1 \text{ m})^2 = 0.05 \text{ kg-m}^2$ and $\alpha = a/r = a/0.1$ m.

Equations for the other two masses are (using $F = ma$)

$$m_1g - T_1 = m_1a \qquad \text{(ii)}$$

$$T_2 - m_2g = m_2a \qquad \text{(iii)}$$

Substituting the values in Equations i, ii, and iii, we get

$$(T_1 - T_2)(0.1 \text{ m}) = (0.05 \text{ kg-m}^2)\left(\frac{a}{0.1 \text{ m}}\right)$$

$$(12 \text{ kg})(9.8 \text{ m/s}^2) - T_1 = (12 \text{ kg})a$$

$$T_2 - (8 \text{ kg})(9.8 \text{ m/s}^2) = (8 \text{ kg})a$$

Solving these equations, we get $a = 1.57 \text{ m/s}^2$, $T_1 = 99$ N, and $T_2 = 91$ N.

EXERCISE 9.5 Solve Example 9.5 if the pulley were a hollow rim (instead of a disk) of the same mass and radius. [*Ans.*: $a = 1.31 \text{ m/s}^2$, $T_1 = 102$ N, and $T_2 = 89$ N.]

9.6. Rotational Kinetic Energy, Work, and Power

We have seen that for the translational quantities s, v, a, F, and m, the corresponding rotational quantities are θ, ω, α, τ, and I, respectively. Keeping these in mind, we can find expressions for kinetic energy, work, and power in rotational motion. These and other rotational analogs are listed in Table 9.2.

Suppose that a rigid body, as shown in Figure 9.15(a), is rotating about an axis with an angular velocity ω. Let us consider a particle of mass m_1 at a distance r_1 from the axis. The velocity v_1 of this particle is given by $v_1 = r_1\omega$, while its kinetic energy will be

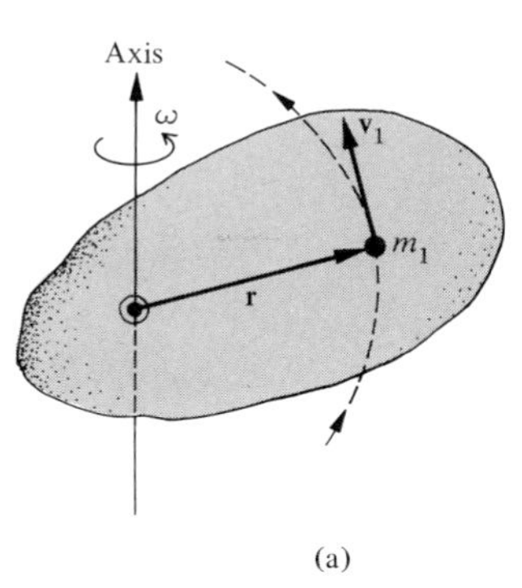

(a)

FIGURE 9.15(a). *Particle of mass m_1 in a rigid body rotating at an angular velocity ω.*

$$K_1 = \tfrac{1}{2}m_1v_1^2 = \tfrac{1}{2}m_1r_1^2\omega^2$$

The total kinetic energy of the body will be the sum of the kinetic energies of all the particles in the body. That is,

TABLE 9.2
Analogy Between Translational and Rotational Quantities

Quantity	Translation	Rotation	Relationship[a]
Displacement	s	θ	$s = r\theta$
Velocity	v	ω	$v_t = r\omega$
Acceleration	a	α	$a_t = r\alpha$
Force	F	τ	$\boldsymbol{\tau} = \mathbf{r} \times \mathbf{F}$
Mass	m	I	$I = \Sigma\, m_i r_i^2$
Newton's second law	$\mathbf{F} \begin{cases} = m\mathbf{a} \\ = \Delta\mathbf{p}/\Delta t \end{cases}$	$\boldsymbol{\tau} \begin{cases} = I\boldsymbol{\alpha} \\ = \Delta\mathbf{L}/\Delta t \end{cases}$	
Work	$\Delta W = F\,\Delta s$	$\Delta W = \tau\,\Delta\theta$	
Kinetic energy	$K_{\text{tran}} = \frac{1}{2}mv^2$	$K_{\text{rot}} = \frac{1}{2}I\omega^2$	
Power	$P = Fv$	$P = \tau\omega$	
Impulse	$F\,\Delta t$	$\tau\,\Delta t$	
Momentum	$\mathbf{p} = m\mathbf{v}$	$\mathbf{L} = I\boldsymbol{\omega}$	$\mathbf{L} = \mathbf{r} \times \mathbf{p}$

[a] Between translational and rotational quantities.

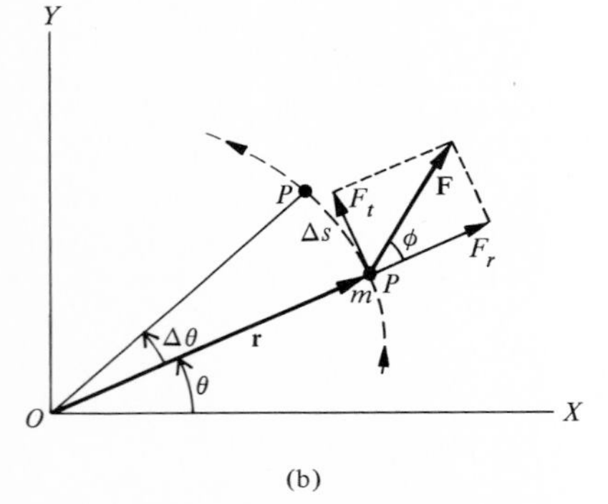

FIGURE 9.15(b). *Rotation of a rigid body through an angle $\Delta\theta$ when a force* **F** *acts on it.*

$$K_{\text{rot}} = \tfrac{1}{2}m_1r_1^2\omega^2 + \tfrac{1}{2}m_2r_2^2\omega^2 + \cdots = \tfrac{1}{2}\left(\sum m_ir_i^2\right)\omega^2$$

The quantity in parentheses is equal to the moment of inertia I about the axis of rotation. Thus, in general, the rotational kinetic energy of a body of moment of inertia I and angular velocity ω is

$$\boxed{K_{\text{rot}} = \tfrac{1}{2}I\omega^2} \tag{9.32}$$

which is analogous to the expression for translational kinetic energy $K = \frac{1}{2}mv^2$.

Let us now consider a particle of mass m at a distance r from the axis of rotation and acted upon by a force **F** as shown in Figure 9.15(b). The tangential component $F_t = F \sin \phi$ of the force F displaces the particle through an angle $\Delta\theta$ so that $\Delta s = r\,\Delta\theta$. Thus, the work done ΔW by the force F_t will be

$$\Delta W = F_t\,\Delta s = F_t r\,\Delta\theta$$

But $F_t r = \tau$; hence,

$$\Delta W = \tau\,\Delta\theta \tag{9.33}$$

or the work done when the body is displaced from θ_i to θ_f is given by

$$\boxed{W = \tau\theta} \tag{9.34}$$

where $\theta = \theta_f - \theta_i$ (which is analogous to $W = Fs$ in linear motion).

Let us see now what is the relation between work done and kinetic energy. Consider the situation shown in Figure 9.15(b). The body has a moment of inertia I about the rotation axis and angular velocity ω_i at $t = 0$. When a force F is applied, its angular velocity changes to ω_f in time t. During this time the angular acceleration is α and the body is displaced through an angle θ. Thus, from Table 9.1,

$$2\alpha\theta = \omega_f^2 - \omega_i^2$$

Substituting for $\alpha = \tau/I$ gives

$$2\frac{\tau}{I}\theta = \omega_f^2 - \omega_i^2$$

Remembering that $\tau\theta = W$ and rearranging, we get

$$\boxed{W = \tfrac{1}{2}I\omega_f^2 - \tfrac{1}{2}I\omega_i^2} \tag{9.35}$$

This is the *work–energy equation for rotational motion*, which is similar to the one in translational motion. That is, the work done in rotational motion is equal to the change in rotational kinetic energy. If the external torque acting on a body is zero, the work done, $W = \tau\theta = 0$, and from Equation 9.35, we get

$$\tfrac{1}{2}I\omega_i^2 = \tfrac{1}{2}I\omega_f^2 = \tfrac{1}{2}I\omega^2 = \text{constant} \tag{9.36}$$

which states that a body keeps on moving with constant angular velocity if no torque acts on it.

Let us go back to Equation 9.33 and divide both sides by Δt. We obtain the rate at which the force F does work, which is the power P. That is,

$$P = \frac{\Delta W}{\Delta t} = \frac{\tau\,\Delta\theta}{\Delta t}$$

Since instantaneous angular velocity $\omega = \Delta\theta/\Delta t$,

$$\boxed{P = \tau\omega} \tag{9.37}$$

This equation is analogous to $P = Fv$ for the linear case.

Of course, pure translational or pure rotational motion is quite common, but there are many cases in which the motion is a combination of both. We illustrate this in the following examples.

EXAMPLE 9.6 A body with a moment of inertia 4 kg-m² is rotating with an angular velocity of 5 rad/s when a torque of 10 N-m is applied for 10 s. Calculate (a) the final angular velocity; (b) the angular displacement; and (c) the change in rotational kinetic energy in this time interval.

From Equation 9.26,

$$\tau = I\alpha \qquad \text{or} \qquad \alpha = \frac{\tau}{I} = \frac{10\text{ N-m}}{4\text{ kg-m}^2} = \tfrac{5}{2}\text{ rad/s}^2$$

$$\omega = \omega_0 + \alpha t = (5\text{ rad/s}) + (\tfrac{5}{2}\text{ rad/s}^2)(10\text{ s}) = 30\text{ rad/s}$$

$$\theta = \omega_0 t + \tfrac{1}{2}\alpha t^2 = (5\text{ rad/s})(10\text{ s}) + \tfrac{1}{2}(\tfrac{5}{2}\text{ rad/s}^2)(10\text{ s})^2 = 175\text{ rad}$$

$$K_i = \tfrac{1}{2}I\omega_i^2 = \tfrac{1}{2}(4\text{ kg-m}^2)(5\text{ rad/s})^2 = 50\text{ J}$$

$$K_f = \tfrac{1}{2}I\omega_f^2 = \tfrac{1}{2}(4\text{ kg-m}^2)(30\text{ rad/s})^2 = 1800\text{ J}$$

or

$$\Delta K = K_f - K_i = 1800 \text{ J} - 50 \text{ J} = 1750 \text{ J}$$

This value agrees with the work done by the torque. That is,

$$W = \tau\theta = (10 \text{ N-m})(175 \text{ rad}) = 1750 \text{ J}$$

EXERCISE 9.6 A flywheel of moment of inertia 8 kg-m^2 is rotating with an angular velocity of 600 rev/min. What opposing torque should be applied so that it will come to a stop in 30 s? What is the change in the rotational kinetic energy? [*Ans.*: $\tau = -16.8$ N-m; $\Delta K = -15{,}775$ J.]

EXAMPLE 9.7 A solid sphere, a solid cylinder, and a ring all having the same mass M and radius R, start from rest and roll down the same inclined plane of height h. In each case find the velocity with which they arrive at the base of the plane. Which body gets to the bottom first?

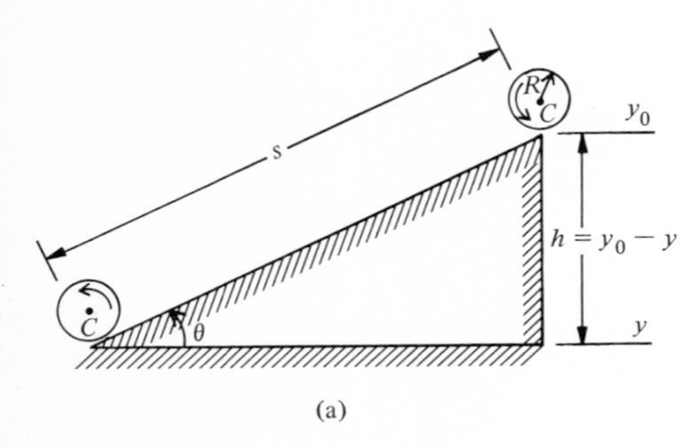

(a)

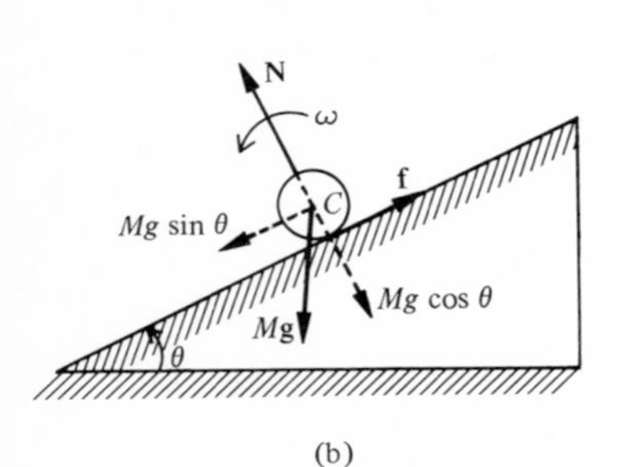

(b)

Ex. 9.7

We can solve this problem by two different methods, either by the energy conservation principle or the dynamical method.

THE ENERGY CONSERVATION PRINCIPLE. Referring to part (a) of the accompanying figure, any one of the three objects rolling down the incline will lose potential energy but will gain translational and rotational kinetic energies. To start with, the initial energy is all potential $E_i = Mgy_0$. let us say that at the bottom the linear velocity is v whereas the angular velocity is ω. Thus, the total energy at the bottom is kinetic and potential,

$$E_f = \tfrac{1}{2}Mv^2 + \tfrac{1}{2}I_c\omega^2 + Mgy$$

where I_c is the moment of inertia about the center of mass. Thus, we may write (frictional forces supply the necessary torque to make the object roll)

$$E_i = E_f$$

$$Mgy_0 = \tfrac{1}{2}Mv^2 + \tfrac{1}{2}I_c\omega^2 + Mgy \quad \text{(i)}$$

Substituting for $\omega = v/R$ and $I_c = Mk^2$, where k is the radius of gyration, after rearranging we get

$$Mg(y_0 - y) = \tfrac{1}{2}M\left(1 + \frac{k^2}{R^2}\right)v^2$$

or

$$v^2 = \frac{2g(y_0 - y)}{1 + (k^2/R^2)} \quad \text{(ii)}$$

where $y_0 - y = h$.

Thus, the velocity of an object rolling down does not depend on the mass of the object, but on mass distribution. That is, using Figure 9.14,

For a sphere: $k^2 = \frac{2}{5}R^2$; hence, $v_s = \sqrt{\frac{10}{7}gh}$

For a cylinder: $k^2 = \frac{1}{2}R^2$; hence, $v_c = \sqrt{\frac{4}{3}gh}$ (iii)

For a ring: $k^2 = R^2$; hence, $v_r = \sqrt{gh}$

That is, $v_s > v_c > v_r$; hence, the sphere translates the fastest, then the cylinder, and the ring the slowest. If the body were not rolling, but only sliding, the final velocity in each case would have been $\sqrt{2gh}$ (the same as for a particle that has a free fall through height h). Thus, the effect of the rotational motion is to slow down the translational motion. This is true because the body has a given amount of initial potential energy which is used in producing both translational as well as rotational kinetic energy.

The Dynamical Method. At any instant, the forces acting on the object are its weight $M\mathbf{g}$, the normal reaction $\mathbf{N}$, and the force of static friction $\mathbf{f}$, as shown in part (b) of the figure. Using Newton's second law, we may write the equations of the translational motion of a body as

$$\begin{aligned} Mg\sin\theta - f &= Ma \qquad \text{for motion along the inclined plane} \\ N - Mg\cos\theta &= 0 \qquad \text{for motion normal to the plane} \end{aligned} \tag{iv}$$

where a is the linear acceleration.

For rotational motion about the center of mass,

$$\sum \boldsymbol{\tau} = I_c\boldsymbol{\alpha}$$

The forces $\mathbf{N}$ and $M\mathbf{g}$ have a zero moment arm because their lines of actions pass through C; hence, they do not produce any rotation about C. The force of friction f has a moment arm R about C, hence, will produce a torque fR. Thus, the equation above may be written

$$fR = I_c\alpha$$

Substituting for $I_c = Mk^2$ and $\alpha = a/R$, we get

$$f = \frac{I_c\alpha}{R} = \frac{Mk^2a}{R^2} \tag{v}$$

Using this in Equation iv and solving for a,

$$a = \frac{g\sin\theta}{1 + (k^2/R^2)} \tag{vi}$$

The speed of the center of mass of the object may be obtained using the relation

$$v^2 = 2as$$

Substituting for a from vi and $\sin\theta = (y_0 - y)/s$, we get

$$v^2 = \frac{2g(y_0 - y)}{1 + (k^2/R^2)} \tag{vii}$$

which is the same equation as we obtained using the energy conservation principle.

EXERCISE 9.7 Repeat Example 9.7 for a hollow sphere (or spherical shell) and a hollow cylinder having the same mass M and radius R, [*Ans.*: $v_s = \sqrt{\frac{6}{5}gh}$, $v_c = \sqrt{gh}$, $v_s > v_c$]

9.7. Angular Momentum

Of the three most important concepts in physics, linear momentum and energy have already been discussed in detail while angular momentum will be introduced in this section. The angular momentum is defined in such a way that the analogy between different quantities in translational and rotational motions is maintained.

Let us consider a particle of mass m moving in the XY-plane with velocity $\mathbf{v}$ and linear momentum $\mathbf{p} = m\mathbf{v}$ at a distance $\mathbf{r}$ from the origin, as shown in Figure 9.16. The angular momentum of the particle about an axis through O and perpendicular to the XY-plane is a vector denoted by $\mathbf{L}$. The magnitude of $\mathbf{L}$ is given by the product of r and the component of $\mathbf{p}$ perpendicular to $\mathbf{r}$. That is, as shown in Figure 9.16(a),

$$L = rp_{\perp} = rp \sin\theta = rmv \sin\theta \tag{9.38}$$

Another way of defining the magnitude of vector $\mathbf{L}$ is shown in Figure 9.16(b). In this case L is equal to the product of the perpendicular distance from O to p, $r_{\perp}$ and the momentum p. That is,

$$L = r_{\perp}p = bp = r\sin\theta\, p = rmv \sin\theta \tag{9.39}$$

The perpendicular distance $b(= r_{\perp} = r\sin\theta)$ is the moment arm.

In terms of the vector cross product, the angular momentum $\mathbf{L}$ is given as

$$\boxed{\mathbf{L} = \mathbf{r} \times \mathbf{p}} \tag{9.40}$$

which means the magnitude of $\mathbf{L}$ is $|\mathbf{L}| = |\mathbf{r}|\,|\mathbf{p}| \sin\theta = rp\sin\theta$, while the direction of $\mathbf{L}$ is given by using the right-hand rule, as explained in Section 9.5. That is, if we grasp the Z-axis with the right hand and curl our fingers from $\mathbf{r}$ toward $\mathbf{p}$ through the smallest angle, the thumb will be in the direction of $\mathbf{L}$. Note from this that a counterclockwise rotation results in a +Z-direction; hence, it is positive, while a clockwise rotation results in a −Z-direction; hence, it is negative.

Examples of objects having angular momentum include the earth rotating about its axis, a satellite going around the earth, the wheels of a car rotating on their axles, and a car going around a curved path. There are numerous other examples.

Analogous to Newton's second law in translational motion, $\mathbf{F} = \Delta\mathbf{p}/\Delta t$, in rotational motion we may write

$$\boxed{\tau = \frac{\Delta\mathbf{L}}{\Delta t}} \tag{9.41}$$

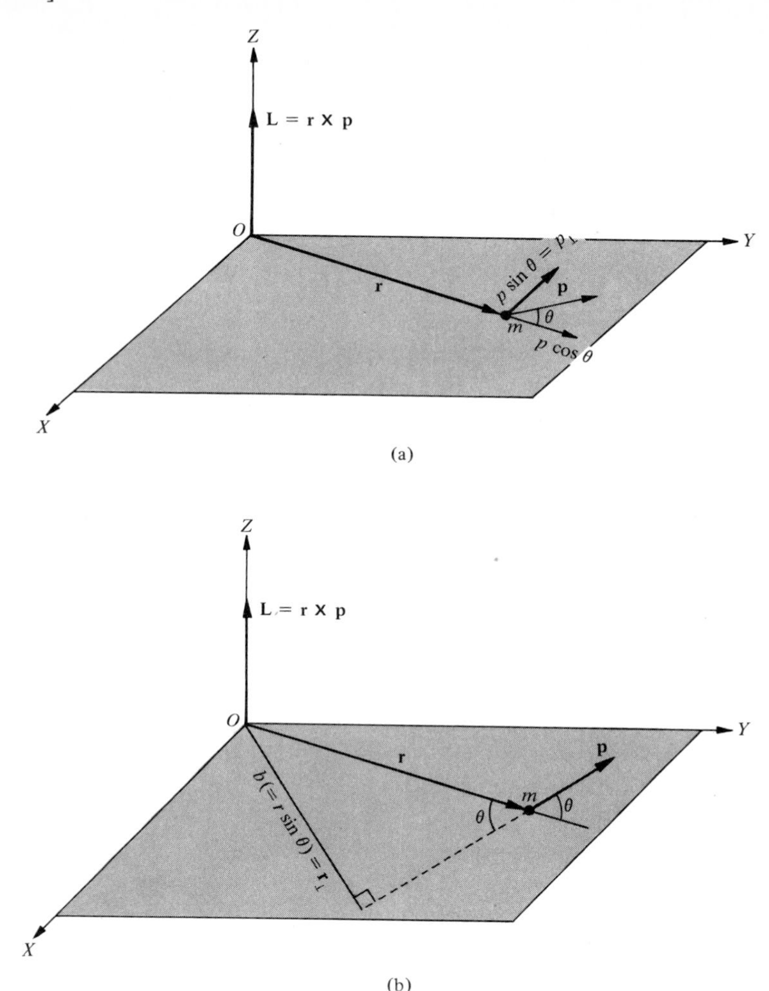

FIGURE 9.16. *(a) Angular momentum* **L** *of a particle of mass m and linear momentum* **p** *at a distance* **r** *from the axis. (b) Relation among* **L, r, p,** *and moment arm b.*

which states that *the time rate of change of angular momentum is equal to the applied torque*. In analogy with $\mathbf{p} = m\mathbf{v}$ we may write

$$\boxed{\mathbf{L} = I\omega} \tag{9.42}$$

Just like the laws of conservation of linear momentum and energy, the law of conservation of angular momentum is one of the most basic laws of physics. If the torque acting on a particle or system of particles is zero, from Equation 9.41 $\tau = 0$ and $\Delta\mathbf{L}/\Delta t = 0$; hence, $\mathbf{L}$ is constant. This leads to the following statement for the

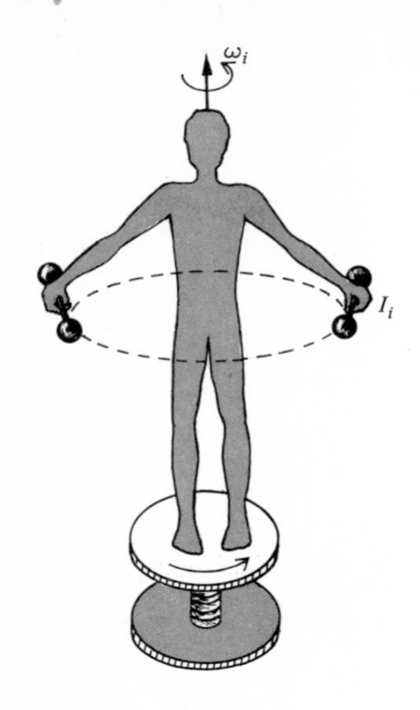

(a)

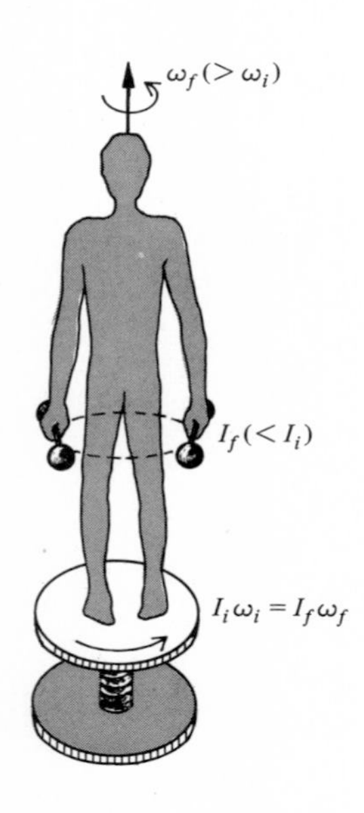

(b)

FIGURE 9.17. *Man with masses in his hands on a rotating platform. Conservation of angular momentum requires that as the man pulls his arms in, the angular velocity must increase.*

Law of Conservation of Angular Momentum: ***If the net external torque acting on a system about a given axis is zero, the angular momentum about that axis will be constant.***

In other words,

$$\mathbf{L} = \text{constant} \qquad \text{if } \boldsymbol{\tau} = 0 \tag{9.43}$$

$\mathbf{L}$ = constant means that if a system has initial angular moment $\mathbf{L}_i$ and if there is a rearrangement of the system without the application of any external torque, the final angular momentum $\mathbf{L}_f$ of the system must satisfy the condition

$$\boxed{\mathbf{L}_i = \mathbf{L}_f \qquad \text{if } \boldsymbol{\tau} = 0} \tag{9.44}$$

Let us now consider a rigid body rotating about the Z-axis in a given inertial reference frame. Thus, using Equation 9.42, we may write

$$\mathbf{L}_z = I\boldsymbol{\omega} \tag{9.45}$$

where $\mathbf{L}_z$ is the component of angular momentum, I the moment of inertia, and $\boldsymbol{\omega}$ the angular velocity, all being along the Z-axis. If the internal parts of the body rearrange themselves, without the application of any external torque the moment of inertia will change. Hence, the angular velocity will have to change so that the angular momentum of the system will remain constant. That is,

$$I_i\omega_i = I_f\omega_f \qquad \text{if } \boldsymbol{\tau}_z = 0 \tag{9.46}$$

where subscripts i and f refer to the initial and final quantities, respectively.

There are numerous examples both in microscopic and macroscopic physics that demonstrate the use and application of the law of conservation of angular momentum. A diver, a circus acrobat, an ice skater, adjusting the spinning motion of a spacecraft—all make use of the principle of the conservation of angular momentum. We shall discuss some of these examples briefly.

When an acrobat just leaves a swing her legs and arms are stretched out and she has a small counterclockwise angular momentum. By pulling her legs and arms in, she decreases her moment of inertia. Her angular velocity must increase in order to conserve angular momentum, since there is no external torque that acts on her. By increasing ω the acrobat can perform a number of somersaults before landing on the net. Before falling on the net, she stretches out again so as to slow down her angular motion.

A skater performing a pirouette on the toe of one skate makes use of the conservation of angular momentum. The situation of a skater is very similar to a man holding two weights in his hands and standing on a rotating table, as shown in Figure 9.17. Initially, when his hands are stretched out, the moment of inertia of the system (platform + man + weights) is I_i and the angular velocity is ω_i, as shown in Figure 9.17(a). Thus, the total angular momentum is $I_i\omega_i$. As the man pulls the weights in, the moment of inertia decreases to I_f; hence, the angular velocity must increase to ω_f so that $I_i\omega_i = I_f\omega_f$, where $I_f < I_i$; hence, $\omega_f > \omega_i$.

A helicopter is provided with a small propeller in its tail. As the rotor of a helicopter starts, the body of the helicopter has a tendency to rotate in the

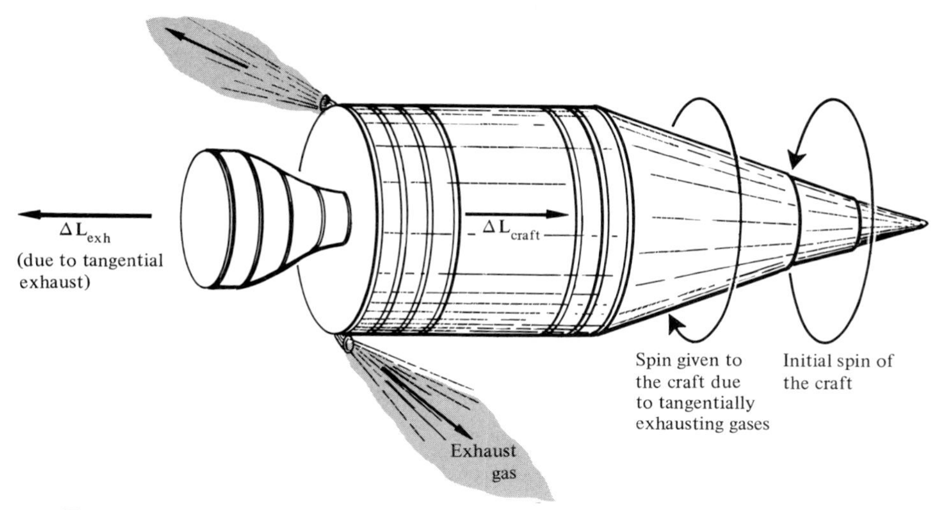

FIGURE 9.18. *The spinning motion of a spacecraft is corrected by means of tangential exhausts.*

opposite direction. This tendency is counterbalanced by the small propeller in the tail.

As another example, suppose that a spacecraft in outer space starts spinning. There is nothing that can be done internally in a spacecraft to stop this motion. But by throwing away mass in the correct manner the rotational motion can be changed. For this purpose the spacecraft is equipped with two tangential nozzles, as shown in Figure 9.18. The exhaust from these nozzles will have angular momentum with respect to the center of the spacecraft. Since there is no external torque acting on the spacecraft, in order to conserve angular momentum the spacecraft and the remaining fuel acquire equal and opposite angular momentum, as shown. That is, as in Figure 9.18,

$$(\Delta\mathbf{L})_{\text{exhaust}} + (\Delta\mathbf{L})_{\text{craft}} = 0 \tag{9.47}$$

Once the spinning motion is corrected, the burning of the fuel is stopped and the exhausts are closed.

EXAMPLE 9.8 A flywheel, when slowed down from 60 rev/min to 30 rev/min, loses 100 J of energy. What is its moment of inertia? What is the change in the angular momentum?

From Equation 9.35, $W = \frac{1}{2}I\omega_f^2 - \frac{1}{2}I\omega_i^2$, where $W = -100$ J.

$$\omega_i = 60 \text{ rev/min} = \frac{(60)(2\pi)}{60}\text{rad/s} = 2\pi \text{ rad/s}$$

$$\omega_f = 30 \text{ rev/min} = \frac{(30)(2\pi)}{60}\text{rad/s} = \pi \text{ rad/s}$$

Therefore, substituting these values.

$$-100 \text{ J} = \tfrac{1}{2}I(\pi \text{ rad/s})^2 - \tfrac{1}{2}I(2\pi \text{ rad/s})^2$$

$$I = \frac{(2)(100)}{3\pi^2} \text{kg-m}^2 = 6.75 \text{ kg-m}^2$$

The change in the angular momentum is

$$\Delta L = I_f\omega_f - I_i\omega_i = I(\omega_f - \omega_i) = \left(\frac{200}{3\pi^2}\right)(\pi - 2\pi)\text{rad/s} = -21.2 \text{ kg-m}^2/\text{s}$$

EXERCISE 9.7 How much energy is needed to speed up a flywheel from 120 rev/min to 240 rev/min if the moment of inertia of the flywheel is 40 kg-m^2? Also calculate the change in angular momentum. [*Ans.*: $E = 9475$ J, $\Delta L = 503$ kg-m^2/s.]

*9.8 Rotation About a Moving Axis (Spinning Tops and Gyroscopes)

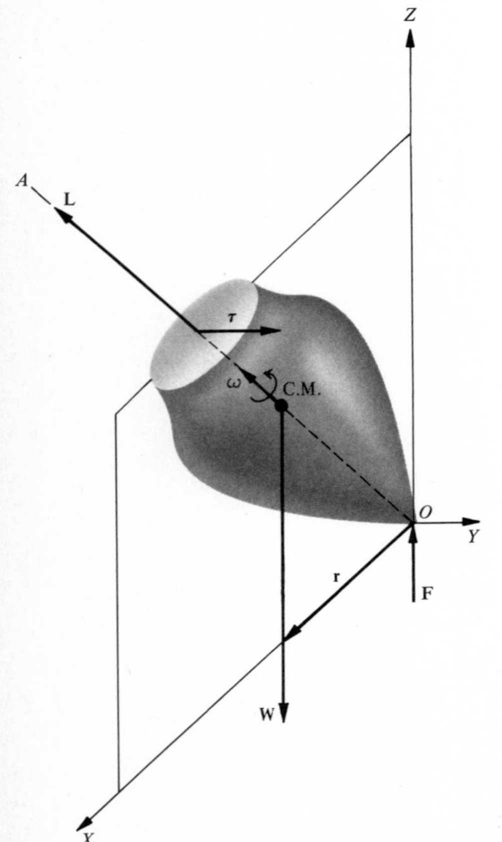

FIGURE 9.19. *Forces and torques acting on a toy gyroscope or top.*

Figure 9.19 shows an object that is not pivoted about its center of mass, but about some other fixed point at O. Ordinarily, such an object will fall in such a way that the center of mass lies directly below the pivot point. On the contrary, the object shown in Figure 9.19 does not fall because it is spinning swiftly about axis OA. *Such an object mounted at a point other than the center of mass and spinning swiftly is a toy gyroscope, more properly called a top.* Let us see why the spinning top does not fall.

Referring to Figure 9.19, we see that the top is pivoted at the origin of the coordinates. There are two forces that act on the top: the weight **W** of the top acting downward through the center of mass, and the reaction force **F** at the point of pivot. The force **F** does not produce any torque about the pivot. The only external torque is that due to the weight **W** and is equal to

$$\tau = Wr \tag{9.48}$$

The direction of τ is along the Y-axis as shown. Ordinarily, this torque will make the top fall, but the angular momentum $\mathbf{L} = I\boldsymbol{\omega}$ (also called the *spinning angular momentum*) of the top is very large and hence a much larger torque than $\boldsymbol{\tau}$ is required to make the top fall.

The torque $\boldsymbol{\tau}$ acting for a time Δt produces a change in angular momentum $\Delta\mathbf{L}$ given by (Figure 9.20)

$$\Delta\mathbf{L} = \boldsymbol{\tau}\,\Delta t \tag{9.49}$$

The direction of $\boldsymbol{\tau}$, as well as that of $\Delta\mathbf{L}$, is perpendicular to the XZ-plane and also perpendicular to **L**. If the top were not spinning, $\mathbf{L} = 0$, the change $\Delta\mathbf{L}$ would make the top fall toward the X-axis.

For a spinning top with an angular momentum **L**, we must add $\Delta\mathbf{L}$ so that the resultant angular momentum is $\mathbf{L} + \Delta\mathbf{L}$, as shown in Figure 9.20. Since $\mathbf{L} >> \Delta\mathbf{L}$, and $\Delta\mathbf{L}$ is always perpendicular to **L**, for all practical purposes the new angular momentum vector has the same magnitude as the old but has different direction. The vector **L** has moved through an angle $\Delta\phi$ about the Z-axis as shown. The direction of $\boldsymbol{\tau}$ is always perpendicular to the plane defined by the axis of the top and the Z-axis. The successive application of $\boldsymbol{\tau}$ over intervals of time causes the repetition of the picture shown in Figure 9.20. The net result is that the **L** vector rotates around the Z-axis on the surface of a cone whose axis is along the Z-axis. For the situation discussed here, the angular momentum vector **L** coincides with the axis of the top; just like **L**, the top itself rotates around the Z-axis. This motion of the top is called *precession* and the top is said to precess about the Z-axis.

In actual practice the motion of the top is much more complicated than stated above. The parameters are hard to adjust so as to have smooth motion. The top may wobble and wander around slightly about the circular path. Depending upon the conditions, the tip of the angular momentum vector follows one of the few paths shown by heavy lines in Figure 9.21.

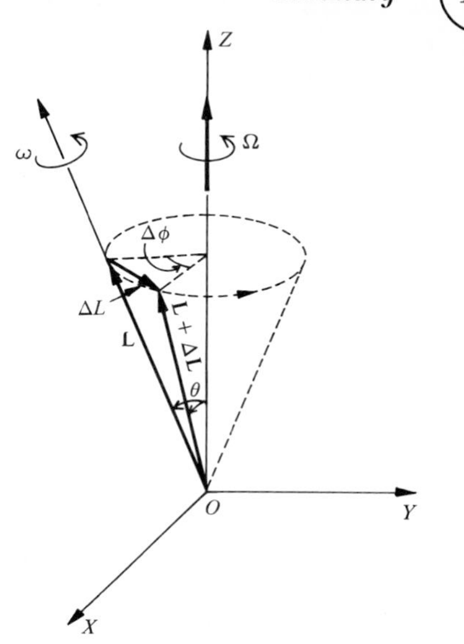

FIGURE 9.20. *Principle of a gyroscope or top.*

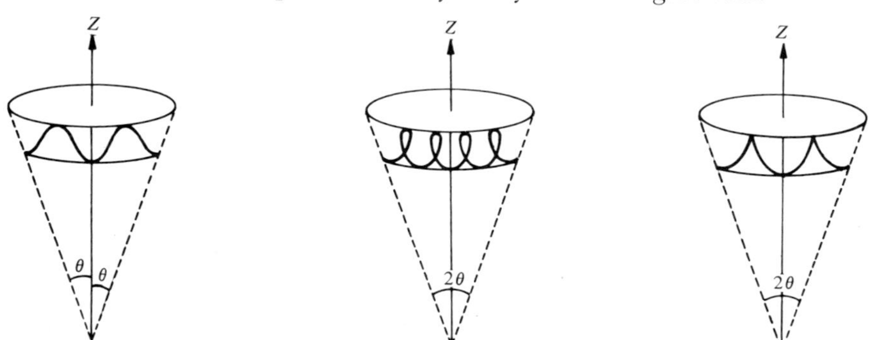

FIGURE 9.21. *The motion of a top usually departs from the circular path. Three such paths are shown here by boldface curves.*

SUMMARY

A rigid body is said to have *rotational motion* if all particles of the body move in circles with their centers on a straight line called the *axis of rotation*. One *radian* is the angle subtended by an arc length equal to the radius of the circle; 2π radians $= 6.28$ rad $= 360°$ $= 1$ rev. There is an analogy between the translational and rotational quantities, as well as the equations of motion (Tables 9.1 and 9.2).

Acceleration of a particle moving on a curved path can be resolved into two components—the *tangential component* a_t, which is due to the *tangential force component* F_t, and a_r, which is due to the *radial force component* F_r. a_t is due to the change in the magnitude of **v**, and a_r is due to the change in the direction of **v**. Also, $v_t = r\omega$ while $a_t = r\alpha$.

A particle moving in a circle with uniform speed has *centripetal acceleration* $a_r = v^2/R = R\omega^2$ and *centripetal force* $F_r = mv^2/R = mR\omega^2$. For a banked road, $\tan\theta = v^2/Rg$. For motion in a vertical circle, the minimum velocity is $v_o = \sqrt{Rg}$.

The torque that produces a rotational motion is $\tau = rF_t = rF\sin\theta = dF$ or, in vector notation, $\boldsymbol{\tau} = \mathbf{r} \times \mathbf{F}$. Also, $\tau = I\alpha$, where $I = \Sigma m_i r_i^2$ and is called the *moment of inertia*. In addition, $I = Mk^2$, where k is the radius of gyration. From the *parallel-axis theorem*, $I_A = I_{cm} + Md^2$.

Rotational kinetic energy is $K_{\text{rot}} = \frac{1}{2}I\omega^2$. Work done $W = \tau\theta$ and power is $P = \tau\omega$. The work–energy equation is $W = \frac{1}{2}I\omega_f^2 - \frac{1}{2}I\omega_i^2$.

Angular momentum is defined as $L = rp_\perp = r_\perp p = rp \sin\theta = rmv \sin\theta$, or in vector notation, $\mathbf{L} = \mathbf{r} \times \mathbf{p}$. Also, $\boldsymbol{\tau} = \Delta\mathbf{L}/\Delta t$, and $\mathbf{L} = I\boldsymbol{\omega}$. If $\boldsymbol{\tau} = 0$, $\mathbf{L}_i = \mathbf{L}_f$, leading to conservation of angular momentum. Tops and gyroscopes are examples of rotation about a moving axis.

QUESTIONS

1. In Figure 9.1, when the body is rotating about ZZ', the path of point P is a circle, as shown. If, in addition, ZZ' has a translational motion along the Z-axis, what will be the path of point P?
2. What are the advantages of using a_r and a_t as components instead of a_x and a_y?
3. A wheel is rotating with constant angular velocity. What types of acceleration will a point on the rim have?
4. A wheel is rotating with constant angular acceleration. What types of acceleration will a point on the rim have?
5. Is there a certain minimum angle for an incline so that a sphere sitting on the surface will roll without slipping? (It is the frictional torque that produces rotation.)
6. Is it possible for a body to accelerate (or decelerate) while having a constant speed?
7. The earth is being pulled by the sun by the gravitational force. Why doesn't it fall into the sun? Explain.
8. Why is it necessary to have frictional force to produce rotation without slipping?
9. Which has the greatest moment of inertia about an axis through the center of mass, if all have the same mass: (1) a ring of radius 10 cm, (2) a flat disk of radius 10 cm, (3) a hollow sphere of radius 10 cm, or (4) a solid sphere of radius 10 cm?
10. Two idential objects are released from the top of an inclined plane with one sliding down without rotating while the other is rotating. Why does the rotating object take longer to reach the bottom?
11. It is well known that the angular velocity of the rotation of the earth is slowly decreasing—the negative angular acceleration is 10^{-3}/s-century. What is the source of the torque that causes this slowing down?
12. You are standing on a platform that is rotating with a constant angular velocity. Are you in equilibrium with the platform? If you jump off the platform, will the angular velocity of the platform increase or decrease? Why? What happens to you if the platform stops suddenly while you are standing on it?
13. For a diver to make two or three turns while in the air, he must pull his legs inward. Why?

PROBLEMS

1. A bicycle wheel has a radius of 0.40 m and a point on its rim has a velocity of 6 m/s. What is the angular velocity?
2. An automobile wheel has a radius of 14 in. What is the angular velocity about its axle when the car is running at 50 mi/h?

3. A phonograph turntable has an angular velocity of (a) 45 rev/min; and (b) 78 rev/min. Calculate these in rad/s and deg/s.
4. A fan rotates at an angular speed of 1200 rev/min. Calculate the angle (in degrees) through which it turns in 2 s. What time does it take to turn through one complete revolution?
5. A wheel of radius 0.5 m turns 120 rev/min. Find the angular and linear velocity of a point at the outer edge and at a point halfway from the center to the outer edge. Calculate the distance traveled in degrees and radians in 1 min in each case.
6. The angular velocity of a point on a disk is 2 rev/s, and after 10 s it is 5 rev/s. Calculate the angular acceleration in rad/s^2.
7. A wheel of 0.5 m diameter starting from rest attains an angular velocity of 36 rad/s in 8 s. Calculate the angular acceleration and the angular distance through which it turns in this time. What is the linear velocity of a point on the rim?
8. An exhaust fan has an angular velocity of 900 rev/min. After the power is cut off, it comes to a stop in 30 s. Calculate the acceleration and the angular distance traveled in this time.
9. Calculate the angular acceleration of a phonograph turntable if, starting from rest, the table reaches an angular velocity of 45 rev/min in one-half a revolution.
10. The flywheel of an automobile has a radius of 20 cm. Starting from rest it reaches an angular velocity of 2000 rev/s in 6 s. Calculate (a) the angular acceleration; (b) the tangential acceleration during the acceleration; and (c) the radial acceleration when rotating at 2000 rev/s.
11. At a given instant a wheel of 60-cm radius has an angular velocity of 4 rad/s and an angular acceleration of 2 rad/s^2. Calculate for a point on the rim: (a) the tangential velocity; (b) the tangential acceleration; (c) the radial (or centripetal) acceleration; and (d) the total acceleration.
12. A wheel of radius 40 cm is rotating about an axis through its center with initial angular velocity of 4 rad/s and a uniform angular acceleration of 2 rad/s^2. After 10 s, calculate the following: (a) the angular velocity; (b) the tangential velocity; (c) the radial and tangential acceleration; (d) the resultant acceleration; and (e) the total angular displacement in degrees and revolutions.
13. A car starts from rest on a curved circular path of radius 800 ft and accelerates at 8 ft/s^2. Calculate the speed at which the centripetal acceleration is equal to the tangential acceleration.
14. The radius of the earth is 6370 km and it rotates about its axis once each sidereal day (= 86,164 s). Calculate the angular velocity in rad/s. Also calculate the linear velocity and the centripetal acceleration of a point on the equator.
15. A car goes around a level curve of radius 100 m at a speed of 15 m/s. What should be the minimum coefficient of friction so that the car would be able to negotiate the curve?
16. What is the safe speed for a car to go around a curve of radius 80 m if the banking angle is 25°?
17. If the car has to go around a curve of radius 250 ft at a speed of 40 mi/h, what should be the banking angle?
18. The coefficient of friction between a bicycle tire and the road is 0.5. Will a

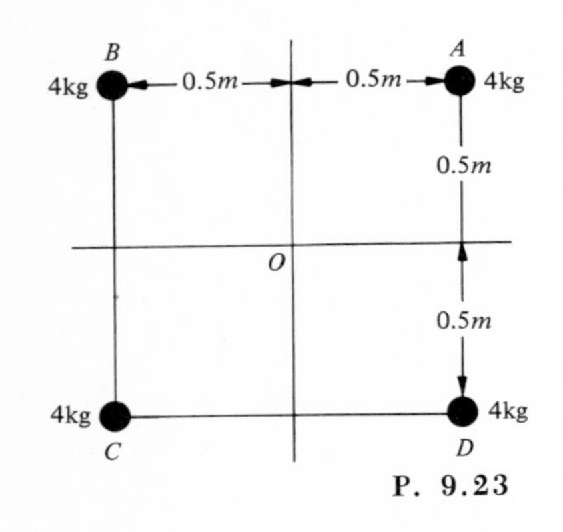

P. 9.23

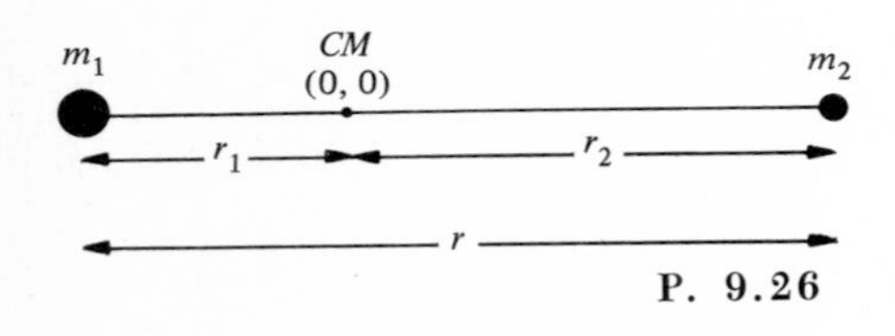

P. 9.26

bicycle rider with a mass of 90 kg going at a speed of 15 m/s be able to negotiate a level curve of radius 100 m? Also, calculate the angle at which the bicycle must be inclined to avoid slipping. (*Hint:* $\tan\theta = f/N$.)

19. A 500-g mass tied with a string is moving in a horizontal circle of radius 1 m and has a uniform angular velocity of 300 rev/min. Calculate (a) the linear velocity; (b) the centripetal acceleration; (c) the centripetal force, and (d) the tension in the string.

20. A force of 10^{-12} N is needed to keep a proton (mass $= 1.67 \times 10^{-27}$ kg) moving in a circle of radius 1 m. What is the velocity of the proton and the direction of the force?

21. A 100 g mass is going in a vertical circle of radius 1 m. What is the minimum speed so that the mass will reach the top of the circle? What is the speed when it is at the bottom?

22. A mass of 500 g is going in a vertical circle of radius 80 cm so that the velocity at the top is 1 m/s. What is the tension and the centripetal force at the top and bottom?

23. For the four masses shown, calculate the moment of inertia about (a) an axis through the center; and (b) an axis through one corner, both perpendicular to the plane of the square.

24. Consider a meter rod of negligible mass with a 2-kg mass at $x = 20$ cm, a 5-kg mass at 50 cm, and an 8-kg mass at 80 cm. Find the moment of inertia about an axis perpendicular to the rod through (a) $x = 0$; and (b) $x = 50$ cm.

25. The moment of inertia of a thin rod of length L and mass M through its center of mass is $ML^2/12$. What is the moment of inertia about an axis through one end? What is the radius of gyration in each case?

26. Show that the moment of inertia of two masses about the center of mass shown is $I = [m_1m_2/(m_1 + m_2)]x^2$. (*Hint:* from CM definition $m_1r_1 = m_2r_2$)

27. What torque will produce an acceleration of 2 rad/s^2 in a body of moment of inertia 500 kg-m^2?

28. A disk of mass 10 kg and radius 20 cm is free to rotate about an axis through its center and perpendicular to the disk. If a force of 50 N is applied tangentially, calculate the angular acceleration.

29. A shaft has a mass of 200 kg and a radius of gyration 1.2 m. If a torque of 1000 N-m is applied when the shaft is at rest, calculate the acceleration and the angular velocity after 5 s.

30. A rotating flat disk has a radius of 30 cm and a mass of 5 kg. A force of 100 N is applied tangentially to the rim of the disk. Calculate the angular acceleration and the angular velocity after (a) 1 s; and (b) 1 revolution.

31. Calculate the torque that should be applied to the turbine of a jet engine of moment of inertia 100 kg-m^2 so that it will have a final angular velocity of 150 rad/s in 20 s. What is the final kinetic energy?

32. A mass of 10 kg connected at the end of a rod of negligible mass is rotating in a circle of radius 30 cm with an angular velocity of 10 rad/s. If this mass is brought to rest in 30 s by a brake, what is the torque that is applied? What is the work done by the applied torque in stopping this mass?

33. A wheel of radius 1 ft weights 96 lb and is rotating at 1200 rev/min. (a) What is its kinetic energy? (b) If a force of 5 lb is applied at the rim to stop the wheel, how long will it take before it comes to rest?

34. An automobile engine with 250 hp has a constant angular velocity of 600 rev/min. Calculate the torque developed.

35. What are the dimensions of (a) angular velocity; (b) angular acceleration; (c) torque; (d) moment of inertia; (e) angular momentum; and (f) rotational kinetic energy?

36. A body with moment of inertia 6 kg-m^2 is rotating at 300 rev/min. What is its angular momentum?

37. A solid disk of diameter 50 cm and mass 2 kg is rotating with an angular velocity of 4 rad/s about an axis through its center and perpendicular to the disk. Calculate its (a) angular momentum; and (b) kinetic energy.

38. A disk of moment of inertia 4 kg-m^2 is rotating at 120 rev/min. If a mass of 1 kg is placed at a distance of 10 cm from the axis, what is the new angular velocity?

39. In Figure 9.17 the moment of inertia of the man is 5 kg-m^2. When holding a 2 kg mass in each hand at a distance of 50 cm from the center, he is rotating at $\frac{1}{4}$ rev/s. If he now drops his hands to 20 cm from the center, what is the new angular velocity?

40. A skater with a moment of inertia 4 kg-m^2 is spinning with 1 rev/s with her hands and one foot extended. When she pulls her hands and foot in, she can spin with 2 rev/s. What is the moment of inertia of the girl with her hands and foot pulled in?

***41.** Show that the angular velocity of precession Ω of a top is given by $\Omega = \frac{\Delta\phi}{\Delta t} = \frac{\Delta L}{L} = \frac{\tau}{I\omega}$ where ω is the spinning angular velocity.

***42.** Using the results of the previous problem, calculate the precessional angular velocity Ω of a top which has a mass of 10 kg, and its center of mass is at a distance of 25 cm. The spinning angular velocity is 10 rev/s, and its radius of gyration about the spinning axis is 10 cm.

***43.** In the above problem what should be the spinning angular velocity so that the precessional angular velocity is 1 rev/s.

10 Vibrational Motion

Vibrational motion is as important as translational and rotational motions and will be the subject of discussion in this chapter. We shall limit ourselves to the case of simple harmonic motion—situations to which Hooke's law is applicable. Uniform circular motion will be shown to be a combination of two simple harmonic motions at right angles to each other. We shall apply the law of conservation of energy to such motion. A few examples of linear and angular simple harmonic motion will be cited. We shall give a brief introduction to damped and forced harmonic motions. Finally, we shall discuss the elastic properties of materials and their relation to Hooke's law.

10.1. Scope of Vibrational Motion

A MOTION that repeats itself in equal intervals of time is called *periodic motion*. If a body in periodic motion executes back and forth motion about a fixed point, the body is said to have *oscillatory or vibratory motion*. The displacement of an object having such a motion can be expressed in terms of sine and cosine functions, which are also called harmonic functions. Because of this, the periodic, oscillatory, or vibratory motion is commonly referred to as *harmonic motion*.

Examples of harmonic motion can be found almost everywhere. A pendulum swinging as in a clock, a balance wheel of a watch, vibrations of strings, and air columns of musical instruments are but a few of the numerous examples of harmonic motion in macroscopic systems. Different types of wave motion, such as sound waves, light waves, radio waves, and other types of electromagnetic waves exhibit single or combinations of harmonic motions. According to modern theories, atoms of the molecules vibrate about their equilibrium positions, and atoms in solids are assumed to be vibrating about fixed positions called lattice points. These are just a few of the numerous examples of harmonic motion in microscopic systems.

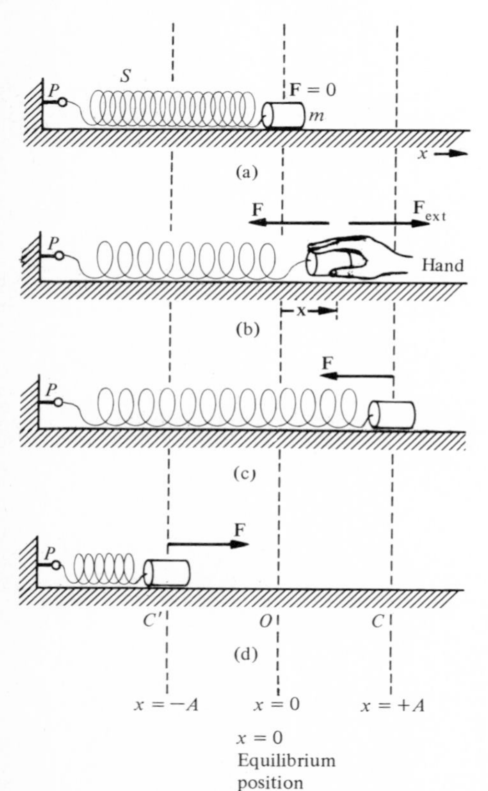

FIGURE 10.1. *A mass m tied to a spring when stretched and released executes simple harmonic motion about the equilibrium position $x = 0$.*

10.2. Hooke's Law and Simple Harmonic Motion

Let us consider a spring S that is tied to a pin P at one end and to a mass m at the other, as shown in Figure 10.1(a). The mass m is lying on a frictionless table surface. The spring in this position is unstretched and is said to be in an equilibrium position, denoted by $x = 0$. Let us stretch this spring by applying an external force $\mathbf{F}_{ext}$ as shown in Figure 10.1(b). According to Hooke's law discussed in Section 6.6, the displacement $\mathbf{x}$ of the spring is directly proportional to this applied force and is in the direction of this force. As soon as the external force is applied, there comes into play another force $\mathbf{F}$, the *restoring force*, which is equal and opposite to the applied force $\mathbf{F}_{ext}$, as shown in Figure 10.1(b). It is this restoring force that brings the system back to its original position when the applied external force is removed. Since $\mathbf{F} = -\mathbf{F}_{ext}$, we may write Hooke's law as

$$\boxed{\mathbf{F} = -k\mathbf{x}} \tag{10.1}$$

The restoring force is directly proportional to the displacement and is in the opposite direction to the displacement, hence the minus sign in Equation 10.1. The term k is the *force constant* defined as the force per unit displacement and its units are newtons per meter or pounds per foot.

Let us investigate the resulting motion when the external force $\mathbf{F}_{ext}$ applied to the spring, as shown in Figure 10.1(b), is removed after the spring has been stretched to the position $x = +A$, as shown in Figure 10.1(c). The restoring force tends to bring the mass m to the equilibrium position, $x = 0$. As the magnitude of the restoring force decreases with decreasing x, the system's (spring and mass m) acceleration also decreases continuously, but the velocity keeps on increasing. When the system reaches $x = 0$, $\mathbf{F} = 0$; hence, the acceleration is zero, but the velocity is maximum. Because of this velocity the mass m keeps moving to the left and the restoring force $\mathbf{F}$ once again comes into play, now acting to the right. Thus, the force $\mathbf{F}$ produces deceleration of m, thereby decreasing the velocity. Eventually, the mass m reaches a point C' to the left where its velocity is zero but its acceleration to the right is maximum; hence, the mass m starts moving to the right, as shown in Figure 10.1(d). It will reach $x = 0$ and then move to the right, reaching a point C shown in Figure 10.1(d), before turning back. If the surface is frictionless and there is no energy loss, the system will keep vibrating between $x = +A$ and $x = -A$ for a long time. The displacement, velocity, and acceleration are shown in Figure 10.2. Such a vibratory motion in which the restoring force is directly proportional to the displacement is called *simple harmonic motion* (SHM). Since $F = ma$, Equation 10.1 may be written $F = -kx = ma$, or

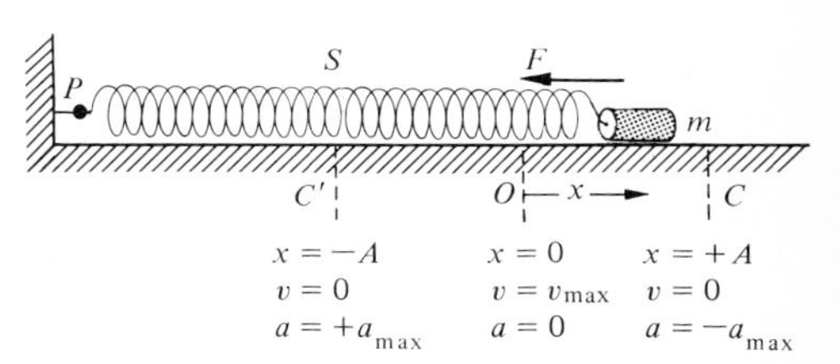

FIGURE 10.2. *Relation among the displacement, velocity, and acceleration in simple harmonic motion.*

$$a = -\frac{k}{m}x \tag{10.2}$$

that is, $a \propto -x$. This enables us to define SHM as follows: If a system vibrates such that its acceleration is directly proportional to the displacement but oppositely directed, the system is said to execute *simple harmonic motion*.

An experimental demonstration of simple harmonic motion is shown in Figure 10.3. A mass m tied to a spring S vibrates vertically up and down. A pointer attached to the mass m leaves a record of its motion on a sheet of paper that is moving at constant speed to the right. As is clear, the motion is about the equilibrium point O and is limited between C and C'. Also, the path marked by the pointer on the paper is sinusoidal; hence, it may be described by the sine and cosine functions.

(Figure 10.3 is on page 190.)

Let us state some definitions to be frequently used.

The *displacement* x is defined as the distance of the vibrating body from the equilibrium position: for example, position x of the mass m from O in Figures 10.1 to 10.3.

The *amplitude* A is the maximum displacement of the vibrating mass from the equilibrium position, that is, $A = x_{max}$ as shown in Figures 10.1 to 10.3, where $|OC| = |OC'| = A$.

A *cycle* or *complete vibration* means one round trip of the vibrating body such as mass m from O to C, C to C', and back to O in Figures 10.1 to 10.3. A

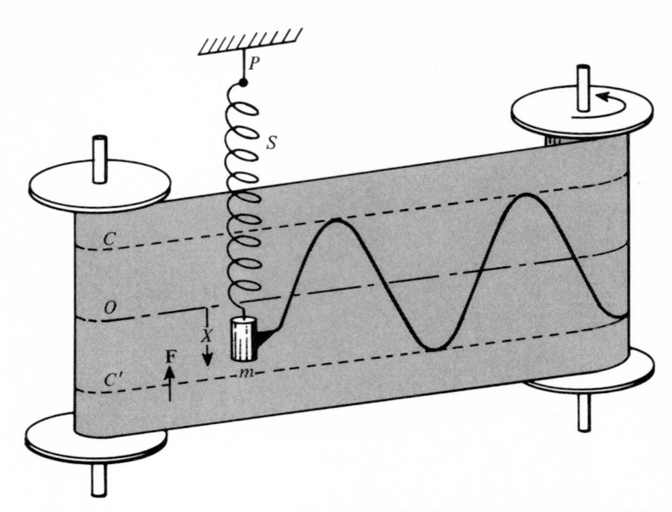

Figure 10.3. *A mass tied to a spring and vibrating in a vertical direction executes simple harmonic motion. A pointer attached to the mass leaves a trace of its displacement on a paper moving to the right with constant speed.*

cycle may be counted from C to C' and back to C, or any similar complete vibration.

The *period* T is the time required to complete one vibration or cycle of motion.

The *frequency* f is the number of complete vibrations of a system in a unit of time. Since T is the time for one vibration, we must have

$$f = \frac{1}{T} \tag{10.3}$$

In the SI system of units, a frequency of 1 cycle per second is called 1 *hertz* (Hz). Thus, 1000 cycles per second will be denoted 1000 Hz or 1 kilohertz (kHz). Furthermore, the frequency f is related to the angular velocity ω by the relation

$$\omega = 2\pi f \tag{10.4}$$

where ω is in radians per second.

10.3. Uniform Circular Motion and Simple Harmonic Motion

Uniform circular motion is a periodic motion. We shall show that it is a combination of two simple harmonic motions. Thus, the geometrical meaning and other properties of simple harmonic motion can be arrived at by considering uniform circular motion.

Consider a point Q in Figure 10.4(a) moving counterclockwise in a circle of radius A and with angular velocity ω. Let us say that at $t = 0$ line OQ makes an angle ϕ with the X-axis, and after a time t it makes an angle $\theta = \omega t + \phi$, as shown in Figure 10.4(a). Let P and P' be the projections of Q on the horizontal

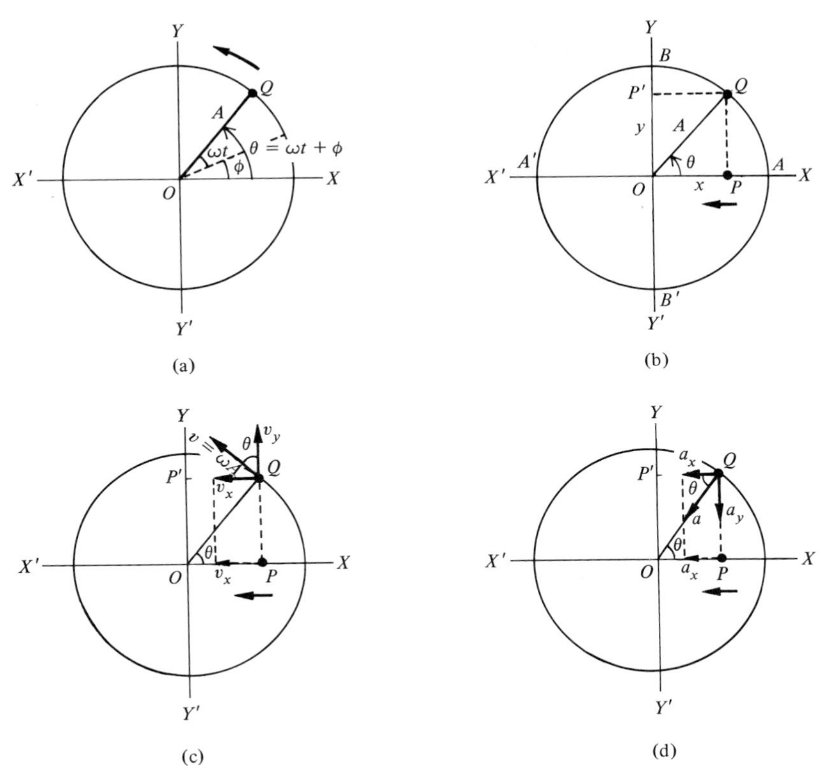

FIGURE 10.4. *Circular motion of point Q is a combination of two simple harmonic motions of points P and P′ along the X- and Y-axes, respectively.*

and the vertical diameters along the X-axis and the Y-axis, respectively, as shown in Figure 10.4(b). Thus,

$$x = A \cos \theta = A \cos (\omega t + \phi) \tag{10.5}$$

$$y = A \sin \theta = A \sin(\omega t + \phi) \tag{10.6}$$

We can now show that as Q moves in a circle, the points P and P' individually execute SHM along their corresponding diameters. The displacements of these two points are given by Equations 10.5 and 10.6. Considering Figure 10.4(c), the tangential velocity of point Q is $v = \omega A$. Thus, when v is resolved along the two axes,

$$v_x = -v \sin \theta = -\omega A \sin (\omega t + \phi) \tag{10.7}$$

$$v_y = v \cos \theta = \omega A \cos (\omega t + \phi) \tag{10.8}$$

which are the velocities of points P and P', respectively. The acceleration of point Q is $a = \omega^2 A$ $(= v^2/A)$ and is directed radially inward [see Figure

10.4(d)]. Resolving a into two components and substituting for $\cos\theta$ and $\sin\theta$ from Equations 10.5 and 10.6, we get

$$a_x = -a\cos\theta = -\omega^2 A\cos(\omega t + \phi) = -\omega^2 x \tag{10.9}$$

$$a_y = -a\sin\theta = -\omega^2 A\sin(\omega t + \phi) = -\omega^2 y \tag{10.10}$$

Multiplying both sides by m, the mass of the particle Q, and remembering that $F_x = ma_x$ and $F_y = ma_y$ gives

$$F_x = -m\omega^2 x \qquad F_y = -m\omega^2 y \tag{10.11}$$

$$F_x = -k_x x \qquad F_y = -k_y y \tag{10.12}$$

where $k_x = k_y = m\omega^2 = k$, a constant for a given value of ω. These equations are similar to $F = -kx$; hence, they represent simple harmonic motion. *Thus, the circular motion of point Q is equivalent to the motions of two points P and P′ which are executing simple harmonic motions perpendicular to each other.* As point Q moves [Figure 10.4(b)] from A to B, B to A', A' to B', and back to A, P moves from A to A' and back to A, while P' moves from O to B, B to B' and B' to O, all completing one vibration in the same *time* period T.

We pointed out earlier that in the case of simple harmonic motion, the variation of displacement with time is sinusoidal. This is quite clear from the discussion above. Both points P and P' execute simple harmonic motion and their displacements x and y versus time t, as given by Equations 10.5 and 10.6, are harmonic. This is further illustrated in Figure 10.5, where x and y are

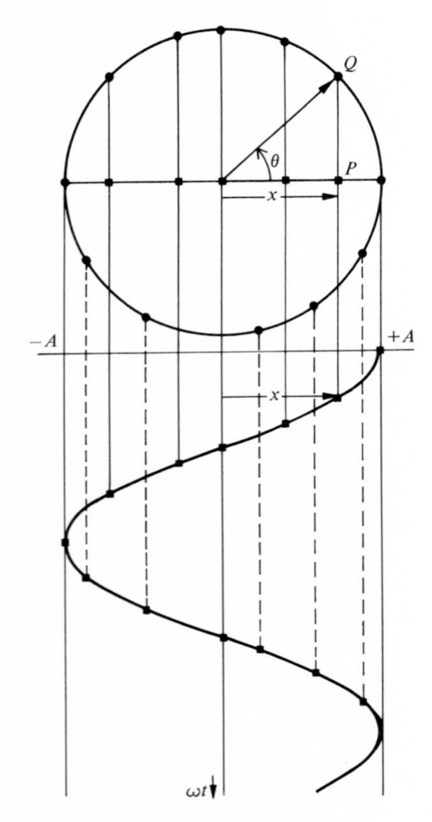

(a) $x = A\cos\theta$
$= A\cos\omega t$

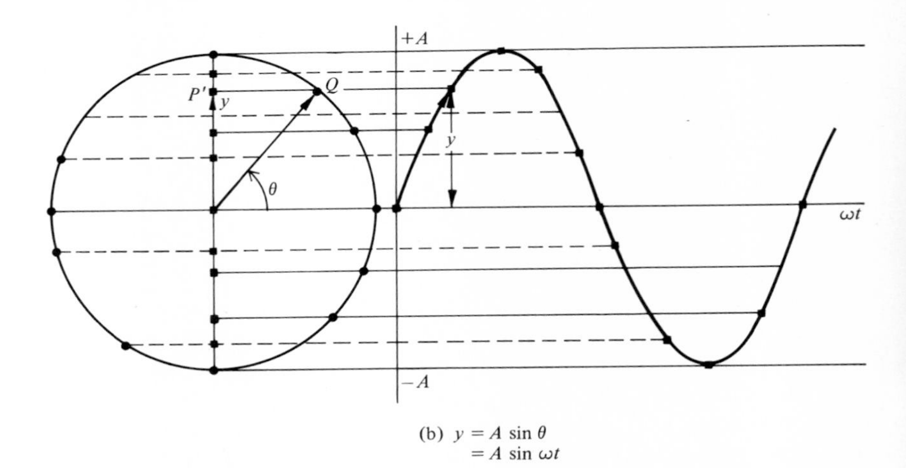

(b) $y = A\sin\theta$
$= A\sin\omega t$

FIGURE 10.5. *As point Q moves in a circle, its projections (a) P along the X-axis and (b) P′ along the Y-axis execute simple harmonic motion.*

plotted as functions of ωt for a special case of $\phi = 0$. The circles in these diagrams are called *reference circles* for simple harmonic motion.

In order to arrive at the characteristic of a simple harmonic motion, let us limit our discussion to point P. In general, any one simple harmonic motion may be represented by Equation 10.5,

$$\boxed{x = A \cos(\omega t + \phi)} \tag{10.13}$$

where A and ϕ are the only unknown constants. A is the *amplitude* and ω is the *angular velocity* in radians per second, $(\omega t + \phi)$ is an angle in radians and is called the *phase angle* or *phase* of the motion, and ϕ is the *initial phase* (phase when $t = 0$) or *phase constant*. Also, from Equations 10.11 and 10.12

$$\boxed{k = m\omega^2} \tag{10.14}$$

or

$$\omega = \sqrt{\frac{k}{m}} \tag{10.15}$$

Since $\omega = 2\pi/T$, the time period of any simple harmonic motion is

$$T = \frac{2\pi}{\omega} = 2\pi\sqrt{\frac{m}{k}} \tag{10.16}$$

and the value of the coordinate x given by Equation 10.13 repeats itself in time T.

The frequency f is given by

$$\boxed{f = \frac{1}{T} = \frac{\omega}{2\pi} = \frac{1}{2\pi}\sqrt{\frac{k}{m}}} \tag{10.17}$$

Note that f, ω, and T are functions only of m (inertial mass) and k (stiffness of the spring) and not of amplitude A. For example, from Equation 10.17, if m is large, T will be large; for a large k (stiff spring), T will be small.

Thus, the problem of SHM is solved, that is, finding the value of x as a function of t from Equation 10.13, provided that we know the values of A and ϕ. The values of these constants depend upon the particular situation under consideration. Let us consider the horizontal vibrating spring in Figure 10.2. If the mass m started oscillating when it was at C, that is, $x = A$ when $t = 0$, on substituting in Equation 10.13, $x = A \cos(\omega t + \phi)$ gives

$$A = A \cos(\omega 0 + \phi_1)$$

that is, $\cos \phi_1 = 1$ or $\phi_1 = 0°$. Thus,

$$x = A \cos \omega t \tag{10.18}$$

The plots of x versus t (and also versus ωt) are shown in Figure 10.6(a).

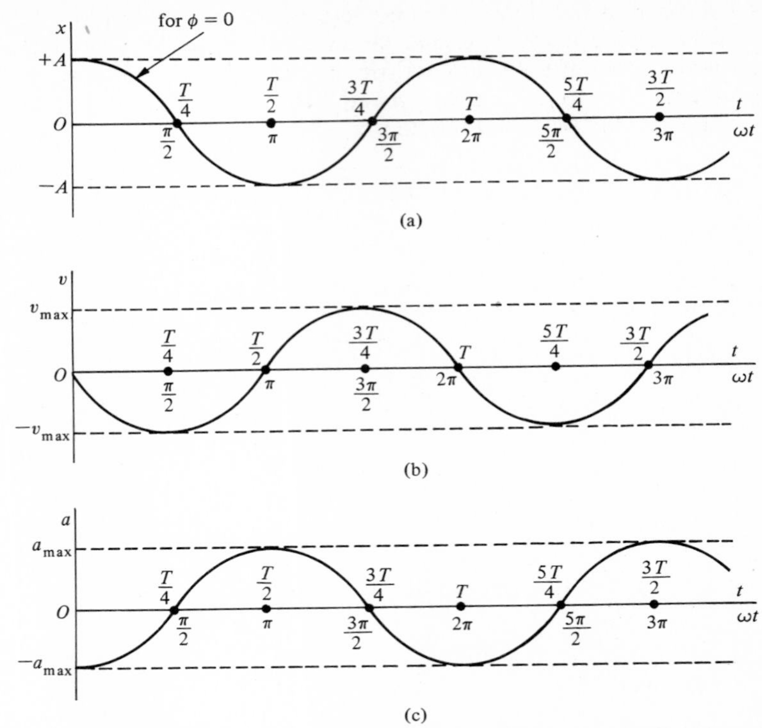

FIGURE 10.6. *Plots of x, v, and a versus t and ωt for simple harmonic motion with the initial phase φ = 0 for the displacement.*

Let us now calculate the velocity and acceleration. Once again consider the motion of point P, which is the projection of the circular motion in Figure 10.4. From Equations 10.7 and 10.9, and defining $v_{max} = \omega A$ and $a_{max} = \omega^2 A$, we get

$$v = -\omega A \sin(\omega t + \phi) = -v_{max} \sin(\omega t + \phi) \tag{10.19}$$

$$a = -\omega^2 A \cos(\omega t + \phi) = -a_{max} \cos(\omega t + \phi) \tag{10.20}$$

Since from Equation 10.13, $\cos(\omega t + \phi) = x/A$ and

$$\sin(\omega t + \phi) = \pm\sqrt{1 - \cos^2(\omega t + \phi)} = \pm\sqrt{1 - \frac{x^2}{A^2}} = \pm\frac{\sqrt{A^2 - x^2}}{A}$$

we may write these equations as

$$v = \pm\omega\sqrt{A^2 - x^2} \tag{10.21}$$

$$a = -\omega^2 x \tag{10.22}$$

That is, for a given x, v can be $+$ (along $+X$-axis) or $-$ (along $-X$-axis). For example, point P in Figure 10.4 may be moving to the right $(+)$ or to the left

(−). Thus, from Equation 10.21, v is maximum when $x = 0$, that is, when the body is passing through the equilibrium position. From Equation 10.22, a is directly proportional to the displacement x and is directed in the direction opposite to that of x, as it should be for simple harmonic motion. For $\phi = 0$, from Equations 10.19 and 10.20, we get

$$v = -v_{max} \sin \omega t \tag{10.23}$$

$$a = -a_{max} \cos \omega t \tag{10.24}$$

The plots of v and a versus t are shown in Figure 10.6(b) and (c), respectively.

EXAMPLE 10.1 The mass attached to the spring in Figure 10.2 is 80 g. When the spring is stretched 10 cm, it requires a force of 0.2 N to hold it motionless. The mass–spring system is set into motion by displacing the mass 20 cm to the right and then letting it go. Calculate (a) the force constant k; (b) the time period T; (c) the frequency f; (d) the angular velocity ω; (e) the amplitude A; (f) the equations of motion for the mass assuming $t = 0$ when $x = x_{max}$; (g) the maximum velocity v_{max} and acceleration a_{max}; and (h) the values of v and a when $x = 10$ cm.

(a) $k = \dfrac{F}{x} = \dfrac{0.2\text{ N}}{0.1\text{ m}} = 2\text{ N/m}$

(b) $T = 2\pi\sqrt{\dfrac{m}{k}} = 2\pi\sqrt{\dfrac{80 \times 10^{-3}\text{ kg}}{2\text{ N/m}}} = \dfrac{2\pi}{5}\text{s} = 1.25\text{ s}$

(c) $f = \dfrac{1}{T} = \dfrac{1}{1.25\text{ s}} = 0.8\text{ Hz}$

(d) $\omega = 2\pi f = 2\pi(0.8/\text{s}) = 5\text{ rad/s}$

(e) $A = x_{max} = 20\text{ cm} = 0.2\text{ m}$

(f) From Equations 10.13, the equation of motion for mass m is

$$x = A \cos(\omega t + \phi)$$

Substituting $x = 0.2$ m when $t = 0$, we get

$$0.2 = 0.2 \cos[5(0) + \phi]$$

that is, $\cos\phi = 1$ or $\phi = 0°$. Therefore, the equation of motion is

$$x = 0.2 \cos 5t\text{ m}$$

Thus, from Equation 10.19, the expression for v is

$$v = -\omega A \sin(\omega t + \phi) = -(5\text{ rad/s})(0.2\text{ m}) \sin 5t = -\sin 5t\text{ m/s}$$

and from Equation 10.20, the expression for a is

$$a = -\omega^2 A \cos(\omega t + \phi) = -(5\text{ rad/s})^2(0.2\text{ m}) \cos 5t = -5 \cos 5t\text{ m/s}^2$$

(g) From Equation 10.19, the value of v_{max} is

$$v_{max} = \omega A = (5\text{ rad/s})(0.2\text{ m}) = 1\text{ m/s}$$

and from Equation 10.20 the value of a_{max} is

$$a_{max} = \omega^2 A = (5 \text{ rad/s})^2(0.2 \text{ m}) = 5 \text{ m/s}^2$$

Note that these values agree with the values calculated in part (f).

(h) To calculate v and a when $x = 10$ cm, we substitute $x = 0.1$ m, $A = 0.2$ m, and $\omega = 5$ rad/s in Equation 10.21,

$$v = \pm\omega\sqrt{A^2 - x^2} = -5\sqrt{(0.2 \text{ m})^2 - (0.1 \text{ m})^2} = -0.87 \text{ m/s}$$

Substituting in Equation 10.22,

$$a = -\omega^2 x = -(5/\text{s})^2(0.1 \text{ m}) = -2.5 \text{ m/s}^2$$

Exercise 10.1 The mass attached to the spring in Figure 10.2 is 50 g. When the spring is stretched 10 cm, it requires a force of 0.5 N to hold it motionless. The mass–spring system is set into motion by displacing the mass 30 cm to the right and letting it go. Calculate (a) k, T, f, ω, and A; (b) the equation of motion for the mass assuming $t = 0$ when $x = x_{max}$; (c) v_{max} and a_{max}; and (d) the values of x, v, and a when $t = 0.25$ s. [*Ans.*: (a) 5 N/m, 0.63 s, 1.6 Hz, 10 rad/s, 0.3 m; (b) $x = 0.3 \cos 10t$ m; (c) 3 m/s, 30 m/s²; (d) −0.24 m, −1.8 m/s, +24 m/s².]

10.4. Energy Conservation in Simple Harmonic Motion

We shall now show that in all simple harmonic motions if frictional (or dissipative) forces are absent, the total mechanical energy, which in this case is the sum of the kinetic and elastic potential energy, is constant. We showed in Section 6.6 that when a spring is stretched through a distance x, the work done is stored as energy called the elastic potential energy U. According to Equation 6.24 and Figure 10.7,

$$U = \tfrac{1}{2}kx^2 \tag{10.25}$$

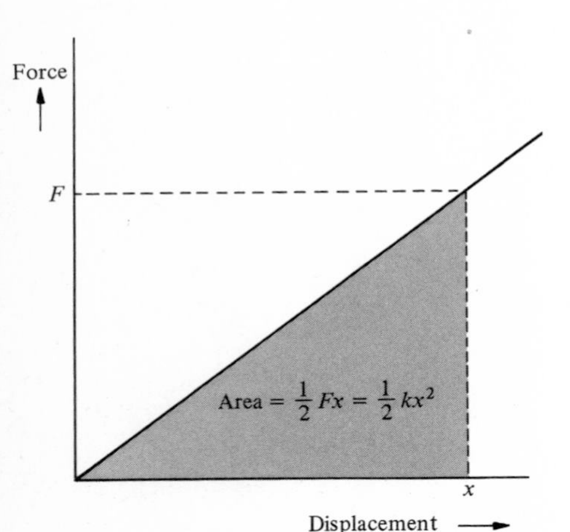

Figure 10.7. *F versus x for a spring. The work done is represented by the shaded area.*

Thus, when the spring has a maximum displacement, say it has been pulled to the extreme right, where $x = A$, the potential energy is maximum U_0 and is equal to

$$U_0 = \tfrac{1}{2}kA^2 \tag{10.26}$$

If the spring is now released, for any value of x between $x = 0$ and $x = A$, the potential energy will be $\frac{1}{2}kx^2$, while the kinetic energy K of the mass m will be $\frac{1}{2}mv^2$, where v is the velocity of the mass. The sum of these two energies is denoted by the total energy E_T and must be equal to the work done when the spring was stretched from $x = 0$ to $x = A$. Thus,

$$\tfrac{1}{2}mv^2 + \tfrac{1}{2}kx^2 = E_T \tag{10.27}$$

or

$$K + U = E_T \tag{10.28}$$

Note that when $x = 0$, the potential energy $U = 0$ and the energy is all kinetic. Thus, if the maximum velocity is v_0 when $x = 0$, the maximum kinetic energy K_0 is

$$K_0 = \tfrac{1}{2}mv_0^2 \tag{10.29}$$

$$E_T = K_0 = U_0 = \tfrac{1}{2}mv_0^2 = \tfrac{1}{2}kA^2 \tag{10.30}$$

Let us see how K and U vary in simple harmonic motion. From Equation 10.13 for $\phi = 0$, we have $x = A\cos\omega t$, and the elastic potential energy U of the spring may be written

$$U = \tfrac{1}{2}kx^2 = \tfrac{1}{2}kA^2\cos^2\omega t = U_0\cos^2\omega t \tag{10.31}$$

where $U_0 = \tfrac{1}{2}kA^2$ is the maximum potential energy of the spring and occurs when $x = A$. Similarly, using Equations 10.19 and 10.14, for $\phi = 0$ the kinetic energy K of the mass may be written as

$$K = \tfrac{1}{2}mv^2 = \tfrac{1}{2}m\omega^2A^2\sin^2\omega t = \tfrac{1}{2}kA^2\sin^2\omega t = K_0\sin^2\omega t \tag{10.32}$$

where K_0 is the maximum energy $\tfrac{1}{2}mv_0^2$ and is equal to $\tfrac{1}{2}kA^2$ (also equal to the maximum potential energy) occurring at $x = 0$. Thus, the total mechanical energy is

$$E_T = K + U = \tfrac{1}{2}kA^2(\sin^2\omega t + \cos^2\omega t) = \tfrac{1}{2}kA^2$$

or

$$\boxed{E_T = \tfrac{1}{2}kA^2 = U_0 = K_0} \tag{10.33}$$

That is, *the total energy (and also the maximum potential or maximum kinetic energy) of a harmonic oscillator is proportional to the square of the amplitude.*

The plots of U, K, and E_T are shown in Figure 10.8. Note that U and K vary sinusoidally with time but E_T remains constant and is equal to the sum of U and

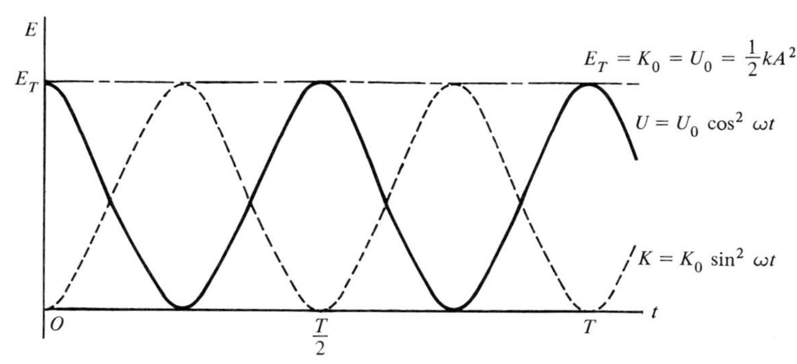

FIGURE 10.8. *Plots of potential energy U, kinetic energy K, and the total energy E_T versus time. Note that the sum of U and K remains constant and is equal to E_T.*

K at any instant. Let us consider another plot, which is closer to a physical situation. Figure 10.9 shows a plot of $U(= \tfrac{1}{2}kx^2)$ versus x, which is a parabola. Draw a horizontal line indicating the value of E_T and drop perpendiculars to the X-axis from points C and C'. Thus, A and $-A$ represent the amplitudes of

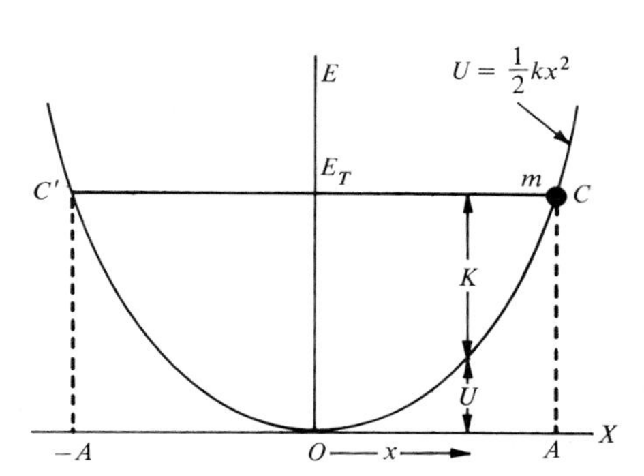

FIGURE 10.9. *Schematic representation of the relation $E_T = U + K$ for a particle in a parabolic potential energy well and executing simple harmonic motion.*

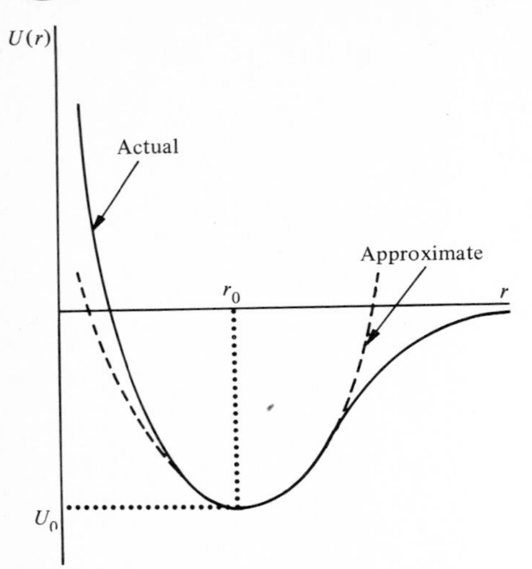

FIGURE 10.10. *$U(r)$ versus r of atoms in molecules and solids, as represented by the continuous potential-energy curve. For small amplitude of vibration, the motion is simple harmonic and $U(r)$ is represented by the dashed potential-energy curve.*

SHM. If the harmonic oscillator is at C or C', its energy is all potential, and it is all kinetic when it is at O. If it is at a displacement x, the potential energy U, is equal to the vertical line between the X-axis and the parabolic curve at x, while the kinetic energy is equal to the vertical line between the parabolic curve and the horizontal line, as shown in Figure 10.9. The situation is like that of a mass m with mechanical energy $\frac{1}{2}kA^2$ after being released from point C and moving on a frictionless parabolic surface between C and C'. We state that the mass m is confined to a *potential-energy well.*

A physical situation corresponding to this occurs both in molecules and in solids. Atoms in molecules and solids under the influence of interatomic forces vibrate about their equilibrium positions. The plot of $U(r)$ versus r in such cases is as shown by the solid line in Figure 10.10. For small amplitudes of atomic oscillations the potential may be approximated by the dashed curve. This curve is similar to the curve in Figure 10.9. Hence, for small oscillations the vibrations of atoms in molecules and in solids are SHM, as if the atoms were tied with springs and vibrating about their equilibrium positions. The distance r_0 between the atoms when in normal or equilibrium position is called the *equilibrium distance.* The minimum energy U_0 is called the *binding energy* or *dissociation energy* and is equal to the energy needed to separate the atoms which are bound in a molecule.

Let us see what happens when energy is supplied to such a bound system: (a) In the case of molecules, when the external energy supplied is less than U_0, no dissociation occurs. But the molecule has more energy than when it is in a normal state. Hence, the molecules are said to be *excited.* These molecules do not remain in the excited states for a long time. In time intervals of less than 10^{-8} s, the molecules give out the extra energy, resulting in what is called molecular spectra (see Chapter 33). (b) The expansion of solids can be explained with the help of Figure 10.10. When a solid is heated, the energy of the atoms is increased, resulting in oscillations that have a larger amplitude, hence a departure from the approximate parabolic potential. As we shall see in Chapter 15, this departure increases the equilibrium distance r_0, leading to the expansion of the solid.

The two examples above are just a small sample of numerous others that correspond to similar physical situations in various branches of physics.

EXAMPLE 10.2 The motion of a 1-kg mass tied to a spring is described by $x = 40 \cos 4t$ cm. Calculate (a) the values of v, a, K, and U when the mass m is halfway from the initial position toward the center; (b) the total energy of the oscillator; and (c) the time to move halfway to the center.

Comparing $x = 40 \cos 4t$ cm with $x = A \cos \omega t$, we get $A = 0.4$ m, $\omega = 4$ rad/s. Since $m = 1$ kg, from Equation 10.14, $k = m\omega^2 = (1 \text{ kg})(4 \text{ rad/s})^2 = 16$ N/m. Also, from Equation 10.19, $|v_{max}| = \omega A = (4 \text{ rad/s})(0.4 \text{ m}) = 1.6$ m/s and from Equation 10.20, $|a_{max}| = \omega^2 A = (4 \text{ rad/s})^2(0.4 \text{ m}) = 6.4 \text{ m/s}^2$.
(a) For halfway, $x = A/2 = 0.4 \text{ m}/2 = 0.2$ m, and from Equations 10.21, 10.22, and 10.27,

$$v = \pm\omega\sqrt{A^2 - x^2} = -(4 \text{ rad/s})\sqrt{(0.4 \text{ m})^2 - (0.2 \text{ m})^2} = -1.39 \text{ m/s}$$

$$a = -\omega^2 x = -(4 \text{ rad/s})^2(0.2 \text{ m}) = -3.2 \text{ m/s}^2$$

$$K = \tfrac{1}{2}mv^2 = \tfrac{1}{2}(1 \text{ kg})(-1.39 \text{ m/s})^2 = 0.96 \text{ J}$$
$$U = \tfrac{1}{2}kx^2 = \tfrac{1}{2}(16 \text{ N/m})(0.2 \text{ m})^2 = 0.32 \text{ J}$$

(b) The total energy is conserved and may be calculated at any instant of the motion. Thus, at $x = 0.2$ m, as calculated in part (a),

$$E = K + U = 0.96 \text{ J} + 0.32 \text{ J} = 1.28 \text{ J}$$

or

$$E = K_{max} = \tfrac{1}{2}mv_{max}^2 = \tfrac{1}{2}(1 \text{ kg})(1.6 \text{ m/s})^2 = 1.28 \text{ J}$$
$$= U_{max} = \tfrac{1}{2}kx_{max}^2 = \tfrac{1}{2}kA^2 = \tfrac{1}{2}(16 \text{ N/m})(0.4 \text{ m})^2 = 1.28 \text{ J}$$

(c) In $x = 40 \cos 4t$, let $x = A/2 = 20$ cm:

$$20 = 40 \cos 4t$$

or $\cos 4t = \tfrac{1}{2}$ or $4t = 60° = \pi/3$:

$$t = \pi/12 \text{ s} = 0.26 \text{ s}$$

Exercise 10.2 The motion of a 2 kg mass tied to a spring is described by $x = 50 \cos 3t$ cm. Calculate v, a, K, and U at $x = 20$ cm. Also, calculate the total energy. What is the time required to move from $x = A$ to $x = \tfrac{1}{4}A$? [*Ans.*: $v = -1.38$ m/s, $a = -1.80$ m/s^2, $K = 1.89$ J, $U = 0.36$ J, $E = 2.25$ J, $t = 0.437$ s.]

10.5. Examples of Linear and Angular Simple Harmonic Motion (Pendulums)

Simple harmonic motion may be linear or angular. A system executing linear simple harmonic motion obeys Hooke's law, Equation 10.1, with its displacement given by Equation 10.5 and the time period given by Equation 10.16, that is, (using x_0 for amplitude instead of A),

$$F = -kx \tag{10.1}$$

$$x = x_0 \cos(\omega t + \phi) \tag{10.5}$$

$$T = 2\pi\sqrt{\frac{m}{k}} \tag{10.16}$$

The equations describing angular simple harmonic motion are completely analogous to these and may be written down by replacing the force F by the torque τ, the linear displacement x by the angular displacement θ, the mass m by the moment of inertia I, and the force constant k by the torsional constant κ. The constant κ is defined as the torque per unit angular displacement; that is, $\kappa = \tau/\theta$. Thus, a system undergoing angular SHM is described by the following equations:

$$\tau = -\kappa\theta \tag{10.34}$$

$$\theta = \theta_0 \cos(\omega t + \phi) \tag{10.35}$$

$$T = 2\pi\sqrt{\frac{I}{\kappa}} \tag{10.36}$$

At present we discuss the simple pendulum and the torsional pendulum to illustrate the use of these equations.

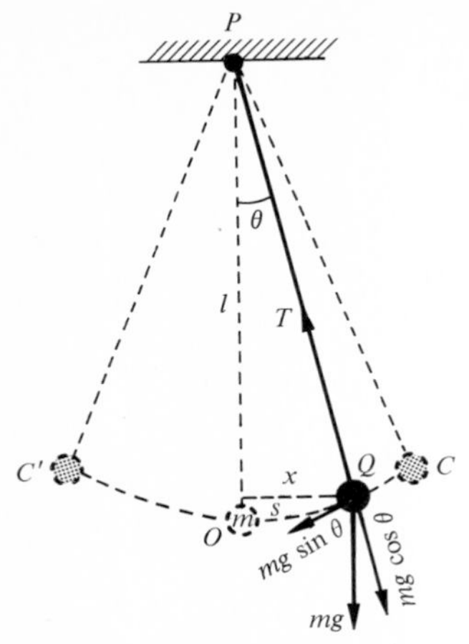

FIGURE 10.11. *Simple pendulum. The tension T is equal to mg cos θ, while the unbalanced component mg sin θ is the restoring force. (Note that T − mg cos θ = mv²/l = 0 only when v = 0 at the ends of a swing.)*

The Simple Pendulum

A simple pendulum (also called an *ideal* or a *mathematical pendulum*) consists of a point mass suspended from a fixed point by means of a weightless inextensible string. Figure 10.11 shows such a pendulum consisting of mass m tied to a light string of length l and suspended from point P. When mass m is displaced from its equilibrium position O to a point C and let go, the mass m swings between C and C'. The period of such a pendulum may be calculated simply.

Let us consider the pendulum in a displaced position Q as shown in Figure 10.11. The two forces acting on mass m are its weight $m\mathbf{g}$ and the tension $\mathbf{T}$ in the string. The weight $m\mathbf{g}$ may be resolved into two components, the radial component $mg\cos\theta$ and the tangential component $mg\sin\theta$. At any instant, $T - mg\cos\theta = mv^2/l$, where mv^2/l is the centripetal force necessary to keep the mass m moving in a circular arc. The tangential component $mg\sin\theta$ is the unbalanced restoring force acting in a direction opposite to that of the displacement θ. Hence, we may write

$$F = -mg\sin\theta \tag{10.37}$$

This is not in the form of Hooke's law. However, if the displacement θ is small, we can justify the approximation $\sin\theta \simeq \theta = s/l \simeq x/l$, where the arc length s is very nearly equal to the linear displacement x. Equation 10.37 takes the form

$$F \simeq -\frac{mg}{l}x \tag{10.38}$$

Comparing with $F = -kx$, we get

$$k = \frac{mg}{l} \tag{10.39}$$

Substituting for k in Equation 10.16 gives the following expression for the time period of a simple pendulum:

$$T = 2\pi\sqrt{\frac{m}{k}} = 2\pi\sqrt{\frac{m}{mg/l}}$$

or

$$\boxed{T = 2\pi\sqrt{\frac{l}{g}}} \tag{10.40}$$

This expression for T, although approximate, is very close to the true value for small amplitudes. For example, for $\theta = 10°$, the true value of T differs from the one given by Equation 10.40 by less than 0.4 percent. (See Problem 10.19.)

Note that the time period of a pendulum is independent of the amplitude. This fact is made use of in a clock pendulum. As the clock runs down, the amplitude of the swings becomes slightly smaller continuously, but the clock still keeps accurate time. The simple pendulum also provides a simple and accurate method for measurement of the acceleration due to gravity g.

The Torsional Pendulum

Figure 10.12 shows one of the many types of torsional pendulums. A disk attached to a wire W at its center O is suspended from point P. When the disk is rotated in a horizontal plane from its equilibrium position OO' to the position OC, the wire will be twisted, and when let go, the disk oscillates between C and C' with an angular amplitude of θ_0. The twisted wire produces a restoring torque on the disk, which tends to return the disk to its original equilibrium position. Thus, Hooke's law takes the form

$$\tau = -\kappa\theta$$

and, using Equation 10.36,

$$T = 2\pi\sqrt{\frac{I}{\kappa}} = 2\pi\sqrt{\frac{\frac{1}{2}MR^2}{\kappa}} \qquad (10.41)$$

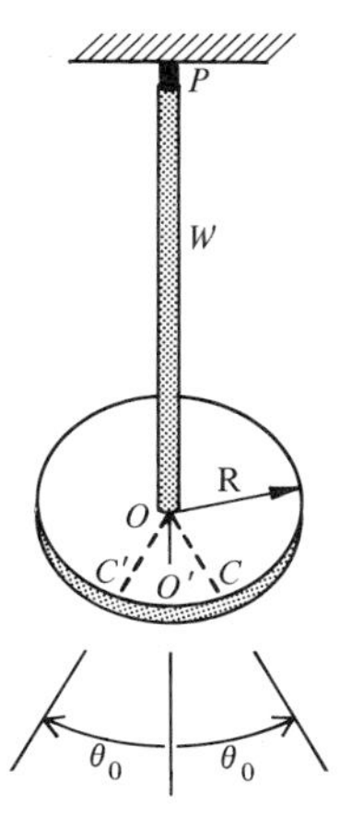

Figure 10.12. *Typical torsional pendulum with a maximum displacement of θ_0 from the equilibrium position OO'.*

where we have used $I = \frac{1}{2}MR^2$, the moment of inertia of the disk.

In actual practice one measures T and, knowing I, calculates the value of the torsional constant κ of the wire. Once κ is known, the wire may be used in electrical equipment such as galvanometers for measuring currents.

As another example, the balance wheel of a watch also exhibits angular harmonic motion. A spiral hairspring provides the restoring torque.

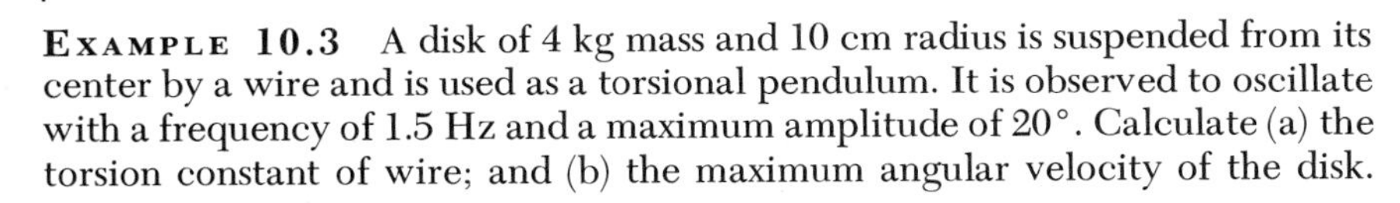

Example 10.3 A disk of 4 kg mass and 10 cm radius is suspended from its center by a wire and is used as a torsional pendulum. It is observed to oscillate with a frequency of 1.5 Hz and a maximum amplitude of 20°. Calculate (a) the torsion constant of wire; and (b) the maximum angular velocity of the disk.

From Equation 10.41,

$$T = \frac{1}{f} = 2\pi\sqrt{\frac{I}{\kappa}} = 2\pi\sqrt{\frac{\frac{1}{2}MR^2}{\kappa}}$$

where $I = \frac{1}{2}MR^2$ is the moment of inertia of the disk. After squaring and rearranging we get

$$\kappa = 4\pi^2 f^2 I = 4\pi^2 f^2(\tfrac{1}{2}MR^2) = 2\pi^2 f^2 MR^2 = 2(3.14)^2(1.5/\text{s})^2(4\text{ kg})(0.1\text{ m})^2$$
$$= 1.78\text{ N-m/rad}$$

Also, from analogy with linear motion $E = \frac{1}{2}kx^2_{max} = \frac{1}{2}mv^2_{max}$, we may write for angular motion,

$$E = \tfrac{1}{2}\kappa\theta^2_{max} = \tfrac{1}{2}I\omega^2_{max}$$

where $\theta_{max} = 20° = \pi/9$ rad:

$$\omega_{max} = \left(\frac{\kappa}{I}\right)^{\frac{1}{2}}\theta_{max} = \left[\frac{1.78\text{ N-m/rad}}{\frac{1}{2}(4)(0.1)^2\text{ kg-m}^2}\right]^{\frac{1}{2}}\left(\frac{\pi}{9}\right) = 3.3\text{ rad/s}$$

EXERCISE 10.3 A certain pendulum is set into oscillation by a torque of 10 N-m, which twists it through 15°. The disk attached to the wire oscillates with a time period of 5 s. Find the moment of inertia of the disk. [*Ans.*: $I = 24.2$ kg-m^2.]

*10.6. Damped and Forced Harmonic Motion

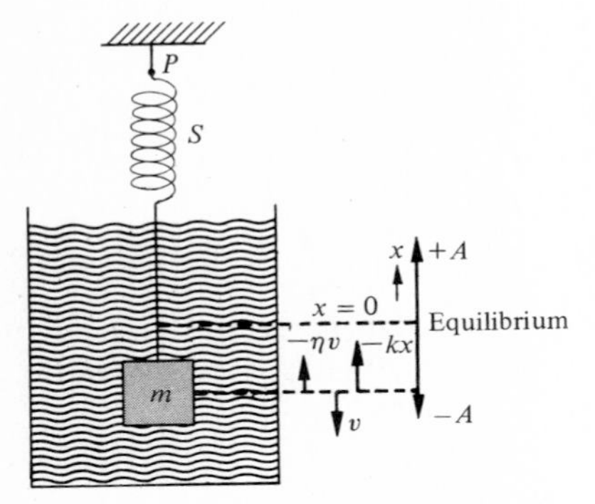

FIGURE 10.13. *A mass tied to a spring and moving in a liquid performs damped harmonic motion.*

In the discussion so far we have assumed that frictional forces are absent. This means that the total energy of the oscillating system remains constant; hence, the body will oscillate with constant amplitude forever. In actual practice, this is not true. There are always some frictional forces present, no matter how small, and the oscillating system must work against this dissipative force. This leads to the decrease in energy, hence a continuous decrease in the amplitude of the swing, ultimately stopping the motion. The motion of the system in the presence of frictional forces is called *damped harmonic motion*. The friction may be due to the internal forces or due to the medium—air, gas, or liquid—in which the body is oscillating. The damping constant η characterizes such frictional force f which is equal to $-\eta v$.

As a particular case, consider a mass m tied to a spring S immersed in a liquid, as shown in Figure 10.13. If there were no friction between mass m and the liquid, the motion would be undamped, as shown in Figure 10.14(a). For small friction the motion is damped, as shown in Figure 10.14(b). For very high friction the motion is overdamped, as shown in Figure 10.14(c), and the system will not complete even one single oscillation.

Since every oscillating system is damped to a certain extent, if no external energy is supplied, the system eventually comes to rest. For example, a child's swing comes to rest unless external forces are applied at proper regular intervals of time to make the swing keep on swinging with a large amplitude. The same is the case with a pendulum or a vibrating mass tied to a spring. For the oscillating system to maintain a constant amplitude, it is necessary to apply an external oscillating force. The external applied force is the *driving force*, the oscillating system is the *driven oscillator*, and the resulting oscillations are

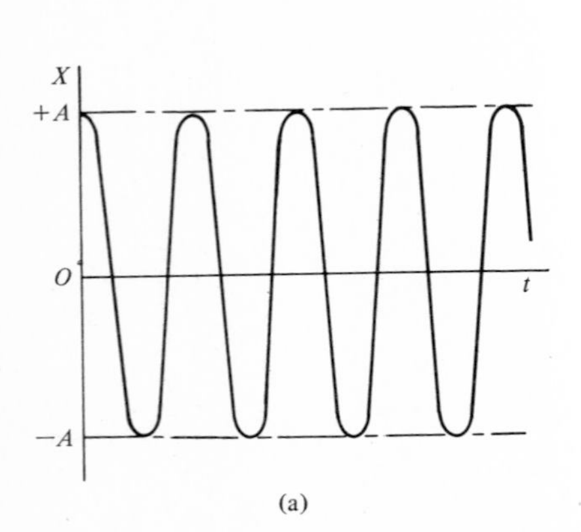

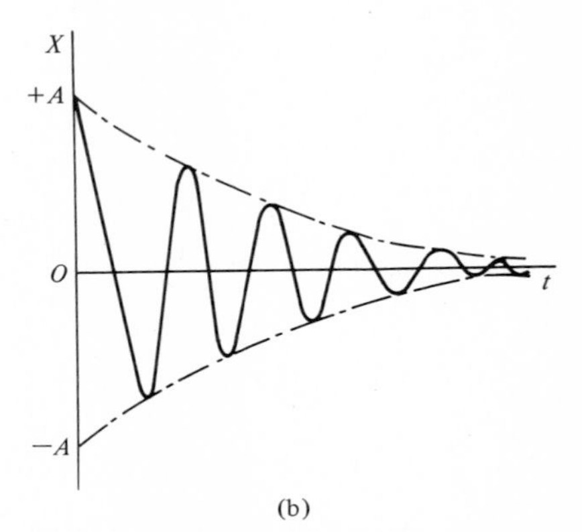

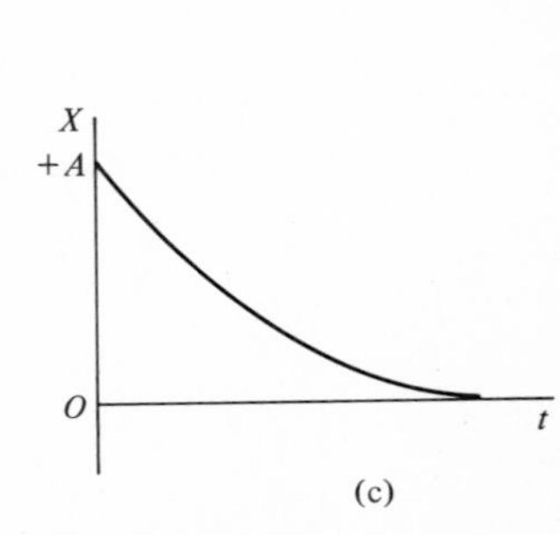

FIGURE 10.14. *Oscillatory motion: (a) undamped motion, (b) damped motion, and (c) overdamped motion.*

forced oscillations. The system oscillates with the frequency of the driving force and not with the natural frequency. We frequently come across examples of driven oscillators in alternating-current systems, acoustical systems, molecular and atomic physics, and so on.

Suppose that there is a driving force F_e of angular frequency ω_e acting on the system and is given by

$$F_e = F_o \cos \omega_e t \tag{10.42}$$

The equation describing the displacement of such a driven oscillator is given by

$$x = A_m \cos (\omega_e t + \phi) \tag{10.43}$$

ω_e is the angular frequency of the driving force, while ω_o $(=\sqrt{k/m})$ is the natural frequency of the oscillator. Note that the system vibrates with frequency ω_e and not ω_o. A_m is the amplitude of the oscillations and its magnitude depends upon ω_o, ω_e, m, and the damping force. In the absence of any damping force, the value of A_m depends on the relative values of ω_e and ω_o. When $\omega_e = \omega_o$, the amplitude A_m will be infinite. In any actual system there is always some damping present, which means that the amplitude of the oscillations will be very large but not infinite, as shown in Figure 10.15. The condition when $\omega_e = \omega_o$ is called *resonance* and the value of the frequency f_e,

$$\boxed{f_e = \frac{\omega_e}{2\pi} = \frac{\omega_0}{2\pi} = \frac{1}{2\pi}\sqrt{\frac{k}{m}}} \tag{10.44}$$

is called the *resonance frequency*. For this value of ω_e the amplitude is larger than at any other frequency for a given system.

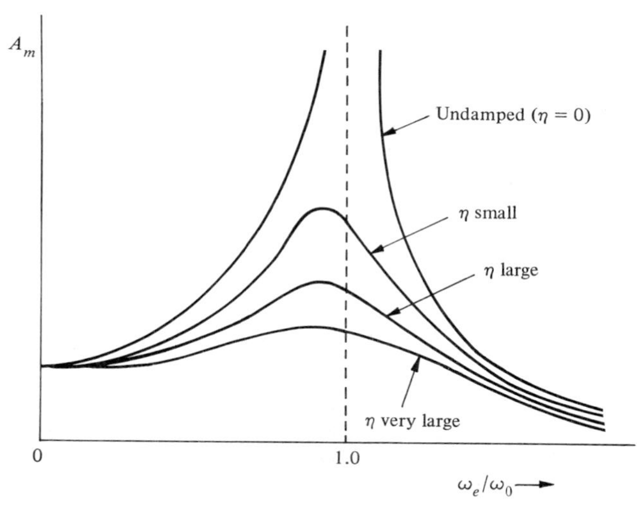

Figure 10.15. *Variation in the amplitude A_m of the forced oscillations versus the frequency ω_e of the applied force for different values of damping constant η. Note the infinitely large value of A_m at resonance ($\omega_e = \omega_0$) for $\eta = 0$.*

The damping force is characterized by a constant η, the damping constant. For any value of η, A_m is maximum when $\omega_e = \omega_o$, and it decreases on either side as ω_e departs from the natural frequency ω_o. If under certain circumstances the amplitude of the oscillating system becomes extremely large, Equation 10.43 does not apply. Under such high amplitudes the system usually breaks down. For example, the reason a column of soldiers is ordered to break

FIGURE 10.16. *The Tacoma Bridge at Puget Sound, Washington, was completed on July 1, 1940. After 4 months the winds from a mild gale blowing down the river valley produced a fluctuating force of frequency equal to the natural frequency of the bridge, resulting in mechanical resonance. There was a steady increase in the amplitude of the oscillating bridge until the bridge was completely destroyed.* (Courtesy of Wide World Photos, Inc.)

step while crossing a bridge is to avoid setting the bridge into forced oscillations of high amplitude. Another example is worth mentioning. The Tacoma Narrows Bridge at Puget Sound, Washington, was completed on July 1, 1940. Four months later, the winds from a mild gale produced a fluctuating force of frequency nearly equal to the natural frequency of the bridge. There was a steady increase in the amplitude of the oscillating bridge, resulting in breaking the main span of the bridge, as shown in Figure 10.16. To avoid such disasters, many other bridges were redesigned to make them aerodynamically stable.

*10.7. Elasticity and Elastic Constants

In most solids atoms are usually arranged in some order. How rigidly these atoms are held about their equilibrium positions depends upon the relative strength of the short-range electrical forces between them. When external forces are applied to such a sample, a distortion results because of the displacement of the atoms from their equilibrium position, and the body is said to be in a state of stress. After the external force is removed, the body returns to its equilibrium position, provided that the applied force was not too great. The ability of the body to return to its original shape is called *elasticity*. Accordingly, while a rubber band is more easily stretched than a metal wire, the metal wire is less plastic (but more elastic) than the rubber band.

Consider a block of some solid material, as shown in Figure 10.17(a) and (b). Suppose that a force **F** is applied to the face, which has an area A. **F** can be resolved into two components, F_n and F_t, where $F_n = F \sin\theta$ is called the normal force and $F_t = F\cos\theta$ is called the tangential force. In general, *stress* is defined as the force per unit area while we define

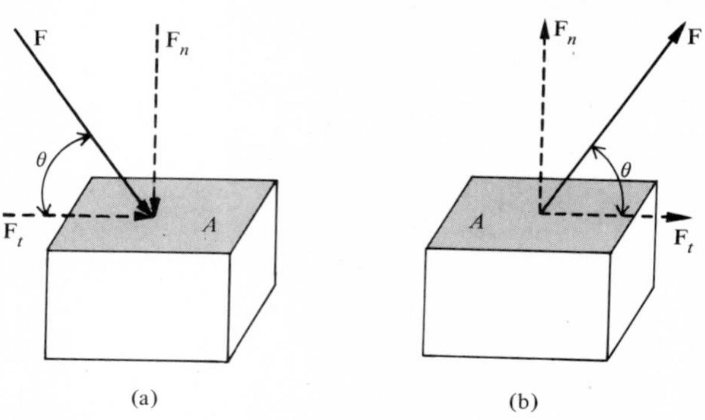

FIGURE 10.17. *Force* **F** *will produce compression in (a) and stretching in (b).*

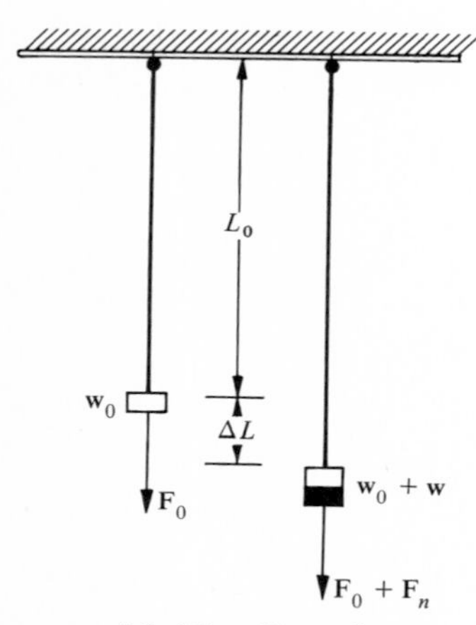

FIGURE 10.18. *Normal stress, producing a longitudinal strain in a thin wire.*

$$\text{normal (tensile) stress} = \frac{F_n}{A} \tag{10.45}$$

$$\text{tangential (shear) stress} = \frac{F_t}{A} \tag{10.46}$$

The units of stress are N/m^2, $dyne/cm^2$, lb/in^2, and so on. It is convenient to study the effect of these two stresses separately.

The effect of stress is to cause distortion or a change in size and shape. *A quantity called strain refers to the relative change in size or shape of the body when under the applied stress.* The normal stress applied to a body can produce a compression or a stretching, as will happen in Figure 10.17(a) and (b), respectively. To be more precise, let us consider a weightless wire that has a cross-sectional area A and length L_0 which is hanging vertically under the action of normal force F_0 (due to the weight w_0, that is, $F_0 = w_0$) as shown in Figure 10.18. Suppose that another weight w is added so that the total normal force is $F_0 + F_n$, while the length changes to $L_0 + \Delta L$. Thus, the normal force F_n acting on an area A has increased the length by ΔL. *The longitudinal (or tensile) strain is defined as the ratio of the change in length to the original length.* That is,

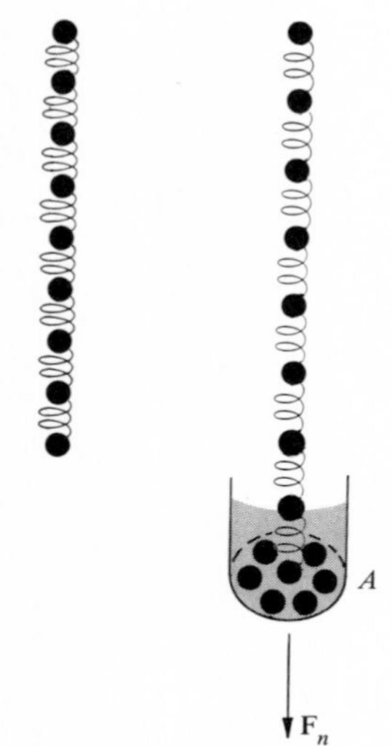

FIGURE 10.19. *Atomic view of longitudinal strain.*

$$\text{longitudinal (tensile) strain} = \frac{\Delta L}{L_0} \tag{10.47}$$

while the normal (tensile) stress is F_n/A.

Note that Figure 10.19 shows schematically what happens on a microscopic scale when a stress is applied. For simplicity, only one line of atoms is shown. It is as if the atoms in a solid are held together by means of springs, and an application of force just stretches these springs from their equilibrium position, thereby causing a change in length.

It is found experimentally that the ratio of stress to strain is a constant for a given material. This is called the *elastic modulus*. The ratio of the longitudinal stress to the longitudinal strain is called *Young's* (or *stretch*) *modulus*, that is,

$$\boxed{Y = \frac{F_n/A}{\Delta L/L_0}} \tag{10.48}$$

Since strain is a dimensionless quantity, the units of Young's modulus are the same as those of stress, that is, N/m^2, $dyne/cm^2$, lb/in^2, and so on. It may be pointed out that transverse strain (change in length perpendicular to the force) is negligible in most cases and will not be considered any further here.

Let us now consider the effect of the tangential or shear stress F_t/A. Such a stress produces a change in the shape of the body, called the shear strain. For example, suppose that a sideways shear or tangential shear is applied at the top of a deck of cards or at the top of a book laying on a table. This often results in a displacement of each card or page relative to its neighbor, the top one being displaced the most while the bottom is not displaced at all, as shown in Figure 10.20.

Thus, when a force F_t is applied to the top surface $ABCD$ of a rectangular block as shown in Figure 10.20(a), it results in a shear strain, as shown in Figure 10.20(b). The effect of a shear or tangential strain is to move plane $ABPQ$ to a new position, $A'B'PQ$, to the right. This is due to the displacement of the atoms to the right, as shown in Figure 10.20(c). According to Figure 10.20(b), the shear strain is defined as

$$\text{shearing strain} = \frac{\Delta x}{h} \tag{10.49}$$

while the shearing stress is F_t/A. Thus, the *shear modulus* or *modulus of rigidity* or *torsion modulus* S is the ratio of the shear stress and shear strain. Thus,

$$S = \frac{F_t/A}{\Delta x/h} \tag{10.50}$$

Usually, Δx is small; hence, $\tan\theta \simeq \Delta x/h$, and we may write

$$\boxed{S = \frac{F_t/A}{\tan\theta}} \tag{10.51}$$

It should be mentioned that when a shearing stress is applied to a fluid, liquid or gas, the fluid simply slips; that is, the fluid cannot sustain a shearing stress. Thus, the shearing modulus is only applicable to solids.

In the case of fluids, forces must be applied normal to the surface to produce a deformation which, in turn, can be used to define the modulus of elasticity. The applied force decreases the volume of the fluid. Suppose that a fluid of volume V is acted upon by a force F_n acting normal to an area A resulting in a change in volume ΔV. The normal force per unit area applied to a fluid is called *pressure*, p. Thus, the normal stress in this case is equal to a change in

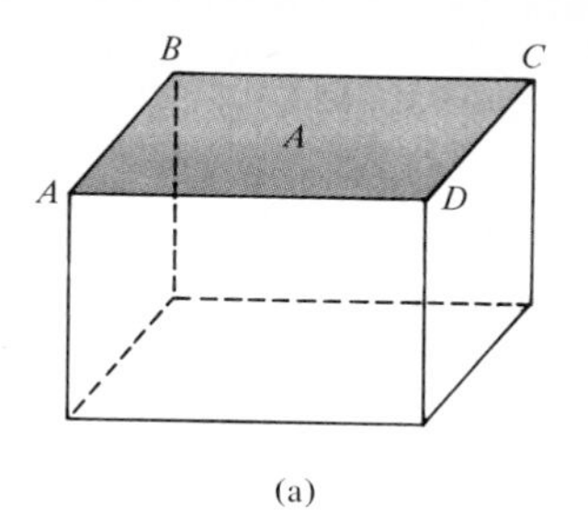

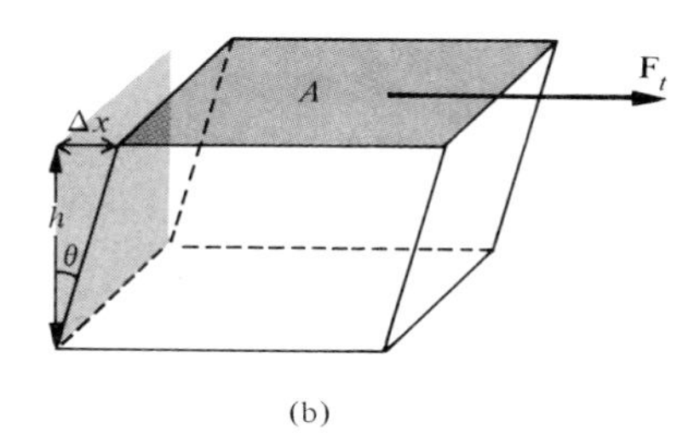

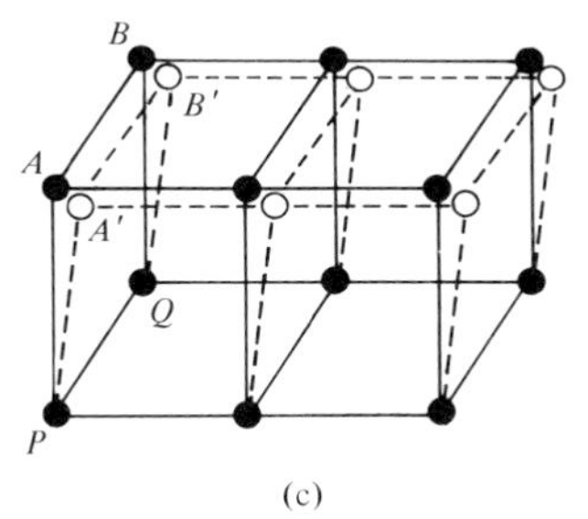

FIGURE 10.20. *A block (a) under no stress, (b) under shearing stress producing shearing strain, and (c) atomic model.*

pressure, Δp, and is called the *volume stress*. In fluids we define volume stress and volume strain as

$$\text{volume stress} = \frac{F_n}{A} = \Delta p \qquad \text{volume strain} = \frac{\Delta V}{V} \tag{10.52}$$

while the *volume elasticity*, also called the *bulk modulus*, is defined as the ratio of volume stress to volume strain, that is,

$$\boxed{B = \frac{\Delta p}{-\Delta V/V} = -V\frac{\Delta p}{\Delta V}} \tag{10.53}$$

The negative sign indicates that as pressure increases, volume decreases. Once again the units of B, S, and Y are all the same, that of stress. The values of these quantities are listed in Table 10.1. The reciprocal of the bulk modulus is denoted by $\beta = 1/B$ and is called the *compressibility*.

TABLE 10.1
Elastic Constants

Material	Y ($\times 10^{10}$ N/m^2)	S ($\times 10^{10}$ N/m^2)	B ($\times 10^{10}$ N/m^2)
Solids			
Aluminum	7	3	7
Bone	1.4	0.6	0.8
Brass	9	3.4	6.1
Copper	11	4.2	14
Glass	5.8	2.4	4.5
Iron	12	4.6	9
Steel	20	8.4	16
Tungsten	36	15	20
Liquids			
Glycerin			0.45
Mercury			2.5
Water			0.02
Gases			
Air, hydrogen, helium			1.01

The definition of the modulus of elasticity comes from the fact that stress is proportional to strain. For example, if we consider Young's modulus, Equation 10.48 may be written as

$$F_n = \frac{YA}{L_0}\Delta L \qquad \text{or} \qquad F_n = k\,\Delta L \tag{10.54}$$

where $k = YA/L_0$. Since Y, A, and L_0 are constants, k is the force constant; hence $F = k\,\Delta L = k\,\Delta x$, which states Hooke's law.

Thus, elongation increases with increasing force. When the force is removed, the wire returns to its original length. The plot of F_n versus ΔL is shown in Figure 10.21. This proportionality relation holds only if stress is less than a certain maximum value. As shown in Figure 10.21, this is reached at point A,

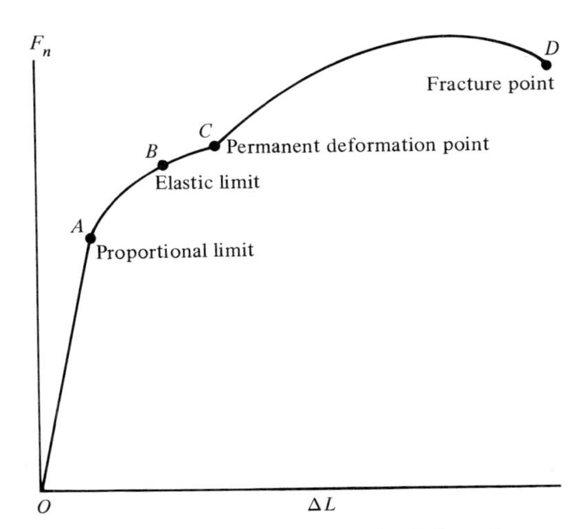

FIGURE 10.21. *Properties of a typical solid under normal stress.*

the *proportional limit* or *yield point*. If the value of the applied stress is between A and B, there is no proportionality between stress and strain, but when the stress is removed, the body does return to its original value. If the applied stress is beyond point C, permanent deformation occurs in the body and eventually, if the applied stress is very high, it will break.

EXAMPLE 10.4 A wire 1 m long and 1 mm in radius stretches 0.6 mm when a load of 20 kg hangs from it. Calculate (a) stress; (b) strain; (c) Young's modulus; and (d) the force constant for such a wire.

$L_0 = 1.0$ m, $\Delta L = 0.6$ mm $= 0.6 \times 10^{-3}$ m, $r = 1$ mm $= 10^{-3}$ m, $A = \pi r^2 = \pi(10^{-3}\text{ m})^2 = \pi \times 10^{-6}\text{ m}^2$, $F_n = mg = 20\text{ kg} \times 9.8\text{ m/s}^2 = 196$ N.

(a) $$\text{Stress} = \frac{F_n}{A} = \frac{196\text{ N}}{\pi \times 10^{-6}\text{ m}^2} = 6.24 \times 10^7\text{ N/m}^2$$

(b) $$\text{Strain} = \frac{\Delta L}{L_0} = \frac{0.6 \times 10^{-3}\text{ m}}{1.0\text{ m}} = 0.6 \times 10^{-3}$$

(c) $$\text{Young's modulus } Y = \frac{\text{stress}}{\text{strain}} = \frac{6.24 \times 10^7\text{ N/m}^2}{0.6 \times 10^{-3}} = 10.4 \times 10^{10}\text{ N/m}^2$$

(d) $$\text{Force constant } k = \frac{YA}{L_0} = \frac{(10.4 \times 10^{10}\text{ N/m}^2)(\pi \times 10^{-6}\text{ m}^2)}{1.0\text{ m}} = 3.27 \times 10^5\text{ N/m}$$

Note that this agrees with the value $k = F/\Delta L = 196\ \text{N}/0.6 \times 10^{-3}\ \text{m} = 3.27 \times 10^5\ \text{N/m}$.

Exercise 10.4 An aluminum wire has a Young's modulus of $7.0 \times 10^{10}\ \text{N/m}^2$. A wire 1 m in length and radius 1 mm, loaded with a mass of 40 kg, is hanging vertically. Calculate (a) stress; (b) change in length; (c) strain; and (d) force constant of such a wire. [*Ans.*: (a) $1.25 \times 10^8\ \text{N/m}^2$, (b) 1.8 mm, (c) 1.8×10^{-3} and (d) $2.2 \times 10^5\ \text{N/m}$.]

Example 10.5 The limiting stress S_b of a typical human bone is $0.9 \times 10^8\ \text{N/m}^2$, while Young's modulus is $1.4 \times 10^{10}\ \text{N/m}^2$. How much energy can be absorbed by two leg bones (without breaking) if each has a typical length of 50 cm and an average cross-sectional area of $5\ \text{cm}^2$?

From Equation 10.25, the potential energy when an elastic material is stretched through $x = \Delta L$ is

$$U = \tfrac{1}{2}kx^2 = \tfrac{1}{2}k(\Delta L)^2 \tag{i}$$

From Equations 10.54 and 10.48,

$$k = \frac{YA}{L_0} \quad \text{and} \quad Y = \frac{F_n/A}{\Delta L/L_0} = \frac{S_b}{\Delta L/L_0} \quad \text{or} \quad \Delta L = \frac{S_b L_0}{Y} \tag{ii}$$

Substituting for k and ΔL in i, we get

$$U = \frac{1}{2}\left(\frac{YA}{L_0}\right)\left(\frac{S_b L_0}{Y}\right)^2 = \frac{1}{2}\frac{L_0 A S_b^2}{Y} \tag{iii}$$

For the bone in question, $L_0 = 2 \times 50\ \text{cm} = 1.0\ \text{m}$, $A = 5 \times 10^{-4}\ \text{m}^2$, $S_b = 0.9 \times 10^8\ \text{N/m}^2$, and $Y = 1.4 \times 10^{10}\ \text{N/m}^2$. Substituting these in iii,

$$U = \frac{1}{2}\frac{(1.0\ \text{m})(5 \times 10^{-4}\ \text{m}^2)(0.9 \times 10^8\ \text{N/m}^2)^2}{1.4 \times 10^{10}\ \text{N/m}^2} = 145\ \text{J}$$

which is a very small amount of energy. Two legs together will be able to absorb 290 J. If $U = 290$ J for a 60 kg man in a vertical fall through height h from $U = mgh$, we get $h = 50$ cm. That is, a man will break his bone if he falls through 50 cm vertically down on a hard floor. In actual practice since one does not land with the leg bones exactly vertical to the force and the muscles absorb a large fraction of the energy, a fall through a much larger height is possible without breaking the leg bones provided that the legs do not lock-in.

Exercise 10.5 A 40 kg boy whose leg bones are $4\ \text{cm}^2$ in area and 50 cm long falls through a height of 2 m without breaking his leg bones. If the bone can stand a stress of $0.9 \times 10^8\ \text{N/m}^2$, calculate (a) Y for the material of the bone, and (b) k. [*Ans.*: (a) $0.21 \times 10^{10}\ \text{N/m}^2$, and, (b) $1.65 \times 10^6\ \text{N/m}$.]

SUMMARY

If a system vibrates such that its acceleration (or force) is directly proportional to the displacement and is directed toward the center, the system is said to execute *simple harmonic motion*. That is, for a system describing SHM, $F = -kx$ and the motion is described by

$$x = A \cos(\omega t + \phi)$$

Also, $f = 1/T = \omega/2\pi = (1/2\pi)\sqrt{k/m}$, $v = \pm\omega\sqrt{A^2 - x^2}$, and $a = -\omega^2 x$.

The potential energy of a spring executing SHM is given by $\frac{1}{2}kx^2$, and the total energy $E_T = \frac{1}{2}mv^2 + \frac{1}{2}kx^2 = K + U$ remains constant in SHM.

For angular SHM the following relations hold true: $\tau = -\kappa\theta$, $\theta = \theta_0 \cos(\omega t + \phi)$, and $T = 2\pi\sqrt{I/\kappa}$. The time period of a simple pendulum is given by $T = 2\pi\sqrt{l/g}$.

* In *damped harmonic motion*, the amplitude of oscillation will depend upon the damping constant η. When the frequency f_e of the applied external force is equal to the natural frequency f_o of the system, it is said to be in *resonance*. At $f_e = f_o$ the amplitude of oscillation is very large.

* The ability of a body to return to its original shape is called *elasticity*. *Stress* is defined as the force per unit area; normal (tensile) stress $= F_n/A$ and tangential (shear) stress $= F_t/A$. The longitudinal *strain* is defined as the ratio of change in length to the original length ($= \Delta L/L_0$). The ratio of the longitudinal stress to the longitudinal strain is *Young's modulus*, $Y = (F_n/A)/(\Delta L/L_0)$. Similarly, the shear modulus is $S = (F_t/A)/\tan\theta$, where $\tan\theta = \Delta x/h$ and the bulk modulus is $B = -\Delta p/(\Delta V/V)$. The force constant k in $F_n = k\,\Delta L$ is YA/L_0.

QUESTIONS

1. Give four examples of systems that execute SHM.
2. In one complete period of SHM, is the average value of (a) velocity; and (b) acceleration zero or nonzero?
3. A mass m is tied to a spring of length L and is vibrating with a period T. If the length of the pendulum is reduced to $L/2$ by cutting the spring, how will it change (a) the spring constant k; and (b) the period T?
4. How will a finite mass (instead of zero mass as assumed in the text) of a spring affect the time period?
5. At what points along the path of a simple pendulum is the tension in the string maximum? Minimum?
6. Why is it hard to find systems with precise SHM? Can you name one?
7. Suppose that a simple pendulum is located in an elevator. If the elevator accelerates upward or downward, how will it affect the time period of the pendulum?
8. Which one has a larger period—a pendulum with a small amplitude or one with a large amplitude, and why?
9. How can you demonstrate experimentally that the displacement-versus-time curve of a simple pendulum is a sine or cosine curve?
***10.** Give some examples of damped and critically damped motion.

*11. Give an example of two materials, one of which is more elastic than the other while the other is more plastic.

*12. When a shear stress is applied, does the lower surface get displaced?

PROBLEMS

1. A force of 10 N is needed to stretch a spring through 20 cm. What is the force constant? What force is needed to stretch it through 25 cm?
2. A horizontal spring stretches 25 cm when a force of 5.0 N is applied to it. Suppose that the same spring hangs vertically. What mass should be attached to it so that the spring will stretch 25 cm? What will be the frequency of oscillation if displaced slightly vertically?
3. The motion of a 100 g mass tied to a spring is described by the equation

$$x = 25 \cos (3t + \pi/4) \text{ cm}$$

Find (a) the angular velocity ω; (b) frequency f; (c) the time period T; (d) the force constant k; (e) the amplitude; and (f) the phase angle.

4. The motion of a 1 kg mass tied to a spring is described by the equation

$$x = 50 \cos (3t + \pi/2) \text{ cm}$$

Calculate (a) ω, f, T, k, and A; (b) the initial phase and the phase angle; and (c) v_{max} and a_{max}.

5. The horizontal spring in Figure 10.2 is attached to a 4-kg mass. When the mass is displaced 10 cm to the right from its equilibrium position, it needs a force of 40 N to hold the mass motionless in that position. The mass is pulled 30 cm and released to execute SHM. Calculate (a) k, T, f, ω, and A; (b) the force on the 4 kg mass when released; (c) v_{max} and a_{max}; and (d) v and a when $x = 10$ cm.
6. In Problem 5, calculate (a) the values of v, a, K, and U when the mass is halfway from the initial position toward the center; and (b) the total energy of the oscillator.
7. The motion of a 2-kg mass tied to a spring is described by $x = 20 \cos 6t$ cm. Calculate the value of v, a, K, and U at $x = 5$ cm and $x = 15$ cm. What is the total energy?
8. A 10 g bullet moving with velocity v_0 strikes a block and becomes embedded in it, as shown. The frequency of the block and the bullet together is 4.0 Hz and its amplitude is 30 cm. If the mass of the block is 990 g, what was the speed of the bullet before the collision?

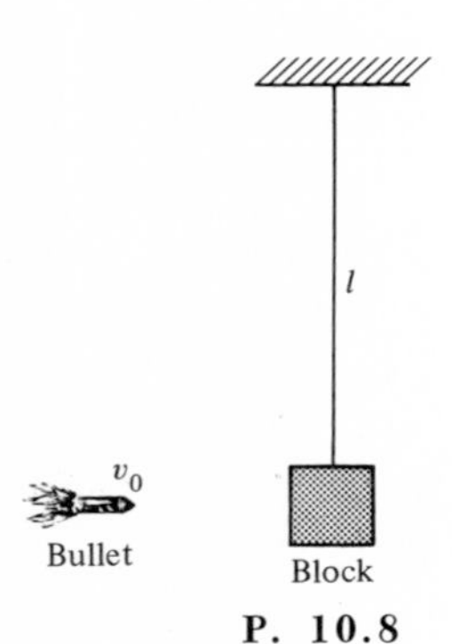

P. 10.8

9. A 1.0 kg mass block tied to a spring is oscillating with a frequency of 2 Hz on a horizontal frictionless surface. The amplitude of the oscillation is 20 cm. What will be the frequency if an additional mass of 1.0 kg is added when the block is at the end point of its oscillation? Assume that the amplitude remains the same.
10. A bullet weighing 0.10 lb with a speed of 400 ft/s is shot vertically upward into a 10.0-lb block which is suspended from the ceiling by means of a spring of force constant 4.0 lb/in. (a) Find the amplitude of the oscillations. (b) What fraction of the original kinetic energy of the bullet is stored in the oscillator? Where is the energy lost? (*Hint:* Calculate the velocity of (m + M) from momentum conservation. Use the relation $K_1 + U_1 = U_2 = \frac{1}{2}kA^2$.)

11. The motion of a block oscillating on a horizontal surface is given by $x = A \cos 10t$. If another block of mass 0.5 kg sitting on the first block slips away when the amplitude of oscillation exceeds 10 cm, what is the coefficient of friction between the two blocks? [*Hint:* Frictional acceleration $a_f = f/m_{\text{top}} \geqslant a(= F/(m_{\text{top}} + m_{\text{bottom}}.)]$

12. A 2-kg mass tied to a spring of force constant 40 N/m is observed to have an amplitude of 0.25 m. If $t = 0$, when $x = +A$, write down the equation of motion of the oscillating mass. What are the values of t, v, and a at $x = +10$ cm? Calculate the kinetic and potential energy at $x = +10$ cm and show that the sum is equal to the total energy of the motion.

13. The motion of a 1 kg mass tied to a spring is described by $x = 20 \cos 6t$ cm. Calculate the time taken by the mass to move from (a) $x = +20$ cm to $x = 10$ cm; and (b) $x = 10$ cm to $x = 0$.

14. A spring tied to a mass of 100 g oscillates vertically with a time period of 2 s and an amplitude of 20 cm. Calculate the time it takes to move 10 cm below the equilibrium position to 10 cm above the equilibrium position. (Remember that $t = 0$ when $y = A$ in $y = A \cos \omega t$.)

15. A spring tied to a mass of 100 g oscillates horizontally on a frictionless surface with a time period of 4 s and an amplitude of 20 cm. Calculate (a) the time it takes for the mass to move from its equilibrium position to a point $x = +8$ cm; and (b) the time it takes to move from $x = +8$ cm to $x = -8$ cm.

16. Show that the frequency of oscillation of the system shown is

$$f = \frac{1}{2\pi}\sqrt{\frac{k_1 + k_2}{m}}$$

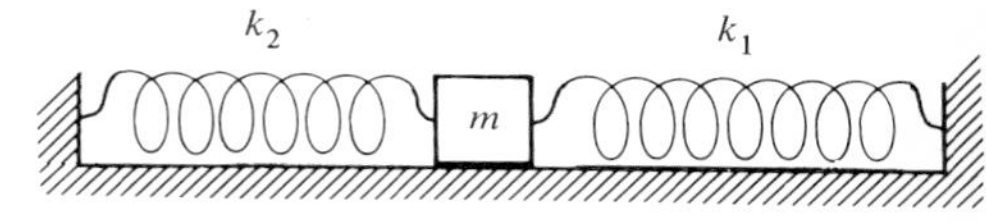

P. 10.16

17. Atoms in a solid can be imagined to be connected to one another by springs. Each atom vibrating can be considered as a simple harmonic oscillator. A typical frequency of such an oscillator at room temperature is $\sim 10^{13}$ vibrations/s. Suppose that the solid under consideration is copper. One mole of copper has a mass of 63.54 g and contains 6.03×10^{23} atoms. Calculate the force constant of a single spring.

18. What should be the length of a simple pendulum so that it will have a time period of (a) 1 s; (b) 2 s; and (c) 4 s?

19. In deriving the expression for the time period of a simple pendulum, we approximated $F = -mg \sin \theta \simeq -mg\theta$. What is the fractional error in the right side of this equation if the maximum deflection is (a) 10°; (b) 20°; and (c) 30°? [*Hint:* Calculate $(\theta - \sin \theta)/\sin \theta$.]

20. In a simple pendulum if we assume that the amplitude is not small, that is, $\sin \theta \neq \theta$, the general expression for the correct time period is given by

$$T_c = 2\pi \sqrt{\frac{l}{g}}\left(1 + \frac{1}{2^2}\sin^2\frac{\theta_m}{2} + \frac{1}{2^2}\frac{3^2}{4^2}\sin^4\frac{\theta_m}{2} + \cdots\right)$$

where θ_m is the maximum angular displacement. Calculate the value of T_c given by this expression for (a) $\theta_m = 10°$; (b) $\theta_m = 20°$; and (c) $\theta_m = 30°$. Compare the value of T_c with T; that is, calculate the fractional percentage $[(T_c - T)/T]$ (100) for the three cases.

21. If the tension in the string of a simple pendulum at the lowest point is 1.25 times the weight of the mass attached to it, what is the angular velocity of such a pendulum?

22. The time period of a simple pendulum on earth is 2 s. What will be its period on the surface of Venus ($g_V = 8.75\ \text{m/s}^2$) and Mars ($g_M = 4.06\ \text{m/s}^2$)?
23. The time period of a simple pendulum for a given l depends on the value of g. Calculate the value of $\Delta T/T$ in terms of $\Delta g/g$.
24. The balance wheel of a watch vibrates with an angular amplitude of π rad and a period of 0.50 s. (a) Write the equations of motion describing the angular displacement, velocity, and acceleration. (b) What are the values of maximum angular velocity and acceleration?
25. A certain torsion pendulum is set into oscillations by a torque of 8.0 N-m which twists it through 20°. The disk attached to the wire oscillates with a time period of 1 s. Find the moment of inertia of the disk.
26. Consider a torsion pendulum of the type shown in Figure 10.12. A thin ring of mass 100 g and radius 10 cm is placed around the torsion wire on the disk. The time period of the system is 1.5 s. The mass of the disk is 2 kg and its radius is 10 cm. Find the torsion constant of the wire.
27. A disk of moment of inertia I_1 suspended from a torsion wire is used as a torsion pendulum and has a period T_1. Another symmetrical object is placed on the disk surrounding the torsion wire. The new time period is T_2. Find the moment of inertia I_2 of the second object about the axis of the torsion wire.

*28. Consider a vertical wire of length 10 m and cross-sectional area 0.2 cm^2. When a load of 50 kg is applied, it stretches 4 mm. Calculate (a) stress; (b) strain; and (c) Young's modulus.

*29. In Problem 28, calculate the value of k by using Equation 10.54 and also by Hooke's law. Show that they agree.

*30. A steel wire 10 m long will break if it stretches more than 1 cm. Calculate the breaking stress. What is the radius of the wire if the applied force is 8000 N?

*31. Suppose that the block shown in Figure 10.20(a) is 0.5 m $\times$ 0.5 m and 1 cm thick and is made of glass. When a force F is applied, the displacement Δx is 0.05 cm. Calculate (a) the stress; (b) the strain; and (c) the force F needed to produce this displacement.

*32. A quantity 10 m^3 of water is subjected to a pressure of $1.05 \times 10^7\ \text{N/m}^2$. What will be the change in the volume of the water?

*33. A fractional change in the volume of oil is 1 percent when a pressure of $2 \times 10^7\ \text{N/m}^2$ is applied. Calculate the bulk modulus and its compressibility.

11 Gravitation and Space Physics

We start with a brief history of planetary motion. The discussion of the universal law of gravitation and Kepler's laws leads to a description of velocity, period, and so on, of the planets in the solar system. The concept of gravitational potential energy will be introduced so as to enable us to discuss satellites, ballistic missiles, and interplanetary travel. We will conclude with a brief description of the fundamentals of rocket propulsion.

11.1. History of Planetary Motion

EVER SINCE the time of the early Greeks (400 B.C.), philosophers, poets, scientists, and others have looked at the sky, sun, moon, stars, and other celestial bodies, and have wondered about the origin, structure, motion, and future of these heavenly objects. Although tremendous progress has been made in the investigations of space and in the technology leading to successful missions of landing men on the moon and bringing them back to earth safely, our understanding of the universe is far from complete. The difficulty involved in exploring microscopic systems such as atoms and nuclei is their small size; the difficulty involved in exploring space physics is just the reverse. The distances between the stars in our own galaxy are so large that to go to these stars will take years even when traveling at the speed of light 3×10^8 m/s or 186,000 mi/s—a speed very high as compared to the speed of rockets of modern technology (which move a few km/s).

The most intriguing part of space physics has been to understand planetary motion. According to the second-century Greek astronomers, the earth was assumed to be the center of the solar system, consisting of the Earth, Moon, Mercury, Venus, Sun, Mars, Jupiter, and Saturn—arranged in order of increasing distances from the earth—and all these "planets" were assumed to be going around the earth in orbits with complex motions. This is the *Ptolemaic* or *geocentric theory* of the planetary system.

Although the Ptolemaic theory of the solar system prevailed up to the sixteenth century, it could not account for the increasing number of facts made available by astronomers. In the sixteenth century, the Polish monk Nicolaus Copernicus (1473–1543) proposed a much simpler theory, the *Copernican* or *heliocentric theory*. According to his theory, the sun is supposed to be at rest at the center of the solar system, while all the other planets revolve around the sun in closed orbits in the following order with increasing distances from the sun—Mercury, Venus, Earth, Mars, Jupiter, and Saturn, and the Moon revolves around Earth as shown in Figure 11.1. Also shown in the figure are planets that were discovered after Copernicus. The difference between the above two theories is that according to Copernicus (heliocentric theory), the sun was considered to be the reference frame; whereas, according to Ptolemy (geocentric theory), the earth was considered to be the reference frame. As we know, Copernicus was correct.

The Danish astronomer Tycho Brahe (1546–1601) spent about 20 years in

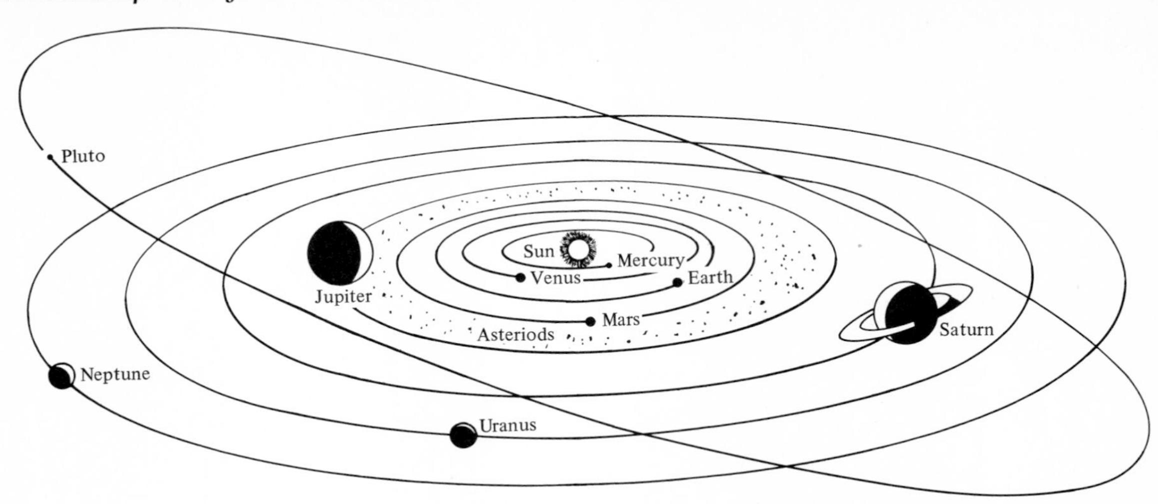

Figure 11.1. *According to the Copernican or heliocentric theory of the solar system the planets move in orbits, with the sun at rest at the center, as shown. The elliptical paths of the planets are according to Kepler's first law.*

the latter half of the sixteenth century compiling observational data about the position of the planets. He was the last great astronomer to make systematic observations without a telescope, and surprisingly his data do not contain errors any larger than $\frac{1}{15}$ of a degree. The fruits of his labor became quite evident when astronomer and mathematician Johannes Kepler (1571–1630),

Table 11.1
The Planets of our Solar System

Name of Planet	Distance[a] from Sun ($\times 10^7$ km)	Mass[b] (earth's mass = 1)	Diameter[a] ($\times 10^3$ km)	Average Density[c]	Revolution Period (yr)	Rotation Period[d]	Average[e] Temperature (K)	Number of Satellites
Mercury	5.79	0.05	4.827	5.4	0.24	59 d	633	0
Venus	10.78	0.82	12.228	5.1	0.62	243 d	463	0
Earth	14.96	1.00	12.759	5.52	1.00	23 h, 56 m, 4.1 s	294	1
Mars	22.85	0.12	6.870	3.97	1.88	24 h, 37 m, 22.6 s	219	2
Jupiter	77.71	317.80	143.201	1.33	11.86	9 h 50 m	173	14
Saturn	142.56	95.2	120.675	0.68	29.46	10 h 14 m	128	10
Uranus	286.4	14.5	48.27	1.00	84.01	10 h 49 m	90	5
Neptune	448.91	17.2	45.052	2.25	164.97	15 h	72	2
Pluto	590.5	~0.1	~6.436	~4	248.4	6.39 d	63	1
Sun		332,776	1,392,000	1.41		25–35 d	5776	

[a] To obtain the distance in miles, multiply by 0.6214.
[b] Mass of earth = 5.98×10^{24} kg.
[c] With respect to water density = 1×10^3 kg/m^3.
[d] d, days; h, hours; m, minutes; s, seconds.
[e] Surface temperature.

who was Tycho Brahe's assistant, spent 20 years analyzing Brahe's data. Kepler, using Brahe's data and Copernicus' heliocentric theory, was able to observe regularities in the positions of the planets and came up with a *kinematical* description of planetary motion in the form of three laws, called *Kepler's laws*, which we shall discuss in Section 5. Once these laws were known, the next thing was to know the *dynamics* of the planetary system or, in modern terminology, the types of interactions that hold these masses in their positions in orbital motion. In 1666, Sir Isaac Newton (1642–1727) sought to find the nature of the forces between the planets and the sun. With the help of Kepler's third law, Newton arrived at the universal law of gravitation. The gravitation law and the laws of motion were sufficient to put most of the data of the planets in order. Table 11.1 shows the latest data about different planets in the solar system.

11.2. Newton's Universal Law of Gravitation

In 1666, Isaac Newton, at the age of 23 years, stated the universal law of gravitation in the following form:

> **Newton's Law of Gravitation:** ***The gravitational force (or interaction) of attraction between any two objects in the universe is directly proportional to the product of their masses and inversely proportional to the square of the distance betweeen them.***

Thus, the force F between two objects of masses m_1 and m_2 separated by a distance r as shown in Figure 11.2 is given by

$$F = G\frac{m_1 m_2}{r^2} \tag{11.1}$$

where G is called the *gravitational constant* and its presently accepted value is

$$G = (6.673 \pm 0.003) \times 10^{-11} \frac{\text{N-m}^2}{\text{kg}^2} \tag{11.2}$$

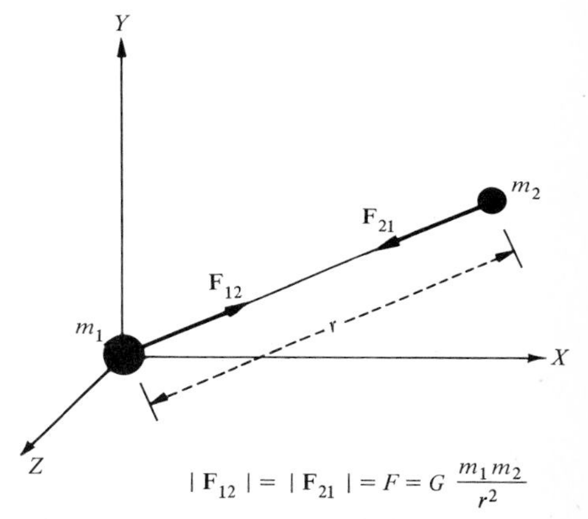

FIGURE 11.2. *According to Newton's third law the gravitational force* $\mathbf{F}_{21}$ *acting on* m_2 *is equal and opposite to the gravitational force* $\mathbf{F}_{12}$ *acting on* m_1.

Newton's first and second laws are laws of motion, whereas the gravitational law is a law of force. According to Newton's third law, the forces must exist in action–reaction pairs. As shown in Figure 11.2, the mass m_2 is attracted by mass m_1 with a gravitational force $\mathbf{F}_{21}$, while m_1 is attracted by mass m_2 with a gravitational force $\mathbf{F}_{12}$. According to Newton's third law, $\mathbf{F}_{12} = -\mathbf{F}_{21}$. Thus,

$$|\mathbf{F}_{12}| = |\mathbf{F}_{21}| = F = G\frac{m_1 m_2}{r^2} \tag{11.3}$$

Since the line of action passes through a fixed point, on the line joining the two masses, that is, the force is directed toward the center of the mass, the gravitational force is a *central force*.

Experimental Determination of G

Even though the universal law of gravitation was given by Newton in 1666, the first reasonably accurate experimental determination of the gravitational constant G was made by Lord Cavendish in 1798. Many improvements were made later by Poynting and Boys. The currently accepted value of G was measured by Heyl and Chizanowski of the U.S. National Bureau of Standards in 1942. We shall briefly describe the method used by Cavendish.

The value of G was determined by Cavendish by measuring the force between two known masses that are separated by a known distance by means of a device called a *torsion balance*, shown in Figure 11.3. Two small spheres

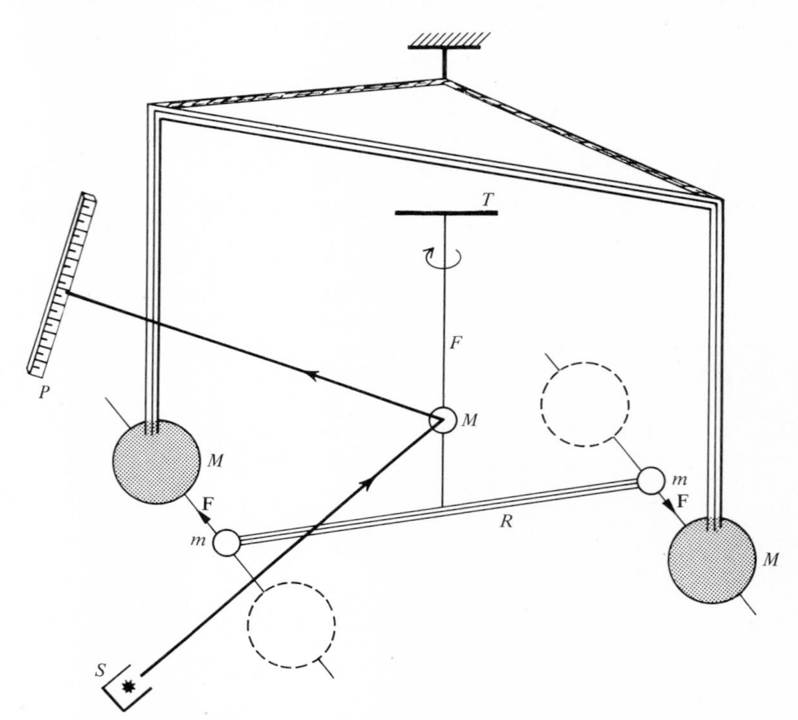

Figure 11.3. *Schematic arrangement of the Cavendish experiment to measure the gravitational force between two masses of ordinary size.*

each of mass m were attached to the ends of a rod R. The rod R was suspended from the center by a very delicate fiber F which was attached to a support T. The twist of the fiber resulting from the horizontal rotational motion of the rod R was magnified by reflecting a beam of light (coming from source S) from a mirror M attached to the fiber and by receiving the beam on a distant scale P.

If two large masses M are brought close to the two small masses m, the forces of attraction will produce a torque that will cause a twist in the fiber. Since the forces and the torque are extremely small, the fiber must have a very small torsion constant; that is, a small torque must produce a large twist. (It should be pointed out that the spherical masses can be treated as point masses.) The torsion constant of the fiber can be calculated by using the system as a torsion pendulum. Knowing the deflection, masses, length of the rod R, and torsion constant, the value of G may be calculated as illustrated in Example 11.1.

EXAMPLE 11.1 In Cavendish's experiment, let each small mass be 20 g and each large mass be 5 kg. The rod connecting the small masses is 50 cm long, while the small and large spheres are separated by 10.0 cm. The torsion constant is 4.8×10^{-8} kg-m²/s² and the resulting angular deflection is 0.4°. Calculate the value of the universal gravitational constant G from these data.

The force on each small sphere is $F = GMm/r^2$, where $M = 5.0$ kg, $m = 20$ g $= 0.2$ kg, and $r = 10$ cm $= 0.1$ m. Each force F has a moment arm of $l/2$ where $l = 50$ cm $= 0.5$ m. Thus, the total torque τ (τ = force $\times$ perpendicular distance) is

$$\tau = \frac{2Fl}{2} = 2G\frac{Mm}{r^2}\frac{l}{2}$$

But the torque, torsion constant κ, and angular displacement θ are related by

$$\tau = \kappa\theta$$

where $\kappa = 4.8 \times 10^{-8}$ kg-m²/s² and $\theta = (0.4°)(2\pi/360°)$ rad $= 0.007$ rad. Thus, combining the equations above,

$$G\frac{Mml}{r^2} = \kappa\theta \qquad \text{or} \qquad G = \frac{\kappa\theta r^2}{Mml}$$

$$G = \frac{(4.80 \times 10^{-8}\ \text{kg-m}^2/\text{s}^2)(0.007\ \text{rad})(0.1\ \text{m})^2}{(5\ \text{kg})(0.02\ \text{kg})(0.5\ \text{m})} = 6.72 \times 10^{-11}\ \text{N-m}^2/\text{kg}^2$$

EXERCISE 11.1 In Cavendish's experiment $M = 10$ kg, $m = 50$ g, and the separation between the two masses is 12 cm, while the length of the rod separating the two masses is 60 cm. The torsion constant of the wire is 5.10×10^{-8} kg-m²/s². Using the value of $G = 6.673 \times 10^{-11}$ N-m²/kg², calculate (a) F; (b) τ; (c) θ; and (d) the linear deflection that will be produced on a scale placed at a distance of 1 m from the mirror. [*Ans.*: (a) $F = 2.32 \times 10^{-9}$ N; (b) $\tau = 1.39 \times 10^{-9}$ N-m; (c) $\theta = 2.73 \times 10^{-2}$ rad $=$ 1.56°; (d) 2.73 cm.]

Once the value of G is determined it is possible to calculate the mass of the earth by knowing the radius of the earth, R_e. This is the reason Cavendish is said to be the first person to weigh the earth. We shall illustrate this point in Example 11.2.

Finally, as pointed out in Chapter 4, the gravitational interaction is the weakest of all the four known types of interactions. This is the reason gravitational interactions are of little consequence for objects of small mass. But this interaction becomes enormous when gravitational forces of attraction between planets are considered. These are the forces that are holding the universe together.

In the macroscopic world, the gravitational law has been verified and seems to hold everywhere. There are speculations that this law may be inadequate in or near the vicinity of very dense celestial objects such as neutron stars. It is postulated that such stars are the result of gravitational collapse and may have densities more than 10^9 times the density of the sun.

11.3. Gravitational Field Strength

According to Equation 11.1, the gravitational force of attraction that acts on mass m due to mass M separated by a distance r is $F = GMm/r^2$, where the direction of $\mathbf{F}$ on m is toward M.

The concept of force may be treated either as an *action-at-a distance* or as a *field concept*. According to the former, the gravitational force between two masses comes into existence instantly when the two masses separated by a distance r come into existence. What this means is that the information about the force is conveyed from one mass to the other with very high (infinite) speed. An alternative approach is to introduce the concept of a gravitational field. The field is assumed to exist in the region surrounding the mass M, and if the mass m, which we may call a "test mass," is brought into this region of space, it interacts with the field, resulting in the force of attraction. It is immaterial whether the field existed before the mass m was brought into position or afterward. The *gravitational field strength* or *intensity* $\mathbf{g}$ (or $\mathbf{I}_g$) at any point in space surrounding the mass M is defined as the force that will act on a unit mass; that is,

$$g = |\mathbf{g}| = \frac{|\mathbf{F}|}{m} = G\frac{M}{r^2} \tag{11.4}$$

Suppose that we consider the mass M (which may be the earth) as shown in Figure 11.4(a). The gravitational field intensity due to this mass varies as $1/r^2$.

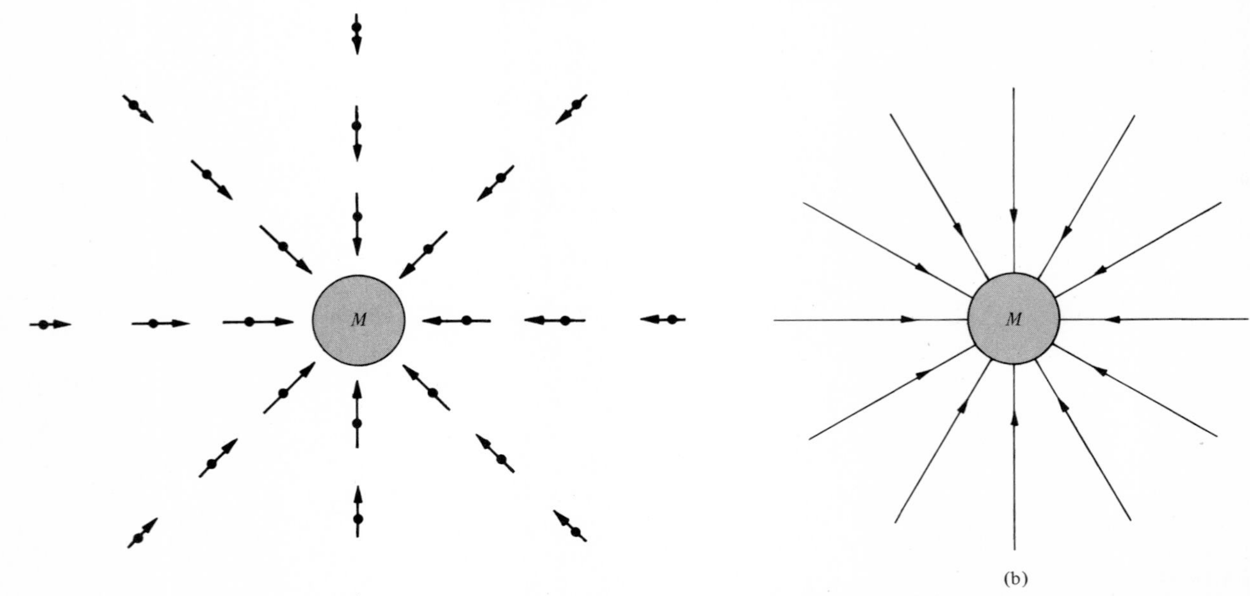

FIGURE 11.4. *(a) The gravitational field strength at different points (dots) surrounding the mass M is proportional to the length of the arrow and is in the direction of the arrow. (b) The gravitational field lines due to a mass M.*

The intensity at these points (dots are the points) is proportional to the length of the arrows at different points. If these arrows are joined together as shown in Figure 11.4(b), the resulting lines are the *field lines*. We shall illustrate this concept in detail when we talk about electric fields in Chapter 19.

11.4. Variation in g

Let us now look into the factors that affect the value of g. Suppose that an object of mass m is on the surface of a perfectly spherical planet of mass M and radius R. The gravitational force F with which the planet pulls the mass m is

$$F = G\frac{Mm}{R^2}$$

If the acceleration due to free fall on the surface of this planet is g_0, we may also write

$$F = mg_0$$

Equating the two equations above gives

$$\boxed{g_0 = G\frac{M}{R^2}} \tag{11.5}$$

Thus, the value of g on the surface of any planet depends on its mass M and radius R. The value of g_e for the planet earth is 9.8 m/s² or 32.2 ft/s². For the moon, which has a mass 1/81.3 of the mass of the earth and a radius 1/3.65 of the radius of the earth, the value of g_m is $(1/6)g_e$, that is, 1.63 m/s² or 5.39 ft/s².

At points above the surface of the planet, the value of the force $F(r)$ on m or the value of $g(r)$ depends upon r, the distance from the center of the planet. Thus,

$$F(r) = mg(r) = G\frac{Mm}{r^2}$$

Substituting the value of $GM = g_0R^2$ from Equation 11.5, we get

$$F(r) = mg(r) = mg_0\left(\frac{R}{r}\right)^2 \propto \frac{1}{r^2} \quad \text{for } r \geqslant R \tag{11.6a}$$

or

$$\boxed{g(r) = g_0\left(\frac{R}{r}\right)^2} \quad \text{for } r \geqslant R \tag{11.6b}$$

Besides the variation in the value of g from planet to planet, there are other factors that cause changes in the value of g. To illustrate our point we shall restrict the discussion to the value of g due to the earth. The same type of remarks will apply equally to the other planets as well. The value of g depends on the following factors:

1. Distance from the center of the earth—as r changes, the value of g changes, as listed in Table 11.2.

TABLE 11.2

A. Variation of g with Altitude at 45° Latitude

Altitude (m)	($\times R_e$)	g (m/s²)
0	1	9.806
100,000	0.0157	9.598
1,000,000	0.157	7.41
12,740,000	2.00	2.4515
380,000,000[a]	59.6546	0.00271

[a] Distance of the moon from the earth.

B. Variation of g with Latitude at Sea Level

Latitude	g(m/s²)	g(ft/s²)
0° (equator)	9.78039	32.0878
20°	9.78641	32.1076
40°	9.80171	32.1578
60°	9.81918	32.3151
80°	9.83059	32.2525
90° (poles)	9.83217	32.2577

2. Rotational motion of the earth—causes centripetal acceleration. This results in the variation in the value of g from the equator to the pole as shown in Table 11.2.

3. Nonspherical shape of the earth—results in the following change in g:

$$(\Delta g)_{\text{shape}} = g_{\text{pole}} - g_{\text{equator}} = 0.018 \text{ m/s}^2$$

EXAMPLE 11.2 Calculate the mass and the average density of the material of the earth if the radius is 6.37×10^3 km.

At the surface of the earth $R = R_e = 6.37 \times 10^6$ m and $g = 9.8 \text{ m/s}^2$. Substituting these values in Equation 11.5 and writing $M = M_e =$ the mass of the earth, we get

$$M_e = \frac{gR_e^2}{G} = \frac{(9.8 \text{ m/s}^2)(6.37 \times 10^6 \text{ m})^2}{6.67 \times 10^{-11} \text{ N-m}^2/\text{kg}^2} = 5.97 \times 10^{24} \text{ kg}$$

Thus, if the earth is assumed to have uniform density $\bar{\rho}$ and a spherical shape of radius R so that its volume V is $(4\pi/3)R^3$, we may write

$$\bar{\rho} = \frac{M_e}{V} = \frac{M_e}{(4\pi/3)R^3} = \frac{5.97 \times 10^{24} \text{ kg}}{(4\pi/3)(6.37 \times 10^6 \text{ m})^3} \simeq 5.5 \times 10^3 \text{ kg/m}^3$$

$$\simeq 5.5 \text{ g/cm}^3$$

The density of the outer layer of the earth is found to be less than 5.5 g/cm^3; hence, the inner layers must have higher densities.

EXERCISE 11.2 The radius of the moon is 1/3.65 of the radius of the earth, while the value of g_m is 1/6 of that on the earth's surface. Calculate the mass and the density of the moon, assuming the moon to be spherical in shape. [*Ans.*: $M_m = 7.46 \times 10^{22}$ kg, $\bar{\rho}_m = 3.35 \text{ g/cm}^3$.]

11.5. Kepler's Laws of Planetary Motion

As pointed out at the beginning of the chapter after years of analyzing the astronomical data of Brahe, Kepler deduced three laws of planetary motion. The first two laws were published in 1609 and the third in 1619. These laws, called Kepler's laws, may be stated as follows:

I. **The Law of Orbits:** ***The planets move in elliptical orbits with the sun at one focus.***
II. **The Law of Areas:** ***A line (or position vector) joining any planet to the sun sweeps out equal areas in equal intervals of time.***
III. **The Law of Periods:** ***The square of the period of revolution of any planet is proportional to the cube of the mean distance of the planet from the sun.***

There is a fundamental difference between Newton's laws of motion and Kepler's laws. Newton's laws are about motion and force, in general, and as such, implicitly involve an interaction between objects; whereas Kepler's laws describe the motion of only a single system, the planetary system, and do not involve interaction. Whereas Newton's laws are *dynamic*, giving relations among force, mass, distance, and time, Kepler's laws are *kinematic*, giving a relation between distance and time. Kepler's laws should apply not only to the solar system but to the moons going around the planets as well as to artificial satellites. In addition, Kepler's laws are valid whenever inverse-square force law is involved.

Kepler's First Law. The gravitational force between any planet and the sun is a central force (a force directed toward or away from the center) and varies inversely as the square of the distance between them. Without going into a derivation, we state that the path of a planet moving under the influence of such a gravitational force is an ellipse with the sun at one focus, as shown in Figure 11.5. (In general, the orbit is a conic section—ellipse, hyperbola, or

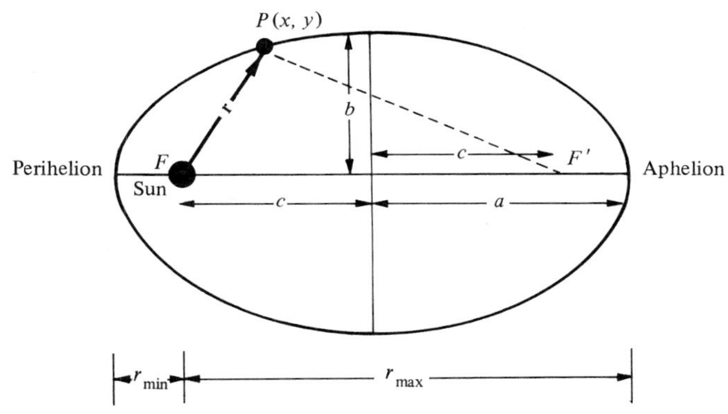

Figure 11.5. *Properties of an elliptical orbit described by a particle moving under a central force.*

parabola. But if the velocity or the energy is just right, it is an ellipse.) An ellipse may be defined as "the locus of points the sum of whose distances from two fixed points is constant." Thus, in Figure 11.5, for any point $P, FP + F'P =$ constant. The fixed points F and F' are the *foci* of the ellipse. The distance a is the *semimajor axis* of the ellipse and the distance b is the *semiminor axis* of the ellipse. Any point (x, y) on the elliptical orbit is given by the equation of the ellipse,

$$\frac{x^2}{a^2} + \frac{y^2}{b^2} = 1 \tag{11.7}$$

In particular for a circular orbit F and F' coincide and $a = b$; hence, $x^2 + y^2 = a^2$, and the focus is the center of the circle.

As shown in Figure 11.6, the gravitational force on the earth due to the sun is directed toward the sun. The closest and furthest positions of the planet from the sun are the *perihelion* and *aphelion* positions, respectively. The corresponding terms used for earth satellites are called *perigee* and *apogee*, respectively.

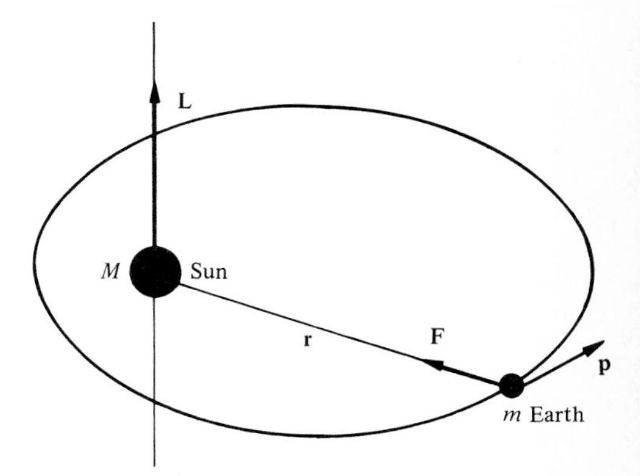

Figure 11.6. *The gravitational force acting on the earth that is directed toward the sun results in an elliptical orbit of the earth.*

Kepler's Second Law. Consider a planet, say the earth, moving around the sun as shown in Figure 11.6. The force F acting on the earth is the gravitational force directed toward the sun. If we take an axis through the center of the sun, this force will not exert any torque about this center because F passes through the center; hence,

$$\tau = \frac{\Delta \mathbf{L}}{\Delta t} = 0 \qquad \text{or} \qquad \Delta L = L_2 - L_1 = 0$$

Thus,

$$L_1 = L_2 = L = \text{constant} \tag{11.8}$$

Therefore, the angular momentum of the earth about an axis through the sun is constant. For the earth moving in an elliptical path for two different points on the orbit, we may write

$$L_1 = L_2 \qquad \text{or} \qquad r_1 p_1 = r_2 p_2 \qquad \text{or} \qquad r_1 m v_1 = r_2 m v_2$$

where r_1 and r_2 are the perpendicular distances from the sun to the mass m at the orbit points 1 and 2, while v_1 and v_2 are the velocities at the two points. Thus,

$$\boxed{r_1 v_1 = r_2 v_2} \tag{11.9}$$

This means that the tangential speed of the planet in the orbit must vary. The speed is least when the planet is at the farthest position (aphelion or apogee) from the force center, and the speed is greatest when it is at the closest position (perihelion or perigee). This should be clear from the fact that when the planet is closest to the force center (at the perihelion position), the force acting on it is maximum and it will be moving the fastest. As it goes away from the force center, it slows down as a result of decreasing force and so is slowest at the aphelion position.

Figure 11.7 shows the velocity of the planet at three different points. From Equation 11.9, the velocity is inversely proportional to the position vector, the distance of the planet from the force center. Thus, $r_1 > r_2 > r_3$ and $v_1 < v_2 < v_3$. Thus we can show that the areas A_1, A_2, and A_3 swept by the

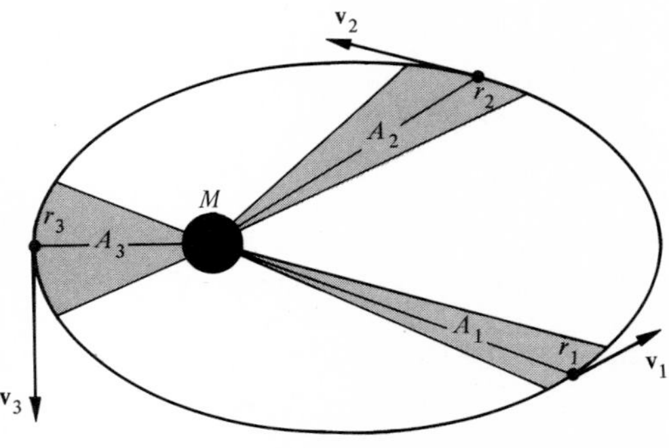

FIGURE 11.7. *The shaded areas A_1, A_2, and A_3 swept out in unit intervals of time are equal; that is, the areal velocity is constant.*

position vectors in unit time are the same, $A_1 = A_2 = A_3$. This fact is stated by saying that the areal velocity, area swept per unit time, $\Delta A/\Delta t$, is constant even though r and v change from point to point. This is Kepler's second law.

Kepler's Third Law. In this law Kepler sought to find the relation between the orbits of different planets in the solar system. Deriving this law by using an elliptical orbit is somewhat complex. Therefore, we make a simplifying assumption that the orbit of the planet is circular with radius r (which is equal to the semimajor axis a) and the sun at the center of the circle as in Figure 11.8. The centripetal force F_c needed to keep the planet of mass m in a circular orbit is the gravitational force F_G exerted on it by the sun of mass M.

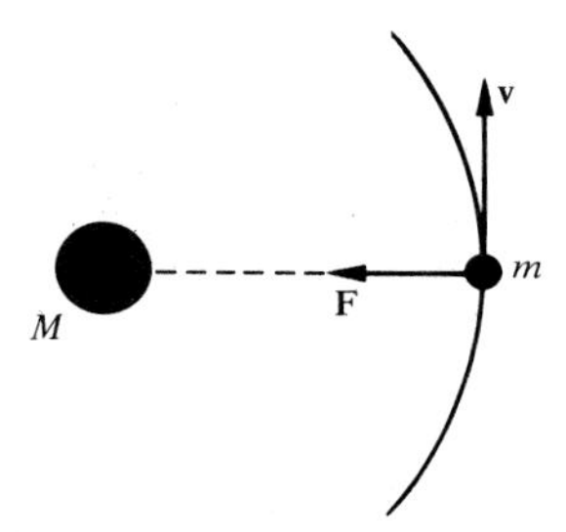

Figure 11.8. *Planet of mass m assumed to be moving in a circular orbit around the sun of mass M.*

Thus,

$$F_c = F_G \quad \text{or} \quad \frac{mv^2}{r} = \frac{GMm}{r^2} \tag{11.10}$$

where v is the speed of the planet, $v = 2\pi r/T$, and T is the period for one revolution. Substituting for v in Equation 11.10,

$$m\left(\frac{2\pi r}{T}\right)^2 \frac{1}{r} = \frac{GMm}{r^2}$$

$$\boxed{T^2 = \frac{4\pi^2}{GM} r^3 = Kr^3} \tag{11.11}$$

which is Kepler's third law. The constant $K = K_S$ for the sun with $M = M_S$ = the mass of the sun is

$$K_S = \frac{4\pi^2}{GM_S} \tag{11.12}$$

The value of this constant for the moon and other satellites around the earth will be

$$K_E = \frac{4\pi^2}{GM_E} \tag{11.13}$$

and for each of the 12 moons going around Jupiter it has a value

$$K_J = \frac{4\pi^2}{GM_J} \tag{11.14}$$

Kepler's third law is illustrated in Figure 11.1 and Table 11.1 for the planets of the sun.

EXAMPLE 11.3 The radius of the orbit of the moon around the earth is 3.84×10^8 m and its time period is 27.3 days. Calculate the mass of the earth from these data.

Applying Equation 11.11 to the earth–moon system with $M = M_e$, we get

$$M_e = \frac{4\pi^2 r^3}{GT^2}$$

where $r = 3.84 \times 10^8$ m, $T = 27.3$ days $= 2.36 \times 10^6$ s, and $G = 6.67 \times 10^{-11}$ N-m²/kg². Substituting these values gives

$$M_e = \frac{4\pi^2(3.84 \times 10^8 \text{ m})^3}{(6.67 \times 10^{-11} \text{ N-m}^2/\text{kg}^2)(2.36 \times 10^6 \text{ s})^2} = 6.02 \times 10^{24} \text{ kg}$$

Compare this value with the one obtained from Equation 11.4; that is, $M = gR^2/G$.

EXERCISE 11.3 The orbital radius of Sputnik I around the earth could be taken to be 6.94×10^6 m while its period was 5.77×10^3 s. Calculate the mass of the earth from these data. How does this value compare with the one calculated in the above example? [*Ans.*: 5.94×10^{24} kg.]

EXAMPLE 11.4 The orbital radius of the earth around the sun is 1.496×10^{11} m (or 1 astronomical unit, A.U.) and its period is 3.156×10^7 s (or 1 earth year). Making use of Kepler's third law, (a) calculate the mass of the sun. (b) If the period of the planet Saturn is 29.6 earth years, what is its orbital radius around the sun?

From Equation 11.11,

$$\frac{T^2}{r^3} = \frac{4\pi^2}{GM} = K = \text{constant}$$

where T and r may be values for any planet, and we shall use those of earth,

$$M = \frac{4\pi^2 r^3}{GT^2} = \frac{4\pi^2(1.496 \times 10^{11} \text{ m})^3}{(6.67 \times 10^{-11} \text{ N-m}^2/\text{kg}^2)(3.156 \times 10^7 \text{ s})^2} = 1.99 \times 10^{30} \text{ kg}$$

Also Equation 11.11 may be written as

$$\frac{T_E^2}{r_E^3} = \frac{T_S^2}{r_S^3}$$

where E stands for the earth and S for Saturn. Substituting the proper values, we get

$$r_S^3 = r_E^3 \frac{T_S^2}{T_E^2} = (1 \text{ A.U.})^3 \left(\frac{29.6}{1}\right)^2$$

$$r_S = 9.57 \text{ A.U.} = (9.57)(1.496 \times 10^{11} \text{ m}) = 1.43 \times 10^{12} \text{ m}$$

Exercise 11.4 If the period of the planet Neptune is 165.0 earth years, what is its orbital radius? [*Ans.*: $r_N = 4.50 \times 10^{12}$ m.]

Perturbation Effect. While considering the orbital path of a planet around the sun, we assumed that the presence of other planets does not affect the motion of the planet under consideration. This was justified because the mass of the sun is very large compared to the mass of any other planet. But it had been recognized by Newton that the force between one planet and another, although very small, will cause significant deviations from the perfect elliptical orbit. These small disturbing forces due to other planets, which may be ignored for most practical purposes, are called *perturbations*. The perturbation can be calculated accurately by means of special techniques that come under the name *celestial mechanics*.

There are two types of effects produced on the planets by this perturbation. It produces (1) rotation of the axis of the ellipse, and (2) the periodic variation or oscillation in the eccentricity of the ellipse from the average value. In fact, the changes are extremely small. For example, in the case of the earth, the rate of advance of the perihelion (rotation of the major axis) is only about 21 arc minutes per century. The total period for the two effects is of the order of 10^5 years.

The perturbation effects have led to very interesting and valuable predictions. At the time of Newton only six planets—Mercury, Venus, Earth, Mars, Jupiter, and Saturn—were known. In 1781 William Herschel in England, with the help of a telescope, discovered the seventh planet, Uranus, which is twice as far from the sun as the sixth planet, Saturn. The orbit of the planet Uranus was calculated assuming the force due to the sun and the perturbations due to the other six planets. Uranus did not seem to follow the orbit precisely and the orbit of Uranus was still perturbed. To account for this perturbation, John Adams in England and Urbain Leverrier in France in the 1840s predicted the presence of an eighth planet and, independent of each other, calculated the exact position of this planet in the sky. Such a planet was discovered at the Berlin Observatory the night the information about its location in the sky was obtained. This eighth planet was named Neptune.

It was the calculations on the orbits of Neptune and Uranus that led an American astronomer, Percival Lowell, to predict still another planet beyond Neptune. After 25 years, in 1930, the ninth planet, Pluto, was discovered at the Lowell Observatory in Arizona. There may be indications of the presence of still another planet, a tenth one, but to date this has not been observed.

11.6. Gravitational Potential Energy and Satellites

The gravitational potential energy of a mass m when raised to a height h from the surface of the earth is mgh. This is true only if g is constant. If h is very large, g is not constant. Thus, the potential energy difference must be calculated from basic principles.

Let a mass m be moved from $r_i = r$ to $r_f = r + \Delta r$ in a gravitational field where the force between r_i and r_f may be approximated to be $F = GMm/r_i r_f$ (we are using $r^2 \approx r_i r_f$). Thus, the work done is given by

$$W = F\,\Delta r = G\frac{Mm}{r_i r_f}(r_f - r_i) = GMm\left(\frac{1}{r_i} - \frac{1}{r_f}\right) \tag{11.15}$$

But the work done is also equal to the potential-energy difference. Thus, if $V(r_i)$ and $V(r_f)$ are the potential energies at r_i and r_f, respectively, we may write

$$W = V(r_f) - V(r_i) \tag{11.16}$$

It is convenient and conventional to define potential energy to be zero at infinity. Thus, substituting $V(r_f) = V(\infty) = 0$ in Equations 11.15 and 11.16, we get

$$0 - V(r_i) = GMm\left(\frac{1}{r_i} - \frac{1}{\infty}\right)$$

Thus, the gravitational potential energy $V(r)$ at any point $r\,[r_i = r$ and $V(r_i) = V(r)]$ is

$$\boxed{V(r) = -\frac{GMm}{r}} \tag{11.17}$$

The negative sign is consistent with the fact that in order for the mass m to be taken to infinity, external work in the amount of GMm/r must be done, so that at $r = \infty$, $V(\infty) = 0$, as defined to start with. Also, if we move the mass m from the surface of the earth to a height $R + h$, the change in potential energy is correctly predicted by Equation 11.17 to be mgh, where h is small.

Now if a mass m is moving with velocity v in a circular orbit of radius r around the earth, its total energy E is given by

$$\text{total energy } E = \text{kinetic energy } K + \text{potential energy } V \tag{11.18}$$

$$\boxed{E = \frac{1}{2}mv^2 + \left(-\frac{GMm}{r}\right)} \tag{11.19}$$

Furthermore, for a circular orbit the centripetal force is equal to the gravitational force. That is, as shown in Figure 11.8, and Equation 11.10,

$$\frac{mv^2}{r} = \frac{GMm}{r^2} \qquad \text{or} \qquad v^2 = \frac{GM}{r}$$

That is,

$$K = \frac{1}{2}mv^2 = \frac{1}{2}\frac{GMm}{r} \tag{11.20}$$

Substituting this in Equation 11.19, we get

$$E = -\frac{1}{2}\frac{GMm}{r} \tag{11.21}$$

which indicates that the total energy E of mass m moving in the gravitational field of mass M is negative. Figure 11.9(a) shows the plots of K, V, and E by using Equations 11.20, 11.17, and 11.21, respectively. For $E < 0$, the orbit of the mass m is an ellipse or a circle, as shown in Figure 11.9(b). Such a closed orbit means that for negative energies, the particle in the system is bound. By supplying extra kinetic energy to the particle, it is possible for the total energy $E = K + V$ to be positive or zero. If the total energy were positive ($E > 0$) or zero ($E = 0$), the orbit will be an open one, that is, a hyperbola ($E > 0$) or a parabola ($E = 0$). The systems with $E > 0$ or $E = 0$ are unbound.

Motion of Satellites. Suppose that a satellite is lifted to a height h from the surface of the earth with the help of a rocket. At P in Figure 11.10 the rocket gives a final thrust to the satellite, resulting in a horizontal velocity v, as shown. Thus, the total energy E of the satellite at point P will be

$$E = \frac{1}{2}mv^2 - \frac{GMm}{R + h}$$

where m is the mass of the satellite, M the mass of the earth, and R the radius of the earth. Depending on the value of v, E will be less than 0, greater than 0, or equal to 0, corresponding to three possible orbits—ellipse, hyperbola, and parabola, respectively, as discussed previously. If $E\,(< 0)$ is too low, the elliptical orbit will intersect the earth and the satellite will fall back. If v is sufficient but E is still < 0, the satellite will keep on moving in a closed

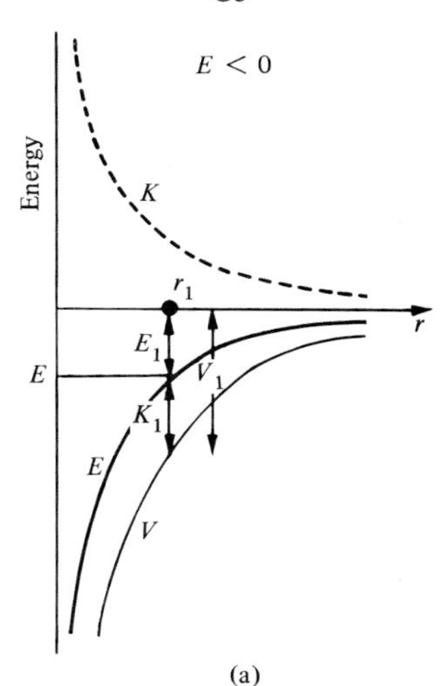

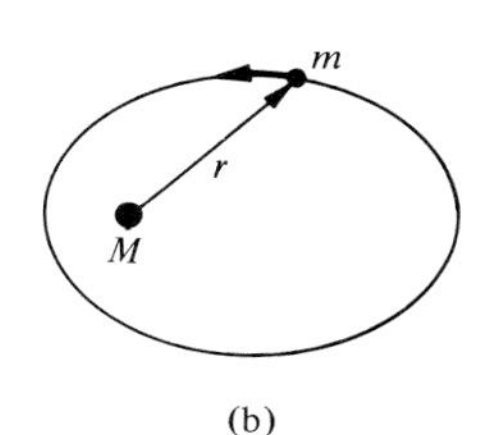

Figure 11.9. *(a) The plots of K, V, and E for a particle with $E < 0$ moving in a bound orbit shown in part (b).*

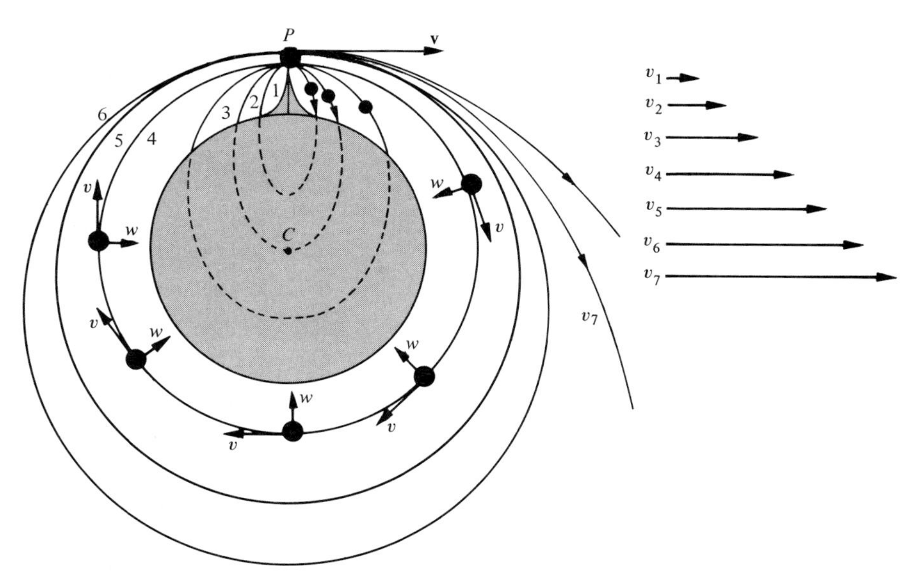

Figure 11.10. *The paths of a satellite for different values of the initial velocity $v = v_1, v_2, \ldots, v_6, v_7$. (Only for circular orbits are **v** and **g** at right angles.)*

elliptical orbit. On the other hand, if v is such that $E \geqslant 0$, the satellite will escape from the earth, as illustrated in Figure 11.10 by orbit 7.

The minimum velocity v_e that is necessary for the satellite to escape from the gravitational pull of the earth is given by substituting $E = 0$ and $r = R$ at the earth's surface in Equation 11.19. Thus,

$$\frac{1}{2}mv_e^2 = \frac{GMm}{R}$$

or

$$\boxed{v_e = \sqrt{\frac{2GM}{R}}} \tag{11.22}$$

Substituting the values of G, M, and R,

$$v_e = \sqrt{\frac{2(6.673 \times 10^{-11}\ \text{N-m}^2/\text{kg}^2)(5.975 \times 10^{24}\ \text{kg})}{6.371 \times 10^6\ \text{m}}} = 1.12 \times 10^4\ \text{m/s}$$

$$= 11.2\ \text{km/s} = 25{,}280\ \text{mi/h} \simeq 7.02\ \text{mi/s}$$

If v is less than this, the object falls back to the earth, and if it is greater than this, the object escapes the gravitational field of the earth and still has some kinetic energy.

Usually, the problem is different. We would like to know what the velocity of a satellite should be so that it is set into an orbit of radius r. From the relation that the gravitational force provides the centripetal force, we get

$$\frac{GMm}{r^2} = \frac{mv^2}{r}$$

or

$$v^2 = \frac{GM}{r} \tag{11.23}$$

But $GM/R^2 = g_R$ or $GM = g_R R^2$, where g_R is the value of g at the surface of the earth. Substituting this in Equation 11.23, we get the velocity v of a satellite going in an orbit of radius r:

$$\boxed{v = R\sqrt{\frac{g_R}{r}}} \tag{11.24}$$

The time period T required by the satellite to complete one revolution is given by

$$T = \frac{2\pi r}{v} = \frac{2\pi r}{R\sqrt{g_R/r}}$$

or

$$\boxed{T = \frac{2\pi r^{\frac{3}{2}}}{R\sqrt{g_R}}} \tag{11.25}$$

Ballistic Missiles. In the case of intercontinental ballistic missiles, the purpose is for an object launched from a certain place to reach a certain designated place on earth after traveling through space. The first step is to launch (with the aid of rockets) a ballistic missile vertically upward (to have less air resistance as compared to launching at an angle). In the second step, when the rocket reaches a precalculated height, it is programmed to automatically turn the missile to a precomputed angle with the vertical in a particular direction. The third stage involves cutting off the rockets when the missile has reached a precalculated velocity v_0. In the fourth step the missile flies freely in an elliptical ballistic trajectory, landing at the desired place. All the steps involved above are programmed and are automatically carried out.

Interplanetary Travel. To throw some light on how interplanetary travel has been accomplished, we use as an example the flight of Apollo 11, the first spacecraft to take men to the moon and bring them back to earth. Apollo 11 consisted of three astronauts, Neil Armstrong, Edwin Aldrin, Jr., and Michael Collins, a mother spacecraft command module, Columbia, and the lunar module, Eagle. First, the Apollo 11 spacecraft was put into an almost circular orbit of radius 119 mi around the earth with a speed of 17,427 mph. Next, its engines were fired to increase the speed of the spacecraft to 24,245 mph toward the moon. Even though this is not equal to the escape velocity, it was enough because once the craft started toward the moon, the gravity of the moon started pulling the spacecraft. All subsequent maneuvering was done by firing retrorockets.

At a distance of 43,500 mi from the moon, the gravitational pull of the moon is equal and opposite that of the earth. Thus, after passing this point, Apollo 11 started accelerating and passed the other side of the moon with a speed of 5645 mph. At this point the total energy is positive; hence, the kinetic energy of the spacecraft is more than the potential energy due to the moon. The spacecraft was slowed down by firing retarding rockets, making the total energy of the spacecraft negative. This put Apollo 11 into an elliptic orbit around the moon, with a maximum altitude of 195 mi and a closest distance of 70 mi.

Now the lunar module Eagle, with Neil Armstrong and Edwin Aldrin, Jr., on board, was separated from the mother ship Columbia, with Michael Collins staying in the mothership. Eagle was slowed down and brought into an orbit at an altitude of 50,000 ft from the surface of the moon. Finally, Eagle was flown to a landing site on the moon. Touchdown on the moon occurred at 4:17:41 P.M. E.D.T. on July 20, 1969, while Neil Armstrong set his foot on the moon at 10:56 P.M. E.D.T. the same day.

The reverse process was followed in coming back to the earth.

Lack of Atmosphere. The expression for the escape velocity can be used to explain why some planets have an atmosphere while others do not. The molecules of gases in the atmosphere are in constant motion with velocities from zero to very large values. Those molecules that have velocities greater than the escape velocity will leave the atmosphere. As the molecules escape, the remaining molecules rearrange their velocities so that their velocity range is still from zero to very large. Thus, the process of escape continues. Of course, the average speed of the molecules also depends on their mass and is given by the expression (we shall see this expression in Chapter 13)

$$v = \sqrt{\frac{3kT}{M}}$$

where k is the Boltzmann constant, equal to 1.38×10^{-23} J/°K; T the temperature (K); and M the molecular mass. This means that light molecules have higher velocities and escape faster. This is the reason why the earth's atmosphere is almost devoid of hydrogen and there is very little helium left. The concentration level of helium has reached equilibrium. As much helium is escaping the earth as is given out by radioactive sources.

The planet Mercury and the earth's moon have smaller escape velocities than that of the earth; hence, they have already lost their atmospheres. Mars has an escape velocity one-sixth that of the earth; hence, it has some atmosphere left. Venus has the same escape velocity as the earth. Other planets have much larger escape velocities; hence, their original atmosphere is still intact, but, for other reasons, their chemical compositions may be different from that of the earth.

Example 11.5 Show that the velocity of a body released at a distance r from the center of the earth when it strikes the surface of the earth is given by

$$v = \sqrt{2GM\left(\frac{1}{R} - \frac{1}{r}\right)}$$

where R and M are the radius and mass of the earth, respectively. Also, show that the velocity with which the meteorites strike the surface of the earth is equal to the escape velocity.

The body of mass m initially at rest has zero velocity and the total energy is only potential:

$$E_i = K_i + V_i = 0 - \frac{GMm}{r} = -\frac{GMm}{r}$$

After the body reaches the earth's surface, its velocity is v; hence, the total energy is

$$E_f = K_f + V_f = \frac{1}{2}mv^2 - \frac{GMm}{R}$$

Equating the above two equations $E_i = E_f$, we get

$$-\frac{GMm}{r} = \frac{1}{2}mv^2 - \frac{GMm}{R}$$

Rearranging and solving for v gives

$$v = \sqrt{2GM\left(\frac{1}{R} - \frac{1}{r}\right)}$$

Since meteorites start at large distances from the earth, $r \rightarrow \infty$; hence, $1/r \rightarrow 0$. Therefore,

$$v = \sqrt{\frac{2GM}{R}} = 1.12 \times 10^4 \text{ m/s} = 25{,}280 \text{ mi/h}$$

which is equal to the escape velocity.

EXERCISE 11.5 Calculate the velocity with which the meteorites are hitting the surface of the moon. [*Ans.*: $v = 2.37 \times 10^3$ m/s.]

EXAMPLE 11.6 Calculate the velocity, acceleration, and time period of an earth satellite that revolves in a circular orbit at a height of 250 miles above the surface of the earth.

Substituting R = radius of earth = 6.37×10^6 m, $g_R = 9.8 \text{ m/s}^2$, and $r = R + 250 \text{ mi} = 6.77 \times 10^6$ m in Equation 11.24,

$$v = R\sqrt{\frac{g_R}{r}} = 6.37 \times 10^6 \text{ m} \sqrt{\frac{9.80 \text{ m/s}^2}{6.77 \times 10^6 \text{ m}}} = 7.66 \text{ km/s} = 4.76 \text{ mi/s}$$

The radial acceleration of the satellite (which is also equal to the value of g at the height of the satellite) is equal to

$$a_R = \frac{v^2}{r} = \frac{(7.66 \text{ km/s})^2}{6.77 \times 10^6 \text{ m}} = 8.67 \text{ m/s}^2$$

The time period T of the satellite is given by

$$T = \frac{2\pi r}{v} = \frac{2\pi(6.77 \times 10^6 \text{ m})}{7.66 \times 10^3 \text{ m/s}} = 5553 \text{ s} = 92.55 \text{ min}$$

EXERCISE 11.6 Calculate the height of a satellite from the surface of the earth so that its radial acceleration will be half the value of g at the surface of the earth. Also, calculate the velocity and the time period of this satellite. [*Ans.*: $h = r - R = 2.64 \times 10^6$ m; $v = 6.64$ km/s; $T = 8.53 \times 10^3$ s.]

* 11.7. Rocket Propulsion

Twenty-five years ago a trip to the moon was to be found only in science fiction books. But rocket technology has developed so rapidly that in the last decade, U.S. astronauts have made several trips to the moon and the duration of these trips has extended to several days. It is through the development of powerful and accurately controlled rockets that such missions have been accomplished. But the amazing part of the whole rocket technology is that it is based on the most simple principles of force and momentum, as explained below.

A rocket is propelled in the forward direction by ejecting mass in the form of gases resulting from the combustion of fuel from itself in the backward direction. Thus, the forward force on the rocket is the reaction to the backward force on the ejected gases (burned-out fuel). The problem is to find the velocity of the rocket at any time after launching or takeoff from the ground.

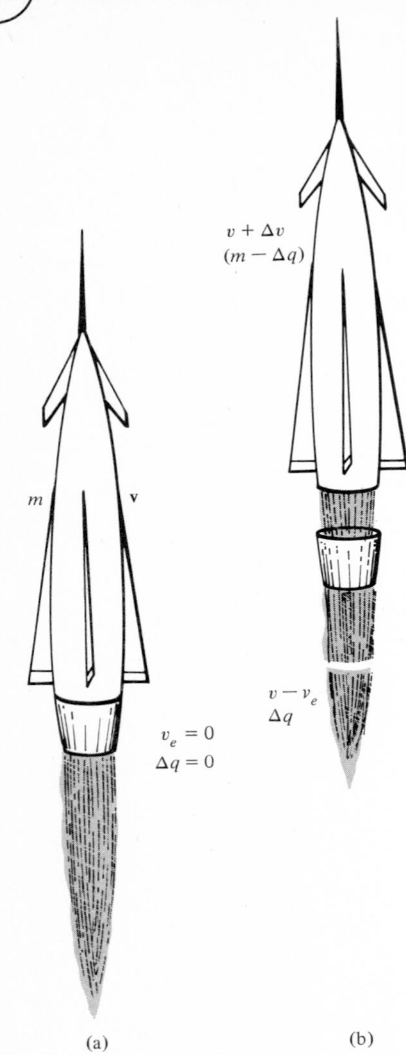

FIGURE 11.11. *Position of the rocket at (a) $t = 0$ and (b) a time Δt later.*

Let us suppose that a rocket is to be launched from the surface of the earth in the upward direction, the direction that we take to be positive. To start with, let the mass of the rocket be $M_0 + m_0$, where M_0 is the mass of the empty rocket and m_0 is the mass of the fuel before the combustion is started. As the fuel starts burning in the combustion chamber of the rocket and is ejected from the rear, the rocket is pushed in the upward direction. Let us say that at one particular instant the mass of the rocket and the fuel is m while the velocity of the rocket with respect to the earth is v, as shown in Figure 11.11(a). The momentum of the system at this time, say $t = 0$, is

$$p_i = mv \tag{11.26}$$

In a short interval Δt, the mass of the gases ejected is, say, Δq, which will increase the velocity of the rocket to $v + \Delta v$, while its mass will change to $(m - \Delta q)$ as in Figure 11.11(b). Let the velocity of the ejected gases be $-v_e$ with respect to the rocket, and $(v - v_e)$ with respect to the ground. Thus, the new momentum of the system will be

$$\begin{aligned} p_f &= (m - \Delta q)(v + \Delta v) + \Delta q(v - v_e) \\ &= mv + m\,\Delta v - v\,\Delta q - \Delta q\,\Delta v + v\,\Delta q - v_e\,\Delta q \end{aligned}$$

Neglecting the product of the two small quantities $\Delta q\,\Delta v$ and noting that the mass of the fuel Δq is equal to the change in the mass of the rocket $-\Delta m$, we get

$$p_f = mv + m\,\Delta v + v_e\,\Delta m \tag{11.27}$$

From Equations 11.26 and 11.27, the change in momentum Δp in time Δt is

$$\Delta p = p_f - p_i = m\,\Delta v + v_e\,\Delta m$$

or

$$\frac{\Delta p}{\Delta t} = F = m\left(\frac{\Delta v}{\Delta t}\right) + v_e\left(\frac{\Delta m}{\Delta t}\right) \tag{11.28}$$

The term $v_e(\Delta m/\Delta t)$ is the *thrust* of the rocket.

The only force acting on the rocket system is the gravitational pull on the rocket, $F = -mg$. Thus, if we neglect any variation in g and also air resistance force, we may write Equation 11.28 as

$$-mg\,\Delta t = m\,\Delta v + v_e\,\Delta m$$

$$\boxed{\Delta v = -g\,\Delta t - v_e\frac{\Delta m}{m}} \tag{11.29}$$

Let us say that it took time t seconds to burn all the fuel and that in this time the velocity of the rocket changes from v_0 at $t = 0$ to v at $t = t$. Under these conditions, Equation 11.29, using integral calculus, becomes

$$v = v_0 - gt + v_e \ln\left(1 + \frac{m_0}{M_0}\right) \tag{11.30}$$

where the value of the logarithmic quantity $\ln(1 + m_0/M_0)$ can be taken from the standard mathematical tables. Comparing this with the equation of a freely falling body $v = v_0 - gt$, we see that it is the last term in Equation 11.30 that determines the final velocity of the rocket. Thus, v depends upon (1) v_e, the speed with which the combustible gases leave the rocket (in order to obtain very high values of v_e, the combustion chamber is maintained at extremely high temperatures), and (2) the ratio m_0/M_0, that is, the ratio of the mass of the fuel to that of the empty rocket. The larger this ratio, the higher will be the final speed of the rocket. It may be pointed out that, in general, $v_0 = 0$.

SUMMARY

According to the *Copernican* or *heliocentric theory*, the sun is the center of the solar system. According to the *universal law of gravitation*, the force between any two objects is given by $F = Gm_1m_2/r^2$. The gravitational field strength at the surface of any planet is given by $g_0 = GM/R^2$, while at any distance r from the center, $g(r) = g_0(R/r)^2$ for $r \geqslant R$.

Kepler's laws describe planetary motion: (I) the planets move in elliptical orbits with the sun at one focus; (II) a line joining any planet to the sun sweeps out equal areas in equal intervals of time; (III) the square of the period of any planet is proportional to the cube of the mean distance. Thus, according to II, $r_1v_1 = r_2v_2$ and $\Delta A/\Delta t =$ constant. According to III, $T^2 = Kr^3$, where $K = 4\pi^2/GM$.

The potential energy of mass m at a distance r from the center of the earth is $V(r) = -GMm/r$ while the total energy is given by $E = \frac{1}{2}mv^2 - GMm/r = -\frac{1}{2}(GMm/r)$.

The *escape velocity* is evaluated from $\frac{1}{2}mv_e^2 = GMm/R$; that is, $v_e = \sqrt{2GM/R} \simeq 11.2$ km/s $\simeq 7.02$ mi/s for the earth. The velocity of a satellite at a distance r is $v = R\sqrt{g_R/r}$.

* In rocket propulsion the change in velocity during a time interval of Δt is $\Delta v = -g\,\Delta t - v_e\,\Delta m/m$, while the final velocity when all the fuel has burned out is $v = v_0 - gt + v_e \ln(1 + m_0/M_0)$.

QUESTIONS

1. Consider a man of mass 100 kg. How will his mass and weight vary as he moves from the earth to the moon and then to Mars? ($g_{Mars} = 4.06$ m/s^2.)
2. How does the value of g vary at different points on the surface of the earth?
3. How will the value of g change if (a) $r > R$ and (b) $r < R$, R being the radius of the earth?
4. What conditions are necessary for the path of a planet around the sun to be an ellipse?
5. Draw the type of orbits that will result if in Figure 11.9 $E = 0$ or $E > 0$.
6. What should be the consequences if the path of the earth around the sun were a parabola or hyperbola?
7. Does the moon orbit around the sun as it does around the earth? Note: the force of gravitational pull by the sun on the moon is twice that of the earth on the moon.
8. There are indications of the presence of a tenth planet. What steps do you think astronomers will take to discover this planet?

9. Can the motions of the 12 moons of Jupiter be described accurately by Kepler's law? Explain why there might be some deviation.
10. Why is there a probability that Mars might have some atmosphere while the moon and Mercury do not?
11. Can we launch a satellite that will move with a given velocity at a desired height from the surface of the earth?
12. Do you think that the presence of a few hundred man-made satellites around the earth will produce perturbation effects on earth (changing the period of the earth around the sun) or on each other?
13. How can we study the shape of the earth by investigating the motion of the satellites?

PROBLEMS

1. Compare the gravitational force between the following pairs of objects: (a) two objects of 1000 kg each separated by a distance of 10 m; (b) the sun and the earth; and (c) the earth and the moon.
2. Compare the gravitational force between the following pairs of objects: (a) two masses each 1 g separated by a distance of 1 cm; and (b) a proton of mass 1.67×10^{-27} kg and an electron of mass 9.1×10^{-31} kg separated by a distance of 0.5×10^{-10} m.
3. Compare the gravitational force between two protons separated by a distance (a) 10^{-13} m; (b) 10^{-15} m; and (c) 10^{-17} m.
4. Calculate the average gravitational force exerted on the planet Uranus by the planet Neptune. How does this compare with the force exerted by the sun on Uranus? (Average distance between Uranus and Neptune is 4.9×10^{12} m.)
5. Find the gravitational force of attraction and the angular deflection in Cavendish's experiment with the following data. The small masses are 50 g each and are separated by a rod 50 cm long. The torsional constant of this system is 8×10^{-8} kg-m²/s². The large masses are 10 kg each. The separation between the large and small mass is 10 cm.
6. In Cavendish's experiment $M = 10$ kg, $m = 50$ g, and the separation between the two masses is 12 cm while the length of the rod separating the two masses is 80 cm. The torsion constant of the wire is 6×10^{-8} kg-m²/s². Calculate the linear deflection produced on a scale placed at a distance of 1 m from the mirror.
7. A man has a mass 100 kg on the surface of the earth. Calculate the gravitational pull on him when he is on the surface of (a) the moon; (b) the sun and (c) Mars.
8. Using the values of G, M, and R in Equation 11.5, show that the values of g at the surface of the earth and moon are 9.8 m/s² and 1.63 m/s², respectively.
9. For the configuration shown, calculate the gravitational field intensity at the center of the square.

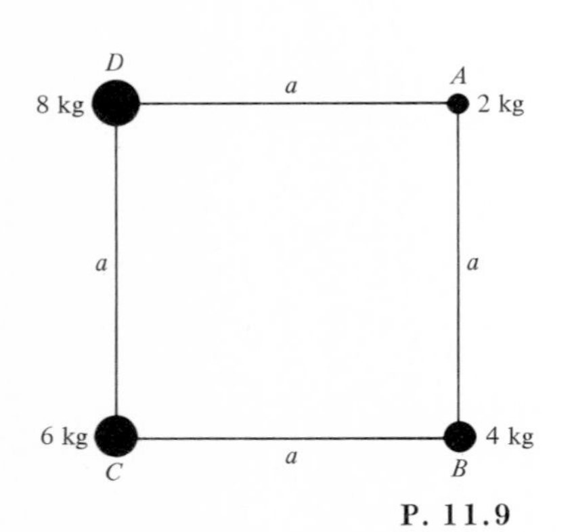

P. 11.9

10. (a) At what distance from the earth will the gravitational field of the earth be equal to the gravitational field of the moon? (b) At what distance from the earth will the gravitational field of the earth be equal to the gravitational field of the sun?

11. Using the expression $g = GM/R^2$, show which planet has the least value of g and which has the largest value of g. Calculate these two extreme values.
12. It is predicted that the sun will eventually collapse into a final state known as a white dwarf. Its radius will be about the radius of the earth while its mass will remain the same. What will be the density and acceleration due to gravity at the surface of such a star?
13. Using Equation 11.6a, make a plot of the weight of mass m as a function of the distance r ($\geqslant R$) from the center of the earth.
14. Using Equation 11.6b, show that for $r = R + h$, where h is small and R is the radius of the earth, $g \simeq g_0(1 - 2h/R)$. [*Hint:* Use $(1 + x)^{-2} \simeq 1 - 2x$, where $x << 1$.]
15. Suppose that we use the same pendulum clock to measure time on the surface of the earth as well as on the surface of the moon. What is the relation between the time intervals on the two places as measured by this clock?
16. Calculate the time period of a pendulum of 1 m length located at $r = R$, $2R$, $10R$, and $100R$, where R is the radius of the earth. Make a plot of T versus r.
17. The earth orbits about the sun in an elliptic orbit. The aphelion distance and the perihelion distance are 1.515×10^{11} m and 1.475×10^{11} m, respectively. The mass of the earth is 5.98×10^{24} kg and its speed at the perihelion is 3.028×10^4 m/s. Calculate the following: the speed of the earth at aphelion, the earth's kinetic and potential energies at perihelion and aphelion.
18. Estimate the magnitude of the areal velocity of the earth around the sun.
19. Assuming the orbit of the earth around the sun to be a circle of radius 1.5×10^{11} m, calculate the mass of the sun. The period of the earth around the sun is 3.16×10^7 s.
20. Calculate the escape velocity from the surface of the moon.
21. Calculate the escape velocity from the surface of the sun. How does this compare with the escape velocity from the surface of the earth?
22. Calculate the height and velocity of a satellite that remains over the same point at all times as seen from the earth. Assume a circular orbit. (This is the case of the communication satellite, Earlybird.)
23. What should be the horizontal velocity of an imaginary satellite that will enable it to circle the earth barely above its surface? Ignore all frictional effects.
24. Calculate the velocity with which a satellite must be ejected at a distance of 200 mi from the surface of the earth so that it will orbit around the earth. What is the period of this satellite?
25. Calculate the period of a satellite at a height equal to the earth's radius. How does this compare with the period given in Example 11.6?
26. Calculate the maximum height an object will reach when fired with velocity v from the surface of the earth.
27. Calculate the escape velocity for molecules that are (a) 100 km; (b) 500 km; (c) 1000 km; and (d) 5000 km above the surface of the earth. Repeat the same for molecules under the influence of the moon and Mars.
28. Calculate the work required to put a satellite of 1000 kg into a circular orbit 100 mi above the earth. How does this work compare with the work required to send the satellite vertically upward to that altitude?
29. Compare the escape velocities from two different planets with the fol-

lowing characteristics: (a) the planets have equal densities but different radii R_1 and R_2; and (b) the planets have equal radii but different masses M_1 and M_2.

***30.** Using Equation 11.29, show that the acceleration a of a rocket is given by $a = -g - (v_e/m)(\Delta m/\Delta t)$.

***31.** A rocket is burning fuel at the rate of 60 g/s. If the velocity of the ejected gases is 60 km/s, calculate the thrust of the rocket.

***32.** The ratio of the mass of the rocket when filled with fuel to that when it is empty is 10. The escape velocity of the gases is 40 km/s and all the fuel burns in 30 s. Calculate the final velocity.

***33.** Calculate the ratio of the mass of the fuel to the empty rocket so that it will achieve a final velocity of 10 km/s in 50 s after launching. The escape velocity of the gases is 400 km/s.

12 Mechanics of Fluids

This chapter is concerned with fluids (liquids and gases) at rest and in motion. Starting with the definition of density and pressure, we briefly describe the physical principles governing static fluids—such as Pascal's principle and Archimedes' principle. Fluid dynamics is described by the continuity equation and Bernoulli's equation. Surface-tension phenomena, such as capillarity, contact angle, and its applications, are described at the end.

12.1. Fluids and Their Scope

THE STUDY of fluids—liquids and gases—has always been of utmost importance from the point of view of everyday life. Flying birds, kites, balloons, flowing streams, construction of dams, weather forecasting, and construction of ships, to name a few, are age-old problems investigated under the name of fluids. The study of fluids has taken on extraordinary importance in the twentieth century because of developments in the air as well as in the sea. This involves the motion of airplanes, jets, rockets, missiles, satellites, and spaceships, on the one hand, and merchant ships and submarines, on the other.

The term *fluid* refers to a substance that can flow, and does not have a shape of its own. Thus, all liquids and gases are fluids. The liquids take the shape of the container in which they are placed, while the gases not only take the shape of the container but also completely fill it. The study of fluids at rest is termed *hydrostatics* (or, more accurately, *fluid statics*), and the study of fluids in motion is termed *hydrodynamics*). A special branch of hydrodynamics dealing with the flow of air and gases is called *aerodynamics*. This chapter will be devoted primarily to the study of liquids.

Density is an important characteristic of all materials. For a homogeneous material, the *density* ρ is defined as mass per unit volume. Thus, if m is the mass of a fluid of volume V, the density ρ is

$$\rho = \frac{m}{V} \tag{12.1}$$

The units of density can be kg/m^3, g/cm^3, or slug/ft^3. It is also sometimes convenient to use *weight density*, which is equal to ρg. Thus, the units are N/m^3, dyn/cm^3, and lb/ft^3. The ratio of the density of a material to that of water is defined as *relative density* (also commonly called by a misleading name, *specific gravity*). Since relative density is a ratio, it is a number without any units. Densities of some commonly used substances are given in Table 12.1.

TABLE 12.1
Densities of Some Materials

Substance	Density (kg/m^3)
Solids	
Alcohol	790
Aluminum	2,700
Brass	8,600
Copper	8,800
Gold	19,320
Ice	920
Iron	7,850
Lead	11,370
Platinum	21,400
Silver	10,500
Steel	7,800
Liquids	
Benzene	920
Glycerin	1,260
Kerosene	820
Mercury	13,600
Water[a]	1,000
Gases[b]	
Air	1.29
Helium	17.85×10^{-2}
Hydrogen	8.99×10^{-2}

[a] To obtain the density in g/cm^3, divide by 1000; for weight density in lb/ft^3, multiply by 62.5×10^{-3}, where 62.5 lb/ft^3 is the weight density of water.
[b] These values are at standard pressure and temperature (see Chapter 13).

12.2. Pressure in a Fluid and Its Measurement

Unlike the case in solids where the force can be applied in any direction with respect to the surface, in liquids the force must be applied at right angles (or normal) to the liquid surface. The reason for this is that fluids at rest cannot sustain a tangential force. If a tangential force is applied to any fluid, the different layers simply slide over one another.

In view of this, it is customary to state pressure acting on the fluid instead of force. The *pressure* p is defined as the magnitude of the normal force acting on a unit surface area. If a force ΔF acts normally on a surface area ΔA, the pressure p is

$$p = \frac{\Delta F}{\Delta A} \tag{12.2}$$

If a constant force of magnitude F acts normally on a surface area A,

$$p = \frac{F}{A} \quad \text{and} \quad F = pA \tag{12.3}$$

The pressure p is a scalar quantity and its units are those of force divided by area: N/m^2, dyn/cm^2, lb/ft^2, bar (1 bar $= 10^6$ dyn/cm^2), and atmospheres (1 atm $= 14.7$ lb/in^2). Another unit, mm Hg, which we shall define more explicitly at the end of this section, is related to the atmosphere as 1 atm $= 760$ mm Hg. The U.S. Weather Service reports atmospheric pressure in millibars (1 bar $= 1000$ millibars).

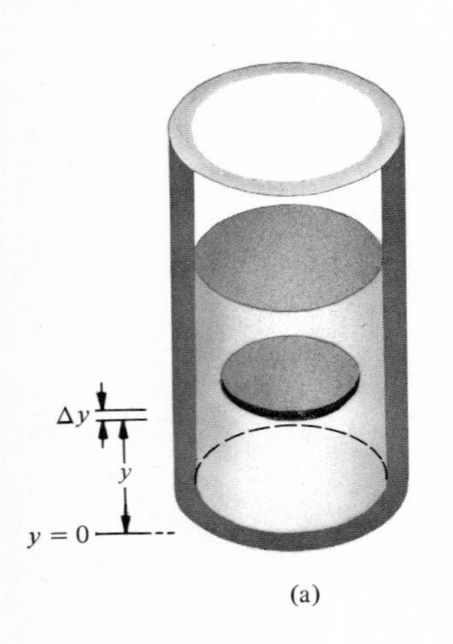

(a)

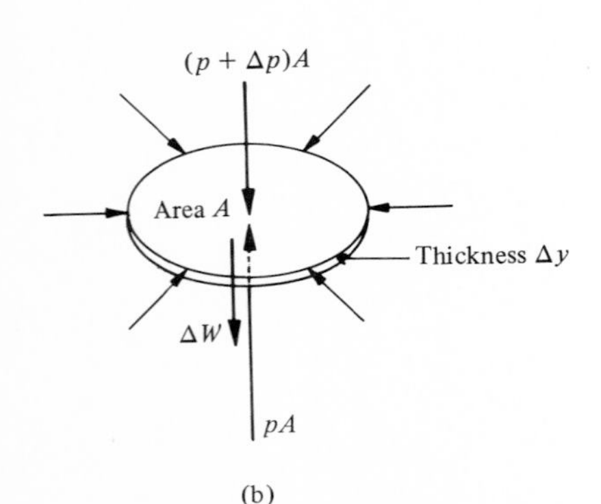

(b)

FIGURE 12.1. *(a) Fluid in a container. (b) Portion of a fluid in equilibrium.*

Figure 12.1(a) shows a fluid in a container that is in equilibrium. Let us consider a small element of fluid in the form of a disk, as shown in Figure 12.1(a) and magnified in (b). The disk is at a height y from the reference level $y = 0$ with thickness Δy and area A. The forces exerted on the disk by the surrounding fluid are perpendicular to the surface at each and every point of the disk.

Since the disk of fluid is in equilibrium, the sum of the x-components of the force ΣF_x and the sum of the y-components ΣF_y must each be zero. There is no net acceleration in the X-direction; hence, $\Sigma F_x = 0$. In the vertical direction let the pressure acting upward on the lower surface of the disk be p and the pressure acting downward on the upper surface of the disk be $(p + \Delta p)$. Thus, the forces acting on the two surfaces must be pA and $(p + \Delta p)A$, respectively. In addition to these two forces, the weight of the disk $\Delta W = \rho(A\,\Delta y)g$ also acts in the downward direction. Thus, for equilibrium from Figure 12.1(b), $\Sigma F_y = 0$ takes the form

$$pA - (p + \Delta p)A - \rho g A\,\Delta y = 0$$

$$\Delta p = -\rho g\,\Delta y \tag{12.4a}$$

$$\frac{\Delta p}{\Delta y} = -\rho g \tag{12.4b}$$

That is, since both ρ and g are constants, a positive increase in y from y to $y + \Delta y$ is accompanied by a decrease in pressure from p to $p - \Delta p$. Also, the pressure gradient defined as $\Delta p/\Delta y$ is constant.

Let the pressure at y_1 be p_1 and at y_2 be p_2. Then $\Delta p = p_2 - p_1$ and $\Delta y = y_2 - y_1$; hence, from Equation 12.4a,

$$p_2 - p_1 = -\rho g(y_2 - y_1) \tag{12.5}$$

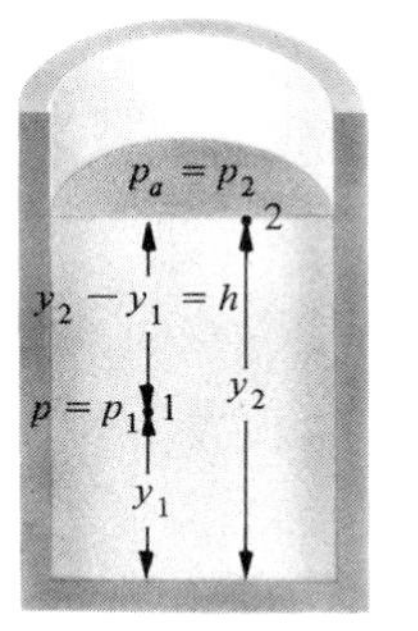

Figure 12.2. *Illustrates the relation $p = p_a + \rho gh$.*

Usually, a liquid has a free surface and the pressure is measured with reference to this level. As shown in Figure 12.2, point 2 is exposed to atmospheric pressure p_a (we shall explain this shortly) while the pressure at point 1 is equal to p, and let $y_2 - y_1 = h$ = depth of the fluid. Thus, from Equation 12.5,

$$p_a - p = -\rho gh$$

$$\boxed{p = p_a + \rho gh} \tag{12.6}$$

That is, the pressure p at a point a distance h below the free surface is equal to the sum of the pressures p_a at the free surface and ρgh. Note that above the free surface the pressure will decrease and will be equal to $p' = p_a - \rho' gh$, where ρ' is the mean density of the fluid, air in this case, above the surface of the liquid.

The atmospheric pressure p_a at any point is numerically equal to the weight of a column of air of unit cross-sectional area extending from that point to the top of the atmosphere. In order to experimentally measure the atmospheric pressure, Evangelista Torricelli (1608–1647) invented a mercury barometer in 1643. It consists of a long glass tube that is filled with mercury and then inverted in a dish of mercury to stand vertically as shown in Figure 12.3. The space above the mercury column is almost empty except for a negligible amount of mercury vapor. This space is called a *Torricelli vacuum*. The weight of the mercury column is supported by the force due to the pressure p_a acting upward on the area at point 1. Thus, applying Equation 12.5 to this situation where $p_1 = p_a$, $p_2 = 0$ and $y_2 - y_1 = h$, we get

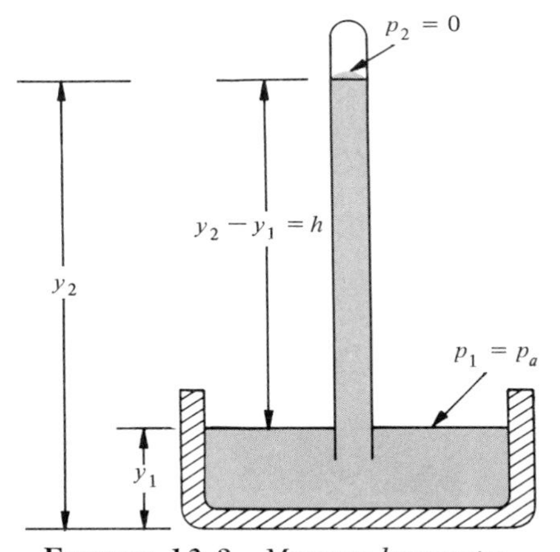

Figure 12.3. *Mercury barometer.*

$$p_a = \rho gh \tag{12.7}$$

where ρ is the density of mercury. At sea level the mercury barometer is found to have a height of about 76 cm. The pressure equivalent to this column of mercury at 0°C is 1 *atmosphere* (atm). It is common practice to state pressure in terms of a length of a mercury column without converting them into proper units of pressure. The pressure exerted by a column of mercury 1 mm high is commonly called 1 *Torr*. A pressure of exactly 1 million dynes per square centimeter (10^6 dynes/cm^2) is 1 *bar*, and 1 *millibar* is equal to 10^{-3} bar; hence, the atmospheric pressure is of the order of 1000 millibars.

The SI unit of pressure is *pascal*, 1 Pa = 1 N/m^2. The atmospheric pressure may be expressed in various units. The density of mercury at 0°C is 13.5950×10^3 kg/m^3, and the value of g is 9.80665 m/s^2. Thus,

$$\begin{aligned} 1 \text{ atm} &= (13.5950 \times 10^3 \text{ kg/m}^3)(9.80665 \text{ m/s}^2)(0.76 \text{ m}) = 1.013 \times 10^5 \text{ N/m}^2 \\ &= 1.013 \times 10^5 \text{ Pa} = 101.3 \times 10^3 \text{ kPa} = 1.013 \times 10^6 \text{ dyn/cm}^2 \\ &= 1.013 \text{ bar} = 1013 \text{ mbar} = 76 \text{ cm Hg} = 760 \text{ mm Hg} = 760 \text{ torr} \\ &= 14.7 \text{ lb/in}^2 = 2116 \text{ lb/ft}^2 \end{aligned} \tag{12.8}$$

Let us go back to Equation 12.6, $p - p_a = \rho gh$, where the pressure p is the *absolute pressure* and the difference $p - p_a$ between absolute pressure and atmospheric pressure is the *gauge pressure* p_g. Thus, the gauge pressure may be either above or below atmospheric pressure. An instrument or gauge that reads pressure below atmospheric is usually called a *vacuum gauge*.

The simplest type of pressure gauge is the open-tube manometer shown in

Figure 12.4. It consists of a U-shaped tube filled with mercury for measuring a high gauge pressure or with a lighter liquid to measure a low gauge pressure.

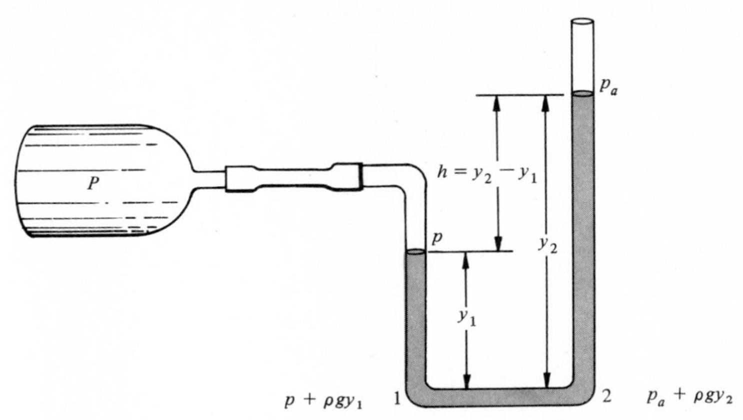

FIGURE 12.4. *Open-tube manometer.*

As shown, one end of the tube is open to the atmosphere while the other end is connected to the container at pressure p to be measured. When equilibrium is reached, the pressure on the left bottom point 1 must be equal to the pressure on the right bottom point 2; that is,

$$p + \rho g y_1 = p_a + \rho g y_2$$

or

$$p - p_a = \rho g(y_2 - y_1) = \rho g h$$

Thus, measuring the height h, knowing the density of the liquid in the manometer and the value of g, the product $\rho g h$ gives the gauge pressure. There are many different designs of instruments used for measuring pressures and gauge pressure, but we shall not go into details.

EXAMPLE 12.1 A gauge pressure of 30.0 lb/in^2 is recommended for air in a tire of an automobile. What is the absolute pressure of air in the tire? Express this pressure in terms of atmospheres and N/m^2.

If p_g, p_a, and p are the gauge, atmosphere, and absolute pressures, respectively, then $p_g = p - p_a$. It is given that $p_g = 30.0$ lb/in^2 and $p_a = 14.7$ lb/in^2. Thus, the absolute pressure p is

$$p = p_a + p_g = (14.7 + 30.0)\ \text{lb/in}^2 = 44.7\ \text{lb/in}^2$$

Noting the relation 1 atm $= 14.7$ lb/in^2 $= 1.013 \times 10^5$ N/m^2, we can calculate

$$p_g = 30.0\ \text{lb/in}^2 = \frac{30.0}{14.7}\ \text{atm} = 2.04\ \text{atm} = (2.04)(1.013 \times 10^5\ \text{N/m}^2)$$

$$= 2.07 \times 10^5\ \text{N/m}^2$$

and

$$p = 44.7 \text{ lb/in}^2 = \frac{44.7}{14.7} \text{ atm} = 3.04 \text{ atm} = (3.04)(1.013 \times 10^5 \text{ N/m}^2)$$
$$= 3.08 \times 10^5 \text{ N/m}^2$$

Exercise 12.1 A gauge pressure of 60 lb/in^2 is recommended for air in a typical 10-speed bicycle. What is the absolute pressure of air in the tire? Express this pressure in terms of atmosphere and N/m^2. [*Ans.*: $p = 5.08 \text{ atm} = 5.15 \times 10^5 \text{ N/m}^2$.]

12.3. Pascal's Principle and Archimedes' Principle

These two principles are necessary consequences of the laws of mechanics and fluid statics.

Pascal's Principle

Consider a fluid in a container such as in Figure 12.5 equipped with a piston at the top. According to Equation 12.6, the pressure p at any point R at a depth h is $p = p_a + \rho gh$. Now if we increase the pressure at the piston by Δp, we find that the pressure at point R is also increased by Δp; that is, the new pressure at depth h is $p + \Delta p$. Such results were generalized by the French scientist Blaise Pascal (1623–1662):

> **Pascal's principle:** ***Pressure applied to an enclosed fluid is transmitted undiminished to every portion of the fluid and the walls of the containing vessel.***

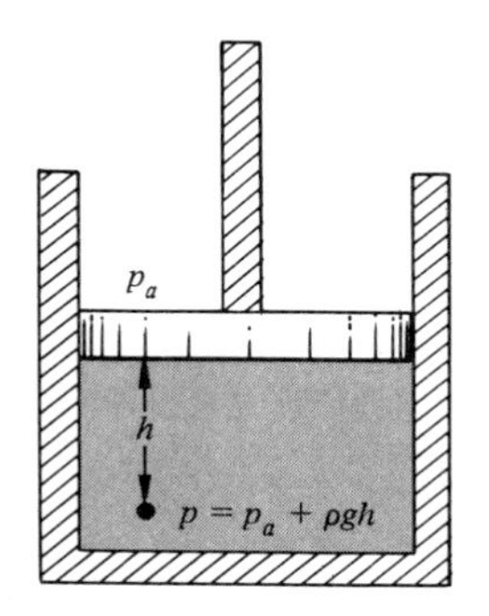

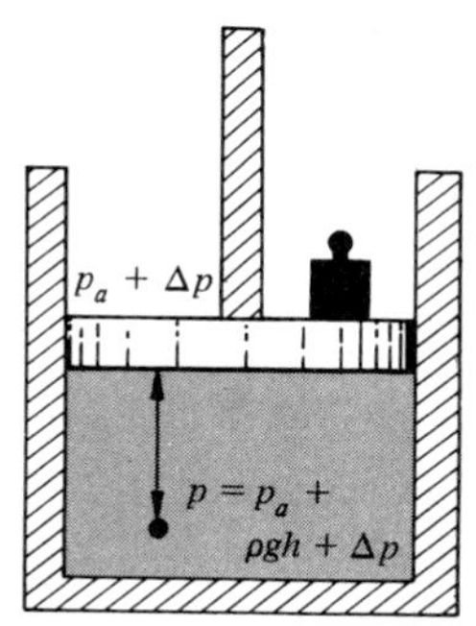

Figure 12.5. *Pressure change Δp is transmitted undiminished.*

To elaborate this point, consider the sketch in Figure 12.6, which illustrates the principle of an hydraulic press. A piston of small cross-sectional area a

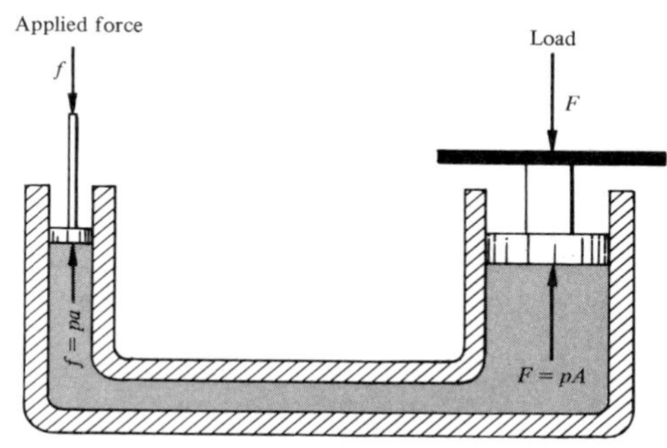

Figure 12.6. *Principle of hydraulic press.*

exerts a force f directly on a liquid such as oil. The pressure $p = f/a$ is transmitted undiminished through the connecting pipe to a larger cylinder, which has a piston of area A. Since the pressure must be the same on both sides,

$$p = \frac{f}{a} = \frac{F}{A} \quad \text{or} \quad F = \frac{A}{a} \times f \tag{12.9}$$

Thus, the force f has been multiplied by a factor of A/a. Hydraulic brakes, car lifts, and dentist chairs are examples that make use of this principle.

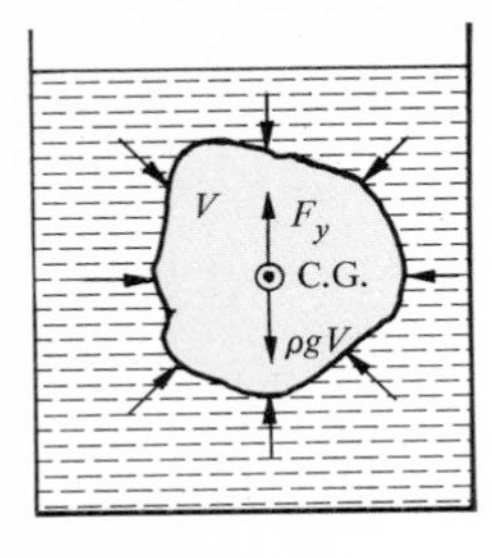

(a)

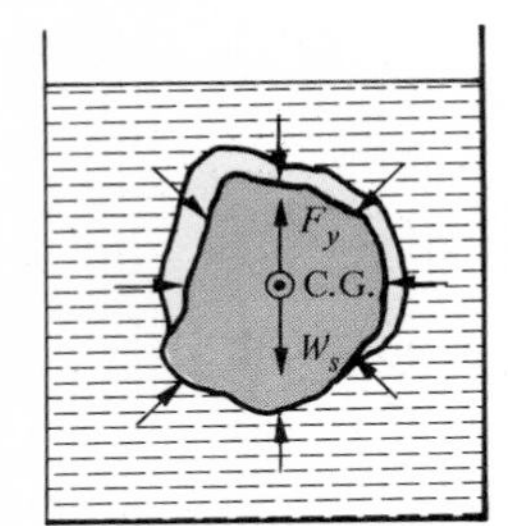

(b)

FIGURE 12.7. *Archimedes' principle. Note that $F_y = F_b$.*

Archimedes' Principle

In Figure 12.7(a), consider an imaginary volume of fluid V enclosed by the boundaries and the forces acting on it. This portion of the fluid, like any other, is in equilibrium; hence, $\Sigma F_x = 0$ and $\Sigma F_y = 0$. The horizontal forces due to the fluid add up to zero, as there is no acceleration in this direction. In the vertical direction the weight of the fluid enclosed by the volume V is $\rho g V$ and acts downward. Thus, there must be a net force $F_y = \rho g V$ exerted by the fluid outside the volume V, which is acting in the upward direction so as to keep the enclosed fluid in equilibrium.

Now suppose that we replace the volume V by a solid of the same shape and volume but of density ρ_s, as in Figure 12.7(b). The force acting on this body due to the fluid is still F_y, while the weight of the solid acting downward is $W_s = \rho_s g V$. The upward force exerted by the fluid on the body is called a *buoyant force* F_b or the *buoyancy* of the immersed body. The buoyant force F_b $(= F_y) = \rho g V$ is equal to the weight of the liquid displaced. Thus, as shown in Figure 12.7(b), the net force F acting on an object immersed in a fluid is

$$F = F_y - W_s = F_b - W_s = \rho g V - \rho_s g V = (\rho - \rho_s) g V \qquad (12.10)$$

According to this equation, a body immersed in the fluid may or may not be in equilibrium. If $F_b = W_s$, the body will be in equilibrium. If $F_b < W_s$, F will act downward and the body will sink. On the other hand, if $F_b > W_s$, F will act upward and the body will rise upward; hence, it will be partially immersed; that is, the body will float.

Experiments with fluids led Archimedes (287–212 B.C.) to the formulation of a principle that may be stated as follows:

> **Archimedes' principle:** ***A body immersed wholly or partially in a fluid is buoyed up with a force in the vertical direction equal to the weight of the volume of the fluid displaced by the body.***

The point through which this vertical force acts when the body is immersed in the fluid is called its *center of buoyancy*. This will coincide with the center of gravity if the solid is homogeneous. If the solid is not homogeneous, the center of gravity may not lie on the line of F_b; hence, there may be a torque that may cause rotation of the body.

EXAMPLES 12.2 To lift an automobile of 2000 kg, a hydraulic pump with a large piston 900 cm² in area is employed. Calculate the force that must be applied to pump a small piston of area 10 cm² to accomplish this.

According to Equation 12.9, $p = f/a = F/A$, where $F = mg = (2000\text{ kg})(9.8\text{ m/s}^2) = 19{,}600\text{ N}$ and $A = 900\text{ cm}^2 = 900 \times 10^{-4}\text{ m}^2$. Therefore,

$$p = \frac{F}{A} = \frac{19{,}600\text{ N}}{900 \times 10^{-4}\text{ m}^2} = 2.18 \times 10^5\text{ N/m}^2$$

If the area of the smaller piston is $10\text{ cm}^2 = 10 \times 10^{-4}\text{ m}^2$, the force f that must be applied is

$$f = pa = (2.18 \times 10^5\text{ N/m}^2)(10 \times 10^{-4}\text{ m}^2) = 218\text{ N}$$

EXERCISE 12.2 The average mass that must be lifted in a dentist chair is 80 kg. If the radius of the larger piston is five times that of the smaller piston, what is the minimum force that must be applied? [*Ans.*: $f = 31.4$ N.]

EXAMPLE 12.3 A piece of alloy has a mass of 250 g in air. When immersed in water it has an apparent weight of 1.96 N and in oil has an apparent weight of 2.16 N. Calculate the density of (a) the metal; and (b) the oil.

Let V be the volume and ρ the density of the alloy. When this is immersed in fluids, there will be different buoyant forces and different apparent weights. Let w, w_w, and w_o be the weights in air, water, and oil, respectively. Thus,

$$w = mg = (0.250 \text{ kg})(9.8 \text{ m/s}^2) = 2.45 \text{ N}$$

$$w_w = 1.96 \text{ N}$$

$$w_o = 2.16 \text{ N}$$

$$\text{weight of water displaced} = w - w_w = (2.45 - 1.96) \text{ N} = 0.49 \text{ N}$$

$$\text{mass of water displaced} = m - m_w = \frac{w - w_w}{g} = \frac{0.49 \text{ N}}{9.8 \text{ m/s}^2} = 0.05 \text{ kg}$$

$$\text{volume } V \text{ of the water displaced} = \frac{\text{mass}}{\text{density}} = \frac{0.05 \text{ kg}}{1000 \text{ kg/m}^3} = 5 \times 10^{-5} \text{ m}^3$$

But this is the volume of the alloy of mass m. Hence, the density of the alloy is

$$\rho = \frac{m}{V} = \frac{0.250 \text{ kg}}{5 \times 10^{-5} \text{ m}^3} = 5000 \text{ kg/m}^3 = 5 \text{ g/cm}^3$$

$$\text{weight of the oil displaced} = w - w_o = (2.45 - 2.16) \text{ N} = 0.29 \text{ N}$$

But this must be equal to $\rho_o gV$, where ρ_o is the density of the oil and V is its volume displaced, which is also equal to the volume of the alloy. That is,

$$\rho_o gV = w - w_o$$

or

$$\rho_o = \frac{w - w_o}{gV} = \frac{0.29 \text{ N}}{(9.8 \text{ m/s}^2)(5 \times 10^{-5} \text{ m}^3)} = 592 \text{ kg/m}^3 = 0.59 \text{ g/cm}^3$$

EXERCISE 12.3 A piece of steel has a weight of 19.6 N in air and a density of 7800 kg/m^3. What is the volume and apparent weight of this piece when completely submerged in (a) water; (b) glycerin (density = 1260 kg/m^3); and (c) kerosene (density = 820 kg/m^3)? [*Ans.*: $V = 2.56 \times 10^{-4}$ m^3, $w_w =$ 17.1 N, $w_g = 16.4$ N; $w_k = 17.5$ N.]

12.4. Fluid Dynamics and the Continuity Equation

A fluid may be divided into infinitesimal volume elements so that each such element may be treated as a fluid particle. In mechanics we have described the motion of a particle by specifying its coordinates (x, y, z) as a function of time t. It is convenient to describe the motion of fluid particles by specifying their density $\rho(x, y, z, t)$ and velocity $v(x, y, z, t)$ at a particular point (x, y, z) in space and at time t as the fluid particles pass through that point. Furthermore, we shall limit our discussion to the motion of those fluids which are (1) incompressible, (2) nonviscous, (3) in steady motion, and (4) without rotational motion. We explain these points briefly.

Flowing liquids are usually treated as *incompressible*, which means that the density ρ is constant. *Viscosity* or internal friction, which is due to the relative motion between two adjacent layers of fluids, is analogous to the friction of a solid moving over another solid surface. In many cases, such as lubricants, viscosity plays an important role. On the other hand, in many fluid flows of interest, the fluid may be treated as *nonviscous*.

When the fluid flow is such that velocity v at any point is constant in time, the flow is said to be *steady*. The velocities might have different magnitudes and directions at different points, but the velocity of any fluid particle passing a given point in space is always the same; it does not change with time. A turbulent flow is an example of a nonsteady velocity.

Finally, the flow of the fluid will be *rotational* if the fluid at any point in space has angular velocity. The rotational motion of fluid can be detected by placing a small paddlewheel in the moving fluid. If the wheel has a linear motion and does not rotate, the flow is irrotational. But if the wheel has both linear as well as rotational motion, the fluid flow is said to be rotational.

An ideal fluid is one that is incompressible and nonviscous. In the following we shall be dealing with an ideal fluid in irrotational, steady flow.

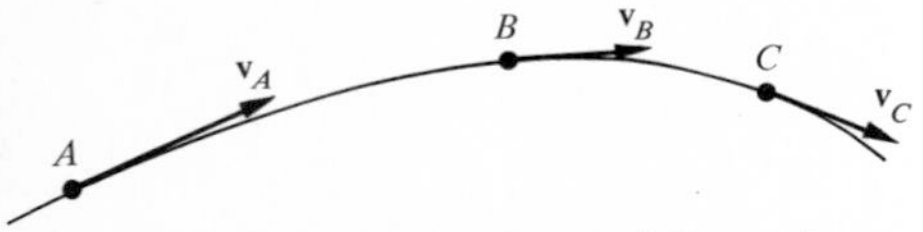

FIGURE 12.8. *Line of flow of a fluid particle.*

The path followed by a fluid particle is called a *line of flow*. Consider, for example, a fluid particle in a steady flow being at points A, B, and C at times t_A, t_B, and t_C with velocities $\mathbf{v}_A$, $\mathbf{v}_B$, and $\mathbf{v}_C$, respectively, as shown in Figure 12.8. Thus, if we trace the path of the particles joining the points A, B, and C, the curve is called a *streamline*. In steady flow the streamlines coincide with the lines of flow and have the following two characteristics: (1) a tangent at any point on the streamline gives the direction of the velocity of the fluid particle at that point; and (2) no two streamlines ever cross one another. If they did, a fluid particle arriving at the cross point could have either one of the two velocities, which would mean that the flow is nonsteady. Thus, in a steady flow, every fluid particle arriving at A must follow the path ABC.

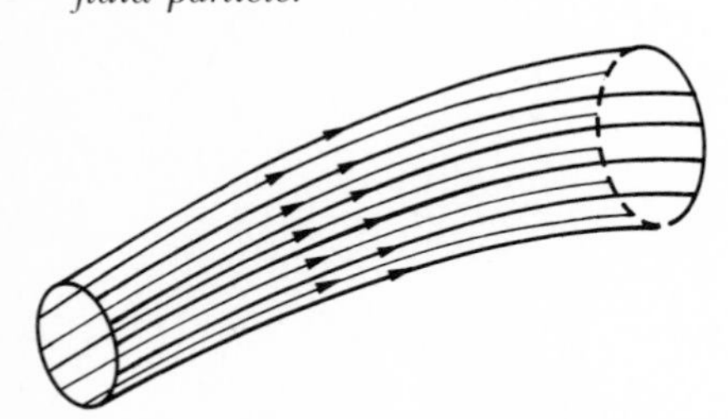

FIGURE 12.9. *Tube of flow enclosing several streamlines.*

Consider an area A shown in Figure 12.9. If we draw streamlines passing the periphery of this area, the lines will enclose a tube called the *tube of flow* or *flow tube*. From the definition, if two fluids are flowing side by side in steady flow, no mixing will take place. That is, the tubes of flow behave like pipes guiding the motion of the fluids; hence avoiding their mixing. The tubes will be straight and parallel if the flow is homogeneous.

To give quantitative meaning to the streamlines, the number of streamlines passing through a unit area perpendicular to the fluid motion are drawn proportional to the velocity at that point. Thus, the streamlines will be drawn closer together where the speed of the fluid is high and far apart where the speed is low. Also, the lines of flow adjust themselves when facing an obstacle in their path. Figure 12.10 shows the rearrangement of the lines of flow or streamlines around an airfoil. Note the existence of both streamline and turbulent flow.

FIGURE 12.10. *Air foil showing streamline flow and turbulence around it. (Reproduced from* Journal of Applied Physics, *August 1943, with permission from the National Advisory Committee for Aeronautics.)*

THE EQUATION OF CONTINUITY The equation of continuity expresses the *law of conservation of mass* in fluid dynamics. Consider a tube of flow for an ideal fluid in steady state, as shown in Figure 12.11. Let us consider areas A_1 and A_2 perpendicular to the streamlines at points A and B, respectively. Let ρ_1 and v_1 be the density and velocity of the fluid particles at point A, and ρ_2 and v_2 be the corresponding quantities at B. As shown in Figure 12.11, the fluid particles crossing at A travel a distance $v_1\,\Delta t$, while those crossing at B travel a distance $v_2\,\Delta t$ in the same time interval, Δt. Thus, the mass of the fluid Δm_1 crossing the area A_1 in time Δt will be approximately

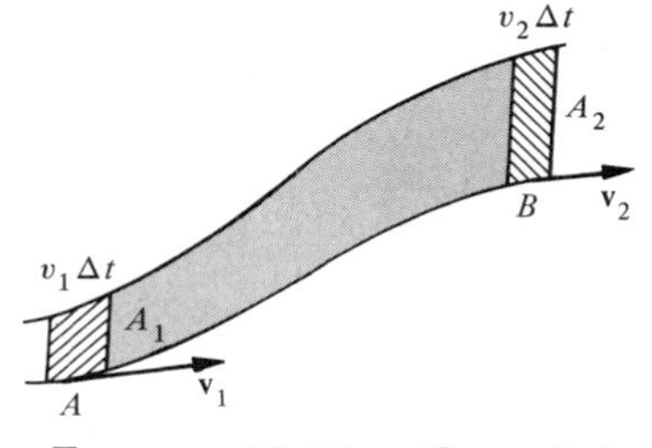

FIGURE 12.11. *Flow of fluid through a tube of varying cross section.*

$$\Delta m_1 = \rho_1\,\Delta V_1 = \rho_1 A_1 v_1\,\Delta t \tag{12.11}$$

where ΔV_1 is the shaded volume at A in Figure 12.11. Thus,

$$\frac{\Delta m_1}{\Delta t} = \rho_1 A_1 v_1 \tag{12.12}$$

where $\Delta m_1/\Delta t$ is the *mass flux* at point A. Similarly, the mass flux at point B is

$$\frac{\Delta m_2}{\Delta t} = \rho_2 A_2 v_2 \tag{12.13}$$

The two mass fluxes must be approximately equal because there are no sources or sinks in the fluid. Thus, Equations 12.12 and 12.13 yield

$$\boxed{\rho_1 A_1 v_1 = \rho_2 A_2 v_2} \tag{12.14}$$

which is the *equation of continuity*. Since an ideal fluid is incompressible, $\rho_1 = \rho_2 = \rho =$ constant and Equation 12.14 takes the form

$$\boxed{v_1 A_1 = v_2 A_2} \tag{12.15}$$

This equation states that as the cross-sectional area of the tube of flow becomes larger, the fluid speed becomes smaller and vice versa. The quantity vA is called the *volume flux* or *discharge rate,* denoted by the symbol Q.

EXAMPLE 12.4 A water pipe entering a home has a diameter of 2 cm and the speed of the water is 0.1 m/s. Eventually, the pipe tapers to a diameter of 1 cm. Calculate (a) the speed of water in the tapered portion; and (b) the amount (the volume and mass) of water that flows per minute across any cross-sectional area.

At the entrance $v_1 = 0.1$ m/s, radius $r_1 = 1$ cm, and $A_1 = \pi r_1^2 = \pi(0.01 \text{ m})^2 = \pi \times 10^{-4} \text{ m}^2$. At the tapered point $r_2 = 0.5$ cm and $A_2 = \pi r_2^2 = \pi(0.005 \text{ m})^2 = \pi \times 2.5 \times 10^{-5} \text{ m}^2$. Substituting these in the continuity equation 12.15, $v_1A_1 = v_2A_2$, then v_2 is given by

$$v_2 = \frac{v_1A_1}{A_2} = \frac{(0.1 \text{ m/s})(\pi \times 10^{-4} \text{ m}^2)}{\pi \times 2.5\ 10^{-5} \text{ m}^2} = 0.4 \text{ m/s}$$

The volume flux or the discharge rate Q is

$$Q = v_1A_1 = v_2A_2 = (0.4 \text{ m/s})(\pi \times 2.5 \times 10^{-5} \text{ m}^2)$$
$$= 3.14 \times 10^{-5} \text{ m}^3\text{/s} = 31.4 \text{ cm}^3\text{/s}$$

The mass flow from Equation 12.12 is

$$\frac{\Delta m_1}{\Delta t} = \rho v_1 A_1 = \rho v_2 A_2 = (1000 \text{ kg/m}^3)(3.14 \times 10^{-5} \text{ m}^3\text{/s})$$
$$= 3.14 \times 10^{-2} \text{ kg/s} = 31.4 \text{ g/s}$$

EXERCISE 12.4 In Example 12.4, what should be the diameter of the tapered portion of the pipe if we want the water to have a speed of 2 m/s? Will the discharge rate still be the same? [*Ans.:* $d = 0.45$ cm, yes.]

EXAMPLE 12.5 In a normal adult, the average speed of the blood through the aorta (which has a radius of 0.9 cm) is 0.33 m/s. From the aorta the blood goes into the major arteries, which are, say, 30 in number, each with a radius of 0.5 cm. Calculate the speed of the blood through the arteries.

The area of the aorta $A_1 = \pi r^2 = \pi(0.9 \times 10^{-2} \text{ m})^2 = 2.5 \times 10^{-4} \text{ m}^2$. Therefore, the blood flow rate is

$$Q = v_1A_1 = (0.33 \text{ m/s})(2.5 \times 10^{-4} \text{ m}^2) = 0.83 \times 10^{-4} \text{ m}^3\text{/s}$$

The area of each artery $= \pi r^2 = \pi(0.5 \times 10^{-2} \text{ m})^2 = 0.78 \times 10^{-4} \text{ m}^2$. If there are 30 arteries, the total area $A_2 = 30(0.78 \times 10^{-4} \text{ m}^2) = 23.4 \times 10^{-4} \text{ m}^2$. From the continuity equation,

$$Q = v_1A_1 = v_2A_2$$

or

$$v_2 = \frac{Q}{A_2} = \frac{0.83 \times 10^{-4}\ \text{m}^3/\text{s}}{23.4 \times 10^{-4}\ \text{m}^2} = 0.035\ \text{m/s}$$

It is clear that $A_2 > A_1$, therefore, $v_2 < v_1$; that is, the blood moves slower in the arteries than in the aorta.

EXERCISE 12.5 Suppose that each major artery in Example 12.5 divides into 60 capillaries, each of radius 0.1 cm. What will be the speed of the blood in the capillary? [*Ans.*: $v = 1.5$ cm/s.]

12.5 Bernoulli's Equation and Its Applications

In 1738, Daniel Bernoulli applied the principle of the work–energy theorem to fluid flow and arrived at what is known as *Bernoulli's equation.* The work done by the resultant force acting on a system is equal to the change in the kinetic and potential energy of the system; that is,

$$\Delta W = \Delta K + \Delta U \tag{12.16}$$

Let us consider an incompressible, nonviscous fluid with a streamline flow through a pipe, as shown in Figure 12.12. At point A the cross-sectional area of the pipe is A_1, fluid velocity v_1, pressure p_1, elevation y_1, and fluid density ρ. The corresponding quantities at B are A_2, v_2, p_2, y_2, and ρ, respectively. Note that the density is the same not only at A and B, but everywhere, because the fluid is incompressible. The forces that work on the system are, as shown in

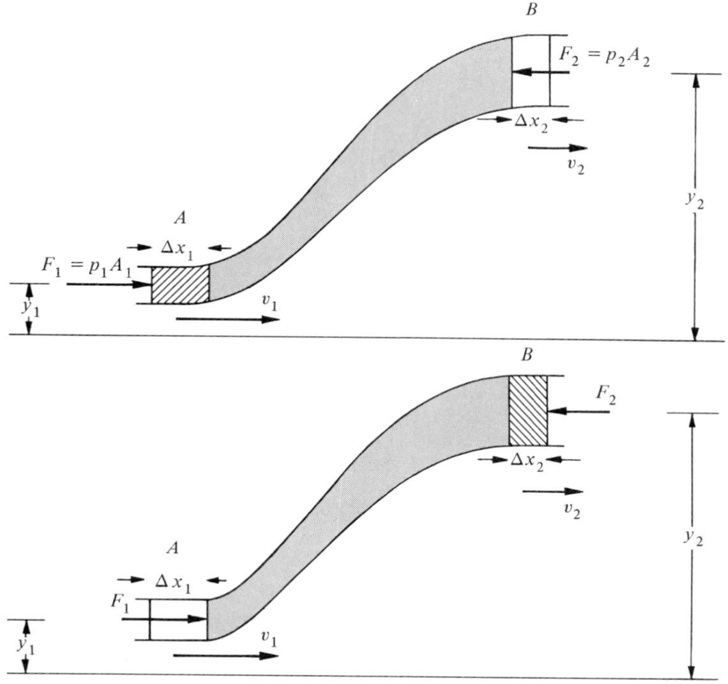

FIGURE 12.12. *Forces acting on a portion of a fluid flowing in a tube.*

Figure 12.12, due to the pressures p_1 and p_2 given by $F_1 = p_1A_1$ acting on the system at A to the right and $F_2 = p_2A_2$ acting on the system at B to the left. Force F_1 is in the direction of the fluid flow; hence, it does *positive* work while F_2, being opposite to the fluid flow, does *negative* work. In time Δt the fluid at point A advances a distance $\Delta x_1 = v_1 \Delta t$ to the right, while at B it advances a distance $\Delta x_2 = v_2 \Delta t$, also to the right. Thus, the net work done by the external forces on the system is

$$\Delta W = F_1 \Delta x_1 - F_2 \Delta x_2 = (p_1A_1)(v_1 \Delta t) - (p_2A_2)(v_2 \Delta t) \tag{12.17}$$

But by using the continuity equation $v_1A_1 = v_2A_2$ in Equation 12.17, we get

$$\Delta W = (v_1A_1 \Delta t)(p_1 - p_2)$$

Also, from Equations 12.12 and 12.13,

$$\Delta m = \rho v_1A_1 \Delta t = \rho v_2A_2 \Delta t \tag{12.18}$$

Substituting for $v_1A_1 \Delta t = \Delta m/\rho$ in the equation above,

$$\Delta W = \frac{\Delta m}{\rho}(p_1 - p_2) \tag{12.19}$$

As the mass Δm moves from point A to B, the change in the kinetic energy is given by

$$\Delta K = \tfrac{1}{2}\Delta m(v_2^2 - v_1^2) \tag{12.20}$$

and the change in the gravitational potential energy is given by

$$\Delta U = \Delta mg(y_2 - y_1) \tag{12.21}$$

Substituting for ΔW, ΔK, and ΔU from Equations 12.19, 12.20, and 12.21 into Equation 12.16, we get

$$\frac{\Delta m}{\rho}(p_1 - p_2) = \frac{1}{2}\Delta m(v_2^2 - v_1^2) + \Delta mg(y_2 - y_1) \tag{12.22}$$

Multiplying by $\rho/\Delta m$ and rearranging the terms yields

$$p_1 + \tfrac{1}{2}\rho v_1^2 + \rho g y_1 = p_2 + \tfrac{1}{2}\rho v_2^2 + \rho g y_2 \tag{12.23}$$

which is Bernoulli's equation. Since points A and B were chosen arbitrarily, we may state *Bernoulli's theorem* as

$$\boxed{p + \tfrac{1}{2}\rho v^2 + \rho g y = \text{constant}} \tag{12.24}$$

This is a statement of the conservation of energy in a steady fluid flow. We shall now briefly discuss some of its applications.

Fluid Statics

Fluid statics is a special case of fluid dynamics. Thus, the law of pressure change with height can be derived from Bernoulli's equation. For the case of a fluid at rest, substituting $v_1 = v_2 = 0$ in Equation 12.23, we get

$$p_1 + \rho g y_1 = p_2 + \rho g y_2$$

or

$$p_2 - p_1 = -\rho g(y_2 - y_1)$$

which is the result already derived in Equation 12.5. Since the units of each term in Equation 12.24 are that of pressure, it is sometimes useful to refer to the term $p + \rho g y$ as the *static pressure* and the term $\frac{1}{2}\rho v^2$ as the *dynamic pressure*.

Speed of Efflux

Consider the situation shown in Figure 12.13 and say $y_1 - y_2 = h$. Let the pressure at 1 be $p_1 = p$ while the velocity at this point is v_1. The pressure at

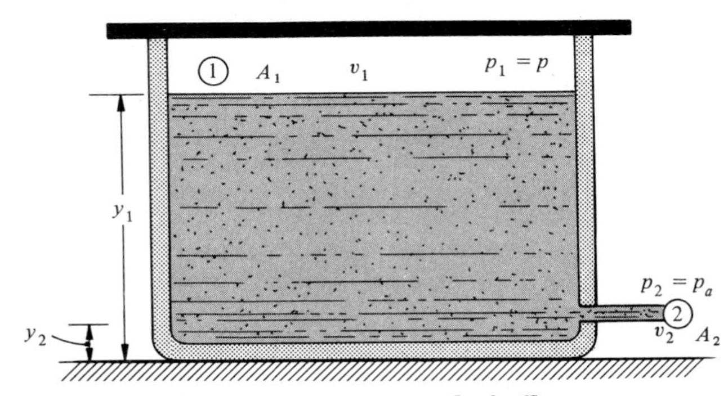

FIGURE 12.13. *Speed of efflux.*

point 2 is $p_2 = p_a$ = atmospheric pressure and the velocity v_2 at this point is called the *speed of efflux*. For this case, Equation 12.23 takes the form

$$p + \tfrac{1}{2}\rho v_1^2 + \rho g h = p_a + \tfrac{1}{2}\rho v_2^2$$

or

$$v_2^2 = v_1^2 + 2\frac{p - p_a}{\rho} + 2gh \tag{12.25}$$

Also, from the continuity equation $v_2 = (A_1/A_2)v_1$ if $A_1 >> A_2$, so that $v_2 >> v_1$, v_1^2 may be neglected as compared to v_2^2. Suppose that the top of the tank in Figure 12.13 is open to the atmosphere; that is, $p = p_a$. Equation 12.25 takes the form

$$v_2 = \sqrt{2gh} \tag{12.26}$$

That is, the speed of efflux is the same as that acquired by a body falling through a height h. This is called *Torricelli's theorem*.

The Venturi Flow Meter

The rate of flow of liquid or gas through a pipe can be measured by applying Bernoulli's equation. Figure 12.14(a) shows a liquid flowing through a horizontal constricted pipe. Let v_1, p_1, and A_1 be the quantities at point 1 and v_2, p_2, and A_2 be the corresponding quantities at point 2. Note that $A_1 > A_2$.

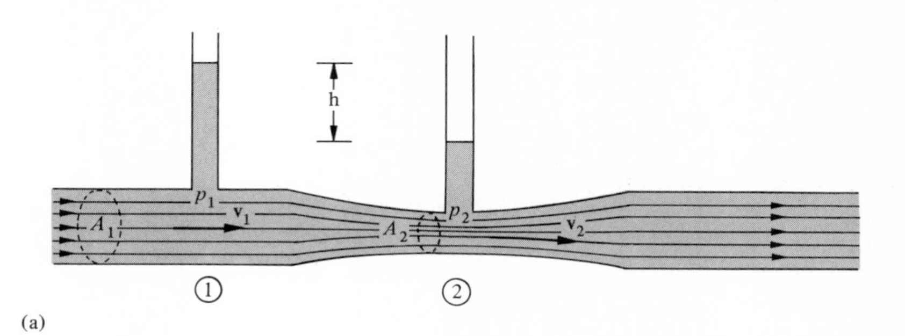

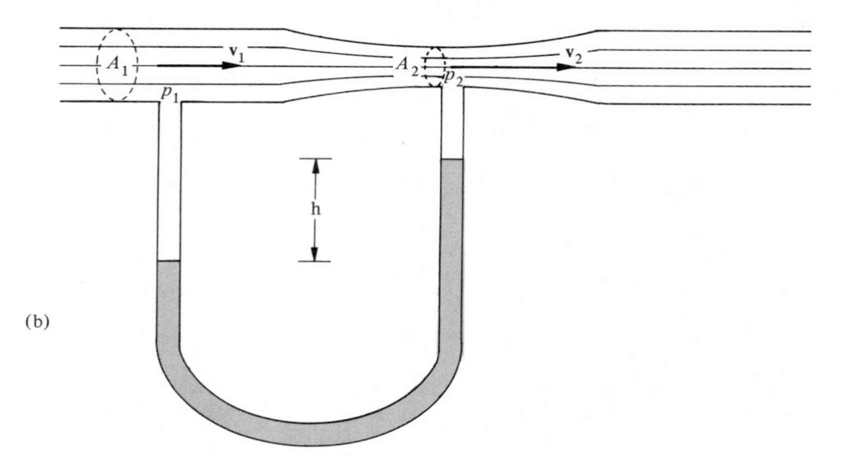

FIGURE 12.14. *Venturi flowmeter.*

Applying Bernoulli's equation, since $y_1 = y_2$, we get

$$p_1 + \tfrac{1}{2}\rho v_1^2 = p_2 + \tfrac{1}{2}\rho v_2^2 \tag{12.27}$$

Substituting for v_2 from the continuity equation $v_2A_2 = v_1A_1$, we get

$$p_1 - p_2 = \frac{\rho v_1^2}{2}\left[\left(\frac{A_1}{A_2}\right)^2 - 1\right] \tag{12.28}$$

Since $A_1 > A_2$, this means that $p_1 > p_2$ and $v_1 < v_2$, as shown in Figure 12.14(a). Solving for v_1 gives

$$v_1 = A_2\sqrt{\frac{2(p_1 - p_2)}{\rho(A_1^2 - A_2^2)}} \tag{12.29}$$

and the flow rate Q (volume/s) is given by

$$Q = v_1 A_1 \tag{12.30}$$

Note that $p_1 - p_2$ in Figure 12.14(a) is equal to ρgh, where h is the difference in heights of the liquid levels in the two tubes. This principle is made use of in the construction of the Venturi flow meter, as shown in Figure 12.14(b).

The Fluid Dynamic Lift

The *fluid dynamic lift* or simply *dynamic lift* is a force that acts on a body because of its motion through a fluid. This is different from the static lift, which is the buoyant force that acts on a body according to Archimedes' principle. The examples of dynamic lift are to be found in a spinning ball, an airplane wing, a hydrofoil, and many others. Let us consider the first two in some detail.

Consider a nonspinning baseball or tennis ball as it moves through the air from right to left. If we look at the motion of the air with reference to the coordinate system fixed on the ball, the flow of air is steady and the streamlines are symmetrical as shown in Figure 12.15(a). As the streamlines approach the

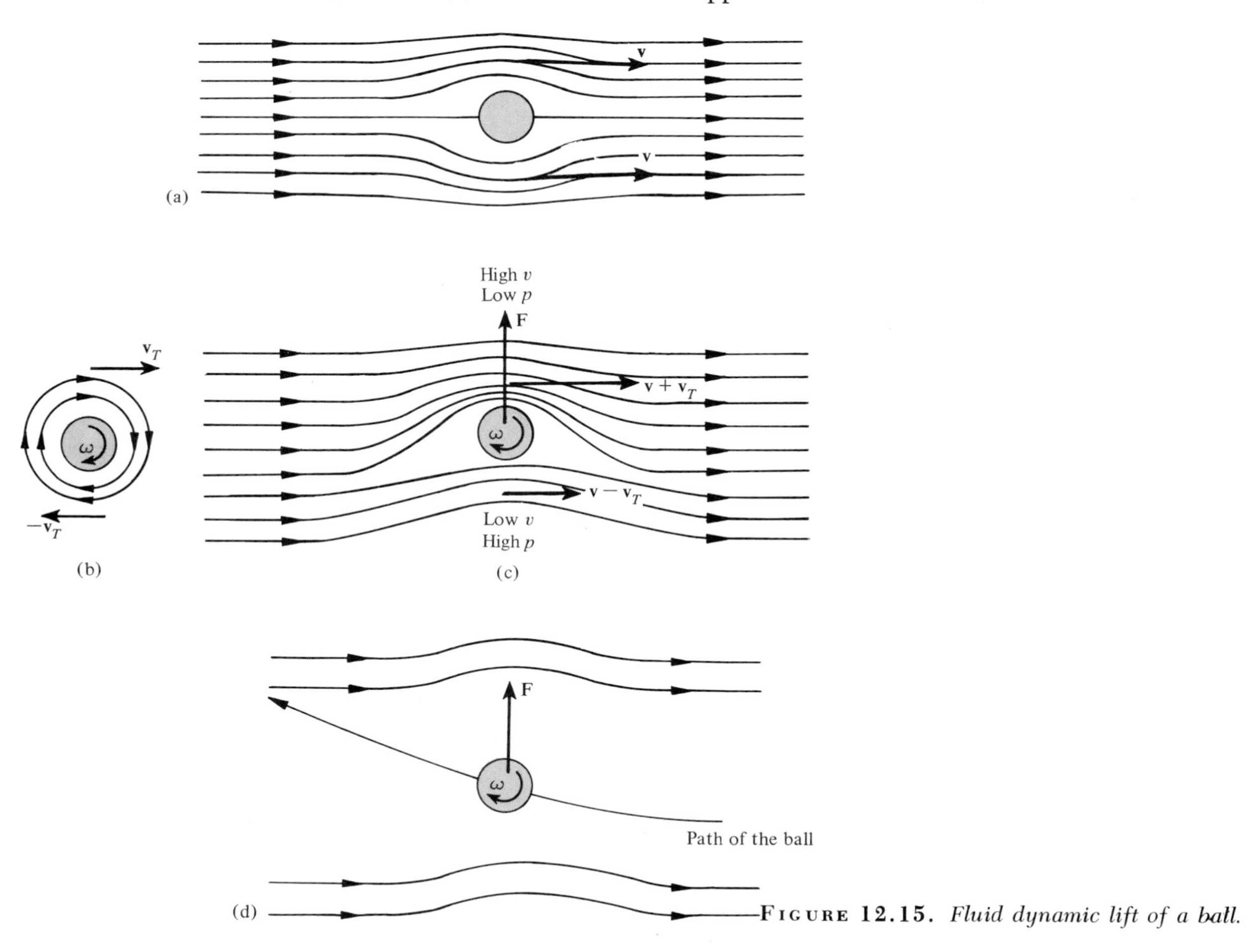

FIGURE 12.15. *Fluid dynamic lift of a ball.*

ball, half bend upward and half downward. Thus, the velocity of the air flow above the ball is equal to the velocity below the ball. According to Bernoulli's principle, the pressure above the ball is equal to the pressure below; hence, there is no net upward or downward force.

Let us now consider a ball that has no translatory motion but is spinning about an axis perpendicular to the plane of the figure, as shown in Figure 12.15(b). Due to friction, air is dragged with the ball and the resulting streamlines, as seen from the spinning ball, are as shown in Figure 12.15(b). Note that the tangential velocity $\mathbf{v}_T$ at the top is to the right, while at the bottom it is to the left.

Let us now consider the case in which the ball has both translatory and spinning motions. The resultant motion will be the sum of the two motions (a) and (b), as shown in Figure 12.15(c). It is clear that the streamlines are much closer above the ball than below the ball because the velocity of air above the ball is $\mathbf{v} + \mathbf{v}_T$, and below the ball is $\mathbf{v} - \mathbf{v}_T$, both to the right. Thus, according to Bernoulli's principle, the pressure above the ball is less than the pressure below the ball. This results in a net force **F** acting upward, as shown in Figure 12.15(c). Thus, as the spinning ball moves through the air, there is an upward lift, as shown in Figure 12.15(d). This is the dynamic lift.

Dynamic lift always occurs when the streamlines that are due to a moving object result in unsymmetrical streamlines crowding on one side and far apart on the other side. As another example, Figure 12.16 shows the streamlines around an airplane wing or airfoil moving from right to left. The airplane wing is designed in such a way that there is a crowding of streamlines above the wing as compared to the streamlines below the wing, thereby producing a net upward force or lift. In addition, the fluid that strikes the lower portion of the wing is deflected downward, and, according to Newton's third law, there will be an upward thrust. Thus, the total lifting force is the combination of the upward lift and the upward thrust.

EXAMPLE 12.6 The cross-sectional area of a water pipe entering the basement is 4×10^{-4} m². The pressure at this point is 3×10^5 N/m² and the speed of the water is 2 m/s. This pipe tapers to a cross-sectional area of 2×10^{-4} m² when it reaches the second floor 8 m above. Calculate the speed and pressure at the second floor.

From the continuity equation the speed of water v_2 at the second floor is

$$v_2 = \frac{v_1 A_1}{A_2} = \frac{(2\text{ m/s})(4 \times 10^{-4}\text{ m}^2)}{2 \times 10^{-4}\text{ m}^2} = 4\text{ m/s}$$

In Bernoulli's equation 12.23,

$$p_1 + \tfrac{1}{2}\rho v_1^2 + \rho g y_1 = p_2 + \tfrac{1}{2}\rho v_2^2 + \rho g y_2$$

where $p_1 = 3 \times 10^5\text{ N/m}^2$, $\rho = 1000\text{ kg/m}^3$, $v_1 = 2\text{ m/s}$, $y_2 - y_1 = 8$ m, and $v_2 = 4$ m/s. Therefore,

$$\begin{aligned} p_2 &= p_1 - \tfrac{1}{2}\rho(v_2^2 - v_1^2) - \rho g(y_2 - y_1) \\ &= (3 \times 10^5\text{ N/m}^2) - \tfrac{1}{2}(1000\text{ kg/m}^3)(16\text{ m}^2/\text{s}^2 - 4\text{ m}^2/\text{s}^2) \\ &\quad - (1000\text{ kg/m}^3)(9.8\text{ m/s}^2)(8\text{ m}) \\ &= (3 \times 10^5 - 0.06 \times 10^5)\text{N/m}^2 - (0.78 \times 10^5\text{ N/m}^2 = 2.16 \times 10^5\text{ N/m}^2 \end{aligned}$$

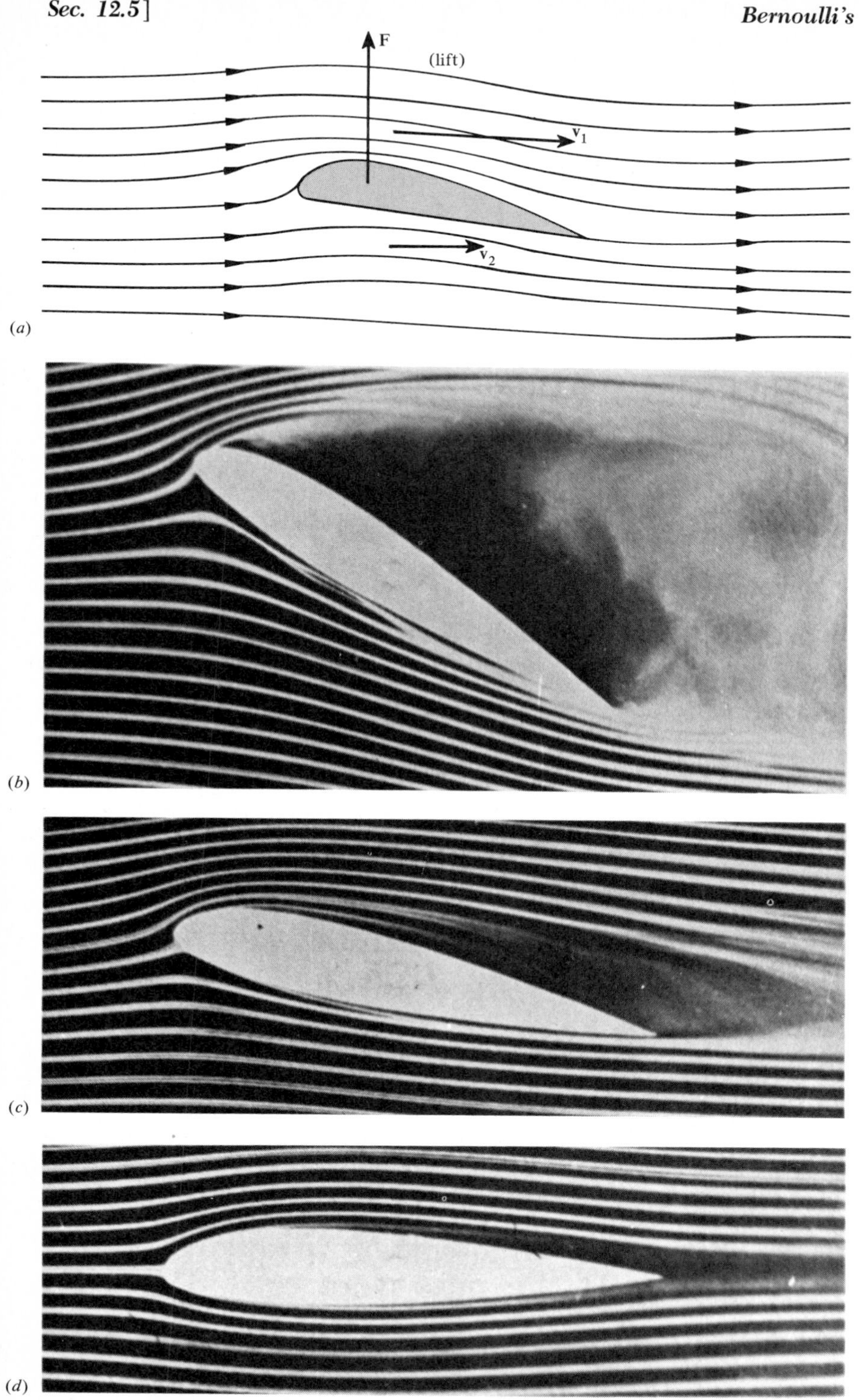

Figure 12.16. *Streamlines around an airplane wing or airfoil. (a) Schematic, (b) Photographs show how the change in the angle of the airfoil results in an increase in turbulence around the foil. (Reproduced from* Journal of Applied Physics, *August 1943, with permission from the National Advisory Committee for Aeronautics.)*

Exercise 12.6 In Example 12.6, to what height can the pipe be extended before the water pressure becomes equal to 1 atm? [*Ans.*: $h = 19.7$ m.]

* 12.6. Surface Tension

When a drop of water is placed on a glass plate, it almost flattens out, whereas a drop of mercury placed on a glass plate more or less retains a spherical shape. Furthermore, if a glass rod is dipped into water and pulled out, the water molecules stick to the glass; hence, it is wet. On the contrary, when the glass rod is dipped into mercury, no mercury molecules stick to the glass and it is completely dry. Also, if we dip a capillary tube into water, the water rises, while if dipped in mercury, the mercury level in the tube depresses.

These interesting phenomena and many others that are commonly observed arise because of an effect called *surface tension*. Surface tension is the result of forces acting on the molecules near the surface or boundaries of fluids and the surface of a fluid and solid. As mentioned earlier, there are attractive forces between the molecules called *molecular forces,* which are of electrical origin. Attractive molecular forces exist not only between molecules of a fluid, but also between the molecules of a fluid and solid in contact. *The force of attraction between like molecules is called cohesion while the force of attraction between unlike molecules is called adhesion.* These attractive forces exist only if the molecules are within a few millionths of a centimeter of each other. In a liquid the molecules are not as close to each other as they are in a solid. This results in a liquid not having a strong-enough cohesive force to have as rigid a structure as a solid. These intermolecular attractive forces are strong enough to give a liquid a definite surface. This means that any change in surface area will involve a change in energy. All fluid surface phenomena can be viewed as either a surface-tension effect or a surface-energy effect, as explained below.

Surface Tension and Surface Energy

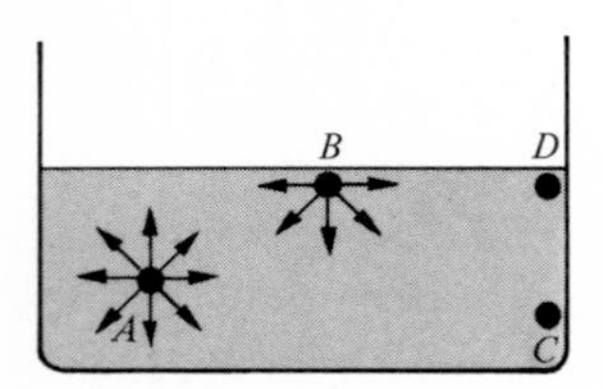

Figure 12.17. *Forces acting on molecules at different positions in a liquid.*

Consider a large volume of a liquid placed in a glass vessel with its surface exposed to air as in Figure 12.17. A molecule such as A which is well inside the liquid is attracted by cohesive forces from all directions by neighboring molecules of the liquid; hence, no net force acts on it. A molecule such as B, which is near the surface of the liquid, is attracted both by the cohesive (liquid–liquid molecules) as well as by adhesive (liquid–air molecules) forces. The adhesive forces between air and liquid molecules are almost negligible as compared to the cohesive forces acting on B. Thus, the molecules at the surface are pulled downward (or inward). That all the surface molecules are in a state of stress or tension will be demonstrated in a simple manner. Furthermore, for molecules such as C and D, there is a strong adhesive force resulting from the liquid–glass molecules. If this adhesive force is stronger than the cohesive force (as in the case of water–glass molecules), the liquid molecules form a thin film on the glass, hence wet it. On the other hand, if the cohesive forces are stronger than adhesive (such as in the case of mercury–glass molecules), the liquid does not wet the glass surface.

Let us now examine the existence of stress or tension in a liquid surface with the help of an arrangement shown in Figure 12.18. A wire is bent into the shape of a U-frame $CABD$ and is provided with a sliding wire PQ, which has a very small weight w_1. The frame is dipped into a soap solution and taken out. The area $PABQ$ is covered with soap film. If w_1 is sufficiently small, the sliding wire PQ is at once pulled close to AB. To hold PQ in equilibrium, an additional weight w_2 must be added, as shown in Figure 12.18. Thus, the total force acting downward is $F = w_1 + w_2$. The magnitude of this force is independent of the area of the soap film; hence, PQ can be set anywhere along the U-frame. Even though the soap film is very thin, it is still many molecular layers thick, and thus there are two surfaces to this film, one in the front and one in the back.

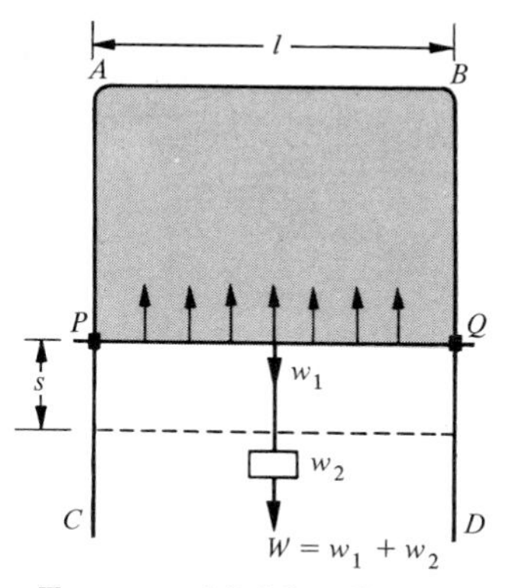

Figure 12.18. *Demonstration of tension in a liquid surface.*

To maintain equilibrium, the downward force F must be balanced by some upward force exerted by the liquid on the sliding wire PQ. The force acting upward is distributed over the whole length of the slider PQ. This force exerted by the liquid is called the *coefficient of surface tension or surface tension T and defined as the force per unit length; the force being parallel to the surface and perpendicular to the length.* Thus, as shown in Figure 12.18, the total force acting upward is $2lT$, the factor of 2 being introduced because there are two surfaces in contact with l. Thus, for equilibrium,

$$F = 2lT \tag{12.31a}$$

$$T = \frac{F}{2l} = \frac{w_1 + w_2}{2l} \tag{12.31b}$$

which states that *surface tension T* is equal to the force per unit length needed to overcome molecular forces. The units of surface tension are N/m or dyn/cm. Values of T are listed in Table 12.2A. Note that the value of T varies with temperature.

Table 12.2
A. Surface-Tension Coefficients

Substance	Temperature (°C)	T (N/m)	T (dyn/cm)
Acetic acid	20	0.0234	23.4
Benzene	20	0.0289	28.9
Carbon tetrachloride	20	0.0268	26.8
Ethyl alcohol	20	0.0216	21.6
Glycerin	20	0.0631	63.1
Mercury	20	0.465	465
Petroleum	20	0.0259	25.9
Soap solution	20	0.025	25
Turpentine	20	0.0289	28.9
Water	0	0.0756	75.6
Water	20	0.0725	72.5
Water	60	0.0662	66.2
Water	100	0.0589	58.9

Still another way of looking at surface tension can be obtained by considering Figure 12.18 in a different way. The system is in equilibrium. Suppose that the slider PQ is pulled down a distance s. The soap film will stretch and increase its surface area by $\Delta A = ls(\text{front layer}) + ls(\text{back layer}) = 2ls$. The work done ΔW by the force $F(= w_1 + w_2)$ is Fs. Thus, the work done per unit area in increasing the area (in the process of stretching) is

$$\frac{\Delta W}{\Delta A} = \frac{Fs}{2ls} = \frac{F}{2l} = T \tag{12.32}$$

Thus, according to this equation we may define *surface tension to be the work per unit area to increase the area.* Therefore, the units of T will be J/m^2 or $ergs/cm^2$, which are equivalent to N/m or dyn/cm, as defined before. This work is stored as potential energy. Thus, *surface tension may also be defined as energy per unit area.*

One of the commonly used methods for measuring the coefficient of surface tension is shown in Figure 12.19. It requires measuring the force F needed to pull a platinum ring through the surface of liquid. The ring forms one arm of a balance while weights are added to the other side. The total length of the liquid surface to be broken is twice the circumference of the ring, because there are two surfaces to the film [see Figure 12.19(b)]. Thus, if r is the radius of the ring, the additional force needed for balance will be

$$F = 2(2\pi r)T = 4\pi rT \tag{12.33}$$

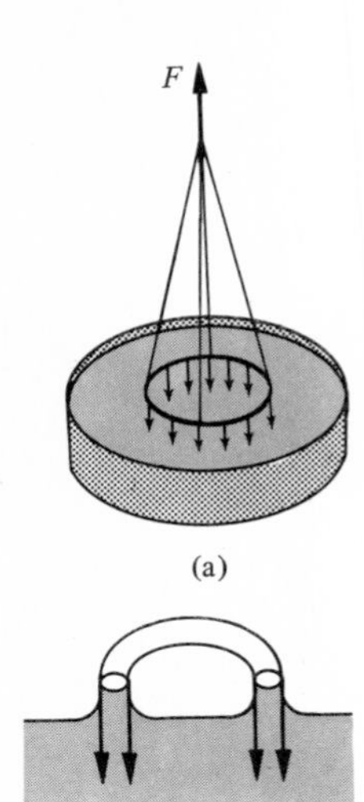

FIGURE 12.19. *Method for measuring surface tension.*

Thus, knowing F and r, T can be calculated. It may be pointed out that surface tension depends upon temperature; hence, temperature must be kept constant in any such measurements.

Capillarity and Contact Angle

If an open tube of very small cross section is dipped into water, the water wets the tube and rises in it, as shown in Figure 12.20(a). On the other hand, if this tube is dipped into mercury, the mercury does not wet the glass tube and the level of the liquid is depressed, as shown in Figure 12.20(b). The smaller the radius of the tube, the greater will be the rise (or depression) of the liquid in it. This phenomenon of rise or depression of liquid in a tube of small radius is called *capillarity*. Furthermore, it may be noticed that the surface of the liquid is curved at the top. This curved liquid surface in the tube is called a *meniscus* and may be curved inward or outward as shown in the two cases. The effects above can be understood as surface-tension phenomena.

First, let us try to understand the shape of the meniscus. Consider a water molecule, such as P, in a glass tube shown in Figure 12.20(c). Three forces act on this molecule: (1) the cohesive force C due to other water molecules pulling P inward, (2) the adhesive force A due to glass molecules pulling P, and (3) the adhesive force due to the air molecules acting on P. The third force is so small that we may, for all practical purposes, neglect it. The forces C and A acting on P are shown in Figure 12.20(c). If A is very large compared to C, the resultant force R will be as shown, and the liquid molecules are pulled by the glass molecules, forming a film, hence wetting the surface.

On the contrary, as demonstrated in Figure 12.20(d), the shape of the meniscus is reversed if the cohesive force C is much greater than the adhesive force A, as in the case of mercury and glass. Not only is the liquid surface pulled inward, but it does not wet the solid surface.

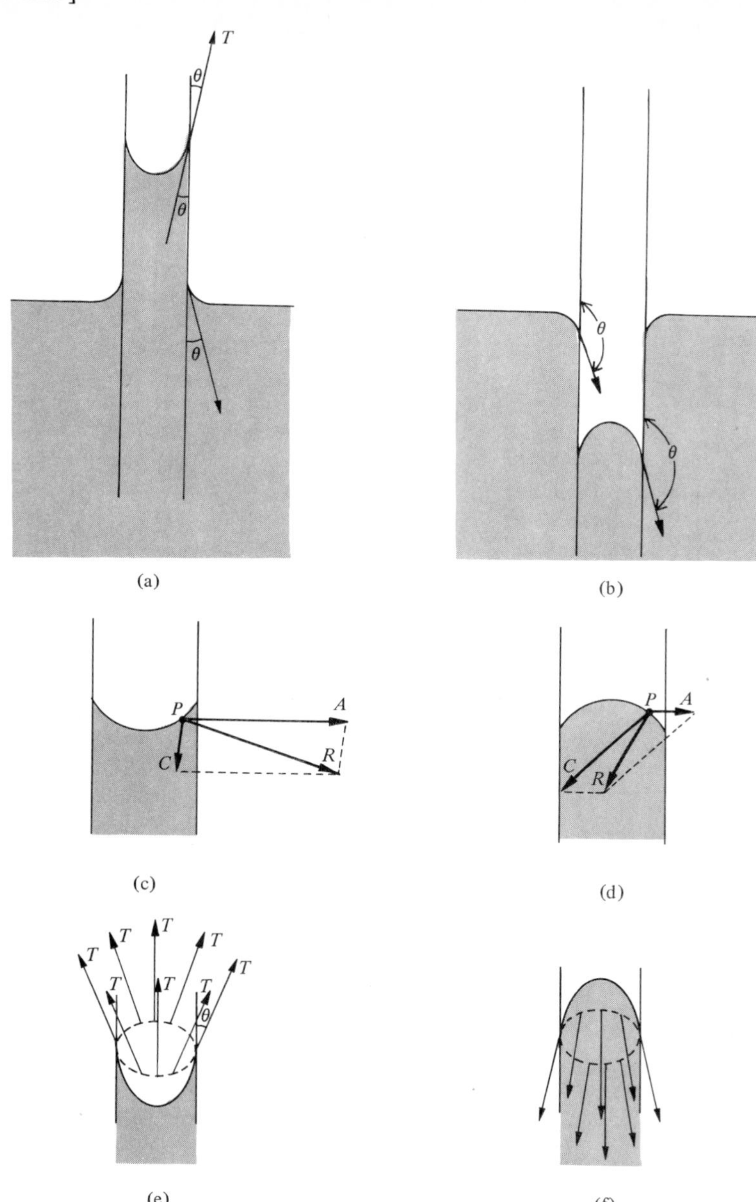

Figure 12.20. *Capillary action and contact angle of water (parts a, c, and e) and mercury (parts b, d, and f).*

Let us go back to Figure 12.20(a). When a glass tube is dipped into water, the molecules on the surface of the glass wall just above the water surface pull the water molecules upward, raising the water level as shown. Thus, the column of water pushed upward is supported by the surface forces. Let us consider the equilibrium of this column of radius r and height h. The surface tension force T per unit length is acting all along the surface of circumference $2\pi r$ as shown in Figure 12.20(e). The *angle of contact*, θ, defined as the angle between the liquid surface and the wall, depends upon the type of liquid, the gas above it, and the material of the contact wall.

Thus, the component of the upward force per unit length is $T \cos \theta$; while the total upward force is $T \cos \theta(2\pi r)$. This force must support a cylinder of liquid of weight $w = (\pi r^2 h)\rho g$, where ρ is the density of the liquid. Thus,

$$T \cos \theta(2\pi r) = \pi r^2 h \rho g$$

or

$$\boxed{h = \frac{2T \cos \theta}{r \rho g}} \tag{12.34}$$

For liquids that wet the solid surface, the contact angle is less than 90°. On the other hand if the liquid does not wet the solid, the angle of contact is greater than 90°. For example, as shown in Figure 12.20(b), for mercury and glass, $\theta = 140°$. For a clear water–glass combination, the contact angle is almost zero. For a water–paraffin combination, $\theta = 107°$; that is, the water does not wet paraffin. Some values of contact angles are listed in Table 12.2(B).

TABLE 12.2
B. Contact Angle

Liquid	Solid Wall	Contact Angle
Water	Glass	0°
Water	Paraffin	107°
Water	Silver	90°
Kerosene	Glass	26°
Mercury	Glass	140°
Methylene iodide	Pyrex, lead glass	29°, 30°

Pressure Difference in Bubbles and Drops

Finally, let us consider surface-tension effects acting on a soap bubble. Let p be the pressure inside the bubble and p_a outside. We want to calculate $p - p_a$. Let r be the radius of this bubble. Imagine the bubble broken into two halves, and consider one-half of it as shown in Figure 12.21. Since there are two surfaces, inner and outer, the total surface tension force is $2(2\pi r)T$. The pressure difference $(p - p_a)$ acts on a cross-sectional area πr^2, the projection of the

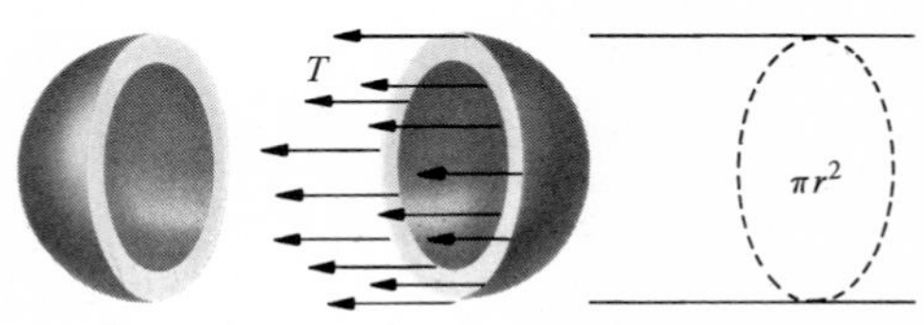

FIGURE 12.21. *Spherical bubble shown broken into two. Horizontal arrows indicate the surface-tension forces acting on the inner and outer surfaces of the circumference.*

spherical bubble; that is, the force is $(p - p_a)\pi r^2$. The surface-tension force must balance this force to maintain equilibrium.

$$(p - p_a)\pi r^2 = 2(2\pi r)T$$

or

$$\boxed{p - p_a = \frac{4T}{r}} \quad \text{for a bubble} \tag{12.35}$$

On the other hand, for a completely filled liquid drop, there is only one surface, and hence Equation 12.35 will take the form

$$\boxed{p - p_a = \frac{2T}{r}} \quad \text{for a drop} \tag{12.36}$$

Applications

We have mentioned some phenomena connected with surface tension, and we come across many others every day. For example, the absorbing power of towels and blotting paper and the rise of oil in the wick of a lamp are all due to capillary action. The dissolving of instant coffee and sugar is due to their wetting properties when coming in contact with the liquids.

In the chemical industry the wetting property is made use of in detergents and waterproofing. When added to liquids, detergent materials decrease the contact angle, thereby increasing the wettability. On the other hand, when a waterproofing material is added to a fabric, it increases the contact angle, making the fabric water-repellent.

Another example is the use of a commercial antiseptic trademarked ST-37 indicating that it has a T of 37 dyn/cm. Such a low value of T prevents the formation of drops that might otherwise block the entrance to skin or a wound.

Example 12.7 Calculate the height to which kerosene will rise in a glass tube of 0.2 mm diameter when dipped in kerosene.

For kerosene the surface tension $T = 0.026$ N/m, contact angle $\theta = 26°$, $\cos 26° = 0.90$, and its density $\rho = 0.82 \times 10^3$ kg/m^3. Since the contact angle θ is less than 90°, the liquid will rise in the glass tube. Substituting these values and $r = 0.1$ mm $= 0.1 \times 10^{-3}$ m in Equation 12.34, the value of h is

$$h = \frac{2T\cos\theta}{r\rho g} = \frac{2(0.026\ \text{N/m})(0.90)}{(0.1 \times 10^{-3}\ \text{m})(0.82 \times 10^3\ \text{kg/m}^3)(9.8\ \text{m/s}^2)}$$
$$= 0.058\ \text{m} = 5.8\ \text{cm}$$

Exercise 12.7 What happens to the level of mercury in a glass tube of diameter 0.2 mm when it is dipped in mercury? For mercury $T = 0.465$ N/m, $\theta = 140°$, and $\rho = 13.6 \times 10^3$ kg/m^3. [*Ans.*: The liquid in the tube is depressed by 5.4 cm.]

SUMMARY

The term "fluid" refers to a substance that can flow and does not have a shape of its own. The study of fluids at rest is termed *hydrostatics* (or fluid statics), and the study of fluids in motion is *hydrodynamics* (or fluid dynamics). Aerodynamics deals with the study of the flow of air or gases.

Density, $\rho = m/V$, is mass per unit volume. *Pressure*, $p = F/A$, is the normal force acting on a unit area. In general, pressure at a depth h below the liquid level is $p = p_a + \rho gh$. One atmosphere $= 1.013 \times 10^5\ \text{N/m}^2 = 101.3 \times 10^3$ kPa. $14.7\ \text{lb/in}^2 = 76$ cm Hg. Gauge pressure $p_g = p - p_a$.

According to *Pascal's principle*, pressure applied to an enclosed fluid is transmitted undiminished to every portion of the fluid, $p = f/a = F/A$. According to *Archimedes' principle*, a body immersed in a fluid is buoyed up with a force in the vertical direction equal to the weight of the volume of the fluid displaced.

An *ideal fluid* is incompressible and nonviscous. An ideal fluid having an irrotational, steady flow is described by the continuity equation $v_1A_1 = v_2A_2$ and by Bernoulli's equation, $p + \frac{1}{2}\rho v^2 + \rho gy =$ constant. The discharge rate is $Q = vA$.

* The force of attraction between like molecules is called *cohesion*, and the force of attraction between unlike molecules is called *adhesion*. The *surface tension* T is defined as the force per unit length or the work per unit area to increase the area. The phenomenon of rise or depression of liquid level in a tube of small radius is called *capillarity*. The relation between T, h, and the contact angle θ is $h = (2T\cos\theta/r\rho g)$, where r is the radius of the tube. The pressure difference in a bubble is $4T/r$ and in a liquid drop is $2T/r$.

QUESTIONS

1. When the molecules are brought together, what factors determine whether the resulting material will be a solid, a liquid, or a gas?
2. How does the buoyant force vary with depth when a stone is immersed in (a) an incompressible fluid; and (b) a compressible fluid?
3. Explain why pressure is a scalar quantity.
4. A ball with compressed air in it floats on the water surface. Will it still float if (a) the air is completely removed; and (b) if the pressure of the gas inside is increased five times?
5. Does the center of gravity coincide with the center of buoyancy of a floating body? If it does not, what happens to the body when immersed in water?
6. A beaker of water is placed on a spring-balance scale. Will there be a change in the scale reading if (a) a piece of wool floats on the water surface; and (b) a piece of metal is immersed in the water without touching the sides of the beaker?
7. If the water is turned off, does the pressure in the pipe decrease or increase?
8. A tennis ball, after being hit, is moving from right to left. If it is spinning clockwise when viewed from below, in which direction will the ball curve?
9. How do the following factors affect the airplane lift: (a) change in altitude; (b) change in wind velocity; and (c) accumulation of ice on the wings?

*10. If the tube in Figure 12.3 is used for measuring atmospheric pressure, is it necessary to correct the height of the mercury column due to the surface tension (capillary action)?

PROBLEMS

1. Calculate the pressure due to (a) a column of mercury 80 cm high; and (b) a column of water 40 ft high.
2. (a) Express the following in atmospheres: 50 lb/in^2, 5.06×10^5 N/m^2, 2026 mbars, and 10.13×10^6 dyn/cm^2. (b) Express the following in N/m^2: 4 atm, 50 lb/in^2, 2026 mbars, and 10.13×10^6 dyn/cm^2.
3. Each of the four tires of an automobile has a 225-cm^2 area in contact with the road. If the mass of the automobile is 2000 kg, calculate the pressure exerted by the automobile on the road.
4. A barometer reads 76 cm Hg at the bottom of a building, 74.8 cm Hg at the top of the building. If the density of air is 1.25 kg/m^3, what is the height of the building?
5. If the pressure at the surface of a lake is 1 atm, how far below the surface will the pressure be 3 atm?
6. A 1-m-thick layer of water floats on a 0.2-m-thick layer of mercury. What is the pressure difference between the top of the water surface and the bottom of the mercury surface?
7. What is the hydrostatic blood pressure difference between the head and the feet of a 2-m-tall man standing straight? The density of blood is 1.06×10^3 kg/m^3.
8. A vacuum gauge reads a pressure of 0.01 atm. Calculate the absolute pressure in atm and N/m^2.
9. A U-tube manometer containing mercury is connected to a gas tank. The difference in the levels of mercury in the two tubes is 25 cm. If the atmospheric pressure is 76 cm Hg, what is the gauge pressure and the absolute pressure, in N/m^2?
10. A submarine is at a depth of 50 m under the sea. Calculate the pressure due to the water and the force on a hatch of 1.2 m $\times$ 0.8 m. The density of sea water is 1.05×10^3 kg/m^3.
11. The radius of a piston of a hydraulic automobile lift is 14 cm. Calculate the gauge pressure needed to lift a car of 1500 kg.
12. An automobile of 2400 lb is to be lifted by a hydraulic pump. What should be the ratio of the areas if the force applied is only 24 lb?
13. In a service station a hydraulic pump is used to lift cars. The radius of the larger piston is 15 cm and that of the smaller piston is 1.5 cm. If a force of 200 N is applied to the smaller piston, calculate the maximum mass of a car that can be lifted.
14. A hydraulic lift is to be used to lift a 4-ton truck by applying a pressure of 8 atm. What should be the radius of the larger piston?
15. The gauge pressure of 3×10^5 N/m^2 must be maintained in the main water pipes of a city. How much work must be done to pump 50,000 m^3 of water at a pressure of 1.0×10^5 N/m^2?
16. An ice cube of volume 0.1 m^3 and density 0.92×10^3 kg/m^3 is immersed in a liquid of density 0.82 kg/m^3. Calculate the buoyant force on the ice.

17. A girl of mass 40 kg is floating almost completely immersed in the water. Calculate her volume and the buoyant force that acts on the girl.
18. An object has a mass of 50 kg in air, and its apparent mass when immersed in water is 45 kg. Calculate (a) the apparent weight; (b) the buoyant force on the object; and (c) the volume of the object.
19. A piece of iron has a mass of 0.12 kg. If its density is 7.85×10^3 kg/m^3, what is its volume? This mass is tied to a spring balance and immersed in water. What will the spring balance read? (This is the apparent weight.)
20. A piece of alloy has a mass of 2 kg in air, 1.75 kg when immersed in water, and 1.50 kg when immersed in oil. Calculate the density of the alloy and of the oil.
21. A cork has a mass of 30 g in air. When tied to a metal immersed in water, the combined weight is 2.10 N. When both metal and cork are immersed in water, the combined weight is 0.82 N. What is the density of the cork?
22. A metal piece when immersed in water, benzene, and glycerin experiences a loss in weight of 2.0 N, 1.82 N, and 2.54 N, respectively. Calculate the relative density (or specific gravity) of benzene and glycerin.
23. What fraction of a cube of floating ice is above water if its density is 0.92×10^3 kg/m^3? What should be the volume of the ice so that when a mass of 10 kg is placed on the ice, it remains floating?
24. Consider a rectangular block of wood 3 m $\times$ 2 m and 0.5 m thick and of density 600 kg/m^3 floating in water. Calculate the depth of the block below the waterline when a man of mass 80 kg sits on it. What is the maximum mass that can be placed on the block and have it still remain afloat?
25. What weight can be supported by a balloon of radius 10 m if it is filled with (a) helium; and (b) hydrogen?
26. Water flowing in a pipe of cross-sectional area 12 cm^2 has a velocity of 10 m/s. What will be the velocity of water if the pipe tapers to a cross-sectional area 4 cm^2? What is the diameter of the pipe at each of the two points?
27. The speed of blood in a major artery of radius 0.35 cm is 0.04 m/s. The artery is divided into 50 capillaries, each of radius 0.1 cm. Calculate the speed of the blood through a capillary.
28. In the previous problem calculate the volume of blood that flows through the artery and each capillary per minute.
29. Consider a tank filled with water to a height of 6 m. A hole of cross-sectional area 1.0 cm^2 is made 4 m below the water level. Calculate the velocity of efflux of water from the hole and the mass of water that is discharged per second. Assume that the tank is open and $A_1 >> A_2$.
30. A tank is filled with water to a height H. A hole is made at h below the water level. At what distance R from the tank will the water strike the ground? Calculate R for $H = 6$ m and $h = 4$ m. Assume that the tank is open and $A_1 >> A_2$.
31. A tank of large cross-sectional area is filled with water to a depth of 8 ft. A hole of area 2 cm^2 is made 6 ft below the water level. Calculate (a) the speed of efflux; and (b) the mass of water that flows out per second.
32. Water stands at a height of 3 m in an enclosed tank with compressed air at the top having a gauge pressure of 3×10^5 N/m^2. A hole of 4 cm^2 is made

at the bottom. Calculate the velocity and the discharge rate of water from the hole.

33. The cross-sectional area of a pipe at point A is twice that at point B, which is 10 m above point A. The speed of water in the pipe at point A is 5 m/s, while the pressure at this point is 1.5×10^5 N/m^2. Calculate the velocity and pressure at point B.

34. What is the absolute pressure and the gauge pressure required in the main-water lines so that the water from a firehose can reach a height of 30 m?

35. Calculate the work done in forcing 10 m^3 of water through a pipe where the pressure difference is 1.8×10^5 N/m^2.

36. Air flows through a pipe of variable cross section. The gauge pressure between two points 1 and 2 is 4×10^5 N/m^2, while the area A_1 is three times A_2. If the area A_2 is 5 cm^2, calculate the speed of the air at point 1. Also, calculate the flow rate Q.

37. The velocity of air at the top of the airplane wing is 220 m/s; under the wing it is 180 m/s. The wing size is 8 m $\times$ 3 m and the density of air is 1.0 kg/m^3. Calculate (a) the pressure difference between the two sides of the wing; and (b) the net lift on the wing.

***38.** In Figure 12.18 the frame is dipped in water and held vertically. What is the minimum weight needed to hold the film in equilibrium if the length PQ is 10 cm? How much work is done if PQ is pulled down 2 cm?

***39.** The frame in Figure 12.18 is dipped in a soap solution and held vertical. The sliding wire PQ is 4 cm. Calculate the weight needed to hold PQ in equilibrium. How much work will be done if PQ is to be pulled another 2 cm? The coefficient of surface tension for soap solution is 0.025 N/m (= 25 dyn/cm = 25 ergs/cm^2).

***40.** A ring of 3 cm diameter is dipped in a liquid and pulled out slowly, as shown in Figure 12.19. If a force of 0.1 N is needed to break the film, calculate the coefficient of suface tension of the liquid.

***41.** Suppose that 64 rain drops combine into a single drop. Calculate the ratio of the total surface energy of the 64 drops to that of a single drop. For water, $T = 0.072$ N/m = 0.072 J/m^2.

***42.** A soap bubble of radius 1 cm expands into a bubble of radius 2 cm. Calculate the increase in total surface energy if T for soap is 25 dyn/cm = 25 ergs/cm^2.

***43.** Calculate the height to which water will rise in a tube of diameter (a) 0.2 mm; (b) 0.8 mm; (c) 2 mm; and (d) 8 mm. Also, calculate the weight of the water raised in each case.

***44.** What should be the diameter of a mercury tube so that when the tube is dipped in mercury, the mercury level in the tube does not drop more than 0.01 mm?

***45.** A tube 0.4 mm in diameter is made of paraffin and dipped in water. Calculate the change in water level if the contact angle is 107° and the surface tension for water is 72 dyn/cm.

***46.** The liquid in a capillar tube of radius 0.1 mm rises to a height of 8 cm. Calculate the surface tension of the liquid if the angle of contact between the liquid and the wall is 0°. The density of the liquid is 750 kg/m^3.

***47.** A glass tube 0.15 mm in diameter is dipped in glycerin whose surface tension is 0.0632 N/m. Calculate the contact angle if the glycerin rises to a height of 14 cm. The density of glycerin is 1250 kg/m^3.

***48.** Calculate the pressure difference between the inside and outside of a soap bubble of radius (a) 4 mm and (b) 8 mm.

***49.** Calculate the excess pressure inside a drop of water and mercury each of radius 2.5 mm and at 20°C.

Part III
Heat and Thermodynamics

13

Concepts of Temperature and Heat—Kinetic Theory

To start with, we shall relate the concepts of heat and temperature with disordered molecular motion. Different temperature scales—Celsius scale, ideal gas scale, and Fahrenheit scale—will be introduced. We shall define an ideal gas and state the gas laws that give the relation between pressure, volume, and temperature of an ideal gas. The kinetic theory of gases, which gives the theoretical derivation of the gas laws and the meaning of temperature and internal energy in terms of atomic and molecular motions, will be discussed in some length.

13.1. Investigation of Disorder Motion

WHEN THE molecules of an object or a system have ordered motion such as shown in Figure 13.1(a), energy is manifested on the macroscopic scale as kinetic and potential energies. The subject of mechanics deals with such ordered motions resulting from the application of external forces. On the other hand, the molecules in Figure 13.1(b) have random or disordered motion. *The study of these individual molecules in disordered motion is called heat.* The energy (kinetic and potential) of these disordered molecules is called *thermal* or *internal energy* of the system. Of course, it is possible that the molecules may have some degree of ordered as well as disordered motion [as shown in Figure 13.1(c)], in which case the total energy is the sum of the mechanical and thermal energies.

A detailed investigation of the subject of heat will involve the study of individual motions of a tremendously large number of molecules. For example, 1 m^3 of gas may have as many as 10^{25} molecules. Thus to write down the equations of motion for all these molecules and to try to solve them is an impossible task, even with the aid of high-speed computers. The subject of heat has evolved under three categories: (1) thermodynamics, (2) the kinetic theory of gases, and (3) statistical mechanics.

Thermodynamics deals with interconversion of thermal energy and mechanical work, and *may be defined as the theory of heat superimposed on the theory of mechanics*. The success of thermodynamics lies in the fact that it finds relations between such macroscopic or bulk quantities as pressure, volume, temperature, internal energy, and mechanical work.

After the bulk properties of heat had been established, scientists started looking into the relations between these quantities and the microscopic properties of atoms and molecules. According to *kinetic theory*, enough simplifying assumptions are made about the molecular motion so that we only need to solve a few equations describing their average behavior. This approach, as we shall see, has been very successful in dealing with gases and in special cases with liquids also.

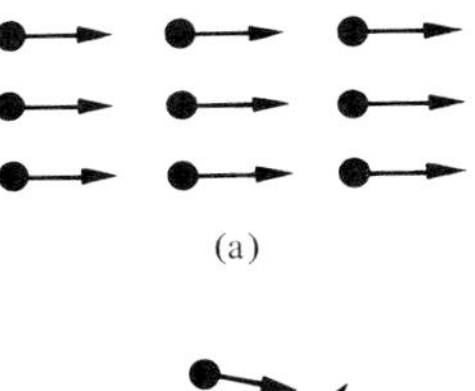

(a)

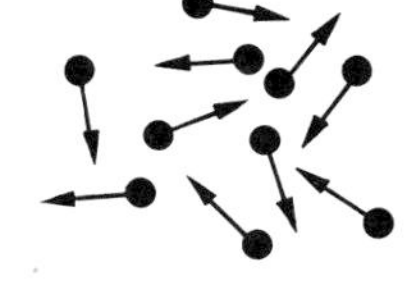

(b)

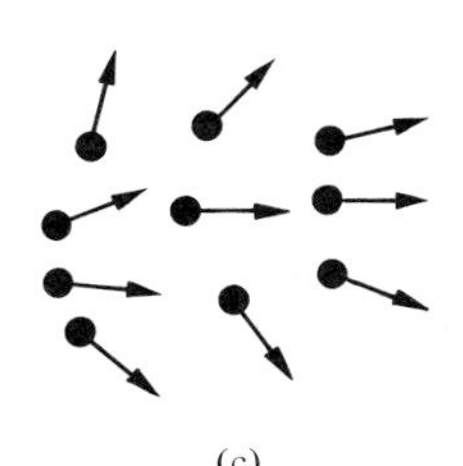

(c)

FIGURE 13.1. *Ordered motion of molecules results in mechanical energy. (b) Disordered motion of molecules results in thermal or internal energy. (c) Partially disordered motion of molecules results in a combination of mechanical and thermal energy.*

Still another approach to the theory of disordered motion is statistical physics or statistical mechanics. According to *statistical mechanics,* one completely ignores the behavior of individual molecules but tries to find the most probable values by making use of statistical methods. This approach, together with quantum mechanics, has been highly successful not only in gases, but in liquids and solids as well. It has played a very significant role in explaining and developing many aspects of modern technology, such as the understanding of transistors.

13.2. Concepts of Heat and Temperature—Zeroth Law

Most of us have some feeling for the meaning of heat and temperature in everyday usage. But we must define these concepts in terms of scientific terminology. Also, a clear distinction must be made between the two easily confused terms—heat and internal (or thermal) energy.

In everyday language we may define *heat as something that produces in us the sensation of warmth while temperature may be defined as the degree of warmth or hotness*. For example, when we sit near a burning fireplace, we feel warmth, and we say that heat coming from the fireplace is being absorbed by our body. Furthermore, we know that the fireplace is much hotter than our body; hence, we say that the fireplace is at a much higher temperature than the temperature of our body. Similarly, when we sit in the summer sun we can feel our body getting warm. Once again we say that heat is being given out by the sun and absorbed by our bodies, and also the temperature of the sun is higher than the temperature of the body. Let us now consider a somewhat different example. Suppose we hold a chunk of ice. Heat flows from hands to the ice and that our hands are at a higher temperature than the ice. As our hands lose heat to the ice, we feel cold because the temperature of our hands has fallen as compared to the rest of the body. Thus, we may conclude: (1) the heat flows from a body at a higher temperature to a body at a lower temperature; and (2) the temperature of the body losing heat decreases while the temperature of the body gaining heat increases.

The discussion above still does not answer in scientific terms the questions: what is heat? And what is temperature? We have already mentioned that bulk or mechanical energy is due to the ordered motion of the atoms and molecules. The higher the disorder of the molecules, the greater the internal energy of the system and the higher its temperature. Suppose that two bodies at different temperatures are brought together. The more highly disordered molecules collide with molecules of lower disorder and transfer energy to them. This process continues until the disorder of the molecules is the same in both bodies; hence, they have the same temperature. *Such energy transfer that takes place due to difference in temperature is called heat* or, more formally: *energy transfer by means of molecular collisions is called heat Q.*

Historically, in the eighteenth and early nineteenth centuries there were three different views held about the nature of heat:

1. The *caloric theory,* according to which heat was considered to be a substance called *caloric*. This substance transferred from hotter to colder bodies when they were brought in contact. But no such substance was ever found.

2. According to some scientists, heat was associated with the random translational motion of molecules.

3. Still another view was that heat is associated with the rotational and vibrational motions of molecules in a fixed position.

In the middle of the nineteenth century, kinetic theory was developed on the assumption of random translational motion of molecules. Its successes clearly established the meaning of both internal energy as well as rotational and vibrational energies, and all three kinds of motion must be taken into consideration in explaining observed facts.

Finally, let us go into more detail about the concept of temperature, T. *Temperature may be defined as a quantity that is proportional to the degree of disordered molecular kinetic energy of a body.* In the case of gases and liquids, it is a measure of the translational kinetic energy of molecules, while in the case of solids (since the atoms are not free to have translational motion), it is a measure of the vibrational kinetic energy of the atoms.

In view of the definition of heat we may add that the magnitude of temperature determines if the heat will flow from or to a body when it is brought into contact with another body. Thus, if heat flows from body A to body B, body A must be at a higher temperature than body B.

Suppose that body A does not lose or gain heat to its surroundings. This is stated by saying that body A is in thermal equilibrium with its surroundings; that is, body A and the surroundings are at the same temperature. Two systems (or bodies) A and B are said to be in *thermal equilibrium* if both are at the same temperature and no heat flow occurs between them when they are brought in contact with each other.

The *zeroth law of thermodynamics* is concerned with the thermal equilibrium conditions of more than two bodies and may be stated as follows:

> **Zeroth Law of Thermodynamics:** ***Suppose that of the three bodies A, B, and C, bodies A and B are separately in thermal equilibrium with body C; then bodies A and B are in thermal equilibrium with one another.***

This statement is so evident that it was not thought necessary to formulate it as a law until the other laws of thermodynamics—first, second, and third—had been formulated; hence the name "zeroth law."

13.3. Temperature Scales and Measurement

The sensation of touch cannot be used to assign quantitative meaning to temperature because it is not always reliable. Just as with length, mass, and time, which are measured in terms of operationally defined quantities such as meter, kilogram and second, respectively, the temperature must also be defined operationally. Thus, we need two things: (1) instruments called thermometers, which measure the relative hotness, that is, the temperatures of different bodies with respect to a certain reference point; and (2) the scales to give relative magnitudes. We may say that temperature is a quantity measured by a thermometer, just as length is a quantity measured by a meter stick. It may be pointed out that temperature is a scalar quantity; hence, it has only magnitude.

The Celsius Scale

To construct a thermometer we must rely on those materials whose properties change with temperature. For example, in general the volume of a gas or liquid increases with increasing temperature. The length of a solid changes with a change in temperature. Similarly, the resistance of an electrical wire changes with changing temperature.

One of the most common thermometers makes use of the expansion of mercury in a glass tube. Mercury is placed in a small spherical bulb attached to a long, thin glass cylindrical tube and sealed at the top. First, the bulb is dipped in a bath of melting ice at normal atmospheric pressure. The mercury level in the glass tube will become stationary when mercury is in thermal equilibrium with the ice bath, that is, when they both have the same temperature. As shown in Figure 13.2(a), this level is marked A and is referred to as the *ice point*. Next, the mercury bulb is put in a bath of boiling water at normal atmospheric pressure. After thermal equilibrium is reached, the mercury level stands at B as shown in Figure 13.2(b), and point B is referred to as the *boiling point*.

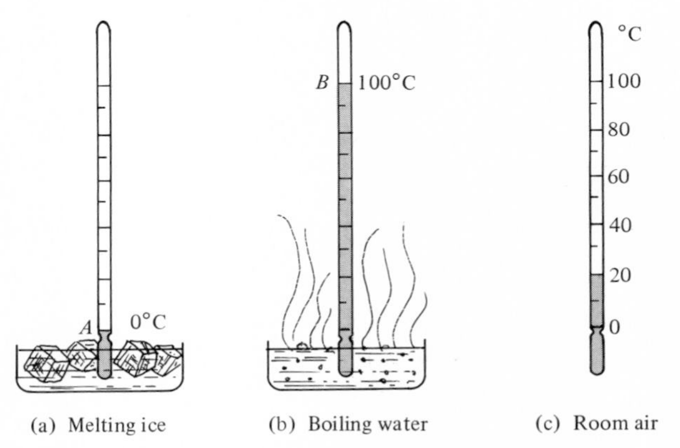

FIGURE 13.2. *Construction of a mercury thermometer.*

The next step is to assign a scale to these two fixed points. On the *Celsius scale*, the ice point is assigned 0°C (zero degrees Celsius) and the boiling point is assigned 100°C (one hundred degrees Celsius), while the length between these two points is divided into 100 equal parts, or a hundred Celsius degrees (100 C°), as shown in Figure 13.2(c). Note that temperature is quoted as °C (degrees Celsius) and a difference in temperature as C° (Celsius degrees). This scale is also called the *centigrade scale*. [The centigrade scale was invented in 1742 by a Swede, Anders Celsius (1701–1744). The Ninth General Conference on Weights and Measures in 1949 decided to name this scale "Celsius" in his honor.] Thus, we have constructed a thermometer that can be used for measuring the temperature by noting the mercury level when the thermometer is in contact with the body and is in thermal equilibrium. The scale can be extended to read temperatures below 0°C and above 100°C.

The mercury used in the construction of the thermometer described above is the *thermometric* substance. What happens if we use something else, say alcohol, as a thermometric substance? We can find the 0°C and 100°C marks by using the ice point and the boiling point, and calibrate with a Celsius scale as before. Suppose that we use the mercury thermometer and the alcohol thermometer to measure the temperature of another body. Of course, the two thermometers agree in their readings at 0°C and 100°C while at temperatures in between, the readings of the two are close but are not exactly equal. The reason for this disagreement is due to the slightly different rates of expansion of different liquids in different temperature regions. Similarly, if we use some other thermometric substance, the temperature readings will be different. We conclude the following: *The temperature measured by any thermometer depends on the thermometric substance used—a very undesirable feature!*

Somehow we must get rid of this drawback in constructing a thermometer, and be able to design a thermometer with a standard scale. This is done by using constant-volume gas thermometers, as we explain next.

Constant-Volume Gas Thermometers and Ideal-Gas Temperature Scale

If we construct a thermometer using a gas, such as H_2, He, N_2, O_2, or air at low density as a thermometric substance, experiments show that there is a very good agreement in their readings in any temperature range. Such thermometers are called *constant-volume gas thermometers*. They consist of a glass bulb G, Figure 13.3, containing a gas, say He. The volume of this gas is kept constant to a fixed point F. The pressure exerted on the gas is measured by use

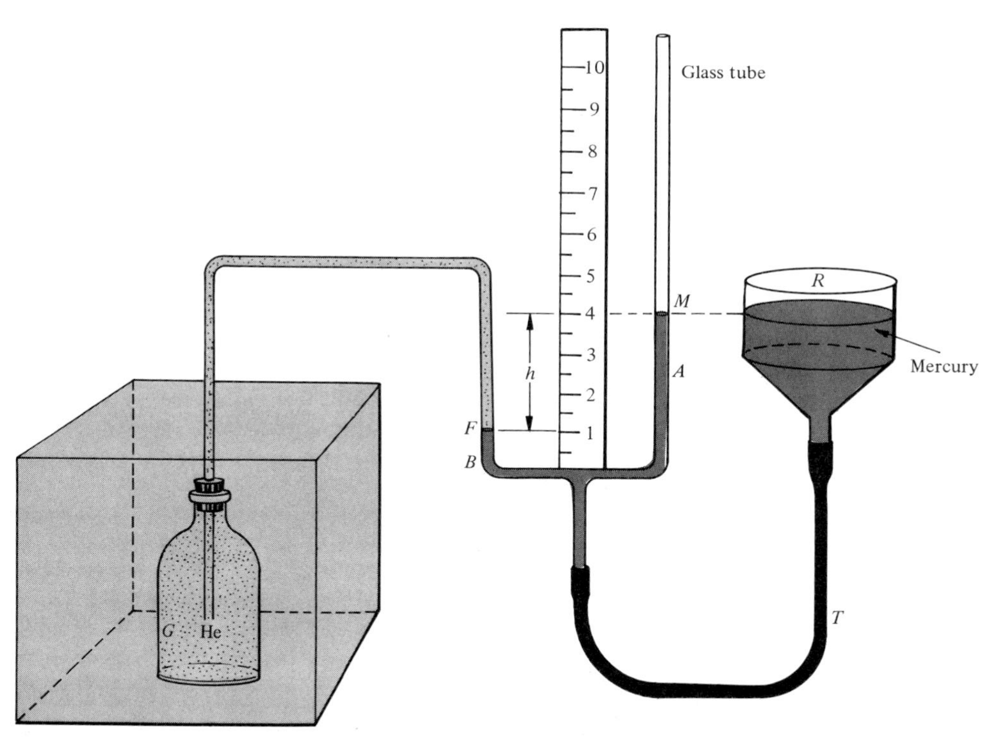

FIGURE 13.3. *Constant-volume gas thermometer.*

of the mercury manometer AB attached as shown. Suppose that the glass bulb is placed in a bath of high temperature T_C. As the temperature of the gas increases, it expands and pushes the mercury down in tube B and up in tube A. The manometer tubes are connected to a mercury reservoir R through the rubber tubing T. By adjusting the height of the reservoir R, the mercury in tube A is brought back to the fixed point F. If the difference in the levels of mercury in the tubes A and B is h, the pressure exerted on the gas is $P = P_0 + \rho gh$, where P_0 is the atmospheric pressure.

Let us consider different values of P for different values of T_C when there are N_1 molecules of gas in the bulb of the thermometer. Make a plot of P (or PV) versus T_C as shown in Figure 13.4. Repeat this experiment with N_2 molecules of gas and then with N_3 molecules of gas. As seen from Figure 13.4, all three plots

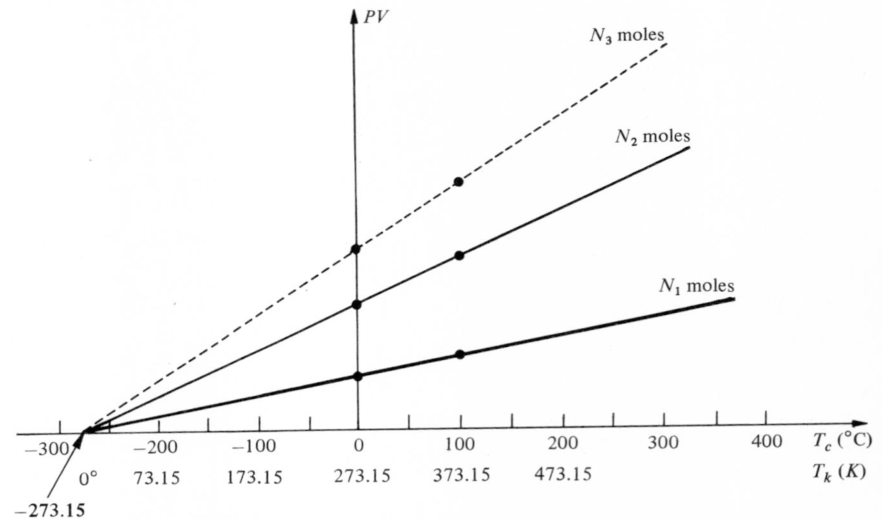

FIGURE 13.4. *P (or PV) versus T for a given gas shows that as $T_c \to -273.15°C$, $P \to 0$.*

are straight lines, although their slopes are different. (All three give the same temperature, even though the densities are different in the three cases.) If we extend all three straight lines, they all meet at one point. This happens at a temperature of $-273.15°C$. The results of Figure 13.4 suggest another linear temperature scale. This scale is called the *ideal gas scale* or *absolute gas scale*, or, as we shall see in Chapter 16, the *Kelvin temperature scale*. Zero temperature on this scale is written as 0°K (zero degrees Kelvin) corresponding to $-273.15°C$, and 0°C and 100°C correspond to 273.15°K and 373.15°K, respectively. Thus, (it is an accepted notation as recommended at the International Conference on Weights and Measures to write K instead of °K)

$$\boxed{T_C = T_K - 273.15\ \text{K}} \tag{13.1}$$

Let us now fill up the gas bulb of the constant-volume thermometer with different gases at different pressures and make a measurement at one particular

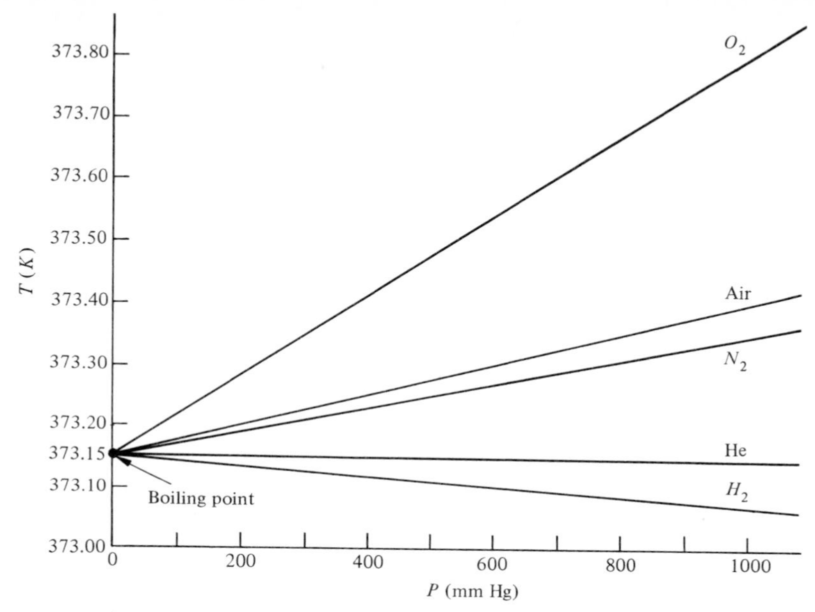

FIGURE 13.5. *Constant-volume thermometers using different gases read the same temperature as long as the pressure (or density) is low.*

temperature, say that of boiling water. The results of such experiments are shown in Figure 13.5. It clearly shows that if He is used as a thermometric substance, it reads the same temperature (horizontal line) independent of its pressure. Also, the results obtained by using other gases do not differ much. Thus we conclude the following: *The temperature measured by a constant-volume gas thermometer is independent of the type of gas as long as the density of the gas is low.* It may be pointed out that most gases liquefy before reaching absolute zero, 0 K. Only He can be used down to 4.2 K and may be extended to ~2 K.

Let us repeat the experiment by surrounding the glass bulb with a bath at a standard temperature T_s. Let the corresponding pressure be P_s. Similarly, when surrounded by an unknown temperature T, the corresponding pressure is P. From Figure 13.4 since the pressure is proportional to the temperature, we may write

$$\frac{T}{T_s} = \frac{P}{P_s} \tag{13.2}$$

Thus, if the standard temperature T_s is known and the pressure P and P_s are experimentally measured, the value of an unknown temperature T can be calculated from Equation 13.2.

Since the zero point is fixed on the absolute gas scale or Kelvin scale, it is important and interesting to note that we do not need two more reference points to construct a temperature scale; only one will suffice. This standard temperature is taken to be the triple point of water. The *triple point of water* is that temperature at which the three phases of water—solid (ice), liquid (water),

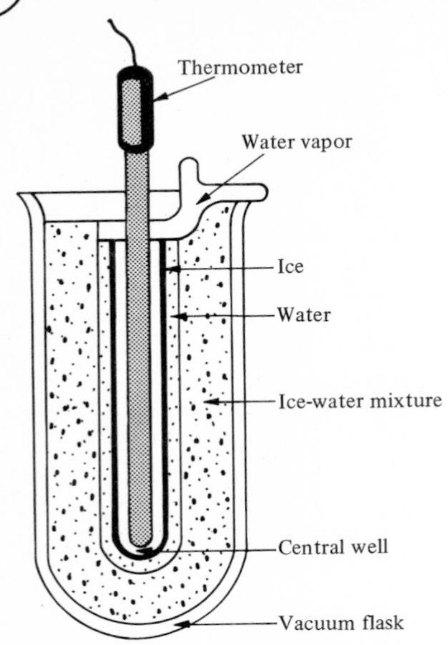

FIGURE 13.6. *Standard triple-point cell at the National Bureau of Standards. (Courtesy of the National Bureau of Standards.)*

and gas (vapor)—coexist in equilibrium. Unlike the melting point of ice or boiling point of water, which vary with atmospheric pressure, the triple point of water can exist only at one pressure. This triple-point pressure is found to be 0.006 atm (or 4.58 mm Hg). This temperature is assigned a value of 273.16 K and not 273.15 K, which is the ice point. Thus, on the Celsius scale the triple point of water will have a temperature of 0.01°C. The reason for assigning 273.16 K to the triple point is that it makes a Celsius degree equal to a Kelvin degree.

To obtain the pressure at the triple point, a triple-point cell very similar to the one shown in Figure 13.6 is used. The gas bulb of the gas thermometer is filled with a very low density gas, almost reaching zero, and is dipped in the triple-point cell. The measured pressure is $P_s = P_t$ and $T_s = T_t = 273.16$ K. Knowing these quantities, Equation 13.2 takes the form (assuming that in the limit $P_t \rightarrow 0$),

$$\boxed{T = (273.16 \text{ K}) \lim_{P_t \to 0} \left(\frac{P}{P_t}\right)} \quad \text{at constant volume} \tag{13.3}$$

Thus, to measure an unknown temperature, we need measure only P by the constant-volume gas thermometer.

EXAMPLE 13.1 A constant-volume gas thermometer reads a pressure of 0.200 atm when in thermal equilibrium with water at the triple point. What will be the reading of this thermometer when in thermal equilibrium with a steam bath?

According to the relation $T/T_s = P/P_s$, where $T_s = 273.16$ K, $P_s =$ 0.200 atm, and $T = 373.15$ K, the pressure P is given by

$$P = \frac{T}{T_s} P_s = \frac{373.15 \text{ K}}{273.16 \text{ K}} (0.200 \text{ atm}) = 0.273 \text{ atm}$$

The difference in the two pressures is $(0.273 - 0.200)$ atm $= 0.073$ atm $=$ 0.073(76 cm Hg) = 5.55 cm Hg. That is, when in thermal equilibrium with steam, the thermometer reads 5.55 cm above the previous mercury level.

EXERCISE 13.1 A constant-volume thermometer reads 0.250 atm when in thermal equilibrium with a body at a temperature of 1000°C. What will be the pressure if the temperature of the body were 110°C? What is the difference in the height of the mercury levels in the two cases? [*Ans.*: $P = 0.075$ atm $=$ 5.72 cm Hg, $\Delta h = 13.3$ cm.]

Different Temperature Scales

There are four temperature scales in general use: (1) the Celsius scale, (2) the Kelvin or absolute or ideal-gas scale, (3) the Fahrenheit scale, and (4) the Rankine scale. The first two, which we have already discussed, are the most commonly used. As stated in Equation 13.1,

$$T_C = T_K - 273.15 \text{ K} \tag{13.1}$$

The Fahrenheit temperature scale was introduced by Gabriel Fahrenheit (1686–1736) and has been in use for a long time in the United States and Great Britain. (Great Britain has recently adopted the Celsius scale in commerce.) According to this scale, water freezes at 32°F and boils at 212°F. The relation between the Celsius temperature T_C and the Fahrenheit temperature T_F is given by

$$\boxed{T_C = \tfrac{5}{9}(T_F - 32)} \quad \text{or} \quad \boxed{T_F = \tfrac{9}{5}T_C + 32} \tag{13.4}$$

Note that a unit temperature interval on the Fahrenheit scale is $\frac{5}{9}$ of the Celsius degree, and zero is also shifted.

Even in the United States there is a gradual trend to change from the Fahrenheit to the Celsius scale. Figure 13.7 shows a clinical thermometer with temperature markings on both scales. A comparison of the three scales—Kelvin, Celsius, and Fahrenheit—is shown in Figure 13.8.

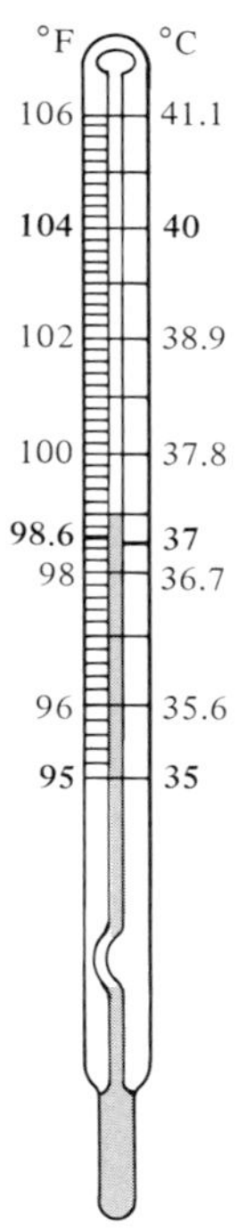

FIGURE 13.7. *Clinical thermometer with both the Fahrenheit and Celsius scales. Small constriction does not let the mercury level drop without applying a jerky motion.*

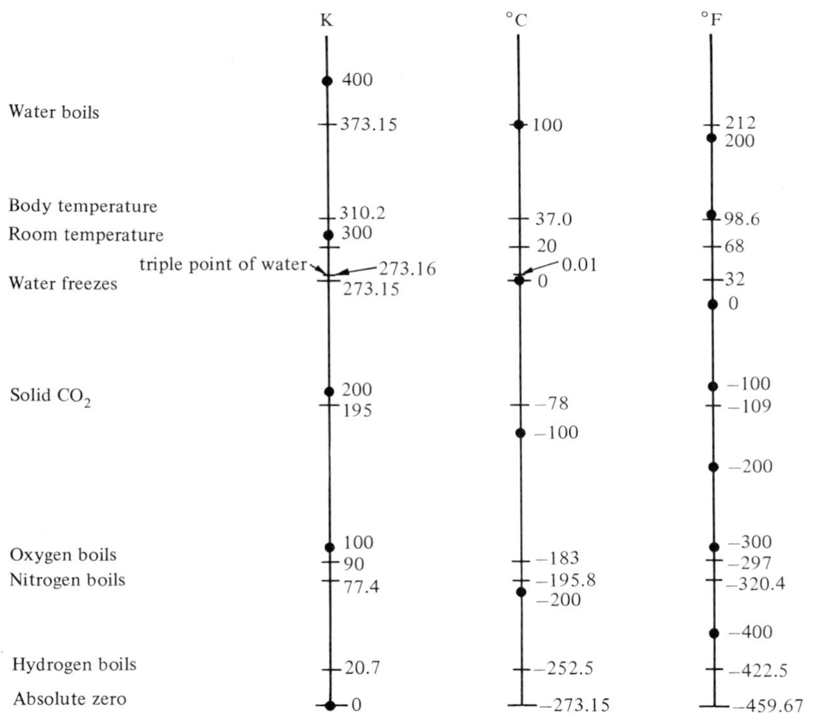

FIGURE 13.8. *Comparison between different temperature scales.*

Finally, the Rankine scale, used primarily in engineering, was proposed by William John MacQuorn Rankine (1820–1872). Its zero agrees with the zero of the Kelvin scale, while the size of the degree is different. The relation between the Rankine scale and Kelvin scale is $T_R = (9/5)\,T_K$. That is, unlike T_C and T_F, the Kelvin and Rankine temperatures are directly proportional.

Example 13.2 A comfortable room temperature in winter or summer is 72°F, while normal body temperature is 98.6°F; that is, $\Delta T_F = 26.6$F°. What are these temperatures and their differences on (a) the Celsius scale; and (b) the absolute-gas scale? (Answer to the nearest degree.)

(a) According to Equation 13.4, $T_C = \frac{5}{9}(T_F - 32)$

for $T_F = 72°\text{F}$ $\quad T_C = \frac{5}{9}(72 - 32)°\text{C} = \frac{5}{9}(40)°\text{C} = 22°\text{C}$

for $T_F = 98.6°\text{F}$ $\quad T_C = \frac{5}{9}(98.6 - 32)°\text{C} = \frac{5}{9}(66.6)°\text{C} = 37°\text{C}$

$$\Delta T_C = (37 - 22)\ \text{C}° = 15\ \text{C}°$$

(b) According to Equation 13.1, $T_C = T_K - 273.15$ K and $T_K = T_C + 273.15$ K.

For $T_C = 22°\text{C}$ $\quad T_K = (22 + 273.15)\text{K} = 295\ \text{K}$

for $T_C = 37°\text{C}$ $\quad T_K = (37 + 273.15)\text{K} = 310\ \text{K}$

$$\Delta T_K = (310 - 295)\text{K} = 15\ \text{K}$$

Exercise 13.2 It is a very hot summer day if the temperature in a typical U.S. city reaches 100°F while on a cold day in winter it may be 32°F (freezing), 0°F (cold), or −20°F (very cold). What are these temperatures on (a) the Celsius scale; and (b) the absolute scale? [*Ans.:* (a) 38°C, 0°C, −18°C, −29°C; (b) 311 K, 273 K, 255 K, 244 K.]

13.4. Ideal Gas and Gas Laws

All through the discussion of heat and thermodynamics we shall be dealing mostly with gases. Most of the definitions, derivations, and laws take a simpler form if we are dealing with an ideal gas instead of a real gas. Hence, we must clearly understand the meaning of an ideal gas and how closely it resembles a real gas.

Ideal Gas

Suppose that we have a container filled with a gas, say hydrogen molecules. Let us consider two particular molecules, 1 and 2, as shown in Figure 13.9. Each hydrogen molecule consists of two nuclei (nuclei are simply protons in the case of hydrogen molecules) and two electrons. Since the electrons are constantly moving with high speed around the nuclei, it is hard to say where exactly these electrons are at a particular time. Hence, these electrons are

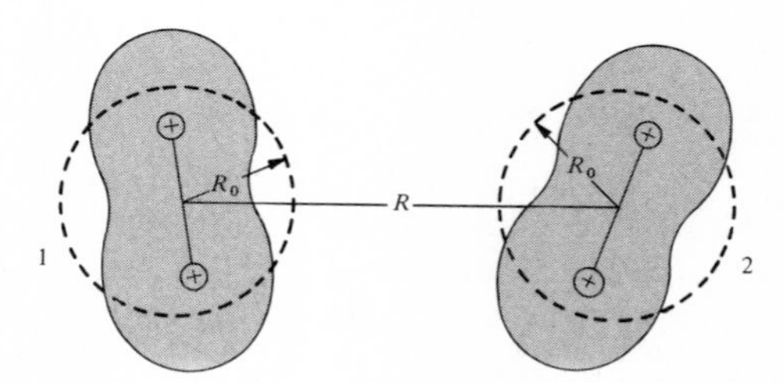

Figure 13.9. *Two hydrogen molecules separated by a distance* R.

shown as a cloud (shaded) of radius R_0 around the center of the two nuclei. Let us now calculate the total energy of the system of these two molecules.

The total energy (excluding the energy that holds the atoms of the molecules together, the binding energy) consists of two parts: (1) the kinetic energy of each of the individual molecules resulting from their random motions, and (2) the energy due to the forces of electrical origin between the two molecules. The energy resulting from this electrical interaction is called *electric potential energy*. In most cases, the separation between the molecules of a gas is much larger than the diameter R_0 of the molecules; hence, the potential energy may be neglected. We define an *ideal gas as one in which the molecules exert no force of electrical origin, and the only force they exert is when they physically collide*. In other words, in an ideal gas the molecules have only kinetic energy and no potential energy. The molecules behave as hard, impenetrable spheres. Most of the gases with low density and well above the condensation temperature behave like an ideal gas.

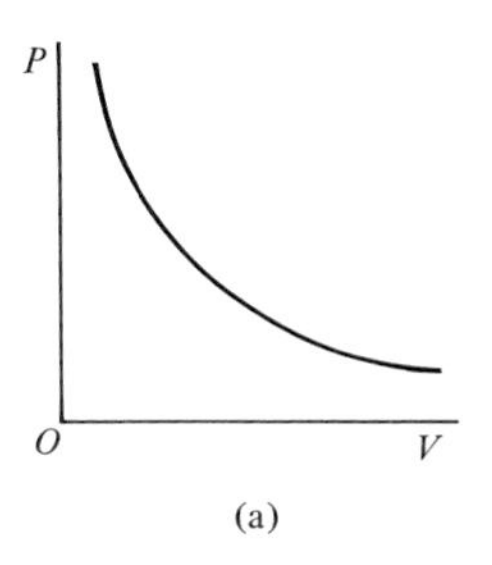

(a)

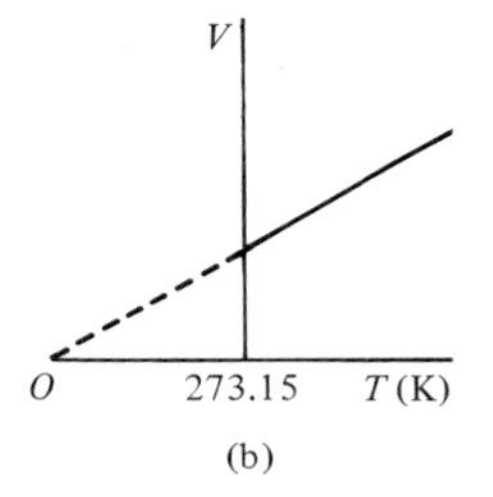

(b)

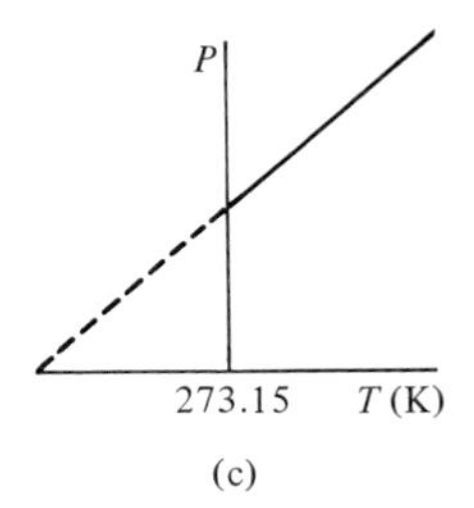

(c)

FIGURE 13.10. *Plots of (a) P versus V, (b) V versus T, and (c) P versus T for an ideal gas.*

Ideal-Gas Laws

We shall now discuss relations between pressure P, volume V, and temperature T for an ideal gas. Take first the relation between the pressure and volume of an ideal gas, which was discovered by an English scientist, Boyle, in about 1660.

> **Boyle's law: *At constant temperature the pressure of a given quantity of gas is inversely proportional to its volume.***

That is,

$$P \propto \frac{1}{V} \qquad \text{or} \qquad PV = \text{constant}$$

or

$$\boxed{P_1V_1 = P_2V_2} \qquad \text{if } T \text{ is constant} \tag{13.5}$$

The value of the constant will be different for different samples of gas. The plot of P versus V is shown in Figure 13.10(a).

The relation between V and the absolute temperature T while keeping pressure P constant was discovered by Jacques Charles in about 1760 and also by Joseph Gay-Lussac. It may be stated as follows.

> **Charles' law: *At constant pressure the volume of a given sample of gas is directly proportional to its absolute temperature T.***

That is,

$$V \propto T \qquad \text{or} \qquad \frac{V}{T} = \text{constant}$$

or

$$\boxed{\frac{V_1}{T_1} = \frac{V_2}{T_2}} \qquad \text{if } P \text{ is constant} \tag{13.6}$$

Once again the constant is different for various samples of gas. The plot of V versus T is shown in Figure 13.10(b).

A third law, which is a relation between P and T at fixed V does not bear any particular name, may be stated as follows.

> **Constant-Volume law:** ***At constant volume the pressure of a given sample of gas is directly proportional to its absolute temperature.***

That is,

$$P \propto T \qquad \text{or} \qquad \frac{P}{T} = \text{constant}$$

or

$$\boxed{\frac{P_1}{T_1} = \frac{P_2}{T_2}} \quad \text{if } V \text{ is constant} \tag{13.7}$$

The plot of P versus T is shown in Figure 13.10(c).

The three laws just described may be combined into one relation, resulting in the ideal-gas law. Suppose that we have a sample of gas with volume V_1, temperature T_1 and pressure P_1. If one of the variables is changed, the others may change also. Suppose that the final volume, temperature, and pressure are V_2, T_2, and P_2. It is a simple matter to show (see Problem 13.18) that

$$\boxed{\frac{P_1V_1}{T_1} = \frac{P_2V_2}{T_2}} \qquad \text{or} \qquad \frac{PV}{T} = \text{constant} \tag{13.8}$$

where P may be in N/m^2, V is in m^3, and T is the absolute temperature. Of course, the value of the constant (in given units) depends on the gas sample; that is, it will be different for different amounts of gases.

Suppose that a given sample of gas has a mass m and its molecular weight is M. The number of moles n in this sample is equal to m/M. In such cases Equation 13.8 may be written as

$$\frac{PV}{T} = nR$$

or

$$\boxed{PV = nRT} \tag{13.9}$$

that is,

$$\frac{P_1V_1}{n_1T_1} = \frac{P_2V_2}{n_2T_2} = R \tag{13.10}$$

where R is called the *universal gas constant* and does not depend on the quantity of gas in the sample. This relation is called the *ideal-gas law* or *general gas equation*. Note that Equation 13.9 is an experimentally confirmed relation as obtained from the plots in Figure 13.4 discussed in the preceding section.

If P is measured in N/m^2, V in m^3, and T in degrees Kelvin, the value of the universal gas constant obtained from Equation 13.9 is

$$R = 8.134 \text{ J/mol-K} \qquad \text{(in SI units)}$$
$$R = 8.134 \times 10^7 \text{ erg/mol-K} \qquad \text{(in CGS units)}$$

Also, the following values of P, V, T, and n are referred to as standard values, while T_0 and P_0 are the standard temperature and pressure (STP).

$$P_0 = 1 \text{ atm} = 1.013 \times 10^5 \text{ N/m}^2 \qquad T_0 = 273 \text{ K}$$
$$V_0 = 22.4 \ \ell \qquad n_0 = 1 \text{ mol}$$

EXAMPLE 13.3 Oxygen gas contained in a cylinder at a pressure of 1 atm and a temperature of 0°C has a mass of 64.06×10^{-3} kg. Suppose that this gas is compressed to half its original volume and the final temperature is 500°C. Find the initial volume and the final pressure of the gas. The mass of an oxygen molecule is 5.32×10^{-26} kg.

In the relation $PV = nRT$, $P = 1 \text{ atm} = 1.013 \times 10^5 \text{ N/m}^2$, $T = 0°\text{C} = 273 \text{ K}$, and $R = 8.134$ J/mol-K. Thus, knowing n, we can calculate V.

$$\text{number of molecules} = \frac{\text{total mass}}{\text{mass of each molecule}}$$
$$= \frac{64.06 \times 10^{-3} \text{ kg}}{5.32 \times 10^{-26} \text{ kg/molecule}} = 12.04 \times 10^{23} \text{ molecules}$$
$$\text{number of moles } n = \frac{\text{number of molecules}}{\text{Avogadro's number}} = \frac{12.04 \times 10^{23} \text{ molecules}}{6.022 \times 10^{23} \text{ molecules/mol}}$$
$$= 2 \text{ mol}$$

Note that we could have obtained the same result by using the relation

$$n = \frac{\text{total wt}}{\text{molecular wt}} = \frac{64.06 \text{ g}}{32.02 \text{ g/mol}} = 2 \text{ mol}$$
$$V = \frac{nRT}{P} = \frac{(2 \text{ mol})(8.314 \text{ J/mol-K})(273 \text{ K})}{1.013 \times 10^5 \text{ N/m}^2} = 4481 \times 10^{-5} \text{ m}^3$$
$$= 44{,}810 \text{ cm}^3 = 44.81 \ \ell$$

Thus, the initial volume $V_1 = 44.81 \ \ell$, while the final volume $V_2 = \frac{1}{2} V_1 = 22.40 \ \ell$. Also, $T_1 = 273$ K, while $T_2 = (273 + 500) = 773$ K. Since $P_1 = 1.013 \times 10^5 \text{ N/m}^2$, the final pressure P_2 from the relation $P_1 V_1/T_1 = P_2 V_2/T_2$ is given by

$$P_2 = \frac{V_1}{V_2}\frac{T_2}{T_1}P_1 = \left(\frac{2}{1}\right)\left(\frac{773 \text{ K}}{273 \text{ K}}\right)(1.013 \times 10^5 \text{ N/m}^2) = 5.74 \times 10^5 \text{ N/m}^2$$
$$= \frac{5.74 \times 10^5 \text{ N/m}^2}{1.013 \times 10^5 \text{ N/m}^2\text{/atm}} = 5.67 \text{ atm}$$

EXERCISE 13.3 Nitrogen gas contained in a cylinder at a pressure of 1 atm and temperature of 0°C has a mass of 112 g. Suppose that this gas is compressed to a pressure of 10 atm and is raised to a temperature of 700°C. Calculate the initial volume and the final volume. [*Ans.*: $V_1 = 8.96 \times 10^{-2} \text{ m}^3 = 89.6 \ \ell$, $V_2 = 3.20 \times 10^{-2} \text{ m}^3 = 32.0 \ \ell$.]

13.5. Kinetic Theory of Gases

To have a better insight into the concepts of temperature and heat (molecular energy) and their interrelations to pressure and volume for an ideal gas, we must investigate the motion of microscopic systems consisting of atoms and molecules. The kinetic theory of gases, developed after making simplifying assumptions about the random motion of the molecules, provides the key answers to this problem.

Basic Assumptions

A very small volume of gas contains an enormous number of molecules moving in all directions in random fashion and colliding with the walls of the container. To derive a quantitative relationship, we must make the following basic assumptions.

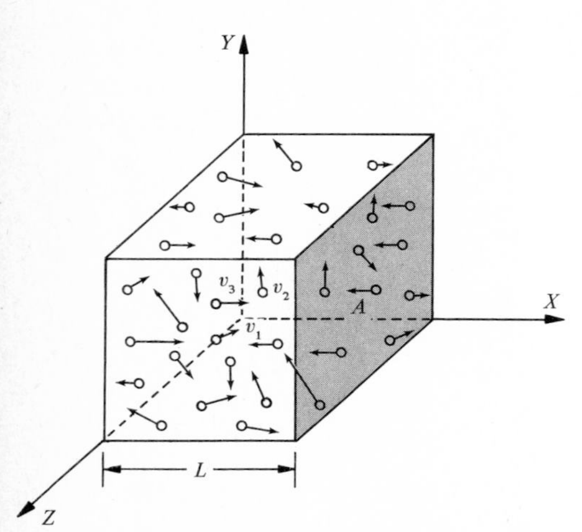

FIGURE 13.11. *Random motion of a large number of molecules in a cubical container of side L.*

RIGID MOLECULES WITH NEGLIGIBLE VOLUME. The gas consists of molecules that may be treated as tiny impenetrable spheres. Even though there is a very large number of molecules, their total volume is negligible as compared to the volume of the container. This implies that the collisions between molecules may be neglected.

RANDOM MOTION OF THE MOLECULES. The molecules do not have any preferred direction of motion. The motion is completely random, as shown in Figure 13.11. This means that if **v** is the average velocity of the molecules, the average component velocities, v_x, v_y, and v_z will each be zero. That is, $\langle v_x \rangle = \langle v_y \rangle = \langle v_z \rangle = 0$. Thus, on the average, as many molecules are moving to the right with given $|v_x|$ as are moving to the left; and so on.

INTERACTION TIME IS SMALL. Molecules travel in straight lines, spending most of the time in free motion and only a small fraction of time in interaction with other molecules.

PERFECTLY ELASTIC COLLISIONS. The collisions between molecules, if any, and between the molecules and the wall of the container, are considered to be perfectly elastic. This means that kinetic energy is conserved in such collisions.

NEWTON'S LAWS ARE APPLICABLE. The motion of molecules and their collisions are governed by Newton's laws of motion.

Derivation

Let us consider a cubical container of side L which has a large number of gas molecules moving in random fashion, as shown in Figure 13.11. The area of each face of the container is $A = L^2$, and the volume of the container is $V = LA = L^3$. Let us concentrate on a single molecule of mass m moving with velocity v_1 with components v_{1x}, v_{1y}, and v_{1z}. We need to calculate the number of times in an interval Δt this molecule collides with the wall on the right (shown shaded) and then bounces back. In the time interval Δt, the molecule will travel a distance $\Delta L = v_{1x}\,\Delta t$ along the X-axis. The distance the molecule travels between successive collisions is $2L$. Thus, the number of collisions f in time Δt with the right wall will be

$$f = \frac{v_{1x}\,\Delta t}{2L} \tag{13.11}$$

Note that the components of velocities v_{1y} and v_{1z} remain unchanged in elastic collisions between molecules and the wall of the container.

The momentum of the molecule before collision is mv_{1x} and after collision is $-mv_{1x}$ (see Figure 13.12). Thus, the change in momentum of the molecule is $-mv_{1x} - (mv_{1x}) = -2mv_{1x}$ while the momentum imparted to the wall is $2mv_{1x}$. Since there are f collisions in time Δt, the momentum imparted to the wall is

$$\Delta p_{1x} = f(2mv_{1x}) = \frac{v_{1x}\,\Delta t}{2L}\,2mv_{1x} = \frac{mv_{1x}^2}{L}\,\Delta t \tag{13.12}$$

The force F_W exerted on the wall by the molecules ($= -F_M$ the force exerted on the molecules by the wall according to Newton's third law) is F_{1x}, given by

$$F_{1x} = \frac{\Delta p_{1x}}{\Delta t} = \frac{mv_{1x}^2}{L} \tag{13.13}$$

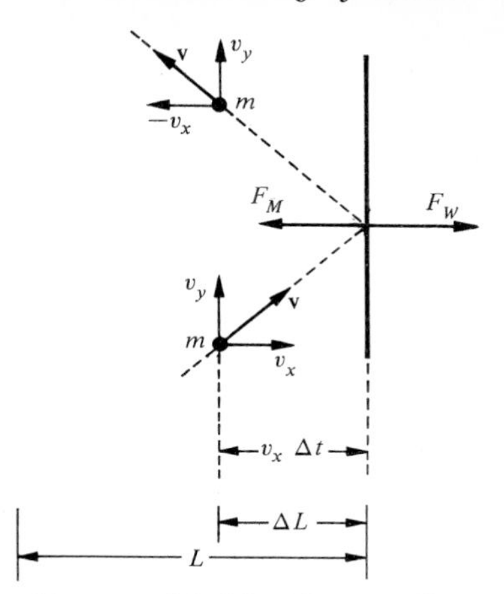

FIGURE 13.12. *Elastic collision between a molecule and the face perpendicular to the X-axis (shown shaded in Figure 13.11). (The v_z-component of the velocity is not shown.)*

Suppose that there are N molecules in the container moving with velocities $v_1, v_2, v_3, \ldots, v_N$. The total force exerted on the right wall of the container by all these molecules will be

$$F_x = \frac{m}{L}(v_{1x}^2 + v_{2x}^2 + \cdots + v_{Nx}^2) = \frac{m}{L}\sum_{i=1}^{N} v_{ix}^2 \tag{13.14}$$

Since the pressure P on the wall by the gas is defined to be the normal force per unit area, we may write

$$P = \frac{F_x}{A} = \frac{F_x}{L^2} = \frac{m}{L^3}\sum_{i=1}^{N} v_{ix}^2 \tag{13.15}$$

According to assumption 2, the average of each component velocity is zero, but the square of these component velocities is not zero. So we define the *mean-square velocity* $\langle v_x^2 \rangle$ of the X-components of the molecular velocity as

$$\langle v_x^2 \rangle = \frac{v_{1x}^2 + v_{2x}^2 + \cdots + v_{Nx}^2}{N} = \frac{\sum v_{ix}^2}{N} \tag{13.16}$$

and the root-mean-square velocity of the X-component $(v_x)_{\text{rms}}$ as

$$(v_x)_{\text{rms}} = \sqrt{\langle v_x^2 \rangle} = \sqrt{\frac{\sum v_{ix}^2}{N}} \tag{13.17}$$

As the name indicates, *root-mean-square* (rms) means the square root of the mean of the squared velocities. Thus, using Equations 13.15 and 13.16,

$$P = \frac{mN}{L^3}\langle v_x^2 \rangle \tag{13.18}$$

Since all component velocities are equally probable, we may write

$$\langle v_x^2 \rangle = \langle v_y^2 \rangle = \langle v_z^2 \rangle \qquad \text{and} \qquad \langle v^2 \rangle = \langle v_x^2 \rangle + \langle v_y^2 \rangle + \langle v_z^2 \rangle = 3\langle v_x^2 \rangle$$

That is,

$$\langle v_x^2 \rangle = \frac{1}{3} \langle v^2 \rangle$$

This is equivalent to saying that only one-third of the total molecules are traveling along the X-axis. Substituting this result in Equation 13.18 yields

$$P = \frac{mN}{L^3} \frac{1}{3} \langle v^2 \rangle \tag{13.19}$$

This relation may be written in two different forms. Since $L^3 = V$ is the volume of the container, we get

$$PV = \tfrac{2}{3} N \langle \tfrac{1}{2} m v^2 \rangle = \tfrac{1}{3} N m v_{\mathrm{rms}}^2 \tag{13.20}$$

or

$$\boxed{P = \frac{2}{3} \frac{N}{V} \left\langle \frac{1}{2} m v^2 \right\rangle = \frac{2}{3} \frac{N}{V} \langle K \rangle} \tag{13.21}$$

which states that

$$\text{pressure} = \frac{2}{3} \begin{bmatrix} \text{number of molecules} \\ \text{per unit volume} \end{bmatrix} \begin{bmatrix} \text{average translational kinetic} \\ \text{energy of a molecule} \end{bmatrix}$$

Thus, the pressure exerted by the gas molecules is directly proportional to the average translational kinetic energy of the molecules. Note that this energy refers to the translational kinetic energy of the molecules with respect to the center of mass of the molecules. Thus, no kinetic energy due to the rotational or vibrational motion of the molecules is included in the expression above. Furthermore, Equations 13.20 and 13.21 reveal that bulk properties of gases, such as pressure and volume, can be expressed in terms of the motion of the individual molecules.

Finally, one may ask this question: Since there will always be fluctuations in the number of molecules hitting the walls of the container, will there be fluctuations observed in the measured pressure or will the pressure remain steady? The answer is that because of the very large number of molecules hitting the wall, the fluctuations as a function of time are negligible; hence, the pressure observed is a steady pressure.

EXAMPLE 13.4 Consider nitrogen gas in a container at standard pressure and temperature. Calculate (a) the root-mean-square speed of the nitrogen molecules; (b) the average kinetic energy of each molecule; and (c) the change in momentum of a molecule in a collision, assuming v perpendicular to the wall.

One mole of the gas at standard pressure ($1 \text{ atm} = 1.013 \times 10^5 \text{ N/m}^2$) and standard temperature ($0°\text{C} = 273 \text{ K}$) has a volume of $22.4 \ \ell = (22.4)(1000 \text{ cm}^3) = 22.4 \times 10^{-3} \text{ m}^3$. The number of molecules in this volume is equal to Avogadro's number, $N_A = 6.02 \times 10^{23}$ molecules/mol, and the weight is equal to the molecular weight, $M = 28 \text{ g} = 28 \times 10^{-3}$ kg of nitrogen molecules. Thus, the mass m of each molecule is

$$m = \frac{M}{N_A} = \frac{28 \text{ g/mol}}{6.02 \times 10^{23} \text{ molecules/mol}} = 4.65 \times 10^{-23} \text{ g} = 4.65 \times 10^{-26} \text{ kg}$$

(a) From the equation $PV = \frac{1}{3}Nmv_{\text{rms}}^2$, where $Nm = N_A m = M = 28 \times 10^{-3}$ kg, $V = 22.4 \times 10^{-3}$ m^3 and $P = 1.013 \times 10^5$ N/m^2, we get

$$v_{\text{rms}} = \sqrt{\frac{3PV}{Nm}} = \sqrt{\frac{3(1.013 \times 10^5 \text{ N/m}^2)(22.4 \times 10^{-3} \text{ m}^3)}{28 \times 10^{-3} \text{ kg}}} = 493 \text{ m/s}$$

(b) $\langle K \rangle = \frac{1}{2}mv_{\text{rms}}^2 = \frac{1}{2}(4.65 \times 10^{-26} \text{ kg})(24.3 \times 10^4 \text{ m}^2/\text{s}^2)$
$= 5.65 \times 10^{-21}$ J.

(c) $\Delta p = 2mv = 2(4.65 \times 10^{-26} \text{ kg})(493 \text{ m/s}) = 4.58 \times 10^{-23}$ kg-m/s.

EXERCISE 13.4 Repeat Example 13.4 for hydrogen gas and compare the results. [*Ans.:* $v_{\text{rms}} = 1842$ m/s, $\langle K \rangle = 5.62 \times 10^{-21}$ J, $\Delta p = 1.22 \times 10^{-23}$ kg-m/s.]

13.6. Kinetic-Theory Interpretation of Temperature and Heat

From kinetic theory, the relation between P and V is

$$PV = \tfrac{2}{3} N \langle \tfrac{1}{2}mv^2 \rangle \tag{13.20}$$

while all experimental evidence indicates that for an ideal gas,

$$PV = nRT \tag{13.9}$$

where n is the number of moles of the gas and T is its absolute or Kelvin temperature. If kinetic theory agrees with the experimental results, the right-hand side of the two equations above must be equal. That is,

$$nRT = \tfrac{2}{3} N \langle \tfrac{1}{2}mv^2 \rangle$$

Substituting for $N = nN_A$ and rearranging, we get

$$T = \frac{2}{3}\frac{N_A}{R} \langle \frac{1}{2}mv^2 \rangle \tag{13.22}$$

Since both N_A and R are constant, the ratio R/N_A is still another constant, called the *Boltzmann constant*, k, given by

$$k = \frac{R}{N_A} = \frac{8.314 \text{ J/mol-K}}{6.022 \times 10^{23} \text{ molecules/mol}} = 1.380 \times 10^{-23} \text{ J/K} \tag{13.23}$$

If we denote the average translational kinetic energy per molecule $\langle \frac{1}{2}mv^2 \rangle$ by $\bar{K}$, we may write Equation 13.22 as

$$T = \frac{2}{3k}\bar{K} \tag{13.24}$$

or

$$\boxed{\bar{K} = \langle \tfrac{1}{2}mv^2 \rangle = \tfrac{3}{2} kT} \tag{13.25}$$

Thus, according to Equation 13.24, we can state: *The temperature of a gas is directly proportional to the average translational kinetic energy of its molecules.* This means that absolute zero temperature is that temperature at which the translational kinetic energy of the molecules of an ideal gas is zero. It is important to note that at $T = 0$, $\bar{K} = 0$ does not mean that the total internal energy of the gas is zero. Even at absolute zero the molecules may have rotational and vibrational kinetic energies. Thus, it is only translational kinetic energy, not rotational or vibrational energies, which defines temperature.

We can draw two important conclusions from Equation 13.25. First, $v_{\text{rms}} = \sqrt{\langle v^2 \rangle} = \sqrt{\frac{3}{m} kT}$ means that

$$v_{\text{rms}} \propto \sqrt{T} \tag{13.26}$$

Second, if a gas contains molecules of two different masses, m_1 and m_2, both at the same temperature, then $\langle \frac{1}{2} m_1 v_1^2 \rangle = \langle \frac{1}{2} m_2 v_2^2 \rangle$; that is,

$$\frac{(v_1)_{\text{rms}}}{(v_2)_{\text{rms}}} = \sqrt{\frac{m_2}{m_1}} \tag{13.27}$$

which says that the velocities of the molecules are inversely proportional to the square root of their masses. This relation is utilized in separating molecules of different gases by the diffusion method. In this method of separation, two mixed gases are contained in a porous container which is placed in an evacuated chamber. The lighter gas molecules have higher speeds, hence leak out of the porous container faster. Over a short interval of time, the amount of gases leaking out of the porous container will be inversely proportional to the square root of their molecular masses. This method is used, for example, in separating the fissionable material ^{235}U from normal uranium, which consists mostly of ^{238}U and only 0.7% of ^{235}U. Experiments with diffusion verify the relation given by Equation 13.27; hence, the kinetic theory of gases is verified as well.

EXAMPLE 13.5 A gas of uranium hexafluoride contains two different isotopes, $^{235}UF_6$ and $^{238}UF_6$, with molecular weights 349 and 352, respectively. The diffusion of this gas through a porous container is used for separating the two isotopes of uranium. Calculate the *enrichment factor* or *ideal separation factor*.

Let m_1 and m_2 be the molecular weights 349 and 352 corresponding to the light and heavy isotopes of uranium. Thus, the ratio of the rms speeds of the two isotopes according to Equation 13.27 will be

$$\frac{(v_1)_{\text{rms}}}{(v_2)_{\text{rms}}} = \sqrt{\frac{m_2}{m_1}} = \sqrt{\frac{352}{349}} = 1.0043$$

That is, the lighter isotope $^{235}UF_6$ will diffuse faster by a factor 1.0043, the enrichment factor.

EXERCISE 13.5 Calculate the ideal separation factor for chlorine isotopes in the form of a gaseous mixture of $^1H^{35}Cl$ and $^1H^{37}Cl$. [*Ans.:* 1.0274]

We now calculate the internal energy of a gas. For any system the energy consists of two parts: (1) energy due to *ordered motion* of the molecules in the system, and (2) energy due to the *disordered* or *chaotic motion* of the molecules of the system. It is the disordered energy, also called *thermal energy,* which we are interested in calculating in the kinetic theory of gases.

The internal energy or thermal energy U of a system containing N molecules is given by

$$U = \sum_{i=1}^{N} K_i$$

where K_i is the kinetic energy of each molecule. If we assume an ideal gas (hard-sphere noninteracting molecules), we can use the expressions derived above and write the equation as

$$U = N\overline{K} = N\langle \tfrac{1}{2}mv^2 \rangle \tag{13.28}$$

Using the results of Equations 13.25 and 13.23, that $\overline{K} = \frac{3}{2}kT$ and $k = R/N_A$, we may write

$$\boxed{U = N\tfrac{3}{2}kT = \tfrac{3}{2}nRT} \tag{13.29}$$

which is the *disordered internal energy or thermal energy* of a system obeying the ideal gas law. Still another way of writing Equation 13.29 is to make use of the gas law, $PV = nRT$. Substituting for $nRT = \frac{2}{3}U$, we get

$$\boxed{PV = \tfrac{2}{3}U} \tag{13.30}$$

which is the relation among P, V, and U, not containing T.

The success of kinetic theory was a great boost for Newton's laws because it extended their domain to infinitesimal particles. Thus, Newton's laws of motion now spread in application from the very small to the very large (stars and planets), thus unifying the structure of mechanics.

EXAMPLE 13.6 Consider 1 mol of oxygen molecules at a temperature of 300 K in a cubical container 10 cm on a side. Calculate (a) the average kinetic energy of a molecule; (b) the total kinetic energy; (c) the average value of the square of the speed; (d) the root-mean-square speed; (e) the average momentum of a molecule; (f) the average force exerted by the molecules on the walls of the container; and (g) the average pressure exerted by the molecules.

(a) The average kinetic energy of a molecule: from Equation 13.25,

$$\overline{K} = \langle \tfrac{1}{2}mv^2 \rangle = \tfrac{3}{2}kT = \tfrac{3}{2}(1.38 \times 10^{-23}\ \text{J/K})(300\ \text{K}) = 6.21 \times 10^{-21}\ \text{J}$$

(b) The total kinetic energy of 1 mol of gas molecules: 1 mol contains $N_A = 6.022 \times 10^{23}$ molecules/mol or $N = 6.022 \times 10^{23}$.

$$U = N\overline{K} = (6.022 \times 10^{23}\ \text{molecules/mol})(6.21 \times 10^{-21}\ \text{J}) = 3740\ \text{J}$$

Also, we may calculate this from the relation

$$U = N\bar{K} = N(\tfrac{3}{2}kT) = \tfrac{3}{2}nRT = \tfrac{3}{2}(1)(8.31\ \text{J/mol-K})(300\ \text{K}) = 3740\text{J}$$

(c) The average value of the square of the speed: from Equation 13.25, $\langle v^2 \rangle = 3kT/m$, where m is the mass of the oxygen molecule given by

$$m = \frac{M}{N_A} = \frac{32.0\ \text{g/mol}}{6.022 \times 10^{23}\ \text{molecules/mol}} \times \left(10^{-3}\frac{\text{kg}}{\text{g}}\right) = 5.31 \times 10^{-26}\ \text{kg}$$

Therefore,

$$\langle v^2 \rangle = \frac{3kT}{m} = \frac{3(1.38 \times 10^{-23}\ \text{J/K})(300\ \text{K})}{5.31 \times 10^{-26}\ \text{kg}} = 23.4 \times 10^4\ \text{m}^2/\text{s}^2$$

(d) The root-mean-square velocity:

$$v_{\text{rms}} = \sqrt{\langle v^2 \rangle} = \sqrt{23.4 \times 10^4\ \text{m}^2/\text{s}^2} = 4.84 \times 10^2\ \text{m/s} = 484\ \text{m/s}$$

(e) The average momentum of a molecule:

$$\bar{p} = mv_{\text{rms}} = (5.31 \times 10^{-26}\ \text{kg})(484\ \text{m/s}) = 2.57 \times 10^{-23}\ \text{kg-m/s}$$

(f) The average force exerted by all the molecules (including a factor of $\frac{1}{3}$ from Equation 13.19):

$$F = \frac{m}{L}\frac{N}{3}(v_{\text{rms}}^2) = \frac{m}{L}\frac{N}{3}\langle v^2 \rangle$$

$$= \frac{5.31 \times 10^{-26}\ \text{kg}}{3(0.10\ \text{m})}(6.022 \times 10^{23})(23.4 \times 10^4\ \text{m}^2/\text{s}^2) = 0.249 \times 10^5\ \text{N}$$

(g) The average pressure:

$$P = \frac{\bar{F}}{A} = \frac{0.249 \times 10^5\ \text{N}}{(0.10\ \text{m})^2} = 24.9 \times 10^5\ \frac{\text{N}}{\text{m}} = \frac{24.9 \times 10^5}{1.013 \times 10^5}\text{atm} = 24.6\ \text{atm}$$

Exercise 13.6 Repeat Example 13.6 considering 1 mol of oxygen gas at 3000 K. [*Ans.:* (a) 6.21×10^{-20} J, (b) 37400 J, (c) $23.4 \times 10^5\ \text{m}^2/\text{s}^2$, (d) 1530 m/s, (e) 8.12×10^{-23} kg-m/s, (f) 2.49×10^5 N, (g) 247 atm.]

SUMMARY

The subject of mechanics of individual molecules in disordered motion is called *heat.* The energy of these disordered molecules is called *thermal* or *internal energy.* The energy transfer that takes place due to the difference in the temperature is called *heat* or heat energy; it is also defined as the energy transfer by means of molecular collisions. *Temperature* is a quantity that is proportional to the degree of disordered kinetic energy of a body. According to the *Zeroth law of thermodynamics:* Of the three bodies *A, B,* and *C,* if *A* and *B* are separately in thermal equilibrium with *C,* then *A* and *B* are in thermal equilibrium with one another.

The relations between T_C, T_K, and T_F are $T_C = T_K - 273.15$ K, $T_C = \frac{5}{9}(T_F - 32)$. The *triple point* of water is that temperature at which the three phases of water—solid, liquid, and gas—coexist in equilibrium. This triple-point temperature is 273.16 K. For a constant-volume gas thermometer, $T = (273.16\text{ K}) \lim_{P_t \to 0} (P/P_t)$.

An *ideal gas* is one in which the molecules exert no force of electrical origin, and the only force they exert is when they collide. The ideal-gas laws are $P_1V_1 = P_2V_2$ (Boyle's law), $V_1/T_1 = V_2/T_2$ (Charles' law), $P_1/T_1 = P_2/T_2$ (constant-volume law), and $P_1V_1/T_1 = P_2V_2/T_2 = nR$ (the gas law). The constant R is 8.314 J/mol-K and STP values are $P_0 = 1\text{ atm} = 1.013 \times 10^5\text{ N/m}^2$, $T_0 = 273\text{ K}$, $V_0 = 22.4\ \ell$, and $n_0 = 1$ mol.

Starting with some basic assumptions, the kinetic theory of gases yields $PV = \frac{2}{3}N\langle\frac{1}{2}mv^2\rangle$. Comparing this with $PV = nRT$ gives $\bar{K} = \langle\frac{1}{2}mv^2\rangle = \frac{3}{2}kT$, where $k = R/N_A = 1.380 \times 10^{-23}$ J/K is the *Boltzmann constant*. Also, $(v_1)_{rms}/(v_2)_{rms} = \sqrt{m_2/m_1}$ and $v_{rms} = \sqrt{3kT/m}$. According to kinetic theory, for disordered internal energy or thermal energy, $U = N\bar{K} = N(\frac{3}{2}kT) = \frac{3}{2}nRT$ and $PV = \frac{2}{3}U$.

QUESTIONS

1. Water always flows from a higher level to a lower level, even though the quantity of water at the higher level may be much smaller than at the lower level. Give a corresponding statement in heat where temperature plays the role of the water level.
2. Could you give an example which will prove that you cannot depend upon the sense of touch to determine the relative hotness of bodies?
3. What determines that a body will remain in thermal equilibrium? Give two examples of systems that are in thermal equilibrium.
4. How do we make use of the zeroth law of thermodynamics in defining temperature?
5. When two systems are in equilibrium, which of the following statements are true and which are not true? (a) They have the same amount of kinetic energy. (b) They are at the same temperature. (c) The molecules have the same amount of heat.
6. Explain why the ideal gas law will not hold if the molecules of a gas are brought very close together.
7. Why does the pressure of a gas in an automobile tire increase if the automobile is driven for a while?
8. Is it possible to observe the fluctuations in pressure due to the collisions of molecules against the walls of the container? Why?
9. Why is the average velocity of the molecules in a gas zero but the average of the square of the velocity is not zero?
10. The expression $PV = nRT$ was derived by assuming a dilute gas. What effect will a high density of the gas have on P and V?
11. In the diffusion process of the separation of ^{235}U and ^{238}U, which isotope will diffuse out and which will be mostly left behind?

PROBLEMS

1. If the most common cooking temperature is about 225°F, convert this temperature into the Celsius scale.

2. Mercury freezes at $-39°C$ and boils at $357°C$. Convert these temperatures into (a) the Fahrenheit scale; and (b) the absolute scale.
3. The ambient temperature in some hot places may reach as high as $120°F$, while in some cold places it may reach as low as $-50°F$. Convert these temperatures into the Celsius scale.
4. The temperature of dry ice is $-78°C$, while nitrogen boils at $-195.8°C$. Convert these temperatures into the Fahrenheit and Kelvin scales.
5. Hydrogen boils at 20.7 K. Convert this temperature into the Celsius and Fahrenheit scale.
6. The melting point of copper is $1083°C$, while its boiling point is $2300°C$. Convert these temperatures into the Kelvin and Fahrenheit scales.
7. A constant-volume gas thermometer, when in thermal equilibrium with water at the triple point, shows a pressure of 0.50 atm. What will be the reading of this thermometer when in equilibrium with an object at $200°C$?
8. If a constant-volume thermometer has to read a temperature difference of the order of 0.01 K, what is the accuracy needed in reading the mercury levels? Let P_s be 0.40 atm.
9. For a constant-pressure thermometer, the volume of gas at the triple point of water is 275 cm^3. What will the temperature of an object be when the thermometer reads a volume of 480 cm^3? (Make use of the relation $T/T_s = V/V_s$.)
10. In Problem 9, if the volume of the gas in the thermometer can be read with an accuracy of 0.01 cm^3, is it possible to read the temperature with an accuracy of 0.01 K?
11. For a constant-pressure thermometer, a gas has a volume of 310 cm^3 at $0°C$ and 615 cm^3 at $100°C$. What is the temperature when the volume is 510 cm^3?
12. A thermometer that uses the resistance of a platinum wire for a thermometric substance is the *platinum resistance thermometer*. Suppose that this thermometer shows a resistance of 1.1 units at the triple point of water. What is the temperature if the thermometer shows a resistance of 15.5 units? [Make use of the relation $T/T_s = R/R_s$; that is, the thermometer has a linear response function.]
13. Originally, a gas in a tank of volume 1.0 m^3 is at a pressure of 1 atm. If this gas is compressed to one-fifth of its original volume, what is the final pressure in (a) atm; (b) N/m^2; and (c) lb/in^2? Assume no change in temperature.
14. Suppose that a cylinder contains 2.0 kg of hydrogen at a pressure of 5 atm. Suppose that the hydrogen is replaced by (a) 10 kg of helium; and (b) 10 kg of oxygen. What will be the new pressures?
15. A gasoline drum is filled with air and sealed at a temperature of $0°C$ and a pressure of 1 atm. If this drum is heated to $50°C$, what will be the new pressure? If the drum can stand a pressure of 10 atm, what is the temperature to which this drum can be heated before it will explode?
16. On a cold morning when the temperature is $0°C$, the gauge pressure of an automobile tire reads 28 lb/in^2 ($= 1.93 \times 10^5$ N/m^2). During the day, if the temperature reaches $38°C$ ($= 100°F$), what is the gauge pressure? (Assume that the volume remains constant.)
17. Initially, a gas contained in a cylinder provided with a movable piston is at atmospheric pressure. Suppose that the volume is reduced to one-fifth of

its original volume by pushing the piston in. Calculate the new pressure under the following circumstances: (a) the temperature of the gas inside the cylinder remains constant; and (b) the temperature of the gas increases from its initial value of 27°C to 65°C.

18. Prove the relation $P_1V_1/T_1 = P_2V_2/T_2$ by the following procedure. A gas of volume V_1 at pressure P_1 is changed to a new pressure P_2 and volume V_1', keeping the temperature T_1 constant. Volume V_1' at pressure P_2 is changed to volume V_2 by increasing the temperature to T_2 while keeping P_2 constant.

19. A balloon containing 5 ft^3 of helium at 20°C has a pressure of 1 atm when on ground level. Suppose that this balloon rises to an altitude where its temperature is −25°C and the pressure is 20 cm Hg. Calculate the new volume.

20. In Problem 19 calculate the amount of helium in the balloon (the molecular weight is 4.00). How much will the weight be if helium is replaced by hydrogen (molecular weight is 2)?

21. The surface of a lake is at a temperature of 27°C and a pressure of 1 atm. The temperature at the bottom of the lake is 10°C. If a bubble of air of 1 mm radius starting from the bottom has a radius of 2 mm when it reaches the top of the lake, what is the depth of the lake? [Volume = $(4\pi/3)R^3$.]

22. Nitrogen is stored in a tank of volume 50 ℓ ($1\ell = 10^{-3}$ m^3) at a pressure of 200 lb/in^2. When some nitrogen is withdrawn from the tank, its pressure drops to 100 lb/in^2 while its temperature drops from 27°C to 17°C. Calculate the amount of nitrogen withdrawn and the amount left in the tank.

23. Of the 10^{14} neutrons incident on a target of 1 cm^2 in 1 s, only 10^6 are absorbed. The energy of each incident neutron is 1 MeV ($=10^6$ eV $= 10^6 \times 1.6 \times 10^{-19}$ J), and its mass is 1.67×10^{-27} kg. Calculate the total (a) change in momentum; and (b) the force imparted by these neutrons to the target.

24. Bullets are fired on a target at a rate of 300 per minute. The mass of each bullet is 100 g and the speed is 200 m/s. Calculate the average momentum and force exerted on the target if (a) the bullets are absorbed by the target; and (b) the bullets bounce back from the target with unchanged speed. What will happen if the target is free to move?

25. A single oxygen molecule contained in a cubical box of 10 cm on a side has perfectly elastic collisions with the walls. The components of the velocity are $v_x = 1500$ m/s, $v_y = 1000$ m/s, and $v_z = 1200$ m/s. Calculate (a) the number of collisions per second; (b) change of momentum in each collision; (c) the average force exerted on each wall; and (d) the average pressure exerted on each wall.

26. Calculate the rms speed of hydrogen molecules (molecular wt = 2) at a temperature of 1000 K and a pressure of 2 atm.

27. Calculate the rms speed of air molecules (that of oxygen and nitrogen) at room temperature (23°C) and at a pressure of 1 atm.

28. A quantity of 0.1 kg of oxygen gas is confined in a volume of 3 m^3 and at a pressure of 3×10^5 N/m^2. Calculate the rms speed of the molecules of this gas and their momentum.

29. Calculate the average kinetic energy of a single oxygen molecule at room temperature. If this molecule is raised 1 m in height, what will be the change in the gravitational potential energy of the molecule? Comparing the two values, can you say that we were justified in neglecting the effect of gravity in our kinetic-theory calculations?

30. What is the ratio of the rms speeds of a molecule at temperatures 3000 K and 30 K?

31. What is the ratio of the rms speeds of molecules of molecular weights 64 and 4?

32. Calculate the temperatures at which the rms speed of the following molecules exceeds the escape velocity from the earth ($= 1.1 \times 10^4$ m/s): (a) hydrogen; (b) nitrogen; and (c) oxygen. Explain why the moon and some other planets, such as Mercury, cannot have an atmosphere.

33. Calculate the rms speed of the hydrogen atoms at the surface of the sun where the temperature is 6000 K. How does this compare with the escape velocity from the sun ($= 615$ km/s)? What is the kinetic energy of the hydrogen atom? (In a hydrogen bomb explosion, the kinetic energy of the protons corresponds to a temperature of 10^8 K.)

34. If to start with, a mixture consists of 0.7 percent and 99.3 percent of two isotopes $^{235}UF_6$ and $^{238}UF_6$, respectively, what will be their relative abundance after diffusion through one porous wall.

35. In the above problem what will be their relative abundance after diffusion through two porous walls.

14 First Law of Thermodynamics—Specific Heat

We want to establish the relations between the heat supplied, work done, and internal energy, and one such relation is the first law of thermodynamics. To investigate the changes in temperature when heat is supplied to a system, we must familiarize ourselves with different types of specific heats. Making use of the principle of equipartition, we shall make theoretical calculations for the specific heats of gases and solids. We shall investigate the phase changes (solid to liquid and liquid to gas) that take place when heat is supplied to a system.

14.1. Heat and Internal Energy

WHEN HEAT energy is added to a system, one of many different things can happen:

1. Internal energy may increase and may appear in the form of (a) translational kinetic energy of the molecules, which is indicated by an increase in temperature; (b) rotational energy of the molecules, (c) vibrational energy of the molecules, and (d) electronic excitations, which will result in the ionization of the atoms if enough energy is transferred.
2. The system may expand, which may or may not result in work being done on the surroundings.
3. As the system absorbs heat energy, it may initiate a chemical reaction, thereby changing the chemical potential of the system.
4. It is possible that the system may accelerate, thereby changing its kinetic energy or potential energy. This is the *mechanical energy.*

If heat is a form of energy, there must be some relation between this and the other forms of energy, such as mechanical energy. This means that the unit of heat should be that of work or energy, that is, the joule. We shall see this to be true, although the conventional unit of heat is called the calorie and is still in use. A *calorie* (cal) as a unit of heat is defined as the amount of heat energy needed to raise the temperature of 1 gram of water through 1 Celsius degree or 1 Kelvin degree (or, to be more precise, from 14.5°C to 15.5°C). The British unit of heat energy is the *British thermal unit* (Btu), defined as the heat energy required to raise the temperature of that amount of water which weighs 1 pound through 1 Fahrenheit degree. Note that the British unit uses the weight and not the mass of water. The relation between the two units is 1 Btu = 252 cal.

14.2. Specific Heats of Solids and Liquids

Suppose that a body absorbs heat energy and its disordered translational kinetic energy increases, which in turn increases the temperature of the body.

The amount of heat needed to raise the temperature of a body through a certain range is different for different bodies. A quantity that describes the ability of a body to absorb heat and increase its temperature is *specific heat*. Suppose that a body of mass m is at temperature T. When an amount of heat ΔQ is supplied to this body, its temperature increases to $T + \Delta T$. The amount of heat ΔQ needed is

$$\boxed{\Delta Q = mc\,\Delta T} \tag{14.1}$$

where c is the specific heat of the material. If we rewrite Equation 14.1 as

$$c = \frac{\Delta Q}{m\,\Delta T} \tag{14.2}$$

we define *specific heat as the amount of heat that must be added to a material of unit mass to raise its temperature through 1 degree*. From Equation 14.2, the unit of c will be cal/g-C°, kcal/kg-C°, or J/kg-C°, while in the British system it will be Btu/lb-F°.

It may be pointed out that specific heat depends not only on the type of material but also on the external conditions under which heat is supplied. Two commonly used specific heats are specific heat at constant pressure, c_p, in which the material is heated while the pressure is kept constant; and specific

TABLE 14.1
Specific Heats of Some Materials

Substance	Temperature Range (°C)	Specific Heat (cal/g-C° or kcal/kg-C°)
Solids		
Aluminum	15–100	0.217
Asbestos	20–100	0.195
Copper	15–100	0.093
Glass	20–100	0.20
Ice		0.50
Iron (steel)	18–100	0.113
Lead	20–100	0.031
Silver	15–100	0.056
Liquids		
Benzene	At 20	0.41
Mercury	0–100	0.033
Methyl alcohol	At 20	0.06
Water	At 15	1.00
Gases (at constant pressure)		
Air	At 50	0.25
Helium	At 20	1.24
Steam	At 110	0.48

heat at constant volume, c_v, in which the material is heated while the volume is kept constant. For gases the values of c_p and c_v are very different. For solids and liquids, the values of c_p and c_v at ordinary pressure and temperature do not differ substantially. Usually only the values of c_p are quoted. Hence, in the following c will stand for c_p when the discussion concerns solids or liquids.

From the definition of a calorie—the amount of heat needed to heat 1 gram of water through 1 Celsius degree—the specific heat of water is $\sim$1 calorie per gram per Celsius degree. The reason for using an approximate sign is because c varies slightly with temperature, as we shall see shortly. Table 14.1 gives the specific heats of many materials at room temperature. It is quite clear that the specific heat of water is much larger than for other materials. Usually, c is approximately constant for most materials over a small range of temperatures. Figure 14.1 shows the plot of variation of specific heat of water with

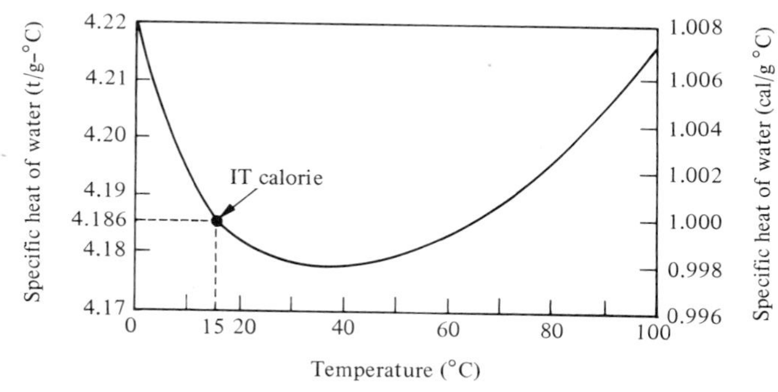

FIGURE 14.1. *Variation of specific heat of water with temperature. Note from the scale how small the variation is.*

temperature. From this we conclude that over a temperature range of 100 degrees, the maximum variation in the specific heat of water is less than 1 percent. In order to be more accurate about the size of the calorie, it is defined as follows:

One calorie is the amount of heat needed to raise the temperature of 1 g of water from 14.5° to 15.5°C. This is the 15°-calorie. When converted into joules,

$$1\ 15°\text{-cal} = 4.186\ \text{J} \tag{14.3}$$

The International Table calorie (IT cal) is identical to the 15°-cal, that is, equal to 4.186 J.

Another useful quantity is the *molar specific heat* or *molar heat capacity*, defined as *the amount of heat needed to raise the temperature of 1 mole of the material by 1C°*. Thus, if a material contains n moles of the substance and needs heat ΔQ to raise its temperature from T to $T + \Delta T$, the molar specific heat C is given

$$\boxed{C = \frac{1}{n}\frac{\Delta Q}{\Delta T}} \tag{14.4}$$

where $n = m/M$, m being the mass of the material and M its molecular weight (the number of grams in 1 mol). Substituting for $m = nM$ in Equation 14.4 and rearranging yields

$$Mc = \frac{1}{n}\frac{\Delta Q}{\Delta T} \qquad (14.5)$$

Comparing Equations 14.2 and 14.5 gives

$$\boxed{C = Mc} \qquad (14.6)$$

that is,

molar specific heat = molecular weight × specific heat

where C is in cal/mol-C°, c in cal/g-C°, and M in g/mol.

Another commonly used term is *heat capacity*, C', which is equal to the product of mass and specific heat:

$$\text{heat capacity} = C' = mc \qquad (14.7)$$

Thus, for a given type of material, its heat capacity will depend on its mass, while molar heat capacity remains constant.

Finally, let us consider the problem of heat balance. Suppose that system 1, of mass m_1, specific heat c_1, and temperature T_1, is brought in contact with system 2, of mass m_2, specific heat c_2, and temperature T_2. If $T_1 > T_2$, heat will flow from system 1 to system 2 until the two are in thermal equilibrium at temperature T. The heat lost by system 1 is $m_1c_1(T_1 - T)$, while the heat gained by system 2 is $m_2c_2(T - T_2)$. The principle of heat balance states that

$$\text{heat lost} = \text{heat gained} \qquad (14.8)$$

Therefore,

$$m_1c_1(T_1 - T) = m_2c_2(T - T_2)$$

Knowing m_1, m_2, c_1, c_2, T_1, and T_2, we can solve for T.

EXAMPLE 14.1 A total of 500 g of lead shot is heated to 100°C and dropped into a 100-g copper can containing 150 g of water at 15°C. If the mixture is stirred well, what is the equilibrium temperature? (The specific heats of copper and lead are 0.093 cal/g-C° and 0.031 cal/g-C°.)

Suppose that the final equilibrium temperature reached is T_f. Thus, the decrease in the temperature of lead will be $(100 - T_f)$C°, while the increase in the temperature of copper and water will be $(T_f - 15)$C°. Heat is lost by lead and gained by copper and water. If we denote the quantities with subscripts L, C, and W for lead, copper, and water, respectively, we may write

$$\text{heat lost} = \text{heat gained}$$

$$m_Lc_L\,\Delta T_L = m_Cc_C\,\Delta T_C + m_Wc_W\,\Delta T_W$$

$$(500\text{ g})(0.031\text{ cal/g-C°})(100 - T_f)\text{C°} = (100\text{ g})(0.093\text{ cal/g-C°})(T_f - 15)\text{C°} + (150\text{ g})(1.0\text{ cal/g-C°})(T_f - 15)\text{C°}$$

$$15.5(100 - T_f) = 159.3(T_f - 15) \qquad \text{or} \qquad 175\,T_f = 3940$$

$$T_f = 22.5\text{°C}$$

Exercise 14.1 To find the temperature of a furnace, the following procedure is used. A 100-g piece of metal with a specific heat of 0.11 cal/g-C° is heated in the furnace. The metal is then dropped into a 150-g copper calorimeter containing 300 g of water at 10°C. The final temperature reached is 18°C. What is the temperature of the furnace? [*Ans.*: $T = 246°\text{C}$.]

14.3. Mechanical Equivalent of Heat

Benjamin Thompson (1753–1814), an American who later became Count Rumford of Bavaria, gave the first conclusive evidence that heat is not a substance. That heat is a form of energy and follows the principle of conservation of energy was established by a British physicist, James Joule (1818–1889). He showed by experiments that when mechanical work is converted into heat, the same amount of work is needed to produce the same amount of heat. Thus, he was able to find the relation between the calorie and the joule. We shall briefly describe his experiment.

Figure 14.2 shows a type of experimental arrangement used by Joule. It consists of a container filled with water and insulated from the outside. A set of

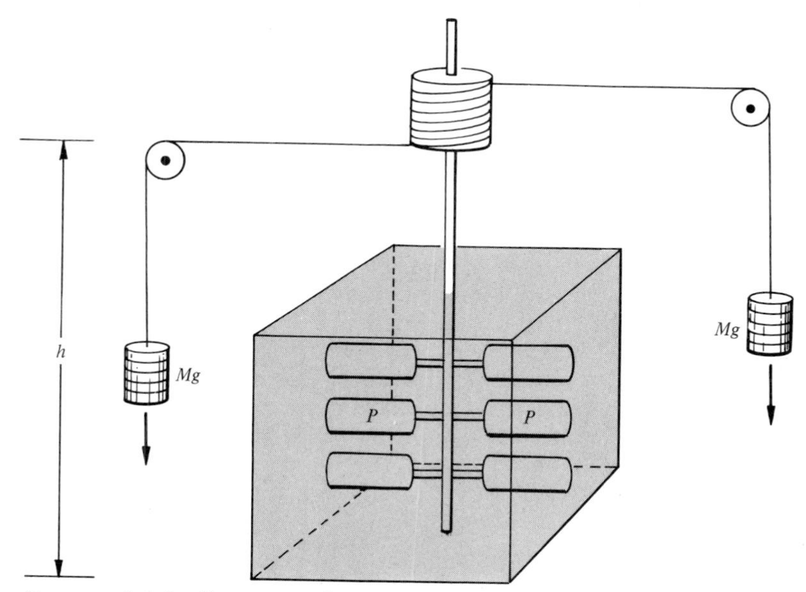

Figure 14.2. *Experimental arrangement used by Joule to establish the relation between mechanical energy and thermal energy.*

paddles P in the water container will rotate when two weights Mg attached to the shaft fall from a height h to the ground. In each fall the total loss in potential energy by the set of two weights is $2Mgh$. If the weights are allowed to fall n times, the total loss in mechanical potential energy will be $2nMgh$. This loss in mechanical potential energy appears as rotational kinetic energy of the paddles. As the paddles are stopped by the water, the rotational kinetic energy is changed into heat, thereby increasing the temperature. Suppose that the mass of water is m, its specific heat is c, and the rise in temperature is ΔT. (Note that the container, as well as the paddles, will be heated. It is assumed that in more accurate calculations, these will be taken into consideration.) Thus, the heat used by the water will be $\Delta Q = mc\,\Delta T$.

From the conservation-of-energy principle, heat energy gained must be equal to the loss in mechanical potential energy. That is,

$$\Delta Q = \Delta W \qquad \text{or} \qquad mc\,\Delta T = 2nMgh \tag{14.9}$$

The left-hand side is in cal, kcal, or Btu; the right-hand side is in joules or ft-lb. For the equation above to be valid, no matter what the magnitudes of ΔQ and ΔW are, the ratio $\Delta W/\Delta Q$ should be constant. This was found to be the case in all experiments. This constant conversion factor is the *mechanical equivalent of heat,* denoted $\mathcal{J}$, and may be defined as the amount of mechanical work that must be done to produce a unit amount of heat. Thus, if W units of work produce Q amount of heat,

$$\boxed{\mathcal{J} = \frac{W}{Q}} \tag{14.10}$$

The most recent value of this constant conversion factor obtained from much more accurate experiments using electrical methods is

$$\begin{gathered}\mathcal{J} = 4.186\ \text{J/cal} = 4186\ \text{J/kcal} = 777.9\ \text{ft-lb/Btu}\\ 1\ \text{cal} = 4.186\ \text{J} \qquad 1\ \text{kcal} = 4186\ \text{J}\\ 1\ \text{Btu} = 252\ \text{cal} = 777.9\ \text{ft-lb}\end{gathered} \tag{14.11}$$

EXAMPLE 14.2 A 15-g lead bullet traveling at 400 m/s is brought to rest when it strikes a target. If all the available energy is used in heating the bullet, what will be the change in its temperature?

The total available kinetic energy is $\frac{1}{2}mv^2$, where m is the mass and v is the velocity of the bullet. In terms of heat units, this is equal to $\frac{1}{2}mv^2/\mathcal{J}$, where $\mathcal{J} = 4186$ J/kcal. Let c be the specific heat of lead and ΔT the increase in the temperature of the bullet. Thus, from the conservation-of-energy principle,

$$\text{mechanical energy lost} = \text{heat energy gained}$$

That is,

$$\frac{W}{\mathcal{J}} = Q \qquad \text{or} \qquad \frac{\frac{1}{2}mv^2}{\mathcal{J}} = mc\,\Delta T$$

$$\Delta T = \frac{1}{2}\frac{v^2}{\mathcal{J}c} = \frac{(400\ \text{m/s})^2}{2(4.186\times 10^3\ \text{J/kcal})(0.031\ \text{kcal/kg-C}^\circ)} = 616\text{C}^\circ$$

EXERCISE 14.2 A 10-g lead bullet strikes a target and comes to rest. All the available kinetic energy is converted into heat and is used up in heating the bullet, without any loss to its surroundings. If the temperature of the bullet changes from 20°C to 400°C, what was the speed of the bullet? [*Ans.:* $v = 314$ m/s.]

14.4. Internal Energy and the First Law

The first law of thermodynamics is a statement of the conservation of energy. The *total internal energy U* of any isolated system of molecules, say a gas confined in a cylinder by means of a piston, is given by the sum of two terms:

$$U = E_k + E_p \tag{14.12}$$

E_k is the internal kinetic energy of the translation, rotation, and vibration of molecules, while E_p is the internal potential energy due to the interaction between different molecules. In the case of an ideal gas, there are no intermolecular forces, and the internal potential energy will be zero. In any case, *if the system is isolated, the total internal energy of the system of particles remains constant*. Thus, just like P, V, and T, we can assign an internal energy U to various states of the system. The value of U can be changed by bringing the system in contact with the surroundings and exchanging energy. This can be accomplished in two different ways: either by doing work W or by heat flow Q. The following sign convention should be kept in mind: The work done *on the system* W_{ext} by an external agent is taken as *positive*, while the work done *by the system* on the surrounding is *negative*; heat Q *absorbed by the system* is considered *positive*, while heat Q *given off by the system* is considered *negative*.

Let us consider a system in state 1 with internal energy U_1. The system absorbs heat Q, and an external agent does work W_{ext} on the system. This changes the system to state 2 with internal energy U_2. Note that the heat Q supplied and the work W_{ext} must be expressed in the same units. Thus, according to the conservation-of-energy principle, the change in the internal energy $U_2 - U_1$ is given by

$$\boxed{U_2 - U_1 = Q + W_{ext}} \tag{14.13}$$

This equation represents the

> **First law of thermodynamics: *The change in internal energy of a system is equal to the heat absorbed by the system plus the external work done on the system.***

A common situation is one in which the system absorbs heat Q and does work W on the surroundings. In this case $W_{ext} = -W$ because the work is done by the system and Equation 14.13 takes the form

$$\boxed{U_2 - U_1 = Q - W} \tag{14.14}$$

Both Equations 14.3 and 14.4 constitute the first law of thermodynamics and are depicted diagrammatically in Figure 14.3. It is Equation 14.14 that we shall be using most frequently.

Suppose that an ideal gas is contained in a cylinder fitted with a movable, frictionless piston of area A, as shown in Figure 14.4. Let P, V, and U_1 be the pressure, volume, and internal energy of the confined gas. For the piston to remain stationary, the external force F applied to the piston must be equal to the force exerted by the gas on the piston; that is, $F = PA$.

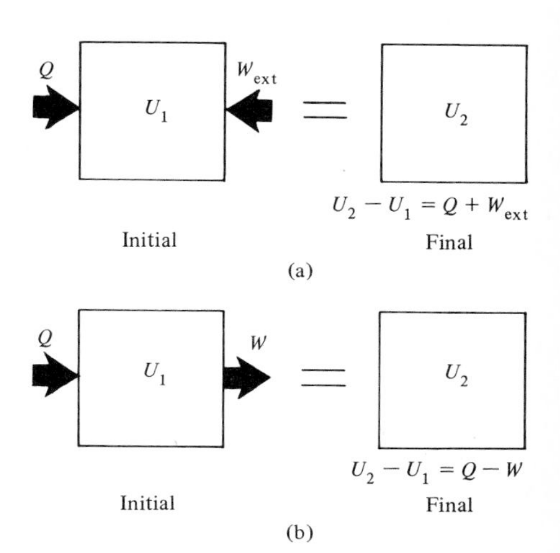

FIGURE 14.3. *First law of thermodynamics.*

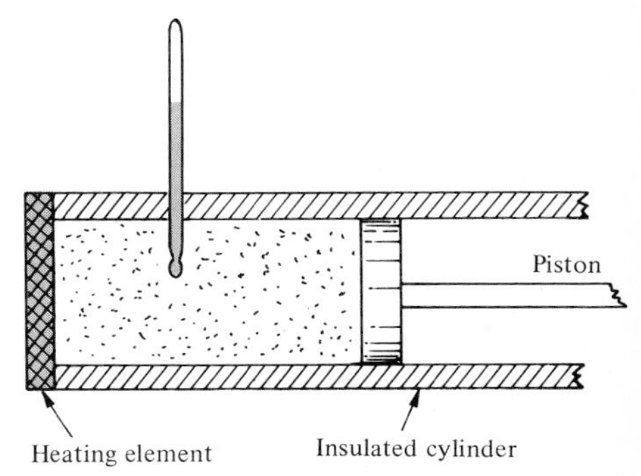

FIGURE 14.4. *Ideal gas confined in a cylinder with a movable piston.*

Let us supply heat ΔQ to the gas. If the pressure is kept fixed, as the temperature increases the gas must be allowed to expand. Suppose that the piston moves through a distance Δx when the temperature increases from T to $T + \Delta T$. The work done by the expanding gas is

$$\Delta W = F\,\Delta x = PA\,\Delta x = P\,\Delta V \tag{14.15}$$

If the new internal energy of the gas is U_2, so that $\Delta U = U_2 - U_1$, the first law may take the form

$$\Delta U = \Delta Q - \Delta W \qquad \text{or} \qquad \Delta U = \Delta Q - P\,\Delta V \tag{14.16}$$

On the other hand, suppose that the piston is pushed in; hence, external work is done on the gas (because ΔV is now negative) and heat ΔQ is also supplied to the gas. The change in internal energy will be

$$\Delta U = \Delta Q + P\,\Delta V \tag{14.17}$$

EXAMPLE 14.3 A quantity of 10 m^3 of a liquid is supplied with 100 kcal of heat and expands at a constant pressure of 10 atm to a final volume of 10.2 m^3. Calculate the work done and change in internal energy of the system.

The change in volume is

$$\Delta V = (V_f - V_i) = (10.2 - 10)\ \text{m}^3 = 0.2\ \text{m}^3$$

The work done from Equation 14.15 is

$$\Delta W = P\,\Delta V = (10 \times 1.013 \times 10^5\ \text{N/m}^2)(0.2\ \text{m}^3) = 2 \times 10^5\ \text{J}$$

$$= 2 \times 10^5\ \text{J} \times \frac{1}{4186\ \text{J/kcal}} = 48\ \text{kcal}$$

From the first law of thermodynamics, Equation 14.14, $\Delta U = \Delta Q - \Delta W$, where $\Delta Q = 100$ kcal and from above $\Delta W = 48$ kcal. Therefore,

$$\Delta U = (100 - 48)\ \text{kcal} = 52\ \text{kcal}$$

That is, of the total 100 kcal supplied, only 52 kcal is used in increasing the internal energy.

EXERCISE 14.3 A total of 5 m^3 of a liquid is supplied with 70 kcal of heat, and it expands at a constant pressure of 1 atm to a final volume of 5.05 m^3. Calculate in joules the work done and the change in the internal energy of the system. [*Ans.:* $\Delta W = 5 \times 10^3$ J; $\Delta U = 2.9 \times 10^5$ J.]

14.5. Equipartition of Energy

In the development of the kinetic theory of gases in Chapter 13, we showed that the average translational kinetic energy of a molecule of an ideal gas in thermal equilibrium at temperature T is $\frac{3}{2}kT$. Since a monatomic gas at high temperature and low density behaves like an ideal gas, kinetic theory should be quite successful in explaining the specific heats of these gases. In the middle of the nineteenth century attempts were made to explain the specific heats of diatomic gases by means of the principle of equipartition of energy, as introduced by Maxwell and explained below.

Let us consider a mixture of gases that are in thermal equilibrium at temperature T and occupy the same volume V. Let m_i be the mass of the ith-type molecule and $\langle v_i^2 \rangle$ be its mean square speed. According to kinetic theory, Equation 13.25, the mean kinetic energy depends only on the temperature, hence

$$\bar{K} = \tfrac{1}{2}m_1 \langle v_1^2 \rangle = \tfrac{1}{2}m_2 \langle v_2^2 \rangle \cdots = \tfrac{3}{2}kT$$

which states that *the mean translational kinetic energy of all the molecules is the same and is independent of their masses.* For example, if the mixture contains both hydrogen and oxygen molecules at temperature T, their masses are in the ratio 2:32, but their mean translational kinetic energies are the same. This also implies that the average speed of the hydrogen molecules must be four times that of the oxygen molecules.

The translational kinetic energy of a molecule may also be written as

$$K_{\text{tran}} = K_x + K_y + K_z = \tfrac{1}{2}m(v_x^2 + v_y^2 + v_z^2) = \tfrac{3}{2}kT$$

of the total energy $\tfrac{3}{2}kT$, we would expect that $\tfrac{1}{2}kT$ is associated with K_x, $\tfrac{1}{2}kT$ with K_y and $\tfrac{1}{2}kT$ with K_z. That is, the total energy $\tfrac{3}{2}kT$ is equally shared among three components, X, Y, and Z. We say that the molecule has 3 degrees of freedom, and each degree of freedom has associated with it an amount of energy $\tfrac{1}{2}kT$.

Each independent quantity (or coordinate) that is required to be specified to determine the energy of a molecule is called a degree of freedom, and with each degree of freedom is associated an amount of energy $\tfrac{1}{2}kT$.

A geometrical point molecule has three independent quantities, v_x, v_y, and v_z, which must be specified to determine the total energy. These quantities are the three free coordinates, and the energy associated with each coordinate is $\tfrac{1}{2}kT$. Thus, we have equipartition (equal sharing) of energy among the three translational degrees of freedom.

Such treatment, when extended to a more general case, takes the following form. According to Boltzmann and Maxwell:

> **The Principle of Equipartition of Energy:** ***If energy is added to a system containing many molecules, it will be distributed equally on the average among all the degrees of freedom of the molecules. Furthermore, if the energy associated with any degree of freedom is proportional to the square of the independent variable specifying the degree of freedom, the average value of energy which it contributes to the total thermal energy is $\tfrac{1}{2}kT$ per degree of freedom.***

Thus, if a system has N molecules and each molecule has f degrees of freedom, the mean energy associated with each molecule is

$$\bar{E} = \frac{f}{2}kT$$

while the total energy of the system is

$$E_T = N\bar{E} = \frac{f}{2}NkT = \frac{f}{2}nRT \tag{14.18}$$

where we have used $Nk = nR$, n being the number of moles and R the universal gas constant. The value of f is different in different situations, as we shall see shortly.

14.6. Specific Heat of Gases

The true test of the kinetic theory of gases and the principle of equipartition of energy should come through its application to the determination of different properties. In this section we apply this theory to the specific heat of gases and see how far we can go without quantum mechanics. We shall see that the theory does well in the case of monatomic gases, but additional use of quantum mechanics is needed in the case of diatomic and polyatomic gases.

As mentioned earlier, heat capacity per unit mass is called *specific heat*, defined as

$$c = \frac{1}{m}\frac{\Delta Q}{\Delta T} \tag{14.19}$$

On the other hand, *molar specific heat* is defined as

$$C = \frac{1}{n}\frac{\Delta Q}{\Delta T} \tag{14.20}$$

where n is the number of moles of the gas.

The gases may be heated in several ways by varying the restrictions imposed on them; hence, the gases will have many specific heats. Two commonly used ones are the specific heat at constant volume C_v, and the specific heat at constant pressure C_p.

Suppose that we have n moles of gas containing N molecules at thermal equilibrium temperature T. As we showed in the preceding section, the total internal energy U (instead of E_T) of such a system is

$$U = E_T = \frac{f}{2}NkT = \frac{f}{2}nRT \tag{14.21}$$

where f is number of the degrees of freedom. Let us now supply heat ΔQ to this system so that its temperature changes from T to $T + \Delta T$ while the volume remains constant. Thus, in the expression for the first law of thermodynamics from Equation 14.16, $\Delta Q = \Delta U - \Delta W$, where ΔW is equal to zero because there is no change in the volume; that is, $\Delta W = P\,\Delta V = 0$. Thus,

$$\Delta Q = \Delta U \qquad \text{at constant } V$$

which means that all the heat energy supplied to the system converts into internal energy. Thus, the molar specific heat capacity at constant volume C_v may be written

$$C_v = \frac{1}{n}\left(\frac{\Delta Q}{\Delta T}\right) = \frac{1}{n}\left(\frac{\Delta U}{\Delta T}\right)$$

From Equation 14.21, $\Delta U = (f/2)\ nR\ \Delta T$. Substituting above,

$$C_v = \frac{1}{n}\left(\frac{(f/2)\ nR(\Delta T)}{\Delta T}\right)$$

or

$$C_v = \frac{f}{2}R \tag{14.22}$$

Let us now calculate the molar specific heat at constant pressure. According to Equation 14.16, if a gas is heated at constant pressure, the amount of heat ΔQ supplied increases its internal energy by ΔU and the gas does work ΔW against the surroundings while the temperature increases from T to $T + \Delta T$. From $PV = nRT$, we get $P\,\Delta V = nR\,\Delta T$; hence,

$$\Delta W = P\,\Delta V = nR\,\Delta T \tag{14.23}$$

The first law of thermodynamics takes the form

$$\Delta U = \Delta Q - P\,\Delta V = Q - nR\,\Delta T \tag{14.24}$$

or

$$\Delta Q = \Delta U + nR\,\Delta T \tag{14.25}$$

Hence, the molar specific heat at constant pressure is given by

$$C_p = \frac{1}{n}\frac{\Delta Q}{\Delta T} = \frac{1}{n\,\Delta T}(\Delta U + nR\,\Delta T) = \frac{1}{n}\frac{\Delta U}{\Delta T} + R$$

By definition, $(1/n)(\Delta U/\Delta T) = C_v$. Thus, we get $C_p = C_v + R$, or

$$C_p - C_v = R \tag{14.26}$$

Substituting for C_v from Equation (14.22) and rearranging

$$C_p = \left(\frac{f}{2} + 1\right)R \tag{14.27}$$

while the ratio of the two specific heats, C_p/C_v, is defined as γ and is given by

$$\gamma = \frac{C_p}{C_v} = \frac{\left(\frac{f}{2} + 1\right)R}{\frac{f}{2}R} = \frac{f + 2}{f} \tag{14.28}$$

Thus, we have found that for an ideal gas,

$$C_v = \frac{f}{2}R \tag{14.22}$$

$$C_p = \left(\frac{f}{2} + 1\right)R \tag{14.27}$$

$$\gamma = \frac{f + 2}{f} \tag{14.28}$$

Note that C_p *is always greater than* C_v because at constant pressure thermal energy has to be supplied not only to increase the internal energy, but the gas must do extra work against the external environment (Equation 14.23). At constant volume no such extra work is needed.

For an ideal monatomic gas in which the molecules are considered geometrical points, the internal energy is all translational kinetic energy; hence, there are three degrees of freedom. That is, for a monatomic gas $f = 3$ and from Equations 14.22, 14.27, and 14.28,

$$C_v = \tfrac{3}{2}R = 1.5\,R, \qquad C_p = \tfrac{5}{2}R = 2.5\,R, \qquad \gamma = \tfrac{5}{3} = 1.66 \quad (14.29)$$

These theoretically predicted values are in perfect agreement with the experimentally measured values for monatomic gases, as shown in Table 14.2. Also, the experimentally measured values of the specific heats of these gases are independent of temperature, hence in agreement with theory.

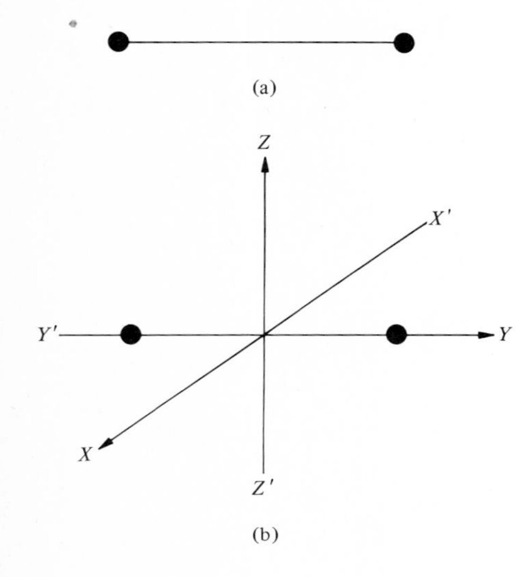

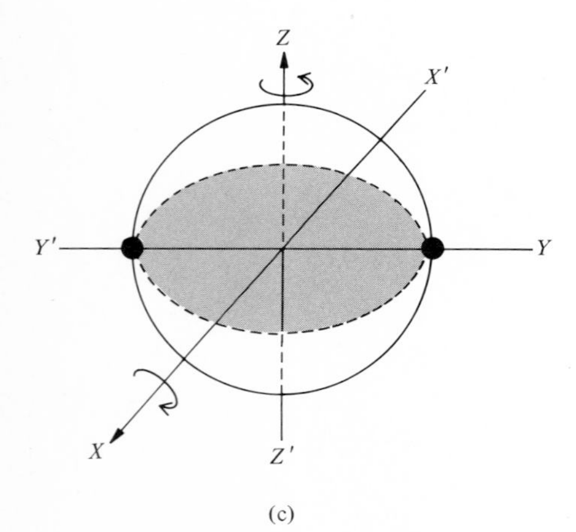

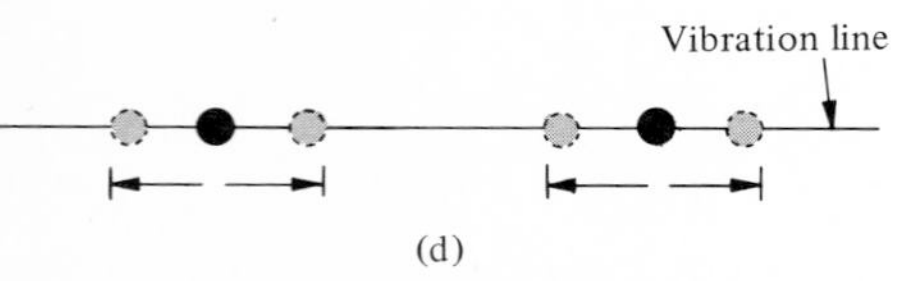

Figure 14.5. *Diatomic molecule: (a) Dumbbell structure, (b) translational motion, (c) rotational motion, and (d) vibrational motion.*

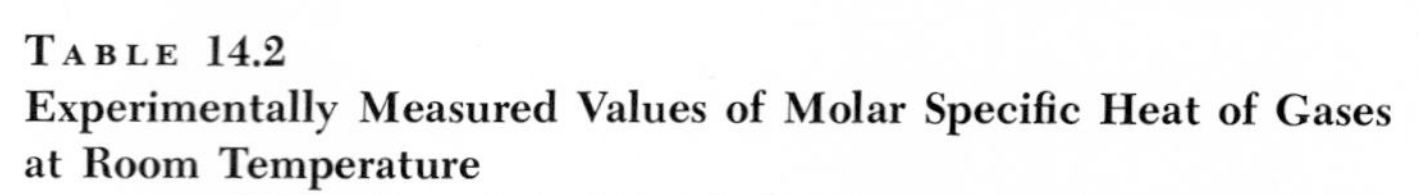

Table 14.2
Experimentally Measured Values of Molar Specific Heat of Gases at Room Temperature
(*Note:* R = 8314 J/kmol-K = 8.314 J/mol-K = 1.99 cal/mol-K)

Gas	C_v/R	C_p/R	γ	$(C_p - C_v)/R$
Monatomic				
Helium (He)	1.519	2.52	1.659	1.001
Neon (Ne)			1.64	
Argon (Ar)	1.509	2.52	1.67	1.008
Krypton (Kr)			1.68	
Xenon (Xe)			1.66	
Diatomic				
Hydrogen (H_2)	2.438	3.42	1.410	0.9995
Oxygen (O_2)	2.504	3.52	1.401	1.004
Nitrogen (N_2)	2.448	3.50	1.404	1.005
Hydrogen chloride (HCl)	2.523	3.58	1.41	1.00
Carbon monoxide (CO)	2.488	3.50	1.404	1.005
Nitric oxide (NO)	2.512	3.52	1.400	1.005
[a] Chlorine (Cl_2)	3.02	4.12	1.36	1.09
[a] Bromine (Br_2)			1.32	
[a] Iodine (I_2)			1.30	
Polyatomic				
Carbon dioxide (CO_2)	3.38	4.40	1.304	1.027
Ammonia (NH_3)	3.42	4.48	1.31	1.06
Methane (CH_4)	3.26	4.27	1.31	1.01

[a] The reason for the high specific heat of these diatomic molecules is explained in the text.

Let us now consider a diatomic molecule having a dumbbell structure as shown in Figure 14.5(a). If we assume that this molecule is a rigid structure, it has mass as well as moment of inertia. Also, it will have both translational kinetic energy and rotational energy. *There are still three translational degrees of freedom,* because the motion of the center of mass must be described by three independent coordinates, as shown in Figure 14.5(b). As shown in Figure 14.5(c), the moment of inertia I_y of this molecule about the Y-axis is almost negligible as compared to its moment of inertia I_x about the X-axis and I_z about the Z-axis. Thus, the rotational kinetic energy of this molecule is given by

$$K_{\text{rot}} = \tfrac{1}{2}I_x\omega_x^2 + \tfrac{1}{2}I_z\omega_z^2$$

where ω_x and ω_z are the respective angular velocities while the component $\frac{1}{2}I_y\omega_y^2$ is negligible because I_y is very small. *Thus, a diatomic molecule has two rotational degrees of freedom.* The situation is shown in Figure 14.5(c).

Finally, if we remove the constraint that the diatomic molecule is a rigid structure, it will have vibrational motion as well. This is assumed to be a simple harmonic motion, as shown in Figure 14.5(d), and the vibrational energy is given by

$$K_{\text{vib}} = \tfrac{1}{2}kx^2 + \tfrac{1}{2}mv^2 \qquad (14.30)$$

where the first term on the right is a potential-energy term while the second term is the kinetic energy. *Thus, a diatomic molecules has two vibrational degrees of freedom.* We may expect diatomic molecules to have three translational, two rotational, and two vibrational, that is, a total of seven degrees of freedom. From the principle of equipartition of energy, the total mean energy $\bar{E}$ for each molecule is

$$\bar{E} = K_{\text{trans}} + K_{\text{rot}} + K_{\text{vib}} = 3(\tfrac{1}{2}kT) + 2(\tfrac{1}{2}kT) + 2(\tfrac{1}{2}kT) = \tfrac{7}{2}kT$$

If there are N molecules, the total energy E_T is

$$E_T = N\bar{E} = \tfrac{7}{2}NkT = \tfrac{7}{2}nRT \qquad (14.31)$$

Using Equations 14.22, 14.27, and 14.28 with $f = 7$, we get

$$C_v = \tfrac{7}{2}R = 3.5\,R, \qquad C_p = \tfrac{9}{2}R = 4.5\,R, \qquad \gamma = \tfrac{9}{7} = 1.29 \qquad (14.32)$$

These values are not in agreement with experimentally measured values for the diatomic molecules, as shown in Table 14.2. The values shown in the table are at room temperature. Typical values of C_v versus T for hydrogen are shown in Figure 14.6. It is quite clear from this figure and Table 14.2 that these experimental values do not agree with the theoretical values given by Equation 14.32. Thus, we must look elsewhere for a theoretical explanation.

Let us have a close look at Figure 14.6. It reveals that C_v for hydrogen molecules at low temperature below 50 K is constant and equal to $\frac{3}{2}R$. According to Equation 14.22, this implies that there are only three degrees of freedom. If we assume that both rotational and vibrational degrees of freedom are *frozen out* at such low temperature, theory and experiment agree at once. As the temperature increases, between 300 K and 700 K, the value of C_v is again constant and equals $\frac{5}{2}R$. If we assume that at this temperature the rotational degrees of freedom thaw out while the vibrational degrees of freedom remain frozen, the predicted value of $\frac{5}{2}R$ agrees with the experiment. If the temperature is increased still further and if the gas does not dissociate, the vibrational degrees of freedom also thaw out and the predicted value of $\frac{7}{2}R$ for C_v agrees with experiment, as shown in Figure 14.5. That the rotational and vibrational degrees of freedom do not contribute (or remain frozen) to the value C_v at all temperature ranges was a mystery for a long time.

(Figure 14.6 is on page 306.)

The variation of C_v with temperature T for diatomic gases was explained for the first time by Einstein in 1907. He made use of the concept of quantization

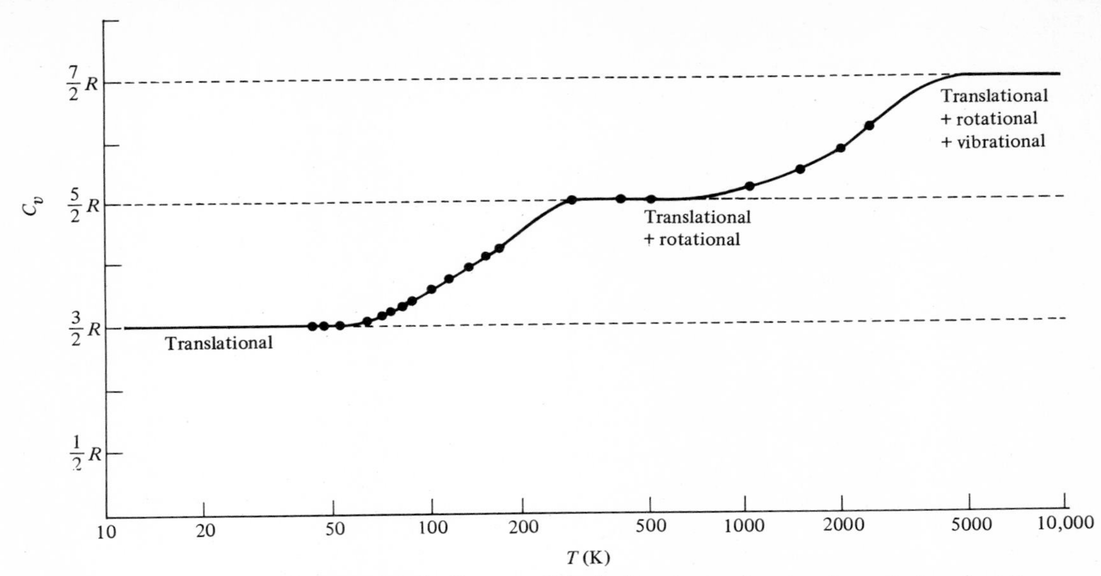

FIGURE 14.6. *C_v versus T for hydrogen gas, showing theoretical explanation (continuous curve) and experimental values (dark circles).*

of energy as introduced by Planck and explained in Chapter 15. According to quantum mechanics, the rotating or vibrating molecules can take only certain minimum discrete energy values and cannot take continuous values. Hence, at low temperatures the conditions are not always such that the principle of equipartition of energy is completely applicable. As the temperature is raised, the minimum discrete energy needed becomes available; hence, rotational, and eventually vibrational, degrees of freedom take part.

Thus, we conclude that at low temperatures, molecules have only translational kinetic energy. As the temperature increases, diatomic molecules have both translational and rotational energy. As the temperature is increased still further, the molecules have vibrational energy also. For example, for a diatomic molecule at room temperature, there is no vibrational excitation, and only translational (three) and rotational (two) degrees of freedom are available. Thus, for $f = 5$, Equations 14.22, 14.27, and 14.28 take the form

$$C_v = 2.5\,R, \qquad C_p = 3.5\,R, \qquad \gamma = \tfrac{7}{5} = 1.40$$

and, as seen from Table 14.2, these theoretical values are in perfect agreement with the experiment. For $T \lesssim 50$ K, $f = 3$ while for $T \gtrsim 3000$ K, $f = 7$, and in between the value of $f = 5$.

The specific heat of the molecules Cl_2, Br_2, and I_2, as shown in Table 14.2, is higher than other diatomic molecules. The reason for this is that these molecules have very low vibrational energy; hence, the vibrational degrees of freedom contribute to the specific heat, even at room temperature.

The situation with polyatomic molecules (molecules with more than two atoms), such as CH_4, CO_2, and NH_3, is much more complicated. But the principle of equipartition coupled with quantum-mechanical principles, can satisfactorily explain the variation of specific heat of gases with temperature.

Example 14.4 The molecular weight of nitrogen is 28. The specific heat at constant volume is 0.174 cal/g-C°. Calculate (a) C_p; (b) C_v; and (c) γ. What can you say about its molecular structure?

From the relation $C_p - C_v = R$ and $C_v = Mc_v = (28 \text{ g/mol})(0.174 \text{ cal/g-C°}) = 4.87$ cal/mol-C° and $R = 1.99$ cal/mol-C°, we get

$$C_p = C_v + R = (4.87 + 1.99) \text{ cal/mol-C°} = 6.86 \text{ cal/mol-C°}$$

$$\gamma = \frac{C_p}{C_v} = \frac{6.86}{4.87} = 1.41$$

From Table 14.2 it is clear that nitrogen is a diatomic gas.

Exercise 14.4 For HCl gas, $C_p - C_v = 1.99$ kcal/kmol-K and $\gamma = 1.41$. Calculate C_p and C_v. [*Ans.*: $C_v = 4.85$ kcal/kmol-K, $C_p = 6.89$ kcal/kmol-K.]

14.7. Specific Heat Capacity of Solids

In the early nineteenth century, Dulong and Petit found experimentally the following law of specific heats of solids.

> **Dulong and Petit law:** ***Molar specific heat at constant volume C_v at high temperature is the same for all solids and is equal to $\sim 3R$ or 6.0 cal/mol-K.***

For solids, C_p increases slightly above room temperature while C_v remains almost constant. The maximum difference between C_p and C_v for solids is between 3 and 5 percent at room temperature. We shall concentrate on C_v only. Figure 14.7 shows plots of C_v versus T for several solids, and there is no question that, except for diamond, solids do obey the Dulong–Petit law. The constant specific heat at high temperatures can be easily explained with the help of classical theory, as we shall discuss below. The sharp increase in C_v

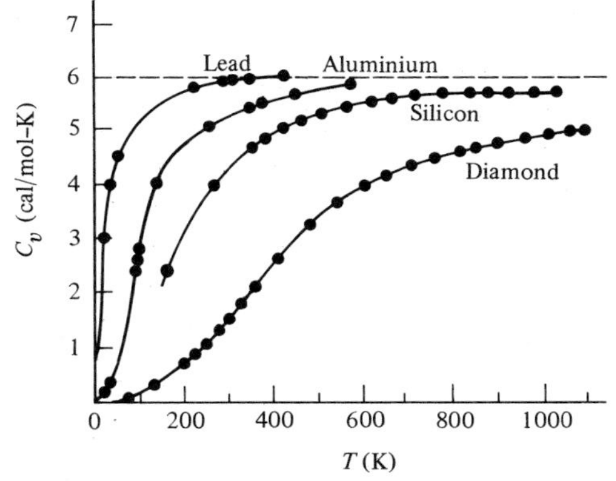

Figure 14.7. *Theoretical (continuous curves) and experimental values (dots) of the specific heat of solids as a function of temperature.*

between 0 K and room temperature can be explained only with the help of quantum mechanics and statistical mechanics

Let us consider the classical interpretation of C_v for solids. Suppose that the solid has N atoms or molecules which are coupled by elastic forces and, because their positions are fixed, cannot have translation or rotational motion. They can only vibrate about their equilibrium position. Since the restoring force is directly proportional to the displacement, the atoms execute simple harmonic motion. Thus, an atom vibrating along a line, say along the X-axis, will have kinetic energy $\frac{1}{2}mv_x^2$ and potential energy $\frac{1}{2}k_x x^2$. This is equivalent to saying that the atom has two degrees of freedom if vibrating only along the X-axis. Since the atoms are allowed to vibrate in any of the three independent directions, they will have a total of six degrees of freedom with energies $\frac{1}{2}mv_x^2$, $\frac{1}{2}k_x x^2$, $\frac{1}{2}mv_y^2$, $\frac{1}{2}k_y y^2$, $\frac{1}{2}mv_z^2$, and $\frac{1}{2}k_z z^2$.

According to the principle of equipartition of energy, the mean energy corresponding to each degree of freedom is $\frac{1}{2}kT$. Since the atoms in the solid have six degrees of freedom, each atom has a mean energy of $6(\frac{1}{2}kT) = 3kT$. The total energy U associated with N atoms is

$$U = 3NkT \tag{14.33}$$

and if the temperature changes to $T + \Delta T$, the change in U is $\Delta U = 3Nk\,\Delta T$. Thus, the specific heat at constant volume C_v is

$$C_v = \frac{1}{n}\frac{\Delta Q}{\Delta T} = \frac{1}{n}\frac{\Delta U}{\Delta T} = \frac{1}{n}3Nk\frac{\Delta T}{\Delta T} = 3\frac{N}{n}k$$

But $Nk = nR$; hence,

$$\boxed{C_v = 3R \simeq 6.0 \text{ cal/mol-K}} \tag{14.34}$$

which is the result found experimentally by Dulong and Petit. Thus, classical theory satisfactorily explains specific heat of solids at high temperatures.

The variation of C_v versus T at low temperature can be explained with the help of quantum mechanics and statistical mechanics. Such a theory of specific heats was developed by Einstein in 1907 and further improved by Debye in 1912. We shall not go into the details of this theory, but Figure 14.7 shows the agreement between theory and experiment.

14.8. Phase Change and Latent Heat

Depending on temperature and pressure, all matter can exist in a solid, liquid, or gaseous state. These three states or forms of matter are also called *phases* of matter. In the solid phase the molecules are rigidly bound together; they vibrate only about their equilibrium positions and cannot wander. In the liquid phase the molecules are loosely bound and can move around within the liquid. In the gaseous or vapor state, these molecules are not bound together at all. It is possible to change the energy of these molecules by supplying heat, which in turn can result in a change of phase. Let us illustrate this point by considering the three phases of H_2O molecules—ice, water, and steam.

To start with, let us have a cylinder half-filled with ice at $-20°$C and a thermometer dipping in it to register the temperature, the pressure being equal

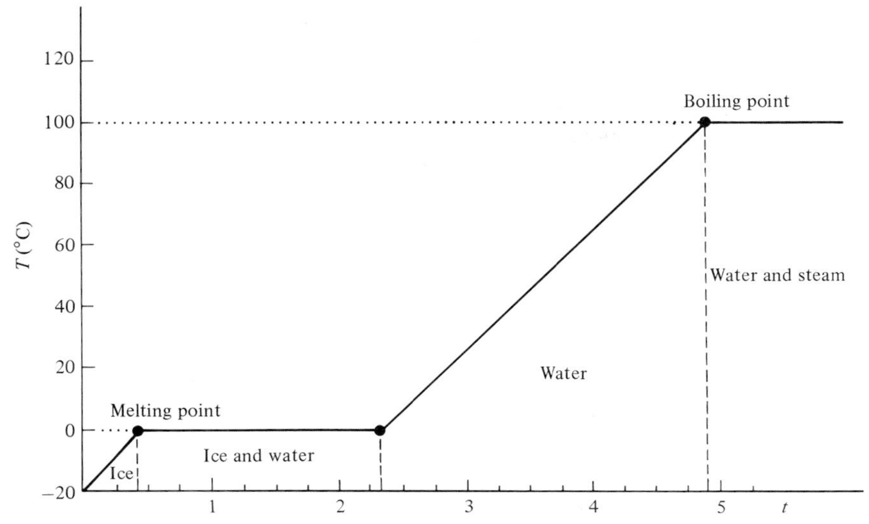

FIGURE 14.8. *Change in water phases as heat is supplied continuously to a sample.*

to atmospheric pressure. Heat the bottom of the cylinder and as time passes, the temperature increases, as indicated by the thermometer. Figure 14.8 shows the plot of the temperature T versus time t. Once the temperature reaches 0°C, the ice starts melting. Even though heat energy is being supplied, the temperature does not change until all the ice has melted and changed into water. This is indicated by the first horizontal line. Thus, a phase change has taken place from a solid to liquid without a change of temperature. The heat supplied is used to increase the energy of the molecules so that they can break lose from their bonds in the solid structure and change into a liquid.

Once all the ice has melted, the temperature once again starts rising, as shown in Figure 14.8. When the water reaches a temperature of 100°C, it starts changing into steam. The temperature does not change, as indicated by the horizontal portion, until all the water has changed into steam. The heat is again being used to break lose the molecules of water into free molecules in the gaseous or steam phase. Once all the water has changed into steam, the temperature starts increasing again. We conclude that the heat supplied can be used to change the phase or to increase the temperature. The reverse of this process—changing steam into water and water into ice—takes place when heat is taken out.

The temperature at which a solid changes into a liquid is called its *melting point;* the temperature at which a liquid changes into a solid is called its *freezing point.* For most crystalline substances the melting and freezing points are the same. As noted above, the meling or freezing point of water is 0°C. As we shall see, the melting point depends upon the pressure. Different materials need different amounts of heat to change phases. The *heat of fusion* L_f of a solid is the amount of heat needed to convert a unit mass of a solid into liquid at the same temperature and pressure. Thus, 1 kg of ice at 0°C and 1 atm pressure needs 80 kcal to change into water. Thus, the heat of fusion of water is 80 kcal/kg. Some melting points and heat of fusions are given in Table 14.3.

The molecules of a liquid have different instantaneous velocities. Some molecules near the surface will have enough energy to escape. The escape of the molecules from the surface is called *evaporation*. If the volume above the liquid surface is enclosed, some of the molecules will return to the liquid, and the process is called *condensation*. The number of molecules escaping and returning keeps on increasing. Eventually, a situation is reached when as many molecules leave the surface as return to it. The space above the liquid is said to have *saturated vapor* and the pressure exerted is called the *saturated vapor pressure*. This pressure depends upon the temperature of the liquid.

As the temperature increases, the rate of evaporation from the surface of the liquid also increases. A point is reached when bubbles of vapor start forming in the volume of liquid and rise to the surface and escape. The temperature at which the bubbles start forming in the liquid is called the *boiling point* of the liquid. Of course, the boiling point is different for different liquids, and for a given liquid it changes with pressure. As the pressure on the liquid is increased, the boiling point also increases, and if the pressure is decreased, the boiling point is decreased. Once boiling starts, the heat is used in changing from the liquid to the vapor phase. The *heat of vaporization* L_v of a liquid is defined as the amount of heat needed to change a unit mass of the liquid to the vapor phase while keeping the temperature and the pressure constant. The heat of vaporization decreases with increasing temperature. The heat of vaporization of water at 100°C and 1 atm pressure is 539 kcal/kg. Some values for different substances are listed in Table 14.3. The process in which the molecules escape directly from a solid into the gas phase without going through the liquid phase is called *sublimation*. As in the case of the heat of fusion and heat of vaporization, we may define *heat of sublimation* as the amount of heat needed to change a unit mass of solid into the vapor phase without a change in temperature.

TABLE 14.3
Latent Heats of Fusion and Vaporization

Substance	Melting Point (°C)	Heat of Fusion, L_f (cal/g)	Boiling Point (°C)	Heat of Vaporization, L_v (cal/g)
Aluminum	660	90	2450	2720
Ammonia	−75	108	−33	327
Copper	1083	32	2567	1211
Gold	1063	15.4	2660	377
Helium	−269.65	1.25	−268.93	5
Hydrogen	−259.65	14	−252.89	108
Lead	327	5.9	1750	206
Mercury	−39	2.8	357	70
Nitrogen	−209.97	6.09	−195.81	48
Oxygen	−218.79	3.30	−182.97	51
Silver	961	21.1	2193	558
Sulfur	119	9.1	444.6	78
Tungsten	3410	44	5900	1150
Uranium	1133	20	3900	454
Water	0	80	100	539

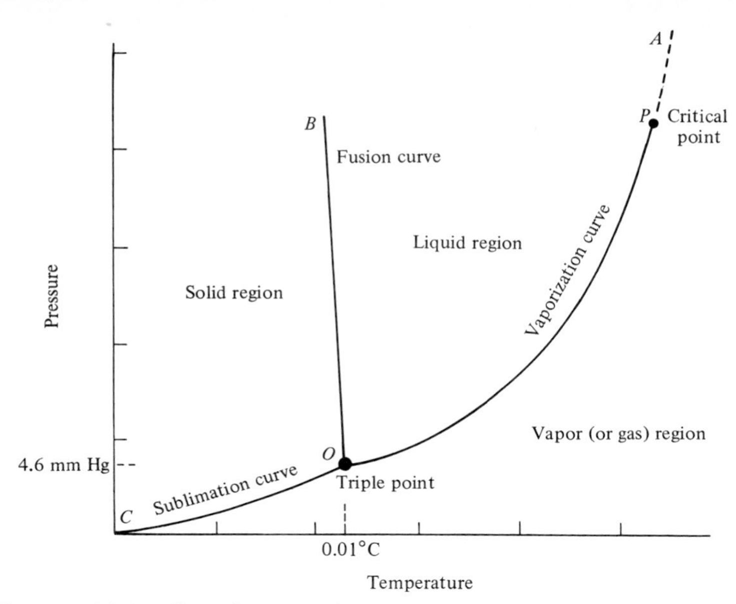

FIGURE 14.9. *Phase diagram of H_2O molecules, indicating various equilibrium states.*

Let us now put all the preceding thoughts together into one picture, called the *phase diagram,* as shown in Figure 14.9. Consider a mixture of ice and water contained in a cylinder. Ordinarily, the ice will melt, the liquid will evaporate, and some of the vapor will condense. We say that the substance is not in equilibrium. On the other hand, if we completely remove the vapor from the cylinder, there are certain values of pressures and temperatures for which ice (solid) and water (liquid) will coexist in equilibrium; that is, they will not change phase. Hence, one phase does not gain mass over the other. Such values of temperature and pressure for ice and water are shown by a line *OB* in Figure 14.9, called the *fusion curve.* Similarly, water and vapor can exist in equilibrium for the values of temperature and pressure indicated by the line *OA*, called the *vaporization curve.* Ice and vapor can exist in equilibrium for different values of temperature and pressure along the line *OC*, which is called the *sublimation curve.* If all three curves are extended, they intersect at a point. This point is called the *triple point,* because for the temperature and pressure corresponding to this point, all three phases (ice, water, and vapor) of H_2O can coexist in equilibrium as shown. For H_2O the triple point corresponds to a temperature of 0.0100°C (273.16 K) and a pressure of 4.6 mm Hg. Since the triple point of H_2O is easily reproduced, it is used as a standard temperature in thermometry, as discussed in Chapter 13. The whole diagram is called the phase diagram of H_2O. Note that the fusion or melting curve *OB* leans slightly toward the left. This implies that as the pressure is increased, the melting point is lowered. Similar interpretations apply to the other curves. Also, different materials have different phase diagrams and contain much information. In Figure 14.9, if we draw a horizontal line, it will indicate that at a given pressure, a substance can exist in different phases, depending upon the temperature. On the other hand, a vertical line will indicate that at a given temperature, a substance can exist in different phases, depending upon the pressure.

It is important to note that the vaporization curve does not extend indefinitely. As the temperature increases, the distinction between the liquid and vapor phases becomes smaller and smaller. Finally, for a specific volume (volume of a unit mass) at a certain temperature and pressure, the distinction between the vapor and the liquid phases disappears. This point, P in Figure 14.9, corresponds to the critical point of H_2O. Different substances have different critical points.

EXAMPLE 14.5 How much heat is needed to change 10 g of ice at −10°C to 10 g of steam at 110°C? (c_{ice} = 0.50 cal/g-C°, c_{steam} = 0.48 cal/g-C°.)

The heat needed to change ice from −10°C to 0°C is

$$\Delta Q_1 = mc_{ice}\,\Delta T = (10\text{ g})(0.50\text{ cal/g-C}^\circ)(10\text{C}^\circ) = 50\text{ cal}$$

The heat needed to melt ice at 0°C is

$$\Delta Q_2 = mL_f = (10\text{ g})(80\text{ cal/g}) = 800\text{ cal}$$

The heat needed to change water from 0°C to 100°C is

$$\Delta Q_3 = mc_{water}\,\Delta T = (10\text{ g})(1.0\text{ cal/g-C}^\circ)(100\text{C}^\circ) = 1000\text{ cal}$$

The heat needed to change water to steam at 100°C is

$$\Delta Q_4 = mL_v = (10\text{ g})(539\text{ cal/g}) = 5390\text{ cal}$$

The heat needed to change steam from 100°C to 110°C is

$$\Delta Q_5 = mc_{steam}\quad \Delta T = (10\text{ g})(0.48\text{ cal/g-C}^\circ)(10\text{C}^\circ) = 48\text{ cal}$$

Therefore, the total heat needed is

$$\begin{aligned} Q &= \Delta Q_1 + \Delta Q_2 + \Delta Q_3 + \Delta Q_4 + \Delta Q_5 \\ &= (50 + 800 + 1000 + 5390 + 48)\text{ cal} = 7288\text{ cal} \end{aligned}$$

EXERCISE 14.5 A quantity of 8000 cal is used to change certain amount of ice at −10°C to steam at 110°C. Calculate the amount of ice. [*Ans.:* 11g.]

EXAMPLE 14.6 How much ice at 0°C should be added to 100 g of water at 20°C so that its final temperature will be 5°C?

Let m_1 grams of ice be needed to raise the temperature to 5°C. The heat lost by ice will be gained by water.

$$\begin{aligned} \text{heat lost by ice} &= m_1L_f + m_1c_1\,\Delta T_1 \\ &= m_1(80\text{ cal/g}) + m_1(1\text{ cal/g-C}^\circ)(5 - 0)\text{C}^\circ = 85\,m_1\text{ cal/g} \end{aligned}$$

$$\begin{aligned} \text{heat gained by water} &= m_2c_2\,\Delta T_2 \\ &= (100\text{ g})(1\text{ cal/g-C}^\circ)(20 - 5)\text{C}^\circ = 1500\text{ cal} \end{aligned}$$

Since

$$\text{heat lost} = \text{heat gained}$$
$$85\, m_1 \text{ cal/g} = 1500 \text{ cal}$$
$$m_1 = \frac{1500}{85} \text{ g} = 17.6 \text{ g}$$

Exercise 14.6 How much steam at 110°C is needed to bring 100 g of water at 20°C to 100°C? $c_{steam} = 0.48$ cal/g-C°. [*Ans.:* 14.7 g.]

SUMMARY

A *calorie* is defined as the amount of heat needed to raise the temperature of 1 g of water through 1 Celsius degree or 1 Kelvin degree. The British unit of heat, 1 Btu = 252 cal.

The *specific heat* c of a material is the amount of heat that must be added to a unit mass to raise its temperature through 1 degree; that is, $\Delta Q = mc\,\Delta T$. The *molar specific heat* C is the amount of heat needed to raise the temperature of 1 mol of the material by 1 C°; that is, $C = Mc$. The *heat capacity* C' of a substance of mass m is $C' = mc$. The principle of heat balance requires that heat lost = heat gained.

The *mechanical equivalent of heat* $\mathcal{J}$ is given by $\mathcal{J} = W/Q = 4186$ J/kcal. According to the *first law of thermodynamics,* the change in internal energy of a system is equal to the heat absorbed by the system plus the external work done on the system; $\Delta U = \Delta Q + \Delta W_{ext}$. If the work is done by the system, $\Delta U = \Delta Q - \Delta W$. If the gas expands at constant pressure, $\Delta W = P\,\Delta V$.

According to the *principle of equipartition* of energy, the energy associated with each degree of freedom is $\frac{1}{2}kT$. Thus, $\bar{E} = (f/2)kT$ and $\bar{E}_T = (f/2)nRT$.

For gases two specific heats are commonly used—specific heat at constant volume C_v, and specific heat at constant pressure C_p. Also, C_p is always greater than C_v. In general $C_v = (f/2)\,R$, $C_p = [(f/2) + 1]R$, and $\gamma = (f + 2)/f$. For monatomic gases $f = 3$, $C_v = 1.5\,R$, $C_p = 2.5\,R$, and $\gamma = 1.66$. For diatomic gases $f = 7$.

According to the *Dulong and Petit law,* the molar specific heat at constant volume C_v at high temperature is the same for all solids and is equal to $\sim 3\,R$ (or 6 cal/mol-K). This can be explained by assigning six degrees of freedom to the atoms in the solids.

The *heat of fusion* is the heat needed for a unit mass to change from a solid to a liquid state at a constant temperature while the *heat of vaporization* is the heat needed to change a unit mass from a liquid to a gaseous state at a constant temperature.

QUESTIONS

1. If water had its maximum density at 0°C, what consequences would it have in nature?
2. What is the advantage of using molar specific heat instead of heat capacity?
3. When a mass m falls from a height h on a block of ice, a part of the ice melts. Why?

4. Why must more work be done when a gas is heated at constant pressure than at constant volume?
5. What is the correspondence between the principle of heat balance and the law of conservation of energy?
6. Can you suggest some experimental arrangement (other than the one in the text) for measuring $\mathcal{J}$?
7. Explain, in words, why the molar specific heat of a diatomic gas is usually larger than that of a monatomic gas.
8. What are the maximum possible degrees of freedom for (a) a monatomic gas; and (b) a diatomic gas?
9. Which has more heat energy—water at 100°C or steam at 100°C? Which will produce more damage as a burn when it comes in contact with bare skin?
10. Why is the heat of vaporization for any substance generally larger than the heat of fusion?
11. In general, why is there an increase in volume when (a) a solid changes into a liquid; and (b) a liquid changes into a gas?
12. Can you explain why ice often melts when a heavy object is placed on it?

PROBLEMS

1. Express the specific heat and molar specific heat of water in kcal/kg-C°, cal/mol-C°, kcal/kmol-C°, J/kg-C°, and J/kmol-C°.
2. Express the specific heat and molar specific heat of copper in kcal/kg-C°, cal/mol-C°, kcal/kmol-C°, J/kg-C°, and J/kmol-C°.
3. How much heat is needed to heat 2 kg of copper from 50°C to 150°C? Express your answer in calories and joules.
4. How much energy is given out when (a) 50 g of water; and (b) 2 liters of water are cooled from 80°C to 20°C? Express your answer in calories, kcal, and joules.
5. How many calories are needed to heat 4 kg of iron from 100°C to 375°C? Express your answer in Btu and joules.
6. A 100-watt heater is used to heat 100 g of water from 20°C to 80°C. How long will it take to do this, assuming no loss of heat to the surroundings or to the container?
7. An electric stove supplies heat at the rate of 50 cal/s. How long will it take to raise the temperature of 500 g of water, placed in an aluminum pan of mass 500 g, from 20°C to 95°C?
8. The waste from a nuclear power plant in the form of heat energy is 8×10^5 kcal/s. How much energy in joules/day is being wasted? This heat appears as a temperature increase of the discharging water. How much water will be needed so that the increase in water temperature does not exceed 5C°?
9. The heat of combustion of natural gas is 13,000 kcal/m^3; that is, 1 m^3 of gas when completely burned produces 13,000 kcal of heat. What is the heat of combustion in units of Btu/ft^3? How much gas must be burned to heat 200 liters ($\approx$53 gal) of water from 10°C to 80°C. Assume that 50 percent of the heat is wasted.
10. The water equivalent (or heat capacity) of a substance is defined as the

product of its mass and specific heat, that is, in heat exchanges it behaves like that amount of water. Calculate the water equivalent of (a) a copper calorimeter of mass 200 g; and (b) a glass container of mass 500 g.

11. A sample of 0.2 kg of an unknown substance is heated to a temperature of 100°C and then dropped into a copper calorimeter of mass 0.10 kg containing 0.50 kg of water at 20°C. After stirring well, the equilibrium temperature reached is 42°C. What is the specific heat of the substance?

12. If 300 g of lead shot heated to 90°C is dropped into an aluminum can containing 200 g of water at 10°C, the final temperature of the water, lead, and can is 12°C. What is the heat capacity and mass of the aluminum can?

13. The water equivalent of a copper calorimeter is 10 g; that is, in heat exchange it behaves like 10 g of water. It contains 100 g of a liquid at 20°C. When a 50-g piece of copper at 80°C is added to this, the equilibrium temperature after mixing is 28°C. What is the specific heat of the liquid?

14. Water splashes down from a height of 80 m. Calculate the rise in temperature if all the potential energy is used up in heating the water.

15. Water at a temperature of 20°C falls through a height of 500 ft. What is its final temperature if all the heat is used in heating the water?

16. An automobile of 2000 kg is traveling at a speed of 80 km/h. Calculate the amount of heat developed if the automobile comes to rest, assuming that 50 percent of the mechanical energy is converted into heat energy.

17. A lead bullet traveling at 400 m/s is brought to rest in a wooden target. Assuming that half of the energy is lost to the target, what will be the rise in temperature of the bullet if its mass is 20 g and specific heat is 0.03 kcal/kg-C°?

18. A lead bullet of mass 20 g traveling at 500 m/s is brought to rest in a 10-kg object that is at a temperature of 30°C. Half of the kinetic energy of the bullet is used in raising the temperature of the object, which has a specific heat of 1 kcal/kg-C°. Calculate the increase in temperature.

19. When a torque of 0.04 N-m is applied to the paddles of a cake mixer, it rotates at 400 rev/min. The rotational kinetic energy so generated is used in heating the cake mix. Calculate (a) the rate at which heat is generated, and (b) the rate of increase of the temperature of the cake mix, which has a mass of 0.8 kg and a specific heat of 0.8 cal/g-C°.

20. One kilogram of water is heated by an electric stove. It takes 10 min for the water at 10°C to reach 90°C. Calculate the rate at which the internal energy (in J/s) increases. Assume no increase in volume.

21. A system absorbs 8×10^4 calories of heat and 6×10^3 joules of work is done on the system. Calculate the increase in internal energy.

22. Suppose that a system absorbs 8 kcal and increases its internal energy by 5 kcal. How much work (in joules) is done on or by the system?

23. Suppose that 2 liters of oxygen at standard temperature and pressure are expanded at constant pressure to double its volume. If the temperature also becomes double, calculate (a) the work done; and (b) the change in internal energy of the system.

24. Use the Dulong and Petit law to calculate the specific heat of copper, silver, gold, and uranium.

25. Calculate the amount of heat needed to raise the temperature of 5 mol of oxygen by 20 K at constant volume.

26. Consider 2 mol of nitrogen at 0°C. Calculate the amount of heat that must be added to (a) double the volume at constant pressure; and (b) double the pressure at constant volume.

27. For a gas $C_p - C_v = 2.14$ kcal/kmol-K and $\gamma = 1.29$. Calculate the values of C_p and C_v. What type of structure has the gas?

28. The molecular weight of chlorine is 71. The specific heat at constant pressure is 0.117 cal/g-K. (a) Calculate C_v. (b) What is the value of γ? What can you say about the structure of this gas?

29. How much heat is needed to melt 1 kg of mercury at −39°C into liquid at 20°C? (The freezing point of mercury is −39°C, the specific heat of liquid mercury is 0.033 cal/g-C°, and heat of fusion is 2.82 cal/g.)

30. Calculate the amount of kcal needed to heat 10 g of mercury from 15°C to 357°C and then to vaporize it at that temperature. (The heat of vaporization of mercury is 65 cal/g and the boiling point is 357°C.)

31. A piece of iron of mass 100 g and at a temperature of 320°C falls on a large block of ice. How much ice will melt?

32. A copper container of mass 100 g contains 200 g of water at 20°C. How much steam at 100°C should be added so that the final temperature will be 30°C?

33. A mass of 100 kg falls through a height of 10 m on a big slab of ice. If all the available energy is used up in melting the ice, calculate the amount of ice melted.

34. In a steam radiator the steam enters at 100°C and leaves as water at 75°C. If on the average 8×10^4 kcal/h is needed to heat a house, what is the total energy, in kcal and joules, needed for 1 month?

35. In a condenser of a power plant, steam enters at 100°C and leaves as water at 60°C. The cooling water enters at 10°C and leaves at 50°C. How much cooling water is being used for each kilogram of steam?

36. An ice skater of mass 50 kg is going in a circle of radius 4 m with a speed of 4 m/s. If the rotational kinetic energy is completely used in melting ice, calculate the rate at which the ice will melt.

37. A quantity totaling 539.5 cal must be supplied to convert 1 g of water at 100°C into steam at 100°C while keeping the pressure at 1 atm. The volume of 1 g of water is 1.00 cm^3, while that of steam is 1671 cm^3. Calculate the heat energy used in (a) doing external work; and (b) increasing the internal energy.

38. Calculate the work done and the change in internal energy when 1 g of water at 0°C is changed into ice at 0°C while the pressure is kept at 1 atm. Note that the latent heat of fusion is 80 cal, the volume of 1 g of water is 1 cm^3, and the volume of ice is 1.087 cm^3.

15

Thermal Properties and Energy Quantization

Thermal expansion plays an important role in everyday life and will be discussed at some length. Conduction, convection, and radiation are the three modes by which heat can be transferred from one point to another and should be clearly understood. The study of radiation has played an important role in physics, and to appreciate this we shall briefly discuss Planck's hypothesis, which led to the quantization of energy—the basis of quantum mechanics.

15.1. Thermal Expansion of Solids and Liquids

EXCEPT FOR a few cases all solids and liquids increase in size as their temperature increases. Solids can change in length, area, or volume while liquids change their volumes. It is experimentally observed that if a solid rod [Figure 15.1(a)] is heated, its increase in length ΔL is proportional to its original length L_0 and change in temperature ΔT (where ΔT is in C° or K):

$$\Delta L \propto L_0 \, \Delta T$$

$$\Delta L = \alpha L_0 \, \Delta T \tag{15.1}$$

or

$$\boxed{\alpha = \frac{\Delta L}{L_0 \, \Delta T}} \tag{15.2}$$

where α is called the *coefficient of linear expansion* and is defined as the increase in length per unit length per degree increase in temperature. The change in α with temperature is negligible; hence, α may be taken as constant in most cases. Since $\Delta L = L - L_0$, where L is the length after heating the rod, Equation 15.1 may be written

$$\boxed{L = L_0(1 + \alpha \, \Delta T)} \tag{15.3}$$

If a solid plate of area A_0 and of negligible thickness is heated, the change in the area is given by

$$\Delta A = \beta A_0 \, \Delta T \tag{15.4}$$

or

$$\boxed{\beta = \frac{\Delta A}{A_0 \, \Delta T}} \tag{15.5}$$

where β is called the *coefficient of area expansion* and is defined as the increase

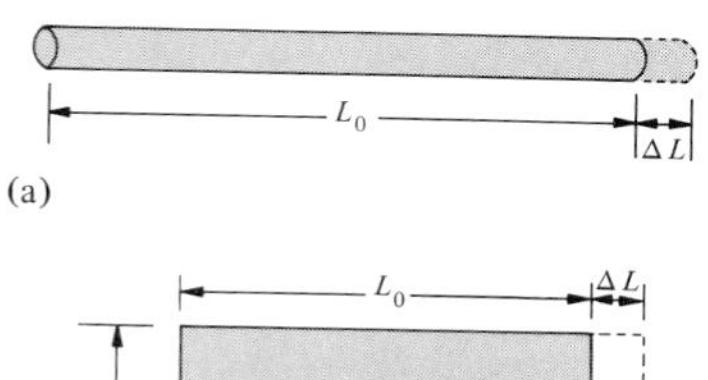

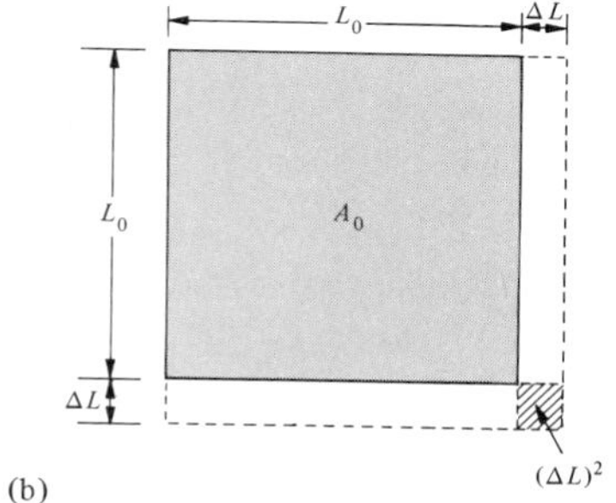

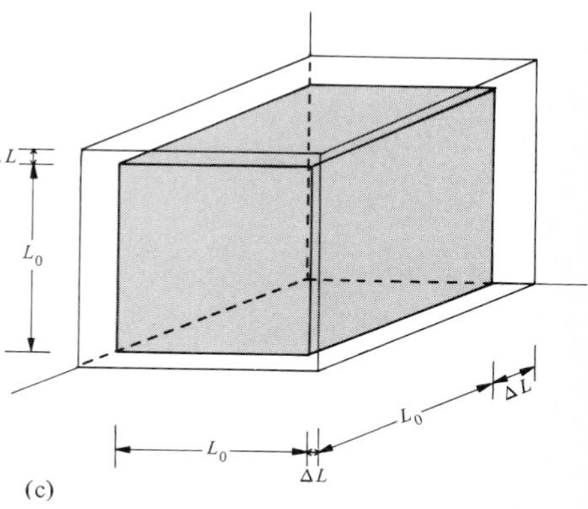

FIGURE 15.1. *Thermal expansion of solids: (a) linear, (b) surface, and (c) volume.*

in area per unit area per degree increase in temperature. Once again, since $\Delta A = A - A_0$, where A is the area after heating, we may write Equation 15.4 as

$$\boxed{A = A_0(1 + \beta\,\Delta T)} \tag{15.6}$$

Similarly, if a solid or liquid of volume V_0 is heated to a new volume V, the change in volume ΔV is given by

$$\Delta V = \gamma V_0\,\Delta T \tag{15.7}$$

or

$$\boxed{\gamma = \frac{\Delta V}{V_0\,\Delta T}} \tag{15.8}$$

where γ is called the *coefficient of volume expansion* and is defined as the increase in volume per unit volume per degree increase in temperature. Again, we may write Equation 15.7 as

$$\boxed{V = V_0(1 + \gamma\,\Delta T)} \tag{15.9}$$

The units of all three coefficients α, β, and γ are reciprocal temperature degrees; that is, $1/\text{C}°$ or $1/\text{K}$ (because the sizes of C° and K are the same). The values of these coefficients are listed in Table 15.1. It may be pointed out that

TABLE 15.1
A. Expansion Coefficients for Solids ($\alpha = \beta/2 = \gamma/3$)

Material	α (per C°)[a]
Aluminum	2.6×10^{-5}
Brass	2.0×10^{-5}
Copper	1.7×10^{-5}
Glass	$0.4 - 0.57 \times 10^{-5}$
Invar (64% Fe + 36% Ni)	0.1×10^{-5}
Platinum	0.9×10^{-5}
Silver	1.93×10^{-5}
Steel	1.2×10^{-5}
Tungsten	0.43×10^{-5}

B. Expansion Coefficients for Liquids

Material	γ (per C°)
Benzene	124×10^{-5}
Ethyl alcohol	75×10^{-5}
Gasoline	95×10^{-5}
Glycerin	49×10^{-5}
Mercury	18.2×10^{-5}

[a] For most materials these values hold between 0 and 100°C.

since the expansion coefficients are extremely small, it is not necessary in the case of solids and liquids to choose V_0 corresponding to any one particular temperature. L_0, A_0, and V_0 refer simply to initial values at temperature T_0 and L, A, and V refer to the final values at temperature T, where $\Delta T = T - T_0$.

The coefficients α, β, and γ are interrelated by the following equation:

$$\boxed{\alpha \simeq \frac{\beta}{2} \simeq \frac{\gamma}{3}} \qquad (15.10)$$

To prove this relation, refer to Figure 15.1. The thermal expansion of A_0 to A is equivalent to both sides changing from L_0 to $L_0(1 + \alpha\,\Delta T)$, while thermal expansion of V_0 to V is equivalent to all three sides, changing from L_0 to $L_0(1 + \alpha\,\Delta T)$. From Figure 15.1(b),

$$A = L^2 = L_0^2(1 + \alpha\,\Delta T)^2 = L_0^2(1 + 2\alpha\,\Delta T + \alpha^2\,\Delta T^2)$$

Since α is a very small quantity, $\alpha^2\,\Delta T^2$ is usually negligible while L_0^2 is equal to A_0. The quantity $\alpha^2\,\Delta T^2$ is the small shaded corner at the right side of Figure 15.1(b). Thus, the equation above may be written as

$$A \simeq A_0(1 + 2\alpha\,\Delta T)$$

Comparing with Equation 15.6 yields

$$\beta \simeq 2\alpha \qquad \text{or} \qquad \alpha \simeq \frac{\beta}{2}$$

Similarly, referring to Figure 15.1(c), we can write

$$V = L^3 = L_0^3(1 + \alpha\,\Delta T)^3 = L_0^3(1 + 3\alpha\,\Delta T + 3\alpha^2\,\Delta T^2 + \alpha^3\,\Delta T^3)$$

Neglecting the last two terms, which are small, and remembering that $L_0^3 = V_0$, we obtain

$$V \simeq V_0(1 + 3\alpha\,\Delta T)$$

Comparison with Equation 15.9 results in

$$\gamma = 3\alpha \qquad \text{or} \qquad \alpha = \frac{\gamma}{3}$$

as stated in Equation 15.10.

The thermal expansion of the material can be understood in terms of the interaction potential between molecules. As discussed in Chapter 10, the interaction $V(r)$ versus r between two molecules varies, as shown in Figure 15.2. At the equilibrium position r_0, the potential between the two molecules is V_0. As the molecules gain kinetic energy, they start vibrating with an amplitude $(r - r_0)$. Since the restoring force is directly proportional to the displacement, the molecules execute simple harmonic motion. Of course, this is true only for small kinetic energy, say K_1, as shown, where the potential is still symmetric. In this case the average separation is r_0, and the time average of the displacement $(r - r_0)$ is also zero.

As the energy of the molecules increases, say to K_2, the molecular system still oscillates back and forth, say between A and B, but the average displace-

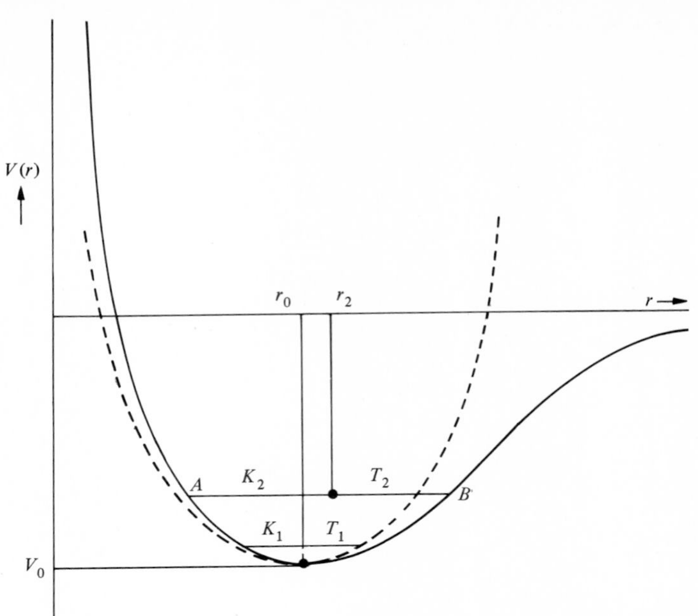

Figure 15.2. *Explanation of thermal expansion in terms of interaction potential* V(r) *versus* r *between two molecules.*

ment is r_2, which is greater than the equilibrium separation r_0. This increase in average displacement from r_0 to r_2 is due to the asymmetry in the interaction potential. As the kinetic energy increases, the average displacement also increases, as shown in Figure 15.2. Since temperature is a direct measure of the kinetic energy, we may state that *the average displacement increases with increasing temperature.* If the potential curve were symmetric about the equilibrium (dotted curve), any increase in temperature, and hence kinetic energy, would have increased the amplitude of the oscillations, but their equilibrium position would have remained the same. This means there would have been no expansion or contraction.

There are many situations in which the expansion of material with a change in temperature must be taken into account. Suppose that the two ends of a rod are fixed so that it cannot expand or contract. If the temperature of the rod is changed, compressive or tensile stresses, called *thermal stress,* will be set up in the rod. If the temperature change is very large, the stress may be large enough to take the rod beyond its elastic limits or breaking strength. This is why in setting railroad tracks enough space is left between the sections so that the tracks will not buckle as a result of the large thermal stresses. Actually, in any large structure subject to large temperature changes, care must be taken to avoid thermal stresses.

Thermal stress, which is equal to the force per unit area, may be calculated in a simple manner. If the two ends of a rod of length L_0 are fixed, when its temperature changes by ΔT, we may write

$$\frac{\Delta L}{L_0} = \alpha \, \Delta T$$

But from the definition of Young's modulus, Equation 10.48,

$$Y = \frac{F/A}{\Delta L/L_0}$$

Substituting for $\Delta L/L_0$ from the equation above and rearranging, we get

$$\text{thermal stress} = \frac{F}{A} = Y\alpha\,\Delta T \tag{15.11}$$

The property that different materials have different coefficient of expansion is utilized in many situations. One example is that of a thermostat that uses a bimetallic strip. A *bimetallic strip* is a device made from two thin strips of different metals (usually steel and brass) that are bonded together, say at temperature T_0, as shown in Figure 15.3(a). If the combination is heated such that $T > T_0$, the two strips expand by different amounts; hence, they must bend as shown in Figure 15.3(b). If $T < T_0$, they will bend the other way, as shown in Figure 15.3(c).

Finally, we should look into some examples in which the material contracts with an increase in temperature. A typical example is water between the temperature range of 0 to 4°C. As the temperature increases from 0°C the volume of water decreases, reaching a minimum volume at 4°C. This is the *anomalous expansion,* where water has a negative coefficient of expansion. Thus, water will have a maximum density at 4°C, while it has less density at all other temperatures, as shown in Figure 15.4. This is the reason why the water in lakes freezes first at the surface. The ice formed remains afloat, because the water under it is at 4°C and is heavier. Ice, being a poor conductor of heat, keeps the water under it from further freezing. Thus, the aquatic life can survive through extreme winter.

Iron at about 1100 K has a slightly negative expansion coefficient. Material such as tellurium, silicon, selenium, and cobalt–iron–chromium alloy have negative temperature coefficients.

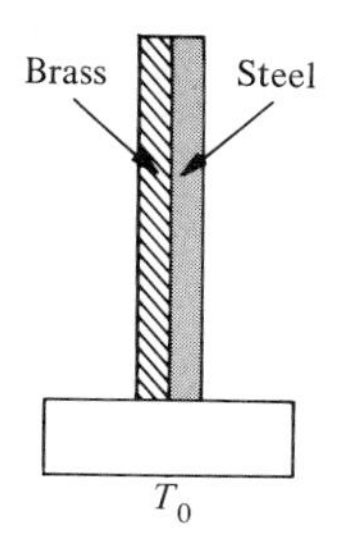

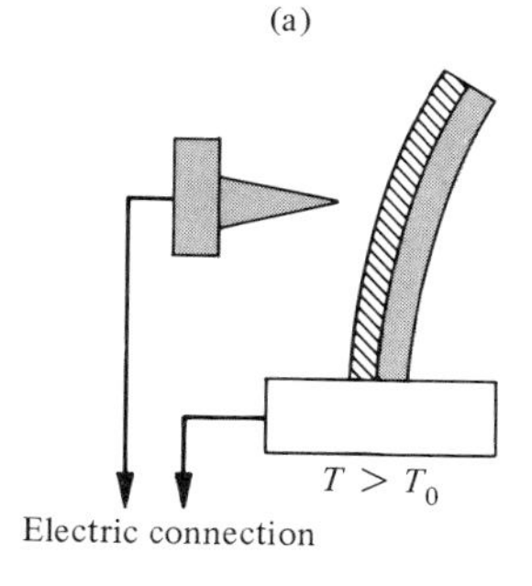

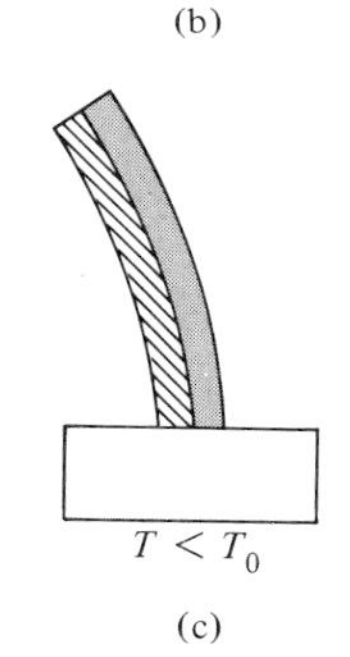

FIGURE 15.3. *Principle of a thermostat.*

EXAMPLE 15.1 A steel rod 10 m long and of cross-sectional area 10 cm² is heated from 0 to 40°C. (a) What will be the increase in length? (b) If the rod is fixed between two rigid supports, what thermal stress and corresponding force must be applied to avoid expansion?

From Equation 15.3,

$$L - L_0 = \Delta L = L_0\,\alpha\,\Delta T$$

$$\Delta L = (10\text{ m})(1.2 \times 10^{-5}/\text{C}^\circ)(40\text{C}^\circ) = 4.8 \times 10^{-3}\text{ m} = 0.0048\text{ m}$$

Therefore, $L = 10.0048$ m.

From Equation 15.11, thermal stress is given by

$$\frac{F}{A} = Y\alpha\,\Delta T = (2 \times 10^{11}\text{ N/m}^2)(1.2 \times 10^{-5}/\text{C}^\circ)(40\text{C}^\circ) = 9.6 \times 10^7\text{ N/m}^2$$

Since $A = 10\text{ cm}^2 = 10 \times 10^{-4}\text{ m}^2$,

$$F = (Y\alpha\,\Delta T)A = (9.6 \times 10^7\text{ N/m}^2)(10 \times 10^{-4}\text{ m}^2) = 9.6 \times 10^4\text{ N}$$

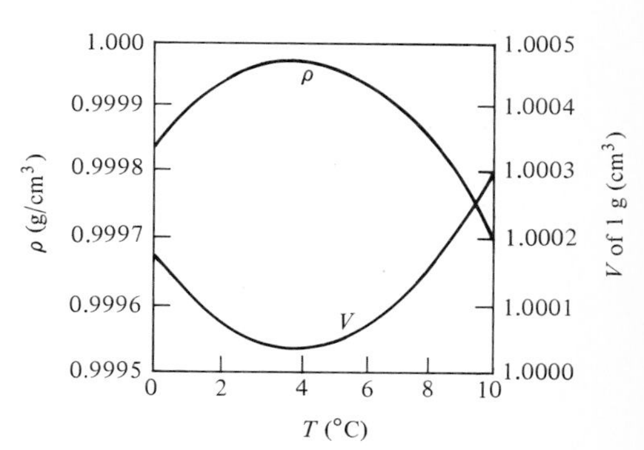

FIGURE 15.4. *Variation of density of water with temperature.*

Exercise 15.1 The main span of the Golden Gate Bridge is 1280 m (= 4200 ft) and is made of steel. If the temperature variations are between $-18°C$ ($\simeq 0°F$) and 60° (= 140°F), what allowance must be made for expansion? If both ends are fixed, what will be the thermal stress? [*Ans.*: $\Delta L = 1.20$ m, $F/A = 1.87 \times 10^8$ N/m^2.]

Example 15.2 A simple pendulum made out of an aluminum wire 1 m long has a period t_1 at temperature 0°C. Calculate the periods t_1 and t_2 at temperatures of 0°C and 40°C and fractional change in period.

The period of a simple pendulum from Equation 10.42 is

$$t_1 = 2\pi\sqrt{\frac{L_1}{g}} = 2\pi\sqrt{\frac{1\ m}{9.8\ \text{m/s}^2}} = 2.0071\ \text{s}$$

Suppose that at temperature of 40°C, the length of the pendulum becomes L_2 and $L_2 = L_1 + L_1\alpha\,\Delta T = 1\ \text{m} + (1\ \text{m})(26 \times 10^{-6}/\text{C°})(40\text{C°}) = 1.00104$ m. Therefore,

$$t_2 = 2\pi\sqrt{\frac{L_2}{g}} = 2\pi\sqrt{\frac{1.00104\ \text{m}}{9.8\ \text{m/s}^2}} = 2.0081\ \text{s}$$

$$\frac{\Delta t}{t_1} = \frac{t_2 - t_1}{t_1} = \frac{0.001\ \text{s}}{2.0071\ \text{s}} = 0.0005$$

Thus, the period increases by 0.001 s, which means that the clock runs slowly; for every 2.0071 s, it loses 0.001 s.

Exercise 15.2 A clock with a steel pendulum has a period of 1 s at 20°C. If the temperature falls to 0°C, (a) what will be the new period; and (b) will the clock run slow or fast and how much over a period of 1 day? [*Ans.*: (a) $t = 0.99988$ s; (b) runs fast by 10.37 s/day.]

Example 15.3 Suppose that one early morning when the temperature is 10°C (= 50°F), a driver of an automobile gets his gasoline tank, which is made of steel, filled with 75 ℓ (= 19.8 gallons, 1 gal = 3.786 ℓ) of gasoline, which is also at 10°C. During the day the temperature rises to 30°C (= 86°F). How much gasoline will overflow? $\alpha_{(\text{steel})} = 12 \times 10^{-6}/\text{C°}$ and $\gamma_{(\text{gasoline})} = 9.5 \times 10^{-4}/\text{C°}$.

Let $V = 75$ ℓ be the original volume of the gasoline and the steel tank. The increase in temperature $\Delta T = (30 - 10)\text{C°} = 20\text{C°}$ increases the volume of both, except that the gasoline expands much more than the steel; hence, it will overflow. The change in volume of the gasoline from Equation 15.8 is

$$\Delta V_g = \gamma\,\Delta T V = (9.5 \times 10^{-4}/\text{C°})(20\text{C°})V = 190 \times 10^{-4} V$$

The change in volume of the steel tank due to expansion is

$$\Delta V_s = \gamma\,\Delta T V = 3\alpha\,\Delta T V = 3(12 \times 10^{-6}/\text{C°})(20\text{C°})V = 7.2 \times 10^{-4} V$$

Thus,

$$\text{overflow} = \Delta V_g - \Delta V_s = (190 - 7.2) \times 10^{-4} V = (182.8 \times 10^{-4})(75\ \ell)$$
$$= 1.37\ \ell$$

That is, 1.37 ℓ will overflow, which is about 1.8 percent.

Exercise 15.3 If you fill the gasoline tank of your car with 20 gallons of gasoline when the temperature is $-10°C$, how much of it will overflow when the temperature reaches 20°C? Can we ignore the expansion of the steel tank? [*Ans.:* Overflow = 0.548 gal = 2.07 ℓ. Neglecting expansion of the tank, overflow = 0.570 gal = 2.16 ℓ.]

15.2. Heat Transfer

Whenever there is a temperature difference between two systems, there will be a transfer of thermal energy from the system at the higher temperature to the system at the lower temperature by one or more of the following three distinct processes.

Conduction

This process of heat transfer takes place in a material medium. The thermal energy is transferred from the molecules of higher translational kinetic energy to the neighboring molecules of lower translational kinetic energy. These, in turn, transfer the energy to the next neighbor, and so on. Thus, the energy is transferred without the actual mass motion of the medium. A typical example is heat transfer in metals, when one end initially is at a higher temperature.

Convection

In this process of heat transfer, the colder portion of the medium moves toward the hotter portion. After being heated it moves away and the colder portion moves in. Thus, the transfer is by the actual mass motion of the medium and takes place in liquids and gases. A typical example is the heating of a liquid by means of a flame or heater.

Radiation

This is the process in which heat transfer takes place by means of electromagnetic radiation. If any object is heated, the vibrating molecules will emit electromagnetic waves (Chapter 27). These waves, when received by a body at lower temperature, will increase the thermal energy of the molecules, hence raise the temperature. Note that no medium is necessary for this process; these waves can travel in a vacuum. A typical example is the heating of the earth by electromagnetic waves coming from the sun.

Of the three processes above, convection is often the slowest, and the rate of heat transfer depends on the motion of the medium, which is typically ~ 1 m/s. As compared to this, conduction takes place at a speed of $\sim 10^4$ m/s, while electromagnetic waves travel at the speed of 3×10^8 m/s—the speed of light. We shall discuss these processes in some detail.

15.3. Thermal Conduction

As pointed out earlier, the conduction method of heat transfer is one in which the molecules of the material transfer translational kinetic energy to

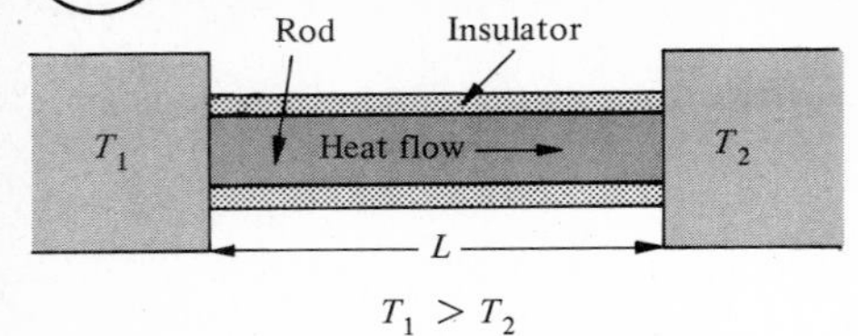

Figure 15.5. *Heat transfer by conduction through a rod of length L.*

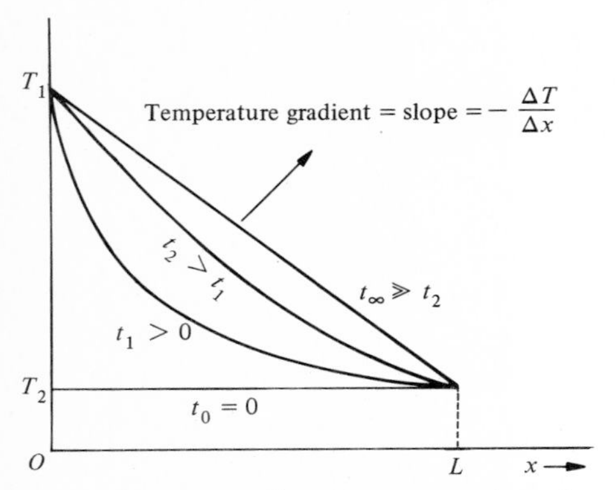

Figure 15.6. *T versus x of a rod after different time intervals.*

neighboring molecules which have lower kinetic energy. For example, as shown in Figure 15.5, the left end of a metal rod of length L is in contact with a hot body of temperature T_1 while the right-hand side is in contact with a cold body of temperature T_2. The molecules at the left end will have very high translational kinetic energy compared to the molecules at the right end. The molecules at the left pass on a portion of their kinetic energy to neighboring molecules; and so on. Thus, the heat energy is transmitted from left to right. As time passes, the temperature at different points along the rod increases, reaching a steady state after a long time, as shown in Figure 15.6. A *steady state* exists if there is no further change in temperature at different points along the length of the rod.

To calculate the amount of heat flow across any area, let us consider Figure 15.7. The amount of heat ΔQ flowing across any cross-sectional area is

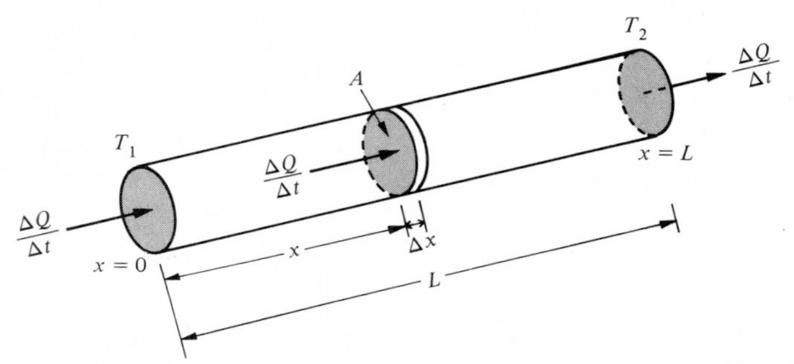

Figure 15.7. *Heat flow across any cross-sectional area of a rod.*

directly proportional to the cross-sectional area A of the rod, to the temperature different ΔT, to the time interval Δt, and inversely proportional to the thickness Δx. That is,

$$\Delta Q \propto \frac{A\,\Delta T}{\Delta x}\,\Delta t$$

or

$$\boxed{\frac{\Delta Q}{\Delta t} = -kA\,\frac{\Delta T}{\Delta x}} \tag{15.12}$$

where k is the constant of proportionality, called the *thermal conductivity* of the material, while $\Delta Q/\Delta t$ is the rate of heat flow across any area and is denoted by H:

$$\text{heat flow rate} = H = \frac{\Delta Q}{\Delta t} \tag{15.13}$$

The quantity $\Delta T/\Delta x$ is called the *temperature gradient* and is defined as the rate of change of temperature T with distance x along the rod.

$$\text{temperature gradient} = \frac{\Delta T}{\Delta x} \tag{15.14}$$

The negative sign in Equation 15.12 is taken for consistency. If we take the direction of heat flow $\Delta Q/\Delta t$ along the direction of increasing x, then for decreasing T, the temperature gradient will be negative, and in order for heat flow to be positive, we must add another negative sign. From Equation 15.12 the expression for k may be written as

$$k = -\frac{\Delta Q/\Delta t}{A\,\Delta T/\Delta x} \tag{15.15}$$

Thus, the units of thermal conductivity k will be cal/s-cm-C°, kcal/s-m-C°, or Btu-in/s-ft^2-F°. The thermal conductivity of different materials is listed in Table 15.2. Metals have very high values of thermal conductivity because, in

Table 15.2
Thermal Conductivities

Substance	k (cal/s-cm-C°)[a]
Air	0.000057
Aluminum	0.49
Brass	0.26
Brick	0.0015
Concrete	0.002
Copper	0.92
Felt, wool	0.0001
Glass	0.0025
Ice	0.004
Lead	0.083
Mercury	0.02
Silver	0.97
Steel	0.11
Wood	0.0001–0.0003

[a] To express k in (kcal/s-m-C°) multiply by 0.1.

addition to the ions participating in conduction, the free valence electrons also contribute to the process. This is not so in the case of thermal nonconductors, which have a very low thermal conductivity. Examples include wood, glass, and ice.

Let us go back to Equation 15.12 and Figure 15.7 and calculate the heat-flow rate in terms of known quantities: temperatures T_1, T_2, length L, area A, and conductivity k. Since in a steady state the same heat will flow across any cross-sectional area, the quantity $H = \Delta Q/\Delta t$ will be constant. Thus, substituting for $\Delta T = T_2 - T_1$ and $\Delta x = L$, Equation 15.12 takes the form

$$\boxed{H = \frac{\Delta Q}{\Delta t} = \frac{kA(T_1 - T_2)}{L}} \tag{15.16}$$

Example 15.4 Two ends of a steel rod 2 m long and having a radius of 0.50 cm are maintained at a steady temperature of 80°C and 20°C, respectively. Assuming no loss of heat to the surroundings, how much heat passes through any cross-sectional area in 1 min?

In Equation 15.16, $L = 2$ m, $T_1 = 80°$C, $T_2 = 20°$C, $r = 0.5$ cm $=$ 5×10^{-3} m, and $\Delta t = 1$ min $= 60$ s. Thus, $A = \pi r^2 = \pi(5 \times 10^{-3}\text{ m})^2 =$ $78.5 \times 10^{-6}\text{ m}^2$, while $k_{(\text{for steel})} = 1.1 \times 10^{-2}$ kcal/s-m-C°. Hence,

$$H = \frac{\Delta Q}{\Delta t} = \frac{kA(T_1 - T_2)}{L}$$

$$= \left(1.1 \times 10^{-2} \frac{\text{kcal}}{\text{s-m-C°}}\right)(78.5 \times 10^{-6}\text{ m}^2)\frac{(80 - 20)\text{C°}}{2\text{ m}}$$

$$= 0.26 \times 10^{-4}\text{ kcal/s}$$

Since $t = 60$ s,

$$Q = (0.26 \times 10^{-4}\text{ kcal/s})(60\text{ s}) = 15.5 \times 10^{-4}\text{ kcal} = 1.55\text{ cal}$$

EXERCISE 15.4 One end of a brass rod 2 m long and having a 1-cm radius is maintained at 250°C. When a steady state is reached, the rate of heat flow across any cross section is 0.50 cal/s. What is the temperature of the other end? [*Ans.*: 128°C.]

15.4. Thermal Convection

As explained earlier, when heat is transferred from one part of the medium to another by actual mass motion of the material, the process is called *thermal convection*. When the temperatures are not too high, the principal method of heat transfer in liquids and gases is thermal convection or simply convection. Examples are hot-air furnace, steam radiator, and hot-water heating system. If the medium is forced to move with the help of a pump or a fan, it is called *forced convection*. If the material moves because of the differences in density of the medium, the process is called *natural* or *free convection*.

In order to understand the process of thermal convection, let us consider the heating of a room by a radiator. The air molecules colliding with the radiator gain translational kinetic energy because the radiator is at a higher temperature than the air molecules. Thus, a thin layer of air next to the radiator is at a higher temperature than the rest of the air; hence, the density of this thin layer is lower. According to Archimedes' principle, buoyant force comes into play and makes this air rise to the top of the room. The next cold layer of air is heated by the radiator and keeps on moving to other parts of the room. Hot air coming in contact with cold walls and the ceiling loses its heat, gets cold, increases in density, and comes back down to the radiator to be reheated. This process of convection continues. Note that heating of a liquid can be explained in a similar fashion. This process can be seen to take place in a liquid very easily by looking through a liquid in a clear container that is being heated.

The mathematical theory of thermal convection is much more complex, and there is no simple equation that may be written as we did for the case of thermal conduction. The heat gained or lost depends on such factors as (1) the geometrical shape of the surface; (2) the characteristics of the fluid, such as its density, specific heat, conductivity, and viscosity; and (3) the type of flow, which may be streamline or turbulent.

As in the case of thermal conduction, we may write an expression for the *convection current* $\Delta Q/\Delta t = H$, defined as the heat crossing any area per unit time, as

$$\boxed{H = hA(T_1 - T_2)} \tag{15.17}$$

where A is the area, T_1 the temperature of the hot body, and T_2 that of the cold body. The term h is the *convection coefficient,* defined as the heat lost or gained by a surface of unit area in unit time per degree change in temperature. That is, the units of h are cal/s-cm²-C° or kcal/s-m²-C°. It is difficult to find a theoretical expression for h; hence, its determination is made for a particular configuration by means of experimentation. For example, the value of h for a horizontal or vertical pipe of diameter D cm with natural convection in air at atmospheric pressure is found to be

$$h = 1.00 \times 10^{-4}(\Delta T/D)^{1/4} \text{ cal/s-cm}^2\text{-C}° \tag{15.18}$$

There is one simple law, discovered by Newton, which gives the relation between the rate of cooling or warming of a given body surrounded by air at a different temperature, and may be stated as follows: *Newton's law of cooling:* The rate at which a body loses heat to the surrounding air is directly proportional to the difference in temperature between the body and the air, provided that the temperature difference is not extremely large. Most heat transfer is by convection; there is a very small contribution from conduction and radiation.

EXAMPLE 15.5 Water at 80°C in a copper pipe 10 m long and 2.6 cm in diameter is surrounded by air at a temperature of 10°C (= 50°F). How much heat is lost by convection in one hour?

For a pipe of 10-m length and radius 2.6 cm/2 = 1.3 cm, the area exposed to air is

$$A = 2\pi rL = 2\pi(1.3 \text{ cm})(1000 \text{ cm}) = 8170 \text{ cm}^2$$

For a horizontal pipe for $\Delta T = (80 - 10)\text{C}° = 70\text{C}°$,

$$h = 1.00 \times 10^{-4}\left(\frac{\Delta T}{D}\right)^{1/4} = 1.00 \times 10^{-4}\left(\frac{70}{2.6}\right)^{1/4}$$

$$= 2.28 \times 10^{-4} \text{ cal/s-cm}^2\text{-C}°$$

From Equation 15.18,

$$H = hA\,\Delta T = (2.28 \times 10^{-4} \text{ cal/s-cm}^2\text{-C}°)(8170 \text{ cm}^2)(70\text{C}°) = 130 \text{ cal/s}$$

In 1 hour (= 3600 s), the heat lost will be

$$Q = H\Delta t = (130 \text{ cal/s})(3600 \text{ s}) = 468{,}000 \text{ cal} = 468 \text{ kcal}$$

EXERCISE 15.5 In Example 15.5, if the pipe has a diameter of 4 cm and the hot water temperature is 90°C, what is the total heat lost in 1 h? [*Ans.:* Q = 765 kcal.]

15.5. Thermal Radiation

As mentioned earlier, thermal radiation is a process in which a body at high temperature loses its heat by giving out electromagnetic waves. When these electromagnetic waves are absorbed by another body, it gains heat energy. The process of thermal radiation can be understood only after one is familiar with the process of production of electromagnetic waves, which will be discussed in Chapter 27. We shall briefly explain it here.

The molecules of a solid are in constant vibrational motion at any temperature. According to electromagnetic theory, whenever charged particles are accelerated or decelerated (and the vibrating particles always have acceleration), they create a disturbance that travels in space and carries energy. This disturbance that carries energy is called a *wave*, and since the energy is of electric and magnetic origin, these waves are called *electromagnetic waves*. These electromagnetic waves, when absorbed by another material, convey this electromagnetic energy to the molecules. This results in an increased energy of the molecules and an increased temperature. These electromagnetic waves travel at the speed of light, that is, 3×10^8 m/s in a vacuum, and do not require any medium to sustain them. Therefore, they can travel in a vacuum. This is exactly the nature of the process called *radiation*, by which heat energy is transferred from the sun to the earth.

Actually, every object is constantly emitting as well as absorbing radiation. The bodies at higher temperature than the surroundings are emitting more radiation than they are absorbing, while bodies at lower temperature are absorbing more than emitting. This is explained by the *Stefan–Boltzmann law*.

> **Stefan–Boltzmann law:** ***The amount of radiation (electromagnetic energy) emitted per unit time from a unit area of a body at absolute temperature T Kelvin is directly proportional to the fourth power of T.***

That is,

$$\boxed{P = e\sigma T^4} \tag{15.19}$$

where P is the power radiated per unit area and may be expressed in SI units as watts/m² or in the CGS system as ergs/cm²-*s*, and σ is a universal constant (the same for all bodies) known as the *Stefan constant*. In SI units it has a value of 5.6699×10^{-8} W/m²-K⁴. The quantity *e*, *emissivity*, depends strictly upon the nature of the surface and has a value between zero and 1. We shall explain this in some detail before closing this section.

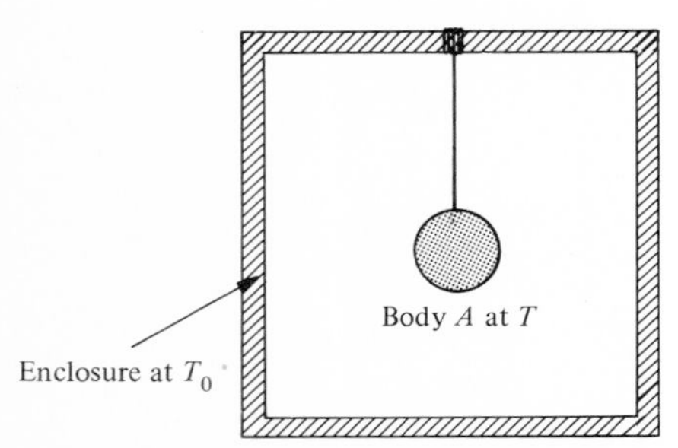

FIGURE 15.8. *Body A at temperature T in thermal equilibrium with its surroundings at temperature T_0.*

A common situation is as follows. Suppose that a body A of emissivity e and at temperature T is surrounded completely in a chamber that is at a temperature T_0, as shown in Figure 15.8 ($T_0 < T$). The body will emit radiation at the rate $e\sigma T^4$, while it will absorb radiation from the chamber at the rate $e\sigma T_0^4$. (Both e's are the same, as will be discussed shortly.) The net rate of loss of energy per unit area by the process of thermal radiation is

$$\boxed{P_{\text{net}} = e\sigma T^4 - e\sigma T_0^4 = e\sigma(T^4 - T_0^4)} \tag{15.20}$$

which is the modified, more complete, form of the Stefan–Boltzmann law. Thus, if the object is hotter than its surroundings, it loses energy, and if it is cooler than the surroundings, it gains energy.

Let us investigate the quantity e, emissivity. According to *Kirchhoff's law*, any object that is a good absorber of radiation of a particular wavelength is also a good emitter of radiation of the same wavelength. That is, a perfect absorber is a perfect radiator, while a poor absorber is a poor radiator. According to thermodynamics, a *blackbody* is one that absorbs all radiation of all wavelength; hence, it is assigned a value of $e = 1$. This also means that a blackbody is not only a perfect absorber, but a perfect radiator; that is, it emits electromagnetic waves of all wavelengths. For other than blackbodies, the value of e is less than 1.

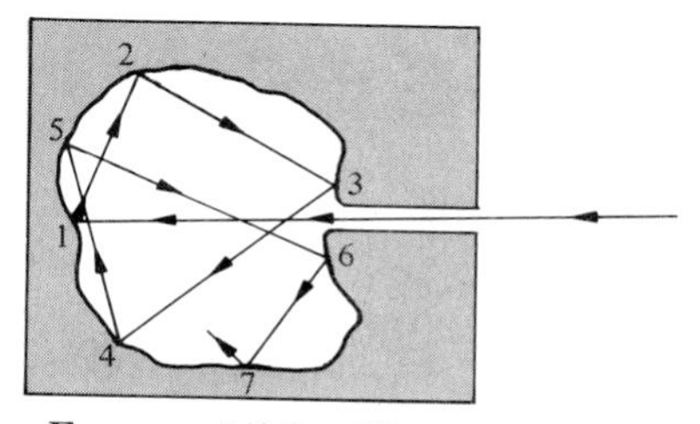

FIGURE 15.9. *Close approximation to a perfect blackbody.*

A blackbody does not mean that its color is black (although a black-colored body may come close to being a blackbody). Consider a hollow container in the form of a cavity with a narrow opening made out of copper, iron, or any other material, as shown in Figure 15.9. The inside of the cavity is painted with lampblack, and such a cavity is a close approximation to a perfect blackbody. Incident radiation that enters the cavity will be reflected from the walls of the cavity and has a very little chance of escaping from the hole. Now suppose that we heat this cavity to a temperature T. If we analyze the radiation escaping from the hole, it is found to contain radiation of all wavelengths. Since $e = 1$ for a blackbody, Equations 15.19 and 15.20 take the form

$$P = \sigma T^4 \tag{15.21}$$

$$P_{\text{net}} = \sigma(T^4 - T_0^4) \tag{15.22}$$

EXAMPLE 15.6 At what rate is the energy radiated by a sphere of radius 5 cm at 3000 K and with an emissivity of 0.30?

According to Equation 15.19, the energy radiated per unit area is $P = e\sigma T^4$ and if the area is A, the total energy radiated will be

$$P_T = PA = e\sigma T^4 A$$

where $e = 0.30$, $\sigma = 5.67 \times 10^{-8}\ \text{W/m}^2\text{-K}^4$, $T = 3000$ K and $A = 4\pi r^2 = 4\pi(5 \times 10^{-2}\ \text{m})^2 = \pi \times 10^{-2}\ \text{m}^2$. Hence,

$$P_T = (0.30)(5.67 \times 10^{-8}\ \text{W/m}^2\text{-K}^4)(3000\ \text{K})^4(\pi \times 10^{-2}\ \text{m}^2) = 43{,}300\ \text{W}$$
$$= 43.3\ \text{kW}$$

EXERCISE 15.6 The average human body may be considered to have a surface area of 1.25 m^2 and a surface temperature of 32°C. Assuming $e = 1$, what is the rate of energy loss by the human body? [*Ans.*: 613 W.]

*15.6. Blackbody Spectrum and Energy Quantization

Let us look closely at the radiation emitted from a blackbody shown in Figure 15.9, which has been heated to a temperature T. If we analyze the radiation coming out of the hole, it will contain radiation of all frequencies (hence, of all wavelengths). We state this fact by saying that the radiation emitted from a blackbody has a *continuous spectrum*. Let us say that $I(\nu)$ is the amount of radiation emitted at frequency ν. Experimental measurements result in a plot of $I(\nu)$ versus ν, as shown in Figure 15.10 for different temperatures.

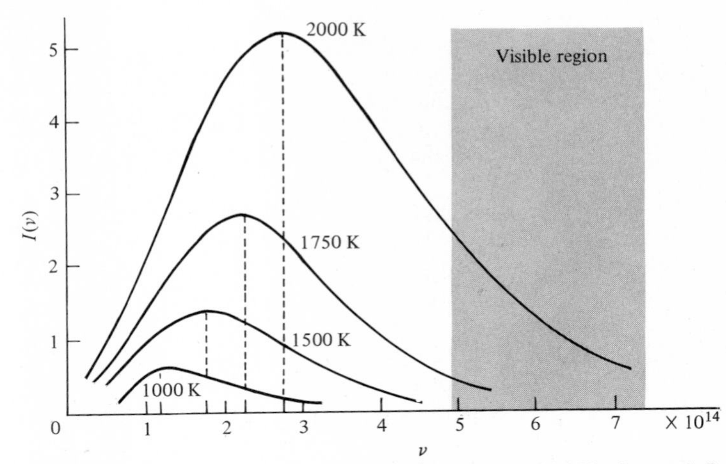

Figure 15.10. *Distribution of radiation from a blackbody at different temperatures. Note that the maximum in each curve shifts toward higher frequencies with increasing T.*

Some simple characteristics of these plots are:

1. The distribution $I(\nu)$ versus ν is a function of the temperature of the blackbody.
2. The total amount of radiation emitted, which is equal to the area under the curve, increases with increasing temperature. Actually, it should be proportional to the fourth power of the temperature, as stated in Equation 15.22. If we measure the area under the curve for $T = 1000$ K and for $T = 2000$ K, we will find that the ratio of the areas is 1:16, while the ratio of the temperatures is 1:2.
3. The position of the maximum peak shifts toward the higher frequencies with increasing temperature. That is,

$$\nu_m \propto T \qquad \text{or} \qquad \frac{\nu_m}{T} = \text{constant } K \tag{15.23}$$

This relation is called *Wien's displacement law*.

Classical theories cannot explain the blackbody radiation spectra exhibited in Figure 15.10. To provide a satisfactory explanation in 1900, the German theoretical physicist Max Planck (1858–1947) announced that by assuming electromagnetic radiation to be emitted or absorbed in bundles of size $h\nu$, where h is Planck's constant, one could correctly predict the blackbody spectrum. Such a bundle of energy is called a *quantum* or *photon* as mentioned in Chapter 2. For a given frequency ν, all the quanta have the same energy, but quanta of high frequency have high energy, and those of low frequency have low energy. The introduction of *Planck's quantization hypothesis* had a profound effect and led to the start of quantum physics, as we shall discuss in Chapter 31.

SUMMARY

The *coefficient of linear expansion* α is defined as the increase in length per unit length per degree increase in temperature, $\alpha = \Delta L/L_0\, \Delta T$; $L = L_0(1 + \alpha\, \Delta T)$. The coefficient of area expansion β and the coefficient of volume expansion γ have similar relations, and $\alpha \simeq \beta/2 \simeq \gamma/3$. Thermal stress due to change in temperature is given by $F/A = Y\alpha\, \Delta T$.

Heat transfer can take place by conduction, convection, or radiation. In *conduction*, molecules of the material transfer translational kinetic energy to neighboring molecules, which have lower kinetic energy. The heat-flow rate is $H = \Delta Q/\Delta t = -kA\, \Delta T/\Delta x = kA(T_1 - T_2)/L$, where k is the coefficient of thermal conductivity and $\Delta T/\Delta x$ is the temperature gradient.

In *thermal convection* the heat transfer takes place by the actual mass motion of the material. The convection current is given by $H = \Delta Q/\Delta t = hA(T_1 - T_2)$, where h is the convection coefficient.

Thermal radiation is a process in which a body at high temperature loses its heat by giving out electromagnetic waves. According to the *Stefan–Boltzmann law*, $P = e\sigma T^4$, where P is the power radiated per unit area, e is emissivity and σ is the Stefan constant. For a perfect blackbody, $e = 1$ and $P = \sigma T^4$ and $P_{net} = \sigma(T^4 - T_0^4)$.

*The blackbody spectrum can be explained by Planck's quantum hypothesis—electromagnetic energy is emitted and absorbed in bundles of size $h\nu$. According to Wien's displacement law, ν_m/T = constant.

QUESTIONS

1. What modification, if any, is needed in Equation 15.3 if the temperature difference ΔT is expressed in (a) the Fahrenheit scale; and (b) the Kelvin scale?
2. When heated through the same temperature, will a solid sphere or a hollow sphere expand more?
3. If a clock has to keep correct time at different temperatures, will it be better to use aluminum or steel?
4. Why is there always some space left between concrete blocks when laying them?
5. When water is heated in a glass pot, why does its level slightly decrease and then increase?
6. Explain why some materials such as rubber have a negative coefficient of linear expansion.
7. When you touch an outside doorknob in winter, it feels colder than the wooden door, and in summer it feels warmer than the wooden door. Why?
8. A hot liquid is added to two cups, one made of glass and the other of aluminum. Occasionally, the glass cup will break while the aluminum cup never breaks. Why?
9. Two spheres are painted, one white and the other black, and heated to incandescence. Which one will look brighter?

PROBLEMS

1. A number of 10-m steel rails are laid in winter when the temperature is 0°C (= 32°F). How much space should be left between the rails to allow for expansion in summer when the temperature may reach as high as 50°C (= 122°F)? If the temperature varies from −40°C (= −40°F) in winter to +40°C (= 104°F) in summer, what is the maximum change in the length of the steel rail?
2. A precise surveying steel tape was calibrated to read 1 m when at 20°C. What will be the percentage error when it is used at (a) 0°C; and (b) 30°C?
3. A 100-ft steel tape is calibrated to read correctly at 20°C. If, on a hot day when the temperature is 100°F (= 37.8°C), the distance measured by the tape is 80.2 ft, what is the true distance at 20°C?
4. A 3-m-long copper wire supports the bob of a simple pendulum. Calculate the change in length if the temperature changes from 15°C to 35°C. Also, calculate the change in the time period.
5. A clock with a steel pendulum has a time period of 2 s at 20°C. If the temperature of the clock rises to 30°C (= 86°F), what will be the gain or loss per day?
6. How closely must the temperature of a brass pendulum be controlled so that it will not gain or lose more than 1 s/day?
7. A square sheet of aluminum is 30 cm on a side at 10°C (= 50°F). Calculate the change in the area when its temperature is 35°C (= 95°F).
8. A copper pipe has a cross-sectional area of 10×10^{-3} m^2 at 20°C. What will be its area when filled with (a) water at 100°C; and (b) steam at 175°C? (*Hint:* use expansion of area.)
9. A certain wheel has a diameter of 1.25 m at 200°C. A steel rim that has a diameter of 1.2495 m at this temperature will slip on the wheel if its diameter is 1.2505 m. To what temperature should the steel rim be heated to accomplish this?
10. An aluminum rivet of 1.215-in diameter is to be fitted in a hole of 1.214-in diameter at 20°C. To what temperature should the rivet be cooled so that it will fit? Can dry ice of temperature −78°C accomplish this? (This method is used to ensure tight fits in airplane construction.)
11. A flask has a volume of 100 cm^3 at 0°C. What is its volume at 20, 50, and 100°C?
12. Consider a cubical block of aluminum 10 cm on an edge at 30°C. What will be its volume if it is immersed in ice water at 0°C?
13. When a liquid is heated in a solid container, usually the expansion of the solid is very small as compared to that of the liquid. The net expansion is called the *apparent expansion*, as we shall illustrate. A glass flask with a volume of 1000 cm^3 is filled with mercury at 0°C. When heated to 100°C, 15 cm^3 of mercury overflows. Calculate the apparent coefficient of volume expansion. If the true volume expansion coefficient is 18.2×10^{-5}/C°, what is the volume expansion coefficient of glass?
14. A number of 10-m-long steel rails are laid in winter when the temperature is 0°C. If no space is left between the rails, what will be the thermal stress when the rails are heated to 40°C in summer? If the cross-sectional area of the rail is 225 cm^2, calculate the force.

15. A copper wire 10 m long is stretched tight at 40°C. Calculate the stress in the wire if cooled to −18°C (≈ 0°F).

16. A steel rod 1 m long and having a 10-cm² cross section is to be cooled from 500°C to 20°C. What force should be applied to prevent it from contracting while cooling? Compare this to the similar case in aluminum.

17. Two rods of iron and copper each 1 m long and of the same cross-sectional area are placed end to end between rigid supports at 20°C. If they are heated to steam temperature, calculate the stress in each.

18. Two metal rods, each of length 50 cm, expand 0.08 cm and 0.04 cm, respectively, when heated from 0 to 100°C. A third rod 50 cm long is made out of these two rods so that it expands 0.06 cm. What is the length of each portion of the composite rod?

19. The coefficient of thermal conductivity k in CGS units is cal-cm/cm²-s-C°. Calculate the multiplication factor to express k in the following units: (a) kcal-m/m²-s-C°; and (b) Btu-in/ft²-h-F°.

20. One end of a copper rod 1 m long and 1 cm in radius is immersed in boiling water at 100°C, while the other end is immersed in ice at 0°C. Calculate the rate of heat flow.

21. In Problem 20 calculate the rate at which the ice will melt. How can you use this method to measure the thermal conductivities of different materials?

22. Water is boiling in a stainless steel pot that has a thickness of 3 mm. The inside temperature is 100°C, while the outside temperature (next to the heating element) is 120°C. How much heat is flowing per minute across a surface area of 400 cm²? Calculate the amount of water that boils away every minute.

23. A boiler made of steel has an inner temperature of 100°C and the outer temperature (next to the heating element) is 150°C. The thickness of the bottom is 1.75 cm. If 0.5 kg of water boils away every minute, what is the surface area of the boiler that is next to the heating element?

24. Consider two rods of lengths L_1 and L_2, each of cross-sectional area A connected as shown. In the steady state one end is at temperature T_1 while the other at T_2. The temperature at the interface is T_x. (a) Calculate T_x and (b) show that the rate of flow of heat is

$$\frac{\Delta Q}{\Delta t} = \frac{A(T_1 - T_2)}{(L_1/k_1) + (L_2/k_2)}$$

Assume no loss to the surroundings. [*Hint:* In the steady state, $\Delta Q_1/\Delta t = \Delta Q_2/\Delta t$, calculate T_x and substitute it in either one of the two expressions for heat flow.]

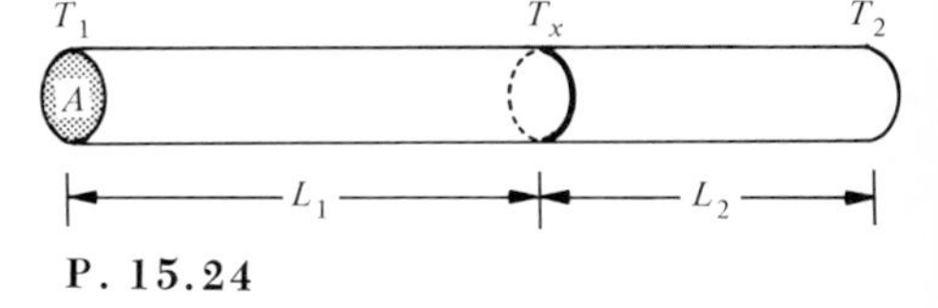

P. 15.24

25. Consider two rods made of copper and steel, each of cross-sectional area 2 cm² and 0.5 m long, connected as shown for Problem 24. One end is at 100°C and the other is 0°C. Calculate the amount of heat flow when in the steady state. (Use the result derived in Problem 24.)

26. A vertical pipe 4 m long and 2.54 cm in diameter carries hot water at 92°C from the basement to the second floor. During the day the temperature of the air surrounding the pipe is 10°C (= 50°F). What is the rate at which the pipe loses heat to the air?

27. In Problem 26 suppose that at night the temperature of the air surrounding the pipe drops to $-20°C$. Calculate the rate of heat loss. Is it possible that over a period of 8 h during the night, there will be enough heat loss to freeze the pipe? (To calculate h, use the average value of ΔT; that is, $(92 - (-20))C°/2 = 56C°$.)

28. Water at 5°C is in a pipe 2 ft long and 2 in in diameter which is surrounded by air at $-15°C$. Calculate the rate of heat loss. How long will it take before the water in the pipe freezes? For calculating h, use $\Delta T = 10C°$.

***29.** Calculate the radiant emittance of a perfect blackbody at a temperature of (a) 3 K; (b) 30 K; (c) 300 K; and (d) 3000 K.

***30.** At what temperature will a blackbody emit double the amount of radiation that it emits at a temperature of 27°C?

***31.** A small sphere that has been painted black and may be treated as a blackbody has a radius of 5 cm and is hung in the middle of an evacuated chamber. The walls of the chamber are maintained at 100°C. What power should be supplied to the copper sphere to maintain its temperature of 200°C?

***32.** A 60-W electric bulb made out of a tungsten filament operates at a temperature of 2450 K, while another bulb has the same power but operates at a temperature of 2850 K. Calculate the ratio of the areas of the filaments in the two cases.

***33.** Calculate the surface area of the tungsten filament of a 100-W electric bulb. The operating temperature of tungsten is 2450 K and its emissivity is 0.3.

***34.** The roof of a typical medium sized home is 18 m $\times$ 9 m (60 ft $\times$ 30 ft) and may be assumed to be completely black and flat. If the sun's radiation strikes normal to the roof, how much energy is received per second? If 15% of this could be converted into useful energy, how much power would be available? Is this a significant amount?

***35.** An average human body may be considered to have a surface area of 1.3 m^2 and surface temperature of 32°C. If the surrounding temperature is (a) 0°C; and (b) 20°C, calculate the rate of radiation emitted by the human body in each case.

16 Disorder-to-Order and Thermodynamics

We shall familiarize ourselves with some of the most important thermodynamical processes. The concept of entropy and the second law of thermodynamics, which limits the efficiency of heat engines (together with the theoretical Carnot engine), will be discussed in detail. We shall see how these concepts lead to the third law of thermodynamics—the unattainability of absolute zero.

16.1. Terminology

WE START with the investigation of the principles that govern the transfer of disordered energy into ordered energy, that is, thermal energy into mechanical energy. *Thermodynamics* deals with the study of this conversion of thermal energy into mechanical energy, and vice versa. The laws of thermodynamics have been found to apply not only to these two forms of energy, but have been extended to the exchange of thermal and other forms of energy such as chemical, electrical, and so on.

A *thermodynamical system* consists of a certain quantity of matter containing a large number of particles. Since a thermodynamical system is macroscopic in nature, it is complex and each particle cannot be treated individually. We shall be concerned primarily with *isolated systems*, which are systems separated from their surroundings in such a way that no energy can be transferred to or from the system. On the other hand, if the system is not isolated, it will be in *thermal contact* with its surrounding, and the energy can flow to or from it. Certain changes may take place in an isolated system, but eventually the system will reach a state in which no further changes will be noticeable. Such an isolated system is said to be in *thermal equilibrium*.

Once a physical system has reached thermal equilibrium, it can be described by means of such quantities as pressure, volume, and temperature. These observable quantities, which describe the state of the system, are called *state variables, thermodynamical variables*, or *thermodynamical coordinates*. Note that the state variables describe the macroscopic system and not the microscopic parts of the system. For example, gas pressure is a macroscopic property that remains constant, even though the forces exerted by individual molecules may fluctuate.

The state variables that describe the thermodynamical system in equilibrium are not all independent. The relation between state variables describing the thermodynamical system in equilibrium is called the *equation of state* for the system. For example, the relation

$$PV = nRT$$

is an equation of state of an ideal gas.

16.2. Thermodynamical Processes

A *thermodynamical process* is said to have taken place when a system changes its state. Thermodynamical principles and methods apply to the systems, which are in thermal equilibrium, and not to the thermodynamical processes. This difficulty can be overcome if we limit our discussion to reversible processes.

A *reversible process* is one in which the changes take place in such a way that the system is always in thermal equilibrium. A reversible process can be achieved if the changes take place very slowly, that is, in a very large number of infinitesimal steps so that at each step the system is only slightly disturbed from its previous equilibrium state. As the name implies, a reversible process can proceed in either direction at every instant and is always in thermal equilibrium. There is no such thing as a truly reversible process because there will always be some frictional processes present, but one can obtain conditions that will be very close to a true reversible process.

A *cyclic process* is one in which the system returns to its initial state. If a cycle consists of reversible processes, it is possible that no observable change is produced either in the system or in its surroundings after the cycle is completed. On the other hand, if a cycle consists (in part or totally) of irreversible processes, the system returns to its initial state, after producing a finite measurable change in its surroundings.

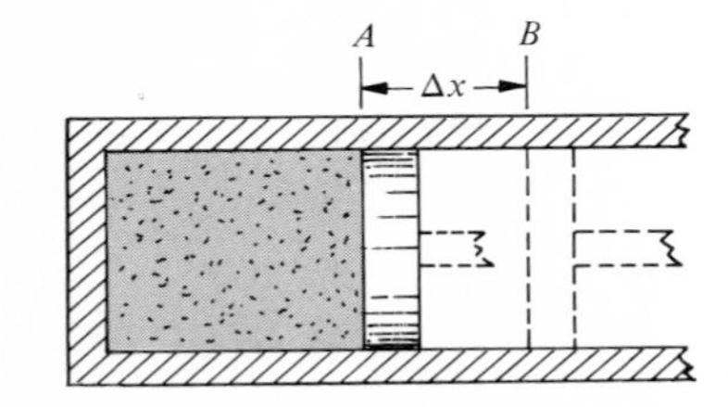

FIGURE 16.1. *Ideal gas enclosed in a cylinder fitted with a movable piston.*

Let us consider the following four processes: (1) isothermal process, (2) adiabatic process, (3) isobaric process, and (4) isochoric process.

Isothermal Process (T-constant)

A reversible isothermal process is one which takes place at a constant temperature so slowly that the system is continuously in equilibrium with the heat source. In general, all coordinates except T will change in this process. According to Equation 13.29, the internal energy of an ideal gas is $U = \frac{3}{2} nRT$; hence, $\Delta U = 0$. Thus, for an ideal gas undergoing an isothermal process, the first law of thermodynamics $\Delta U = \Delta Q - \Delta W$ takes the form

$$\boxed{\Delta Q = \Delta W} \quad \text{if } T \text{ is constant} \tag{16.1}$$

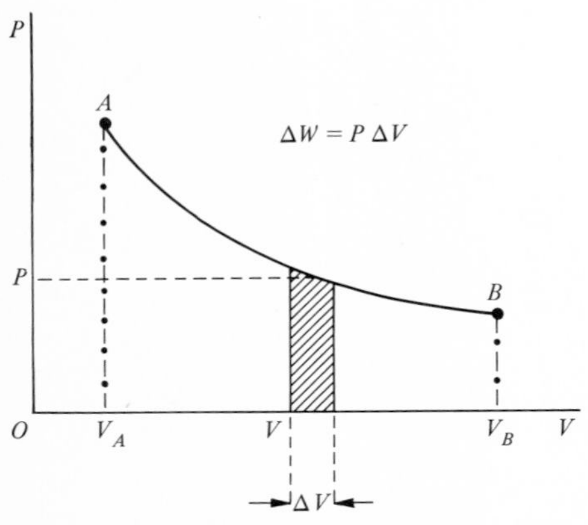

FIGURE 16.2. *P versus V plot for an isothermal process. Shaded area is equal to the work done in a small isothermal expansion ΔV.*

Let us see what happens when an ideal gas undergoes an isothermal change. Figure 16.1 shows an ideal gas confined in a cylinder with a movable piston. The cylinder is completely insulated from its surroundings except the bottom, which can be insulated or connected to a heat source. To start with, let the piston be at position A so that the gas pressure is P_A and volume V_A. Suppose that we supply heat Q to this ideal gas. In order to keep the temperature constant, the gas must expand and push the piston outward so that the system does work W on its surroundings. Since temperature is constant, $PV = nRT$ reduces to $PV =$ constant, which is an equation of a hyperbola. The plot of P versus V is shown in Figures 16.2 and 16.3, where AB is an isothermal expansion curve.

Since the process is reversible, the system will change in exactly the opposite way in going from B to A as it did in going from A to B. The curve BA shown in Figure 16.4 is an isothermal compression. This process is achieved by pushing the piston inward as in Figure 16.1. The result will be external work done on the system. Since the temperature has to remain constant, there can be no increase in internal energy. The result of the isothermal compression is to increase the pressure (from P_B to P_A) and decrease in volume (from V_B to V_A), as shown in Figure 16.4. From the above we conclude that in an isothermal volume expansion, the work W_{AB} is done by the system, while in isothermal volume compression, the work W_{BA} $(= -W_{AB})$ is done on the system. Let us calculate this work.

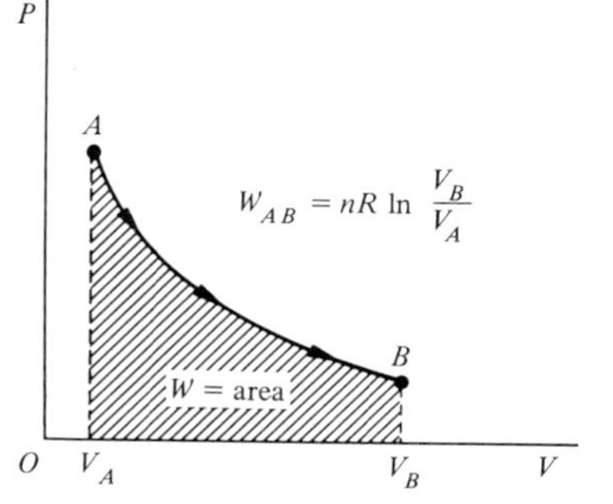

Figure 16.3. *Work done in an isothermal expansion is equal to the area under the PV curve.*

Consider the isothermal expansion of gas in the cylinder shown in Figure 16.1. Suppose that the piston moves through a distance Δx when the gas expands isothermally. If the cross-sectional area of the piston is A and the external pressure acting is P, the force exerted by the system will be $F = PA$. The work ΔW done when the piston moves through an infinitesimal distance Δx is given by

$$\Delta W = F\,\Delta x = PA\,\Delta x$$

But $A\,\Delta x = \Delta V$; therefore,

$$\Delta W = P\,\Delta V \tag{16.2}$$

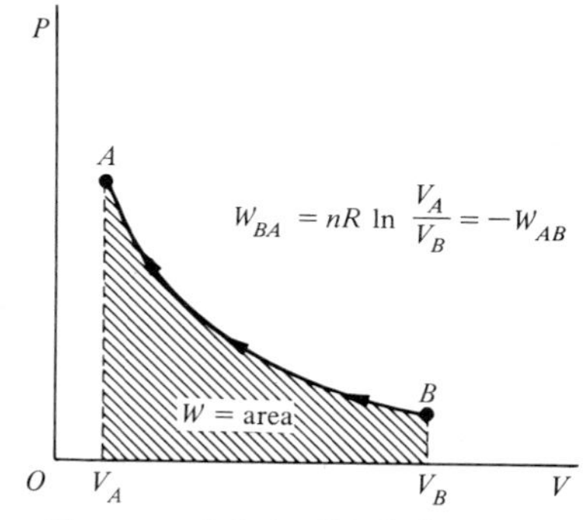

Figure 16.4. *Work done in an isothermal compression is equal to the area under the PV curve.*

On a PV diagram this infinitesimal amount of work is equal to the shaded area shown in Figure 16.2. Thus, the finite amount of work done W when the gas expands isothermally from point A to B is

$$W = \Delta W_1 + \Delta W_2 + \cdots$$

or

$$\boxed{W = \sum \Delta W = \sum P\,\Delta V} \tag{16.3}$$

As drawn in Figure 16.3, the work is equal to the area under the P versus V curve between points A and B as shown by the shaded area.

We can evaluate the sum in Equation 16.3 for the case of an ideal gas if we assume that $PV = nRT$. Substituting for P in Equation 16.3, we get (remembering that T is a constant)

$$W = nRT \sum \frac{\Delta V}{V} \tag{16.4}$$

It can be shown that $\Sigma\,\Delta V/V = \ln V_B/V_A$, where ln is the natural logarithm. Therefore, the total work done in going from the initial volume V_A to the final volume V_B is

$$W_{AB} = nRT \sum_A^B \frac{\Delta V}{V} = nRT \ln \frac{V_B}{V_A} \tag{16.5}$$

This work is equal to the heat supplied to the system as it expands.

The case of isothermal compression is shown in Figure 16.4. The work done when the gas is compressed from volume V_B to volume V_A is given by

$$W_{BA} = nRT \ln \frac{V_A}{V_B} = -W_{AB}$$

Once again the work done is equal to the area under the P-versus-V curve. The work done depends not only on the initial and final state coordinates, but also on the paths depending upon how P varies with V.

Adiabatic Process (Q = *constant*)

A *reversible adiabatic process* is one that takes place with no flow of heat into or from the system, and the process takes place sufficiently slowly that the working substance is continuously in equilibrium. Thus in Equation 14.14 the first law of thermodynamics $\Delta U = \Delta Q - \Delta W$ for $\Delta Q = 0$ takes the form

$$\boxed{\Delta U = -\Delta W} \quad \text{if } Q \text{ is constant} \tag{16.6}$$

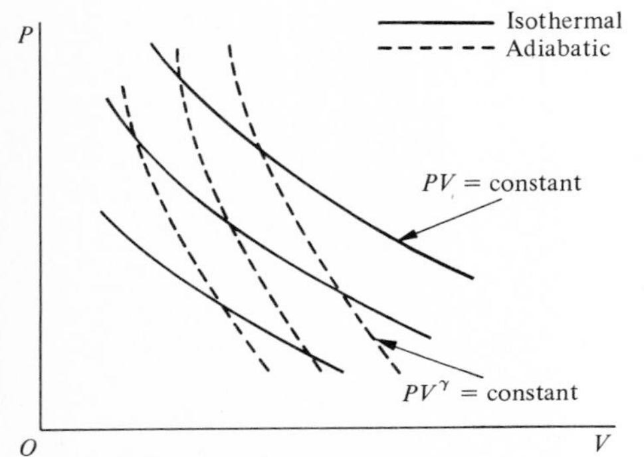

FIGURE 16.5. *Plots of isothermals (solid curves) and adiabatics (dashed curves) on a PV diagram for an ideal gas. Note that the adiabatics are steeper than isotherms.*

According to Equation 16.6, in an adiabatic process the work is done at the expense of the internal energy of the system. Thus, if the system expands adiabatically, the system does work and ΔW is positive; hence, ΔU is negative and the system's *temperature decreases*. On the other hand, if the system is *compressed* adiabatically, the work must be done on the system and ΔW is negative; hence, ΔU is positive and the system's *temperature increases*. An ideal adiabatic system is obtained by completely insulating the system from its surroundings. In Figure 16.1 the cylinder and piston must be completely insulated or suspended in a vacuum.

In an adiabatic process all three variables, P, V, and T, change and the ideal-gas equation, $PV = nRT$, is not a very useful relation. The plots of P versus V for adiabatic processes are shown in Figure 16.5 together with the plots of isothermal processes for comparison. It is quite clear that the adiabatic curves are steeper than the isothermal curves. These adiabatic plots are represented by the following relation:

$$\boxed{\begin{array}{c} P_1V_1^\gamma = P_2V_2^\gamma \\ \text{or} \quad PV^\gamma = \text{constant} \end{array}} \tag{16.7}$$

where $\gamma(=C_p/C_v)$ is the ratio of the specific heat of a gas at constant pressure to the specific heat at constant volume. Equation 16.7 is an equation of state for an adiabatic process.

Finally, let us calculate the work done by an ideal gas in an adiabatic expansion. Suppose that an ideal gas undergoes an infinitesimal expansion. In the expression $\Delta U = \Delta Q - \Delta W$, $\Delta Q = 0$, ΔW is the amount of work done by the gas while $\Delta U = nC_v\, \Delta T$ is the decrease in the internal energy of the system due to a decrease in temperature ΔT of the system (see Section 14.6). Thus, $\Delta U = -\Delta W$ takes the form

$$\Delta W = -nC_v\, \Delta T$$

where $\Delta T = T_B - T_A = -(T_A - T_B)$. Therefore,

$$\boxed{\Delta W = nC_v(T_A - T_B)} \tag{16.8}$$

Once again, ΔW, the work done, is always equal to the area under the P-versus-V curve.

Isobaric Process (P = constant)

An *isobaric process* is one in which the pressure remains constant. Since $\Delta P = 0$ and $\Delta W = P\,\Delta V$, Equation 14.14, $\Delta U = \Delta Q - \Delta W$, for the first law takes the form

$$\boxed{\Delta U = \Delta Q - P\,\Delta V} \quad \text{if } P \text{ is constant} \tag{16.9}$$

Isochoric Process (V = constant)

An *isochoric process* is one in which the volume remains constant. If $\Delta V = 0$, $\Delta W = P\,\Delta V = 0$; that is, no work is done. The first law, $\Delta U = \Delta Q - \Delta W$, takes the form

$$\boxed{\Delta U = \Delta Q} \quad \text{if } V \text{ is constant} \tag{16.10}$$

Equation 16.10 implies that all the heat absorbed by the system is used in increasing the internal energy of the system.

To summarize our discussion of this section, Figure 16.6 shows plots of P versus V for all four processes. It may be pointed out once again that the quantities of heat Q and work W depend upon the path and are different for different paths. On the other hand, U is independent of the path; and like P, V, and T, U is a state variable.

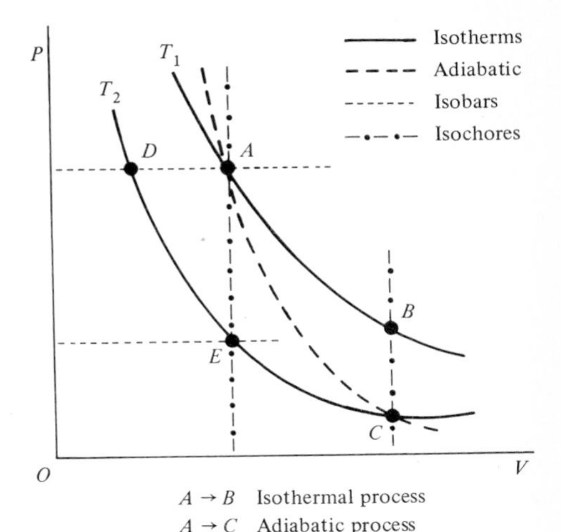

FIGURE 16.6. *Different thermodynamical processes on a PV diagram.*

EXAMPLE 16.1 A 2 kg quantity of oxygen expands isothermally to double its original volume at 0°C. Calculate the work done by the gas and the heat that must be supplied.

For oxygen, since $m = 2$ kg and $M = 32$ kg/kmol, $n = m/M = \frac{2}{32}$ kmol; $T = 273$ K and $V_B = 2V_A$. Substituting in Equation 16.5,

$$W_{AB} = nRT \ln (V_B/V_A) = (\tfrac{2}{32}\text{ kmol})\left(8310\frac{\text{J}}{\text{kmol-K}}\right)(273\text{ K}) \ln 2$$
$$= (14.18 \times 10^4\text{ J})(\ln 2) = (14.18 \times 10^4\text{ J})(0.693) = 9.83 \times 10^4\text{ J}$$

A positive W means that this amount of work is done by the gas. In order to maintain a constant-temperature heat in this amount must be supplied. In units of kcal, since 1 kcal = 4186 J,

$$Q = \frac{9.83 \times 10^4\text{ J}}{4186\text{ J/kcal}} = 23.5\text{ kcal}$$

EXERCISE 16.1 Repeat Example 16.1 for the case in which oxygen compresses to half its original volume. [*Ans.:* $W = -9.83 \times 10^4$ J, $Q = -23.5$ kcal.]

16.3. Entropy—Microscopic and Macroscopic Definitions

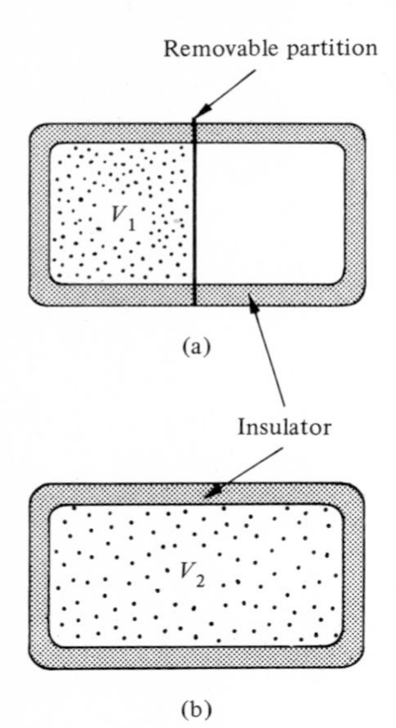

Figure 16.7. *Localization of molecules (a) in a volume V_1 and (b) in a volume $V_2 = 2V_1$.*

To have a better understanding of thermodynamical processes and the second and third laws of thermodynamics, it is worthwhile at this point to introduce the concept of entropy. This will give us an opportunity to look at different processes not only from the macroscopic point of view, but to see their relations to microscopic quantities.

Even though thermal energy is disordered energy, it is not a measure of disorder. A quantity that denotes the amount of disorder is called *entropy*, S. The total energy is always conserved in any process, but the total entropy always increases (or remains the same) in any process. As we shall see, it is the principle of increase of entropy that allows a complete conversion of mechanical energy to thermal energy, but not vice versa. The principle of entropy increase is one of the basic laws of physics and is applicable to many processes throughout chemistry, engineering, biology, and other fields.

In short, we may say that *entropy* is a measure of molecular disorder and in any process the entropy increases or remains constant; that is, the disorder increases or remains constant. Let us illustrate these points by means of a few simple examples.

Consider a large number of molecules of a gas confined in an insulated cylinder fixed with a removable partition as shown in Figure 16.7(a). To start with, all the molecules are confined in a volume $V_1 = V$. We say that the molecules are localized within a volume V. Suppose now that the partition is removed. The molecules, instead of staying confined to volume V, now occupy the whole volume $V_2 = 2V$, as shown in Figure 16.7(b), and are less localized than they were before the partition was removed. Actually, the degree of localization is a measure of disorder. As the system increases in volume, the disorder increases, and we say that the entropy increases. Also, there is almost *no* chance that all the molecules by themselves will collect in the original volume, V_1. That is why we made the statement "entropy or disorder always increases or remains constant."

Let us consider H_2O molecules in ice. These molecules have a certain degree of order—molecules of a solid cannot move around freely, but move slightly about their equilibrium positions. Suppose that the ice is exposed to room temperature ($\sim 20°C$), where it absorbs heat from the surroundings and changes into water. The molecules of water can now move around anywhere in the whole volume and are no longer that well localized. We state this by saying that "in going from a solid to a liquid phase, the disorder has increased or the entropy has increased." Once again there is almost *no* chance that heat will flow out of the water into the surroundings and change this water into ice, because that would mean less disorder or a decrease in entropy.

Finally, it is an experimental fact that heat is always transferred from a hot body to a cold body. The reverse has never been observed. In 1865, Clausius realized that there must be some quantity which determines the transfer of heat. He named this quantity *entropy* (a Greek word meaning transformation), which measures the potentiality of a system for heat-energy transformation. Just like gravitational potential, which measures the capability of a system to do work, entropy measures the potential of a thermodynamical system to transfer heat energy.

Entropy—Microscopic View

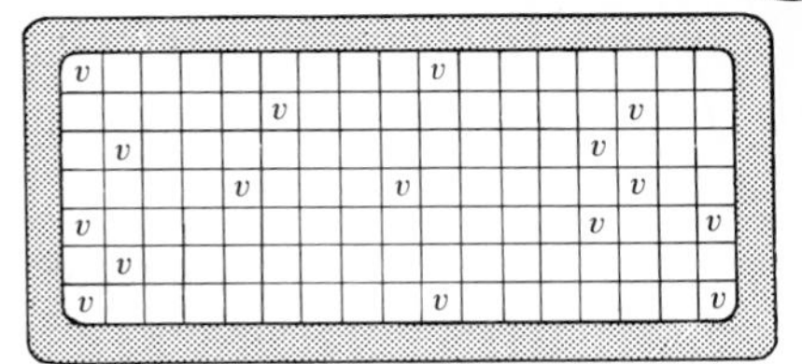

FIGURE 16.8. *A large volume V divided into a large number of small cells of volume v each.*

Unlike macroscopic systems, in microscopic systems we talk in terms of probabilities for things to happen. Suppose in a container of volume V there is only one molecule. This molecule has an equal probability of being found anywhere in this volume. Its probability of being found in a unit volume will be $1/V$. Suppose that this volume V is divided into small cells of volume v each, as shown in Figure 16.8. The probability of the molecule being found in this volume v will be v/V. Now, let us assume that there are two noninteracting molecules (i.e., assuming an ideal gas) in volume V. Each will have a probability of (v/V) of being found in different cells of volume v. The probability of finding *both* of them *simultaneously in one particular cell of volume* v will be $(v/V)(v/V) = (v/V)^2$. Similarly, the probability of finding three such molecules simultaneously in one cell will be $(v/V)^3$; and so on. Let us say in volume V there are n moles of gas with a total of nN_0 molecules, where N_0 is Avogadro's number. Thus, the probability of finding all these molecules simultaneously in one particular cell of volume v will be $(v/V)^{nN_0}$. Let

$$P = \left(\frac{v}{V}\right)^{nN_0} \tag{16.11a}$$

where P is called the probability that all the nN_0 molecules will be found in an extremely small volume v instead of the very large volume V available to them. Since nN_0 is a very large quantity and v/V is a small fraction, the value of P will be very, very small, but not zero. Since probability is proportional to disorder, according to statistical mechanics, as developed by Ludwig Boltzmann in 1877, it is found convenient to define entropy S by the relation

$$S = k \ln P \tag{16.11b}$$

where $\ln P$ is the logarithm of P (P can be calculated with the help of the methods of statistical mechanics) and k is the Boltzmann constant.

Entropy—Macroscopic View

The definition of entropy given by Equation 16.11 is based on a molecular viewpoint, hence is more fundamental. But the earlier macroscopic definition of entropy given by Rudolph Clausius in 1865 is equally important because of its simplicity and practical importance. According to Clausius, the change in entropy ΔS (not the absolute value of entropy) when a system changes from one equilibrium state to another is defined as

$$\boxed{\Delta S = \frac{\Delta Q}{T}} \tag{16.12}$$

where ΔS is equal to the amount of heat ΔQ added to any part of the system divided by the absolute temperature T at which heat is added. If heat is added to the system, ΔQ is positive and entropy increases, while if the system loses heat, ΔQ is negative and entropy decreases. Note that the units of S are J/K or cal/K. We shall illustrate the use of Equation 16.12 by the following example. Suppose that a body A at temperature T_1 is in contact with body B at temperature T_2, where $T_1 > T_2$. Let body A lose heat ΔQ and body B gain the same amount. The change in the entropy of body A will be

$$\Delta S_1 = -\frac{\Delta Q}{T_1}$$

while in body B,

$$\Delta S_2 = +\frac{\Delta Q}{T_2}$$

Thus, the total change in the entropy of the system will be

$$\Delta S = \Delta S_1 + \Delta S_2 = \Delta Q\left(\frac{1}{T_2} - \frac{1}{T_1}\right) \tag{16.13}$$

For ΔS to be positive, T_1 must be greater than T_2, which implies that heat must flow from a body of higher temperature to a body of lower temperature. The reverse is not possible, because that leads to a decrease in entropy.

EXAMPLE 16.2 Calculate the change in entropy when 10 kg of ice at 0°C is changed into water at 2°C.

There are two processes involved in this change. First, the heat must be supplied to melt the ice from 0°C ice to 0°C water, and then heat is supplied to raise the temperature of water from 0°C to 2°C. Heat ΔQ_1 needed to melt ice at 0°C (=273 K) is

$$\Delta Q_1 = mL = (10.0 \text{ kg})(80 \text{ kcal/kg}) = 800 \text{ kcal}$$

The change in entropy ΔS_1 is

$$\Delta S_1 = \frac{\Delta Q_1}{T} = \frac{800 \text{ kcal}}{273 \text{ K}} = 2.93 \text{ kcal/K}$$

Heat ΔQ_2 needed to heat water from 0°C (=273 K) to 2°C (=275 K) is

$$\Delta Q_2 = mc\,\Delta T = (10 \text{ kg})(1 \text{ kcal/kg-C°})(2\text{C°}) = 20 \text{ kcal}$$

In this process the temperature does not remain constant; it changes from 273 K to 275 K. Because this difference in temperature is very small, we may use either temperature in calculating the entropy change. We shall use the average of the two, that is, 274 K. (In more accurate work, integral calculus must be used.) Thus, the change in entropy is

$$\Delta S_2 = \frac{\Delta Q_2}{T} = \frac{20 \text{ kcal}}{274 \text{ K}} = 0.07 \text{ kcal/K}$$

Therefore, the total increase in the entropy is

$$\Delta S = \Delta S_1 + \Delta S_2 = (2.93 + 0.07) \text{ kcal/K} = 3.00 \text{ kcal/K}$$

EXERCISE 16.2 Calculate the change in entropy when 10 kg of water at 98°C is changed to steam at 100°C. [*Ans.:* $\Delta S = 14.6$ kcal/K.]

16.4. Entropy and the Second Law of Thermodynamics

A restriction imposed on different processes by the first law of thermodynamics is that only those processes take place which conserve energy. Does this mean that as long as energy is conserved, any thermodynamical process will take place? The answer is no. The second law of thermodynamics imposes further restrictions on the processes taking place. Before we formally state this law, let us consider a few examples.

When two bodies at different temperatures are brought together, heat always flows from the body at a higher temperature to the body at a lower temperature. The reverse is not possible, even though the total amount of energy is always conserved. Similarly, a ball rolling on the floor always comes to rest when all its mechanical energy has been lost and changed into heat energy. The reverse process, in which a ball sitting on a floor will gather heat from the floor and start moving does not happen. Similarly, gases mix but never unmix by themselves. A drop of ink thrown into a beaker of water will dissipate throughout the fluid, but the dissolved ink never (or in the language of statistical mechanics—has a very small probability) separates itself from the rest of the water.

A careful examination of the processes described above reveals that in the final state the system is more disordered than it was in the initial state. Of course, the increase in disorder means an increase in the entropy of the system. This essential asymmetry of nature (order $\rightarrow$ disorder is possible while the reverse is not) is expressed by the second law of thermodynamics.

The Second Law in Terms of Entropy: ***For an isolated system, only those processes can take place for which the entropy of the system increases or remains constant.***

That is,

$$\Delta S \geqslant 0 \tag{16.14}$$

$\Delta S > 0$ for irreversible processes, $\Delta S = 0$ for reversible processes while $\Delta S < 0$ is not possible. A part of the system may have a decrease in entropy, but the overall effect is $\Delta S \geqslant 0$. Before the development of statistical mechanics, the second law of thermodynamics was stated in the following two alternative forms:

Rudolf Clausius's Statement of the Second Law: ***It is impossible to construct a device that operating in a cycle will produce no effect other than the transfer of heat from a cooler body to a hotter body;*** **in other words,** ***the spontaneous flow of heat from a colder body to a hotter body is impossible.***

Lord Kelvin and Max Planck's Statement of the Second Law: ***It is impossible to construct a device operating in a cycle for the sole purpose of extracting heat from a reservoir and changing it entirely into an equal amount of work without rejecting a part of the heat.***

According to this statement, an engine or any other device, when converting heat into mechanical work, cannot convert all of it into work. A part of the heat must be rejected to a cooler reservoir, the exhaust. In other words, Kelvin–Planck's statement of the second law says that it is impossible to devise a machine that would have an efficiency of 100 percent for the conversion of heat into work, even though the first law will be satisfied. The machine that violates the second law is called a perpetual motion machine of the second kind, and the conclusion from the second law is the following: *It is impossible to devise a perpetual motion machine of the second kind.* (The perpetual-motion machine of the first kind is one that violates the first law of thermodynamics, that is, a machine that will do more work than the heat supplied to it.)

Thus, from the discussion of the first and second laws of thermodynamics, we may conclude that these laws are a big disappointment to the seekers of perpetual motion machines of the first and second kind. *The first law says that you cannot win, and the second law adds that you cannot break even;* that is, you always get less work done than the amount of total heat added.

16.5. The Carnot Cycle and Heat Engines

General Principles

A heat engine is a device that transforms thermal energy (disordered energy) into mechanical work (ordered energy) by repeating the same thermodynamical processes over and over again. At first sight it may appear that such a device is not possible because it leads to a decrease in entropy (disorder → order), but this is not so. In calculating entropy changes we must include the source that supplies the heat and the exhaust (or the environment) to which the heat is rejected. The net effect, after taking everything into consideration, is that the net entropy increases or remains constant.

Basically, any heat engine consists of a working substance, a hot reservoir, and a cold reservoir. The *working substance* (a gas or liquid) is the system that undergoes a cyclic thermodynamical process. The *cyclic process* means that the working substance returns to its original state at regular time intervals. Hence, over a complete cycle the internal energy of the working substance is unchanged, $\Delta U = 0$. Furthermore, the processes are *reversible,* which means that the working substance can retrace the process back and forth by slightly changing the state variables and is continuously in equilibrium. The *hot reservoir* is a source of very large heat capacity at high temperature, from which the working substance draws heat. This loss of heat by the hot reservoir does not change its temperature.

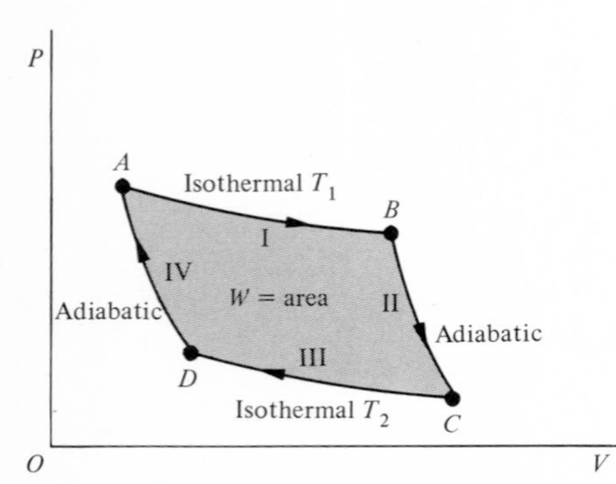

FIGURE 16.9. *PV diagram of a Carnot engine. The net work done is equal to the shaded area.*

Let us see how thermodynamical processes can be used to convert heat into mechanical work. In Figures 16.3 and 16.4 we showed that on a *PV* diagram the work done by or on the system is equal to the area under the curve that represents the process. In Figure 16.3 heat Q is absorbed by the system that is changed into work W_{AB} done by the gas. In Figure 16.4 heat Q_2 is given out by the gas when work W_{BA} is done on the gas. Since $W_{BA} = -W_{AB}$ and $Q_2 = -Q_1$, no net work is done in such an isothermal expansion and compression cycle.

Suppose that the working substance goes through the cyclic process shown in Figure 16.9, which consists of four paths I → II → III → IV. Along path I ($A \rightarrow B$) the working substance absorbs heat Q_1 from the hot reservoir at temperature T_1 and is used in doing work by the expanding gas on the surroundings (such as pushing the piston out). Along path III ($C \rightarrow D$) the working substance rejects heat Q_2 ($<Q_1$) to the cold reservoir at temperature T_2 ($<T_1$) and compresses back to state D. The heat Q_2 given out is the result of the external work done on the gas. The two isothermal paths I and III are connected by an adiabatic expansion path II and an adiabatic compression path IV. The work done in these two processes (II and IV) cancel each other. Thus, the net work W done by the gas is equal to $Q_1 - Q_2$. This partial transformation of heat ($Q_1 - Q_2$) into work W by a heat engine is illustrated schematically in Figure 16.10. This is the flow diagram of a heat engine. The engine takes in heat Q_1 from the source converts part of it into work W and rejects heat Q_2 to the sink. The relative cross section of the three pipes connected to the engine represents the relative amounts of Q_1, Q_2, and W ($=Q_1 - Q_2$). *The efficiency η of such an engine is defined as the fraction of the total heat that is converted into mechanical work.* Thus,

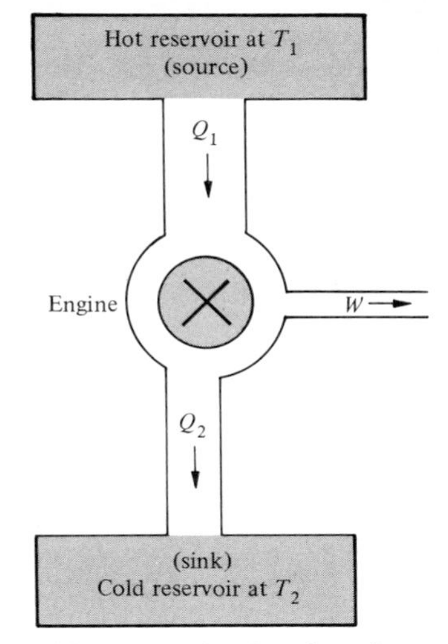

FIGURE 16.10. *Flow diagram of a Carnot engine.*

$$\eta = \frac{\text{output}}{\text{input}} = \frac{\text{work done}}{\text{heat supplied}} = \frac{W}{Q_1} \tag{16.15}$$

But W from the first law is $Q_1 = Q_2 + W$ or $W = Q_1 - Q_2$. Therefore,

$$\boxed{\eta = \frac{W}{Q_1} = \frac{Q_1 - Q_2}{Q_1} = 1 - \frac{Q_2}{Q_1}} \tag{16.16}$$

The source loses heat Q_1 at temperature T_1 and its entropy decrease is $-Q_1/T_1$, while the sink gains heat Q_2 at temperature T_2, so that its increase in entropy is Q_2/T_2. Since according to the second law the net change in entropy ΔS must increase, we may write

$$\Delta S = \frac{Q_2}{T_2} - \frac{Q_1}{T_1} \geqslant 0 \tag{16.17}$$

or from this equation,

$$\frac{Q_2}{Q_1} \geqslant \frac{T_2}{T_1} \tag{16.18}$$

Thus, combining Equation 16.18 with Equation 16.16,

$$\eta \leqslant 1 - \frac{T_2}{T_1} \tag{16.19}$$

For an ideally perfect engine that will be free of dissipative effects such as friction, it will be a truly cyclic process and the entropy change ΔS will be zero. In this case the efficiency will be maximum and $\leqslant$ is replaced by an $=$ sign. That is,

$$\boxed{\eta_{\text{max}} = \frac{W_{\text{max}}}{Q_1} = 1 - \frac{T_2}{T_1}} \tag{16.20}$$

Thus, from Equation 16.20, the efficiency will be 100 percent only if $T_2 = 0$ K, that is, if the sink is at absolute zero temperature. On the other hand, if

$T_1 = T_2$, $\eta = 0$; that is, for a heat engine to operate the source must be at a higher temperature than the sink.

The Carnot Cycle

The ideas discussed above were first introduced in 1824 by Sadi Carnot, who developed a particular type of theoretical cycle, called the Carnot cycle. A *Carnot cycle* on a PV diagram is bounded by two isothermals and two adiabatics, as shown in Figure 16.9. A *Carnot engine* is an ideal reversible engine which operates between two reservoirs, source and sink, and the working substance follows the Carnot cycle. Although theoretical, the Carnot engine (1) has a maximum efficiency for the two temperature limits between which it operates and is independent of the working substance, and (2) no engine can be more efficient than a Carnot engine. Points (1) and (2) together comprise *Carnot's theorem.* A detailed study of the Carnot engine can help us understand practical engines, such as steam, gasoline, and diesel engines.

According to Carnot's theorem, point (1), the efficiency η of a Carnot engine is

$$\eta = 1 - \frac{T_2}{T_1} \tag{16.21}$$

and depends only on the temperatures of the hot and cold reservoir. For example, in the case of a steam engine, where $T_1 = 100°\text{C} = 373\text{ K}$ and $T_2 \simeq 27°\text{C} = 300\text{ K}$, the efficiency of this engine will be

$$\eta = 1 - \frac{300}{373} = 0.20 \quad \text{or} \quad 20\%$$

This is the best efficiency a steam engine can have while working between these temperature limits.

The only way the Carnot engine can be made 100 percent efficient is to have an exhaust temperature at 0 K; that is, $T_2 = 0\text{ K}$, $\eta = 1$. This fact can be used to define absolute zero (see Section 16.7).

The Kelvin Temperature Scale

We have shown above that if an ideal gas is used as a working substance in a Carnot engine, the ratio of the heat rejected to the heat absorbed is equal to the ratio of the temperatures of the two reservoirs; that is,

$$\frac{Q_2}{Q_1} = \frac{T_2}{T_1}$$

where T_1 and T_2 are the absolute temperatures defined by the ideal gas thermometer.

We can use the relation above to define another temperature scale, called the *thermodynamic* or *Kelvin temperature scale*. Suppose that the heat sink is assigned a temperature of the triple point of water; that is, $T_2 = T_t = 273.16\text{ K}$ and $Q_2 = Q_t$. Then the temperature of the heat source, that is, $T_1 = T$ and $Q_1 = Q$, will be given by

$$\frac{Q}{Q_t} = \frac{T}{T_t}$$

or

$$T = 273.16 \text{ K} \left(\frac{Q}{Q_t}\right) \tag{16.22}$$

Thus, knowing the heat Q going into the engine and heat Q_t rejected by the engine, one can measure T of the hot reservoir. The importance of the Kelvin temperature scale lies in the fact that the Carnot engine is independent of the working substance; hence, the *Kelvin scale*, given by Equation 16.22, is a truly absolute temperature scale.

The ideal-gas temperature scale, which was defined by Equation 13.3,

$$T = 273.16 \text{ K} \lim_{P \to 0} \left(\frac{P}{P_t}\right) \quad \text{at constant } V$$

is found to be completely identical with the absolute Kelvin temperature scale defined by Equation 16.22.

EXAMPLE 16.3 An ideal engine operates by taking in steam from a boiler at a temperature of 327°C and exhausting it at a temperature of 27°C. The engine runs at 500 revolutions per minute and the heat taken in is 600 kcal in each revolution. Calculate (a) the Carnot efficiency of the engine; (b) the work done in each revolution; (c) the heat rejected in each revolution; and (d) the power output of this engine.

$$T_2 = 27°\text{C} = (273 + 27) \text{ K} = 300 \text{ K}$$
$$T_1 = 327°\text{C} = (273 + 327) \text{ K} = 600 \text{ K}$$
$$Q_1 = 600 \text{ kcal} = 600(4.18 \times 10^3 \text{ J}) = 2.51 \times 10^6 \text{ J}$$

(a) From Equation 16.21, the Carnot efficiency is

$$\eta = 1 - \frac{T_2}{T_1} = 1 - \frac{300 \text{ K}}{600 \text{ K}} = 0.5 = 50\%$$

(b) $\eta = W/Q_1$; $W = \eta Q_1 = (0.5)(600 \text{ kcal}) = 300 \text{ kcal} = 1.25 \times 10^6$ J. This is the work done in 1 revolution.
(c) $Q_2 = Q_1 - W = 600 \text{ kcal} - 300 \text{ kcal} = 300 \text{ kcal} = 1.25 \times 10^6$ J.
(d) The power output is work done per second. The engine runs at 500 revolutions per minute; therefore,

$$P = \frac{W}{t} = \frac{(500)(1.25 \times 10^6 \text{ J})}{60 \text{ s}} = 10.42 \times 10^6 \text{ W} = 10.42 \times 10^3 \text{ kW}$$

EXERCISE 16.3. Calculate the quantities (a) to (d) in Example 16.3 if the exhaust temperature were 127°C instead of 27°C. [*Ans.:* $\eta = 0.33$, $W =$ 200 kcal, $Q_2 = 400$ kcal, $P = 6.97 \times 10^3$ kW.]

16.6. The Refrigerator and the Heat Pump

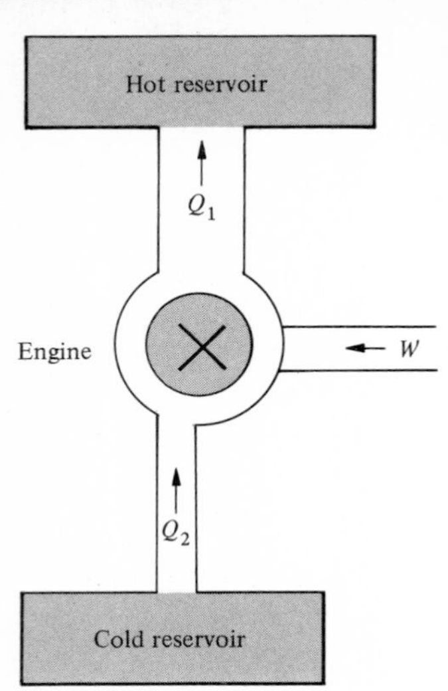

Figure 16.11. *Heat-flow diagram of a refrigerator.*

A heat engine operating in reverse transfers heat from a cold reservoir to a hot reservoir with the help of external work which is also converted into heat and rejected to the hot reservoir, as shown by the heat flow diagram in Figure 16.11. A heat engine operating in reverse is called a *refrigerator* or a *heat pump*. In a refrigerator the interest is in the amount of heat removed from a cold reservoir, while in a heat pump the interest lies in the amount of heat rejected to the hot reservoir. Note that both these devices cause heat to flow "uphill" from a low-temperature reservoir to a high-temperature reservoir at the cost of supplying external mechanical work to the system.

Figure 16.11 shows that heat Q_2 from a cold reservoir at temperature T_2 is transferred to a hot reservoir at temperature T_1 with the help of external mechanical work W. $Q_2 + W$ is the amount of heat rejected to the hot reservoir equal to Q_1. That is, from the first law of thermodynamics,

$$Q_1 = Q_2 + W \tag{16.23}$$

Instead of efficiency in refrigerators, we use the quantity *coefficient of performance E*, which is defined as the ratio of what we want to what we pay.

$$E = \frac{\text{desired output}}{\text{required input}} = \frac{Q_2}{W}$$

If it is truly a Carnot cycle, $Q_2/Q_1 = T_2/T_1$; hence,

$$\boxed{E = \frac{Q_2}{W} = \frac{Q_2}{Q_1 - Q_2} = \frac{T_2}{T_1 - T_2}} \tag{16.24}$$

which is the coefficient of performance of an ideal refrigerator. T_2 may be the temperature of the refrigerator, while T_1 is the exhaust temperature, room temperature in this case. The actual coefficient E is between 2 and 7, depending upon the irreversible processes such as friction.

Example 16.4 A refrigerator freezes 10 kg of water at 0°C into ice at 0°C in a time interval of 30 min. Assume the exhaust temperature (the room temperature) to be 22°C. If the cost of electrical energy supplied to the refrigerator is 3 cents/kWh, what will be the cost of making 10 kg of ice? Also, calculate the minimum amount of power needed to accomplish this.

The amount of heat that must be extracted from the water to convert it into ice is

$$Q_2 = mL_f = (10\text{ kg})(80\text{ kcal/kg}) = 800\text{ kcal}$$

Since $T_2 = 0°\text{C} = 273\text{ K}$ and $T_1 = 22°\text{C} = 295\text{ K}$, the minimum amount of work that must be done from Equation 16.24 is

$$\begin{aligned} W &= Q_2[(T_1/T_2) - 1] = 800\text{ kcal}\,[(295\text{ K}/273\text{ K}) - 1] \\ &= (800\text{ kcal})(0.081) = (64.8\text{ kcal})(4.18 \times 10^3\text{ J}) = 271 \times 10^3\text{ J} \end{aligned}$$

Since 1 kWh = $(1 \times 10^3 \text{ J/s})(3600 \text{ s}) = 3.6 \times 10^6$ J,

$$W = (0.271 \times 10^6)\left(\frac{1 \text{ kWh}}{3.6 \times 10^6 \text{ J}}\right) = 0.075 \text{ kWh}$$

Since the cost is 3 cents/kWh, the total cost will be (0.075 kWh)(3 cents/kWh) = 0.23 cent. The work must be done in a $\frac{1}{2}$ h; therefore, the minimum power needed to run the refrigerator is

$$P = \frac{W}{t} = \frac{271 \times 10^3 \text{ J}}{30 \times 60 \text{ s}} = 151 \text{ W}$$

Of course, we have assumed that it is an ideal refrigerator, that the water is at 0°C and no heat is used by the surroundings. The actual cost is many times higher than this.

Exercise 16.4 Suppose that the actual refrigerator is only 50 percent as efficient as an ideal one, and the temperature of the water is 23°C instead of 0°C. What will be the cost of converting 10 kg of water into ice at −10°C? What is the power needed to accomplish this is 30 min? [*Ans.*: cost = 0.50 cents, P = 336 W.]

16.7. Absolute Zero and the Third Law of Thermodynamics

We showed that the efficiency of a Carnot engine is

$$\eta = \frac{W}{Q_1} = \frac{Q_1 - Q_2}{Q_1} = 1 - \frac{T_2}{T_1}$$

As mentioned earlier, the efficiency will be 100 percent only if $T_2 = 0$. This is the *absolute zero temperature*. If T_2 were less than zero, that is, negative temperature, the efficiency would be greater than 100 percent. This is impossible because it is incompatible with the law of conservation of energy. Hence, the lowest possible temperature is absolute zero. The next question to ask is the following: Is it possible to achieve absolute zero by any physical processes? With the development of the latest experimental techniques for liquefying and freezing of various gases, very low temperatures, such as 1 K, have been reached. By using the special techniques of adiabatic demagnetization, a temperature as low as 10^{-3} K has been reached while the extension of this method has made it possible to reach 10^{-6} K (=0.000001 K). There is enough evidence to make the following statement about the unattainability of absolute zero.

> **Third Law of Thermodynamics:** ***It is impossible to reach the absolute zero of temperature in any physical process.***

Furthermore, in the expression $\eta = 1 - (Q_2/Q_1)$, the smallest possible value of Q_2 is zero, and the temperature at which this will happen is absolute zero. In other words, absolute zero is the temperature of that reservoir to which the Carnot engine will reject no heat. Since there is no transfer of heat at absolute zero, there will be no change in entropy as well. Thus absolute zero can be reached only by a truly reversible process (in an irreversible process, there will be a change in entropy). There is no physical process that is truly reversible (there will always be some change in entropy); hence, it is not possible to attain an absolute zero temperature.

16.8. Thermal Pollution

Many undesirable effects that result from heat waste (heat discharged to the environment) from heat engines or other devices is called *thermal pollution.* The main sources of thermal pollution are the power plants used for the production of electricity. More than 85 percent of the power plants are run by steam engines, which utilize fossil fuel (coal and oil), about 12 percent of the electricity is produced by water power in hydroelectric plants, and about 3 percent by nuclear power plants using uranium 235 as fuel. The basic workings of a plant producing thermal pollution is shown schematically in Figure 16.12.

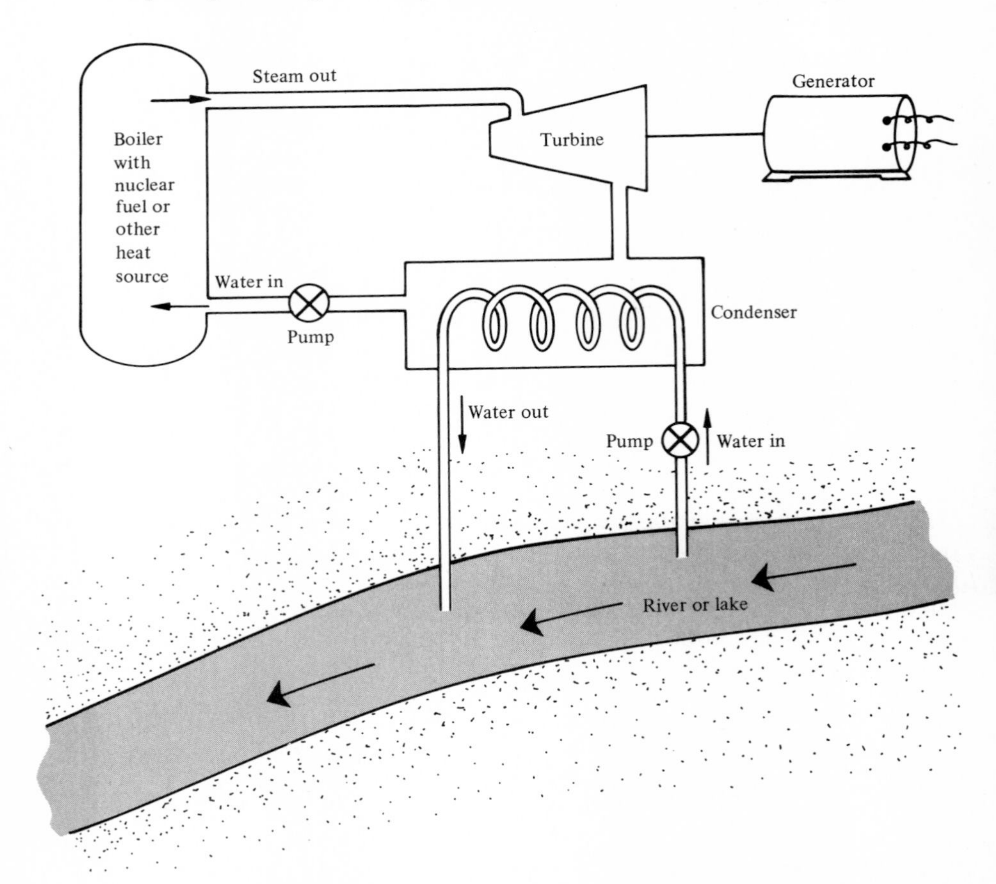

FIGURE 16.12. *Basic workings of a power plant producing thermal pollution.*

Water at a low temperature and pressure enters a boiler where it is changed into steam at a very high temperature and pressure. The heat in the boiler is provided by the direct burning of fossil fuel or by using nuclear fission (described in Chapter 34). The steam enters a turbine, where heat energy is converted into mechanical energy, thereby causing a shaft to rotate. Then, mechanical energy is converted into electricity by a generator (described in Chapter 25). The rejected or exhaust steam at a much lower temperature goes to a condenser. This steam must be brought to still a lower temperature before being fed back to the boiler for recycling. For this purpose, water taken from a nearby source (a lake, river, or stream) is used. This water, after going through the condenser, is rejected back to the water source but at a higher temperature. It is this heat that is called *thermal waste* or *thermal pollution*. In a typical power plant, the difference between the ingoing and outgoing water from the condenser may be from 10 to 12C°. Also, since the efficiency of most engines is ~34 percent, only one-third of the heat energy is used for power production, while the other two-thirds is a complete waste from the engineering point of view and a hazard from society's point of view. Both are interested in reducing this waste. Of course, we know from the second law of thermodynamics that it is not possible to reduce this loss to zero (that will happen if the engine is 100 percent efficient—an impossibility).

The ingoing water used for cooling has a temperature of ~85°F and the outgoing water has a temperature ~100°F, while the flow rate in a typical power plant is ~10,000 gal/s (or ~40 m^3/s). This leads to an increase in temperature of the water in the stream, river, or lake, where the heated water is dumped. This rise in temperature has many diverse effects. It reduces the amount of oxygen in the water, which is essential for aquatic life and also for decomposition of organic matter. This increased temperature speeds up the metabolism of fish and eventually inhibits reproduction. It also stimulates the growth of bacteria and the blue-green algae, while normal vegetation is adversely affected. This may lead to the unbalancing of plant–fish–insect food chain.

There have been many suggestions for avoiding the discharge of thermal waste into water. Some of these are the use of cooling ponds, where the water is left to stand until it cools off. Another method involves the use of cooling towers, in which heat is used for evaporating water or giving heat directly to the air. Some of these methods have been put to work, but the problem is far from being solved.

SUMMARY

Thermodynamics deals with the conversion of thermal energy into mechanical energy. An isolated system with no thermal contact with the surroundings is said to be in *thermal equilibrium*. A *thermodynamical process* is said to have taken place when a system changes its state. A *reversible process* is one in which the changes take place in such a way that the system is always in thermal equilibrium. A *cyclic process* is one in which the system returns to its initial state.

An *isothermal process* is one that takes place at constant temperature and $\Delta Q = \Delta W$. Also, work done is $W = nRT \Sigma \Delta V/V = nRT \ln V_B/V_A$. A *reversi-*

ble adiabatic process is one that takes place with no flow of heat into or from the system, $\Delta U = -\Delta W$. For adiabatic processes, PV^{γ} = constant, where $\gamma = C_p/C_v$ and $\Delta W = nC_v(T_A - T_B)$. An *isobaric process* is one that takes place at constant pressure, and $\Delta U = \Delta Q - P\,\Delta V$. An *isochoric process* is one that takes place at constant volume, and $\Delta U = \Delta Q$.

The *entropy* is a measure of molecular disorder, and in any process it always increases or remains constant. The change in entropy is $\Delta S = \Delta Q/T$.

The *second law of thermodynamics* states: (1) For an isolated system, only those processes take place for which the entropy increases or remains constant, or (2) in the *Clausius form*, it is impossible to construct a device that operating in a cycle will produce no other effect than to transfer heat from a cooler body to a hotter body, or (3) in the *Kelvin form*, it is impossible to construct a device operating in a cycle that will extract heat from a reservoir and change it entirely into work.

A *heat engine* is a device that transfers thermal energy into mechanical work by repeating the same thermodynamical processes over and over again. The efficiency η of an engine is $\eta = W/Q_1$ = work done/heat supplied.

For a Carnot engine, $\eta = 1 - T_2/T_1$, and no engine can be more efficient than the Carnot engine. In a Carnot engine, $Q_2/Q_1 = T_2/T_1$, and this can be used to define a thermodynamic or Kelvin temperature scale $T = 273.16\text{ K}(Q/Q_t)$.

A heat engine operating in reverse is called a *refrigerator*, and its coefficient of performance is $E = Q_2/W = T_2/(T_2 - T_1)$.

According to the *third law of thermodynamics*, it is impossible to reach a temperature of absolute zero in any physical process.

QUESTIONS

1. To apply the laws of thermodynamics, the system must be in thermal equilibrium. Why?
2. Give two examples each of reversible and irreversible processes.
3. Will the velocity of the gas molecules increase or decrease in the following processes: (a) isothermal; (b) adiabatic; (c) isobaric; and (d) isochoric?
4. Why are the slopes of all the isotherms and the adiabatics in Figure 16.5 negative?
5. Under what conditions can you add heat to a system without changing its temperature?
6. Why does the temperature drop in an adiabatic expansion process?
7. Give two examples (other than the one discussed in the text) in which disorder increases.
8. If work is done by friction, will the entropy increase or decrease?
9. Why is it not possible to convert a given amount of heat completely into work? Is the reverse possible? Why.
10. Give an example of processes that are permitted by the law of conservation of energy but are forbidden by the second law of thermodynamics.
11. Is it possible, according to the second law of thermodynamics, to construct an engine that will be free of thermal pollution?
12. How can you increase the efficiency of an engine by changing its temperature?

13. What steps do you suggest should be taken to improve the low efficiencies of different engines?
14. "If an engine is 100 percent efficient, its exhaust must be at absolute 0 K." Explain. Is it possible to construct such an engine?
15. Is it wise and practical to cool a kitchen by keeping the refrigerator door open? Explain.

PROBLEMS

1. How much work is done if air at 0°C and 1 atm expands isothermally from a volume of 1.5 m^3 to 2.0 m^3?
2. A quantity of 1 kg of nitrogen gas at 20°C expands isothermally to three times its original volume. Calculate the work done by the gas and the heat that must be supplied.
3. Gas at 6 atm pressure is confined in a cylinder fitted with a movable piston of 5 cm in radius. The gas is heated so that the piston moves a distance of 6 cm while the pressure remains constant. Calculate the amount of work done by the gas during this expansion.
4. Calculate the amount of work done in changing 1 kg of water at 100°C into 1 kg of steam at 100°C and 1 atm pressure. One kg of steam has a volume of 1.671 m^3.
5. A monatomic gas expands adiabatically to double of its volume. What is the ratio of the final pressure to its initial pressure?
6. A diatomic gas at 1 atm has a volume of 2 m^3 and expands adiabatically to a volume of 2.5 m^3. What is the final pressure?
7. A quantity of 0.2 kg of oxygen is compressed adiabatically from 20°C to 100°C. Calculate the work done in J and kcal.
8. Heat in the amount of 250 cal is supplied to oxygen gas at constant pressure of 1 atm and it expands from 700 cm^3 to 800 cm^3. Calculate the change in internal energy.
9. Calculate the change in entropy when 5 kg of water at 100°C is changed into steam at 102°C (specific heat of steam is 0.48 cal/g-C°).
10. Calculate the change in entropy when 5 kg of water at 96°C is changed into steam at 104°C.
11. Calculate the changes in entropy when 1 kg of (a) ice at −10°C changes to ice at −8°C; (b) water at 49°C changes to water at 51°C; and (c) steam at 150°C changes to steam at 152°C. How do you explain the difference in the magnitudes of the entropy?
12. A quantity of 4 kg of dry ice (CO_2) sublimates at a temperature of −78°C and a pressure of 1 atm. Calculate the change in entropy if the latent heat of sublimation is 138 kcal/kg.
13. The melting point of copper is 1083°C and the latent heat of fusion is 32 kcal/kg. Calculate the change in entropy of copper when it changes from solid to liquid.
14. If 10 kg of water at 90°C is mixed with 5 kg of water at 80°C, what is the change in entropy of the system?
15. If 1 kg of ice is mixed with 50 kg of water at 10°C, what will be the total change in entropy?

16. A total of 500 kcal is extracted from a hot reservoir at 100°C and rejected to a cold reservoir at 0°C. Calculate the change in the entropy of each reservoir and the total change in entropy. Is the process reversible?

17. If 500 kcal is extracted from a hot reservoir at 100°C with 100 kcal changed into mechanical work and the remainder rejected to a cold reservoir at 0°C, calculate the change in entropy of (a) the hot reservoir; (b) the cold reservoir; and (c) the whole system. Is the process reversible?

18. A total of 400 kcal is extracted from a hot reservoir at 500 K. Of these, how many kcal will be rejected to a reservoir at 300 K so that the process will be reversible?

19. Calculate the efficiency of a Carnot engine that takes in 200 kcal and rejects 160 kcal in each cycle. If the temperature of the hot reservoir is 500 K, what is the temperature of the cold reservoir?

20. Calculate the efficiency of a heat engine that takes in 2 kcal of heat and does 8×10^2 J of mechanical work.

21. The temperatures of the hot and cold reservoirs of a heat engine are at 227°C and 47°C. The engine absorbs 2 kcal from the hot reservoir and rejects 1.8 kcal to the cold reservoir. Calculate (a) the work done; (b) the Carnot efficiency; and (c) the actual efficiency.

22. A Carnot engine has an efficiency of 30 percent while operating at a low-temperature reservoir at 127°C. What is the temperature of the hot reservoir? The efficiency can be increased either by decreasing T_2 or by increasing T_1. Suppose that we want to increase the efficiency to 50 percent. Calculate (a) T_2 if T_1 is kept constant; and (b) T_1 if T_2 is kept constant. Which of the two alternatives do you prefer?

23. A steam engine works with reservoirs at temperatures of 207°C and 37°C while a gasoline combustion engine has temperatures of 1550°C and 475°C. Which of the two engines is more efficient?

24. A heat engine is working at 300 revolutions per minute. In each revolution it does 300 ft-lb of work and rejects heat-energy equivalent to 1200 ft-lb. Calculate (a) the efficiency of the engine; (b) the heat intake of the engine in each revolution; and (c) the power output both in hp and kW.

25. A gasoline engine has a combustion temperature of 1475°C and exhaust temperature of 350°C. The engine is doing useful work at the rate of 20 kW and losing to friction in the engine at the rate of 5 kW. Calculate (a) the Carnot efficiency; (b) the rate at which heat energy is taken in; and (c) the actual efficiency of the engine.

26. The overall efficiency of a steam engine is 12 percent. The power of the engine is 1000 hp (1 hp = 0.75 kW) and it operates between the temperatures of 212°F and 104°F. Calculate (a) the Carnot efficiency; (b) the heat drawn from the source; (c) the heat rejected to the exhaust; and (d) the fraction of the energy lost.

27. An eight-cylinder engine has a piston with a cross-sectional area of 80 cm^2. The length of the stroke is 15 cm, during which the average gauge pressure is 8×10^5 N/m^2. How many revolutions per minute should the engine make to develop 100 kW? What is this power, in units of hp?

28. A power plant is being constructed that uses steam at a temperature of 207°C in its turbine. There is a choice of exhausting into the atmosphere at a temperature of 20°C or into lake water at a temperature of 10°C. Calculate the efficiency in the two cases. Does it make much difference?

29. A Carnot refrigerator works between temperature limits of 0°C and 27°C (hot kitchen temperature). Suppose that 10 kg of water at 0°C are converted into ice at 0°C. Calculate (a) the heat ejected; and (b) the energy supplied to the refrigerator.

30. Do Problem 29 if the water was initially at 20°C.

31. A Carnot refrigerator takes 5 kcal in each cycle from a freezer compartment at a temperature of −25°C and ejects to a room at a temperature of 27°C. Calculate the amount of work done; and the heat ejected to the room.

32. In Problem 31, suppose that the actual efficiency of the engine is 50 percent of the Carnot engine efficiency. If 300 kcal must be removed every minute from the freezer compartment, what is the power supplied by the electric motor of the refrigerator?

33. Refrigerator *A* works between −10°C and +27°C while refrigerator *B* works between −20° and +17°C, both removing 2000 J from the freezer. Which of the two is the better refrigerator?

Part IV
Wave Motion and Sound

17
Wave Motion

Wave phenomena include both mechanical waves and electromagnetic waves. In this chapter we limit ourselves to the study of mechanical waves, longitudinal as well as transverse. We shall derive expressions for the wave velocity as a function of the properties of the medium in which they travel. We conclude the chapter with a brief introduction to interference and diffraction of sound waves.

17.1. Wave Phenomena

WE OBSERVE that if we throw a pebble into a pond of still water, a few circular ripples, called *waves* or *pulses*, move outward on the surface of the water. Let us dip one end of a stick into the water while its other end is connected to a vibrator. The vibrating stick produces a continuous train of circular waves, as shown in Figure 17.1. These waves are a form of disturb-

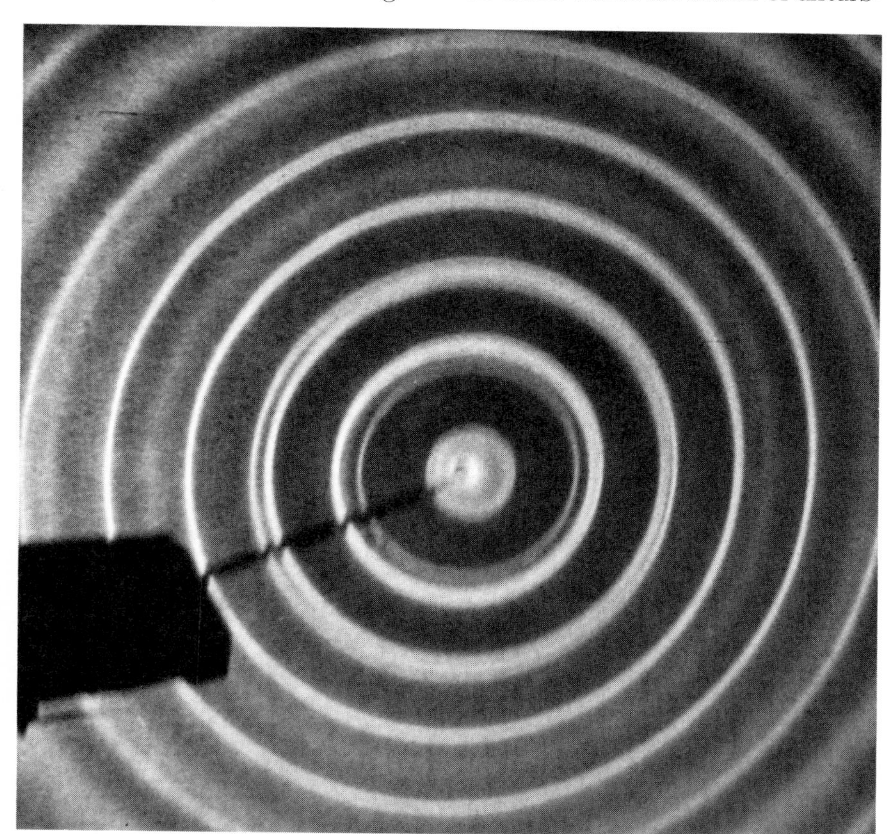

FIGURE 17.1 *Instantaneous photograph of the shadows of circular waves on the water surface produced by a vibrating stick.* (From *PSSC Physics*, 2nd ed., 1965, copyright by Education Development, Inc. Reprinted by permission of D. C. Heath & Company.)

ance that travel outward, and no portion of the medium (water in this case) is transported from one part of the medium to another, even though it may look otherwise. This may be verified by placing a small piece of cork or wood on the surface of the water. As the waves reach the cork, it starts moving up and down, and also back and forth. As soon as the waves have passed by, it stops.

Water waves are just one of the several kinds of wave motion that we come across everyday. Wave motion appears not only in almost every branch of physics, but in numerous natural phenomena. One common characteristic of all wave motion is that *it provides a mechanism for transfer of energy from one point to another without physical transfer of any material between the points.* Examples of wave motion are radio waves, light waves (from the sun, a candle, a light bulb, or any other light source), sound waves, waves on strings, ocean waves, and so on. All these wave motions can be divided into two groups.

Electromagnetic Waves

Electromagnetic waves is a name given to all types of light waves, sun rays, x-rays, gamma rays, microwaves, and radiowaves, as listed in Chapter 27. These waves require no medium to sustain them (as water waves need water as a medium) and can travel through the vacuum. For example, rays from the sun travel through the free space before reaching us on the surface of the earth. Electromagnetic waves are a subject of detailed study in Chapter 27.

Mechanical Waves

These are sound waves in air, water waves, waves in solids and liquids, and so on. As a matter of fact all types of waves that are produced in deformable or elastic media are called *mechanical* or *elastic waves*. Note that, unlike electromagnetic waves, mechanical waves need a medium to sustain them. This and the next chapter will be devoted to the study of such waves.

The mechanical wave motion involves (1) the vibrator, which produces displacement; (2) some sort of disturbance, which travels from the source into the medium; and (3) the medium to support the disturbance. The disturbance or wave motion carries energy from the source into the medium. There are two properties of the medium—elasticity and inertia—that determine the characteristics of wave motion. When the vibrator is in motion, it displaces a portion of the elastic medium, causing it to oscillate about an equilibrium position. The disturbance is transmitted from one portion of the medium to the next; hence, the kinetic and potential energy of the vibrating portion of the medium is passed to successive portions of the medium. In short, *mechanical wave motion* is an energy-carrying disturbance that is produced by the repeated periodic motion of the elastic medium and is passed on from one portion of the medium to the next.

17.2. Type and Range of Elastic Waves

When a continuous train of waves is produced, the wave motion traveling outward from the source causes oscillating motions of the particles of the medium. The particles of the medium continuously keep vibrating about their equilibrium position. The motion of the particles will be simple harmonic if the source producing the initial displacement is vibrating with simple harmonic motion and if the amplitudes are small. The wave motion may be classified as a transverse wave or longitudinal wave, depending on the nature of the motion executed by the particles of the medium.

> ***Transverse Waves* are those in which the particles of the medium vibrate at right angles to the direction of propagation of the wave motion. *Longitudinal Waves* are those in which the particles of the medium vibrate back and forth about their equilibrium position along the direction of propagation of the wave motion.**

Wave motion in a stretched string is a typical example of transverse waves. To see how these waves are propagated, let us consider the following situation. Figure 17.2(a) shows a long stretched string tied at both ends. The portion ABC of the string is distorted by holding C to its initial position while B is pulled upward. If we now release B and C, B starts moving down toward the equilibrium position, while the portion at C starts moving upward. This process continues and results in a pulse traveling to the right, as shown in Figure 17.2(c), (d), and (e). This is a transverse pulse. If the string would have been displaced downward, a pulse would still travel to the right, except that it would have been inverted. To produce a continuous train of waves, the left end of the string can be attached to a vibrator which executes simple harmonic motion of amplitude A, time period T, and frequency $f = 1/T$, as shown in Figure 17.3(a). This will result in a continuous train of transverse waves traveling to the right, as shown in Figure 17.3(b). It is quite clear that each

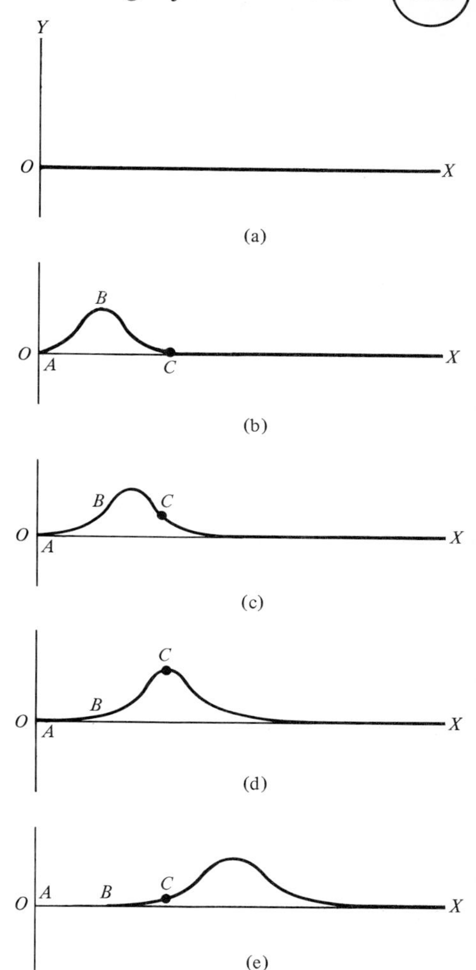

FIGURE 17.2 *Transverse pulse traveling to the right in a string is produced by pulling the string at B while holding C, and then releasing both B and C, as in part (b).*

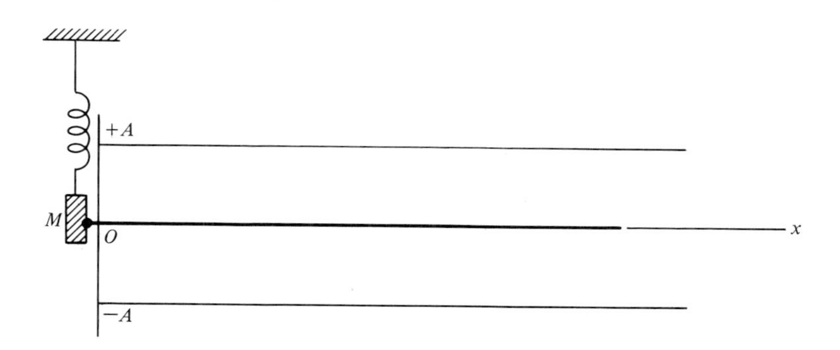

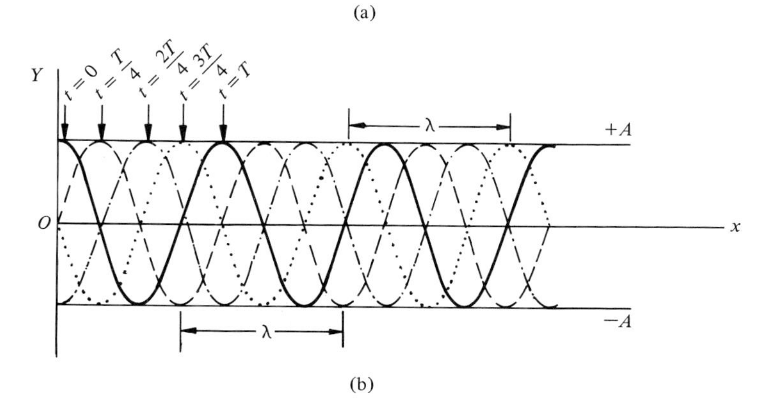

FIGURE 17.3 *(a) Spring–mass system having a vibrator that produces transverse sinusoidal waves traveling to the right in a long string. (b) Wave forms produced by the motion of the string as the mass M moves from +A to O to −A to O and back to +A.*

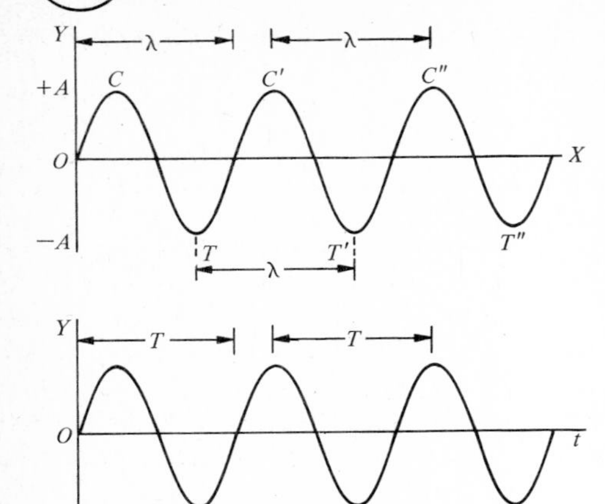

FIGURE 17.4 *(a) Displacement of different particles of strings at any instant of time. (b) Displacement of one particle of the string at different times.*

portion of the string executes simple harmonic motion, but all these motions are not in phase. Figure 17.4(a) is a graph showing the displacement of different particles of the string at one instant of time, while Figure 17.4(b) shows a graph of the displacement of one particle of the string at different times. Thus, the wave is a sinusoidal wave because the plots of the displacement y versus x and also y versus t are sine (or cosine) curves. The work done by the vibrator on the string to cause displacement is transmitted along the string as potential and kinetic energy of the up-and-down motion of the particles of the string.

Referring to Figure 17.4(a), the points of maximum positive displacement ($x = +A$), such as C, C', C'', . . . , are called *crests* while the points of maximum negative displacement ($x = -A$), such as T, T', T'', . . . , are called *troughs*. The wavelength λ is the distance between two successive crests or two successive troughs. More generally, the *wavelength* is the distance between any two successive points that are in the same phase. Two points in the same phase means that they are precisely in the same state of disturbance at the same time. Two particles undergoing simple harmonic motion are said to be *in phase* if the difference in their phase angles is 0° or multiples of 360°; otherwise, they are *out of phase*. Any two crests and any two troughs are in phase, whereas a crest and a trough are 180° out of phase with respect to each other.

We can now find a relation between the wave speed v, the time period T, wavelength λ, and the frequency f. By definition, a wave disturbance travels a distance of one wavelength in one time period. Thus,

$$v = \frac{\text{distance traveled}}{\text{time taken}} = \frac{\lambda}{T}$$

since $f = 1/T$,

$$\boxed{v = \frac{\lambda}{T} = f\lambda} \tag{17.1}$$

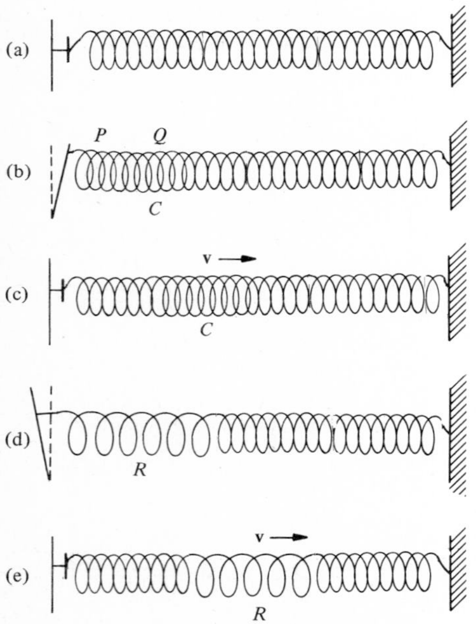

These relations hold good for any type of sinusoidal wave motion, transverse or longitudinal. Note that as shown in Figure 17.4(b), T is also the time in which a particle of the medium completes one vibration.

To understand longitudinal wave motion, consider a helical spring in equilibrium position and attached to a vibrator, as shown in Figure 17.5(a). As the vibrator moves to the right, it compresses the spring as shown in Figure 17.5(b). The region PQ is called *condensation*. As the vibrator moves back to the equilibrium position, the compression of the spring moves to the right of Q; that is, the condensation moves to the right. This process continues and a region of condensation moves to the right as shown in Figure 17.5(c). Similarly, as the vibrator moves to the left of the spring, a condition called *rarefaction* exists, as shown in Figure 17.5(d). A region of rarefaction can be made to travel to the right as shown in Figure 17.5(e).

FIGURE 17.5 *(a) Coil spring in equilibrium; production of longitudinal pulses of condensation as in parts (b) and (c); pulses of rarefaction as in parts (d) and (e).*

Thus, the spring attached to a vibrator that vibrates between $x = +A$ and $x = -A$, as shown in Figure 17.6, sends waves traveling to the right, which consist of condensations and rarefactions. Note that these are longitudinal waves because the particles of the medium, the coils of the spring in this case, move back and forth along the direction of propagation of the wave motion.

The distance between two successive condensations or rarefactions is called the *wavelength*. The formula $v = f\lambda = \lambda/T$ of Equation 17.1 also holds in this case. Note that any two condensations are in phase, whereas a condensation and a rarefaction are completely out of phase (180° out of phase). It should be kept clearly in mind that the velocities of the particles of the medium, moving up and down or back and forth are quite different from the velocity v of the wave motion. In the case of a longitudinal wave, the displacement of any particle is still represented by y, even though it is along the horizontal direction. The maximum displacement, called the *amplitude*, is $y = \pm A$, and the plots of y versus x and y versus t shown in Figure 17.4 for transverse waves hold in this case as well.

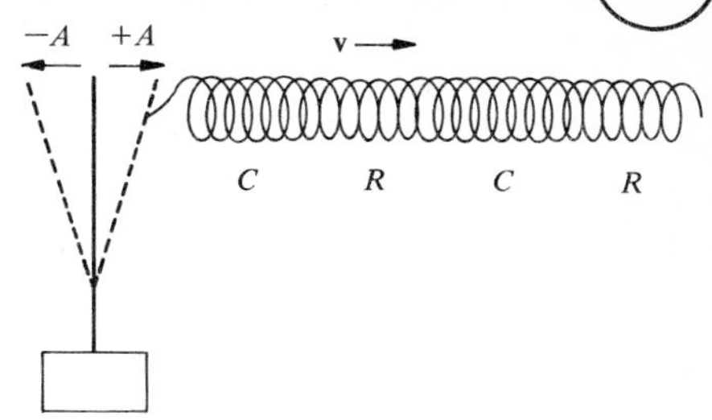

FIGURE 17.6 *Production of a continuous sinusoidal longitudinal wave train traveling to the right in a long coil spring.*

Let us now discuss one of the most important of all wave motions, sound waves in air. In air or gas the only type of wave motion possible is longitudinal wave motion. The formation of condensations and rarefactions can be demonstrated with the help of Figure 17.7. Suppose that a thin layer of powder is

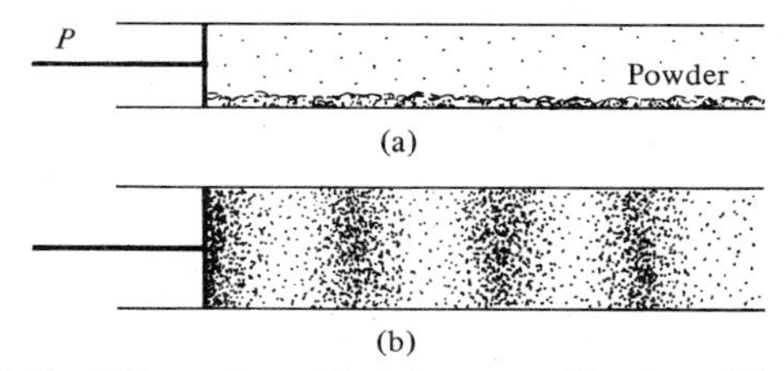

FIGURE 17.7 *When piston Pin (a) is moved back and forth, the powder in the tube redistributes into condensations and rarefactions as shown in part (b) at a particular instant.*

placed all along the inside of a long tube which is provided with a plunger on the left. The spacing of the dots indicates the density of the air molecules throughout the tube before starting the disturbance. The plunger is set into back-and-forth periodic motion. This sends waves of condensation and rarefaction in the tube, as shown in Figure 17.7(b) for one particular instant. Of course, these condensations and rarefactions move to the right with time as the piston moves back and forth. At a condensation the molecules are crowded together, resulting in higher pressure than normal, and at a rarefaction the molecules are farther apart, resulting in lower pressure than normal. The presence of powder makes the condensations and rarefactions visible. The horizontal displacement y of the molecules from their equilibrium positions versus x or t is the same as is shown in Figure 17.4 for transverse waves.

It may be pointed out that in a sound wave the amplitude of oscillation of air molecules at any point is very small. For example, at a frequency of 1000 Hz, a very loud sound causes a displacement amplitude of the order of 1/100 mm, but it is sufficient to hurt the ear. Also, the corresponding pressure variation is about 28 N/m^2, which is only 0.03 percent of normal atmospheric pressure.

In solids, both types of vibrations, longitudinal as well as transverse, are possible, but the speed of the longitudinal waves is always greater than the transverse waves. Water waves, discussed in Section 17.1, also consist of both types of vibrations, although the transverse waves are predominant.

As we shall see later, the speed of sound in air is 331 m/s at 0°C; that is, sound travels about 1 km in 3 s (or 1 mi in 5 s). The speed of sound increases by 0.6 m/s for each Celsius degree increase in temperature. The speed of sound in different media is different. For example, in water at 15°C the speed of sound is 1450 m/s.

As shown in Figure 17.8, elastic (or mechanical) vibrations have a very wide range of frequencies—from below 1 hertz (Hz) to $\gtrsim 10^{13}$ Hz. The human ear

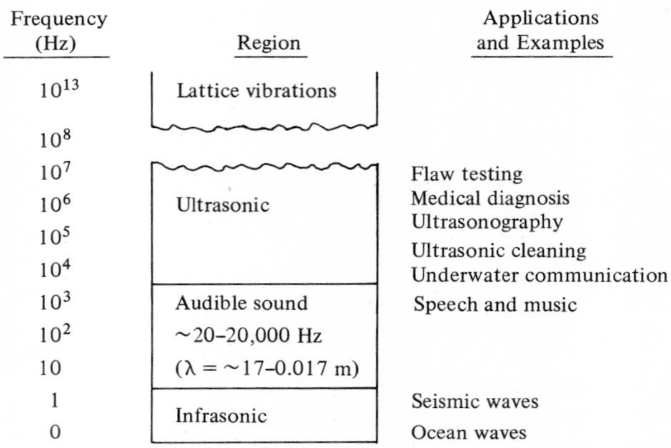

Frequency (Hz)	Region	Applications and Examples
10^{13}	Lattice vibrations	
10^8		
10^7	Ultrasonic	Flaw testing
10^6		Medical diagnosis Ultrasonography
10^5		Ultrasonic cleaning
10^4		Underwater communication
10^3	Audible sound	Speech and music
10^2	~20–20,000 Hz	
10	(λ = ~17–0.017 m)	
1	Infrasonic	Seismic waves
0		Ocean waves

FIGURE 17.8 *Range of mechanical waves.*

responds only to frequencies between 20 and 20,000 Hz. This is the *audible range* of sound. Since $v = 331$ m/s at 0°C, the corresponding wavelengths of the audible range are between 17 and 0.017 m. Sound waves include not only the audible range, but all mechanical waves as well. The waves of frequencies above the audible range are called *ultrasonic* (2×10^4 to $\sim 10^{10}$ Hz). The waves of frequencies below the audible range are called *infrasonic* (<1 to 15 Hz).

EXAMPLE 17.1 A message from ship A to ship B is sent simultaneously by two methods, using sound waves in air as well as water. The message through air reaches 6.0 s later than the message through water. What is the distance between the two ships? The speed of sound in air is 340 m/s and in water is 1480 m/s.

Let the distance between the two ships be x. If the message in water takes t seconds, it will take $(t + 6.0)$ seconds in air. Using the relation $x = vt$,

In water: $x = (1480 \text{ m/s})t$ (i)

In air: $x = (340 \text{ m/s})(t + 6.0)$ (ii)

Equating the two,

$$(1480 \text{ m/s})t = (340 \text{ m/s})(t + 6)$$
$$t = 1.8 \text{ s}$$

Substituting in Equation i

$$x = 2664 \text{ m}$$

EXERCISE 17.1 In Example 17.1, what will be the time interval between the two messages if the two ships are (a) 1 km apart; and (b) 6 km apart? [*Ans.:* (a) 2.3 s; (b) 13.6 s.]

17.3. Equation Representing a Traveling Wave

Let us consider a long string stretched in a horizontal direction along the X-axis. Suppose that the left end of this string is attached to a harmonic vibrator. This will make the waves travel from the left end of the string to the right. At any particular instant, say $t = 0$, the wave train on the string may be represented by a sine curve, as shown in Figure 17.9. The equation of this sinusoidal wave is (solid curve)

$$y = A \sin \frac{2\pi}{\lambda} x \qquad \text{at } t = 0 \tag{17.2}$$

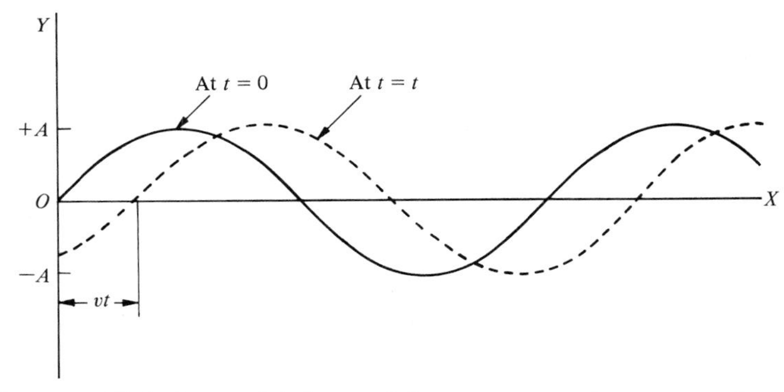

FIGURE 17.9 *Sinusoidal shape of the string at* $t = 0$. *In time* t *the sine wave has moved to the right a distance* $x = vt$.

Note that $y = 0$ where $x = 0, \lambda/2, 2\lambda/2, 3\lambda/2, 4\lambda/2, 5\lambda/2, \ldots$, and $y = \pm A$ where $x = \lambda/4, 3\lambda/4, 5\lambda/4, \ldots$. Since the waves are traveling to the right, the displacement at any point will vary from instant to instant, as shown in Figure 17.3(b). As Figure 17.9 indicates, the original curve at $t = 0$ is displaced laterally in time t through a distance vt and is shown by the dashed curve. To make the curve at time t coincide with the curve at $t = 0$, the whole curve will have to be displaced along the $-X$-axis through a distance $-vt$. We say that the path difference between the two wave trains at $t = 0$ and $t = t$ is $-vt$. Thus, the general equation giving the displacement y in terms of both x and t will be otained by replacing x by $(x - vt)$ in Equation 17.2. That is,

$$\boxed{y = A \sin \frac{2\pi}{\lambda}(x - vt)} \tag{17.3}$$

According to Equation 17.1, $v = \lambda/T$. Substituting for v in Equation 17.3, we get

$$y = A \sin 2\pi \left(\frac{x}{\lambda} - \frac{t}{T}\right) \tag{17.4}$$

This equation states that at any given instant t, y has the same value at x, $x + \lambda$, $x + 2\lambda$, . . . , and that at a given position x, y has the same value at t, $t + T$, $t + 2T$, ... Equation 17.4 may be written in a compact form by introducing the quantities *wave number* k and *angular frequency* ω. Thus, if

$$k = \frac{2\pi}{\lambda} \qquad \text{and} \qquad \omega = \frac{2\pi}{T} = 2\pi f \tag{17.5}$$

for a wave traveling to the right, Equation 17.4 may be written

$$y = A \sin (kx - \omega t) \tag{17.6}$$

while the wave traveling to the left is given by

$$y = A \sin(kx + \omega t)$$

At this point it is important to note that the wave velocity or phase velocity v discussed above is different from the speed v_p of a particle at a distance x from the origin. This particle speed may be found in a manner similar to that used for simple harmonic motion, making use of a reference circle as described in Chapter 10.

EXAMPLE 17.2 The equation of a transverse traveling wave is given by

$$y = 0.05 \sin 2\pi(0.5x - 10t)$$

where x and y are in meters and t in seconds. Calculate (a) the amplitude; (b) the wavelength; (c) the time period; (d) the frequency; and (e) the speed of the waves.

Comparing with Equation 17.6, $y = A \sin (kx - \omega t)$, where $k = 2\pi/\lambda$ and $\omega = 2\pi/T$, we get

(a)
$$A = 0.05 \text{ m} = 5 \text{ cm}$$

(b)
$$kx = 2\pi(0.5x) \qquad \text{or} \qquad k = \pi$$

$$\frac{2\pi}{\lambda} = 2\pi(0.5) \qquad \lambda = 2 \text{ m}$$

(c)
$$\omega t = 2\pi(10)t \qquad \text{or} \qquad \frac{2\pi}{T} = 20\pi$$

$$T = 0.1 \text{ s}$$

(d)
$$f = \frac{1}{T} = 10 \text{ Hz}$$

(e) $v = f\lambda = (10/\text{s})(2\text{ m}) = 20\text{ m/s}$

EXERCISE 17.2 The equation of a transverse traveling wave is given by

$$y = 0.01 \sin 2\pi(0.05x - 1000t)$$

where x and y are in meters and t in seconds. Calculate (a) the amplitude; (b) the wavelength; (c) the time period; (d) the frequency; and (e) the speed. [*Ans.*: (a) 0.01 m; (b) 20 m; (c) 0.001 s; (d) 1000 Hz; (e) 20 km/s.]

17.4. Transverse Waves in Strings

In this section we shall derive an expression for the velocity v of transverse waves in a string in terms of the properties of the medium. Let us consider a string that is under tension T. When in equilibrium, the string is straight; otherwise, it has a sinusoidal transverse wave form as shown in Figure 17.10(a).

Instead of a continuous train of sinusoidal waves, consider a single pulse shown in Figure 17.10(b) of such a shape that the small element of length Δl is an arc of a circle of radius R, as shown in Figure 17.10(c). Let us observe this pulse from the point of view of an observer traveling with the pulse so that the pulse will seem to remain stationary while the particles of the string will continue to pass through the pulse.

Consider length Δl of the string, which is acted upon by tension T at each end as in Figure 17.10(c). Let μ be the mass per unit length of the string. The mass $\Delta m = \mu \Delta l$ of the segment Δl is moving with speed v in a circular arc of radius R. Thus, the centripetal force that acts on this mass Δm is

$$F = \frac{\Delta m v^2}{R} = \frac{\mu \Delta l\, v^2}{R} \tag{17.7}$$

This force is provided by the vertical components of the tensions at the two ends of the segment of length Δl. Thus, from Figure 17.10(c) (assuming that θ is small, $\sin\theta \approx \theta$),

$$F = 2T\sin\theta \simeq 2T\theta = 2T\frac{\Delta l/2}{R} \tag{17.8}$$

The horizontal components of the tensions $(+T\cos\theta)$ and $(-T\cos\theta)$ cancel each other. Equating the two equations above,

$$\frac{\mu \Delta l\, v^2}{R} = 2T\frac{\Delta l/2}{R}$$

that is,

$$\boxed{v = \sqrt{\frac{T}{\mu}}} \tag{17.9}$$

Thus, the velocity v of the transverse waves depends upon two factors—the elastic factor T and the inertial factor μ.

We have confined our discussion to the *plane-polarized waves,* that is, waves in which the vibrations of the particles of the medium are confined to one

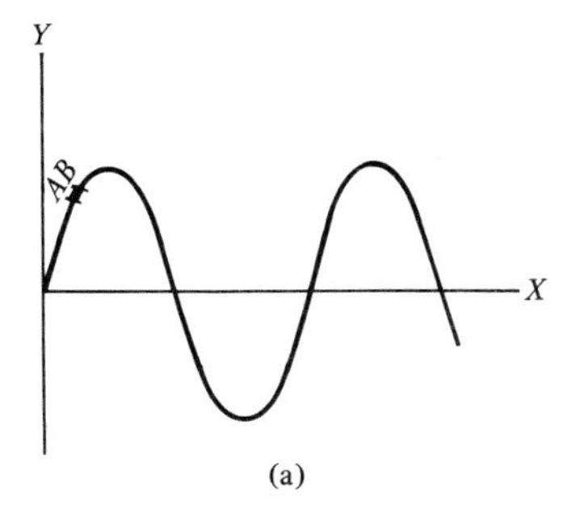

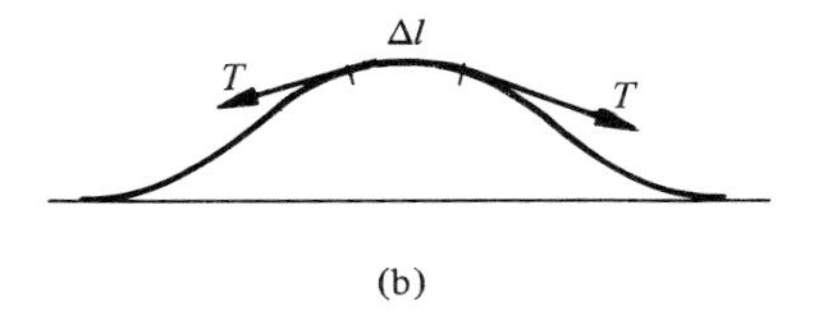

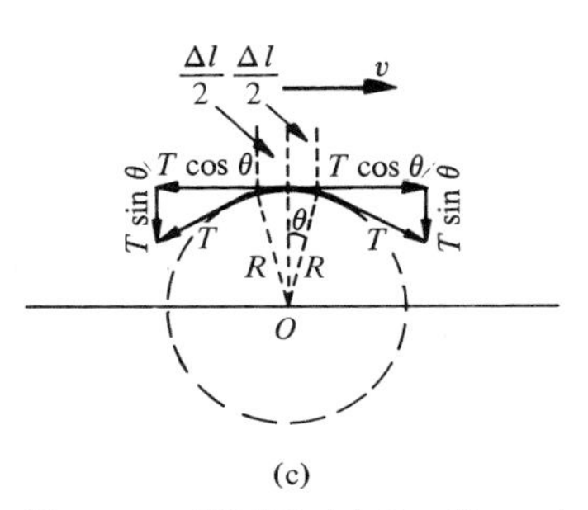

FIGURE 17.10 *(a) Small portion AB of a string that is in transverse motion. (b) Transverse pulse in motion. (c) Tensions acting on Δl, which is a small element of a circle of radius R.*

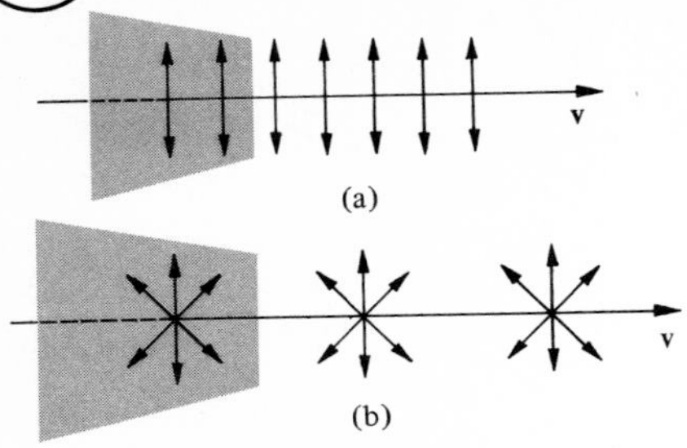

FIGURE 17.11 *Vibration of particles in (a) a plane-polarized wave and (b) an unpolarized wave in three dimensions.*

plane, usually the vertical plane, as in the case shown in Figure 17.11(a). The plane-polarized transverse waves in strings are a special case. The velocity of the unpolarized transverse waves in a three-dimensional medium such as a solid, as shown in Figure 17.11(b), is given by

$$v = \sqrt{\frac{S}{\rho}} \tag{17.10}$$

where S is the shear modulus and ρ the density of the medium.

EXAMPLE 17.3 A string 3 m long has a mass of 7.5 g. It is tied at one end, runs horizontally, goes over a pulley, and a mass of 10 kg hangs vertically at the other end. If this string is vibrating, what will be the velocity of the transverse waves?

From Equation 17.9, $v = \sqrt{T/\mu}$, where $\mu = m/L = 7.5 \times 10^{-3}\text{ kg}/3\text{ m} = 2.5 \times 10^{-3}\text{ kg/m}$ and $T = Mg = (10\text{ kg})(9.8\text{ m/s}^2) = 98\text{ N}$. Therefore,

$$v = \sqrt{\frac{T}{\mu}} = \sqrt{\frac{98\text{ N}}{2.5 \times 10^{-3}\text{ kg/m}}} = 198\text{ m/s}$$

EXERCISE 17.3 In Example 17.3 what should be the magnitude of the vertically hanging mass so that the velocity of the transverse waves becomes double the velocity above? [*Ans.*: 42 kg.]

17.5. Longitudinal Waves in a Gas

Sound waves are longitudinal in nature and may be regarded as elastic waves in a gas. Owing to the periodic motion of the source of sound, the gas moves back and forth, resulting in a change in density. The density corresponds to a change in pressure. The pressure changes generate a gas disturbance, creating condensations and rarefactions (Figure 17.7) that propagate in all three directions. For simplicity we shall limit our discussion to the transmission of longitudinal waves in one direction.

Let us now calculate the velocity of this pressure disturbance in the gas. As in the case of transverse waves, the velocity of longitudinal waves also depends upon two factors, the elastic factor and the inertial factor. If the elastic factor is represented by the elasticity E of the medium and the inertial factor by mass density ρ, the velocity v of the longitudinal wave is given by

$$v = K\sqrt{\frac{E}{\rho}}$$

where K is a constant, which is found to be unity both experimentally as well as theoretically. Thus, in general, we may write

$$\boxed{v = \sqrt{\frac{\text{elastic modulus}}{\text{density}}}} \tag{17.11}$$

Depending upon the type of medium, this expression will take different forms. If we are dealing with longitudinal waves traveling along the length of a long solid rod, the velocity is given by

$$v = \sqrt{\frac{Y}{\rho}} \quad \text{(in solids)} \tag{17.12}$$

where Y is Young's (or stretch) modulus.

On the other hand, if longitudinal waves are traveling in a liquid or gas in which the waves spread out in all directions, Equation 17.11 takes the form

$$v = \sqrt{\frac{B}{\rho}} \quad \text{(in fluids)} \tag{17.13}$$

where B is the bulk modulus.

Note that sound waves are longitudinal waves, and the term "sound wave" is used for such mechanical waves in any medium.

EXAMPLE 17.4 Calculate the speed of mechanical waves in railroad tracks made of steel. (Young's modulus for steel is 2×10^{11} N/m^2; its density is 7.8×10^3 kg/m^3.) If the frequency of the source is 240 Hz, what is the wavelength of the waves in steel?

From Equation 17.12,

$$v = \sqrt{\frac{Y}{\rho}} = \sqrt{\frac{2 \times 10^{11}\ \text{N/m}^2}{7.8 \times 10^3\ \text{kg/m}^3}} = 5064\ \text{m/s}$$

In the relation $v = f\lambda$ where $f = 240$ Hz,

$$\lambda = \frac{v}{f} = \frac{5064\ \text{m/s}}{240/\text{s}} = 21.1\ \text{m}$$

EXERCISE 17.4 What is Young's modulus for brass if the velocity of mechanical waves in it is 3068 m/s and its density is 8500 kg/m^3? What is the wavelength of the wave in brass if the frequency of the source is 240 Hz? [*Ans.*: $Y = 8 \times 10^{10}$ N/m^2, $\lambda = 12.8$ m.]

Let us now apply Equation 17.13 to calculate the velocity of sound waves in air. In order to calculate B, we must consider whether compressions and rarefactions are isothermal or adiabatic processes. To start with, let us assume that during all compressions and rarefactions, the temperature remains constant; that is, the process is isothermal. In such cases, $B = P$, the pressure of the air. Replacing B by P, atmospheric pressure, Equation 17.13 becomes

$$v = \sqrt{\frac{P}{\rho_0}} \tag{17.14}$$

Substituting the values of P and ρ_0 (at 0°C),

$$v = \sqrt{\frac{1.013 \times 10^5\ \text{N/m}^2}{1.3\ \text{kg/m}^3}} = 279\ \text{m/s at } 0°\text{C}$$

This value is about 16 percent less than the measured value of 331 m/s at 0°C. This discrepancy can be removed by the following considerations.

We assumed above that compressions and rarefactions taking place due to the sound-wave disturbance were isothermal. But this is not true because there is not enough time for the heat to flow "out" of compressions and "into" rarefactions so that the temperature will remain constant. On the other hand, it will be a justifiable assumption to say that the total heat content remains constant. In this case we should use an adiabatic equation of state; that is,

$$PV^{\gamma} = \text{constant} \tag{17.15}$$

where γ is the ratio of the specific heat at constant pressure to that at constant volume. Thus, B in Equation 17.13 must be replaced by $B_{\text{adiabatic}}$ $(= B_{\text{ad}})$. Under these circumstances, we state without proof that $B_{\text{ad}} = \gamma P$, where $\gamma = 1.40$ for monatomic gases. Hence, we may write

$$\boxed{v = \sqrt{\frac{\gamma P}{\rho_0}}} \tag{17.16}$$

Substituting the values of γ, P, and ρ_0, we obtain

$$v = \sqrt{\frac{(1.40)(1.013 \times 10^5\ \text{N/m}^2)}{1.3\ \text{kg/m}^3}} = 330\ \text{m/s}$$

which is in excellent agreement with the experimental value.

Finally, let us eliminate P and ρ_0 and calculate the velocity of sound in air as a function of temperature. From the ideal gas law, $PV = nRT$, and substituting $P = nRT/V$, $\rho_0 = m/V$, and $n = m/M$ (where m is the mass of the gas and M is its molecular weight so that n is the number of moles) in Equation 17.16, we get

$$v = \sqrt{\frac{\gamma RT}{M}} \tag{17.17a}$$

Let v_0 be the velocity at temperature T_0; that is,

$$v_0 = \sqrt{\frac{\gamma RT_0}{M}} \tag{17.17b}$$

Dividing Equation 17.17b by Equation 17.17a,

$$\boxed{\frac{v}{v_0} = \sqrt{\frac{T}{T_0}}} \tag{17.18}$$

Using the relations $T = (273 + t)$ K and $T = 273$ K, where t is the temperature in Celsius degrees,

$$\frac{v}{v_0} = \sqrt{\frac{273 + t}{273}} = \left(1 + \frac{t}{273}\right)^{1/2} = 1 + \tfrac{1}{2}\left(\frac{t}{273}\right) + \cdots$$

where we have used the binomial expansion. If $t \ll 273$, we may neglect the higher terms and write

$$\frac{v}{v_0} = \left(1 + \frac{t}{546}\right) \quad \text{or} \quad v = v_0 + \frac{v_0}{546}t \tag{17.19}$$

In air $v_0 = 331$ m/s; hence,

$$v = (331 + 0.61\,t) \text{ m/s} \tag{17.20}$$

That is, the speed of sound in air increases by 0.61 m/s with each Celsius-degree increase in temperature. From Equation 17.17 it is clear that the velocity is directly proportional to the square root of the absolute temperature. Also it is clear that v is independent of pressure if the temperature remains constant.

The speeds of sound waves in some solids, liquids, and gases are listed in Table 17.1.

TABLE 17.1
Speed of Sound in Various Media

Medium	Temperature (°C)	Speed m/s	Speed ft/s
Gases			
Air	0	331.36	1,087.1
Air	20	344	1,129
Air	100	366	1,201
Air	1000	700	2,297
Hydrogen	0	1,286	4,200
Oxygen	0	317.2	1,041
Steam	100	404.8	1,328.1
Liquids			
Alcohol, methyl	25	1,143	3,750.2
Mercury	25	1,450	4,753.5
Seawater	25	1,532.8	5,029.1
Turpentine	25	1,326	4,350.6
Water	25	1,493.2	4,899.2
Solids			
Aluminum	20	5,100	16,700
Copper	20	3,560	11,680
Glass	20	5,550	18,050
Iron	20	5,130	16,830
Lead	20	1,322	4,340
Nickel	20	4,973	16,326
Vulcanized rubber	0	54	177

* 17.6. Superposition—Interference and Diffraction

So far we have been dealing with a single wave traveling in an elastic medium. There are many situations in which we are concerned with two or more waves traveling in the same medium, and we are interested in finding the resultant displacement at different points. Since these waves travel simultaneously and independently of each other, the resultant displacement at any point is simply their vector addition. Suppose that two waves produce displacements y_1 and y_2 at a certain point x. The resultant displacement y at x will be simply $y_1 + y_2$. This is the *superposition principle* and may be stated as follows.

> **Superposition principle:** ***When two or more waves travel simultaneously through a portion of a medium, each wave acts independently as if the other were not present. The resultant displacement at any point and time is the vector sum of the displacements of the individual waves.***

The superposition of two wave trains of the same frequency and the same amplitude traveling in the same direction results in *interference phenomena.* Mathematically, we may calculate the displacement at different points by adding the two waves, say

$$y_1 = y_0 \sin(kx - \omega t) \tag{17.21}$$

and

$$y_2 = y_0 \sin(kx - \omega t - \phi) \tag{17.22}$$

where y_0 is the amplitude and ϕ is the phase angle difference between the two waves at the point under consideration. Thus, the resultant displacement is given by

$$y = y_1 + y_2 = y_0 \sin(kx - \omega t) + y_0 \sin(kx - \omega t - \phi)$$

Using the relation

$$\sin\theta_1 + \sin\theta_2 = 2\cos\left(\frac{\theta_1 - \theta_2}{2}\right)\sin\left(\frac{\theta_1 + \theta_2}{2}\right)$$

where $\theta_1 = (kx - \omega t)$ and $\theta_2 = (kx - \omega t - \phi)$, we get

$$y = \left(2y_0 \cos\frac{\phi}{2}\right)\sin\left(kx - \omega t - \frac{\phi}{2}\right)$$

or

$$\boxed{y = A_m \sin\left(kx - \omega t - \frac{\phi}{2}\right)} \tag{17.23}$$

This resultant wave has the same frequency ω, but the resultant amplitude is $A_m = 2y_0\cos(\phi/2)$. This amplitude is not constant, but varies with the value of ϕ at the point under consideration. Suppose that $\phi = 0°$ (or $360°$), in which case $\cos(\phi/2) = \cos 0° = 1$; hence, $A_m = 2y_0$. The resultant amplitude is doubled, that is, the crest of one wave coincides with the crest of the other, and likewise the troughs. The waves are said to be in phase and yield *constructive interference,* as shown in Figure 17.12(a) for $t = 0$ at different values of x. On the other hand, if $\phi = 180°$, $\cos(\phi/2) = \cos 90° = 0$; hence $A_m = 0$. That is, the crest of one wave coincides exactly with the trough of the other. The waves are said to be out of phase and the result is *destructive interference,* as shown in Figure 17.12(b). Of course, for ϕ between $0°$ and $180°$ the amplitude of the resultant wave will be in between the two cases, as shown in Figure 17.12(c). Thus, for constructive interference $\cos(\phi/2) = \pm 1$ or $\phi = 0, 2\pi, 4\pi, 6\pi, \ldots$, while for destructive interference $\cos(\phi/2) = 0$; that is, $\phi = \pi, 3\pi, 5\pi, \ldots$.

The situation above is beautifully demonstrated by means of transverse water waves produced in a ripple tank as shown in Figure 17.13. The bright

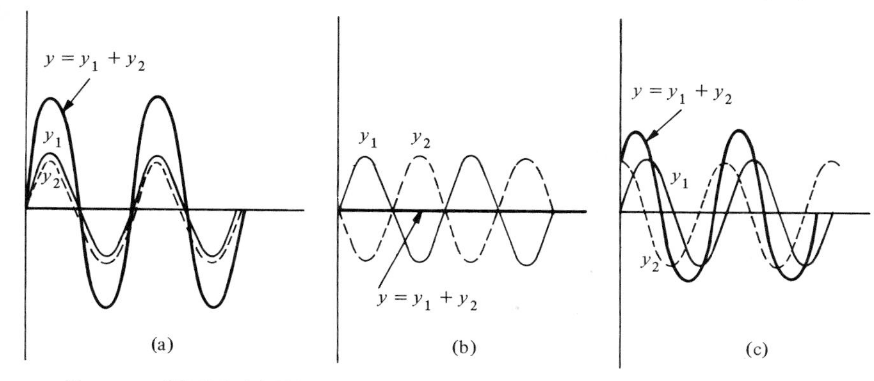

FIGURE 17.12 *(a) Waves arriving in phase at a point produce constructive interference. (b) Waves arriving 180° out of phase at a point produce destructive interference. (c) Situation that falls between parts (a) and (b).*

FIGURE 17.13 *The overlapping waves from two vibrating sources produce an interference pattern. The crests and troughs on the water surface show up as bright and circular bands.* (From *PSSC Physics,* 2nd ed., 1965, copyright by Education Development Center, Inc. Reprinted by permission of D. C. Heath & Company.)

circles correspond to crests and the dark circles to troughs. The interference patterns due to two waves produced by two vibrating rods are clearly visible. The nodal lines join points of destructive interference and antinodal lines join points of constructive interference.

Figure 17.14 shows a way of observing an interference pattern between two longitudinal sound waves. Two sources of sound S_1 and S_2 emit sinusoidal waves of the same amplitudes and frequencies and are initially in phase. This is achieved, for example, by connecting two identical loudspeakers to the same electric oscillator as shown. The sources emit longitudinal waves. The continuous circular arcs represent positions of maximum condensations, while the broken circular arcs represent positions of maximum rarefactions. The distance between two successive solid arcs or broken arcs is equal to the wavelength. By a suitable detector, one can observe maxima and minima of sound waves.

In Figure 17.14 we see that there are regions of overlap between the waves from the two sources. Since the waves have the same frequency and amplitude and are traveling almost in the same direction, interference will result, owing

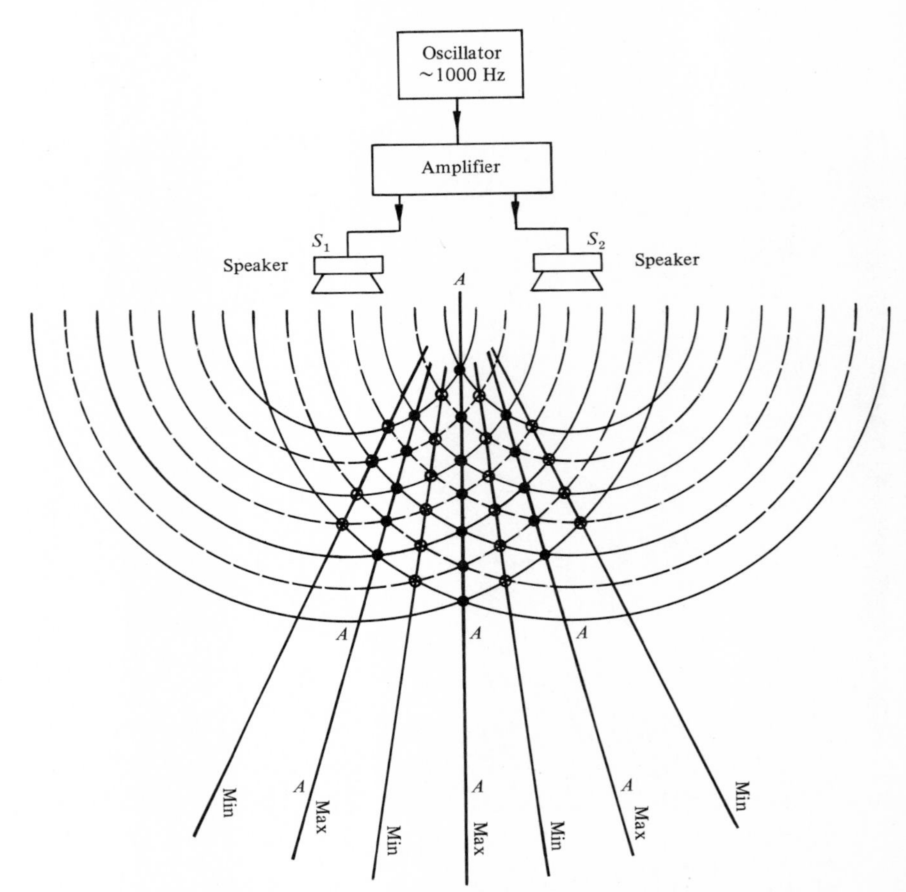

FIGURE 17.14 *Interference resulting from two sound sources.*

to phase differences. At some points, condensations from two sources arrive together, resulting in constructive interference. At other points rarefactions arrive together also, resulting in constructive interference. At still other points, condensations of one wave arrive at the same time as rarefactions of another, as shown in Figure 17.14, resulting in destructive interference. Thus, along line AA the waves are in phase, condensations due to two waves are together, and so are rarefactions. The amplitude of the resultant wave along the line AA is twice that of the wave from either source.

The surfaces represented by circular arcs are parts of spheres called *wave fronts*. As is clear the wave fronts are surfaces formed by joining points having the same phase. The wave fronts emitted by a point source are spherical, as shown in Figure 17.15(a), but if we look at small portions of spheres at distances far away from the source, the wave fronts look plane, as shown in Figure 17.15(b). A line drawn perpendicular to the wave fronts is called a *ray* and represents the direction of propagation of the wave.

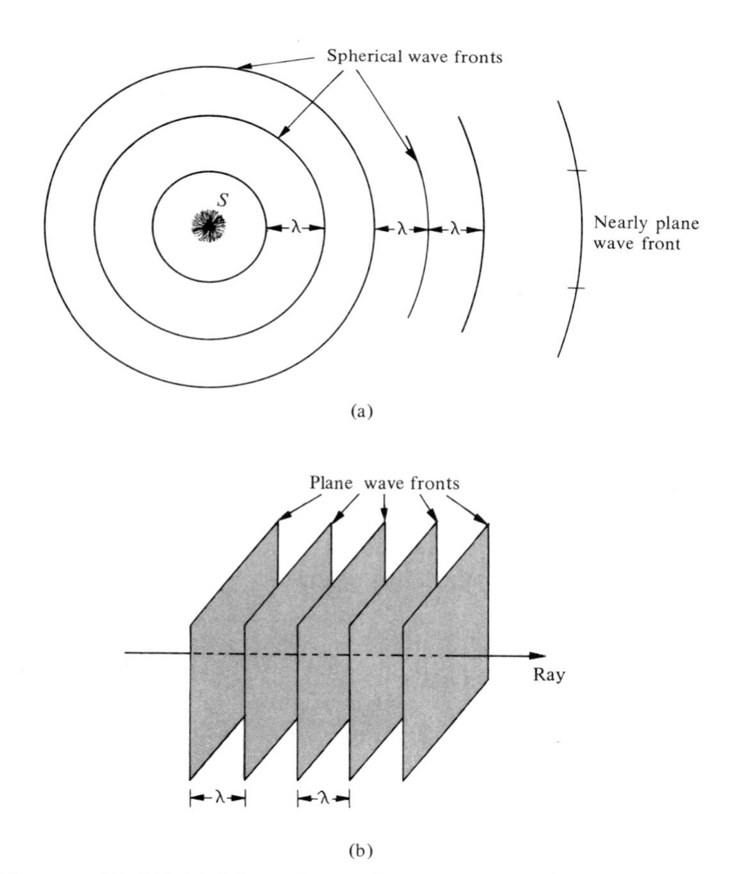

FIGURE 17.15 *(a) Spherical wave fronts emanating from a sound source* S. *(b) Plane wave fronts perpendicular to the ray, the direction of wave propagation.*

Before closing this section we discuss briefly the phenomenon of diffraction that is closely related to the superposition principle. Suppose that sound waves with plane wave fronts are incident on a small aperture. After passing through the aperture, the waves undergo spreading and fan into a cone. This phenomenon, by which the energy of sound waves is spread or waves are bent around edges, is called *diffraction*. If the aperture is very large, such as a window or door, the diffraction spreading is very small. That is, as the aperture becomes larger, the spreading becomes smaller. For sound of wavelength λ striking a circular aperture of diameter D, the waves on the other side of the aperture spread into a cone of vertex angle θ (for small angles, $\sin\theta \simeq \theta$), given by

$$\theta \simeq \frac{\lambda}{D} 100 \text{ degrees} \tag{17.24}$$

FIGURE 17.16(a) [*Figure continued on the facing page.*]

Figure 17.16 illustrates the diffraction of water waves for various D. More specific examples of the superposition principle—interference in space (standing waves) and interference in time (beats)—will be discussed in Chapter 18.

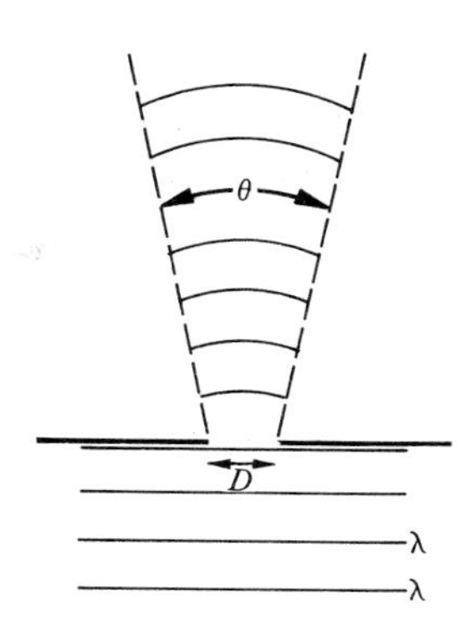

(b)

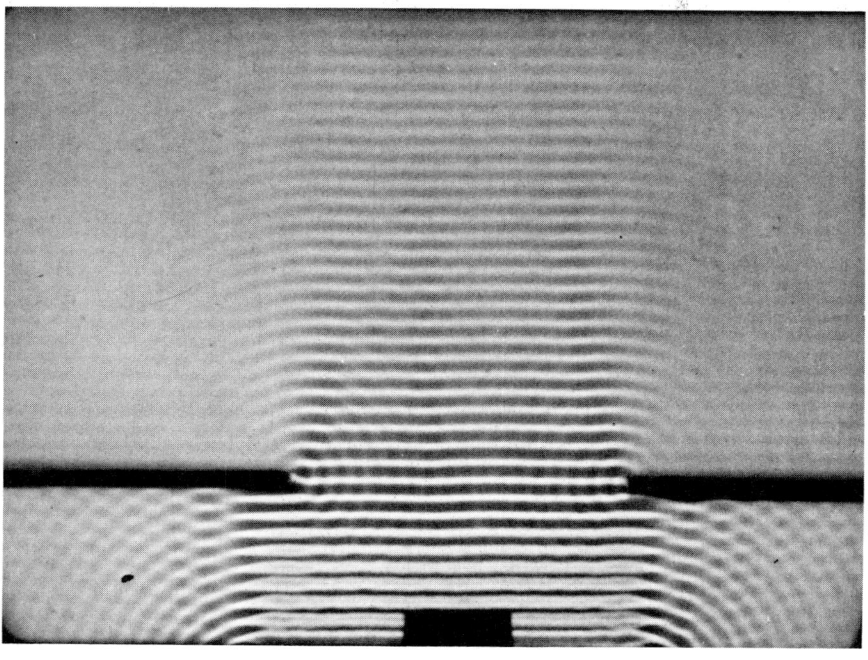

(c)

FIGURE 17.16 *Diffraction of plane water waves of wavelength* λ *through an aperture of various sizes D.* (From the *Project Physics Course*, Holt, Rinehart and Winston, Inc., New York, 1970.)

SUMMARY

Wave motion provides a mechanism for transfer of energy from one point to another without physical transfer of any material medium between the points. In *transverse waves* the particles of the medium vibrate at right angles to the direction of propagation, while in *longitudinal waves* the particles of the medium vibrate back and forth about their equilibrium position along the direction of propagation of the wave motion. *Wavelength* λ is the distance between any two successive points that are in the same phase, $v = \lambda/T = f\lambda$.

A general wave motion is described by the equation $y = A \sin [(2\pi/\lambda)(x - vt)] = A \sin [2\pi(x/\lambda - t/T)]$. Defining wave number $k = 2\pi/\lambda$ and angular frequency $\omega = 2\pi/T = 2\pi f$, we may write $y = A \sin (kx - \omega t)$.

The velocity of the transverse waves is $v = \sqrt{T/\mu}$ in a string and that of the longitudinal waves is $v = \sqrt{Y/\rho}$ in solids and $v = \sqrt{B/\rho}$ in fluids. For velocity of sound in air, $v = \sqrt{\gamma P/\rho_0} = \sqrt{\gamma RT/M}$ and $v/v_0 = \sqrt{T/T_0}$ or $v = (331 + 0.62t)$ m/s.

*According to the superposition principle, when two or more waves travel simultaneously through a portion of a medium, the resultant displacement at any point and time is the vector sum of the displacements of the individual waves, resulting in *constructive interference* ($\phi = 0, 2\pi, 4\pi, \ldots$) and *destructive interference* ($\phi = \pi, 3\pi, 5\pi, \ldots$). The phenomenon by which the energy of the sound waves is spread, or the bending of waves around edges, is called *diffraction;* $\theta \simeq (\lambda/D)100$ degrees.

QUESTIONS

1. Give two examples which will show that it is the energy disturbance and not the medium that is transported by mechanical waves.
2. Give physical reasoning proving that sound waves in gases must be longitudinal and not transverse.
3. Which of the two methods of energy transport is better (a) by wave motion; and (b) by actual motion of the particles?
4. What experiment would you perform to distinguish whether the energy is being carried by a wave or a particle in a medium such as air, water, or a string?
5. Why is the speed of sound independent of its wavelength?
6. From Equation 17.16, if $v = \sqrt{\gamma P/\rho_0}$, then why is the velocity independent of pressure and not of temperature? Explain.
7. How do the temperature, density, pressure, and heat change in a longitudinal wave in a small volume located at (a) a compression; and (b) a rarefaction?

* 8. Give two examples of constructive and destructive interference in space.

* 9. How can you tell whether the incoming wave is plane or spherical?

*10. Give two everyday examples of diffraction of sound waves.

PROBLEMS

1. Suppose that waves are traveling at a speed of 960 m/s through a medium. If 60 waves pass by any point in one second, what are the wavelength and the time period?
2. Waves traveling from one medium to another do not change their frequency, but their wavelength changes. Suppose that a sound disturbance has a wavelength of 30 cm in air, where the wave velocity is 340 m/s. What will be the wavelength of this disturbance in helium, water, and steel if the speed of sound waves in these media is 970 m/s, 1450 m/s, and 5000 m/s, respectively?
3. A man standing in a valley between two parallel mountains fires a gun and hears echoes at intervals of 2 s and 3.5 s. What is the distance between the two mountains and the location of the gunner with respect to the mountains?
4. The human ear can respond to frequencies in the range 20 to 20,000 Hz. Calculate the corresponding wavelengths for these frequencies for sound waves in air and water.
5. An observer far away from a railroad station hears the train starting. The sound arrives both from the steel rails and through the air with a time difference of 3 s. How far away is the railroad station from the observer?
6. The fifth crest on the water surface is 35 m from the center of the disturbance and the ninth crest is 63 m away. It takes 3 s to travel from the fifth to the ninth crests. Calculate (a) the wavelength; (b) the speed of the disturbance; and (c) the frequency of the disturbance.
7. Calculate the number of compressions and rarefactions that pass an observer in 5 min if the velocity of the sound waves is 340 m/s and the wavelength is 3 m. What are the distances between compressions and rarefactions?
8. A cork on the water surface moves up and down completing five vibrations in 4 s. The waves (crests or troughs) travel from the cork to the shore, which is 20 m away, in 10 s. Calculate the speed, frequency, and wavelength. What is the distance between a crest and a trough? How many crests reach the shore in 1 min?
9. A displacement of a certain particle is described by $y = 0.03 \sin (2\pi/5)t$. Calculate (a) the amplitude; (b) the time period; and (c) the frequency. Make a plot of y versus t for $t = 1, 2, \ldots, 10$ s. Note y is in meters.
10. A displacement of a certain particle is given by $y = 0.05 \text{ m} \cos (\pi/10)t$. Calculate (a) the amplitude; (b) the time period; and (c) the frequency. Make a plot of y versus t for $t = 0.01, 0.02, \ldots, 0.1$ s.
11. The equation of a traveling wave is given by

$$y = 0.1 \sin [2\pi(0.01x - 100\, t)]$$

where x and y are in meters and t is in seconds. Calculate (a) the amplitude; (b) the wavelength; (c) the period; (d) the frequency; and (e) the speed.
12. In Problem 11 make the following plots: (a) y versus x for $t = 0$; (b) y versus x for $t = T/2$; and (c) y versus t for $x = 0$; and (d) y versus t for $x = \lambda/2$.

13. Repeat Problems 11 and 12 for a traveling wave given by

$$y = 0.01 \cos [2\pi(100t - 0.05x)]$$

14. A steel wire of length 3 m and of mass 75 g is under a tension of 2000 N and is set into vibration. Calculate the speed of propagation of transverse waves in this wire. If the tension is increased while keeping the length the same, will it increase or decrease the velocity?
15. A wire of linear mass density 0.01 kg/m is attached at one end to an electrically driven tuning fork of frequency 120 Hz while the other end goes over a pulley and supports a mass of 2 kg. Calculate the speed of transverse waves in the string and the wavelength.
16. Consider a string of length 5 m and mass 50 g which is vibrating with a frequency of 60 Hz under a tension of 500 N. Calculate the speed of the transverse waves and the wavelength. If the tension in the string is doubled, how will it change the speed and the wavelength if the frequency remains the same?
17. Calculate the speed of sound in brass if its stretch modulus is 9×10^{10} N/m^2 and its density is 8500 kg/m^3.
18. Young's modulus of brass is 9×10^{10} N/m^2 and of steel is 20×10^{10} N/m^2 while their densities are 8500 kg/m^3 and 7800 kg/m^3, respectively. What is the ratio of the speed of sound waves in the two media?
19. The density of metal A is 4 times that of B while Young's modulus of A is double that of B. What is the ratio of the speed of sound waves in the two metals?
20. Show that Equation 17.11 is dimensionally correct.
21. A metal wire having a diameter of 1 mm and a density of 7×10^3 kg/m^3 and Young's modulus of 24×10^{10} N/m^2 is fixed between two rigid supports with zero tension at 40°C. Suppose that the temperature is decreased to (a) 20°C; and (b) 0°C. What will be the speed of the transverse waves in the two cases provided that the coefficient of linear expansion α is 1.5×10^{-5}/C°? Note that $F/A = \alpha Y \Delta t$.
22. Calculate the speed of sound waves in an atmosphere of helium at 0°C and 1 atm pressure. Note that 4 g of helium under standard pressure and temperature has a volume of 22.4 ℓ. For helium, $\gamma = 1.67$.
23. Calculate the bulk modulus of air at 0°C from the fact that the speed of sound in air is 331.5 m/s. The density of air is 1.3 kg/m^3.
24. What will be the speed of sound in mercury if its bulk modulus is 2.7×10^{10} N/m^2 and its density is 13.6 g/cm^3?
25. The speed of sound in air at 0°C is 331 m/s. At what temperature will the speed of sound be (a) 350 m/s; (b) 400 m/s; and (c) 663 m/s? Calculate $\Delta v/\Delta T$ is each case.
26. Calculate the speed of sound waves at 30°C in (a) hydrogen (b) helium and (c) oxygen. How do they compare with respect to their masses?
27. Make a plot of the speed of sound in air versus temperature between -20 and $+50$°C. Calculate $\Delta v/\Delta T$.
28. Aluminum and steel pipes each 100 m long and tied parallel to each other are struck at one end. An observer at the other end receives three sounds, which are due to the three longitudinal waves in air, aluminum, and steel,

respectively. Calculate the time interval between the three sounds. The speed of sound in air is 340 m/s and the stretch moduli of aluminum and steel are 7×10^{10} N/m^2 and 20×10^{10} N/m^2, while their densities are 2700 kg/m^3 and 7800 kg/m^3, respectively.

***29.** Two waves of the same frequency and amplitude $y_0 = 0.01$ m arrive at a point differing in phase by 90°. What is the amplitude of the resultant wave at this point?

***30.** Two waves arriving at a point are described by (x and y in m and t in s)

$$y = 0.05 \sin [2\pi - 10t)]$$

and differ in phase by 60°. Describe the resultant wave at this point.

***31.** Consider the diffraction of sound waves in air through an aperture of 10 mm. If the diffracted beam is concentrated in a cone with a vertex angle of 5°, what is the wavelength of the incident waves?

***32.** Suppose that waves of wavelength 1 m are incident on an aperture of 5 m. Calculate the angle of diffraction. What is the value of θ if λ is 2 m?

18 Sound Waves

In this chapter we are interested in investigating the characteristics of sound waves and their application to more practical situations. To start with, we discuss sound waves in strings, air columns, rods, and plates which are commonly applicable to musical instruments. We shall discuss the difference between musical sound and noise. This will be followed by the relation between frequencies and the velocities of the source and the observer—the Doppler effect. We shall conclude the chapter with the discussion of shock waves, application of sonic waves, and noise pollution.

18.1. Stationary Waves

STATIONARY or *standing waves* are formed when two waves of the same frequency and speed travel in the opposite directions. We apply the superposition principle to calculate the resultant wave motion, as explained below.

Let us consider two transverse wave trains of the same frequency, amplitude, and speed traveling in a string in opposite directions, say along the $+X$-axis and $-X$-axis. We may represent these waves as

$$y_1 = y_0 \sin (kx - \omega t) \tag{18.1}$$

and

$$y_2 = y_0 \sin (kx + \omega t) \tag{18.2}$$

and the resultant displacement y may be written as

$$\begin{aligned} y = y_1 + y_2 &= y_0 \sin (kx - \omega t) + y_0 \sin (kx + \omega t) \\ &= y_0 [(\sin kx \cos \omega t - \sin \omega t \cos kx) + (\sin kx \cos \omega t + \sin \omega t \cos kx)] \end{aligned}$$

That is,

$$y = (2y_0 \sin kx) \cos \omega t$$

or

$$\boxed{y = A_x \cos \omega t} \tag{18.3}$$

where

$$\boxed{A_x = 2y_0 \sin kx} \tag{18.4}$$

Equation 18.3 represents a wave motion of angular frequency ω and amplitude A_x. In a traveling wave all the particles not only vibrate with the same frequency but they have the same amplitudes as well. On the other hand, in standing waves given by Equation 18.3, all the particles still vibrate with the same frequency, but particles with different values of x have different amplitudes, as given by Equation 18.4. Thus, amplitude A_x will be maximum at those positions of x for which $2y_0 \sin kx$ is maximum. That is, $\sin kx = 1$ or $kx = \pi/2$, $3\pi/2$, $5\pi/2$, ... Since $k = 2\pi/\lambda$,

$$x = \frac{\lambda}{4}, \frac{3\lambda}{4}, \frac{5\lambda}{4}, \ldots \tag{18.5}$$

These points of maximum amplitude are called *antinodes* and are a half-wavelength apart from each other. In between the positions of maximum amplitudes are the points where the amplitude A_x has a minimum value of zero. This is possible only if in Equation 18.4, $\sin kx = 0$ or $kx = \pi, 2\pi, 3\pi, \ldots$, or, since $k = 2\pi/\lambda$,

$$x = \frac{\lambda}{2}, \frac{2\lambda}{2}, \frac{3\lambda}{2}, \frac{4\lambda}{2}, \ldots \tag{18.6}$$

The points of minimum amplitudes are called *nodes* and are a half-wavelength apart. The distance between any two adjacent nodes or antinodes is a half-wavelength. The points above are clearly illustrated by means of an example below and Figure 18.1.

In Figure 18.1(a) a wave of wavelength λ, frequency f, and amplitude y_0 is

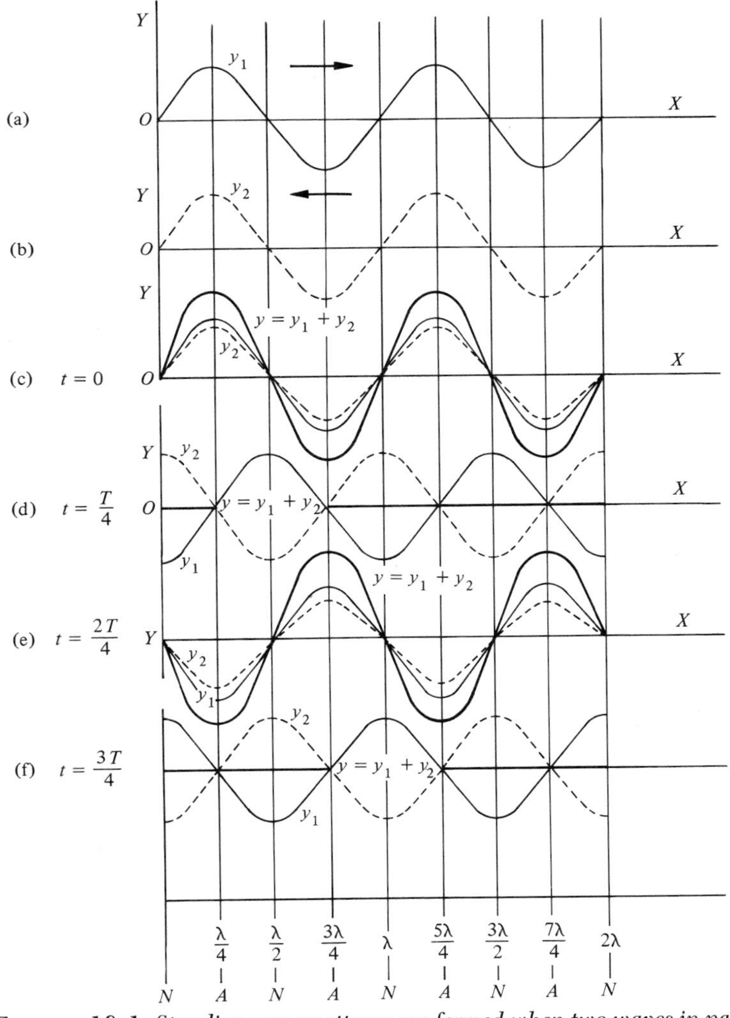

FIGURE 18.1 *Standing-wave patterns are formed when two waves in parts (a) and (b) of the same frequency and amplitude travel in opposite directions.*

traveling to the right while another wave of the same λ, f, and y_0 is traveling to the left, as in Figure 18.1(b). These two waves are superposed at $t = 0$, as shown in Figure 18.1(c). After $t = T/4$, that is, after a fourth of a period, the wave in (a) has moved $\frac{1}{4}$ wavelength to the right, while the wave in (b) has moved $\frac{1}{4}$ wavelength to the left. Thus, the resultant obtained by the superposition principle after $t = T/4$ is as shown in Figure 18.1(d). Note that the resultant of these two waves is zero, as shown by the heavy horizontal line. We can obtain the resultant amplitudes after each successive period of $T/4$ as shown in Figure 18.1(c) to (f) from $t = 0$ to $t = T$. The resultant in each of these cases, as shown by the heavy lines, is quite different from the traveling waves shown in (a) and (b), and we may observe the following characteristics.

The points at $x = \lambda/2, 2\lambda/2, 3\lambda/2, \ldots$ given by Equation 18.6 and shown in Figure 18.1 are nodes denoted by N, while points at $x = \lambda/4, 3\lambda/4, 5\lambda/4, \ldots$, given by Equation 18.5 and shown in Figure 18.1, are antinodes denoted by A. At the nodes N (which are in between the antinodes A), the amplitude of vibrations is zero; that is, the particles at N are at rest. The particles in between the nodes have different amplitudes but the same frequency. Since particles at the nodes are not in motion, the energy cannot be transmitted from one region to another across the nodes. Thus, these waves are called *standing* or *stationary waves*. The energy of each particle in the standing wave (in strings in this case) remains in that very particle as it executes simple harmonic motion. (Thus a string may be thought of as a collection of a very large number of oscillators vibrating side by side.) The energy of each particle on the string alternates between the vibrational kinetic energy and elastic potential energy, but this energy is not transported to the left or right of the string, and remains standing in the string. Note that in standing waves the positions of nodes and antinodes remain fixed.

Longitudinal standing waves may also be formed from two longitudinal traveling waves of the same λ, f, and y_0, which are traveling in opposite directions, as we shall discuss later. It is important to note that no standing-wave interference pattern will be formed if two traveling waves have slightly different wavelengths.

The transverse or longitudinal waves in most cases do not travel indefinitely. For example, waves in strings or in air columns do meet the boundaries. The points at the fixed and rigid boundaries cannot be set into vibration; hence, they must be displacement nodes. If the boundary is free, it must be a displacement antinode.

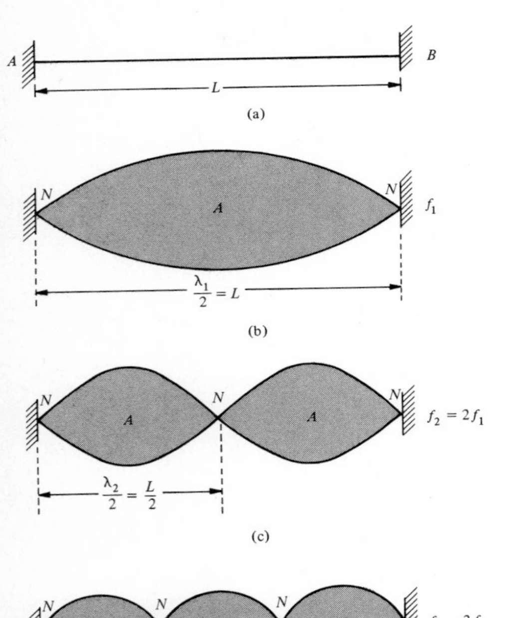

FIGURE 18.2 *Formation of standing waves in a string fixed at both ends. Shaded areas represent the limits within which different portions of the string can vibrate.*

18.2. Vibrations of Strings and Air Columns

Let us investigate in some detail the formation of different modes of transverse standing waves in strings fixed at both ends and the longitudinal standing waves in air pipes.

Vibrations of Strings Fixed at Both Ends

Figure 18.2(a) shows a string of length L fixed at both ends A and B. This string can be put into transverse vibrations of different frequencies. Any vibrational motion may be regarded as a superposition of certain simple wave motions called *normal modes* of vibrations, as discussed below.

Remembering that the two fixed ends of the string must be nodes, the first mode of standing waves of vibration is as shown in Figure 18.2(b) and is produced by displacing the string (or plucking the string) at the middle and letting it go. The waves reaching at the ends are reflected, hence result in the formation of standing waves. The midpoint of the string is an antinode. Since the distance between the two nodes is a half-wavelength, as we see from Figure 18.2(b),

$$\frac{\lambda_1}{2} = L \qquad \text{or} \qquad \lambda_1 = 2L$$

If v is the velocity of the wave on the string, the frequency f_1 is given by

$$f_1 = \frac{v}{\lambda_1} = \frac{1}{2L} v \tag{18.7}$$

This is the lowest possible natural frequency with which the string can vibrate and is called the *fundamental frequency*. The next possible mode of vibration of this string is as shown in Figure 18.2(c), in which case

$$\frac{\lambda_2}{2} = \frac{L}{2} \qquad \text{or} \qquad \lambda_2 = \frac{2L}{2}$$

and the frequency f_2 of this mode, called the *first overtone*, is given by

$$f_2 = \frac{v}{\lambda_2} = \frac{2}{2L} v \tag{18.8}$$

This mode is produced by holding the middle of the string with one hand and plucking at the middle of either half, then releasing both hands. Similarly, the frequency f_3 of the next higher mode, called the *second overtone*, shown in Figure 18.2(d), is given by

$$f_3 = \frac{v}{\lambda_3} = \frac{3}{2L} v \tag{18.9}$$

In general, the frequency f_n of the $(n - 1)$th overtone is given by

$$\boxed{f_n = \frac{n}{2L} v = nf_1} \qquad n = 2, 3, 4, \ldots \tag{18.10}$$

Thus, the frequencies of the overtones are integral multiples of the fundamental frequency $f_1 = v/2L$. The overtones, together with the fundamental note, are said to form a *harmonic series*. The fundamental with frequency f_1 is called the *first harmonic*. The first overtone of frequency $f_2 = 2f_1$ is called the *second harmonic;* the second overtone of frequency $f_3 = 3f_1$ is called the *third harmonic;* . . . .

Note that the patterns formed in Figure 18.2 are standing-wave patterns. This means all the points along the line AB move up and down with the amplitudes restricted by the shaded areas. To the eye standing waves appear as loops, but the multiflash photograph shown in Figure 18.3 reveals the actual position of the string at different instants.

FIGURE 18.3 *Multiflash photograph of a vibrating string, showing its position at different instances.* (Courtesy of the National Film Board of Canada.)

As we saw in Section 17.4, the velocity of waves in a string is given by

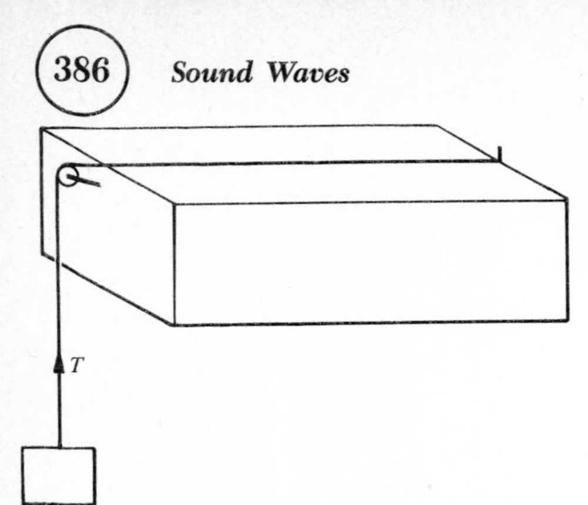

Figure 18.4 *The string can be made to vibrate with different frequencies by adjusting the tension in the string by means of weight* W.

$v = \sqrt{T/\mu}$, where T is the tension in the string and μ is the mass per unit length of the string. By using the arrangement of Figure 18.4, the tension in the string can be adjusted to the desired value by means of the vertical weights W, giving $T = W$. Thus, the frequency of the nth harmonic or $(n - 1)$th overtone (using Equation 18.10) may be written

$$\boxed{f_n = \frac{n}{2L}\sqrt{\frac{T}{\mu}}} \tag{18.11}$$

The string usually vibrates with more than one mode simultaneously. It may be pointed out that no matter how complex the resultant motion of the string is, it can always be represented as a superposition of the various normal modes of vibration.

Example 18.1 A string 2 m long has a mass of 4 g and is under a tension of 3000 N. Calculate (a) the velocity of the transverse waves in the string and (b) the frequencies of the fundamental and the first overtone.

From Equation 18.11,

$$f_n = \frac{n}{2L}v = \frac{n}{2L}\sqrt{\frac{T}{\mu}}$$

where $L = 2\ m$, $T = 3000\ N$, $\mu = m/L = 4 \times 10^{-3}\ \text{kg}/2\ \text{m} = 2 \times 10^{-3}\ \text{kg/m}$.

(a)
$$v = \sqrt{\frac{T}{\mu}} = \sqrt{\frac{3000\ \text{N}}{2 \times 10^{-3}\ \text{kg/m}}} = 1225\ \text{m/s}$$

(b) For the fundamental note $n = 1$,

$$f_1 = \frac{n}{2L}v = \frac{1}{2(2\ \text{m})}(1225\ \text{m/s}) = 306/\text{s} = 306\ \text{Hz}$$

while for the first overtone $n = 2$,

$$f_2 = \frac{n}{2L}v = \frac{2}{2(2\ \text{m})}(1225\ \text{m/s}) = 612/\text{s} = 612\ \text{Hz}$$

Note that the corresponding wavelengths are

$$\lambda_1 = v/f_1 = (1225\ \text{m/s})/(306/\text{s}) = 4\ \text{m}$$
$$\lambda_2 = v/f_2 = (1225\ \text{m/s})/(612/\text{s}) = 2\ \text{m}$$

Exercise 18.1 A wire of length 80 cm has a mass of 8 g and is under a tension of 500 N. Calculate the frequencies and the wavelengths of the first four modes of vibration. [*Ans.*: 140 Hz, 280 Hz, 420 Hz, 560 Hz; 160 cm, 80 cm, 53 cm, 40 cm.]

Vibrations of Air Columns

The vibrations of air in an organ pipe are a good example of longitudinal waves. The pipe may be open at both ends, as shown in Figure 18.5(a), in

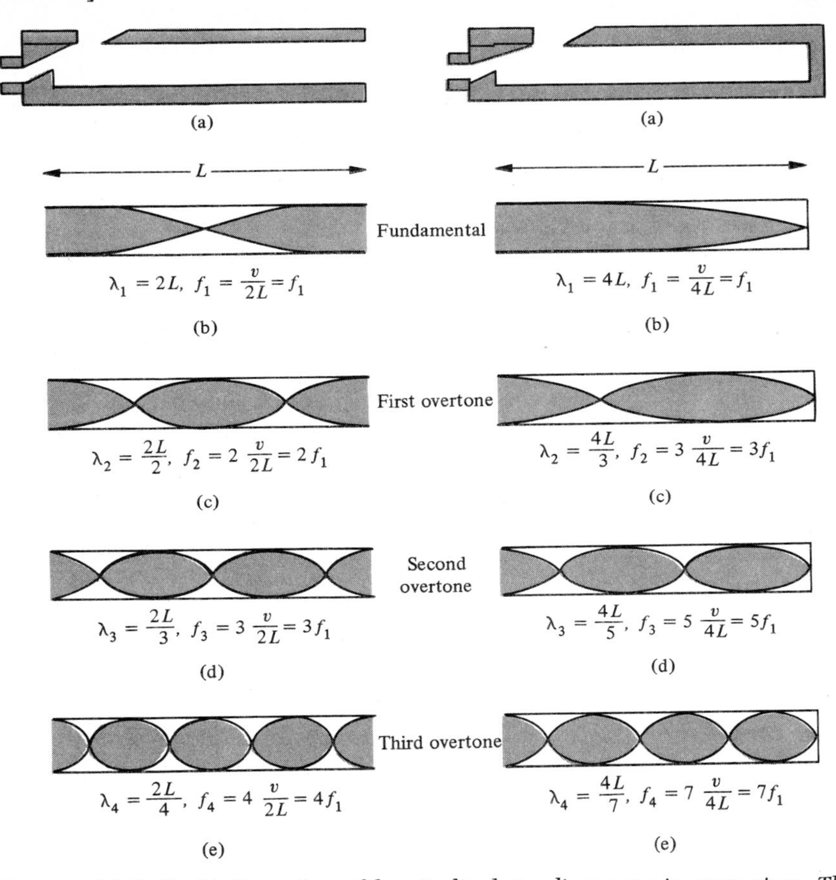

FIGURE 18.5 (Left) *Formation of longitudinal standing wave in open pipes. The frequencies of the harmonic series are* f_1, $2f_1$, $3f_1$ . . . .

FIGURE 18.6 (Right) *Formation of longitudinal standing waves in closed pipes. The frequencies of the harmonic series are* f_1, $3f_1$, $5f_1$ . . . .

which case it is called an *open pipe*, or it may be closed at one end, as shown in Figure 18.6(a), and called a *closed pipe*. By blowing air from the left, the longitudinal standing waves may be formed. Remembering that the closed end is always a displacement node while the open end is a displacement antinode, different modes of vibrations of open and closed pipes may be formed as shown in Figure 18.5 and Figure 18.6, respectively. These figures show the fundamental notes and the overtones in each case. As in the case of standing waves in strings, the fundamental and overtones can exist at the same time. Even though in Figures 18.5 and 18.6 we have shown the displacement of different particles by means of transverse displacements for convenience, the actual displacements are longitudinal; that is, molecules of air move back and forth.

From Figure 18.5 it is quite evident that if the frequency of the fundamental note in an open pipe is f_1, the frequencies of the overtones are $2f_1$, $3f_1$, $4f_1$

. . . . That is, *the frequencies of the natural modes of vibration in an open pipe are given by the integer multiples of the fundamental note.*

In the case of a closed pipe as shown in Figure 18.6, if the frequency of the fundamental note is f_1, the frequencies of the overtones are $3f_1$, $5f_1$, $7f_1$, ... That is, *the frequencies of the natural modes of vibration in a closed pipe are given by the odd-integer multiples of the fundamental note.*

We can extend our discussion of vibrations of air columns and strings to the formation of stationary vibrations in rods and plates. By clamping a rod at the middle and striking it at one end, the rod is set into transverse vibrations as shown in Figure 18.7(a). The ends are free to vibrate, hence are antinodes, while the middle point, which is held fixed by a clamp, is a node. This is the fundamental mode of vibration and is similar to the fundamental mode in the case of an open pipe. The rod can be clamped at different points to obtain other modes of vibrations, as shown in Figure 18.7.

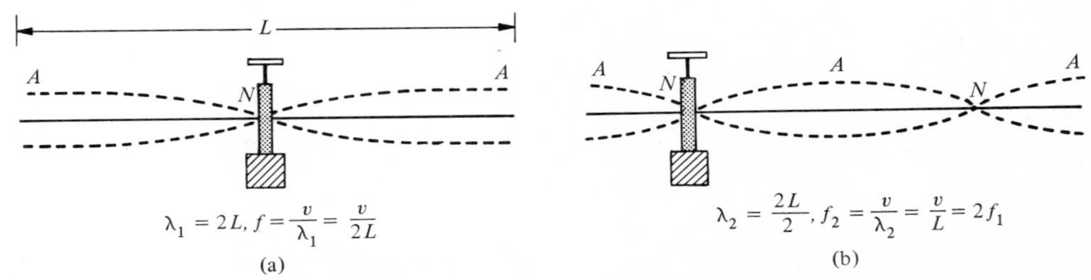

FIGURE 18.7 *Two different modes of vibration of a clamped rod.*

EXAMPLE 18.2 Consider an open pipe of length 40 cm. Calculate the frequency of the first harmonic. How many harmonics can be heard by a person with a normal hearing range (20 to 20,000 Hz)?

For an open pipe from Figure 18.5, $\lambda_1 = 2L = 2(0.40\ \text{m}) = 0.80\ \text{m}$

$$f_1 = \frac{v}{\lambda_1} = \frac{344\ \text{m/s}}{0.80\ \text{m}} = 430/\text{s} = 430\ \text{Hz}$$

In $f_n = nf_1$, trying $n = 47$ gives $f_{47} = (47)(430/\text{s}) = 20{,}210/\text{s}$ and $n = 46$ gives $f_{46} = (46)(430/\text{s}) = 19{,}780/\text{s}$. Therefore, in the range 20 to 20,000 Hz we will hear 46 harmonics.

EXERCISE 18.2 Consider a closed pipe of length 40 cm. Calculate the frequency of the first harmonic. How many harmonics can be heard by a person with a normal hearing range (20 to 20,000 Hz). [*Ans.:* $f_1 = 215$ Hz, 93 harmonics.]

A typical example of transverse vibrations in two dimensions is that of a drumhead. A drumhead is a circular sheet of stretched flexible membrane with its outside perimeter held stationary. This situation is equivalent to replacing a one-dimensional string by a two-dimensional sheet. The analysis can be made simpler by replacing the circular sheet by a rectangular sheet. We shall not go into detailed analysis but simply show the result in Figure 18.8.

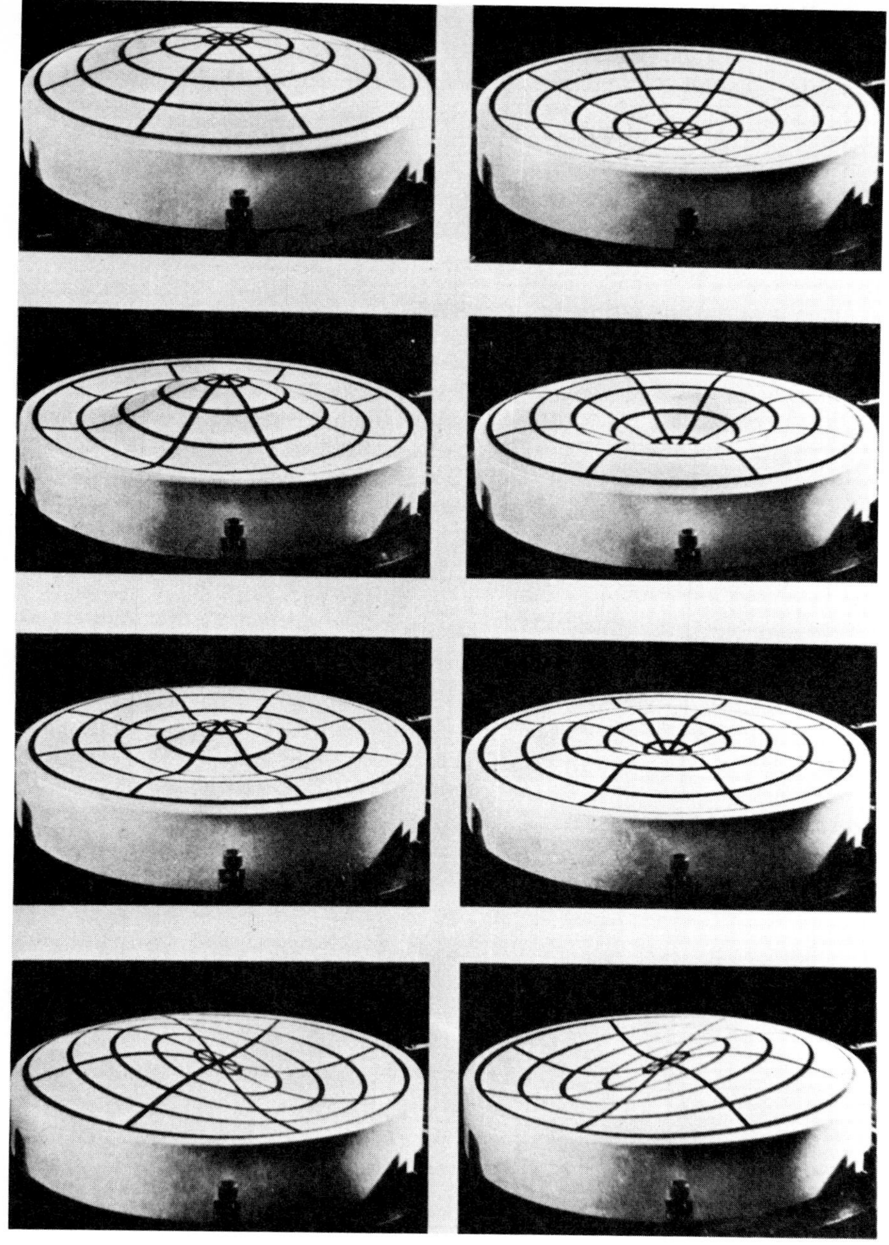

FIGURE 18.8 *Eight different modes of vibration of a marked drumhead or membrane.* (Courtesy of The National Film Board of Canada.)

18.3. Resonance

The phenomenon of resonance was discussed in connection with vibrational motion in Chapter 10. Resonance is not limited to mechanical systems only, but occurs in sound, electromagnetism, optics, atomic, molecular, and nuclear systems as well. Let us discuss resonance in sound.

Consider a source of sound which instead of emitting waves is absorbing energy when sound waves from some other source are incident on it. The body absorbing this energy is set into *forced oscillations* at the frequency of the incident wave called the *driving frequency*. The amplitude of this forced oscillation is very small unless the driving frequency is near the natural frequency of the body. If the driving frequency is exactly equal to the natural frequency of the system, the system is set into oscillation with a very large amplitude. The phenomenon resulting in a large response at certain driving frequencies is called *resonance*, and the corresponding frequencies are called *resonant frequencies*. In short, we may say: *whenever a system capable of oscillating is acted on by a series of impulses having frequency equal to one of its natural frequencies, the system responds by setting itself into oscillation with a relatively large amplitude. This phenomenon is called resonance, and the system is said to resonate with the applied impulses.*

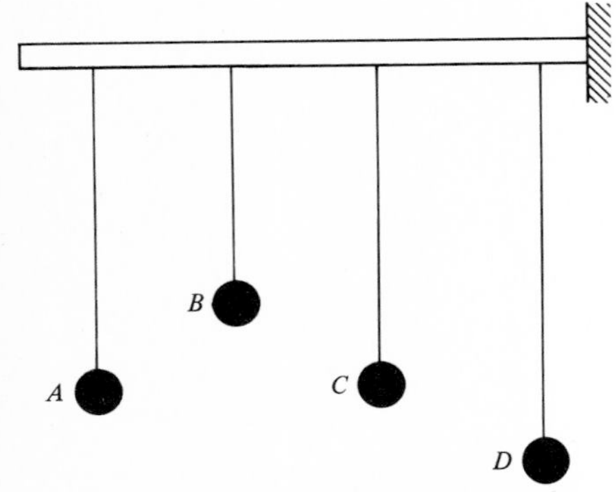

FIGURE 18.9 *When pendulum A is vibrating, it makes pendulum C resonate while pendulum B and D are set into small forced oscillations.*

For example, let us suspend many pendula of different lengths from a semi-rigid support, as shown in Figure 18.9. Suppose that pendulum *A* is set into oscillation. After a short time the rest of the pendula will be set into oscillations as well as they receive longitudinal waves from *A*, but pendulum *C*, which has the same length as *A*, will have a much larger amplitude than any other. This is because the frequency of pendulum *A*, which is the driving frequency, is equal to the natural frequency of pendulum *C*.

Consider another example of resonance due to longitudinal sound waves, as shown in Figure 18.10. A tuning fork *A* mounted on a sound box is set into vibration. If the frequency of the tuning fork *B* mounted on another sound box

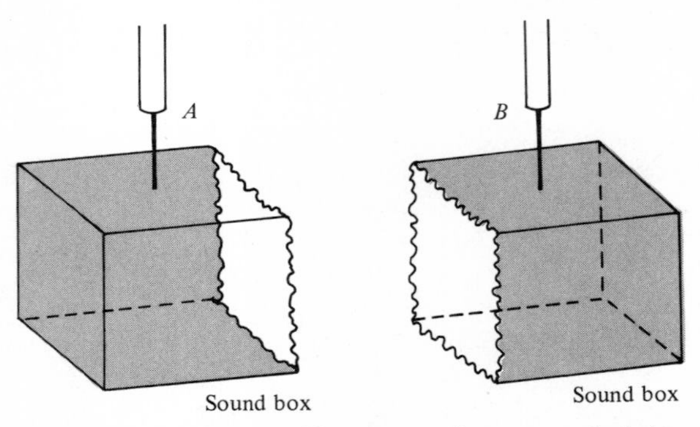

FIGURE 18.10 *Tuning fork B will resonate when tuning fork A is vibrating, because both have the same frequency.*

is the same as that of *A*, the tuning fork *B* will be set into vibrations as it receives longitudinal waves coming from *A*. If a small piece of wax is attached to the tuning fork *B*, its frequency is changed and it will not resonate; that is, it will not respond to the incident impulses with large amplitude.

The phenomenon of resonance can also be demonstrated clearly by considering vibrations of strings. Suppose that in Figure 18.4 the right end of the string is connected to a vibrator of frequency f and wavelength λ. Whenever the length of the string is an integral multiple of $\lambda/2$, it will be set into resonance. Thus, the condition for resonance in a string is (see Figure 18.4)

$$L = n\frac{\lambda}{2} \qquad \text{where } n = 1, 2, 3, \ldots$$

or

$$f = \frac{v}{\lambda} = \frac{v}{2L/n} = \frac{n}{2L}v = \frac{n}{2L}\sqrt{\frac{T}{\mu}} \tag{18.12}$$

This implies that if the frequency of the vibrator is equal to any one of the normal frequencies of the string given by Equation 18.12, the string will resonate; that is, it will vibrate with a very large amplitude.

An example of electrical resonance is the tuning of a radio. By turning the dial, the frequency of the oscillator in the receiving circuit is changed. If this frequency is equal to the frequency of a particular broadcasting station, the receiving circuit is set into an oscillation of large current amplitude even though the incoming signal is weak.

18.4. Loudness, Pitch, and Quality

For the most part, loudness depends on the intensity, pitch on the frequency, while quality of sound depends on a combination of frequencies, overtones, and their relative intensities. Together these three quantities determine the characteristics of sound—musical or otherwise. Before trying to understand these, let us define intensity.

The *intensity* of a sound wave is equal to the energy crossing per unit area per second, the area being normal to the direction of propagation. Since energy per second is power, we may define intensity as power transferred through a unit area normal to the direction of propagation. Thus, the unit of intensity I is watts/m^2 (=W/m^2).

Loudness

The psychological sensation of *loudness* in the human ear *is intimately connected with the intensity* of the incident sound wave. There is no simple relation, in general, between the loudness of a sound and its intensity. But for a pure tone of given frequency, *the loudness increases with increasing frequency* if intensity increases with increasing frequency. It is not possible to measure loudness in terms of physical quantities, because it depends upon the response of the ear and the judgment of the individual. For example, two or more observers can easily agree that two sounds are almost equally loud, but they will rarely agree that one sound is twice as loud as another. The loudness of the two sounds may be harder to compare if the frequencies of the two sound sources differ widely.

Hearing capabilities of different individuals are sensitive to a different range of intensities and frequencies, as shown in Figure 18.11. As can be seen, the ear is sensitive to sound of frequencies anywhere from 20 Hz to about 20,000 Hz and is most sensitive to frequencies between 2000 and 3000 Hz. At a frequency

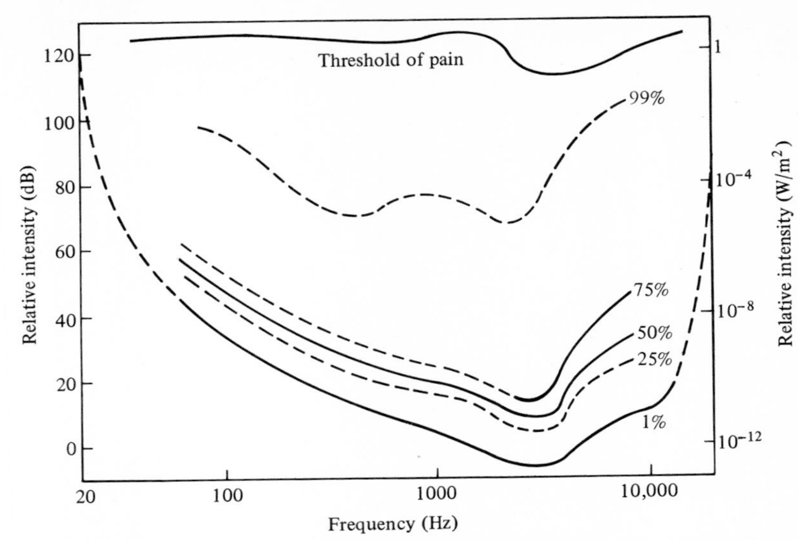

FIGURE 18.11 *Percentage of different individuals whose ear responds to sounds of various intensities and frequencies.*

of about 1000 Hz, the ear can detect sounds varying from the hardly audible low intensity of 10^{-12} W/m^2 to very high intensities of 1W/m^2, as shown in Table 18.1. Because of this very enormous range of intensity to which the ear responds, it is not convenient to use a linear scale for measuring intensity levels. It is found that a scale proportional to the logarithm of the intensity corresponds approximately to the response of the ear. In order to use a

TABLE 18.1
Comparative Sound Levels Due to Various Sources

Source or Situation	Intensity (dB)	Intensity (W/m^2)	Effect on Ear
Thunderclap	120	1	Painful
Rocket	115	4×10^{-1}	
Artillery	110	10^{-1}	Deafening
Airplane	100	10^{-2}	
Elevated train	90	10^{-3}	
Busy street traffic	80	10^{-4}	Very loud
Automobile (noisy)	70	10^{-5}	
Conversation	60	10^{-6}	Loud
Quiet automobile	50	10^{-7}	
Average office or home	40	10^{-8}	Average
Quiet office or home	30	10^{-9}	
Average whisper	20	10^{-10}	Faint
Rustle of leaves	10	10^{-11}	
Threshold of audibility (zero loudness)	0	10^{-12}	Hardly audible

logarithmic scale, we must define the intensity level I in terms of some other standard intensity level I_0 so that I/I_0 is dimensionless; hence, we may take the logarithm of this ratio. Thus, the difference L in intensity levels of two sound waves of intensity I and I_0 is defined in units of bel as

$$L = \log\left(\frac{I}{I_0}\right) \quad \text{bel} \tag{18.13}$$

The unit of intensity level, *bel*, was given this name in honor of the inventor of the telephone, Alexander Graham Bell (1847–1922). This unit is very large; hence, one-tenth of a bel, known as a *decibel*, is used more frequently. Thus, the intensity level of sound in decibels (dB) is given by

$$L = 10\log\left(\frac{I}{I_0}\right) \quad \text{dB} \tag{18.14}$$

It was agreed by the International Conference to use an intensity of 10^{-12} W/m^2 as a standard threshold of audibility I_0. According to this, the intensity level of an average sound, which has a loudness of 10^{-6} W/m^2, will be

$$L = 10\log\frac{10^{-6}}{10^{-12}} = 10\log 10^6 = 60 \text{ dB}$$

above the threshold, while a painfully loud sound of intensity 1 W/m^2 will correspond to an intensity level of

$$L = 10\log\frac{1}{10^{-12}} = 10\log 10^{12} = 120 \text{ dB}$$

These and other sound levels are listed in Table 18.1. Note that the intensity-level difference between two sounds of intensities I_1 and I_2 will be

$$L_1 - L_2 = 10\log\frac{I_1}{I_0} - 10\log\frac{I_2}{I_0} = 10\log\frac{I_1/I_0}{I_2/I_0}$$

That is,

$$\boxed{L_1 - L_2 = 10\log\frac{I_1}{I_2} \quad dB} \tag{18.15}$$

Also, it is found that the intensity of the sound level must be doubled before an observer can respond to the change in intensity and say that the sound is definitely louder. Thus, doubling the intensity corresponds to a change in intensity level of about 3 dB.

Pitch

Pitch is another psychological property of sound that is related to frequency. Just like loudness it is a subjective quantity and cannot be measured by means of instruments. *The pitch refers to that characteristic of sound sensation that enables one to classify a note as a high note or low note.* Unless the sound is extremely loud, there is a one-to-one correspondence between pitch and frequency. The higher the frequency, the higher the pitch. Thus, as we go up the scale one octave, say on a piano keyboard, we double the frequency, and the pitch also increases. Thus, we may say that *for a pure note of constant intensity, the pitch increases with increasing frequency.*

Quality (*or Timbre*)

The musical quality, whether it pleases or displeases, is quite a subjective quantity and depends on the psychological response of the ear and the brain to such quantities as loudness, pitch, and tone quality. Two of these quantities have been discussed above; tone quality will be explained below.

If a musical instrument could be made that had only the fundamental frequency and constant intensity, it will not produce a pleasant music. *The tone quality of any musical sound is determined by the presence of the number of overtones and their relative intensities.* For example, a clarinet, piano, and violin all could be tuned to have the same fundamental frequency, say for middle C. But we will have little difficulty in distinguishing between the sounds produced by these three instruments. It is because all have their characteristic mixtures of overtones and their relative intensities as shown in Figure 18.12.

In contrast to musical sounds shown in Figure 18.12, *noise* has a waveform that consists of a series of random displacements without any regularity.

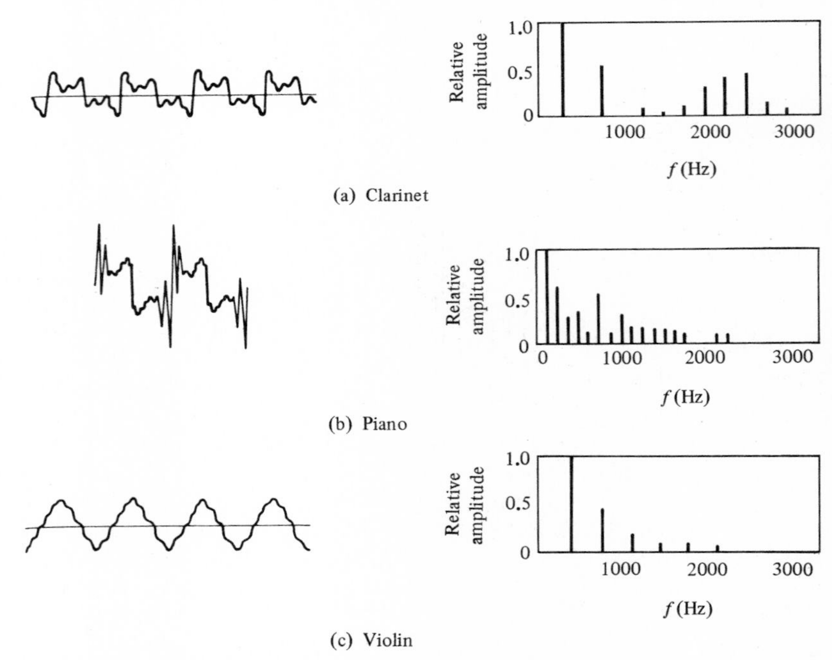

FIGURE 18.12 *Sound spectra—the characteristic mixtures of overtones and their amplitudes—of some musical instruments.*

EXAMPLE 18.3 The intensity of sound at point A a distance of 10 m from the source is 10^{-4} W/m^2, while the intensity at point B at 1000 m from the source is 10^{-8} W/m^2. Calculate the intensities at points A and B in dB. What is the difference between the intensity levels at these two points?

$$I_A = \text{intensity at point } A = 10^{-4}\ \text{W/m}^2$$
$$I_B = \text{intensity at point } B = 10^{-8}\ \text{W/m}^2$$

From Equation 18.14, $L = 10 \log (I/I_0)$ dB, where $I_0 = 10^{-12}\ \text{W/m}^2$. Therefore,

$$L_A = 10 \log \frac{I_A}{I_0} = 10 \log \frac{10^{-4}\ \text{W/m}^2}{10^{-12}\ \text{W/m}^2}$$
$$= 10 \log 10^8 = 80 \log 10 = 80\ \text{dB}$$
$$L_B = 10 \log \frac{I_B}{I_0} = 10 \log \frac{10^{-8}\ \text{W/m}^2}{10^{-12}\ \text{W/m}^2}$$
$$= 10 \log 10^4 = 40 \log 10 = 40\ \text{dB}$$
$$L_A - L_B = (80 - 40)\ \text{dB} = 40\ \text{dB}$$

or from Equation 18.15,

$$L_A - L_B = 10 \log \frac{I_A}{I_B} = 10 \log \frac{10^{-4}\ \text{W/m}^2}{10^{-8}\ \text{W/m}^2}$$
$$= 10 \log 10^4 = 40 \log 10 = 40\ \text{dB}$$

Exercise 18.3 In Example 18.3, if the intensity at point C 100,000 m from the source is $10^{-12}\ \text{W/m}^2$, what is the intensity at point C in dB? What is the difference in intensity levels between (a) A and C; and (b) B and C? [*Ans.:* $L_C = 0$ dB; (a) $L_A - L_C = 80$ dB; (b) $L_B - L_C = 40$ dB.]

* 18.5. The Ear

We already mentioned the range of frequencies and intensities to which a human ear responds. Let us investigate its structure and workings. When sound waves impinge on an ear, certain nerves which are very sensitive to pressure variations in sound waves lead to the sensation of hearing. The human ear may be divided into three parts—the outer ear, middle ear, and inner ear—as shown in Figure 18.13. The sound waves collected at the outer ear are transmitted

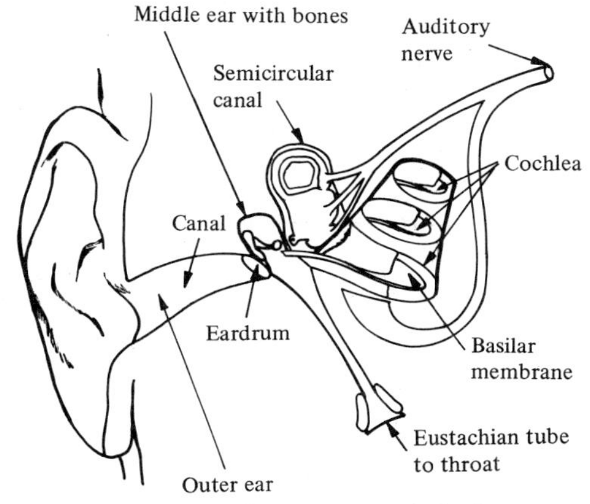

Figure 18.13 *Different parts of the human ear.*

through the ear canal, the *auditory canal* (~2.7 cm long), to the ear drum, the *tympanic membrane,* which is set into vibrations. The ear canal is like a closed pipe vibrating with a fundamental note of frequency ~3200 Hz. The middle ear consists of three bones: the *hammer, anvil,* and the *stirrup*. These bones transmit the vibrations from the vibrating drum to an oval window connected to the inner ear. The important components of the inner ear are the *cochlea, basilar membrane,* and three *semicircular canals*. The cochlea, looking like a snail shell, is filled with fluid, which may be set into vibrations by the vibrations of the oval window. The pressure variations in the vibrations of the fluid are ~60 times larger than those of the incident sound waves on the ear drum. (This is because the area of the ear drum is ~20 times that of the oval window.)

The cochlea, a spiral of about 2.5 turns, is divided along its length by the basilar membrane, which is connected to ~30,000 nerve endings. It is here the sound energy is converted into electrical energy and conveyed to the brain as nerve impulses for interpretation. The semicircular canals do not play a vital part in hearing but are essential for keeping one's balance.

The human ear is a very sensitive device, and the faintest sound detected by the ear has pressure variations of $\sim 2 \times 10^{-5}\ \text{N/m}^2$ (corresponding to the amplitude of $\sim 10^{-11}$ m). The sensation of pitch depends upon the position where resonance occurs along the basilar membrane, while the amount of stimulation of the basilar membrane fibers determines the loudness.

18.6. Interference in Time—Beats

The application of superposition to two waves of exactly equal frequencies and amplitudes traveling in opposite directions in the same region of space leads to the formation of standing waves, as we discussed in the case of strings and air columns. This is the phenomenon of *interference in space*. In this section we shall consider *interference in time*. When two waves of slightly different frequencies traveling through the same region of space are observed at a given point, the result is a sound with pulsating intensity called *beats*. If condensations (or crests) from both the waves reach the ear of the listener at the same time, the amplitudes of the two waves are added resulting in a loud sound. Since the frequencies of the two wave trains are slightly different, some time later a condensation (or crest) from one wave reaches at the same time as a rarefaction (or trough) from the other, resulting in little or no sound to be heard by the ear. Thus, the listener hears a periodic rise and fall in the loudness of the sound, called *beats*. The number of beats heard per second is equal to the difference in the frequencies of the two wave trains, as shown below and demonstrated in Figure 18.14.

Consider a point in space through which two waves of slightly different frequencies f_1 and f_2 are passing. For simplicity, let us assume that their amplitudes y_m are equal. That is (using $\omega = 2\pi f$),

$$y_1 = y_m \cos 2\pi f_1 t \tag{18.16}$$

and

$$y_2 = y_m \cos 2\pi f_2 t \tag{18.17}$$

We want to find the resultant of these two waves at a given point as a function of time. Thus, the net displacement will be

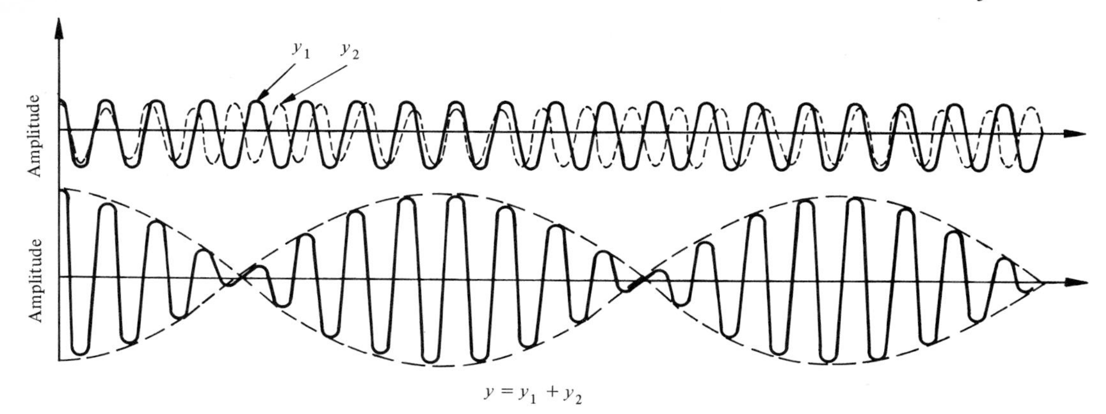

FIGURE 18.14 *Formation of beats when two waves of slightly different frequencies travel in the same direction.*

$$y = y_1 + y_2 = y_m (\cos 2\pi f_1 t + \cos 2\pi f_2 t) \tag{18.18}$$

since

$$\cos\theta_1 + \cos\theta_2 = 2\cos\frac{\theta_1 - \theta_2}{2}\cos\frac{\theta_1 + \theta_2}{2} \tag{18.19}$$

We may write Equation 18.18 as

$$y = \left[2y_m \cos 2\pi \left(\frac{f_1 - f_2}{2}\right) t\right] \cos 2\pi \left(\frac{f_1 + f_2}{2}\right) t$$

or

$$y = A_m \cos 2\pi \bar{f} t \tag{18.20}$$

where $\bar{f}$ is the average frequency,

$$\bar{f} = \frac{f_1 + f_2}{2} \tag{18.21}$$

and A_m is the amplitude modulation factor,

$$A_m = 2y_m \cos 2\pi \left(\frac{f_1 - f_2}{2}\right) t \tag{18.22}$$

Thus, the resulting wave motion given by Equation 18.20 has a frequency $\bar{f}$, which is the average of the frequencies of the two waves, and has an amplitude given by Equation 18.22. The amplitude of the resultant wave is not constant, but, as given by Equation 18.22, it varies with frequency f_m, given by

$$f_m = \frac{f_1 - f_2}{2} \tag{18.23a}$$

Thus, if f_1 and f_2 are very close so that $f_1 - f_2$ is small, the frequency of variation in the amplitude will be small as well. (The ear cannot respond to very rapid, high f_m changes in loudness.) A beat, which is maximum in the amplitude, will be heard whenever

$$\cos 2\pi \left(\frac{f_1 - f_2}{2}\right) t = \pm 1 \tag{18.23b}$$

That is, twice in each cycle there will be maxima. But the amplitude variation frequency is $f_m = (f_1 - f_2)/2$. Hence, the number of beats will be twice this value, that is, $2f_m = f_1 - f_2$, and the beat frequency is equal to the difference in the two frequencies.

$$\boxed{\text{Beat frequency} = f_1 - f_2} \tag{18.24}$$

Figure 18.14 illustrates the case in which the waves are of frequencies $f_1 = 8$ Hz and $f_2 = 10$ Hz. The beat frequency will be 2, as shown in Figure 18.14. Thus, we may say that the beats are fluctuations in amplitude when two waves of slightly different frequencies interfere in time at a given point. Beats between two different tones can be easily detected by the ear if the beat frequency does not exceed seven. Musicians tune two strings to the same frequency by tightening one of them until the frequencies of the two are the same; hence, no beats are heard.

Example 18.4 Two sources of sound waves are emitting waves of wavelengths 5.0 m and 5.50 m. If the sound velocity is 340 m/s, what is the number of beats that will be produced?

Let $\lambda_1 = 5.0$ m, $\lambda_2 = 5.50$ m, and f_1 and f_2 be the corresponding frequencies. Since $v = 330$ m/s, from the relation $v = f\lambda$,

$$f_1 = \frac{v}{\lambda_1} = \frac{340 \text{ m/s}}{5.0 \text{ m}} = 68/\text{s}$$

$$f_2 = \frac{v}{\lambda_2} = \frac{340 \text{ m/s}}{5.5 \text{ m}} = 62/\text{s}$$

$$\text{number of beats} = f_2 - f_1 = (68 - 62)/\text{s} = 6/\text{s}$$

Exercise 18.4 Two sources of sound waves in water are emitting waves of wavelengths 5.0 m and 5.50 m. If the speed of sound in water is 1480 m/s, what is the number of beats that will be produced? [*Ans.:* 27/s.]

18.7. The Doppler Effect

It is a matter of common experience that whenever there is a relative motion between a source of sound and an observer (listener), the pitch (frequency) measured by the observer is different as compared to the measured pitch when both of them are at rest. For example, when standing at a railroad platform, we observe that the pitch of the whistle from an approaching locomotive is higher than the pitch from the receding locomotive. Similarly, we notice an abrupt drop in the pitch of the sound from the horn of a car when it passes by with a great speed.

The observed change in the pitch due to the relative motion between the source and the observer is called the *Doppler effect*. This effect was first analyzed in 1842 by the German-born Austrian physicist Christian Johann Doppler (1803–1853). Doppler observed this effect in sound waves, but pointed out that it should be observed in light waves as well. Actually, the Doppler effect applies to all wave motion in general. For simplicity we shall analyze the Doppler effect for sound waves. Also, we shall limit our analysis to the case in which the source and the observer move in a straight line relative to each other; that is, we are dealing with the *longitudinal Doppler effect.*

Figure 18.15 shows spherical wave fronts diverging from a source S, which is at rest and emits sound waves with frequency f_S. If v is the velocity of the

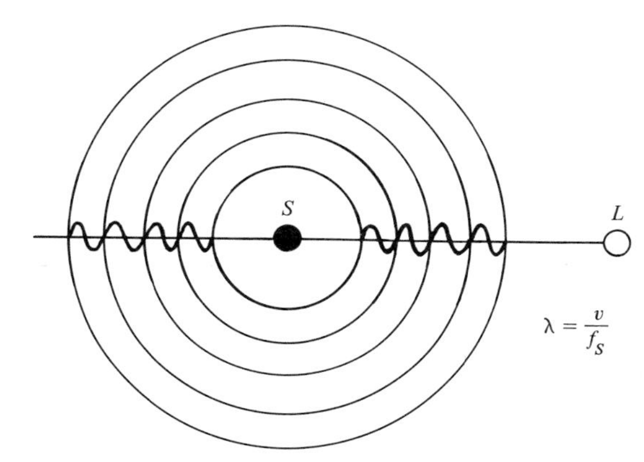

FIGURE 18.15 *Stationary source of sound emitting waves in all directions. The listener L at rest will hear sound with the same frequency as that of the source.*

sound waves with respect to stationary air, the wavelength of these waves will be $\lambda = v/f_S$ in all directions as measured by a listener L and the frequency will be f_S. Now let us assume that S is moving with velocity v_S and L with velocity v_L, both along the $+X$-axis as shown in Figure 18.16. The purpose is to calculate the frequency f_L as measured by the observer L. To start with suppose that L is at rest and the source S is at 0 (or 1) when $t = 0$. As the source moves to the right and is at positions 1, 2, 3, 4, . . . , it emits waves resulting in spherical wave fronts 1′, 2′, 3′, 4′, . . . , respectively. As is clear from this figure, the waves are closely spaced in the direction in which the source is moving and are widely spaced in the opposite direction. Thus, to a listener L at rest in front of the source, the waves look shorter while if the listener is behind the source, the waves look longer. We can calculate these changed wavelengths in a simple manner.

(Figure 18.16 is on page 400.)

Suppose that in time t the source S has moved from 0 to a, as shown in Figure 18.16. The distance $0a = v_St$, whereas

$$ab = vt - v_St = (v - v_S)t$$

$$ac = vt + v_St = (v + v_S)t$$

In time t, the source emits f_St waves. These waves are crowded in front of the source in distance ab while behind the source they are spread out over a distance ac. Thus, the wavelength observed by the listener L in front of the

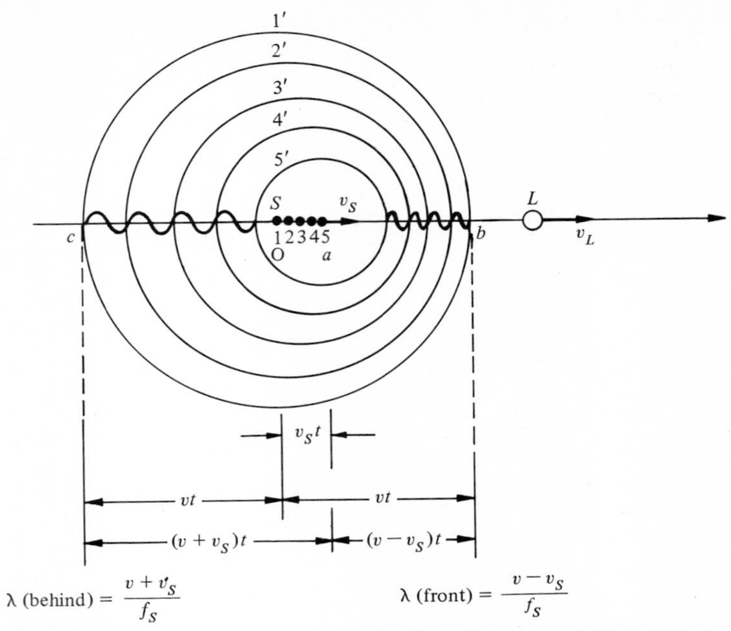

FIGURE 18.16 *Doppler effect—the wavelength as measured by the listener when the source* S *is in motion.*

source will be

$$\lambda(\text{front}) = \frac{ab}{f_S t} = \frac{(v - v_S)t}{f_S t} = \frac{v - v_S}{f_S} \tag{18.25}$$

while behind the source the wavelength observed will be

$$\lambda(\text{behind}) = \frac{ac}{f_S t} = \frac{(v + v_S)t}{f_S t} = \frac{v + v_S}{f_S} \tag{18.26}$$

Thus, to the listener L at rest in front of the source, the wavelengths are shorter (and frequency higher), while behind the source the wavelengths are longer (and frequency lower). The change in wavelength in front and back of the source is beautifully demonstrated by the Doppler effect in a liquid, as shown in Figure 18.17. The source is a vibrating stick.

Note that the frequency in the front $= v/\lambda = vf_S/(v - v_S)$, while the frequency in the rear $= vf_S/(v + v_S)$.

If the listener is in motion, the wavelengths, and hence the frequencies, will be further changed. As shown in Figure 18.16, the listener L is moving to the right in front of the source with velocity v_L. The relative speed of the waves with respect to the listener will be $v - v_L$. Thus, the frequency f_L at which the listener L will receive these waves is

$$f_L = \frac{v - v_L}{\lambda(\text{front})} = \frac{v - v_L}{(v - v_S)/f_S} = f_S\left(\frac{v - v_L}{v - v_S}\right)$$

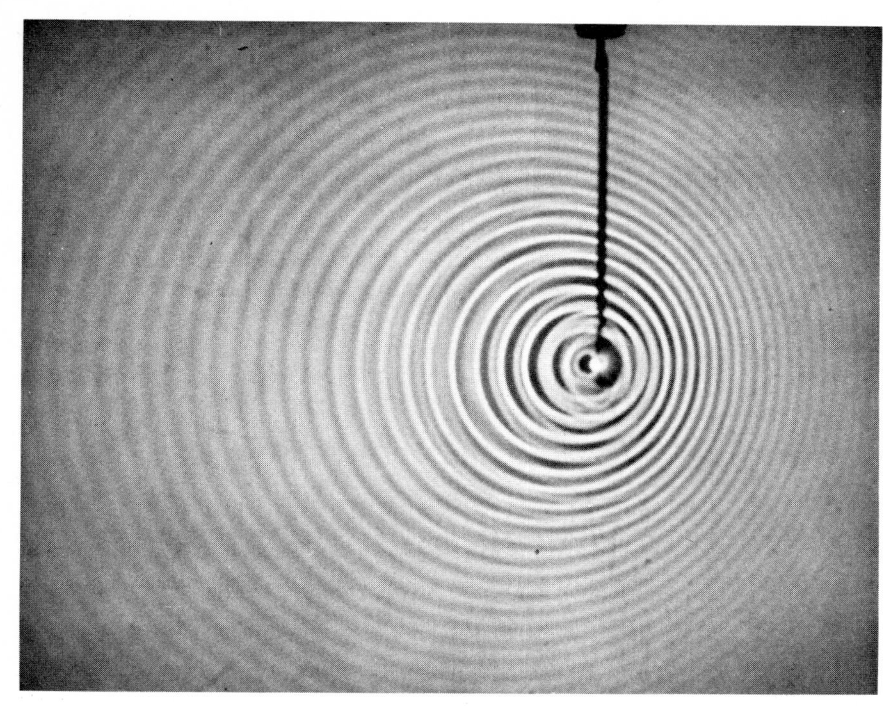

FIGURE 18.17 *Doppler effect in a liquid. The source is a vibrating stick moving to the right.* (Courtesy of Education Development Center, Inc., Newton, Mass.)

or

$$\boxed{\frac{f_L}{v - v_L} = \frac{f_S}{v - v_S}} \qquad \text{for} \quad \overset{S}{\bullet} \longrightarrow \quad \overset{L}{\circ} \longrightarrow \tag{18.27}$$

This result is much more general and may be applied to many other situations. Let us consider some special situations.

Suppose that the listener is approaching the source, in which case v_L in Equation 18.27 should be replaced by $-v_L$; that is,

$$f_L = f_S\left(\frac{v + v_L}{v - v_S}\right) \qquad \text{for} \quad \overset{S}{\bullet} \longrightarrow \quad \longleftarrow \overset{L}{\circ} \tag{18.28}$$

On the other hand, if the source was moving to the left—receding, v_S in Equation 18.27 should be replaced by $-v_S$; that is,

$$f_L = f_S\left(\frac{v - v_L}{v + v_S}\right) \qquad \text{for} \quad \longleftarrow \overset{S}{\bullet} \quad \overset{L}{\circ} \longrightarrow \tag{18.29}$$

Furthermore, if both the source and the observer were moving to the left, v_L and v_S in Equation 18.27 must be replaced by $-v_L$ and $-v_S$, respectively. Therefore,

$$f_L = f_S\left(\frac{v + v_L}{v + v_S}\right) \quad \text{for} \quad \overset{S}{\longleftarrow \bullet \longleftarrow} \quad \overset{L}{\circ} \tag{18.30}$$

As pointed out earlier, the Doppler effect is observed not only in sound waves, but also in electromagnetic waves, such as light waves, radio waves, and so on. But the above equations do not hold for electromagnetic waves. This is because the electromagnetic waves can travel through a vacuum without the need for any medium and their speed is always constant ($c = 3 \times 10^8$ m/s) and is independent of the relative speed of the source and the observer. Also, we do not talk about the relative motion of the source or the observer with respect to the medium, only relative to each other. Besides the longitudinal Doppler effect, electromagnetic waves exhibit the transverse Doppler effect as well.

Example 18.5 A train S traveling at a speed of 20 m/s (~45 mi/h) and blowing a whistle with a frequency of 240 Hz is approaching a train L which is at rest. Assuming the speed of sound to be 340 m/s, calculate the following: (a) Wavelengths in air (1) in front and (2) behind the train S. (b) Frequencies measured by a listener in train L while train S is (1) approaching and (2) receding from train L. (c) If train L starts moving with a speed of 10 m/s, what will be the frequencies heard by a passenger in train L if both trains were (1) approaching and (2) receding.

(a) The situation is $\overset{S}{\bullet} \longrightarrow \overset{L}{\circ}$, with $v_S = 20$ m/s and $v_L = 0$.

(1) From Equation 18.25

$$\lambda(\text{front}) = \frac{v - v_S}{f_S} = \frac{(340 - 20)}{240} = 1.33 \text{ m}$$

(2) From Equation 18.26

$$\lambda(\text{behind}) = \frac{v + v_S}{f_S} = \frac{(340 + 20)}{240} = 1.5 \text{ m}$$

(b) (1) The situation is $\overset{S}{\bullet} \longrightarrow \overset{L}{\circ}$, with $v_L = 0$ and from Equation 18.28,

$$f_L = f_S\left(\frac{v}{v - v_S}\right) = 240\left(\frac{340}{340 - 20}\right) = 255 \text{ Hz}$$

(2) The situation is $\longleftarrow \overset{S}{\bullet} \; \overset{L}{\circ}$, so that $v_L = 0$ while the source speed is $-v_S$. From Equation 18.29,

$$f_L = f_S\left(\frac{v}{v + v_S}\right) = 240\left(\frac{340}{340 + 20}\right) = 227 \text{ Hz}$$

(c) (1) The situation is $\overset{S}{\bullet} \longrightarrow \quad \longleftarrow \overset{L}{\circ}$, where v_L is negative. From Equation 18.27 or 18.28,

$$f_L = f_S\left(\frac{v + v_L}{v - v_S}\right) = 240\left(\frac{340 + 10}{340 - 20}\right) = 263 \text{ Hz}$$

(2) The situation is $\overset{S}{\longleftarrow \bullet} \quad \overset{L}{\circ \longrightarrow}$, where v_S is negative. From Equation 18.27 or 18.29,

$$f_L = f_S\left(\frac{v - v_L}{v + v_S}\right) = 240\left(\frac{340 - 10}{340 + 20}\right) = 220 \text{ Hz}$$

Exercise 18.5 A train passes a stationary observer at a railroad crossing at a speed of 30 m/s and is blowing a whistle with a frequency of 800 Hz. Calculate the wavelengths in front and behind the source. What is the change in frequency as measured by a stationary observe as the train approaches and recedes from him? What is the percentage change in frequency? [*Ans.*: λ(front) $= 38.8$ cm, λ(behind) $= 46.3$ cm, $\Delta f = 142$ Hz, $\Delta f/f_0 = 17.8\%$.]

18.8. Shock Waves—Sonic Boom

The Doppler effect is concerned with sources moving with *subsonic speeds*, that is, speeds lower than the speed at which sound waves propagate through a medium. The propagation of spherical wave fronts with respect to the source positions was shown in Figure 18.16. Sources that move faster than the speed of sound are said to have *supersonic speeds*. In such cases, the sources advance more than the wave fronts, and the situation is as shown in Figure 18.18. In this

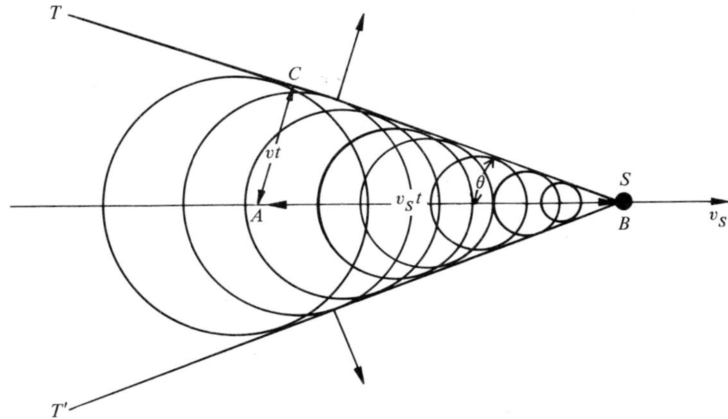

Figure 18.18 *Formation of wave fronts when the source is moving faster than the speed of sound.*

case, energy emitted is unable to move in front of the source, hence is concentrated on the sides. In time t the source S moves a distance AB, while the wave front emitted by S moves a distance AC. The figure shows that as the source moves out, the spherical wave fronts overlap behind. The surface tangents ST and ST' drawn to all these wave fronts result in a cone whose axis is the line of motion of the source. The aperture θ of the cone is

$$\sin\theta = \frac{AC}{AB} = \frac{vt}{v_S t} = \frac{v}{v_S} \tag{18.31}$$

Thus, the resultant wave motion is a *conical wave* that propagates in the direction indicated by the arrows. Along the tangents to these wave fronts, the wave crests reinforce each other, leading to a concentration of energy. The concentration of energy or pressure discontinuity as it travels outward is called a *shock wave.*

Note that it is not necessary for an object to have a vibrating source of sound in order to create a shock wave. The only condition necessary is that the object moves with a speed greater than the speed of sound in that medium. The ratio of the speed of the source to that of the speed of sound, that is, v_s/v, is called the *Mach number* (named for Ernst Mach (1838–1916). Thus, shock waves (also called Mach waves) are generated by objects moving with a Mach number greater than 1.

Figure 18.19 shows two examples of shock waves. In (a) the shock wave is produced by a bullet moving in air with a speed of more than Mach 2. In (b) the shock waves are produced by a vibrating reed moving with speed greater than the speed of propagation of water waves. Another example of a shock wave is the *sonic boom,* which is produced by a distant supersonic aircraft and is clearly heard by an observer at rest on the ground. The energy carried by a conical shock wave from such a craft is sometimes sufficient to crack glass windows. The abrupt change in the air pressure may easily exceed 2 pounds per square foot.

The counterpart of a sonic boom in light (or electromagnetic waves) is Cerenkov radiation, discovered by the Russian physicist P. A. Cerenkov (1904–). Accordingly, whenever a particle moves through a medium with a speed greater than the speed of light in that medium, it emits blue Cerenkov radiation in the transparent medium, exactly in the form as observed from the shock wave in Figure 18.18. By measuring the cone angle at which the bluish radiation is emitted, the speed of the particle can be easily calculated.

18.9. Sonic Applications and Noise Pollution

As shown in Figure 17.1, the sonic spectrum consists of a wide range of mechanical vibrations. So far we have been discussing the audible range (20 to 20,000 Hz) of the sonic spectrum. In this section we shall discuss the other two ends of the sonic spectrum.

Sounds with frequencies greater than 20,000 Hz are called *ultrasonic.* The human ear cannot detect such high-frequency sounds, but dogs and bats, for example, can. Contrary to the audible sound waves, ultrasonic sound waves can be obtained in the form of a well-defined narrow beam. This is because diffraction effects are directly proportional to the wavelength; that is, $\sin\theta = 100°(\lambda/D)$. Thus, for a fixed aperture width D, if the frequency of the waves is large, λ will be small; hence, θ will be small. Applications involve the use of a well-defined beam of ultrasonic waves.

Communication

Ultrasonic waves in the range 20,000 to 100,000 Hz are commonly used by a ship's sonar to locate the position of other ships or submarines and to measure

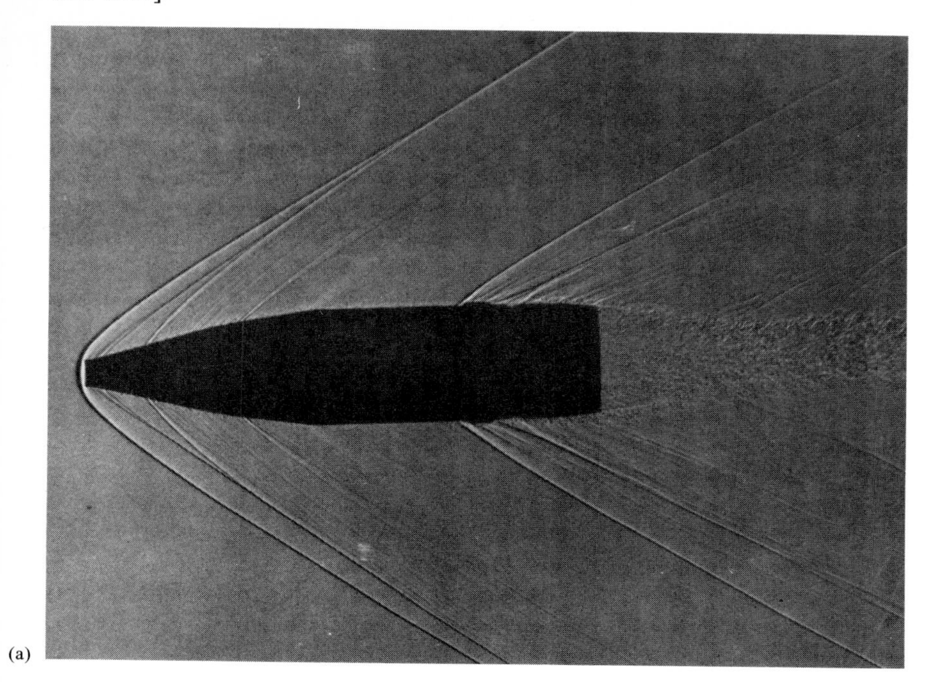

(a)

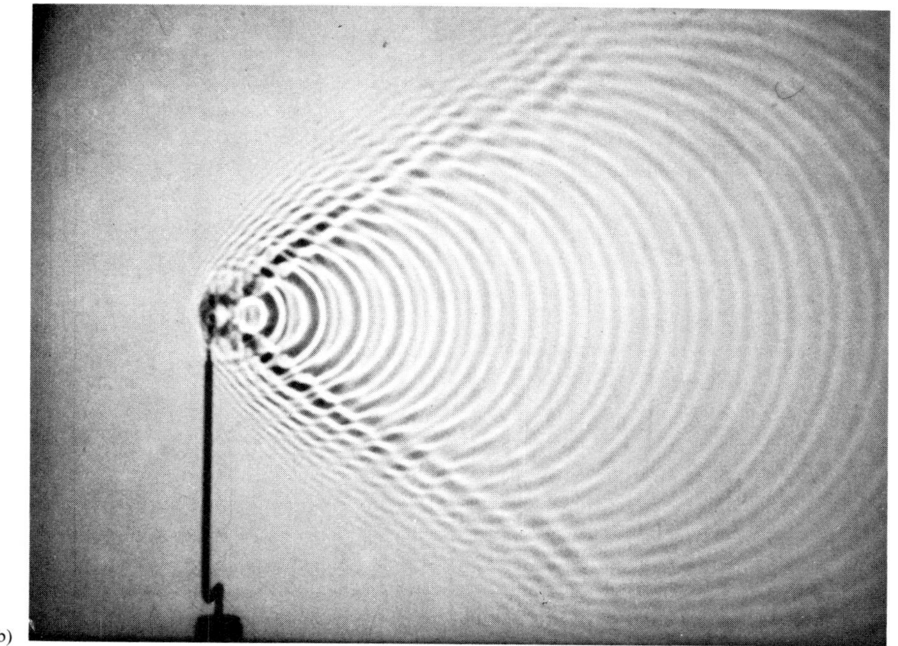

(b)

FIGURE 18.19 *Examples of shock produced by (a) a bullet in air* courtesy of U.S. Army Ballistic Research Laboratories Aberdeen Proving Ground, MA. *(b) a vibrating reed moving to the right while touching a water surface:* courtesy of Education Development Center, Inc., Newton, Mass.

the depth of the sea. A transmitter sends ultrasonic waves which after being reflected from the object under investigation are received by the transmitter, which works as a detector as well. The time elapsed between the emission and the reception of the ultrasonic waves determines the location of the objects, or depth of the sea. This is the *echo principle*. Ship-to-ship communication also uses ultrasonic waves. Another advantage of using ultrasonic waves in the applications above is that these waves cannot be heard without the aid of special instruments.

Industrial

Cleaning and sterilization by means of ultrasonic waves is another application. The cleaning is done by the method called *cavitation* or *cold boiling*. An instrument that needs cleaning but whose parts cannot be reached directly may be placed in a liquid. The ultrasonic waves passing through the liquid produce tiny bubbles where the rarefaction of the ultrasonic wave reaches. When the compression of the wave reaches these bubbles, the bubbles are compressed until they implode (explode inward). This leads to the creation of several small localized shock waves. These shock waves blast away any dirt or contamination from the unreachable portions. Usually, frequencies in the range 20,000 to 30,000 Hz are used for this purpose.

Still another application of ultrasonic waves in industry is in plastic welding. Application of small pressures and ultrasonic vibration to two similar surfaces produces sufficient thermal energy to bond the surfaces together.

Medical

Ultrasonic waves are finding their applications in medical technology for diagnostic purposes. The advantage of the ultrasonic waves over x-ray photography lies in the fact that ultrasonic waves can distinguish between soft tissues and liquids, whereas x-rays cannot. Ultrasonic waves can distinguish between cysts and tumors. The usual method used is *ultrasonography*. It involves sending an ultrasonic wave, for example, to the brain, liver, kidneys, and so on, and looking at the reflected or transmitted waves. The observations are made on an oscilloscope after the ultrasonic waves have been converted into electrical pulses by means of transducers.

The other end of the spectrum is the infrasonic with frequencies less than 20 Hz. Seismic waves have such low frequencies. The useful applications of infrasonic vibrations are still in the initial process of investigation.

In some instances, infrasonic waves have proved dangerous. It is found that at very low frequencies of 5 to 10 Hz, certain organs of the body tend to resonate, leading to a vibration-induced sickness. This resonating of one organ leads to rubbing against another, thereby producing noticeable ill effects, even at low power levels.

Excessive exposure to intense sound, ultrasonic and infrasonic, for long duration and at high intensity levels has created a widespread concern that the noisy environment of modern man may be detrimental to his health. Constant exposure to noise in industry, automobile traffic, and sonic booms from aircrafts is not only distracting but may affect hearing (leading to deafness) and produce physiological effects leading to nervous disorders. Even music, which is so pleasing to the ear, can be hazardous. Medical studies have revealed that certain teenage rock musicians have developed boilermaker's ears. That is, their hearing is on a par with that of the average 65-year-old person. There still remains much to be done to investigate noise pollution and its control in modern industrialized society, which is exposed to higher and higher noise pollution levels every day.

SUMMARY

Stationary or standing waves are formed when two waves of the same frequency and speed travel in opposite directions. The resultant displacement is $y = A_x \cos \omega t$, $A_x = 2y_0 \sin kx$. For antinodes $x = \lambda/4, 3\lambda/4, \ldots$, and for nodes $x = \lambda/2, \lambda, \ldots$.

In strings the fundamental frequency is $f_1 = v/2L$, while for the $(n - 1)$th overtone, $f_n = n(v/2L)$, $n = 2, 3, \ldots$. The overtones together with the fundamental tone form a harmonic series. The frequency of the nth harmonic in the string is $f_n = (n/2L)\sqrt{T/\mu}$.

For open pipes, $f_n = f_1, 2f_1, 3f_1, 4f_1, \ldots$, and for closed pipes $f_n = f_1, 3f_1, 5f_1, \ldots$.

Whenever a system capable of oscillating is acted on by a series of impulses having frequencies equal to one of the natural frequencies, the system responds by setting itself into oscillation with a relatively large amplitude—this phenomenon is called *resonance.*

The *loudness* is intimately connected with the intensity of sound and usually increases with the frequency. The difference in the intensity levels of two sounds is $L_1 - L_2 = 10 \log (I_1/I_2)$ dB. The *pitch* refers to that characteristic of sound sensation that enables one to classify a note as a high note or a low note. For a pure note of constant intensity, the pitch increases with increasing frequency. The *tone quality* of any musical sound is determined by the presence of the number of overtones and their relative intensities.

When two waves of slightly different frequencies traveling through the same region of space are observed at a given point, the result is a sound with pulsating intensity called *beats* (interference in time). The frequency of the beats is $f_1 - f_2$.

The observed change in the pitch due to the relative motion between the source and the observer is called the *Doppler effect*. In general, $f_L/(v - v_L) = f_S/(v - v_S)$.

The concentration of an energy or pressure discontinuity as it travels outward is called a *shock wave*. The aperture θ of the conical wave is $\sin \theta = v/v_S$. The ratio v_S/v is called the *Mach number.*

Ultrasonic waves find several applications in communication, industry, and medicine.

QUESTIONS

1. If a violin string is (a) tightened; and (b) loosened, what effect will it have on the frequencies of the harmonic series?
2. Give an example of a musical instrument of the following type: (a) open pipe; (b) closed pipe; (c) string; (d) rod; and (e) plate.
3. How would you locate the position of nodes and antinodes in a vibrating drum?

4. Name two musical instruments whose overtones are not harmonics.
5. Why are marching soldiers ordered to break step while crossing a bridge?
6. What factor determines the change in the quality of sound from person to person?
7. Give two examples of constructive and destructive interference in time, that is, beats.
8. Is there any energy loss when the waves interfere to produce beats?
9. When the waves interfere, what effect has this on the velocity of the individual waves and their total energy?
10. Consider a source and a listener. What can cause (a) the crowding of waves; and (b) the thinning of waves near the listener?
11. Suppose that (a) a source is moving toward you; or (b) you are moving toward a source. What effect will the wind velocity have on the frequency heard by you?
12. Is it necessary to have a vibrating source to produce shock waves? Give two examples in which shock waves are produced.
13. Explain how a sonic boom from an aircraft can break a window?
14. What is the danger from ultrasonic and infrasonic rays to the human ear, eyes, and other parts of the body?

PROBLEMS

1. An iron wire of length 1.5 m and linear mass density 0.004 kg/m is under a tension of 420 N. Calculate the speed of the transverse waves and the frequencies of the first three harmonics.
2. A string fixed at two ends, when plucked at the center and let go, vibrates with a frequency of 400 Hz. With what other possible frequencies can the string be made to vibrate?
3. What are the wavelengths and frequencies of the first three harmonics and overtones for a stretched string of length 1 m and fundamental frequency of 400 Hz?
4. Calculate the length of a stretched string which when under a tension of 200 N will produce a fundamental note of 500 Hz. The linear mass density of the string is 0.004 kg/m.
5. A certain violin string 0.50 m long is tuned to middle C, which has a frequency of 264 Hz, by adjusting the tension in it. Calculate the value of this tension. A 1-m-long wire has a mass of 0.60 g.
6. A piano wire has a length of 1.4 m and has a fundamental frequency of 330 Hz. Calculate the tension in the string and the velocity of the transverse wave. The linear mass density of the wire is 0.01 kg/m.
7. A stretched string with its supports 50 cm apart is vibrating in a fundamental mode with a frequency of 110 Hz. The mass of the string is 25 g. Calculate the tension in the string. What will be the frequencies of the first two overtones?
8. Calculate the ratio of the fundamental frequencies of two strings under the conditions: (a) Same length, material, and diameter, but one has four times the tension of the other. (b) Same length, material, and tension, but one has double the diameter of the other. (c) Same material, diameter, and tension,

but one has double the length of the other. (d) Same length, diameter, and tension, but the linear density of one is 4 times that of the other.

9. An open organ pipe is 80 cm long. Calculate the wavelengths and frequencies of the following: (a) fundamental note; (b) second harmonic; (c) second overtone; (d) third harmonic; and (e) third overtone.
10. Repeat Problem 9 for a closed pipe of length 80 cm.
11. A 1-m-long stretched string of mass 5 g and a 1-m-long open pipe are placed side by side. What should be the tension in the string so that the second harmonic of the string is in resonance with the fundamental of the open pipe?
12. At 0°C an open pipe resonates at a frequency of 240 Hz. What will be the resonant frequency at 30°C?
13. To what different possible frequencies will (a) an open pipe; and (b) a closed pipe resonate if each pipe has a length of 0.6 m and the temperature is 20°C?
14. A wire of linear density 0.01 kg/m is under a tension of 400 N. Its fundamental frequency is in resonance with the first overtone of a closed pipe of length 0.80 m. Calculate the length of the wire? What is the speed of the transverse waves?
15. A glass tube is dipped into water and the length of the air column is changed by moving the tube vertically. The air column resonates when a tuning fork of unknown frequency is brought to the top and length of the air column is either $l_1 = 12$ cm or $l_3 = 36$ cm. If the speed of sound in air is 340 m/s, what is the unknown frequency? (Note that the air column behaves like a pipe closed at the lower end.)
16. Standing waves formed in an air column are of a frequency of 880 Hz and produce nodes that are 20 cm apart. Calculate the speed of the longitudinal waves in this gas.
17. A steel rod 1 m long is clamped at the middle as shown in Figure 18.7(a). The rod can be set into longitudinal vibrations by stroking it. Calculate the frequencies of the fundamental and first two overtones. Draw schematic wave forms as in the case of pipes. What is the ratio of the frequencies?
18. A steel rod 1 m long is clamped 25 cm from one end. The rod is set into longitudinal vibrations by stroking it. Calculate the frequencies of the fundamental and first two overtones. Draw schematic wave forms as in the case of pipes. What is the ratio of the frequencies?
19. Relative to the arbitrary intensity level of 10^{-12} W/m^2, what is the intensity level in dB of a sound with an intensity of 10^{-7} W/m^2? (What difference will it make in the value of dB if the intensity was expressed in W/cm^2?)
20. The intensity level of a classroom is 50 dB if only one person is talking. What will be the intensity level if 100 persons are talking simultaneously?
21. The intensity of the threshold of hearing is 10^{-12} W/m^2. Calculate the intensity of (a) a whisper that has a sound level of 20 dB; and (b) a painful noise that has a noise level of 120 dB. What is the ratio of the intensity in the two cases?
22. The intensity of 10 violins is 10 times that of a single violin. What is the change in dB in the sound level if all 10 instead of one are playing simultaneously?

23. The intensity of sound B is 500 times that of sound A. What is the difference in their intensity level in dB?
24. The intensity of the threshold of hearing is 10^{-12} W/m^2. The intensity of another sound is 30 dB higher than this. Calculate the intensity of this source.
25. The power output of a double speaker at a frequency of 60 Hz is 50 W. At a frequency of 90 Hz, the intensity of the sound is 20 dB higher. What is the power output at 90 Hz?
26. The power output of a hi-fi system is so high that the intensity of sound is 100 dB above the threshold of hearing. Calculate the amount of energy falling on an ear of area 5 cm^2. How much is this energy in microjoules? How long will it take to accumulate 1 J of energy?
27. Two sources are emitting sounds of wavelengths 9 ft and 10 ft, respectively, while the velocity of sound is 1080 ft/s. How many beats will be heard per second?
28. Using a 270-Hz tuning fork and sound from a certain piano note, a piano tuner hears six beats per second. Using a 260-Hz tuning fork, he hears four beats per second. What is the frequency of the piano note?
29. A tuning fork of unknown frequency produces four beats per second with a standard source of frequency 240 Hz. (a) What are the possible values of the unknown frequency? (b) A small piece of wax attached to a prong of the tuning fork increases the number of beats per second. What is the true frequency of the tuning fork?
30. Two strings are vibrating with a frequency of 612 Hz. How much tension should be changed in one of the strings so that four beats per second will be heard? [*Hint:* Use Binomial theorem for $(1 + \Delta T/T)^{1/2}$.]
31. Two strings are in resonance. The tension in one string is decreased slightly so that it has a frequency of 508 Hz and produces seven beats per second when sounded together with the other string. What is the initial frequency of the two strings?
32. Two open pipes of length 80 and 82 cm produce fundamental notes simultaneously. What is the beat frequency?
33. Two closed pipes of length 40 and 41 cm produce fundamental notes simultaneously. What is the beat frequency?
34. Two identical closed pipes are in tune at 20°C and produce a note of frequency of 840 Hz. If one of the pipes is cooled to 0°C, how many beats per second will be produced? (Ignore thermal expansion of the pipes.)
35. Two open pipes are in tune at 24°C and produce a note of frequency of 420 Hz. If one of the pipes is heated to 34°C, how many beats per second will be produced? (Ignore thermal expansion of the pipes.)
36. Calculate the frequency heard by a listener at rest when a source moving with a velocity of 25 m/s and emitting a sound with a frequency of 800 Hz is (a) approaching; and (b) receding from the listener.
37. Suppose that a source at rest is emitting a sound having a frequency of 800 Hz. Calculate the frequency observed when a listener moving with a velocity of 25 m/s is (a) approaching; and (b) receding from the source.
38. A source is emitting a sound with a frequency of 400 Hz. A listener hears a sound with a frequency of 420 Hz. What is the speed and direction if (a) the source is in motion; and (b) the listener is in motion?

39. A source is emitting a sound with a frequency of 400 Hz. A listener hears a sound with a frequency of 380 Hz. What is the speed and direction if (a) the source is in motion; and (b) the listener is in motion?

40. Consider cars *A* and *B* each moving with a speed of 30 m/s and a police car moving with a speed of 35 m/s in between the two cars and all are in a straight line. If a siren on the police car is emitting a note of frequency of 1200 Hz, what is the frequency heard by the passengers in cars *A* and *B*? What are the wavelengths in front and behind the police car?

41. The airport ground control is emitting sound waves of frequency 1000 Hz. What is the speed of an approaching jet if the pilot measures the frequency of the sound to be 1200 Hz?

42. How fast should a sound-emitting source be moving (a) toward you; and (b) away from you, so that there will be a change of 20 percent in frequency?

43. A train is traveling at a speed of 90 km/h and whistling at a frequency of 1000 Hz. Calculate the following quantities. (a) The wavelengths of the sound waves in front and behind the train. (b) The frequency of the sound waves as heard by a stationary listener in front of and behind the train. (c) The frequency heard by another passenger traveling with 70 km/h parallel to the train and approaching or receding from the train.

44. An automobile moving away from you at a speed of 88.5 km/h (55 mi/h) toward a cliff blows a horn with a frequency of 480 Hz. (a) Calculate the wavelengths and frequencies of the sound waves that come directly from the horn and those reflecting from the cliff. (b) What is the speed, wavelength, and frequency of the reflected waves as heard by the driver?

45. Draw a sketch of the shock wave produced by a supersonic jet traveling at twice the speed of sound. Also, calculate the vertex angle of the cone formed by the shock wave.

46. A bullet traveling through air sends a shock wave in a cone of vertex angle of 45°. What is the speed of the bullet?

47. A supersonic jet is traveling with a speed of 1.4 times the speed of sound at an altitude of 8 km. Calculate the vertex angle of the cone of the shock wave. How long after the plane passes overhead will the shock wave reach the ground?

48. A ship is moving at a speed of 10 m/s. If the wave front of the bow wave makes an angle of 15° with the ship, what is the speed of the water waves?

Part V
Electricity and Magnetism

19 Electrostatic Interactions

We are interested in investigating electrostatic interactions between point charges. To do this, we shall familiarize ourselves with the force law for point charges—Coulomb's law—and apply it to different situations. The concept of a force field in general, and an electric field in particular, which has played an important role in modern theories, will be introduced.

19.1. Scope of Electric Interactions

BESIDES mechanics, an equally important branch of physics is electricity and magnetism. Although some discoveries in the field of electricity go back to 600 B.C. and William Gilbert (1540–1603), it was not until the middle of the eighteenth century that a real beginning of understanding was made by the work of Benjamin Franklin (1706–1790) and Charles Augustin de Coulomb (1738–1806). Much deeper understanding in the field of electricity was achieved through the work of Michael Faraday (1791–1867), James Clerk Maxwell (1831–1879), H. A. Lorentz (1853–1928), Heinrich Hertz (1857–1894), and André Marie Ampère (1775–1836), just to name a few scientists of the nineteenth century involved in such investigations.

The study of electrical interactions may be divided into (1) *electrostatics*, which deals with charges at rest, and (2) *electrodynamics*, which includes current electricity and electromagnetism. *Current electricity* is the study of charges in motion, while *electromagnetism* is the study of electric and magnetic fields produced by the moving charges. The two branches—electricity and magnetism—developed quite independently of each other until 1820, when Hans Christian Oersted (1777–1851) demonstrated a close connection between the two by showing that a current-carrying wire produces a magnetic field and hence deflects a magnetic compass needle when it is placed near the wire. Most of the interactions between charges at rest can be expressed mathematically by means of Coulomb's law, but complete sets of relations giving the connection between the charges at rest, charges in motion, electric fields, and magnetic fields were derived theoretically by Maxwell in 1865. His results are summarized in four equations, called Maxwell's equations, as we shall see in Chapter 27. These are the basic equations of *classical electromagnetic theory*. These equations play the same role in electricity and magnetism as Newton's laws play in mechanics. Although many changes have taken place in the theories of physics through the introduction of the theories of relativity and quantum mechanics, the laws of classical electromagnetic theory as developed by Maxwell remain unaffected in their basic form.

Furthermore, we shall be dealing with electromagnetic interactions or forces. These electromagnetic interactions are much stronger than gravitational interactions and weak interactions, but not as strong as the nuclear interactions. The forces resulting from electromagnetic interactions are responsible for the formation of stable chemical systems such as atoms, molecules, and the composition of the planets. The gravitational forces are responsible for the stability of the planets in their orbits.

19.2. The Two Types of Charges

As early as 600 B.C., it was known to Thales of Miletus that an amber rod when rubbed with a woolen cloth acquired the property of attracting light objects such as pith balls, straw, and so on, when placed near them. The amber is said to have acquired an *electric charge*. In Greek the word *elektron* means amber, hence the terminology. Further experiments revealed that any object when rubbed with a different object acquires an electric charge. Later detailed investigations led to the conclusion that there were two different types of charges, as we explain below.

Rub a glass rod with a piece of silk. Touch the glass rod for a short time to the pith balls A and B shown in Figure 19.1(a). These balls are tied to silk threads and hang vertically. The balls A and B fly apart as shown in Figure 19.1(b). We say that the two balls have acquired electric charges or are electrified with the same type of charge and repel each other. Now repeat the experiment by rubbing a hard-rubber, amber, or ebonite rod with a fur, wool, or a cat skin. The two balls will still fly apart; that is, they will repel as shown in Figure 19.1(c). Let us start the experiment over again with two pith balls hanging vertically as shown in Figure 19.1(a). Touch the pith ball A with a glass rod that has been rubbed with silk, and touch pith ball B with a hard-rubber rod that has been rubbed with fur. This time we will notice that balls A and B attract each other as shown in Figure 19.1(d).

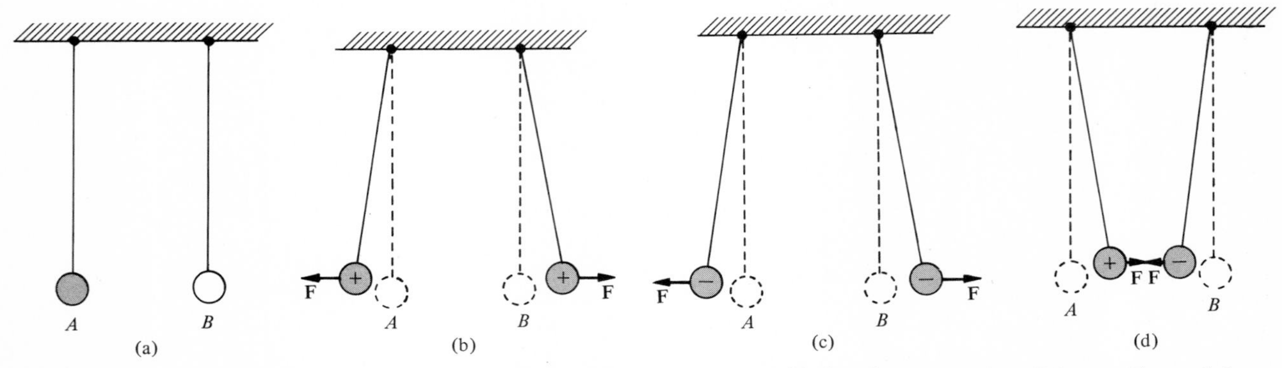

FIGURE 19.1. *(a) Two uncharged pith balls A and B. (b) Balls A and B, with positive electric charges, repel. (c) Balls A and B, with negative electric charges, repel. (d) Ball A, with a positive charge, and ball B, with a negative charge, attract each other.*

We may say that in Figure 19.1(b) the charges on A and B are alike and there is a force of repulsion between them. The same is true in Figure 19.1(c); that is, the charges on A and B are alike, although different from that in Figure 19.1(b), hence there is a repulsion.

In Figure 19.1(d) the type of charge on A is different from that on B, and we notice that two unlike charges have a force of attraction. Benjamin Franklin (1706–1790) was the first physicist to name the kind of electric charge appearing on a glass rod (rubbed with silk) as *positive charge*, and the kind of electric charge appearing on hard-rubber (rubbed with fur) as *negative charge*. Thus, we may conclude from the observations above that *like charges repel each other while unlike charges attract each other.*

Further experiments to prove the existence of the two types of charges can be performed with the help of an *electroscope* shown in Figure 19.2. It consists of two strips L_1 and L_2 of thin aluminum or gold foil fastened to a metal rod R and connected to the knob K at the top. The rod R is supported and also separated from the frame F by means of wood fittings W. When a charged body touches the knob of the electroscope, the leaves diverge, indicating that the electroscope is charged. (In some electroscopes there is only one leaf that moves, while the other remains stationary.)

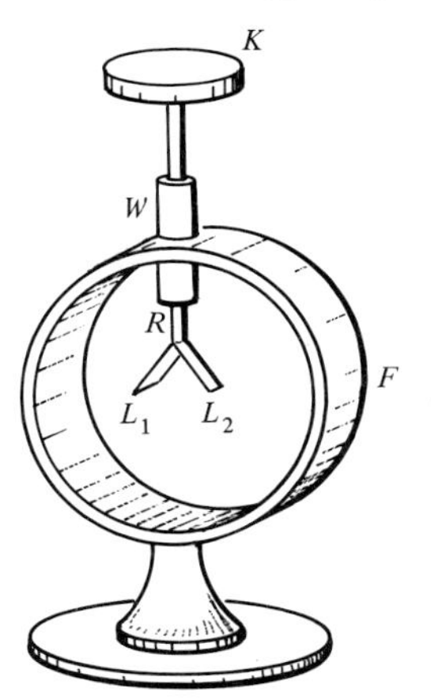

FIGURE 19.2. *Typical electroscope.*

Figures 19.3(a) and (b) show that when a glass rod rubbed with silk touches the knob of the electroscope, the leaves diverge. After the rod is removed, the leaves remain diverged; that is, the electroscope is still charged. If the electroscope is touched with a hand or a metallic wire, the leaves collapse, indicating that the electroscope is uncharged as shown in Figure 19.3(c). (This is because charge flows through the hand to the ground.) Now touch the electroscope knob with an ebonite rod that has been rubbed with fur. The leaves diverge, as shown in Figure 19.3(d). Now remove the rod. The leaves remain diverged, indicating that the electroscope is charged as shown in Figure 19.3(e). Next, touch the knob of the charged electroscope with a charged glass rod. The leaves converge. This indicates that the charge on the glass rod is opposite to that of the ebonite rod.

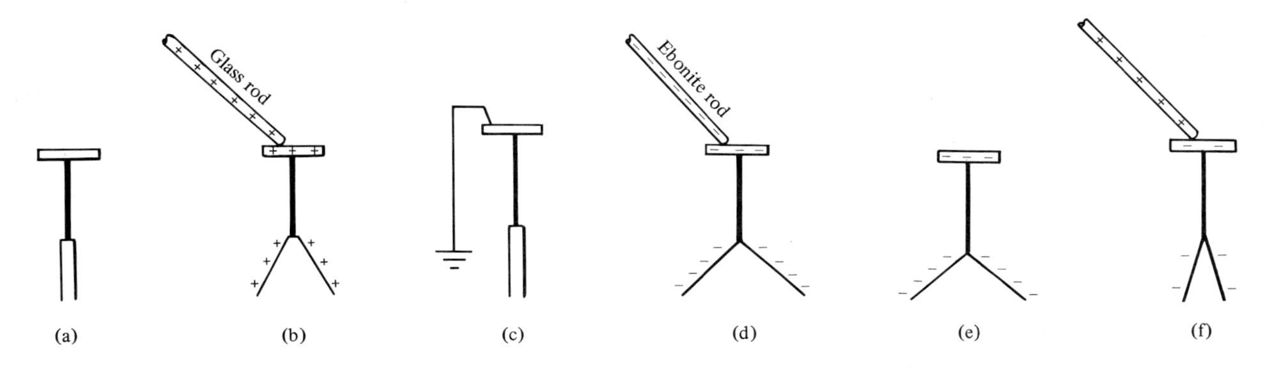

FIGURE 19.3. *(a) Uncharged electroscope. (b) Positively charged electroscope. (c) Grounded electroscope. (d) and (e) Negatively charged electroscope. (f) The decrease in the divergence of the electroscope leaves when a positively charged glass rod touches the knob.*

The charging process of different objects can be easily understood in terms of atomic theory. Each object consists of atoms that are electrically neutral. In each atom there are as many positive charges as there are negative. The negative charge is due to the electrons, while the positive charge is due to the charge of the protons. As two objects are rubbed together, one object loses electrons while the other gains electrons; that is, there is a transfer of electrons from one object to the other. The object that gains electrons becomes negatively charged, while the object that loses electrons has an excess of positive charges, hence is positively charged.

If we hold a metallic rod in one hand and rub it with any material, the metallic rod cannot be charged. The reason for this is that the electrons can easily flow through the metal to our hand and then to the ground. However, if the metallic rod has a handle made of wood or glass and is rubbed while holding it by the handle, the rod will become charged. The handle prevents the flow of electrons from the metal rod to the hand; hence to the ground. Materials such as wood and glass which do not allow a free flow of electrons through them are called *insulators*. Metals and other materials that allow the free flow of electrons through them are called *conductors*. There are other materials which are not as good conductors as metals, but are not good insulators as are wood and glass. Such materials are called *semiconductors*. Examples are germanium and silicon.

In order to investigate quantitatively the forces between the charged bodies, it is necessary to have bodies with relatively known amounts of charge. In such cases we find the method of charging by induction very useful. Consider two metallic spheres A and B mounted on insulating stands and placed touching each other, as shown in Figure 19.4(a). Now bring a positively charged glass

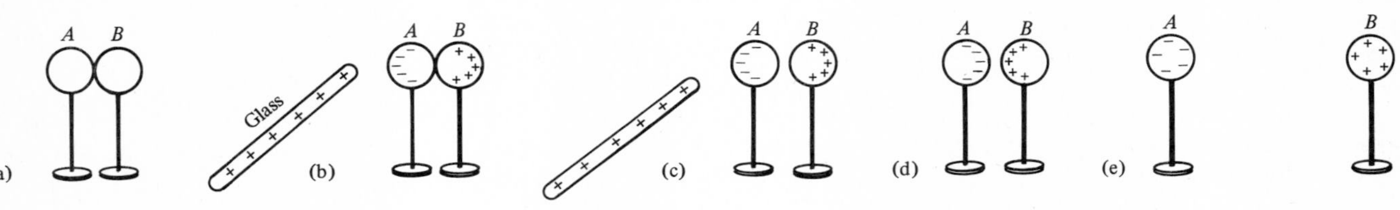

FIGURE 19.4. *Producing two equal and opposite charges by the method of induction.*

rod next to sphere A without touching it. Since the electrons will be attracted by the positive charges on the glass, some of the electrons in the two spheres rearrange themselves and accumulate in that part of sphere A, which is near the glass rod. The positive charges are not as mobile as the electrons, even though there is some tendency for them to be pushed to the right because of repulsion. The net result is that sphere A has excess negative charge while sphere B has excess positive charges, as shown in Figure 19.4(b). The next step is to separate spheres A and B slightly, as shown in Figure 19.4(c), followed by removing the glass rod as shown in Figure 19.4(d). Note the redistribution of the charges so that spheres A and B will attract. Finally, spheres A and B are taken far apart, as shown in Figure 19.4(e). The net result is that the negative charge on A is equal in magnitude to the positive charge on B. Thus, we have two equal and opposite charges. If sphere A is now touched to another uncharged metallic sphere C of the same size, sphere A will give half its charge to sphere C. This process can be repeated for further subdividing the charges.

In Figure 19.3(b) and (d), the electroscope was charged by touching the charged bodies with the knob. The electroscope may also be charged by induction as shown in Figure 19.5. The steps are self-explanatory if we follow the procedure in Figure 19.4.

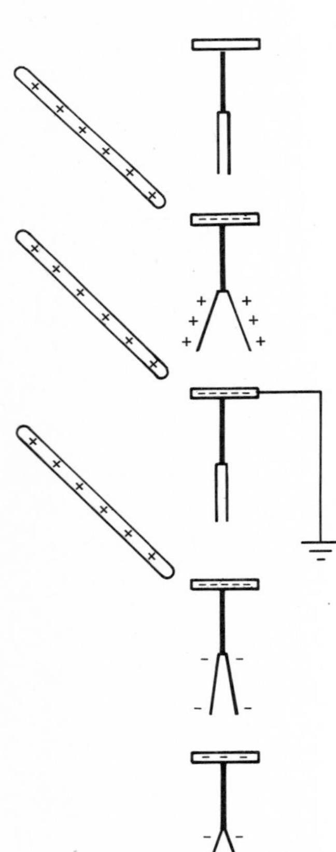

FIGURE 19.5. *Charging an electroscope by induction.*

19.3. Coulomb's Law and Units of Charge

Coulomb's Law

In the previous section we concluded that like charges repel each other and unlike charges attract each other. The next thing we would like to investigate is the quantitative nature of these forces; that is, we would like to know the magnitudes and directions of such forces of repulsion or attraction. The answers to those questions were given by the French scientist Charles Augustin de Coulomb (1736–1805) in 1784. According to the conclusions he drew from his experimental data;

Coulomb's law: ***the force between two charges is directly proportional to the product of the magnitudes of the charges and inversely proportional to the square of the distance between them.***

Thus, as shown in Figure 19.6, if q_1 and q_2 are the two charges separated by a distance r, the force F between them is

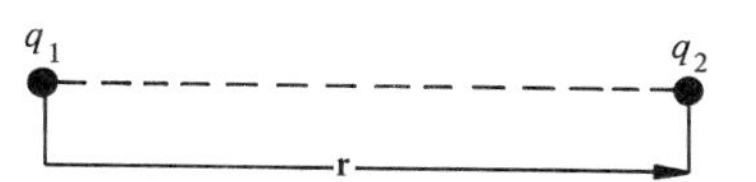

FIGURE 19.6. *Two point charges, q_1 and q_2, separated by a distance r.*

$$F \propto \frac{q_1 q_2}{r^2}$$

or

$$\boxed{F = k\frac{q_1 q_2}{r^2}} \tag{19.1}$$

where k is the constant of proportionality, and its value depends upon the types of units used. It is assumed that q_1 and q_2 are point charges.

The experimental apparatus used by Coulomb consisted of a torsion balance of the type shown in Figure 19.7 and is similar to the one used by Cavendish for measuring the gravitational constant G as discussed in Chapter 11. The torsion balance consists of a horizontal insulating rod R carrying two metallic spheres a and a' at its ends suspended from the middle with a fiber supported from F. A small mirror M is attached to the fiber, and a beam of light L reflected from this mirror falls on the scale S. Another insulating rod P, carrying a small sphere b, equal in size to a, can be brought in the vicinity of a as shown in Figure 19.7. Sphere b is charged by rubbing, as explained in the previous section. Sphere b, when touched to sphere a, shares its charge, so that both have equal charges, $q_1 = q_2 = q$. The charge on b can be further subdivided into $q/2$, $q/4$, $q/8$, . . . , by touching it with spheres of equal sizes. By keeping the distance between a and b fixed, one finds that the twist in the fiber, which is measured by the deflection on the scale S, is directly proportional to the magnitudes of the two charges. Similarly, by keeping the magnitudes of the charges on a and b fixed, but varying the distance between them, he found that the force is inversely proportional to the square of the distance.

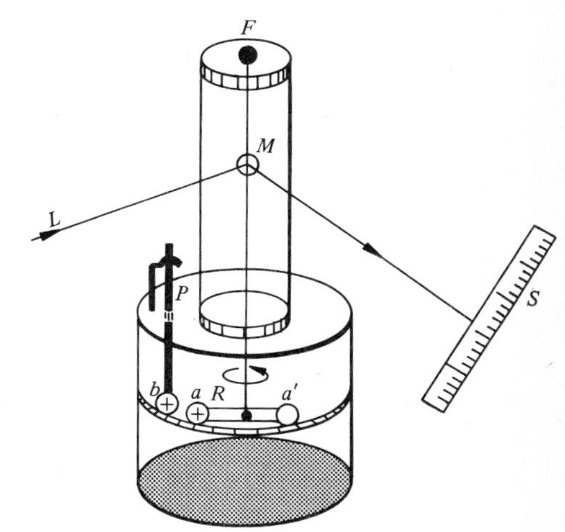

FIGURE 19.7. *Schematic of a torsion balance used by Charles de Coulomb to verify his inverse-square law of force between electrical charges.*

Of course, one can ask: Does the electric force follow the exact inverse square law, or is it only approximate? More precise and detailed investigations in 1936 by Plimpton and Lawton of the U.S. National Bureau of Standards revealed that the value of the power of r lies between the following limits 1.999999999 and 2.000000002, which is very close to 2.

Equation 19.1 gives the magnitude of the force between two charges but does not say anything about the direction of the force. Suppose that we denote the force on charge q_1 due to charge q_2 as $\mathbf{F}_{12}$, and that on charge q_2 due to charge q_1 as $\mathbf{F}_{21}$. Then according to Newton's third law,

$$\mathbf{F}_{12} = -\mathbf{F}_{21} \tag{19.2}$$

If q_1 and q_2 are like charges, they will repel each other and the direction of $\mathbf{F}_{12}$ and $\mathbf{F}_{21}$ will be as shown in Figure 19.8(a). On the other hand, if q_1 and q_2 are unlike charges, the forces are attractive and their directions are as shown in Figure 19.8(b).

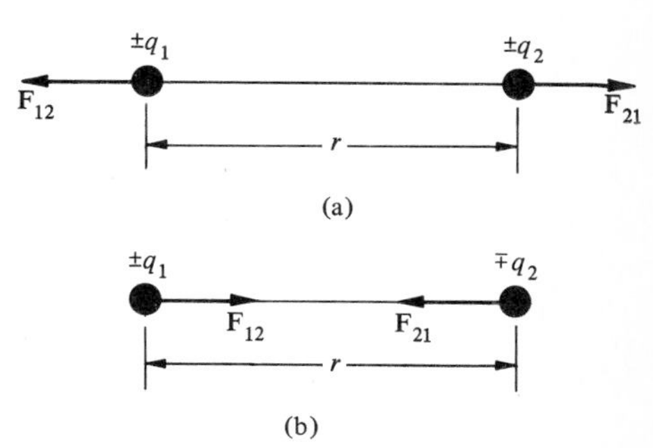

FIGURE 19.8. *(a) Repulsive forces between like charges and (b) attractive forces between unlike charges.*

Units of Charge

In order to know the value of the proportionality constant k in Equation 19.1, we must know the units of r, F, and q. Since the units of r and F are already known, we define the units of q. There are three different ways of doing this: (1) CGS or electrostatic units (esu), (2) electromagnetic units (emu), or (3) SI (or rationalized MKS) units.

To avoid confusion between the esu and emu a single system that is applicable to both electricity and magnetism is desirable. The SI incorporates such a unit.

In the SI system, the basic unit is the unit of current called the ampere (A), while the unit of charge is called the coulomb (C), a derived unit defined in terms of the ampere. The ampere is the rate of flow of charge across any cross-sectional area, and will be defined precisely in terms of electrical force in Chapter 24. We may define the coulomb in terms of the ampere as follows: The *coulomb* (C) is that quantity of charge that flows across any cross section of a wire in 1 second when there is a steady current of 1 ampere. The SI, after incorporating the ampere (A) as the unit of current, is also referred to as the MKSA system.

In Equation 19.1, if we take $F = 1$ N, $q_1 = q_2 = 1$ C, and $r = 1$ m, the value of the constant k turns out to be

$$k = 8.98742 \times 10^9 \text{ N-m}^2/\text{C}^2 \simeq 9 \times 10^9 \text{ N-m}^2/\text{C}^2 \tag{19.3}$$

Many equations of electricity and magnetism take a much simpler form if the constant k is expressed in terms of the quantity ϵ_0 by the following relation:

$$\boxed{k = \frac{1}{4\pi\epsilon_0} \simeq 9 \times 10^9 \text{ N-m}^2/\text{C}^2} \tag{19.4}$$

where ϵ_0 is called the *permittivity constant* of free space or vacuum, and its value is

$$\epsilon_0 = 8.85418 \times 10^{-12} \text{ C}^2/\text{N-m}^2 \tag{19.5}$$

Now we may write Coulomb's law as

$$\boxed{F = \frac{1}{4\pi\epsilon_0}\frac{q_1 q_2}{r^2}} \tag{19.6}$$

Using the value of $1/4\pi\epsilon_0$ given by Equation 19.4, we may define a *coulomb* to be that amount of charge which when placed at a distance of 1 m from an equal and similar charge repels it with a force of 8.98742×10^9 N. The introduction of 4π in k $(= 1/4\pi\epsilon_0)$ is referred to as *rationalization*. Thus, Equation 19.6 is referred to as Coulomb's law in the rationalized MKS system of units.

As we shall investigate in Chapter 32, the charge on an electron is

$$q = -e = -1.60207 \times 10^{-19} \text{ C}$$

Thus, it will take $1/1.60207 \times 10^{-19} = 6.24 \times 10^{18}$ electrons to have 1 C of charge.

Electric forces, just like any other vector quantities, obey the principle of superposition; that is, the resultant force is equal to the vector sum of the individual forces acting on the charged particles.

Comparison with Gravitational Force

There are many similarities and differences in gravitational and Coulomb forces. Both (1) follow the inverse-square law, (2) are central and, (3) are conservative.

There are two great differences between the two types of forces. First, as we noted earlier, electrical forces are much stronger than gravitational forces. For example, the force of attraction between an electron and a proton is 10^{39} times as large as the gravitational force between them as illustrated in Example 19.2. Second, there are two types of electrical charges—positive and negative, but there is only one type of mass known. Note, however, in electricity that like charges repel, while in gravitation, like masses attract.

EXAMPLE 19.1 Two tiny identical metal balls of 10 g each carrying equal positive charges are suspended from 1-m-long strings. These balls repel each other and then come to equilibrium making an angle of 37° with the vertical as shown in the accompanying figure. Calculate the charge on each ball.

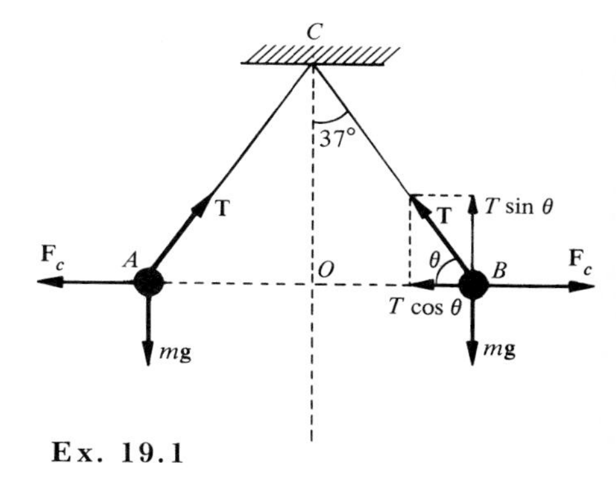

Ex. 19.1

Let q be the charge on each ball. Since both the balls are in equilibrium, we can consider either of them, say ball B on the right. There are three forces acting on it: the tension **T**, the weight of the ball $m\mathbf{g}$ acting vertically downward, and the force of electrical repulsion $\mathbf{F}_c$. Thus, for equilibrium,

$$\sum \mathbf{F} = \mathbf{T} + m\mathbf{g} + \mathbf{F}_c = 0$$

Resolving **T** into two components $T \sin\theta$ and $T\cos\theta$, we may write the equilibrium condition as

$$\sum F_y = T\sin\theta - mg = 0 \qquad T\sin\theta = mg$$

$$\sum F_x = F_c - T\cos\theta = 0 \qquad T\cos\theta = F_c$$

Dividing one by the other, we get $\sin\theta/\cos\theta = mg/F_c$. Since $\sin\theta/\cos\theta = \tan\theta$,

$$\tan\theta = \frac{mg}{F_c}$$

Substituting for $F_c = kq^2/r^2$, where $k = 1/4\pi\epsilon_0$,

$$\tan\theta = mgr^2/kq^2 \qquad \text{or} \qquad q^2 = \frac{mgr^2}{k\tan\theta}$$

In this equation $m = 10\text{ g} = 10 \times 10^{-3}\text{ kg}$, $g = 9.8\text{ m/s}^2$, $\theta = 90° - 37° = 53°$, so $\tan 53° = 1.327$, $k = 9 \times 10^9\text{ N-m}^2/\text{C}^2$, and $r = AB = 2OB = 2CB\cos\theta = 2CB\cos 53° = 2(1\text{ m})(0.6) = 1.2\text{ m}$. Thus,

$$q^2 = \frac{(10 \times 10^{-3}\ \text{kg})(9.8\ \text{m/s}^2)(1.2\ \text{m})^2}{(9 \times 10^9\ \text{N-m}^2/\text{C}^2)(1.327)} = 11.8 \times 10^{-12}\ \text{C}^2$$

$$q = 3.44 \times 10^{-6}\ \text{C} = 3.44\ \mu\text{C}$$

where $1\ \mu\text{C} = 1$ microcoulomb $= 10^{-6}$ C.

EXERCISE 19.1 Two tiny aluminum-foil balls of mass 10 g each are suspended from the same point with strings that are 50 cm long. When equal charges are given to these balls, they repel and then come to equilibrium with their centers 50 cm apart. Calculate (a) the charge on each ball; (b) T; and (c) F_c. [*Ans.*: (a) $q = 1.25\ \mu\text{C}$; (b) $T = 0.113$ N; (c) $F_c = 0.056$ N.]

EXAMPLE 19.2 According to Bohr's model of the hydrogen atom, an electron is going around a proton in a circular orbit of radius 0.53×10^{-10} m. (a) Calculate the electrical force between the electron and the proton. (b) Calculate the gravitational force between the electron and the proton. (c) Calculate the gravitational force between the earth and the sun, assuming that the earth is going around the sun in a circular orbit of radius 1.5×10^{11} m. (d) Compare the results of (a), (b), and (c).

(a) The electrical force between a proton and an electron, according to Coulomb's law, is

$$F_e = \frac{1}{4\pi\epsilon_0}\frac{q_p q_e}{r^2} = \frac{(9 \times 10^9\ \text{N-m}^2/\text{C}^2)(1.6 \times 10^{-19}\ \text{C})(1.6 \times 10^{-19}\ \text{C})}{(0.53 \times 10^{-10}\ \text{m})^2}$$

$$= 8.2 \times 10^{-8}\ \text{N}$$

(b) The gravitational force between the proton and the electron, according to Newton's law of gravitation, is

$$F_g = G\frac{m_p m_e}{r^2} = \frac{(6.67 \times 10^{-11}\ \text{N-m}^2/\text{kg}^2)(1.67 \times 10^{-27}\ \text{kg})(9.11 \times 10^{-31}\ \text{kg})}{(0.53 \times 10^{-10}\ \text{m})^2}$$

$$= 3.6 \times 10^{-47}\ \text{N}$$

(c) The gravitational force between the sun and the earth is

$$F = G\frac{m_s m_e}{r^2} = \frac{(6.67 \times 10^{-11}\ \text{N-m}^2/\text{kg}^2)(2 \times 10^{30}\ \text{kg})(5.98 \times 10^{24}\ \text{kg})}{(1.5 \times 10^{11}\ \text{m})^2}$$

$$= 3.5 \times 10^{22}\ \text{N}$$

(d) It is quite evident from the calculations above in (a) and (b) that the gravitational force is negligible compared to the electrical force when dealing with microscopic systems such as atoms and molecules; that is, using the values calculated in (a) and (b) the ratio $F_e/F_g \approx 10^{39}$. On the other hand, as is clear from (c), the gravitational forces are of utmost importance when dealing with such objects as the sun, moon, earth, and so on. Because the planets on the whole are electrically neutral, there is no electrical force between them.

Exercise 19.2 Suppose that the force between the sun and the earth was of electrical origin. Assume that the charges on the sun and the earth are equal in magnitude but opposite in sign. Calculate this charge and compare it with the charge of an electron. [*Ans.*: $q = 2.96 \times 10^{17}$ C $= 1.85 \times 10^{36}$ e.]

19.4. The Concept of Electric Field

Newton's universal gravitation law and Coulomb's law enable us to calculate the magnitudes as well as the directions of the gravitational and electrical forces, respectively. These laws are limited to describing only interactions of two point charges (or masses). On the contrary, the field concept is a powerful tool to handle distributed and continuous charges and their interactions. But there are still no answers to the following more fundamental and basic questions: (a) what are the origins of these forces? (b) how are the forces transmitted from one mass to another or from one charge to another?

The answer to (a) is unknown; the existence of forces is accepted de facto. But ever since the formulation of the force laws, there have been theories proposed to explain (b). Most scientific thought has been divided along two lines: (1) the action-at-a-distance effect, and (2) the field effect or field theory. In the following we limit ourselves to the discussion of Coulomb or electrical forces, but the same remarks apply equally to gravitational forces.

Action-at-a-distance means that the force between two charged bodies is conveyed directly and instantaneously (with no time delay) between the two bodies. Accordingly, the force between two charges is considered to be a one-step process. Even though it is an experimental fact that the forces can be conveyed across empty space, the idea of action-at-a-distance has never been a satisfying thought to scientists. There must be some agent or intermediate medium that transmits force, or change in force, from one point to another.

The concept of a field theory was introduced by Michael Faraday (1791–1867) and is becoming more and more useful and convenient. According to the *field concept,* the interaction between two charges q and q_0 separated by a distance r is explained in the following manner. The charge q produces an electric field in the space surrounding it. This field exists whether there are other charges present in the space or not. But the presence of this field cannot be detected until another charge q_0 is brought into the field. The field of charge q interacts with the test charge q_0 to produce an electrical force. Thus, the interaction between q and q_0 is a two-step process: (1) the charge q produces a field, and (2) the field interacts with charge q_0 to produce a force **F** on q_0. These two steps are illustrated in Figure 19.9. The two processes are reversible. For example, charge q_0 may produce a field, and if a charge q is brought in the field of q_0 at a distance of r, it feels a force $-\mathbf{F}$.

We know from Coulomb's law that the magnitude of the force **F** depends upon the separation distance r between the two charges. In terms of the field concept, we say that the strength of the field is very high near charge q and decreases as $1/r^2$. We may define the *electric field strength* or *electric field intensity* **E** due to a charge q at a distance r from it to be

$$\mathbf{E} = \frac{\mathbf{F}}{q_0} \tag{19.7}$$

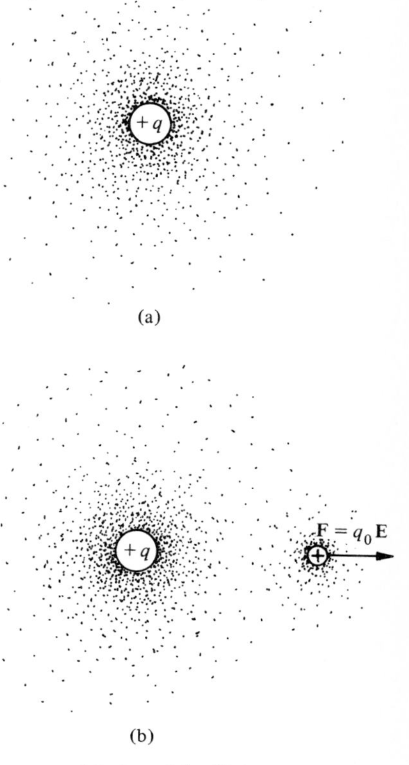

Figure 19.9. (*a*) *Dots surrounding the positive charge indicate the presence of the electric field. The density of the dots is proportional to the strength of the electric field at different points.* (*b*) *Interaction of the field with the charge* q_0.

where $\mathbf{F}$ is the force on q_0 due to q given by Coulomb's law. *Thus, the electric field strength* $\mathbf{E}$ *at any point surrounding the charge* q *is defined as the force per unit positive charge in the field.* Also, $\mathbf{E}$ is a vector quantity and its direction is that of the force that will act on a positive charge in the field. The density of points in Figure 19.9(a) indicates the relative strength of the field. As shown in Figure 19.9(b), we may write $\mathbf{F} = q_0\mathbf{E}$.

There are inherent difficulties in the definition of $\mathbf{E}$ as given by Equation 19.7. The test charge q_0 when brought in the field of the charge q will interfere with this field; hence, the measurements will not yield the true value of $\mathbf{E}$. To overcome this difficulty, we make the charge q_0 infinitesimally small so that it will not alter the field in any fashion. Thus, we may define electric field strength $\mathbf{E}$ as

$$\boxed{\mathbf{E} = \lim_{q_0 \to 0} \frac{\mathbf{F}}{q_0}} \tag{19.8}$$

We may point out that the electrical field represented by $\mathbf{E}$, like the gravitational field represented by $\mathbf{g}$, is a *vector field*. The quantity $\mathbf{E}$ can be defined at all points in space surrounding the charge q. Thus, $\mathbf{E}$ may be regarded as a continuous distributed electrical property over a region of space surrounding the charge. When we can define a quantity over all points throughout a region of space, we say there exists a *field* in that region of space.

At a distance r from a point charge q, the force on charge q_0 is kqq_0/r^2; hence, the magnitude of the electric intensity $\mathbf{E}$ is

$$\boxed{E = \frac{F}{q_0} = \frac{kq}{r^2}} \tag{19.9}$$

The units of E are newtons/coulomb ($=$ N/C).

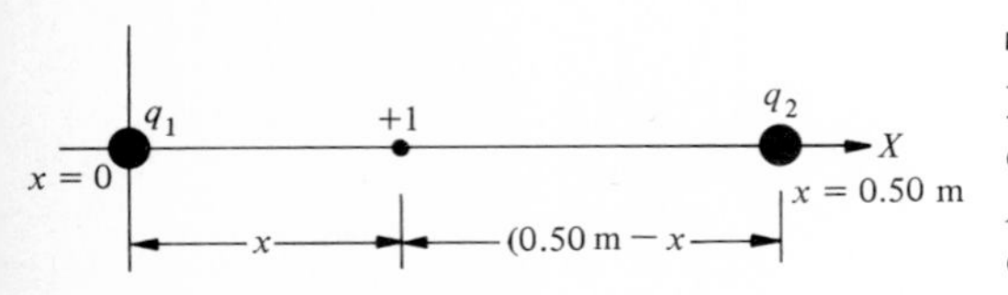

Ex. 19.3

EXAMPLE 19.3 A charge $q_1 = +5\,\mu$C is placed at the origin while a charge $q_2 = +10\,\mu$C is placed at a distance of 50 cm from the origin along the X-axis, as shown. Find a point where the electric intensity due to these two charges will be zero.

If we place a unit positive charge at any point to the right of charge q_2, the forces on this charge will be to the right, due to both q_1 and q_2; hence, the electric intensity cannot be zero at such points. The same is true for a unit positive test charge to the left of q_1. But if we place a unit positive charge between these two charges as shown in the accompanying figure, the forces act in opposite directions; hence, the electric intensity can be zero. Let this point be at a distance x from q_1, hence at a distance $(0.5\text{ m} - x)$ from q_2. Thus, for the resultant E to be zero,

$$E = E_1 + E_2 = 0$$

That is,

$$k\frac{q_1}{x^2} - k\frac{q_2}{(0.5 - x)^2} = 0$$

$$q_1(0.5 - x)^2 = q_2x^2$$

Substituting for q_1 and q_2 and taking the square root,

$$0.5 - x = \pm\sqrt{2}\,x$$

$$x = \frac{0.5}{1 + \sqrt{2}} = \frac{0.5}{2.414} = 0.21\text{ m} \qquad \text{or} \qquad x = \frac{0.5}{1 - \sqrt{2}} = -1.21\text{ m}$$

As explained above, $x = -1.21$ is not possible. Hence, $x = 0.21$ m.

EXERCISE 19.3 If in the accompanying figures, the charge $q_1 = +10\,\mu\text{C}$ while charge $q_2 = -5\,\mu\text{C}$, find the position of the point where the total intensity due to these two charges will be zero. [*Ans.:* $x = 1.71$ m; that is, 1.21 m to the right of q_2.]

EXAMPLE 19.4 Consider point charges $q_1 = +2\,\mu\text{C}$, $q_2 = +4\,\mu\text{C}$, $q_3 = +6\,\mu\text{C}$, and $q_4 = -6\,\mu\text{C}$ placed at four corners of a square as shown in the figure. Each diagonal is of length 2 m. (a) Calculate the electric intensity at the center of the square. (b) What will be the force on a charge of $+5\,\mu\text{C}$ placed at the center?

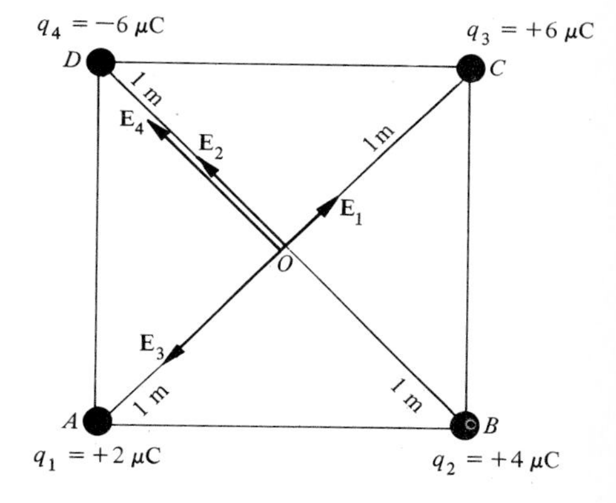

(a)

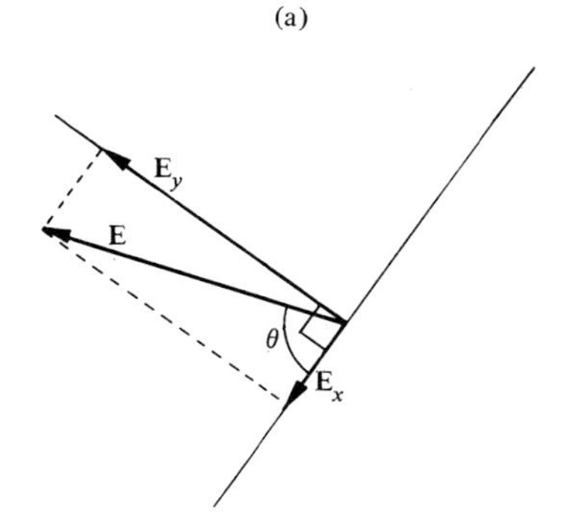

(b)

Ex. 19.4

The forces acting on a unit positive charge placed at the center of the square are the intensities E_1, E_2, E_3, and E_4 with their directions as shown. Note that $r_1\,(= OA) = r_2\,(= OB) = r_3\,(= OC) = r_4\,(= OD) = 1$ m. Thus,

$$E_1 = k\frac{q_1}{r_1^2} = 9 \times 10^9\text{ N-m}^2/\text{C}^2\,\frac{2 \times 10^{-6}\text{ C}}{(1\text{ m})^2} = 18 \times 10^3\text{ N/C}$$

$$E_2 = k\frac{q_2}{r_2^2} = 9 \times 10^9\text{ N-m}^2/\text{C}^2\,\frac{4 \times 10^{-6}\text{ C}}{(1\text{ m})^2} = 36 \times 10^3\text{ N/C}$$

$$E_3 = k\frac{q_3}{r_3^2} = 9 \times 10^9\text{ N-m}^2/\text{C}^2\,\frac{6 \times 10^{-6}\text{ C}}{(1\text{ m})^2} = 54 \times 10^3\text{ N/C}$$

$$E_4 = k\frac{q_4}{r_4^2} = 9 \times 10^9\text{ N-m}^2/\text{C}^2\,\frac{6 \times 10^{-6}\text{ C}}{(1\text{ m})^2} = 54 \times 10^3\text{ N/C}$$

The directions of E depend upon the signs of the charges. As in part (b), let

$$\sum E_x = E_3 - E_1 = (54 - 18) \times 10^3\text{ N/C} = 3.6 \times 10^4\text{ N/C}$$

$$\sum E_y = E_2 + E_4 = (36 + 54) \times 10^3\text{ N/C} = 9.0 \times 10^4\text{ N/C}$$

Therefore,

$$E = \sqrt{\left(\sum E_x\right)^2 + \left(\sum E_y\right)^2}$$

$$= \sqrt{(3.6)^2 + (9.0)^2} \times 10^4\text{ N/C} = 9.7 \times 10^4\text{ N/C}$$

As shown in part b, the resultant intensity E makes an angle θ with OA given by

$$\tan\theta = \frac{\sum E_y}{\sum E_x} = \frac{9.0}{3.6} = 2.5 \qquad \text{or} \qquad \theta = 68.2^\circ$$

The force on charge $q_0 = +5\,\mu\text{C}$ placed at O will be given by

$$F = q_0E = (5 \times 10^{-6}\ \text{C})(9.7 \times 10^4\ \text{N/C}) = 0.485\ \text{N}$$

Exercise 19.4 In Example 19.4, if $q_1 = q_2 = 2\,\mu\text{C}$, $q_3 = q_4 = -2\,\mu\text{C}$, what is the electric intensity at the center? What will be the intensity if $q_1 = q_2 = q_3 = q_4 = 2\,\mu\text{C}$? [*Ans.:* $E = 5.09 \times 10^4$ N/C toward the midpoint of the line joining q_3 and q_4. $E = 0$.]

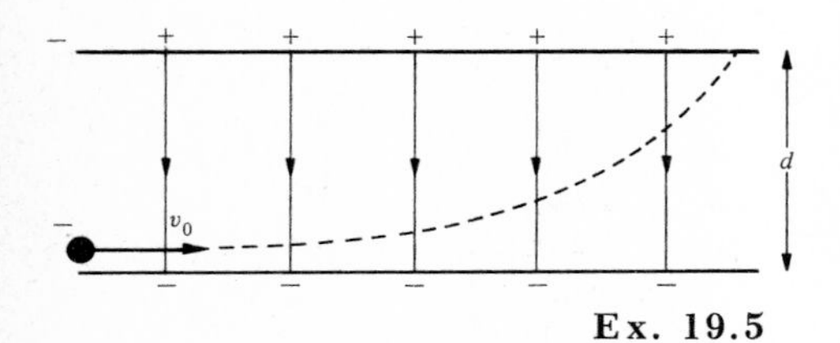

Ex. 19.5

Example 19.5 Suppose that an electron with a velocity v_0 is ejected between two horizontal plates separated by a distance d as shown in the accompanying figure. The electric field between the plates has a constant value E. When the electron hits the upper plate, what is (a) its velocity; (b) the time it takes to reach the upper plate; and (c) the horizontal distance it travels during this time?

The force acting on the electron will be upward and is

$$F = qE = eE \tag{i}$$

Since $F = ma$ and acts along the Y-axis,

$$a_y = \frac{F}{m} = \frac{eE}{m} \tag{ii}$$

The initial vertical velocity is zero. Therefore, from the relation $v_y^2 = v_{0y}^2 + 2a_yy$ and substituting for a_y, $v_{0y} = 0$ and $y = d$, we get

$$v_y^2 = 2\left(\frac{eE}{m}\right)d \qquad \text{or} \qquad v_y = \sqrt{\frac{2eEd}{m}} \tag{iii}$$

The time it takes to travel this vertical distance d can be found from the relation $y = v_{0y}t + \frac{1}{2}a_yt^2$, where $y = d$, $v_{0y} = 0$, and $a_y = eE/m$. Therefore,

$$d = \frac{1}{2}\left(\frac{eE}{m}\right)t^2$$

or

$$t = \sqrt{\frac{2md}{eE}} \tag{iv}$$

Since the horizontal velocity remains constant, $v_{0x} = v_0$, the horizontal distance x traveled is

$$x = v_{0x}t = v_0\sqrt{\frac{2md}{eE}} \tag{v}$$

Exercise 19.5 In Example 19.5, if $d = 10$ cm, $v_0 = 10^2$ m/s, and $E = 1000$ N/C, calculate a_y, v_y, t, and x. [*Ans.:* $a_y = 1.76 \times 10^{14}$ m/s^2, $v_y = 5.91 \times 10^6$ m/s, $t = 3.36 \times 10^{-8}$ s, $x = 3.36 \times 10^{-6}$ m.]

19.5. Electric Field Lines

The concept of electric field lines (or electric lines of force) was introduced by Michael Faraday (1791–1867). The *electric field lines are imaginary,* but they afford a very convenient way of pictorial visualization of the electric field.

Consider a charge $+q$ as shown in Figure 19.10(a). The arrows indicate the electric intensity at different points (black dots) surrounding the charge. The size of any arrow is proportional to the electric intensity at that point while the direction of the arrow indicates the direction of the force on a unit positive charge. If these arrows are joined together, they result in radially outward lines emanating from charge $+q$ as shown in Figure 19.10(b). These lines are *electric field lines or lines of force.* There is no limit to the number of lines that can be drawn from a given charge. But once we assign a certain number of lines coming out of the $+q$ charge, these lines spread out radially and their density becomes less and less as we move out. Of course, Figure 19.10(b) is only a two-dimensional case, while in a three-dimensional case it will look like needles sticking out of a spherical pin cushion.

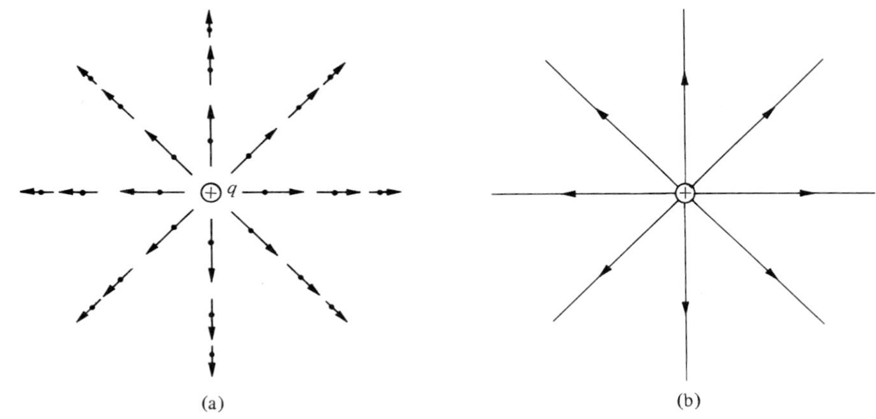

Figure 19.10. *(a) Electric intensity at different points surrounding a point charge $+q$. (b) Electric field lines emanating from a charge $+q$.*

The field lines due to a negative charge $-q$ may be drawn as shown in Figure 19.11. In this case when a unit positive charge is placed in space, it will experience a force directed toward the charge $-q$. Thus, the field lines in this case are directed toward the negative charge, as shown in Figure 19.11(b).

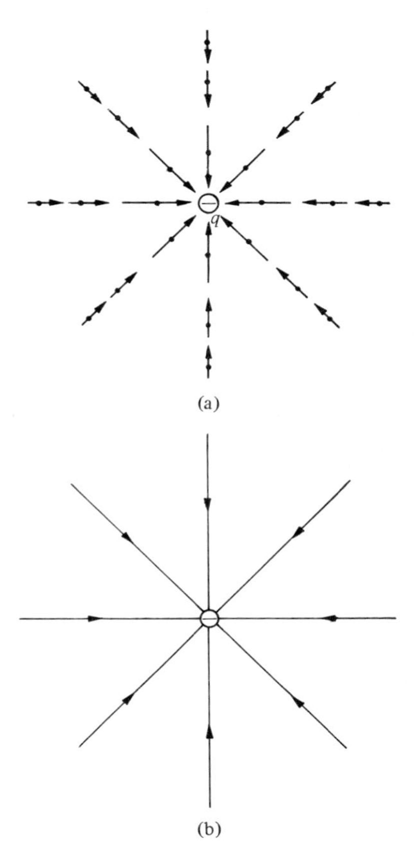

Figure 19.11. *(a) Electric intensity at different points surrounding a point charge $-q$. (b) Electric field lines converging to a point charge $-q$.*

These ideas of electric field lines from point charges can be put together to draw the lines from different combinations of charges such as shown in Figure 19.12. In Figure 19.12(a) two charges $+q$ and $-q$ are placed at a certain distance. The field lines start from a positive charge and end on a negative charge. The electric intensity at any point such as P is a combination

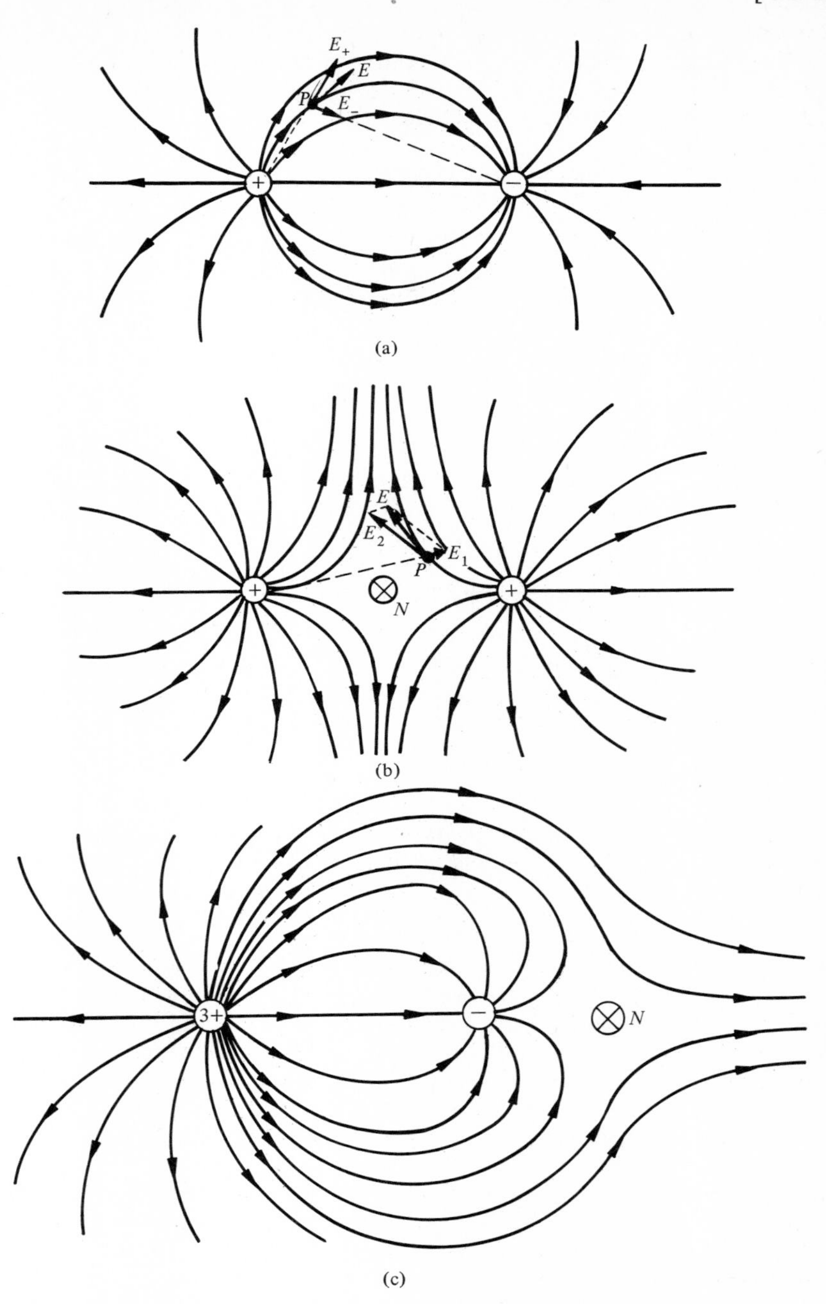

Figure 19.12. *Electric field lines due to different charge configurations. N indicates the point where the net electric field is zero.*

of intensities $\mathbf{E}_+$ and $\mathbf{E}_-$ due to $+q$ and $-q$ charges, respectively. Thus, the net intensity at P is $\mathbf{E}$, which is in the direction of the tangent to the field line at P. These ideas may be extended to draw field lines from other charge configurations, as shown in Figures 19.12(b) and (c). Note that the points, like N in Figures 19.12(b) and (c), where the resultant intensity is zero, are called *neutral points*. Also, the field lines starting from a positive charge will extend to infinity if there is no negative charge in the vicinity. Lines closing in on a negative charge, if not coming from a positive charge, are supposed to come from infinity.

Let us now consider two special cases. Figure 19.13(a) shows an infinite plate that has a uniform positive charge on it. The electric field lines are parallel, have uniform density, and extend to infinity. If the plate was finite in size as shown in Figure 19.13(b), the lines will be almost parallel; hence, the field will be uniform in the middle region. At the edges the field is not uniform, and we call this a *fringing field*. Figure 19.14(a) shows two infinitely large parallel plates separated by a very small distance. The upper plate has a positive charge, while the lower plate has a negative charge. The field lines are as shown, and the field is uniform. Also, the direction of $\mathbf{E}$ is the direction in which the force acts on a unit positive charge. Figure 19.14(b) shows the field for two parallel plates of finite size. There is a fringing field at the edges; that is, the field is nonuniform near the edges.

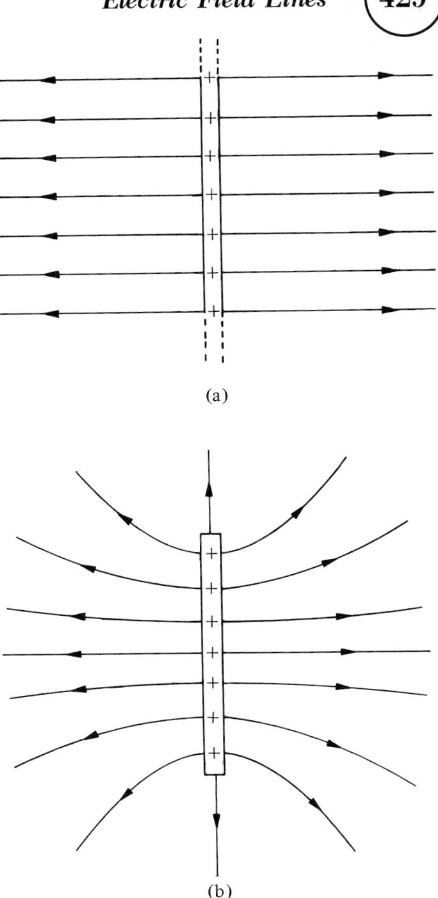

FIGURE 19.13. *Electric field lines due to a uniform positive charge on (a) infinite plate and (b) finite plate.*

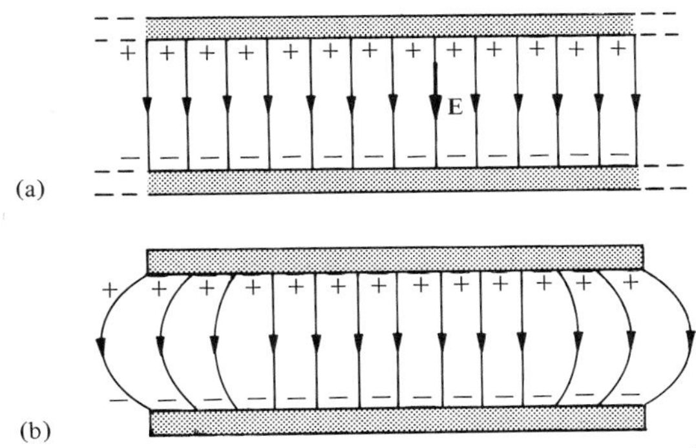

FIGURE 19.14. *Electric field lines due to oppositely charged parallel plates of (a) infinite size and (b) finite size.*

We are now in a position to summarize the properties of electric field lines.

1. Electric field lines originate from positive charges and end on negative charges.
2. The tangent to a field line at any point gives the direction of the electric field at that point as illustrated in Figure 19.15.
3. Field lines are drawn so that the number of lines crossing per unit cross-sectional area perpendicular to the field lines is proportional to the magnitude of the electric field in that region. Thus, the lines are closer where the field is strong, the lines are farther apart where the field is weak, and the lines are parallel and equidistant where the field is uniform.
4. No two field lines cross each other. This is because E has only one value at any one given point. If the lines cross, E could have more than one value.

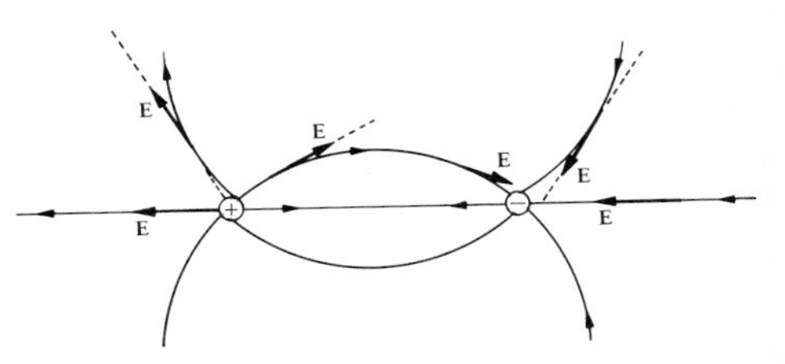

FIGURE 19.15. *Tangent drawn to the field line gives the direction of the electric intensity at that point.*

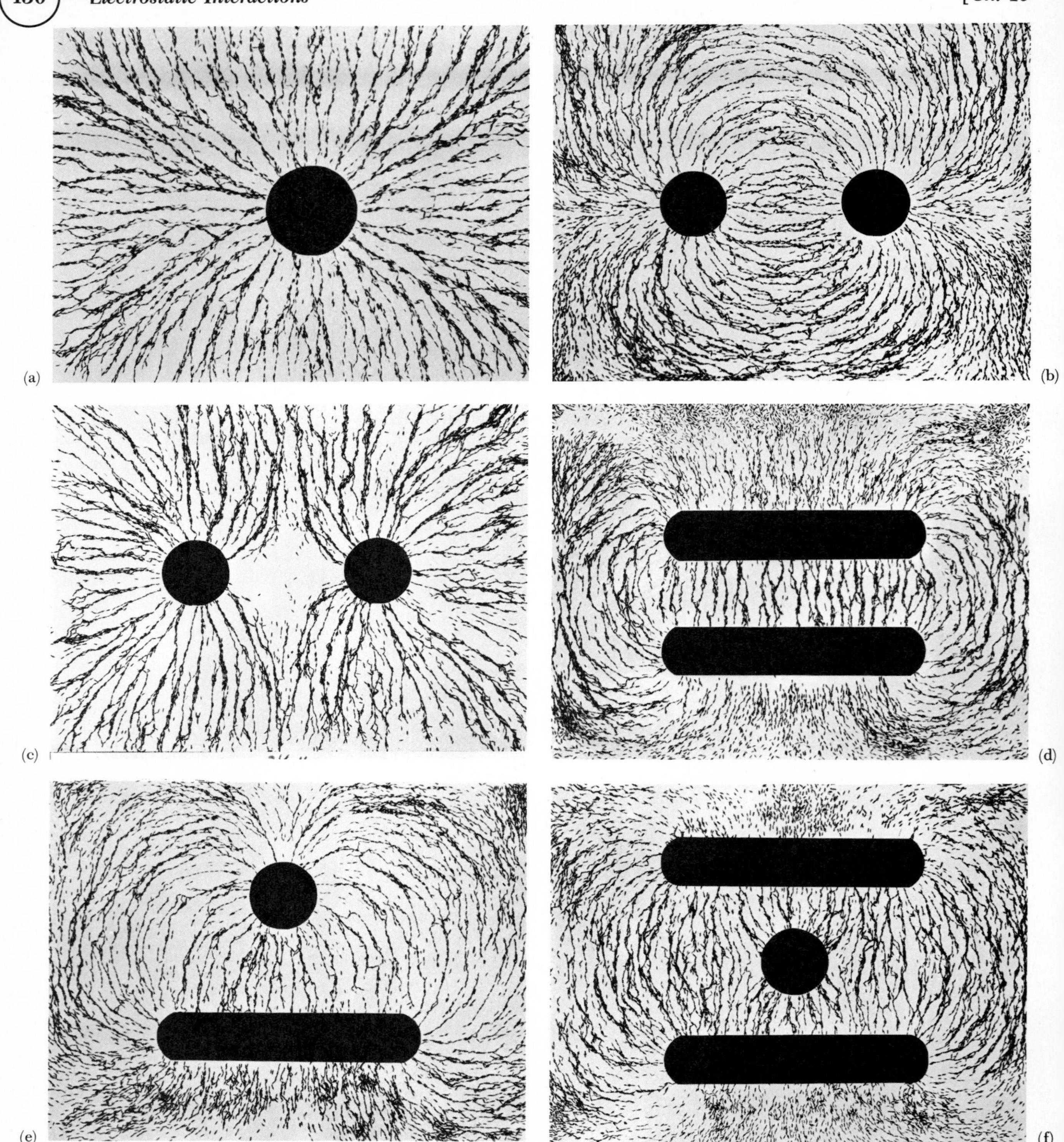
(a)
(b)
(c)
(d)
(e)
(f)

5. A small positive test charge released in such an electric field will follow a path along the field line on which it is placed.

Figure 19.16 shows patterns of electric field lines obtained from different charge configurations. The patterns were obtained by using grass seeds. The preceding ideas will be put into quantitative form by introducing the concept of electric flux in the next section.

Let us consider another interesting property of electric field lines. Suppose that a charge $+q$ is placed near a metal plate as shown in Figure 19.17. The

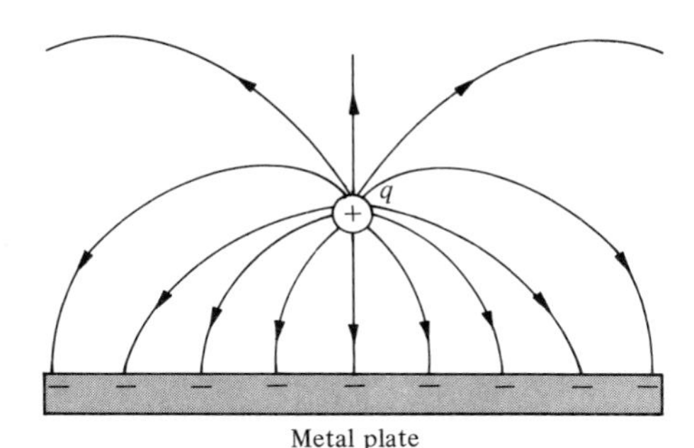

FIGURE 19.17. *Electric field lines are always perpendicular to the surface of a conductor under electrostatic conditions.*

positive charge will attract the negative charges (electrons) in the metal plate, resulting in the motion of these charges until some of them reach that surface of the metal plate which is near the $+q$ charge. This means that the charges are at rest now—in a state of electrostatic equilibrium. Thus, the field lines starting from $+q$ charge will end on the negative charges of the metal plate. Furthermore, these lines are always perpendicular to the metal surface as shown in Figure 19.17. *The electric field lines are always perpendicular to the surface of a metal or conductor under electrostatic conditions*. This statement can be proved as follows.

Suppose field lines were not perpendicular to the metal surface. This means that **E** will make an angle, say θ, as shown in Figure 19.18. Resolve **E** into two components, $E_\perp$ and $E_\parallel$. Component $E_\parallel$ will act on the electrons in the metal and cause them to move, thereby producing an electric current in the metal. But this is contrary to the condition that the charges in the metal are under electrostatic equilibrium. Thus, E must be perpendicular to the surface of the conductor.

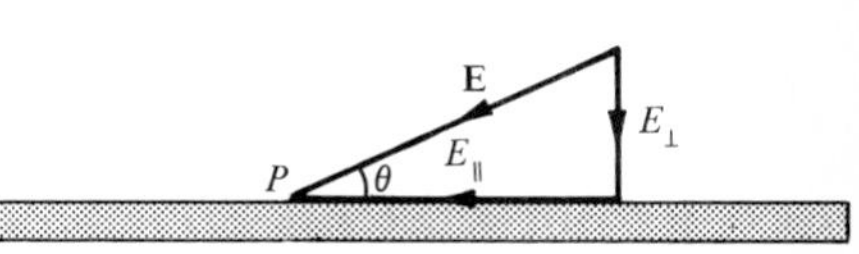

FIGURE 19.18. *If* **E** *were not perpendicular to the metal surface, the component parallel to the surface* $E_\parallel$ *will produce electric current.*

[OPPOSITE] FIGURE 19.16. *Electric field lines due to (a) a positive charge, (b) two spheres carrying equal and opposite charges, (c) two spheres carrying equal and similar charges, (d) two oppositely charged parallel plates, (e) a sphere and a plate carrying opposite charges, and (f) a sphere and two parallel plates, oppositely charged.* (Courtesy of Dr. Oleg D. Jefimenko and Electret Scientific Company, Star City, West Va.)

19.6. Electric Flux—Quantitative Aspects

Figure 19.19 shows a three-dimensional representation of the electric field lines due to a uniform electric field intensity E pointing to the right. In order to

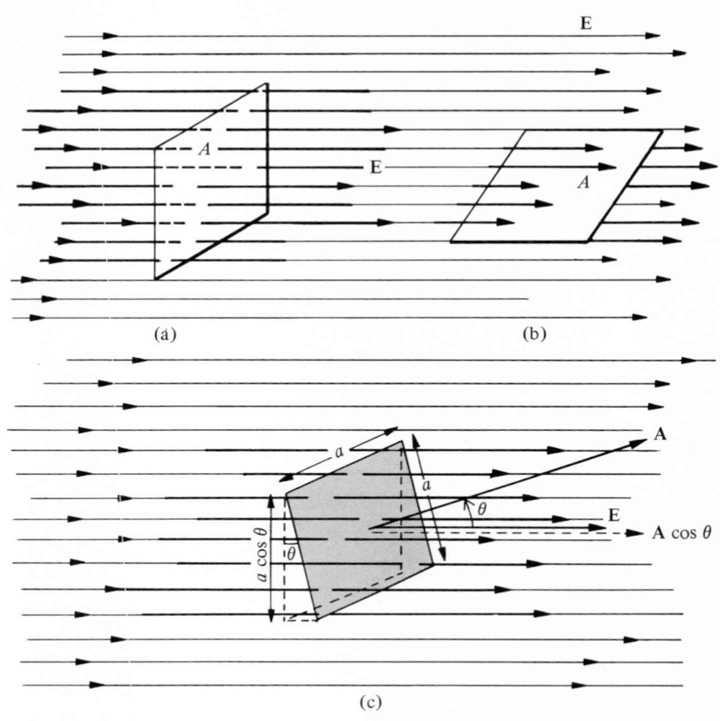

FIGURE 19.19. *Relation among Φ, E, and A, showing that $\Phi = EA \cos \theta$.*

give a quantitative meaning to field lines (or lines of force), a unit area is held perpendicular to E at a given point so that the field lines pass through this area. Suppose that at a point P, E is equal to 5 N/C. This means that if a 1-m^2 area is held perpendicular to E at point P, five field lines will pass through this unit area. *Electric flux* Φ through any surface area is defined as the number of lines that pass through that area. In order to find the relation between Φ, **E** and area A we consider Figure 19.19.

In Figure 19.19(a) area A is held perpendicular to the electric field; hence, $EA_{\perp}$ lines pass through. The flux Φ in this case is given by

$$\Phi = EA_{\perp} \tag{19.10}$$

where $A_{\perp}$ denotes that A is held perpendicular to the field lines. In Figure 19.19(b), A is held parallel to the field lines and, as is obvious, no lines cross this area. Hence, the flux Φ in this case is given by

$$\Phi = EA_{\parallel} = 0 \tag{19.11}$$

where $A_{\parallel}$ indicates that A is held parallel to the field lines. Figure 19.19(c) shows the case where A is neither perpendicular nor parallel to the lines, but makes an angle θ, as shown. To calculate the flux we must know what fraction of A is perpendicular to the field lines. As is clear from Figure 19.19(c), an area that will be perpendicular to the lines is shown dashed and is equal to $A \cos\theta$. Thus, the flux through this area A is given by

$$\Phi = EA_\theta = EA\cos\theta \tag{19.12}$$

where A_θ ($= A\cos\theta$) means that area A is inclined at an angle θ as shown. The units of Φ in all cases are N-m^2/C.

The results derived in the preceding three equations may be combined into a single equation if we make use of the dot or scalar product of two vectors. Before we can do this, the area A must be given a vector representation. The area is denoted by a vector **A**. The direction of this vector is perpendicular to the plane of the area as shown in Figure 19.19(c), and the magnitude of this vector is equal to the magnitude of the area A itself. Thus, the flux Φ is given by the scalar or dot product of the vectors **E** and **A**. That is,

$$\boxed{\Phi = \mathbf{E}\cdot\mathbf{A} = EA\cos\theta} \tag{19.13}$$

where θ is the angle between the vectors **E** and **A** as shown in Figure 19.19(c). For $\theta = 0°$ and $\theta = 90°$, Equation 19.13 reduces to Equations 19.11 and 19.12, respectively. It is important to note that θ is the angle between **E** and **A**.

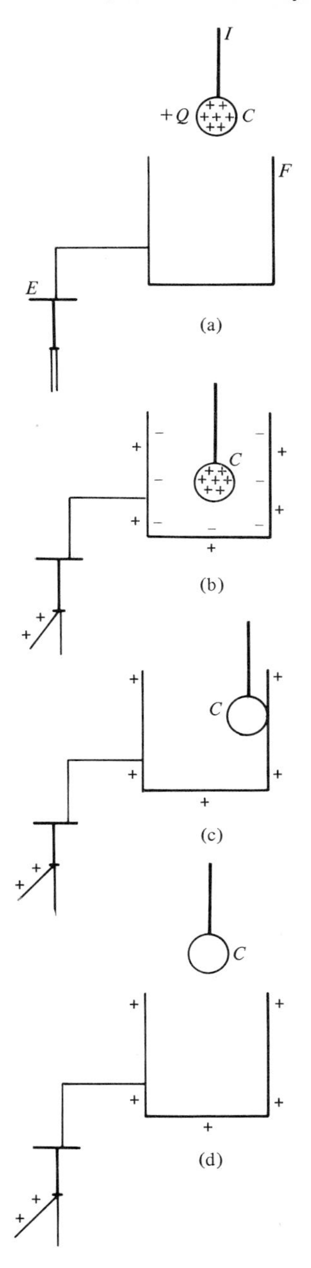

FIGURE 19.20. *Steps in the Faraday ice-pail experiment to prove that (1) the induced charge is equal and opposite to the inducing charge and (2) charge resides on the outer surface of a conductor.*

19.7. An Insulated Conductor—Faraday's Ice-Pail Experiment

In Figure 19.20(a), F is a metallic container (Faraday originally used an ice pail) connected to an electroscope E. C is a spherical charged conductor with an insulated handle I carrying a charge $+Q$. If this is lowered into F without touching, as shown in Figure 19.20(b), it induces a $-Q$ charge on the inner surface of F and $+Q$ on the outer surface of F. Some of this positive charge goes to the electroscope E giving rise to a divergence of a single leaf electroscope. If the charged conductor C is pulled out of F, the container returns to its original state and the leaf of the electroscope converges.

On the other hand, if the conductor in Figure 19.20(b) is allowed to touch the inside of pail F as shown in Figure 19.20(c), the charge $+Q$ on C and $-Q$ on F cancel each other leaving only $+Q$ on the outer surface of F. If C is now pulled out of the pail as in Figure 19.20(d), it is found that the leaves remain diverged and unaffected while C is found to have no charge. (We can test this by touching C to another electroscope.) Let us now sample the charge on the inside and outside of container F using a test sphere with a subsequent transfer to another electroscope. This easily shows that there is no charge on the inside, the only charge is on the outside. Thus, these experiments prove the following: (1) *the induced charge is equal to the inducing charge, and* (2) *the excess charge resides on the outer surface of a conductor.*

* 19.8. Field of an Electric Dipole

A system containing two equal and opposite charges separated by a certain distance constitutes an *electric dipole*. Figure 19.21(a) shows an electric dipole

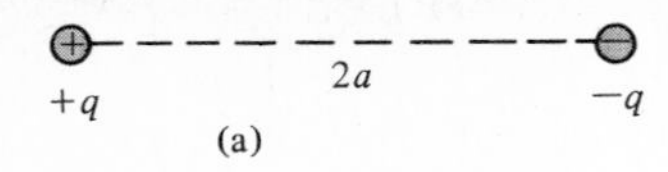

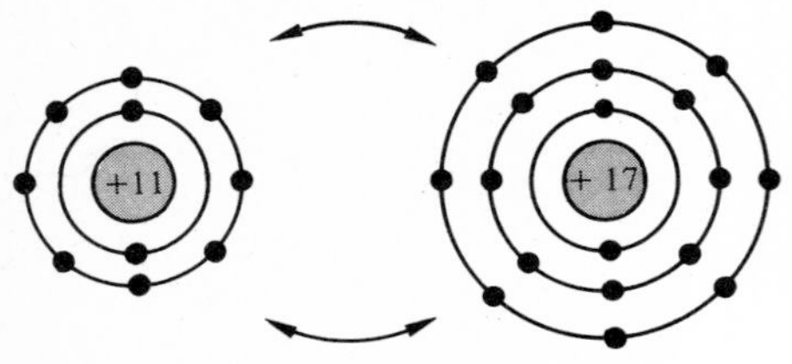

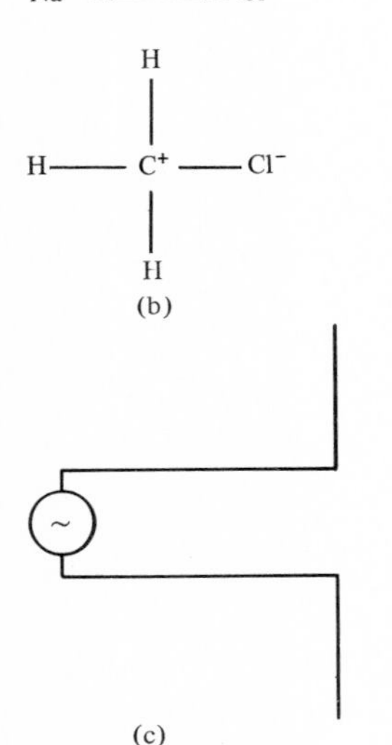

FIGURE 19.21. *Examples of simple electric dipoles.*

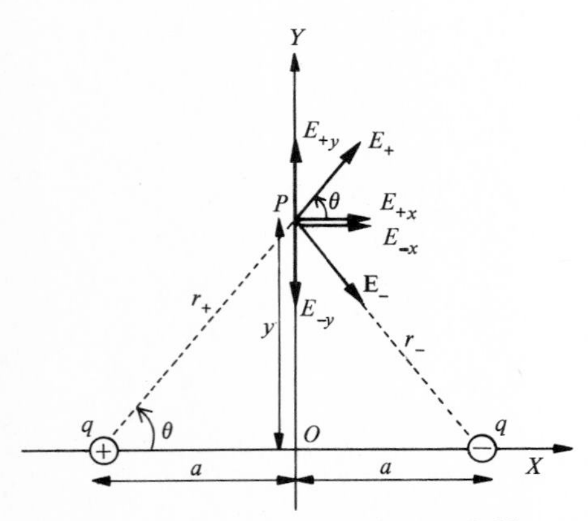

FIGURE 19.22. *Electric field at point P due to an electric dipole.*

containing charges $+q$ and $-q$, which are separated by a distance $2a$. Figure 19.21(b) shows the two most common molecules which are equivalent to simple electric dipoles. Figure 19.21(c) shows a simple radio antenna which is equivalent to an oscillating dipole. These are just a few of many more examples of electric dipoles.

We are interested in finding the electric strength **E** at a point P resulting from an electric dipole shown in Figure 19.22. The point P is at a distance y on the perpendicular bisector of the line joining the two charges. Thus, the total field **E** at P is

$$\mathbf{E} = \mathbf{E}_+ + \mathbf{E}_- \tag{19.14}$$

where

$$E_+ = E_- = \frac{1}{4\pi\epsilon_0}\frac{q}{a^2 + y^2} \tag{19.15}$$

If we resolve $\mathbf{E}_+$ and $\mathbf{E}_-$ into components as shown in Figure 19.22, we find that E_{-y} cancels E_{+y} while E_{-x} adds to E_{+x}. Thus, the magnitude of the total intensity at P may be written

$$E = E_+ \cos\theta + E_- \cos\theta = 2E_+ \cos\theta \tag{19.16}$$

But $\cos\theta = a/\sqrt{a^2 + y^2}$; therefore,

$$E = \frac{1}{4\pi\epsilon_0}\frac{2qa}{(a^2 + y^2)^{3/2}} \tag{19.17}$$

The direction of E is parallel to the X-axis. Usually, the product $2qa$ is denoted by p, where p is called the *electric dipole moment;* hence, E is written as

$$\boxed{E = \frac{1}{4\pi\epsilon_0}\frac{p}{(a^2 + y^2)^{3/2}}} \tag{19.18}$$

In a special case in which $y >> a$, $(a^2 + y^2) \approx y^2$ and

$$E = \frac{1}{4\pi\epsilon_0}\frac{p}{y^3} \tag{19.19}$$

which means that the electric field strength due to an electric dipole varies as the inverse cube of the distance $(\propto 1/y^3)$ and not as the inverse square of the distance, as in the case of a point charge. Thus, the dipole field decreases much more rapidly as compared to the field of a point charge. **E** at a point along the line joining the two charges can be calculated in a similar manner for the case $x >> a$ (see Problem 19.30).

SUMMARY

There are two types of charges, negative and positive. Like charges repel each other and unlike charges attract each other. Materials that do not allow free flow of electrons through them are called *insulators*. Metals and other materials that allow the flow of electrons are called *conductors*. In between these two types are materials called semiconductors.

According to *Coulomb's law,* the force between two charges is directly proportional to the product of the magnitude of the charges and inversely proportional to the square of the distance between them; that is, $F = kq_1q_2/r^2$ where $k = 1/4\pi\epsilon_0 \simeq 9 \times 10^9$ N-m^2/C^2. The SI unit of charge is the *coulomb* (C), defined as the quantity of charge that flows across any cross section of a wire in 1 second when there is a steady current of 1 ampere. The gravitational forces are negligible as compared to electrical forces.

The concept of electric field was introduced by Faraday. The *electric field strength* **E** at any point surrounding the charge q is defined as the force per unit positive charge in the field $\mathbf{E} = \mathbf{F}/q_0$, or $E = kq/r^2$. The units of E are N/C. The electric field lines, which are imaginary, always start from positive charges and end on to negative charges. The tangent to the field lines at any point gives the direction of the electric field at that point.

The *electric flux* Φ through any surface area is defined as the number of lines that pass through that area. $\Phi = \mathbf{E} \cdot \mathbf{A} = EA \cos \theta$, where the direction of **A** is a vector perpendicular to the plane of the surface area. The units of Φ are N-m^2/C.

According to Faraday, (1) the induced charge is equal to the inducing charge, and (2) the excess charge resides on the outer surface of a conductor.

* A system consisting of two equal and opposite charges separated by a certain distance is an *electric dipole*. The dipole moment $p = 2qa$, and at a point on the Y-axis the field due to the dipole is $E = (1/4\pi\epsilon_0)[p^2/(a^2 + y^2)^{3/2}]$.

QUESTIONS

1. Can you think of an experiment (other than the one in the text) to demonstrate that like charges repel each other while unlike charges attract each other?
2. How will you prove that spheres A and B in Figure 19.4 have equal and opposite charges?
3. An electrified rod attracts pieces of paper. After a while these pieces fly away. Why?
4. Why is it difficult to electrify objects on a humid day?
5. How will you proceed to identify the + and − terminals of an unmarked battery?
6. Show that "repulsion is the sure test of electrification." Use this to show whether a body is (a) charged—negatively or positively; or (b) uncharged.
7. Can you think of some consequences if the force F between two charges were (a) $\propto r$; (b) $\propto 1/r^5$; and (c) $\propto 1/q_1q_2$.
8. Suppose that somewhere in space there exist two types of masses just like electrical charges. What do you predict about the gravitational interaction when objects from earth reach there?
9. What will happen to the orbits of the planets if the planets were not electrically neutral?
10. In your opinion which concept, action-at-a-distance or field theory, is better in explaining the force between two charges? Why?
11. Describe the force or forces that act on a positive charge placed between two parallel plates with (a) equal and similar charges; and (b) equal and opposite charges.

12. Can you devise a simple experiment to show that the charge resides on the outer surface of a conductor?
13. Give some examples of electric dipoles.

PROBLEMS

1. How many electrons are needed to produce a total charge of (a) 1 C; and (b) 1 μC?
2. Calculate the distance between two charges of 5 μC each so that the force between them will be 0.1 N.
3. If the force of repulsion between two electrons is equal to the weight of either electron, what is the distance between them?
4. If the force of repulsion between two protons is equal to the weight of either proton, what is the distance between them?
5. An oxygen nucleus has eight protons; that is, it has a charge of $+8e$. What is the force on an electron in the oxygen atom that is at a distance of 10^{-11} m from the nucleus?
6. Find the ratio of the electrical force to the gravitational force between two α-particles. The mass of an α-particle is 4 times the mass of a proton, while its charge is twice that of the proton.
7. Two spheres of mass 10 g each are 10 cm apart. How much charge should be given to each sphere so that the force of electrical repulsion is balanced by the force of gravitational attraction between the two spheres?
8. Three identical charges, $q_1 = q_2 = q_3 = 5\ \mu$C, are placed along the X-axis at $x = 0$, 0.5, and 1.5 m, respectively. What is the magnitude and the direction of the force that acts on q_2?
9. A charge $+q$ is placed at the origin while a charge $+4q$ is placed at a distance of 2 m from the origin along the X-axis. Calculate the position of the point where the resultant force will be zero.
10. Repeat Problem 9 if (a) $+q$ is replaced by $-q$; and (b) $+4q$ is replaced by $-4q$.
11. Three charges, $q_1 = +5\ \mu$C, $q_2 = -10\ \mu$C, and $q_3 = -5\ \mu$C, are placed along the X-axis at $x = 0$, 2, and 4 m, respectively. Calculate the force on (a) q_1 due to q_2 and q_3; and (b) q_2 due to q_1 and q_3.
12. Four equal charges, each of $+5\ \mu$C, are placed at each of the four corners of a square of side 0.5 m. Calculate the force on any one of the four charges.
13. Point charges of $+2\ \mu$C each are placed at three corners of a square whose side is 0.4 m. If a point charge of $+1\ \mu$C is placed at the fourth corner, what will be the force on the 1 μC charge?
14. Show that in Example 19.5 the effect of the gravitational pull on the electron is negligible as compared to the electrostatic force.
15. An electron is placed in a uniform electric field of 2000 N/C. Calculate the force and the acceleration that acts on this electron. What are the values if the electron is replaced by a proton?
16. A particle of charge $+10 \times 10^{-9}$ C feels a force of 10^{-8} N downward. What are the magnitude and direction of the electric field?
17. Calculate the magnitude and direction of the electric field at a point between two parallel horizontal plates if (a) a proton experiences a force of 2×10^{-16} N in the upward direction; and (b) if an electron experiences the same force in the downward direction.

18. An electron is released between two parallel plates where there is a uniform electric field of 10^4 N/C. Calculate (a) the electrical force that acts on it; (b) the acceleration it acquires; (c) the velocity after traveling 5 cm; and (d) the kinetic energy it acquires.
19. Suppose that an electron is shot into an electric field along the X-axis with a velocity of v_0. Show that after a time t the coordinates of the electron are (see Figure P. 19.19)

$$y = -\left(\frac{eE}{2mv_0^2}\right)x^2$$

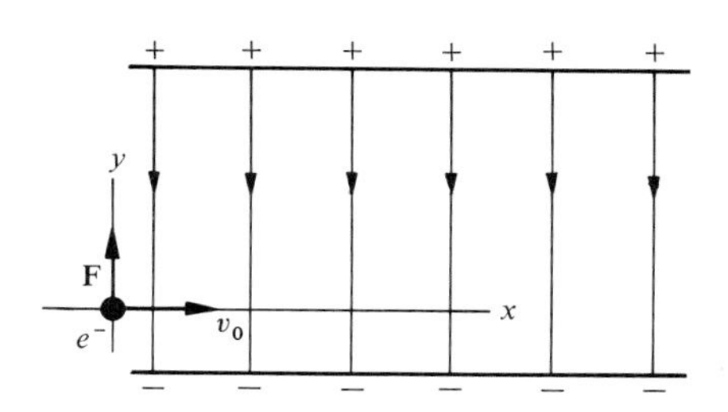

P. 19.19

20. Using the results of Problem 19, calculate the value of E so that $x = 0.5$ m, $y = 0.10$ m, and v_0 is (a) 10^4 m/s; (b) 10^6 m/s; and (c) 10^7 m/s.
21. An electron is released from the bottom plate A as shown. Calculate (a) the force on the electron; (b) the acceleration of the electron; (c) the velocity of the electron when it reaches the plate B; (d) the time it takes to travel from A to B; and (e) the kinetic energy it acquires. E is 10^4 N/C.

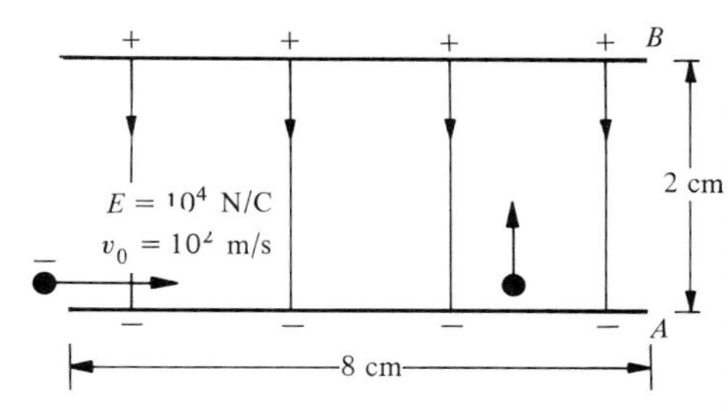

P. 19.21

22. An electron with an initial horizontal velocity of $v_0 = 10^7$ m/s is released between two parallel plates as shown in Example 19.5. After 10^{-8} s, (a) what is the vertical component of the velocity; (b) the resultant velocity; (c) the value of x; and (d) the value of y? Let the value of E be 10^4 N/C.
23. Two charges, $q_1 = -1\ \mu$C and $q_2 = +4\ \mu$C, are separated by a distance of 5 m. Find the position of a point along the line joining the two charges where the net electric intensity will be zero.
24. The force on a charge of $+5\ \mu$C is found to be 0.5 N along the $+X$-axis. What is the magnitude and direction of the electric intensity at this point?
25. Two charges, $q_1 = +100\ \mu$C and $q_2 = +200\ \mu$C, are separated by a distance of 4 m. What is the net electric intensity at a point half way between the charges?
26. Three charges, each of $+10\ \mu$C, are placed at three of the four corners of a square. Calculate the electric intensity (a) at the fourth corner; and (b) at the center of the square. A side of the square is 0.5 m.
* 27. Show that the force on a charge q_0 shown is given by

$$F = 2k\frac{q_0 q}{r^2}\sin\theta = 2k\frac{q_0 q}{r^3}(r^2 - a^2)^{1/2}$$

(a) Calculate F for $q_0 = +2\ \mu$C, $q = +10\ \mu$C, $r = 1$ m, and $a = 10$ cm. (b) Calculate F when $r = a$; and (c) when $r >> a$ (say $r = 10$ m and $a = 10$ cm).

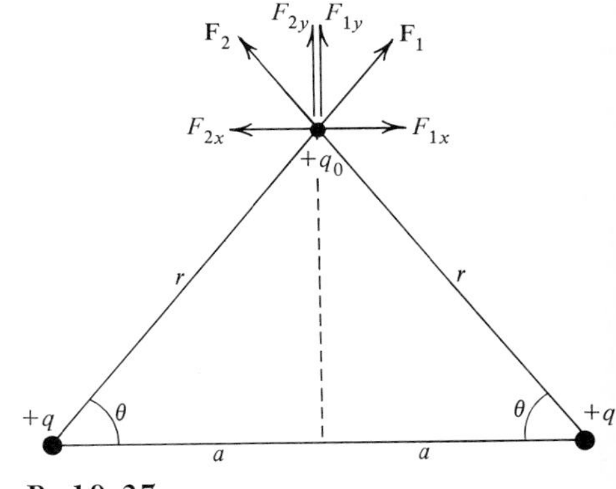

P. 19.27

* 28. Consider the electric dipole shown in Figure 19.22. Let $q_1 = +5\ \mu$C, $q_2 = -5\ \mu$C, and the distance between them be 25 cm. Calculate the electric intensity at the following points: (a) 10 cm to the left of charge q_1; (b) 10 cm to the left of charge q_2; and (c) 25 cm from each of the two charges.
29. An oil drop of mass 0.01 g between two parallel plates is in equilibrium under the action of two forces: (1) its weight acting downward, and (2) an electric force due to a uniform electric intensity of 1000 N/C between the plates. What are the magnitude and sign of the electric charge on the oil drop and the direction of the electric field?
30. Find the electric field along the axis of a dipole for $x >> a$.

20
Electric Potential and Capacitance

For practical purposes the concept of electric field must be translated into the concepts of electric potential and electrical potential energy. These quantities are analogous to gravitational potential and gravitational potential energy. Another quantity that plays an important role in electrical circuits is the capacitance—the ability to store a quantity of charge—and its dependence on the dielectrics. Putting these different concepts together, we shall calculate the electric energy stored in space.

20.1. Electric Potential Energy and Potential Difference

LET US consider a particle with a positive charge q_0 that is allowed to move in an electric field produced between two oppositely charged parallel plates, as shown in Figure 20.1(a). The positive charge will freely move from B to A, gaining kinetic energy. On the other hand, if the charged particle is moved from A to B, an external force must be applied to make the charge move uphill against the electric field.

Let us impose a condition that as the charge is moved from B to A, it is always maintained in equilibrium; that is, it is moving with uniform velocity. This is achieved by applying a force $\mathbf{F}$ which is equal and opposite to $q_0\mathbf{E}$ at every point along its path, as shown in Figure 20.1(b). The force $\mathbf{F}$ keeps the charge q_0 from accelerating when moving from B to A, and avoids its deceleration if the charge moves from A to B. The situation is analogous to the gravitational field shown in Figure 20.2. As in the gravitational field, the work done by the force in the electric field is stored as potential energy. Just like the gravitational field, the electric field is conservative; that is, the work done is independent of the path, as we shall show in the next section.

Thus, we may conclude that a charged particle placed in an electric field has potential energy because of its interaction with the electric field. Let W_{AB} be the work done by the force in carrying a positive charge q_0 from point A to point B while keeping the charge in equilibrium. *The change in potential energy ΔU of charge q_0 is defined to be equal to the work done by the force in carrying the charge q_0 from one point to the other.* That is,

$$\Delta U = W_{AB} \qquad \text{or} \qquad U_B - U_A = W_{AB} \tag{20.1}$$

where U_A and U_B designate the potential energy at points A and B, respectively.

To describe the electric field in terms of a quantity that will be independent of the magnitude of the charge q_0, we introduce the concept of electric potential difference.

The *potential difference* ΔV between two points A and B in an electric field is defined as the work done in carrying a unit positive charge from A to B while keeping the charge in equilibrium. That is,

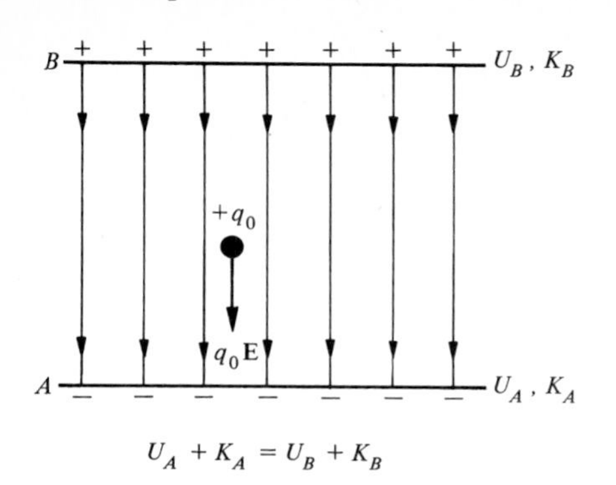

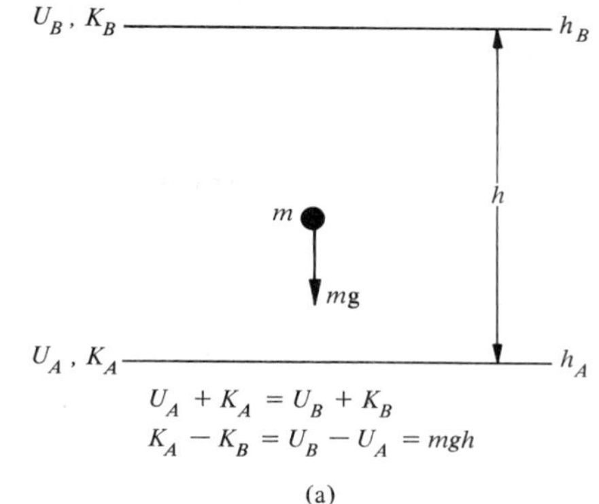

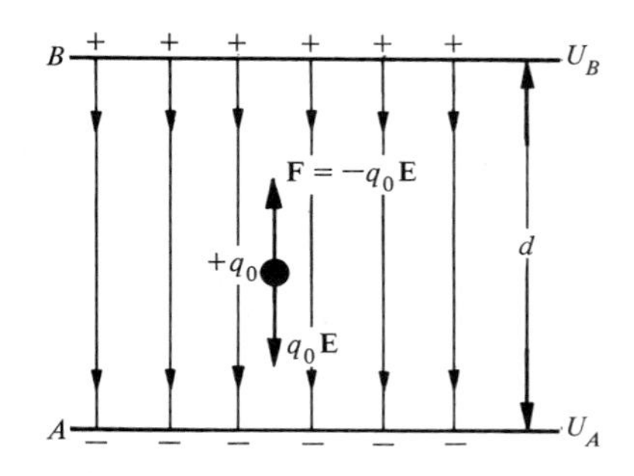

$W_{AB} = Fd = q_0Ed$
$W_{AB} = U_B - U_A = q_0(V_B - V_A)$

(b)

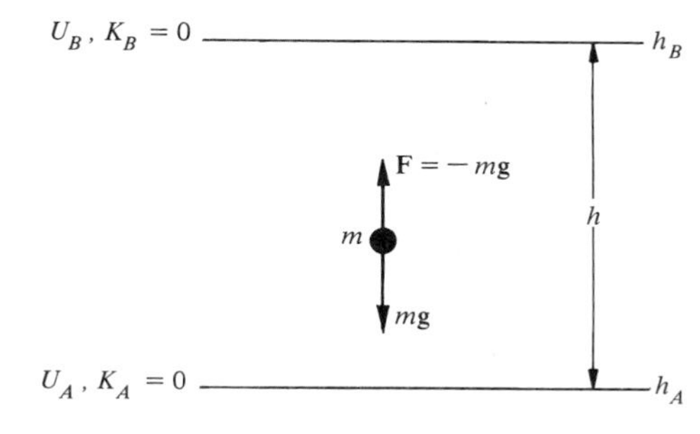

$W_{AB} = U_B - U_A = Fh = mgh$

(b)

FIGURES 20.1 *and* 20.2. *The analogy between the electric field (Figure 20.1) and the gravitational field (Figure 20.2): (a) accelerated motion and (b) equilibrium motion.*

$$\Delta V = V_B - V_A = \frac{W_{AB}}{q_0} \tag{20.2}$$

where V_A and V_B are defined as the electric potentials at points A and B, respectively. From above equations, the relation between the potential at a point, change in potential energy ΔU, and potential difference ΔV is

$$U_A = q_0 V_A, \qquad U_B = q_0 V_B$$

and

$$\Delta U = q_0\,\Delta V = W_{AB} \tag{20.3}$$

The unit of work is the joule (J) and that of charge is the coulomb (C). From Equation 20.1 it is clear that the unit of potential energy will be the same as that of work; that is, the joule is the unit of potential energy. According to Equation 20.2 or 20.3, the unit of potential difference is the joule per coulomb. Since potential difference is used extensively, it is convenient to define a new unit, which has been given the name volt:

$$1 \text{ volt} = \frac{1 \text{ joule}}{1 \text{ coulomb}} \tag{20.4}$$

That is, a potential difference of 1 *volt* is said to exist between two points if 1 joule of work is required to move a positive charge of 1 Coulomb from one point to the other while keeping the charge in equilibrium.

In order to talk in terms of the absolute potential or simply potential at a point in an electric field, we must designate some point as the *reference point* or *standard point*. This reference point is arbitrarily taken to be at infinity and is assigned a potential of zero. This is justified because any charge configuration will produce a zero field at infinity. Thus, in Equation 20.2 if we take A to be at infinity so that $V_A = 0$, the electric potential at B will be $V_B = W_{\infty B}/q_0$, or, dropping the subscript,

$$\boxed{V = \frac{W}{q_0}} \tag{20.5}$$

which states that the *electric potential* at any point in an electric field is equal to the work done in bringing a unit positive charge from infinity to the point under consideration. Note that potential at a point is still a potential difference between the potential at that point and the potential at infinity. Since $U = q_0V$, we may define the potential energy of charge q_0 at any point in the field to be equal to the amount of work done in bringing the charge q_0 from infinity to that point. We may emphasize that both potential and potential energy are scalar quantities, because W and q_0 are scalars.

In the discussion above we have assumed that the charged particle, which is moving, is in equilibrium. If no external force is applied to maintain this equilibrium, the charged particle will change its kinetic energy as well as its potential energy as it moves from point to point. Thus, if a particle of mass m and charge q_0 has potential energy U_A and kinetic energy K_A at point A and U_B and K_B at B, the conservation of energy requires (in the absence of any external force acting on the charge)

$$\boxed{U_A + K_A = U_B + K_B} \tag{20.6a}$$

Since $U_A = q_0V_A$ and $U_B = q_0V_B$, we may write

$$q_0(V_A - V_B) = K_B - K_A \tag{20.6b}$$

Of course, this is true only if the electric field is conservative, and we shall show this to be the case in the next section.

Example 20.1 Calculate the potential difference between two points A and B if it requires 8×10^{-4} J of external work to move a charge of $+4\,\mu$C from A to B. Which point is at a higher potential?

According to Equation 20.2, $\Delta V = V_B - V_A = W_{AB}/q_0$, where $q_0 = +4\,\mu C = +4 \times 10^{-6}$ C and $W_{AB} = 8 \times 10^{-4}$ J. Therefore,

$$\Delta V = V_B - V_A = \frac{8 \times 10^{-4}\text{ J}}{4 \times 10^{-6}\text{ C}} = 200\frac{\text{J}}{\text{C}} = 200\ V$$

Thus, the potential difference between A and B is 200 V and positive. Hence, B is at a higher potential than A.

Exercise 20.1 Calculate the amount of work done in carrying a charge of $+2\,\mu$C from A to B if A is at a potential of -50 V and B is at $+20$ V. [*Ans.:* $W_{AB} = 1.4 \times 10^{-4}$ J.]

20.2. Electric Potential and Electric Intensity

Let us now show how the potential difference between two points is related to the electric intensity. As a special case, let us consider the situation shown in Figure 20.1(b). The electric field between the two charged plates is uniform; hence, the electric intensity E is constant. The potential difference between A and B is given by Equation 20.2:

$$V_B - V_A = \frac{W_{AB}}{q_0} \tag{20.2}$$

where $W_{AB} = Fd = -q_0Ed$ (the negative sign is needed because F must be applied opposite to q_0E so as to keep it in equilibrium). Thus, substituting above gives

$$V_B - V_A = -Ed, \qquad E = -\frac{V_B - V_A}{d} = \frac{V_A - V_B}{d} \tag{20.7}$$

Of course, this is a very special case in which the electric field is uniform. If the field were nonuniform, **E** as well as V will change everywhere. But we can derive a more general relation as explained below.

Let points A and B be separated by an infinitesimal distance Δs so that the potential difference between them is ΔV and the field E is almost constant. If the work done is ΔW, we may write Equation 20.2 as

$$\Delta V = \frac{\Delta W}{q_0} = \frac{F\,\Delta s}{q_0}$$

Substituting once again for $F = -q_0E$ and rearranging, we get

$$E = -\frac{\Delta V}{\Delta s} \tag{20.8}$$

The negative sign in Equation 20.8 is introduced so that E will be positive when $\Delta V/\Delta s$ is negative. This is keeping in mind the fact that the positive charge moves from a high potential point to a low potential point, that is, downhill. The quantity $\Delta V/\Delta s$ is the rate of change of potential with distance, called the *potential gradient*. Thus, according to Equation 20.8, *in an electric field the component of electric intensity in any direction is equal to the negative of the potential gradient in that direction.*

The unit of electric field strength E which we have used so far is newton/coulomb (N/C). The definition of the volt provides us with another unit of E. Since from Equation 20.7, $E = (V_A - V_B)/d$, the unit of E may be defined as volt/meter (V/m). That V/m is equivalent to N/C is shown below:

$$1\,\frac{\text{volt}}{\text{meter}} = 1\,\frac{\text{joule/coulomb}}{\text{meter}} = 1\,\frac{\text{newton-meter}}{\text{coulomb-meter}} = 1\,\frac{\text{newton}}{\text{coulomb}}$$

Thus, we shall use V/m or N/C as the unit of E.

Let us now derive an expression for the potential at a point in the field of a point charge. For this case E in the relation $\Delta V = -E\,\Delta s$ is no longer constant. Since the field is produced by a point charge q, the electric intensity E at a distance r from q will be $E = (1/4\pi\epsilon_0)(q/r^2)$. The field is radial, as shown in Figure 20.3. Let us consider two very close points A and B at distances r_A and r_B from q so that $\Delta s = r_B - r_A$ and $r^2 \approx r_A r_B$. Thus, $\Delta V = -E\,\Delta s$ takes the form

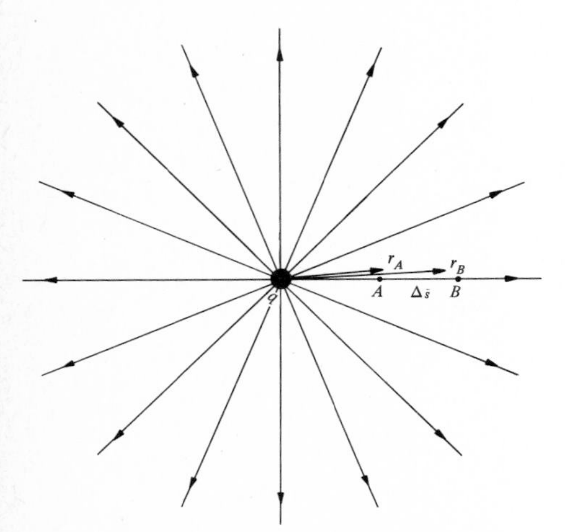

FIGURE 20.3. *Electric field surrounding a point charge used for calculating potential at a point.*

$$\Delta V = V_B - V_A = -\frac{1}{4\pi\epsilon_0}\frac{q}{r^2}\Delta s = -\frac{1}{4\pi\epsilon_0}\frac{q}{r_A r_B}(r_B - r_A)$$

That is, the potential difference between any two points is

$$\Delta V = V_B - V_A = \frac{q}{4\pi\epsilon_0}\left(\frac{1}{r_B} - \frac{1}{r_A}\right) \tag{20.9}$$

The potential $V(r)$ at any point a distance r from the point charge q may be obtained by using the substitution $V_B = 0$ at $r_B = \infty$ and $V_A = V(r)$ when $r_A = r$. That is,

$$\boxed{V(r) = \frac{1}{4\pi\epsilon_0}\frac{q}{r}} \tag{20.10}$$

Even though $V(r)$ is called the potential at a point, actually $V(r)$ is equal to the potential difference between r and ∞.

Finally, we may point out that the superposition principle is applicable. The potential produced by any charge is not affected by the presence of other charges. Thus, if a charge distribution consists of N discrete charges q_1, $q_2, \ldots, q_N$, the potential at a point P, located at distances $r_1, r_2, \ldots, r_N$, respectively, from these charges, may be obtained by adding algebraically the potentials due to the individual charges. That is, the potential at P is given by

$$V = V_1 + V_2 + V_3 + \cdots + V_N$$

or

$$V = \sum_{i=1}^{N} V_i = \frac{1}{4\pi\epsilon_0}\sum_{i=1}^{N}\frac{q_i}{r_i} \tag{20.11}$$

where V_i is the potential at point P produced by charge q_i, which is at a distance r_i from P. The potential, unlike field strength, is a scalar quantity; hence, a simple addition given by Equation 20.11 is possible.

EXAMPLE 20.2 Consider charges $q_1 = 2 \times 10^{-7}$ C, $q_2 = 3 \times 10^{-7}$ C, and $q_3 = -4 \times 10^{-7}$ C placed at three corners of a square of side $a = 1$ m as shown in the accompanying figure. Calculate (a) the potential at the center P of the square; and (b) the potential at the fourth corner D of the square. (c) What is the potential difference between points P and D? (d) If a charge of 2 μC is moved from D to P, what will be the change in its potential energy?

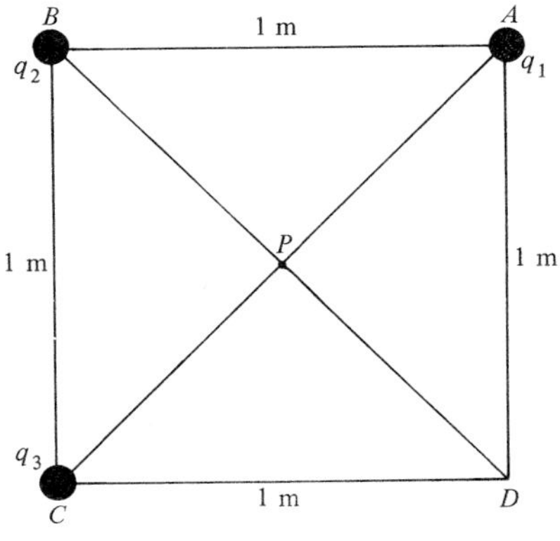

Ex. 20.2

(a) Using the general expression

$$V = \sum V_i = \frac{1}{4\pi\epsilon_0}\left(\frac{q_1}{r_1} + \frac{q_2}{r_2} + \frac{q_3}{r_3}\right)$$

the potential at point P for which $r_1 (= AP) = r_2 (= BP) = r_3 (= CP) = a/\sqrt{2} = 1/\sqrt{2}$ m $= 0.71$ m is given by

$$V_P = 9 \times 10^9 \frac{\text{N-m}^2}{\text{C}^2}\left[\frac{2 \times 10^{-7}\text{ C} + 3 \times 10^{-7}\text{ C} - 4 \times 10^{-7}\text{ C}}{0.71\text{ m}}\right] = 1268\text{ V}$$

(b) The potential at D is given by (where BD $= \sqrt{2}$ m)

$$V_D = 9 \times 10^9 \frac{\text{N-m}^2}{\text{C}^2}\left[\frac{2 \times 10^{-7}\text{ C}}{1\text{ m}} + \frac{3 \times 10^{-7}\text{ C}}{\sqrt{2}\text{ m}} - \frac{4 \times 10^{-7}\text{ C}}{1\text{ m}}\right] = 109\text{ V}$$

(c) The potential difference between points P and D is

$$V_P - V_D = 1268\text{ V} - 109\text{ V} = 1159\text{ V}$$

(d) The change in potential energy ΔU is given by

$$\Delta U = U_P - U_D = q(V_P - V_D) = 2 \times 10^{-6}\text{ C }(1159\text{ V}) = 2.32 \times 10^{-3}\text{ J}.$$

EXERCISE 20.2 Consider charges $q_1 = 1 \times 10^{-7}$ C, $q_2 = 2 \times 10^{-7}$ C, $q_3 = 3 \times 10^{-7}$ C, and $q_4 = 4 \times 10^{-7}$ C placed at the four corners of a square of side 1 m. Calculate the potential at the center of the four charges. If a charge of 5×10^{-7} C is placed at the center of the square, what is its potential energy? How much work will have to be done to move this charge from the center of the square to infinity? [*Ans.*: $V = 1.27 \times 10^4$ V, $U = 6.35 \times 10^{-3}$ J, and $W = -6.35 \times 10^{-3}$ J.]

20.3. Equipotential Lines, Surfaces, and Volumes

A graphical representation of an electric field was drawn in terms of electric field lines in Figure 19.11. Some sort of graphical representation for electrostatic potential is also desirable.

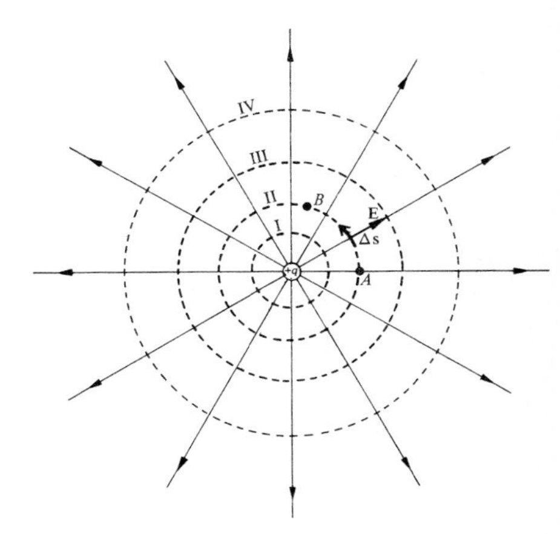

FIGURE 20.4. *The dashed-line circles indicate the equipotential lines surrounding a charge $+q$. Note that the radial electric field lines are perpendicular to the equipotential lines.*

Consider a point charge $+q$, as shown in Figure 20.4. The radial lines shown represent the electric field strength **E**. Let us calculate the potential difference between A and B which lie on the same circle, II. Since the force **F** (= q**E**) is perpendicular to the displacement the work done, $\Delta W = q\Delta V = 0$ (because

$\theta = 90°$, $\cos 90° = 0$; see Chapter 6). We can arrive at the same result as follows. In Equation 20.9,

$$V_B - V_A = \frac{q}{4\pi\epsilon_0}\left(\frac{1}{r_B} - \frac{1}{r_A}\right)$$

If we substitute $r_A = r_B$, we get

$$V_B - V_A = 0 \qquad \text{or} \qquad V_A = V_B$$

which means that points A and B are at the same potential. For $\Delta V = 0$, $\Delta W = q\,\Delta V = 0$; that is, no work is done.

As is clear, no work will be done in moving a test charge from any point on circle II to any other point on the same circle. Thus, all points on circle II are at the same potential. We call this circle an *equipotential line*. Actually, this circle is only a portion of an *equipotential surface* of a sphere of radius equal to the radius of circle II. Other equipotential lines are shown by dashed circles in Figure 20.4. One could draw equipotential lines and surfaces at each and every point, but we have drawn only a few representative circles. Another example of equipotential lines is shown in Figure 20.5.

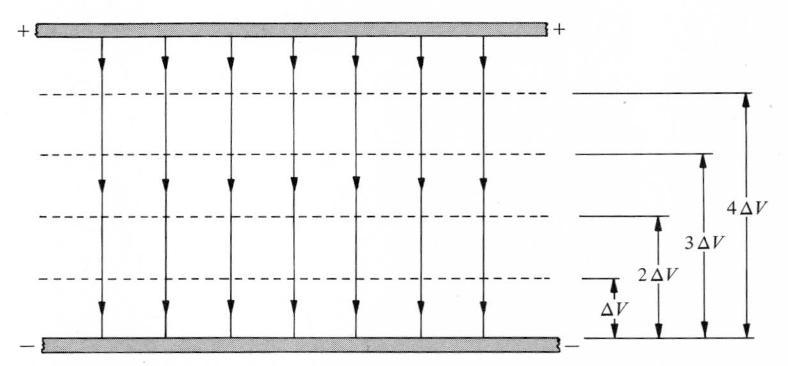

FIGURE 20.5. *The dashed lines are the equipotential lines due to two oppositely charged parallel plates; the continuous lines are the electric field lines.*

It is essential to note that equipotential lines are always drawn so as to be perpendicular to the electric field lines, since no work can be done in moving a test charge along them. Thus, the field lines and equipotential lines (and surfaces) form a set of mutually perpendicular lines. In general, the field lines are curved and so are the equipotential lines. If the field is uniform as between the parallel plates, the field lines are parallel straight lines while the equipotential surfaces are planes perpendicular to the field lines, as shown by the dashed lines in Figure 20.5.

Without proof we state that since charge resides on the outer surface of a conductor, **E** inside the conductor is zero. Therefore, the potential difference between any two points inside the conductor is zero. That is, all the points in the inner region are at the same potential. We state this fact by saying that the interior region of any charged conductor is an *equipotential volume*. That is, no work will be done in moving charges in such a volume.

20.4. The Electron Volt

We have stated that when a particle of charge q moves from a point where the potential is V_A to a point where the potential is V_B, the change in potential energy ΔU of the particle is

$$U = q(V_B - V_A) = q\,\Delta V$$

If no external force acts on the charge to maintain the equilibrium, this change in potential energy appears in the form of a change in kinetic energy. Suppose that the particle carries the charge $q = \pm e = \pm 1.60 \times 10^{-19}$ C, which is the charge on a proton or an electron. Thus, the energy ΔK acquired by the particle will be

$$\Delta K = q\,\Delta V = e\,\Delta V = (1.60 \times 10^{-19}\ \text{C})(\Delta V)$$

Furthermore, assume that $\Delta V = 1$ Volt; hence,

$$\Delta K = (1.60 \times 10^{-19}\ \text{C})(1\ \text{V}) = 1.60 \times 10^{-19}\ \text{J}$$

To eliminate the factor 1.60×10^{-19}, we introduce another unit of energy, called the *electron volt* (eV) and define it as follows: One *electron volt* (eV) is the amount of energy acquired or lost by an electron as it traverses a potential difference of 1 volt. That is,

$$\boxed{1\ \text{eV} = 1.60 \times 10^{-19}\ \text{J}} \tag{20.12}$$

This is a very convenient unit and may be used for potential energy, kinetic energy, or any other form of energy. For example, if a particle carrying a charge of $2e$ falls through a potential difference of 1 V, it will acquire an energy of 2 eV. On the other hand, if a particle of charge $2e$ falls through a potential difference of 100 V, its energy change will be $\Delta U = q\,\Delta V = (2e)(100\ \text{V}) = 200$ eV. In terms of joules this will be equal to $200(1.6 \times 10^{-19}\ \text{J}) = 3.2 \times 10^{-17}$ J. Thus, when a particle of charge $|q| = ne$, where n is an integer, falls through a potential difference of V volts, it acquires an energy of (nV)eV.

In terms of the amount of energy particles can acquire after being accelerated through modern high-energy accelerators, an eV is a very small unit. The bigger units commonly used are

$$1\ \text{kilo-electron volt} = 1\ \text{keV} = 10^3\ \text{eV}$$
$$1\ \text{million electron volts} = 1\ \text{MeV} = 10^6\ \text{eV}$$
$$1\ \text{giga-electron volt} = 1\ \text{GeV} = 10^9\ \text{eV}$$

One giga-electron volt is also called 1 billion electron volts and is denoted 1 BeV.

Example 20.3 A proton placed in a uniform electric field of 5000 N/C directed to the right is allowed to go a distance of 10 cm from A to B. Calculate (a) the potential difference between those two points; (b) the work done; (c) the change in potential energy of the proton; (d) the change in kinetic energy of the proton; and (e) its velocity.

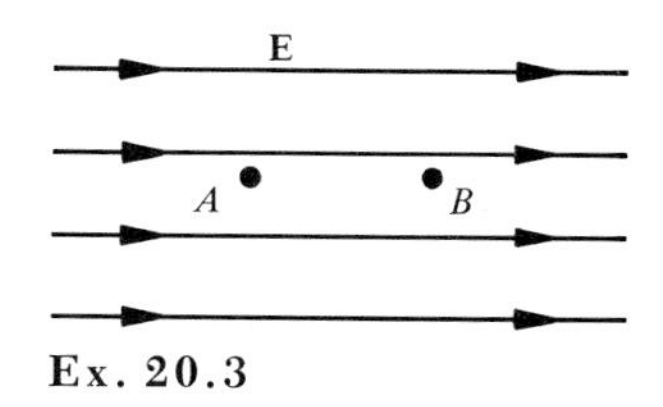

Ex. 20.3

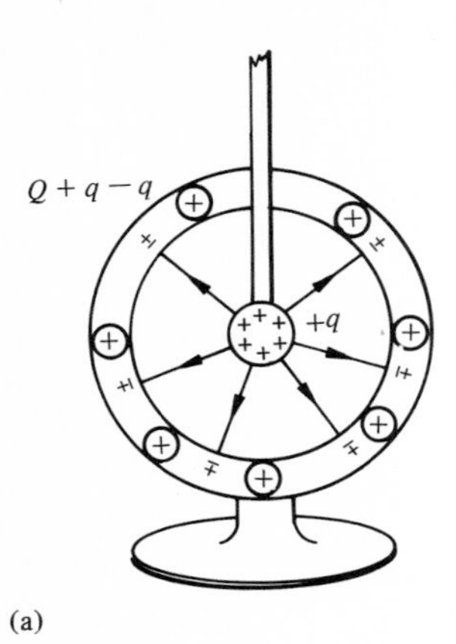

(a)

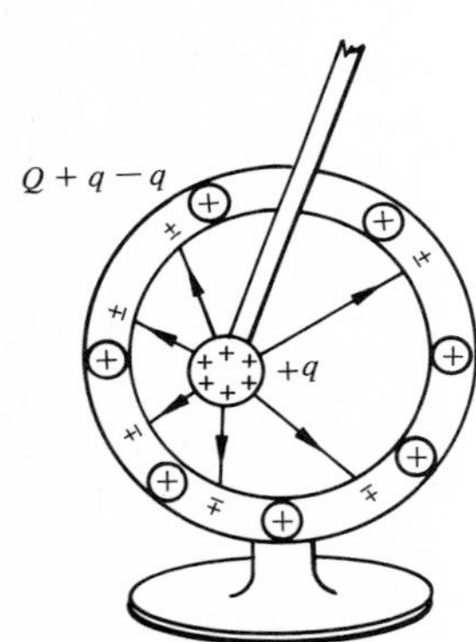

(b)

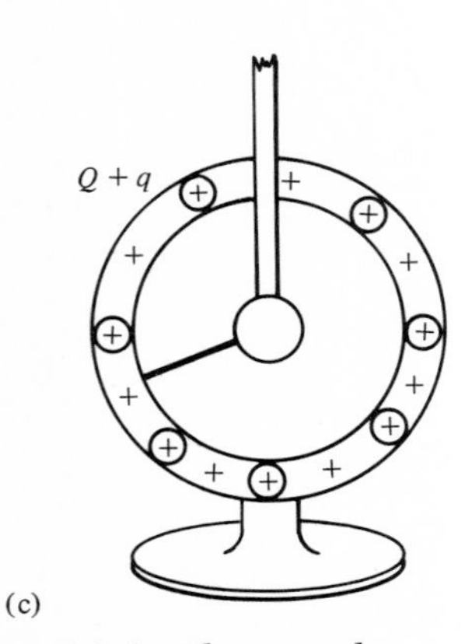

(c)

FIGURE 20.6. *The setup demonstrates the principle of operation of an electrostatic generator.*

(a) According to Equation 20.8, $E = -\Delta V/\Delta s$ or

$$\Delta V = V_B - V_A = -E\,\Delta s = -(5000\ \text{N/C})(0.1\ \text{m}) = -500\ \text{V}$$

The minus sign indicates that B is at a lower potential than A. Thus, $|\Delta V| = 500$ V.

(b) According to Equation 20.2, the work done is

$$W = q_0|\Delta V| = (1.602 \times 10^{-19}\ \text{C})(500\ \text{V}) = 801 \times 10^{-19}\ \text{J}$$

or

$$W = q_0|\Delta V| = e\,|\Delta V| = e(500\ \text{V}) = 500\ \text{eV}$$

(c) From Equation 20.3, the change in potential energy is

$$\Delta U = q_0|\Delta V| = \Delta W = 500\ \text{eV}$$

(d) Also from Equation 20.6, the change in kinetic energy is given by

$$K_B - K_A = -q_0(V_B - V_A) = -q_0\,\Delta V$$

But $K_A = 0$ while $K_B = \frac{1}{2}mv^2$. Therefore,

$$K_B = \tfrac{1}{2}mv^2 = q_0|\Delta V| = e(500\ \text{V}) = 500\ \text{eV}$$

(e) $$v^2 = \frac{2q_0|\Delta V|}{m} = \frac{2(1.602 \times 10^{-19}\ \text{C})(500\ \text{V})}{1.67 \times 10^{-27}\ \text{kg}} = 9.59 \times 10^{10}\ \text{m}^2/\text{s}^2$$

or

$$v = 3.097 \times 10^5\ \text{m/s}$$

If we assume that the mass does not remain constant with velocity, we must use the relativistic relation for kinetic energy. That is,

$$K_B = mc^2 - m_0c^2 = m_0c^2\left(\frac{1}{\sqrt{1-(v^2/c^2)}} - 1\right) = \Delta U$$

where $\Delta U = 500$ eV and $m_0c^2 = 940$ MeV $= 940 \times 10^6$ eV. Further calculation shows that $v = 3.0945 \times 10^5$ m/s, which is very close to the value calculated above. Hence, it is not necessary to use relativistic mechanics for this problem.

EXERCISE 20.3 Find the kinetic energy and the velocity of an electron that starts from rest and falls through a potential difference of 10^5 V. Is it necessary to use the relativistic treatment? (Note: for the electron $m_0c^2 = 0.51$ MeV.) [*Ans.:* $K = 0.1$ MeV $= 1.6 \times 10^{-14}$ J, $v = 1.64 \times 10^8$ m/s, yes.]

20.5. Electrostatic Generators

In order to understand the working of an electrostatic generator—the Van de Graaff generator—refer to Figure 20.6(a). Suppose a big metal shell of radius R carries a charge $+Q$. A small sphere of radius r carrying charge $+q$ is brought into the big sphere. The small sphere will induce $-q$ charge on the inner surface (the surface nearest to the charge) and $+q$ charge on the outer surface (the surface farthest from the charge) of the bigger sphere. If the small sphere is moved around, the electric field lines may rearrange, but the total charge distributed remains unchanged, as shown in Figure 20.6(b). Now suppose that the small sphere is connected to the big sphere by means of a conducting rod as shown in Figure 20.6(c). The positive charge on the small sphere cancels the induced negative charge on the bigger sphere. Thus, the net effect is that the charge on the bigger sphere is increased from Q to $Q + q$. The process above can be repeated and the charge on the bigger sphere can be increased more and more by bringing a small charged sphere repeatedly inside. Also, as the charge on the sphere of radius R increases, its potential also increases.

An electrostatic generator that makes use of this principle was designed by Robert J. Van de Graaff in 1930 and is called a Van de Graaff generator. Instead of charged bodies being inserted into a hollow conductor one after another, charge is carried in continuously by means of a belt B made of insulated fabric running between the two rollers R and R', as shown in Figure 20.7.

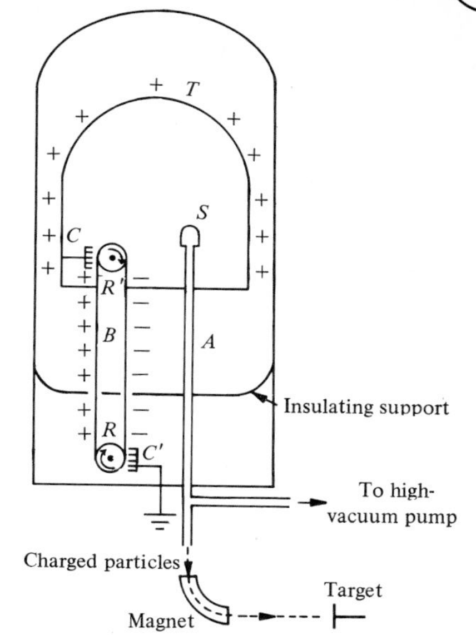

Figure 20.7. *A typical Van de Graaff accelerator.*

The roller R is run by hand or by a small electric motor. The rollers R and R' are covered with different materials so that when the belt B makes contact with R′, it acquires a negative charge, and when in contact with R, it acquires positive charge. The belt moving upward on the left carries positive charge. At the sharp points C the field is very high, and causes ionization of the air between the belt and the sharp points, thereby making it conducting. This causes the flow of the positive charge from the belt to the sphere T.

As the belt moves downward on the right, it gains negative charge by rubbing against the roller R'. Negative charge from the belt is removed by a sharp point C', which is connected to the ground. Removing negative charge is equivalent to adding positive charge. The net effect is that both sides of the belt contribute to the increase of the positive charge on the sphere T.

The potential on sphere T keeps on increasing as the belt keeps on moving. A limit is reached when the insulation breaks down or if the charge is drained off through a different path. A tube A connected to the high-potential terminal T may be used for accelerating charged particles (from source S) as they fall from a high potential at the top to the ground potential at the bottom of the tube. A voltage up to 10 MV can be achieved with a modified form of this design.

20.6. Capacitor and Capacitance

A device used for storing electric charge is called a *capacitor* or *condenser*. Usually, it consists of two neighboring conductors that have equal and opposite charges. When we say that the charge of a capacitor is Q, it means that one conductor has $+Q$ charge while the other has $-Q$. Figures 20.8 and 20.9 illustrate two typical situations. The two conductors of a capacitor are connected to the two terminals of a battery where the conductors acquire equal and opposite charges. A typical capacitor is capable of storing a large amount of charge (hence, a large amount of energy) in a small space. The capability of any such arrangement to store charge is called its *capacitance*, C. The space in the capacitor may be filled with air or any other insulator. In this section we shall assume that it is filled with air.

It is found experimentally that for a capacitor the potential difference V between the two conductors is directly proportional to the charge Q. The constant ratio of the charge Q to the potential difference V is defined as the *capacitance* C of a capacitor; that is,

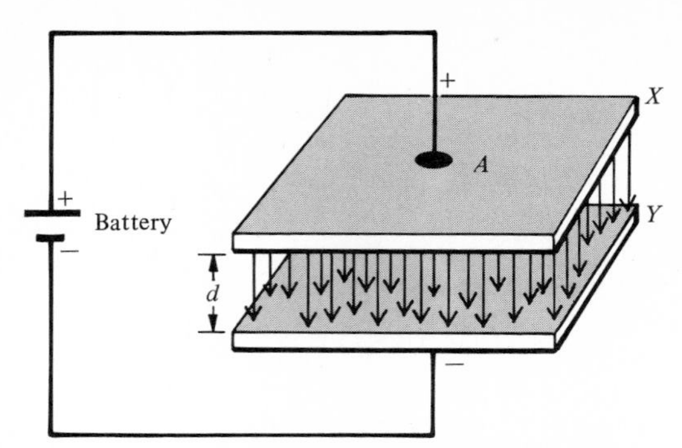

FIGURE 20.8. *System of two conductors forming a parallel-plate capacitor.*

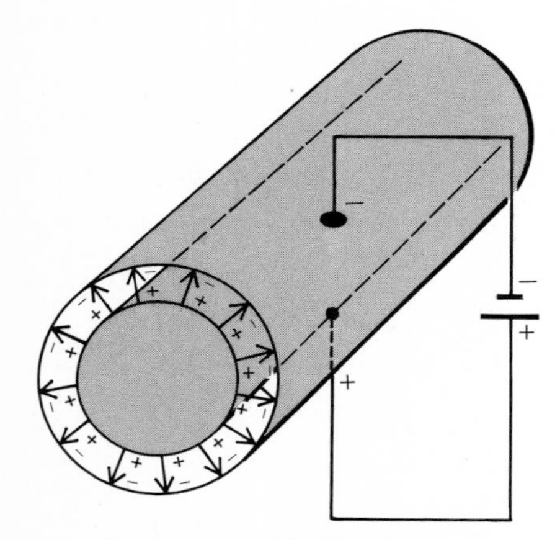

FIGURE 20.9. *Cylindrical capacitor.*

$$\boxed{C = \frac{Q}{V}} \tag{20.13}$$

In SI units Q is in coulombs, V in volts; hence, C will be in coulombs per volt. This unit of C, coulomb/volt, has been named the *farad* (F), in honor of Michael Faraday. This is a very large unit; hence, smaller units called a microfarad (μF), where 1 μF $= 10^{-6}$ F, and a picofarad (pF), where 1 pF $= 10^{-12}$ F $= 10^{-6}$ μF, are commonly used.

A capacitor of fixed capacitance is represented by the symbols shown in Figures 20.10(a) and (b); a capacitor of variable capacitance is shown in Figure 20.10(c). Let us now consider capacitors with some special configurations and calculate their capacitances.

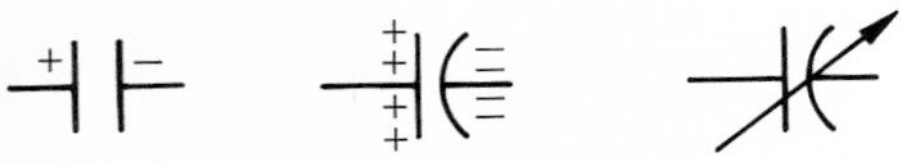

FIGURE 20.10. *Symbols used for representing capacitors of (a) and (b) fixed capacitance and (c) variable capacitance.*

First, consider the parallel-plate capacitor shown in Figure 20.8. Let $+Q$ be the charge on plate X and $-Q$ on plate Y. The area of each plate is A while the separation between the two plates is d. The electric field that is confined in the space between the two plates is uniform and is shown in Figure 20.8. The electric field lines start from X and end on Y. Note that the process of *charging* a capacitor means supplying equal and opposite charges to the two plates. This can be achieved by connecting the plates to the opposite poles of a battery. This is also equivalent to removing a $+Q$ charge from plate Y and supplying it to plate X, thus leaving plate Y with a $-Q$ charge or alternatively taking charge $-Q$ from plate X to plate Y. After the capacitor is charged, suppose that the two plates are connected by means of a wire to provide a conducting path for the flow of charge. The plates of the capacitor lose charge and become neutral. The process is called *discharging* and the capacitor is said to be *discharged*.

The capacitance C of a parallel-plate capacitor in air or in vacuum is experimentally found to be directly proportional to the area A of either plate and inversely proportional to the separation distance d between the plates. That is,

$$C \propto \frac{A}{d}$$

or

$$C = \epsilon_0 \frac{A}{d} \tag{20.14}$$

where ϵ_0 is the permittivity of free space and its value in SI units, as discussed in Chapter 19, is $\epsilon_0 = 8.85 \times 10^{-12}\ \mathrm{C^2/N\text{-}m^2}$, where C stands for coulomb. Since $k = 1/4\pi\epsilon_0 = 9 \times 10^9\ \mathrm{N\text{-}m^2/C^2}$, we may write Equation 20.14 as

$$C = \frac{1}{4\pi k}\frac{A}{d} \tag{20.15}$$

The expression for C given in Equation 20.14 can be derived theoretically, but we shall not go into the derivation. It must be emphasized, however, that capacity is a function of the geometry and material of the capacitor.

Variable capacitors are commonly used in tuning circuits of radio and TV receivers. A typical one is shown in Figure 20.11. It consists of a number of fixed metal plates connected together to form one conductor while a second set

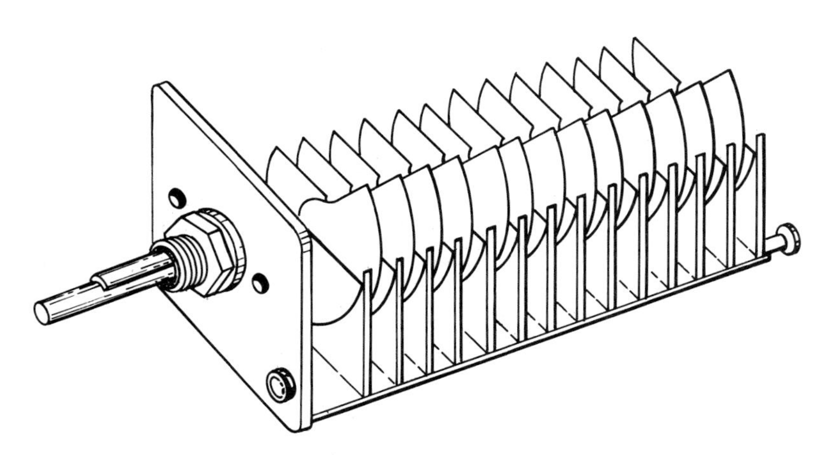

Figure 20.11. *Variable capacitor.*

of plates, which are movable, are connected together to form the other conductor. The two sets of plates are interleaved as shown and the effective area can be changed by rotating the shaft connected to the movable set of plates. Thus, the effective value of C is changed. Besides parallel-plate capacitors, spherical and cylindrical capacitors are also used commonly in laboratories.

Commercial-type capacitors are much more compact, and a few of them are shown in Figure 20.12. Figure 20.12(a) shows a tubular capacitor consisting of plates of thin metal foil separated by a sheet of paper or plastic film and rolled up into a small package. The whole unit is then encapsulated in plastic

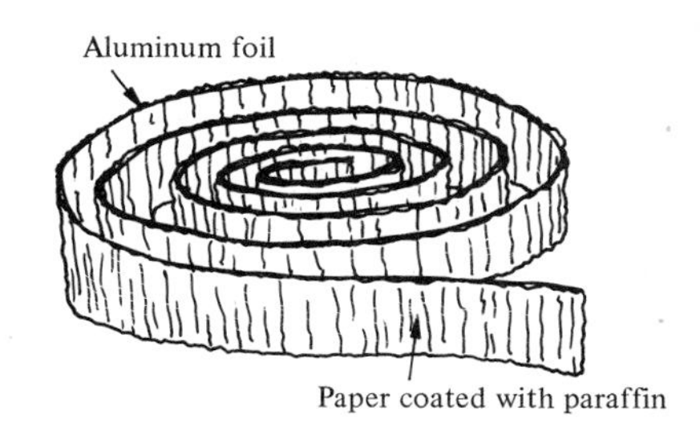

(a) Tubular capacitor

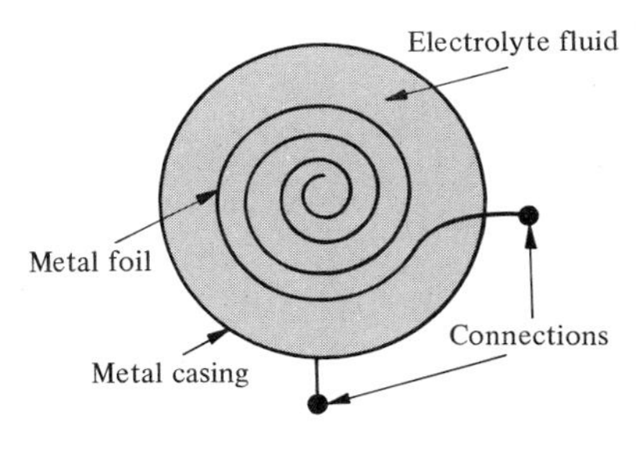

(b) Electrolyte capacitor

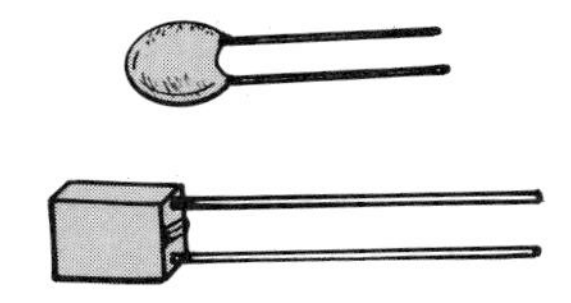

(c) Miniature capacitors

Figure 20.12. *Commercial-type capacitors: (a) tubular, (b) electrolytic, and (c) miniature.*

film. Figure 20.12(b) shows another compact capacitor, called an *electrolytic capacitor.*

20.7. Dielectrics

In the previous discussion we have assumed that the space between the plates of a capacitor is filled with air or is a vacuum, essentially making no difference in the magnitude of the capacitance. Suppose that a nonconducting material such as glass, wood, paper, wax-impregnated paper, or plastic is brought in between the plates. The effect of inserting such an insulator between the plates of the capacitor is to increase the capacitance of the capacitor. Such insulators are called *dielectric materials* or simply *dielectrics.* Let us demonstrate the effect of dielectrics by the following experiment.

Consider a charged capacitor whose plates are connected to a voltmeter (or an electroscope) as shown in Figure 20.13(a). The deflection of the meter is a measure of the potential difference between the plates. Now insert a dielectric

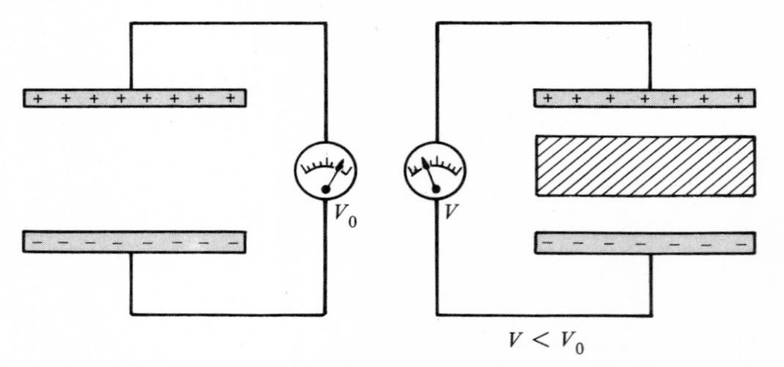

FIGURE 20.13. *Effect of a dielectric on the capacitance of a capacitor.*

material between the plates of the capacitor. As shown in Figure 20.13(b), the reading drops, indicating a decrease in the potential difference between the plates. From the definition $C = Q/V$, since V decreases while Q remains constant, the value of C increases. That the charge Q remains unaltered can be shown by removing the dielectric. We notice that the needle will come to the original position, thereby indicating the initial values of V and C.

Let C_{med} be the capacitance of a capacitor when the space between the plates is filled with a dielectric, and C_{vac} when the plates are separated by vacuum or air. *The ratio of Capacitance with and without the insulator is called the dielectric constant K of the insulator.* That is,

$$\boxed{K = \frac{C_{\text{med}}}{C_{\text{vac}}}} \tag{20.16}$$

The values of K are listed in Table 20.1. Thus, for a parallel-plate capacitor with air between the plates, $C = \epsilon_0 A/d$, while with a dielectric between the plates its value is

$$C = K\epsilon_0 \frac{A}{d} = \epsilon \frac{A}{d} \tag{20.17}$$

Table 20.1
Dielectric Constant K and Dielectric Strength

Material	K	Dielectric Strength V/m
Air (1 atm)	1.0006	3×10^6
Ammonia (liquid)	22–25	
Bakelite	5–18	
Benzene	2.284	
Germanium	16	
Glass	4.8–10	30×10^6
Mica	3–7.5	200×10^6
Paraffined paper	2	40×10^6
Plexiglas	3.40	
Rubber	2.94	21×10^6
Teflon	2.1	
Transformer oil	2.1	$(5\text{–}15) \times 10^6$
Vacuum	1	
Water (distilled)	78.5	3×10^6

where ϵ, called the *permittivity of the dielectric,* is given by

$$\boxed{\epsilon = K\epsilon_0} \tag{20.18}$$

The presence of a dielectric not only increases the capacitance of the capacitor, but also makes it possible to stand much higher voltages. It increases the voltage value at which the breakdown will take place, that is, the value of the voltage beyond which the space between the two plates becomes conducting, hence shorting and discharging the capacitor. The breakdown voltages or dielectric strengths for different materials are listed in Table 20.1.

Let us now explain how the presence of a dielectric increases the capacitance. This effect is due to the polarization of the dielectric material. The dielectric consists of atoms and molecules that are electrically neutral on the average. When such material is placed in an electric field, the electrons are attracted by the positively charged plate of the capacitor and the nuclei by the negatively charged plate. The electrons in the insulator are not free, but it is possible that the electrons and the nuclei can undergo slight displacements. This creates an electric field E_i that is opposite the original field E, as shown in Figures 20.14(a) and (b). This reduces the effective charge on the conductor without altering the original charge. The net result is that the total effective field is decreased as shown in Figure 20.14(c). Thus, from the expression $E = V/d$, V will also be decreased. If the capacitor were connected to a source or battery, extra charges would flow to the plates, so that the initial value of the voltage is restored. (For comparison the case of a metal plate between the plates of a capacitor is shown in Figure 20.14(d). The field inside the metal plate is zero.)

When the center of the charges of the electron cloud and that of the nuclei in an atom or a molecule do not coincide, the material is said to be *polarized.*

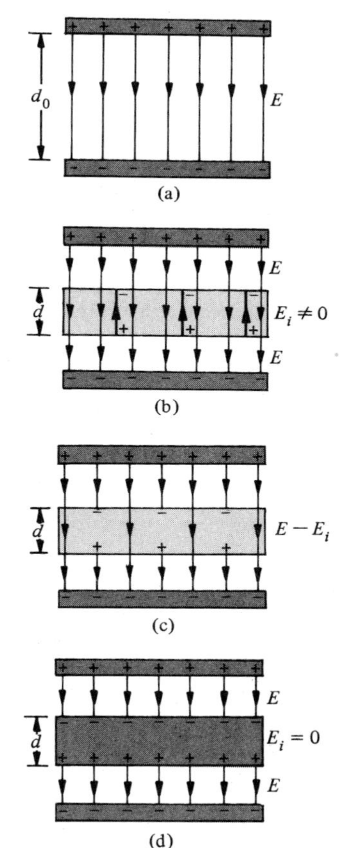

Figure 20.14. *The electric field in the space between two parallel plates of a capacitor when the space has: (a) air; (b) and (c) an insulator, and (d) a metal plate. (b) shows that the field lines E_i are opposite to E, thereby reducing the total effective field as shown in (c). The field inside the metal plate is zero as shown in (d).*

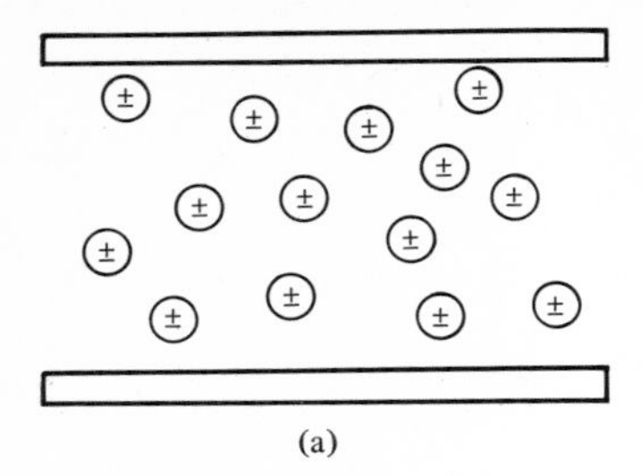

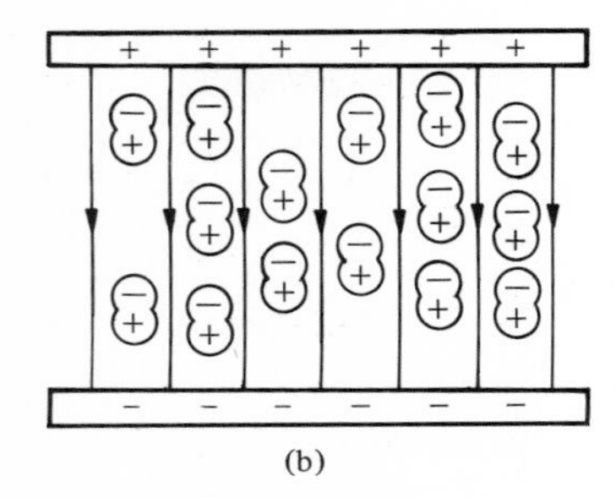

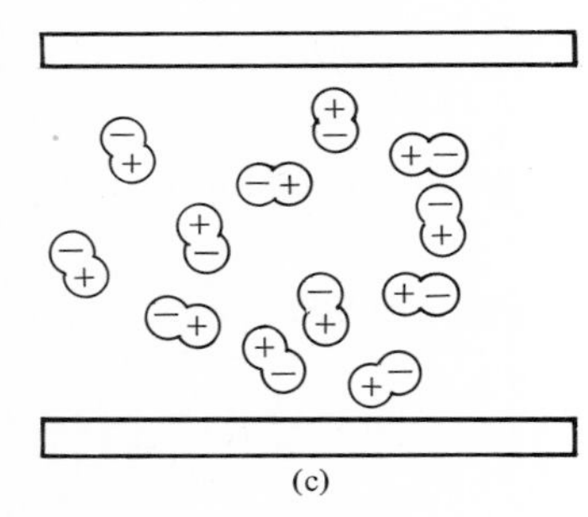

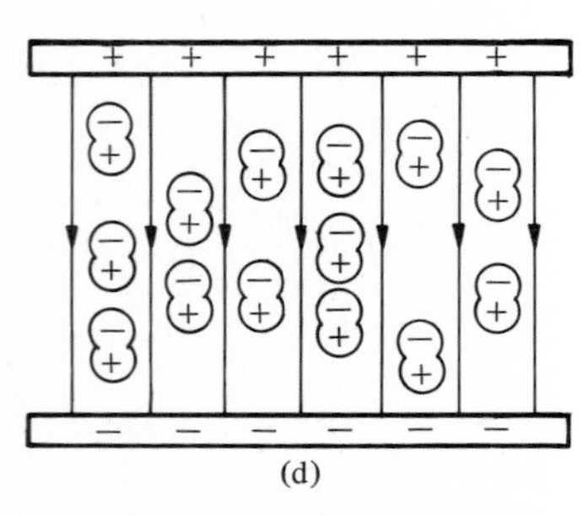

FIGURE 20.15. *Polarization produced by an externally applied electric field: (a) and (b) for symmetrical charge distribution and (c) and (d) for randomly oriented charges.*

This polarization arises from two different situations. If the charge distribution is symmetrical, as in Figure 20.15(a), the presence of an electric field simply displaces the charges, as shown in Figure 20.15(b). On the other hand, if the molecules already have effective negative and positive charges displaced permanently but are randomly oriented, the presence of the electric field aligns them as shown in Figures 20.15(c) and (d). The net effect in both cases is to form layers of oppositely charged surfaces on the plates of the capacitor. These two new layers produce an electric field in the direction opposite to the original field. This decreases the net value of E, hence that of V, thereby increasing C.

20.8. Capacitors in Series and in Parallel

There are many situations in electrical circuits where more than one capacitor is used. Two of the most common methods of combining them are (1) capacitors in series, and (2) capacitors in parallel. We wish to find an equivalent capacitor that has the same capacitance as the combination.

Capacitors in Series

Figure 20.16 shows three capacitors of capacitance C_1, C_2, and C_3 connected in series. One characteristic of a series combination (as explained below) is that *when the capacitors are charged in series, each capacitor acquires the same charge.*

Suppose that we apply a potential difference of V between the points A and B. Say that the left plate of capacitor C_1 acquires a $+Q$ charge, so that the right plate will have a charge of $-Q$. This $-Q$ charge must come from the left plate of capacitor C_2 by a transfer of electrons. This makes the left plate of capacitor C_2 positive with a charge of $+Q$. This process continues as shown in Figure 20.16.

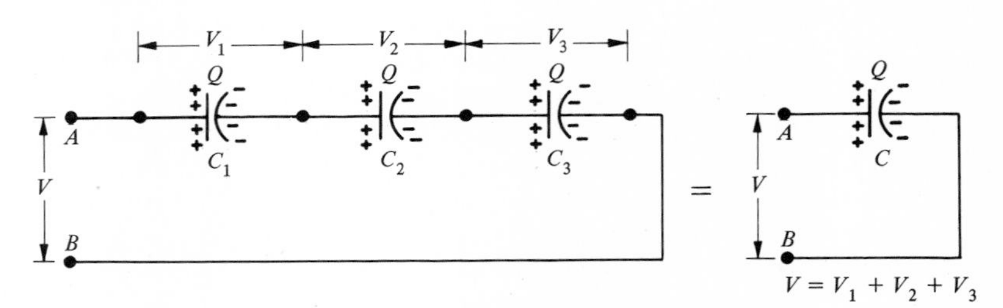

FIGURE 20.16. *Capacitors in series.*

The potential difference V must be equal to the sum of the potential drops V_1, V_2, and V_3 across the three capacitors. Thus,

$$V = V_1 + V_2 + V_3$$

where $V_1 = Q/C_1$, $V_2 = Q/C_2$, and $V_3 = Q/C_3$. Thus,

$$V = \frac{Q}{C_1} + \frac{Q}{C_2} + \frac{Q}{C_3}$$

Let C_{eq}^s be the equivalent of the three capacitors in series. This, too, must have the same charge Q as the other capacitors. Hence,

$$V = \frac{Q}{C^s_{eq}}$$

Comparing the equations above, we get

$$\frac{1}{C^s_{eq}} = \frac{1}{C_1} + \frac{1}{C_2} + \frac{1}{C_3}$$

or, in general, for more than three capacitors

$$\boxed{\frac{1}{C^s_{eq}} = \frac{1}{C_1} + \frac{1}{C_2} + \frac{1}{C_3} + \cdots} \qquad (20.19)$$

One of the interesting features of the series combination is that the applied potential difference is shared among all the capacitors. This makes it possible to apply a voltage in excess of any one single capacitor rating.

Capacitors in Parallel

Figure 20.17 shows three capacitors C_1, C_2, and C_3 arranged in a parallel combination. When a potential difference of V is applied across A and B, all

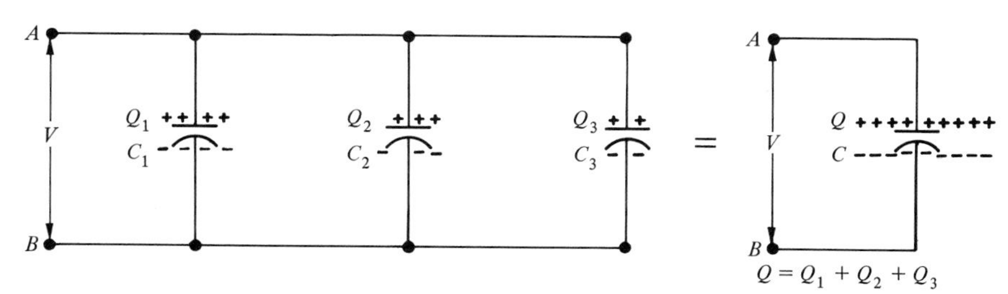

FIGURE 20.17. *Capacitors in parallel.*

capacitors have equal potential differences V across them but have different amounts of charge Q_1, Q_2, and Q_3 (because C_1, C_2, and C_3 are different). If Q is the total charge, we can write

$$Q = Q_1 + Q_2 + Q_3$$

where $Q_1 = C_1V$, $Q_2 = C_2V$ and $Q_3 = C_3V$. Thus, substituting above yields

$$Q = (C_1 + C_2 + C_3)V$$

Let C^p_{eq} be the equivalent capacitance of this combination, as shown in Figure 20.17. Thus,

$$Q = C^p_{eq}\, V$$

Comparing the two equations above gives

$$C^p_{eq} = C_1 + C_2 + C_3$$

or, in general, for more than three capacitors,

$$C_{eq}^{p} = C_1 + C_2 + C_3 + \cdots \tag{20.20}$$

The equivalent capacitor has a higher value than any one of the single capacitor when in a parallel combination. This is contrary to the series combination, where the equivalent capacitor has a lower value of capacitance than any single capacitor.

EXAMPLE 20.4 Two capacitors $C_1 = 6\,\mu\text{F}$ and $C_2 = 12\,\mu\text{F}$ are first connected (a) in series and then (b) in parallel. In each case the external source of voltage is 1000 V. Calculate in each case: the total capacitance, the potential drop across each capacitor, and the charge on each capacitor.

(a) Let us first consider the capacitors in series. In this case $C_{eq} = C$ is

$$\frac{1}{C} = \frac{1}{C_1} + \frac{1}{C_2} = \frac{1}{6\,\mu\text{F}} + \frac{1}{12\,\mu\text{F}} = \frac{1}{4\,\mu\text{F}}$$

or

$$C = 4\,\mu\text{F}$$

When the capacitors are in series, the charge on each capacitor is the same and is given by

$$Q = CV = (4\,\mu\text{F})(1000\text{ V}) = 4000\,\mu\text{C} = 0.004\text{ C}$$

The potential difference across each capacitor is different and is given by

$$V_1 = \frac{Q}{C_1} = \frac{4000\,\mu\text{C}}{6\,\mu\text{F}} = 666.7\ V$$

$$V_2 = \frac{Q}{C_2} = \frac{4000\,\mu\text{C}}{12\,\mu\text{F}} = 333.3\text{ V}$$

(b) When the two capacitors are connected in parallel, the total equivalent capacitance $C_{eq} = C$ is given by

$$C = C_1 + C_2 = 6\,\mu\text{F} + 12\,\mu\text{F} = 18\,\mu\text{F}.$$

The potential difference across each one of them is the same; that is, 1000 V, but the charges on the capacitors are different. Thus,

$$Q_1 = C_1V = 6\,\mu\text{F}(1000\text{ V}) = 6000\,\mu\text{C}$$
$$Q_2 = C_2V = 12\,\mu\text{F}(1000\text{ V}) = 12000\,\mu\text{C}$$

EXERCISE 20.4 Repeat Example 20.4 for the case in which there are three capacitors $C_1 = 5\,\mu\text{F}$, $C_2 = 10\,\mu\text{F}$, and $C_3 = 20\,\mu\text{F}$. [*Ans.*: (*a*) $C = 2.86\,\mu\text{F}$, $V_1 = 572$ V, $V_2 = 286$ V, $V_3 = 143$ V, $Q_1 = Q_2 = Q_3 = 2860\,\mu\text{C}$; (b) $C = 35\,\mu\text{F}$, $V_1 = V_2 = V_3 = 1000$ V, $Q_1 = 5000\,\mu\text{C}$, $Q_2 = 10{,}000\,\mu\text{C}$, $Q_3 = 20{,}000\,\mu\text{C}$.]

20.9. Energy Stored in a Capacitor and Energy Density

The process of charging a capacitor is equivalent to transferring charge from one plate of the capacitor to the other. At any stage of the charging there is a potential difference between the plates. Thus, to transfer charges, work must be done due to this potential difference. This work done is stored as energy in the capacitor, which may be calculated as follows.

Let ΔW be the work done in transferring a charge of ΔQ coulombs against a potential difference V so that $\Delta W = V\,\Delta Q$. Suppose that the final charge on the capacitor is Q_0, while the initial potential difference is zero and the final V_0. Thus, the average potential difference during the process of transferring charge Q_0 is $\frac{1}{2}(0 + V_0) = \frac{1}{2}V_0$. Thus, the total work done is $W_0 = \frac{1}{2}V_0Q_0$.

We may arrive at the same result by a slightly different approach. Suppose at any instant during the process of charging the capacitor, the charge is Q and the potential difference is V. If C is the capacitance, $Q = CV$. As V increases, so does Q. The plot of V versus Q is a straight line, as shown in Figure 20.18. Thus, the work done in transferring charge ΔQ against a potential difference V is $\Delta W = V\,\Delta Q$, and this is equal to the area of the crosshatched strip. The total work done, W, for the final values of V_0 and Q_0 is equal to the shaded area; that is,

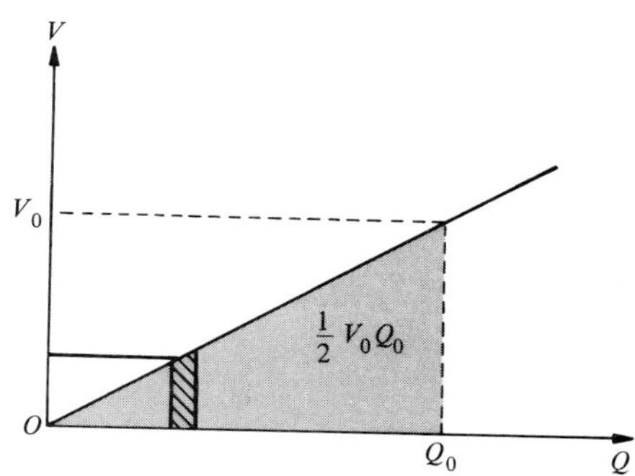

FIGURE 20.18. *The work done in charging a capacitor is equal to the area under the V-versus-Q plot.*

$$W_0 = \tfrac{1}{2}V_0Q_0$$

Using the relation $Q_0 = CV_0$ and dropping the zero subscripts, we may write

$$\boxed{W = \frac{1}{2}VQ = \frac{1}{2}CV^2 = \frac{1}{2}\frac{Q^2}{C}} \qquad (20.21)$$

A stretched spring is a mechanical analog of an electrically charged capacitor. When a spring of spring constant k is stretched through a distance x, the work done that is stored in the form of potential energy is $W = \frac{1}{2}kx^2$. Comparing this equation with $W = \frac{1}{2}(Q^2/C)$ yields $1/C$ to be equivalent to k and Q to be equivalent to x. Also a plot of V versus Q is analogous to the plot of F versus x as shown in Figure 6.10.

Of course, the work done in charging the capacitor is stored as energy. We may look upon this energy as the potential energy of the charges, or as the energy stored in the form of a localized electric field in the space between the plates of the capacitor. It is usually convenient and useful to take the second alternative. Let us calculate this stored energy in the special case of a parallel-plate capacitor in vacuum for which

$$C = \epsilon_0\frac{A}{d} \qquad \text{and} \qquad E_0 = \frac{V}{d}$$

Substituting for C and $V = E_0d$ in $W = \frac{1}{2}CV^2$, we get

$$W = \frac{1}{2}CV^2 = \frac{1}{2}\left(\epsilon_0\frac{A}{d}\right)(E_0d)^2 = \frac{1}{2}\epsilon_0E_0^2Ad \qquad (20.22)$$

where Ad is the volume of the space between the two plates. Thus, we define the *energy density* u_E to be energy per unit volume given by

$$u_E = \frac{\text{total energy}}{\text{volume}} = \frac{W}{Ad} \tag{20.23}$$

Substituting for W from Equation 20.22, we get

$$\boxed{u_E = \tfrac{1}{2}\epsilon_0 E_0^2} \tag{20.24}$$

where we have assumed that the space between the two plates is vacuum or air. If it were filled with a dielectric material of dielectric constant K $(= \epsilon/\epsilon_0)$, in such cases the expression above takes the form

$$u'_E = \tfrac{1}{2}K\epsilon_0 E^2 = \tfrac{1}{2}\epsilon E^2 \tag{20.25}$$

EXAMPLE 20.5 The parallel plates of a capacitor are 5 cm by 4 cm each and are separated by a distance of 0.5 cm. To start with the potential difference between the plates is 2000 V, but when the space between the plates is filled with a dielectric, the potential difference drops to 1000 V. Calculate the following quantities: (a) the initial capacitance C_0; (b) the initial charge Q_0 on each plate; (c) the final capacitance C when filled with a dielectric; (d) the dielectric constant K of the dielectric; (e) the permittivity ϵ of the dielectric; (f) the initial electric field E_0; (g) the final electric field E; (h) the initial energy stored in the capacitor; and (i) the final energy stored in the capacitor.

(a) $A = 5 \times 10^{-2}\text{ m} \times 4 \times 10^{-2}\text{ m} = 20 \times 10^{-4}\text{ m}^2$, $d = 0.5 \times 10^{-2}$ m.

$$C_0 = \epsilon_0 \frac{A}{d} = \left(8.85 \times 10^{-12}\frac{\text{C}^2}{\text{N-m}^2}\right)\frac{20 \times 10^{-4}\text{ m}^2}{0.50 \times 10^{-2}\text{ m}} = 3.54 \times 10^{-12}\text{ F}$$
$$= 3.54\text{ pF}$$

(b) $$Q_0 = C_0 V_0 = (3.54 \times 10^{-12}\text{ F})(2000\text{ V}) = 7.08 \times 10^{-9}\text{ C}$$

(c) After the space is filled with a dielectric, the charge on the plates is still the same; that is, $Q = Q_0$, while the potential difference drops to 1000 V.

$$C = \frac{Q}{V} = \frac{7.08 \times 10^{-9}\text{ C}}{1000\text{ V}} = 7.08 \times 10^{-12}\text{ F} = 7.08\text{ pF}$$

(d) $$K = \frac{C}{C_0} = \frac{7.08\text{ pF}}{3.54\text{ pF}} = 2 \quad \text{or} \quad K = \frac{V_0}{V} = \frac{2000\text{ V}}{1000\text{ V}} = 2$$

(e) $$\epsilon = K\epsilon_0 = 2(8.85 \times 10^{-12}\text{ C}^2/\text{N-m}^2) = 17.70 \times 10^{-12}\text{ C}^2/\text{N-m}^2$$

(f) $$E_0 = \frac{V_0}{d} = \frac{2000\text{ V}}{0.5 \times 10^{-2}\text{ m}} = 4 \times 10^5\text{ V/m}$$

(g) The electric field after the space is filled with a dielectric

$$E = \frac{V}{d} = \frac{1000\text{ V}}{0.5 \times 10^{-2}\text{ m}} = 2 \times 10^5\text{ V/m}$$

(h) $u_E = \frac{1}{2}\epsilon_0 E_0^2 = \frac{1}{2}\left(8.85 \times 10^{-12} \frac{\text{C}^2}{\text{N-m}^2}\right)(4 \times 10^5 \text{ V/m})^2$
$= 0.71 \text{ J/m}^3$

(i) $u'_E = \frac{1}{2}\ \epsilon E^2 = \frac{1}{2}\left(17.70 \times 10^{-12} \frac{\text{C}^2}{\text{N-m}^2}\right)(2 \times 10^5 \text{ V/m})^2$
$= 0.35 \text{ J/m}^3$

Note that u'_E is not equal to u_E. How do you account for this apparent loss of energy?

Exercise 20.5 Keeping the separation between the plates the same as in Example 20.5 and using the same dielectric and the charge, what should be the area of the plates so that the electric energy stored per unit volume is 1.0 J/m^3? [*Ans.*: $A = 12 \text{ cm}^2$.]

SUMMARY

The *change in potential energy* ΔU of charge q_0 is equal to the work done by the force in carrying the charge from one point to another, $\Delta U = \Delta W$. The *potential difference* ΔV between two points A and B in an electric field is defined as the work done in carrying a unit positive charge from A and B while keeping the charge in equilibrium; $\Delta V = V_B - V_A = W_{AB}/q_0$ and $\Delta U = q_0\,\Delta V$. A potential difference of 1 *volt* is said to exist between two points if 1 joule of work is required to move a positive charge of 1 coulomb from one point to the other while keeping the charge in equilibrium; 1 volt = 1 joule/1 coulomb.

The *electric potential* at any point in an electric field is equal to the work done in bringing a unit positive charge from infinity to the point under consideration, $V = W/q_0$. The conservation of energy requires that $U_A + K_A = U_B + K_B$ or $q(V_A - V_B) = K_B - K_A$.

The quantity $\Delta V/\Delta s$ is called the *potential gradient.* In an electric field the component of *electric intensity* in any direction is equal to the negative of the potential gradient in that direction, $E = -\Delta V/\Delta s$. The unit of E, N/C, is equivalent to V/m. The potential at a point at a distance r from a charge q is $V(r) = (1/4\pi\epsilon_0)(q/r)$ and $V = \Sigma V_i$. The lines joining points of equal potential are equipotential lines with similar meaning for equipotential surfaces and volumes.

One *electron volt* (eV) is the amount of energy acquired or lost by an electron as it traverses a potential difference of 1 volt; $1 \text{ eV} = 1.60 \times 10^{-19}$ J.

A device used for storing electric charge is called a *capacitor* or *condenser.* The capacitance C is equal to the ratio Q/V. The unit of capacitance is the *farad* (F) = coulomb/volt. For a parallel-plate capacitor, $C = \epsilon_0 A/d$. The ratio of capacitance with and without the insulator is called the *dielectric constant* K of the insulator, $K = C_{\text{med}}/C_{\text{vac}}$ and $\epsilon = K\epsilon_0$.

For *capacitors in series* $1/C_{\text{eq}}^{s} = 1/C_1 + 1/C_2 + 1/C_3 + \cdots$, and in *parallel,* $C_{\text{eq}}^{p} = C_1 + C_2 + C_3 + \cdots$.

The work done in charging a capacitor is $W = \frac{1}{2}VQ = \frac{1}{2}CV^2 = \frac{1}{2}Q^2/C$. The *energy density* u_E is defined as the energy per unit volume $u_E = \frac{1}{2}\epsilon_0 E_0^2$ in air or vacuum. In the presence of a dielectric $u'_E = \frac{1}{2}\epsilon E^2 = \frac{1}{2}K\epsilon_0 E^2$.

QUESTIONS

1. Write electrical field statements analogous to the following in a gravitational field: (a) water flows from a higher level to a lower level; (b) water always maintains its level; (c) the total mass is conserved; (d) when a body falls through a height h, it loses potential energy mgh and gains kinetic energy.
2. Draw equipotential lines for the following configurations: (a) two equal and opposite charges separated by a distance; (b) two equal and similar charges separated by a distance; and (c) a positive charge at a distance d from a negatively charged flat plate.
3. If instead of zero volts the surface of the earth was assigned a potential of 100 V, how will it affect Equation 20.10 for the potential at a point due to a point charge?
4. Suppose that you follow an electric field line due to a positive point charge. Do E and the potential increase or decrease?
5. Under what conditions at a given point will the following situations occur: (a) $E = 0$, $V \neq 0$; and (b) $V = 0$, $E \neq 0$?
6. How can you tell which plate of a parallel-plate capacitor is charged positively and which is charged negatively?
7. Consider a parallel-plate capacitor. Draw the electric field lines and potential lines if the dielectric between the plates is (a) air; (b) Teflon; and (c) water.
8. What changes will take place in the potential between the plates, the charge on the plate, and the capacitance of the system if a dielectric plate is inserted between the plates of a capacitor.
9. Name some everyday examples in which a capacitor is used for storing charge, hence energy.
10. Is the work done on the system or by the system when a metal plate is (a) brought between; or (b) withdrawn from the parallel plates of a capacitor?
11. Suppose that you have two capacitors of equal value. Should you connect them in series or in parallel so that a maximum energy will be stored in them at constant V?

PROBLEMS

1. Calculate the potential difference between two points if 2 J of work must be done to move a 4 mC (1 mC = 1 millicoulomb = 10^{-3} C) charge from one point to another.
2. If 8 J of work is done in moving a positive charge of 0.1 C from A to B, what is the potential difference between the two points? Which point is at a higher potential?
3. A charge of $+10\ \mu$C is moved from a point where the potential is -10 V to a point where the potential is $+100$ V. Calculate the work in J and eV.
4. Two parallel plates are 4 cm apart and are at potentials 300 V and -100 V, respectively. (a) What is the magnitude and direction of the electric field between the two plates? (b) What are the potentials at distances of (1) 1 cm, (2) 2 cm, and (3) 3 cm from the -100 V plate?
5. The electric field between two parallel plates is 2000 V/m and is directed

downward. If the potential at some point P is 100 V, what will be the potential at a point (a) 2 cm above point P; and (b) 2 cm below point P?

6. Two horizontal parallel plates are separated by 4 mm and are at a potential difference of 2800 V. Calculate the potential gradient and the electric intensity.
7. To move an electron from point A to B, 8×10^{-15} J of work must be done. Calculate the potential difference between the points. What is the separation between the two points if the magnitude of the electric field strength is 50,000 V/m? What is the direction of the field? Which of the two points is at a higher potential?
8. A potential difference between two plates separated by a distance d is 5000 V. The maximum field that can exist between the plates is 3×10^6 N/C before the air ionizes, hence becomes conducting. What is the minimum distance between the two plates?
9. Two charges $+10\ \mu C$ and $-5\ \mu C$ are 10 cm apart. Calculate the potential at a point P at a distance of 1 m from either charge. Also, calculate the potential energy of a $+2$-μC charge placed at point P.
10. Calculate the amount of work done in moving an electron in the field of a 50-μC charge from a point 6 m to a point 1 m from the charge. What is the potential difference between the two points?
11. A unit positive charge has to be brought from infinity to a point between two charges, 20 μC and 10 μC separated by a distance of 50 cm. How much work will be required?
12. Charges of 6 μC, 8 μC, and $-10\ \mu C$ are placed at three corners of a square of 10-cm sides. Calculate the potential at the fourth corner.
13. A charge $q_1 = +4\ \mu C$ is placed at the origin, a charge $q_2 = +4\ \mu C$ is placed at $x = 20$ cm, and a charge $-4\ \mu C$ is placed at $y = 20$ cm. Calculate the potentials at point A located at $x = 10$ cm, $y = 0$ and at point B located at $x = 0$, $y = 10$ cm. What is the potential difference between A and B? Calculate the work in carrying a charge of $+1\ \mu C$ from A to B.
14. Consider point charges $+q$ at $x = a$ and $+q$ at $x = -a$. Show that the potential at any point along the Y-axis is given by

$$V = k\frac{2q}{\sqrt{y^2 + a^2}}$$

What is the potential at the origin?

15. Consider four charges, each $+q$, placed at the four corners of a square of side a. (a) Calculate the potential at the center of the square, (b) Repeat part (a) if the charges at the second and fourth corners are replaced by $-q$ each.
16. Calculate the speed of (a) an electron; and (b) a proton if each is accelerated through a potential difference of 400 V.
17. A particle accelerator is used for accelerating positively charged particles through a potential difference of 10 MV ($= 10 \times 10^6$ V). Calculate the kinetic energy in eV and joules and the speed of (a) protons; and (b) α-particles.
18. The potential at point A is 600 V and that at B is zero. An electron is ejected with a kinetic energy of 1000 eV from point A. What will be its kinetic energy when it reaches B? What should be the potential at B so

that when the electron reaches there, it has zero kinetic energy? What happens to the electron after that?

19. The electrons in an electron beam in a TV tube are accelerated starting from rest through a potential difference of 20,000 V. Calculate their energy in (a) eV; and (b) in joules. Also calculate their final speed using (c) classical mechanics; and (d) relativistic mechanics.

20. A certain potential difference is maintained between two parallel plates A and B. An electron leaves plate A with an initial speed of 6×10^6 m/s and when it reaches plate B its speed is 2×10^6 m/s. What is the potential difference between the two plates, and which point is at a higher potential?

21. Through what potential difference should (a) an electron; and (b) a proton be accelerated so that their gain in kinetic energy is 1 percent of their rest energy? What are their speeds?

22. An electron is accelerated through a potential difference of 10^9 V. Calculate the final speed of the electron (a) from the classical expression; and (b) from the relativistic expression. What is the kinetic energy and the total energy of the electron?

23. Two parallel plates separated by a distance of 5 cm are maintained at a potential difference of 2000 V. An electron is released from the negative plate at the same time a proton is released from the positive plate. Calculate (a) their velocities; and (b) their kinetic energies when they strike the opposite plate. (c) At what distance from the plates do they pass each other?

24. A parallel-plate capacitor in air has a capacitance of 0.01 μF while the charge on each plate is 4 μC. Calculate the potential difference between the two plates.

25. A capacitor of 50 μF has a charge of 0.1 C on each plate. How much more charge should be added to increase the potential difference by 200 V?

26. A capacitor of 0.1 μF is to be constructed by using tin foil as plates and wax paper of thickness 0.1 mm and dielectric constant 3 as the dielectric material. What should be the area of each foil?

27. Show that it is impractical to make a parallel-plate capacitor of 1 F in which the plates are separated by 1 cm and air is the dielectric. (*Hint:* Calculate the area of each plate.)

28. The maximum breakdown potential of a 100 μF capacitor is 1000 V. What is the maximum charge that can be placed on each plate?

29. The dimensions of the plates of a parallel-plate capacitor are 8 cm by 8 cm, and they are separated by a distance of 2 mm. Calculate the capacitance if (a) air is between the plates; and (b) glass fills the space between the plates. The dielectric constant of glass is 5.

30. A parallel-plate capacitor of 20 pF has mica ($K = 6$) as a dielectric. What is the minimum thickness of the mica if the capacitor has to stand a voltage of 8000 V? The area of each plate is 10^{-2} m^2.

31. Three capacitors of capacitance 10 μF, 20 μF and 60 μF are connected (a) in series; and (b) in parallel. Calculate the total capacitance in each case. Which arrangement gives the maximum and minimum values?

32. (a) A capacitor of 5 μF and another of 20 μF are connected in series and are charged by a battery of 100 V. Calculate the potential difference and charge on each capacitor. (b) Repeat part (a) if the capacitors are connected in parallel.

33. Calculate the equivalent capacitance of the network between points A and B shown where $C_1 = 15\ \mu F$, $C_2 = 5\ \mu F$, and $C_3 = 10\ \mu F$.

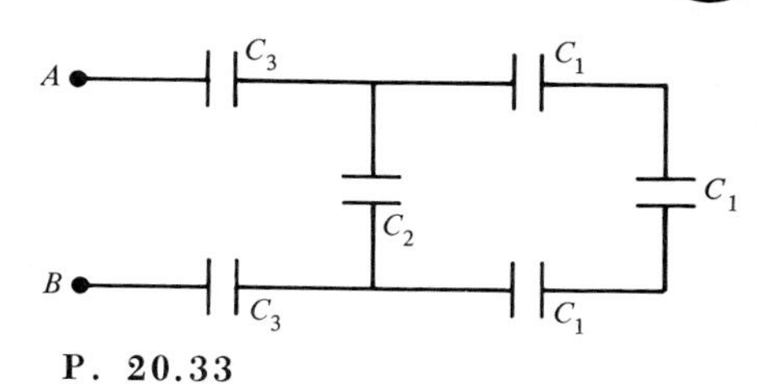

P. 20.33

34. A potential difference of 120 V is applied between A and B as shown. Calculate the total capacitance and the charge on each capacitor. Calculate the potential difference across each capacitor.

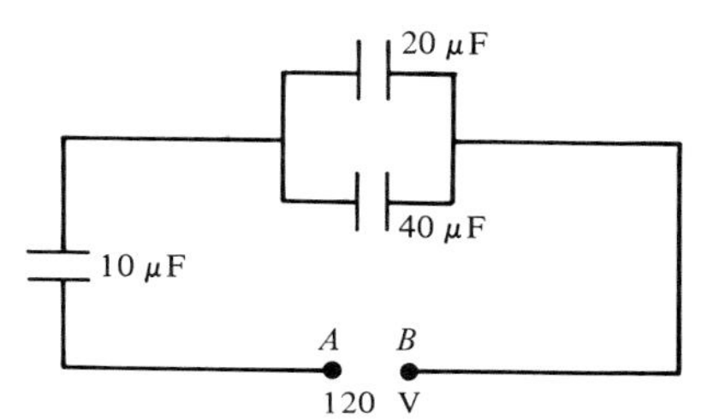

P. 20.34

35. Consider the network shown. Calculate (a) the total net capacitance; and (b) the potential difference across each capacitor.

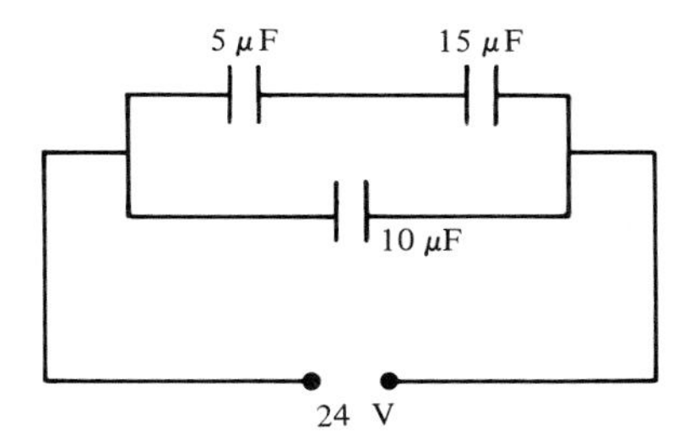

P. 20.35

36. A 20-μF capacitor with a charge of 500 μC is connected in parallel with an uncharged capacitor of 10 μF. Calculate the final potential difference and the charge on each.
37. The potential difference between the plates of a capacitor is 1000 V and each plate carries a charge of 250 μC. Calculate the capacitance and energy stored in the capacitor.
38. A 5-μF capacitor carries a charge of 10 mC. Calculate the potential and the energy stored in the capacitor.
39. A parallel-plate capacitor of 1 μF with a dielectric can stand a potential of 10,000 V. Calculate the maximum charge and maximum energy stored.
40. Consider a parallel-plate capacitor with plates of area 500 cm^2 and separated by a sheet of mica 0.3 mm thick. The capacitor is charged to 600 V. The dielectric constant of mica is 6. Calculate the capacitance and the energy density.
41. Calculate the energy stored in a capacitor of 100 μF and charged to a potential of 400 V. This capacitor is discharged by connecting the plates by a metallic wire. Suppose 25 percent of the stored energy goes into light energy while the rest of it goes into heating the wire. Calculate this heat energy in calories.
42. Calculate the total energy stored and the energy density in a parallel-plate capacitor. The plates each have an area of 2 cm^2 and are separated by a distance of 5 mm. The plates are connected to a source of 100 V, and the space between the plates is filled with air. How would the results be altered if the space between the plates were filled with a dielectric material of dielectric constant 5?
43. Consider a parallel-plate capacitor each plate being 20 cm by 20 cm and separated by 2 mm. The dielectric constant of the material between the plates is 5. The plates are connected to a voltage source of 500 V. Calculate (a) the capacitance, (b) the charge on either plate; (c) the energy density; and (d) the total energy stored in the capacitor. How do these values change if the dielectric is replaced by air?
44. Consider a parallel-plate capacitor with plates 20 cm by 20 cm and filled with mica which has a dielectric constant of 3 and a breakdown field of 100 kV/m. Calculate the maximum charge that each plate can be given. If the separation between the plates is 2 mm, what will be the maximum energy density?
45. Energy stored in a capacitor can be used by a flashbulb to take a picture. If we need at least 80 J of energy over a very short time, what is the magnitude of the capacitor if 20 V batteries are used?
46. Two capacitors, 5 μF and 10 μF, are connected in series to a 240-V source. (a) Calculate the charge, the potential difference, and the energy stored in each capacitor. (b) Repeat if the capacitors are connected in parallel.

21
Current Electricity

Now we pursue the study of charges in motion—current electricity. Starting with the sources that produce current, we shall look into the characteristics of the materials that affect the current through them. After getting familiar with the terminology, we investigate different relations between potential difference, current, resistance, electric field, and so on, leading to the statement of Ohm's law. We shall calculate the energy and power loss by moving charges in simple circuits.

21.1. Sources of Electromotive Force

A CONTINUOUS flow of charges across any cross-sectional area of a conductor is referred to as *current electricity*. If the average motion of the charges is always in the same direction, it is referred to as *direct current*. Consider two parallel plates A and B, A being positively charged while B is negatively charged, as shown in Figure 21.1. Thus, A is at a higher potential than B. Suppose that the two plates are connected by means of a

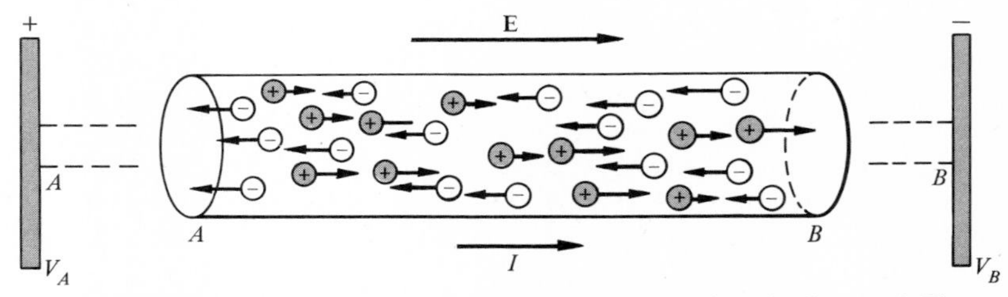

FIGURE 21.1. *Flow of charge carriers in a material with electric field E. Direction of current I is that of E.*

conductor, say a copper wire. An electric field is established from A to B inside the conductor, but this field does not last long. The electrons from plate B start flowing through the conductor to plate A. In a very short time, enough charge is transferred so that the two plates are at the same potential and the electric field in the copper wire becomes zero. There is no more flow of charges. During the short time interval the charges are flowing, a *transient current* is said to exist.

It is quite clear from the example above that in order to maintain a steady direct current, the potential difference between the two plates must be maintained at a constant value. This can be done only if the two plates are connected to some device such as a dry-cell battery or an electric generator which converts some other form of energy into electrical energy, thereby maintaining the required potential difference. This is done by transporting positive charges from a lower potential to the higher potential. Such a device, which converts nonelectrical energy into electrical energy, is called a source of *electromotive force* (emf)—a totally misleading name because the source refers to energy and not to force. It should have been called *electromotive energy* (eme). Since emf is in common use, we will stick with it and represent the emf of a source by $\mathcal{E}$. We define a *source emf* $\mathcal{E}$ equal to the work done in carrying 1 coulomb of charge through the source. Suppose that q coulombs require an amount of work W joules; then $\mathcal{E}$ in volts in given by

$$\boxed{\mathcal{E} = \frac{W}{q}} \tag{21.1}$$

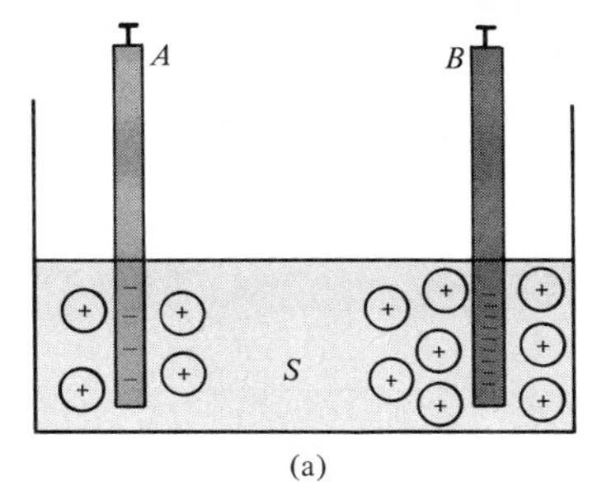

Thus, if $\mathcal{E} = 1.5$ V, it means that 1.5 J of work must be done to move a charge of 1 coulomb through the source.

There are many sources of emf. A few examples are: (1) batteries or cells convert chemical energy into electrical energy, (2) electric generators convert mechanical energy into electrical energy, and (3) thermocouples convert heat energy into electrical energy. Some of these sources will be discussed in this text while at present we confine ourselves to explaining how chemical energy is converted into electrical energy. It may be pointed out that it was the Italian physicist Alessandro Volta (1745–1827) who discovered this process of energy conversion, in 1800.

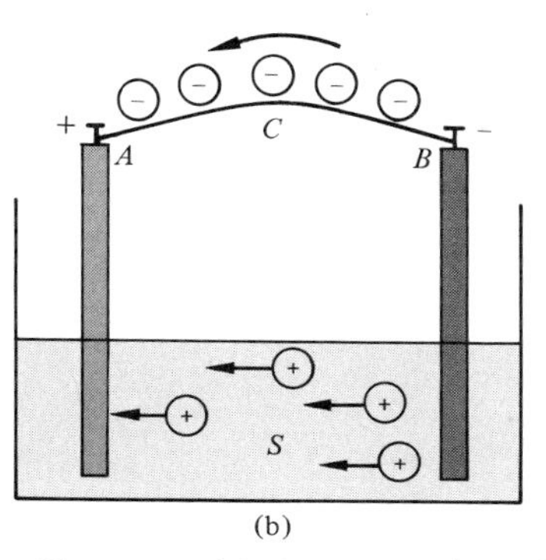

FIGURE 21.2. *Two disimilar metal rods immersed in a dilute acid solution (a) before connecting A and B and (b) after connecting A and B by a conductor C.*

A simple chemical battery consists of two dissimilar metal rods A and B immersed in a dilute acid solution S, as shown in Figure 21.2(a). Let us concentrate on rod A. Like most metals, rod A dissolves in acid. When this happens, each atom leaves an electron on the rod while the atom itself enters the solution as a positive ion. As dissolving continues, more and more electrons are left on rod A, and the positive ions go into solution. Eventually, rod A becomes so negative that as many positive ions go into solution as are attracted back by the rod. We can say that *equilibrium* has been reached. Rod A, being negative, is at a lower potential with respect to the solution. Similar things happen to rod B. When equilibrium is reached, rod B is not only at a lower potential with respect to the solution S, but also with respect to rod A (because rod B dissolves faster in acid and has even more negative charge than rod A) as shown in Figure 21.3. Thus, there exists a potential difference between rods A and B, with A being at a higher potential than B. That is, A is a positive terminal while B is a negative terminal, as illustrated in Figure 21.3.

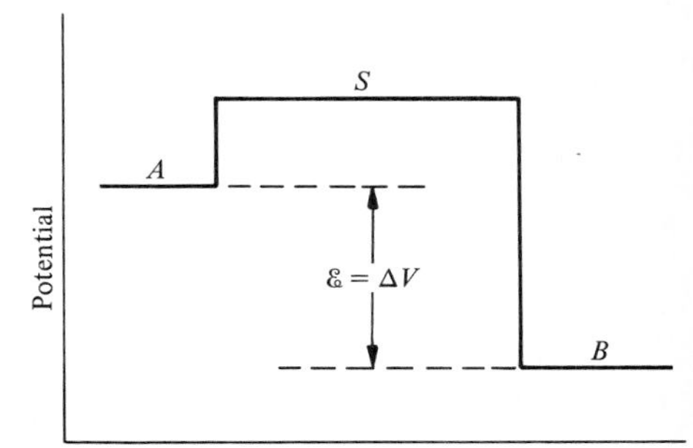

FIGURE 21.3. *Potential of metal rods A and B with respect to the potential of solution S.*

Outside the solution the electric field is from A to B. If terminals A and B are connected by a wire [Figure 21.2(b)], the positive charges will flow from A (the higher potential) to B (the lower potential) while the electrons will flow from B to A. Since in metals, the positive ions are fixed, only the electrons flowing from B to A contribute to the electric current. At the same time, as the electrons leave B, the rod becomes less negative. Some of the positive ions leave rod B and go into the solution. The arrival of electrons at rod A through the wire helps attract the positive ions in the solution which get deposited on rod A. Thus, the metal rod B keeps on dissolving while rod A, because of the positive ions depositing on it, gains mass. If rod A gets completely covered with the ions coming from rod B, A and B will be at the same potential; hence, there will be no current between them. Thus, rod A must be cleaned occasionally.

A common battery of this type uses Zn and Pb plates in a sulfuric acid solution, although any two different metal plates should work. A dry-cell battery consists of a carbon rod and a zinc can separated by a thick moist paste of ammonium chloride and manganese dioxide. Commercial batteries are much more involved and we shall not discuss them here.

21.2. Electric Current and Current Density

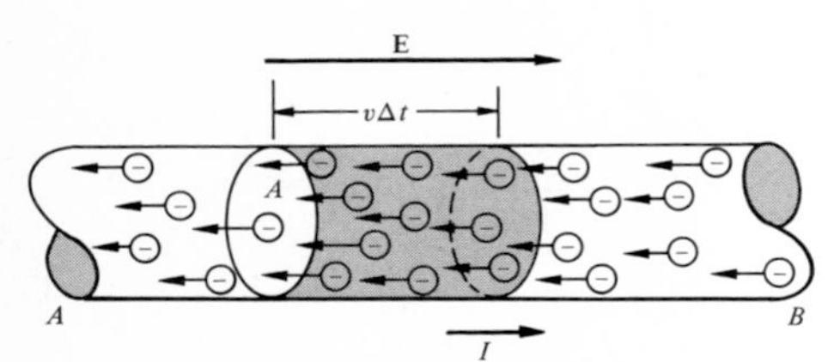

FIGURE 21.4. *Charge carriers in a metallic conductor under the influence of* **E**.

Let us consider a portion AB of a material connected to a voltage source as shown in Figure 21.1. An electric field **E** will be established inside the material. If there are free charges inside the material, the positive charges will drift from left to right—that is, in the direction of **E**, while the negative charges will drift from right to left, that is, in the direction opposite **E**.

If the material AB is a metallic conductor, the charge carriers are free electrons, as shown in Figure. 21.4; hence, the current is due to these drifting electrons. On the other hand, if the material AB is an electrolyte such as a salt–water solution, there are no free electrons, but the charge carriers are both positive and negative ions, drifting in opposite directions. If the material AB is an ionized gas, such as in a mercury vapor lamp, the positive charge carriers are positive ions while the negative charge carriers are free electrons. Note that if the material AB is an insulator, there will be no flow of charges; hence no current.

There is another class of materials, called semiconductors. The charge transfer in semiconductors is described by means of electrons as well as the motion of holes which behave almost like positive charges. We shall discuss this in Chapter 33.

In Figure 21.1 the positive charges moving to the right increase the positive charges on the right side of the conductor. The effect of the electrons moving to the left will be to decrease the negative charge on the right side of the conductor, or equivalently to increase the positive charge on the right. Thus, for all practical purposes, as far as current is concerned, a positive charge moving in one direction is equivalent to a negative charge moving in the opposite direction.

In order to define current quantitatively, we must state its magnitude and its sense of direction. *The conventional current direction is that in which the positive charges will drift, that is, in the direction of* **E**. Thus, in Figure 21.1 or 21.4, even though the negative charges move to the left, the direction of current is to the right. *The magnitude of the current across any area is defined as the net charge flowing across the area per unit time.* Thus, if a net charge ΔQ flows across an area in a time interval Δt, the current I across the area is

$$\boxed{I = \frac{\Delta Q}{\Delta t}} \tag{21.2}$$

where I has a sense of direction but is not a vector quantity. It is a scalar, as is clear from Equation 21.2. The SI unit of current is 1 *ampere* (1 A). It is defined as that current in which the charges flow across any cross section at the rate of 1 C/s. The definition of ampere in terms of magnetic forces will be given in Chapter 24. This unit was named in honor of the French scientist André Marie Ampère (1775–1836), whose contributions to the field of electricity and magnetism will also be discussed in Chapter 24. One ampere is a large unit; hence, smaller units that are commonly used are

$$1 \text{ milliampere} = 1 \text{ mA} = 10^{-3} \text{ A}$$
$$1 \text{ microampere} = 1 \mu\text{A} = 10^{-6} \text{ A}$$

The instrument used for detecting very small currents is called a *galvanometer* or, in its modified form, an *ammeter.*

Let us now express current I in terms of quantities such as drift velocity and the charge on the charge carrier. For simplicity let us consider a metallic conductor in which the free electrons will be the charge carriers, as shown in Figure 21.4. These electrons are ordinarily in random thermal motion. Typical speeds of electrons in metals are $\sim 10^{-4}$ m/s. The motion is so random that in a given time as many electrons cross any cross-sectional area to the left as to the right; hence, there is no net current. These electrons constantly collide with the atoms (or ions) of the metal and change direction as shown in Figure 21.5 by the continuous line. However, when an electric field is applied from one end of the conductor to the other, there will be a drift velocity superimposed on these electrons in the direction opposite **E**. Thus, the electrons' motion will be slightly biased toward the left, and they follow the dashed path shown in Figure 21.5(b). Note that, owing to random motion, in seven collisions the electron in the absence of an electric field moves from position 1 to X, while the effect of the electric field has been to move the electron to its final position, X'. Thus, the drift distance is only XX'. This explains why the drift velocity v is of the order of a fraction of a centimeter per second. It is this drift velocity that is used in calculating current.

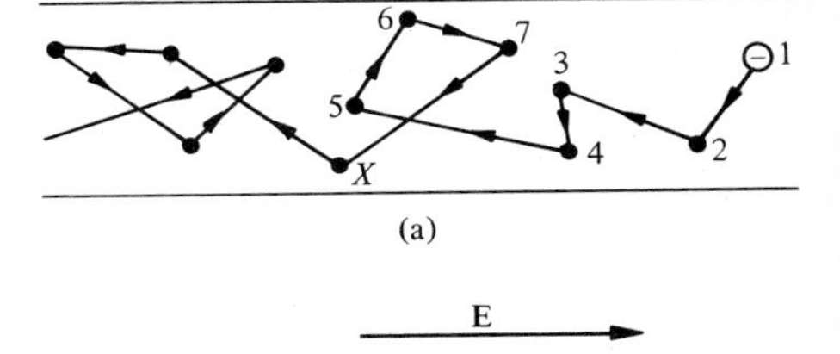

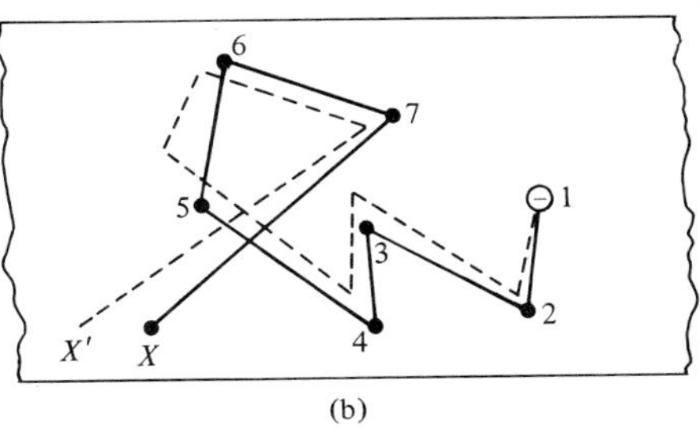

FIGURE 21.5. *Random motion of an electron in a conductor (a) when there is no electric field and (b) a portion of part (a) in the presence of an electric field results in the dashed-line path.*

Suppose that there are n free electrons per unit volume, each with a charge q and drift velocity v. In time Δt each electron will travel a distance $v \, \Delta t$. Thus, all the electrons within the shaded cylinder of length $L = v \, \Delta t$, as shown in Figure 21.6, will flow across the cross-sectional area A of the cylinder in time Δt. Since the volume of such a cylinder is $v \, \Delta t \, A$, the total number of negative charge carriers will be $nv \, \Delta t \, A$, each having a charge q. Thus, the charge ΔQ crossing the cross section A in time Δt will be $\Delta Q = qnv \, \Delta t \, A$, or the current I will be

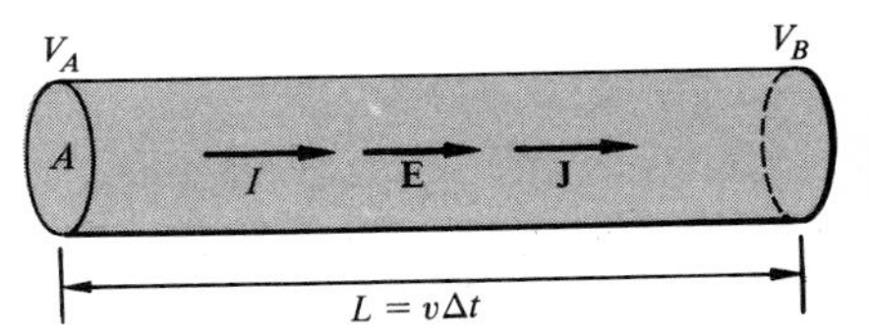

FIGURE 21.6. *Current in a conductor of length L and cross-sectional area A.*

$$I = \frac{\Delta Q}{\Delta t} = nqvA$$

In general, if the conductor contains different types of charge carriers both positive and negative with different charge densities and moving with different drift velocities, the current I is given by

$$I = n_1 q_1 v_1 A + n_2 q_2 v_2 A + n_3 q_3 v_3 A + \cdots$$

or

$$I = A \sum_i n_i q_i v_i \tag{21.3a}$$

For convenience we shall drop the summation sign; it will be used when needed in a particular situation. We write Equation 21.3a as

$$I = nqvA \tag{21.3b}$$

Thus, I depends upon A—a geometrical factor. A quantity independent of A is called *current density* J and defined as the current per unit cross-sectional area; that is,

$$J = \frac{I}{A} = nqv \tag{21.4}$$

Actually, current density is a vector quantity. The current density vector **J** is defined as

$$\mathbf{J} = nq\mathbf{v} \tag{21.5}$$

For positive charge carriers the direction of **v** is the same as that of **E**. Thus, the direction of **J** is the same as that of **E** for positive charge carriers. Suppose that the charge carriers are negative, say electrons. The direction of **v** will be opposite to **E**. Since q is negative, the direction of **J** will be opposite to that of **v**; that is, **J** has the same direction as **E**. *Thus, whether the charge carriers are positive or negative, the current density vector* **J** *is always parallel to* **E**. Also, the direction of motion of positive charge carriers is parallel to **J**, while those of negative charge carriers is parallel to $-\mathbf{J}$.

Example 21.1 Consider a copper wire of 3 mm² cross-sectional area in which there is a current of 6 A. Calculate (a) the number of free electrons per m³; (b) the total charge that crosses any cross-sectional area in 1 h; (c) the current density; and (d) the drift velocity.

(a) Copper has a density $\rho = 8.93 \times 10^3$ kg/m³, an atomic mass $A = 63.54$ u, while Avogadro's number is $N_A = 6.022 \times 10^{26}$/kmol. Thus, the number of electrons per unit volume (since there is one free electron associated with each atom) is

$$n = \rho\frac{N_A}{A} = 8.93 \times 10^3 \text{ kg/m}^3 \frac{6.022 \times 10^{26}/\text{kmol}}{63.54 \text{ kg/kmol}} = 0.85 \times 10^{29}/\text{m}^3$$

(b) $$Q = It = 6 \text{ A} \times 1 \text{ h} = 6 \text{ C/s} \times 3600 \text{ s} = 2.16 \times 10^4 \text{ C}$$

(c) From Equation 21.4,

$$J = \frac{I}{A} = \frac{6 \text{ A}}{3 \text{ mm}^2} = \frac{6 \text{ A}}{3 \times 10^{-6} \text{ m}^2} = 2 \times 10^6 \text{ A/m}^2$$

(d) From Equation 21.4, $J = nqv$. Since $q = e$.

$$v = \frac{J}{nq} = \frac{2 \times 10^6 \text{ A/m}^2}{(0.85 \times 10^{29}/\text{m}^3)(1.6 \times 10^{-19} \text{ C})} = 1.47 \times 10^{-4} \text{ m/s}$$

Note how small this drift velocity is.

Exercise 21.1 Repeat Example 21.1 for the case of an aluminum wire of the same cross-sectional area. Which of the two wires, Cu or Al, would you prefer for electrical conduction? The atomic mass of Al is 26.98 kg/mol and its density is 2700 kg/m³. [*Ans.:* (a) $n = 6.026 \times 10^{28}$ electrons/m³; (b) $Q = 2.16 \times 10^4$ C; (c) $J = 2 \times 10^6$ A/m²; (d) $v = 2.08 \times 10^{-4}$ m/s. Cu is preferred because of the greater density of free electrons.]

21.3. Resistivity and Resistance

When an object falls vertically through a viscous medium, in spite of the fact that there is a gravitational force that accelerates the object, because of the resistance offered by the medium the object reaches a terminal velocity and subsequently moves without any acceleration. Similarly, in conductors when an electric field **E** is applied, there is a net force $q\mathbf{E}$ on the charge, hence an acceleration. But over a very short distance, because of the resistance of the medium to the flow of the charges (as a result of the collisions) the charges achieve a terminal velocity, that is, the drift velocity we have referred to in the previous section. In this section we investigate how the flow of charge is related to the characteristics of the conductor and the applied field **E**.

The current density **J** depends upon the value of **E** and the type of conductor. We define the *resistivity* ρ of a conductor as the ratio of the electric intensity to the current density. Thus,

$$\rho = \frac{E}{J} \tag{21.6}$$

In vector form we may write

$$\boxed{\mathbf{J} = \frac{\mathbf{E}}{\rho} = \sigma\mathbf{E}} \tag{21.7}$$

[This is Ohm's law, in general form, if σ is constant (see Section 21.4).] That is,

$$\sigma = \frac{1}{\rho} \tag{21.8}$$

where σ is called the *conductivity* of the material and is equal to the reciprocal of the resistivity ρ. Thus, from Equation 21.7, the greater the resistivity of a material, the larger the value of the electric intensity **E** needed to produce a given current density, and vice versa. Thus, a perfect insulator will have infinite resistivity (or zero conductivity). The resistivities of different material at 20°C are listed in Table 21.1. Note the change in ρ from insulators to metals by a factor of the order of 10^{22}. Also, the resistivities of the semiconductors are between those of the conductors and insulators.

To calculate ρ from Equation 21.6, we must know the values of E and J. But the values of E and J are not directly measurable. Thus, we must express ρ in some other form. Let us consider a cylindrical conducting wire of length L and cross-sectional area A. When a potential difference of $V = V_A - V_B$ is established between the two ends, it results in a field **E** inside the wire and establishes a current I as shown in Figure 21.6. From the definition of E and J,

$$E = \frac{V}{L} \quad \text{and} \quad J = \frac{I}{A}$$

Substituting these in Equation 21.6 and rearranging,

$$\rho = \frac{E}{J} = \frac{V/L}{I/A}$$

TABLE 21.1
Resistivity of Different Materials at 20°C and their Mean Temperature Coefficient

Material	ρ (Ω-m)	$\bar{\alpha}$ (per C°)
Conductors		
Aluminum	2.8×10^{-8}	0.0039
Brass	7×10^{-8}	0.0020
Constantan	49×10^{-8}	0.00002
Copper	1.79×10^{-8}	0.0039
Gold	2.27×10^{-8}	0.0034
Iron	11×10^{-8}	0.0052
Manganin	44×10^{-8}	0.0000
Mercury	96×10^{-8}	0.00089
Nichrome	100×10^{-8}	0.0004
Platinum	11×10^{-8}	0.0039
Silver	1.6×10^{-8}	0.0038
Tungsten	5.6×10^{-8}	0.0045
Semiconductors		
Carbon	3.5×10^{-5}	−0.0005
Germanium	0.60	−0.048
Silicon	2300	−0.075
Insulators		
Amber	5×10^{14}	
Glass	10^{10}–10^{14}	
Lucite	$>10^{13}$	
Mica	10^{11}–10^{15}	
Quartz	7.5×10^{16}	
Wood	$>10^{13}$	

or

$$\frac{V}{I} = \rho \frac{L}{A} \tag{21.9}$$

In this particular situation ρ, A, and L are all fixed; hence, the ratio V/I will be constant. The ratio of the potential difference V to the current I is called the *resistance* R. That is,

$$\boxed{R = \frac{V}{I} = \rho \frac{L}{A}} \tag{21.10}$$

The quantities V, I, L, and A are directly measurable; hence, R, as well as ρ, may be easily calculated.

From Equation 21.10, if V is measured in volts and current I in amperes, the resistance is given in ohms (Ω). That is,

$$1 \text{ ohm} = \frac{1 \text{ volt}}{1 \text{ ampere}} \tag{21.11}$$

From Equations 21.10 and 21.11, we get the *units of resistivity* ρ as ohm-m (Ω-m); those of conductivity σ are 1/ohm-m.

It is necessary to point out that the three microscopic quantities $\mathbf{E}$, ρ, and $\mathbf{J}$ which are related by the equation $\mathbf{E} = \rho\mathbf{J}$ are not directly measurable by instruments, while the corresponding macroscopic quantities V, R, and I are measurable and given by the relation $V = RI$.

Finally, let us investigate the temperature dependence of resistivity ρ of different materials. The variation of resistivity with temperature for three different types of materials (a) a metallic conductor (b) a semiconductor and (c) a superconductor are shown in Figure 21.7.

The resistivity of a metallic conductor increases slightly with temperature and is nonlinear as shown in Figure 21.7(a). But over a few hundred degrees, the variation of ρ with T may be assumed to be linear as shown for the case of copper in Figure 21.8. Thus, over a limited range of temperature, the variation of ρ with T may be expressed by the equation

$$\rho = \rho_0[1 + \bar{\alpha}(T - T_0)] \tag{21.12}$$

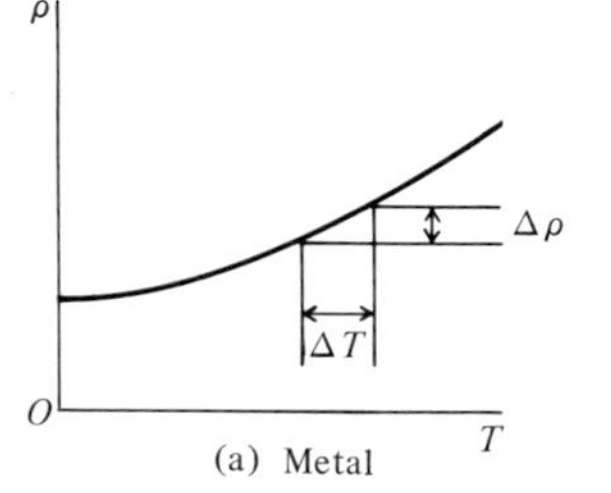

(a) Metal

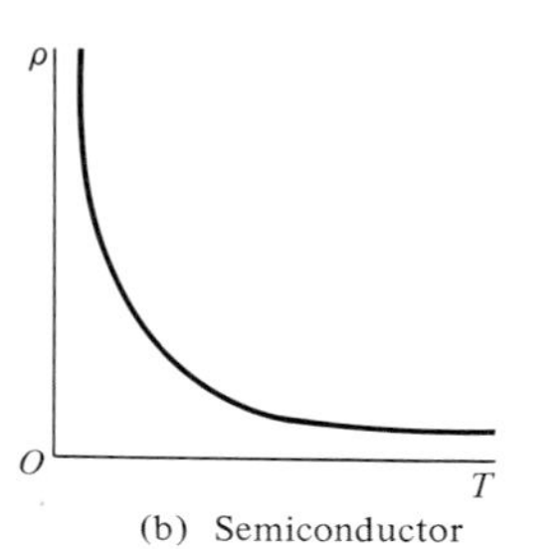

(b) Semiconductor

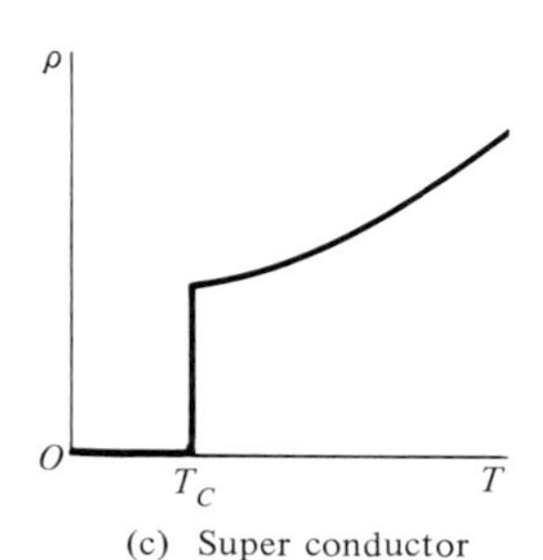

(c) Super conductor

FIGURE 21.7. *Variation of resistivity with temperature for various materials.*

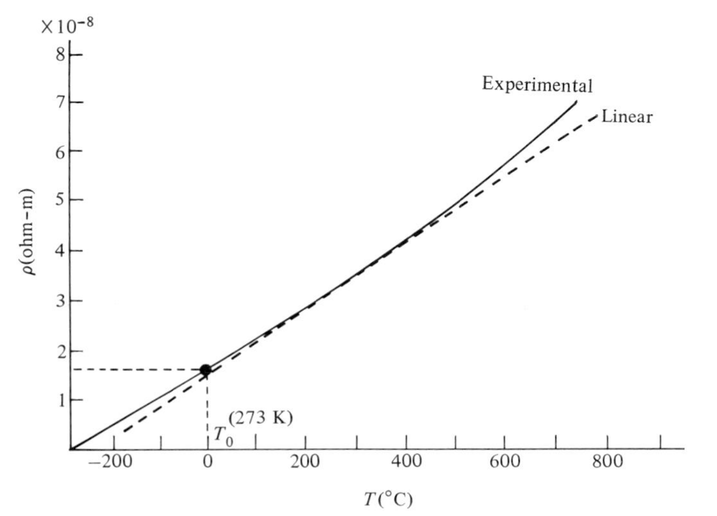

FIGURE 21.8. *Experimental variation of R versus T for copper conductor. Note the linear variation over a limited temperature range.*

where ρ_0 is the resistivity at a reference temperature T_0 (which may be, say, 300 K or 27°C), ρ is the resistivity at temperature T, and $\bar{\alpha}$ is a *mean temperature coefficient of resistivity.* $\bar{\alpha}$ means that the value used in Equation 21.12 is the mean value of α at two or three different temperatures within the range of interest. Let us rewrite Equation 21.12 as

$$\bar{\alpha} = \frac{1}{\rho_0}\frac{\rho - \rho_0}{T - T_0} = \frac{1}{\rho_0}\frac{\Delta\rho}{\Delta T} \tag{21.13}$$

Thus, $\bar{\alpha}$ may be defined as *the fractional increase in resistivity per degree increase in temperature.* (It should be noted that T is in Kelvins and ΔT may be in C° or K.) Thus, for a conductor of fixed length L and cross-sectional area A, we may use the relation $R = \rho L/A$ and rewrite Equation 21.12 as

$$\boxed{R = R_0[1 + \bar{\alpha}(T - T_0)]} \tag{21.14}$$

For a given specimen of wire this change in ρ or change in R with temperature may be used to measure the temperature. A platinum resistance thermometer is an example. Table 21.1 lists the values of $\bar{\alpha}$ for different materials. Notice that carbon has a negative temperature coefficient of resistivity, the metal constantan has a very small positive coefficient while manganin has almost zero.

The resistivity of semiconductors is controlled by two factors: (1) the impurities present, and (2) the temperature variation. A small amount of impurities of a proper foreign material in a semiconductor can decrease the resistivity by a very large factor. Similarly, the resistivity of a semiconductor decreases very rapidly with increasing temperature, as shown in Figure 21.7(b).

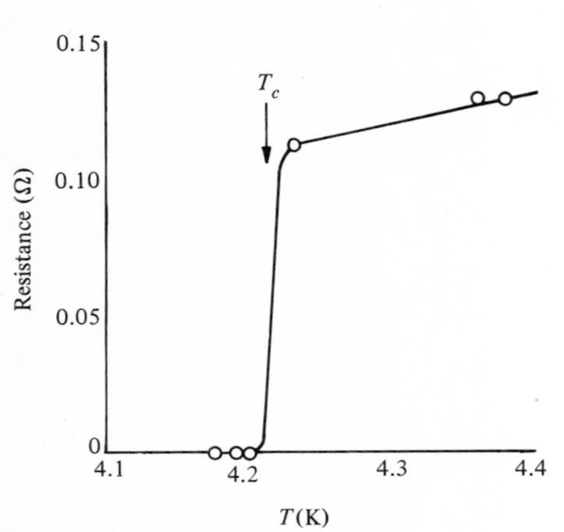

Figure 21.9. *Experimental values of R versus T for mercury at low temperatures, showing the superconducting phenomenon.* [From H. K. Onnes, *Leiden Commun.* **124C** (1911).]

So far we have been dealing with the cases where the resistivity of a metal conductor decreases with decreasing temperature. In 1911 Kamerlingh Onnes measured the resistivity of mercury in the neighborhood of 4 K, the plot of which is shown in Figure 21.9. It is clear from this plot that the resistance of mercury drops very abruptly almost to zero below 4.2 K. Since then many metals and even many poor conductors such as lead are found to have abnormally low resistivity (or very high conductivity), similar to that for Hg in Figure 21.9. This property is called *superconductivity,* while the material under such conditions is said to be in the *superconducting state.*

The temperature below which the material is in a superconducting state is called the *transition* or *critical temperature* T_c. Table 21.2 lists such temperatures for some materials. One of the outstanding features of the superconductor

Table 21.2
Some Superconducting Materials with their Transition Temperatures

Element	T_c(K)	Compound	T_c(K)
Indium	0.14	$AnSb_2$	0.58
Titanium	0.40	$PdSb_2$	1.25
Aluminum	1.19	NiBi	4.25
Mercury	4.15	NbN	16.0
Lead	7.18	Nb_3Sn	18.05
Technetium	11.2	$Nb_3Al_{0.8}Ge_{0.2}$	20.05

is the following. Suppose that we have a ring made of a superconducting material kept below the transition temperature. Once the current is started and the battery is switched off, because of the extremely small resistivity the current will continue indefinitely (without needing an emf source). Different experiments performed reveal that the current does not decrease appreciably over periods of years.

EXAMPLE 21.2 Calculate the resistance of a wire 20 m long that has a radius of 2 mm and resistivity of $5.5 \times 10^{-8}\ \Omega$-m. If the two ends of the wire are connected to a source of 24 V, what are the values of I, J, and E?

The cross-sectional area of the wire is $A = \pi r^2 = \pi(2 \times 10^{-3}\ \text{m})^2 = 4\pi \times 10^{-6}\ \text{m}^2$. Substituting the values of ρ, L, and A in Equations 21.10, we get

$$R = \rho\frac{L}{A} = 5.5 \times 10^{-8}\ \Omega\text{-m}\frac{20\ \text{m}}{4\pi \times 10^{-6}\ \text{m}^2} = 8.75 \times 10^{-2}\ \Omega$$

From Equation 21.10, $R = V/I$ or

$$I = \frac{V}{R} = \frac{24\ \text{V}}{8.75 \times 10^{-2}\ \Omega} = 274\ \text{A}$$

Therefore,

$$J = \frac{I}{A} = \frac{274\ \text{A}}{4\pi \times 10^{-6}\ \text{m}^2} = 21.8 \times 10^6\ \text{A/m}^2$$

while from Equation 21.6,

$$E = \rho J = 5.5 \times 10^{-8}\ \Omega\text{-m} \times 21.8 \times 10^6\ \text{A/m}^2 = 1.2\ \text{V/m}$$

EXERCISE 21.2 To be safe we do not want the current in the wire used in Example 21.2 to exceed 100 A, while keeping the length the same but changing the cross-sectional area. For this case calculate R, A, J, and E. [*Ans.:* $R = 0.24\ \Omega$, $A = 4.58 \times 10^{-6}\ \text{m}^2$, $J = 2.18 \times 10^7\ \text{A/m}^2$, $E = 1.2\ \text{V/m}$.]

21.4. Linear and Non-linear Circuit Elements

Figure 21.10 shows an experimental arrangement for measuring the resistance R of a metallic conductor. Voltmeter V measures the potential difference across the resistor R. Ammeter A measures the current through the resistor R. The voltage V across R and the current I through R can be varied by means of a source B of variable voltage. The plots of V versus I for three different resistors are shown in Figure 21.11. It is clear from these plots that for a given resistor the ratio V/I, called resistance, R, is constant. From such experiments the German physicist Georg Simon Ohm (1787–1854) concluded:

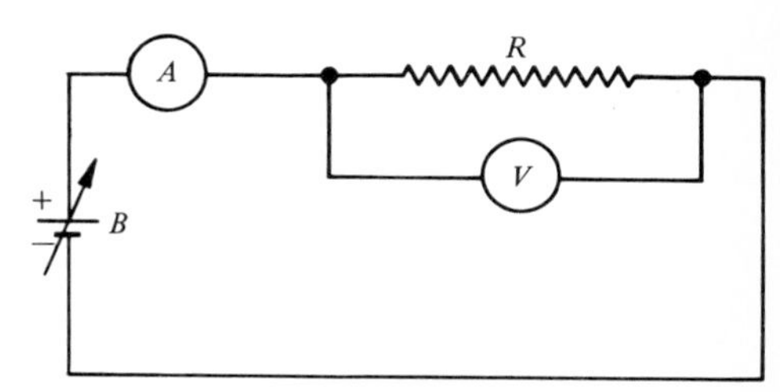

FIGURE 21.10. *Arrangement for measuring the resistance R of a metallic conductor.*

The resistance R of a metallic conductor at a given temperature is the same no matter what the value of the potential difference is across the terminals of the conductor. This is a statement of *Ohm's law*, which may be written mathematically as

$$R = \frac{V}{I} \quad \text{or} \quad V = IR \tag{21.15}$$

(Note that this is a special case of Equation 21.7.) The product IR, which is equal to V, is referred to as the IR drop across the conductor. If the direction of V is changed, the direction of I will change, too, but the magnitude of R will remain constant.

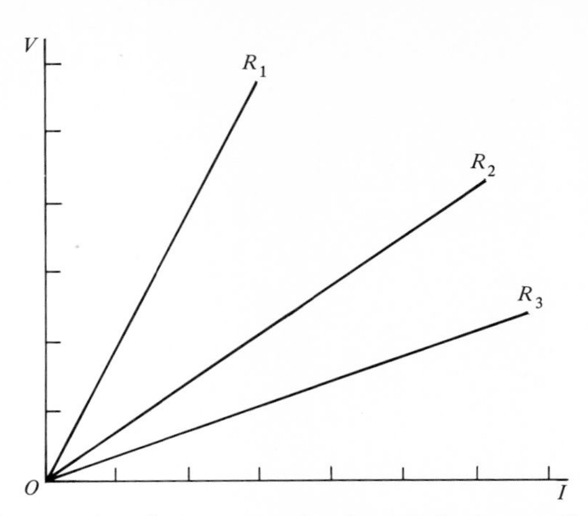

FIGURE 21.11. *Plot of V versus I for three different metallic resistors.*

Any resistor that obeys Ohm's law (i.e., the plot of V versus I is a straight line at a given temperature) is said to be an *Ohmic circuit element*. The Ohmic circuit element is also referred to as a *linear circuit element*. The most common example of a linear circuit element is that of a copper or other metal wire used commonly in electrical circuits.

Ohm's law is a special case. Many conductors do not obey Ohm's law. This means that their current varies nonlinearly with the applied voltage. Figure 21.12 shows three different circuit elements which are nonohmic because of the nonlinearity of the V-versus-I plots. The nonohmic circuit elements are also called *nonlinear circuit elements*. A thermistor is a tiny bead of semiconducting material (with current variation as shown in Figure 21.12(c)) that has a large negative temperature coefficient of resistivity $\bar{\alpha}$. It is a very sensitive temperature device. Such modern devices as vacuum tubes, transistors, and crystal rectifiers are nonlinear circuit elements.

We may remind ourselves that even though nonlinear circuit elements do not obey Ohm's law, the definition of resistance R given by $R = V/I$ still holds for any particular values of V and I.

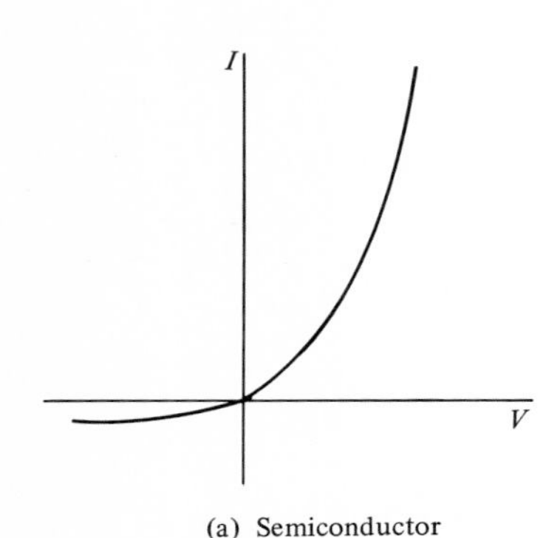

(a) Semiconductor

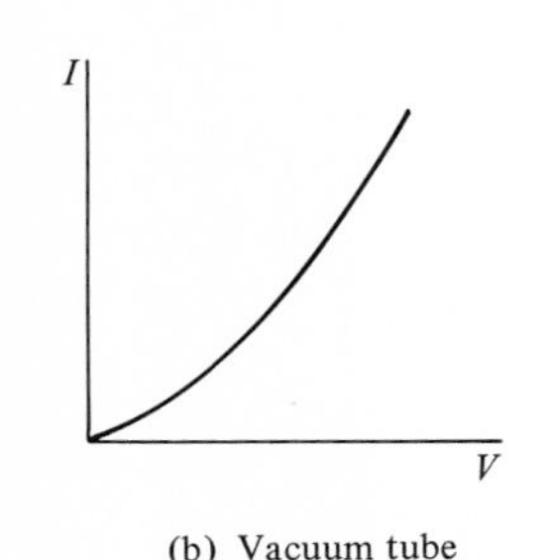

(b) Vacuum tube

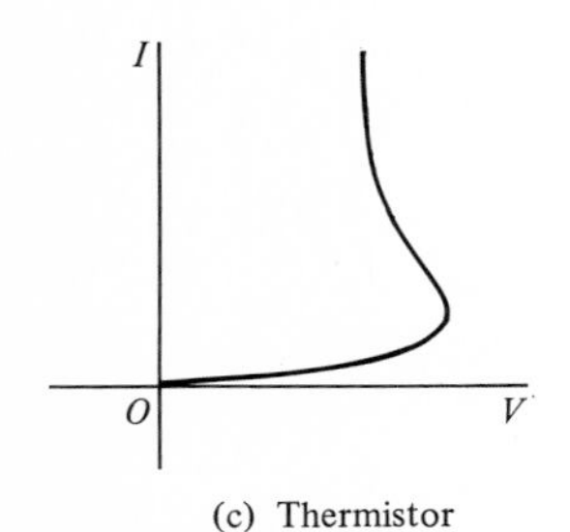

(c) Thermistor

FIGURE 21.12. *Examples of nonlinear circuit elements.*

EXAMPLE 21.3 The resistance of a tungsten wire at room temperature (20°C) is 1.50 Ω. What will be its resistance at 0°C, 100°C, and 1000°C?

The temperature coefficient of resistivity of tungsten at 20°C is $\bar{\alpha} = 0.0045/\text{C}°$. Substituting in Equation 21.4, $R = R_0[1 + \bar{\alpha}(T - T_0)]$, the values of $R_0 = 1.50\ \Omega$, $T_0 = 20°\text{C}$, and $T = 0°\text{C}$, we get

$$R = 1.50\ \Omega\ [1 + (0.0045/\text{C}°)(0 - 20)\text{C}°] = 1.365\ \Omega$$

At $T = 100°\text{C}$: $R = 1.50\ \Omega\ [1 + (0.0045/\text{C}°)(100 - 20)\text{C}°] = 2.04\ \Omega$

At $T = 1000°\text{C}$: $R = 1.50\ \Omega\ [1 + (0.0045/\text{C}°)(1000 - 20)\text{C}°] = 8.12\ \Omega$

We have assumed a linear increase in resistance with temperature. $\bar{\alpha}$ may not be the same in different temperature ranges.

Exercise 21.3 At what temperature will the resistance of the tungsten wire in Example 21.3 be double its resistance at 20°C? [*Ans.*: $T = 242°C$.]

21.5. Energy and Power in Electric Circuits

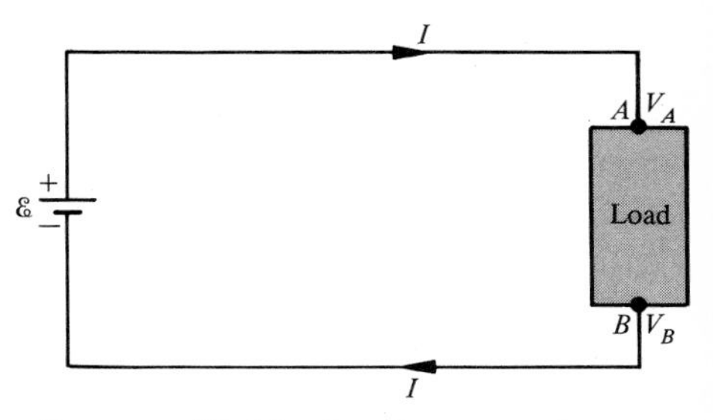

Figure 21.13. *Load connected to a source of emf.*

Energy and Power

Consider a source of emf ℰ connected to a load AB through conducting wires of negligible resistance, as shown in Figure 21.13. There is a potential difference $V_{AB} = V_B - V_A$ across the load and a steady current I in the circuit directed from A to B. The load in the shaded box may be a resistor, a storage battery, a motor, and so on.

Suppose that in a time interval Δt a quantity of positive charge $\Delta q = I\,\Delta t$ enters the terminal A while the same charge leaves the terminal B. Thus, the charge Δq passing through the load experiences a change of *potential energy* ΔU given by

$$\Delta U = V_B\,\Delta q - V_A\,\Delta q = (V_B - V_A)\,\Delta q$$

or

$$\Delta U = V_{AB}\,\Delta q = V_{AB} I\,\Delta t \tag{21.16}$$

If the potential V_A at A is greater than V_B at B, the charge Δq will lose potential energy in going through AB. Thus, there is an *energy input* by the charge to the portion AB of the circuit. On the other hand, if V_A is less than V_B, the charge passing through AB will gain potential energy. This implies that there will be an *energy output* from this portion AB of the circuit.

The rate of energy transfer between the circulating charge and the circuit is called the power, P. Thus, from Equation 21.16,

$$\boxed{P = \frac{\Delta U}{\Delta t} = V_{AB} I} \tag{21.17}$$

which is a general expression for the electrical power input or output to or from the load between A and B. In Equation 21.17, the unit of V is the volt and the unit of I is the ampere, which yield the unit of power to be the joule/s or watt, as shown below.

$$\text{(volt)(ampere)} = \left(\frac{\text{joule}}{\text{coulomb}}\right)\left(\frac{\text{coulomb}}{\text{second}}\right) = \left(\frac{\text{joule}}{\text{second}}\right) = \text{watt}$$

If the portion of the circuit AB contains a motor, the potential energy appears in the form of mechanical energy, while if it is a storage battery that is being charged (for this purpose the positive terminal of the storage battery will be connected to the positive terminal of the source), the potential energy will be used as stored chemical energy in the battery. On the other hand, if the circuit AB is a resistor, the potential energy appears as heat energy in the resistor, as discussed below.

Joule's Law of Heating

Let us assume that the load AB in Figure 21.11 is purely resistive. In this case the circulating charge loses potential energy in going from A to B; hence, there will be an energy input to the resistor. The potential energy lost by the circulating charges appear as thermal energy in the resistor. The temperature of the resistor rises until the heat flowing out of the resistor equals the energy input to the circuit. Thus, we say that the potential energy is being dissipated in the resistor, and the process is called *Joule heating*.

We can explain Joule heating from the microscopic point of view by considering the motion of electrons in a metallic resistor. After the initial acceleration the electrons colliding with the atoms settle down to a steady drift velocity. Any further loss in potential energy does not appear as the increased kinetic energy of the electrons. In collisions between electrons and atoms, this energy is transferred to the atoms of the crystal, thereby increasing the amplitude of thermal vibrations of the crystal lattice. This leads to an increase in temperature.

If the circuit is an ohmic circuit element, $V_{AB} = IR$; hence, the power dissipated P_R in a resistor according to Equation 21.17 takes the form

$$\boxed{P_R = I^2R} \tag{21.18}$$

which is known as *Joule's law*.

Unless it is our intention to convert electrical energy into heat energy, the power loss as given by Equation 21.18 is undesirable. This power loss is called the *I^2R loss* or the *Joule heating loss*. Since every electrical circuit contains resistance, there will always be some I^2R loss. This loss becomes quite significant when electricity is transferred from a generating station to homes. To keep this loss to a minimum, besides using a wire of very low resistance, it is necessary and desirable to use very low current I. But to keep the power supplied $P = VI$ constant, if I is low, V must be high. This means that from a generating station, the electric power is transmitted at low current and high voltage. Because high voltage is dangerous, for safety reasons when this electric power is received in homes and businesses it is changed into high current and low voltage. Change from high current and low voltage into low current and high voltage is accomplished by means of a step-up transformer; the reverse is accomplished by means of a step-down transformer. We shall discuss these transformers in Chapter 26.

Source Power Conversion

The purpose of a source of emf is to convert nonelectrical energy into electrical potential energy. The rate at which this energy is converted is the power supplied by the source, and may be calculated as shown below.

Suppose in time Δt the charge Δq is transported within the source from the low-potential terminal to the high-potential terminal. If $\mathcal{E}$ is the emf, the work done ΔW by the nonelectrical force will be

$$\Delta W = \mathcal{E}\,\Delta q$$

Then the rate at which the source does work, called power P_S, is given by

$$P_S = \frac{\Delta W}{\Delta t} = \mathcal{E}\frac{\Delta q}{\Delta t}$$

Since $\Delta q/\Delta t = I$,

$$\boxed{P_S = \mathcal{E}I} \qquad (21.19)$$

If the source is a storage battery or a dry cell, the chemical energy is converted into electrical potential energy according to Equation 21.19. On the other hand, if the source is a generator or a dynamo, it is the mechanical energy that is converted into electrical potential energy.

Normally, the conventional current direction within the source is from a low potential to a high potential. If the current in the source is made to go in the backward direction, that is, from the high potential terminal to the low potential terminal by connecting the source to another source of higher emf, the electrical potential energy will be converted into nonelectrical energy. Such is the case when the storage battery is being charged so that the electrical potential energy will be converted into chemical potential energy. Similarly, a dynamo working backward is an electric motor, because electrical potential energy is being converted into mechanical energy.

EXAMPLE 21.4 Consider the circuit shown in the accompanying figure, consisting of a source of emf $\mathcal{E}$ with an internal resistance $r = 0.1\ \Omega$ and an electric appliance L with resistance R that operates at 300 W. The voltage V is 110 V while the current through the circuit is 5 A. Calculate the following quantities: (a) the power output from the source; (b) the power used in the joule heating of R; (c) the magnitude of R; (d) the power loss in the source itself; and (e) the emf of the source.

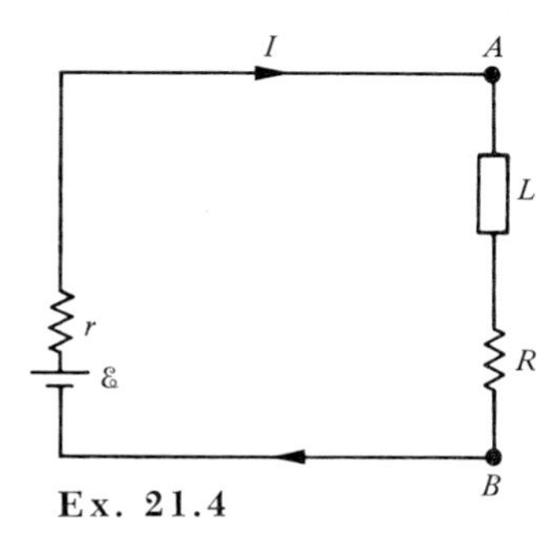

Ex. 21.4

(a) The power output from the source which is delivered to R and L is

$$P = V_{AB}I = 110\text{ V} \times 5\text{ A} = 550\text{ W}$$

(b) Of the 550 W, 300 W ($= P_A$) is used by L. Therefore, P_R power loss in Joule heating of R is

$$P_R = P - P_A = 550\text{ W} - 300\text{ W} = 250\text{ W}$$

(c) From the relation $P_R = I^2R$,

$$R = \frac{P_R}{I^2} = \frac{250\text{ W}}{(5\text{ A})^2} = 10\ \Omega$$

(d) Joule heating loss in the internal resistance r of the source is

$$P_r = I^2r = (5\text{ A})^2(0.1\ \Omega) = 2.5\text{ W}$$

(e) Thus, the total power provided by the source, or rate at which the source does work, is

$$P_S = P_A + P_R + P_r = 300\text{ W} + 250\text{ W} + 2.5\text{ W} = 552.5\text{ W}$$

From the relation $P_S = \mathcal{E}I$, the emf of the source is

$$\mathcal{E} = \frac{P_S}{I} = \frac{552.5\ \text{W}}{5\ \text{A}} = 110.5\ \text{V}$$

EXERCISE 21.4 Suppose in Example 21.4 that $V_{AB} = 100$ V, $R = 8\ \Omega$, $I = 6$ A, and $r = 0.2\ \Omega$. Calculate the power used by the appliance and its resistance, P_r, P_S, and $\mathcal{E}$. [*Ans.:* $P_A = 312$ W, $P_R = 288$ W, $P_r = 7.2$ W, $P_S = 607.2$ W, $\mathcal{E} = 101.2$ V.]

SUMMARY

The source emf $\mathcal{E}$ is equal to the work done in carrying 1 C of charge through the source; $\mathcal{E} = W/q$. Current is the net charge flowing across the area per unit time, $I = \Delta Q/\Delta t$. The direction of current is that in which the positive charge will drift, that is, in the direction of **E**. The unit of current is the *ampere*. The *current density* **J** is defined as the current per unit cross-sectional area, $J = I/A = nqv$.

The *resistivity* ρ of a conductor is the ratio of the electric intensity to the current density. Thus, $\rho = E/J$ and $J = E/\rho = \sigma E$, where $\sigma = 1/\rho$ is called the *conductivity* of the material. The ratio of the potential difference V to the current I is called the *resistance* R. That is, $R = V/I = \rho L/A$. The unit of resistance is the *ohm* (Ω), while the unit of resistivity ρ is in *ohm-m*.

The variation of ρ with temperature T is given by $\rho = \rho_0[1 + \bar{\alpha}(T - T_0)]$ or $R = R_0\,[1 + \bar{\alpha}(T - T_0)]$, where $\bar{\alpha}$ is the *mean temperature coefficient of resistivity*. In the neighborhood of 4 K, many metals exhibit abnormally low resistivity. This is the *superconducting state*, and the property is called *superconductivity*. The *critical temperature* T_c is the temperature below which the material is superconducting.

The resistance of a metallic conductor is $R = V/I$, which is the statement of *Ohm's law*. Any resistor that obeys Ohm's law is a *linear circuit element*. For a nonlinear circuit element, the plot of V versus I is not linear.

The rate of energy transfer between the circulating charge and the circuit is called the *power*, P; $P = \Delta U/\Delta t = V_{AB}I$. The unit of power, J/s, is called the *watt*. The process of dissipation of potential energy in the resistor is called *Joule heating*, and $P_R = I^2R$ is known as *Joule's law*. The rate at which the source does work called power is $P_S = \mathcal{E}I$.

QUESTIONS

1. When charge carriers reach the terminals of the source of an emf, what happens to their velocity or kinetic energy?
2. When a potential difference is maintained between two ends of an insulator, what effect has it on the charges in regard to the velocity, force, and energy?
3. How do you measure drift velocity and distinguish it from the random motion of the charge carriers?
4. In a gas where there are both positive and negative charge carriers, what are the relative directions of **v**, **J**, and the force **F** on the charge carries if the direction of **E** is from left to right?
5. Are collisions between the charge carriers and the atoms in a conductor elastic or inelastic? What types of collisions contribute to Joule heating?

6. Should the conductivity of a conductor decrease or increase with temperature? Explain.
7. Can you use a resistance thermometer in the vicinity of absolute zero? Explain.
8. Suppose that there is a current I in (a) a conductor; and (b) a semiconductor. What effect will an increase or decrease in temperature have on the current?
9. Why are thermistors used as automatic temperature-control switching devices?
10. Give an example of each of the following systems in which electrical energy is converted into (a) mechanical energy; (b) heat energy; and (c) chemical energy. Also, give an example in which (d) chemical energy is converted into electrical energy; and (e) mechanical energy is converted into electrical energy.

PROBLEMS

1. Calculate the emf of a source if the work done by the nonelectrical forces on a charge of 1.5 C is 30 J.
2. Considering the electrons as charge carriers in a battery of 12 V, calculate the change in potential energy in terms of joules as the electrons migrate from one pole to another.
3. If there are 10^{18} electrons flowing across any cross section of a wire in 1 min, what is the current in the wire?
4. There is a current of 5 mA in a wire. How much charge passes any cross section of the wire in 10 s? If this current is due to the flow of electrons, how many electrons pass through the wire in 10 s?
5. An automobile battery is rated at 120 ampere-hours, that is, 1 A for 120 h, 10 A for 12 h, and so on. If a starter in the automobile draws a current of 400 A, how long could the starter be kept on before the battery runs down? How much is the total charge that goes through the starter in this time?
6. An electron going around a proton in Bohr-type orbit makes 6.6×10^{15} revolutions per second. What is the current in such an orbit caused by the electron motion?
7. A gas between two electrodes is ionized by applying a very high potential difference between two electrodes. The gas ionizes and every minute 5×10^{20} electrons, and 2×10^{20} positive ions move in opposite directions. Calculate the current and its direction.
8. A 0.5-m-wide belt of a Van de Graaff machine runs at 20 m/s. How much charge in C/s must be sprayed on the belt to give a current of 5 mA?
9. A metal wire 1 mm in diameter contains 3×10^{28} electrons/m^3, and a charge of 100 C/h crosses any cross-sectional area. Calculate the current and current density in the wire and the drift velocity of the electrons.
10. Consider a copper wire of 1 mm diameter in which there is a current of 5 A. Calculate (a) the number of free electrons per m^3; (b) the total charge that crosses any cross-sectional area in 1 h; (c) the current density; and (d) the drift velocity.
11. Calculate the resistance of a wire 10 m long that has a diameter of 2 mm and resistivity of 2.63×10^{-8} Ω-m.
12. A copper wire in household wiring has a cross section of 0.32×10^{-6} m^2

(called 22-gauge wire). The resistivity of such a wire is $1.72 \times 10^{-8}\ \Omega$-m. What is the resistance of such a wire that is 30 m long?

13. A wire has a resistance of 200 Ω. What will be the resistance of a wire of the same material if it is 2 times as long and its cross section is 2 times as large?

14. Consider a copper wire of length 30 m and cross-sectional area $2 \times 10^{-6}\ m^2$. The IR drop across this wire is 3 V. Calculate I, J, and E.

15. Calculate the resistance of a copper plate of length 15 cm and rectangular cross section of 3 cm × 0.5 cm.

16. A 22-gauge copper wire has a diameter of 0.64 mm. Calculate the resistance of such a wire 5 km long at 0°C.

17. What is the resistance of a wire made of (a) copper; (b) tungsten; (c) iron; and (d) constantan if each is 1 m long and has a diameter of 1 mm?

18. A 10-m-long wire of 1 mm diameter has a resistance of 10 Ω. What should be the length of a wire of the same material having a 2-mm diameter so that it has a resistance of 5 Ω.

19. A copper wire of length L and cross-sectional area A is to be replaced by an aluminum wire of the same length and resistance but of different cross-sectional area A_1. Calculate A_1 in terms of A.

20. Show that the relation $R = \rho L/A$ may also be written as $R = \rho d L^2/m = \rho m/dA^2$, where ρ, L, A, m, and d are the resistivity, length, cross-sectional area, mass, and density, respectively, of a wire.

21. A total of 50 g of copper is stretched into a wire 100 m long at 0°C. What is the resistance of the wire? (*Hint:* Find A from $m = dV$, d = density.)

22. A wire of 50 Ω has a certain length. If it is stretched to 4 times its length while keeping its volume the same, what is the new resistance?

23. If 1 g of gold, 1 g of copper, and 1 g of aluminum are drawn into uniform wires of length 10 m each, calculate the resistance of each. What is the ratio of the resistance to the diameter in each case?

24. An electrical system operating at 12 V draws a current of 0.05 A. What is the resistance of the system?

25. Calculate the resistance of the heating element of a toaster if a current of 5 A passes through it when connected across 120 V.

26. A voltmeter reads 24 V across a resistor and an ammeter reads a current of 2 A through it. Calculate R. What will be the current through the resistor if the voltage changes to 12 V?

27. A wire carries a current of 100 A and has a resistance of 120 μΩ/m. What is the potential difference across 1 m?

28. A copper wire at 0°C has a resistance of 10 Ω. (a) What will be the resistance when this wire is immersed in steam? (b) At what temperature will its resistance become double, that is, 20 Ω?

29. A certain wire has a resistance of 10 Ω at 0°C and 10.2 Ω at 200°C. Calculate the temperature coefficient of resistance. What is the value of this resistance at 70°C?

30. Calculate the resistance of a carbon rod at 100°C which has a resistance of 0.032 Ω at 0°C.

31. Consider a copper wire and a tungsten wire each 1 m long and of 1 mm diameter. Compare their resistances at 0°C and 100°C.

32. Calculate the resistance of a tungsten wire at 20°C which has a resistance of 10 Ω at 1000°C. (First calculate the resistance at 0°C.)

33. The resistance of constantan wire at room temperature (20°C) is 100 Ω. What will be its resistance at (a) 0°C; (b) −100°C; (c) +100°C; and (d) 1000°C?

34. A heater is rated at 1200 W when the current is 5 A. Calculate (a) the resistance of the heater; and (b) the voltage across the heater.

35. Suppose that a current of 5 A is drawn from a battery to obtain 1200 J of work over a time interval of 1 min. Calculate (a) the available power; and (b) the voltage across the battery.

36. Calculate the current supplied by the 12-V battery of an automobile to a starter that develops a power of 2 kW (1 kW = 1.34 hp).

37. Calculate the cost of lighting a 100-W bulb every night for 6 h for 1 year at the rate of 3 cents/kWh.

38. The maximum energy a certain 100,000-Ω resistor can dissipate is 2 W; that is, the power rating is 2 W. What is the maximum voltage that can be applied to this resistor? What is the maximum voltage if the power rating is 5 W?

39. How much current is drawn from a 12-V automobile battery to operate a 1.0-hp (=746-W) starter? How much energy is used up if it takes 5 s to start the automobile?

40. A motor is designed to operate at 120 V and 5 A. If its efficiency is 50 percent, calculate the energy lost in cal/s.

41. A $\frac{1}{2}$-hp (1 hp = 746 W) motor draws a current of 4.00 A from a line source of 110 V. What is its efficiency?

42. The headlights of an automobile are left on by mistake. This results in a 12-V battery going dead in 4 h. If the two lamps have a total power of 400 W, what was the total chemical energy in the battery that was available for conversion into electrical energy?

43. A 720-W heater operates from a 120-V line voltage. Calculate (a) the resistance of the heater; and (b) the current drawn by the heater. What fluctuations in power take place if the line voltage fluctuates between 110 and 130 V?

44. A 2500-W heater has to be made out of a constantan wire that can carry a maximum current of 5 A. What is the length of this wire if its diameter is 0.8 mm?

45. The current through a hot-water heater is 20 A when connected to a voltage of 240 V. If the voltage drops to 110 V, what will be the current? How much longer will it take to heat the same amount of water through the same temperature interval at the lower voltage?

46. An electric iron draws 10 A of current from a 120-V line voltage. Half of the energy is radiated while the other half heats the iron. If the mass of the iron is 1.5 kg and its specific heat is 0.1 kcal/kg-C°, how long will it take to heat the iron from 20 to 110°C?

47. A typical 100-W lamp operates at 120-V line voltage. Calculate (a) the current through the filament; (b) the resistance of the filament; and (c) the number of electrons through the filament/second.

48. A commercial 12-V battery has a resistance of 0.01 Ω. It has to be charged by connecting it to a 120-V line voltage. Show the circuit diagram for this purpose. If it is charged for 6 h, drawing a current of 10 A, what is the cost of the electrical energy used at the rate of 3 cents/kWh?

22
Electrical Circuits and Instruments

Simple electrical circuits can be solved by applying Ohm's law. For circuits that are not too complicated, we can use the method of conservation of energy or the definition of potential difference between two points in a circuit. Ultimately we must depend upon the more advanced method of Kirchhoff's rules, as will be discussed here.

We shall see the application of these rules not only to solve circuit problems, but also to many electrical instruments as well. The measurement of different electrical quantities is done by using a voltmeter, ammeter, Wheatstone bridge, potentiometer, or other instruments. These will be discussed in this chapter. In the end we shall discuss charging and discharging of a capacitor in a RC series circuit.

22.1. Single-Loop Circuit

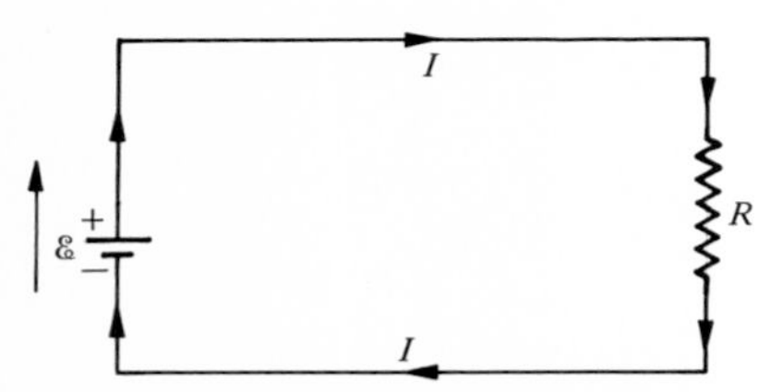

FIGURE 22.1. *Single-loop circuit containing R and ℰ.*

TO START we shall consider a simple single-loop circuit, and find the relation between the current, potential drop, and emf. Figure 22.1, containing a source of emf $\mathcal{E}$ and resistance R, is a single-loop circuit. According to energy conservation, the electrical power P_S supplied by the source must be equal to the power dissipated, P_R, as thermal energy (Joule heat) in the resistor. That is, if the current in the circuit is I,

$$P_S = P_R \qquad \text{or} \qquad \mathcal{E}I = I^2R \tag{22.1}$$

Thus,

$$\boxed{\mathcal{E} = IR, \qquad I = \frac{\mathcal{E}}{R}} \tag{22.2}$$

which is a *single-loop equation.*

We have assumed that the source has zero resistance, the connecting wires have negligible resistance, and R is the only resistance present in the circuit. In reality this is not true for the source resistance, although it is often true for the resistance of the connecting wires. The source always has some resistance, say r, however small, as shown in Figure 22.2. The shaded block implies that r is not separate from the source, r being inside the source itself and is called the *internal resistance*. Some power P_r must be lost as thermal energy in the resistor r as well. Thus, Equation 22.1 will take the following form:

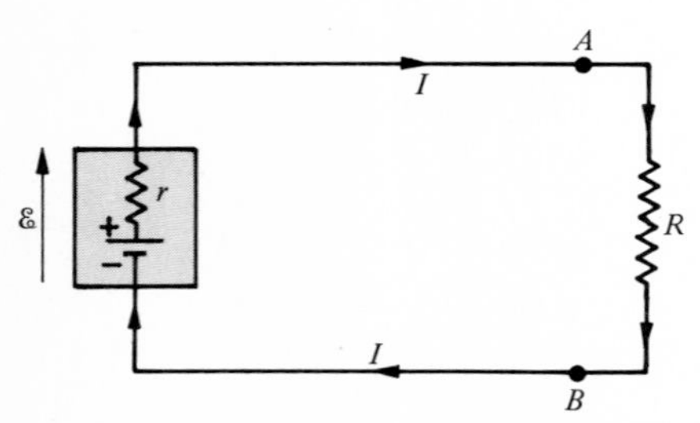

FIGURE 22.2. *Simple single-loop circuit containing R, ℰ, and r (internal resistance).*

$$P_S = P_R + P_r \qquad \text{or} \qquad \mathcal{E}I = I^2R + I^2r$$

Thus,

$$\boxed{\mathcal{E} = I(R + r), \qquad I = \frac{\mathcal{E}}{R + r}} \tag{22.3}$$

The application of the energy conservation principle is not always as simple as discussed above. An alternative is to make use of the definition of electrical potential. Accordingly, in the direct-current circuits, the potential at any point in the circuit must have the same value at any time. This is only possible if the following is true: *In a complete traversal of a circuit, the algebraic sum of all the potential differences encountered must add up to zero.* This is actually the second of the two Kirchhoff's rules which we shall discuss in Section 3.

Before we use this rule, we must decide on a convenient sign convention stated below and demonstrated in Figure 22.3. Let us keep in mind that in each of the four cases shown, point A is at a higher potential than the corresponding

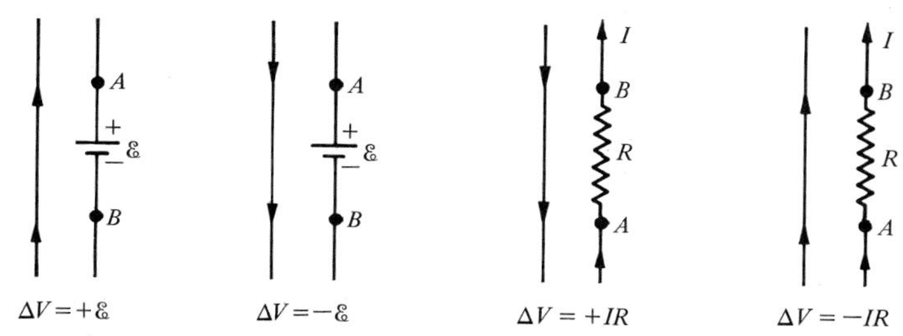

FIGURE 22.3. *Sign convention while traversing a circuit.*

point B. This is because the positive terminal is at a higher potential than the negative. The direction of the current is from a higher to lower potential.

1. If we travel through a source from − to +, the change in potential $\Delta V = +\mathcal{E}$, and in the opposite direction from + to −, $\Delta V = -\mathcal{E}$.
2. If we travel through a resistor in the direction opposite that of the current, the change in potential $\Delta V = +IR$, and in the same direction as the current, $\Delta V = -IR$. (This change is also called the IR drop.)

Let us apply these rules to the simple circuit in Figure 22.2. Starting with point A, where the potential is V_A, going clockwise we encounter resistor R, across which the potential change will be $-IR$. Next we meet the source where the potential drop will be $+\mathcal{E}$, while across r it will be $-Ir$. Finally, we arrive at A. The sum of the potentials above must be zero. That is,

$$-IR + \mathcal{E} - Ir = 0$$

or

$$\mathcal{E} = I(R + r) \tag{22.4}$$

which is the result already derived in Equation 22.3.

This method has another advantage. Suppose that we want to find the potential difference between points A and B in Figure 22.2. Let V_A and V_B be the potentials at points A and B. Starting from A and going clockwise to point B, we may write

$$V_A - IR = V_B$$

or

$$\Delta V = V_A - V_B = IR$$

Substituting for I from Equation 22.3, we get

$$\boxed{\Delta V = \mathcal{E}\frac{R}{R + r}} \tag{22.5}$$

We can arrive at the same result if we start from A and follow the circuit in a counterclockwise direction. Thus,

$$V_A + Ir - \mathcal{E} = V_B$$
$$V_A - V_B = \mathcal{E} - Ir$$

Substituting for $I = \mathcal{E}/(R + r)$ and simplifying, we get the same value of ΔV as given by Equation 22.5, which is called the *terminal voltage* V_t.

Example 22.1 A typical 12-V automobile battery has a resistance of 0.012 Ω. What is the terminal voltage of this battery when the starter draws a current of 100 A? Also, calculate R, P_S, P_R, and P_r.

Since $\mathcal{E} = 12$ V, the terminal voltage V_{AB} from Equation 22.4 may be written

$$V_t = IR = \mathcal{E} - Ir = 12\text{ V} - (100\text{ A})(0.012\ \Omega) = (12 - 1.2)\text{ V} = 10.8\text{ V}$$

Also, from this relation

$$R = \frac{\mathcal{E} - Ir}{I} = \frac{10.8\text{ V}}{100\text{ A}} = 0.108\ \Omega$$

To calculate P,

$$P_r = I^2r = (100\text{ A})^2(0.012\ \Omega) = 120\text{ W}$$
$$P_R = I^2R = (100\text{ A})^2(0.108\ \Omega) = 1080\text{ W}$$
$$P_S = \mathcal{E}I = (12\text{ V})(100\text{ A}) = 1200\text{ W}$$

Note that $P_S = P_R + P_r$, as it must be.

Exercise 22.1 A typical 12-V automobile battery of internal resistance 0.01 Ω is connected to an external circuit of resistance 0.10 Ω. Calculate the terminal voltage, current in the circuit, P_r, P_R, and P_S. [*Ans.:* $V_t = 10.9$ V, $I = 109$ A, $P_r = 119$ W, $P_R = 1188$ W, $P_S = 1308$ W.]

22.2. Series and Parallel Circuits

Let us now analyze circuits containing series and parallel combinations of resistors and sources of emf.

Series Circuits

Figure 22.4 contains three resistors, R_1, R_2, and R_3, which are connected end to end and are said to be in series. The characteristic describing this series

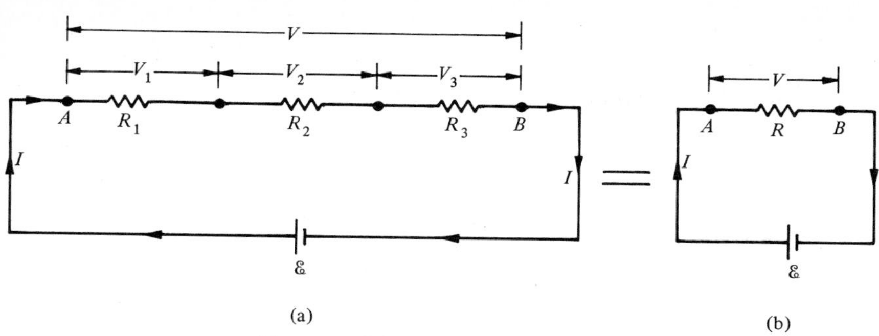

FIGURE 22.4. *Resistors in series.*

circuit is that there is the same current I through each resistor. Let V be the potential difference between A and B. Let V_1, V_2, and V_3 be the potential drops across the resistors R_1, R_2, and R_3, respectively, so that

$$V = V_1 + V_2 + V_3 \tag{22.6}$$

where $V_1 = IR_1$, $V_2 = IR_2$, and $V_3 = IR_3$. Let R be the resistance that is equivalent to this series combination as far as external connections are concerned, so that, as shown in Figure 22.4(b), $R = V/I$ or $V = IR$. Substituting the values of V, V_1, V_2, and V_3 in Equation 22.6, we get

$$IR = IR_1 + IR_2 + IR_3$$

or

$$R = R_1 + R_2 + R_3$$

If there were more than three resistors in series, a single resistance equivalent to this combination will be

$$\boxed{R = R_1 + R_2 + R_3 + \cdots} \tag{22.7}$$

That is in a series circuit a single equivalent resistance R is equal to the sum of the individual resistors.

In order to connect sources of emf in series, the positive terminal of one source is connected to the negative terminal of the next one, as shown in Figure 22.5. A common example is a flashlight, where two or three dry cells are connected in series. The law of conservation of energy requires that a single source of emf which must be equivalent to several sources of emf connected in series, as shown in Figure 22.5, is given by

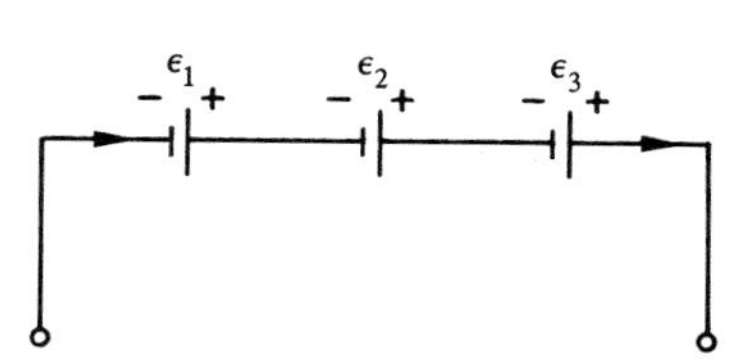

FIGURE 22.5. *Sources of emfs in series.*

$$\boxed{\mathcal{E} = \mathcal{E}_1 + \mathcal{E}_2 + \mathcal{E}_3 + \cdots} \tag{22.8}$$

where $\mathcal{E}$ is equal to the total work done in moving a unit positive charge through all the sources.

Finally, Figure 22.6 shows three resistors R_1, R_2, and R_3, which are in series, and three sources of emf $\mathcal{E}_1$, $\mathcal{E}_2$, and $\mathcal{E}_3$, which are in series also. In this case the equivalent resistance is $R = R_1 + R_2 + R_3$ and the equivalent emf is $\mathcal{E} = \mathcal{E}_1 + \mathcal{E}_2 + \mathcal{E}_3$. The current through the circuit will be

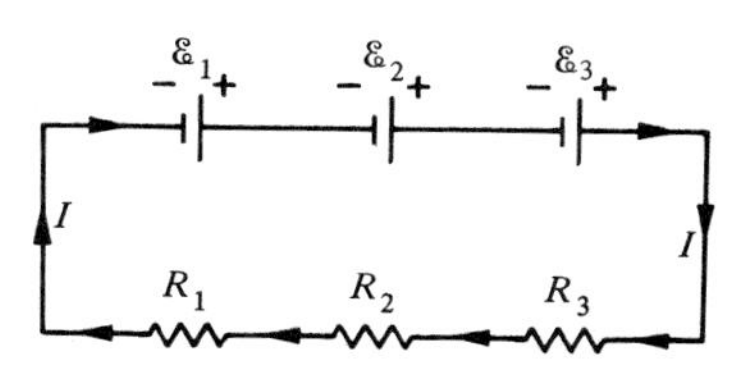

FIGURE 22.6. *Resistors and sources of emfs in series.*

$$I = \frac{\mathcal{E}}{R} = \frac{\mathcal{E}_1 + \mathcal{E}_2 + \mathcal{E}_3}{R_1 + R_2 + R_3} \tag{22.9}$$

Parallel Circuits

Figure 22.7 shows three resistors, R_1, R_2, and R_3, connected in parallel. The characteristic describing resistors in parallel is that there is the same potential

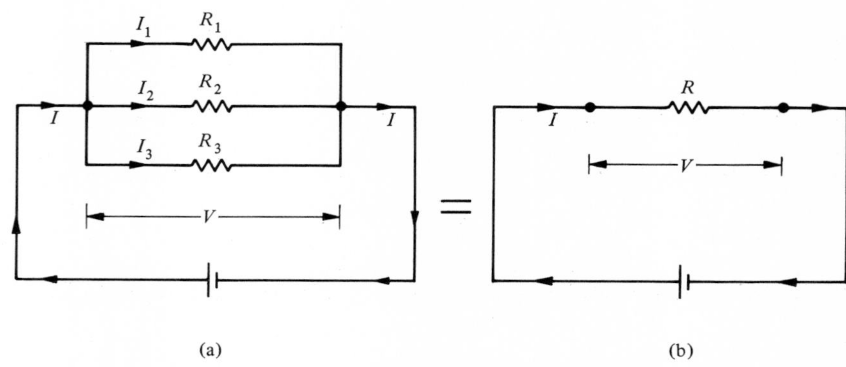

FIGURE 22.7. *Resistors in parallel.*

difference V across each of the resistors. The current I in the circuit must be equal to the sum of the currents I_1, I_2, and I_3 in the three resistors. Thus,

$$I = I_1 + I_2 + I_3 \tag{22.10}$$

where $I_1 = V/R_1$, $I_2 = V/R_2$, and $I_3 = V/R_3$. Let R be the resistance which is equivalent to the three resistors in parallel so that, as shown in Figure 22.7(b), $I = V/R$. Substituting the values of I, I_1, I_2, and I_3 in Equation 22.10,

$$\frac{V}{R} = \frac{V}{R_1} + \frac{V}{R_2} + \frac{V}{R_3}$$

or

$$\frac{1}{R} = \frac{1}{R_1} + \frac{1}{R_2} + \frac{1}{R_3}$$

In general,

$$\boxed{\frac{1}{R} = \frac{1}{R_1} + \frac{1}{R_2} + \frac{1}{R_3} + \cdots} \tag{22.11}$$

That is, the reciprocal of the single equivalent resistor R is equal to the sum of the reciprocals of the resistances R_1, R_2, R_3, . . . , which are connected in parallel.

Figure 22.8(a) shows how to connect sources of emf in parallel combination—all the positive terminals are connected together and all the negative terminals are connected together. All sources of emf in parallel should always be of the same voltage (otherwise, some will behave as sources while others behave as sinks). Thus, the voltage of the system is equal to the voltage across any one of them, as shown in Figure 22.8(b). Thus the total current is equal to

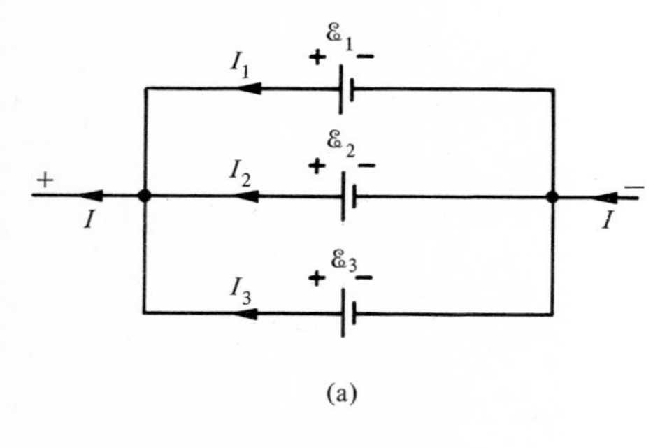

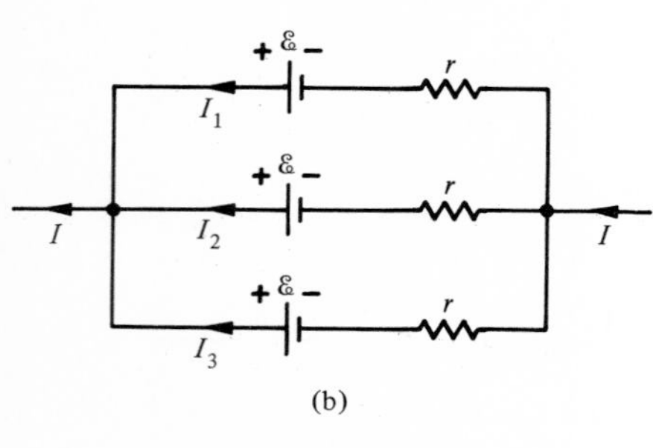

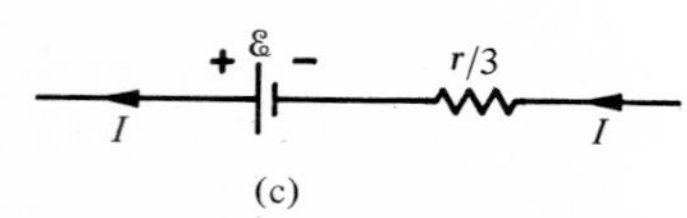

FIGURE 22.8. *Sources of emfs in parallel.*

the sum of the currents through each of the sources, that is, $I = I_1 + I_2 + I_3$ while $\mathcal{E} = \mathcal{E}_1 = \mathcal{E}_2 = \mathcal{E}_3$. Thus, using this combination it is possible to increase the current without changing the emf. Also if each has an internal resistance r, the net resistance will be $r/3$, as shown in Figure 22.8(c).

Example 22.2 For the network shown in the accompanying figure, calculate the total resistance and current I_1.

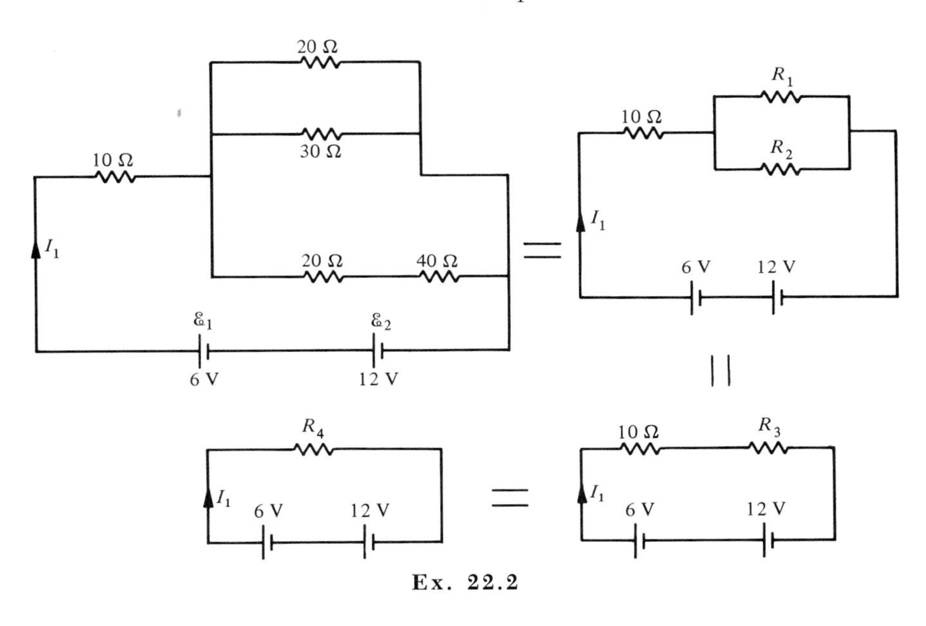

Ex. 22.2

Since the 20-Ω and 30-Ω are in parallel, equivalent resistance R_1 is:

$$\frac{1}{R_1} = \frac{1}{20} + \frac{1}{30} = \frac{5}{60} = \frac{1}{12} \qquad \text{or} \qquad R_1 = 12\ \Omega$$

The 40 Ω and 20 Ω are in series; therefore, their combined resistance R_2 is

$$R_2 = 40\ \Omega + 20\ \Omega = 60\ \Omega$$

Now R_1 and R_2 are in parallel. Therefore, the combined resistance R_3 is

$$\frac{1}{R_3} = \frac{1}{12} + \frac{1}{60} = \frac{6}{60} = \frac{1}{10} \qquad \text{or} \qquad R_3 = 10\ \Omega$$

R_3 and 10 Ω are in series. Hence, the combined resistance R_4 is

$$R_4 = R_3 + 10\ \Omega = 10\ \Omega + 10\ \Omega = 20\ \Omega$$

The combined voltage of the sources $\mathcal{E}_1$ and $\mathcal{E}_2$ is

$$\mathcal{E} = \mathcal{E}_1 + \mathcal{E}_2 = 6\text{ V} + 12\text{ V} = 18\text{ V}$$

Thus, the whole circuit reduces to an equivalent circuit containing a net resistance of 20 Ω and a source of 18 V. Thus, I_1 will be

$$I_1 = \frac{V}{R} = \frac{18\ \text{V}}{20\ \Omega} = 0.9\ \text{A}$$

Exercise 22.2 Suppose that in Example 22.2, the 20-Ω and 30-Ω resistors which are in parallel are replaced by a single resistor R. What should be the value of R so that the current I_1 will be exactly 1 A? [*Ans.:* $R = 9.23\ \Omega$.]

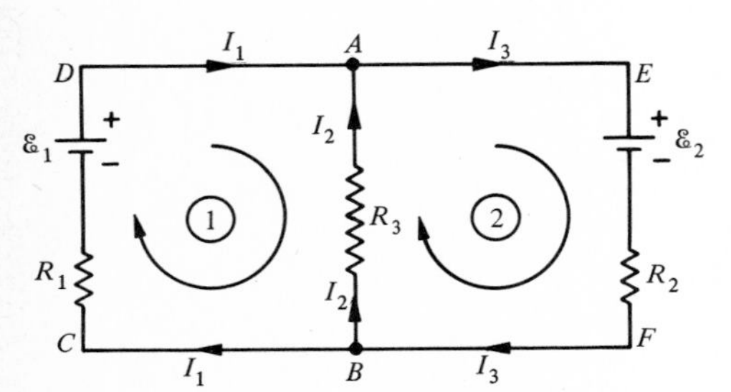

Figure 22.9. *Simple multiloop circuit.*

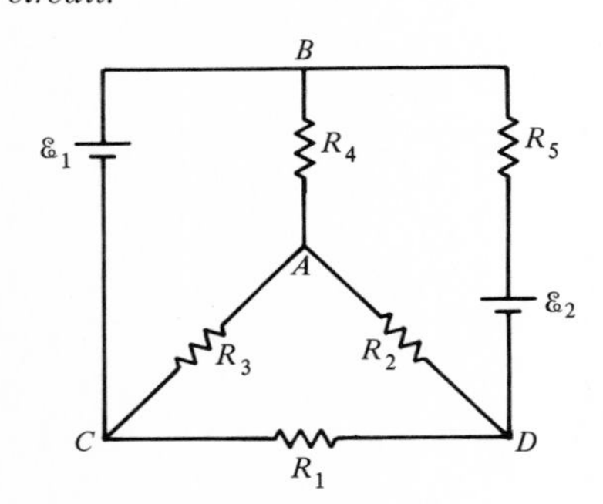

Figure 22.10. *Complex multiloop circuit.*

22.3. Multiloop Circuits (Kirchhoff's Rules)

There are many circuits that cannot be solved by methods discussed previously. These circuits may be as simple as the one shown in Figure 22.9 or as complex or more as the one shown in Figure 22.10. Such circuits may be analyzed by means of a systematic procedure based upon two rules devised by Gustav Robert Kirchhoff (1824–1887), known as Kirchhoff's rules. These rules are a generalization of the methods discussed in Section 22.1.

Kirchhoff's First Rule (Junction Theorem): ***The algebraic sum of the currents at any junction, or branch point, is equal to zero.***

That is,

$$\sum I = 0 \tag{22.12}$$

A junction or branch point in an elecrical network is a point where three or more conductors are connected, such as point A or B in Figure 22.9 or point A, B, C, or D in Figure 22.10. This rule is simply a consequence of the *law of conservation of charge* and the fact that in a steady-state current, no charge can accumulate at the junction. In other words, the junction rule states that *the sum of the electric currents entering any junction must equal the sum of the electric currents leaving that junction.* That is, Equation 22.12 takes the form

$$\left(\sum I\right)_{\text{in}} = \left(\sum I\right)_{\text{out}} \tag{22.13}$$

Kirchhoff's Second Rule (Loop Theorem): ***The algebraic sum of the potential differences encountered in going around a closed loop is equal to zero.***

In general, the potential differences encountered are the sources of emf and the IR drops across resistors. Hence, we may write this rule as

$$\sum \mathcal{E} + \sum IR = 0 \tag{22.14}$$

$\mathcal{E}$ is equal to the amount of work done by the source on a unit positive charge while the IR is equal to the work performed by a unit positive charge when going through the resistor. Thus, the loop theorem is a direct consequence of the *law of energy conservation.* Note that $\mathcal{E}$ and IR are taken as positive or negative according to the convention established in Figure 22.3.

To explain the use of the Kirchhoff rules, let us consider the electrical network shown in Figure 22.9. $\mathcal{E}_1$, $\mathcal{E}_2$, R_1, R_2, and R_3 are given. We want to find the currents in different branches of this network assuming them to be I_1, I_2, and I_3, as shown in Figure 22.9. In this network there are two junctions, A and B. We apply the junction rule to point A, which yields

$$I_1 + I_2 = I_3 \tag{22.15}$$

If the direction of any one of these currents is assumed to be wrong, the final calculations will yield a negative value for that current. Applying the junction rule to point B does not yield anything new. In general, if there are n junction points, one applies the junction rule to only $(n - 1)$ points. The application of the junction rule to the nth point does not lead to an independent equation.

Before applying the loop rule, we must decide arbitrarily to traverse the loops clockwise or counterclockwise. We choose a clockwise direction. Now imagine the network to be separated into a number of simple loops. Thus, in Figure 22.9 we have two loops—loop 1 being $ABCDA$ and loop 2, $AEFBA$. (There is a third loop, $DAEFBCD$, but it does not yield any new information; it is simply a combination of the other two loops.) Remember that: (1) If the resistor is traversed in the direction of the current, the change in potential is $-IR$; otherwise, it is $+IR$. (2) If a source of emf is traversed from $-$ to $+$, the change in potential is $+\mathcal{E}$; otherwise it is $-\mathcal{E}$. Thus, for loop 1 we may write

$$I_2R_3 - I_1R_1 + \mathcal{E}_1 = 0 \tag{22.16}$$

while for loop 2 we may write

$$-\mathcal{E}_2 - I_3R_2 - I_2R_3 = 0 \tag{22.17}$$

Solving Equations 22.15, 22.16, and 22.17 yields the values for the currents I_1, I_2, and I_3. We shall illustrate this in the following example.

Example 22.3 Calculate the currents through all the resistors in the accompanying figure.

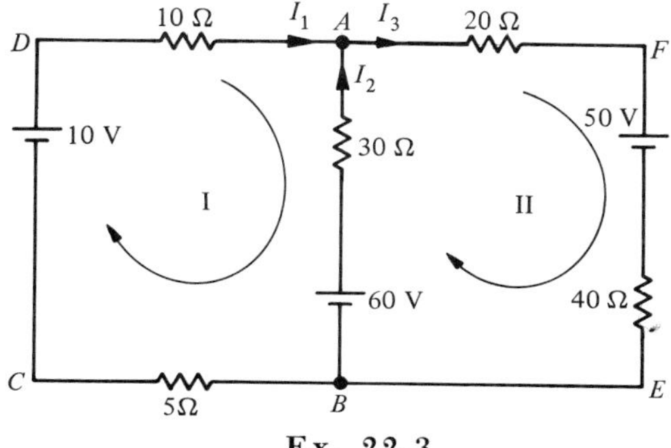

Ex. 22.3

If we indicate I_1, I_2, and I_3 as three possible currents, that will take care of the currents through all the resistors. We arbitrarily assign directions to these currents. Let us apply the junction rule to point A.

$$I_1 + I_2 = I_3 \tag{i}$$

If we know I_1 and I_2, we need not solve for I_3, which is equal to $I_1 + I_2$.

Applying the loop rule to loop I, we start from A and go clockwise.

$$30\,I_2 - 60 - 5\,I_1 + 10 - 10\,I_1 = 0$$

or

$$30\,I_2 - 15\,I_1 - 50 = 0 \tag{ii}$$

Applying the loop rule to loop II, we start from A:

$$-20I_3 - 50 - 40I_3 + 60 - 30I_2 = 0$$

$$30I_2 + 60I_3 - 10 = 0 \tag{iii}$$

Thus, we have three Equations—i, ii, and iii—and three unknowns—I_1, I_2, and I_3. To solve these, first let us substitute $I_3 = I_1 + I_2$ from Equation i into Equation iii:

$$30I_2 + 60(I_1 + I_2) - 10 = 0$$

or

$$60I_1 + 90I_2 - 10 = 0 \tag{iv}$$

Multiplying Equation ii by 3, we get

$$90I_2 - 45I_1 - 150 = 0 \tag{v}$$

Subtracting Equation v from iv,

$$105I_1 + 140 = 0 \qquad \text{or} \qquad I_1 = -1.33 \text{ A}$$

Substituting this in Equation v,

$$90I_2 - 45(-1.33) - 150 = 0 \qquad \text{or} \qquad I_2 = 1 \text{ A}$$

and

$$I_3 = I_1 + I_2 = -1.33 \text{ A} + 1 \text{ A} = -0.33 \text{ A}$$

The negative signs for I_1 and I_3 mean that the actual directions of I_1 and I_3 are opposite to those shown.

Exercise 22.3 With what voltage battery should we replace the 10-V battery in the figure of Ex. 22.3 so that the current I_1 will be 2 A? [*Ans.*: $\mathcal{E} = 13.3$ V.]

22.4. Direct-Current Measuring Instruments

Instruments such as an ammeter, voltmeter, Wheatstone bridge, or potentiometer are used for the purpose of measuring currents, voltages, resistances, and emf. One of the basic instruments for making electrical measurements is a galvanometer. We shall discuss the working and construction of a galvanometer in detail in Chapter 23. At this point we give a very brief discussion and show how to modify a galvanometer to use it as an ammeter or a voltmeter.

A *galvanometer* consists of a coil that is freely pivoted and placed in a magnetic field. A pointer attached to the coil moves over a calibrated scale as shown in Figure 22.11. When the current passes through the coil, it interacts with the magnetic field, thereby producing a deflection of the coil. The deflection is directly proportional to the current. Two basic quantities that describe a particular galvanometer are the resistance R_G of its coil, and the current I_G which going through the coil will produce a full-scale deflection. The galvanometer may also be used for measuring potential differences across two points. This is possible because the coil is a linear conductor; hence, according to Ohm's law, the current is proportional to the potential difference between the terminals of the coil. Thus, the deflection will be proportional to the potential difference, and the maximum deflection will correspond to a voltage drop of $V_G = R_G I_G$.

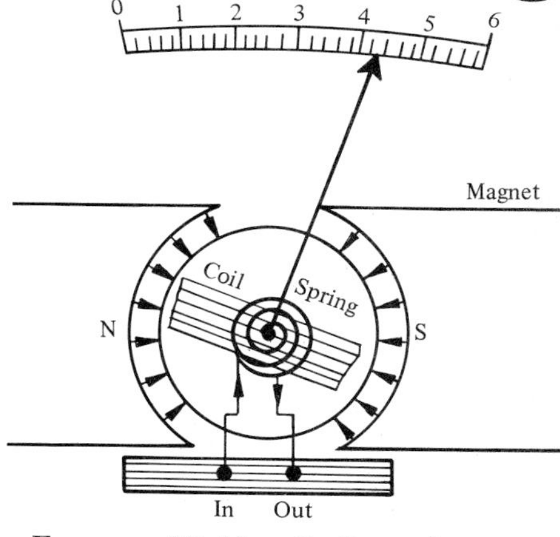

Figure 22.11. *Outline of a typical galvanometer.*

A typical galvanometer has a resistance ranging between 10 and 100 Ω, while a current of a few milliamperes produces a full-scale deflection. Let us say that $R_G = 50\ \Omega$ and $I_G = 2$ mA; therefore, $V_G = R_G I_G = 50\ \Omega \times 2 \times 10^{-3}\ \text{A} = 0.1$ V. This means that the galvanometer may be used as a current-measuring device for currents up to 2 mA, or it may be used as a potential-difference-measuring device up to a maximum range of 0.1 V. Obviously, these ranges are too small for most practical purposes. Modified forms of a galvanometer, called ammeter and voltmeter, are used for measuring current and voltage difference, respectively, and are described below. Also a current-measuring device must be inserted in series into the circuit in which the current has to be measured, whereas the voltage-measuring device must be connected across the points between which the potential difference is to be measured.

The Ammeter

An *ammeter* is a modified galvanometer. If the galvanometer without modification is inserted in the circuit to measure the current, the galvanometer resistance will change the current in the circuit. An ideal current-measuring device should have zero resistance. Thus, to convert a galvanometer into an ammeter, we have to do two things: (1) increase the range, and (2) make the effective resistance as close to zero as possible. This may be done as follows.

Suppose we want to make an ammeter that will read 10 A ($= I_0$) full scale from the galvanometer, which reads full scale for current $I_G = 2$ mA and has a resistance $R_G = 50\ \Omega$. This means that a path must be provided for the extra current $I_{sh}\ (= I_0 - I_G)$. This can be achieved by putting an extra resistor R_{sh}, called a *shunt*, parallel to the resistance R_G, as shown in Figure 22.12. The value of the shunt R_{sh} must be adjusted so that

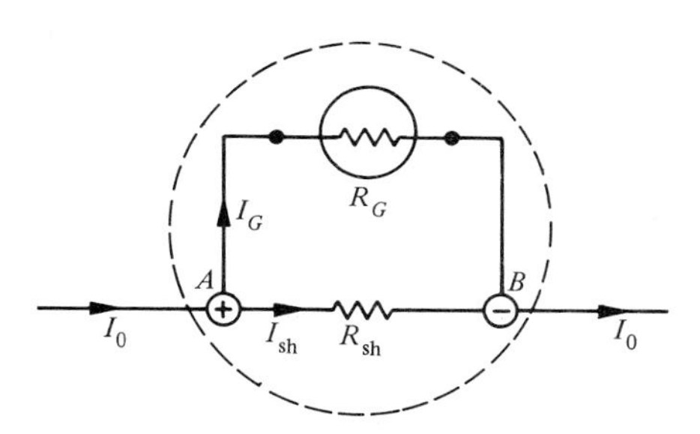

Figure 22.12. *Circuit showing conversion of a galvanometer into an ammeter, using a shunt resistance.*

$$I_0 = I_G + I_{sh} \tag{22.18}$$

Since R_G and R_{sh} are parallel, the potential difference across them is the same. That is, from Figure 22.12,

$$V_{AB} = R_G I_G = R_{sh} I_{sh}$$

or

$$\boxed{R_{sh} = \frac{I_G}{I_{sh}} R_G = \frac{I_G}{I_0 - I_G} R_G} \tag{22.19}$$

Thus, if $I_G = 2$ mA $= 0.002$ A and $R_G = 50\ \Omega$, we get

$$R_{\text{sh}} = \frac{0.002\ \text{A}}{(10 - 0.002)\text{A}}(50\ \Omega) = 0.010\ \Omega$$

Thus to change a galvanometer into a 10-A ammeter, a wire that will have a resistance of 0.010 Ω (and carry a current of 10 A with negligible change in temperature) is connected in parallel to the galvanometer resistance R_G.

Also, the total resistance R of this ammeter is given by

$$\frac{1}{R} = \frac{1}{R_G} + \frac{1}{R_{\text{sh}}} = \frac{1}{50} + \frac{1}{0.010} \qquad \text{or} \qquad R \simeq 0.01\ \Omega$$

which is very small compared to 50 Ω. Thus, in the process of converting a galvanometer into an ammeter, not only have we increased the range, we also have decreased the effective resistance to almost zero, as required for an ideal ammeter.

A point of caution: An ammeter is always connected in series in the branch of the circuit in which the current is to be measured. If connected in parallel, because of the small resistance it will draw a very large current (producing a short, that is, a path of zero or negligible resistance) and may burn out quickly.

The Voltmeter

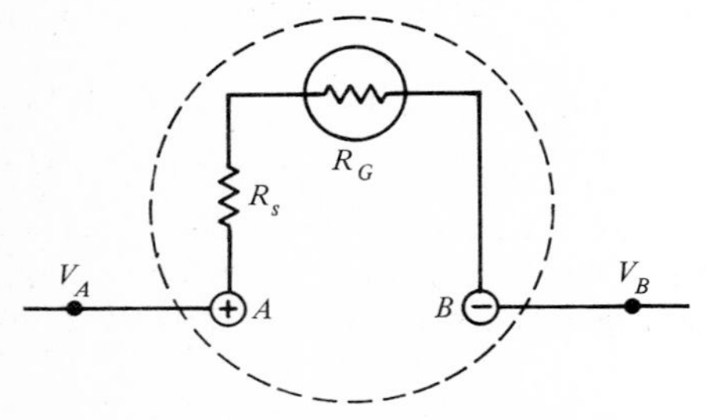

FIGURE 22.13. *Circuit showing conversion of a galvanometer into a voltmeter, using a high resistance in series.*

Suppose that we want to convert a galvanometer with $R_G = 50\ \Omega$, $I_G = 2$ mA, and $V_G = R_G I_G = 0.1$ V into a voltmeter capable of measuring $V_0 = 100$ V for a full-scale deflection. To achieve this, resistance R_s is added in series to the resistance R_G as shown in Figure 22.13. To find the value of the series resistance R_s, we impose the condition that the potential drops across R_s and R_G should add to V_0. Remembering that the current through the voltmeter will be I_G for full-scale deflection, we may write

$$V_0 = R_G I_G + R_s I_G$$

or

$$\boxed{R_s = \frac{V_0}{I_G} - R_G} \tag{22.20}$$

which for the case under consideration yields.

$$R_s = \frac{100\ \text{V}}{0.002\ \text{A}} - 50\ \Omega = 49{,}950\ \Omega$$

and the total equivalent resistance is

$$R = R_s + R_G = 49{,}950 + 50 = 50{,}000\ \Omega = 50\ \text{k}\Omega$$

As pointed out earlier, a voltmeter is connected in parallel to the circuit element across which the potential difference is to be measured. Since the voltmeter provides a second path, some current may branch off from the main circuit under consideration. To avoid this, the resistance of the voltmeter is made as high as possible so that the current in this second path will be as small as possible. The voltmeter described above has a resistance of 50,000 Ω for

100 V, or it is a 50,000 Ω/100 V = 500 Ω/V voltmeter. Typical voltmeters are rated higher than 10^4 Ω/V, while electronic voltmeters are 10^5 Ω/V.

22.5. Measurement of Resistance

We now describe three commonly used methods for measuring resistances: (1) the voltmeter–ammeter method, (2) the Wheatstone bridge method, and (3) the ohmmeter method.

Voltmeter-Ammeter Method

A straightforward method is to use Ohm's law, $R = V/I$. Suppose that we want to measure resistance R shown in Figure 22.14(a), which is connected to a source of emf, $\mathcal{E}$. We need to measure the current I through the resistor and the potential drop V across the resistor. To measure I and V we connect an ammeter and voltmeter in either one of two ways shown in Figure 22.14(b) or (c). Either method would give the same value of R if both the ammeter and the voltmeter were ideal (zero resistance and infinite resistance, respectively). But this is not so, and each method will need correction to get the right value of R.

Let us assume that the voltmeter resistance $R_V >>> R$. In this case not much current will go through the voltmeter and arrangement (b) or arrangement (c) will be suitable for measuring R. On the other hand, if $R_V \approx R$, the current in the voltmeter will be comparable to the current in R. But to use $R = V/I$, the ammeter must read only the current through the resistor R. Hence, arrangement (c) will be suitable but not (b). If the resistance R_A of the ammeter is such that $R_A <<< R$, there is no appreciable potential drop across R_A; hence, either arrangement (b) or arrangement (c) is suitable. On the other hand, suppose that $R_A \approx R$, for which only arrangement (b) is suitable, because in (c) the voltmeter will read the potential difference not only across R, but across both R and R_A. In cases not covered by these situations, proper corrections must be made.

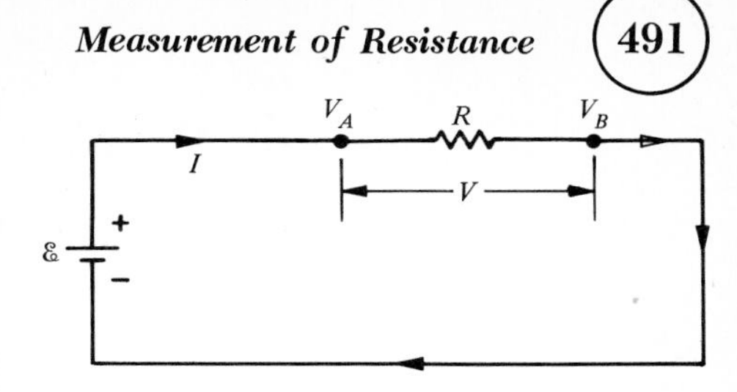

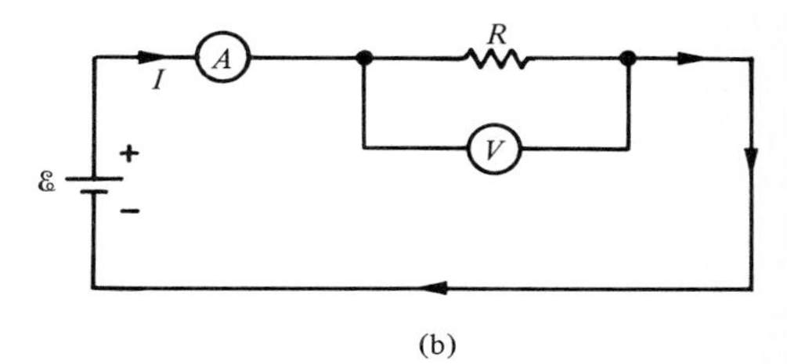

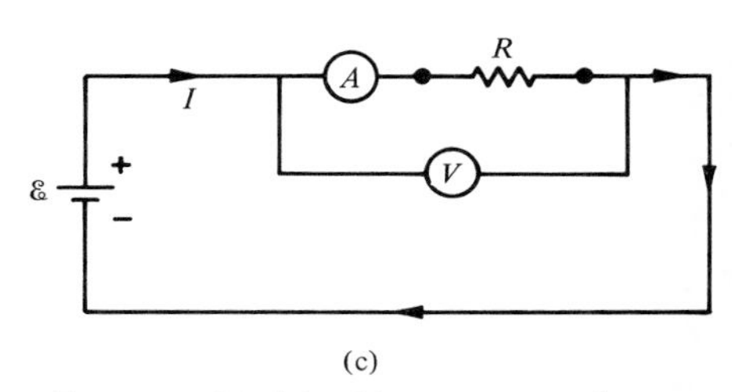

FIGURE 22.14. *Measurement of a resistance, using a voltmeter and an ammeter.*

The Wheatstone Bridge Method

One of the most accurate methods of measuring resistance, devised in 1843 by the English scientist Charles Wheatstone (1802–1875), is called the *Wheatstone bridge method,* shown in Figure 22.15. It consists of fixed resistors R_1 and R_2, and a variable resistance R inserted in three sides of the bridge while the unknown resistance R_x is connected in the fourth side, AB. A very sensitive galvanometer is connected across B and C while a battery with a key K is connected across AD. The variable resistance R is adjusted until there is no current through the galvanometer when key K is closed. Under such conditions current I_x continues past junction B and current I_R continues across junction C. Thus, B and C are at the same potential.

Let us apply Kirchhoff's second law to loops $ABCA$ and $BDCB$. Remembering that the current through the galvanometer is zero, we get

$$-I_x R_x + I_R R = 0$$
$$-I_x R_1 + I_R R_2 = 0$$

Rearranging and dividing one by the other, we get

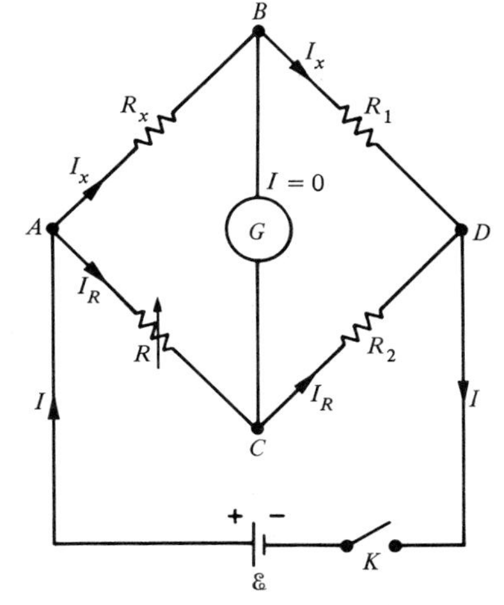

FIGURE 22.15. *Wheatstone bridge method of measuring unknown resistance.*

$$R_x = R\frac{R_1}{R_2} \tag{22.21}$$

Thus, knowing R, R_1, and R_2, R_x may be calculated. Usually, the values of R_1 and R_2 are fixed so that $R_1/R_2 = 0.001,\ 0.01,\ 0.1,\ 1,\ 10,\ 100, \ldots$. The accuracy with which R_x can be measured depends on the accuracy of R and the sensitivity of the galvanometer G.

EXAMPLE 22.4 Calculate an unknown resistance, using the voltmeter–ammeter method. The voltmeter reads 20 V and the ammeter 4 A. The same resistance is now measured by the Wheatstone method. If the variable resistance R is 2.45 Ω and R_1/R_2 is 2, what is the unknown resistance?

According to Ohm's law, the unknown resistance is $R = V/I = 20\text{ V}/4\text{ A} = 5\ \Omega$. From Equation 22.21, for the Wheatstone bridge

$$R_x = R(R_1/R_2) = (2.45\ \Omega)(2) = 4.90\ \Omega$$

EXERCISE 22.4 In the Wheatstone bridge method, if $R_1/R_2 = 4$ and the variable resistance R is 2.50 Ω, what is the unknown resistance? If this resistance were measured by using the voltmeter–ammeter method with the voltmeter reading 5 V, what will the ammeter read? [*Ans.:* $R_x = 10\ \Omega$, $I = 0.5$ A.]

The Ohmmeter

A rapid but not very precise method of measuring resistance is by use of the *ohmmeter*. It consists of a galvanometer G, a resistor R_s, and a source $\mathcal{E}$, all connected in series as shown in Figure 22.16. The scale of the galvanometer is calibrated to read resistance directly in the following manner.

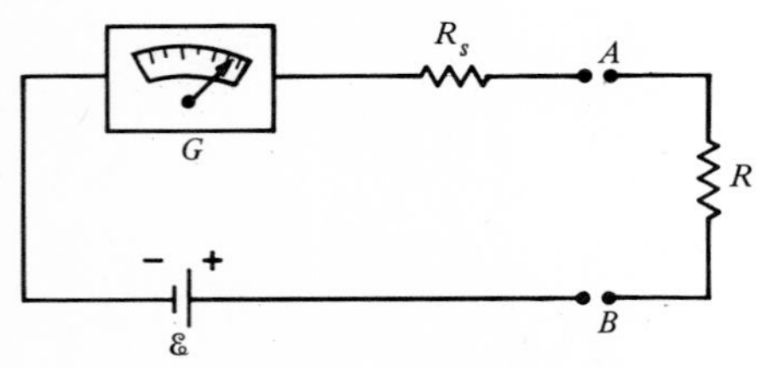

FIGURE 22.16. *Circuit diagram of an ohmmeter.*

The value of resistance R_s is such that when the path AB is short-circuited (resistance between A and B is equal to zero), the galvanometer has full deflection. On the other hand, when the path between A and B is an open circuit, that is, resistance $= \infty$ between A and B, the galvanometer shows zero deflection. Thus, the galvanometer is set to read resistance between 0 and ∞ when an unknown resistance R is connected across AB.

22.6. The Potentiometer

A voltmeter put across a source does not measure its emf, but its terminal voltage $V_t = \mathcal{E} - Ir$, where r is the internal resistance of the source. Unless $r = 0$, which is not possible, there will always be some current no matter how small and $V_t \neq \mathcal{E}$. In order to measure $\mathcal{E}$ directly, another null-type instrument commonly used is the *potentiometer*.

The principle of the potentiometer is based on comparing an unknown potential difference with a standard source of voltage. A schematic diagram of a potentiometer is shown in Figure 22.17. $\mathcal{E}$ is called the *working battery* and is a source of current I through the resistor AB. The resistance AB is a wire of uniform thickness called the *slide wire*. The connecting point P can be moved back and forth; hence, a variable potential difference can be achieved between A and P. The standard source of voltage $\mathcal{E}_s$ and the unknown voltage source $\mathcal{E}_x$ may be connected one at a time through switch S and galvanometer G to the contact point P.

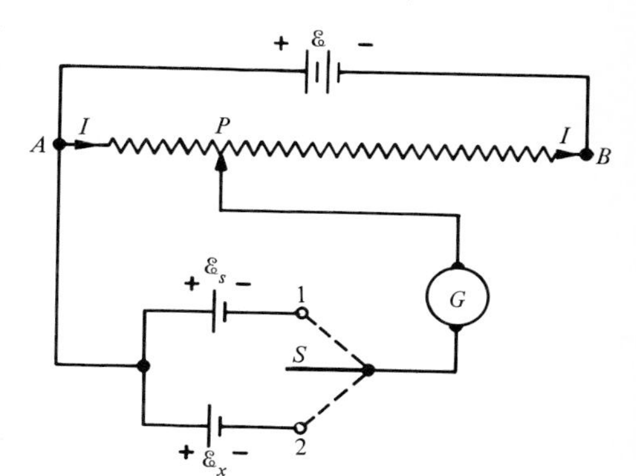

FIGURE 22.17. *Potentiometer method for measuring emf.*

To start with, switch S is thrown to position 1 so that the standard cell of voltage $\mathcal{E}_s$ is connected. The contact point P is moved until there is no current through the galvanometer G. Suppose that the length $AP = l_s$ of the wire has resistance R_s. Hence, the loop equation applied to R_s and $\mathcal{E}_s$ gives

$$-IR_s + \mathcal{E}_s = 0 \qquad \text{or} \qquad \mathcal{E}_s = IR_s$$

Now the switch is thrown to position 2 and the contact point P is moved until once again there is no current through the galvanometer. Suppose in this case that the null point is obtained when the length of the wire $AP = l_x$ is such that its resistance is R_x. Once again the loop equation yields

$$-IR_x + \mathcal{E}_x = 0 \qquad \text{or} \qquad \mathcal{E}_x = IR_x$$

From the preceding two equations we get

$$\mathcal{E}_x = \mathcal{E}_s \frac{R_x}{R_s} \tag{22.22}$$

Since the resistance of the slide wire is proportional to its length, Equation 22.22 may be written

$$\boxed{\mathcal{E}_x = \mathcal{E}_s \frac{l_x}{l_s}} \tag{22.23}$$

Thus, knowing $\mathcal{E}_s$ and measuring l_x and l_s directly, we can calculate an unknown $\mathcal{E}_x$. (Note that the current I does not enter the final result.)

EXAMPLE 22.5 For a standard cell of 2 V the null point is obtained if the length of the potentiometer wire is 40 cm. What is the emf of an unknown battery if the null point is obtained at 60 cm?

According to Equation 22.23, $\mathcal{E}_x = \mathcal{E}_s\, l_x/l_s$, where $\mathcal{E}_s = 2$ V, $l_s =$ 40 cm $= 0.4$ m, and $l_x = 60$ cm $= 0.6$ m. Therefore,

$$\mathcal{E}_x = (2\text{ V})(0.6\text{ m}/0.4\text{ m}) = 3\text{ V}$$

EXERCISE 22.5 If in Example 22.5 the battery had an emf of 2.5 V, what is the length of the potentiometer wire for the null point; that is, what is l_x? [*Ans.*: $l_x = 50$ cm.]

---

* 22.7. The *RC* Series Circuit

In the circuits considered so far, the currents and voltages remained constant with time. We now introduce a simple circuit containing a resistor, capacitor, and battery in which the currents and voltages are functions of time. Such an arrangement is shown in Figure 22.18. We can analyze this system by energy conservation or by applying the loop rule. We will do the latter.

To start with, switch S is closed through ab, and the direction of the current is as shown. As the current starts, the capacitor starts getting charged, with the left plate becoming positively charged and the right plate negatively charged.

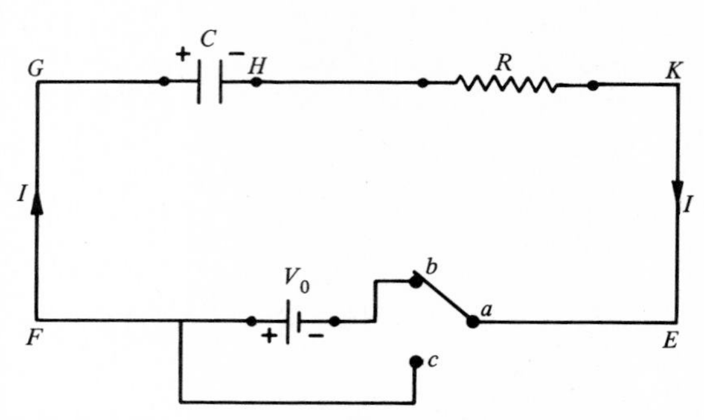

FIGURE 22.18. *Arrangement for charging (by connecting across ab) and discharging (by connecting across ac) a capacitor in an RC circuit.*

Let the current through the circuit at any instant be I, the voltage across the capacitor be $V_C = Q/C$, and the voltage across the resistor be $V_R = IR$. Applying Kirchhoff's loop rule to $EFGHKE$ and going clockwise, we get

$$V_0 - V_C - V_R = 0 \qquad \text{or} \qquad V_0 - \frac{Q}{C} - IR = 0$$

After rearranging, we may write

$$I = \frac{V_0}{R} - \frac{Q}{RC} \tag{22.24}$$

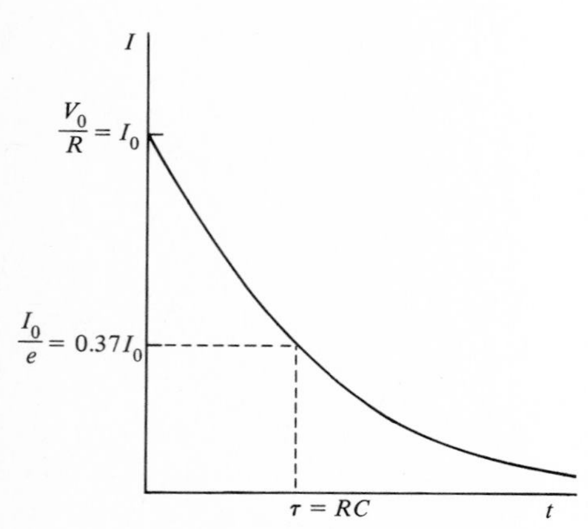

FIGURE 22.19. *Current I versus time t in an RC circuit while a capacitor is being charged.*

where I and Q are related through the definition of current $I = \Delta Q/\Delta t$. Substituting this in the equation above, we can find the variation in I versus t or Q versus t. Since this involves the use of calculus, we shall not go into such details but simply point out the most important features and quote the results.

Initially, the charge on the capacitor is zero and the current from Equation 22.24 will be maximum, that is, at $t = 0$, $Q = 0$ and $I = I_0 = V_0/R$. As time passes, the charge on the capacitor increases, reaching a maximum value $Q_0 = CV_0$, after which the battery cannot charge the capacitor anymore. Substituting this in Equation 22.24 yields $I = 0$; that is, the current in the circuit is zero and so is the potential drop across the resistor R. Thus, the current in a RC series circuit is maximum when the battery is initially connected, decreases as time passes, and eventually reaches zero. This variation in I with t is as shown in Figure 22.19 and is represented by

$$I = I_0 e^{-t/RC} \tag{22.25}$$

where $I_0 = V_0/R$. Similarly, to start with, the charge on the capacitor is zero and as time passes, it reaches a maximum value of $Q_0 = CV_0$. The behavior of Q versus t is as shown in Figure 22.20, and is represented by

$$Q = Q_0(1 - e^{-t/RC}) \tag{22.26}$$

where $Q_0 = CV_0$.

Thus, as seen from Figures 22.19 and 22.20 as well as from Equations 22.25 and 22.26, the current I in an RC circuit decreases exponentially with time while the charge Q on the capacitor increases exponentially. That is, the capacitor is being charged. It is also clear that how rapid these changes take place depend upon the value of the product RC. The quantity RC is called the *characteristic time* or *time constant* or *capacitive time constant,* denoted by τ. Thus, the equations above may be written

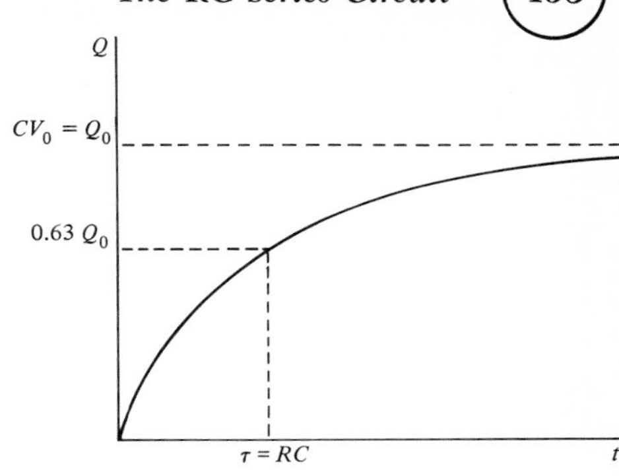

Figure 22.20. *Charge Q on a capacitor versus time t in an RC circuit while the capacitor is being charged.*

$$\boxed{I = I_0e^{-t/\tau}} \tag{22.27}$$

and

$$\boxed{Q = Q_0(1 - e^{-t/\tau})} \tag{22.28}$$

Let us consider three special cases: (1) If $t = 0$, $e^{\circ} = 1$, $I = I_0$, and $Q = 0$, as it should be. (2) If $t = \infty$, $e^{-\infty} = 0$, $I = 0$, and $Q = Q_0$; that is, it takes an extremely long time before the capacitor is fully charged and Q is exactly equal to $Q_0\ (= CV_0)$ and the current I reduces to exactly zero. (3) If $t = \tau = RC$, that is, t is equal to one time constant, $e^{-1} = 0.37$ and $(1 - e^{-1}) = 0.63$. The values of I and Q change to $I = I_0/e = 0.37I_0$ and $Q = Q_0(1 - e^{-1}) = 0.63Q_0$, as shown in Figures 22.19 and 22.20, respectively. Thus, we may define the *time constant* τ to be that time in which the maximum value of I_0 decreases to I_0/e. Although in one time constant a capacitor gets charged to 63 percent of its full value, it takes a few time constants to be charged close to saturation. Of course, the value of τ depends upon the individual values of R and C, which can be adjusted as desired.

So far we have been talking about charging a capacitor. Let us now see what happens if the capacitor is fully charged to Q_0 and the switch S in Figure 22.18 is connected through ac so that the battery V_0 will be out of the circuit. The positive charge on the capacitor starts to flow from capacitor C through the circuit to the negative plate of the capacitor. Thus the direction of the current is opposite in this case. The current continues until the capacitor is completely discharged. In Equation 22.24 if $V_0 = 0$,

$$I = -\frac{Q}{RC} \tag{22.29}$$

Once again the relation $I = \Delta Q/\Delta t$ yields

$$I = I_0e^{-t/RC} = I_0e^{-t/\tau} \tag{22.30}$$

and

$$Q = Q_0e^{-t/RC} = CV_0e^{-t/\tau} \tag{22.31}$$

where $I_0 = V_0/R$ and $Q_0 = CV_0$. The plot of Q versus t while the capacitor is discharging is shown in Figure 22.21. In one time constant the capacitor is left with only 37 percent of its initial charge. The I-versus-t plot in discharging is similar to the case of charging except that the direction of the current is reversed.

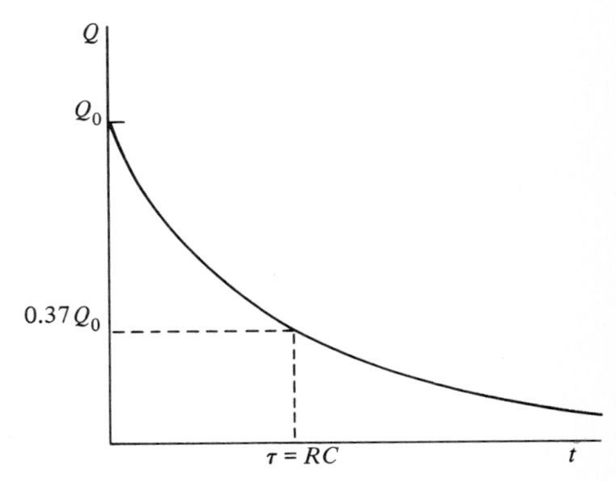

Figure 22.21. *Q versus t while a capacitor is discharging in an RC circuit.*

Thus, if the switch S in Figure 22.18 is continuously alternated between b

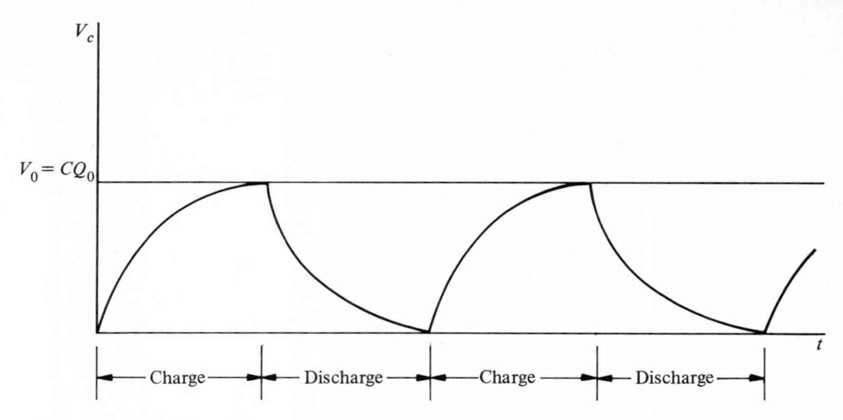

Figure 22.22. *Alternate charging and discharging of a capacitor in a RC circuit.*

and *c*, remaining in each position for a few time constants, the capacitor will alternately charge and discharge, as shown in Figure 22.22. The processes can be demonstrated by connecting the terminals of the capacitor to an oscilloscope (Chapter 26). The screen of the scope will exhibit V_C $(= CQ)$ versus t and will look similar to Figure 22.22. The *RC* series circuit plays an important role in electronics, automobile starters, rectifiers, and many other fields.

Example 22.6 In a *RC* series circuit $R = 5 \times 10^6\,\Omega$, $C = 0.1\,\mu$F, and $V_0 = 10$ V. Calculate τ, Q, and I when $t = \tau$.

From the definition of the time constant,

$$\tau = RC = (5 \times 10^6\,\Omega)(0.1 \times 10^{-6}\,\text{F}) = 0.5\,s$$

Also, $I_0 = V_0/R = 10\,\text{V}/5 \times 10^6\,\Omega = 2 \times 10^{-6}\,\text{A}$ and $Q_0 = CV_0 = 0.1 \times 10^{-6}\,\text{F} \times 10\,\text{V} = 10^{-6}\,\text{C}$. From Equations 22.25 and 22.26,

$$I = I_0 e^{-t/\tau} = I_0 e^{-1} = (0.2 \times 10^{-5}\,\text{A})(0.37) = 0.74\,\mu\text{A}$$

$$Q = Q_0(1 - e^{-t/\tau}) = Q_0(1 - e^{-1}) = (10^{-6}\,\text{C})(0.63) = 0.63\,\mu\text{C}$$

Exercise 22.6 In a *RC* series circuit $R = 1 \times 10^6\,\Omega$ and $C = 0.05\,\mu$F and $V_0 = 24$ V. Calculate τ, Q, and I when $t = 2\tau$. [*Ans.:* $\tau = 0.05$ s, $Q = 1.04\,\mu$C, $I = 3.25\,\mu$A.]

SUMMARY

For a single-loop circuit with a source of emf $\mathcal{E}$ and resistance R, $P_S = P_R$, or $I = \mathcal{E}/R$. If there is an internal resistance r, $P_S = P_R + P_r$ and $I = \mathcal{E}/(R + r)$. In a complete traversal of a circuit, the algebraic sum of all the potential differences encountered must add up to zero. If we travel through a source from $-$ to $+$, $\Delta V = +\mathcal{E}$, and in a resistor if we travel in a direction opposite the current, $\Delta V = IR$. The terminal voltage is $V_t = \mathcal{E} - Ir$.

For *resistors in series* $R = R_1 + R_2 + R_3 + \cdots$, and for *sources in series* $\mathcal{E} = \mathcal{E}_1 + \mathcal{E}_2 + \mathcal{E}_3 + \cdots$. For *resistors in parallel* $1/R = 1/R_1 + 1/R_2 + 1/R_3 + \cdots$. All sources in parallel have the same voltage, that is, $\mathcal{E} = \mathcal{E}_1 = \mathcal{E}_2 = \mathcal{E}_3$ and $I = I_1 + I_2 + I_3$.

According to *Kirchhoff's first rule* (junction theorem), the algebraic sum of the currents at any junction is equal to zero, $\Sigma I = 0$ or $(\Sigma I)_{in} = (\Sigma I)_{out}$. *Kirchhoff's second rule* (loop theorem), the algebraic sum of the potential differences in going around a closed loop is equal to zero, $\Sigma \mathcal{E} + \Sigma IR = 0$.

To convert a galvanometer into an *ammeter*, a small resistance called a shunt resistor R_{sh} is placed parallel to the resistance R_G so that $R_{sh} = [I_G/(I_0 - I_G)]R_G$.

To convert a galvanometer into a voltmeter, a large resistor R_s is connected in series with the galvanometer resistance R_G so that $R_s = (V_0/I_G) - R_G$.

Resistance R can be measured by (1) the voltmeter–ammeter method, making use of the equation $R_x = V/I$; or (2) by the Wheatstone bridge method, so that $R_x = R(R_1/R_2)$.

The *potentiometer* is used for measuring an unknown emf $\mathcal{E}_x = \mathcal{E}_s\,(l_x/l_s)$.

* In a *RC series circuit* the time constant $\tau = RC$ and is defined as that time in which the current reduces to $I_0/e\,(= 0.37\,I_0)$ while the charge reaches $0.63\,Q_0$. Both current and charge vary exponentially with time during charging as well as discharging.

QUESTIONS

1. Is the principle of energy conservation always applicable to electrical circuits? Explain.

2. Why are dry cells in a flashlight always connected in series?

3. When resistors are connected in parallel, the equivalent resistance is smaller. Does this mean that the total I^2R loss in the resistors is also reduced?

4. Is it possible to have a situation in which the terminal voltage will be greater than the emf of the battery?

5. Under what conditions is the emf of a battery equal to the terminal voltage? What is physically happening in this situation?

6. What happens if two batteries of different emf are connected in parallel? Discuss in terms of the current and power loss.

7. In measuring an unknown resistance using a Wheatstone bridge, does the voltage of the battery used make any difference?

8. An emf of a battery is measured by using a voltmeter and a potentiometer. Which gives a higher reading? Which is the correct reading, and why?

* **9.** Could you use the arrangement of a RC circuit as a timing device?

***10.** How will you use an RC series circuit to find the unknown value of (a) R; and (b) C?

***11.** Would you prefer to change the value of R or C to (a) increase the time constant; and (b) decrease the time constant?

***12.** What happens to the energy stored in the capacitor while the capacitor is discharging?

***13.** Under what conditions can energy supplied in charging a capacitor be taken back without any loss? Is it possible in practice?

PROBLEMS

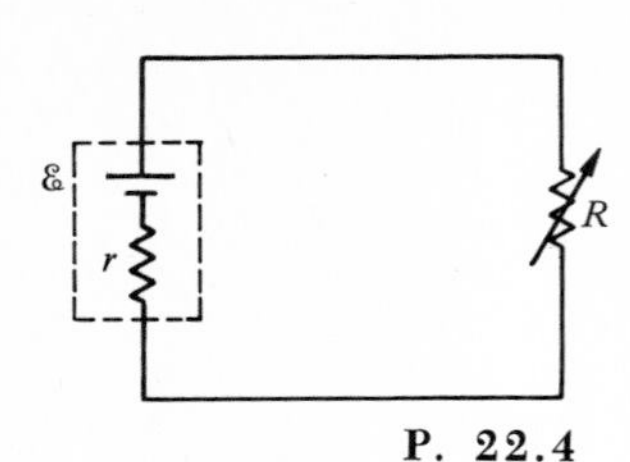

P. 22.4

1. Calculate the current through a single-loop circuit if $\mathcal{E} = 120$ V, $R = 1000\ \Omega$, and the internal resistance $r = 0.01\ \Omega$.
2. An external resistor $R = 100\ \Omega$ is connected to a 12-V battery having an internal resistance of $r = 0.1\ \Omega$. Calculate I, $P_{\mathcal{E}}$, P_R, and P_r.
3. A typical 12-V automobile battery has a resistance of 0.015 Ω. What is the terminal voltage of this battery when the starter draws a current of 150 A?
4. Consider a source of emf $\mathcal{E}$ with an internal resistance r connected to a variable resistor R as shown. (a) Calculate the power P supplied to R in terms of $\mathcal{E}$, r, and R. (b) Make a plot of P versus R for $R = \frac{1}{4}r, \frac{2}{4}r, \frac{3}{4}r$. From this graph find the value of R in terms of r for which P is maximum. Show that the maximum value is $\mathcal{E}^2/4r$.
5. An automobile battery of 12 V has an internal resistance of 0.04 Ω and when fully charged has available a charge in the amount of 150 A-h. To charge this battery it is connected to a 120-V dc source through a variable resistance R so as to limit the charging current to 10 A. (a) Show the connections for charging the battery; (b) the value of R; (c) the electrical power converted into chemical potential energy; and (d) the cost of charging the battery at the rate of 3 cents/kWh, neglecting the waste through R.
6. A 12-V battery with an internal resistance of 0.008 Ω is short-circuited by connecting its terminals to a wire of very low resistance, say $2 \times 10^{-4}\ \Omega$. Calculate the current through the circuit. Suppose that the wire is made of copper and has a mass of 1 g. How much energy is dissipated every second in the Joule heating of this wire? How long will it take before the wire melts? (The melting point of copper is 1083°C and its heat of fusion is 49.0 cal/g.)
7. Consider three 12-Ω resistors. What is the total resistance when these are connected (a) in series; and (b) in parallel?
8. Consider N resistors, each of $R\ \Omega$. Show that the total net resistance is NR if they are connected in series and R/N if connected in parallel.
9. How many 5-Ω resistors should be connected in parallel so that the total resistance will be 1 Ω?
10. Show that when two or more resistors are connected in parallel, the total resistance is smaller than the smallest of all the resistors.
11. Four 10-Ω resistors are connected as shown. What is the net resistance in each case?
12. Calculate the net resistance between A and B in the accompanying figure.

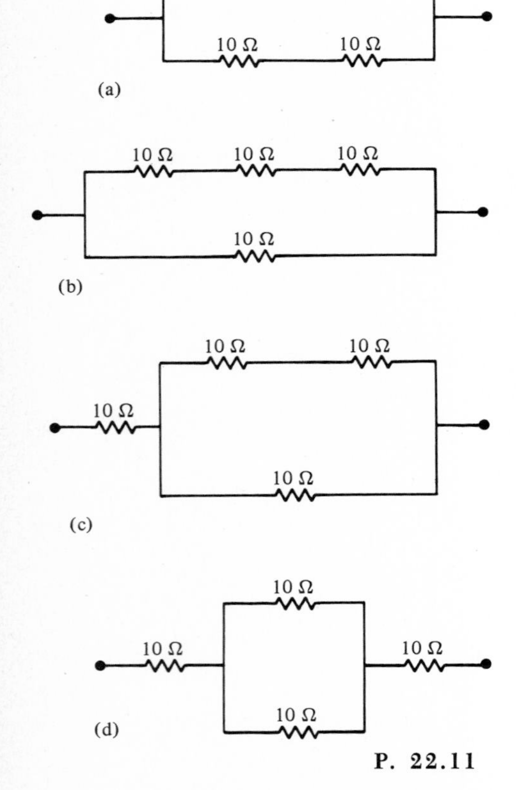

P. 22.11

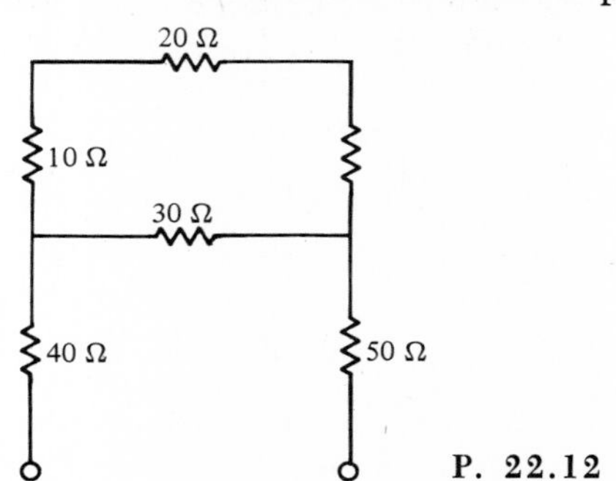

P. 22.12

13. Two resistors of 15 Ω and 30 Ω are connected in parallel. What should be the value of R to be connected in series with the other two so that the net resistance will be 20 Ω?
14. Calculate the net resistance of a circuit if (a) a 10-Ω resistor is connected in series with a parallel combination of two resistors, each 15 Ω. (b) A 10-Ω resistor is connected in parallel with a series combination of two resistors, each 15 Ω.
15. What should be the value of R so that the total resistance of the network shown is $2R$?

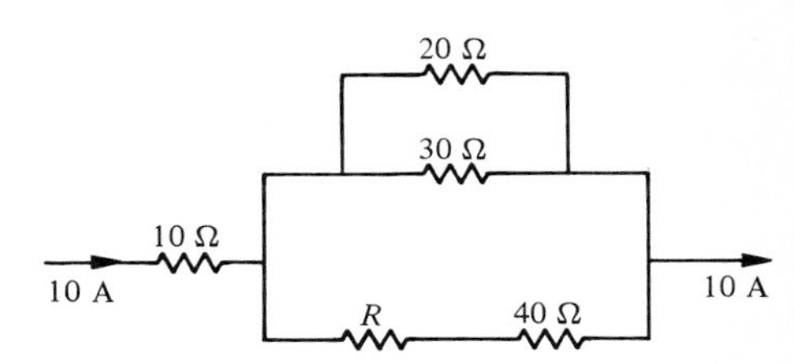

P. 22.15

16. Four cells each of 1.5 V and having an internal resistance of 0.01 Ω are connected (a) in series; and (b) in parallel. Calculate the current in each case.
17. Each cell of a storage battery has an internal resistance of 0.2 Ω and an emf of 2 V. Four of these are connected in (a) series; and (b) in parallel with a 10-Ω resistor. Calculate the current and the potential difference across the resistor in each case.
18. A 100-Ω resistor, when connected to the terminals of a battery, results in a current of 1.5 A. If the resistor is replaced by 150 Ω, the current is 1 A. Calculate the emf and the internal resistance of the battery.
19. Three resistors, 20 Ω, 30 Ω, and 60 Ω, are connected in parallel. If the current through the 30-Ω resistor is 2 A, calculate the current through the other two resistors.
20. In the arrangement shown, $\mathcal{E} = 1.5$ V, $r = 0.01\ \Omega$, $R_1 = 10\ \Omega$, and $R_2 = 20\ \Omega$. Calculate the currents I_1, I_2, I_3, and I_4.

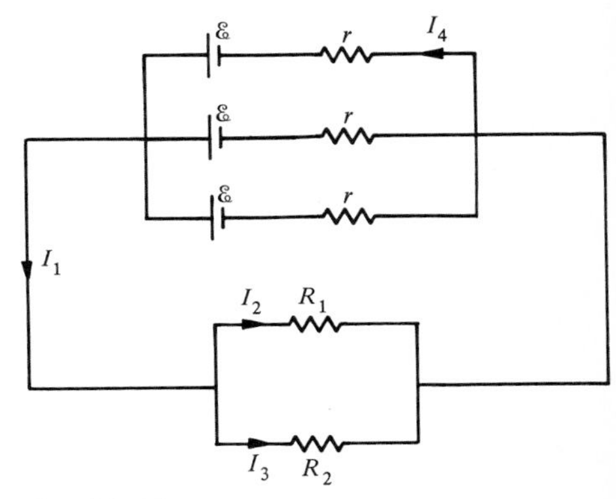

P. 22.20

21. In the circuit shown, calculate the potential difference (a) across AB; (b) across AC; (c) across AD; and (d) across BD.

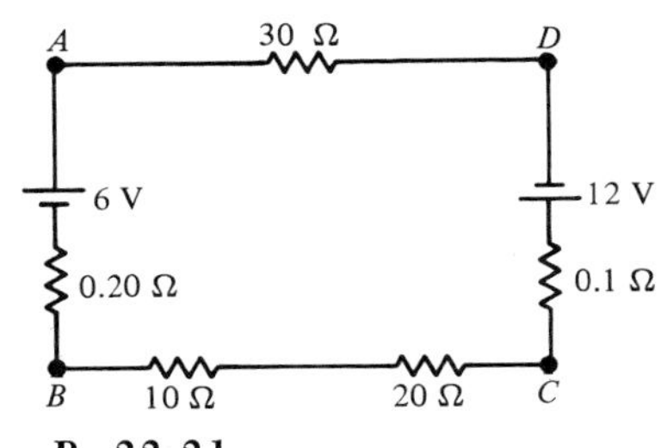

P. 22.21

22. A 12-V battery has an internal resistance of 0.1 Ω. A 1000-Ω resistor and another 20,000-Ω resistor are connected in series with the battery. What is the current through the circuit? Suppose that a voltmeter of $5 \times 10^4\ \Omega$ is connected across the 1000-Ω resistor. Calculate the current through this resistor and the voltmeter.
23. Calculate the combined resistance of two bulbs 50 W (120 V) and 100 W (120 V) when they are connected (a) in series; and (b) in parallel.
24. Two 60-W bulbs are connected with a 120-V line source (a) in series; and (b) in parallel. Calculate the total power used in each case.
25. The total power consumed by two equal resistors connected in series is 100 W. Calculate the power consumed if these are connected in parallel.
26. Consider the arrangement shown. Calculate (a) the current through each resistor; (b) the power lost as heat in each resistor; and (c) the power lost by the source. Does this agree with the sum in part (b)?

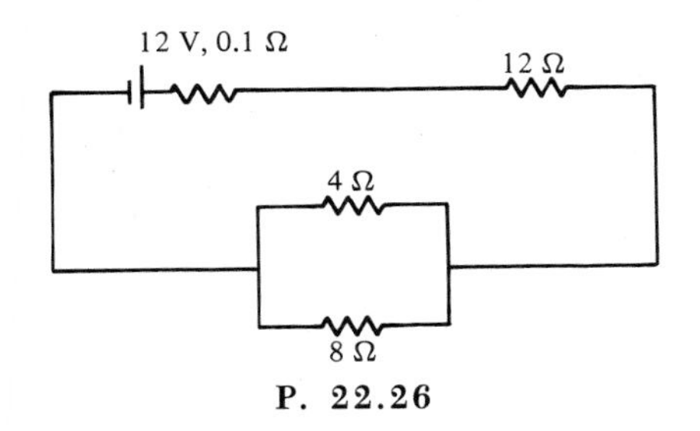

P. 22.26

27. Calculate I_1, I_2, and I_3 in the figure.
28. Calculate $\mathcal{E}_1$ and $\mathcal{E}_2$ in the figure.
29. Calculate the current in the 2-Ω, 3-Ω, and 5-Ω resistors in the figure.

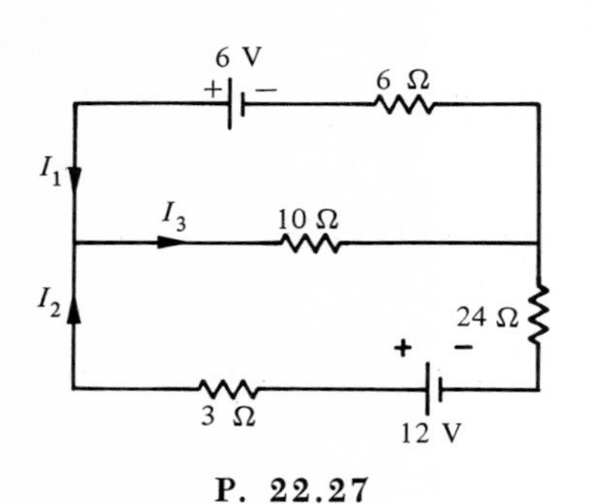

P. 22.27

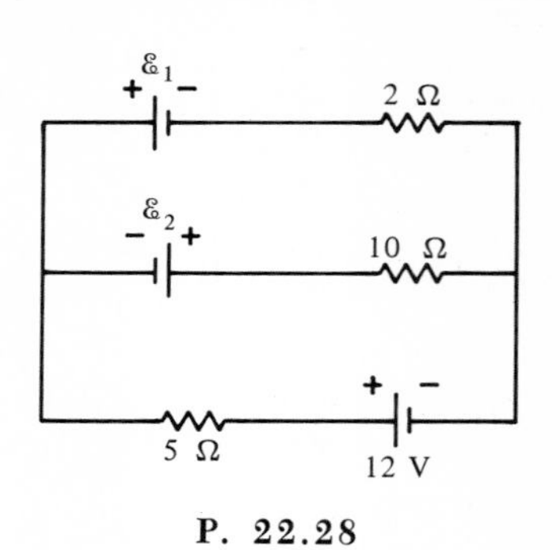

P. 22.28

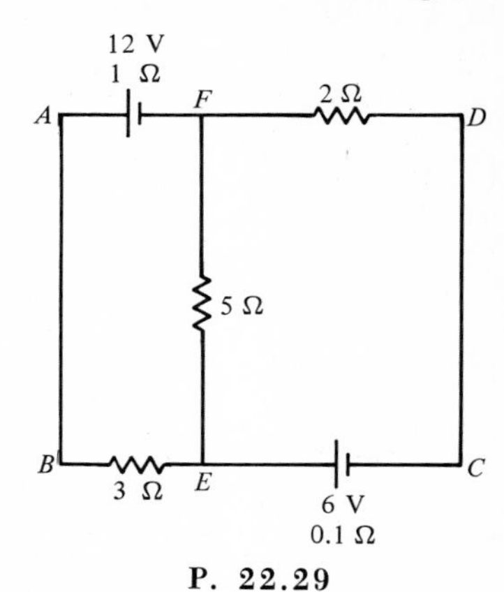

P. 22.29

30. Calculate $\mathcal{E}_1$, I_2, and I_3 in the figure.
31. Calculate I_1, I_2, and I_3 in the figure.

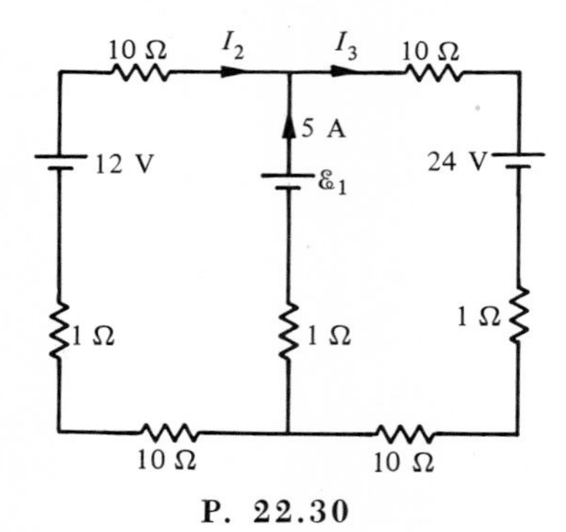

P. 22.30

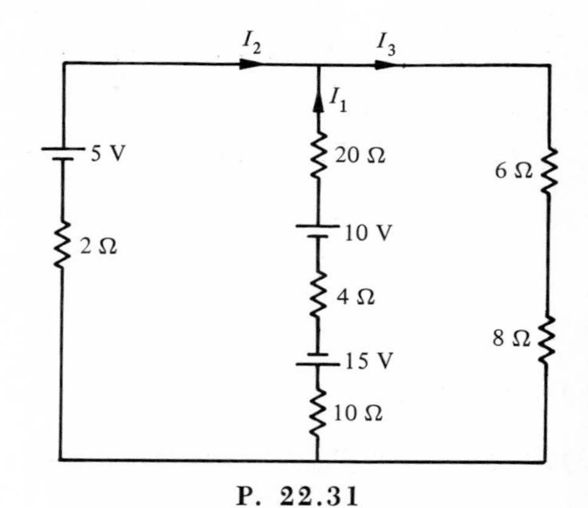

P. 22.31

32. A galvanometer with 10 Ω resistance reads full scale with a 2-mA current. Show how you will convert it into a 2-A full-scale deflection ammeter.
33. The resistance of a galvanometer coil is 10 Ω and reads full scale with a current of 1 mA. What should be the values of the resistances R_1, R_2, and R_3 to convert this galvanometer into a multirange ammeter of 100, 10, and 1 A. (See accompanying figure.)

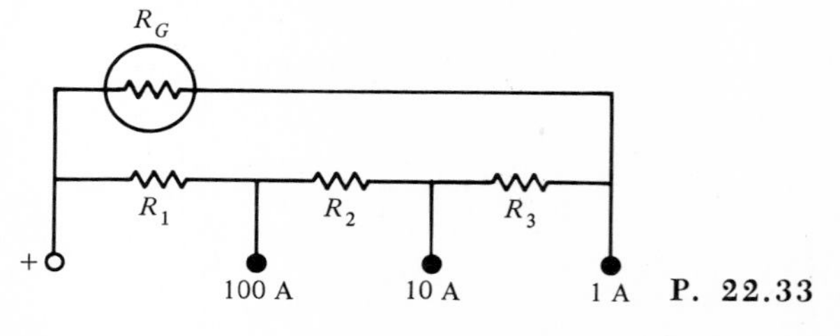

P. 22.33

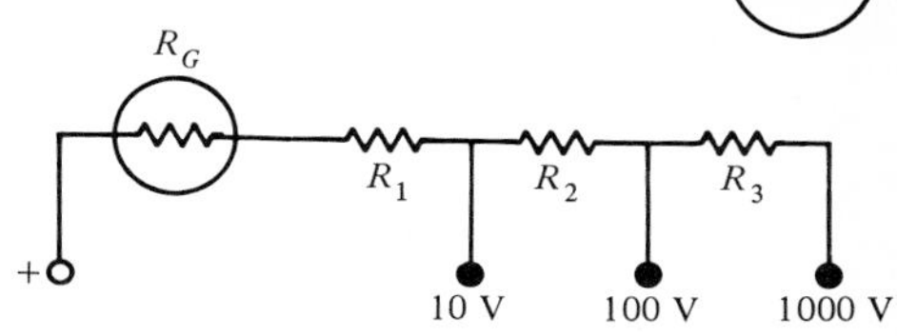

P. 22.34

34. The resistance of a galvanometer coil is 10 Ω and reads full scale with a current of 1 mA. What should be the values of the resistances R_1, R_2, and R_3 to convert this galvanometer into a multirange voltmeter of 10, 100, and 1000 V? (See accompanying figure.)

35. The resistance of a galvanometer coil is 50 Ω and reads full scale with a current of 2 mA. Show in a diagram how to convert this galvanometer into (a) an ammeter reading 5 A full scale; and (b) a voltmeter reading 200 V full scale.

36. Show that according to the arrangement in Figure 22.14(b), the unknown resistance is given by

$$R = \frac{V}{I - (V/R_V)}$$

while according to the arrangement in Figure 22.14(c), it is given by

$$R = \frac{V}{I} - R_A$$

V and I are the voltmeter and ammeter readings, and R_A and R_V are their respective resistances.

37. When an ammeter is connected to the terminals of a 12-V battery, it reads a current of 5 A. If the resistance of the ammeter and the wires is 0.2 Ω, what is the internal resistance of the battery?

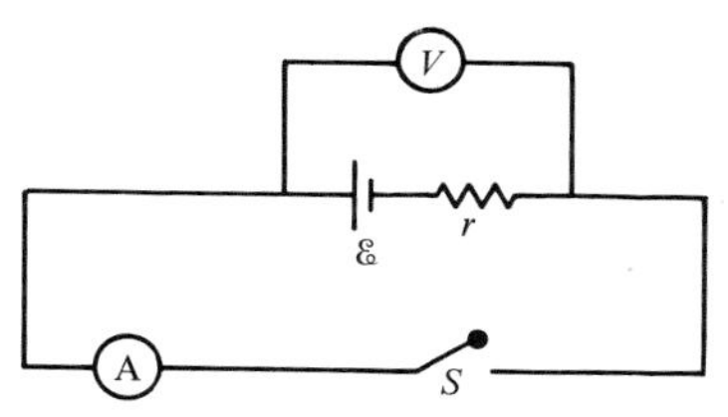

P. 22.38

38. Calculate the internal resistance r of the battery provided that the voltmeter reads 6.30 V when switch S is open and 5.8 V when S is closed. The ammeter reads 5A when switch S is closed. (See Figure P. 22.38.)

39. A galvanometer with 50-Ω resistance reads full scale with a 10-mA current. What should be the cross section of a 10-cm-long copper bar so that when used as a shunt, it converts the galvanometer into a 5-A ammeter?

40. In a Wheatstone bridge, $R_1 = 100\ \Omega$, $R_2 = 200\ \Omega$, and $R_3 = 35.6\ \Omega$. Calculate the fourth resistance if the bridge is balanced.

41. A standard 3-V cell balances a potentiometer wire at 36 cm. What is the voltage of an unknown cell if it balances at 54 cm?

***42.** In an RC circuit, if $R = 10 \times 10^6\ \Omega$ and $C = 0.1\ \mu$F, what is the time constant? If C is changed to 0.5 μF, what should be the value of R so that the time constant remains unchanged?

***43.** Consider an RC circuit with $R = 10^6\ \Omega$ and $C = 0.2\ \mu$F. If $V_0 = 20$ V, calculate τ, also I and Q when $t = \tau$.

***44.** Repeat Problem 43 for the case $t = 2\tau$, 3τ, and 4τ. Make a plot of I versus t and Q versus t. What do you conclude?

***45.** Consider a RC circuit with $R = 5 \times 10^6\ \Omega$ and $C = 0.01\ \mu$F. Calculate the time in which (a) current $I = I_0/2$; and (b) charge $Q = Q_0/2$.

***46.** Consider an RC circuit with $R = 5 \times 10^6\ \Omega$ and $C = 0.2\ \mu$F connected to a battery of 10 V. After the capacitor is fully charged, the battery is removed and the capacitor is discharged. Calculate I and Q after $t = 2\tau$. What happens to the energy stored in the capacitor?

23 Magnetic Interactions

We shall summarize the characteristics of the magnetic fields produced by permanent magnets or magnetic material and show that all magnetic fields are the result of charges in motion. We shall look into the forces that act on the charges moving in magnetic fields, and the torques that act on currents in magnetic fields. A few important applications of such magnetic interactions will be given.

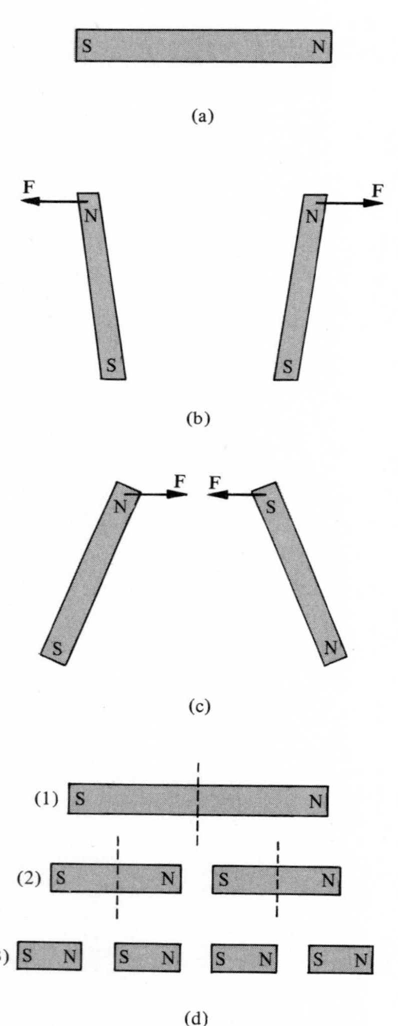

FIGURE 23.1 *Some basic characteristics of magnets.*

23.1. Magnetic Phenomena

MANY centuries ago men in the ancient city of Magnesia in Asia Minor discovered that chunks of certain iron ore (containing iron oxide, Fe_3O_4) such as lodestone, which was found locally, have the property of attracting small pieces of iron. This property of attraction by a natural ore of iron was named *magnetism*. Later it was found that this property is exhibited not only by iron, but also by cobalt, nickel, and many compounds of these metals. These are called *magnetic materials*. As early as 121 A.D., it was known to the Chinese that an iron bar, when brought near a natural magnet, itself acquired this property of magnetism, and retained it for a long time. Furthermore, it was well known that a bar magnet or a magnetized needle, when freely suspended, always points approximately in the north–south direction. This characteristic of magnetic materials was utilized in navigation at least as far back as the eleventh century.

Let us briefly summarize the characteristics of magnetic materials. A magnetized body is called a *magnet*. The property of attracting small pieces of iron seems to be concentrated in certain regions of the magnetic material, called *magnetic poles*. Let us consider a bar magnet, shown in Figure 23.1(a). This bar (or any magnetic body) has two magnetic poles, marked north N (or +) and south S (or −). The north pole is one that points north when the magnet is freely suspended in air. When two magnets are brought together, they will repel or attract each other according to the following rule.

The magnetic interaction between like poles results in a repulsive force, and the magnetic interaction between unlike poles results in an attractive force. *That is, like poles repel each other and unlike poles attract each other.* This is demonstrated in Figures 23.1(b) and (c).

The earth itself behaves like a huge magnet with its north pole approximately at the geographic south pole and the south pole approximately at the geographic north pole. This explains why a magnetic bar or magnetic needle when suspended in air points approximately in the direction of geographic north-south.

Finally, it may be pointed out that it is not possible to separate the north and south poles of a magnet. For example, if the bar magnet shown in Figure 23.1(d1) is broken into two pieces, two new magnets are obtained, as shown in Figure 23.1(d2). Each new magnet has both a north and a south pole. This process may be repeated as shown in Figure 23.1(d3). This means that *magnetic monopoles do not exist.* This is unlike the electric interaction, where electric monopoles are readily available. Recent experiments have failed to show evidence of magnetic monopoles.

The next natural question to ask is this: is this magnetic interaction part of any of the four interactions mentioned earlier—(1) gravitational interaction, (2) electromagnetic interaction, (3) nuclear interaction, and (4) weak interaction? As a matter of fact, magnetic interactions are not independent but are part of the electromagnetic or Coulomb interactions. We have already seen that electrical charges at rest produce electrostatic interactions. Electrical charges in motion produce magnetic interactions. As we shall show in this and the following chapters, currents produce magnetic fields, while the magnetic properties of magnetic materials are due to the currents produced by the motion of the electrons in the atoms of the magnetic materials.

That the magnetic interaction is the result of electric current was shown in 1820 by a Danish physics professor, Hans Christian Oersted (1777–1851). He showed that a magnetic needle (or compass needle) was deflected when it was placed in the vicinity of a wire carrying an electric current. Thus, a connection between electricity and magnetism was discovered accidentally by Oersted for the first time. As we shall see, the understanding of magnetic fields came through the works of Michael Faraday (1791–1867), J. B. Biot (1774–1862). F. Savart (1791–1841), André Marie Ampère (1775–1836), H. F. E. Lenz (1804–1865), and Joseph Henry (1797–1878).

Just as electrostatic and gravitational interactions are represented by means of fields by introducing the concept of field lines, so are electromagnetic interactions. The space surrounding a magnet is said to have a *magnetic field.* The magnetic field may be represented by magnetic field lines (or magnetic lines of force). Consider the bar magnet shown in Figure 23.2(a). If we place

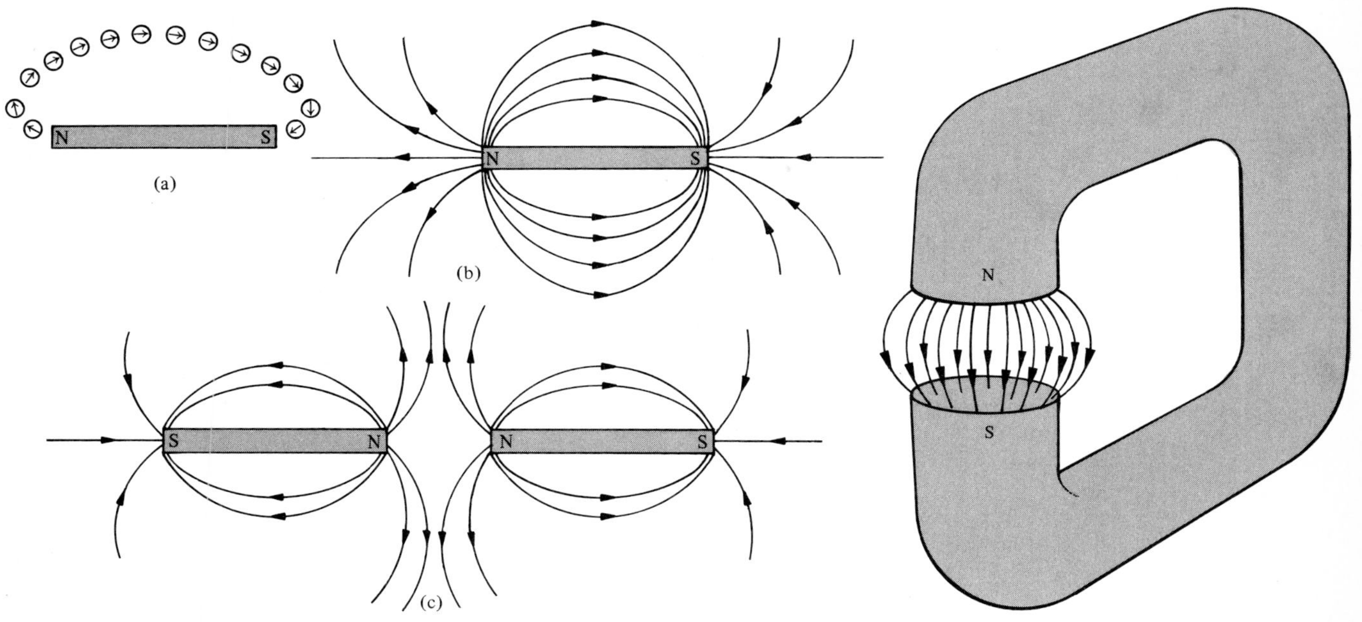

FIGURE 23.2 *Magnetic field lines for various configurations.*

several compass needles in the space surrounding the magnet, the needles point in the direction of the magnetic field. The path that the magnetic needle follows is a *magnetic field line*. The magnetic field lines for different configurations of magnetic poles are shown in Figure 23.2. As shown the field lines start from a north pole and end on a south pole, the direction being from north to south. That like poles repel each other and unlike poles attract each other can be demonstrated by drawing field lines as shown in Figures 23.2 and 23.3. In Figure 23.3 the field lines have been demonstrated by sprinkling iron filings in the space surrounding the magnetic poles.

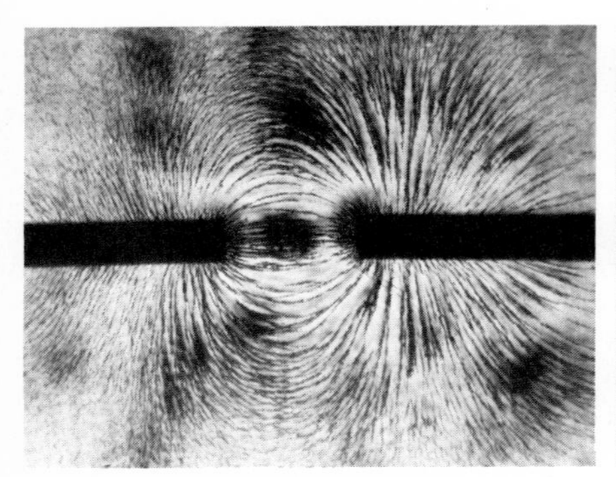
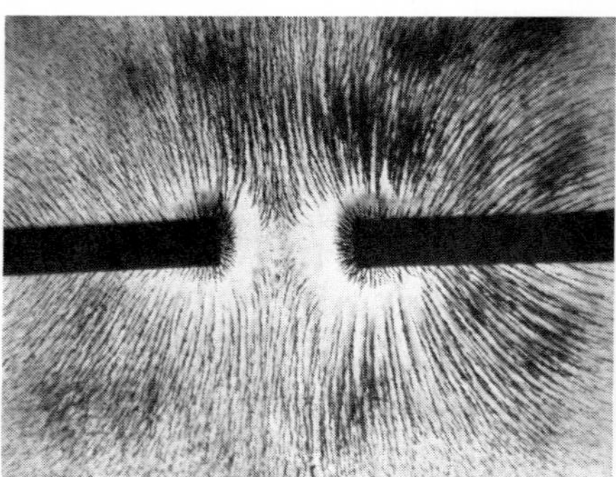
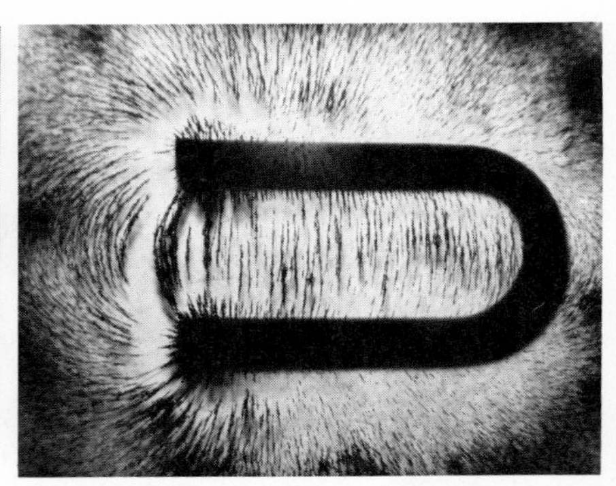

FIGURE 23.3 *Demonstration of magnetic field lines formed by sprinkling iron filings in the space surrounding the magnetic poles.*

The conclusions derived by H. C. Oersted in 1820 that a moving charge produces a magnetic field in the space surrounding it can be demonstrated in a simple manner, as shown schematically in Figure 23.4(a). Consider a conducting wire W carrying current I and a number of magnetic needles lying in the plane PP', which is perpendicular to the conducting wire. If there is no current in W, all the needles point north, as they should. The moment the current is switched on, the needles line up, as shown in Figure 23.4(b), forming a closed circle around the wire. This circular path is the magnetic field line (or line of force). There are many more magnetic field lines encircling this wire in a plane perpendicular to the wire, as shown in Figures 23.4(c) and (d). A common notation for giving direction to these circular lines is the following. The symbol • or ⊙ means an arrow coming toward the reader; that is, the field line is coming toward the reader. The symbol × or ⊗ means an arrow going away from the reader; that is, the field line is going away from the reader. The symbols • and × are supposed to represent the tip and tail of an arrow.

To find the direction of the magnetic field lines surrounding a current-carrying wire, we make use of the *right-hand rule*. According to this rule, if we grasp the wire with our right hand in such a way that our thumb points in the direction of the current, our fingers will encircle the wire in the same direction as the magnetic field lines. This rule is demonstrated in Figure 23.5. In general, the wire will not be straight; it may be curved or coiled. The right-hand rule may be applied to small portions of the wire at a time; hence, the field lines may be drawn for the whole space surrounding current-carrying wires.

(Figure 23.5 is on page 506.)

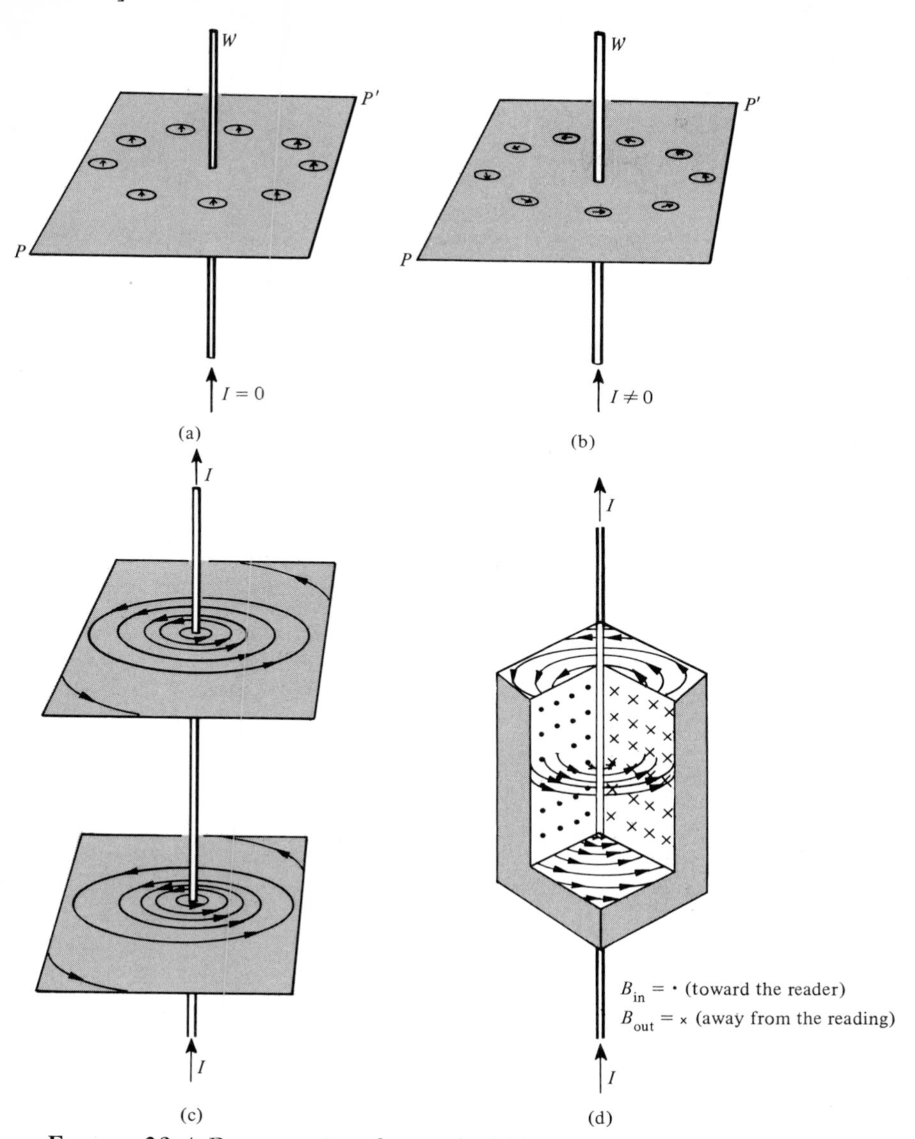

FIGURE 23.4 *Demonstration of magnetic field surrounding a wire carrying current I (that is, moving charges).*

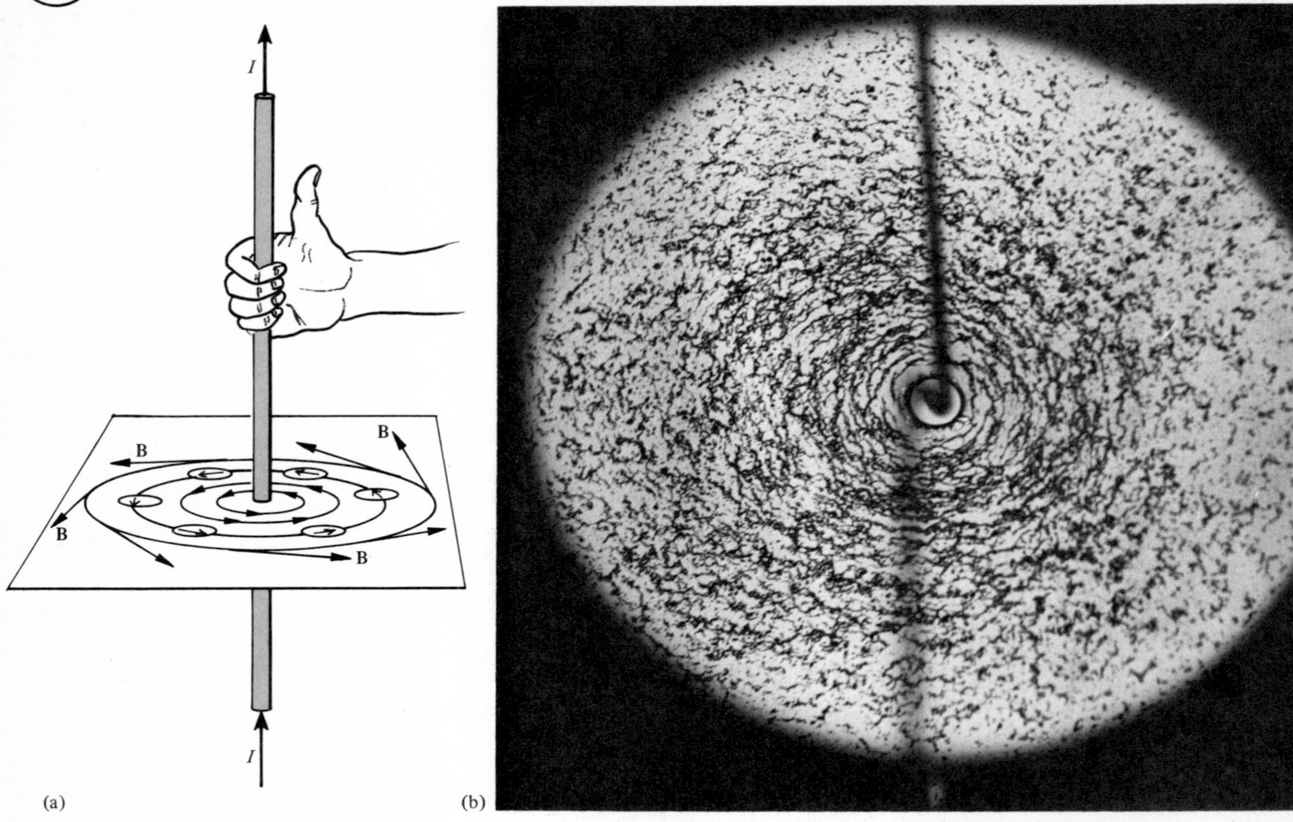

FIGURE 23.5 *(a) Right-hand rule for the direction of magnetic field lines and (b) magnetic field lines formed by iron filings sprinkled around a current-carrying wire.*

23.2. Magnetic Flux

In analogy with other interactions, the magnetic interaction can be described as a two-step process. First, the moving charges, the sources, produce a magnetic field. Second, this magnetic field interacts with other moving charges, called objects, to produce a magnetic force on the objects.

Just like the electrostatic field strength vector **E**, the electric intensity, we associate with each point in space where there is a magnetic field a vector quantity **B** the *magnetic field* at that point. This vector quantity **B** is also called the *magnetic induction* or *magnetic flux density*. Let us investigate the relation between **B** and the magnetic field lines. Similar to electric field lines, the magnetic field lines, also called *lines of induction* or *magnetic flux lines*, are defined as:

A magnetic field line is a line the tangent to which at any point gives the direction of ***B*** *at that point.*

To represent magnetic fields, the lines are drawn by keeping the following points in mind. (1) Each induction line is a closed curve in space, always encircling some electric current (unlike the lines of **E** which begin on positive charges and end on negative charges). (2) Each induction line is provided with an arrow. The tangent to a line of induction at any point gives the direction of **B** at that point. (3) The lines of induction are drawn in such a way that the number of lines crossing per unit cross-sectional area is proportional to the magnitude of **B** at that point. If B is large, the lines are crowded together, and if B is small, the lines are farther apart. These points are illustrated in Figures 23.4 and 23.5.

The magnetic flux Φ_B across a surface may be defined in exact analogy with the electric flux Φ_E. Thus, if a normal **n** drawn to the surface area A makes an angle θ with the direction of the magnetic induction **B**, the flux Φ_B through area A is given by

$$\Phi_B = BA \cos\theta \tag{23.1}$$

or, in vector notation,

$$\boxed{\Phi_B = \mathbf{B} \cdot \mathbf{A} = BA \cos\theta} \tag{23.2}$$

If **B** is uniform and parallel to A, that is, **B** is perpendicular to **n**, $\theta = 90°$, $\cos 90° = 0$, and $\Phi_B = 0$. On the other hand, if **B** is normal to the area A, that is, **B** is parallel to **n**, $\theta = 0°$, $\cos 0° = 1$, and the flux through this area will be

$$\Phi_B = BA \tag{23.3}$$

If the area ΔA is at right angles to the lines of induction and flux $\Delta\Phi$ goes through it, we may write (we need not carry the subscript B with Φ)

$$\boxed{B = \frac{\Delta\Phi}{\Delta A}} \tag{23.4}$$

The SI unit of A is m^2, and if we know the units of B (or Φ), we can find the units of Φ (or B). The unit of B will be defined in the next section. It will suffice to say here that the SI unit of flux Φ is given the name *weber* (Wb), in honor of Wilhelm Weber (1804–1890). Hence, according to Equation 23.4, the unit of B will be weber/m^2. We will discuss these units in detail in the next section.

Example 23.1 A coil of area 0.2 m^2 is placed in a uniform magnetic field of 0.04 Wb/m^2. Calculate the flux through the coil if it is held with its plane (a) parallel; and (b) perpendicular to the direction of the magnetic field.

(a) When the plane of the coil is parallel to **B**, the normal to the coil will make an angle $\theta = 90°$ and $\cos 90° = 0$. Thus, from Equation 23.2,

$$\Phi = BA \cos\theta = (0.04\ \text{Wb/m}^2)(0.2\ \text{m}^2)(\cos 90°) = 0$$

(b) If the plane of the coil is perpendicular to **B**, its normal will be parallel to **B**. That is, $\theta = 0°$ and $\cos 0° = 1$. From Equation 23.2,

$$\Phi = BA \cos\theta = (0.04\ \text{Wb/m}^2)(0.2\ \text{m}^2)(\cos 0°) = 0.008\ \text{Wb}$$

This is the maximum value of the flux that can be enclosed by this coil.

EXERCISE 23.1 In what direction should we hold a circular coil of radius 20 cm so that the magnetic flux due to a uniform magnetic field of 4×10^{-5} Wb/m² pointing south–north is (a) minimum; (b) maximum? Also, calculate the magnitude of Φ in each case. [*Ans.*: (a) axis of coil is perpendicular to **B** (i.e., vertical), $\Phi_B = 0$; (b) axis of coil is parallel to **B** (i.e., horizontal, pointing south–north), $\Phi_B = 5 \times 10^{-6}$ Wb.]

23.3. Magnetic Force on a Moving Charge

Consider a uniform magnetic field **B** pointing in the Y-direction. Suppose that a positive charge q is moving with a velocity **v** in the XY-plane as shown in Figure 23.6. The moving charge will experience a magnetic force **F**. By using different charges, moving with different velocities, we measure the force **F** and draw the following conclusions.

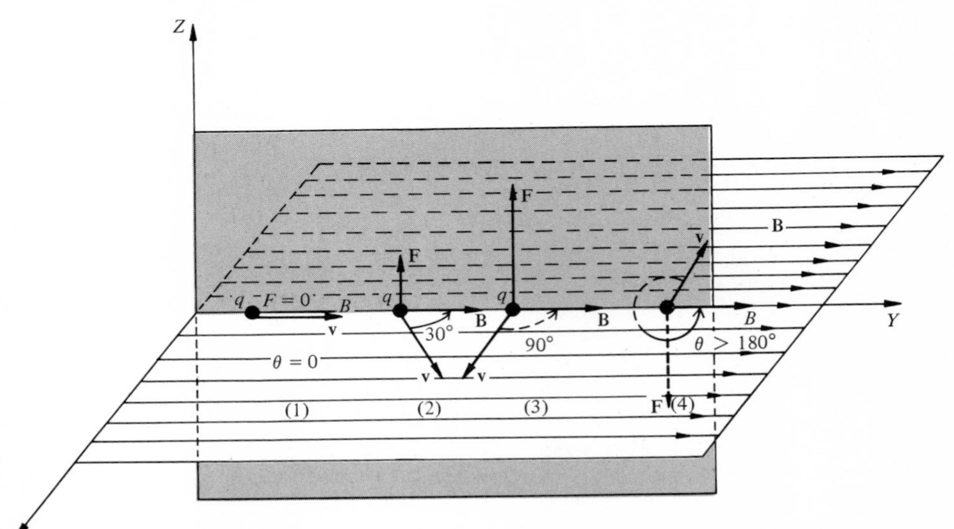

FIGURE 23.6 *Force* **F** *experienced by a moving charge q in a magnetic field* **B**.

The magnitude of the force **F** on the moving charge q exerted by the magnetic field **B** is directly proportional to the magnitudes of the charge q, velocity **v**, magnetic field **B**, and $\sin\theta$, where θ is the angle between **v** and **B**. Since in SI units the constant of proportionality is unity,

$$F = qvB\sin\theta \tag{23.5}$$

The direction of the magnetic force **F** is quite interesting. The force **F** is always perpendicular both to the magnetic field **B** and to the velocity **v** of the charge q. That is, **F** is perpendicular to the plane defined by **v** and **B**. Thus, remembering the definition of a vector product (see Chapter 9) we may write Equation 23.5 as

$$\boxed{\mathbf{F} = q\mathbf{v} \times \mathbf{B}} \tag{23.6}$$

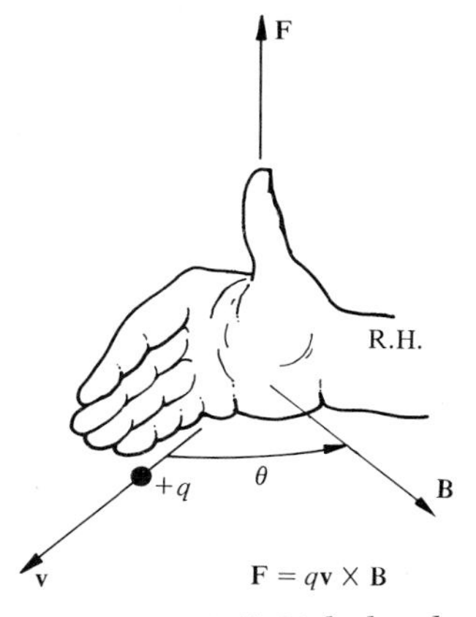

FIGURE 23.7 *Right-hand rule for finding the direction of the deflecting magnetic force* **F.**

The direction of the force **F** can be found from the given directions of **v** and **B** by using the right-hand rule illustrated in Figure 23.7. According to this rule, if the fingers of the right hand are stretched along the direction of **v** and then curled toward **B** through the smallest angle, the thumb will point in the direction of **F**, as shown in Figure 23.7. The meaning of the preceding equations and the right-hand rule are made clear in Figure 23.6.

To start with, since **F** is always perpendicular to **v** (hence, the force does no work), it does not change the magnitude of the velocity; it simply changes the direction of the velocity. That is, **F** is a deflecting force. As shown in Figure 23.6, if **v** and **B** are in the XY-plane, the deflecting force **F** will be along the $+Z$-axis if $0° < \theta < 180°$ and **F** will be along the $-Z$-axis if $180° < \theta < 360°$. For example, as shown in Figure 23.6:

If $\theta = 0°$	$\sin 0° = 0$	$F = 0$
If $\theta = 30°$	$\sin 30° = 1/2$	$F = qvB/2$
If $\theta = 90°$	$\sin 90° = 1$	$F = qvB$
If $\theta = 180°$	$\sin 180° = 0$	$F = 0$
If $\theta = 210°$	$\sin 210° = -1/2$	$F = -qvB/2$
If $\theta = 270°$	$\sin 270° = -1$	$F = -qvB$

It is clear that **F** has a maximum positive value $F = qvB$ when $\theta = 90°$ and points in the $+Z$-axis, while for $\theta = 270°$ it still has the maximum magnitude but the direction of **F** is along the $-Z$-axis.

The fact that $\mathbf{F} = 0$ for $\theta = 0°$ can be used in finding the direction of the magnetic field **B.** The direction of **B** is that direction in which the moving charge will experience no deflecting magnetic force.

We are finally in a position to define the magnitude of **B** in terms of known quantities. According to Equation 23.5, for $\theta = 90°$ we may write

$$B = \frac{F}{qv} = \frac{\text{newton}}{\text{coulomb-meter/s}} = \frac{\text{newton}}{\text{ampere-meter}}$$

Thus, B is defined as the force per unit charge per unit velocity provided that v is perpendicular to B, or the unit of B will be newton/ampere-meter, which is given the special name webers/meter^2; that is,

$$\frac{\text{weber}}{\text{meter}^2} = \frac{\text{newton}}{\text{ampere-meter}} \quad \text{or} \quad \frac{\text{Wb}}{\text{m}^2} = \frac{\text{N}}{\text{A-m}}$$

It is clear from this that since $\Phi = BA$, the unit of flux is a weber. The unit of B, Wb/m^2, has also been given the name *tesla* (T) in honor of the Yugoslavian-born American engineer Nicholas Tesla (1856–1943). We shall use Wb/m^2. Sometimes when the magnetic fields are small, the CGS unit, the *gauss* (G), given by the following relation, is also used:

$$1\,\mathrm{T} = 1\,\frac{\mathrm{Wb}}{\mathrm{m}^2} = 10^4\,\mathrm{G}$$

To indicate how large a field 1 $\mathrm{Wb/m^2}$ is, we give a few examples. The earth's magnetic field is $\sim 10^{-6}$ T = 0.01 G. The field between the pole faces of permanent magnets may be ~ 0.1 T = 1000 G. Typical electromagnets in laboratories have fields on the order of 1 $\mathrm{Wb/m^2}$ = 1 T = 10,000 G, and superconducting magnets may have fields as high as ~ 10 $\mathrm{Wb/m^2}$ = 10 T = 100,000 G = 100 kG.

Finally, magnetic forces, just like any other vector quantities, obey the principle of superposition; that is, the resultant force is equal to the vector sum of the individual forces acting on a moving charged particle. When the charged particle q is moving with velocity $\mathbf{v}$ in a region where there is an electric field $\mathbf{E}$ and magnetic field $\mathbf{B}$, the total force is the vector sum of the electric force $q\mathbf{E}$ and the magnetic force $q\mathbf{v} \times \mathbf{B}$. That is,

$$\mathbf{F} = \mathbf{F}_e + \mathbf{F}_m$$

or

$$\mathbf{F} = q\mathbf{E} + q\mathbf{v} \times \mathbf{B} \tag{23.7}$$

This force $\mathbf{F}$ is called the *Lorentz force*. Once again we remind ourselves that only the electric force does work, while no work is done by the magnetic force, which is simply a deflecting force.

Example 23.2 An electron is moving horizontally through a vertical magnetic field of 0.01 $\mathrm{Wb/m^2}$ with one-half the speed of light. Calculate the force acting on the electron and its acceleration.

The force acting on the electron according to Equation 23.5 is $F = qvB\sin\theta$, where $q = e = 1.6 \times 10^{-19}$ C, $v = \frac{1}{2}c = \frac{1}{2}(3 \times 10^8\ \mathrm{m/s}) = 1.5 \times 10^8$ m/s, $B = 0.01$ $\mathrm{Wb/m^2}$, and θ between v and B is 90°, $\sin 90° = 1$. Thus,

$$F = (1.6 \times 10^{-19}\ \mathrm{C})(1.5 \times 10^8\ \mathrm{m/s})(0.01\ \mathrm{Wb/m^2})\sin 90° = 2.4 \times 10^{-13}\ \mathrm{N}$$

From the relation $F = ma$, the acceleration a of the electron is

$$a = \frac{F}{m} = \frac{2.4 \times 10^{-13}\ \mathrm{N}}{9.11 \times 10^{-31}\ \mathrm{kg}} = 0.26 \times 10^{18}\ \mathrm{m/s^2}$$

(If we make use of the relativistic mechanics, $a = 0.23 \times 10^{18}\ \mathrm{m/s^2}$.)

Exercise 23.2 A charge of 1 μC is moving with a speed of 10^3 m/s through a magnetic field so that the maximum force acting on it is 5×10^{-6} N. Calculate the magnitude and direction of the magnetic induction. [*Ans.:* $B = 5 \times 10^{-3}$ $\mathrm{Wb/m^2}$, perpendicular to the direction of motion of the charge.]

23.4. Motion of a Charge in a Magnetic Field

To start with, let us consider the motion of a charged particle in a uniform magnetic field, that is, in a magnetic field having the same magnitude and direction for **B** at all points. For simplicity let us divide our discussion into three parts: the charged particle injected into the field with (1) **v** parallel to **B**, (2) **v** perpendicular to **B**, and (3) **v** making an angle θ with **B**.

If the charged particle is injected into the field along the direction of **B**, then from the relation $F = qvB \sin\theta$, $\theta = 0°$; hence, $F = 0$. As shown in Figure 23.8, the positively charged particle will simply move to the right in a straight line with constant velocity while the negatively charged particle will move to the left. For $\theta = 180°$, $F = 0$ also, but the positive charge now moves to the left and the negative to the right.

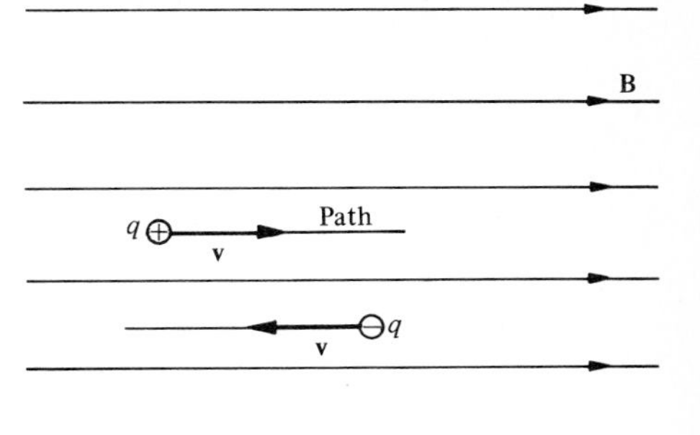

Figure 23.8 *Motion of a charged particle in a magnetic field injected parallel or antiparallel to* **B**.

Let us now consider a particle of mass m and charge q moving with velocity **v** in a direction perpendicular to the magnetic field **B** as shown in Figure 23.9(a). The force on charge q according to Equation 23.5 is $F = F_m =$

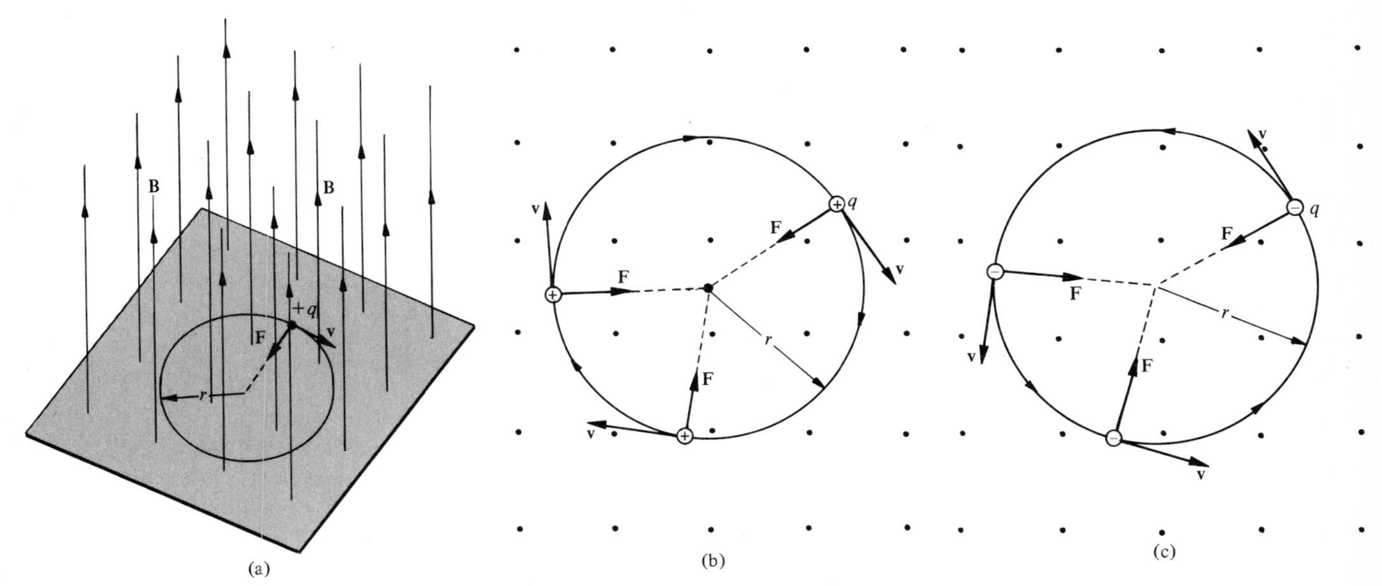

Figure 23.9 *Motion of a charged particle in a magnetic field injected perpendicular to the magnetic field* **B**.

$qvB \sin 90° = qvB$, and is always perpendicular to the velocity. Thus, the effect of force F_m is to change the direction of the velocity without changing its magnitude. This results in a uniform circular motion for the particle with centripetal acceleration v^2/r and centripetal force $F_c = mv^2/r$. This centripetal force is equal to the magnetic force F_m, $F_c = F_m$. That is,

$$\boxed{\frac{mv^2}{r} = qvB} \tag{23.8}$$

or

$$r = \frac{mv}{qB} \tag{23.9}$$

which gives the radius of the circle described by the charged particle. The situation for a positively charged particle is shown in Figures 23.9(a) and (b) and for a negatively charged particle in Figure 23.9(c). In both cases the field is coming out of the paper, that is, pointing toward the reader. As shown, the positive charge moves in a clockwise direction and a negative charge counterclockwise. The momentum of the particle may be written as

$$\boxed{p = mv = qBr} \tag{23.10}$$

where for high energies or fast-moving particles $m = m_0/\sqrt{1 - (v^2/c^2)}$.

From Equation 23.9 it is clear that the radius r is directly proportional to the momentum $p = mv$ and inversely proportional to the magnetic field B. Thus, if the particle has a large energy, its momentum p, and hence its radius r, will be large as well. Also, if the two particles have momentum, but a different sign of charge, they will be deflected in opposite directions.

Let us rewrite Equation 23.9 in a slightly different form. Since $\omega = v/r$, we get

$$\omega = 2\pi f = \frac{qB}{m} \tag{23.11}$$

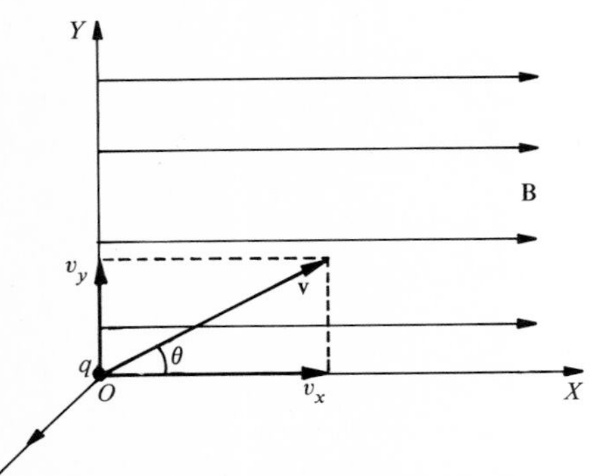

Figure 23.10 *Charged particle q moving with velocity* **v** *injected into a magnetic field making an angle θ with* **B**.

That is, the angular velocity is independent of r, and depends only on the charge-to-mass ratio q/m and the field B. As we shall see later, ω is the *cyclotron frequency*.

Finally, let us consider the case in which **v** is neither parallel nor perpendicular to **B**, but makes an angle θ, as shown in Figure 23.10. The problem is simplified by resolving **v** into two components, $v_x = v \sin\theta$, which is parallel to **B**, and $v_y = v \cos\theta$, which is perpendicular to **B**. Thus, v_x will make the particle move along the X-axis with uniform velocity while v_y makes the particle move in circular motion in the plane perpendicular to the X-axis, that is, in the YZ-plane. The combination of these two motions (uniform motion along the field **B** and circular motion perpendicular to **B**) results in a path that is a helix, as shown for a positively charged particle in Figure 23.11.

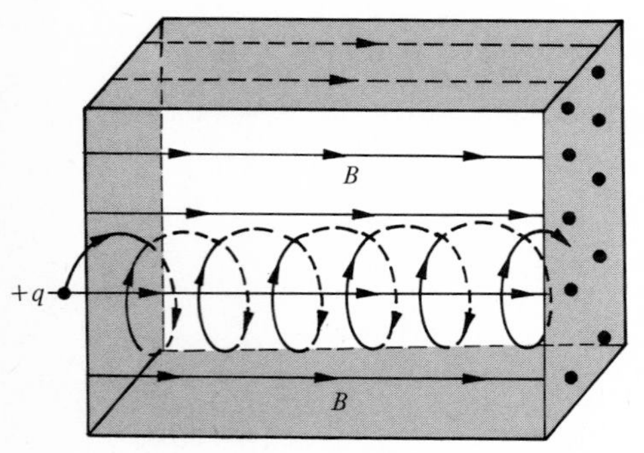

Figure 23.11 *Path of a charged particle making an angle θ with* **B** *is a helix.*

An interesting case is the motion of a charged particle in a nonuniform magnetic field which results in the trapping of a charged particle. The nonuniform field works as a reflector at the two ends. The *Van Allen radiation belt* is an example of trapped cosmic-ray charged particles, mostly electrons and protons. The trapping happens because of the inhomogeneous magnetic field of the earth. The inner belt extends from 800 km to 4000 km (600 mi to 2500 mi) above the earth's surface, and the outer belt extends to 60,000 km (37,500 mi). The only way particles can get out of this trap is by colliding with other ions.

Example 23.3 An electron is moving horizontally through a vertical field of 0.01 Wb/m^2 with a velocity of 1.5×10^8 m/s. Calculate the radius of the electron orbit and the cyclotron frequency. (Ignore relativistic effects.)

From Equation 23.9 the radius of the orbit is

$$r = \frac{mv}{qB} = \frac{(9.11 \times 10^{-31}\text{ kg})(1.5 \times 10^{8}\text{ m/s})}{(1.6 \times 10^{-19}\text{ C})(0.01\text{ Wb/m}^2)} = 8.54 \times 10^{-2}\text{ m} = 8.54\text{ cm}$$

From Equation 23.11, the cyclotron frequency ω is

$$\omega = \frac{qB}{m} = \frac{(1.6 \times 10^{-19}\text{ C})(0.01\text{ Wb/m}^2)}{9.11 \times 10^{-31}\text{ kg}}$$

$$= 1.76 \times 10^{9}\text{ rad/s} = \frac{1.76 \times 10^{9}}{2\pi}\text{ rev/s}$$

$$= 0.28 \times 10^{9}\text{ rev/s}$$

Exercise 23.3 What will be the radius if the electron is replaced by a proton? Is the magnitude of r reasonable? What value of B must you use so that the radius of the proton orbit is 20 cm? [*Ans.:* $r = 157$ m, $B = 7.83$ Wb/m^2.]

23.5. The Cyclotron

The first cyclotron was constructed in 1932 by E. Lawrence and M. Stanley. A sketch of a typical cyclotron in Figure 23.12 reveals that it consists of two hollow metal chambers D_1 and D_2 which are in the form of D's. The "dees" are placed with the open sides parallel and facing each other, with a small gap between them. An alternating electric field (i.e., the field changes direction) is produced in this gap by connecting the dees to a high-frequency alternating-voltage source. We may remind ourselves that because of the electrical shielding, there cannot be any field inside the dees. A strong electromagnet with its pole faces above and below the dees provides a strong magnetic field in the direction shown. Source S, which provides protons or deuterons for accelerating, is placed at the center.

In order to understand the basic working of a cyclotron at one particular instant, let the face of D_1 be positive and that of D_2 be negative. An ion of mass m and charge $+q$ will be accelerated by the electric field (between the gap) toward D_2. Once inside the dee, there is no electric field, but the ion feels the magnetic field and goes into a circular path. The frequency of the alternating-voltage source is such that when the ion comes out of D_2, the field has a reversed direction; that is, D_2 is now positive and D_1 is negative. Thus, the ion is now accelerated in the gap toward D_1. This process continues and the ion is repeatedly accelerated while going across the gap. This means that as the velocity of the ion becomes higher, the magnetic field makes it move in larger and larger circular orbits. After achieving a desirable velocity, the ions can be drawn out of the chamber by means of a deflecting magnetic field.

Let us calculate the kinetic energy acquired by the ion. Equating the magnetic field with the centripetal force, from Equation 23.8 we obtain

$$v = qBr/m \tag{23.12}$$

and

$$f = \frac{\omega}{2\pi} = \left(\frac{1}{2\pi}\right)\left(\frac{q}{m}\right)B \tag{23.13}$$

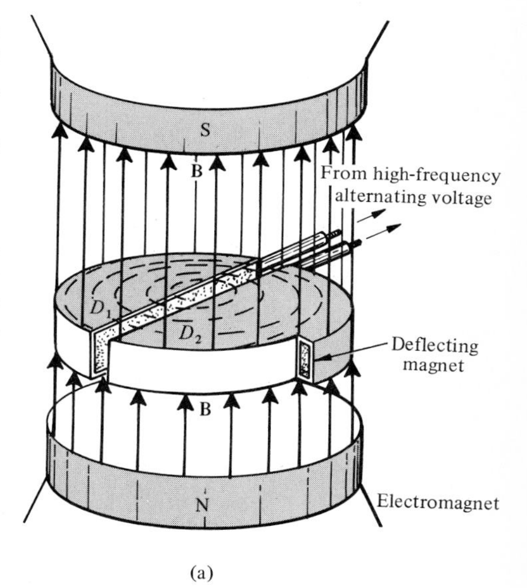

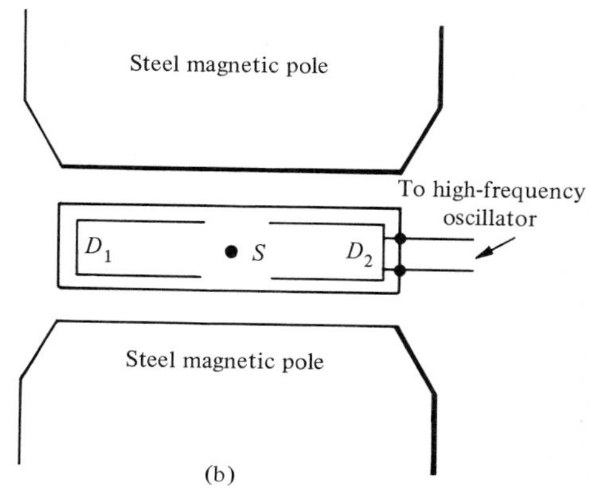

Figure 23.12 *Typical cyclotron: (a) outline sketch and (b) side view.*

If R is the radius of the dees, the ion will attain maximum velocity v_{max} when $r = R$. Thus, the maximum kinetic energy of the ion will be, using $v = v_{max}$ in Equation 23.12,

$$K_{max} = \frac{1}{2}mv_{max}^2 = \frac{1}{2}\frac{(qBR)^2}{m} \tag{23.14}$$

A typical alternating-voltage source has a frequency of ~10 to 12 megahertz and the voltage applied to the dees is ~200 kV while $B \sim 2$ Wb/m². The maximum kinetic energy of protons in some cyclotrons is ~55 MeV.

The calculations above do not hold if the ions reach speeds in the relativistic region. Taking relativistic corrections into account, cyclotrons under construction will provide much higher energies. For light particles such as electrons, relativistic corrections become appreciable at much lower energies, making the cyclotron unsuitable.

Example 23.4 The radius of a cyclotron dee is 0.4 m and the magnetic induction is 1.5 Wb/m². What is the maximum energy of a beam of protons?

Substituting for m = mass of proton = 1.67×10^{-27} kg, $q = 1.6 \times 10^{-19}$ C, $B = 1.5$ Wb/m², and $R = 0.4$ m in Equation 23.14:

$$K_{max} = \frac{1}{2}\frac{q^2B^2R^2}{m} = \frac{1}{2}\frac{(1.6 \times 10^{-19}\ \text{C})^2(1.5\ \text{Wb/m}^2)^2(0.4\ \text{m})^2}{1.67 \times 10^{-27}\ \text{kg}}$$

$$= 2.76 \times 10^{-12}\ \text{J} = 2.76 \times 10^{-12}(1/1.602 \times 10^{-19}\ \text{J/eV})$$

$$= 17.2 \times 10^6\ \text{eV} = 17.2\ \text{MeV}$$

The frequency of the alternating voltage applied to the dees of such a cyclotron, according to Equation 23.13, is

$$f = \frac{1}{2\pi}\left(\frac{q}{m}\right)B = \frac{1}{2\pi}\left(\frac{1.6 \times 10^{-19}\ \text{C}}{1.67 \times 10^{-27}\ \text{kg}}\right)(1.5\ \text{Wb/m}^2) = 2.29 \times 10^7\ \text{Hz}$$

$$= 22.9\ \text{MHz}$$

Exercise 23.4 Repeat Example 23.4 if the proton is replaced by a deuteron. [*Ans.*: $K_{max} = 1.38 \times 10^{-12}$ J = 8.61 MeV, $f = 11.4$ MHz.]

23.6. Magnetic Force on an Electric Current

Let us consider a wire of length L carrying a current I placed in a magnetic field **B** as shown in Figure 23.13. The current is due to the charge carriers, and these charge carriers are in a magnetic field. Thus, there results sideways deflecting magnetic forces on these charges, thereby leading to the deflection of the current-carrying wire. Most charges in the wire are at rest or in random motion, hence experience no steady resultant force. The resultant force is the average of the forces acting on the charge carriers moving with drift velocity v.

Suppose that there are n charge carriers per unit volume, each with a charge q and drift velocity **v**. A wire of cross-sectional area A and length ΔL has a volume $A\ \Delta L$ and contains $N = nA\ \Delta L$ charge carriers. Let us assume that the

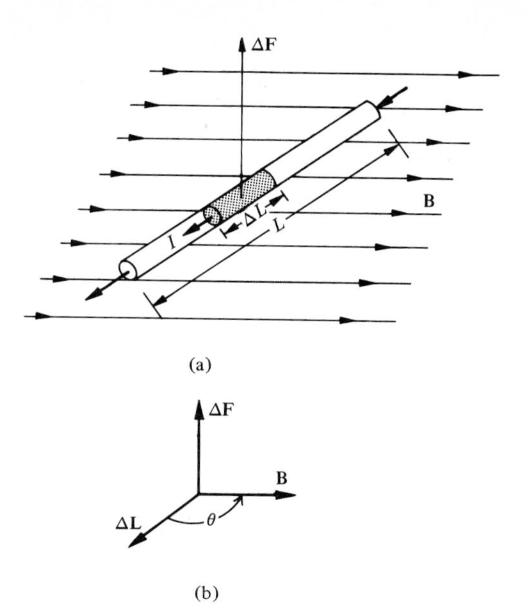

FIGURE 23.13 *(a) Conductor of length L carrying current I in a magnetic field* **B.** *(b) Directions of* **ΔL**, **B**, *and* **ΔF**.

wire is perpendicular to the field. Since the force on each charge, according to Equation 23.5, is qvB, the force ΔF on a length of wire ΔL and containing N electrons (or charge carriers) will be

$$\Delta F = NqvB \tag{23.15}$$

The charges moving with velocity v will take time $\Delta t = \Delta L/v$ to go through the length ΔL. This means that the charges Nq in the shaded length will take time Δt to completely cross through this length ΔL. Thus, by definition the current I is given by

$$I = \frac{\Delta Q}{\Delta t} = \frac{Nq}{\Delta L/v} = \frac{Nqv}{\Delta L}$$

or

$$Nqv = I\,\Delta L \tag{23.16}$$

Substituting this in Equation 23.15,

$$\Delta F = I\,\Delta LB \tag{23.17}$$

If the wire makes an angle θ with the magnetic field, the force on length ΔL will be given by

$$\Delta F = I\,\Delta LB \sin\theta \tag{23.18}$$

which in vector notation is written as

$$\mathbf{\Delta F} = I\,\mathbf{\Delta L} \times \mathbf{B} \tag{23.19}$$

The direction of this magnetic force $\Delta\mathbf{F}$ acting on a length of wire $\Delta\mathbf{L}$ is given by the right-hand rule, hence is perpendicular to the plane containing $\Delta\mathbf{L}$ and **B**, as shown in Figure 23.13(b).

If we consider a wire of length L carrying current I in a magnetic field **B**, the magnetic **F** will be given by

$$\mathbf{F} = I\,\mathbf{L} \times \mathbf{B}$$

or

$$F = ILB \sin\theta \tag{23.20}$$

where **L** is in the direction of current I and θ is the angle that **L** makes with **B**, same as in Figure 23.13. If $\theta = 90°$, the force is given by

$$F = ILB \qquad \text{for } \theta = 90° \tag{23.21}$$

Thus, if a wire 1 m long, carrying a current of 1 A, is placed in a transverse ($\theta = 90°$) magnetic field of 1 Wb/m^2, the resulting deflecting magnetic force will be 1 N. The sign of the charge carriers is of no consequence. In the case above the positive charges move with drift velocity v to the left and negative charges move with drift velocity v to the right. Both types of charge have deflecting forces acting upward.

EXAMPLE 23.5 A wire 10 cm long and carrying a current of 5 A is placed in a magnetic field of 1 Wb/m^2 making an angle of $\theta = 0°, 45°, 90°, 135°$, and $180°$ with the magnetic field. Calculate the force on the wire in each case.

From Equation 23.20,

$$F = ILB \sin\theta = (5\text{ A})(0.1\text{ m})(1\text{ Wb/m}^2)\sin\theta = (0.5 \sin\theta)\text{ N}$$

For $\theta = 0°$: $F = (0.5 \times 0)\text{ N} = 0$
For $\theta = 45°$: $F = (0.5 \times 0.707)\text{ N} = 0.35\text{ N}$
For $\theta = 90°$: $F = (0.5 \times 1)\text{ N} = 0.5\text{ N}$
For $\theta = 135°$: $F = (0.5 \times 0.707)\text{ N} = 0.35\text{ N}$
For $\theta = 180°$: $F = (0.5 \times 0)\text{ N} = 0$

EXERCISE 23.5 A wire 12 cm long and carrying a current of 2 A is placed perpendicular to a uniform magnetic field. If a force of 0.8 N acts on it, calculate the value of the magnetic induction B. [*Ans.*: $B = 33.3\text{ Wb/m}^2$.]

23.7. Magnetic Torque on a Current Loop and Galvanometer

Suppose that a rectangular loop of wire of length a and width b, carrying current I, is placed in a uniform magnetic field **B** as shown in Figure 23.14(a). The magnetic field is along the X-axis, while the normal **n** to the plane of the coil makes an angle θ with **B** as shown in Figure 23.14(b). We want to calculate the torque on this loop, say for convenience, about the Z-axis.

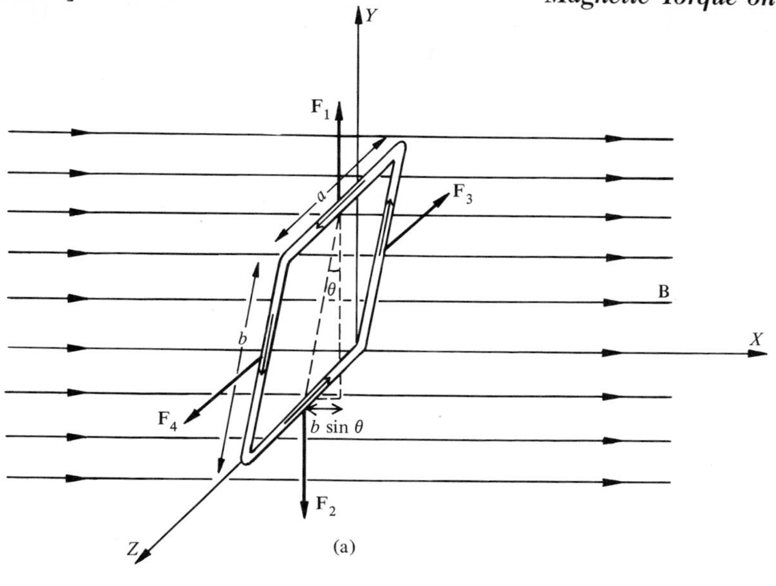

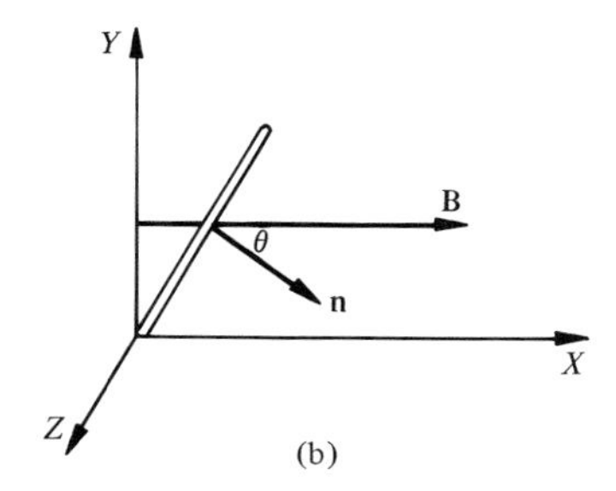

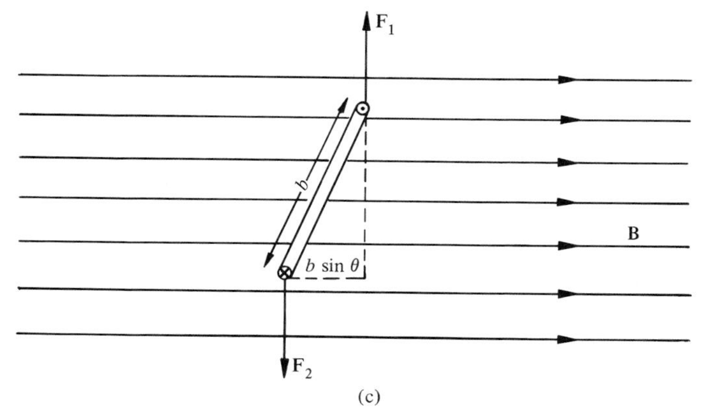

FIGURE 23.14 *Current-carrying rectangular conducting loop in a magnetic field.*

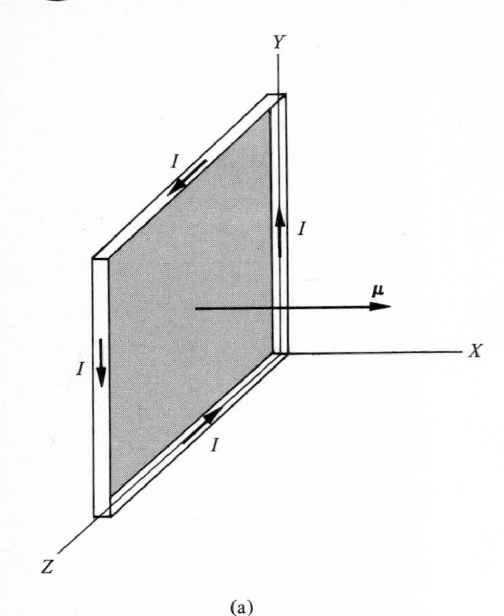

(a)

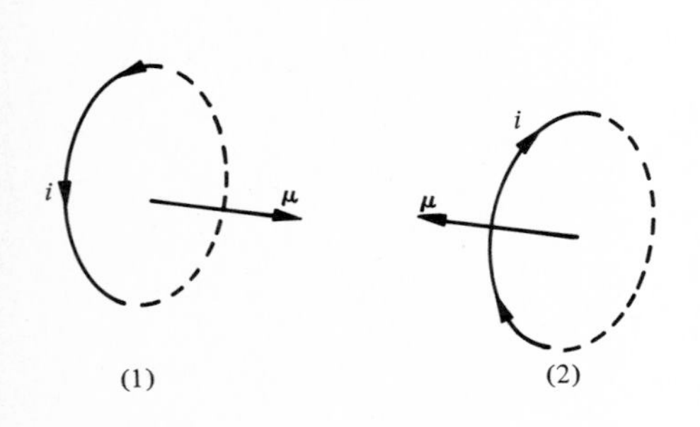

(1) (2)

(b)

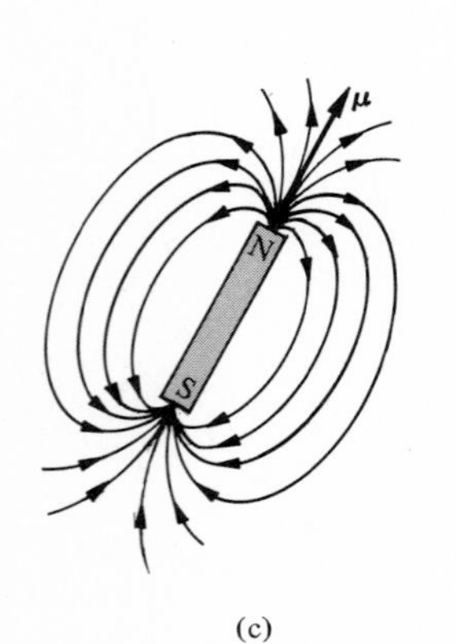

(c)

FIGURE 23.15 *Examples of magnetic dipoles.*

There are four forces, F_1, F_2, F_3 and F_4, acting on four sides of the loop. The forces F_3 and F_4 are exactly equal and opposite and are also in the same line of action. Hence, F_3 and F_4 will produce merely stretching of the loop but no torque and no rotation. The forces F_1 and F_2 are exactly equal in magnitude, but have different lines of action. If we calculate the torque about the Z-axis, the moment arm for F_2 will be zero, while the moment arm for F_1 will be $b \sin \theta$, as shown in Figure 23.14(c). Thus, the net torque τ about the Z-axis will be

$$\tau = F_1 b \sin \theta \tag{23.22}$$

But F_1 $(= F_2)$, according to Equation 23.21, is $F_1 = IaB$. Therefore,

$$\tau = (IaB)(b \sin \theta)$$

Since the area of the loop $A = ab$, we may write the torque τ on a single-turn loop to be

$$\tau = IAB \sin \theta \tag{23.23}$$

(The same value of τ is obtained if we took torques about any other axis as long as it is parallel to the Z-axis.) If there are N turns in the loop, the total magnetic torque will be

$$\boxed{\tau_N = N\tau = NIAB \sin \theta} \tag{23.24}$$

Even though we have assumed that the loop is rectangular, the results above hold for any shape plane loop as long as A represents the area. In what follows, we shall use Equation 23.23, it being understood that IA must be replaced by NIA if there are N turns in the loop. Also, the effect of the leads taking the current into and out of the loop is negligible if the leads are twisted together.

Let us go back to Equation 23.23. The product of the current in a loop and its area is defined as the *magnetic dipole moment* μ. That is,

$$\boxed{\mu = IA} \tag{23.25}$$

The units of μ are ampere-meter^2 ($=$ A-m^2). Also, for N loops, $\mu = NIA$. Thus, Equation 23.24 for the torque may be written

$$\tau = \mu B \sin \theta \tag{23.26}$$

As a matter of fact, a rectangular loop, a circular loop, or a loop of any other shape with electric current in it, a bar magnet, and a compass needle, as shown in Figure 23.15, are all examples of magnetic dipoles. The two faces of a plane loop behave like north and south poles, respectively. If we look at the top face of the loop and the current is in a counterclockwise sense, the top face is the north-seeking pole. If the current is clockwise, the top face is the south-seeking pole. Also, the magnetic dipole moment μ is a vector quantity; hence, it must be assigned a direction. As shown in Figure 23.15, the magnetic dipole moment $\boldsymbol{\mu}$ must lie along the axis of the loop. In Equation 23.26 both $\boldsymbol{\mu}$ and $\mathbf{B}$ are vector quantities, and so is the torque $\boldsymbol{\tau}$. Hence, from the definition of cross product, the relation of Equation 23.26 takes the form where

$$\boxed{\begin{array}{c}\tau = \boldsymbol{\mu} \times \mathbf{B} \\ \hline \tau = \mu B \sin\theta\end{array}} \tag{23.27}$$

As described in Section 22.4, a galvanometer consists of a coil that is freely pivoted and placed in a magnetic field. A pointer attached to the coil moves over a calibrated scale as shown in Figure 22.11; and is reproduced in Figure 23.16 with its top view. When a current passes through the coil, it interacts with the magnetic field, producing a deflection of the coil. Also, the coil is attached to an elastic helical spring which provides a restoring torque to the coil. The deflection is directly proportional to the current, as we shall explain next.

Suppose that a coil with N turns and a cross-sectional area A is placed in a magnetic field B. If there is a current I in the coil, the torque that acts on the coil, according to Equation 23.24, is $\tau_i = NIAB \sin\theta$. In order to make this torque independent of angle θ, the pole faces of the magnet in Figure 23.16 are cylindrically shaped so that the magnetic field B is radial (along the radii), as shown. For this radial field the angle θ is 90° for all orientations of the coil. Thus, for $\theta = 90°$, $\sin 90° = 1$ and

$$\tau_i = NIAB \tag{23.28}$$

As the coil starts rotating, the restoring torque due to the elastic spring also comes into play. According to Hooke's law, the restoring torque τ_r is proportional to the angular displacement ϕ (ϕ is the deflection of the pointer); that is,

$$\tau_r = \kappa\phi \tag{23.29}$$

where κ is the torsional constant of the spring. Thus, the coil will stop rotating and will come to an equilibrium at an angular deflection when $\tau_i = \tau_r$; that is,

$$NIAB = \kappa\phi$$

or

$$\boxed{I = \left(\frac{\kappa}{NAB}\right)\phi \propto \phi} \tag{23.30}$$

That is, the current is directly proportional to the deflection. Since $\phi/I = NAB/\kappa$, for a small current to produce a large deflection, κ should be very small. Thus, for very sensitive galvanometers, the spring is made of material for which κ is extremely small.

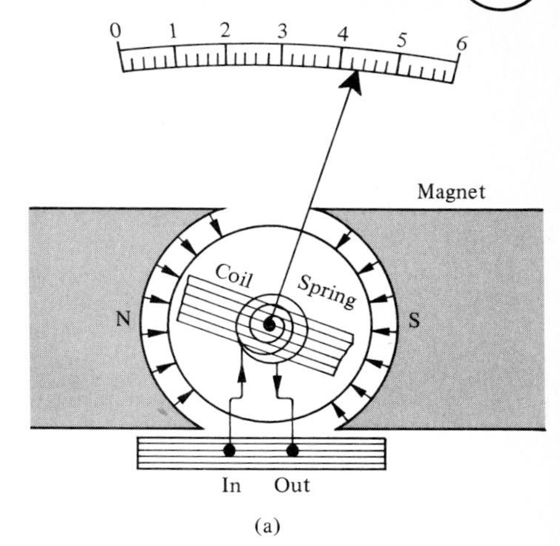

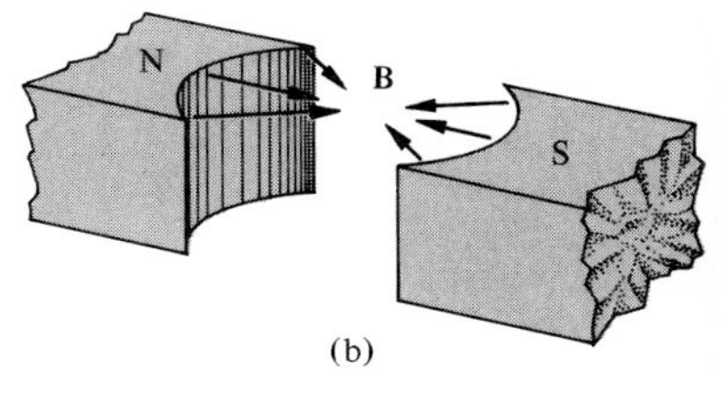

FIGURE 23.16 *(a) Outline of a typical galvanometer. (b) The top view of the magnetic poles and the field.*

EXAMPLE 23.6 A rectangular loop of size 12 cm by 10 cm with 50 turns and a current of 3 A is placed in a uniform magnetic field of 0.40 Wb/m². Calculate (a) the magnetic moment of the loop; and (b) the torque on the loop if the plane of the coil is (1) perpendicular and (2) parallel to the magnetic field **B**.

(a) The area of the coil is $A = 12 \text{ cm} \times 10 \text{ cm} = 120 \text{ cm}^2 = 1.20 \times 10^{-2} \text{ m}^2$, $I = 3$ A, and $N = 50$. Therefore,

$$\mu = NIA = 50(3\text{ A})(1.20 \times 10^{-2}\text{ m}^2) = 1.8\text{ A-m}^2$$

(b) From Equation 23.27,

$$\tau = \mu B \sin\theta = (1.8\text{ A-m}^2)(0.40\text{ Wb/m}^2)\sin\theta = 0.72\sin\theta\text{ N-m}$$

(1) When the plane of the coil is perpendicular to **B**, the normal is parallel to **B**. That is, $\theta = 0°$:

$$\tau = 0.72\sin 0° = 0$$

(2) When the plane of the coil is parallel to **B**, the normal is perpendicular to **B**. That is, $\theta = 90°$.

$$\tau = 0.72\sin 90°\text{ N-m} = 0.72\text{ N-m}$$

EXERCISE 23.6 A coil of 10 turns and area 5 cm² has a magnetic moment of 4×10^{-8} A-m² and experiences a maximum torque of 2×10^{-8} N-m when placed in a uniform magnetic field. Calculate I in the coil and the magnetic induction B. [*Ans.:* $I = 8 \times 10^{-6}$ A $= 8\mu A$, $B = 0.5$ Wb/m².]

SUMMARY

Like magnetic poles repel, whereas unlike magnetic poles attract. The space surrounding a magnet is said to have a *magnetic field*. The search for magnetic *monopoles* has been futile.

A *magnetic field line* is a line the tangent to which at any point gives the direction of the *magnetic field* (magnetic induction or magnetic flux density) **B** at that point. Also, the *magnetic flux* Φ through an area A is $\Phi = \mathbf{B} \cdot \mathbf{A} = BA\cos\theta$, and $B = \Delta\Phi/\Delta A$. The unit of Φ is the weber (Wb).

The *magnetic force* that acts on a charge moving in a magnetic field is $\mathbf{F} = q\mathbf{v} \times \mathbf{B}$ or $F = qvB\sin\theta$. The units of B are 1 Wb/m² = 1 N/A-m = 1 T = 10^4G.

For a charge moving in a uniform magnetic field $mv^2/r = qvB$, $p = mv = qBr$ and $\omega = v/r = qB/m = 2\pi f$. This principle is utilized in the working of a cyclotron. A charged particle in a cyclotron of radius R attains $K_{max} = \frac{1}{2}(qBR)^2/m$.

The magnetic force on a conductor of length L carrying current I in a magnetic field **B** is $\mathbf{F} = I\mathbf{L} \times \mathbf{B}$ or $F = ILB\sin\theta$.

The magnetic torque on a loop of area A is $\tau = IAB\sin\theta = \mu B\sin\theta$, where $\mu = IA$ = magnetic dipole moment. A bar magnet and current-carrying coils are examples of *magnetic dipoles*. The units of μ are A-m². If there are N turns, $\mu = NIA$. Also, $\boldsymbol{\tau} = \boldsymbol{\mu} \times \mathbf{B}$. This principle is utilized in the working of a galvanometer.

QUESTIONS

1. Give two methods by which you would identify the north and south poles of a bar magnet.
2. What do we mean by a negative magnetic flux and positive magnetic flux?
3. How would you distinguish between fields produced by an electric dipole and a magnetic dipole?

4. Can you suggest some configuration of permanent magnets so that the magnetic field over a certain space will be reasonably uniform?
5. If a magnetic monopole were discovered, can you describe some characteristic that it would possess? How would it change the characteristics of a magnetic field?
6. Suppose that a charge q is moving in a uniform magnetic field with a velocity v. Why is there no work done by the magnetic force that acts on charge q?
7. Draw the earth's magnetic field lines and show the path of a positively charged particle that enters (a) the north pole at $\sim 0°$; and (b) the equator at $\sim 90°$.
8. Can you draw a nonuniform magnetic field in which if a charged particle enters, it will be trapped?
9. Why isn't the cyclotron used for accelerating electrons?
10. In a galvanometer, a coil is placed in the magnetic field that is due to a permanent magnet. Why couldn't we place the coil in the magnetic field of a bigger current-carrying coil?
11. What should be the orientation of a coil carrying current in a magnetic field so that there will be no net force?

PROBLEMS

1. A circular coil of radius 0.25 m is placed with its plane perpendicular to a magnetic field of 0.004 Wb/m^2. Calculate the magnitude of the magnetic flux through the coil. What would be the flux if the coil were held with its plane parallel to the direction of the magnetic field?
2. A rectangular coil of sides 10 cm by 6 cm is placed in a uniform magnetic field of 5×10^{-4} Wb/m^2 pointing horizontally toward the west. In what direction should we hold this coil so that the flux through the coil will be maximum? Calculate the magnitude of the flux.
3. A coil of area 2 m^2 is placed in a magnetic field of 0.05 Wb/m^2 so that the flux through the coil is maximum. Calculate this flux. If the coil is rotated through an angle of (a) 30°; (b) 45°; (c) 60°; (d) 90°; and (e) 180°, what will be the flux in each case?
4. A cube of 10-cm sides is placed in a uniform magnetic field of 0.2 Wb/m^2 pointing along the X-axis, as shown. Calculate the flux through each face of the cube as well as through the area $ACGE$.

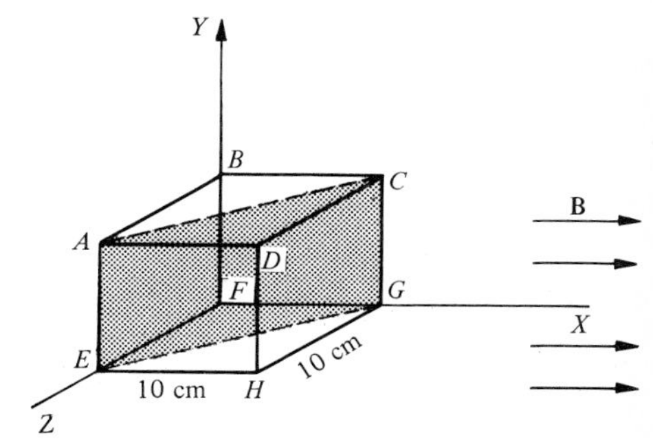

P. 23.4

5. Prove the following relations: 1 Wb = 1 J-s/C = 1 Ω-C and 1 T = 1 V-s/m^2 = 1 Ω-C/m^2.
6. An electron is moving toward the east with a speed of 1.0×10^8 m/s in a uniform magnetic field of 0.8 Wb/m^2 directed upward. Find the magnitude and direction of the magnetic force acting on the electron. If the directions of the electron motion and the magnetic field are interchanged, what is the magnitude and direction of the force?
7. In a television tube an electron is moving with a speed of 1.5×10^8 m/s in a magnetic field of intensity 0.0001 T. Calculate the force acting on the electron and the acceleration of the electron.
8. A proton is moving with a speed of 3×10^7 m/s in a magnetic field of 1.0 T. What is the maximum force and acceleration that can act on the

proton? What will be its relative direction? How does this value change if the proton is replaced by an electron?

9. Charges q_1, q_2, and q_3 are electrons at G, B, and E moving with velocities 2×10^8 m/s along the X, Y, and Z axes, respectively, in a magnetic field acting to the right as shown in figure of problem 4. Calculate the forces on each of the three charges if B is 0.2 Wb/m^2.
10. A particle of mass 0.1 g carrying a charge of 5×10^{-8} C is moving horizontally toward the east with a speed of 5×10^4 m/s. What should be the magnitude and direction of a uniform magnetic field so that it will balance the gravitational pull on the particle and keep it moving horizontally?
11. An electric field of 8×10^5 V/m is acting horizontally toward the east while a magnetic field of 0.2 Wb/m^2 is acting vertically downward. A proton with a speed of 10^7 m/s is ejected at right angles to both the E and B fields, that is, in the direction of (a) north; and (b) south. Calculate the electric force, magnetic force, and the net force in each case.
12. In Problem 11 what should be the direction and the speed of the proton so that the net force on the proton is zero? (This is one of the methods for finding the speed of charged particles, that is, using a cross field and adjusting E or B so the net deflection is zero.)
13. What should be the velocity of an electron and the direction of the electric and magnetic fields so that when E is 8×10^4 V/m and B is 2×10^{-2} Wb/m^2, the net deflection of the electron is zero? Draw a clear diagram showing the orientation of **v**, **E**, and **B**. What happens if E or B is switched off?
14. A proton and an electron both moving with a speed of 10^5 m/s enter a uniform magnetic field of 2 Wb/m^2 in a plane perpendicular to the field. Describe the motion in each case.
15. What is the magnitude of a uniform magnetic field if an electron moving in a plane perpendicular to the field makes one circular revolution in 2×10^{-9} s?
16. Calculate the radius, momentum, time period, and angular frequency of (a) a proton; (b) a deuteron; and (c) an α-particle each moving with a tangential velocity of 10^6 m/s in a magnetic field of 1.2 Wb/m^2.
17. A uniform magnetic field of 2 Wb/m^2 is used in a cyclotron to accelerate protons. The radius of the cyclotron is 0.32 m. Calculate (a) the frequency f; (b) the maximum velocity of the protons; and (c) the maximum kinetic energy of the protons.
18. Repeat Problem 17 using deuterons (a deuteron nucleus consists of a proton and a neutron). Also, calculate the time a deuteron will take to complete one revolution.
19. Protons in a cyclotron describe a circle of radius 0.40 m before they are ejected. The applied frequency is 10 MHz (1 MHz $= 1 \times 10^6$ cycles/s). Calculate the magnetic field, the kinetic energy, and the speed of the protons just before ejection.
20. Repeat Problem 19 using deuterons. Through what potential difference will the deuteron have to be accelerated to acquire the same energy and velocity?
21. Calculate the magnitude and direction of a magnetic field so that an electron entering at A with a speed of 10^6 m/s will describe a circle of

radius 10 cm. If this keeps on moving in a circle, what are the frequency and the time period?

22. A cyclotron is to be constructed to accelerate deuterons to acquire an energy of 20 MeV using a magnetic field of 2.0 Wb/m^2. Calculate the frequency and the size of the dees to be used.

23. Calculate the magnitude of force that acts on a 15-cm-long wire carrying a current of 4 A and placed perpendicular to a uniform magnetic field of 2 T.

24. A copper wire is carrying a current of 2 A from east to west. What is the direction and magnitude of the magnetic field so that a force of 0.01 N per unit length acts upward on the wire?

25. A copper wire of density 8.9×10^3 kg/m^3 has a current density of 4.0×10^6 A/m^2 from west to east. Calculate the magnitude and direction of the magnetic field so that the force per unit length will balance the gravitational push downward.

26. A square coil with one side 10 cm long and having 10 turns with current of 1A in each is placed in a uniform magnetic field of 0.05 Wb/m^2. Calculate the torque if the angle between the normal to the coil and the direction of the magnetic field is (a) 0°; (b) 30°; (c) 45°; (d) 60°; and (e) 90°.

27. A coil of 0.1 m $\times$ 0.1 m and of 200 turns carrying a current of 1 mA is placed in a uniform magnetic field of 0.10 Wb/m^2. Calculate the maximum torque that acts on the coil. What is the magnetic moment of the coil?

28. In Problem 27, what should be the position of the coil so that the torque acting on it is (a) zero; and (b) half the maximum value?

29. A circular loop, a square, and a rectangular loop all have the same perimeter. Which will have the maximum magnetic moment and torque when placed in a uniform magnetic field?

30. A coil of 10 turns and area 5 cm^2 has a magnetic moment of 4×10^{-8} $A\text{-}m^2$ and experiences a maximum torque of 2×10^{-8} N-m when placed in a uniform magnetic field. Calculate the current in the coil and the magnetic field intensity B.

31. A galvanometer coil has an area of 8 cm^2 and has 60 turns and is placed in a field of 0.02 Wb/m^2. The torsion constant of the galvanometer's hairspring is 8×10^{-8} N-m/deg. Calculate the deflection when a current of 1 mA passes through the coil.

24 Sources of Magnetic Fields

In this chapter we seek an analytical relation between current and the magnetic field (magnitude as well as direction) it produces. For this we introduce Ampère's law and the Biot and Savart law. This leads us to the definition of an ampere—the basic SI unit of current. Finally, we shall explain the classification of magnetic materials in terms of the magnetic field produced by a current due to the motion of the electrons in the atoms.

24.1. Magnetic Fields Produced by Currents

IN THIS section we want to calculate the magnetic field at different distances from current sources. Most of the discussion can be summarized under two fundamental laws: (1) Ampère's law, and (2) the Biot and Savart law.

Ampère's Law

As is the case for all basic laws of physics, Ampère's law cannot be derived. Its validity is based on the correctness of the results which it predicts. Furthermore, Ampère's law is especially useful for calculating magnetic fields in situations that are highly symmetrical.

Suppose that current I is enclosed by an arbitrary curved path C. The value of the magnetic induction $\mathbf{B}$ will be different at different points along C. Let there be a constant magnetic induction B_1 over a small element of length Δl_1 on the curved path and, similarly, B_2 for Δl_2, B_3 for Δl_3, . . . , while θ_1 is the angle between B_1 and Δl_1, θ_2 between B_2 and Δl_2, ... We may now state

> **Ampère's law:** ***The sum of the products of the quantities*** $B_i\,\Delta l_i \cos\theta_i$ ***over any closed path is equal to the product of the constant*** μ_0 ***and the net current enclosed by the path.***

That is,

$$B_1\,\Delta l_1 \cos\theta_1 + B_2\,\Delta l_2 \cos\theta_2 + \cdots = \mu_0 I$$

or

$$\sum_{\substack{\text{closed}\\\text{path}}} B_i\,\Delta l_i \cos\theta_i = \mu_0 I \tag{24.1}$$

In vector notation, the left-hand side may be written as a dot product:

$$\sum_{\substack{\text{closed}\\\text{path}}} \mathbf{B}_i \cdot \Delta l_i = \mu_0 I \tag{24.2}$$

The constant μ_0 is known as the *permeability of free space,* and its value in SI units is

$$\mu_0 = 4\pi \times 10^{-7}\ \text{Wb/A-m} \tag{24.3}$$

Thus, as shown in Figure 24.1, if currents I_1, I_2, I_3, and I_4 are enclosed by an arbitrary path C, Ampère's law takes the form

$$\sum_{\substack{\text{closed}\\\text{path}}} \mathbf{B}_i \cdot \Delta \boldsymbol{l}_i = \mu_0(I_1 - I_2 - I_3 + I_4)$$

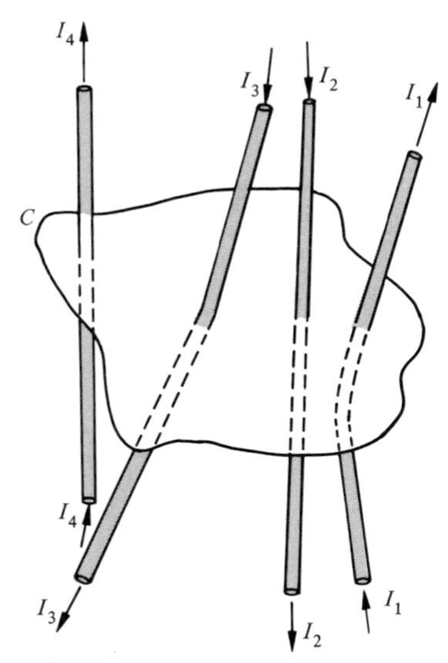

FIGURE 24.1 *An arbitrary path C encloses different current-carrying conductors.*

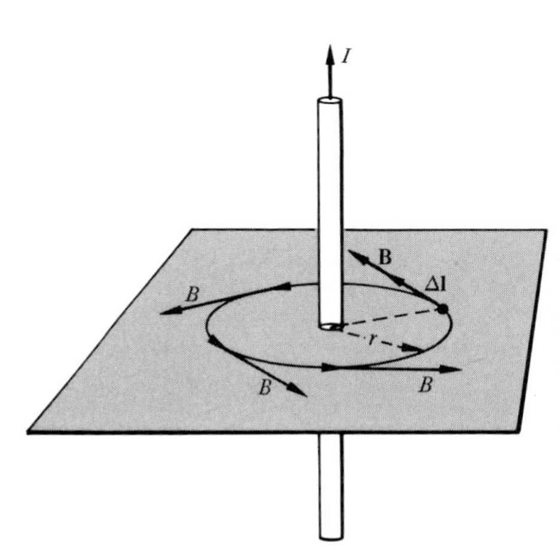

FIGURE 24.2 *Application of Ampère's law to a highly symmetrical situation.*

The left side is not easy to evaluate unless the path is symmetrical.

Let us apply Ampère's law to a symmetrical case as shown in Figure 24.2. Suppose that we want to calculate B at a distance r from a conductor which is carrying current I. Draw a circle of radius r in a plane perpendicular to the conductor as shown. This circle may be considered as a magnetic field line and the tangent at any point on this circle gives the direction of **B** at that point. Since **B** and $\Delta \boldsymbol{l}$, as shown, are in the same direction, $\mathbf{B} \cdot \Delta \boldsymbol{l} = B\,\Delta l \cos 0° = B\,\Delta l$, and the magnitude of **B** is the same at every point on the circle. We may write Equation 24.1 as

$$B(\Delta l_1 + \Delta l_2 + \cdots) = \mu_0 I$$

But the quantity in parentheses is equal to the circumference $2\pi r$. That is,

$$B(2\pi r) = \mu_0 I$$

or

$$\boxed{B = \frac{\mu_0 I}{2\pi r}} \tag{24.4}$$

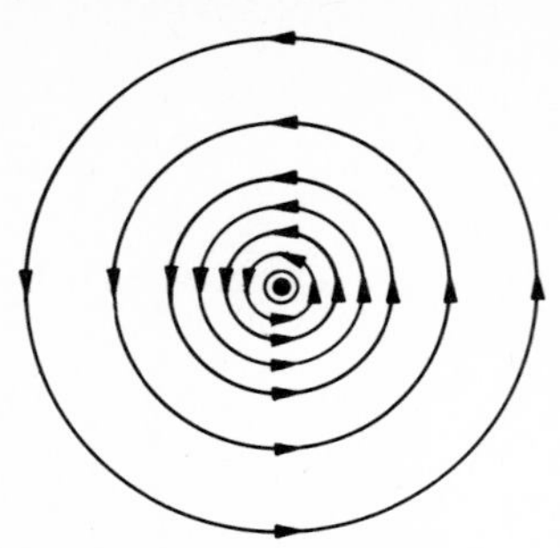

Figure 24.3 *Magnetic induction lines due to a single conductor carrying current I out of the page.*

The resulting magnetic field lines are as shown in Figure 24.3. The field at any point at a distance r from the conductor for a given current I is inversely proportional to r.

Example 24.1 A long, straight wire 2 mm in diameter is carrying a current of 5 A. Calculate the magnetic induction B at distances of 0.1 m, 0.2 m, and 0.4 m.

Substituting for $\mu_0 = 4\pi \times 10^{-7}$ Wb/A-m, $I = 5$ A, and $r = 0.1$ m, 0.2 m, and 0.4 m in Equation 24.4,

$$B = \frac{\mu_0 I}{2\pi r} = \frac{(4\pi \times 10^{-7}\ \text{Wb/A-m})(5\ \text{A})}{2\pi r} = \frac{10^{-6}}{r}\ \text{Wb/m}$$

$$\text{If } r_1 = 0.1\ \text{m:}\quad B_1 = \frac{10^{-6}}{0.1\ \text{m}}\ \text{Wb/m} = 10 \times 10^{-6}\ \text{Wb/m}^2$$

$$\text{If } r_2 = 0.2\ \text{m:}\quad B_2 = \frac{10^{-6}}{0.2\ \text{m}}\ \text{Wb/m} = 5 \times 10^{-6}\ \text{Wb/m}^2$$

$$\text{If } r_3 = 0.4\ \text{m:}\quad B_3 = \frac{10^{-6}}{0.4\ \text{m}}\ \text{Wb/m} = 2.5 \times 10^{-6}\ \text{Wb/m}^2$$

Thus as r increases in the ratio 1 : 2 : 4, B decreases in the ratio 4 : 2 : 1.

Exercise 24.1 Repeat Example 24.1 if instead of one wire, there are two wires together: (a) one carrying a current of 10 A in one direction and the other a current of 5 A in the opposite direction; and (b) both carrying currents in the same direction. [*Ans.:* (a) $(10, 5, 2.5) \times 10^{-6}$ Wb/m^2; (b) $(30, 15, 7.5) \times 10^{-6}$ Wb/m^2.]

The Biot-Savart Law

One of the difficulties with Ampère's law is that it can be used successfully for calculating magnetic fields only in those cases where the symmetry of the current distribution is high enough to easily evaluate the sum $\Sigma\, \mathbf{B} \cdot \Delta \mathbf{l}$. Even in the case of a simple coil carrying current, Ampère's law becomes impractical. It does not fail, but simply becomes difficult to evaluate. In such cases we must make use of the Biot–Savart law, as given by the French physicists J. B. Biot (1774–1862) and F. Savart (1791–1841).

Consider an arbitrary current distribution AA' carrying current I, as shown in Figure 24.4. First, let us calculate the field ΔB produced at point P due to a small current element ΔL. The point P is at a distance r from ΔL and θ is the angle r makes with the element ΔL (or with the tangent TT' drawn to ΔL), as shown. According to the *Biot–Savart law*, the field at P is given by

$$\boxed{\Delta B = \frac{\mu_0}{4\pi}\frac{I\,\Delta L \sin\theta}{r^2} = K'\frac{I\,\Delta L \sin\theta}{r^2}} \qquad (24.5)$$

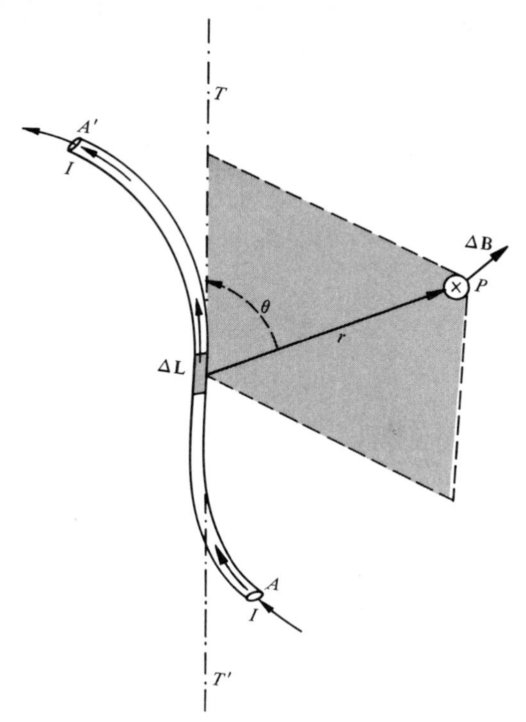

FIGURE 24.4 *Magnetic field* $\Delta\mathbf{B}$ *at a distance* $\mathbf{r}$ *from an arbitrarily shaped current-carrying conductor.*

where $K' = \mu_0/4\pi$ is similar to the case in Coulomb's law in electrostatics, where $k = 1/4\pi\epsilon_0$. As usual, the direction of ΔB is given by the right-hand rule. In the case shown in Figure 24.4, ΔB is perpendicular to the plane containing ΔL and r (shaded area) and is into the page. The total field B at P due to the entire current distribution may be found by summing the quantities $\Delta\mathbf{B}_i$. That is,

$$\mathbf{B} = \sum_i \Delta\mathbf{B}_i = \Delta\mathbf{B}_1 + \Delta\mathbf{B}_2 + \Delta\mathbf{B}_3 + \cdots \qquad (24.6)$$

To demonstrate the use of the Biot–Savart law, we calculate the magnetic field at point P, which is at the center of a current-carrying circular loop of radius R, as shown in Figure 24.5. Let us divide the circular path into small elements $\Delta L_1, \Delta L_2, \ldots$. According to Equation 24.5, the field at point P due to a small element ΔL_i carrying current I is given by

$$\Delta B_i = K' \frac{I\,\Delta L_i \sin\theta_i}{R^2}$$

Every element ΔL_i is at the same distance $r = R$ from the center P and also the angle between every element ΔL_i and r is 90°; that is, $\sin\theta_i = \sin 90° = 1$. Thus,

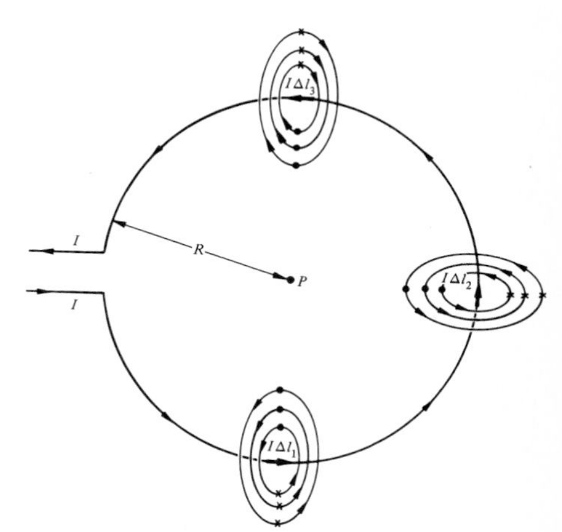

FIGURE 24.5 *Magnetic induction at the center of a current-carrying circular loop.*

$$\Delta B_i = K' \frac{I\,\Delta L_i}{R^2}$$

The total magnetic field B at P is

$$B = \Delta B_1 + \Delta B_2 + \cdots = K' \frac{I}{R^2}(\Delta L_1 + \Delta L_2 + \cdots)$$

But the sum of all the elements in the parentheses is equal to the circumference, $2\pi R$. Remembering that $K' = \mu_0/4\pi$,

$$B = \frac{\mu_0}{4\pi}\frac{I}{R^2}(2\pi R)$$

That is,

$$\boxed{B = \frac{\mu_0 I}{2R}} \tag{24.7}$$

To find the direction of B, we apply the right-hand rule. Hold the wire in your right hand with the thumb pointing in the direction of I through ΔL; the direction of the curling fingers gives the path of the magnetic field. B at any point will be tangent to these lines. The field inside the loop points out of the paper, while outside the loop it points into the paper, as shown.

EXAMPLE 24.2 Calculate the magnetic field at the center of a coil of radius 0.1 m and carrying a current of 10 A. At what distance from a straight wire carrying a current of 10 A will the magnetic field be the same as in the case of the loop?

From Equation 24.7,

$$B = \frac{\mu_0 I}{2R} = \frac{4\pi \times 10^{-7}\ \text{Wb/A-m} \times 10\ \text{A}}{2 \times 0.1\ \text{m}} = 2\pi \times 10^{-5}\ \text{Wb/m}^2$$

For a straight wire from Equation 24.4, $B = \mu_0 I/2\pi R$, where $B = 2\pi \times 10^{-5}$ Wb/m^2, $I = 10$ A. Thus,

$$r = \frac{\mu_0 I}{2\pi B} = \frac{4\pi \times 10^{-7}\ \text{Wb/A-m} \times 10\ \text{A}}{2\pi \times 2\pi \times 10^{-5}\ \text{Wb/m}^2} = 0.032\ \text{m}$$

This distance is 3.2 cm, compared to the radius of the loop, which is 10 cm.

EXERCISE 24.2 A loop carrying a current of 20 A produces a field of $2\pi \times 10^{-5}$ Wb/m^2 at the center. What should be the current in a straight conductor so that it produces the same field ($2\pi \times 10^{-5}$ Wb/m^2) at a distance equal to the radius of the loop? [*Ans.:* $I = 62.8$A.]

24.2. Two Parallel Conductors—Definition of Ampere

One week after the news of Oersted's experiment in 1820 reached Paris, Ampère showed that two long parallel conductors carrying currents in the same direction, as shown in Figure 24.6(a), attract each other. The force between these two conductors is purely magnetic, although, apparently, no magnets are involved. We calculate the magnitude of this force as follows.

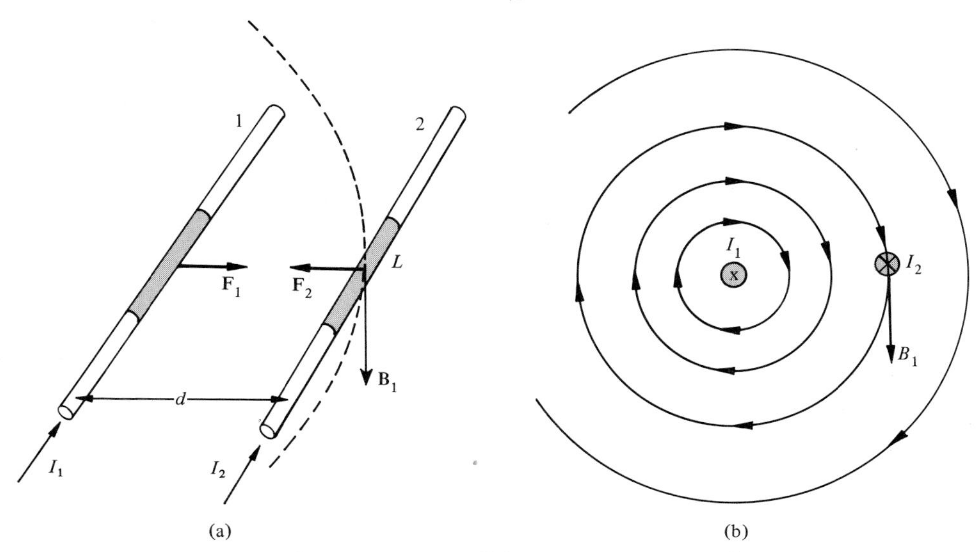

FIGURE 24.6 *Force between two parallel current-carrying conductors.*

Consider two long parallel conducting wires separated by a distance d and carrying currents I_1 and I_2 in the same direction, as shown in Figure 24.6(a). The current I_1 in wire 1 will produce a magnetic field B_1 around itself, as shown by the lines of induction in Figure 24.6(b). At the site of wire 2 the magnetic field B_1 is directed vertically downward, and its magnitude is, from Equation 24.4,

$$B_1 = \frac{\mu_0 I_1}{2\pi d} \tag{24.8}$$

Wire 2, carrying current I_2, finds itself in an external field of magnetic induction B_1. Thus, the length L of wire 2 will experience a sideways deflecting magnetic force F_2 given by (using the relation $\mathbf{F} = I\mathbf{L} \times \mathbf{B}$)

$$F_2 = I_2 L B_1 \tag{24.9}$$

According to the right-hand rule, the force F_2 lies in the plane of the two wires and points to the left, as shown in Figure 24.6(a). Substituting the value of B_1 from Equation 24.8 into Equation 24.9, we get

$$F_2 = \frac{\mu_0 I_1 I_2 L}{2\pi d} \tag{24.10}$$

or the force per unit length $\Delta F/\Delta L$ experienced by wire 2 is given by

$$\frac{\Delta F}{\Delta L} = \frac{F_2}{L} = \frac{\mu_0 I_1 I_2}{2\pi d} \tag{24.11}$$

Starting with wire 2 and going through similar reasoning shows that wire 1 is attracted to wire 2 with equal force per unit length. (The interchange of subscripts 1 and 2 in Equation 24.11 leaves the expression for the force unchanged.) Thus, as shown in Figure 24.6(a), the forces $\mathbf{F}_1$ and $\mathbf{F}_2$ which the two wires exert on each other are equal and opposite. Following this method, we can show that if the currents were antiparallel in the two wires, there will be a repulsion between them.

Equation 24.11 is a statement of the fundamental law of force between currents. There is a close similarity between this and the force law between static point charges. The force between the static charges is conveyed through the intermediary of the electric field, while the force between two currents is conveyed through the intermediary of the magnetic field. It is true that Faraday believed that the force is a direct interaction of the action-at-a-distance type, but later Maxwell clearly established that all electric or magnetic interactions are due to the intermediary, "the field." That is, a charge or a current produces a field, which in turn interacts with other charges or currents in this field to produce force.

The attraction between two current-carrying conductors is used to define, in principle, the unit of current, the ampere. The official definition of the ampere as given by the International Committee on Weights and Measures is stated: "The *ampere* is that constant current which, if maintained in two straight parallel conductors of infinite length, of negligible circular cross section, and placed 1 meter apart in vacuum, would produce between these two conductors a force equal to 2×10^{-7} newton per meter of length."

Substituting these values in Equation 24.11, we get

$$\frac{\Delta F}{\Delta L} = \frac{\mu_0 I^2}{2\pi d} = \frac{(4\pi \times 10^{-7}\ \text{Wb/A-m})(1\ \text{A})^2}{(2\pi)(1\ \text{m})} = 2 \times 10^{-7}\ \text{N/m}$$

as expected from the definition of the ampere.

Once the ampere is defined, the SI unit of charge, the *coulomb,* is defined in terms of the ampere and the second from the relation $Q = It$; that is, 1 coulomb = (1 ampere)(1 second). As stated earlier, the *coulomb* is defined to be that electric charge which passes any cross section in a wire in 1 second when there is a current of 1 ampere in the wire.

The primary measurements of current are made with a current balance. The current balance used at the National Bureau of Standards is shown in Figure 24.7. It consists of a coil of precisely known size and shape placed between two other coils. As shown in the lower portion of the balance, the two outer coils are fixed while the inner one is hung from the arm of a balance. The coils exert forces on one another which can be measured by adding weights on the balance pan. In this way, current can then be defined in terms of measured forces.

FIGURE 24.7 *Current balance used at the National Bureau of Standards.* (Courtesy of the National Bureau of Standards.)

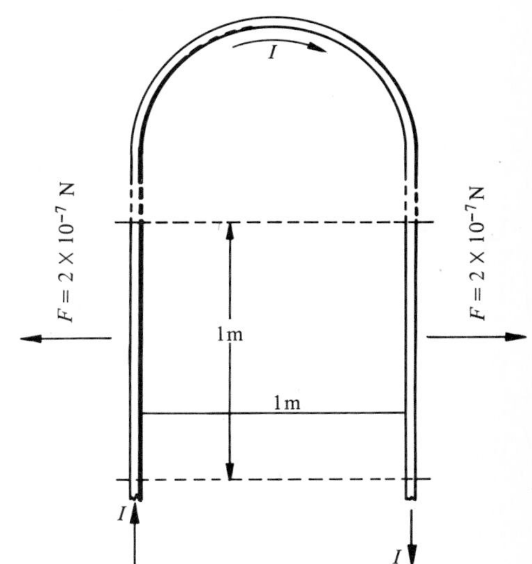

EXAMPLE 24.3 Two parallel conducting wires each carry a current of 50 A. The force per unit length between these wires is equal to the force of the gravitational pull on a wire 1 m long. If the mass of 1 m of wire is 1 g, calculate the distance d between the two parallel wires.

Substitute $I_1 = I_2 = I = 50$ A and $\Delta F/\Delta L = mg = (1 \times 10^{-3}$ kg/m)(9.8 m/s^2) $= 9.8 \times 10^{-3}$ N/m in Equation 24.11:

$$\frac{\Delta F}{\Delta L} = \frac{\mu_0 I^2}{2\pi d}$$

$$9.8 \times 10^{-3}\ \text{N/m} = \frac{(4\pi \times 10^{-7}\ \text{Wb/A-m})(50\ \text{A})^2}{2\pi d}$$

or

$$d = \frac{4\pi \times 10^{-7}\ \text{Wb/A-m} \times 2500\ \text{A}^2}{2\pi \times 9.8 \times 10^{-3}\ \text{N/m}} = 5.1 \times 10^{-2}\ \text{m} = 5.1\ \text{cm}$$

EXERCISE 24.3 What should be the current in each of the two parallel wires separated by a distance of 0.1 m so that the force per unit length between them is equal to the force of gravitation per unit length? The mass of 1 m of the wire is 1 g. [*Ans.*: $I = 70$ A.]

24.3. Magnetic Field Due to a Solenoid and a Toroid

The magnetic field produced by a single permanent bar magnet is not uniform, as shown in Figure 23.2. In order to obtain a more uniform magnetic field we could bend a bar magnet in the shape of a U, as shown in Figure 23.2. Of the many methods that can be used for obtaining a uniform magnetic field, one of them is the use of a solenoid, as described below.

The Solenoid

A solenoid is constructed from a long wire wound in a close-packed helix around the surface of a cylindrical form. The field produced from current I in this solenoid can be understood by first examining the field due to a single loop or a very compact coil, as shown in Figure 24.8(a), which shows the side view as well as the top view. The magnetic field lines encircle the current-carrying wires while at the center of the loop the lines are straight. A cylindrical stack of N such current-carrying loops can be thought of being a solenoid with the field being N times as large as that due to a single loop.

A solenoid of fine, very loosely wound turns is shown in Figure 24.8(b). The resulting magnetic field lines are as indicated. For a more tightly wound solenoid, the situation is as shown in Figure 24.8(c). The following observations may be made from the form of the magnetic field lines. Inside the solenoid the magnetic lines are directed lengthwise along the axis of the solenoid, resulting in a uniform magnetic field for a tightly wound long solenoid. As the magnetic field lines come out of the solenoid, they spread out widely before coming back around to enter the other end. This results in a very weak magnetic field outside the coil. Also, in between the turns the field lines are in the opposite directions, owing to the adjacent turns, and the field is canceled out.

B may be calculated by applying Ampère's law to a point such as P. Without proof we state that for a solenoid having n turns per unit length B inside the solenoid is

$$\boxed{B = \mu_0 n I} \tag{24.12}$$

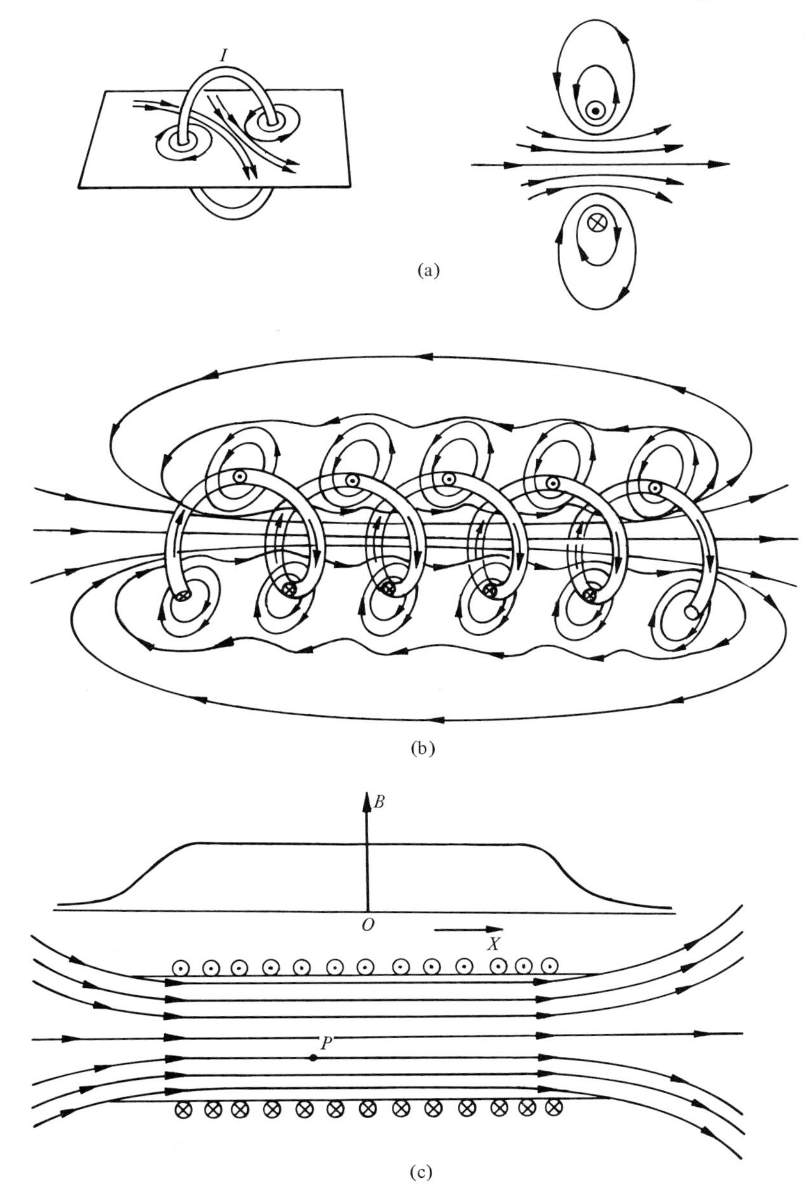

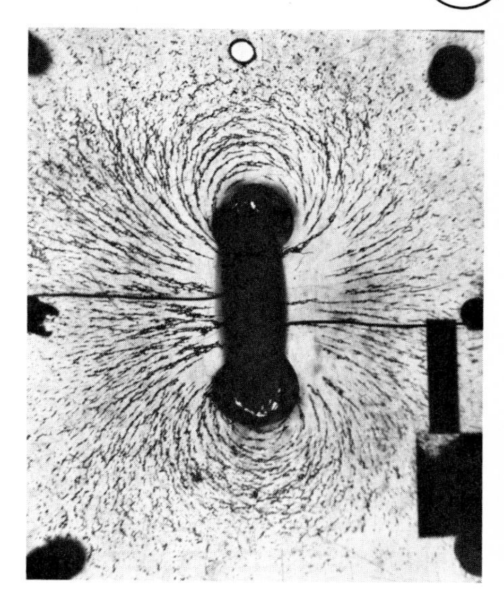

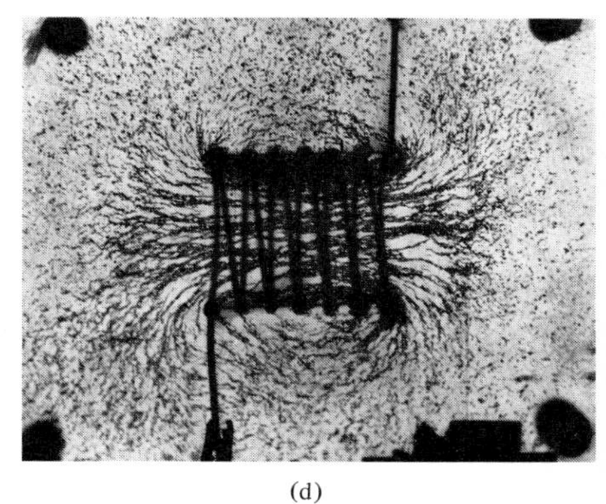

FIGURE 24.8 *(a) Field due to a single loop. (b) Field of a solenoid. (c) Field of a tightly packed solenoid. (d) Actual field lines.*

Thus, B inside a long solenoid is independent of the position of the point inside the solenoid and depends upon I and the number of turns per unit length (and not the total number of turns in the solenoid). For $n = 100$ turns/cm and $I = 1$ A, the field $B \simeq 0.0125$ Wb/m^2 $= 125$ G. Solenoids are available that can produce field up to $\sim$2000 G.

The fields produced by permanent magnets are very weak. Electromagnets can produce fields of the order of ~100,000 G. Solenoids using ferromagnetic materials can also produce large fields. The major problem is the removal of heat produced due to the resistance of the wire. This problem has been solved by using superconducting material in the solenoids instead of ordinary conducting materials. Superconducting solenoids are available that can produce fields from 60 to 125 kG (6 to 12.5 Wb/m^2). Of course, magnets made out of superconducting materials must be operated within a few degrees of absolute zero temperature.

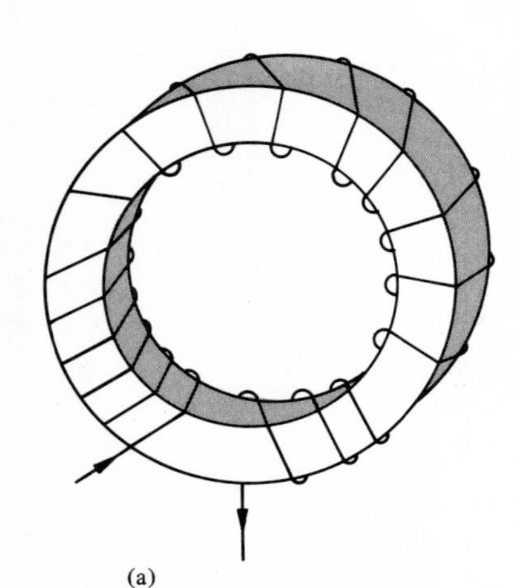

The Toroid

A long solenoid bent into a circle so that it looks like the shape of a doughnut as shown in Figure 24.9(a) is called a *toroid*. We can evaluate the magnetic field at different points by applying Ampère's law. There are three different regions, I, II, and III, as shown in Figure 24.9(b). In region I there is no current enclosed by the dashed-line path; hence, $\mathbf{B} \cdot \Delta \boldsymbol{l}$ must be zero. Also, for this path, $\mathbf{B}$ is perpendicular to $\Delta \boldsymbol{l}$, hence $\mathbf{B} \cdot \Delta \boldsymbol{l} = 0$. Since the circumference of the dashed-line path is not zero, the field $\mathbf{B}$ must be zero. Similarly, in region III the net current enclosed by a dashed-line path (not shown) is zero, hence B must be zero, too.

The field in region II is not zero. By using the right-hand rule, the lines of induction inside the toroid are found to form concentric circles. Let us calculate $\mathbf{B}$ at a point such as P which is at r from the center of the toroid. Draw a circular path, shown dashed, which passes through P. By symmetry the magnetic field $\mathbf{B}$ is tangent to the path and is the same at every point at a distance r. If there are N turns in the toroid and the current in each is I, Ampère's law,

$$B(\Delta L_1 + \Delta L_2 + \cdots) = \mu_0 NI$$

takes the form

$$B(2\pi r) = \mu_0 NI$$

or

$$\boxed{B = \frac{\mu_0 NI}{2\pi r}} \tag{24.13}$$

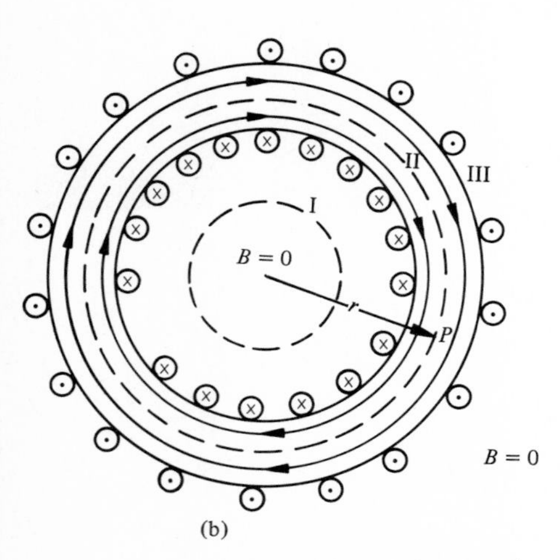

FIGURE 24.9 *Toroid.*

Thus, inside the toroid the field varies as $1/r$. On the other hand, if the cross-sectional area of the toroid is very very small compared to r, we can neglect any variation in r. Considering $2\pi r$ to be the circumference of the toroid, $N/2\pi r$ will be a constant equal to the number of turns per unit length n. In this case, Equation 24.13 takes the form $B = \mu_0 nI$, which is the same as for a long solenoid.

EXAMPLE 24.4 Calculate the magnetic field due to a narrow solenoid 50 cm long with 4000 turns and a current of 2 A. If this solenoid is made into a toroid, what will be the radius and the magnetic field?

$N = 4000$, $L = 50$ cm $= 0.5$ m, $n = N/L = 4000/0.5$ m $= 8000$/m, and

$I = 2$ A. Substituting these in Equation 24.12, the field due to the solenoid will be $B = \mu_0 nI = (4\pi \times 10^{-7}$ Wb/A-m)(8000/m)(2 A) $= 0.02$ Wb/m². If this solenoid is changed into a toroid, $2\pi r = L$ or $r = L/2\pi = (50 \text{ cm}/2\pi) =$ 7.96 cm. Also, if we substitute $r = L/2\pi$ in Equation 24.13, the field due to the toroid will be

$$B = \frac{\mu_0 NI}{2\pi r} = \frac{\mu_0 NI}{2\pi(L/2\pi)} = \frac{\mu_0 NI}{L} = \mu_0 nI$$

which is the same as the field due to the solenoid, that is, 0.02 Wb/m².

EXERCISE 24.4 A solenoid produces a magnetic field of 0.1 Wb/m² when there is a current of 4 A and the total number of turns is 6000. What is the length of the solenoid? [*Ans.*: $L = 30.2$ cm.]

24.4. Magnetism in Matter—Atomic Dipole

We have ample evidence suggesting a close relationship between charges in motion and the resulting magnetic fields. It was André Marie Ampère who from his investigations of magnetic fields of currents suggested that all magnetic effects may be due to currents. Scientists assumed that the presence of microscopic current loops within matter produces fields of natural and permanent magnets.

According to the atomic theory of matter, electrons in every atom are moving in orbits of radius $\sim 10^{-10}$ m. This is the *orbital motion* of the electrons and the electrons are said to have orbital *angular momentum*. In addition to this, each electron is spinning about its own axis, resulting in a *spin angular momentum* or *intrinsic angular momentum*. These two motions of the electrons (shown in Figure 24.10) lead to two types of magnetic dipole moments, as explained below. Thus, every material will have some magnetic properties because of the atomic properties associated with its atoms. In most materials the currents corresponding to the two motions of the electrons are completely randomly oriented in all directions so that the magnetic fields created by these current loops cancel out. Such materials do not show any magnetic effects on the macroscopic scale. There are, however, other materials that will show magnetic effects on a large scale.

Consider an electron of mass m and charge $-e$ moving with a velocity v in a circular orbit (Bohr orbit) of radius r, as shown in Figure 24.10(a). The circulating charge is equivalent to a loop of wire carrying a current i given by

$$i = \frac{e}{T} = \frac{e}{2\pi r/v} = \frac{ev}{2\pi r} \tag{24.14}$$

where T is the period of time the electron takes to make its circular motion, which is equal to $2\pi r/v$. The angular momentum L of the electron as shown is

$$L = rp = rmv \tag{24.15}$$

According to Equation 23.25, the magnetic dipole moment μ_l resulting from the current in this loop is

$$\mu_l = \text{current} \times \text{area enclosed by the loop} = iA$$

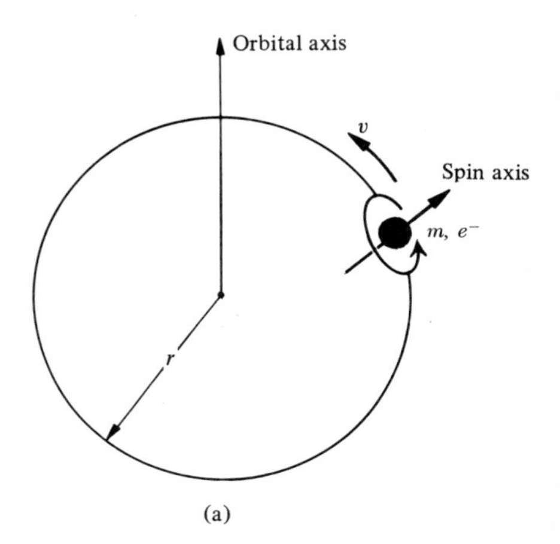

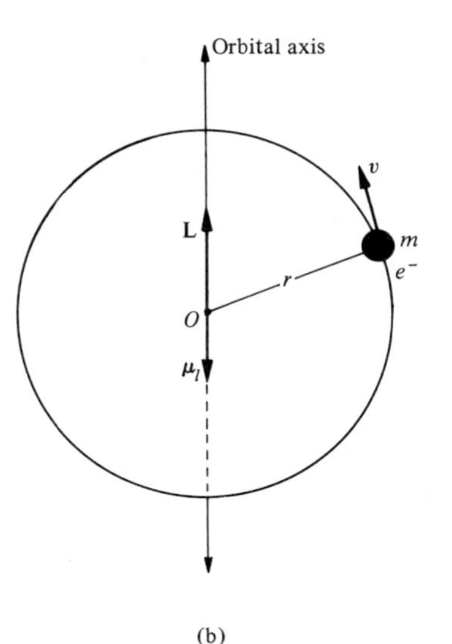

FIGURE 24.10 (*a*) *Orbital and spin motions of an electron in an atom.* (*b*) *Angular momentum and magnetic dipole moment due to orbital motion of an electron.*

The direction of the dipole is perpendicular to the plane of the orbit, as shown. The direction of $\boldsymbol{\mu}_l$ is opposite that of the angular momentum **L** because of the negative charge of the electron. The area A of the loop is πr^2. Thus, substituting for i and A in the equation above, we get

$$\mu_l = iA = \frac{ev\pi r^2}{2\pi r} = \frac{evr}{2} \tag{24.16}$$

We cannot use such a simple picture to calculate the magnetic moment μ_s due to the spin angular momentum of the electron. Typical values of μ_l and μ_s according to classical theory are found to be

$$\mu_l \simeq \mu_s \simeq 9.1 \times 10^{-24} \text{ A-m}^2 \tag{24.17}$$

EXAMPLE 24.5 The electron in a hydrogen atom moves in a circular orbit of radius 0.53 Å with a velocity of 2.2×10^6 m/s. Calculate the magnetic dipole moment of this electronic loop.

According to Equation 24.14,

$$i = \frac{ev}{2\pi r} = \frac{(1.6 \times 10^{-19} \text{ C})(2.2 \times 10^6 \text{ m/s})}{2\pi(0.53 \times 10^{-10} \text{ m})} = 1.06 \times 10^{-3} \text{ A}$$

while $A = \pi r^2 = \pi(0.53 \times 10^{-10} \text{ m})^2 = 0.88 \times 10^{-20} \text{ m}^2$. From Equation 24.16,

$$\mu_l = iA = (1.06 \times 10^{-3} \text{ A})(0.88 \times 10^{-20} \text{ m}^2) = 0.93 \times 10^{-23} \text{ A-m}^2$$

EXERCISE 24.5 Calculate the magnetic dipole moment if the electron were moving in an orbit of radius 2.12 Å (1 Å = 10^{-10} m) and has a velocity of 1.1×10^6 m/s. [*Ans.:* $\mu_l = 1.87 \times 10^{-23}$ A-m^2.]

24.5. Properties of Magnetic Materials

Most electrical equipment, such as motors, generators, and transformers, contain material such as iron or its alloy. The main purpose of this is to increase the magnetic flux in the region of interest without increasing the current. It is true that the presence of most materials in the region of the magnetic field does not appreciably change the field, but materials such as iron, cobalt, and nickel make a tremendous change.

To investigate the changes in magnetic fields that will result due to the presence of different materials, we consider a toroid inside which the magnetic flux density **B** is uniform throughout. If there is no material inside the toroid (except for vacuum or air), the magnitude of the magnetic induction B_0 is given by Equation 24.13. We can fill the interior of the toroid with different materials and measure the resulting field B for the same current i. (The situation is similar to the electrostatic case, where in order to demonstrate the effect of a dielectric on the electric field the space between the parallel plates is filled with such a dielectric.)

The flux density in the toroid can be measured by winding a secondary coil S as shown in Figure 24.11 and connecting this coil to a ballistic galvanometer [a device in which deflection is proportional to the charge due to the current in the circuit for a short time, $i \propto \Delta Q\,(= i\,\Delta t)$]. When the current in the primary (the toroid in this case) is turned on or off, there is a current in the secondary coil due to a flux change (see Chapter 25). The deflection of the ballistic galvanometer is directly proportional to the change in flux through the secondary coil. This change in flux is a direct measure of the flux density B.

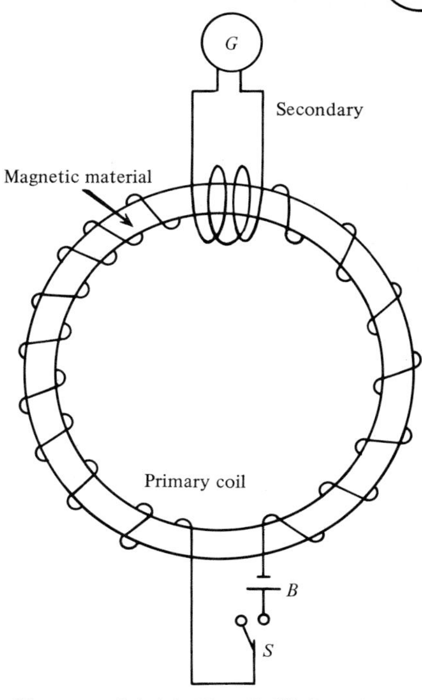

FIGURE 24.11 *Toroid filled with material under investigation for its magnetic properties.*

It is found that if the toroid is filled with air, oil, water, plastic, wood, or one of many other materials, the magnetic field inside the toroid does not change appreciably (no more than 1 part in 10^4). But there is a special class of material such as iron, cobalt, nickel, and their alloys with which if the toroid is filled, they produce a very large flux density. Let us denote B_0 as the flux density inside the toroid when it is empty (vacuum or air) and B as the flux density when it is filled with some other material even though the current in the toroid is the same. It is found that all material can be classified into the following three categories, according to their behavior in the magnetic field:

$$\text{diamagnetic material:} \quad \frac{B}{B_0} < 1$$

$$\text{paramagnetic material:} \quad \frac{B}{B_0} > 1$$

$$\text{ferromagnetic material:} \quad \frac{B}{B_0} >> 1$$

In other words, we may write

$$\boxed{B = k_m B_0} \tag{24.18}$$

where k_m, the *relative permeability* of the material, is defined as the ratio of B to B_0. In the case of diamagnetic and paramagnetic material, k_m is only slightly different from 1; that is, for

$$\text{diamagnetic material:} \quad k_m \simeq 1.00000 - 0.00001$$

$$\text{paramagnetic material:} \quad k_m \simeq 1.00000 + 0.00001$$

$$\text{ferromagnetic material:} \quad k_m \gtrsim 100$$

The values of k_m for different materials are listed in Table 24.1. For diamagnetic and paramagnetic materials the change from B_0 to B is of the order of only 1 part in 10^5, while for ferromagnetic material B may be anywhere between $100B_0$ and $1000B_0$ or more.

Since for nonferromagnetic materials, the value of k_m differs very slightly from unity, it is more convenient to introduce a new constant called *magnetic susceptibility*, χ_m, defined as

$$\boxed{\chi_m = k_m - 1 = \frac{B}{B_0} - 1} \tag{24.19}$$

TABLE 24.1
Values of Permeabilities and Susceptibilities of Various Materials

Material	k_m	χ_m
Vacuum	1.00000	0
Air	1.0000004	0.00000004
Aluminium	1.000021	0.000021
Oxygen (gas)	1.0000018	0.0000018
Bismuth	0.99983	−0.00017
Copper	0.9999904	−0.0000096
Lead	0.999984	−0.000016
Mercury	0.999972	−0.000028
Cobalt	250	249
Iron	5000	4999
Nickel	600	599
Steel	2000	1999

k_m and χ_m are both defined in terms of B/B_0, hence are dimensionless. From Equations 24.18 and 24.19 and the values of χ_m given in Table 24.1, it is quite clear that for diamagnetic materials χ_m is negative, for paramagnetic materials χ_m is positive, and for ferromagnetic materials χ_m is a very large positive number. Also, a negative χ_m means that the presence of a diamagnetic material decreases the value of B slightly, a paramagnetic material increases the value of B slightly, and a ferromagnetic material increases the value of B by a very large factor.

It may also be pointed out that the relative permeability k_m of a magnetic material is related to the permeability μ of the material and the permeability μ_0 of free space by the relation

$$\boxed{k_m = \frac{\mu}{\mu_0}} \tag{24.20}$$

Thus, for vacuum $\mu = \mu_0$; hence, $k_m = 1$.

Another important quantity that we shall come across in our discussion is the *magnetic field strength, H*. It is related to magnetic field induction B by the relations

$$B = \mu H \qquad \text{and} \qquad B_0 = \mu_0 H \tag{24.21a}$$

that is,

$$\boxed{H = \frac{B_0}{\mu_0} = \frac{B}{\mu}} \tag{24.21b}$$

This relation is true only in simple situations.

As we shall see in Chapter 25, whenever any material is placed in an external field, there is always an induced magnetic field in the direction opposite that of the externally applied field. This induced magnetic field is always negligible compared to the magnetic field due to permanent magnetic dipoles, except in the case of diamagnetic materials. In diamagnetic materials the outer electronic shell is closed, pairing off all the electrons (e.g., in copper, lead, and the noble gases); hence, there is no net magnetic moment due to permanent atomic dipoles. In diamagnetic materials there will be a small induced magnetic field in a direction opposite the direction of the applied field.

Except for diamagnetic materials, magnetic properties of solids are due to the magnetic moments of permanent atomic dipoles in the solids as shown in Figure 24.12. In addition to paramagnetic and ferromagnetic materials, there are two more special materials—antiferromagnetic and ferrimagnetic. In paramagnetic materials all but a few electrons are completely paired off. The magnetic field is the result of the permanent magnetic dipole moments of such unpaired electrons.

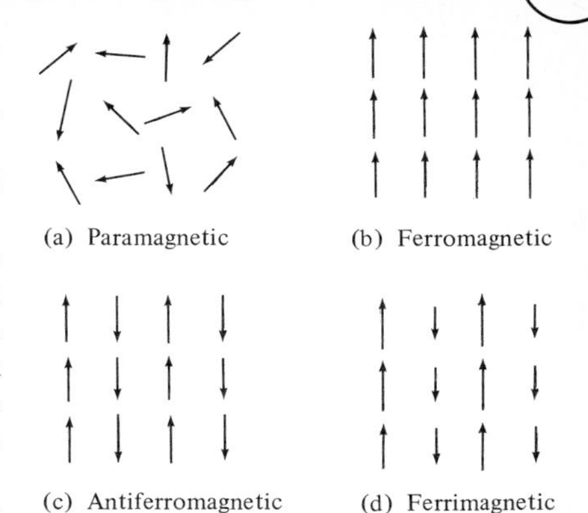

FIGURE 24.12 *Arrangement of atomic dipoles in various materials: (a) paramagnetic, (b) ferromagnetic, (c) antiferromagnetic, and (d) ferrimagnetic.*

Ferromagnetism is due to the interaction between and the alignment of the neighboring permanent atomic dipoles in atoms that result from unpaired electrons in the outer electronic shells. (In paramagnetic materials such interactions are negligible.) Ferromagnetic material such as iron, nickel, and cobalt, which are among the transition elements, have a peculiar electronic structure. That is, the outermost shell contains electrons even though the inner shell next to the outermost is still not filled. This results in an unusually large contribution of electron spin magnetic moments, resulting in large atomic magnetic dipole moments.

The interaction between different atoms in the neighborhood is very complex and is called the *exchange interaction*. The exchange interaction, or exchange coupling, according to modern physics is a quantum-mechanical phenomenon and will not be discussed here. But this interaction is strong enough to cause the neighboring atoms in small regions to align themselves with their magnetic moments in the same direction. These small regions in which all the atoms have their magnetic moments aligned are called *domains*. These domains for a bar magnet are shown schematically in Figure 24.13, where each arrow represents the total magnetic moment of the domain. As is seen from this figure, because of the random orientation, there may be no net overall magnetic field associated with the domains. These domains are usually very small but may be as large as a fraction of a millimeter in length.

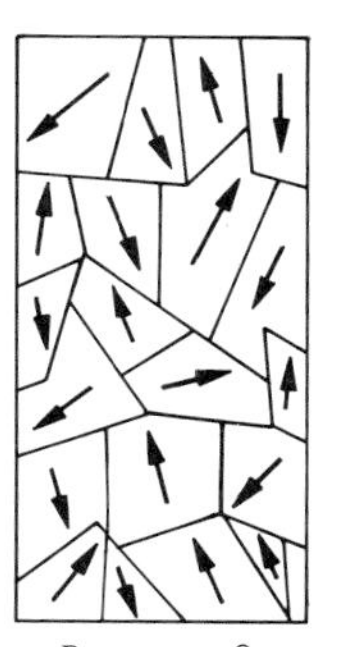

FIGURE 24.13 *The magnetic moments of domains (represented by arrows) in a bar magnet are randomly oriented when $B_{ext} = 0$.*

Let us now apply an external magnetic field on a ferromagnetic material, say iron. This results in a considerably large alignment of the atomic magnetic moments in the direction of the applied field. The size of the domains grow by the motion of the domain boundaries or walls. At low external fields the atoms are partially aligned, but as the field increases, so does the alignment and the size of the domains. As shown in Figure 24.14, at very high fields all the atoms are aligned and there is no further increase in the magnetization; hence, no more increase in the total magnetic field (internal + external). The examples of magnetic domains are shown in Figure 24.15.

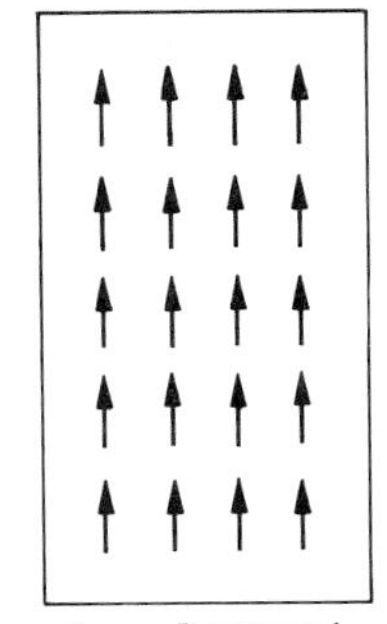

FIGURE 24.14 *Complete alignment of magnetic domains of a bar magnet under a large B_{ext}.*

As the temperature of the material is raised, it increases its thermal energy. At very high temperatures the thermal energy may be high enough to cause all the atoms to break loose from their neighbors, hence losing alignment. The temperature at which the atomic dipoles of the ferromagnetic material lose their alignment is called the *Curie temperature*.

In antiferromagnetic materials the domains of equal magnitudes are oppositely aligned, as shown in Figure 24.12(c), while in ferrimagnetic materials the

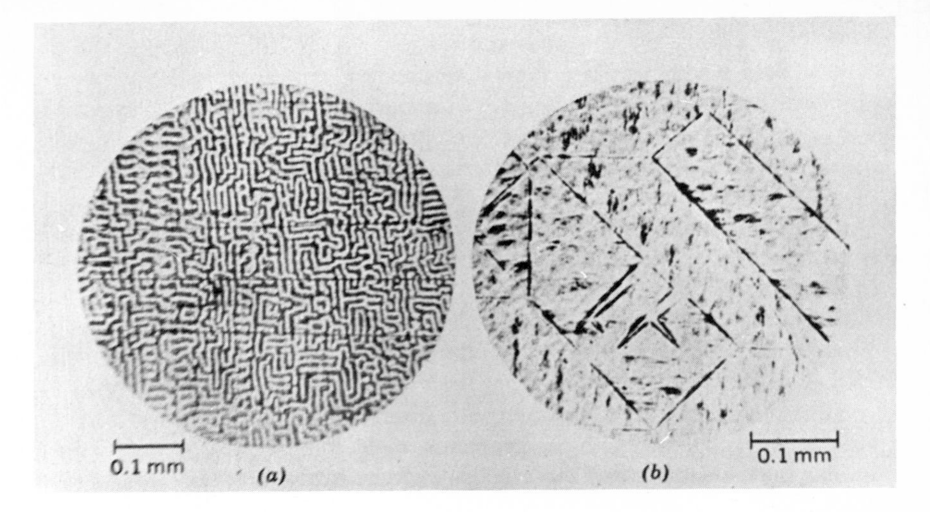

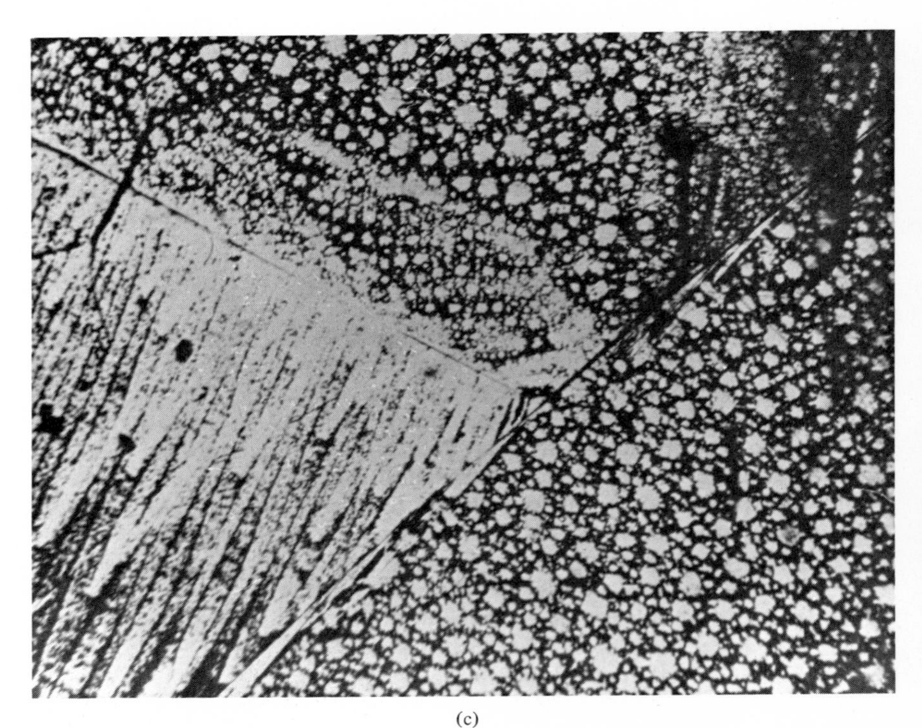

(c)

Figure 24.15 *Examples of ferromagnetic domains visible under the microscope. (a) Maze domain pattern, and (b) true domain pattern after removal of 28 μ of the surface layer. (c) Domain and domain walls in a ferromagnetic Si-Fe crystal.* (Parts (a) and (b) from S. Chikazumi and S. H. Charap, *Physics of Magnetism*, John Wiley & Sons, Inc., New York, 1964. Part (c) from Walter J. Moore, *Seven Solid States*, W. A. Benjamin, Inc., Menlo Park, California.)

domains of unequal magnitudes are oppositely aligned, as shown in Figure 24.12(d).

Let us consider once again a toroid filled with some ferromagnetic material as a core. To start with, the core is unmagnetized when the current in the toroid winding is zero. As the current I is increased, the fields B and H (or B_0) both increase, where $B = \mu H$ and $H = B_0/\mu_0$. The plot of the magnetic intensity B as a function of the magnetizing field H ($\propto I$) is called a *magnetizing curve*. Figure 24.16 is such a curve for a typical sample of iron. The maximum value reached is called the *saturation field*, and at this point B is $\sim 10^3$ as large as B_0.

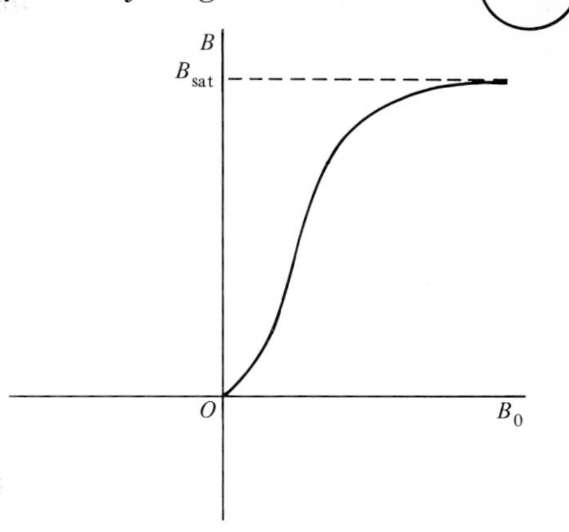

FIGURE 24.16 *Magnetization curve—the plot of B (the field in the presence of a ferromagnetic material) versus B_0 (no ferromagnetic material present).*

As shown in Figure 24.17, after reaching point A the magnetic core is saturated and B does not increase with H (or I). Suppose now that H is reduced to zero (by reducing I to zero). As is clear from this figure, B does not reduce to zero. This is because once the magnetic domains have been aligned by magnetization, they try to keep themselves aligned, even though the external field H is zero. This remaining field, $B = B_r$, corresponding to point R is called the *retentivity* of the material. To reduce B to zero an external magnetic field H must be applied in the opposite direction. This happens when $H = H_c$, corresponding to point C. The external field H_c for which B is zero is called the *coercive force*.

The core specimen can be magnetized to saturation in the opposite direction by making H more negative until it reaches point D. If the value of H is reversed (by reversing the current), then B versus H follows the path $DEFA$. The plot $OARCDEFA$ of B versus H in which the core is magnetized in one direction and then in the opposite direction is called a *hysteresis curve* of the core specimen. The area of hysteresis loop is a measure of energy lost to heat per unit volume of the core material in completing each cycle. Figure 24.18 shows two different hysteresis curves corresponding to different alloys. Curve (a) is much broader than the one in Figure 24.17 and has a much higher retention of magnetism after the external field is switched off. Materials with such characteristics are used for making permanent magnets. Of course, such materials are not suitable for use in transformers and electromagnets. Owing to a very large loop area, there will be too much energy loss in each cycle. On the other hand, materials with hysteresis loops such as the one shown in Figure 24.18(b) have very small loop areas and are quite suitable for use in

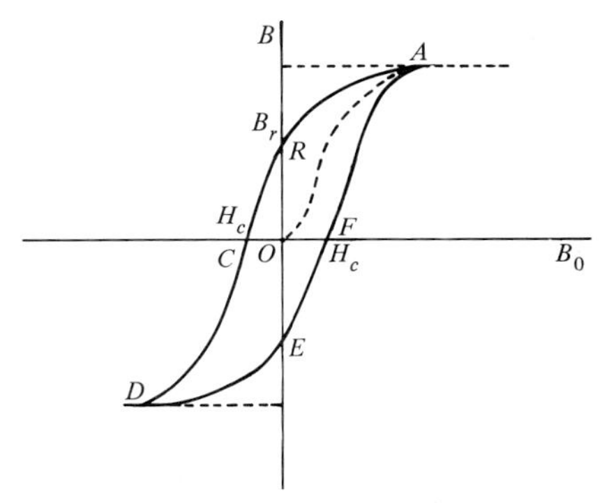

FIGURE 24.17 *Hysteresis loop for a typical iron sample.*

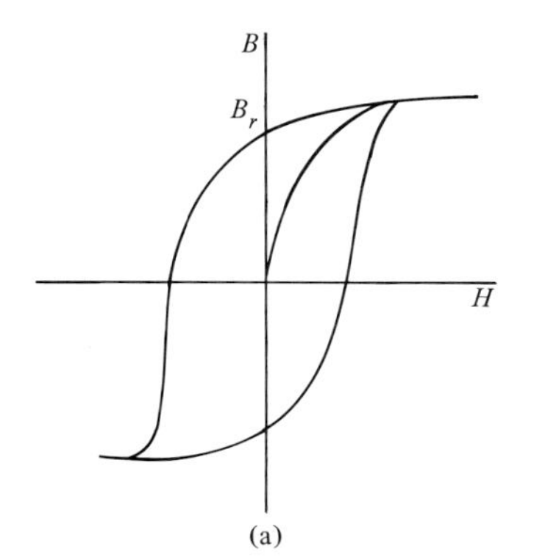

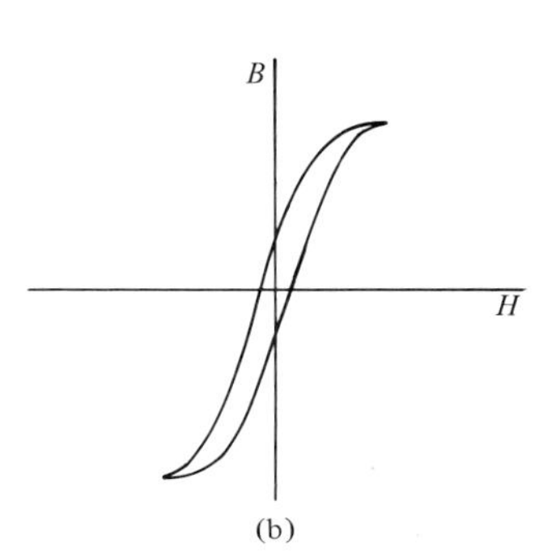

FIGURE 24.18 *Hysteresis loops for materials (a) used for permanent magnets and (b) used for transformers and electromagnetics.*

transformers and electromagnets. In addition to less energy loss, the retention of magnetization is very small. The hysteresis properties of materials have been utilized in magnetic core computer memories, magnetic tapes, and so on.

SUMMARY

According to *Ampère's law*, the sum of the product of the quantities $B_i \, \Delta l_i \cos\theta_i$ over any closed path is equal to the product of the constant μ_0 and the net current enclosed by the path. That is, $\Sigma \, B_i \, \Delta l_i \cos\theta_i = \mu_0 I$, where $\mu_0 = 4\pi \times 10^{-7}$ Wb/A-m. B at a distance from a straight conductor is $B = \mu_0 I/(2\pi r)$. Ampère's law is applicable only to symmetrical cases. In other situations we can make use of the *Biot–Savart law*, according to which $B = (\mu_0/4\pi)(I \, \Delta L \sin\theta/r^2)$. Making use of this, the magnetic field at the center of a circular coil is found to be $B = \mu_0 I/2R$.

If two parallel wires carrying currents I_1 and I_2 are separated by a distance d, the force per unit length on either wire is $\Delta F/\Delta L = F/L = \mu_0 I_1 I_2/(2\pi d)$. If $I_1 = I_2 = 1$ A and $d = 1$ m, then $F/L = 2 \times 10^{-7}$ N/m.

The field B inside a solenoid of n turns per unit length is $B = \mu_0 nI$. The B due to a toroid is $B = \mu_0 NI/2\pi r$.

The orbital motion of an electron in an atom results in a *magnetic dipole moment* given by $\mu_l = iA = (e/T)(\pi r^2) = evr/2 \simeq 9.1 \times 10^{-24}$ A-m^2. The magnetic moment due to the spinning motion $\mu_s \simeq \mu_l$.

The presence of material inside a solenoid or toroid changes the magnetic field from B_0 to B. *Relative permeability* $k_m = B/B_0$. For diamagnetic materials $k_m < 1$, for paramagnetic materials $k_m > 1$, and for ferromagnetic materials $k_m >> 1$. Also, $k_m = \mu/\mu_0$ and magnetic susceptibility $\chi_m = k_m - 1 = B/B_0 - 1$. The relative field strength H is defined as $H = B_0/\mu_0 = B/\mu$.

Magnetic properties of the materials can be explained in terms of permanent magnetic dipole moments. The small regions in which all the atoms have their magnetic moments aligned are called *domains*. The alignment of these domains in the direction of the applied field results in *ferromagnetism*. The temperature at which the atomic dipoles of the material lose their alignment is called the *Curie temperature*. The plot of B versus H forming a closed loop is called a *hysteresis loop*.

QUESTIONS

1. Give two examples of situations where Ampère's law can be applied to calculate B.
2. In your opinion which is more fundamental—Ampère's law or the Biot–Savart law, and why?
3. Which is a stronger magnetic field: the center of a coil of radius R carrying current I, or a point at a distance R from an infinitely long wire carrying current I?
4. Will the force per unit length between the two wires shown in Figure 24.6(a) still be equal and opposite if (a) the two wires are of unequal length; and (b) the two wires are not parallel?
5. Draw the magnetic forces that act on the two conductors shown in Figure 24.6(a) that are (a) parallel but carry unequal current; and (b) not parallel but carry equal current.

6. Devise a simple (although not very precise) laboratory experiment to measure forces of attraction and repulsion between two straight current-carrying conductors.
7. Can you use a current-carrying coil as a compass? Explain how.
8. Arrange two current-carrying coils so that the resulting magnetic field will be similar to that of the permanent bar magnet shown in Figures 23.3(a) and (b). Draw the magnetic field lines in each case.
9. What should be the relative orientation of a current-carrying coil in the field of a solenoid so that there will be (a) a force but no torque; and (b) a torque but no net force?
10. Show that at large distances the magnetic field lines due to the orbital motion of an electron in an atom are identical to a permanent magnet.
11. Describe the behavior of a diamagnetic needle, paramagnetic needle, and ferromagnetic needle in (a) a uniform magnetic field; and (b) a nonuniform magnetic field.

PROBLEMS

1. A long straight wire is carrying a current of 2 A. Calculate the magnetic field intensity B at a distance of 0.1 m from it lying on a line at right angles to the wire.
2. A wire along the floor carries a current of 10 A. What is the magnitude of B at a distance of 2 m from it?
3. A long straight wire carries a current of 20 A. At what distance from the wire will the magnetic field be (a) 0.1 T; (b) 0.01 T; and (c) 0.001 T?
4. Calculate the magnetic field B at points P and Q shown. The current I_1 in A is 20 A and I_2 in B is 10 A, and the currents are parallel to each other.

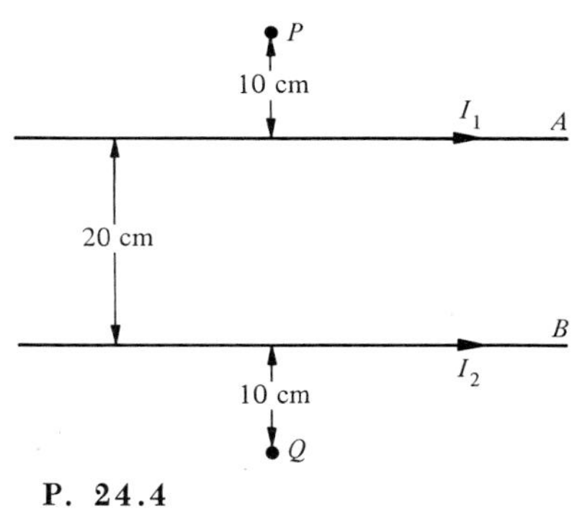

P. 24.4

5. Repeat Problem 4 if the currents in A and B are in opposite directions.
6. Consider two parallel wires A and B. Calculate the magnetic field at point P (at mid point) if the current in A is I_1 and in B is I_2; both I_1 and I_2 are parallel to each other and distance d apart. What is the value of B if I_1 and I_2 are in opposite directions?
7. Find the magnetic field at point P if $a = 10$ cm and $b = 20$ cm as shown, where I_1 is 10 A and I_2 is 5 A for (a) I_1 parallel to I_2; (b) I_1 antiparallel to I_2.

P. 24.7

8. An electron in an atom revolves in a circle of radius 0.5×10^{-10} m with a velocity of 2.2×10^6 m/s. Calculate the current due to this motion and then calculate the magnetic field at the center, that is, at the site of the nucleus.
9. Consider a segment of a wire 1 cm in length carrying a current of 10 A. Calculate the magnetic field intensity at a point 50 cm from it, making an angle of 30° with the wire.
10. Consider a segment of a wire 2 cm long carrying a current of 20 A. Calculate the magnetic field at a point P a distance of 1 m at an angle of (a) 0°; (b) 45°; (c) 90°; (d) 135°; and (e) 180° from the wire.
11. Consider a coil of 50 turns and radius 5 cm which carries a current of 10 A. Calculate the magnetic field intensity at the center of this coil.
12. What should be the current in a circular loop having a radius of 25 cm so that it will produce a magnetic field of 0.001 Wb/m^2 at the center?

13. What should be the current in a loop of radius 10 cm so that the magnetic field produced at the center will be equal to the earth's magnetic field ($= 5 \times 10^{-5}$ T)?
14. If the Biot–Savart law is written in the form $B = K'I/r^2$, what are the dimensions of K′?
15. Calculate the force per meter length between two wires separated by 0.2 m and carrying a current of 5 A and 10 A in (a) the same direction; and (b) in opposite directions.
16. Two long, straight, parallel wires are separated by a distance of 20 cm and carry equal currents. If the force of attraction between them is 0.2 N/m, calculate the currents.
17. Consider two parallel, straight, long wires P and Q 10 cm apart carrying currents of 10 A and 20 A respectively, in the same direction. (a) Calculate B and the force per unit length at wire Q due to wire P. (b) Calculate B and the force per unit length at wire P due to wire Q. Are the forces attractive or repulsive?
18. Repeat Problem 17 if the currents in the two wires are in opposite directions.
19. Calculate the force per unit length on a long wire carrying a current of 5 A and placed at the mid-point and parallel to two parallel wires separated by 0.40 m and carrying currents of 10 A and 20 A, respectively, in the same direction.
20. A 40-cm-long solenoid has 8000 turns of copper wire. If the current in the windings of the solenoid is 20 A, what is the magnitude of the magnetic field produced at the center?
21. What current should pass through a solenoid that is 0.5 m long with 10,000 turns of copper wire so that it will have a magnetic field of 0.4 Wb/m^2?
22. Calculate the number of turns per unit length in a solenoid so that a current of 10 A will produce a magnetic field of 0.05 Wb/m^2 at the center. If the solenoid is 0.6 m long, what is the total number of turns in the solenoid?
23. Consider two concentric solenoids, the inner one having 10 turns per centimeter while the outer one has 20 turns. What will be the field at the center of these two solenoids if a current of 2 A in the inner one and 4 A in the outer one are (a) in the same direction; and (b) in opposite directions?
24. Two concentric solenoids have a length of 50 cm, with the inner one having 500 turns and the outer one having 1500. If a current of 2 A is in the inner solenoid, what should be the current in the outer solenoid so that the magnetic field at the center will be zero?
25. To cancel the effect of the earth's magnetic field (tangential component), a solenoid with a large diameter should produce a field equal and opposite to that of the earth's. If the tangential component of the earth's magnetic field is 2×10^{-5} Wb/m^2, what should be the current in a coil of 20 turns per centimeter? How should this coil be located with respect to the earth's magnetic field?
26. Consider a toroid with 500 turns and a radius of 20 cm. If there is a current of 5 A, what will be the magnetic field inside the toroid?
27. Consider a toroid with 1000 turns and a radius of 25 cm. What should be the current so that the field B in the toroid is 1.0 Wb/m^2?

28. Using the value of $i = ev/2\pi r$ as given by Equation 24.14, where $r = 0.53 \times 10^{-10}$ m and $v = 2.2 \times 10^6$ m/s, calculate the value of the magnetic field produced at the center of the electronic orbit in the hydrogen atom. (Use $B = \mu_0 i/2r$.)
29. Consider a solenoid of length 1.0 m having 8000 turns. If there is a current of 5 A in the solenoid, what will be the field at the center if the solenoid is placed in (a) air; and (b) filled with iron, assuming that $\mu = 80\ \mu_0$? If the radius of the solenoid is 4 cm, what is the flux through the solenoid in each case?
30. Consider a toroid of 1000 turns and a radius of 25 cm. What is the field B in the toroid if there is a current of 2 A? What will be the field when the toroid is filled with iron for which $\mu = 100\ \mu_0$?
31. In Problem 30, if the toroid is filled with iron, what should be the current if the magnetic field inside the toroid is (a) 0.01 T; and (b) 1 T?
32. Calculate the magnetic moment due to an electron moving in an orbit of radius 0.53×10^{-10} m having a velocity of 2.2×10^6 m/s.
33. From the data of Problem 29, calculate k_m and χ_m.
34. From the data of Problem 30, calculate k_m and χ_m.

25
Time-Dependent Magnetic Fields

We have seen how charges at rest produce static electric fields and moving charges produce static magnetic fields. This suggests the possibility that magnetic effects might produce electric effects. Now we investigate the laws of electromagnetic induction discovered by Faraday and others. According to Faraday, the time-varying magnetic field produces an electric field, hence currents. We introduce the concept of induction and apply the laws of electromagnetic induction to such situations as motors and generators. We shall calculate the energy stored in a magnetic field and discuss LR *and* LC *circuits.*

25.1 Laws of Electromagnetic Induction—Faraday's Law and Lenz's Law

HANS Christian Oersted (1777–1851) showed that an electric current gives rise to magnetic effects. Michael Faraday (1791–1867) suspected an inverse effect; that is, perhaps a magnet could be made to produce electric current. His continued efforts were finally successful in 1831, when he discovered that if a magnet was moved in the vicinity of a coil as shown in Figure 25.1(a), a current was induced in the coil as indicated by the meter. Production of an electric current in the coil implies the presence of an emf. Thus, we may say that a changing magnetic field produces an induced emf. Using the concept of magnetic field lines and magnetic flux, Faraday arrived at a relation between the changing flux and the induced emf. This relation, known as Faraday's law of electromagnetic induction, will be stated shortly. Almost simultaneously and independently, Joseph Henry (1797–1878) working at the Albany Academy in the United States, made the same discoveries about the laws of electromagnetic induction, but his results were reported a few months later. Hence, Faraday's law is also called the Faraday–Henry law, and may be stated as follows:

> **Faraday's law:** ***An induced emf is produced in any closed circuit if there is a varying magnetic flux. The magnitude of this induced emf is equal to the negative of the time rate of change of the magnetic flux through the circuit.***

Thus, if there are N turns in a coil, Faraday's law may be written, mathematically, as

$$\boxed{\mathcal{E} = -N\frac{\Delta\Phi}{\Delta t} = -\frac{\Delta(N\Phi)}{\Delta t}} \tag{25.1}$$

where the induced emf $\mathcal{E}$ is in volts if the rate of change of flux $\Delta\Phi/\Delta t$ is in Wb/s. The quantity $N\Phi$ in Equation 25.1 is a measure of the *flux linkage* in a particular device. The meaning of the negative sign will become clear later

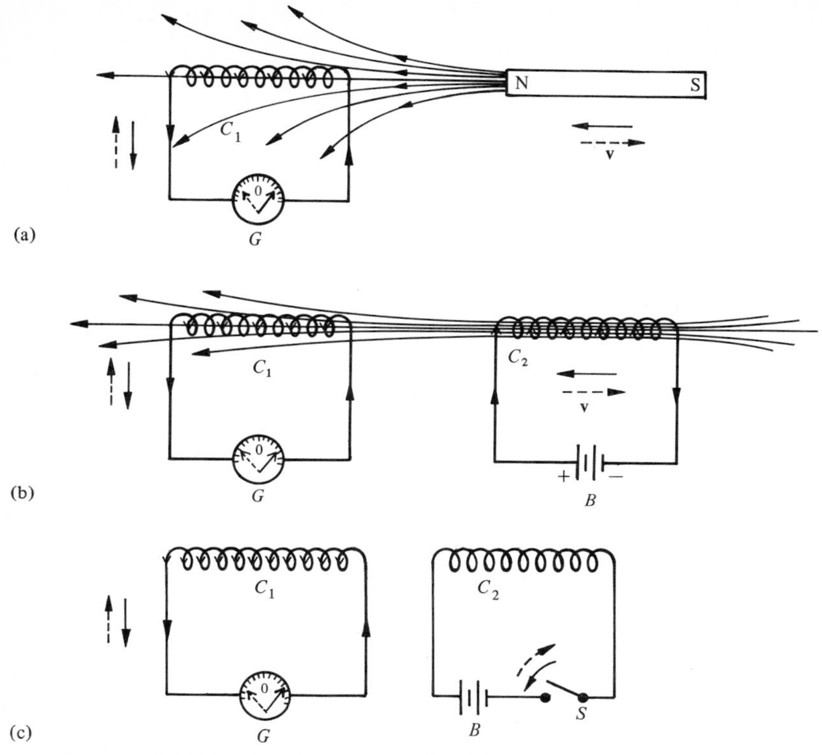

FIGURE 25.1 *Production of induced emf due to changing magnetic fields by (a) moving a magnet, (b) moving a current-carrying coil, and (c) starting and stopping a current in a coil.*

when we discuss Lenz's law, which gives the direction of the induced emf.

As explained in Chapter 23, according to Equation 23.2, the number of magnetic field lines or magnetic flux $\Delta\Phi$ through an area ΔA is given by

$$\Delta\Phi = \mathbf{B} \cdot \Delta\mathbf{A} = B\,\Delta A \cos\theta \tag{25.2}$$

Thus, the change in flux may be achieved by changing (1) magnetic field B; (2) the area, ΔA; or (3) θ, the relative orientation of B and ΔA. Figure 25.1 shows three different configurations that may be used to produce a changing magnetic field B, hence to produce an induced emf in the circuit. (We shall encounter other methods of changing flux in Section 25.3.)

In Figure 25.1(a) the permanent magnet is moved toward a coil C_1. The changing magnetic flux through the coil induces emf in the coil as indicated by the measuring device G. On the other hand, if the magnet is moved in the opposite direction, the current is again observed but in the opposite direction.

In Figure 25.1(b) the permanent magnet is replaced by a second coil C_2, connected to a source of emf. The current in the coil C_2 produces a magnetic field as shown. If coil C_2 is moved toward or away from coil C_1, it results in a changing magnetic field across coil C_1. This results in an induced emf, hence current in coil C_1. The direction of the current is opposite in the two cases.

In Figure 25.1(c) a switch S is added in the circuit containing coil C_2. A current is observed in coil C_1 when the switch S is closed or opened. This is because the changing current in coil C_2 results in a changing magnetic field. Once again the direction of current is opposite in the two cases.

The direction of the induced current in the conductor is given by Lenz's law. A German scientist, H. F. E. Lenz (1804–1864), without knowledge of the work of Faraday and Henry, duplicated many of their experiments. It is convenient to find the direction of the induced current by the following law:

> **Lenz's law:** ***The direction of the induced current is always such as to oppose the cause that produces it.***

Lenz's law can be arrived at by the application of the conservation-of-energy principle. The same result can also be arrived at by a detailed application of Faraday's law.

Let us apply Lenz's law to the case shown in Figure 25.2. In Figure 25.2(a) we show a single loop. As the permanent magnet moves toward the loop, the

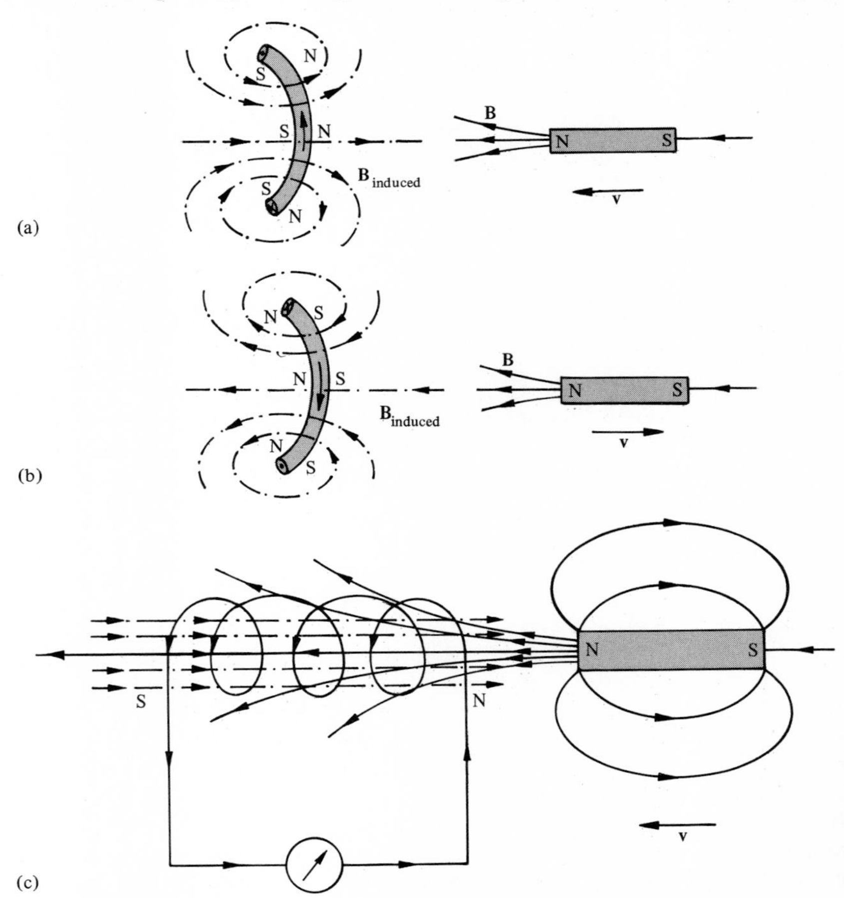

FIGURE 25.2 *Direction of the induced current and* B_{induced}, *illustrating Lenz's law.*

lines of induction through the loop will increase; hence, **B** will increase, pointing to the left. To reduce this effect, the current induced in the loop should be in such a direction so that $\mathbf{B}_{\text{induced}}$ will point to the right as shown. From the right-hand rule this is possible only if the current in the loop is in a counterclockwise direction, as shown.

Similarly, if the magnet is being withdrawn from the loop, as shown in Figure 25.2(b), the field **B** across the loop will decrease. To reduce this decrease in **B**, $\mathbf{B}_{\text{induced}}$ should point to the left. This is possible only if the induced current is in a clockwise direction, as shown in Figure 25.2(b).

Note that it is the *rate of increase or decrease* of flux and not the absolute value of the flux that determines the *magnitude of the induced* emf. It is whether the flux change is an *increase or decrease* (and not the rate) which determines *the direction of the induced current*. If **B** across a loop is increasing, $\mathbf{B}_{\text{induced}}$ will oppose **B**; if **B** is decreasing, $\mathbf{B}_{\text{induced}}$ will help **B**, as illustrated in Figures 25.2(a) and (b). Figure 25.2(c) shows the case in which a permanent magnet is approaching a coil of many turns. Using Lenz's law, we should be able to predict the direction of the induced current in all three cases illustrated in Figure 25.1. In Figures 25.2(a) and (b), if the south pole of the magnet is approaching or receding from the loop, the direction of the induced current will be opposite. The negative sign in Faraday's law is the implication of Lenz's law.

EXAMPLE 25.1 A circular loop of radius 5 cm is placed with its plane perpendicular to **B**. If the magnetic induction **B** changes from 0.1 Wb/m² to 0.5 Wb/m² in 0.025 s, calculate the induced emf.

From Equation 25.1, $\mathcal{E} = \Delta\Phi/\Delta t$, and from Equation 25.2, $\Phi = BA\cos\theta$, where θ, the angle between the normal to the plane of the coil and **B**, is 0°; that is, $\cos\theta = \cos 0° = 1$. The value of A remains constant equal to

$$A = \pi r^2 = \pi(0.05\text{ m})^2 = 0.008\text{ m}^2$$

Thus,

$$\mathcal{E} = \frac{\Delta\Phi}{\Delta t} = \frac{\Delta(BA)}{\Delta t} = A\frac{\Delta B}{\Delta t} = 0.008\text{ m}^2\frac{\Delta B}{\Delta t}$$

But

$$\frac{\Delta B}{\Delta t} = \frac{0.5\text{ Wb/m}^2 - 0.1\text{ Wb/m}^2}{0.025\text{ s}} = 16\text{ V/m}^2$$

hence,

$$\mathcal{E} = (0.008\text{ m}^2)(16\text{ V/m}^2) = 0.128\text{ V}$$

EXERCISE 25.1 What will be the induced emf in Example 25.1 if the normal to the plane makes an angle θ with **B** such that (a) $\theta = 30°$; (b) $\theta = 60°$; and (c) $\theta = 90°$? [*Ans.:* (a) 0.111 V; (b) 0.064 V; (c) 0.]

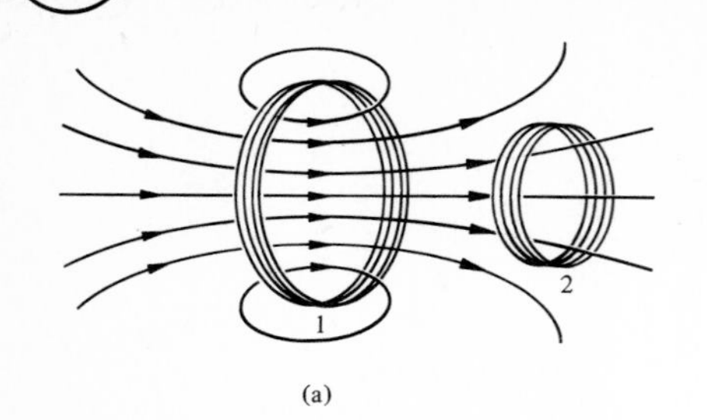

(a)

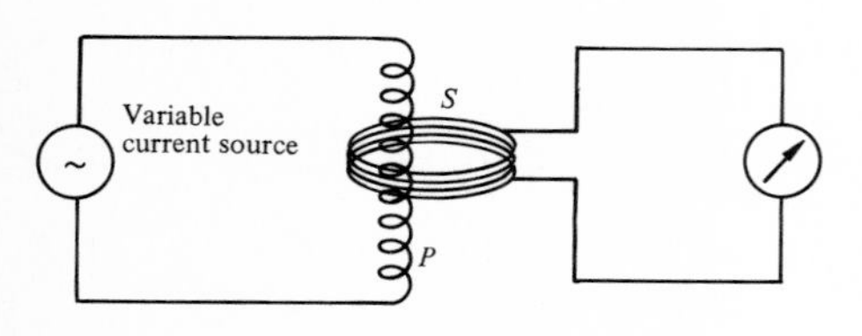

(b)

FIGURE 25.3 *Two different arrangements having mutual inductance due to flux linkages.*

25.2. Mutual Inductance and Self-Inductance

Mutual Inductance

Let us consider two coils, 1 and 2, as shown in Figure 25.3(a). These coils have tightly packed N_1 and N_2 turns, respectively. Suppose that the current i_1 in coil 1 produces a magnetic flux Φ_{21} at the sight of coil 2; the flux linkage with coil 2 will be $N_2\Phi_{21}$. This flux linkage is directly proportional to the current i_1 in coil 1 (provided no magnetic materials are present). That is,

$$N_2\Phi_{21} \propto i_1$$

or

$$N_2\Phi_{21} = M_{21}i_1 \tag{25.3}$$

The quantity M_{21} is called the *coefficient of mutual inductance*. The mutual inductance is a purely geometrical factor, and depends on the number of turns in each coil, their shapes, and their arrangement.

If the current i_1 changes with time, it will produce a varying magnetic field, hence a varying flux linkage. According to Faraday's law, there will be an induced emf $\mathcal{E}_2$ in coil 2 given by

$$\mathcal{E}_2 = -\frac{\Delta(N_2\Phi_{21})}{\Delta t} \tag{25.4a}$$

substituting for $N_2\Phi_{21}$ from Equation 25.3, we get

$$\mathcal{E}_2 = -\frac{\Delta(M_{21}i_1)}{\Delta t} = -M_{21}\frac{\Delta i_1}{\Delta t} \tag{25.4b}$$

or

$$M_{21} = -\frac{\mathcal{E}_2}{\Delta i_1/\Delta t} \tag{25.5}$$

From this relation we may define *mutual inductance* as the induced emf in coil 2 per unit rate of change of current in coil 1.

If the roles of coil 1 and coil 2 are interchanged, the induced emf $\mathcal{E}_1$ in coil 1 due to a changing current $\Delta i_2/\Delta t$ in coil 2 is given by (in analogy with Equation 25.4b)

$$\mathcal{E}_1 = -M_{12}\frac{\Delta i_2}{\Delta t} \tag{25.6}$$

If the rates of change of currents $\Delta i_1/\Delta t$ and $\Delta i_2/\Delta t$ are the same, it is found that induced emf's $\mathcal{E}_1$ and $\mathcal{E}_2$ are equal. This implies that $M_{12} = M_{21} = M$, as it should be, because inductance is a purely geometrical factor. Thus, from Equation 25.3,

$$\boxed{M = \frac{N_1\Phi_{12}}{i_2} = \frac{N_2\Phi_{21}}{i_1}} \tag{25.7}$$

and

$$\boxed{\mathcal{E}_1 = -M\frac{\Delta i_2}{\Delta t}, \qquad \mathcal{E}_2 = -M\frac{\Delta i_1}{\Delta t}} \tag{25.8}$$

The SI unit of mutual inductance M from Equation 25.8 is the *volt per ampere per second,* and this is given the name *henry* (H), in honor of Joseph Henry.

$$1 \text{ henry} = \frac{1 \text{ volt}}{1 \text{ ampere/second}} \tag{25.9}$$

Figure 25.3(b) is another arrangement in which there is a flux linkage. The coil in which there is a varying current is called the *primary coil P*; the coil that has an induced emf is called the *secondary coil* S. The designation of the two coils is interchangeable.

Self-Inductance

It is not necessary to have two coils to show inductive effects. An induced emf appears in a single coil if the current in this coil is changed. This inductive effect is called *self-induction,* and the induced emf is called a *self-induced emf.* For example, in Figure 25.4 the current i in the coil sets up a magnetic field. If the current is changed by moving the sliding contact S, it will result in a

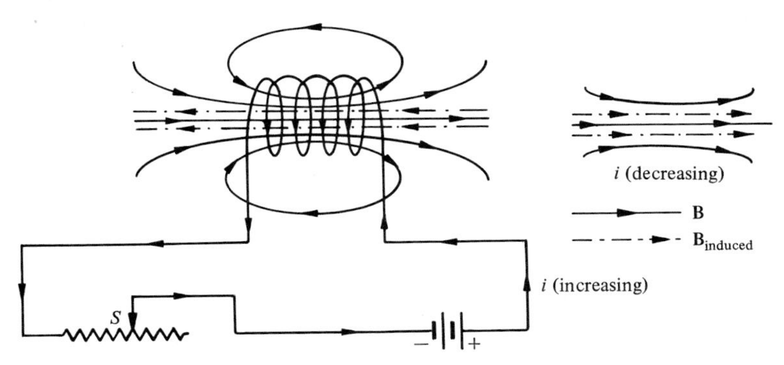

FIGURE 25.4 *Arrangement resulting in self-inductance due to flux linkage.*

changing magnetic field, therefore a changing magnetic flux. This, according to Faraday's law, will result in a self-induced emf. Thus, if there are N turns in the coil, and current i produces magnetic flux Φ, the flux linkage $N\Phi$, in analogy with Equation 25.3, will be

$$N\Phi = Li \tag{25.10}$$

or

$$\boxed{L = \frac{N\Phi}{i}} \tag{25.11}$$

where L is called the *coefficient of self-induction.* Thus, from Faraday's law, the self-induced emf $\mathcal{E}$ may be written

$$\mathcal{E} = -\frac{\Delta(N\phi)}{\Delta t} = -\frac{\Delta(\text{Li})}{\Delta t} = -L\frac{\Delta i}{\Delta t}$$

or

$$L = -\frac{\mathcal{E}}{\Delta i/\Delta t} \tag{25.12}$$

In analogy with definition for mutual inductance, we may define *self-inductance* of a coil in a circuit to be the self-induced emf per unit rate of change of current. As with mutual inductance, the SI unit of self-induction is the *henry*.

Any circuit containing a coil will have self-inductance, so the coil is called an *inductor* and is represented in the circuit by a coil-like symbol. Once again, by using Lenz's law we can find the direction of the induced field or current.

EXAMPLE 25.2 Calculate the self-inductance of a solenoid of radius 3 cm, length 30 cm, and 3000 total number of turns.

Let l and A be the length and cross-sectional area of the solenoid and n be the number of turns per unit length so that the total number of turns $N = nl$. If i is the current in the solenoid, $B = \mu_0 ni$, and the flux $\Phi = BA$. Substituting these in Equation 25.11,

$$L = \frac{N\Phi}{i} = \frac{(nl)(\mu_0 niA)}{i} = \mu_0 n^2 lA$$

In this case $n = 3000/0.30 \text{ m} = 10{,}000 \text{ turns/m}$, $A = \pi r^2 = \pi(0.03 \text{ m})^2$, $l = 0.3 \text{ m}$, while $\mu_0 = 4\pi \times 10^{-7} \text{ Wb/A-m}$. Thus,

$$L = (4\pi \times 10^{-7} \text{ Wb/A-m})(10{,}000/\text{m})^2(0.3 \text{ m})(\pi)(0.03 \text{ m})^2 = 0.11 \text{ H}$$

The induced emf $\mathcal{E}$ in the solenoid can be calculated from $\mathcal{E} = L(\Delta i/\Delta t)$. Suppose that i changes from 0 to 5 A in 0.2 s:

$$\mathcal{E} = (0.11 \text{ H})\left(\frac{5 \text{ A}}{0.2 \text{ s}}\right) = 2.75 \text{ V}$$

EXERCISE 25.2 Consider a solenoid of length 50 cm and radius 4 cm. What should be the number of turns per unit length so that the self-inductance will be 0.5 H? What will be the induced emf if i changes from 0 to 5 A in 0.2 s? [*Ans.*: $n = 12{,}600$ turns/m, $\mathcal{E} = 12.5$ V.]

25.3. Motional EMF—Generators and Motors

According to Faraday's law given by Equation 25.1, changing magnetic flux produces induced emf. This induced emf implies the existence of a local electric field. The electromagnetic induction leading to an induced emf can be achieved without necessarily changing **B** with time. This can be done by having a relative motion between the field **B** and the conductor in which the emf is induced. This is called the *motional emf.* Let us consider two simple cases in which the magnetic field is steady (time-independent) while the conductor is in motion.

Moving Conductor

Let us consider a U-shaped conductor $NRSM$ in the XY-plane, as shown in Figure 25.5. The conductor PQ moves with velocity **v** parallel to the Y-axis, as shown. Thus, $PQRS$ forms a closed circuit enclosing an area that changes as PQ moves. A uniform magnetic field **B** is applied perpendicular to the plane of this system; that is, **B** is parallel to the Z-axis. Thus, any charge q in conductor PQ

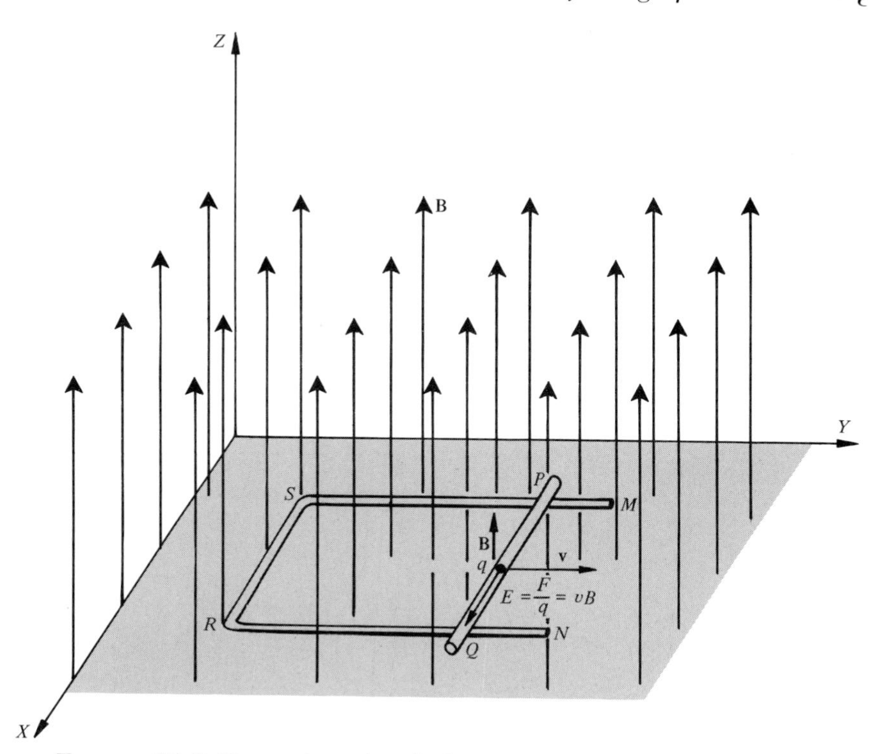

Figure 25.5 *Current is produced when a conductor moves in a magnetic field.*

moving with velocity **v** in a magnetic field **B** will experience a force **F** according to Equation 23.5. Since **v** and **B** are perpendicular, sin 90° = 1, and we get

$$F = qvB \tag{25.13}$$

According to the right-hand rule, this force is acting along PQ, that is, parallel to the X-axis. Dividing both sides by q, $F/q = vB$. But by definition F/q is equal to the electric field. Thus, we introduce an electric field strength **E** pointing along PQ (as shown in Figure 25.5) and given by

$$E = \frac{F}{q} = vB \tag{25.14}$$

It is important to note that F and E are the induced electromagnetic force and field, respectively. The existence of an electric field E between P and Q means that we can calculate the potential difference V between these two points. Knowing that $PQ = l$, we may write

$$V = V_P - V_Q = El = vBl \tag{25.15}$$

On the other hand, the other sections QR, RS, and SP of the conductor which are at rest, have no net forces exerted on them. This means that the emf $\mathcal{E}$ along the closed circuit $PQRS$ is equal to V. That is,

$$\mathcal{E} = vBl \tag{25.16}$$

This emf $\mathcal{E}$ exists as long as the conductor is in motion. If the conductor stops, $v = 0$; hence, $\mathcal{E} = 0$.

The same result can be arrived at directly by relating the emf to the change in the magnetic flux enclosed by the loop $PQRS$. If the length $RQ = x$ and $RS = l$, the magnetic flux Φ enclosed by the loop $PQRS$ will be

$$\Phi = Blx \tag{25.17}$$

Since x is changing with time, the rate of change of flux is given by

$$\frac{\Delta\Phi}{\Delta t} = \frac{\Delta(Blx)}{\Delta t} = Bl\frac{\Delta x}{\Delta t} = Blv \tag{25.18}$$

where we have used $v = \Delta x/\Delta t$, which is the velocity of the conductor PQ. Combining Equations 25.16 and 25.18,

$$\boxed{\mathcal{E} = \frac{\Delta\Phi}{\Delta t} = vBl} \tag{25.19}$$

Except for a minus sign, Equation 25.19 is a statement of Faraday's law. Thus, we have been able to produce induced emf by moving a conductor instead of varying the magnetic field, that is, by changing the magnetic flux enclosed by the conducting loop.

If the resistance of the circuit is R (which changes as the position of the conductor PQ changes), the induced current i in the circuit will be

$$\boxed{i = \frac{\mathcal{E}}{R} = \frac{vBl}{R}} \tag{25.20}$$

The direction of this induced current is clockwise.

Rotating Coils

This is the case in which the orientation of **B** and **A** changes. When a coil rotates in a magnetic field, the magnetic flux passing through the coil varies with time, hence leads to an induced emf in the coil. Let us consider the situation shown in Figure 25.6. At the particular instant shown, the magnetic field **B** makes an angle θ with the normal to the loop. Thus, the flux through the loop $PQRS$ will be

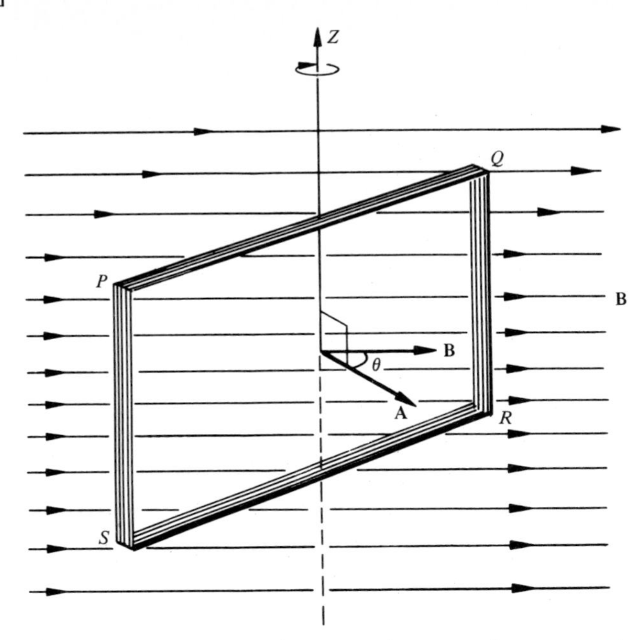

FIGURE 25.6 *Rotating conducting coil in a uniform magnetic field.*

$$\Phi = BA \cos \theta \tag{25.21}$$

where A is the area $PQRS$ of the loop. If the coil is rotating with angular frequency ω, $\theta = \omega t$, and we may write Equation 25.21 as

$$\Phi = BA \cos \omega t \tag{25.22}$$

In analogy with the motion of the X-component of a particle in circular motion as in Chapter 10, where $x = A_0 \cos \omega t$ and $v_x = -\omega A_0 \sin \omega t$, we may write

$$\frac{\Delta\Phi}{\Delta t} = -\omega BA \sin \omega t$$

If there are N turns in the coil, the induced emf may be written as

$$\mathcal{E} = -N\frac{\Delta\Phi}{\Delta t} = \omega ABN \sin \omega t \tag{25.23}$$

Since ω, A, B and N are all constant, let $\mathcal{E}_0 = \omega ABN$ = maximum induced emf and

$$\boxed{\mathcal{E} = \mathcal{E}_0 \sin \omega t = \mathcal{E}_0 \sin 2\pi f t} \tag{25.24}$$

where $\omega = 2\pi f$ and f is the frequency. Thus, the induced emf $\mathcal{E}$ is sinusoidal in time and takes the form shown in Figure 25.7. Also $\mathcal{E} = \mathcal{E}_0$ = maximum when $\theta = 90°$ and the flux through the loop is zero. On the other hand, $\mathcal{E} = 0$ when $\theta = 0°$ while the flux through the loop is maximum. Once more, this empha-

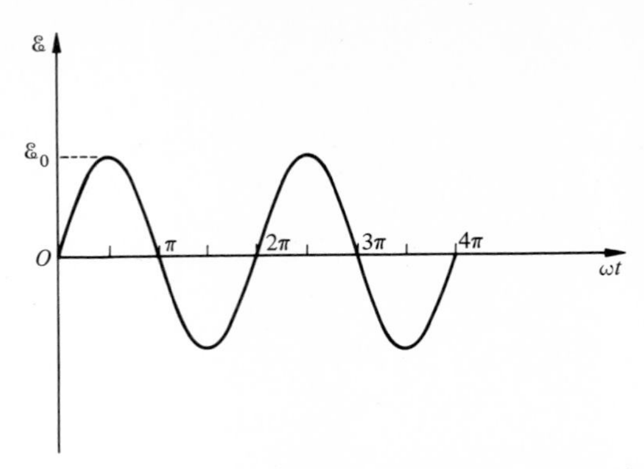

FIGURE 25.7 *$\mathcal{E}$ versus t in a conducting coil rotating in a uniform magnetic field.*

sizes that the induced emf depends not on the flux through the loop, but on the rate of change of flux through the loop. This is a typical type of alternating potential difference and we shall discuss it in detail in Chapter 26.

Generator

The principle of a rotating coil in a magnetic field, discussed above, is made use of in electric generators and electric motors. Figure 25.8 illustrates the working of a prototype ac generator. The mechanical energy is supplied to

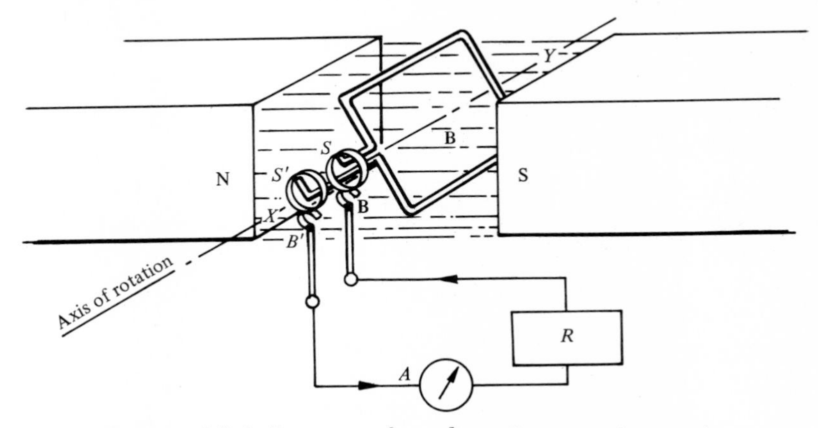

FIGURE 25.8 *Prototype of an alternating-current generator.*

keep the coil rotating at constant speed in the magnetic field. The generator converts this mechanical energy into electrical potential energy. As usual the power converted is given by:

$$\text{power supplied} = \text{power converted} = (\text{generator emf})(\text{current})$$

or

$$\boxed{P = \mathcal{E}i} \tag{25.25}$$

The rotating coil is connected to the slip rings S and S′, which rotate with the coil. Stationary brushes B and B' against the rings S and S′ make connections to an external circuit, giving an alternating current through the load R. Note that the resulting current, $i = \mathcal{E}/R$, is sinusoidal because $\mathcal{E}$ is sinusoidal. The potential difference V across B is usually written

$$V = V_0 \sin 2\pi ft \tag{25.26}$$

while the current is

$$i = i_0 \sin 2\pi ft \tag{25.27}$$

where $i_0 = V_0/R$.

By having an arrangement called the *split-ring* or *commutator*, it is possible to have a unidirectional, though fluctuating, induced emf. This is achieved by the split-ring reversing the connections to the external circuit every half turn corresponding to the positions when the emf in the coil reverses sign. This is the prototype of a dc generator. Commercial dc generators have many coils and commutator segments, so the output voltage is practically constant. Such generators are too complicated to discuss here.

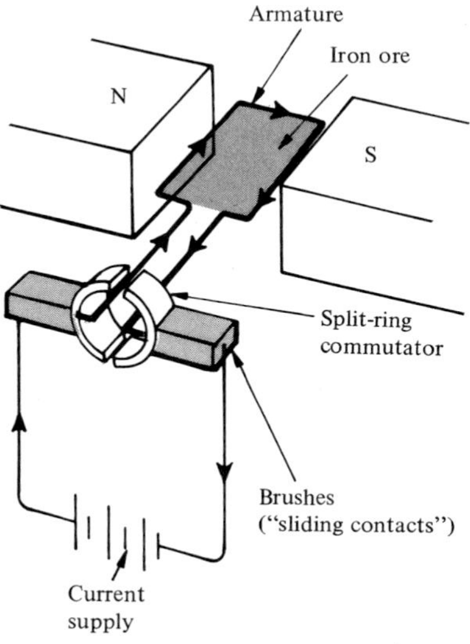

Figure 25.9 *Prototype of an electric motor.*

Electric Motor

The physical construction of an electric motor is similar to that of an electric generator, as shown in Figure 25.9. A direct-current (dc) motor may also be treated as a galvanometer coil that is continuously turning about its axis. As shown in Figure 25.9, a motor has a coil, called the *armature,* wound on an iron core. The armature leads are connected to a split-ring, the commutator, which slides on the sliding contacts, called *brushes*. The brushes are connected to a current supply. As the current goes through the coil, the coil rotates as explained below.

Consider the two parallel portions of the coil AB and CD (each of length L) each carrying current I in the directions shown and placed in a magnetic field B shown in Figure 25.10. The force $F = ILB$ experienced by each portion is perpendicular to both the length of the wire and the magnetic field. The force on AB is downward and on CD is upward, as shown in Figure 25.10. The torque due to these two forces produces rotational motion. As the coil rotates, the brushes make contact with the opposite segments of the commutator, reversing the direction of the current in the coil. This reversal of current is necessary to maintain the same direction of the force, hence that of the torque. This simple arrangement does not result in a smooth-running motor. To achieve this, the armature is made of a number of partial windings, which are brought out to a many-segmented commutator.

Before closing this section, let us briefly discuss back emf, which plays an important role both in electric motors and in generators. When the conductors in the armature of a motor are rotating in a magnetic field, according to Lenz's law, an emf is induced which opposes the current in the conductors. This induced emf due to the generator action in a motor is called a *back emf*. Thus, if there is a lamp in a series circuit with a motor, it will be bright when the motor is not running (no back emf). When the armature starts running, the back emf reduces the current in the circuit, dimming the lamp. The faster the armature rotates, the higher the back emf, and the less the current in the circuit.

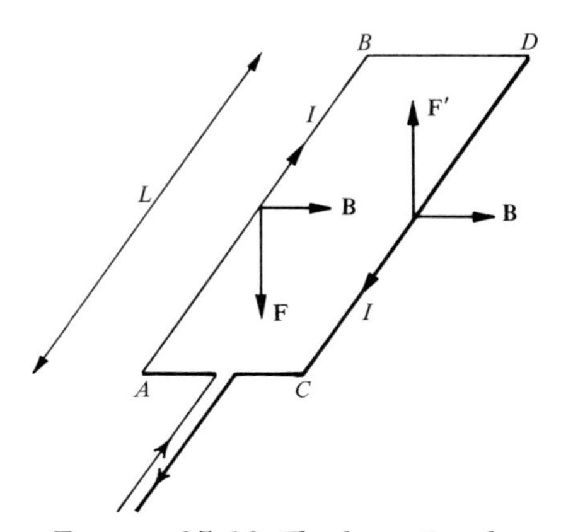

Figure 25.10 *The forces F and F′ on conductors AB and CD produce a torque that causes the rotation of the coil.*

Example 25.3 In Figure 25.5, let $l = 0.1$ m, $v = 0.2$ m/s, $B = 1$ Wb/m^2, and the resistance of the loop at an instant be $R = 0.2\ \Omega$. Calculate the motional emf, the current in the loop, and the power needed to move the loop.

According to Equation 25.16, the motional emf $\mathcal{E}$ is

$$\mathcal{E} = vBl = (0.2 \text{ m/s})(1 \text{ Wb/m}^2)(0.1 \text{ m}) = 0.02 \text{ V}$$

and the current in the loop is

$$i = \frac{\mathcal{E}}{R} = \frac{0.02 \text{ V}}{0.2\ \Omega} = 0.1 \text{ A}$$

The power needed to move the loop is $P = Fv$, where F is the force on the loop and is given by

$$F = iBl = (0.1 \text{ A})(1 \text{ Wb/m}^2)(0.1 \text{ m}) = 0.01 \text{ N}$$

Therefore,

$$P = Fv = (0.01 \text{ N})(0.2 \text{ m/s}) = 0.002 \text{ W}$$

This is equal to the product $\mathcal{E}I$.

Exercise 25.3 At what velocity should we move the loop in Example 25.3 so that the motional emf will be 0.1 V? What is i, F, and P in this case? [*Ans.*: $v = 1$ m/s, $i = 0.5$ A, $F = 0.05$ N, $P = 0.05$ W.]

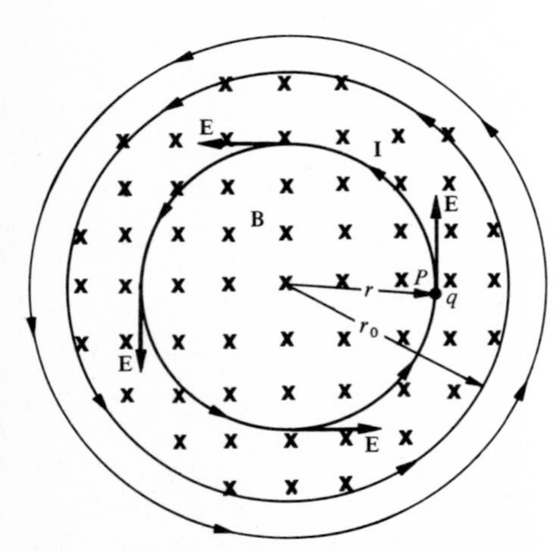

Figure 25.11 *Electric field E associated with a changing magnetic field.*

25.4. Induced EMF and Electric Field—Betatron

According to Faraday's law, a changing magnetic flux enclosed by a conducting loop produces an induced emf. This induced emf $\mathcal{E}$ forces the charges into motion, thereby resulting in an induced current in the loop. Thus, if R is the resistance of a loop placed in a varying magnetic flux as shown in Figure 25.11, the induced current i in the loop is

$$i = \frac{\mathcal{E}}{R} \qquad \text{where } \mathcal{E} = -\frac{\Delta\Phi}{\Delta t} \tag{25.28}$$

Let us elaborate on this by considering a uniform magnetic field of induction **B** pointing into the plane of the paper and confined to a space of radius r_0, as shown in Figure 25.11. Suppose that **B** is increasing in magnitude at a constant rate. Consider a particular circular path I of radius r, which may be an actual physical loop or an imaginary path. From the symmetry of the problem it is clear that the electric field **E** induced at various points of the path must be tangent to the path. This means that the induced electric field lines that are produced by the changing magnetic field are concentric circles in this particular case. Thus, the force on charge q is qE, while the work done in going around the closed circular path will be $qE(2\pi r)$. This must also be equal to $q\mathcal{E}$. Thus,

$$q\mathcal{E} = qE(2\pi r) \tag{25.29}$$

Since $2\pi r = l =$ total path length, Equation 25.29 combined with Equation 25.28 yields

$$\mathcal{E} = El = -\frac{\Delta\Phi}{\Delta t} \tag{25.30}$$

That is, a changing magnetic field produces an electric field proportional to the rate of change of magnetic flux. The principle of induction has many applications, such as in the betatron, search coil, and eddy currents. We shall discuss only the betatron.

Betatron. The *betatron*, also called an *induction accelerator*, is used for accelerating electrons to high energies and is based on the following principle: a changing magnetic field produces an electric field (as shown in Figure 25.11) which accelerates the electrons; at the same time, the magnetic field produces the necessary radial forces for the electrons to move in a circular orbit of constant radius. The machine was conceived in 1928, but Donald W. Kerst (1911–   ) of the University of Illinois designed and constructed the first betatron in 1941, and accelerated electrons to energies of 2 MeV.

A schematic of a typical betatron is shown in Figure 25.12. A pulse of electrons is injected into a high-vacuum glass or ceramic annular tube, a doughnut-shaped chamber or toroidal vacuum tube. This chamber is placed between the pole faces of a large electromagnet. Alternating current (see Chapter 26 for details) at about a frequency of 60 Hz, applied to the primary coil of the electromagnet with circular pole faces, produces a changing magnetic flux through the plane of the toroidal chamber. The changing flux reaches its maximum value in one-fourth of a time period, that is, in $\frac{1}{4}(1/60) = 1/240$ s. Thus, only one-quarter of the cycle is usable for accelerat-

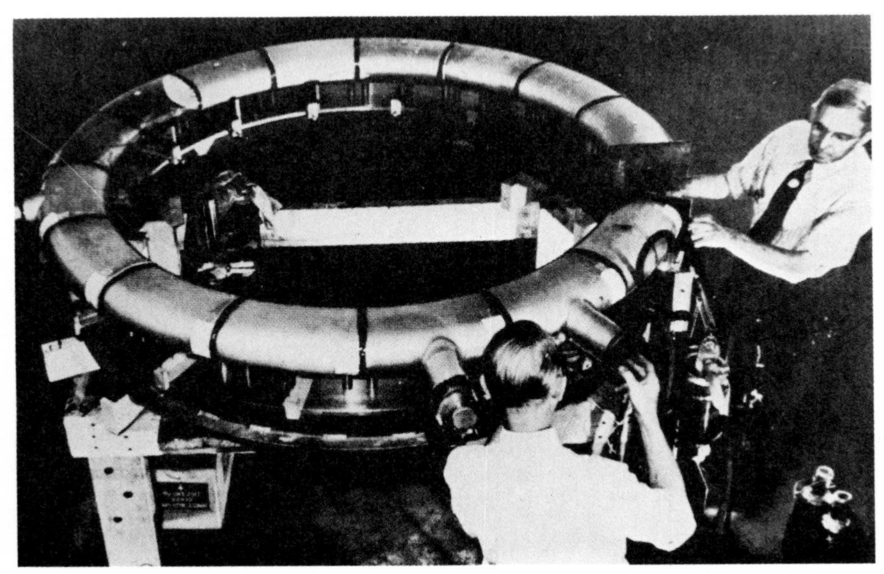

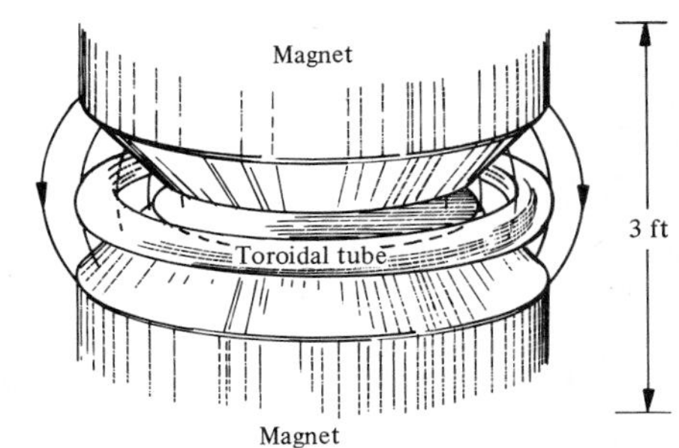

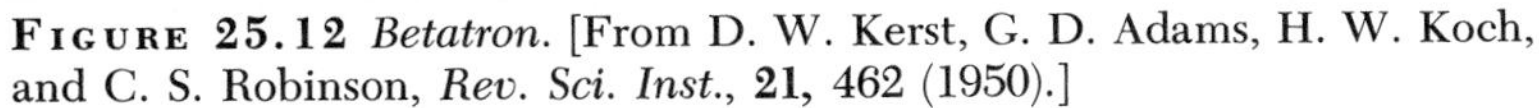
FIGURE 25.12 *Betatron.* [From D. W. Kerst, G. D. Adams, H. W. Koch, and C. S. Robinson, *Rev. Sci. Inst.*, **21**, 462 (1950).]

ing electrons. The electrons are accelerated by the tangential electric field **E** produced by the increasing magnetic flux within the orbit. The electrons travel thousands of revolutions in a circle of fixed radius and build up high energies without the need for insulation. The electrons are guided, kept in a circular orbit, or focused toward it, by a specially shaped magnetic guide field. In each revolution the energy gained is $e\mathcal{E}$, and for n revolutions the energy is $ne\mathcal{E}$.

Many betatrons have been constructed for accelerating electrons up to 400 MeV. The electrons are made to strike an external target, resulting in a beam of x-rays. These machines are frequently used in industry and medical technology, where the x-rays are employed for radiography and cancer treatment.

25.5. Energy in a Magnetic Field

In Chapter 20 we showed that energy can be stored in an electric field. Specifically, we showed that the work done in charging a capacitor is stored as energy in the electric field established in the space between the plates of the capacitor. The electrical energy per unit volume, in vacuum, was found to be

$$u_E = \tfrac{1}{2}\epsilon_0 E^2 \tag{20.24}$$

where E is the electric field strength at the point in question.

In analogy to the situation above the energy can be stored in a magnetic field. We consider an inductor connected to a battery. Suppose that at a given instant the current in the inductor is i and it is changing at the rate $\Delta i/\Delta t$. The induced emf $\mathcal{E}$, according to Equation 25.12, is given by

$$\mathcal{E} = -L\frac{\Delta i}{\Delta t} \tag{25.31}$$

where L is the inductance of the coil. This emf opposes the current in the inductor. Thus, the battery must do work against this emf to maintain a current in the inductor. Since the induced emf $\mathcal{E}$ is equal to the potential difference V between the two ends of the coil, the instantaneous power P supplied to the coil by the battery is

$$P = iV = i\mathcal{E} = iL\frac{\Delta i}{\Delta t} \tag{25.32}$$

Since $P = \Delta W/\Delta t$, the work done in time Δt is

$$\Delta W = P\,\Delta t = iL\,\Delta i$$

When the current in the inductor starts from $i = 0$ at $t = 0$ and reaches a value $i = I$ in time t, the total work done by the battery is obtained (as in the case of a capacitor or spring) by replacing i by the average value $(0 + I)/2 = I/2$ and Δi by $(I - 0) = I$. That is,

$$W = \tfrac{1}{2}LI^2$$

This work is stored as energy in the magnetic field in the space surrounding the inductor. Thus, the total stored magnetic energy U_B in any inductor of inductance L is given by

$$\boxed{U_B = \tfrac{1}{2}LI^2} \tag{25.33}$$

To find the general expression for the magnetic energy stored in a unit volume, that is, the magnetic energy density U_B, we must consider a particular inductor. The self-inductance of a long solenoid of length l and cross-sectional area A as given in Example 25.2 is $L = \mu_0 n^2 lA$. The field in the solenoid is $B = \mu_0 nI$ or $I = B/\mu_0 n$. Substituting for L and I in Equation 25.33, we get

$$U_B = \frac{1}{2}LI^2 = \frac{1}{2}(\mu_0 n^2 lA)\left(\frac{B}{\mu_0 n}\right)^2$$

or

$$U_B = \frac{1}{2\mu_0} B^2 lA \tag{25.34}$$

lA is the volume enclosed by the solenoid, and for a long solenoid almost all the magnetic field is confined to this volume. Thus, the *energy density* of the magnetic field u_B defined as the magnetic energy per unit volume in vacuum, that is, $u_B = U_B/lA$ is given by

$$\boxed{u_B = \frac{1}{2}\frac{B^2}{\mu_0}} \tag{25.35}$$

(Compare this with the electric energy density, $u_E = \frac{1}{2}\epsilon_0 E^2$.) Even though Equation 25.35 has been derived for the special case of a long solenoid, the expression for u_B is more general and holds for all cases except in the presence of magnetic materials.

EXAMPLE 25.4 Calculate the total energy U_B and the energy density u_B of the magnetic field stored in a solenoid 0.5 m long, having 5000 turns and a current of 10 A. The radius of the solenoid is 4 cm.

The magnetic field inside the solenoid given by Equation 24.12 is $B = \mu_0 nI$. Therefore,

$$u_B = \frac{1}{2}\frac{B^2}{\mu_0} = \frac{1}{2}\frac{(\mu_0 nI)^2}{\mu_0} = \frac{1}{2}\mu_0 n^2 I^2 = \frac{1}{2}\left(4\pi \times 10^{-7}\frac{\text{Wb}}{\text{A-m}}\right)\left(\frac{5000}{0.5\text{ m}}\right)^2(10\text{ A})^2$$
$$= 2\pi \times 10^3\text{ J/m}^3 = 6.28 \times 10^3\text{ J/m}^3$$

The volume of the solenoid is

$$V = lA = l\pi r^2 = (0.5\text{ m})\pi(0.04\text{ m})^2 = 0.0025\text{ m}^3$$

Therefore,

$$U_B = u_B V = (6.28 \times 10^3\text{ J/m}^3)(0.0025\text{ m}^3) = 15.7\text{ J}$$

EXERCISE 25.4 In Example 25.4, how will the value of u_B change if (a) I is doubled; (b) the length is doubled; and (c) the radius is doubled? [*Ans.*: (a) increases by a factor of 4; (b) decreases by a factor of 4; (c) is unchanged.]

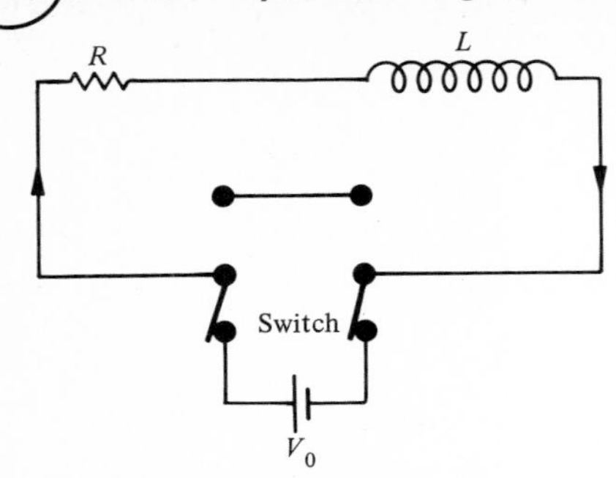

FIGURE 25.13 *Circuit containing an inductor L and a resistor R.*

*25.6. *LR* and *LC* Circuits

LR Circuit

An inductor is simply loops of wire. All wires have some resistance no matter how small, except in the case of a superconducting coil. Thus, there is no such thing as a pure inductor (unless it is a superconductor; Chapter 33). Each inductor has a self-inductance L and resistance R. This resistance is physically a part of L, but we show it separately in Figure 25.13. Many electrical circuits contain some sort of inductor. We are interested in investigating i versus t when a battery of voltage V_0 is connected in a circuit containing an inductor L and resistance R (which may be simply the resistance of the inductance coil or, in addition, any other resistance in the circuit).

Let us apply Kirchhoff's loop rule to the circuit shown in Figure 25.13. As i increases, an emf is induced across the inductor equal to $\mathcal{E} = -L\,\Delta i/\Delta t$, where the minus sign indicates that the induced emf opposes V_0 of the battery. Thus keeping the sign convention in mind, we may write

$$V_0 - iR - L\frac{\Delta i}{\Delta t} = 0 \tag{25.36}$$

Let us multiply both side by i and rewrite as

$$iV_0 = i^2R + Li\frac{\Delta i}{\Delta t} \tag{25.37}$$

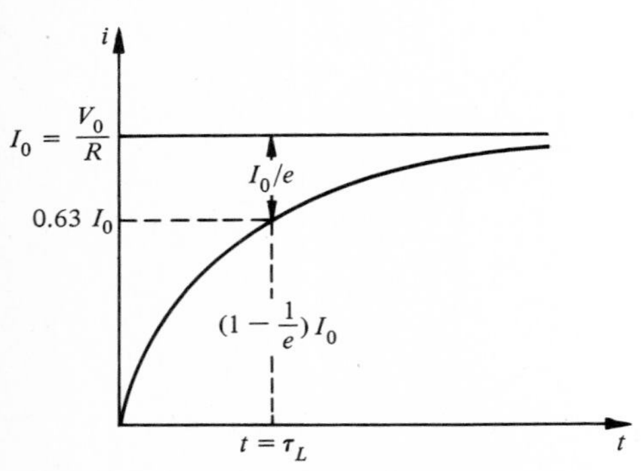

FIGURE 25.14 *Current i versus t in an inductor after a battery has been connected.*

where iV_0 is the rate at which the battery is supplying energy to the circuit, i^2R is the rate at which the energy is being dissipated as heat across R, and $\text{Li}(\Delta i/\Delta t)$ is the rate at which the energy is being used across the inductor and is stored in the magnetic field. Solving the equation above means finding the value of i versus t. Without going into details, we state that

$$i = \frac{V_0}{R}(1 - e^{-(R/L)t}) \tag{25.38}$$

The plot of i versus t of this equation is shown in Figure 25.14. It clearly indicates that the current reaches its maximum value $I_0 = V_0/R$ when $t = \infty$. Also, if we define τ_L to be the *inductive time constant* given by

$$\boxed{\tau_L = \frac{L}{R}} \tag{25.39}$$

we may write Equation 25.38 as

$$\boxed{i = I_0(1 - e^{-t/\tau_L})} \tag{25.40}$$

We can show that the unit of τ_L is time; that is,

$$\tau_L = \frac{L}{R} = \frac{\text{henry}}{\text{ohm}} = \frac{\text{volt/(A/s)}}{\text{ohm}} = \frac{\text{ohm-s}}{\text{ohm}} = \text{s}$$

where we have used the fact that $L = \mathcal{E}/(\Delta i/\Delta t)$. Let us calculate the value of i in terms of I_0 when $t = \tau_L$ in Equation 25.40. Thus,

$$i = I_0(1 - e^{-1}) = I_0\left(1 - \frac{1}{e}\right) = I_0(1 - 0.37) = 0.63I_0$$

which states that the *inductive time constant* τ_L is that time in which the current in the inductive circuit reaches 0.63 (63 percent) of its maximum saturation value. This is clearly seen from Figure 25.14.

Suppose that after the maximum current I_0 *has been established in the* circuit, the switch in Figure 25.13 is thrown to the other side so that the battery is removed from the circuit. Under such circumstances Equation 25.36 for the circuit (which has been short-circuited) takes the form

$$iR + L\frac{\Delta i}{\Delta t} = 0 \qquad (25.41)$$

which may be solved by using the initial condition that $i = I_0$ at $t = 0$. The calculations will yield

$$i = I_0 e^{-(R/L)t} = I_0 e^{-t/\tau_L} \qquad (25.42)$$

Thus, the current decreases exponentially as shown in the plot of i versus t in Figure 25.15. Note that the energy needed to maintain the current in the circuit after the battery has been removed is provided by the energy stored in the magnetic field of the inductor.

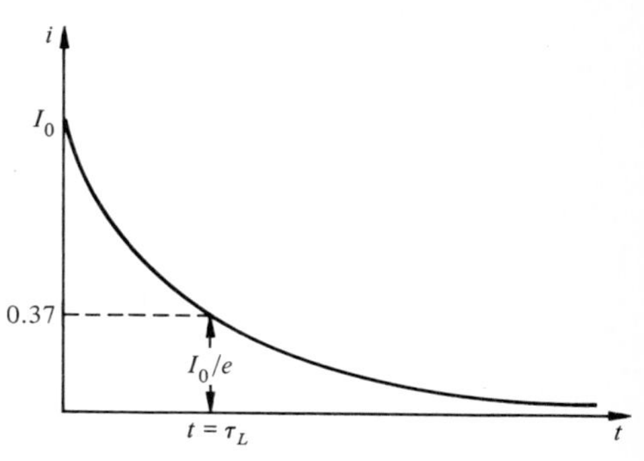

FIGURE 25.15 *Current i versus t in an inductor after the battery has been replaced by a short-circuit.*

Assuming a pure inductor, a resistor, and a battery in a circuit, Figure 25.16 shows the plot of V_R versus t and V_L versus t and the sum of the two versus t

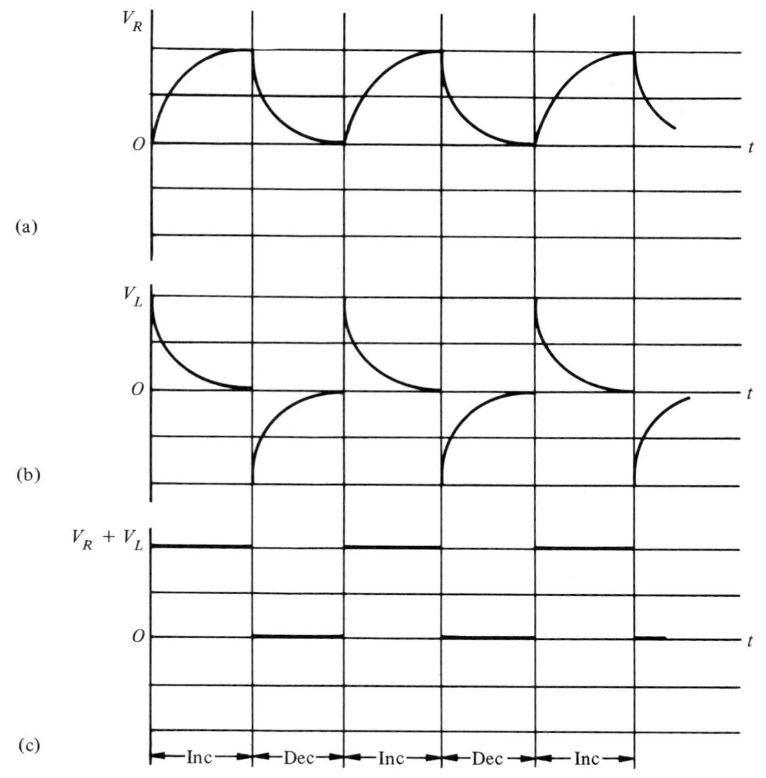

FIGURE 25.16 *Variation of (a) V_R versus t, (b) V_L versus t, and (c) $V_R + V_L$ versus t in an LR circuit for both increasing (inc) and decreasing (dec) current.*

for both increasing current with the battery present and decreasing current without the battery. What do you conclude from this figure about the conservation of energy?

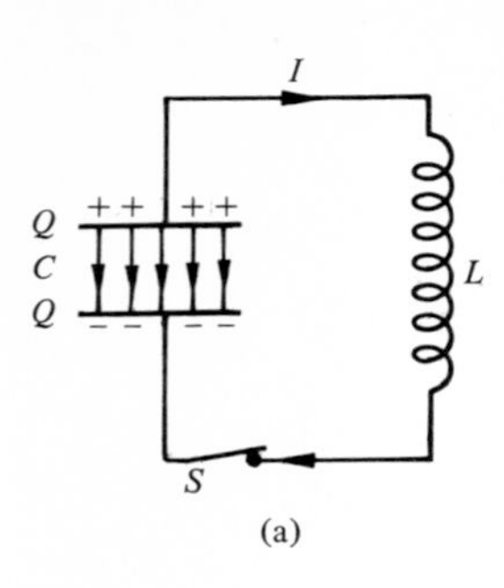

(a)

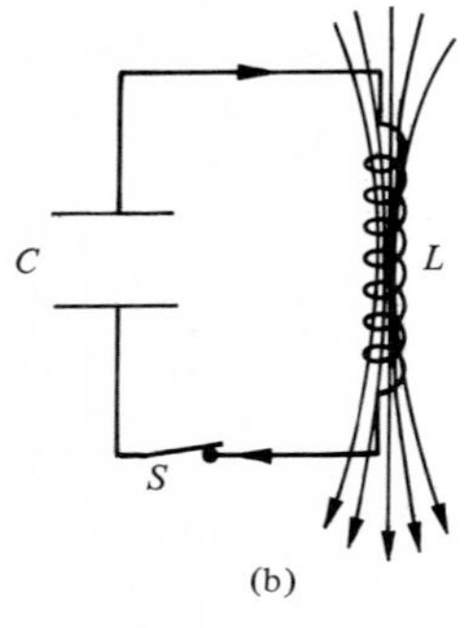

(b)

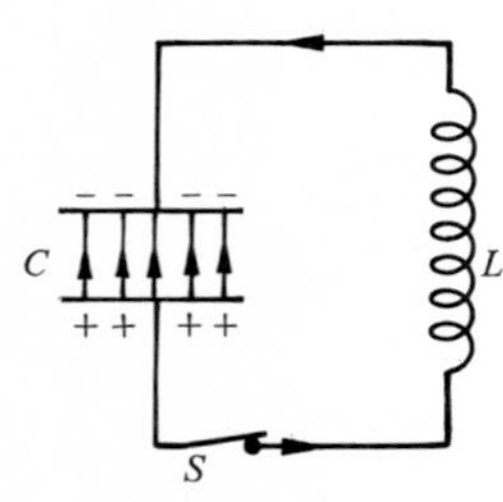

(c)

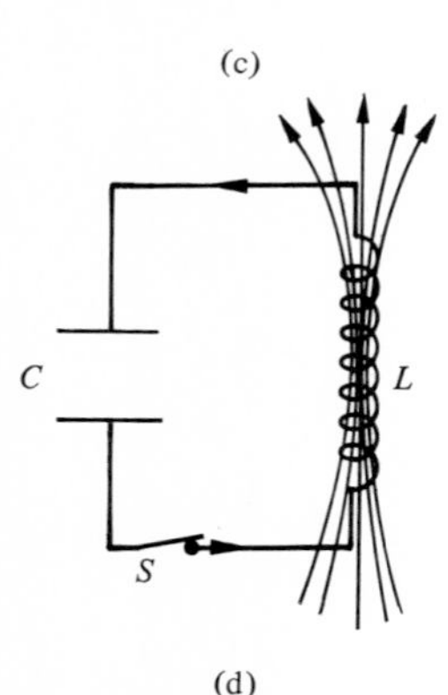

(d)

FIGURE 25.17 *Circuit containing an inductor and a capacitor.*

LC Circuit

Figure 25.17 shows a circuit containing a capacitor C and a pure inductor L while $R = 0$. To start with, the capacitor is fully charged, with charge Q as shown in Figure 25.17(a). If the switch S is closed, a current starts in the circuit, discharging the capacitor. As the current in the circuit increases, there is an induced emf in the inductor which results in establishing a magnetic field in the space around the inductor. After the capacitor is completely discharged, as shown in Figure 25.17(b), the magnetic field established is maximum. The decreasing magnetic field now induces an emf in the same direction as the original current. The current continues until the magnetic field disappears and the capacitor is fully charged with the opposite polarity, as shown in Figure 25.17(c). The process now continues in the opposite direction. In the absence of any energy loss, the process continues indefinitely. These repeated variations in the electric and magnetic fields are called *electrical oscillations* (Chapter 26).

Electrical oscillations in the LC circuit are analogous to the mechanical oscillations of a mass tied to a spring and vibrating without any friction. Let us show this analogy more closely from energy considerations. At any instant when the current in the LC circuit is i and the capacitor charge is q, the total energy is

$$U = U_E + U_B = \frac{1}{2}\frac{q^2}{C} + \frac{1}{2}Li^2 \tag{25.43}$$

Even though the values of U_E and U_B change with time, the total energy U remains constant. Similarly, if a mass m is tied to a spring of spring constant k, the total energy is the sum of the potential energy $U_P = \frac{1}{2}kx^2$ and the kinetic energy $U_K = \frac{1}{2}mv^2$. That is, the total energy

$$U = U_P + U_K = \tfrac{1}{2}kx^2 + \tfrac{1}{2}mv^2 \tag{25.44}$$

Comparing the preceding two equations, we get the following analogy between the two systems:

Mechanical	x	m	k	$v(=\Delta x/\Delta t)$	$m(\Delta v/\Delta t)$
Electrical	q	L	$1/C$	$i(=\Delta q/\Delta t)$	$L(\Delta i/\Delta t)$

Also, the frequency of oscillation, f_M, for the spring-mass system is

$$f_M = \frac{1}{2\pi}\sqrt{\frac{k}{m}}$$

and the corresponding frequency, f_E, of the LC oscillations is

$$f_E = \frac{1}{2\pi}\sqrt{\frac{1}{LC}} \tag{25.45}$$

f_E is the frequency of the electrical oscillations mentioned above.

The analogy is more than just formal. Just as the inertial mass of a particle has a tendency to maintain a constant velocity v, similarly the inductance L of an inductor shows a tendency to maintain current i in the circuit. The restoring constant k of a spring is analogous to the restoring constant $1/C$ of a capacitor. Similarly, the mechanical force $m\,\Delta v/\Delta t$ corresponds to the induced emf $L(\Delta i/\Delta t)$.

An electrical circuit containing a resistance R in addition to L and C is equivalent to a mechanical mass–spring system in a viscous medium. The dissipative force $F = bv$ in mechanical systems is equivalent to Ri in electrical systems. As in the mass–spring system, which will vibrate forever if there were no viscous forces, in electrical systems the LC circuit will oscillate indefinitely if R were zero, which is not possible. There is always an i^2R loss and the system stops if no extra energy is supplied.

The correspondence between the two systems is so good that the behavior of many mechanical systems can be investigated before actual construction by studying analogous electrical circuits.

SUMMARY

According to *Faraday's law*, an induced emf is produced in any closed circuit if there is a time-varying magnetic flux, and the magnitude of this induced emf is equal to the negative of the time rate of change of the magnetic flux through the circuit. That is, $\mathcal{E} = -N\,\Delta\Phi/\Delta t = -\Delta(N\Phi)/\Delta t$. The quantity $N\Phi$ is *flux linkage*. Since $\Phi = BA\cos\theta$, the change in the flux is achieved by changing B, A, or θ. According to *Lenz's law*, the direction of the induced current is always such as to oppose the cause that produces it.

The coefficient of *mutual inductance* M is defined as $M = N_1\,\Phi_{12}/i_2 = N_2\,\Phi_{21}/i_1$ or $\mathcal{E}_1 = -M\,\Delta i_2/\Delta t$. That is, M is equal to the induced emf in coil 1 per unit rate of change of current in coil 2. The unit of M is volt per ampere per second, called a *henry* (H). Similarly, the *self-inductance* L is defined as $L = N\,\Phi/i$ or $\mathcal{E} = -L\,\Delta i/\Delta t$. That is, L is equal to the self-induced emf per rate of change of current, and its unit is also the henry.

For a *moving conductor* of length l and velocity v, the motional emf is given by $\mathcal{E} = \Delta\Phi/\Delta t = vBl$ and $i = \mathcal{E}/R = vBl/R$. For a *rotating coil*, the motional emf is $\mathcal{E} = \mathcal{E}_0 \sin\omega t$, where $\mathcal{E}_0 = \omega ABN$. The principle of a rotating coil in a magnetic field is used in the construction of electric generators and electric motors.

A changing magnetic field produces an electric field proportional to the rate of change of the magnetic flux. That is, $\mathcal{E} = El = -\Delta\Phi/\Delta t$. This principle is used in the working of a betatron, a machine used for accelerating electrons to high energies.

The total stored magnetic energy U_B in an inductor is $U_B = \frac{1}{2}LI^2$. The energy density of the magnetic field u_B defined as magnetic energy per unit volume in vacuum is given by $u_B = \frac{1}{2}B^2/\mu_0$.

* In an LR circuit with a source of voltage V_0, the current increases as $i = I_0(1 - e^{-t/\tau_L})$, where $\tau_L = L/R$ is the *time constant*, and the current decreases as $i = I_0 e^{-t/\tau_L}$.

* Electrical oscillations in the LC circuit are analogous to the mechanical oscillations of a mass tied to a spring and vibrating without any friction.

The system has total energy $U = U_E + U_B = \frac{1}{2}q^2/C + \frac{1}{2}Li^2$, and frequency $f_E = (1/2\pi)\sqrt{1/LC}$.

QUESTIONS

1. Give two examples other than those in the text to demonstrate induced emf.
2. Give qualitative reasoning showing that Lenz's law can be derived from Faraday's law.
3. Suppose that we want to increase the mutual inductance of two coils. What steps do you think can be taken to do this?
4. What steps do you suggest to reduce the self-inductance of a solenoid?
5. Give an example of a system in which the induced current is produced by changing (a) B; (b) A; and (c) θ, the orientation between **B** and **A**.
6. Explain the following statement: "An electric motor is an electric generator operating in reverse."
7. What are the similarities and differences between the electric field produced (a) between the parallel plates of a capacitor; and (b) by the changing magnetic flux?
8. Can you think of some reason why a betatron is used for accelerating electrons and not heavy charged particles such as protons?
9. Is it possible to have a large resistor with no inductance or a large inductor with no resistance?
10. In electrical circuits why (a) are wires kept as small as possible; and (b) are wires carrying currents in opposite directions wrapped around each other?

*12. List as many electrical quantities as possible analogous to a mass–spring system oscillating in a viscous medium, especially the quantities corresponding to the applied force $m\,\Delta v/\Delta t$ and the restoring force $-kx$.

PROBLEMS

1. One hundred turns of copper wire are wound on a wooden cylinder of radius 0.2 m. If a magnetic field through the coil is changed at a uniform rate from a value of 0.2 Wb/m^2 to 0.6 Wb/m^2 in 0.1 s, calculate the emf between the two ends of the wire.
2. Calculate the magnitude of the emf induced in a coil of 500 turns and having a cross-sectional area of 0.25 m^2 if the magnetic field through the coil changes from 0.1 Wb/m^2 to 0.4 Wb/m^2 in 0.06 s.
3. Calculate the magnitude of the emf induced in a coil of 200 turns and of radius 5 cm if the magnetic field through the coil reduces from 0.10 Wb/m^2 to zero in 0.05 s.
4. A coil of size 10 cm by 5 cm with 100 turns is placed parallel to the earth's magnetic field, which is 5×10^{-5} Wb/m^2. If in 0.05 s the coil is turned so that it is perpendicular to the field, calculate the magnitude of the induced emf.
5. A circular loop of 50 turns and radius 10 cm is placed in a magnetic field with its plane perpendicular to the magnetic induction **B**. When the coil is turned in 0.01 s so that its plane is parallel to the field, an emf of 0.05 V is induced. What is the magnitude of **B**?
6. Consider a coil of area 25 cm^2 and 1000 turns placed with its plane perpendicular to the magnetic field. Suppose that the field changes from

$0.1\ Wb/m^2$ to $1.0\ Wb/m^2$ in 0.1 s. If the resistance of the coil is $0.01\ \Omega$, calculate the current through the coil. What is the total charge that flows in 0.1 s?

7. A square loop of side 10 cm and 500 turns is placed perpendicular to a uniform magnetic field of $0.1\ Wb/m^2$. If it is pulled out of the field in 0.05 s, calculate the magnitude of the induced emf.
8. A coil of area $0.2\ m^2$ is placed with its plane perpendicular to a uniform magnetic field of $0.04\ Wb/m^2$. Calculate the flux through the coil. (a) If this coil were turned very rapidly in 0.1 s through 90°, what will be the change in the flux? (b) If the coil were turned through 180° in 0.2 s, what will be the change in the flux? Calculate the average induced emf in each case.
9. A 0.5-m-long solenoid P with 50 turns per centimeter is carrying a current of 2 A and has a cross-sectional area of $0.01\ m^2$. This is the primary coil P. If the current in P is reduced to zero in 0.01 s, what will be the induced emf in the secondary coil S which has 10 turns.
10. Show that 1 Wb/s equals 1 V.
11. The coefficient of mutual inductance between two coils is 0.01 H. If the current in one coil changes at the rate of 500 A/s, calculate the emf induced in the second coil.
12. One coil is wrapped around another coil. If the current in the inner coil changes at the rate of 100 A/s, an emf of 5 V is induced in the outer coil. Calculate the mutual inductance of the coil.
13. When the current in one coil changes from 5 A to 25 A in 0.005 s, an emf induced in a nearby coil is 60 V. What is the mutual inductance of the system?
14. In a coil the current changes from 0 to 10 A in 0.001 s and results in an average induced emf of 150 V. Calculate the self-inductance of the coil.
15. A coil has a self-inductance of 10 H. At what rate should the current change in the coil so that there is an induced emf of 120 V?
16. A potential difference of 100 V is suddenly applied to a coil of self-inductance 0.01 H and a resistance of $1.5\ \Omega$. Calculate the following: (a) the initial rate of increase of the current; (b) the current in the coil when the rate of increase is 200 A/s; and (c) the final current in the coil.
17. If the current through a coil of 50 turns and of inductance 0.2 H is 3 A, what is the flux linkage of the coil?
18. A horizontal wire 10 cm long is moving with a speed of 2 m/s through a magnetic field of $0.1\ Wb/m^2$. Calculate the emf induced in the wire.
19. In Figure 25.5, if $B = 0.8\ Wb/m^2$, $l = 1$ m, and $v = 2$ m/s, calculate the motional emf in the rod and the potential difference between the two ends. Also, calculate the change in flux in (a) 1 s; and (b) 2 s.
20. In Problem 19 if the resistance of the closed circuit is $0.5\ \Omega$, what is the magnitude of the current in the circuit due to the induced emf? What force is required to maintain the rod in motion?
21. Calculate the emf induced across a 2.0-m-long bumper of a car which is moving with a velocity of 30 m/s in the magnetic field of the earth, which has a vertical component equal to $5 \times 10^{-5}\ Wb/m^2$ downward.
22. A coil has a self-inductance of 4 H and a current of 10 A. Calculate the amount of energy stored in the inductor. If the coil is 10 cm long and has a radius of 2 cm, what is the energy density?

23. If energy of 100 J is stored in an inductor when the current in the inductor changes from 0 to 5 A, what is the inductance of the coil?
24. The magnetic field at the center of a solenoid 1 m long and having a 0.05 m radius is 0.1 Wb/m^2. Calculate the energy stored in the magnetic field and the energy density.
*25. Show that the units of $\sqrt{L/C}$ are that of resistance.
*26. Show that the units of L/R and RC are those of time.
*27. An inductor of inductance 2 H and resistance 10 Ω is connected to a battery of 12 V of negligible internal resistance. Write an equation describing the current versus time. Calculate the value of the final steady-state current.
*28. A 10-H inductor of resistance 100 Ω is connected to the terminals of a 12-V battery. Calculate the final steady-state current in the inductor. Write an equation describing i versus t. Calculate the value of i for $t = 0, 0.1, 0.25, 0.45, 0.65, 0.85$, and 1 s and make a plot of it.
*29. In Problem 28, if the inductor is short-circuited, it will discharge. Write an equation describing i versus t and also make a plot of i versus t.
*30. Show that the units of $1/\sqrt{LC}$ are the same as that of frequency, that is, 1/s.
*31. Show that the Equations 25.43 and 25.44 are dimensionally correct.
*32. Calculate the frequency of a circuit containing an inductor of 10 H and a capacitor of (a) 10 μF; and (b) 10 μμF.
*33. What should be the value of a capacitor C in an LC circuit containing an inductor of 2 H so that the oscillation frequency will be 3×10^6 Hz?

26

AC Circuits and Electronics

We shift our emphasis to the study of sources of alternating voltages, hence, alternating currents (ac). The frequencies of such sources may vary from less than ~50 Hz for the usual power sources to ~10^{10} (=10 GHz) for radar use and microwave communication. Investigating the behavior of resistance, inductance, and capacitance in ac circuits prepares us to look into the many diverse uses of these circuit elements and ac sources. Before closing this chapter a brief introduction to transformers and electronics will be made.

26.1. Average and Root-Mean-Square Values

IN THE previous chapters we mostly used steady sources of voltages and currents, called direct current (dc). The voltage and current remain constant with time as shown in Figure 26.1. More and more power circuits in use are being supplied with alternating voltages, hence carry alternating currents (ac).

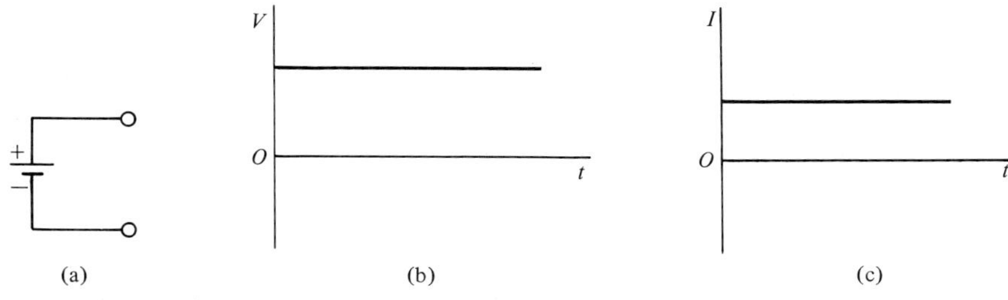

FIGURE 26.1. *Symbolic representation of direct current: (a) source, (b) voltage versus time, and (c) current versus time.*

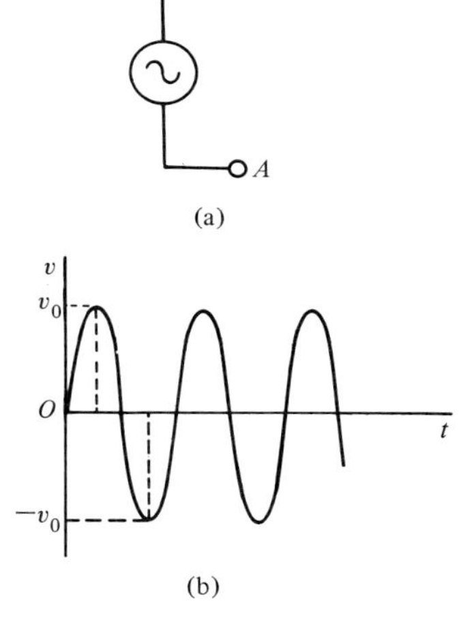

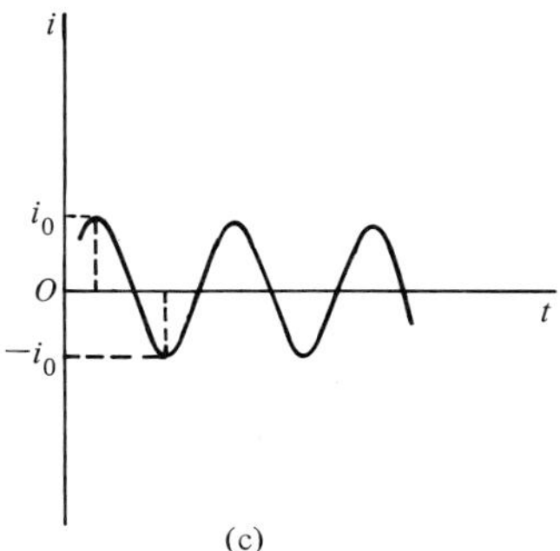

FIGURE 26.2. *Symbolic representation of alternating current: (a) source, (b) voltage versus time, and (c) current versus time. v_0 and i_0 are the voltage and current amplitudes, respectively.*

The circuits containing such sources of power are called alternating-current circuits or ac circuits. The frequencies of the ac power sources range from less than ~50 Hz ($f = 50$ cycles/s) for usual power sources to ~10 GHz (1 GHz = 10^9 cycles/s) for radar use and microwave communication.

The most typical sources of alternating potentials and currents are sinusoidal, in the form shown in Figures 26.2(b) and (c) respectively. We may write the equations for voltage and current as

$$v = v_0 \sin \omega t = v_0 \sin 2\pi f t \tag{26.1}$$

$$i = i_0 \sin \omega t = i_0 \sin 2\pi f t \tag{26.2}$$

where v and i are the instantaneous values of voltage and current across the terminals A and B of the ac source ⊙ shown in Figure 26.2(a). v_0 is called the *voltage amplitude* and i_0 the *current amplitude,* and these are the maximum values of v and i, as shown in Figure 26.2. ω is the angular frequency (rad/s), which is equal to $2\pi f$, where f is the frequency (cycles/s or Hz). ω, f, and the period T are related as

$$\omega = 2\pi f = \frac{2\pi}{T}$$

It is clear from Figure 26.2 that v or i are positive as often as they are negative; hence, the average value of v or i over a cycle will be zero. This means that if a current or voltage measuring device is placed between the terminals A and B of the ac source in Figure 26.2(a), these meters will read zero on the average. If the pointer of the meter can follow the fluctuations in i and v, we may be able to read the peak values of i_0 and v_0. This is possible only if the frequency is very small. Most ac measuring instruments avoid this difficulty because their working principle is based on the square of the currents and voltages. Thus, instead of using the average values of i and v, these instruments use the average values of i^2 and v^2, as we explain below. From Equation 26.1, $i^2 = i_0^2(\sin 2\pi ft)^2$. The average value of $(\sin 2\pi ft)^2$ (or any $\sin^2\theta$) is $\frac{1}{2}$. And the average value of i^2, denoted by $(i^2)_{av}$, is $(i^2)_{av} = i_0^2/2$.

We define the *root-mean-square current* i_{rms}, denoted by I, as the square root of $(i^2)_{av}$. Since $(i^2)_{av} = i_0^2/2$,

$$I \equiv i_{rms} \equiv \sqrt{(i^2)_{av}} = \frac{i_0}{\sqrt{2}} = 0.707 i_0 \qquad (26.3)$$

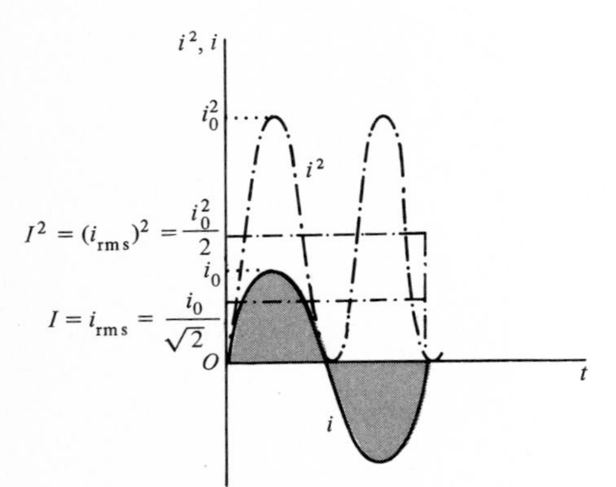

FIGURE 26.3. *Plots of i and i^2 versus t. As shown, $(i^2)_{av} = (i_{rms})^2 = I^2$, which is not zero.*

These values are demonstrated in Figure 26.3. Similarly, the *root-mean-square voltage* V can be shown to be

$$V \equiv v_{rms} \equiv \sqrt{(v^2)_{av}} = \frac{v_0}{\sqrt{2}} = 0.707 v_0 \qquad (26.4)$$

The root-mean-square values are also called *effective values*, and ammeters and voltmeters are calibrated to read these values. Thus, when we say that the alternating-current voltage is 120 V, we mean that the rms value V is 120 V while the peak value of the voltage v_0 is given by $V = v_0/\sqrt{2}$ or $v_0 = \sqrt{2}V = \sqrt{2} \times 120\ \text{V} \simeq 169.7$ V. Thus, the alternating voltage in this case fluctuates between $+169.7$ and -169.7 V.

The advantage of using the rms values I and V is that Ohm's law can be used for devices in ac circuits provided that we replace the instantaneous values i and v by the rms values I and V. Thus, we may write the average power loss across a resistor R in an ac circuit as

$$P_{av} = (iv)_{av} = \frac{i_0}{\sqrt{2}}\frac{v_0}{\sqrt{2}} = i_{rms}v_{rms} = IV$$

or

$$P_{av} = (i^2R)_{av} = (i^2)_{av}R \equiv (i_{rms})^2R = I^2R$$

That is,

$$P_{av} = IV = I^2R \qquad (26.5)$$

EXAMPLE 26.1 If the effective or rms values of current and voltage are 5 A and 100 V, what are their maximum values? What power loss by heating will take place across a 10-Ω resistor?

From Equations 26.3 and 26.4,

$$I = \frac{i_0}{\sqrt{2}} \qquad \text{or} \qquad i_0 = \sqrt{2}\, I = 1.414 \times 5 \text{ A} = 7.07 \text{ A}$$

$$V = \frac{v_0}{\sqrt{2}} \qquad \text{or} \qquad v_0 = \sqrt{2}\, V = 1.414 \times 100 \text{ V} = 141.4 \text{ V}$$

That is, the instantaneous value of i will fluctuate between $+7.07$ and -7.07 A, and the instantaneous value of V will fluctuate between $+141.4$ and -141.4 V. From Equation 26.5,

$$P_{av} = I^2 R = (5 \text{ A})^2\, 10\ \Omega = 250 \text{ W}$$

Note that a dc current of 5 A through a 10-Ω resistor will produce the same power loss.

EXERCISE 26.1 AC current passing through a 25-Ω resistor results in a power loss of 3600 W. Calculate the rms value of the current and the maximum value of the current. [*Ans.:* $I = 12$ A, $i_0 = 16.97$ A.]

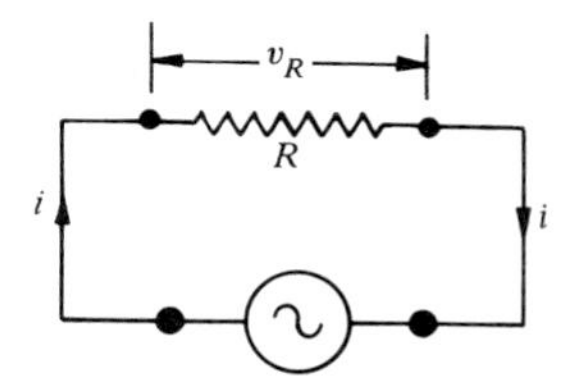

FIGURE 26.4. *Circuit containing a source of alternating current and a resistor.*

26.2. AC Circuits with Resistance, Capacitance and Inductance

Let us investigate the behavior of each individual component in the ac circuit.

AC Circuit with Resistance

Let us consider a circuit containing a resistor R and source of alternating current (or voltage) as shown in Figure 26.4. Let the ac source supplying the current through the circuit be

$$i = i_0 \sin \omega t \tag{26.6}$$

The instantaneous potential difference v_R across the resistor is

$$v_R = iR = i_0 R \sin \omega t = v_{0R} \sin \omega t \tag{26.7}$$

where

$$v_{0R} = i_0 R \tag{26.8}$$

Comparing Equations 26.6 and 26.7, we find that the potential difference across the resistor varies sinusoidally and also is *in phase* with the current; that is, both i and v_R reach their maximum values at the same time, as shown in Figure 26.5. Different scales are used in plotting i and v_R.

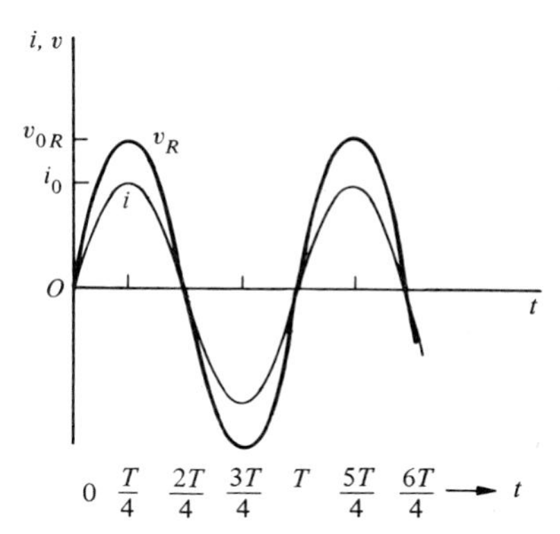

FIGURE 26.5. *Plots of v_R and i versus t, showing that the voltage across the resistor and the current in the circuit are in phase.*

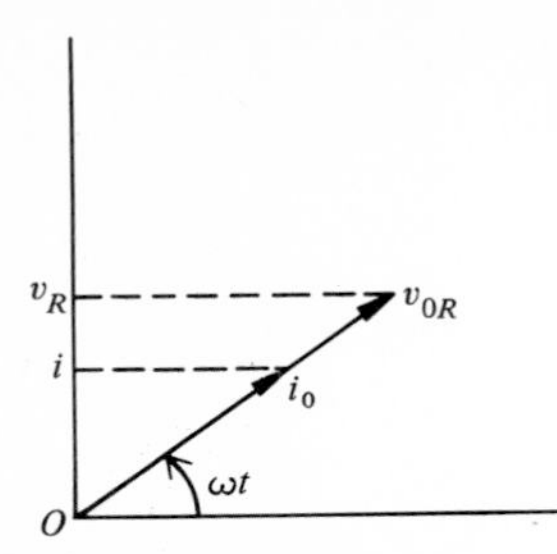

FIGURE 26.6. *Rotor diagram for current i_0 and voltage v_{0R} versus ωt.*

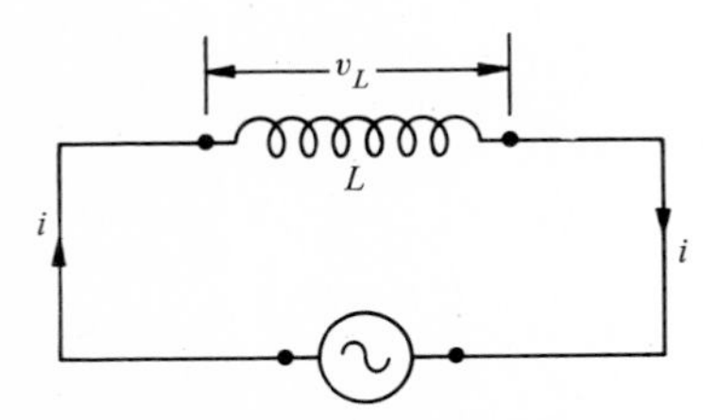

FIGURE 26.7. *Circuit containing a source of alternating current and an inductor.*

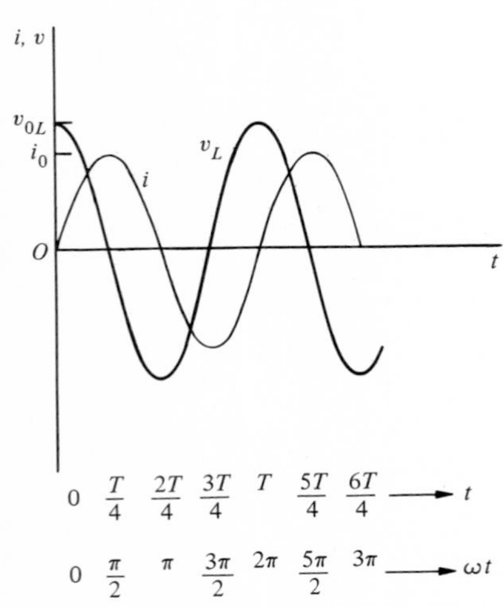

FIGURE 26.8. *Plots of v_L and i versus t, showing that v_L leads i by $\pi/2$ (or 90°).*

To find the rms value of the instantaneous voltage, using Equation 26.7 we get

$$(v_R)_{\text{rms}} = R(i_0 \sin \omega t)_{\text{rms}} = R(i)_{\text{rms}}$$

or

$$\boxed{V_R = RI} \tag{26.9}$$

where V_R and I are the rms values of voltage and current. Equation 26.9 is Ohm's law for an ac circuit provided that we use rms values.

A convenient way of expressing the relation between the instantaneous values of v_R, i and the maximum values of v_{0R} and i_0 at different times is by means of a phasor or rotor diagram, shown in Figure 26.6. The values of i_0 and v_{0R} are shown as vectors rotating counterclockwise. The projections of these on the vertical axis gives the instantaneous values of i and v_R, as shown.

AC Circuit with Inductance

An ac circuit containing a pure inductor is shown in Figure 26.7. Even though an inductor always has some resistance, we assume that it is small and may be neglected. Once again let the current through the circuit be

$$i = i_0 \sin \omega t \tag{26.10}$$

The voltage drop v_L across the inductor is (from Equation 25.12)

$$v_L = L\frac{\Delta i}{\Delta t} \tag{26.11}$$

where L is the inductance of the inductor. The value of $\Delta i/\Delta t$ may be obtained either by using calculus or by comparing Equation 26.10 with Equation 10.6, $y = A \sin \omega t$, and noting that from Equation 10.8, $v_y = \omega A \cos \omega t$, we may write $\Delta i/\Delta t = \omega i_0 \cos \omega t$. Thus,

$$v_L = \omega L i_0 \cos \omega t = v_{0L} \cos \omega t \tag{26.12}$$

where

$$v_{0L} = (\omega L)i_0 \tag{26.13}$$

The plots of i and v_L versus t using Equations 26.10 and 26.12 shown in Figure 26.8 indicate that the voltage drop v_L across the inductor is not in phase with the current. The voltage drop v_L across a pure inductor is $\pi/2$ (or 90°) or $\frac{1}{4}$ cycle ahead of the current i. This means that when the current i is zero, v_L has its maximum value, as shown in Figure 26.8.

The relation between v_L, i, v_{0L}, and i_0 versus ωt may be illustrated by means of a rotor diagram, as shown in Figure 26.9. Unlike resistance, in the case of inductance the v_{0L} vector is drawn making an angle of 90° ahead of the current i_0. The instantaneous values of i and v_L are obtained by projecting i_0 and v_{0L} on the vertical axis as shown.

From Equation 26.12 the rms value of the instantaneous voltage v_L is

$$(v_L)_{\text{rms}} = \omega L(i_0 \cos \omega t)_{\text{rms}}$$

or

$$V_L = \omega L I \tag{26.14}$$

The quantity ωL behaves like a resistance and is called the *inductive reactance*, X_L:

$$X_L = \omega L = 2\pi f L \tag{26.15}$$

where the units of X_L are ohms, and unlike R its magnitude increases with increasing frequency f. Thus, if $f = 0$, that is, for dc through the circuit, v_L or V_L across the inductor will be zero. Using Equation 26.15, we may write Equation 26.14 as

$$V_L = X_L I \tag{26.16}$$

where V_L and I are the rms values. Equation 26.16 is the form of Ohm's law for an inductor in an ac circuit, with X_L taking the place of resistance. The reactance X_L increases with increasing inductance. An ideal inductor offers no resistance to the flow of direct currents, but the resistance to the alternating currents increases with increasing f.

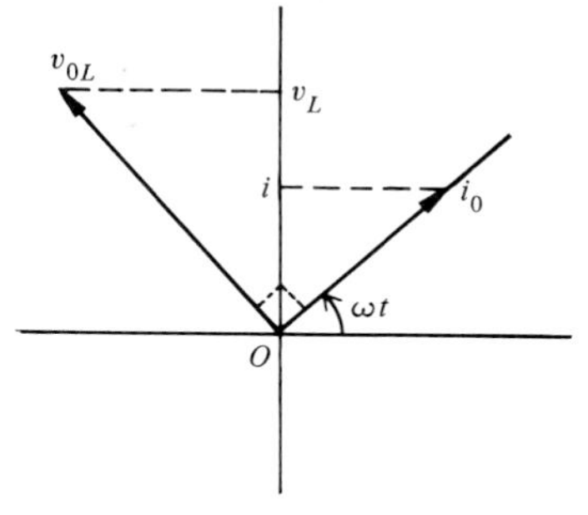

FIGURE 26.9. *Rotor diagram for current i_0 and voltage v_{0L} versus ωt.*

AC Circuit with Capacitance

An ac circuit containing a capacitor is shown in Figure 26.10. Let us say that the current through the circuit at any instant is given by

$$i = i_0 \sin \omega t \tag{26.17}$$

The current entering into one plate of the capacitor must be equal to the current leaving the other plate so that the charges on the two plates are equal and opposite. The instantaneous voltage v_C across the capacitor is

$$v_C = \frac{q}{C} \tag{26.18}$$

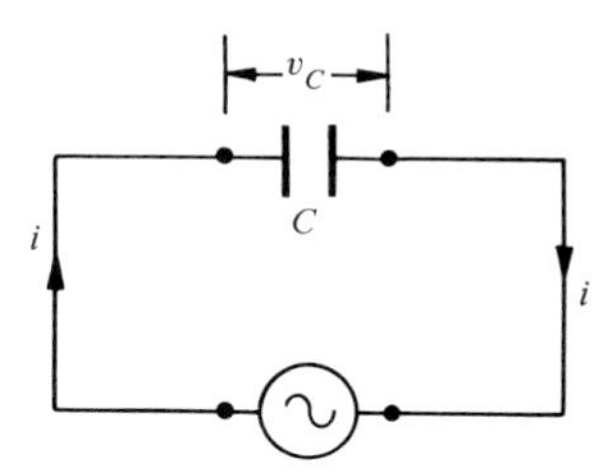

FIGURE 26.10. *Circuit containing a source of alternating current and a capacitor.*

where q is the instantaneous charge on the capacitor of capacitance C. To calculate v_C we must calculate q from Equation 26.17 by using the following relation:

$$\frac{\Delta q}{\Delta t} = i = i_0 \sin \omega t$$

This equation is satisfied (can be proven by using integral calculus in analogy with Equations 10.5 and 10.7) if we assume q to be

$$q = -\frac{1}{\omega} i_0 \cos \omega t$$

Substituting q in Equation 26.18,

$$v_C = -\frac{1}{\omega C} i_0 \cos \omega t = -v_{0C} \cos \omega t = v_{0C} \sin(\omega t - 90°) \tag{26.19}$$

where

$$v_{0C} = \left(\frac{1}{\omega C}\right) i_0 \tag{26.20}$$

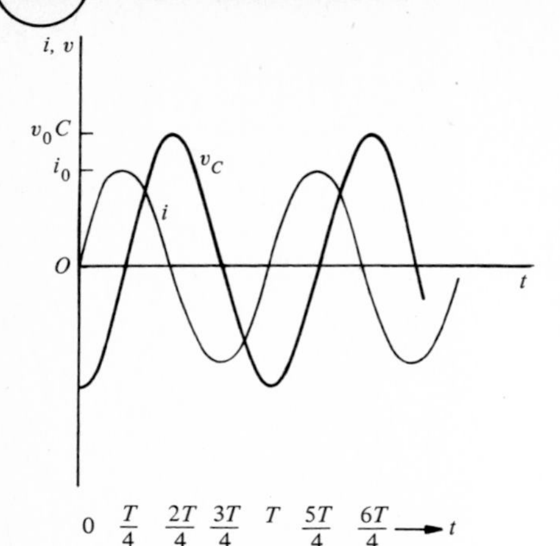

FIGURE 26.11. *Plots of v_C and i versus t, showing that v_C lags current i by $\pi/2$ (or 90°).*

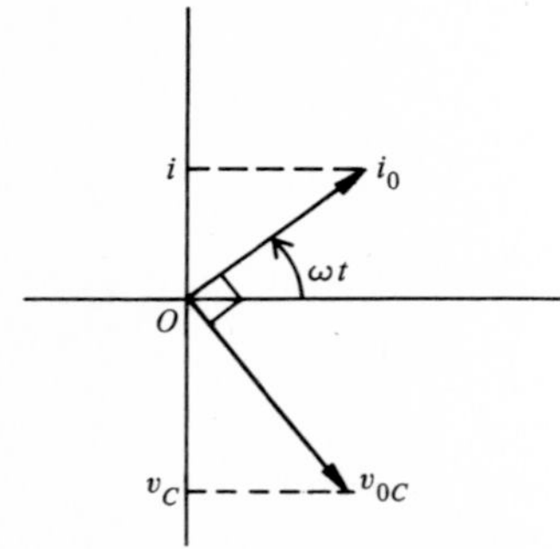

FIGURE 26.12. *Rotor diagram for current i_0 and voltage v_{0C} versus ωt.*

The plots of i and v_C versus t in Figure 26.11 indicate that the voltage v_C across the capacitor is not in phase with the current. The voltage v_C across the capacitor *lags* behind the current by $\pi/2$ (or 90°) or $\frac{1}{4}$ cycle as shown in Figure 26.11, which is the same as saying the current leads the voltage by $\pi/2$ (or 90°).

The relation between v_C, i, v_{0C}, and i_0 versus t may be illustrated by means of a rotor diagram, shown in Figure 26.12. Note the phase relationship between v_{0C} and i_0; that is, v_{0C} is drawn in such a way that it lags i_0 by 90° (or i_0 leads v_{0C} by 90°). Projections of these rotors along the vertical axis yield the instantaneous values.

As before, we can find the rms value of v_C by using Equation 26.19 or Equation 26.20

$$(v_C)_{\text{rms}} = \frac{1}{\omega C}(-i_0 \cos \omega t)_{\text{rms}}$$

or

$$V_C = \frac{1}{\omega C} I \tag{26.21}$$

The quantity $1/\omega C$ behaves like a resistance and is called the *capacitive reactance*, X_C:

$$\boxed{X_C = \frac{1}{\omega C} = \frac{1}{2\pi f C}} \tag{26.22}$$

where the units of X_C are ohms, and unlike R, X_C is not constant. It decreases with increasing f and C. Thus, for alternating currents of large frequencies, the capacitor is an easy bypass (i.e., the current goes through it easily) while for dc for which $f = 0$, it has infinite resistance; hence, no dc passes through a capacitor. Using Equation 26.22, we may write Equation 26.21 as

$$\boxed{V_C = X_C I} \tag{26.23}$$

where V_C and I are the rms values. Equation 26.23 is the form of Ohm's law for a capacitive ac circuit.

EXAMPLE 26.2 Calculate the resistance or reactance and the voltage across each of the following for an ac source with an effective current of 5 A and a frequency of 60 Hz, 600 Hz and 6000 Hz (a) a resistor of 10 Ω (b) an inductor of 1 H (c) a capacitor of 1 μF. Make plots of R, X_L and X_C versus ω.

(a) The resistance R is independent of frequency; hence, remains 10 Ω. From Equation 26.9 the voltage V_R also remains constant.

$$V_R = RI = (10\ \Omega)(5\ \text{A}) = 50\ \text{V}$$

(b) From Equation 26.15, $X_L = \omega L = 2\pi f L = 2\pi f(1\ \text{H})$, and from Equation 26.16, $V_L = IX_L = (5\ \text{A})X_L$. Therefore, for

$f = 60/\text{s}$	$600/\text{s}$	$6000/\text{s}$
$X_L = 377\ \Omega$	$3770\ \Omega$	$37{,}700\ \Omega$
$V_L = 1885\ \text{V}$	$18{,}850\ \text{V}$	$188{,}500\ \text{V}$

(c) From Equation 26.22,

$$X_C = \frac{1}{\omega C} = \frac{1}{2\pi f C} = \frac{1}{2\pi \times 10^{-6}\ \text{F}}\frac{1}{f}$$

and from Equation 26.23, $V_C = IX_C = (5\ \text{A})X_C$. Therefore, for

$f = 60/\text{s}$	$600/\text{s}$	$6000/\text{s}$
$X_C = 2650\ \Omega$	$265\ \Omega$	$26.5\ \Omega$
$V_C = 13{,}250\ \text{V}$	$1325\ \text{V}$	$132.5\ \text{V}$

The plots (not to scale) of R, X_L, and X_C versus ω are shown in the accompanying figure. Note that X_L increases with ω, X_C decreases, and R remains constant.

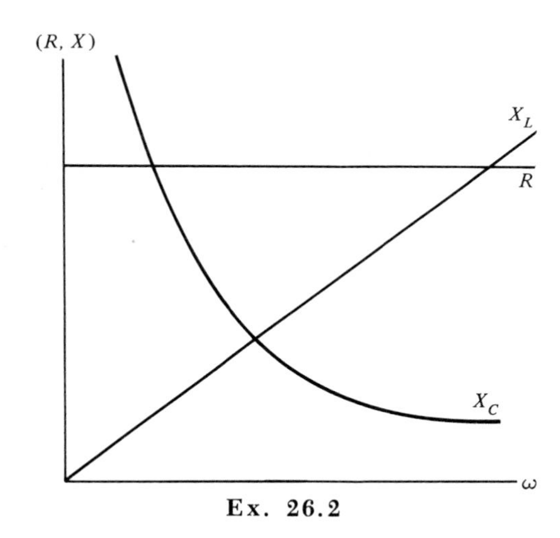

Ex. 26.2

Exercise 26.2 Repeat Example 26.2 for $f = 6/\pi$ Hz and $60{,}000/\pi$ Hz. From the data of this exercise and the example, you can find the value of f_0 for which $X_L = X_C$ (the point where the plot of X_L versus ω crosses X_C versus ω)? [*Ans.*: $f = 6/\pi\ \text{Hz} \to X_L = 12\ \Omega$, $V_L = 60\ \text{V}$, $X_C = 8.33 \times 10^4\ \Omega$, $V_C = 4.17 \times 10^5\ \text{V}$; $f = 6000/\pi\ \text{Hz} \to X_L = 1.2 \times 10^5\ \Omega$, $V_L = 6 \times 10^5\ \text{V}$, $X_C = 8.33\ \Omega$, $V_C = 41.7\ \text{V}$; $f_0 = 500/\pi\ \text{Hz} \approx 159\ \text{Hz}$.]

26.3. The *RLC* Series Circuit

A more general case of an ac circuit is the one that contains all three elements: resistance, inductance, and capacitance, as shown in Figure 26.13. Let the instantaneous current i through any point in the circuit be given by

$$i = i_0 \sin \omega t \tag{26.24}$$

v_R, v_L, and v_C are the instantaneous values of the voltages across R, L, and C, respectively. Applying Kirchhoff's loop rule to the closed circuit, we may write

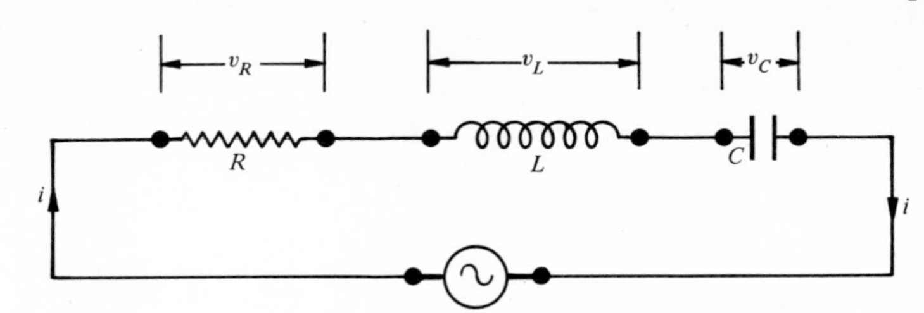

FIGURE 26.13. *AC circuit containing resistance R, inductance L, and capacitance C.*

$$v = v_R + v_L + v_C \tag{26.25}$$

which states that the instantaneous voltage v across the generator is equal to the sum of the instantaneous voltages across the three elements. Let v_0, v_{0R}, v_{0L}, and v_{0C} be the values of the maximum voltage drops across the source, R, L, and C, respectively. As we shall show, because of the phase differences between i and v, $v_0 \neq v_{0R} + v_{0L} + v_{0C}$ even though Equation 26.25 for instantaneous value does hold.

In order to find v as a function of t, we make use of the results discussed in the previous section. From Equations 26.7, 26.12, and 26.19,

$$v_R = v_{0R} \sin \omega t = i_0 R \sin \omega t$$

$$v_L = v_{0L} \cos \omega t = i_0 X_L \cos \omega t \qquad \text{where } X_L = \omega L$$

$$v_C = -v_{0C} \cos \omega t = -i_0 X_C \cos \omega t \qquad \text{where } X_C = \frac{1}{\omega C}$$

The plots of v_R, v_L, and v_C together with i are shown in Figure 26.14. Substituting these in Equations 26.25, we get

$$v = i_0[R \sin \omega t + (X_L - X_C) \cos \omega t] \tag{26.26}$$

or, after multiplying and dividing by $\sqrt{R^2 + (X_L - X_C)^2}$,

$$v = i_0 \sqrt{R^2 + (X_L - X_C)^2}\left[\frac{R}{\sqrt{R^2 + (X_L - X_C)^2}} \sin \omega t + \frac{X_L - X_C}{\sqrt{R^2 + (X_L - X_C)^2}} \cos \omega t\right] \tag{26.27}$$

Let

$$\cos \phi = \frac{R}{\sqrt{R^2 + (X_L - X_C)^2}} \quad \text{and} \quad \sin \phi = \frac{X_L - X_C}{\sqrt{R^2 + (X_L - X_C)^2}} \tag{26.28}$$

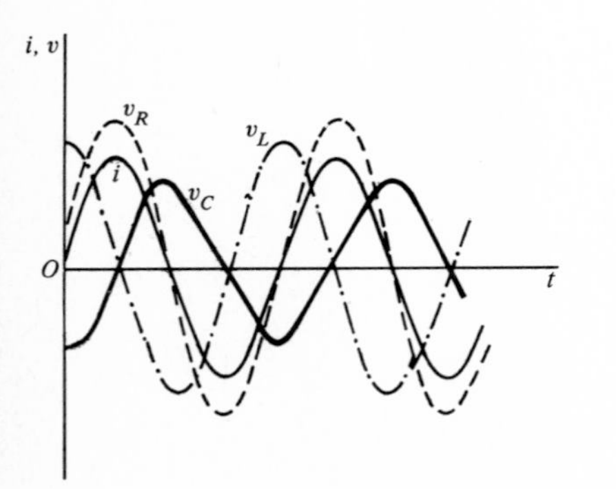

FIGURE 26.14. *Plots of i, v_R, v_L, and v_C versus t and ωt.*

Thus, Equation 26.27 takes the form

$$v = i_0 \sqrt{R^2 + (X_L - X_C)^2}\,(\sin \omega t \cos \phi + \cos \omega t \sin \phi)$$

or

$$v = i_0 \sqrt{R^2 + (X_L - X_C)^2} \sin (\omega t + \phi) \tag{26.29}$$

where, from Equation 26.28, dividing $\sin \phi$ by $\cos \phi$,

$$\boxed{\tan\phi = \frac{X_L - X_C}{R}} \tag{26.30}$$

Equation 26.29 may also be written as

$$\boxed{v = v_0 \sin(\omega t + \phi)} \tag{26.31}$$

where

$$\boxed{v_0 = i_0 Z} \tag{26.32}$$

and

$$\boxed{Z = \sqrt{R^2 + (X_L - X_C)^2}} \tag{26.33}$$

with $X_L = \omega L$ and $X_C = 1/\omega C$. In this general case, Z has taken the place of resistance and is called the *impedance* of the circuit.

Comparison of Equation 26.31 with Equation 26.24 indicates that voltage v and current i are out of phase by an angle ϕ. The value of ϕ is determined by Equation 26.30 and depends on the values of X_L, X_C, and R, that is, on the values of L, C, R, and f. If $X_L > X_C$, ϕ will be positive and the voltage leads the current. If $X_L < X_C$, ϕ is negative, hence current leads the voltage.

We can simplify the discussion by means of a rotor diagram as shown in Figure 26.15. We draw i_0 and v_{0R} in line, making an angle ωt with the horizontal line. v_{0L} is always 90° ahead of i_0 and v_{0C} lags behind i_0 by 90°, as shown. Since v_{0L} and v_{0C} are always in line, the resultant of these two is $v_{0L} - v_{0C}$, as shown. Hence, the resultant voltage v_0 may be found by the vector addition of v_{0R} and $(v_{0L} - v_{0C})$. The resultant v_0 leads i_0 by an angle ϕ if $v_{0L} > v_{0C}$ (or $X_L > X_C$). On the other hand, if $v_{0C} > v_{0L}$ (or $X_C > X_L$), v_0 will lag behind i_0 by an angle ϕ. The instantaneous values of i and v may be obtained by projecting i_0 and v_0 on the vertical axis. Thus, as we have shown in Figure 26.15,

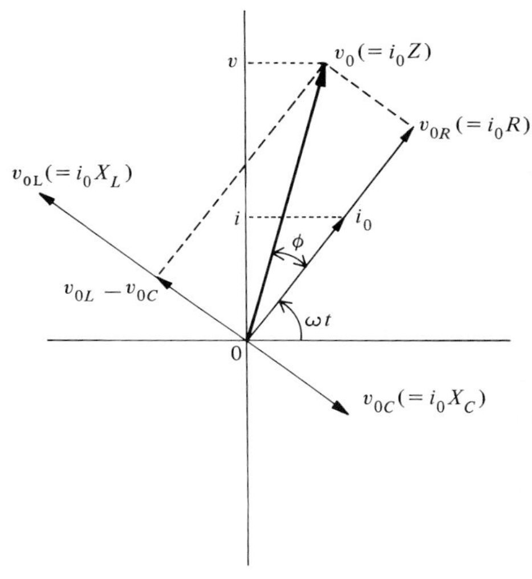

FIGURE 26.15. *Use of the rotor diagram to evaluate v, i, and ϕ for an RLC series circuit.*

$$v_0^2 = v_{0R}^2 + (v_{0L} - v_{0C})^2 \tag{26.34a}$$

or

$$v_0 = \sqrt{(i_0R)^2 + (i_0X_L - i_0X_C)^2} = i_0\sqrt{R^2 + (X_L - X_C)^2} = i_0 Z \tag{26.34b}$$

where

$$Z^2 = R^2 + (X_L - X_C)^2 \tag{26.35}$$

and

$$\tan\phi = \frac{v_{0L} - v_{0C}}{v_{0R}} = \frac{i_0(X_L - X_C)}{i_0 R} = \frac{X_L - X_C}{R} \tag{26.36}$$

Using these equations and Figure 26.15, we may express relations between v_0, v_{0R}, v_{0L}, and v_{0C} as in Figure 26.16(a), and between Z, R, X_L, and X_C as in Figure 26.16(b). Note that in Figure 26.16(a) we may replace v_0, v_{0R}, v_{0L}, and v_{0C} by their respective rms values V, V_R, V_L, and V_C.

To find the rms value of the voltage v, we start with Equations 26.29 and 26.33:

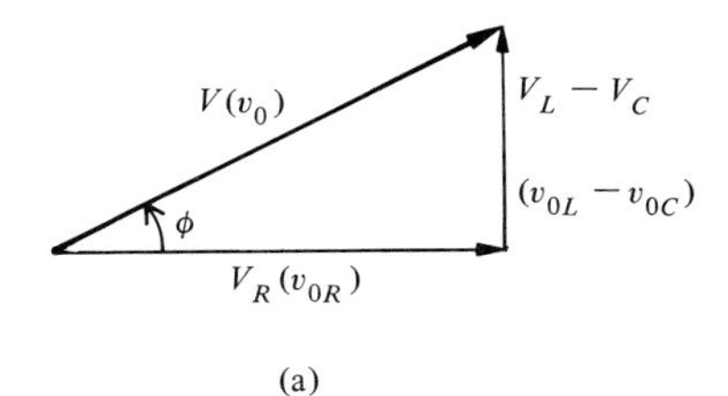

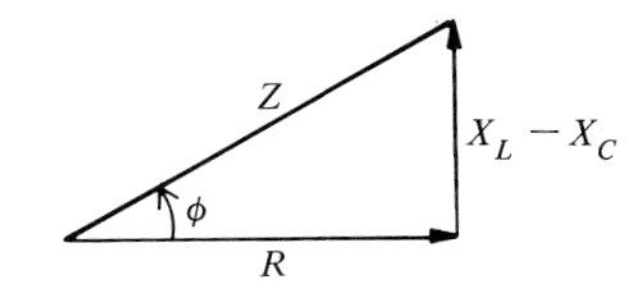

FIGURE 26.16. *(a) Relation between v_{0R}, v_{0L}, v_{0C}, v_0, and ϕ. (b) Relation between R, X_L, X_C, Z, and ϕ.*

$$v_{rms} = Z[i_0 \sin(\omega t + \phi)]_{rms}$$

Thus,

$$V = ZI \qquad \text{or} \qquad I = \frac{V}{Z} \tag{26.37}$$

where I and V are the rms values while Z, which has units of V/A or ohms, takes the place of the resistance. Thus, Equation 26.37 represents Ohm's law for this general case.

Finally, let us talk about the instantaneous power P given to the circuit. From Equations 26.24 and 26.31,

$$P = vi = v_0 \sin(\omega t + \phi) i_0 \sin \omega t \tag{26.38}$$

or

$$P = i_0 v_0 \sin^2 \omega t \cos \phi + i_0 v_0 \cos \omega t \sin \omega t \sin \phi$$

The average value of $\sin^2 \omega t$ is $\frac{1}{2}$, while that of $\sin \omega t \cos \omega t$ is zero. Therefore, the average power P_{av} is

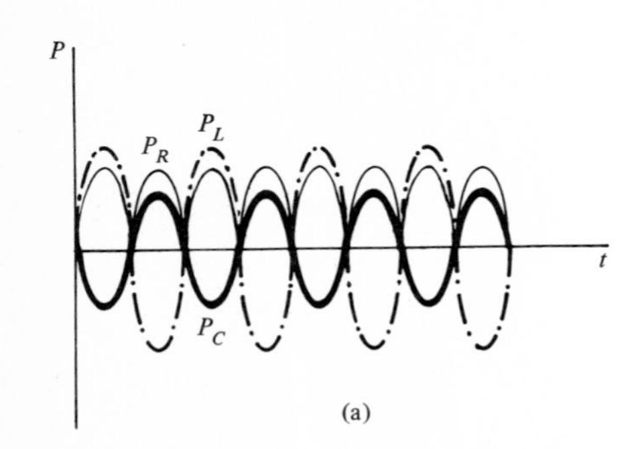

$$P_{av} = \frac{i_0 v_0}{2} \cos \phi = \frac{i_0}{\sqrt{2}} \frac{v_0}{\sqrt{2}} \cos \phi = i_{rms} v_{rms} \cos \phi$$

that is,

$$\boxed{P_{av} = IV \cos \phi} \tag{26.39}$$

Figure 26.17(a) shows the plots of power consumed in individual components of the circuit containing R, L, and C while Figure 26.17(b) shows the net instantaneous power P consumed in the circuit as a function of time, together with instantaneous values of i and v through the circuit. This figure also illustrates the phase angle ϕ between v and i; that is, v and i reach their maximum values ϕ degrees apart. From Figure 26.17(a) it is clear that the average values of P_L and P_C are zero. That is, the power consumed by L and C over any complete cycle is zero. Thus, Equation 26.39 must represent power consumed by the resistor only. That is, from Figure 26.16(a),

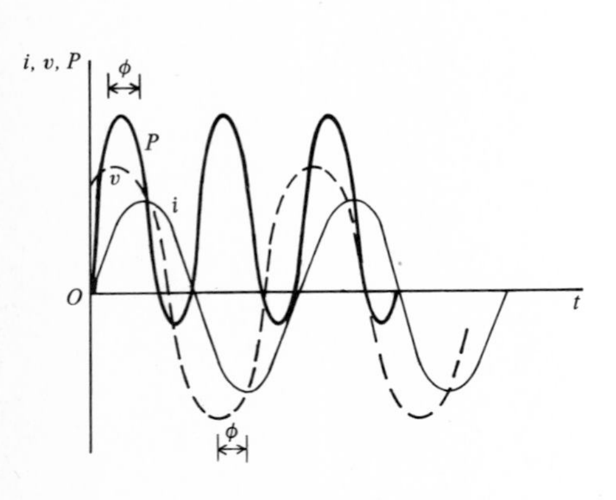

FIGURE 26.17. *(a) Plots of P_R, P_L, and P_C versus t in an RLC series circuit. (b) Plots of i, v, and P versus t in an RLC series circuit. Also illustrated is the phase angle ϕ between i and v.*

$$\cos \phi = \frac{v_{0R}}{v_0}$$

Substituting in Equation 26.39,

$$P_{av} = IV \cos \phi = \frac{i_0}{\sqrt{2}} \frac{v_0}{\sqrt{2}} \frac{v_{0R}}{v_0} = \frac{i_0}{\sqrt{2}} \frac{v_{0R}}{\sqrt{2}} = IV_R = I(IR)$$

That is,

$$P_{av} = I^2 R \tag{26.40}$$

as it should be. Thus, all the energy supplied to the circuit by the source is dissipated as heat energy in the resistor at the average rate of I^2R, where I is the rms value.

In Equation 26.39, $P_{av} = IV \cos \phi$, where $\cos \phi$ is called the *power factor*. For a pure resistance circuit $v_{0R} = v_0$ (also $Z = R$); hence, $\phi = 0$ and $P_{av} = IV$. On the other hand, if $R = 0$ and the circuit contains L, C, or both,

$\phi = 90°$, where $\cos\phi = 0$; hence, $P_{av} = 0$. Also, Equation 26.39 holds not only for a series circuit, but for any ac circuit.

Let us try to understand why $(P_L)_{av}$ and $(P_C)_{av}$ are zero. During a portion of a cycle when both i and v_L are in the same direction, the energy source supplies power to the inductor, thereby increasing the magnetic flux in the inductor. But during the part of a cycle when i and v_L are in opposite directions, the energy stored in the inductor is transferred back to the circuit. Thus, a pure inductor takes in and then gives back to the circuit equal amounts of energy alternatively. Energy goes into establishing a magnetic field, and then the magnetic field collapses and supplies back the energy.

In the case of a capacitor, during the charging the external source supplies energy to the capacitor which is stored in the form of an electric field. In the next half of the cycle the capacitor discharges and supplies back the energy to the circuit and the process continues. There is no energy loss in the form of heat for a pure inductor or a pure capacitor.

Example 26.3 Consider an RLC circuit with $R = 10\ \Omega$, $L = 0.1$ H, and $C = 50\ \mu$F, as shown in part(a) of the figure. The external ac voltage applied is 100 V and a frequency of 60 Hz. Calculate (a) the impedance across each component and the total impedance; (b), the current in the circuit; (c) voltage across each component; and (d) the phase angle.

(a) $X_L = \omega L = 2\pi f L = 2\pi \times (60/\text{s})(0.1\ \text{H}) = 37.7\ \Omega$

$$X_C = \frac{1}{\omega C} = \frac{1}{2\pi f C} = \frac{1}{2\pi(60/\text{s})(50 \times 10^{-6}\text{F})} = 53.1\ \Omega$$

$$Z = \sqrt{R^2 + (X_L - X_C)^2} = \sqrt{(10)^2 + (37.7 - 53.1)^2}\ \Omega = 18.36\ \Omega$$

(b) The current in the circuit is given by

$$I = \frac{V}{Z} = \frac{\mathcal{E}}{Z} = \frac{100\ \text{V}}{18.36\ \Omega} = 5.45\ \text{A}$$

(c)
$$V_{ab} = V_R = IR = 5.45\ \text{A} \times 10\ \Omega = 54.5\ \text{V}$$
$$V_{bc} = V_L = IX_L = 5.45\ \text{A} \times 37.7\ \Omega = 205.5\ \text{V}$$
$$V_{cd} = V_C = IX_C = 5.45\ \text{A} \times 53.1\ \Omega = 289.4\ \text{V}$$

(d) From Equation 26.36,

$$\tan\phi = \frac{V_L - V_C}{V_R} = \frac{X_L - X_C}{R} = \frac{37.7\ \Omega - 53.1\ \Omega}{10\ \Omega} = -1.54 \text{ or } \phi = -57°$$

The negative sign means that the voltage lags the current in the circuit. The rotor or phasor diagrams for the voltages and the impedances are shown in parts (b) and (c), respectively. Note that

$$\mathcal{E} = V_{ad} \neq V_{ab} + V_{bc} + V_{cd}$$

but

$$V_{ad} = \sqrt{V_{ab}^2 + (V_{bc} - V_{cd})^2} = \sqrt{(54.5)^2 + (205.5 - 289.4)^2}\ \text{V} = 100\ \text{V}$$

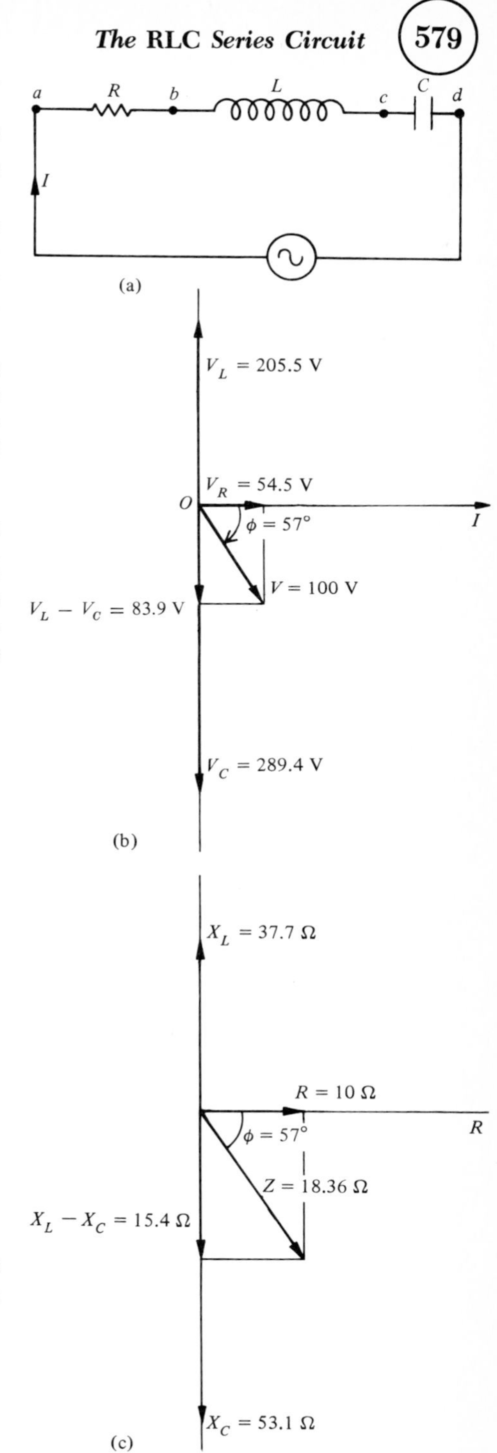

Ex. 26.3

Exercise 26.3 Repeat Example 26.3 if C is changed to 100 μF. [*Ans.:*

(a) $X_L = 37.7\ \Omega$, $X_C = 26.5\ \Omega$, $Z = 15.0\ \Omega$; (b) $I = 6.67$ A; (c) $V_R = 66.7$ V, $V_L = 251.5$ V, $V_C = 176.8$ V; (d) $\phi = 48.2°$.]

26.4. The *RLC* Series Resonance Circuit

In the previous section we discussed the relationship between the quantities i, v, R, X_L, X_C, Z, P, and ϕ in an ac circuit with a source of fixed angular frequency ω ($=2\pi f$). Let us now investigate these quantities as the frequency f of the source is varied continuously from almost zero to a very large value. According to Equation 26.37, the relation between the rms values I and V in an ac circuit is

$$I = \frac{V}{Z} = \frac{V}{\sqrt{R^2 + (X_L - X_C)^2}} \tag{26.41}$$

where $X_L = \omega L = 2\pi f L$ and $X_C = 1/\omega C = 1/2\pi f C$. Figure 26.18 shows the plots of R, X_L, X_C, and Z versus ω. As ω increases, so does X_L, but X_C decreases with increasing ω while R remains constant. Thus, the value of the impedance Z depends upon ω. According to Equation 26.41, the current I in the circuit will be maximum when Z is minimum. Since R is constant, Z will be minimum when X_L will be equal to X_C. Let us say that this happens at a frequency ω_0, as shown in Figure 26.18.

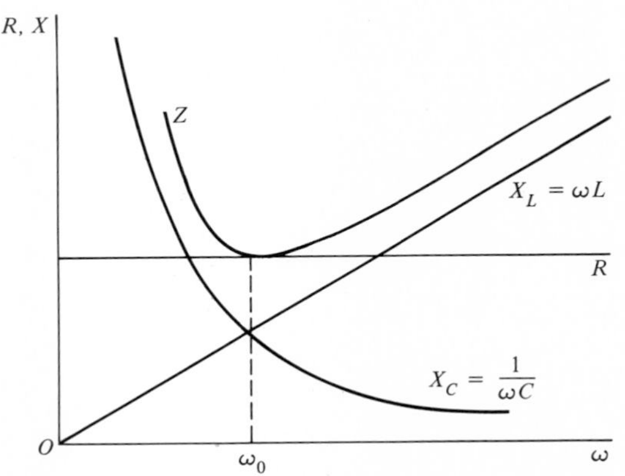

FIGURE 26.18. *R, X_L, X_C, and Z versus ω in an RLC series circuit.*

$$X_L = X_C \qquad \text{or} \qquad \omega_0 L = \frac{1}{\omega_0 C}$$

That is,

$$\boxed{\omega_0 = \sqrt{\frac{1}{LC}} \qquad \text{or} \qquad f_0 = \frac{1}{2\pi}\sqrt{\frac{1}{LC}}} \tag{26.42}$$

and the maximum current is

$$I_0 = \frac{V}{Z} = \frac{V}{R} \qquad \text{when } f = f_0 \tag{26.43}$$

The frequency f_0 is called the *resonant frequency*. At this frequency the current in the circuit is maximum, as given by Equation 26.43. (At f_0 the source transfers energy at a maximum rate, and the source and circuit are said to be in resonance.) The plots of I versus ω for different values of R are shown in Figure 26.19. No matter what the value of R, I is maximum when $\omega = \omega_0$. Also, when $R = 0$, I is infinitely large.

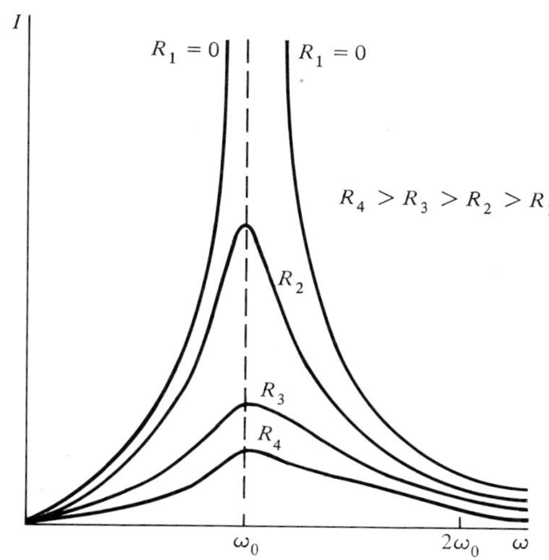

FIGURE 26.19. *Variation of current in an RLC series circuit as a function of frequency. Note that current is maximum at resonant frequency ω_0 when $X_L = X_C$.*

The electrical problem above is analogous to a mechanical system consisting of a mass tied at the end of a spring. When an external force is applied to such a system, the amplitude of the vibrations depends upon the frequency of the driving force, and results in curves similar to ones shown in Figure 26.19, as discussed in Chapter 10. Such a mechanical system resonates when the frequency of the driving force is equal to the natural frequency of the system; that is,

$$f_0 = \frac{1}{2\pi}\sqrt{\frac{k}{m}} \tag{26.44}$$

where m is the mass and k is the spring constant. Comparing Equation 26.44 with Equation 26.42 yields that L and $1/C$ in the electrical circuit are equivalent to m and k, respectively, in the mechanical system.

EXAMPLE 26.4 In Example 26.3 if the ac source has a variable frequency, at what frequency will resonance take place?

From Equation 26.42 at resonance $X_L = X_C$, or $2\pi fL = 1/2\pi fC$, or

$$f = \frac{1}{2\pi}\sqrt{\frac{1}{LC}} = \frac{1}{2\pi}\sqrt{\frac{1}{(0.1\ \text{H})(50 \times 10^{-6}\ \text{F})}} = 71.2\ \text{Hz}$$

EXERCISE 26.4 In Example 26.4, what will be the resonance frequency if (a) C is changed to 100 μF; (b) L is changed to 1 H; and (c) both C and L are changed to 100 μF and 1 H. [*Ans.*: (a) 50.3 Hz; (b) 22.5 Hz; (c) 15.9 Hz.]

26.5. Transformers

There are many situations where the available voltage of the ac source is either too high or too low. For example, ordinary appliances use 110 V ac, heavy appliances and industrial equipment use 220 V, and electric toy trains, transistor radios, and many other electrical units use much lower voltage, say from 6 to 24 V. A *transformer* is a device that changes the effective value of the emf. If the transformer changes the input voltage to a higher output voltage, it is called a *step-up transformer;* if the output voltage is smaller than the input voltage, it is called a *step-down transformer.*

A typical transformer as shown in Figure 26.20 consists of two coils, the primary and secondary coils, which are electrically insulated from each other

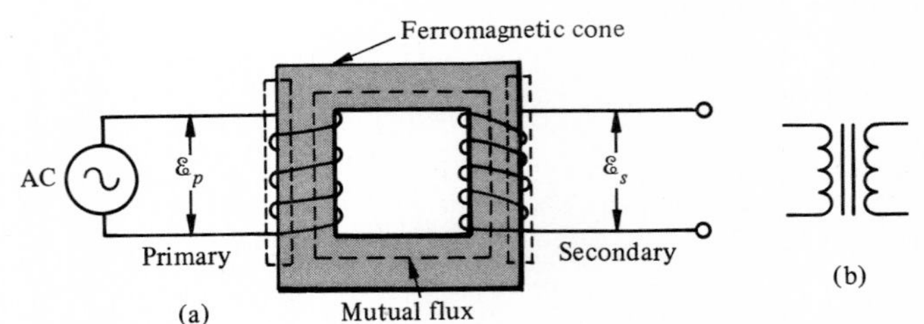

FIGURE 26.20. *(a) Schematic of a typical transformer and (b) its symbolic representation.*

and wound on the same iron core (ferromagnetic material). The working of this transformer is based on the principle that an emf is produced by a changing flux. Suppose that ac power is supplied to the primary coil. A varying current in this coil produces a varying magnetic flux confined to the ferromagnetic core. Thus, the secondary coil encloses a varying flux. Suppose that the rate of change of flux experienced by each of the N_s turns of the secondary coil is $\Delta\Phi/\Delta t$. Thus, the emf $\mathcal{E}_s$ produced in the secondary coil will be

$$\mathcal{E}_s = N_s \frac{\Delta\Phi}{\Delta t} \qquad (26.45a)$$

The magnetic flux and the rate of change of flux is the same through both the coils. Thus, if the primary has N_p turns, the emf $\mathcal{E}_p$ in the primary coil must be given by

$$\mathcal{E}_p = N_p \frac{\Delta\Phi}{\Delta t} \tag{26.45b}$$

From the two equations above

$$\frac{\mathcal{E}_s}{\mathcal{E}_p} = \frac{N_s}{N_p} \tag{26.46}$$

Thus, the ratio of the emf's in the secondary coil to the primary coil is equal to the ratio of their turns in the coils. If $N_s > N_p$, $\mathcal{E}_s > \mathcal{E}_p$ and it is a step-up transformer. If $N_s < N_p$, $\mathcal{E}_s < \mathcal{E}_p$; hence, it is a step-down transformer.

In ac circuits it is more meaningful to talk about the effective value of the voltage; hence, we replace $\mathcal{E}_s$ and $\mathcal{E}_p$ by V_s and V_p, respectively, in the equation above:

$$\boxed{\frac{V_s}{V_p} = \frac{N_s}{N_p}} \tag{26.47}$$

The power source in the primary circuit must be an ac source. If a dc source is inserted, there will be a changing flux in the core only for a very short interval when the primary circuit is closed or opened.

Most of the transformers are very efficient; that is, 90 to 99 percent. This means that the power input in the primary is equal to the power output at the secondary. Very little power is wasted. Thus, from the energy conservation principle, we may write

$$\text{power input } (P_p) = \text{power output } (P_s) \tag{26.48}$$

or

$$V_pI_p = V_sI_s \tag{26.49}$$

that is,

$$\boxed{\frac{V_s}{V_p} = \frac{I_p}{I_s}} \tag{26.50}$$

Combining with Equation 26.47,

$$\boxed{\frac{I_p}{I_s} = \frac{N_s}{N_p}} \tag{26.51}$$

That is, *the currents in the primary and secondary are inversely proportional to the number of turns*. Also, Equation 26.50 implies that in either coil, if the voltage increases, the current decreases, and vice versa, so that $P_p = P_s$.

Let us now give a few examples of the use of transformers. At a power-generating station, ac electricity is changed from low voltage–high current to high voltage–low current before sending it to various locations. This is necessary because every transmission line has some resistance. The Joule heating loss, which is equal to I^2R, will be much smaller if I is low, hence the advantage of transferring power at low current and high voltage. These wires, called *high-tension wires*, carry power which has a voltage of several hundred thousand. Before the power is supplied to a residential home, the voltage, for safety reasons, must be stepped down to 110 or 220 V by means of a step-down transformer. A voltage of 110 V is used for electric lamps, small appliances, and so on, while air-conditioning units, furnaces, and heavy appliances use 220 V. Small transistor radios, tape recorders, and calcultors use still smaller voltages, which are achieved by using step-down transformers, usually installed in the unit itself.

EXAMPLE 26.5 A power station has to transmit 10,000 W. It can be done either by sending 100 A at 100 V or 10 A at 1000 V. If the resistance of the transmission line is 0.25 Ω, calculate the power loss in each case. What transformer is needed to convert the lower voltage into the higher voltage?

When $I = 100$ A, the power loss by heating is

$$P_R = I^2R = (100\text{ A})^2(0.25\ \Omega) = 2500\text{ W}$$

When $I = 10$ A, the power loss by heating is

$$P_R = I^2R = (10\text{ A})^2(0.25\ \Omega) = 25\text{ W}$$

Thus, in the former case the generator will have to supply power 10,000 W + 2500 W = 12,500 W; while in the latter case, we have 10,000 W + 25 W = 10,025 W. There is much less loss by heating at low current and high voltage.

From Equation 26.47, where $V_s = 1000$ V and $V_p = 100$ V,

$$\frac{N_s}{N_p} = \frac{V_s}{V_p} = \frac{1000\text{ V}}{100\text{ V}} = 10$$

That is, the secondary of the transformer should have 10 times as many turns as the primary in order to convert 100 V into 1000 V before transmitting.

EXERCISE 26.5 A power station has to transmit 50 kW of power. It can be done either at 100 A and 500 V or 10 A and 5000 V. If the transmission line has a resistance of 0.2 Ω, calculate the power loss by heating in each case. Which one will you prefer and what transformer will you use at the generating station to convert a lower voltage to a higher one? [*Ans.: P* = 2000 W or 20 W; 10 A at 5000 V is preferred, $N_s/N_p = 10$, Step-up transformer.]

26.6. Vacuum-Tube Electronics

It is unnecessary to emphasize the importance of electronics in this modern age. All the different technologies—nuclear, space, energy, computer, aerospace, chemical, and communication—could not have reached such high perfection without electronics. The advantages to mankind from such technical development are numerous and obvious. In fact, we can say that "electronics" is now a household word.

The development of the field of electronics can be divided into two stages—vacuum-tube electronics and solid-state (or semiconductor) electronics.

Although solid-state electronic devices have replaced the majority of vacuum-tube devices, it is worthwhile to understand the basic principles involved in vacuum tubes. In this section we shall be limited to the discussion of two simple devices: (1): a diode vacuum tube as a rectifier, and (2) a triode vacuum tube as an amplifier. The solid-state devices will be discussed in Chapter 33.

Diode Vacuum Tube as Rectifier

As shown in Figure 26.21, a diode vacuum tube consists of two conductors, called *electrodes*, sealed into a tube. There is a very low vacuum inside the tube and leads from the two electrodes come out of the tube. One electrode, called the *cathode* (or *filament*) is heated either directly, as in Figure 26.21(a), or indirectly by means of a heater, as shown in Figure 26.21(b). The heating of the cathode results in the emission of electrons. These electrons are attracted by the second electrode, called the *anode* (or *plate*), which is maintained at a high positive potential. This flow of electrons from the cathode to the plate inside the vacuum tube leads to a current in the external circuit. Thus, these electrons in motion are the charge carriers.

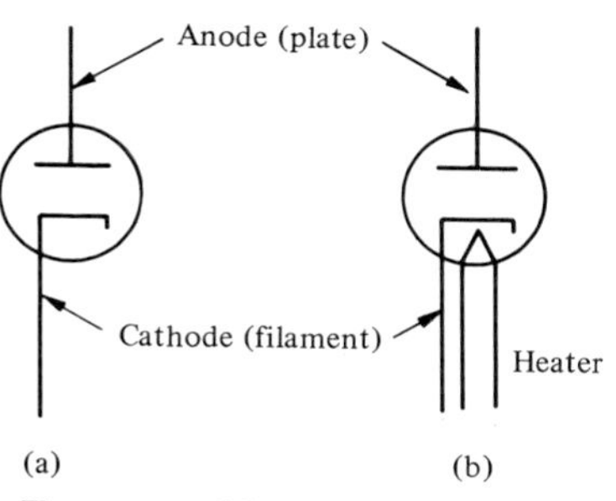

FIGURE 26.21. *Vacuum-tube diode: (a) directly heated and (b) indirectly heated.*

There are many electrical applications which require a dc source while the only thing available is an ac source. In such situations the alternating current must be changed into direct current or pulsating direct current. The process of converting ac into dc is called *rectification*, while a device that accomplishes this is called a *rectifier*.

Figure 26.22(a) shows a schematic of a rectifier that uses a diode vacuum tube. The ac input is such that in the first half of a cycle, the voltage is positive;

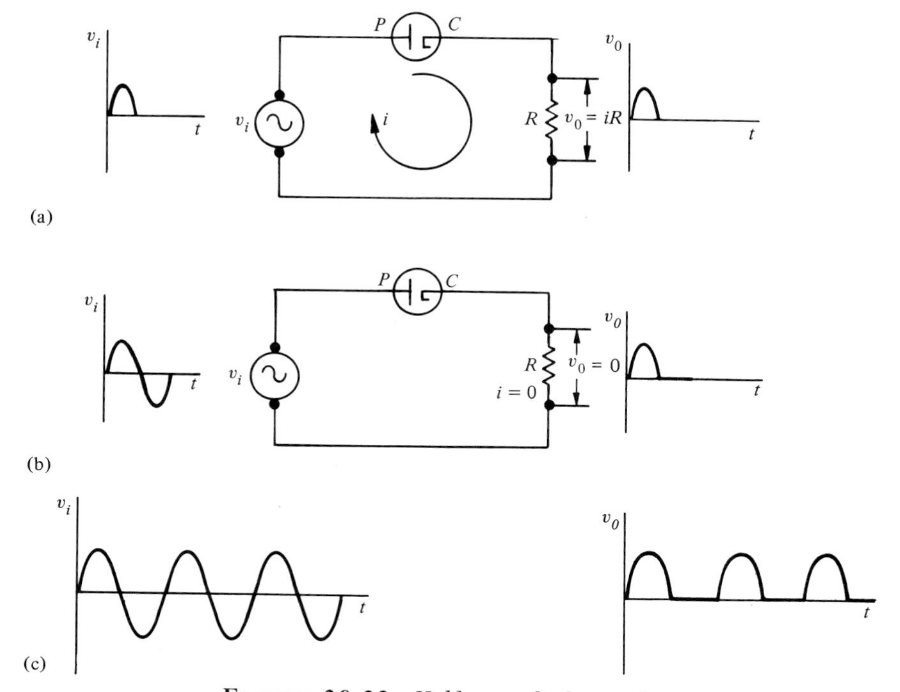

FIGURE 26.22. *Half-wave diode rectifier.*

hence, plate P is at a positive potential. This means that the electrons emitted from the cathode C (which has been heated indirectly by a heater, not shown) are attracted by plate P. Thus, there will be a flow of electrons in the external circuit, and we say that the tube is conducting. The conventional current direction is clockwise, as shown. Since there is a current i in the resistor R, there is a potential drop of $v_0 = iR$ across this resistor. The plot of the output voltage v_0 versus time t is as shown. The output voltage v_0 is an exact replica of the input voltage v_i.

In the next half of a cycle, the applied voltage v_i is negative and P is at a negative potential with respect to the cathode C. The electrons emitted by C are repelled by the plate P, and the tube is in a nonconducting stage and $i = 0$; hence, $v_0 = 0$, as shown in Figure 26.22(b). The tube becomes conducting once again in the next half-cycle and nonconducting in the next following half. This process continues and the output consists of a unidirectional pulsating voltage, as shown in Figure 26.22(c). Thus v_0 is always zero or positive and not negative. Such pulsating dc is rectified only for one-half of each cycle, and the device performing this act is called a *half-wave rectifier*.

In a half-wave rectifier, since the tube conducts only half of the time, the other half of the time the applied voltage is wasted. To avoid this loss, it is desirable to have a device called a *full-wave rectifier* (Figure 26.23). The transformer T is the source of an alternating emf and is tapped at the center as

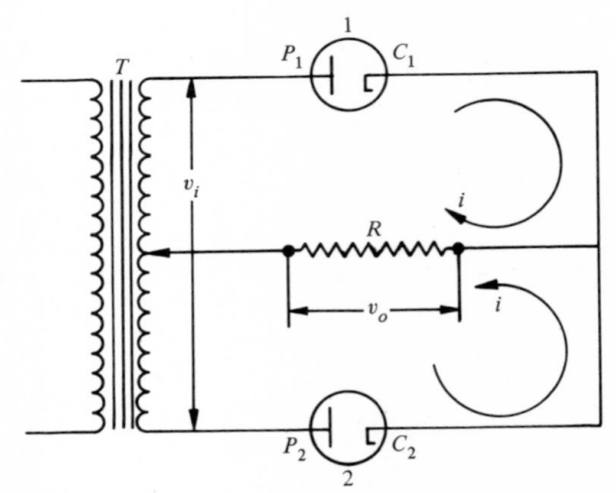

FIGURE 26.23. *Schematic of a full-wave rectifier.*

shown, with one diode in the upper half and another in the lower half. In this arrangement the upper end of the transformer is positive, while the lower end is negative. Hence, tube 1 is conducting and tube 2 is nonconducting. The current in the resistor R is from right to left. In the next half of the cycle, the lower end of the transformer is positive while the upper end is negative. Thus, tube 2 is conducting while tube 1 is nonconducting. The current in the resistor R is once again from right to left. Figure 26.24(a) shows the input voltage v_i while the output voltage v_0 $(=iR)$ across R is shown in Figure 26.24(b). As is clear, we have rectification during each half cycle—or full-wave rectification.

One of the shortcomings in the arrangement above is that the output is pulsating dc. If a much smoother output is needed, a device called a *filter* is

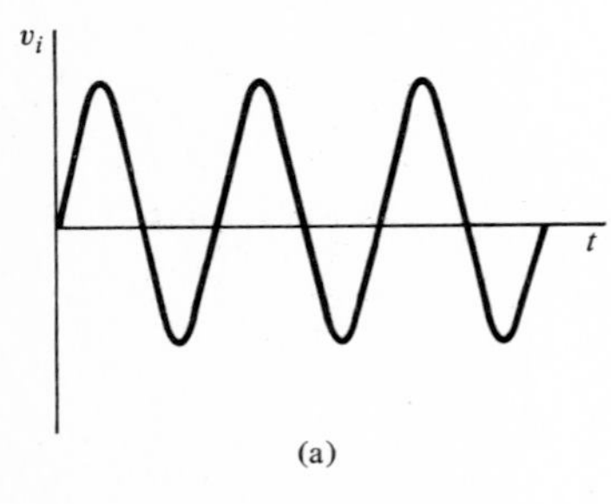

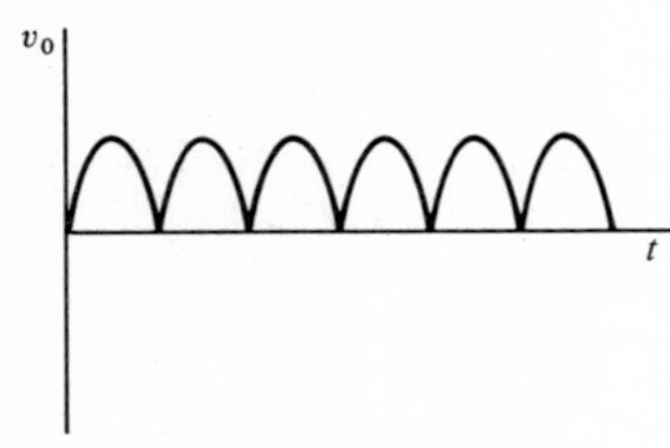

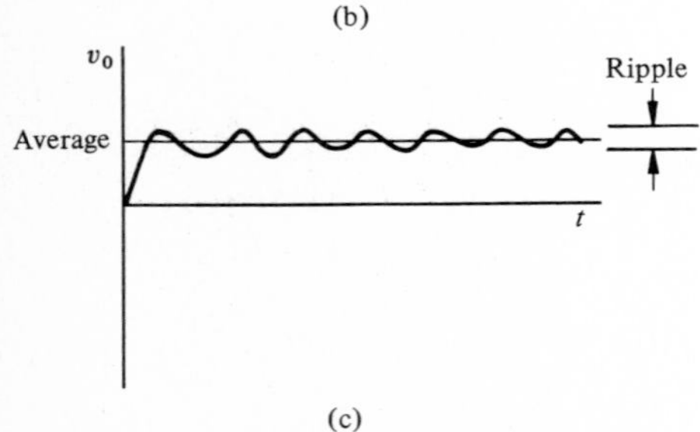

FIGURE 26.24. *Alternating-current (a) input, (b) output of a full-wave rectifier, and (c) filtered output.*

used. The simplest form of filtering is to put a capacitor in parallel with the resistor R in these devices. The result is the output waveform shown in Figure 26.24(c).

Triode Vacuum Tube as an Amplifier

As the name indicates, the *triode* is a thermionic tube with three electrodes sealed in it, as shown in Figure 26.25(a). The third electrode is called a *grid* and is placed closer to the cathode than to the plate. The grid usually consists of a mesh of cross-wires surrounding the cathode, as shown in Figure 26.25(b). As in a diode, the cathode in a triode may be heated indirectly by the heater H.

The plate P is maintained at a high, constant, positive voltage (with respect to the cathode, which is at zero potential) while the grid is supplied with a small negative voltage, v_g. The grid G plays a very important role by controlling the flow of electrons from the cathode C to the plate P. When $v_g = 0$, the triode behaves like a simple diode and the tube conducts with a constant current in the circuit. When a varying negative voltage is applied to the grid as shown in Figure 26.25(b), the resulting plate current is as shown in Figure 26.26 by the boldface parabolic curve. We see when v_g is very negative, the plate current I_p is zero. That is, the grid G repels all the electrons emitted from the cathode C, and none reach the plate P. As the grid voltage v_g is made less negative, the current I_p increases very rapidly, as shown in Figure 26.26. The I_p versus v_g curve is called the plate current–grid voltage characteristic of the triode. By making use of this characteristic, the triode is used as an amplifying device, as illustrated in Figures 26.26 and 26.27.

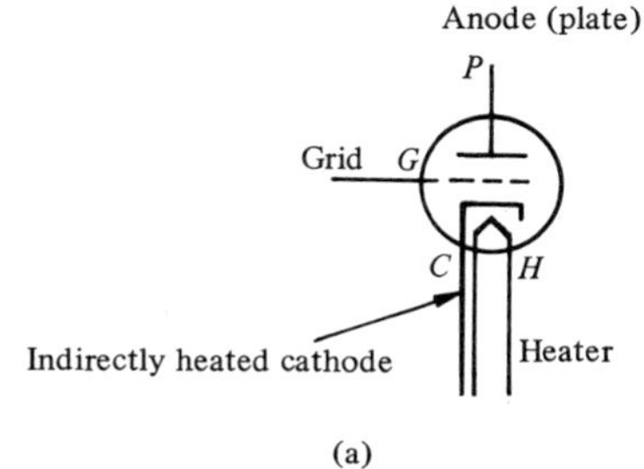

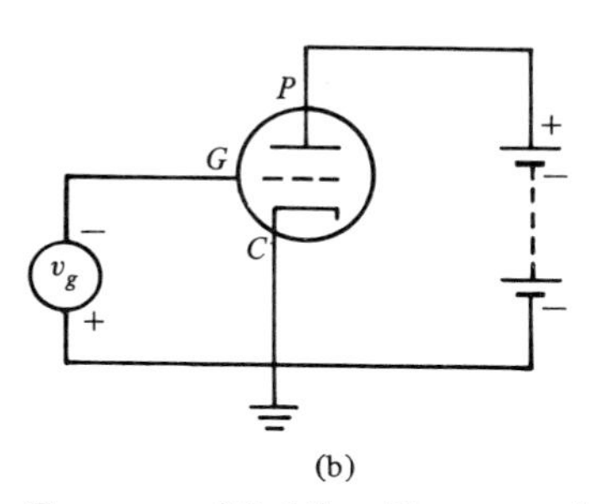

FIGURE 26.25. *Vacuum-tube triode: (a) schematic and (b) voltage connections.*

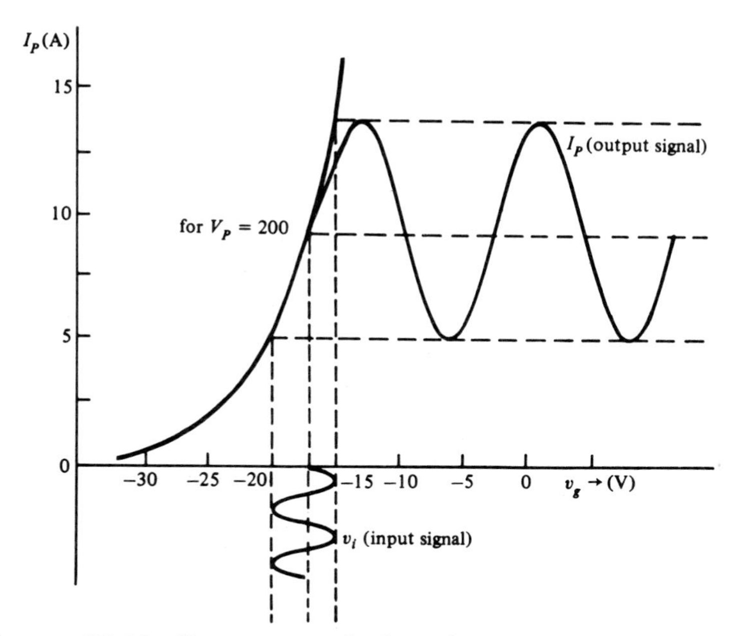

FIGURE 26.26. *Plate current-grid voltage characteristic of a triode, illustrating its use as an amplifier.*

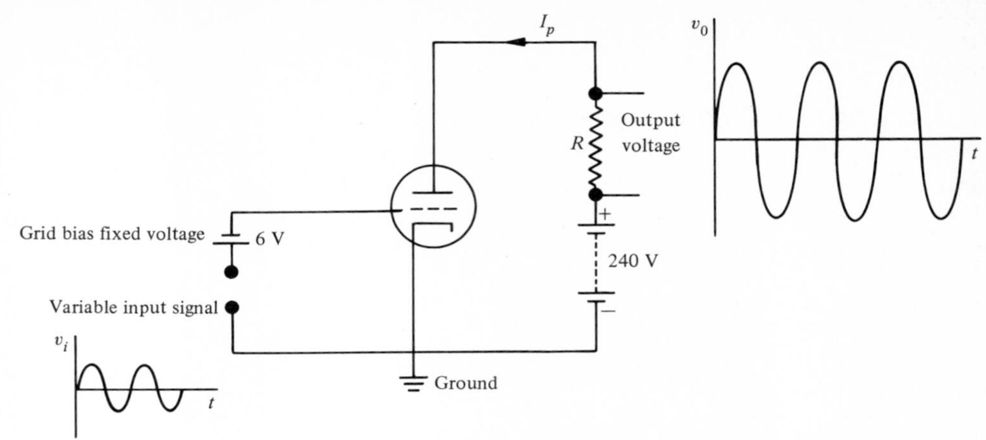

FIGURE 26.27. *Schematic of a triode used as an amplifier.*

Figure 26.27 shows an arrangement for using a triode as an amplifier. A weak input ac signal is applied at the grid. As the value of v_i varies, the current I_p in the circuit and in the resistor R also varies, as shown in Figure 26.27. The output voltage is $v_0 = I_pR$, that is, I_p is multiplied by the load resistor R to obtain v_0. The output is an amplified signal and the device is called a *voltage amplifier*. Depending upon the arrangement, v_0 may be as large as 2 or 3 times v_i or 1000 times v_i. The importance of the amplifier should be clear. The voltage-amplification circuits are used for signal amplification in radios, TV, and numerous other electronic devices.

26.7. The Cathode-Ray Oscilloscope

The *cathode-ray oscilloscope* (CRO) is an essential part of any small or large electronic and electromagnetic laboratory. Because of its usefulness, we shall describe it briefly. As shown in Figure 26.28, the cathode-ray tube (CRT) essentially consists of three parts: an electron gun, deflecting fields, and a fluorescent screen. The entire tube is highly evacuated.

Actually, the electron gun consists of a heater, cathode, control grid, focusing anode, and accelerating anode. The cathode, heated by a heater, emits electrons. Because the anode is maintained at a high positive potential with respect to the cathode the electrons are accelerated toward the anode, and come out as a well-defined beam through a hole in the anode striking the fluorescent screen. The *control grid* regulates the number of electrons that reach the anode, hence the brightness of the spot on the screen. The function of the *focusing electrode* is to make sure that all those electrons leaving the cathode in slightly different directions are brought back in the beam (or eliminated) so that they will hit the screen at the same spot.

A sharp, well-defined, and controlled beam coming out of the anode passes between two sets of plates provided with an appropriate potential difference. The first set provides the vertical deflection of the beam, while the second set provides the horizontal deflection. This deflected beam is received on the screen as shown. If the deflecting plates are provided with a periodically varying potential difference, the electric field between the plates changes

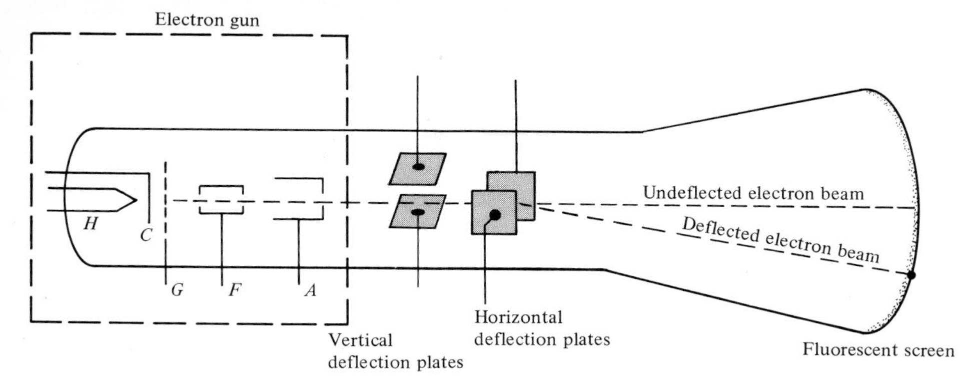

FIGURE 26.28. *Schematic of a typical oscilloscope.*

accordingly, which then affects the electron beam passing through the plates. Thus, when the beam strikes the screen, the spot on the screen moves up and down, and back and forth, representing the changes in potential that were applied to the plates. The intensity of the spot, as mentioned earlier, is controlled by the controlling grid.

This principle of the cathode-ray oscilloscope or tube (CRT) is utilized in television tubes as well, except that magnetic deflecting forces are used instead of electric forces.

SUMMARY

A typical ac voltage and current are represented by $v = v_0 \sin \omega t$ and $i = i_0 \sin \omega t$, where $\omega = 2\pi f$. The *rms* or *effective* values are given by $I = i_{\text{rms}} = i_0/\sqrt{2} = 0.707 i_0$ and $V = v_{\text{rms}} = v_0/\sqrt{2} = 0.707 v_0$, while the *average power* is $P_{\text{av}} = IV = I^2R$.

For an ac circuit with resistance, the current and voltage are always *in phase* and $V_R = RI$. For an ac circuit with inductance, the voltage v_L across a pure inductor is always 90° *ahead* of the current, and $V_L = X_L I$, where $X_L = \omega L$ is the *inductive reactance*. For an ac circuit with a capacitor, the voltage across the capacitor always *lags* behind the current by 90° and $V_C = X_C I$, where $X_C = 1/\omega C$ is the *capacitive reactance*.

For an *RCL* series circuit, $i = i_0 \sin \omega t$ and $v = v_0 \sin(\omega t + \phi)$, where $v_0 = i_0 Z$ with $Z = \sqrt{R^2 + (X_L - X_C)^2}$ is the *impedance* and ϕ is the *phase angle* given by $\tan \phi = (X_L - X_C)/R$. Also, we may write *Ohm's law* for an ac circuit as $I = V/Z$.

The *average power* is $P_{\text{av}} = IV \cos \phi = IV_R = I^2R$, where $\cos \phi$ is called the *power factor*. For a series resonance $X_L = X_C$ and $f_0 = (1/2\pi)\sqrt{1/LC}$, where f_0 is the *resonant frequency*.

The *transformer* can be step-up or step-down. In any transformer, power input = power output, $V_p I_p = V_s I_s$, and $V_s/V_p = I_p/I_s = N_s/N_p$; that is, the currents in the primary and secondary are inversely proportional to the number of turns.

A *diode* vacuum tube consists of two conductors, a cathode and an anode, sealed into a tube. The process of converting ac into dc is called *rectification,* while a device that accomplishes this is called a *rectifier,* such as a diode. A triode is a thermionic tube with three electrodes—cathode, grid, and anode. A triode may be used as a signal amplifier. The *cathode-ray oscilloscope* consists of an electron gun, deflecting fields, and a fluorescent screen.

QUESTIONS

1. Give some examples of the use of ac sources.
2. List some advantages of using ac sources over dc sources.
3. Is it possible to have a pure inductor so that there will be no energy loss in the form of heat?
4. Is it possible to have a pure capacitor? If there is an energy loss, in what form does it happen?
5. What advantages or disadvantages does a 220-V 50-Hz ac source have over a 110-V 60-Hz ac source?
6. What principle is utilized in making ac measuring instruments? Can you suggest some other way of doing this?
7. How does a decrease or increase in the area of the parallel plates of a capacitor affect the frequency of an *LC* circuit in a radio receiving set?
8. Give three examples of transformers that are in use at your place of residence or work.
9. What would happen if the primary of a transformer is connected to a dc source?
10. Is it necessary to use the linear portion of a triode characteristic when using a triode as an amplifier as shown in Figure 26.26?
11. Is it possible to use a triode as a rectifier and a diode as an amplifier?

PROBLEMS

1. A sinusoidal ac source has a rms value of 110 V and a frequency of 60 Hz. Calculate the instantaneous value of the voltage at $t = 2/600$, $4/600$, $5/600$, $6/600$, $8/600$, and $10/600$ s after it passes through zero.
2. A sinusoidal ac source has a rms value of 10 A and a frequency of 300 Hz. Calculate the instantaneous value of the current at $t = 2/3000$, $4/3000$, $5/3000$, $6/3000$, $8/3000$, and $10/3000$ s after it passes through zero.
3. A 12-Ω resistor is connected to an ac source of 60 Hz that has a voltage (rms) of 110 V. Calculate the maximum voltage, maximum current, the rms current, and the total power dissipated in the resistor.
4. A 20-Ω resistor is connected across an ac source that has an amplitude of 160 V and a frequency of 60 Hz. Calculate the rms voltage, rms current, and the power loss by heating.
5. When an alternating current passes through a 15-Ω resistor, the power loss by heat is 225 W. What are the rms and maximum values of current and voltage? What dc current will produce the same loss as this ac?
6. An ac generating power plant generates 50,000 W and transmits it through

a power line of 1.5 Ω resistance. Find the power loss in the line if V is (a) 1000 V; and (b) 5000 V.

7. Calculate the reactance of a 2 H inductor at a frequency of (a) 50 Hz; (b) 200 Hz; and (c) 800 Hz.
8. Calculate the frequency at which the reactance of a 0.6 H inductor will be (a) 18 Ω; (b) 180 Ω; and (c) 1800 Ω.
9. Calculate the reactance of a 1 μF capacitor at a frequency of (a) 50 Hz; (b) 200 Hz; and (c) 800 Hz.
10. Calculate the frequency at which the reactance of a 2 μF capacitor will be (a) 0.1 Ω; (b) 10 Ω; and (c) 100 Ω.
11. Calculate the reactance of a 1 H inductor and 1 μF capacitor at a frequency of 60 Hz.
12. A 0.1 H inductor is connected across an ac source that has an rms value of 110 V and a frequency of 60 Hz. What is the current amplitude in this case?
13. A 1 μF capacitor is connected across an ac source that has an rms value of 110 V and a frequency of 60 Hz. What is the current amplitude in this case?
14. An inductor of 0.2 H and a 20 Ω resistor are connected in series with an ac source of 110 V and of frequency 60 Hz. Calculate the impedance, current, and the potential difference across the resistor and the inductor. Draw a rotor (or phasor) diagram.
15. A capacitor of 1 μF and a 50 Ω resistor are connected in series with an ac source of 110 V and 60 Hz. Calculate the impedance, current, and the potential difference across the resistor and the capacitor. Draw the rotor diagram.
16. An inductor of 0.5 H and a capacitor of 10 μF are connected in series with an ac source of 110 V and 60 Hz. Calculate the impedance, current, and the potential difference across the inductor and the capacitor. Draw the rotor (or phasor) diagram.
17. An inductor of 1.0 H which has a resistance of 2 Ω and another resistor of 50 Ω are connected in series with an ac source of 220 V and 50 Hz. Calculate the impedance, current, and the potential difference across the inductor (including its resistance) and the resistor. Draw the rotor (or phasor) diagram.
18. Consider an *RLC* series circuit with $R = 100\ \Omega$, $L = 0.8$ H, and $C = 20\ \mu$F and an ac source of 220 V, 50 Hz. Calculate (a) the impedance across each component and the total impedance; (b) the current in the circuit; (c) the voltage across each component; and (d) the phase angle. Also, draw the rotor diagram for the impedance and the voltage.
19. Repeat Problem 18 if the frequency of the ac source is 500 Hz.
20. Repeat Problem 18 if $R = 10\ \Omega$, $L = 0.1$ H, and $C = 1\ \mu$F. The ac source is 50 V, 200 Hz. Also, calculate the power factor.
21. A 20 Ω resistor and an inductor are connected in series with an ac source of 110 V, 60 Hz. If the current in the circuit is 2 A, calculate the magnitude of the inductor and phase angle between the current and the voltage. Calculate the power dissipated in the resistor. What is the power factor?
22. A 50 Ω resistor and 5 μF capacitor are connected to an ac source of 50 V, 500 Hz. Calculate the impedance, current, and the phase angle. Calculate the power dissipated in the resistor and the power factor.

23. A 100 μF capacitor and a 0.1 H inductor are connected in series with an ac source of variable frequency. At what frequency will resonance take place?
24. A 10 μF capacitor and an inductor are connected to an ac source of frequency 2000 Hz. What is the value of the inductor to produce a series resonance with the capacitor?
25. An FM radio station broadcasts at a frequency of 100 MHz (1 MHz = 10^6 Hz). Calculate the value of the capacitance needed to produce a resonance with an inductor of 0.01 H.
26. An AM radio station broadcasts at a frequency of 1600 kHz (1 kHz = 10^3 Hz). Calculate the value of the capacitance needed to produce a resonance with the inductor of 0.01 H.
27. The primary of a certain transformer has 100 turns while the secondary has 1200 turns. If the voltage across the primary is 5 V and the current is 1 A, what will be the voltage and current in the secondary?
28. The primary of a transformer has 20 turns while the secondary has 1000 turns. If the voltage across the primary is 3 V, what is the voltage across the secondary? If the secondary has a resistance of 0.5 Ω, what is the current in the primary and secondary circuits?
29. In Problem 28 calculate the power in the primary and secondary.
30. A voltage of 110 V ac is supplied to a transistor radio which needs only 10 V input. What type of transformer will you use?
31. A voltage of 120 V ac is supplied to a calculator which uses only 24 V. What type of transformer will you recommend? If the resistance of the calculator is 10 Ω, what is the current in the secondary and primary circuits of the calculator?
32. A power station is generating electricity at 100 V and 500 A. Before transmitting, the voltage must be raised to 5000 V. What type of transformer will you use? If the resistance of the transmission line is 0.5 Ω, what will be the net power delivered at the receiving station? How much loss would have occurred due to heating if the power was transmitted at 100 V?
33. The 2200 V ac received in homes must be changed to 110 V at 20 A. What type of transformer will you recommend for this purpose? If the circuit has a resistance of 0.05 Ω, how much power is lost in heating?
34. A radio signal, after going through a triode amplifier, has an amplitude of 0.1 V. If the amplification factor for the triode is 50, what is the amplitude of the radio signal before amplification?
35. The amplitude of a radio signal received is 1 mV. A triode has an amplification of 20. How many triodes will have to be used to produce an amplified signal of 0.2 V?

27 Electromagnetic Waves

A theory describing the relation between charges, currents, and electric and magnetic fields was given by James Clerk Maxwell in 1864 in the form of four equations, called the Maxwell equations, which constitute classical electromagnetic theory. This theory predicts that accelerated charges produce electromagnetic waves, which travel at the speed of light. It also indicates that light waves are electromagnetic. In 1887 Heinrich Hertz produced and detected these waves. Just like other waves, electromagnetic waves transport energy. Besides discussing these concepts, we shall look at the whole range of electromagnetic waves. Finally, methods for measuring the speed of light will be outlined.

27.1. Displacement Currents

ACCORDING to James Clerk Maxwell (1831–1879) electric and magnetic fields are interrelated. Furthermore, equations describing electric and magnetic fields should be identical and symmetrical. It was this pursuance of symmetry that led Maxwell to introduce the concept of displacement currents. Before going into Maxwell's field equations, let us briefly discuss displacement currents.

Consider a circuit shown in Figure 27.1(a), containing a capacitor C. The current through the external circuit can be varied by changing the position of the sliding contact P of a variable resistance R. A_1 and A_2 are the two current-measuring devices. If P is stationary at one point, no current is indicated by A_1 and A_2. Once the slider is in motion, A_1 and A_2 indicate currents that are equal. One would think that because the capacitor is an insulator, there will be no current through it. On the other hand, as the slider moves, the capacitor is being charged, and the electric field **E** between the plates of the capacitor increases continuously. This changing electric field strength $\Delta E/\Delta t$ will be equivalent to some sort of current through the capacitor and is called the *Maxwell displacement current*, i_D. Also, the displacement current i_D must be equal to the current i in the external circuit. Because of the current through the wire, there is a magnetic field surrounding the wire. Even though there is no real current across the gap of the capacitor, the magnetic field **B** around the capacitor plates is the same as that around the conducting wires, as shown in Figure 27.1(b).

The relation between the displacement current and $\Delta E/\Delta t$ may be easily determined. Let the charge on the plates of the capacitor at any instant be Q, the area of the plates be A, and the permittivity be ϵ. Since $C = \epsilon_0 A/d$ and $V = Ed$,

$$Q = CV = \left(\frac{\epsilon_0 A}{d}\right)(Ed) = \epsilon_0 AE \qquad (27.1)$$

or

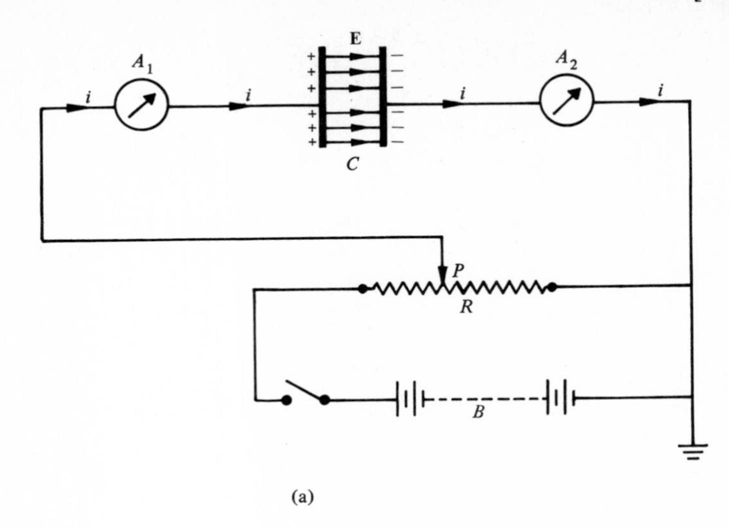

(a)

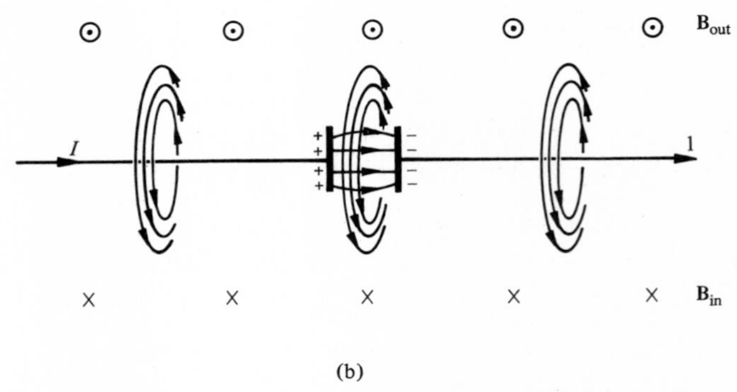

(b)

FIGURE 27.1. *(a) Experimental arrangement for establishing the concept of displacement currents. (b) Equivalent representation of part (a). The changing electric field between the plates of a capacitor produces a magnetic field.*

$$E = \frac{Q}{\epsilon_0 A} \tag{27.2}$$

and

$$\frac{\Delta E}{\Delta t} = \frac{1}{\epsilon_0 A}\frac{\Delta Q}{\Delta t} = \frac{i}{\epsilon_0 A} \tag{27.3}$$

where $i = \Delta Q/\Delta t$. Denoting $i = i_D$, the displacement current is

$$\boxed{i_D = \epsilon_0 A \frac{\Delta E}{\Delta t}} \tag{27.4}$$

EXAMPLE 27.1 A parallel-plate capacitor consists of two circular plates with radius $R = 10$ cm separated by $d = 0.5$ mm. The capacitor is being charged at a uniform rate by applying a changing potential difference between the two plates. Calculate the displacement current for the capacitor. Assume that the field is due to the displacement current only, and that the rate at which the electric field between the plates changes is 5×10^{13} V/m-s.

Since $E = V/d$, Equation 27.4 takes the form

$$i_D = \epsilon_0 A \frac{\Delta E}{\Delta t} = \frac{\epsilon_0 A}{d} \frac{\Delta V}{\Delta t} \tag{i}$$

We may use either result to calculate i_D. Substituting the values of $\epsilon_0 = 8.85 \times 10^{-12}$ C^2/N-m^2, $\Delta E/\Delta t = 5 \times 10^{13}$ V/m-s, and $A = \pi R^2 = \pi(0.1 \text{ m})^2$ in Equation i, we obtain

$$i_D = (8.85 \times 10^{-12} \text{ C}^2/\text{N-m}^2)(\pi \times (0.1 \text{ m})^2)(5 \times 10^{13} \text{ V/m-s}) = 13.9 \text{ A}$$

EXERCISE 27.1 A parallel capacitor consists of two circular plates each of radius $R = 2$ cm separated by a distance $d = 0.1$ mm. Calculate the displacement current i_D, if $\Delta V/\Delta t = 5 \times 10^{13}$ V/s. [*Ans.*: $i_D = 5.56 \times 10^3$ A.]

*27.2. Electromagnetic Field Equations

Maxwell's aim was to put all the laws and the equations describing electric and magnetic fields and their interactions with charges and currents into one unified set of equations. His investigations led to a set of four fundamental equations called the *electromagnetic field equations* or *Maxwell field equations*. These equations had many important consequences and predictions, as we shall describe shortly.

The first of these is Gauss's law for electrostatics, which states the relation between the electric field and the charges that product it. According to Equation 19.13, the electric flux $\Delta\Phi$ through an area ΔA where the electric field is E, is

$$\Delta\Phi = \mathbf{E} \cdot \Delta\mathbf{A} = E\,\Delta A \cos\theta \tag{27.5}$$

Suppose that the net charge q is enclosed by an arbitrary volume of surface area A. Let this area be divided into small elements of area. Let $\mathbf{E}_1, \mathbf{E}_2, \ldots$ be the electric intensities at areas $\Delta\mathbf{A}_1, \Delta\mathbf{A}_2, \ldots$ and $\theta_1, \theta_2, \ldots$ be the angles between $\mathbf{E}_1$ and $\Delta\mathbf{A}_1$, $\mathbf{E}_2$ and $\Delta\mathbf{A}_2, \ldots$, respectively. Thus, the total electric flux Φ_E through this arbitrary area is

$$\Phi_E = E_1\,\Delta A_1 \cos\theta_1 + E_2\,\Delta A_2 \cos\theta_2 + \cdots = \sum \mathbf{E} \cdot \Delta\mathbf{A}$$

According to *Gauss's law for electrostatics*, the total flux through a closed surface is equal to the net charge enclosed by the surface divided by the permittivity ϵ_0, that is, the first Maxwell equation is given by

$$\boxed{\Phi_E = \sum \mathbf{E} \cdot \Delta \mathbf{A} = q/\epsilon_0} \tag{27.6}$$

In particular, if $q = 0$, $\Phi = 0$. Also, Coulomb's law is a special case of Gauss's law. For example, suppose that a charge q is placed at the center of a sphere of radius r. From the symmetry of the spherical situation $\Sigma\, \mathbf{E} \cdot \Delta \mathbf{A} = \Sigma\, E\, \Delta A \cos 0° = E\, \Sigma\, \Delta A = E(4\pi r^2) = q/\epsilon_0$ or $E = (1/4\pi\epsilon_0)(q/r^2)$. The force F on a charge Q on the surface of the sphere will be $F = QE = (1/4\pi\epsilon_0)(qQ/r^2)$, which is Coulomb's law.

The second Maxwell equation is Gauss's law for magnetism. This equation can be written in analogy to Equation 27.6. We replace E by B, Φ_E by Φ_B. Since a magnetic monopole does not exist, any enclosed volume will contain equal and opposite magnetic poles, leading to a net magnetic pole strength zero, hence reducing the right side of Equation 27.6 to zero for magnetism. Thus, Gauss's law for magnetism is

$$\boxed{\Phi_B = \sum \mathbf{B} \cdot \Delta \mathbf{A} = 0} \tag{27.7}$$

which is the second Maxwell equation.

The third and fourth Maxwell equations can be arrived at by considering Ampère's and Faraday's laws. According to Ampère's law in electricity from Equation 24.2,

$$\sum \mathbf{B} \cdot \Delta \boldsymbol{l} = \mu_0 i \tag{27.8}$$

while according to Faraday's law in magnetism from Equations 25.1 and 25.30,

$$\sum \mathbf{E} \cdot \Delta \boldsymbol{l} = -\frac{\Delta \Phi_B}{\Delta t} \tag{27.9}$$

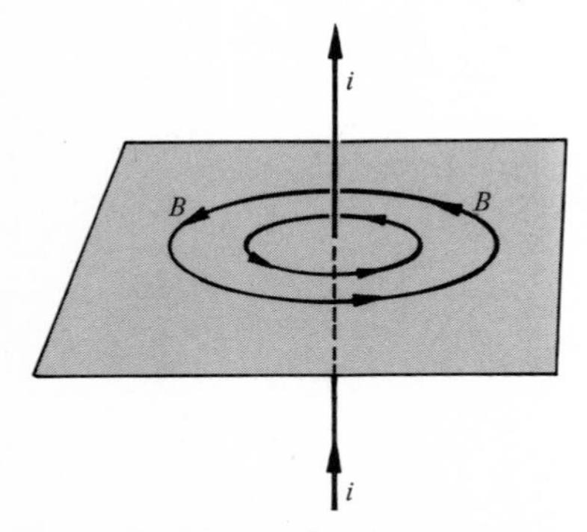

Figure 27.2. *The current in a conductor produces a magnetic field that circles the current.*

Let us see if these two equations are symmetrical. According to Ampère's law, Equation 27.8, there is a magnetic field that circles a current in a conductor as shown in Figure 27.2. According to Faraday's law, Equation 27.9, an electric field circles a changing magnetic flux as shown in Figure 27.3. Even though there seems to be a similarity between these two laws, a close look reveals that there is no exact correspondence between them. Firstly, Ampère's law contains a term $\mu_0 i$, while Faraday's law does not. But the absence of this term in Faraday's law is justified because free magnetic poles do not exist.

Secondly, Faraday's law contains a rate of change of magnetic flux term $\Delta\Phi_B/\Delta t$ while Ampère's law does not. The correspondence between the two laws will be perfect if there existed a term of the form $\Delta\Phi_E/\Delta t$ on the right-hand side of Equation 27.8. This means that a changing electric flux produces a magnetic field. That is, Ampère's law takes the form

$$\sum \mathbf{B} \cdot \Delta \boldsymbol{l} = \mu_0 i + \epsilon_0 \mu_0 \frac{\Delta \Phi_E}{\Delta t} \tag{27.10}$$

The quantity $\epsilon_0\mu_0$ is introduced to be consistent with the units. The introduction of the quantity $\Delta\Phi_E/\Delta t$ should not be surprising. Since $\Phi_E = EA$,

$\Delta\Phi_E/\Delta t = A\,\Delta E/\Delta t$, which when compared to Equation 27.4, reveals that $\Delta\Phi_E/\Delta t$ is proportional to the displacement current i_D. Thus, the above modification of Ampère's law constitutes the third Maxwell equation.

Equations 27.6, 27.7, 27.10, and 27.9 form a set of four Maxwell equations, stated as

$$\text{I. } \Phi_E = \sum \mathbf{E}\cdot\Delta\mathbf{A} = q/\epsilon_0 \tag{27.6}$$

$$\text{II. } \Phi_B = \sum \mathbf{B}\cdot\Delta\mathbf{A} = 0 \tag{27.7}$$

$$\text{III. } \sum \mathbf{B}\cdot\Delta l = \mu_0 i + \epsilon_0\mu_0 \frac{\Delta\Phi_E}{\Delta t} \tag{27.10}$$

$$\text{IV. } \sum \mathbf{E}\cdot\Delta l = -\frac{\Delta\Phi_B}{\Delta t} \tag{27.9}$$

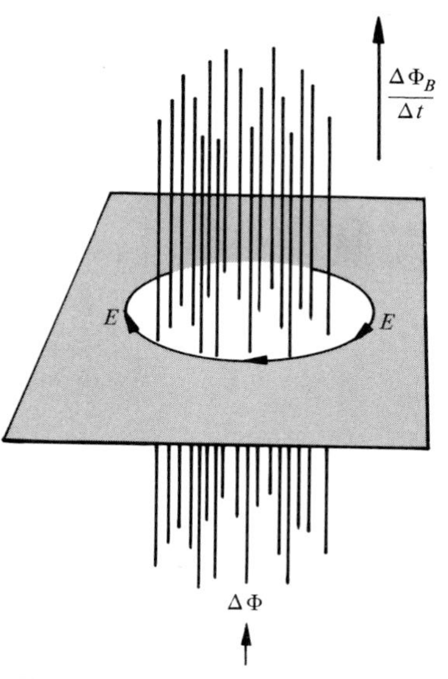

FIGURE 27.3. *Changing the magnetic flux produces an electric field that encircles the flux as shown.*

We now discuss some of the important consequences of these equations. In the Maxwell equation III, $\epsilon_0\mu_0 \simeq 1.1 \times 10^{-17}$ s²/m², which is a very small quantity. Thus, the term $\epsilon_0\mu_0\,\Delta\Phi_E/\Delta t$ will not contribute appreciably unless $\Delta\Phi_E/\Delta t$ is a very large quantity; that is, the electric flux must be changing very rapidly. Thus, this displacement current term will not be detected unless $\Delta\Phi_E/\Delta t$ is extremely large.

These equations not only completely describe electromagnetic phenomena, but also predict that an accelerated charge should generate an electromagnetic disturbance or an electromagnetic wave that propagates with a speed c, given below. The following two important results are also a consequence of the Maxwell equations:

$$\frac{B}{\mu_0} = H = \sqrt{\frac{\epsilon_0}{\mu_0}}\,E \tag{27.11}$$

and

$$c = \frac{1}{\sqrt{\epsilon_0\mu_0}} \tag{27.12}$$

After substituting for ϵ_0 and μ_0, we get

$$\frac{E}{H} = \sqrt{\frac{\mu_0}{\epsilon_0}} = \sqrt{\frac{4\pi \times 10^{-7}\ \text{Wb/A-m}}{8.85 \times 10^{-12}\ \text{C}^2/\text{N-m}^2}} = 377\ \Omega \tag{27.13}$$

which clearly indicates that not only the E and B fields are interrelated, but their ratio also remains constant at any point. Also, from Equation 27.12,

$$c = \frac{1}{\sqrt{\epsilon_0\mu_0}} = \frac{1}{\sqrt{(8.85 \times 10^{-12}\ \text{C}^2/\text{N-m}^2)(4\pi \times 10^{-7}\ \text{Wb/A-m})}}$$
$$= 2.9979 \times 10^8\ \text{m/s}$$

which is the speed of light. Hence, electromagnetic waves propagate at the same speed as visible light, indicating that light may consist of electromagnetic waves.

27.3. Electromagnetic Radiation from an Accelerated Charge

Maxwell's theory predicts that when an electric charge is accelerated, it emits electromagnetic radiation, a form of disturbance, that travels through space as explained below.

Consider a charge q that can move along the Z-axis as in Figure 27.4. Suppose that at $t = 0$ the charge is at rest at the origin O. The electric field

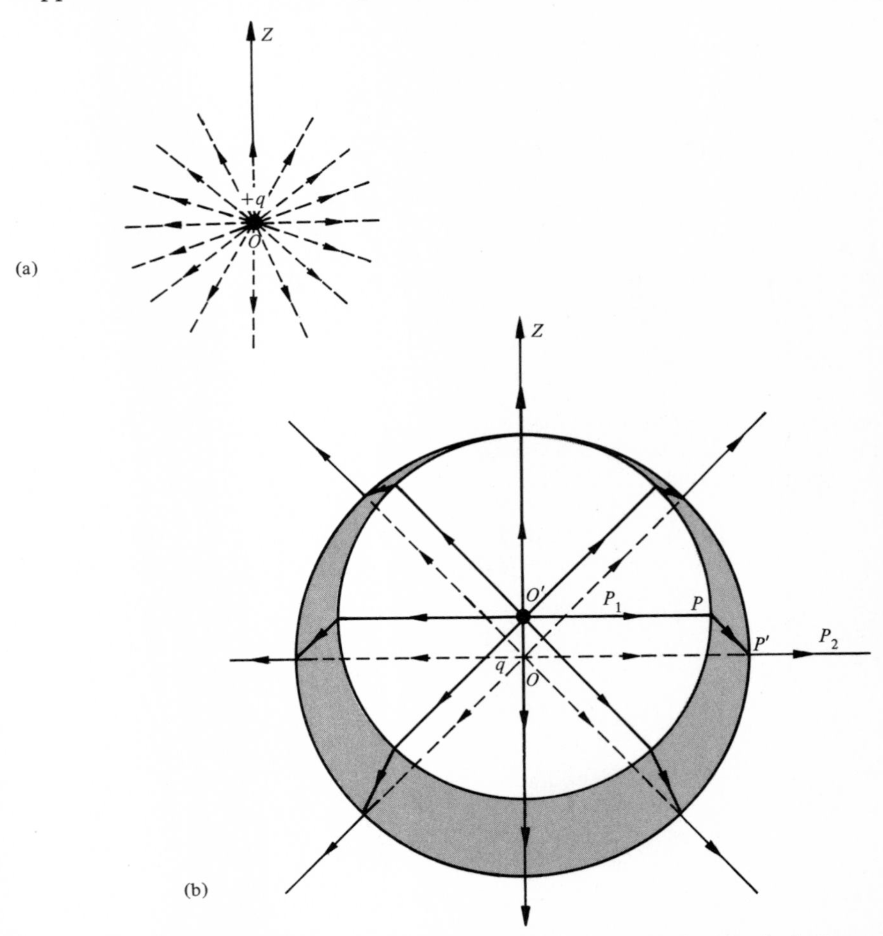

FIGURE 27.4. *(a) Field lines due to a stationary charge. (b) The dashed lines are the radial electric field lines due to a charge q at rest at O; the continuous lines are electric field lines due to the charge after being accelerated from O to O′. The shaded area includes the electromagnetic disturbance, or wave, resulting from the acceleration of the charge.*

lines will extend radially outward as shown in Figure 27.4(a) by the dashed lines. These lines would be similar even if the charge is moving with a uniform velocity v provided that $v < c$ and $v^2 << c^2$, where c is the speed of light. But suppose that during its motion the charge is accelerated before it reaches O', as in Figure 27.4(b). In such cases the field lines must adjust themselves so that they will continue to be centered at the charge and point radially outward. Such lines centering at O' are shown as continuous lines ending on the inner sphere in Figure 27.4(b). The news of the change in motion that took place between O and O' cannot reach all parts of the field at the same time, but we may say that it travels at speed c. For example, to an observer at P_1 the field lines have already adjusted and are coming from the charge at O'. To an observer at P_2 no change in motion has yet occurred and the field lines are still coming from the charge as if it were at O. The kink between PP' caused by the change in motion is the one that has to travel outward to readjust the field lines. This kink or disturbance, which at this particular time is confined to the shaded area, travels at a speed c and is called the *electromagnetic wave*. The disturbance or electric field intensity represented by PP' can be resolved into two components, the radial and the transverse. The radial component, which varies as $1/r^2$, vanishes at large distances as compared to the transverse component, which varies as $1/r$. Thus, the electromagnetic waves are transverse because the radial component becomes negligible.

Note that in the discussion above we have talked about the electric field only. A moving charge produces a magnetic field, and an accelerated charge will produce a changing magnetic field. Thus, the electric kink PP' is also accompanied by a magnetic kink. These two kinks or disturbances traveling together constitute an electromagnetic wave.

27.4. Production of Electromagnetic Waves

Even though Maxwell's theory was advanced in 1865, it was Heinrich Hertz who in 1888 actually demonstrated the production and detection of electromagnetic waves by electrical means using an oscillator that is equivalent to charges having variable acceleration. Figure 27.5 shows the experimental arrangement used by Hertz. Two large spheres, S and S′, are attached to two large metal plates P and P', respectively. The spheres are connected to an

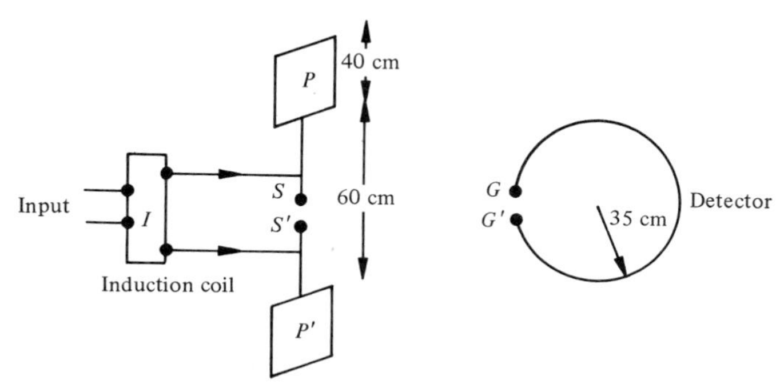

FIGURE 27.5. *Experimental arrangement used by Hertz to demonstrate the production and detection of electromagnetic waves.*

induction coil I. By causing a discharge between the spheres, the current distribution in the plates is such that a wave is generated with its electric vector parallel to the gap. The radiation thus produced was detected by Hertz by using an open wire loop which acted as a resonator. The electromagnetic wave reaching the gap of the detector had an electric field strong enough to establish a high potential difference between the gap GG' and cause a spark. When the detector gap is at right angles to the source gap, no electromagnetic radiation is detected; the sensitivity is maximum when the two gaps are parallel. This clearly demonstrates the state of polarization, leading to the conclusion that the electromagnetic waves are transverse; that is, the electric vector is perpendicular to the propagation direction.

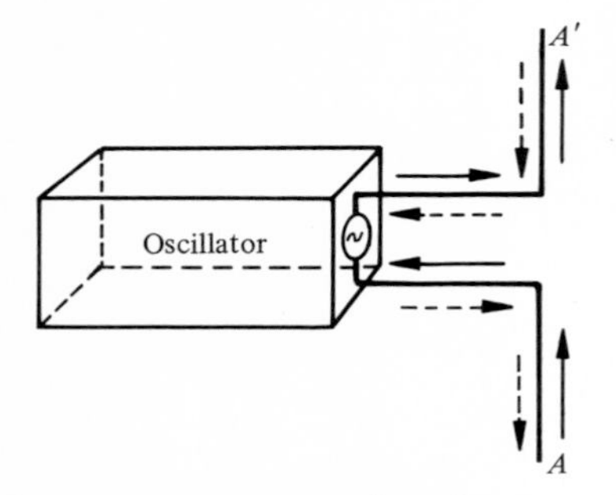

FIGURE 27.6. *Simple radio antenna that produces dipole radiation.*

We shall now explain in some detail the qualitative features of the propagation of electromagnetic radiation from a dipole antenna. A simple dipole antenna AA' is shown in Figure 27.6. The antenna is connected to an alternating-voltage generator. The charges accelerate in one direction $\rightarrow$ then in the other direction $\leftarrow$ alternatively in the arms of the antenna. Such a situation is equivalent to an oscillating electric dipole because the charges keep on changing their positions, as shown in Figure 27.7. The *dipole moment* at any instant is equal to either charge multiplied by the distance between the two

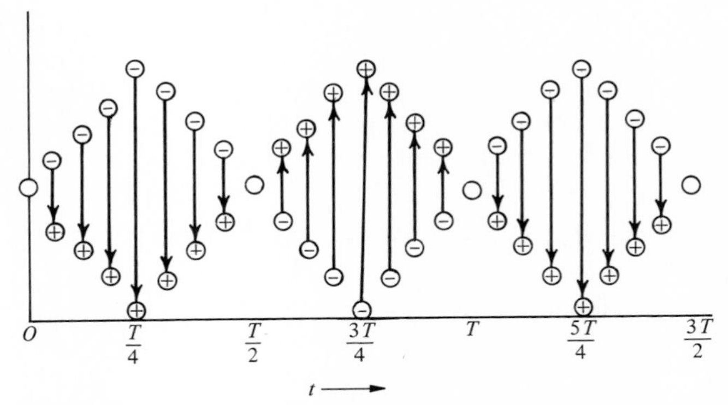

FIGURE 27.7. *Positions of the charges representing an oscillating electric dipole as a function of time. The length of the vertical line is equal to the dipole moment at different times.*

charges. Figure 27.7 shows that at $t = 0$ the two charges are together with no distance between them; hence, the dipole moment is zero. For $t > 0$, the charges separate and the dipole moment increases, reaching a maximum at $t = T/4$, where T is the time period of the oscillation. The variation in the dipole moment with time is shown by the vertical lines in Figure 27.7. These variations are simple harmonic and can be used to explain the generation of electromagnetic waves.

To start with, the dipole moment is zero, as shown in Figure 27.8(a). As the two charges separate, the dipole moment increases from zero and the electric field lines start from the + charge and end on the − charge, as shown in Figure 27.8(b). As the separation increases, reaching a maximum as shown in Figure 27.8(c), the dipole moment is maximum and the field lines are as shown.

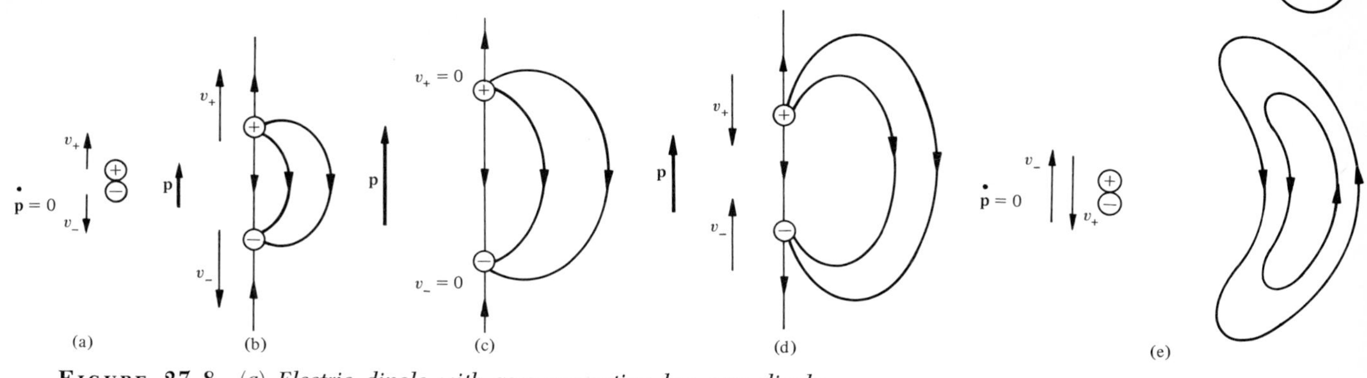

FIGURE 27.8. *(a) Electric dipole with zero separation has zero dipole moment and no electric field. (b) Simple electric dipole with its associated electric field. (c) Dipole with maximum separation, hence maximum dipole moment. (d) Electric field lines breaking away from a collapsing dipole. (e) Collapsed dipole. The breakaway lines form closed loops. For the next half of the cycle the direction of* **p**, *and the directions of the field lines, will be reversed.*

After $t = T/4$ the dipole starts returning to zero; hence, the dipole moment starts decreasing. At the point when the dipole moment is zero, the electric field lines, instead of continuing to spread, break away from the dipole and keep on propagating into space. This may be justified as follows.

As the lines break away from the dipole, they do not have + and − charges to begin and end on. But the lines of **E** must be closed. This is accomplished when the breakaway lines combine with lines from the previous half-cycle and form closed loops, some enclosed by others, as shown in Figures 27.8(d) and (e). Note that for convenience we have drawn lines only on the right side. The electric field lines on the left are the mirror image of those on the right. The complete picture with both electric and magnetic (explained below) is shown in Figure 27.9. At large distances the electric field at any point varies as shown in Figure 27.10.

So far we have talked about an electric field, but there is a magnetic field which also radiates from the antenna. As the charges at the end of the antenna change, there results a current in the antenna. When the two ends of the antenna are joined, it behaves like a wire carrying current. From the right-hand rule the magnetic field will be perpendicular to the page, that is, along the Z-axis. This magnetic field at any point oscillates with the same frequency as the frequency of the oscillating current in the antenna.

From the discussion above, we conclude that in the simple case of an oscillating dipole, the electromagnetic wave propagates along the X-axis, the variation in E is along the Y-axis, and the variation in B is along the Z-axis, as shown in Figure 27.11. That is, **E** and **B** are perpendicular to each other and to the direction of propagation.

The simplest type of electromagnetic wave is a monochromatic, plane, linearly polarized wave traveling in a vacuum. The term *monochromatic* means that the field strength at a point varies with time according to a sine or cosine function. As shown in Figure 27.11, for the wave propagating along the X-axis

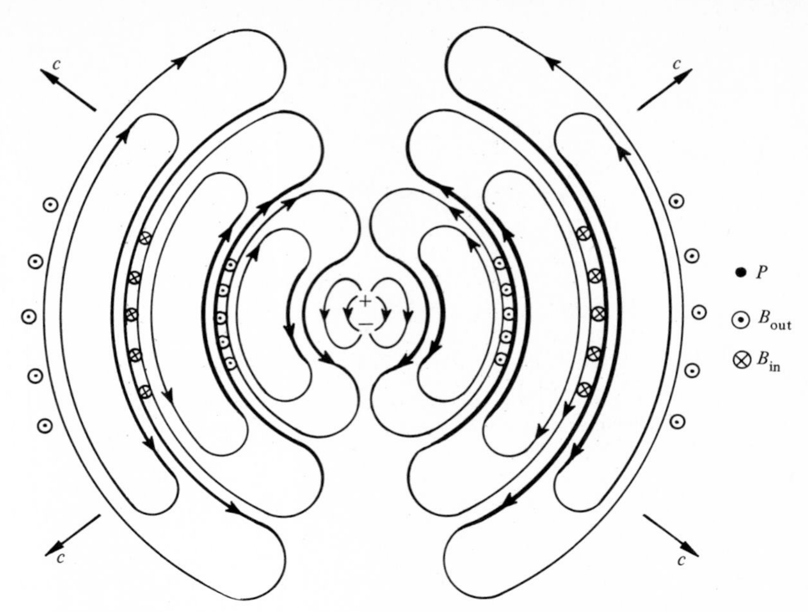

FIGURE 27.9. *Instantaneous radiation pattern of the electric and magnetic fields from an oscillating electric dipole.*

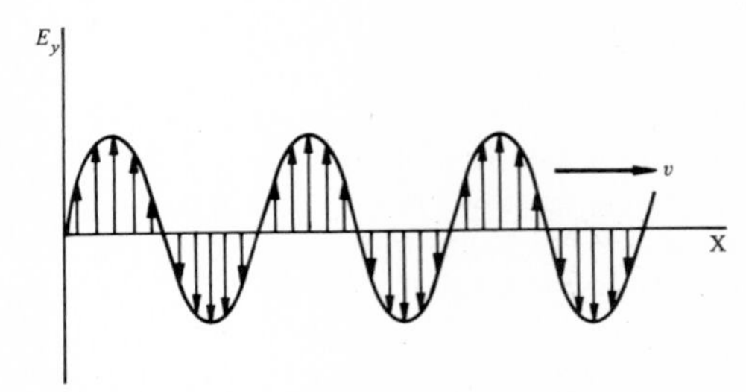

FIGURE 27.10. *At large distances from the dipole antenna, the electric field at any point varies sinusoidally as the signal travels from the antenna to the point under consideration.*

(making use of the relation $v = f\lambda = \lambda/T$),

$$E = E_y = E_0 \sin\left[2\pi\left(ft - \frac{x}{\lambda}\right)\right] = E_0 \sin\left[2\pi\left(\frac{t}{T} - \frac{x}{\lambda}\right)\right] \quad (27.14)$$

and

$$B = B_z = B_0 \sin\left[2\pi\left(ft - \frac{x}{\lambda}\right)\right] = B_0 \sin\left[2\pi\left(\frac{t}{T} - \frac{x}{\lambda}\right)\right] \quad (27.15)$$

where E_0 and B_0 are the amplitudes of the electric and the magnetic field strengths, which are positive. The term *plane* means that the wave always travels in the same direction, say the X-direction, as shown in Figure 27.11, and at any instant the value of **E** is the same at all points in any plane perpendicu-

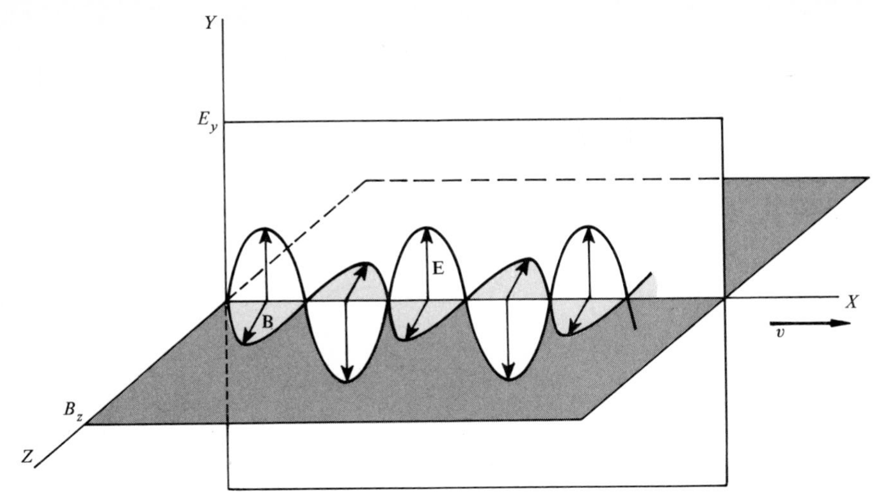

Figure 27.11. *Electromagnetic wave propagating along the X-axis while the amplitudes of the electric and magnetic field are along the Y-axis and Z-axis, respectively. This is a monochromatic, plane, linearly polarized electromagnetic wave.*

lar to the propagation direction. If the wave is traveling in the X-direction, the YZ-plane is perpendicular to the direction of propagation. The term *linearly polarized* means that **E** lies in one direction only. Because the electromagnetic wave is transverse, **E**, as well as **B**, must be perpendicular to the direction of propagation. Because the wave is propagating in the X-direction, we can take the direction of **E** to be the Y-axis, and **B** will then be along the Z-axis. The plane of polarization is conventionally defined as the plane that contains **E** and the direction of propagation.

So far we have been talking about Maxwell's equations, and in this section we have showed how to represent an electromagnetic wave. Just like Newton's second law in mechanics, there is one equation that is the net result of putting everything together. Such an equation is called Maxwell's equation in differential form, and Equations 27.14 and 27.15 are the solutions of such an equation, but we shall not discuss this further.

27.5. Energy Flow in Electromagnetic Waves—Poynting Vector

The electromagnetic waves or disturbances transport electric and magnetic energies from one point to another. In order to calculate the amount of energy carried (the energy flow) by the electromagnetic wave, we must know separately the energies due to the electric and magnetic fields.

As discussed in Chapter 20, the quantity u_E, the *electric field energy density*, is defined as the electric energy per unit volume and is given by

$$u_E = \tfrac{1}{2}\epsilon_0 E^2 \tag{27.16}$$

Similarly, as shown in Chapter 25, the quantity u_B, the *magnetic field energy density*, is defined as the magnetic energy per unit volume and is given by (remembering that $B = \mu_0 H$)

$$u_B = \frac{1}{2}\frac{B^2}{\mu_0} = \frac{1}{2}\mu_0 H^2 \tag{27.17}$$

In the case of an electromagnetic field where both E and B are present, we can define the *energy density* u of the electromagnetic field as the energy per unit volume and may write

$$u = u_E + u_B = \frac{1}{2}\epsilon_0 E^2 + \frac{1}{2}\frac{B^2}{\mu_0} = \frac{1}{2}\epsilon_0 E^2 + \frac{1}{2}\mu_0 H^2 \tag{27.18}$$

A quantity of more interest than energy density is *energy intensity* S, defined as the amount of energy crossing per unit area per time. Let us calculate S for the case of electromagnetic radiation. Consider an element of area ΔA in a plane perpendicular to the direction of propagation of an electromagnetic wave, as shown in Figure 27.12. Energy crossing the area ΔA in time Δt will be equal to the energy contained in a cylinder of volume $\Delta A\, c\, \Delta t$, where $c\, \Delta t$ is

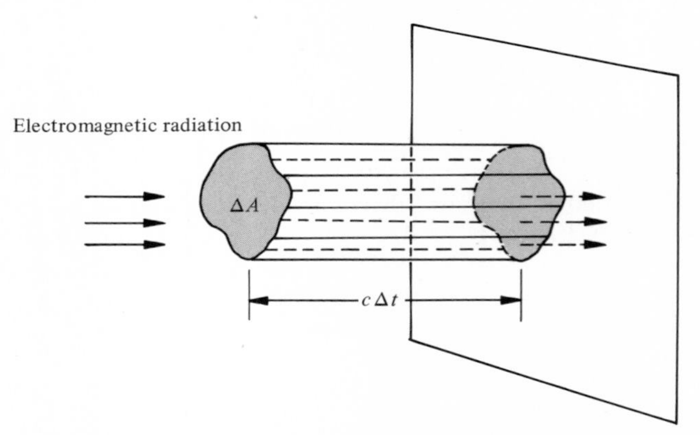

FIGURE 27.12. *A cylinder of length $c\,\Delta t$ and of cross-sectional area ΔA contains electromagnetic radiation that passes through the area ΔA in time Δt.*

the length of the cylinder and ΔA is the cross-sectional area of the cylinder. If u is the energy density, then $u\, \Delta A\, c\, \Delta t$ is the amount of energy that passes through an area ΔA in time Δt. Thus, we may find S, the energy intensity, to be

$$S = \frac{\text{total energy}}{\text{area} \times \text{time}} = \frac{u\, \Delta A\, c\, \Delta t}{\Delta A\, \Delta t} = cu \tag{27.19}$$

Because $u = u_E + u_B$, we may write

$$\boxed{S = cu = c(\tfrac{1}{2}\epsilon_0 E^2 + \tfrac{1}{2}\mu_0 H^2)} \tag{27.20}$$

It is conventional and useful to write Equation 27.20 in different forms by making use of the following relations that come from Maxwell's equations, as mentioned in Section 27.2:

$$\frac{B}{\mu_0} = H = \sqrt{\frac{\epsilon_0}{\mu_0}}\,E \quad \text{and} \quad c = \frac{1}{\sqrt{\epsilon_0\mu_0}} \tag{27.21}$$

Combining Equations 27.20 and 27.21, we get

$$u = \frac{1}{c}EH \tag{27.22}$$

Therefore, Equation 27.20 becomes

$$\boxed{S = cu = EH = \frac{1}{\mu_0}EB} \tag{27.23}$$

Let us consider a special case in which the wave is propagating along the X-axis, E is along the Y-axis, and B along the Z-axis. Thus,

$$S_x = E_y H_z = \frac{1}{\mu_0}E_y B_z \tag{27.24}$$

where S_x is the amount of energy crossing per unit area per second in the X-direction, as shown in Figure 27.13.

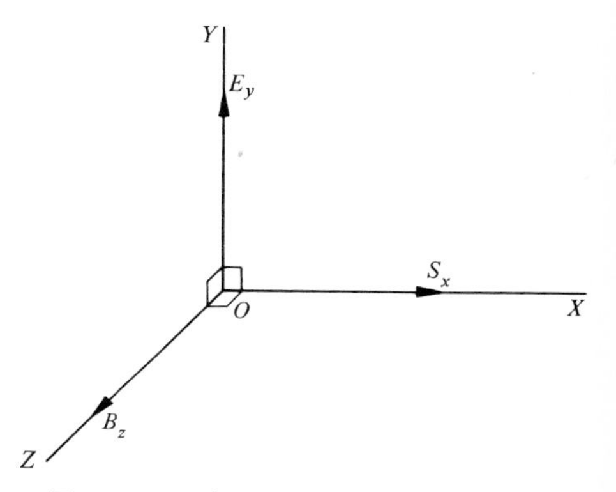

Figure 27.13. *For an electromagnetic wave propagating along the X-axis, the direction of E and B are along the Y- and Z-axis, respectively.*

In general, the energy intensity is a vector quantity with components S_x, S_y, and S_z of **S**, where **S** is called the *Poynting vector* and **S** is perpendicular to both **E** and **H** while **E** and **H** are perpendicular to each other. The Poynting vector **S** gives the intensity or instantaneous flux, which in the SI system is W/m^2. Because the average values of E^2 is $\frac{1}{2}E_0^2$ and H^2 is $\frac{1}{2}H_0^2$, we may write the average value of the Poynting vector S_{av} as

$$\boxed{S_{av} = \frac{1}{2}E_0H_0 = \frac{1}{2\mu_0}E_0B_0} \tag{27.25}$$

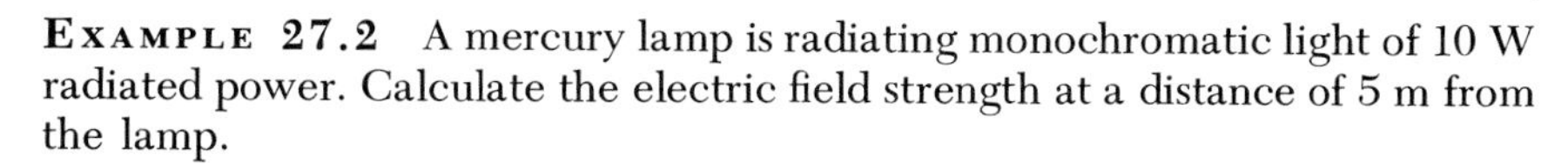

Example 27.2 A mercury lamp is radiating monochromatic light of 10 W radiated power. Calculate the electric field strength at a distance of 5 m from the lamp.

Let us assume that the total power radiated is P_0. If this power is radiated uniformly in all directions, the power passing through a unit area at a distance r from the source will be given by

$$S_{av} = \frac{P_0}{4\pi r^2}$$

where S_{av} is the intensity or the mean power radiated per unit area. Also, from Equation 27.25 and using Equation 27.21,

$$S_{av} = \frac{1}{2}E_0H_0 = \frac{1}{2}E_0\left(\sqrt{\frac{\epsilon_0}{\mu_0}}\,E_0\right) = \frac{1}{2}c\epsilon_0E_0^2$$

Equating the two equations,

$$\frac{1}{2}c\epsilon_0 E_0^2 = \frac{P_0}{4\pi r^2} \qquad \text{or} \qquad E_0 = \frac{1}{r}\sqrt{\frac{P_0}{2\pi c\epsilon_0}}$$

Substituting for $r = 5$ m, $P_0 = 10$ W, $c = 3 \times 10^8$ m/s, and $\epsilon_0 = 8.85 \times 10^{-12}$ C²/N-m²,

$$E_0 = \frac{1}{5 \text{ m}}\sqrt{\frac{10 \text{ W}}{2\pi(3 \times 10^8 \text{ m/s})(8.85 \times 10^{-12} \text{ C}^2/\text{N-m}^2)}} \simeq 4.9 \text{ V/m}$$

EXERCISE 27.2 Assuming an isotropic distribution, calculate the electric and magnetic field strengths at a distance of 1 m from a 100-W lamp bulb. [*Ans.*: $E_0 = 77.4$ V/m, $H_0 = 0.205$ A/m.]

EXAMPLE 27.3 A monochromatic plane-polarized electromagnetic wave is traveling eastward. The wave is polarized vertically with electric field strength directed alternately up and down. Calculate E, H, B, and the Poynting vector S provided that the amplitude of the electric field strength is 0.05 V/m and the frequency is 6 MHz.

Let the wave be propagated along the X-axis, the polarization direction is along the Y-axis, so that $E_x = E_z = 0$ and $E_y \neq 0$. The magnetic field is confined along the north–south direction, that is, along the Z-axis, so that $H_x = H_y = 0$ and $H_z \neq 0$. The frequency of the wave is $f = 6 \times 10^6$ Hz; hence, the period T is

$$T = \frac{1}{f} = \frac{1}{6 \times 10^6/\text{s}} = 1.67 \times 10^{-7} \text{ s}$$

and the wavelength is

$$\lambda = \frac{c}{f} = \frac{3 \times 10^8 \text{ m/s}}{6 \times 10^6/\text{s}} = 50 \text{ m}$$

Thus, from Equation 27.14, E_y is given by

$$E_y = E_0 \sin 2\pi\left(\frac{t}{T} - \frac{x}{\lambda}\right) = 0.05 \sin(3.76 \times 10^7 t - 0.126x) \text{ V/m}$$

The amplitude of the magnetic field H_0 is given by

$$H_0 = \sqrt{\frac{\epsilon_0}{\mu_0}}E_0 = \frac{1}{376.7\ \Omega}(0.05 \text{ V/m}) = 1.33 \times 10^{-4} \text{ A/m}$$

Therefore,

$$H_z = 1.33 \times 10^{-4} \sin(3.76 \times 10^7 t - 0.126x) \text{ A/m}$$

and $B_z = \mu_0 H_z = 4\pi \times 10^{-7} H_z$ or we could use the relation

$$B_z = \sqrt{\epsilon_0\mu_0}\, E_y = E_y/c$$

$$= 1.67 \times 10^{-10} \sin(3.76 \times 10^7 t - 0.126x) \text{ Wb/m}^2$$

The Poynting vector **S**, in this case, has only X-component, that is, the wave is traveling in the eastward direction and is given by

$$S_x = E_y H_z = E_0 H_0 \sin^2 2\pi \left(\frac{t}{T} - \frac{x}{\lambda}\right)$$
$$= 6.65 \times 10^{-6} \sin^2(3.76 \times 10^7 t - 0.126x) \text{ W/m}^2$$

while from Equation 27.24,

$$(S_x)_{av} = (E_y H_z)_{av} = \tfrac{1}{2} E_0 H_0 = 3.325 \times 10^{-6} \text{ W/m}^2$$

EXERCISE 27.3 A laser produces an intense beam of light with a cross section of radius 2.00 mm. The light emitted is monochromatic of wavelength 6328 Å at a power of 10.0 mW. What are the effective (rms) values of the electric and magnetic fields of this beam? [*Ans.*: 547 V/m and 1.45 A/m.]

27.6. The Range of Electromagnetic Waves

At the time Maxwell established his electromagnetic theory, the only well-known electromagnetic waves were visible light waves while the existence of ultraviolet and infrared was barely established. By the close of the nineteenth century, x-rays and gamma rays had also been discovered. At present we know, in addition to visible rays, electromagnetic waves include x-rays, gamma rays, radio waves, and microwaves, just to name a few (Figure 27.14). All these waves have very different frequencies f, hence different

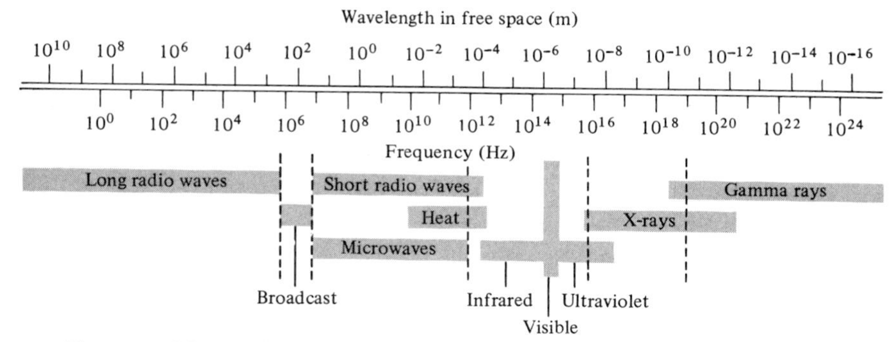

FIGURE 27.14. *Range of electromagnetic waves. Note the very narrow range representing visible light.*

wavelengths λ, but they all travel with the same speed, $c = 3 \times 10^8$ m/s in vacuum, where

$$c = f\lambda \tag{27.26}$$

As mentioned earlier, all electromagnetic waves are transverse in nature. The arrangement of wavelengths or frequencies shown in Figure 27.14 is also termed the *electromagnetic spectrum*. It is worthwhile to notice that visible light forms a very small portion of the whole electromagnetic spectrum. Since the range of wavelengths changes drastically from region to region, different units are used for wavelength measurements in different regions. For example, 1 angstrom $= 1$ Å $= 10^{-10}$ m; 1 X unit $= 1$ X U $= 10^{-13}$; and so on.

As shown earlier, the speed of light or any electromagnetic wave in vacuum according to Equation 27.12 is

$$c = \frac{1}{\sqrt{\epsilon_0 \mu_0}} = 3 \times 10^8 \text{ m/s} \tag{27.12}$$

The speed v of light in any other medium of permittivity ϵ and permeability μ is

$$v = \frac{1}{\sqrt{\epsilon\mu}} = \frac{1}{\sqrt{K\epsilon_0 K_m \mu_0}} = \frac{c}{\sqrt{KK_m}} \tag{27.27}$$

where K is the dielectric constant of the medium while K_m is the relative permeability. The magnetic behavior for most substances (except for ferromagnetic materials) differs slightly from that in vacuum; hence, $K_m \simeq 1$. Thus, from the equation above,

$$\boxed{v = \frac{c}{\sqrt{K}}} \tag{27.28}$$

As we shall see in Chapter 28, $c/v = \sqrt{K} = n$, where n is the refractive index of the medium in which the wave is traveling.

27.7. The Speed of Light

The extremely large magnitude of the speed of light makes its accurate measurement difficult. It was not until the early part of this century, with the development of sophisticated instruments, that the precise value of c became available.

Before 1600 it was thought that light traveled with an infinite speed. The first attempt to measure c was made by Galileo (1563–1642). He stationed himself on one hill and his assistant on another hill about 2 miles away. Each was provided with a lantern and a cover. Galileo uncovered his lantern and his assistant, seeing the light, uncovered his lantern. When Galileo saw the light, he noted the time. Hence, knowing the distance ($\sim$4 mi for the round trip) and the time, he thought he could calculate the speed of light. But the time elapsed was so small that no results were obtained (the time to travel 4 mi is 0.00002 s, which is far smaller than the reaction time of the two experimenters.)

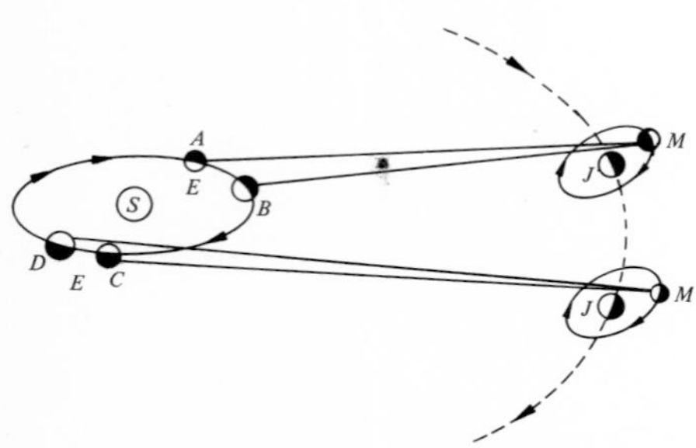

Figure 27.15. *Roemer's method for measuring the speed of light by noticing the eclipses of Jupiter's moon.*

The first measurement of the speed of light was made in 1676 by the Danish astronomer Olaf Roemer (1644–1710) from astronomical observations. (There is still some question as to who was the first to measure the speed of light.) The method involves making measurements on one of the 12 small satellites (or moons) of Jupiter. (Four of these moons are easily visible with the aid of an ordinary telescope.) The plane of the orbits of these satellites is the same as the plane of the earth and Jupiter's orbits. This means that standing on earth one can observe the eclipse of Jupiter's moon, as shown in Figure 27.15. Roemer was involved in measuring the period of revolution ($\sim$42 h) of the innermost moon by measuring the time between successive eclipses by Jupiter. He found discrepancies in the times of the successive onset of the eclipses. He observed that as the earth was receding from Jupiter (Figure 27.15), the successive measurements of the period were somewhat longer than the average. On the other hand, when the earth was approaching Jupiter (Figure 27.15), the successive periods became shorter and shorter; that is, the eclipse appeared earlier and earlier.

Roemer correctly concluded that these discrepancies were due to the time required for the light to travel the changing distance between Jupiter and the earth. The situation is as shown in Figure 27.15. Roemer found that it took $\sim$1300 s for light to travel the distance of the diameter of the earth's orbit, which is $\sim 3 \times 10^{11}$ m. Accordingly, the speed of light was found to be $c \approx 3 \times 10^{11}\ \text{m}/1300\ \text{s} \approx 2.3 \times 10^{8}$ m/s.

The terrestrial method for measuring the speed of light was initiated by the French scientist Fizeau in 1849. This was later improved by Foucault, who replaced Fizeau's tooth wheel by a rotating mirror. Using the Focault method, precise measurements were made by the American physicist Albert A. Michelson (1852–1931). He made the first measurement in 1878. The experiment as carried out during 1923–1927 is demonstrated in Figure 27.16. A light ray

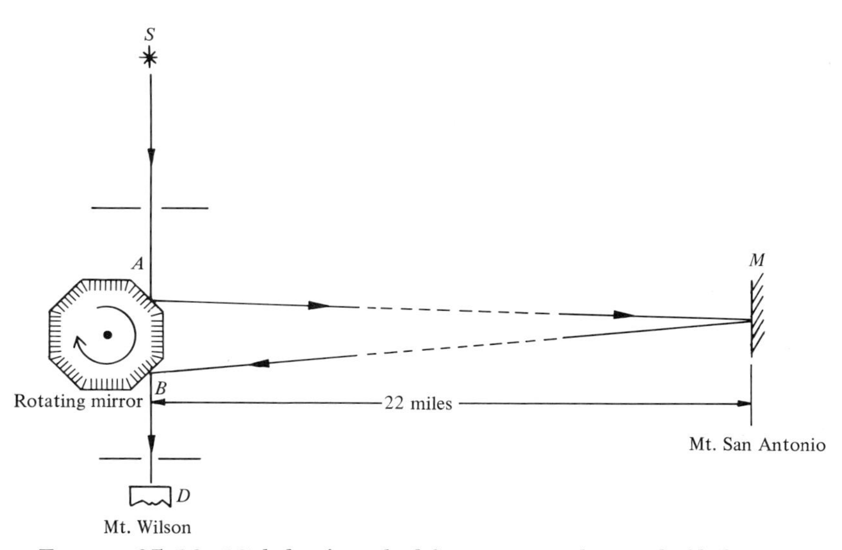

FIGURE 27.16. *Michelson's method for measuring the speed of light using a rotating mirror.*

from a source S was reflected from side A of an eight-sided mirror located at Mt. Wilson. The beam traveled 22 miles and was reflected back from a fixed mirror M located at Mt. San Antonio to side B of the eight-sided mirror and into a detector D. The mirror is then made to rotate. Michelson adjusted the speed so that when the light beam traveled from A to M and back to B the mirror rotated through 1/8 of a revolution. This way the detector was continuously illuminated. Thus, knowing the speed of rotation and the distance of the round trip (known with an accuracy of better than 2 cm), he calculated c. His results gave the speed of light a value of 2.99798×10^{8} m/s.

Since the speed of light c is one of the fundamental physical constants, keen interest in its accurate determination continued. The modern most precise techniques involving the use of laser beams for the measurement of the speed of light yields the following value:

$$c = 2.99791458 \times 10^8 \text{ m/s}$$

EXAMPLE 27.4 In the Michelson experiment discussed above, what was the speed (rev/s) of the eight-sided mirror if the measured value of the speed of light is 2.99798×10^8 m/s?

Since the round-trip distance between the two hills is 2×22 mi $= 2 \times 22 \times 1609$ m $= 70{,}796$ m, the time Δt taken by the light beam will be

$$\Delta t = \frac{\Delta x}{c} = \frac{70{,}796 \text{ m}}{2.99798 \times 10^8 \text{ m/s}} = 0.23615 \times 10^{-3} \text{ s}$$

In this time interval the mirror rotates through 1/8 of a revolution. Therefore, the time T for 1 revolution is

$$T = 8 \times 0.23615 \times 10^{-3} \text{ s} = 1.8892 \times 10^{-3} \text{ s}$$

Hence, the number of revolutions/s, f, is

$$f = \frac{1}{T} = \frac{1}{1.8892 \times 10^{-3} \text{ s}} = 529.32 \text{ rev/s}$$

EXERCISE 27.4 Suppose that one uses a 16-sided mirror in Michelson's experiment and obtains the speed of light to be 2.997915×10^8 m/s. What is the rotational speed of the mirror? [*Ans.:* 264.66 rev/s.]

SUMMARY

Maxwell introduced the concept of *displacement current,* $i_D = \epsilon_0 A\, \Delta E/\Delta t$, that is, a changing electric field produces current.* The basic concepts and laws of electricity and magnetism were put together into a set of four fundamental equations, called Maxwell's equations or electromagnetic field equations. These are (I) Gauss's law for electrostatics, $\Sigma\, \mathbf{E} \cdot \Delta \mathbf{A} = q/\epsilon_0$; (II) Gauss's law for magnetism, $\Sigma\, \mathrm{B} \cdot \Delta \mathbf{A} = 0$; (III) a modified form of Ampère's law, $\Sigma\, \mathbf{B} \cdot \Delta \boldsymbol{l} = \mu_0 i + \epsilon_0\mu_0\, \Delta\Phi_E/\Delta t$; and (IV) Faraday's law, $\Sigma\, \mathbf{E} \cdot \Delta \boldsymbol{l} = -\Delta\Phi_B/\Delta t$.

One of the predictions of Maxwell's equations was that an accelerated charge emits *electromagnetic waves* which are transverse in nature. This was confirmed by Hertz in 1888. Other predictions include $E/H = \sqrt{\mu_0/\epsilon_0} = 377\ \Omega$ and $c = 1/\sqrt{\epsilon_0\mu_0}$ = the speed of light, indicating that light waves are also electromagnetic waves. A monochromatic, plane, linearly polarized wave traveling in vacuum is represented by $E_y = E_0 \sin[2\pi(t/T - x/\lambda)]$ and $B_z = B_0 \sin[2\pi(t/T - x/\lambda)]$.

The *energy density* of an electromagnetic field $u = u_E + u_B = \frac{1}{2}\epsilon_0 E^2 + \frac{1}{2}B^2/\mu_0 = \frac{1}{2}\epsilon_0 E^2 + \frac{1}{2}\mu_0 H^2$. The *energy intensity* S, defined as the amount of energy crossing per unit area per unit time, is given by $S = cu = EH = EB/\mu_0$. **S** is called the *Poynting vector* and $S_{av} = \frac{1}{2}E_0 H_0 = E_0 B_0/2\mu_0$.

The first measurement of the speed of light was made in 1676 by Olaf Roemer, who made measurements on one of the moons of Jupiter. The terrestrial methods were initiated by the French physicist Fizeau in 1849. The first accurate measurement made by Michelson during 1923–1927 yielded a value of $c = 2.997998 \times 10^8$ m/s.

QUESTIONS

1. Does a varying current in a conductor produce displacement current? Explain.
2. Why, according to Ampère's law, doesn't a slowly varying electric field affect the magnetic field B?
3. How do electromagnetic waves differ from mechanical sound waves?
4. Do electromagnetic waves always travel in straight lines?
5. Which charged particle—electron or proton—will radiate more electromagnetic energy both having the same acceleration?
6. If electromagnetic waves were longitudinal, what would be the relative directions of **E**, **B**, and **S**?
7. Do the amount of charge and the magnitude of the acceleration of the charge affect the speed with which the electromagnetic disturbance travels? What quantities are affected by these factors?
8. In Figure 27.9, in a small region far away from the dipole, draw the electric and magnetic field lines. How does this field differ from the field near the dipole?
9. Which of the following quantities determine the amount of light needed to read a book: E, E^2, B, B^2, P, or S?
10. Suppose that the antenna of a transmitting station has a dipole structure. In what direction will you orient the antenna of your radio to receive a strong signal? How do you find this direction? What are the corresponding directions of **E**, **B**, and **S**?
11. The visible light wavelength range is what fraction of (a) the total infrared range; and (b) the ultraviolet region?
12. What consequences, if any, can you think of if the electromagnetic waves did travel with a speed greater than the speed of light?

PROBLEMS

1. Show that for a parallel-plate capacitor the displacement current i_D is given by $i_D = C\,\Delta V/\Delta t$, where C is the capacitance.
2. For a parallel-plate capacitor of 1 μF, what should be the value of $\Delta V/\Delta t$ so that the displacement current will be 1 mA?
*3. Suppose that the existence of magnetic monopoles becomes a fact. Would you have to modify Maxwell's equations because of this?
*4. Show that the ratio of $E_{0y}/B_{0z} = c$.
5. If the electric amplitude of the wave is 5 V/m, what is the magnetic amplitude of this wave?
6. If the velocity of an electromagnetic wave in a medium is 2×10^8 m/s and the electric amplitude of the wave is 5 V/m, what will be the magnetic amplitude?

7. A plane monochromatic linearly-polarized light wave is traveling eastward. The wave is polarized with E directed vertically up and down. Write expressions for E, H, and B provided that $E_0 = 0.1$ V/m and f is 20 megahertz.
8. Repeat Problem 7 for visible light (of $\lambda = 5890$ Å) with $E_0 = 0.01$ V/m and traveling westward. Draw a diagram showing **E, H,** and the propagation direction.
9. A sodium lamp is radiating light of 50 W radiated power. Calculate the electric field strength at a distance of 3 m.
10. If $E_0 = 12$ V/m at a distance of 4 m from an electric lamp, what should be the power of this lamp?
11. Neutrons in a neutron beam are traveling with a speed of 3×10^6 cm/s. If the neutron flux is 10^{10} neutrons/cm^2-s, calculate the neutron density.
12. Consider a parallel-plate capacitor with plate dimensions of 5 cm by 5 cm and a separation of 1 mm. If a potential difference of 500 V is applied to the plates, calculate the total energy stored in the capacitor.
13. A laboratory magnet produces a uniform field of 0.1 Wb/m^2 between 2 in by 2 in pole faces and separated by 2 in. What is the energy density of the magnetic field at this strength?
14. Dry air at room temperature can withstand a maximum electric field of 3×10^6 V/m. What is the maximum energy density in air, treating air as a vacuum?
15. A radio signal of frequency 5 MHz, when received by an antenna, has an intensity of 10^{-5} W/m^2. Calculate E_0, H_0, B_0 and E, H, B as a function of time.
16. Consider a plane, monochromatic, linearly polarized electromagnetic wave traveling in the X-direction and polarized vertically; that is, $E_x = E_z = 0$ and $E_y \neq 0$ while $B_x = B_y = 0$ and $B_z \neq 0$. If the amplitude of the electric field strength is 0.05 V/m and the frequency is 5 megahertz, calculate (a) E_y; (b) B_0; (c) B; (d) H_0; (e) H; (f) S; and (g) S_{av}.
17. If the electromagnetic radiation from the sun reaches the earth's surface at an average rate of 1400 W/m^2, calculate E_0 and B_0.
18. If the average energy received on the surface of the earth is 2.2 cal/cm^2-s, calculate the magnitudes of the electric and magnetic vectors.
19. Suppose that the intensity of the solar radiation on the surface of the earth is 2×10^3 J/m^2-s. Assume that all the radiation from the sun has one wavelength. Calculate the amplitude of the electromagnetic wave on the surface of the earth.
20. A radio signal received by an antenna has a maximum electric field intensity of 10^{-4} V/m. Calculate (a) the maximum flux density of the magnetic field and (b) the magnitude of the Poynting vector of such a wave.
21. A radio receiver antenna that is 2 m long is oriented along the direction of the electromagnetic wave and receives a signal of intensity 5×10^{-16} W/m^2. Calculate the maximum instantaneous potential difference across the two ends of the antenna.
22. Calculate the time that it takes light to travel across the diameter of the earth's orbit around the sun. [Take the diameter of this orbit to be $2(1.49 \times 10^8$ km).]

23. In Michelson's experiment, if the distance between the two hills is 30 mi, what will be the rotational speed of the eight-sided mirror if the speed of light is found to be 2.99798×10^8 m/s?

24. Suppose in Michelson's experiment that the eight-sided mirror was rotating at a speed of 31,680 rev/min. What will be the speed of light, assuming the distance between the two hills to be 22 miles?

Part VI
Optics

28
Geometrical Optics

This and the following two chapters will be devoted to the study of optics. After briefly discussing the theories of light, we shall make use of Huygens' principle in arriving at the laws of reflection and refraction. A relation between wavelength, speed of light, and refractive index of different media will be given. These laws will be applied to internal reflection: deviation and dispersion by prisms. A further application of these laws to the image formation by mirrors and lenses will conclude the chapter.

28.1. Nature and Theories of Light

By THE beginning of the seventeenth century, many properties of visible light were well known. Among those were the rectilinear propagation of light, the laws of reflection and refraction, and the dispersion of white light into different colors by a prism. In an effort to explain these properties, the *corpuscular theory* of light was put forth by Newton in 1675. According to this theory, light consists of tiny particles emitted by a source. These corpuscles, when received by the eye, produced a sensation of visibility. This theory could easily explain the rectilinear propagation of light, such as the formation of shadows of objects in the path of light, the occurrence of eclipses, and so on.

In 1678, Newton's contemporary, the Dutch physicist Christian Huygens (1629–1695), suggested the *wave theory* of light. The wave theory could explain almost all the then-known properties, except for the rectilinear propagation of light. Primarily because of Newton's great reputation, the wave theory was abandoned in favor of the corpuscular theory for the next 100 years.

A major step was taken in 1801–1804 when Thomas Young was able to produce an interference pattern between two light beams (just as in the case of two sound beams) and explain it with the help of the wave theory. This was further confimed by Fresnel, who used the wave theory to explain diffraction patterns.

The wave nature of light was further confirmed by Maxwell's development of equations for the electromagnetic field and their subsequent verification by Hertz in 1887 (as discussed in Chapter 27). It was shown that the velocity with which electromagnetic waves travel through vacuum is the same as the velocity of light ($c \approx 3 \times 10^8$ m/s). These and other experimental evidence led to the conclusion that light waves are transverse electromagnetic waves. As was shown in Figure 27.14, visible light forms a very small fraction of the known spectral range of electromagnetic waves.

The wavelength of light is in the range ~0.0000004 to 0.0000007 m. It is convenient to express wavelengths in terms of a smaller unit, the nanometer (1 nm) or the angstrom (1 Å), where

$$1 \text{ nm} = 10^{-9} \text{ m} = 10^{-7} \text{ cm}$$
$$1 \text{ Å} = 10^{-10} \text{ m} = 10^{-8} \text{ cm}$$

Thus, the visible range of light is ~400 nm (4000 Å) to ~700 nm (7000 Å). The relation between the wavelength λ, the frequency ν, and the velocity c given by $\nu = c/\lambda$ holds for the whole range of electromagnetic radiation. The frequency range of the visible spectrum is $\sim 7.8 \times 10^{14}$ Hz to 4×10^{14} Hz. The colors of different parts of visible light depend upon the wavelength range:

violet	~3800–4500 Å	$\sim(7.8\text{–}6.6) \times 10^{14}$ Hz
blue	~4500–5000 Å	$\sim(6.6\text{–}6.0) \times 10^{14}$ Hz
green	~5000–5600 Å	$\sim(6.0\text{–}5.4) \times 10^{14}$ Hz
yellow	~5600–6000 Å	$\sim(5.4\text{–}5.0) \times 10^{14}$ Hz
orange	~6000–6400 Å	$\sim(5.0\text{–}4.7) \times 10^{14}$ Hz
red	~6400–7500 Å	$\sim(4.7\text{–}4.0) \times 10^{14}$ Hz

In 1890 physicists thought that the wave theory of light was here to stay. But to everyone's surprise it started showing signs of weakness after only 10 years (as will be discussed in Chapter 31), leading to the introduction of quantum theory by Max Planck in 1900. In our study of optics (Chapters 28 to 30) we shall limit ourselves to the use of the wave theory of light as proposed by Huygens and Maxwell. We shall construct the quantum theory of light later in Chapter 31.

28.2. Huygens' Principle—Wave Fronts and Rays

Before stating Huygens' principle, we must explain what we mean by a wave front. A source of light emits electromagnetic waves which propagate in space with different points of vibration having different phases. A *wave front* is defined as the locus of all the points that are in the same phase of vibration of the physical quantity. For example, in the case of water waves (Chapter 17), any sphere drawn with its center at the source is a wave front. Of course, all the points on the sphere will not only be in the same phase but will have the same displacement. In the case of electromagnetic waves, a sphere with its center at the source will be a wave front with all the vibrations in phase and the electromagnetic field being equal at every point. Three such spherical wave fronts (WF) are shown in Figure 28.1.

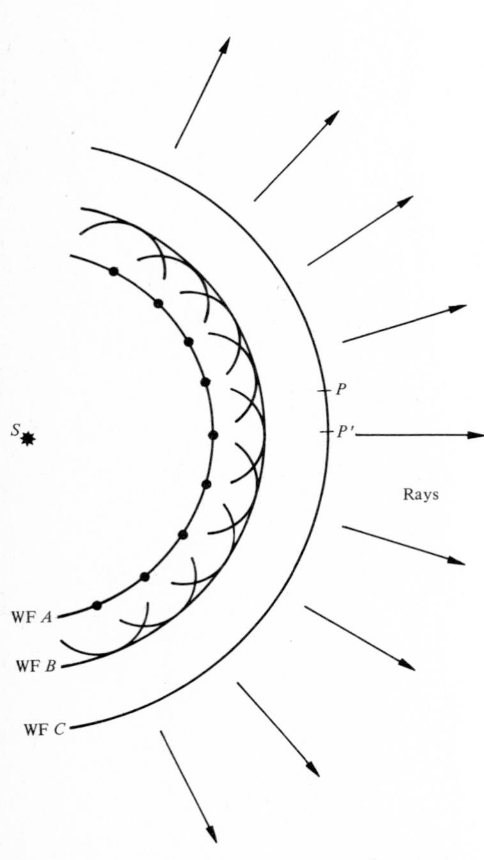

Figure 28.1. *Huygens' principle, showing spherical wave fronts and rays.*

The use of Huygens' principle tells us how the wave fronts propagate in space from point to point. According to

> **Huygens' principle:** ***Every point on a wave front may be considered as a source that produces secondary wavelets. These wavelets propagate in the forward direction with a speed equal to the speed of the wave motion.***

Let us consider a wave front A produced by a source of light S in Figure 28.1. Consider the dots on this wave front as secondary sources, each producing a spherical wave. This is done by drawing hemispheres of radii $c\,\Delta t$, where c is the speed of light and Δt is the time during which the wave propagates from one wave front to the other. If we join the points on these spherical surfaces which are in phase, the result is a new wave front, B. Similarly, we can construct a wave front C from wave front B. The process can be repeated. There are an infinitely large number of wave fronts, but we have just shown three of them.

If the speed of light is the same in all directions, the spherical wave front drawn in space in Figure 28.1 remains spherical. This means that the energy of the wave is transmitted equally in all directions as the waves propagate. The direction of the energy flow, that is, the direction in which the waves travel, is called a *ray*, and some are shown in Figure 28.1. A *beam* of light consists of many waves traveling parallel to each other in a given direction; that is, it is a bundle of rays. A beam is *monochromatic* if all the waves have the same wavelength. It is not possible to get a true monochromatic beam, but a beam spread over, say, ± 5 Å is fairly close to being monochromatic.

A medium in which the speed of the wave is the same in all directions is an *isotropic medium*. In an isotropic and homogeneous medium, the rays are always perpendicular to the wave fronts at every point. Such isotropic media are glass, water, Lucite, air, and many others. There are many crystals which are transparent to light, but the speed of propagation of the wave is different in different directions. Such media are called *anisotropic*. Typical examples are quartz and calcite. In an anisotropic medium, the rays are not always perpendicular to the wave front at every point.

Let us consider a spherical wave front which is far away from a source S. If we consider a small portion of this wave front, it will be an almost plane surface, such as PP' in Figure 28.1. Such a wave front is called a *plane wave front*. These plane wave fronts (plane surfaces parallel to each other and the rays that are perpendicular to those planes) are demonstrated in Figures 28.2(a) and (b). Figure 28.2(b) shows the configuration of the **E** and **B** fields and the propagation direction (vector **S**) in a plane electromagnetic wave.

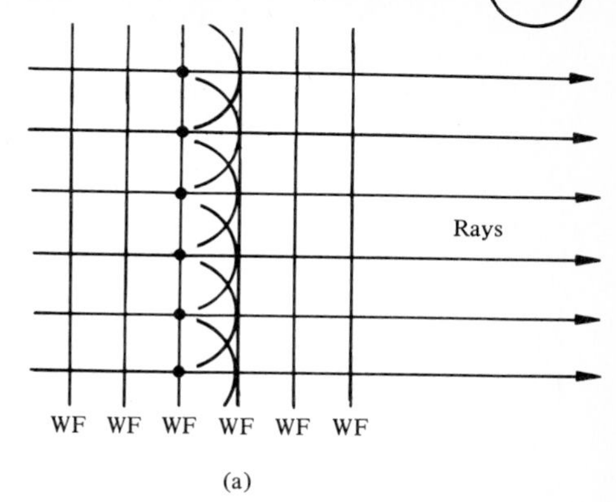

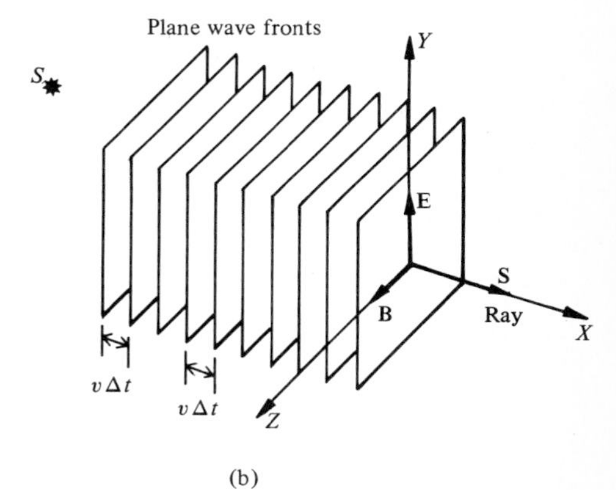

FIGURE 28.2. *Plane wave fronts and rays according to Huygens' principle: (a) two-dimensional view and (b) three-dimensional view, showing electric and magnetic vectors, and propagation direction.*

28.3. Laws of Reflection and Refraction

Consider that a beam of light traveling in a homogeneous medium strikes an object or comes across a second medium. Depending upon the type and smoothness of the surface, the beam may be deflected or transmitted into the medium, or both. The process of deflecting a beam is called *reflection* while the transmission of the beam is called *refraction*. If the surface is well polished like a mirror, better than 90% of the beam may be reflected. The process of reflection and refraction follows certain rules which we shall now discuss.

Laws of Reflection

Consider a ray of light directed toward a smooth surface or a mirror, as shown in Figure 28.3. The ray AO is called an *incident ray*, the ray OB that bounces back is called a *reflected ray*, and a line ON drawn perpendicular to the surface at the point O where the incident ray hits the surface is called a *normal*. The angle $AON = \theta_i$ between the incident ray and the normal is called the *angle of incidence*, and the angle $BON = \theta_r$ between the reflected ray and the normal is called the *angle of reflection*. Experiments in reflection have yielded the following:

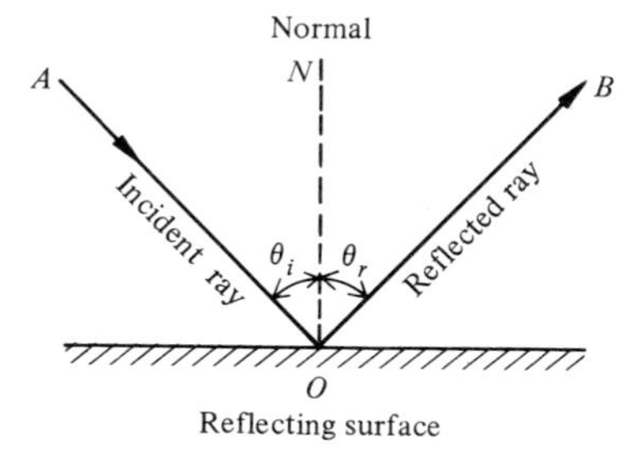

FIGURE 28.3. *Reflection at a smooth surface.*

Laws of Reflection:

Law 1: ***The incident ray, the reflected ray, and the normal to the surface all lie in the same plane.***

Law 2: ***The angle of incidence is always equal to the angle of reflection.***

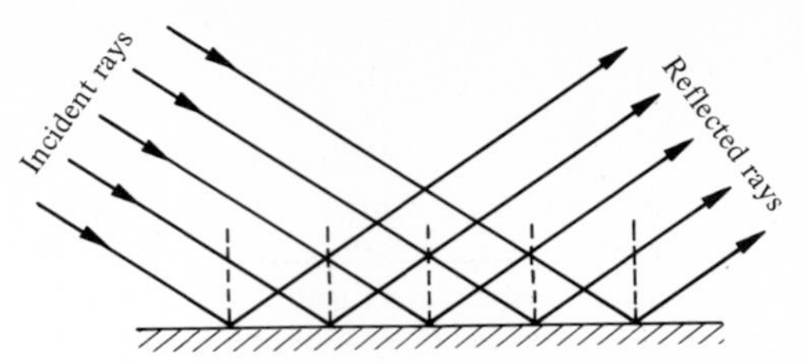

FIGURE 28.4. *Regular or specular reflection.*

A beam of light consisting of many rays, as shown in Figure 28.4, is incident on a smooth reflecting surface. Each ray will follow the laws of reflection, and all the rays in the reflected beam will be parallel as shown. Such a reflection is called a *regular* or *specular reflection*. On the other hand, if the reflecting surface is uneven, as shown in Figure 28.5, the reflected rays will not be parallel, but will be scattered in different directions. Such a reflection from a

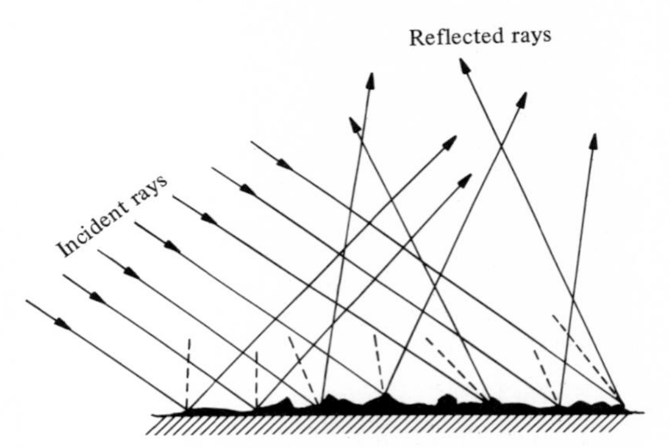

FIGURE 28.5. *Diffuse reflection.*

rough surface is called a *diffuse reflection*. How do we decide whether the surface is smooth or rough? If successive elevations on the surface between adjacent regions are less than a quarter of a wavelength of the incident light, the surface is said to be *smooth* or *polished*. A surface may act like a polished surface for a long wavelength of light, but may be rough for a short wavelength.

An interesting example of the two types of reflections is observed while driving a car at night. When the road is dry, the surface is rough. The light from the headlights is scattered in different directions (diffused) and reaches the driver, making the road visible. On the other hand, on a rainy night the road is covered with water, making a smooth surface. The reflection taking place is specular, and no reflected light reaches the driver, making it difficult to see the road clearly.

Laws of Refraction

Whenever a ray of light travels from one transparent medium to another, it deviates from its original path as shown in Figure 28.6. This bending of light rays at the interface of two media is called *refraction*. As an example of refraction, a stick dipped obliquely into water always looks bent at the surface. In Figure 28.6 the bent ray OC in the second medium is called the *refracted ray* and the angle $CON' = \theta_R$ which the refracted ray OC makes with the normal NON' is called the *angle of refraction*. The refracted ray will bend toward the normal when going from a rarer to a denser medium, as in Figure 28.6(a), or away from the normal when going from a denser to a rarer medium, as in Figure 28.6(b). The only exception to this is that when the ray is at normal incidence, that is, $\theta_i = 0$, no bending of the refracted ray takes place. Light rays traveling from one medium to another follow the laws given below.

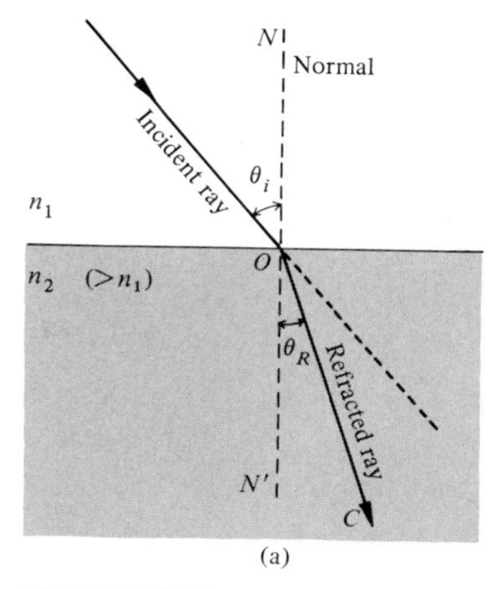

(a)

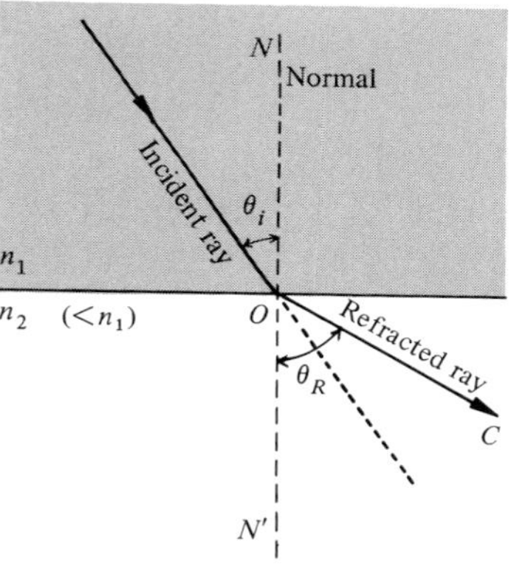

(b)

Figure 28.6. *Refraction of light. (a) From a rarer to a denser medium, and (b) from a denser to a rarer medium.*

Laws of Refraction:

Law 1: ***The incident ray, the normal, and the refracted ray all lie in the same plane.***

Law 2: ***The ratio of the sine of the angle of incidence to the sine of the angle of refraction is always constant (a different constant for different media).***

$$\frac{\sin\theta_i}{\sin\theta_R} = \text{constant} = n_{21} \tag{28.1}$$

Equation 28.1 was discovered experimentally in 1621 by the Dutch mathematician Willebrord Snell (1591–1627) and is known as *Snell's law*. The constant n_{21} in Equation 28.1 is called the *refractive index* of medium 2 with respect to medium 1 when a ray travels from medium 1 to medium 2.

A theoretical explanation of reflection and refraction was given by Huygens, who made use of the wave theory. Whenever a wave travels from one medium to another, the velocity of the wave changes. The velocity of any wave in vacuum is a maximum equal to $c = 3 \times 10^8$ m/s. The velocity of this wave in another medium will be less than c, say equal to v. We define the refractive index n of this medium to be

$$n = \frac{\text{speed of light in vacuum}}{\text{speed of light in medium}} = \frac{c}{v} \tag{28.2}$$

Since v is less than c, n is greater than 1. The refractive indexes of different media (with respect to vacuum) are listed in Table 28.1. Since the refractive index of air is almost unity (= 1.00029), we could define the refractive indexes of different media with respect to air instead of vacuum. A medium with a high value of n is optically dense as compared to a medium with a lower n value. Suppose that the velocity of light in medium 1 is v_1 and that in medium 2 is v_2. Thus, the corresponding refractive indexes n_1 and n_2 are

$$n_1 = \frac{c}{v_1} \quad \text{and} \quad n_2 = \frac{c}{v_2} \tag{28.3}$$

Dividing the second equation by the first and rearranging

$$\boxed{n_1 v_1 = n_2 v_2} \tag{28.4}$$

In Equation 28.1, if we replace n_{21} by n_2/n_1, θ_i by θ_1 and θ_R by θ_2, we get

TABLE 28.1
Index of Refraction
(Solids and Liquids at 20°C;
Gases at 0°C and 1 atm)

Substance	Index
Air	1.000293
Benzene	1.501
Carbon dioxide	1.00045
Carbon tetrachloride	1.461
Crown glass	1.52
Diamond	2.419
Ethyl alcohol	1.361
Flint glass	1.66
Fused silica	1.458
Hydrogen	1.000132
Ice	1.31
Water	1.333

$$\frac{\sin \theta_1}{\sin \theta_2} = \frac{n_2}{n_1}$$

or

$$\boxed{n_1 \sin \theta_1 = n_2 \sin \theta_2} \tag{28.5}$$

which is the general statement of *Snell's law.*

The speed of light waves may be different in different media, but the frequency ν (instead of f used previously) cannot change as the waves propagate from one medium to another. This means that for the relation $\nu = c/\lambda$ to hold good, its wavelength must change. Suppose that a wave motion of frequency ν, wavelength λ_1, and velocity v_1 passes from medium 1 to medium 2, where the corresponding quantities are ν, λ_2, and v_2. Let the refractive index of the two media be n_1 and n_2, respectively. Thus,

$$v_1 = \nu\lambda_1 \qquad \text{and} \qquad v_2 = \nu\lambda_2 \tag{28.6}$$

Dividing one by the other, and using Equation (28.3),

$$\frac{\lambda_2}{\lambda_1} = \frac{v_2}{v_1} = \frac{c/n_2}{c/n_1} = \frac{n_1}{n_2}$$

or

$$\boxed{\lambda_1 n_1 = \lambda_2 n_2} \tag{28.7}$$

Suppose that medium 1 is a vacuum; then $n_1 = 1$ and $\lambda_1 = \lambda_0$. From Equation 28.7 we get $\lambda_2 = \lambda_0/n_2$ or, in general,

$$\boxed{n = \frac{\lambda_0}{\lambda}} \tag{28.8}$$

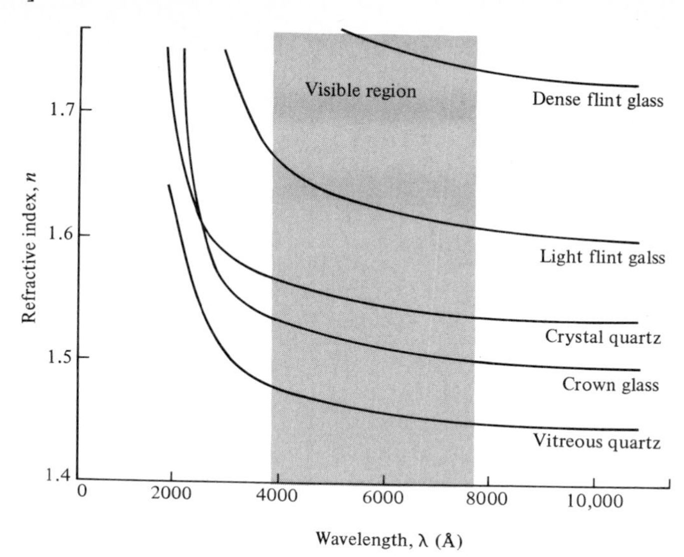

FIGURE 28.7. *Variation in the refractive index versus wavelengths for various materials.* (From E. Hecht and A. Zajac, *Optics*, Addison-Wesley Publishing Company, Inc., Reading, Mass., 1974.)

Thus, according to this equation, the refractive index depends upon the wavelength; that is, n varies with the color of light, even though the variation is very small. For example, the refractive index of glass for red, green, and violet light is 1.51, 1.52, and 1.53, respectively. Figure 28.7 shows the variation in the refractive index versus wavelength for different materials. The shaded area is the visible region.

Finally, we should say a few words about the intensity of the reflected and refracted beams. Suppose that a beam of light traveling in air is incident on a glass surface. Both reflection and refraction will take place. If the angle of incident is 0°, that is, at normal incidence, the ray travels undeviated into the second medium and the smallest percentage of light is reflected. As the angle of incidence increases, the intensity of the reflected beam increases, and becomes almost greater than 95 percent at $\theta \gtrsim 90°$, that is, at grazing incidence.

EXAMPLE 28.1 Light of a certain color has a frequency of 6×10^{14} Hz. Calculate its wavelength and number of waves per centimeter in vacuum. What is its speed and wavelength in glass of $n = 1.50$?

For vacuum,

$$\lambda_0 = \frac{c}{\nu} = \frac{3 \times 10^8 \text{ m/s}}{6 \times 10^{14} \text{ /s}} = 0.5 \times 10^{-6} \text{ m} = 5000 \text{ Å}$$

The number of wavelengths per centimeter is given by

$$\frac{1 \text{ cm}}{\lambda_0} = \frac{10^{-2} \text{ m}}{0.5 \times 10^{-6} \text{ m}} = 2 \times 10^4 = 20{,}000$$

For glass,

$$v = \frac{c}{n} = \frac{3 \times 10^8 \text{ m/s}}{1.50} = 2 \times 10^8 \text{ m/s}$$

From Equation 28.8, $n = \lambda_0/\lambda$ or for glass,

$$\lambda = \frac{\lambda_0}{n} = \frac{0.5 \times 10^6 \text{ m}}{1.50} = 0.3333 \times 10^{-6} \text{ m} = 3333 \text{ Å}$$

EXERCISE 28.1 Consider a light of wavelength 4000 Å. Calculate its frequency and the number of waves in a 2-cm-thick layer of air. Also, calculate its wavelength in glass of refractive index 1.66. How many waves are there in a 2-cm-thick glass? [*Ans.*: $\nu = 7.5 \times 10^{14}$ Hz; 5×10^4 waves; $\lambda_{\text{glass}} = 2410$ Å; 8.30×10^4 waves.]

Wave Theory

We shall now make use of Huygens' wave theory to derive the laws of reflection and refraction. To derive the laws of reflection, consider Figure 28.8. The wave front AA' is perpendicular to the incident rays. As ray 2 travels from A' to B, ray 1 from point A will spread out in a sphere of radius $AB' = A'B = v_1 t$. The line BB' is drawn so that it is tangent to the sphere at B'. Thus, BB' is a new wave front of the reflected beam. From geometry it is clear that $AA' = BB'$. Consider the right-angle triangles $B'BA$ and $A'AB$. These triangles are congruent because $BB' = AA'$ and AB is common to both of them. Therefore, $\angle B'BA = \angle A'AB$ or $90° - \theta_i = 90° - \theta_r$. That is, $\theta_i = \theta_r$.

Let us now consider a general case in which the plane wave front of a ray of light is incident on a surface separating two media of refractive index n_1 and n_2, as shown in Figure 28.9. It is assumed that $n_2 > n_1$; that is, n_2 is optically denser; hence, light waves will travel slower in medium 2 than in 1. Figure 28.9 also shows the reflected and refracted wave fronts. Since $v_2 < v_1$, the refracted wave fronts separated by distance $v_2 \Delta t$ are closer than the reflected wave fronts separated by distances $v_1 \Delta t$.

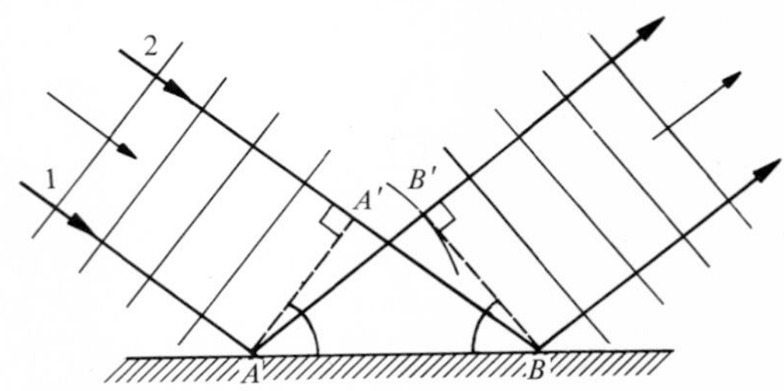

FIGURE 28.8. *Use of Huygens' principle to derive the laws of reflection.*

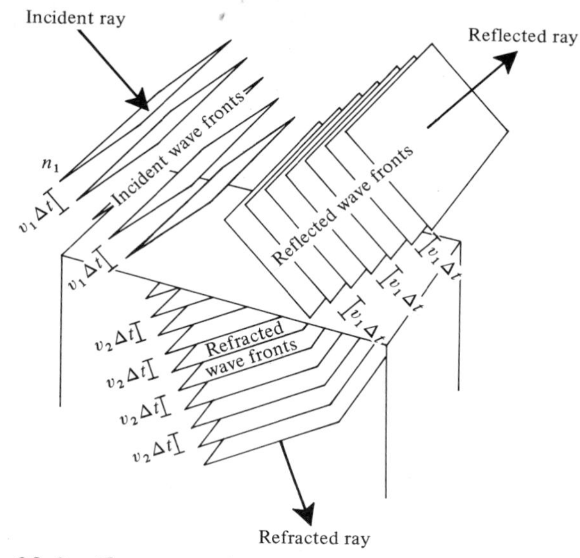

FIGURE 28.9. *Plane wave fronts incident on a surface separating two media.*

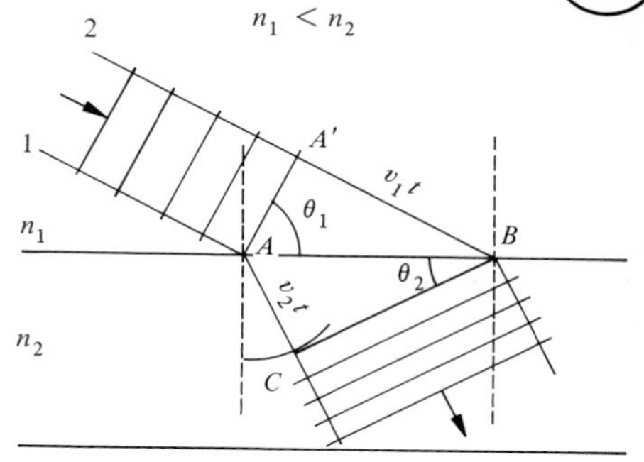

FIGURE 28.10. *Use of Huygens' principle to derive the laws of refraction.*

To derive Snell's law of refraction, let us refer to Figure 28.10. Consider a plane wave front AA'. In time t ray 2 travels a distance $A'B = v_1 t$. During the same time, ray 1 enters medium 2 and travels with velocity $v_2 < v_1$. Draw a sphere of radius $AC = v_2 t$ with A as the center. Draw line BC so that it is tangent to the sphere at C. Thus, BC is a new wave front traveling in medium 2. From the two right triangles $AA'B$ and ACB, we get

$$\sin \theta_1 = \frac{v_1 t}{AB} \quad \text{and} \quad \sin \theta_2 = \frac{v_2 t}{AB}$$

Dividing one by the other and using the relation $v = c/n$,

$$\frac{\sin \theta_1}{\sin \theta_2} = \frac{v_1}{v_2} = \frac{c/n_1}{c/n_2} = \frac{n_2}{n_1}$$

or

$$n_1 \sin \theta_1 = n_2 \sin \theta_2$$

which is Snell's law.

EXAMPLE 28.2 A ray of light making an angle of 30° with the normal enters a glass plate 4 cm thick having a refractive index 1.50 as shown in the accompanying figure. Calculate the distance the light ray travels in the glass before it leaves the plate. What angle does the emerging ray make with the normal?

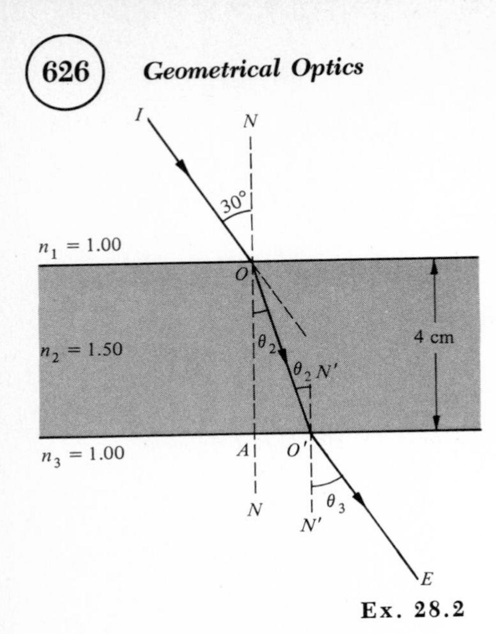

Ex. 28.2

Applying Snell's law, Equation 28.5, at the interface at O:

$$\sin\theta_2 = \left(\frac{n_1}{n_2}\right)\sin\theta_1 = \left(\frac{1.00}{1.50}\right)\sin 30^\circ = \frac{1}{1.5}0.50 = 0.33$$

or

$$\theta_2 = 19.5^\circ$$

From triangle OAO', $(OA/OO') = \cos\theta_2$, or

$$OO' = \frac{OA}{\cos\theta_2} = \frac{4\text{ cm}}{\cos 19.5^\circ} = \frac{4\text{ cm}}{0.94} = 4.26\text{ cm}$$

Applying Snell's law once again at the interface at O',

$$\sin\theta_3 = \left(\frac{n_2}{n_3}\right)\sin\theta_2 = \left(\frac{1.50}{1.00}\right)\sin 19.5^\circ = (1.5)(0.33) = 0.50$$

or

$$\theta_3 = 30^\circ$$

That is, ray $O'E$ is parallel to IO.

Exercise 28.2 Repeat Example 28.2 if the glass plate is replaced by water ($n = 1.33$). Also, calculate AO'. [*Ans.:* $\theta_2 = 22.1^\circ$, $OO' = 4.32$ cm, $\theta_3 = 30^\circ$, $AO' = 1.62$ cm.]

28.4. Total Internal Reflection and Apparent Depth

Total Internal Reflection

So far we have been discussing refraction for the case in which light travels from an optically rarer to a denser medium. Let us consider the reverse situation. That is, light is traveling from an optical denser medium such as water or glass into an optical rarer medium such as air. The situation is as shown in Figure 28.11(a), where S is a source of light in a medium of refractive index n_1 such that $n_1 > n_2$. From Snell's law, $n_1 \sin\theta_1 = n_2 \sin\theta_2$ if $n_1 > n_2$, $\sin\theta_2$ must be greater than $\sin\theta_1$. In this case the angle of refraction θ_2 is always greater than the angle of incidence θ_1. As shown in Figure 28.11(a), as the angle of incidence increases, the angle of refraction also increases. When the angle of refraction is 90°, as shown for ray 4, the refracted ray just grazes along the surface, separating the two media. Any further increase in the angle of incidence will turn the refracted ray back into the same medium as for ray 5. This ray is said to be *totally internally reflected*. Thus, total internal reflection takes place if the angle of incidence is greater than a certain angle, called the critical angle θ_c, defined below.

When a ray of light travels from a denser to a rarer medium, the *critical angle*, θ_c, is that angle of incidence for which the corresponding angle of refraction is 90°. The situation is shown in Figure 28.11(b). Substituting $\theta_1 = \theta_c$, $\theta_2 = 90^\circ$, $\sin 90^\circ = 1$ in Snell's law,

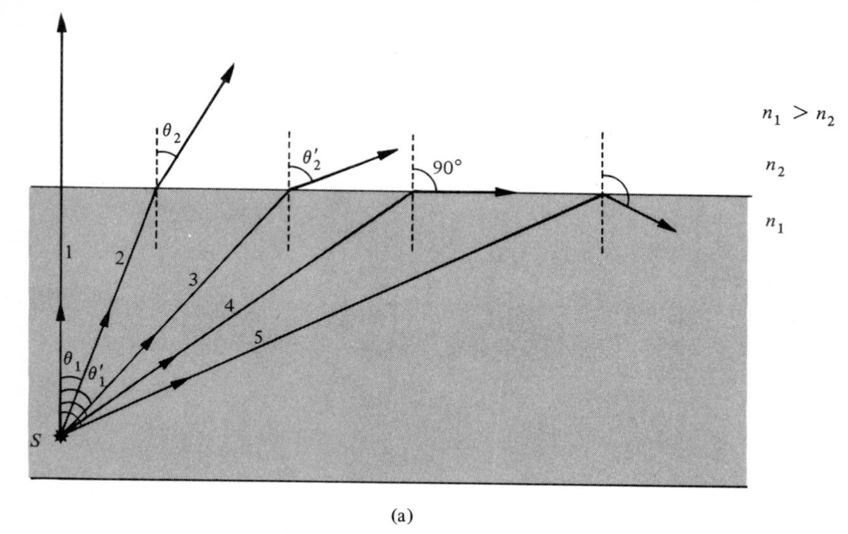

(a)

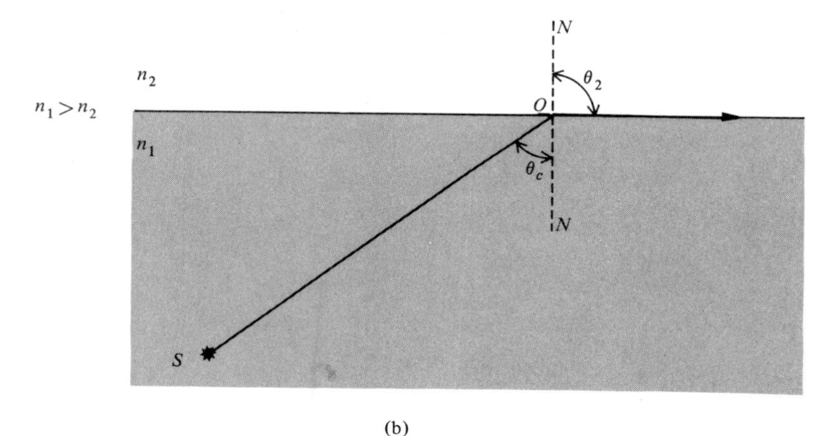

(b)

FIGURE 28.11. *(a) Refraction from an optically denser medium to an optically rarer medium, illustrating internal reflection and (b) the critical angle.*

$$\boxed{\sin\theta_c = \frac{n_2}{n_1}} \tag{28.9}$$

For a water–air system, $n_1 = 1.33$, $n_2 = 1$; hence, $\sin\theta_c = 1/1.33 = 0.75$ and $\theta_c \approx 49°$. For a glass–air system, $n_1 = 1.50$, $n_2 = 1$; $\sin\theta_c = 1/1.50 = 0.67$ or $\theta_c \approx 42°$.

An experimental situation demonstrating total internal reflection is shown in Figure 28.12. A source of light is placed at the bottom of a glass prism. The light is divided into several beams by using a slit system. For angles smaller

Figure 28.12. *Total internal reflection and the critical angle.* (Courtesy of Education Development Center, Inc., Newton, Mass.)

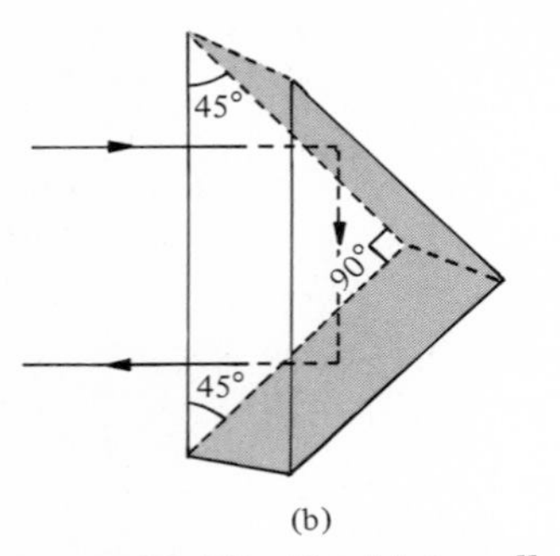

Figure 28.13. *Total internally reflecting prisms: (a) beam reflected through 90° and (b) beam reflected through 180°.*

than the critical angle, when the beam strikes the surface, part of it is refracted and part reflected back into the glass. If the angle is greater than the critical angle, all of the beam is totally internally reflected, with no refraction at all. This principle is utilized in colored water fountains. Once the colored light gets into the water stream, as long as the angle of incidence is greater than the critical angle, the colored beam will remain in the stream.

Since the critical angle for a glass–air system is 42°, a 45°–45°–90° glass prism is utilized in turning a light beam through 90° or 180° as shown in Figure 28.13. Such a prism is used in optical instruments. There are many advantages of using a prism rather than a metallic surface (or mirror) for this purpose. First, the light is totally reflected (no absorption or refraction), while no metallic surfaces will be 100 percent efficient in reflection. Second, unlike metallic surfaces, prisms do not tarnish. A little loss of light by reflection when the light enters or leaves the prism can be avoided by coating the surface with a nonreflecting film.

Figure 28.14 shows a light pipe made of a transparent material such as Lucite. Once the light enters such a pipe, the angle of incidence is greater than

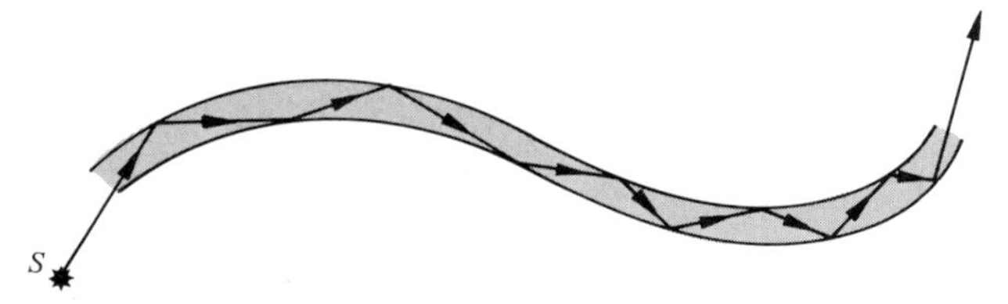

FIGURE 28.14. *Light pipe made of a transparent material such as Lucite.*

the critical angle; hence, the light is totally reflected. As long as the curvature of the pipe is not too great, the light once entering the pipe will be successively totally internally reflected until it comes out of the other end.

Putting thousands of thin fibers together results in a flexible light pipe. Such pipes made of fibers can be used for transmitting light, as well as images, as shown in Figure 28.15. Presently *fiber optics* is an active research field. A fiber pipe, together with a laser beam, is used in medical applications, such as in

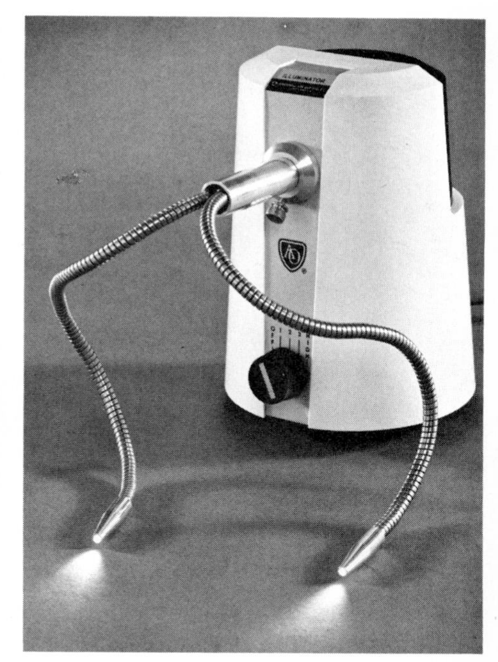

FIGURE 28.15. *Demonstrating a flexible light pipe made by putting thousands of thin fibers together.* (Courtesy of American Optical Corporation.)

surgery, where lamps producing too much heat cannot be used. Inaccessible parts of the body, such as under the skin and in the lungs, can be examined by looking at the images transmitted by using extremely thin fiber pipes.

Apparent Depth

When we look into a clear pond of water or a swimming pool, the bottom seems to be raised; that is, the apparent depth seems to be less than the actual depth. This is easily understood by the phenomenon of refraction. Suppose that an object is placed at O in the water. Ray 1 travels vertically upward without bending. When ray 2 is at point A', it bends away from the normal, as shown in Figure 28.16. When ray 2 is produced backward, it meets ray 1 at I. Thus, someone looking at the object at O from the position B of ray 2 will see the

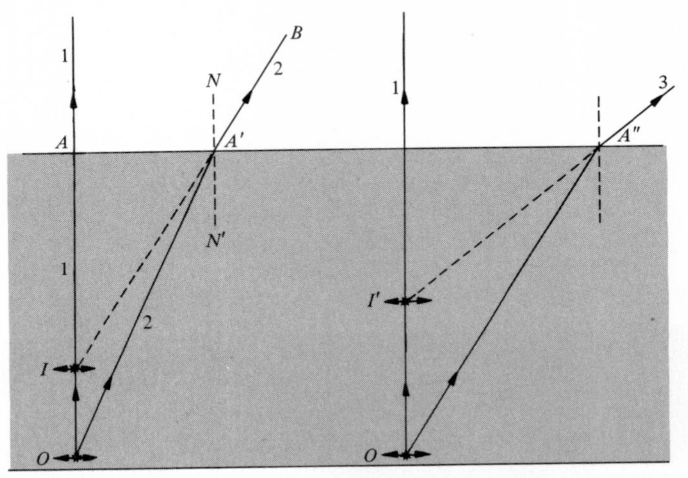

FIGURE 28.16. *Apparent depth of an object in water.*

image of O to be at I. Similarly, if a person is located at such a position that ray 3 enters his eye, the image will be located at I'. The object seems to be raised because the rays seem to diverge from a higher point I'. By definition of the index of refraction,

$$n = \frac{\sin NA'B}{\sin OA'N'} = \frac{\sin AIA'}{\sin AOA'} = \frac{AA'/IA'}{AA'/OA'} = \frac{OA'}{IA'}$$

If we look at the object very close to the normal incidence, we may write $OA'/IA' = OA/IA$. Therefore,

$$n = \frac{OA}{IA} = \frac{\text{actual depth}}{\text{apparent depth}}$$

or

$$\boxed{\text{apparent depth} = \frac{1}{n} \times \text{actual depth}} \tag{28.10}$$

Thus, for water, $n = 4/3$, $1/n = 3/4$; hence, the depth of the pool will look to be 3/4 of its actual depth.

Before closing this section we may point out that an optical phenomenon such as a mirage in a desert (shown in Figure 28.17) is due to the continuous variation in the refractive index of hot air above the earth surface.

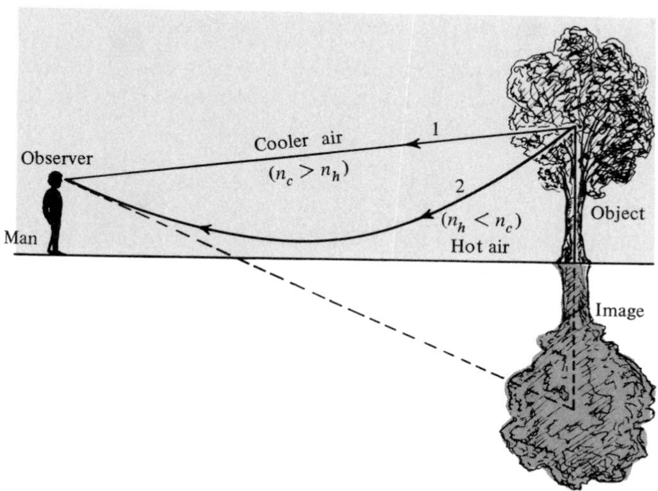

FIGURE 28.17. *A mirage is observed when the air near the ground is hotter than the air above. Thus, ray 2, traveling from a denser to a rarer medium, suffers total internal reflection; and when produced backward forms a virtual inverted image.*

EXAMPLE 28.3 A source of light is located at a corner inside a 3-m-deep pool as shown in the accompanying figure. If you were standing 4 m away on the side of the pool, will the light be totally reflected? If you were standing directly above the source of light, what is the apparent depth of the light source?

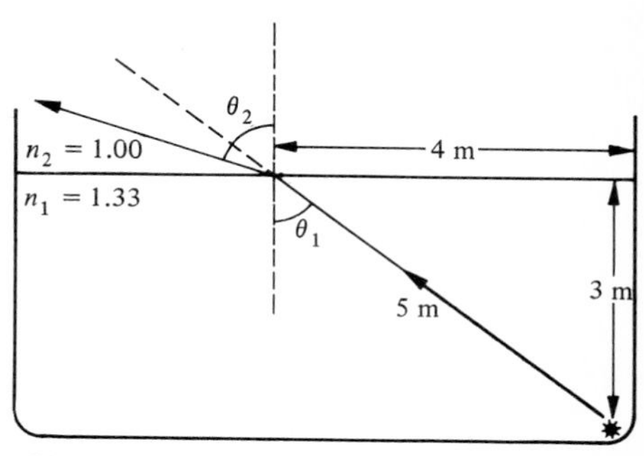

Ex. 28.3

From Equation 28.9,

$$\sin\theta_c = \frac{n_2}{n_1} = \frac{1.00}{1.33} = 0.75 \qquad \text{or} \qquad \theta_c = 49°$$

From the figure, $\sin\theta_1 = 4/5 = 0.8$ or $\theta_1 = 53°$. Since $\theta_1 > \theta_c$, there will be a total internal reflection. Also, from Equation 28.10,

$$\text{apparent depth} = \frac{1}{n} \times \text{actual length} = \frac{1}{1.33} \times 3\text{ m} = 2.26\text{ m}$$

EXERCISE 28.3 If water in Example 28.3 were replaced by a material of (a) $n = 1.2$; and (b) $n = 1.6$, will there be total internal reflection? Also, calculate the apparent depth in each case. [*Ans.:* (a) $\theta_c = 56.4°$; $\theta_1 < \theta_c$, no internal reflection. Apparent depth = 2.5 m; (b) $\theta_c = 38.7°$; $\theta_1 > \theta_c$, total internal reflection, apparent depth = 1.9 m.]

28.5. Deviation and Dispersion by a Prism

A prism is a wedge-shaped figure bounded by two plane surfaces. We briefly mentioned its use for total reflection, and this section is devoted to two other aspects, deviation and dispersion.

Deviation

Consider a monochromatic beam of light incident on a face of a prism, as shown in Figure 28.18. At a point O where the ray enters the prism, the angle of incidence is i while the angle of refraction is r. The ray bends toward the normal NN. When the ray emerges from the second surface, the angle of incidence is r', the angle of refraction (or emergence) is i', and the emergent ray is bent away from the normal. The angle D through which the ray is deviated from its initial path is called the *angle of deviation* (Figure 28.18). As the angle i is changed, the angle D also changes. For a certain particular value

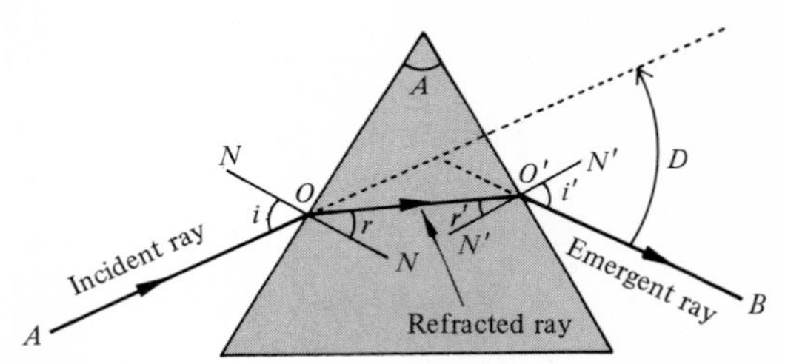

FIGURE 28.18. *Deviation of a beam of light through a prism.*

of i, say i_m, the value of D is minimum, say D_m. For i greater or smaller than i_m, D is larger than D_m. Thus, D_m is called the *angle of minimum deviation.*

It is simple to show that when the prism is set in a position of minimum deviation, $i = (A + D_m)/2$ and $r = A/2$, where A is the angle of the prism. Thus, applying Snell's law at a point on the first surface, the refractive index n of the glass is given by

$$n = \frac{\sin i}{\sin r} = \frac{\sin [(A + D_m)/2]}{\sin (A/2)} \tag{28.11}$$

Also, the angle of minimum deviation will be different for different wavelengths. Knowing the angle A of the prism and measuring D_m for different colors (or wavelengths), we can calculate n versus λ for different material. Such plots are already shown in Figure 28.7.

Dispersion

Let a narrow beam of white light be incident on a prism, as shown in Figure 28.19. After passing through the prism, the beam is split into a band of colors called a *spectrum*. The splitting of white light into its component colors

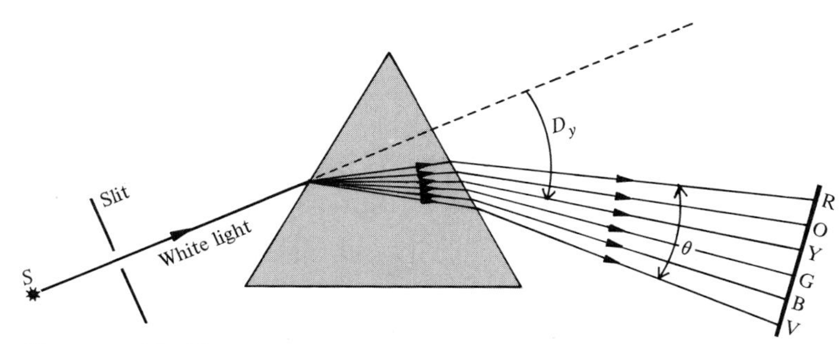

FIGURE 28.19. *Dispersion of a beam of white light through a prism.*

is called *dispersion*. As shown in Figure 28.19, violet light is deviated the most and red the least. The red light travels faster in the prism than violet, and hence its refractive index (1.51 in glass) in less than that of violet (1.53). Thus, from Equation 28.11, the deviation of red light (which has a longer wavelength) is less than that of violet. This explanation holds for other colors in between. In order to see the dispersion, we must use a very well defined narrow beam of light. If a broad beam is used, all the colors will overlap because of the small dispersion. Dispersion is measured in terms of the angle of deviation per unit wavelength, that is, $\Delta D_m/\Delta\lambda$. The mean dispersion for white light is referred to in terms of yellow light. It is possible to combine two different prisms in such a way that if light of different wavelengths falls on the system, it produces (1) deviation with no net dispersion, or (2) dispersion with no net deviation.

Such phenomenon as the formation of a rainbow is due to the combined effects of refraction, internal reflection, and dispersion of sunlight by rain droplets, as illustrated in Figure 28.20. These droplets behave like tiny prisms and cause splitting of white light into different colors.

EXAMPLE 28.4 Calculate the angle of minimum deviation for a 60° prism made out of flint glass of refractive index 1.66.

Since $n = 1.66$, $A = 60°$, and $\sin(A/2) = \sin 30° = \frac{1}{2}$, from Equation 28.11,

$$\sin(A + D_m)/2 = n \sin(A/2) = 1.66 \times \tfrac{1}{2} = 0.83$$

$$(A + D_n)/2 = 56° \qquad \text{or} \qquad D_m = 2(56°) - A = 112° - 60° = 52°$$

EXERCISE 28.4 A 60° prism is made out of a material with a refractive index of 1.84. What is the angle of minimum deviation in this case? [*Ans.:* 74°.]

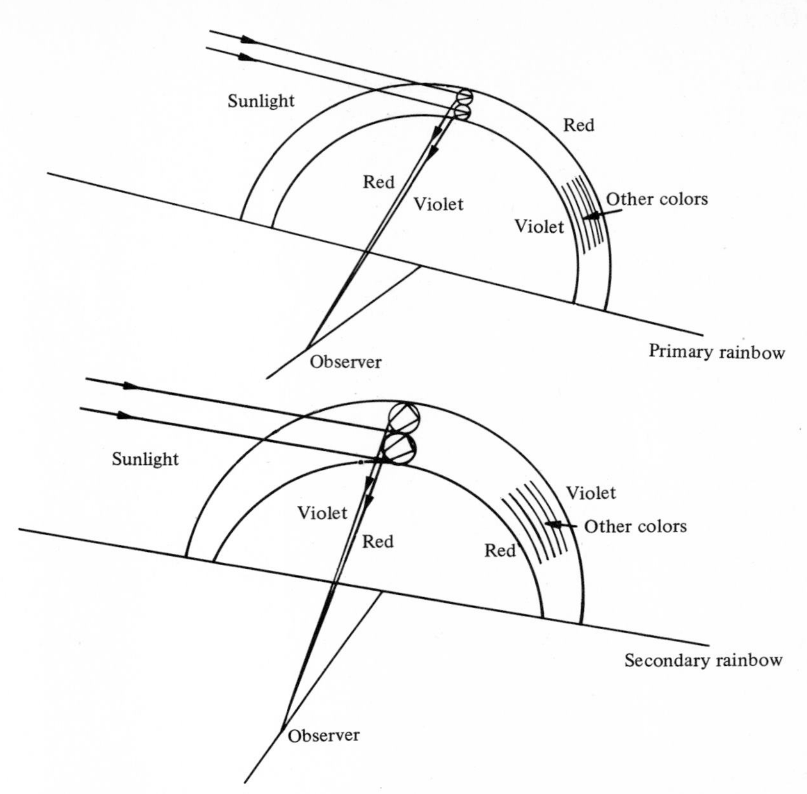

Figure 28.20. *Formation of primary and secondary rainbows. Small circles are water droplets in air working as a prism, causing dispersion of sunlight. (Of a large collection, only two drops are shown in each case.)*

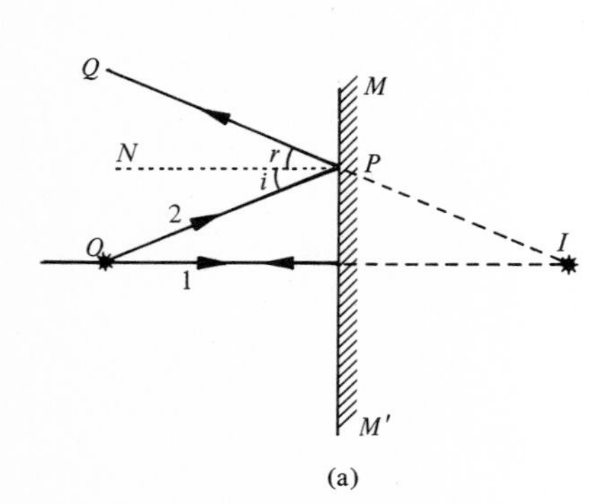

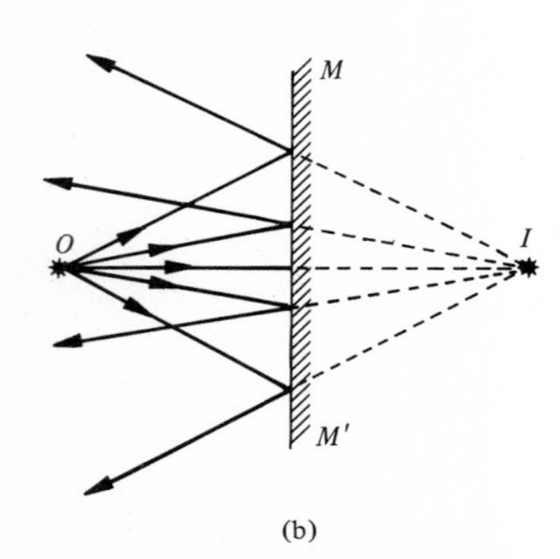

Figure 28.21. *Image formation of a point object by a plane mirror.*

28.6. Image Formation by Mirrors

When we stand in front of a mirror, what we see in the mirror is an image of ourselves. The position and size of the image depends upon the type of mirror—plane or curved—and can be calculated by using the laws of reflection.

Plane Mirror

Consider an object O placed in front of a mirror MM', Figure 28.21(a). Two light rays 1 and 2 starting from O traveling toward the mirror are reflected from it. Ray 1, which is normal to the mirror, is reflected directly opposite its incident path. Ray 2, striking the mirror at P, makes an angle i with the normal PN drawn at P and the reflected ray PQ makes an angle r with this normal such that $\angle r = \angle i$. The two reflected rays seem to meet at I, when they are produced backward (dashed lines). The point I is called the image of the object O. Since the rays do not actually meet but just seem to meet, the image is a *virtual image*. A minimum of two rays is needed to locate an image as shown in Figure 28.21(a). But there are many more rays that leave the object. If we

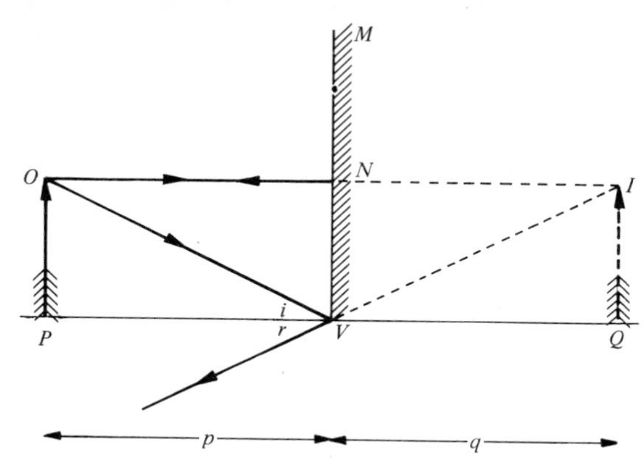

FIGURE 28.22. *Image formation of a finite-size object by a plane mirror.*

apply the laws of reflection, all the rays seem to diverge from the image at point I, as shown in Figure 28.21(b).

The laws of reflection can be applied to locate the image of a finite-sized object as shown in Figure 28.22. It is clear that triangles OPV and IQV are congruent, and so are triangles OPV and INV; hence, $NI = NO$. *Thus, the image formed in the mirror is as far behind the mirror as the object is in front of it.* Also, since $IQ = OP$, the size of the image is equal to the size of the object. We define the *linear (lateral or transverse) magnification* as

$$\text{linear magnification } M = \frac{\text{image size}}{\text{object size}} = \frac{IQ}{OP} \tag{28.12}$$

Since $IQ/OP = q/p$ and $q = p$,

$$\boxed{M = \frac{q}{p} = 1} \tag{28.13}$$

The value $M = 1$ indicates that the image size = object size.

Finally, let us consider another important property of the image formed by a plane mirror. If you stand in front of a mirror and raise your right hand, the image raises its left hand. Your heart is on your left, but the heart of your image is on the right. This is called a left–right *reversal* or *inversion* and the image has inversion. This is demonstrated in Figure 28.23. Of the three coordinate axes, X, Y, and Z the images of Y and Z remain unchanged, but the $+X$-axis in the mirror becomes the $-X$-axis. That is, if you are facing along the $+X$-axis, your image will face along the $-X$-axis. Thus, we may conclude: *The image formed by a plane mirror is as far beyond the mirror as the object is in front; it is erect, virtual, and has inversion and unit magnification.*

Curved Mirrors

A spherical mirror is a segment of a spherical shell, as shown in Figure 28.24(a). If the inner surface of this mirror is polished, as in Figures 28.24(a) or (b), it is a *concave mirror.* If the outer surface is polished, as in Figure 28.25,

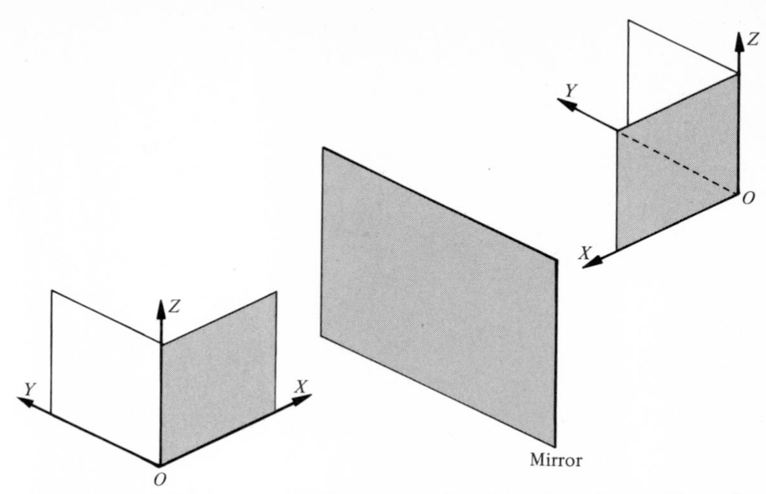

Figure 28.23. *Image by a plane mirror has inversion; that is, if you face along the +X-axis, your image faces along the −X-axis.*

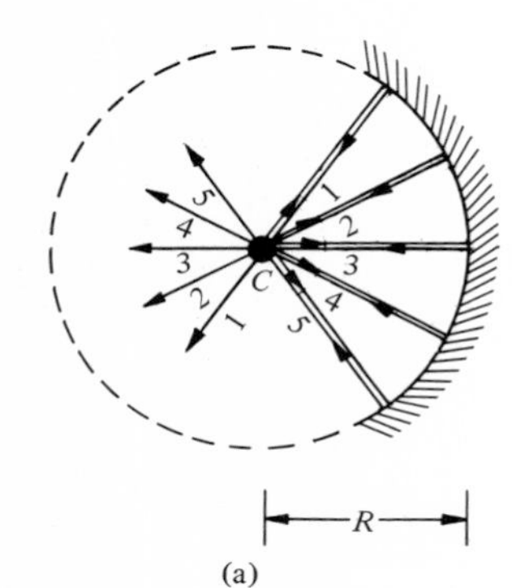

it is a *convex mirror*. The center of the sphere from which the mirror was cut is called the *center of curvature*, *C*, of the mirror. The radius of the sphere is called the *radius of curvature*, *R*, of the mirror. The middle point of the mirror surface is the *vertex*, *V*. The line *CV* through the vertex and the center of curvature is called the *principal axis*. These points are illustrated in Figures 28.24 and 28.25.

Suppose that we place a source of light at the center of curvature *C* of a concave mirror, as shown in Figure 28.24(a). Every ray of light falling on this

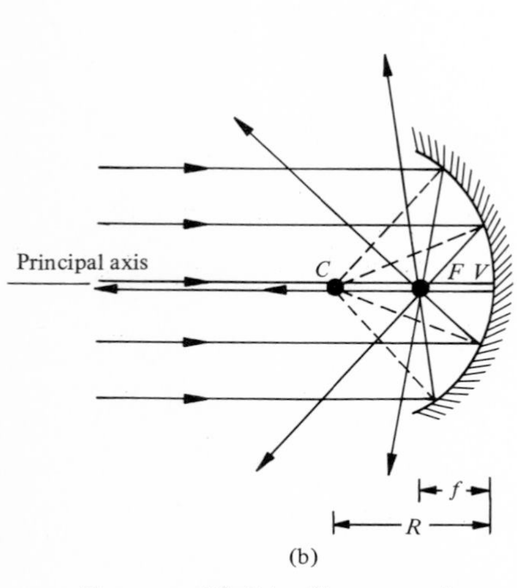

Figure 28.24. *Concave mirror.*

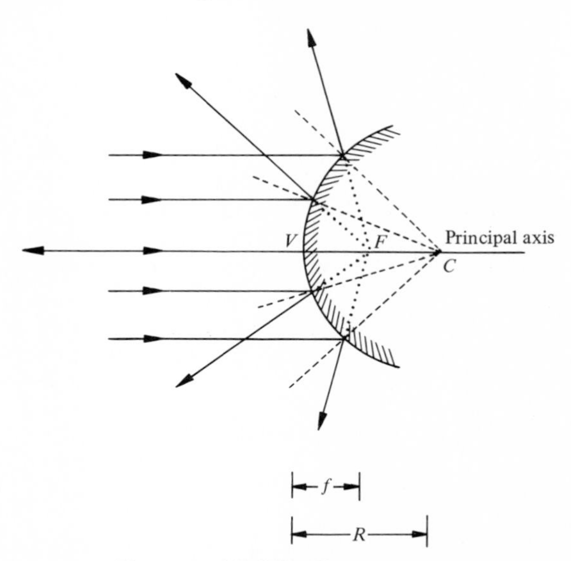

Figure 28.25. *Convex mirror.*

mirror will be normal at the point where it strikes the mirror. Hence, each and every ray is reflected back as shown. Suppose that we send many rays of light that are parallel to the principal axis toward a concave mirror, as shown in Figure 28.24(b). These parallel rays after being reflected from the concave mirror converge to a common point F, called the *principal focus*. In the case of a convex mirror, the parallel rays after reflection diverge, as shown in Figure 28.25. But if these reflected rays are projected backward, they seem to diverge from a point called the principal focus F, as shown in Figure 28.25. Because of these characteristics, the concave mirror is also called a *converging mirror* (Figure 28.24), while a convex mirror is called a *diverging mirror* (Figure 28.25). In each case the distance between the vertex and the principal focus is called the *focal length, f*.

We shall now show that the focal length f is equal to half the radius of curvature R. Consider a ray AB parallel to the principal axis incident on a concave mirror, as shown in Figure 28.26. After reflection this ray will go through the principal focus F. The line BC is normal to the mirror at point B; hence, $\angle i = \angle r$. Since AB is parallel to CF, $\angle \theta = \angle i = \angle r$. Therefore, the triangle CFB is isosceles and $CF = FB$. If B is not too far away from V, FB will be almost equal to FV. Thus, $FV = CF$; that is, F is halfway between V and C. Therefore,

$$\boxed{f = \frac{R}{2}} \tag{28.14}$$

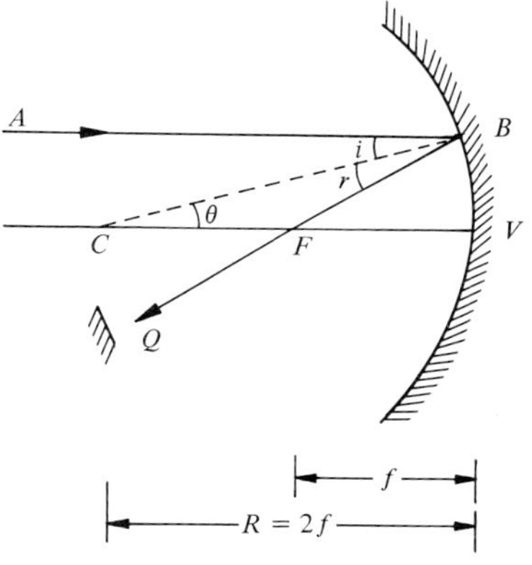

Figure 28.26. *Relation between f and R, f = R/2.*

This relation holds good for a convex mirror as well.

Let us place a plane mirror at Q normal to the ray BF. This ray will be reflected back and will retrace the path along which it came, that is, $QFBA$. This is called the principle of *reversibility*.

Suppose that an object OP is placed in front at a distance p from the center of a concave mirror, as shown in Figure 28.27(a). Let the image be formed at a distance q from the center of the mirror. In order to locate the image graphically the following three rays starting from any point of the object are used.

1. Ray 1 parallel to the principal axis will go through the focal point of a concave mirror after reflection (or appear to be coming from a focal point of a convex mirror).
2. Ray 2, which goes through the focal point (or appears to proceed in that direction) is parallel to the principal axis after reflection from the mirror.
3. Ray 3, going through the center of curvature (or appearing to), is normal to the mirror surface, hence is reflected back along the same path.

The point where these three rays intersect is the image of the corresponding point on the object from where the rays started. It is not necessary to use all three rays; two are enough; the third may be used to confirm the result. The use of this graphical method is well demonstrated in five different cases in Figure 28.27. When the object is brought from infinity toward a concave mirror no closer than the principal focus, the image is always inverted and real.

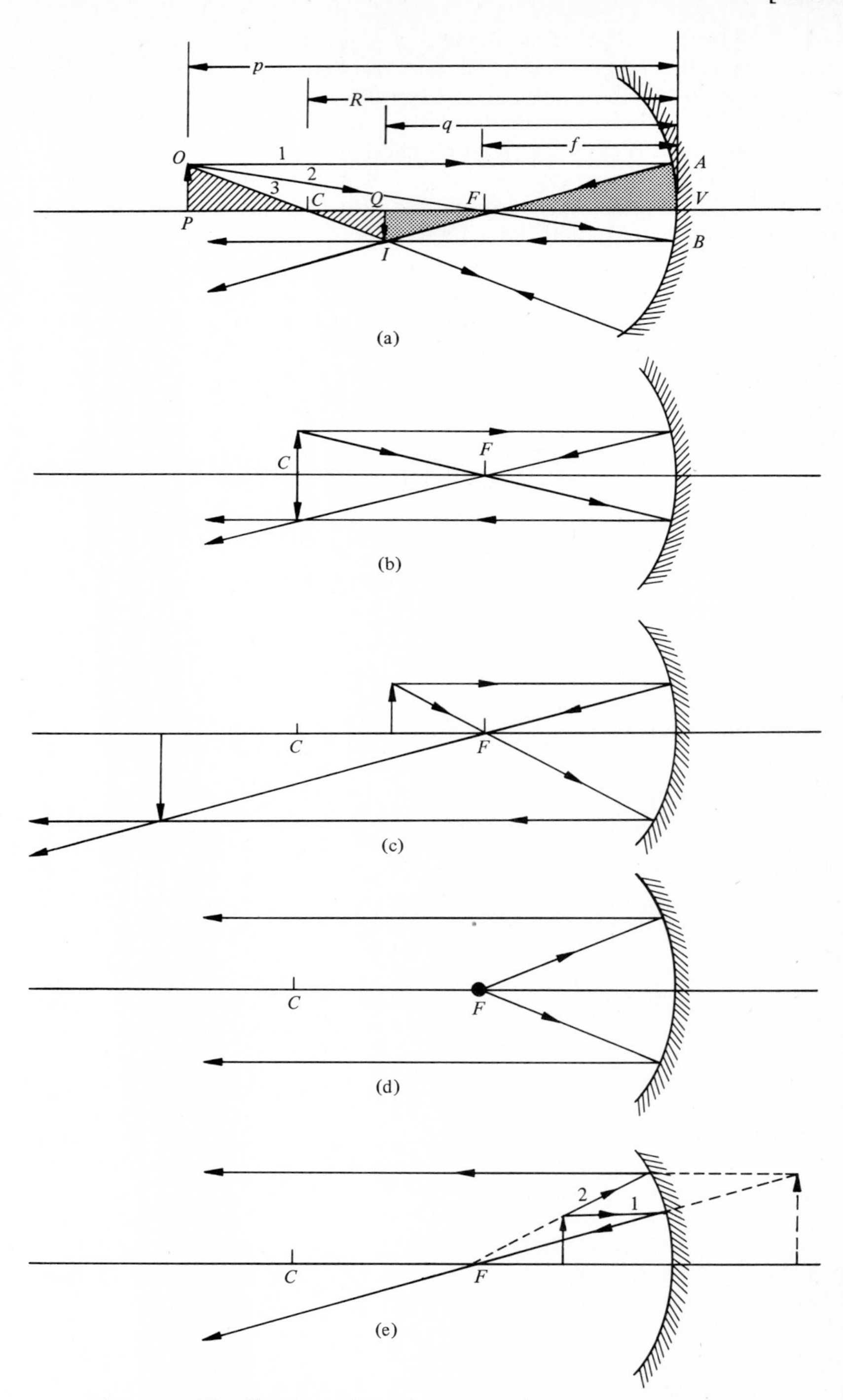

FIGURE 28.27. *Formation of an image by a concave mirror.*

Object location	*Image location*
$\infty > p > R$	$R > q > f$
$p = R$	$q = R$
$R > p > f$	$\infty > q > R$
$p = f$	$q = \infty$

The last case in Figure 28.27(e) is different from the rest. The object is between the vertex and the principal focus. As shown, rays 1 and 2 both diverge after reflection. But if extended backward, they do intersect. Thus, the image formed is virtual and erect.

In many situations it is not convenient to use a graphical method, but an analytical method is preferable which gives a relation between p, q, and f. Referring to Figure 28.27(a), let us assume that AV is almost a straight line, so that it is equal to OP. From triangles FAV and FQI, and $AV = OP$,

$$\frac{AV}{IQ} = \frac{OP}{IQ} = \frac{VF}{QF} = \frac{f}{q - f} \tag{28.15}$$

From triangles OPC and IQC,

$$\frac{OP}{IQ} = \frac{PC}{QC} = \frac{p - R}{R - q} = \frac{p - 2f}{2f - q} \tag{28.16}$$

Equating the two equations above,

$$\frac{f}{q - f} = \frac{p - 2f}{2f - q}$$

or

$$qf + pf = pq$$

Dividing both sides by pqf, we get

$$\boxed{\frac{1}{p} + \frac{1}{q} = \frac{1}{f}} \tag{28.17}$$

which is the *mirror formula*. Thus, when $p = \infty$, $q = f$, and when $p = f$, $q = \infty$, as it should be.

Next we would like to find the *linear* (lateral or transverse) *magnification M* defined as the ratio of the size of the image to the object size. Thus, from Figure 28.27(a) and Equation 28.15,

$$M = \frac{\text{image size}}{\text{object size}} = \frac{IQ}{OP} = \frac{q - f}{f} = \frac{q}{f} - 1$$

Substituting for $1/f$ from Equation 28.17, we get

$$M = q\left(\frac{1}{p} + \frac{1}{q}\right) - 1 = \frac{q}{p} + 1 - 1 = \frac{q}{p}$$

or adding a minus sign, which will indicate that the image is inverted, the magnification is given by

$$M = -\frac{q}{p} \tag{28.18}$$

The preceding ideas can be extended to the case of a convex mirror if we use the following convention: *For a typical case p, q and f are each positive on the side from which light is incident on a converging mirror forming a real image of a real object. Any deviation from this we must add a negative sign.*

Thus for a diverging convex mirror f (on the opposite side of the incident light) will be negative. In Figure 28.27(e) q (to the right of the mirror) is negative because the image is virtual and when this value is substituted in Equation 28.18, M is positive, denoting an erect image, as it should. Figure 28.28 shows the formation of an image in the case of a convex mirror. In

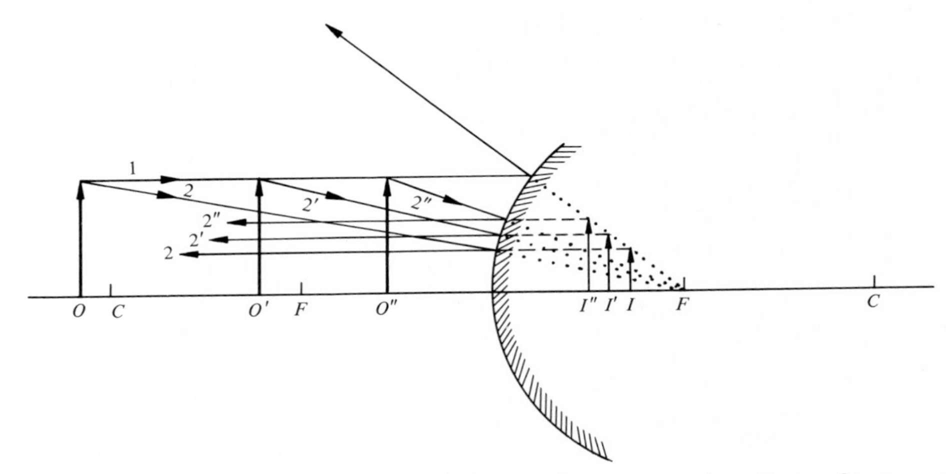

FIGURE 28.28. *Formation of an image by a convex mirror for an object located at O, O′, and O″.*

this case no matter what the location and size of the object, the image is always smaller, virtual, and erect. This property of a convex mirror is utilized in rear-view mirrors in some vehicles. In large stores it is used to keep watch on various portions of the store, because the reflection in a small convex mirror can correspond to a large region.

EXAMPLE 28.5 A 3-cm-high object is located a distance 50 cm from a concave mirror having a radius of curvature of 40 cm. Calculate the position, size, and the nature of the image.

Substituting $p = 50$ cm and $f = R/2 = 40\text{ cm}/2 = 20$ cm in Equation 28.17,

$$\frac{1}{p} + \frac{1}{q} = \frac{1}{f} \quad \text{or} \quad \frac{1}{q} = \frac{1}{20\text{ cm}} - \frac{1}{50\text{ cm}} = \frac{3}{100\text{ cm}}$$

$$q = \frac{100 \text{ cm}}{3} = 33\tfrac{1}{3} \text{ cm}$$

That is, the image is $33\frac{1}{3}$ cm from the mirror and is real. The magnification M, from Equation 28.16, is

$$M = \frac{\text{image size}}{\text{object size}} = -\frac{q}{p} = \frac{(100/3) \text{ cm}}{50 \text{ cm}} = -\frac{2}{3}$$

$$\text{image size} = -\frac{2}{3}(\text{object size}) = -\frac{2}{3}(3 \text{ cm}) = -2 \text{ cm}$$

The minus sign indicates that the image is inverted. Thus, the image is $33\frac{1}{3}$ cm from the mirror, 2 cm in size, inverted, and real. The ray diagram is similar to Figure 28.27(a).

Exercise 28.5 An object 3 cm in height is located a distance 10 cm from a concave mirror of radius of curvature 40 cm. Calculate the position, size, and the nature of the image. [*Ans.:* $q = -20$ cm, $M = 2$, image size = 6 cm, image is virtual and erect.]

28.7. Image Formation by Lenses

Lenses are objects made of optically transparent material, usually glass of such forms as to converge or diverge a beam of light when it passes through them. We may also consider lenses to be a combination of prisms, as shown in Figures 28.29(a) and (b). In (a) the rays, after passing through the prisms,

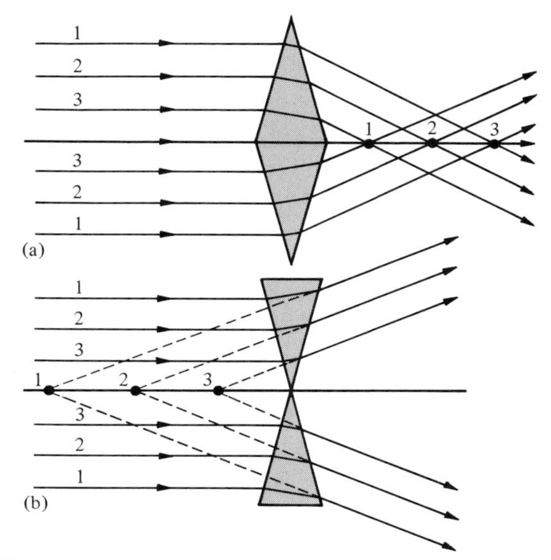

Figure 28.29. *(a) Two prisms placed base to base converge incident parallel rays of light. (b) Two prisms placed vertex to vertex diverge incident parallel rays of light.*

converge, while in (b) they diverge. Even though these parallel rays may come from the same object, they converge at different points such as 1, 2, and 3 in Figure 28.29(a), or seem to diverge from different points, as in Figure 28.29(b).

In order to obtain sharp images by bringing different rays to focus at the same position, the lenses are made in the form of symmetrical spherical surfaces, as shown for the converging lens in Figure 28.30 and the diverging lens in Figure 28.31. In the converging or convex or positive lens shown in Figure 28.30(a), all the rays are brought to focus at one point F, called the *principal focus*. The two surfaces forming this lens are portions of spheres

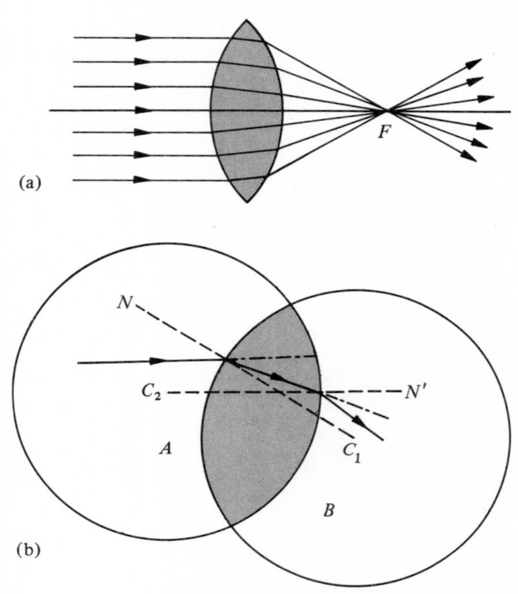

FIGURE 28.30. *Converging (or convex) lens.*

whose centers of curvatures are C_1 and C_2, as shown in Figure 28.30(b). Also shown in this figure is the path a ray follows through the lens. The ray entering the surface A bends toward the normal C_1N, and when leaving the surface B it bends away from the normal C_2N'.

Figure 28.31(a) shows a diverging concave or negative lens. As shown, the incident rays, after passing through the lens, if produced backward meet at point F, its principal focus. Once again a concave lens is formed out of two spherical surfaces, as shown in Figure 28.30(b). This figure also shows the path of an individual ray through the lens.

There are mainly three different types of convex (or converging) lenses and three concave (or diverging) lenses, as shown in Figure 28.32(a) and (b), respectively. A physical distinguishing factor between these lenses is that the convex lenses are thicker at the center as compared to their outer edges, while the concave lenses are thinner at the center as compared to their outer edges. In the following discussion we shall limit ourselves to thin lenses, convex or concave. This implies that the thickness of the lens is much smaller than the distance between the lens and the focal point.

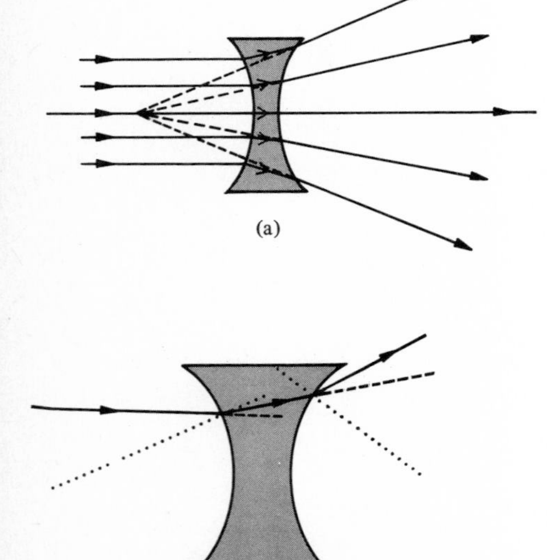

FIGURE 28.31. *Diverging (or concave) lens.*

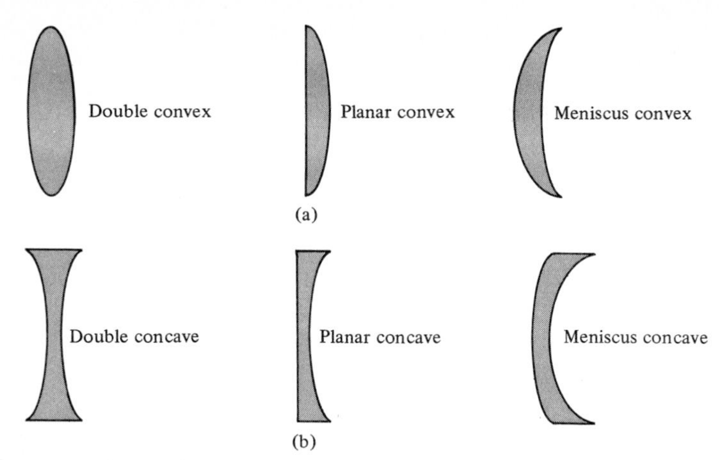

FIGURE 28.32. *Different types of (a) converging lenses and (b) diverging lenses.*

To start with, let us consider the image formation by a convex lens. To do this we must be familiar with the terms defined below and shown in Figure 28.33. The *principal axis* is a line that passes through the center of the lens and connects the centers of curvatures of the two surfaces of the lens. The *optical center* is a point on the principal axis through which the incident rays pass without deviation. For a thin lens this coincides with the geometrical center of the lens. The *principal focus F* is a point on the principal axis where the incident rays parallel to the principal axis are brought to focus. This focal point is denoted by F while a symmetrical point on the opposite side of the lens is denoted by F'. Thus, there are two focal points, F and F', unlike the case of

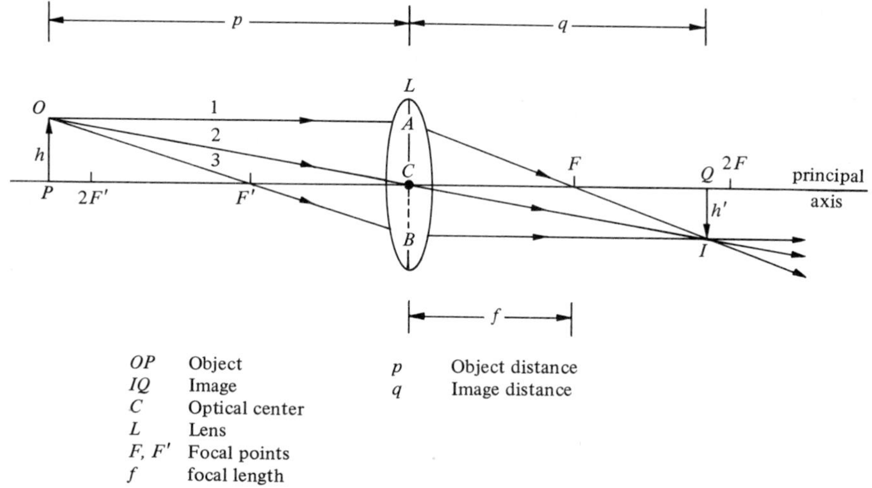

FIGURE 28.33. *Three rays used for locating the position and size of an image formed by a convex lens.*

mirrors, where there is only one. The *focal length* f is the distance along the principal axis between the optical center of the lens and its principal focus, F or F'.

To locate the position and size of the image, we can follow the graphical method or the analytical method. The graphical method is shown in Figures 28.33 to 28.35. It uses the following three rays to locate the image of each point of the object. It is not necessary to use all three rays—two will do.

1. Ray 1, parallel to the principal axis, passes through the focal point F of a converging (or convex) lens after refraction through the lens, or appears to come from the focal point F of a diverging (or concave) lens.
2. Ray 2 passes through the center of the lens undeviated. A small displacement is negligible because the lens is thin.
3. Ray 3, after passing through the focal point F', hits the convex lens surface and emerges parallel to the principal axis. In the case of a concave lens the ray appears to proceed toward F'.

Using these three rays the six possible cases of image formation are shown in Figure 28.34. Note that in the first five cases, the images are inverted and real, while in the last case, when the object is between F' and the center of the lens, the image is erect and virtual, as shown. The nature and size of the image changes as the object is brought from infinity toward the center of the lens. Figure 28.35 shows the image formations in the case of a concave lens.

Let us now derive the lens equation so that we can calculate the position and size of the image analytically. Refer back to Figure 28.33. Let p and q be the distances of the object and the image from the center. From similar triangles ACF and QIF, noting that $AC = OP$,

$$\frac{OP}{IQ} = \frac{AC}{IQ} = \frac{CF}{QF} = \frac{f}{q-f} \tag{28.19}$$

Similarly, from similar triangles OPC and IQC, we get

$$\frac{OP}{IQ} = \frac{CP}{CQ} = \frac{p}{q} \tag{28.20}$$

Equating the two equations above,

$$\frac{f}{q-f} = \frac{p}{q} \qquad \text{or} \qquad qf + pf = pq$$

Dividing both sides by pqf, we get the *lens formula*,

$$\boxed{\frac{1}{p} + \frac{1}{q} = \frac{1}{f}} \tag{28.21}$$

which is identical to the mirror equation. Although we have derived this equation for a converging lens, it can be used for a diverging lens also if the following sign convention is used: *f is positive for a converging lens and*

FIGURE 28.34. *Six different cases of image formation by a convex lens.*

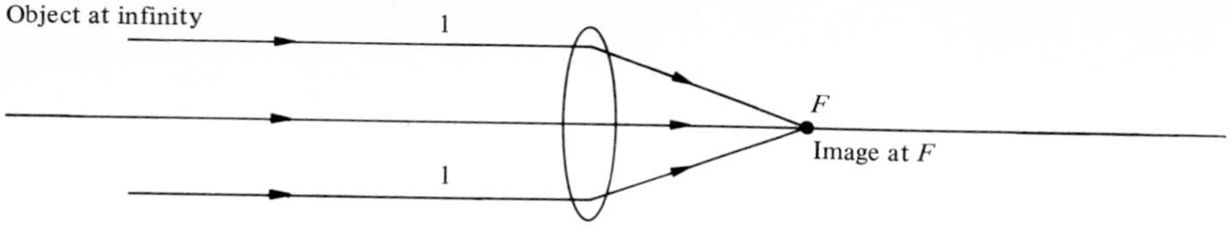

Object at infinity

1

F

Image at F

1

(a) Telescope

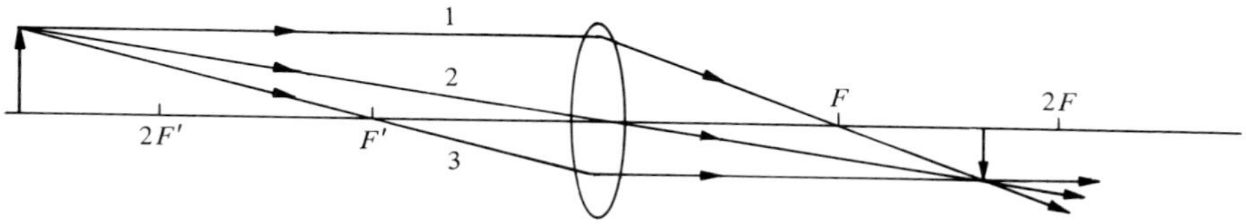

1

2

F

2F

2F′

F′

3

(b) Cameras

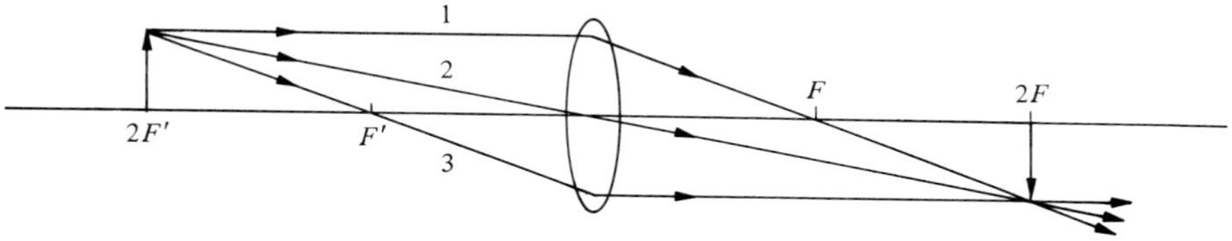

1

2

F

2F

2F′

F′

3

(c) Copy machine

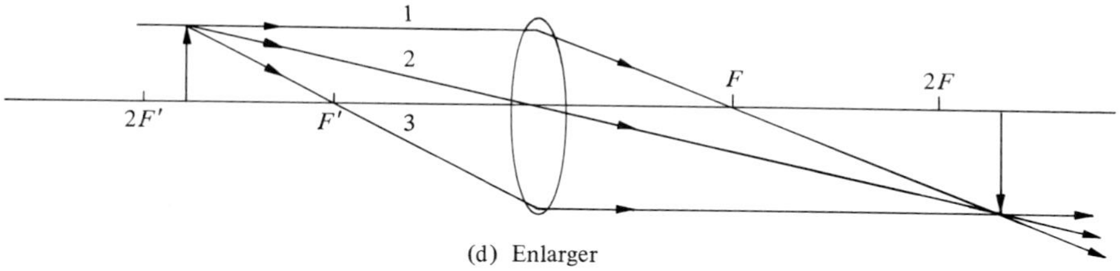

1

2

F

2F

2F′

F′

3

(d) Enlarger

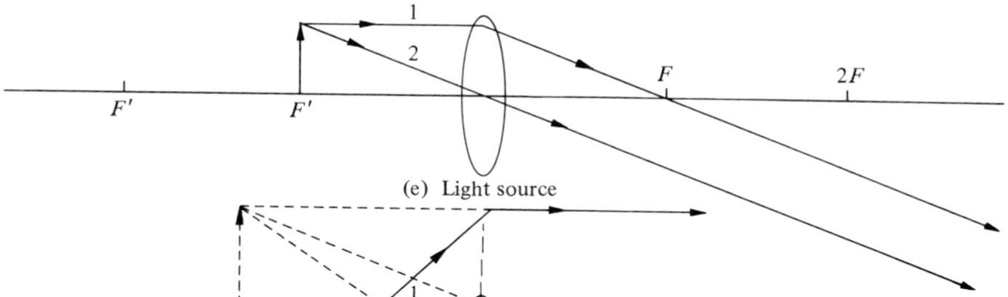

1

2

F

2F

F′

F′

(e) Light source

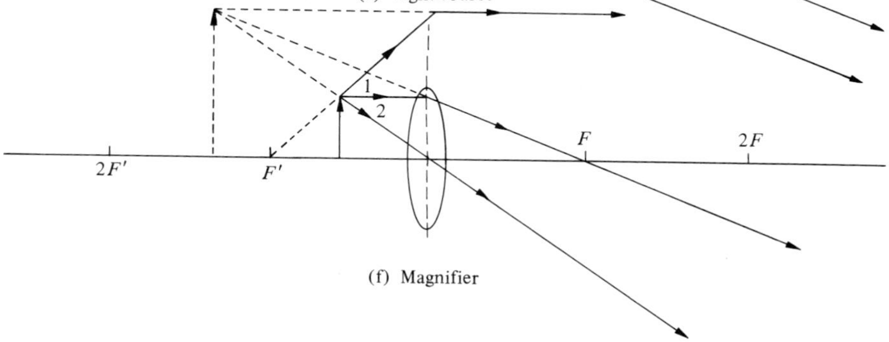

1

2

F

2F

2F′

F′

(f) Magnifier

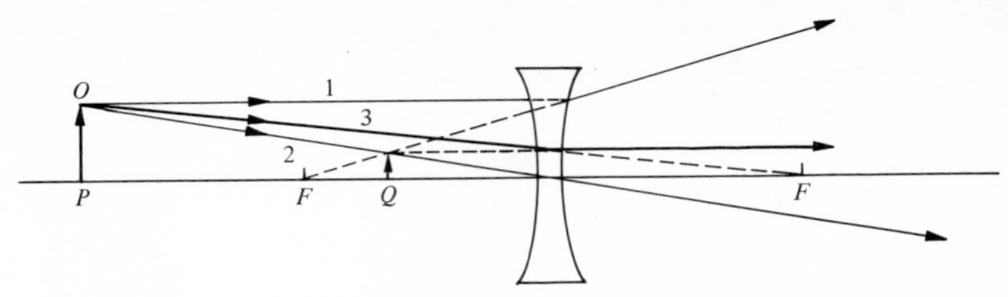

FIGURE 28.35. *Image formation by a concave lens. The image formed is always virtual, erect, and smaller.*

negative for a diverging lens; p is positive for a real object and negative for a virtual object; q is positive for a real image and negative for a virtual image.

Finally, let us calculate the linear (lateral or transverse) magnification M. Using Figure 28.33 and Equation 28.20, we get

$$\frac{\text{size of image}}{\text{size of object}} = \frac{h'}{h} = \frac{IQ}{OP} = \frac{q}{p}$$

where h' is the height of the image and h that of the object. Thus, we define it to be

$$\boxed{M = \frac{h'}{h} = -\frac{q}{p}} \tag{28.22}$$

The minus sign is included to distinguish between upright or erect and inverted objects and images. Upright objects and images are positive, inverted ones are negative.

The lens formula, Equation 28.21, can be extended to the case when more than one lens system forms the image. In such cases the image formed by the first lens acts as an object for the second and so on, as illustrated in Example 28.8.

So far we have assumed a thin symmetrical lens. Suppose that the two surfaces of a lens have completely different radii of curvature, and the index of refraction of the material between them is n. Let R_1 and R_2 be the radii as shown in Figure 28.36. In such cases the effective focal length f is given by

$$\boxed{\frac{1}{f} = (n-1)\left(\frac{1}{R_1} - \frac{1}{R_2}\right)} \tag{28.23}$$

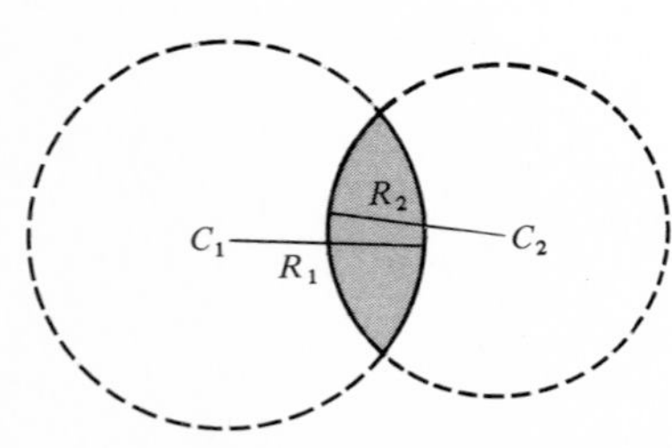

FIGURE 28.36. *Radii of curvature of two lens surfaces.*

This is called the *lens maker's formula*. Remember that R_1 and R_2 are positive for convex surfaces and negative for concave surfaces (when measured from the incident ray side), and Equation 28.23 is true only for thin lenses.

The quantity $1/f$ is used quite often and if f is small, $1/f$ will be large, and vice versa. $1/f$ is called the *power* of a lens of focal length f. Thus, a shorter-focal-length lens is more powerful. (Power is not used in the same sense as work and power.) This *power* is a measure of the ability of a lens to change the curvature of the incident wave front. The power of a lens is usually measured in *diopters*, defined as

$$D = \text{power in diopters} = \frac{1}{\text{focal length in meters}} \tag{28.24}$$

Thus, if $f = 10$ cm $= 0.1$ m, $D = 1/0.1$ m $= 10$ diopters. If $f = 20$ cm $= 0.2$ m, $D = 1/0.2$ m $= 5$ diopters; and so on.

EXAMPLE 28.6 An object 1 cm in size is placed at a distance of 50 cm from a converging lens of focal length 20 cm. Calculate the position, size, and the nature of the image.

Since $p = +50$ cm, $f = +20$ cm, and $h = 1$ cm, from Equation 28.21,

$$\frac{1}{p} + \frac{1}{q} = \frac{1}{f} \quad \text{or} \quad \frac{1}{q} = \frac{1}{20\text{ cm}} - \frac{1}{50\text{ cm}} = \frac{3}{100\text{ cm}} = \frac{1}{33.3\text{ cm}}$$

$q = +33.3$ cm and the image is located a distance 33.3 cm to the right of the lens. The positive sign means that it is a real image. From Equation 28.22,

$$M = \frac{h'}{h} = -\frac{q}{p} = -\frac{33.3\text{ cm}}{50\text{ cm}} = -\frac{2}{3}$$

$$h' = Mh = \left(-\frac{2}{3}\right)(1\text{ cm}) = -\frac{2}{3}\text{ cm}$$

Thus, the image is only $\frac{2}{3}$ cm in size (two-thirds of the object). The negative sign means that the image is inverted. The final image is real, inverted, and smaller.

EXERCISE 28.6 An object 5 mm in size placed 40 cm in front of a convex lens forms a real image at a distance of 60 cm from the lens. Calculate the focal length and the magnification. [*Ans.:* $f = 24$ cm, $M = -1.5$.]

EXAMPLE 28.7 One surface of a glass lens is convex with a radius of curvature 1.5 m, while the other is concave with a radius of curvature 1 m. Calculate the focal length and the power of the lens if the refractive index of glass is 1.5.

Substituting $R_1 = +1.5$ m, $R_2 = -1.0$ m, and $n = 1.5$ in Equation 28.23,

$$\frac{1}{f} = (n-1)\left(\frac{1}{R_1} - \frac{1}{R_2}\right) = (1.5-1)\left(\frac{1}{1.5\text{ m}} - \frac{1}{-1.0\text{ m}}\right) = \frac{5}{6\text{ m}}$$

$$f = \frac{6}{5}\text{ m} = 1.2\text{ m}$$

Therefore, from Equation 28.24,

$$D = \frac{1}{f\,(\text{in meters})} = \frac{1}{1.2\text{ m}} = 0.625\text{ diopters}$$

EXERCISE 28.7 A lens is made of glass of refractive index 1.52 with two surfaces of radii of curvature 40 cm and 20 cm. What will be the focal length and power if two surfaces are combined to form (a) a double convex lens; and (b) a double concave lens? [*Ans.*: (a) 0.256 m, 3.9 diopters; (b) −0.256 m, −3.9 diopters.]

EXAMPLE 28.8 An object 4 cm high is placed at a distance of 40 cm from a converging lens of focal length 15 cm. A diverging lens of focal length −10 cm is placed a distance of 30 cm from the first lens as shown. Find the position, size, and nature of the final image.

For the first lens, $p_1 = +40$ cm, $f_1 = +15$ cm, and in Equation 28.21,

$$\frac{1}{p_1} + \frac{1}{q_1} = \frac{1}{f_1} \quad \text{or} \quad \frac{1}{q_1} = \frac{1}{15\text{ cm}} - \frac{1}{40\text{ cm}} = \frac{5}{120\text{ cm}} = \frac{1}{24\text{ cm}}$$

Therefore, $q_1 = +24$ cm and the image is real:

$$M_1 = \frac{h_1'}{h_1} = -\frac{q_1}{p_1} = -\frac{24\text{ cm}}{40\text{ cm}} = -\frac{3}{5}$$

$$h_1' = M_1 h = \left(-\frac{3}{5}\right)(4\text{ cm}) = -2.4\text{ cm}$$

The image is smaller and inverted (see the figure).

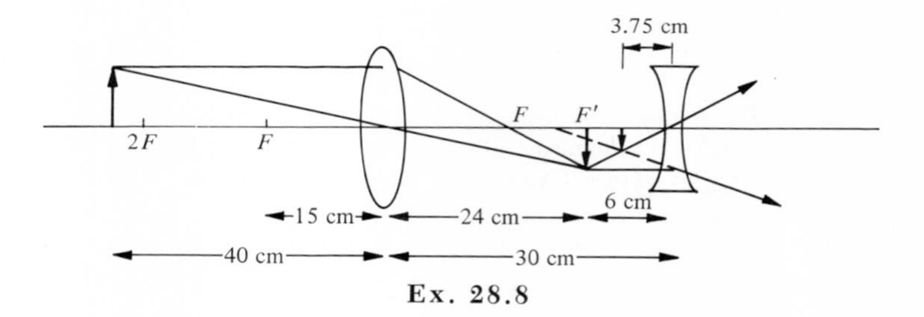

Ex. 28.8

For the second lens, $p_2 = (30 - 24)$ cm $= +6$ cm and $f_2 = -10$ cm:

$$\frac{1}{p_2} + \frac{1}{q_2} = \frac{1}{f_2} \quad \text{or} \quad \frac{1}{q_2} = -\frac{1}{10\text{ cm}} - \frac{1}{6\text{ cm}} = -\frac{4}{15\text{ cm}}$$

Therefore, $q_2 = -15/4 = -3.75$ cm and the image is virtual:

$$M_2 = \frac{h_2'}{h_2} = -\frac{q_2}{p_2} = -\left(\frac{-3.75\text{ cm}}{+6\text{ cm}}\right) = +\frac{5}{8}$$

Since $h_2 = h_1' = -2.4$ cm,

$$h_2' = M_2 h_2 = \left(+\frac{5}{8}\right)(-2.4\text{ cm}) = -1.5\text{ cm}$$

and the image is inverted. Thus the total magnification is

$$M = M_1 M_2 = \left(-\frac{3}{5}\right)\left(+\frac{5}{8}\right) = -\frac{3}{8}$$

Hence, if $h = 4$ cm,

$$h' = Mh = \left(-\frac{3}{8}\right)(4 \text{ cm}) = -1.5 \text{ cm}$$

That is, the final image is inverted and reduced in size, even though the second lens forms an erect image. (The first lens forms an inverted image while the second lens keeps it inverted.) Thus, the final image is virtual, inverted, and reduced in size and to the left of the concave lens, as shown.

Exercise 28.8 In Example 28.8, suppose that $f_1 = +30$ cm, and the second lens is placed a distance of 150 cm from the first lens. Repeat the example. [*Ans.:* final image is inverted, virtual, 3 cm high, located 7.5 cm in front of the second lens.]

SUMMARY

The wave theory of light started showing signs of weakness by the close of the nineteenth century. According to *Huygens' principle:* Every point on a wave front may be considered to be a source of secondary wavelets that propagate in the forward direction. A beam is monochromatic if all the waves have the same wavelength. A medium in which the speed of the wave is the same in all directions is an *isotropic medium;* if the speed is different in different directions, it is *anisotropic*.

The two *laws of reflection* are: (1) the incident ray, reflected ray and the normal all lie in the same plane; and (2) the angle of incidence is always equal to the angle of reflection. According to the *laws of refraction:* (1) the incident ray, the normal, and the refracted ray all lie in one plane; and (2) the ratio of the sine of the angle of incidence to the sine of the angle of refraction is constant, or $n_1 \sin\theta_1 = n_2 \sin\theta_2$, which is called *Snell's law*. Also, $n = c/v = \lambda/\lambda_0$ and $n_1\lambda_1 = n_2\lambda_2$.

When a ray of light travels from a denser to a rarer medium, the critical angle θ_c is that angle of incidence for which the corresponding angle of refraction is 90°, $\sin\theta_c = n_2/n_1$. For $\theta_i > \theta_c$ there is *total internal reflection*. The apparent depth = $(1/n)$(actual depth).

A prism is a wedge-shaped figure bounded by two surfaces. The *deviation* is given by $n = \sin[(A + D_m)/2]/\sin(A/2)$. The splitting of white light into its component colors is called *dispersion*.

The image formed by a *plane mirror* is as far beyond the mirror as the object is in front; it is erect, virtual, and has inversion and unit magnification. The following relations with proper sign conventions hold good for both concave and convex mirrors and lenses. $f = R/2$, $1/p + 1/q = 1/f$, and linear magnification M = image size/object size = $-q/p$.

If a lens is made of two surfaces of radii of curvatures R_1 and R_2, the focal length is given by the relation $1/f = (n - 1)(1/R_1 - 1/R_2)$. The power of a lens is a measure of its ability to change the curvature of the incident wave front, D = power in diopters = 1/focal length in meters.

QUESTIONS

1. Are Huygens' principle and Snell's law applicable to all electromagnetic waves, such as x-rays, radio waves, heat waves, and so on?
2. As light propagates from one medium to another, why does its wavelength change but not its frequency?
3. Explain the formation of shadows by using (a) Newton's corpuscular theory; and (b) Huygens' wave theory.
4. An object is placed between two parallel vertical mirrors. Explain the formation of an extremely large number of images.
5. Which conservation law or laws do you think are the basis for the laws of reflection and refraction?
6. While using totally reflecting prisms, why is light always made to enter normally and leave normally?
7. Why does dispersion happen only in refraction and not in reflection?
8. Can you use a prism to measure the variation of n with wavelength?
9. What is the difference between a real and a virtual image? Can a virtual image be (a) photographed; and (b) projected on a screen?
10. A mirror forms a real, equal, and inverted image next to the object. How can you distinguish between the image and the object by just looking at them?
11. For a rearview side mirror of a car, is it better to use a convex or a concave mirror? Why?
12. Give a quick and easy method of measuring the focal length of (a) a concave mirror; and (b) a convex lens, using only a pencil and a meter stick.
13. A convex lens is used to focus the sun's rays on a piece of paper to start a fire. Can this be done by using a concave lens?

PROBLEMS

1. By drawing a ray diagram, show that in order to see a full image of an object, a plane mirror should be half the size of the object.
2. A 10-cm plane mirror held vertically 25 cm from the eye is completely filled by the image of a building 100 m away. Calculate the height of the building.
3. An object is placed exactly in the middle of two plane mirrors placed at right angles to each other. By drawing a ray diagram, calculate the total number of images that are formed.
4. For a light beam incident on a mirror, if the mirror is rotated through an angle θ, show that the reflected beam is rotated through an angle 2θ.
5. Find the speed and wavelength of yellow light in glass of refractive index 1.52 if its wavelength in vacuum is 5890 Å.
6. What is the wavelength and speed of light of frequency 5×10^{14} Hz in benzene having a refractive index of 1.5?
7. The wavelength range of visible light is 4000 to 7000 Å in air and its speed is 3×10^8 m/s. Calculate the range of the wavelengths and the speed of visible light in a plastic of refractive index 2.00.
8. The wavelength of a certain color light in vacuum is 5500 Å. If the speed of light in water is three-fourths its speed in vacuum, calculate the

wavelength in water. Also, calculate the number of wavelengths per centimeter in air and in water.

9. If the speed of light in flint glass having a refractive index of 1.66 is 1.88×10^8 m/s, what is the speed in diamond having a refractive index of 2.42?
10. At an air–liquid interface, a ray of light makes an incident angle equal to 60° while the angle of refraction is 45°. Calculate the refractive index and the speed of light in the liquid.
11. A ray of light enters from air into water ($n = 1.33$), making an angle of 30° with the surface of the water. Calculate the angle of refraction.
12. What should the angle of incidence θ be in air so that the angle of refraction in a liquid ($n = 1.60$) is $\theta/2$? Is it possible with water ($n = 1.33$)?
13. Show that the lateral displacement shown in the accompanying figure is $d = t \sin(\theta_i - \theta_R)/\cos\theta_R$, t being the thickness of the plate.
14. In Problem 13 calculate the value of d for $t = 5$ cm and $n = 1.42$ for (a) $\theta_i = 30°$; and (b) $\theta_i = 60°$.
15. A source of light is placed 10 cm below a water–air surface. Calculate the angles of refraction for the incident rays, which make angles of 10°, 30°, and 40° with the normal. Draw a clear diagram.
16. Calculate the critical angle for (a) a water–air interface; (b) an ice–water interface; and (c) a glass–water interface.
17. A fish in a pond appears to be 1 m below the surface of the water. What is the actual depth of the fish?
18. If a glass plate 2 in thick is placed over a printed page, how much do the letters seem to be raised if the refractive index of glass is 1.5?
19. A ray of light starting from the bottom of a pool 10 ft deep strikes the water surface a distance of 12 ft from a point directly above the light source. The refracted ray just grazes along the water surface. Calculate the refractive index of the water in the pool.
20. For a certain 60° glass prism the angle of minimum deviation is found to be 42°. What is the refractive index of the material of the prism?
21. The refractive index of glass for yellow light is 1.51 and for blue light is 1.54. Calculate the angle of dispersion for the two colors in a 60° prism.
22. A beam consisting of wavelengths 4800 Å and 6600 Å is incident on a flint glass, making an angle of 30° with the surface. After refraction, what is the angular separation between the two rays if the refractive index of glass for the two wavelengths is 1.67 and 1.65?
23. Show that if the angle A of the prism is small, the angle of minimum deviation is given by $D_m = (n - 1)A$.
24. The apex angle of a prism made out of a certain glass is 15°. If the angle of minimum deviation is 9.5°, what is the refractive index of the material? (Use the formula given in Problem 23.)
25. Two prisms made out of glass of refractive index 1.5 and 1.66 are placed in such a way that there is no net deviation. If the prism with refractive index 1.5 has $A = 10°$, what is the value of A for the other prism? (Use the formula given in Problem 23.)
26. An object of height 5 mm is placed a distance of 45 cm from a concave mirror of focal length 10 cm. Calculate the position and the size of the image.

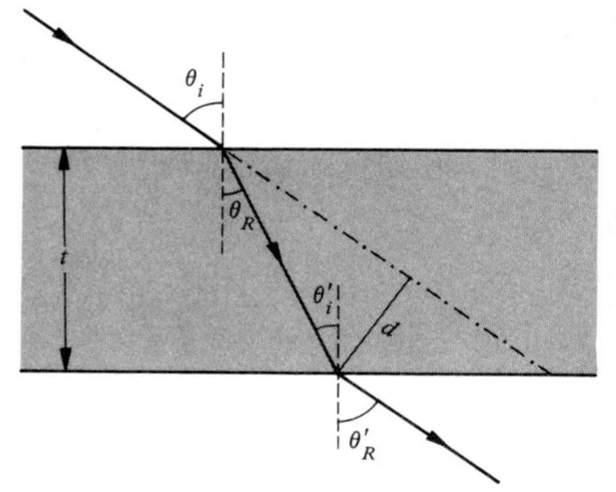

P. 28.13

27. A building 60 m high (20 stories) is 300 m from a concave mirror of radius of curvature of 1 m. Find the position and size of the image.
28. An object of height 4 mm is placed a distance of 45 cm from a concave mirror. The image is formed a distance of +20 cm from the mirror. Calculate the focal length of the mirror and the size of the image.
29. A concave mirror has a focal length of 10 cm. Where should we place an object so that the size of the image formed is half the size of the object?
30. The distance between a concave mirror and screen is 5 m. A 1-cm object placed between them produces an image 5 cm in size on the screen. Calculate f, R, and the position of the object.
31. A man uses a concave mirror for shaving and stands about 45 cm from it. If the desired linear magnification is 3, what should be the radius of curvature of the mirror? What is the nature of the image?
32. To obtain a linear magnification of 4 of a cavity, a dentist holds a concave mirror a distance of 2 cm from the tooth. What is the focal length of the mirror?
33. Where in front of a concave mirror of focal length 30 cm must an object be placed so that it produces (a) a real image; and (b) a virtual image, each with a magnification of 2?
34. An object is placed a distance 30 cm from a convex mirror of 20-cm focal length. Calculate the position, nature, and magnification of the image.
35. An object 5 cm tall is placed 20 cm from a spherical mirror, resulting in a virtual, enlarged image 10 cm tall. Calculate the focal length of the mirror and the position of the image. What type of mirror is this?
36. A converging lens has a focal length of 20 cm. If an object is placed at a distance of 10, 20, 30, 40, and 50 cm, calculate the position, nature, and magnification of the image.
37. An object 1 cm in height is placed at a distance of 0.80 m from a double convex lens of focal length 0.2 m. Calculate the location and size of the image formed. Draw a ray diagram.
38. A diverging lens of 10-cm focal length is placed at a distance of 30 cm from an object of 3 mm in size. Calculate the location and size of the image.
39. Calculate the focal length and power of a combination of two lenses, a converging lens of focal length of +40 cm and a diverging lens of focal length of −30 cm both placed together.
40. A converging lens of focal length +20 cm forms an erect, virtual, and three-times-magnified image of an object. Calculate the position of the object and the image. Draw a ray diagram.
41. A screen is placed a distance of 1 m from a converging lens of focal length 0.25 m. Calculate the position and size of the object if the image formed on the screen is 4 cm high.
42. A magnifying lens of power +4 diopters is held a distance 10 cm from a printed page. If the size of a letter printed on the page is 2 mm, what is its size when viewed through the lens?
43. A certain lens made of glass ($n = 1.5$) has a power of +5 diopters. What is its focal length?
44. A converging lens is made of glass of refractive index 1.52. The front surface has a radius of curvature of +20 cm and the back surface has

-10 cm. Find the final focal length and the power of the lens. Calculate the focal length and the power if the two surfaces are interchanged.

45. The radii of curvature of the two surfaces of a thin lens are (a) $+20$ cm and $+30$ cm; (b) -20 cm and -30 cm; (c) -20 cm and $+30$ cm; and (d) $+20$ cm and -30 cm. The refractive index of the material is 1.50. Calculate the focal length of the lens.

46. An object 1 cm high is placed a distance of 20 cm from a lens of focal length $+15$ cm. A second lens of focal length $+20$ cm is placed 30 cm away from the first lens. Calculate the position, nature, and size of the final image.

47. Two double convex lenses, each with a focal length $+20$ cm, are separated by a distance of 30 cm. An object 4 cm high is placed at a distance of 50 cm. Find the position, size, and nature of the image.

48. A double convex lens of focal length 30 cm is placed 60 cm in front of a double concave lens of focal length 20 cm. If an object 5 cm high is placed 80 cm in front of the convex lens, calculate the position, size, and nature of the image.

49. Consider an object 4 cm in height placed a distance of 30 cm from a convex lens of focal length 20 cm. On the other side of the lens a mirror of focal length $+30$ cm is placed 100 cm away from the lens. Calculate the final position, nature, and size of the image.

29 Physical Optics

In Chapter 28 the emphasis was on ray optics, even though we made an occasional reference to Huygen's principle. There are three important phenomena—interference, diffraction, and polarization—which cannot be explained with the help of ray optics (or corpuscular theory). The subject of interference, diffraction, and polarization is called physical optics *or* wave optics *and should be explained by using the wave theory of light. Our aim in this chapter is to elaborate on these points. The establishment of the wave theory of light came in 1801 when the English scientist Thomas Young performed the interference experiment. Very shortly, diffraction was explained by Fresnel and Fraunhofer while the transverse nature of light explained polarization experiments.*

29.1 Conditions for Interference

THE PHENOMENA of interference and diffraction of water waves and sound waves were discussed briefly in Chapter 17. It was pointed out that the superposition of two wave trains of the same frequency and the same amplitude traveling in the same direction results in *interference phenomena.* Furthermore, at any point if the waves are *in phase,* the result is *constructive interference* (increase in the resultant amplitude), and if the waves are *out of phase* the result is *destructive interference* (decrease in the resultant amplitude). In the case of water waves, the interference leads to a change in the amplitude of the vibrating particle; in sound waves it is the pressure that changes, and for light waves the interference will lead to a change in the amplitudes of the electric and magnetic field vectors.

If the true nature of light is wave motion, it should be possible to observe the interference phenomenon. We may remind ourselves that the wavelength of light is much smaller [$\sim$4000 to 7000 Å (or 4 to 7 $\times$ 10^{-7} m)] than the wavelength of sound waves (0.017 to 17 m), creating several difficulties. To observe the interference of light waves, the following conditions should be achieved: (1) the sources should be monochromatic, (2) the sources should be coherent, and (3) the principle of linear superposition should be applicable. We explain these terms next.

A monochromatic source of light will give light of only one wavelength. It is impossible to get a truly monochromatic source of light. But by using filters it is possible to get a source that gives light within a very narrow band of wavelengths, say $\pm$5 Å. Such a source is called *quasi* (almost)-*monochromatic* and is quite satisfactory in most cases. (Note that interference will take place even if the sources contain more than one wavelength, but the interference pattern will not be well defined.)

To produce an interference pattern, we must have two sources, producing two wave trains. These two waves must always be produced with a constant phase difference between them. This will produce a stationary interference pattern in space. If the phase difference between these two wave trains changes continuously, the interference pattern changes continuously. If two ordinary light sources are used, the phase changes in each source are so rapid and haphazard (light is emitted in all directions randomly) that *no* interference pattern is observed at all. Thus, we must use *coherent sources*—sources that maintain a constant phase difference between the wave trains they emit. If no constant phase difference is maintained, they are *incoherent sources*. In order to have two coherent beams, a single source of light is split into two by the method of wavefront splitting or amplitude splitting. The alternative to this is to use two laser beams. A laser emits a unidirectional and almost coherent beam, as we shall discuss in Chapter 33.

Suppose that two waves produce certain displacements, y_1 and y_2, at a certain point x. The resultant displacement y at x will be simply $y_1 + y_2$. This is the

Linear Superposition Principle: ***When two or more waves travel simultaneously through a portion of a medium, each wave acts independently, as if the other were not present. The resultant displacement at any point and time is the vector sum of the displacements of the individual waves.***

This principle is applicable only if the amplitudes and the intensities of the waves are not very large. The phenomena that involve high-intensity beams come under the heading *nonlinear optics*, a topic that will not be discussed in this text.

Let us now see how interference takes place. Consider two waves traveling along the X-axis,

$$E_1 = E_0 \sin(kx - \omega t) \tag{29.1}$$

and

$$E_2 = E_0 \sin(kx - \omega t - \delta) \tag{29.2}$$

where E_0 is the electric amplitude, δ the phase difference between the two waves at the point under consideration, $\omega = 2\pi\nu$ the angular frequency, ν the frequency, and $k = 2\pi/\lambda$ is the propagation constant (or wave number), while λ is the wavelength. The resultant of these, as explained in Chapter 17, is

$$E = E_1 + E_2$$

or

$$E = E_m \sin\left(kx - \omega t - \frac{\delta}{2}\right) \tag{29.3}$$

where

$$E_m = 2E_0 \cos\left(\frac{\delta}{2}\right) \tag{29.4}$$

As before, this amplitude E_m is not constant but varies with the value of δ at the point under consideration. Suppose that $\delta = 0$ or $360°$, in which case $\cos\delta/2 = \pm 1$; hence, $E_m = \pm 2E_0$. The resultant amplitude is doubled; that is, the crest of one wave coincides with the other, and likewise for the troughs. The waves are said to be *in phase* and yield *constructive interference*, as shown in Figure 29.1(a), for $t = 0$ at different values of x. On the other hand, if $\delta = 180°$, $\cos\delta/2 = \cos 90° = 0$; hence, $E_m = 0$. That is, the crest of one coincides exactly with the trough of the other. The waves are said to be out of

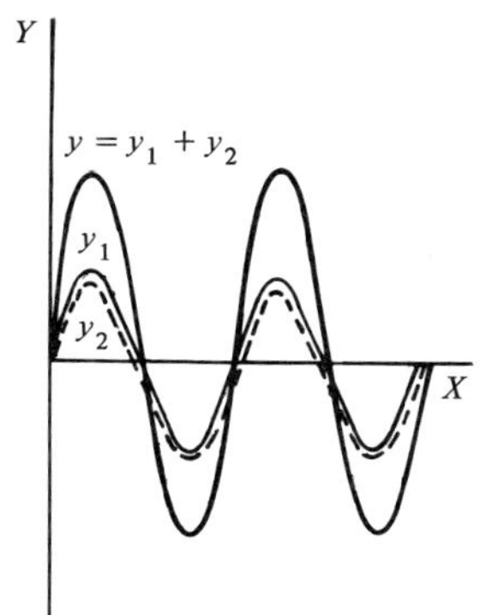

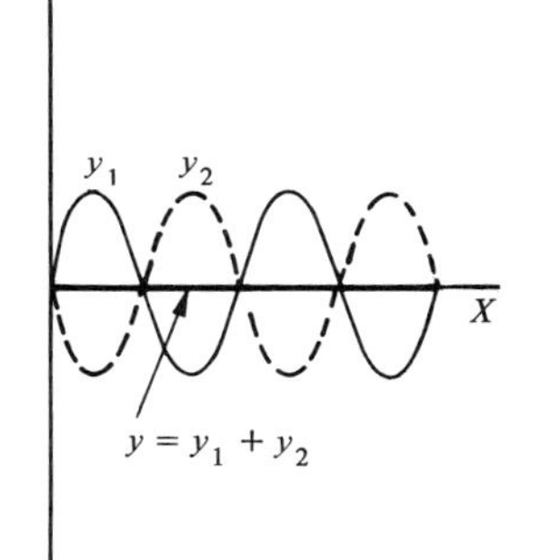

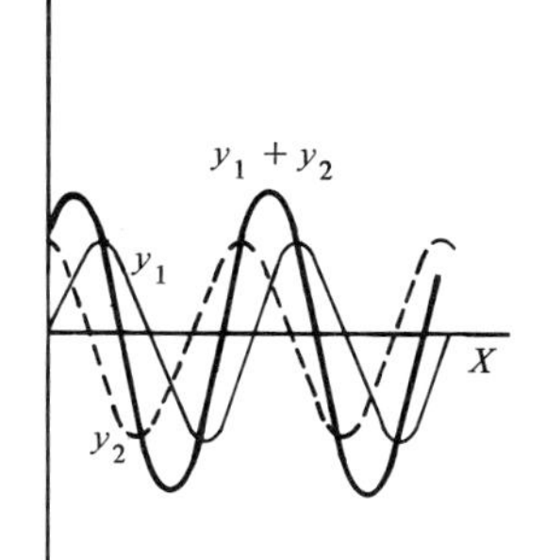

FIGURE 29.1. *(a) Waves arriving in phase produce constructive interference. (b) Waves arriving 180° out of phase produce destructive interference. (c) A situation that falls between parts (a) and (b).*

phase and result in *destructive interference,* as shown in Figure 29.1(b). Of course, for δ between $0°$ amd $180°$ the amplitude of the resultant wave will be between the two cases, as shown in Figure 29.1(c). Thus,

$$\begin{aligned} \cos \delta/2 &= \pm 1 \qquad \delta = 0, 2\pi, 4\pi, \ldots \qquad \text{constructive} \\ \cos \delta/2 &= 0 \qquad \delta = \pi, 3\pi, 5\pi, \ldots \qquad \text{destructive} \end{aligned} \tag{29.5}$$

The remarks above apply to the magnetic field vector B as well.

We may point out further that when the two wave trains differ in phase by 2π, the path difference in their waves is λ as shown in Figure 29.1. Thus if the phase difference is δ, the corresponding path difference d will be

$$\frac{\lambda}{d} = \frac{2\pi}{\delta}$$

or

$$\boxed{d = \frac{\lambda}{2\pi}\delta} \tag{29.6}$$

Thus, the conditions for constructive and destructive interference may be written in terms of the path difference by using Equations 29.5 and 29.6. Hence, for constructive interference $d = 0, \lambda, 2\lambda, 3\lambda, \ldots$ and for destructive interference $d = \lambda/2, 3\lambda/2, 5\lambda/2, \ldots$ or, in general, if m is an integer,

$$d = m\lambda \qquad \text{constructive interference} \tag{29.7}$$

$$d = (m + \tfrac{1}{2})\lambda \qquad \text{destructive interference} \tag{29.8}$$

29.2. Young's Double-Slit Experiment (Wave Front—Splitting Interferometer)

In 1801, the English physicist Thomas Young performed an experiment that led to the establishment of the wave theory of light, as suggested by Huygens. The outline of his experimental arrangement, called *Young's double-slit experiment,* is shown in Figure 29.2(a). A screen A with a narrow rectangular slit S_0 is placed in front of a source S. The cylindrical wave fronts emerge on the other side of screen A. These wave fronts arrive at screen B, which has two narrow slits, S_1 and S_2. As shown, S_1 and S_2 are on the same cylindrical wave front; hence, these two points are in phase; that is, S_1 and S_2 behave as coherent sources producing cylindrical wave fronts. These wave fronts produce interference in space, as shown in Figure 29.2(b). When these wave fronts are received on a screen, the resulting interference pattern, consisting of alternating dark and bright vertical lines called *fringes,* is as shown in Figure 29.2(c).

Before we calculate the conditions for the position of maxima and minima on the screen, it is worthwhile to point out the following. The location of point O on the center of the screen C is such that if light consists of rays (and not waves), the point O would be dark because it is in the shadow of the opaque portion of the screen B. But it is not so. The waves do bend around the edges (this phenomenon is called *diffraction*) and make point O the brightest.

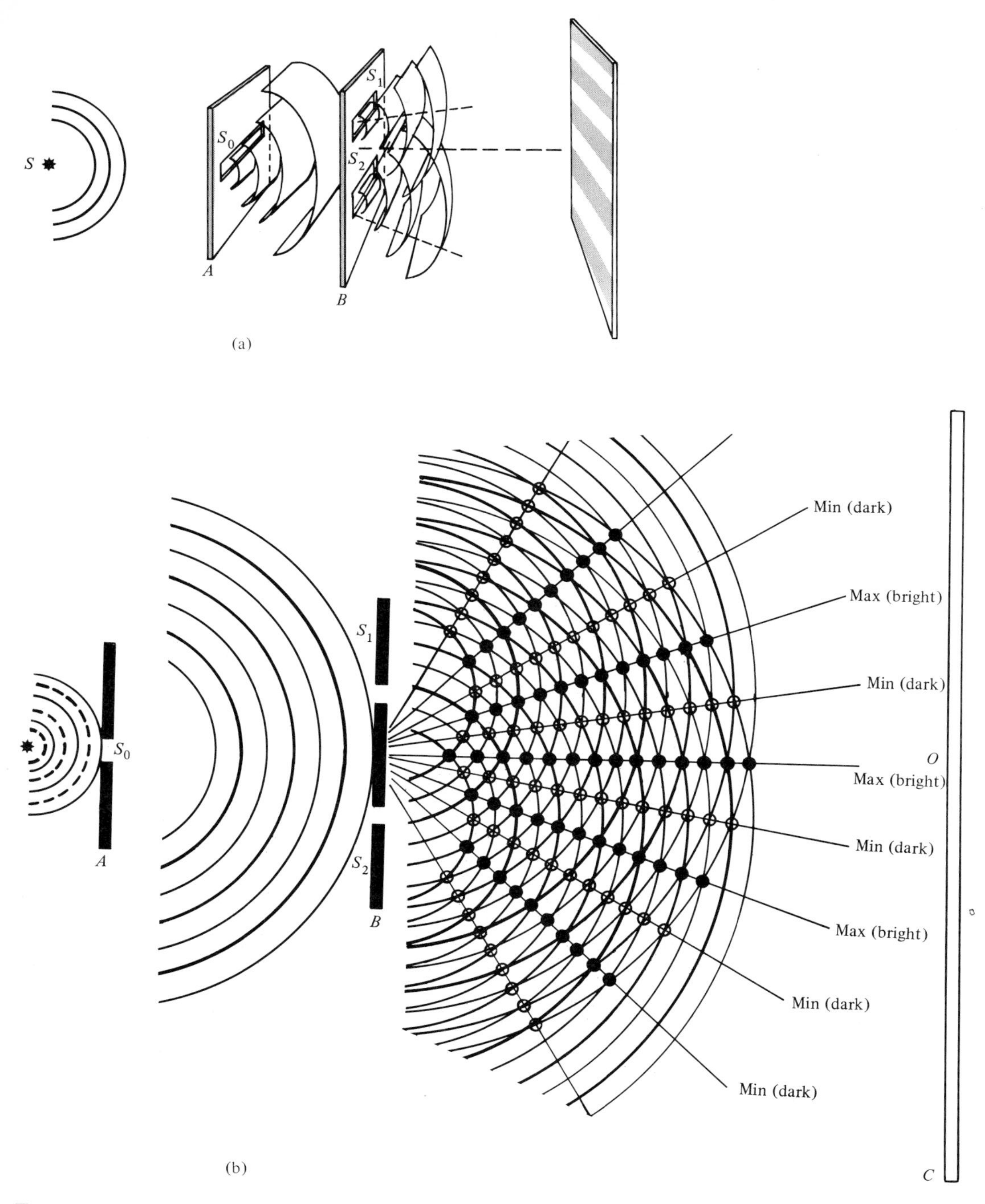

Figure 29.2. *(a) Schematic of Young's double-slit experiment. (b) Schematic of interference in space produced by cylindrical wave fronts.* [See page 658 for parts (c) and (d).]

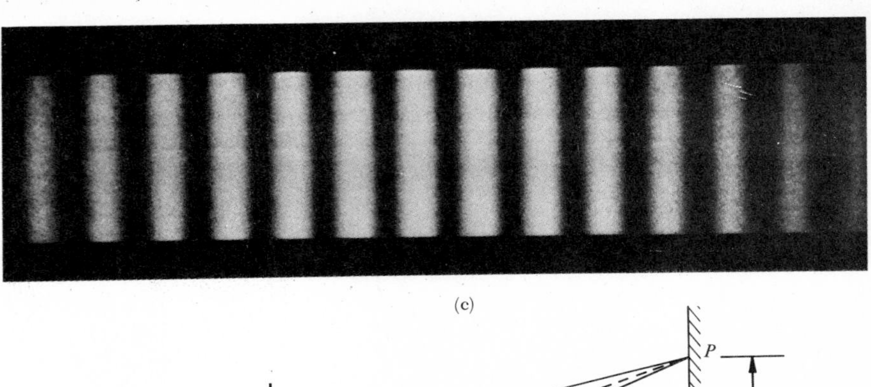

(c)

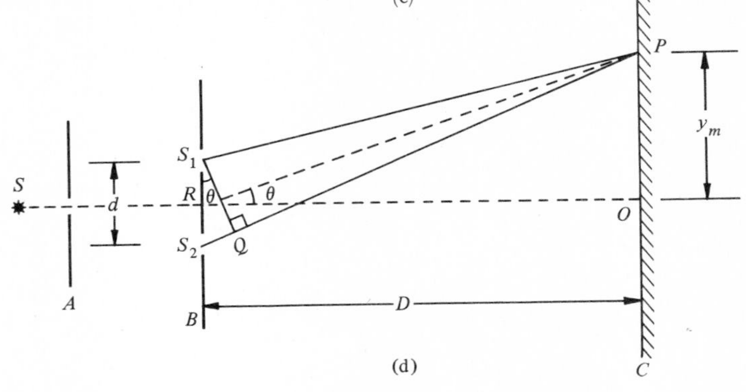

(d)

Figure 29.2 (Cont.). *(c) Photograph of an actual interference pattern. (d) Schematic to calculate the interference condition.* [Part (c) from M. Cagnet, M. Françon, and J. C. Thrierr, *Atlas of Optical Phenomena*, Springer-Verlag, Berlin, 1962.]

Even though we have used rectangular slits, Young used circular slits and white sunlight instead of a monochromatic source of light. The fringes obtained were circular and of mixed colors. Typical dimensions of the experimental setup in Figure 29.2(d) are the following. The distance between screen A (containing the source slit S_0) and screen B (containing the source slits S_1 and S_2) is between 10 and 100 cm. The distance D between screens B and C is between ~1 and 5 m. The slits S_1 and S_2 are about 0.1 mm to 0.2 mm wide, while the distance d between S_1 and S_2 is ~1 mm.

In order to calculate the positions of maxima and minima of intensities on the screen C, refer to Figure 29.2(d). Let us consider point P. The waves arriving at point P travel distances S_1P and S_2P; hence, the path difference between the two is $S_2P - S_1P$. Draw S_1Q perpendicular to S_2P. Since the distance $D >> d$, we can assume that $S_1P \simeq QP$ and also $S_2Q = d \sin\theta$. The path difference between the two rays arriving at P is

$$S_2P - S_1P = S_2Q = d \sin\theta \tag{29.9}$$

Thus, for P to be a bright fringe, the path difference must be an integral multiple of wavelengths as given in Equation 29.7; that is,

$$\boxed{d \sin\theta = m\lambda \qquad m = 0, 1, 2, \ldots \quad \text{for maxima}} \tag{29.10a}$$

At point O, $S_1P - S_2P = 0$, the path difference is zero; hence, O will be a bright fringe corresponding to $m = 0$, and this is called the *zeroth-order fringe.* In between the bright fringes are the dark fringes given by

$$\boxed{d \sin \theta = (m + \tfrac{1}{2})\lambda \qquad m = 0, 1, 2, \ldots \quad \text{for minima}} \tag{29.10b}$$

Let us now calculate the position of point P from the central point O. Let point P be at the position of the mth fringe. According to Figure 29.2(d),

$$\tan \theta_m = \frac{OP}{OR} = \frac{y_m}{D} \qquad \text{or} \qquad y_m = D \tan \theta_m \tag{27.11a}$$

Usually, θ_m is very small ($\sim 10°$) and we may write $\tan \theta_m \simeq \sin \theta_m$; $y_m = D \sin \theta_m$. Substituting for $\sin \theta_m = m\lambda/d$ from Equation 29.10a, we get

$$\boxed{y_m = m\lambda \frac{D}{d} \qquad \text{or} \qquad \lambda = \frac{y_m}{m} \frac{d}{D}} \tag{29.11b}$$

By knowing d and D, measuring y_m, and counting m from O, we can calculate the wavelength λ.

If white light is used, each wavelength will produce its own interference pattern. There will be a white spot at the center surrounded by a few colored fringes. Away from the center there will be overlapping with no clear fringes.

We may notice that Young's double-slit experiment utilizes the principle of wave-front splitting to produce coherent sources. The wave-front incident on screen B was split into two by slits S_1 and S_2. There are many other examples in which interference patterns are produced by this method. A few of these are Fresnel's double mirror, Fresnel's biprism, and Lloyd's mirror. We shall not discuss these here.

EXAMPLE 29.1 Consider a double-slit arrangement in which the slits are 0.015 mm apart and the screen is 0.75 m away. The incident light has a wavelength of 6000 Å. Calculate the angular and the linear separation of the second-order bright fringe from the central maximum. What is the separation between the two bright fringes in the second order; one on either side of the central maximum?

From Equation 29.10, $d \sin \theta = m\lambda$, where $d = 0.015 \text{ mm} = 0.015 \times 10^{-3}$ m, $m = 2$, and $\lambda = 6000 \text{ Å} = 6000 \times 10^{-10}$ m. Therefore,

$$\sin \theta = \frac{m\lambda}{d} = \frac{2(6000 \times 10^{-10} \text{ m})}{0.015 \times 10^{-3} \text{ m}} = 0.08, \qquad \theta = 4.6°$$

From Equation 29.11, $y_m = m\lambda(D/d)$, where $m = 2$, $\lambda = 6000 \times 10^{-10}$ m, $D = 0.75$ m, and $d = 0.015 \times 10^{-3}$ m. Therefore,

$$y_2 = 2(6000 \times 10^{-10} \text{ m}) \frac{0.75 \text{ m}}{0.015 \times 10^{-3} \text{ m}} = 0.06 \text{ m} = 6.0 \text{ cm}$$

Thus, the angular distance from the central maximum is 4.6°, while the linear distance is 6.0 cm. The separation between the two second-order bright fringes will be double this; that is, the angular separation $= 2 \times 4.6° = 9.2°$ and the linear separation $= 2 \times 6 \text{ cm} = 12$ cm.

EXERCISE 29.1 Calculate the angular separation and the linear separation in Example 29.1 for the following cases: (a) λ is changed to 4000 Å; (b) m is changed to 4; and (c) d is changed to 0.03 mm. [*Ans.*: (a) 6.1°, 8 cm; (b) 18.4°, 24 cm; (c) 4.6°, 6 cm.]

29.3. Michelson's Interferometer (Amplitude—Splitting Interferometer)

Michelson's interferometer, which utilizes the principle of amplitude splitting, is one of the most precise instruments for measuring different wavelengths. Figure 29.3(a) shows the basic outline of a Michelson interferometer while Figure 29.3(b) shows the type of fringes observed. A beam of monochromatic light of wavelength λ, from source S after passing through lens L, falls on a half-silvered glass plate P that splits the beam into two beams. The part being transmitted is called beam 1, and the part being reflected is called beam 2. These beams, after being reflected from mirrors M_1 and M_2, respectively, arrive at plate P and are then transmitted into the telescope T and to the eye at E. Beam 2 goes through the plate P three times while beam 1 goes through only once. So to make the optical paths of the two beams equal, a compensating plate CP is placed in the path of beam 1. When beams 1 and 2 arrive at the telescope T, they produce interference fringes of the type shown in Figure 29.3(b).

We emphasize that the optical path is defined as

$$\text{optical path} = \text{refractive index} \times \text{length} = nL \tag{29.12}$$

where L is the distance of the mirrors from plate P and n is the refractive index of the medium. Thus, if the optical path of the two beams arriving at E are exactly equal (i.e., if the two beams are in phase), a bright fringe will be produced (constructive interference). If one mirror, say M_1, is moved a distance $\lambda/4$, this will produce a path difference of $\lambda/2$ ($\lambda/4$ each way, before and after reflection from mirror M_1) between the two beams. The two beams are then out of phase and it results in a dark fringe at E (destructive interference). Thus, by moving mirror M_1, we can make the fringes move past a reference mark and count the fringes.

There will be a shift of one complete fringe if mirror M_1 is moved through a distance of $\lambda/2$. If there is to be a shift of N fringes, the mirror must be moved through a distance x so that

$$\boxed{x = N\frac{\lambda}{2} \qquad \text{or} \qquad \lambda = \frac{2x}{N}} \tag{29.13}$$

Thus, measuring x accurately by a micrometer screw and counting the number of fringes, we can calculate λ. Since N can be as large as 1000 fringes, x will be also sufficiently large to be measured accurately; hence, λ can be calculated precisely. This instrument is so sensitive that it is possible to measure a shift of 1/20 of a fringe. It is the precise nature of this instrument that has made it very useful.

This instrument, as mentioned in Chapter 1, has been used to define a

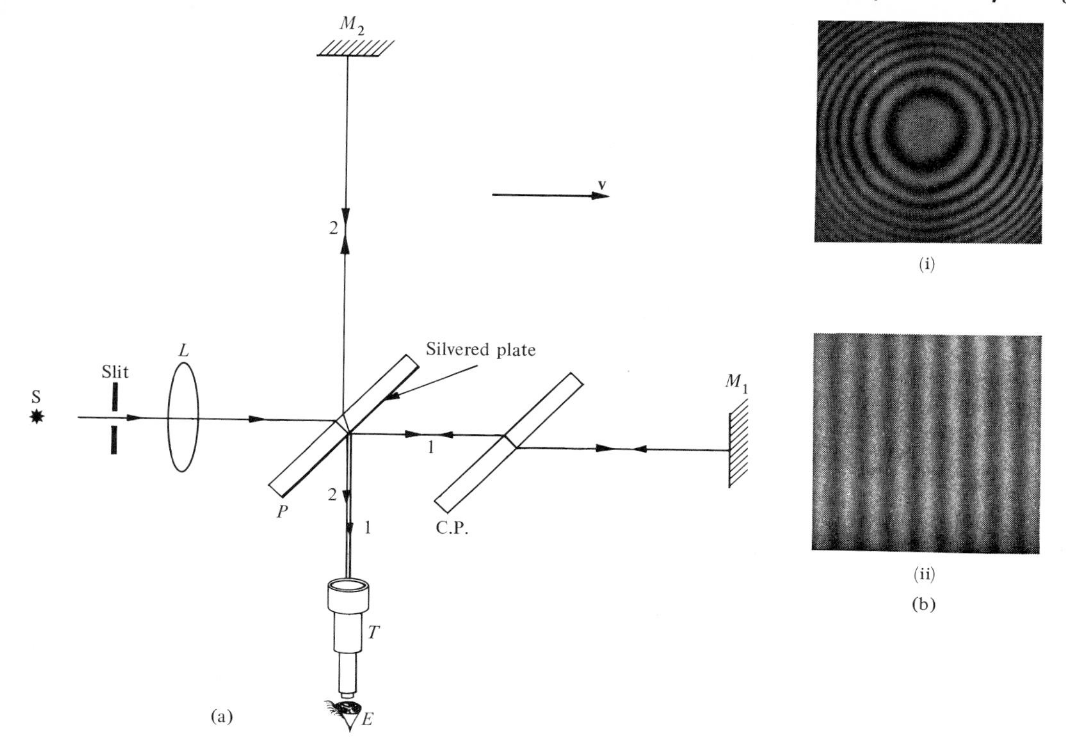

FIGURE 29.3. *Michelson interferometer: (a) outline, (b) Fringes obtained are circular as in (i) if the two mirrors are exactly at right angles, otherwise the fringes are curved or straight as in (ii).*

standard meter. The meter is defined as 1,650,763.73 wavelengths of the orange-red light of krypton 86, which has a wavelength of $\lambda_{air} = 6056.16$ Å and $\lambda_{vac} = 6057.802105$ Å. This number of wavelengths was found by measuring the length of the former standard meter (distance between two scratches on a bar of platinum–iridium located near Paris) in terms of this wavelength by using a Michelson interferometer.

Another use of this instrument can be found in the accurate measurement of the refractive index of glass, gases, or any other materials in the form of a thin film. Suppose that beam 1 of the interferometer is surrounded by a gas of refractive index n. Thus, the total optical path of beam 1 in a round trip is $2nL$, while that of beam 2 is still $2L$. Thus, the path difference between the two beams is

$$2nL - 2L = 2L(n - 1)$$

Suppose that this change in path makes m fringes move across the pointer. This shift corresponds to a path difference of $m\lambda$. Thus,

$$\boxed{2L(n - 1) = m\lambda} \qquad (29.14)$$

Knowing λ, L, and m, n can be calculated. Refractive indices of gases and thin films can be calculated to the accuracy of the fourth decimal place.

Of course, we cannot ignore the fact that this interferometer was used in one of the most important experiments, the Michelson–Moreley experiment, in finding the relative motion of the earth with respect to the ether, which resulted in a negative result, as discussed in Chapter 8.

EXAMPLE 29.2 One arm of a Michelson interferometer contains a cell 5 cm long which is parallel and in the path of one of the beams. Suppose that the air in the cell is completely evacuated. Calculate the number of fringes that will shift. The wavelength of the light being used is 6000 Å and the refractive index of air is 1.00028.

Substituting $L = 5$ cm, $\lambda = 6000 \times 10^{-8}$ cm, $n = 1.00028$ in Equation 29.14, where we use L to be the length of the cell.

$$m = \frac{2L(n-1)}{\lambda} = \frac{2(5\text{ cm})(1.00028 - 1)}{6000 \times 10^{-8}\text{ cm}} = 46.7 \simeq 47 \text{ fringes}$$

If we wanted to produce the same shift by moving one of the mirrors instead of evacuating the tube, we use Equation 29.13, $x = N(\lambda/2)$, where $N = m = 47$, $\lambda = 6000 \times 10^{-8}$ cm. Therefore,

$$x = N\frac{\lambda}{2} = 47\left(\frac{6000 \times 10^{-8}\text{ cm}}{2}\right) = 0.141 \times 10^{-2}\text{ cm} = 0.0141\text{ mm}$$

Thus, the mirror has to be moved only 0.0141 mm to produce a shift of 47 fringes.

EXERCISE 29.2 Suppose in Example 29.2 that after evacuation, the cell is filled with a gas of refractive index 1.0012. Calculate the shift in the number of fringes that will be observed. What change in the position of one of the mirrors will produce the same shift? [*Ans.*: $m = 200$, $x = 0.06$ mm.]

29.4. Interference in Thin Films

There are many examples of interference produced in thin films. Colored fringes formed in soap bubbles and in thin layers of oil on water or other liquids, are just two of the many examples.

To start with, consider a thin film of thickness t, shown in Figure 29.4. Let monochromatic light of wavelength λ be incident on the upper surface. A part of this ray AB is reflected as ray BC and the rest of it is transmitted as ray BD. At D a part of this is again reflected, which eventually emerges as ray EF. The rays BC and EF are superimposed to produce interference fringes. Even though we have drawn only one incident ray, in a practical situation there are a large number of rays close together, and also the film is very thin. There are two important cases of this situation.

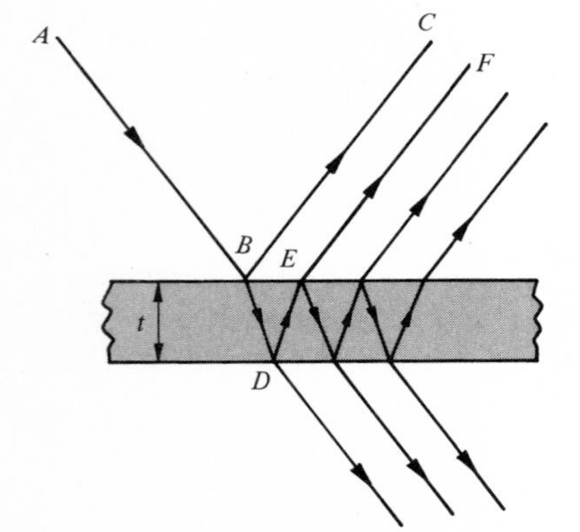

FIGURE 29.4. *Interference produced by a thin film by splitting the incident beam into multiple beams.*

First, there is the case in which the two surfaces of the thin film are parallel and the incident light is normal to these surfaces [Figure 29.5(a)]. The result is a formation of circular fringes. The second case is a wedge-shaped film such as

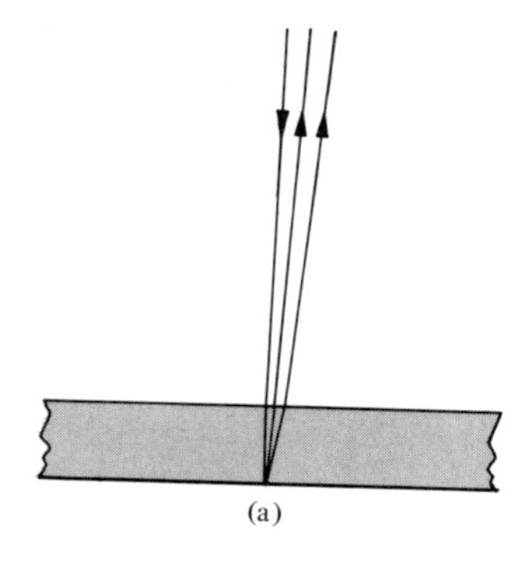

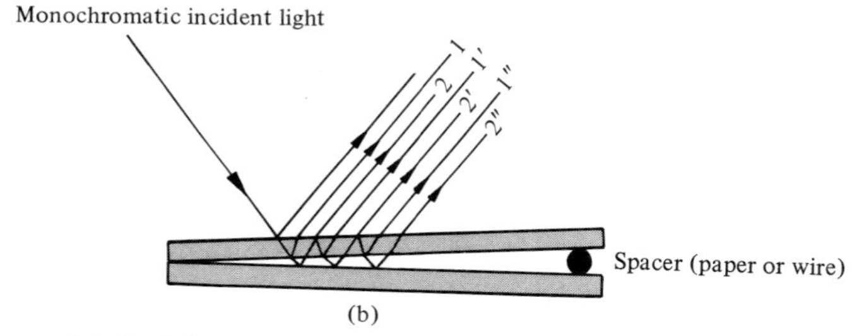

FIGURE 29.5. *Schematic of the apparatus used for producing interference fringes by a thin film: (a) for normal incident to thin parallel film and (b) wedge-shaped film.*

the air film between two smooth glass surfaces, as shown in Figure 29.5(b). The wedge may be produced by placing a thin sheet of paper or a small-diameter wire between two flat surfaces. In this case the reflected light from the upper and lower surfaces of the air film, such as ray 1 and 2 or 1′ and 2′ or 1″ and 2″, when emerging from the top is superimposed to produce interference fringes. The fringes in this case are straight lines. Also, the fringe at the point where the film thickness is zero is a dark fringe even though the optical path lengths are the same. The reason for this is the following rule.

When a light wave traveling in one medium is reflected from the surface of a second medium which has a greater refractive index, the reflected ray suffers an additional phase change of 180° (or path difference of $\lambda/2$). No phase change takes place if the second medium has a lower refractive index. The conditions of constructive and destructive interference will interchange if there is an additional phase difference of 180°.

A common situation is the one shown in Figure 29.6(a), where there is an air film formed between a convex lens and a plane surface of glass. This arrangement is due to Newton and the corresponding fringes formed in Figure 29.6(b) are called *Newton's rings*. Note that once again the central fringe is a dark one, as explained in the case above.

Two common applications are (1) testing the flatness of an optical surface, and (2) monitoring the application of a thin film on glass to produce nonreflecting glass. In optical instruments such as in an astronomical telescope it is necessary to produce surfaces which are optically flat; that is, the variation in thickness is less than a quarter of a wavelength of the light to be used. When such a surface

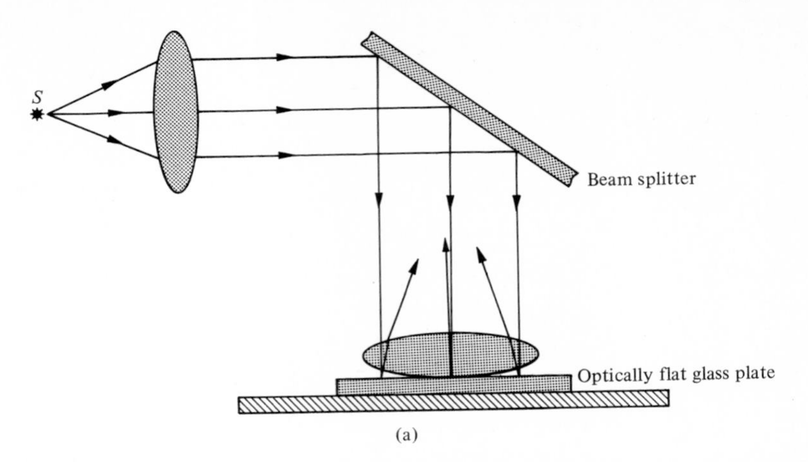

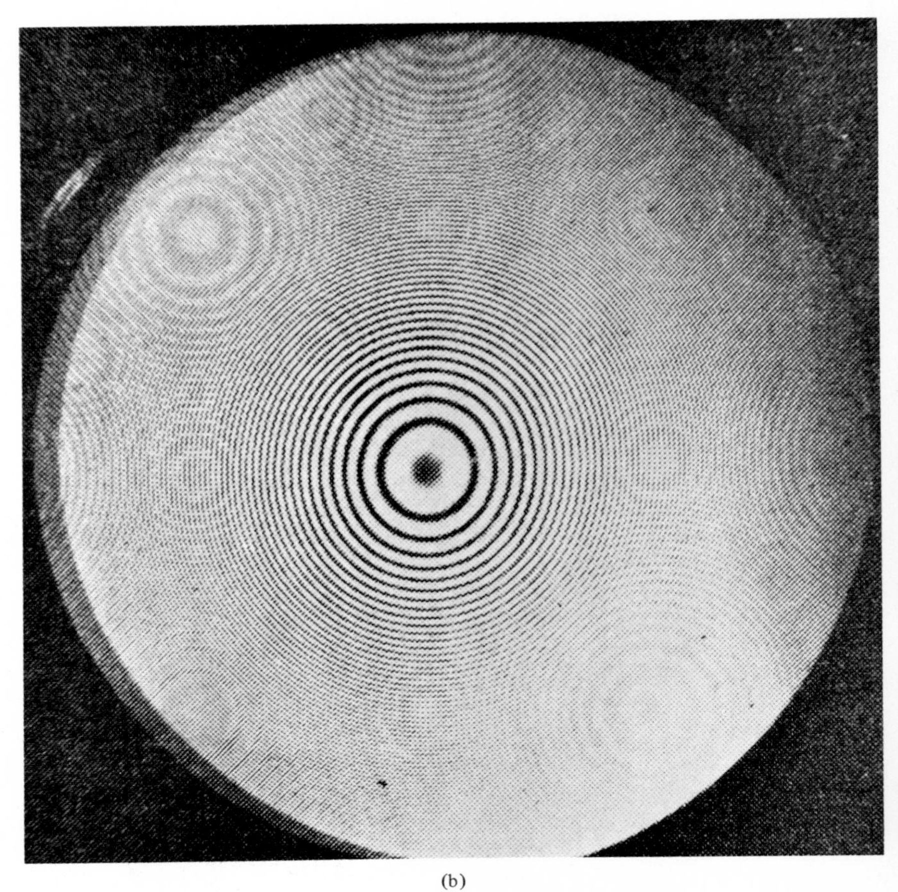

Figure 29.6. *(a) Experimental arrangement for obtaining Newton's fringes (or rings). (b) Photographs of Newton's rings.* (Courtesy of Bushnell Optical Company, a division of Bausch & Lomb.)

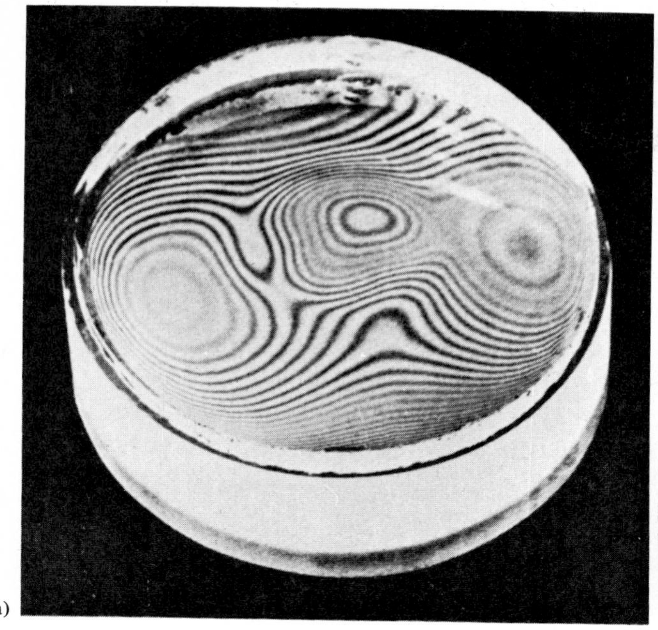

FIGURE 29.7. *Interference fringes obtained from (a) a surface that is not optically flat (it has three high spots) and (b) a surface that is optically flat.* (Courtesy of Bushnell Optical Company, a division of Bausch & Lomb.)

is viewed with a normal incident light, the fringes formed are as shown in Figure 29.7(a) for a surface that is not optically flat, and Figure 29.7(b) when the surface is optically flat.

To produce nonreflecting surfaces, a film of hard transparent material that has an optical thickness of a quarter of a wavelength ($\lambda/4$) of the light to be used and has a refractive index less than that of glass is deposited on the glass as shown in Figure 29.8. The light reflected from the upper and lower surfaces differ in a path difference of $\lambda/2$, leading to destructive interference. Also, the refractive index is such that each surface reflects almost an equal amount of light.

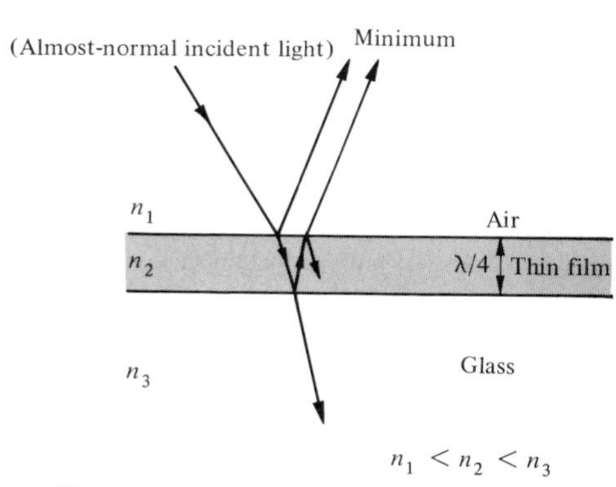

FIGURE 29.8. *Application of a thin coating on glass so as to produce nonreflecting glass.*

EXAMPLE 29.3 Calculate the position of the fifth dark and bright fringe formed by the air wedge shown in the accompanying figure. The bottom glass plate is 12 cm long and the space thickness SS' is 0.005 cm, while the wavelength of the light used is 5800 Å.

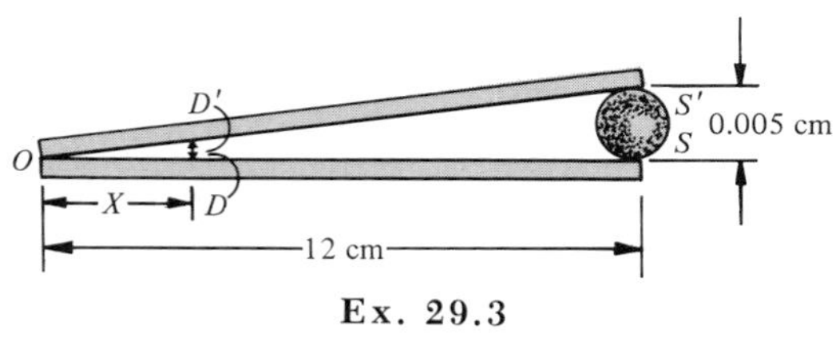

Ex. 29.3

Let us consider point O, where the air film thickness is zero. Because of the additional phase difference of π on reflection (as explained in the text), the point O is the position of the dark fringe. Suppose that the next dark fringe occurs at x cm from O. From triangles ODD' and OSS',

$$\frac{x}{12 \text{ cm}} = \frac{DD'}{SS'}$$

where DD' must be equal to $\lambda/2$ (for position of the dark fringe), $DD' = \lambda/2 = 5800 \text{ Å}/2 = 0.2900 \times 10^{-4}$ cm, and $SS' = 0.005$ cm. Therefore,

$$x = \frac{DD'}{SS'} \times 12 \text{ cm} = \frac{0.2900 \times 10^{-4} \text{ cm} \times 12 \text{ cm}}{0.005 \text{ cm}} = 0.0696 \text{ cm}$$

The fifth dark fringe will occur at

$$5 \times 0.0696 \text{ cm} = 0.3480 \text{ cm}$$

while the sixth dark fringe will occur at

$$6 \times 0.0696 \text{ cm} = 0.4176 \text{ cm}$$

In between these two is the position of the fifth bright fringe; that is,

$$\frac{(0.3480 + 0.4176) \text{ cm}}{2} = 0.3828 \text{ cm}$$

Exercise 29.3 In Example 29.3, what is the total number of bright and dark fringes. What is the position of the nth dark and nth bright fringe? [*Ans.:* 172 fringes; dark: $x_n = n \times (0.0696 \text{ cm})$; bright: $x_n = (n + \frac{1}{2})(0.0696 \text{ cm})$.]

29.5. Diffraction

Let us consider three types of waves—sound waves, water waves, and light waves. We have already shown in Chapter 17 that sound waves and water waves exhibit the phenomenon of *diffraction*, that is, bending of the waves around the edges of an obstacle placed in their path. For example, suppose that you speak from one room to another through a hole in the wall. A person standing in the other room does not necessarily have to stand in front of the hole to hear your voice. He can hear regardless of where he stands, but the intensity level will not be the same everywhere. Similarly, we showed in Figure 17.17 that when water waves reach a slit, they spread on the other side outside the width of the slit (or obstacle).

Now let us return to light waves. Ordinarily, when light shines on a slit placed in front of a screen as in Figure 29.9(a), we expect a sharp patch of light and a shadow as shown. This is very well explained if we assume the rectilinear propagation of light—the corpuscular theory. But a very close examination of the light distribution on the screen reveals dark and bright fringes near the edges, as shown in Figure 29.9(b). As a matter of fact, Figures 29.9(c) and (d) reveal that whenever an obstacle is placed in the path of light, instead of a sharp shadow we get bright and dark fringes near the edges. This

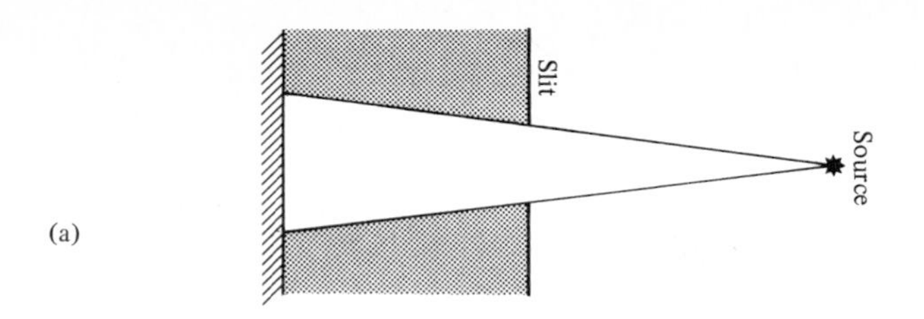

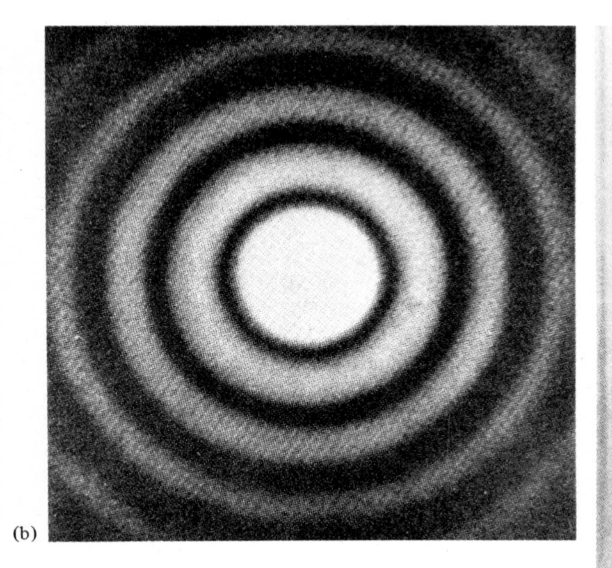

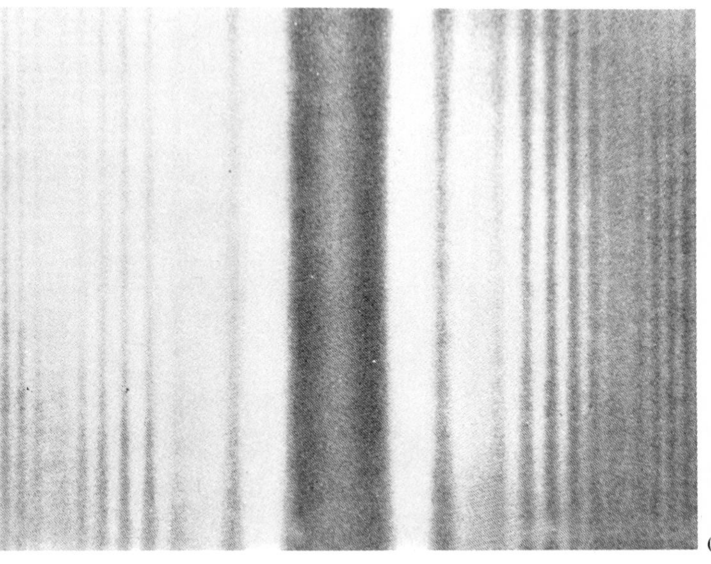

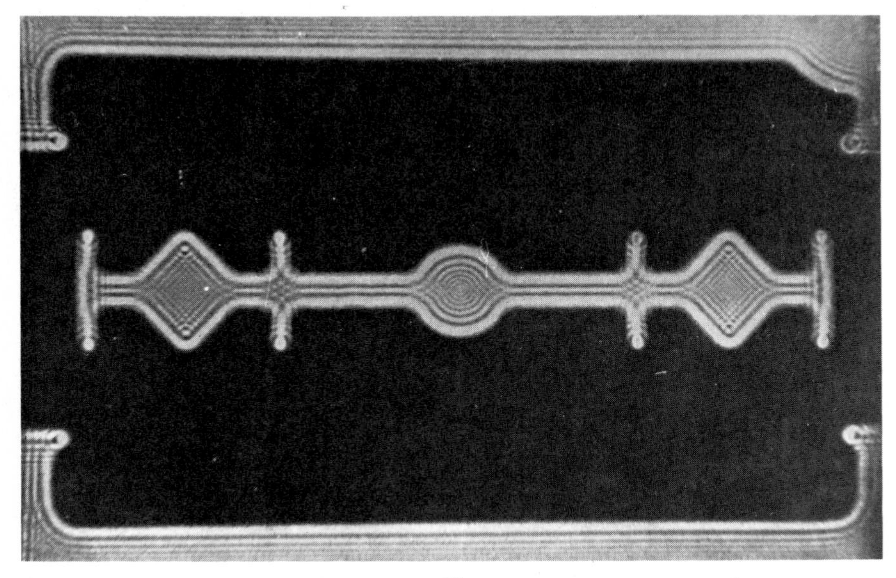

FIGURE 29.9. *Patch of light due to a circular slit a few millimeters wide, as in part (a). The dark and bright fringes in the geometrical shadow of: (b) a circular aperture, (c) a mechanical pencil, and (d) a razor blade are due to diffraction.* (Parts (b) and (c) courtesy of Bushnell Optical Company, a division of Bausch & Lomb. Part (d) from *Optics* by F. W. Sears, Addison-Wesley Publishing Co., Inc., Reading, Mass. 1949. Reprinted with permission.)

can be explained by assuming that light waves bend around an obstacle—and this phenomenon is called diffraction. As a matter of fact, we may define *diffraction* as the spreading of waves into the geometrical shadow of an obstacle. The distribution of light intensity resulting in dark and bright fringes (i.e., with alternate maxima and minima) is called a *diffraction pattern*.

The diffraction effects due to light waves are small, but they are always there. The diffraction effect is large if the wavelength is relatively large as compared to the size of the obstacle, and it is small if the wavelength is small. Whenever there is an obstacle, opaque or transparent, in the path of light, certain portions of the wave fronts will change in amplitude and phase; that is, diffraction occurs. The various parts of the wave fronts that propagate beyond the obstacle interfere, thereby producing a diffraction pattern. Actually, there is no significant difference between an interference pattern and a diffraction pattern. Interference is usually referred to as the superposition of only a few (two, three, or four) waves, whereas diffraction is concerned with the superposition of a very large number of waves.

Before discussing how diffraction patterns are formed, we must briefly explain the distinction between Fresnel and Fraunhofer diffraction. *Fresnel diffraction* (named after Augustin Jean Fresnel, 1788–1827) is one in which the incident waves, as well as the diffracted waves, which interfere to produce diffraction patterns, have spherical wave fronts. The situation is shown in Figure 29.10(a), where the source and the screen are placed at finite distances from the obstacle (slit) and no lenses are used. This is also called *near-field diffraction*. On the other hand, in *Fraunhofer diffraction* (named after Joseph von Fraunhofer, 1787–1826) the incident waves as well as the diffracted rays can be approximated to have plane wave fronts. This can be achieved by placing the source and the screen far away from the obstacle or by using lenses, as shown in Figure 29.10(b), so that the source and the screen are at the focal planes. Fraunhofer diffraction is also called *far-field diffraction*. Owing to the very nature of the wave fronts, Fraunhofer diffraction (planar wave fronts) is much simpler to analyze than Fresnel diffraction (spherical wave fronts). Therefore, we shall limit ourselves mainly to Fraunhofer diffraction.

Diffraction by a Single Slit

(Figures 29.11 and 29.12 are on page 670.)

Consider a single slit of width w as shown in Figure 29.11(a). The plane wave fronts are incident on this slit from the left. Each point on the incident wave front is a source of secondary Huygens wavelets. These wavelets then interfere to produce the diffraction pattern. It is simpler to deal with rays instead of wave fronts, as shown in Figure 29.12. In this figure we have drawn only nine rays. In actual practice there are a very large number of them, each differing in phase from the adjacent ones. The diffracted rays drawn in Figure 29.12 are those that are headed toward P, and make an angle α with the normal to the slit.

It is important to note that this single slit is much wider than the slits used in Young's double-slit experiment. In Young's double-slit experiment each slit was so narrow that a bundle of waves coming out of a slit were almost completely in phase with each other. This is not so in this single slit.

In order to calculate the conditions of maxima and minima of the intensity of the light on the screen, we may refer to Figure 29.12. Let us consider waves 1 and 5. These waves are in phase when in the wave front AB. After these

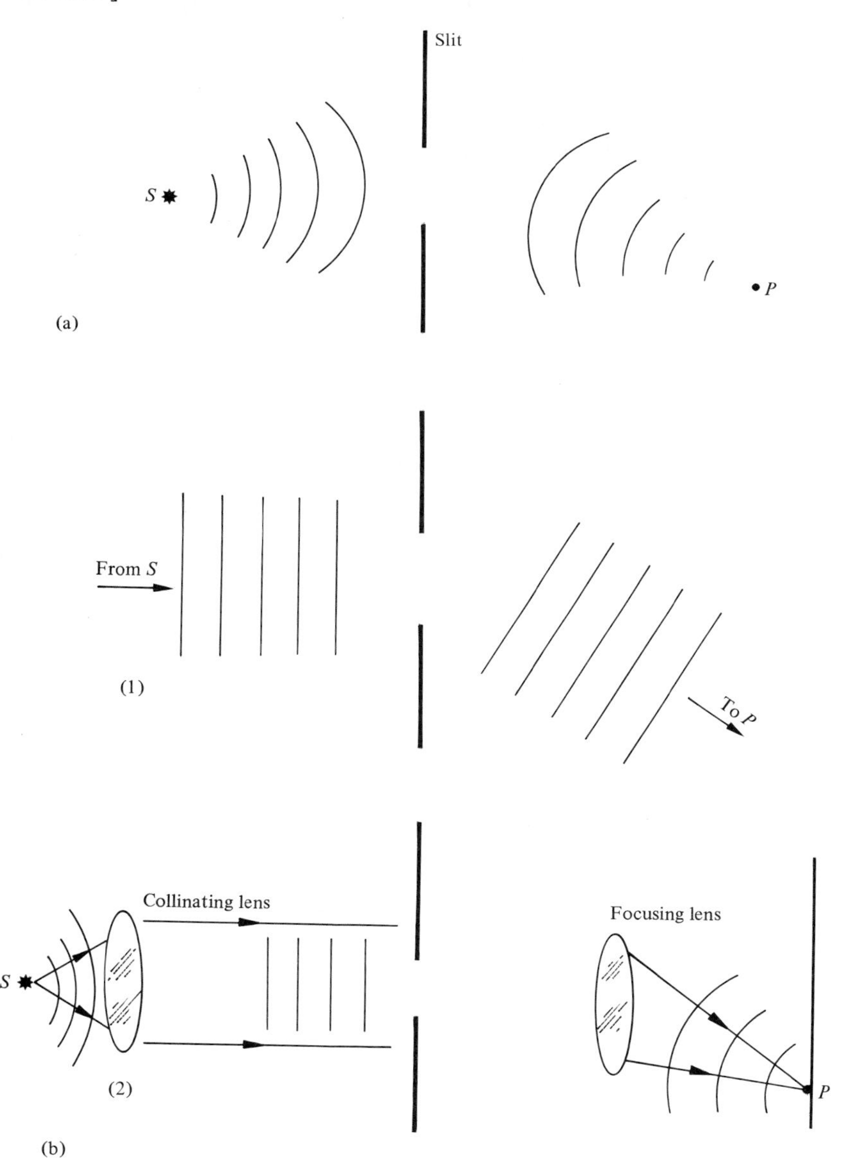

Figure 29.10. *(a) Conditions for producing Fresnel diffraction. (b) (1) and (2) are conditions for producing Fraunhofer diffraction.*

reach the wave front AC, wave 5 would have an additional path difference ab, say equal to $\lambda/2$. Thus, when these two rays reach point P, they will interfere destructively. Similarly, as shown, each pair 2 and 6, 3 and 7, 4 and 8 differ in path by $\lambda/2$ (or phase of 180°), and they will interfere destructively. But the path difference is $ab = (w/2)\sin\alpha$. Hence, the condition for destructive interference is

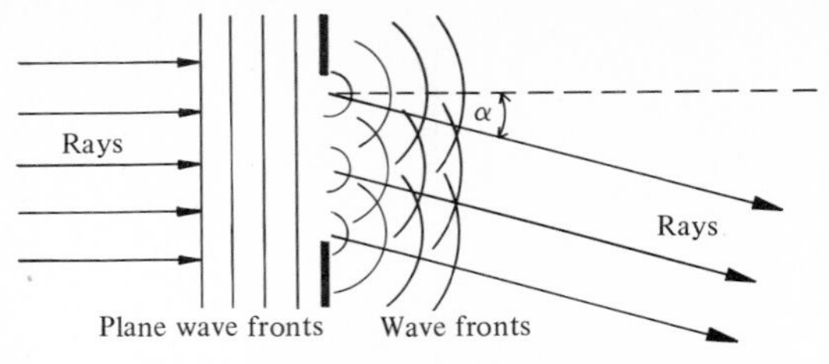

Figure 29.11. *Diffraction from a single-slit, wave-front representation.*

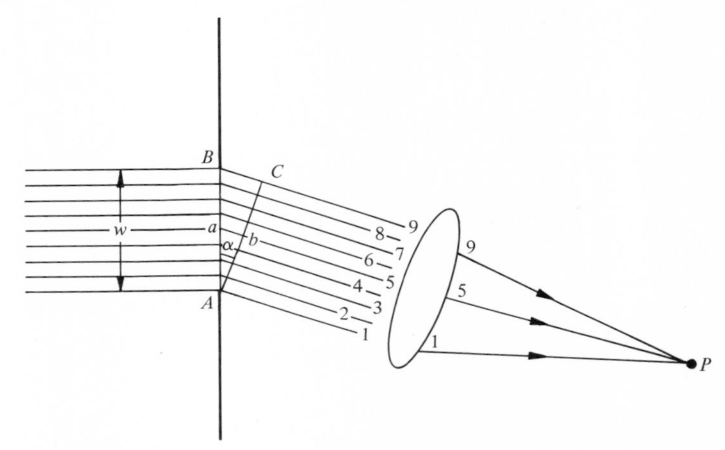

Figure 29.12. *Condition for minima for single-slit diffraction.*

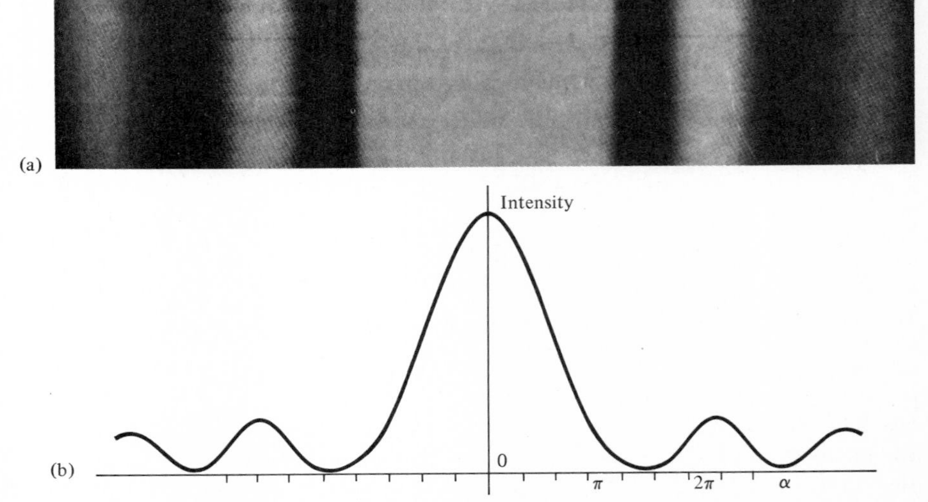

Figure 29.13. *Diffraction pattern due to a single slit: (a) Photograph (b) Intensity plot.* (From M. Cagnet, M. Françon, and J. C. Thrierr, *Atlas of Optical Phenomena,* Springer-Verlag, Berlin, 1962.)

$$ab = \frac{w}{2} \sin \alpha = \frac{\lambda}{2} \qquad \text{or} \qquad w \sin \alpha = \lambda$$

This is the condition for the first minimum on either side of the central maximum as shown in Figure 29.13. But the waves arriving at the central point O from the upper half of the slit are in phase with the waves arriving from the lower half of the slit, thereby, forming a central maxima. In general, the conditions for different orders of minima on either side of the central maximum are given by

$$w \sin \alpha = n\lambda \qquad n = 1, 2, 3, \ldots \quad \text{for minima} \tag{29.15}$$

Substituting $n = 1, 2, 3, \ldots$ we can calculate α; hence, the position of different minima is as shown in Figure 29.13. In between the minima are the maxima. Figure 29.13 shows the actual photograph of such a diffraction pattern. The relative intensities of these dark and bright fringes are shown by means of a plot below the photograph.

Diffraction by a Circular Aperture

We showed a circular aperture in Figure 29.9(a), with the resulting diffraction pattern in Figure 29.9(b). Instead of a bright spot on the screen, we obtain dark and bright fringes. This is an important situation because most optical instruments have circular apertures. No matter how improved the instrument is, a diffraction pattern always results—a bright image of a bright source with additional dark and bright fringes surrounding it. The central bright disk in this pattern is called *Airy's disk*. The diffraction pattern limits the ability of an instrument to resolve or separate the images of two neighboring objects. (We shall discuss this in Chapter 30.)

Diffraction Pattern by a Straightedge and a Circular Obstacle

Figure 29.14(a) shows the configuration for obtaining a diffraction pattern by a straightedge and Figure 29.14(b) shows the diffraction pattern. According to the rectilinear propagation of light, there should be a uniform illumination above the line AA', and complete darkness below it. But this is not true, as shown in Figure 29.14(b). The illumination does extend a short distance below and above AA', but instead of being uniform, a series of dark and bright fringes appear parallel to the edge. These are all explainable with the help of the wave theory.

Finally, let us consider a circular aperture, obstacle, or wire, shown in Figure 29.15. The diffraction pattern of the aperture on either end is surrounded by dark and bright fringes, each edge behaving as a straightedge. The most important feature of this pattern is that the central point O is a bright fringe, even though it is very narrow and not as bright as the two edges, the reason being that light arriving at O from the two points A and B is in phase. This feature is quite contrary to the rectilinear propagation of light but easily explainable by the wave theory.

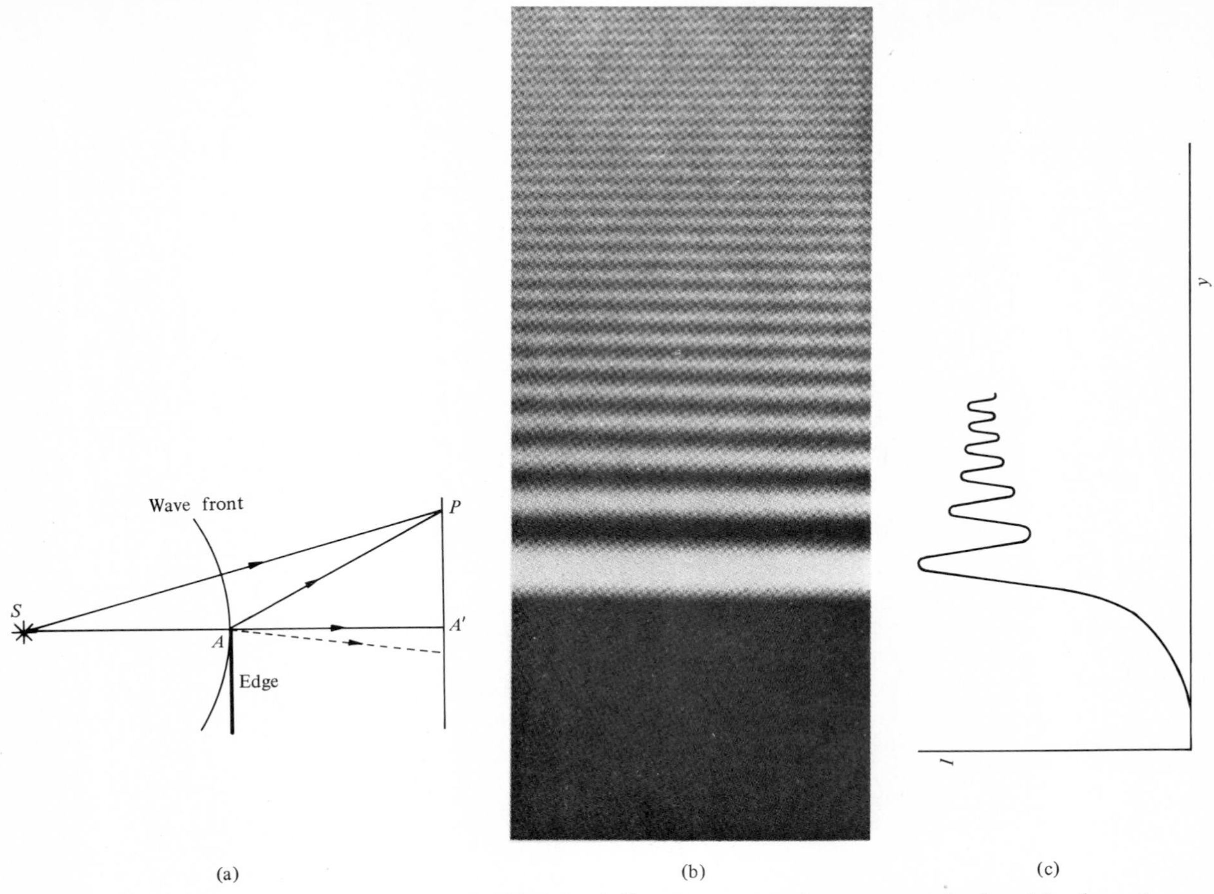

Figure 29.14. *Diffraction pattern due to a straightedge: (a) schematic, (b) photograph of the fringes, and (c) the intensity distribution.*

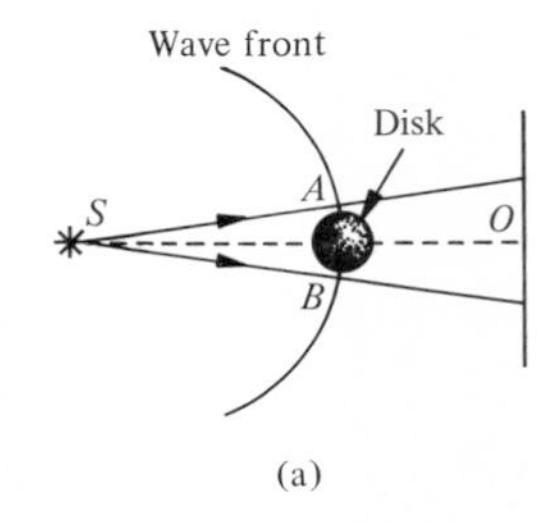

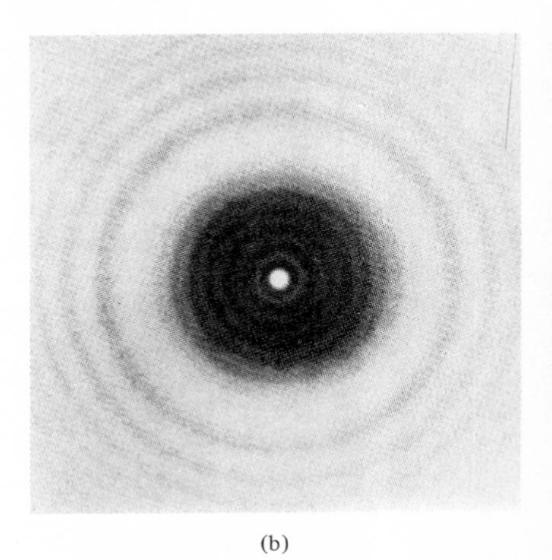

Figure 29.15. *Diffraction pattern due to a circular obstacle: (a) schematic and (b) photograph of the fringes.* (From M. Cagnet, M. Françon and J. C. Thrierr, *Atlas of Optical Phenomena,* Springer-Verlag, Berlin, 1962.)

29.6. Diffraction Grating

An array of a large number of parallel slits, all of the same width and spaced at equal intervals, is said to form a *diffraction grating*. In earlier diffraction gratings thin wires of thickness from 0.005 to 0.1 mm and separated by the same order of distances were used as the grating. Later techniques involved drawing a series of very fine equidistant parallel slits on a glass plate by means of a fine diamond pen. The light cannot go through the lines drawn by the diamond while the spacing between the lines is transparent to the light. Such mechanical gratings may have as many as 12,000 lines per centimeter to produce a diffraction of visible light. The spacing d between adjacent slits is called the *diffraction element* and is $d = 1/N$, where N is the number of slits in one unit length of a diffraction grating.

Suppose that a beam of light is incident normally on a diffraction grating G, as shown in Figure 29.16. We may calculate the condition for interference at any point on the screen. Let the slits in the diffraction grating, just as in Young's double-slit experiment, be considered so narrow (a special case) that a band of waves coming out of any one of these slits are in phase with each other (unlike the single slit discussed previously, which is quite wide). Figure 29.16 shows the secondary Huygens wavelets that start from each slit. These wavelets interfere with each other, producing destructive or constructive interference, depending upon whether they are out of phase or in phase at a given point.

Suppose that a beam of parallel rays coming out of the slit are brought to focus at point O on the screen by means of a convex lens. The point O will be a bright spot. On the other hand, let us see what happens if the point under

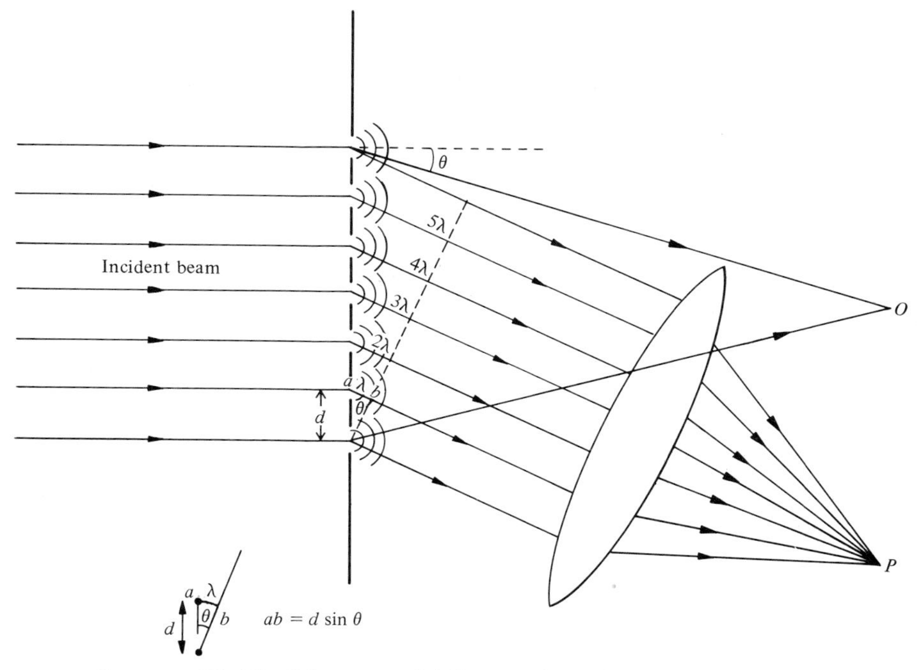

FIGURE 29.16. *Schematic of diffraction by a diffraction grating.*

consideration is P so that the diffracted rays make an angle θ with the normal to the grating, as shown. Suppose that at point P the light from one slit is one wavelength behind the light from the adjacent slit; for example, ray 2 is λ behind ray 1 if $ab = \lambda$, and so on. This means that the rays from one slit are in phase with the rays from the adjacent slit, and they will reinforce each other, producing constructive interference. Thus, $ab = d \sin \theta = \lambda$ is the condition for the first maximum. If the path differences between the rays from the adjacent slits are 2λ, 3λ, 4λ, . . . , we still will get constructive interference although at different angles. Thus, in general, the condition for constructive interference is

$$d \sin \theta = m\lambda \qquad m = 0, 1, 2, 3, \ldots \quad \text{for maxima} \tag{29.16}$$

where m is called the *order* of the diffraction pattern; $m = 0$ is the central bright spot; $m = 1$ is the first-order maximum; $m = 2$ is the second order maximum; and so on.

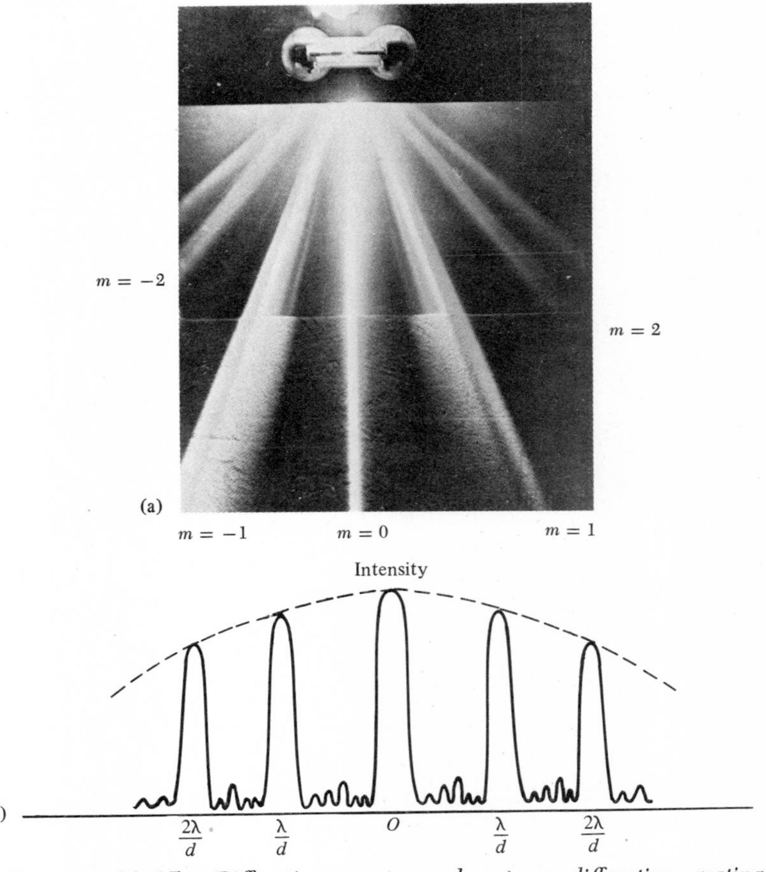

FIGURE 29.17. *Diffraction spectrum due to a diffraction grating: (a) photograph, and (b) intensity plot.* [Part (a) courtesy of Klinger Educational Products Corp., 83-40 Parsons Blvd., Jamaica, N.Y. 11432.]

If we use a monochromatic source of light, we get bright fringes at regular intervals. Knowing d and m and measuring θ, the value of λ can be measured very accurately.

Another outstanding feature of the diffraction grating is the number of different orders (m) that are obtained. In a prism when white light is used, it is dispersed and we get one spectrum. If white light is used in a grating, the diffraction produces not only the dispersion of different colors, but also more than one spectrum, one on each side of the central maximum for each m. The value of m is limited by d, λ, and $\sin\theta$ (the maximum $\sin\theta = \pm 1$). Figure 29.17(a) shows the diffraction spectrum obtained by using a diffraction grating while Figure 29.17(b) shows the intensity plot. Furthermore, it is important to note that the interference of radiation from different slits produces the principal interference pattern. The maxima we have been discussing are very sharp, bright, and narrow. They are separated by λ/d as shown in Figure 29.17 and are called the *principal maxima*. In between these consecutive principal maxima there are $N_t - 1$ minima, where N_t is the number of slits. Of course, in between each pair of minima, there are very faint maxima called subsidiary maxima as shown.

An interesting demonstration of diffraction is shown in Figure 29.18. A

FIGURE 29.18. *Chalk dust displays the diffracted rays of laser light.* [From Dr. David K. Walker, *The Physics Teacher,* **11** (1973).]

grating with more than 30 slits per mm is placed in the path of a laser beam. The diffraction rays are made visible by chalk dust obtained by tapping two dusty erasers against each other. The rays are scattered by the chalk dust, making the diffracted rays visible.

EXAMPLE 29.4 A grating 2.5 cm long has 15,000 lines, and the normal incident light used has a wavelength of 6060 Å. Calculate the angular positions for the first-, second-, and third-order maxima.

$d = \dfrac{2.5 \text{ cm}}{15{,}000} = 0.000167 \text{ cm} = 1.67 \times 10^{-4} \text{ cm}, \quad \lambda = 6060 \text{ Å} = 6060 \times 10^{-8} \text{ cm} = 0.6060 \times 10^{-4} \text{ cm}$. Therefore, from Equation 29.16, $d \sin\theta = m\lambda$:

$$\text{For } m = 1: \quad \sin\theta_1 = \frac{m\lambda}{d} = \frac{1(0.6060 \times 10^{-4}\text{ cm})}{1.67 \times 10^{-4}\text{ cm}} = 0.363 \qquad \theta_1 = 21.3°$$

$$\text{For } m = 2: \quad \sin\theta_2 = 2(0.363) = 0.726 \qquad \theta_2 = 46.5°$$

$$\text{For } m = 3: \quad \sin\theta_3 = 3(0.363) = 1.089$$

Since $\sin\theta_3 > 1$, the third order will not be visible.

When $\theta = 90°$, $\sin\theta = 1$; from the relation $d\sin\theta = m\lambda$, if $m = 3$, we get

$$\lambda = \frac{d\sin\theta}{m} = \frac{(1.67 = 10^{-4}\text{ cm})(1)}{3} = 0.5566 \times 10^{-4}\text{ cm}$$

That is, only for $\lambda \leq 5566$ Å can one observe the third order with this grating.

Exercise 29.4 In the above example calculate the positions of the first three maxima if $\lambda = 5000$ Å. Is it possible to observe the third order maximum? [*Ans.:* $\theta_1 = 17.4°$, $\theta_2 = 36.8°$, $\theta_3 = 63.9°$; yes.]

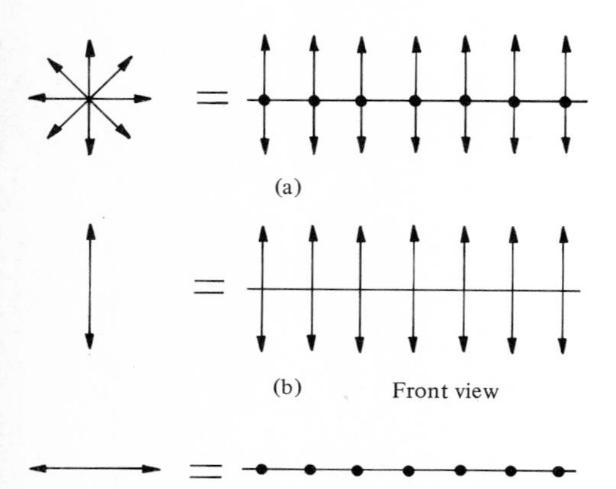

Figure 29.19. *Schematic representation of (a) unpolarized ordinary light, (b) plane-polarized (or linearly polarized) light with the electric vector vibrating vertically, and (c) plane-polarized (or linearly polarized) light with the electric vector vibrating horizontally.*

29.7. Polarization

In previous sections we have established the wave nature of light by discussing its application to explain the phenomena of interference and diffraction. However, this does not establish that light waves are transverse in nature. That is, the electric and magnetic vectors, which are at right angles to each other, are also perpendicular to the direction of propagation of the waves. The phenomenon of polarization establishes the transverse nature of light and of electromagnetic waves in general. Since the interaction of electromagnetic waves with matter for the most part depends upon the orientation of the electric vector in space, we shall discuss polarization with reference to the vibrating electric vector only.

Let us consider a beam of ordinary light. It consists of a large number of waves, each with its own plane of vibration. All directions of vibration are equally probable and are always perpendicular to the ray (the direction of propagation). Such a beam of light is said to be an *unpolarized light beam,* and looking head-on, is represented schematically as in Figure 29.19(a). By making such an unpolarized beam go through a polarizing device (to be discussed shortly), called a *polarizer,* it is possible that the transmitted beam will have electric vibrations only in certain directions. The resulting light beam is said to be *polarized,* as shown in Figures 29.19(b) and (c). If the electric vibrations are vertical, denoted by a vertical arrow as shown in Figure 29.19(b) then the light beam is said to be *vertically plane-polarized.* If the electric vibrations are horizontal, as shown by the dots in Figure 29.19(c), the light beam is said to be *horizontally plane-polarized light.* Either component is referred to as simply *plane* or *linearly polarized.*

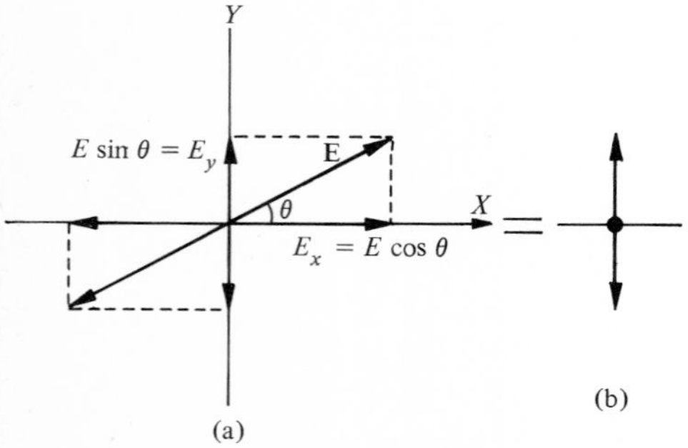

Figure 29.20. *Single electric vector resolved into two linearly polarized components: (a) horizontal component $E_x = E\cos\theta$ and (b) vertical component $E_y = E\sin\theta$.*

Any vibration in an unpolarized light beam can be decomposed into two components, one horizontal and the other vertical, as shown in Figure 29.20(a). Thus, an *unpolarized light may be resolved into two plane-polarized light beams at right angles to each other,* represented as shown in Figure 20.20(b); the arrow indicates the vertical vibrations and the dots the horizontal vibrations. In an ordinary light source, each atom or molecule emits vibrations which are individually linearly polarized. Since there is an enormously large number of molecules in any given sample, the total number of vibrations that are randomly oriented result in an unpolarized light beam. The statements above can be demonstrated by means of two simple equivalent experiments, one dealing with mechanical waves and the other with light waves.

Consider a transverse wave on a string as shown in Figure 29.21. The vibrations are at right angles to the propagation of the wave. Suppose that a wooden piece with a slit P is placed in its way. If this slit is parallel to the direction of the transverse vibrations, the wave passes through the slit undisturbed. If the slit is at right angles, the wave is not transmitted at all. This means that if the vibrations are in all directions in a plane perpendicular to the direction of propagation, only those parallel to the slit P are transmitted through, as shown in Figure 29.21(a). The slit P is called the polarizer and the wave transmitted through it is said to be polarized. Now suppose that a second

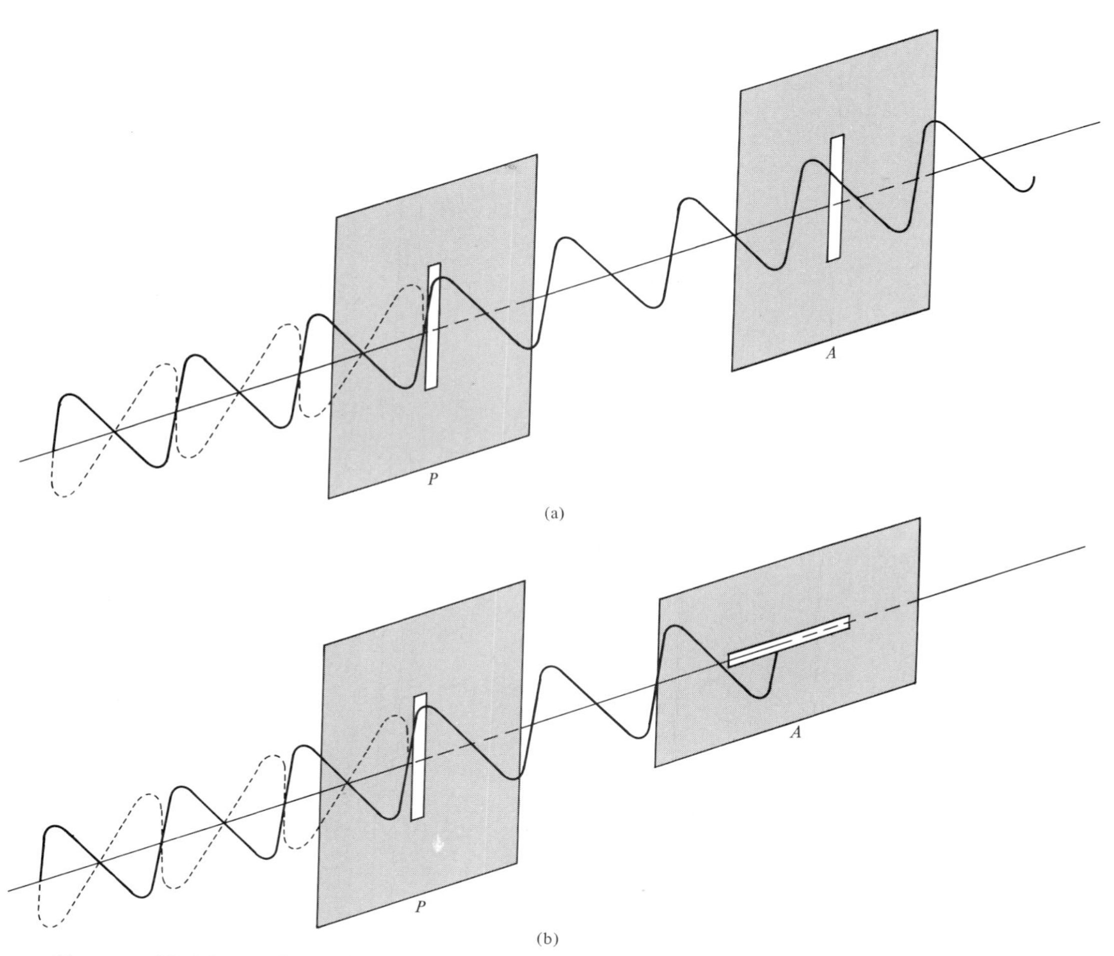

FIGURE 29.21. *Mechanical experiment (using transverse waves on strings) to demonstrate polarization.*

slit A is placed in the path of the beam, as shown in Figure 29.21. If the slit A is parallel to P, the polarized wave is transmitted through it. If A is at right angles to P, no wave is transmitted. The slit A works as an analyzer. For no transmission the slits are said to be *crossed*. On the other hand if the waves were longitudinal, the position of the slit would have not made any difference, and the wave would not have been polarized.

Let us now extend this argument to the case of light waves. The function of the slit is performed by a polarizing disk or sheet, which we shall discuss shortly. Suppose that the source S gives natural unpolarized light, as shown in Figure 29.22. When this falls on a polarizer P, only those components that are parallel to a certain direction in the polarizer, called the *optic axis*, are transmitted; the rest are stopped. When this transmitted light is viewed by a photocell, its intensity is less than the intensity of the incident beam, but the rotation of the polarizer P does not change the transmitted intensity. The transmitted beam is polarized, as can be detected by placing in the path of the beam a second polarizing disk A, called an *analyzer*. If the analyzer A is rotated, the light reaching the photocell is maximum for a certain position. This happens when the axes of P and A are parallel. On the other hand, if the analyzer A is rotated 90° further, the polarizing disks A and P are crossed and the photocell reads zero intensity. Further rotation to 180° (90° more from the zero intensity position) brings maximum intensity again. While for angles between 0 and 90°, the intensity varies from maximum to zero.

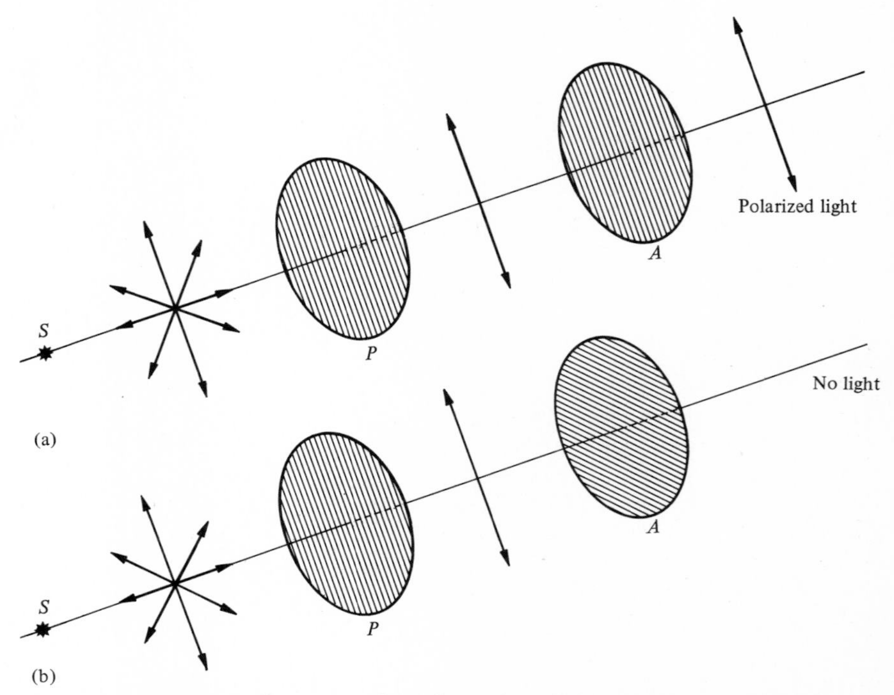

FIGURE 29.22. *Light passing through a Polaroid sheet P is plane-polarized. The second sheet of Polaroid A works as an analyzer. Light may or may not pass through it, depending upon its orientation.*

Thus, by the method above we can detect whether or not the beam is polarized. Suppose that the beam coming out of the polarizer P was partially polarized. If we now rotate the analyzer A, for certain positions the photocell will read a maximum intentisy I_{max}, while at another position, 90° from the previous position, it reads a minimum intensity I_{min}. We can now define the percentage of polarization P of the beam as

$$P = \frac{I_{max} - I_{min}}{I_{max} + I_{min}} \times 100\% \tag{29.17}$$

Suppose that the angle between the optic axes of the polarizer P and the analyzer A is θ. The linearly polarized beam of light coming out of P makes an angle θ with the optical axis of A. Let us resolve **E** (the electric field vector) into two components, $E \cos \theta$ and $E \sin \theta$. Only $E \cos \theta$, which is parallel to the optic axis, is transmitted through. Since the intensity is equal to the square of the amplitude, the transmitted intensity I is given by

$$I = E^2 = E_0^2 \cos^2 \theta$$

or

$$I = I_0 \cos^2 \theta \tag{29.18}$$

where I_0 is the maximum light transmitted. The relation above was discovered by Etienne Louis Malus in 1809, and is called *Malus' law*. Thus, in general, when a polarized light of intensity I_0 falls on an analyzer A, the transmitted intensity I, given by Malus' law, depends upon the angle θ which the **E** vector makes with the optic axis of the analyzer.

29.8. Production and Applications of Polarization

Polarization by Reflection

It is found that when a beam of unpolarized light is reflected from a dielectric medium such as glass or water, the reflected light, depending upon the angle of incidence θ_i, is unpolarized, partially polarized, or completely plane-polarized. If the angle of incidence θ_i is 0° (normal incidence) or 90° (the grazing incidence), the reflected beam is unpolarized.

For example, consider a beam of unpolarized light incident on a glass plate as shown in Figure 29.23(a). The incident beam is shown decomposed into two components. When this beam of light falls on a reflecting surface, the component parallel to the surface is preferentially reflected. Thus, the reflected and refracted beams will be partially polarized, as shown in Figure 29.23(a).

Now the angle of incidence θ_i is changed to such a value that the angle between the reflected and the refracted beam is 90°, as shown in Figure 29.23(b). In this situation the reflected beam consists only of the parallel vector component, while the refracted ray consists of both the perpendicular and parallel components. Thus, when the angle of incidence is equal to the *polarizing angle*, $\theta_i = \theta_p$, the reflected beam is completely plane-polarized, while the refracted ray is partially polarized. The relation between the

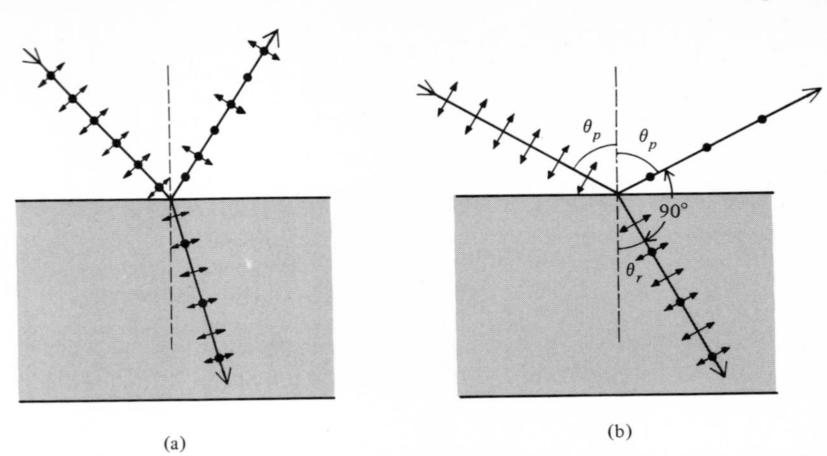

FIGURE 29.23. *(a) Incident light is partially polarized by reflection and refraction at any angle. (b) The reflected beam is completely plane-polarized when the incident angle is equal to the polarizing angle.*

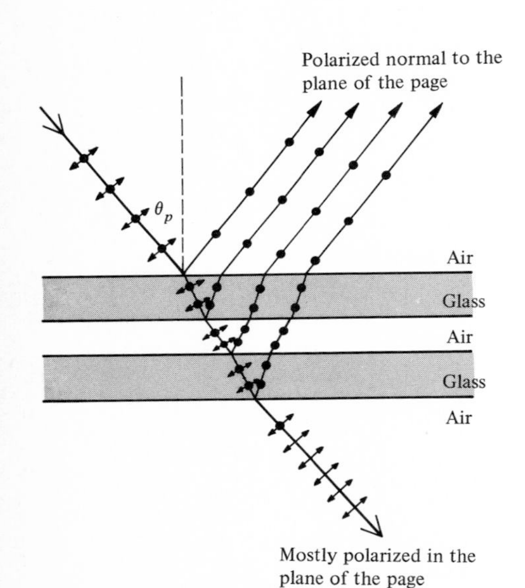

FIGURE 29.24. *Use of a stack of thin glass plates to enhance the intensity of the polarized reflected beam (only two plates shown).*

polarizing angle θ_p and the refractive index n can be arrived at easily. Referring again to Figure 29.23(b), $\theta_p + \theta_r + 90° = 180°$ or $\theta_r = 90° - \theta_p$. That is, $\sin\theta_r = \sin(90° - \theta_p) = \cos\theta_p$. From Snell's law,

$$n = \frac{\sin\theta_p}{\sin\theta_r} = \frac{\sin\theta_p}{\cos\theta_p} = \tan\theta_p$$

$$\boxed{n = \tan\theta_p} \qquad (29.19)$$

This relation is called *Brewster's law*, and the angle θ_p is called *Brewster's angle*.

At the polarizing angle, even though the reflected light is completely linearly (or plane-) polarized, it is very weak (intensity is low). Most of the light intensity goes into the refracted beam. To enhance the intensity of the polarized reflected beam, several thin plates are kept parallel to each other, as shown in Figure 29.24. Each plate reflects a small amount of the incident beam, eventually leading to a reflected, completely plane polarized beam of appreciable intensity. The sunlight reflected from the surface of a lake or paved road also contains partially polarized light according to this phenomenon, but involving only one surface.

Polarization by Double Refraction

We know that when a ray of light enters a medium such as glass, it travels with a single definite speed which is independent of the direction of propagation. But this is not true always. There are crystals, such as calcite (also called Iceland spar), which although homogeneous, are anisotropic. *Anisotropic* means that the speed of light is different in different directions in the crystal. Such crystals are called *doubly refracting* or *birefringent*. Calcite and quartz are examples of such anisotropic crystals; glass and water are examples of isotropic media.

We now demonstrate the anisotropy of the double-refracting crystals. Let us put a mark on a paper and look at it through a calcite crystal. We will observe two images. If we examine these images with the help of an analyzer (Polaroid sheet), we will find that the images disappear at two different positions, separated by a rotation of 90° of the analyzer. From this we conclude that each image is produced by plane-polarized light with the planes of vibrations at right angles to each other. Thus, when light enters the crystal, it splits into two plane-polarized beams traveling in different directions. This phenomenon is known as *double refraction*. Instead of one set of spherical Huygens wave fronts, in doubly refracting crystals there are two sets. The one with its electric vibrations perpendicular to the page is a spherical wave front which is called an *ordinary wave front* and is represented by dots as shown in Figure 29.25. The other wave front with its vibrations in the plane of the page is elliptical and is called an *extraordinary wave front*, represented by lines as shown in Figure 29.25.

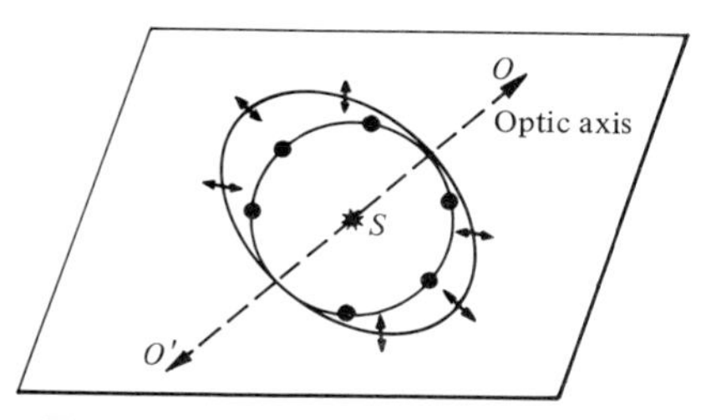

FIGURE 29.25. *Point source S in a doubly refracting crystal (such as calcite) produces two wave fronts—ordinary and extraordinary.*

In the case above, the speed of the extraordinary vibrations is greater than the ordinary vibrations, as indicated by the circle being inside the ellipse. There is one direction, OO', as shown in Figure 29.25, in which both ordinary and extraordinary rays travel with the same speed. This direction is called an *optic axis*. Since an optic axis is a direction, any line parallel to OO' is also an optic axis. For any other direction, because of different speeds, double refraction takes place and the paths of the two wave fronts are different. That the beam splits up into two beams is demonstrated in Figure 29.26. Suppose that light is incident normally on a crystal. Two images are formed on the screen. If the crystal is rotated using the incident beam as the axis, the image corresponding to the ordinary wave front remains fixed while the image corresponding to the extraordinary wave front goes around the first image in a circle as shown.

Thus, the double-refracting crystals can be used to produce plane-polarized light beams. To accomplish this, in the last 150 years many prisms of such crystals have been constructed, such as the Nicol prism, Wollaston prism, Rochon prism, and many others. Because of different refractive indices for the ordinary and extraordinary rays, one goes through undisturbed while the other is totally internally reflected, as is illustrated in Figure 29.26.

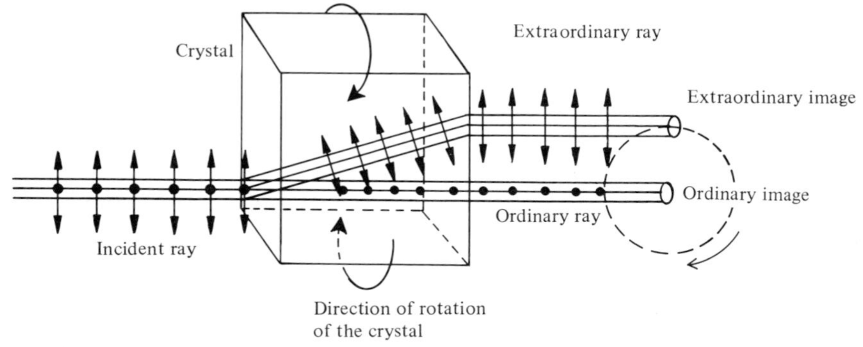

FIGURE 29.26. *An incident beam is split in two by a calcite crystal. If the crystal is rotated, the extraordinary image rotates around a fixed ordinary image.*

Polarization by Selective Absorption

Some doubly refracting crystals, such as tourmaline, possess a property called *dichroism*. Such crystals show selective absorption; that is, one polarized component is completely absorbed while the other is transmitted.

Polarization Using Polaroid Sheets

The doubly refracting prisms are difficult and costly to make and their size is limited. Large sheets of polarizing material (having a transmission axis) became available in 1938 when Edwin H. Land invented a material called Polaroid. Polaroid is manufactured by stretching a plastic sheet containing long hydrocarbon chains (as polyvinyl alcohol, PVA) so as to align the molecules. The sheet is then dipped in an iodine solution (or ink containing iodine). The iodine gets attached to the long hydrocarbon chains, thereby providing mobile electrons. The component of the incident vibrating electric vector which is parallel to this chain is absorbed by these mobile electrons (the electrons are set into a back-and-forth motion). At right angles to the direction of the chain, absorption cannot take place; this provides a transmission axis.

These Polaroid sheets do not polarize all wavelengths. Some extreme red and ultraviolet do get through unpolarized.

Polarization by Scattering

Polarization produced by scattering can be explained with the help of the following considerations. The electric vector of the incident electromagnetic wave interacts with the electrons of the medium and sets them into oscillation with their line of vibration parallel to the incident electric vector. A vibrating electron behaves like an oscillating dipole and emits radiation in all directions except along its own line of vibration (see Chapter 27). The resultant wave in the medium is the sum of the incident wave and the waves emitted from the oscillating dipole. The intensity and the percentage of polarization of the resultant light depends upon the wavelength of the incident light and the location of the observer with respect to the oscillating dipoles as explained below.

When rays of light fall on air molecules or dust, the amount of light scattered depends upon the wavelength λ and the relative size of the wavelength and the scattering target particle size d. If $d \ll \lambda$, the shorter wavelengths are scattered much more than the longer wavelengths. (The amount of scattering is proportional to $1/\lambda^4$.) Thus, if white light from the sun is incident on air molecules in the atmosphere, blue light is scattered much more than red light ($\lambda_{blue} < \lambda_{red}$). Because of the blue scattered light we receive, the sky looks blue. For the same reasons, at sunrise and sunset the sun appears redder. The white light goes through a far greater thickness of air; therefore, most of the blue has been scattered, and the light reaching an observer will be rich in red.

Let us consider a few applications of the phenomenon of polarization. One most common occurrence is sunlight reflected from smooth surfaces such as roads and lakes. The reflected light, the glare, is linearly polarized in the horizontal plane. This glare can be avoided or eliminated by using sunglasses made out of a Polaroid sheet with its optic (or transmission) axis vertical. Thus, the horizontally polarized light cannot go through.

Another application is connected with the study of optical activity. When a beam of light is made to pass through certain crystals and liquids, the direction of vibration of the transmitted polarized light is found to be rotated. The phenomenon of the rotation of the plane of polarization is called *optical activity*. The amount of rotation of the plane of polarization by certain liquids, such as sugar solutions, depends upon the density of the molecules in the liquid. This method is used commercially for measuring the amount of sugar present in a given sample; the rotation of the plane of polarization is directly proportional to the concentration of sugar.

Another useful application of the phenomenon of polarization is in optical stress analysis. Suppose that a polarizer P and an analyzer A are in a crossed position. No light comes out of A. If a doubly refracting crystal is placed between them, some light will come out of A. Some substances, such as glass and plastics, under great stress become doubly refracting. Hence, when such an object is placed between P and A, light will come out of A. The pattern of the interference fringes obtained reveals the strong and weak points in the structure. Non-transparent objects may be analyzed by reflected light.

EXAMPLE 29.5 A ray of light is incident on a surface of benzene of refractive index 1.50. If the reflected light is linearly polarized, calculate the angle of refraction.

From Equation 29.20, $n = \tan\theta_p$, where $n = 1.50$.

$$\tan\theta_p = 1.50 \qquad \text{or} \qquad \theta_p = 56.3°$$

That is, the incidence angle which will be equal to the polarizing angle is 56.3°. The angle of refraction θ_r is given by

$$\theta_r = 90 - \theta_p = 90° - 56.3° = 33.7°$$

EXERCISE 29.5 What should be the angle of incidence so that light reflected from a glass ($n = 1.66$) surface is linearly polarized? What is the corresponding angle of refraction? [*Ans.*: 58.9°, 31.1°.]

SUMMARY

The superposition of two wave trains of the same frequency and the same amplitude traveling in the same direction results in *interference phenomena*. The resultant amplitude is $E_m = 2E_0 \cos(\delta/2)$. For *constructive interference* the waves are in phase and $d = m\lambda$, while for *destructive interference* the waves are out of phase and $d = (m + \frac{1}{2})\lambda$, where $m = 0, 1, 2, 3, \ldots$ and $d = (\lambda/2\pi)\delta$.

In Young's double-slit experiment the condition for constructive interference is $d \sin\theta = m\lambda$ while a point on this bright fringe is given by $y_m = (m\lambda)(D/d)$. In a Michelson interferometer to produce a shift of N fringes, the mirror must be moved through a distance $x = N(\lambda/2)$. If one beam of the interferometer goes through a gas of refractive index n, then $2L(n - 1) = m\lambda$.

Two important cases of interference in thin films are (1) two surfaces of thin film are parallel and the incident light is normal to these surfaces and (2) the film is wedge-shaped. While considering interference, the following rule holds: When a light wave traveling in one medium is reflected from the surface of a second medium which has a greater refractive index, the reflected ray suffers an additional phase change of 180° (or a path difference of $\lambda/2$).

We may define *diffraction* as the bending of light waves around an obstacle or as the spreading of waves into the geometrical shadow of an obstacle. The distribution of light intensity resulting in dark and bright fringes is called a *diffraction pattern*. Diffraction may be Fresnel diffraction, in which the wave fronts are spherical, or Fraunhofer diffraction, in which the wave fronts are plane. The condition for an intensity minimum in a diffraction pattern by a single slit is $w \sin \alpha = n\lambda$. In a diffraction pattern by a circular aperture, the central bright disk is called *Airy's disk*.

An array of a large number of parallel slits is said to form a *diffraction grating*. The conditions for maxima are $d \sin \theta = m\lambda$ where $m = 0, 1, 2, \ldots$.

Polarization is possible because light waves (and electromagnetic waves in general) are transverse. If all directions of vibrations are equally probable, the beam is *unpolarized*. For a *plane-polarized* beam, the electric vibrations are either vertical or horizontal. An unpolarized beam may be represented by two plane-polarized light beams at right angles to each other. The *percentage of polarization* $P = [(I_{max} - I_{min})/(I_{max} + I_{min})] \times 100\%$. According to *Malus' law*, the intensity of the beam transmitted is $I = I_0 \cos^2 \theta$.

A beam can be polarized by reflection from a dielectric medium. When the angle of incidence is equal to the *polarizing angle* θ_p, the reflected beam is completely polarized, and $\tan \theta_p = n$ is called *Brewster's law*. A doubly refracting crystal such as calcite, which though homogeneous is *anisotropic*—speed of light is different in different directions—can produce polarization. When a light beam falls on such crystals, it splits into beams—ordinary and extraordinary. The direction in which the ordinary and extraordinary beams travel with the same speed is called the *optic axis*.

Polarization can be produced by selective absorption using Polaroid sheets. Polaroid sheets can be used as polarizers and analyzers. Polarization can also be produced by scattering.

QUESTIONS

1. Is it possible to distinguish between two points in an interference pattern that correspond to $E_m = 2E_0$ and $E_m = -2E_0$?
2. Two light beams of the same wavelengths but one with double the amplitude of the other and 180° out of phase are traveling in the same direction. Draw the plots of the individual and the resultant waves.
3. What effect has the presence of vacuum, air, and water on an interference pattern of (a) sound waves; and (b) light waves?
4. Why is it hard to obtain an interference pattern using thick films?
5. How does the dispersion of sun light by a prism differ from diffraction by a grating?
6. What effect does the width of the slit have on the diffraction pattern of (a) a single slit; and (b) a double slit?
7. Which is easier to obtain, the diffraction of sound waves or light waves? Why?
8. In a diffraction grating pattern, is it possible that the first order of one color may coincide with the second order of another color?

9. What effect does the number of slits per centimeter have on the diffraction pattern of a grating?
10. How will you determine whether a given sheet of material is (a) an ordinary dielectric; (b) a Polaroid sheet; or (c) a doubly refracting material?
11. You are given Polaroid sheets. How will you construct sun glasses to remove the glare due to sunlight reflected from a road?
12. Given a Polaroid sheet, a protractor, and a ruler, how will you determine the refractive index of a sheet of glass?

PROBLEMS

1. Calculate the angular and the linear distance from the central maximum to the second bright fringe in a double-slit experiment for which $d = 0.002$ m, $D = 1.5$ m, and $\lambda = 6000$ Å (red light).
2. Consider a double-slit arrangement in which the slits are 2 mm apart, the screen is 2 m away, and the incident light has a wavelength of 5500 Å. Calculate the angular and linear separation of the fourth-order maximum from the central maximum.
3. In a double-slit experiment the two slits are 0.25 mm apart and the screen is 2 m away. The linear distance between the two second-order maxima is 20 mm. Calculate the wavelength of the incident light.
4. In a double-slit experiment, the slits are 0.2 mm apart and placed 1 m from a screen. The light incident on the slits has a wavelength of 5500 Å. Calculate the angular and linear separation between the second and third bright fringes.
5. In Young's double-slit experiment, mercury light of wavelength 5461 Å is used and the screen on which the fringes are formed is placed 1 m away from the double slit. The linear separation between two fifth-order fringes (one on each side of the central maximum) is 1.2 cm. Calculate the separation d between the two slits.
6. In a double-slit experiment yellow sodium light of wavelength 5893 Å is used. If the screen is at a distance of 1 m and the separation for the first-order maxima is 0.3 cm, calculate the distance d between the two slits.
7. Consider a double-slit experiment in which the slits are 0.025 cm apart and the screen is 2 m away. The incident light consists of blue light of wavelength 4360 Å and green light of wavelength 5460 Å, both from mercury. Calculate the fringe separation of these two (a) in the first order; and (b) in the second order.
8. Typically, one can see about 10 fringes on each side of the central maximum in a double-slit experiment using a 5000-Å wavelength. Calculate the order of magnitude of the spacing between the two slits.
9. Monochromatic light of wavelength 5461 Å is used in a Michelson interferometer. One of the mirrors is moved through a distance of 0.2 mm. Calculate the shift in the number of fringes.
10. When one of the mirrors in a Michelson interferometer is moved through a distance of 0.18 mm, there is a fringe shift of 725 fringes. Calculate the wavelength of the light used in the interferometer.
11. A Michelson interferometer uses krypton 86 light of wavelength 6060 Å.

How much must the movable mirror be moved so that there will be a shift of 2000 fringes?

12. A piece of plastic 0.01 mm thick, when placed in one arm of a Michelson interferometer, produces a fringe shift of 210 fringes. If the wavelength of the monochromatic light used is 6000 Å, what is the refractive index of the plastic?

13. One arm of a Michelson interferometer has a cell r cm long, and the light used has a wavelength of 4862 Å. Calculate the fringe shift if this cell is evacuated. Calculate the fringe shift if this evacuated cell is filled with a liquid of refractive index 1.18.

14. Calculate the fringe shift observed when a plastic 1 mm thick of refractive index 1.38 is used in one arm of the Michelson interferometer. The wavelength of the light used is 5890 Å.

15. A film of soap is 0.00006 cm thick and has a refractive index of 1.36. For what wavelengths of normal incident light is there no reflection?

16. What thickness of coating of a material with a refractive index 1.35 is needed to make a glass prism surface nonreflecting for light of wavelength 6000 Å?

17. Calculate the thinnest film of a material with a refractive index of 1.42 needed to cause destructive interference by reflection of light of wavelength 4400 Å. If white light is incident on such a film, what colors will be absorbed and what colors are left?

18. A glass lens made out of a refracting material of refractive index 1.66 has a nonreflecting coating of material with refractive index 1.30. What is the thickness of the coating to be nonreflecting for light of wavelength 4000 Å?

19. A glass surface of refractive index 1.50 has a coating of magnesium fluoride of refractive index 1.25. What is the thickness of the coating so that the reflected green light of wavelength 5000 Å produces destructive interference?

20. What is the thickness of the oil film ($n = 1.40$) floating on water so that yellow light of wavelength 5900 Å is completely eliminated from the reflected light by destructive interference?

21. A film of oil ($n = 1.40$) is formed on a glass surface ($n = 1.50$). Calculate the thickness of the film to eliminate green light of wavelength 5000 Å from the reflected light.

22. Light is incident from air on a thin film of refractive index 1.30 on glass of refractive index 1.50. Suppose that the thickness of the film is 10^{-6} cm. What wavelength of light will be eliminated by the destructive interference of the reflected light?

23. A nonreflecting material (magnesium fluoride, MgF_2) used as a coating on glass has a refractive index of 1.38. If the incident light has a wavelength of 5500 Å, calculate the thickness of the coating.

24. When light of wavelength 5893 Å falls on a single slit of width 3×10^{-4} cm, calculate the position of the first dark and bright band.

25. When light of wavelength 5500 Å falls perpendicularly on a single slit, the two dark fringes on either side are separated by 45°. Calculate the width of the slit.

26. For an incident light of wavelength 6000 Å, what should be the width of a single slit so that the first diffraction minimum will occur at an angle of (a) 10°; (b) 20°; and (c) 30°?

27. Monochromatic light of wavelength 5800 Å is incident normally on a grating with 4000 lines per centimeter. Find the angular deviation for the first two orders.

28. Consider a diffraction grating with 5000 lines per centimeter. If normally incident monochromatic light is diffracted through 22° in the first order, what is the wavelength?

29. Monochromatic light of wavelength 5500 Å is incident on a grating with 6000 lines per centimeter. How many orders can be observed on each side of the center, and what are their angular positions?

30. Using a grating with 5000 lines per centimeter, calculate the longest wavelength that can be used to observe a third-order spectrum.

31. Using monochromatic light of wavelength 5800 Å, a grating produces a spectrum in the second order at 18°. Calculate the number of lines per centimeter on the grating.

32. A grating 2 cm long has 12,000 lines, and a screen on which the spectrum is received is placed 1 m away. Find the angular width and the linear width for the first-order visible spectrum. (The range of the visible spectrum is 4000 to 7000 Å.)

33. Suppose that light consisting of two monochromatic wavelengths λ_1 and λ_2 is incident on a diffraction grating of element d. The spectrum consists of doublets in each order. Show that the angular separation $\Delta\theta$ in the nth order is given by $\Delta\theta = n(\Delta\lambda/d)$, where $\Delta\lambda = \lambda_1 - \lambda_2$. (Assume θ to be small so that $\sin\theta \simeq \theta$.)

34. Calculate the separation between the sodium doublet lines ($\lambda_1 = 5890$ Å and $\lambda_2 = 5896$ Å) in the first, second, and third orders. The grating is 2 cm long and has 8000 lines. (Use the result derived in Problem 33.)

35. In Problem 34 what should be the value of d so that the angular separation in the first order between the two sodium lines is at least $\frac{1}{2}$°? What linear separation will this produce on a screen 50 cm away?

36. What fraction of the incident light is transmitted through an analyzer if it is set to transmit vibrations at 40° to the incident plane-polarized beam?

37. A polarizer and an analyzer are crossed. Suppose that the analyzer is rotated through an angle of 30°. Calculate the percentage of the plane-polarized light that is transmitted from the polarizer. What percentage is transmitted through the analyzer?

38. A polarizer and an analyzer are crossed. A third Polaroid is placed between the two at different angles. Calculate the effect on the transmitted intensity by rotation of the third Polaroid.

39. Find the polarizing angle for the following material: diamond ($n = 2.419$), flint glass ($n = 1.66$), and ethyl alcohol ($n = 1.361$).

40. The sunlight reflected from the surface of a lake is completely polarized. What is the angle the sun makes with the horizontal?

41. A ray of light with an angle of incidence of 58° when reflected from a flint glass surface is plane-polarized. What is the refractive index of the glass?

42. A glass plate has a refractive index of 1.52. Calculate the Brewster angle when the plate is (a) in air; (b) in water ($n = 1.33$); and (c) in carbon tetrachloride ($n = 1.461$).

43. The specific rotation for a sugar solution in a 10-cm tube of a polarimeter is 50°/(g/cm^3). If the total rotation is 112°, what is the concentration of sugar?

30
Optical Instruments

The theories of geometrical and physical optics have been utilized in the construction of many useful and common optical instruments, such as the camera, microscope, telescope, and spectroscope, just to name a few. Before discussing these optical instruments, we shall briefly discuss some of the defects in lenses and methods of removing them. At the end we discuss the limitations to the improvement of the image imposed by the diffraction effects.

30.1. Lens Aberrations

In the simple theory of image formation by mirrors and lenses, it is assumed that an object is located very close to the principal axis, and each point object produces a corresponding point image. That is, the lens produces a perfect or an ideal image of an object. But this is really not true. No matter how smooth and exact the surface upon which the image is formed may be, the actual image formed departs from the ideal image. There are certain inherent characteristics of the lens which lead to the formation of defective images when the laws of reflection and refraction are applied.

The departures of an actual image from that of the ideal or perfect image (as predicted by the simple theory) are called *aberrations*. The aberrations may be classified into two groups: chromatic aberration and monochromatic aberrations. *Chromatic aberration* results from the inability of a lens to bring to a focus in one plane all rays of different colors. Those aberrations that still exist, even if monochromatic light were used in image formation, are called *monochromatic aberrations*. Monochromatic aberrations include spherical aberration, astigmatism, coma, curvature of field, and distortion. Of all these aberrations, only three will be discussed briefly in the following: (1) spherical aberration, (2) astigmatism and (3) chromatic aberration.

Spherical Aberration

Let us refer to Figure 30.1(a), in which rays parallel to the optical axis are incident on a converging lens. Such a converging lens has a shorter focal length for rays that are incident on the outer edges and a longer focal length for the rays near the center of the lens or close to the principal axis. Thus, rays 1 and 2 coming from a point object at infinity converge to a point F_1 closer to the lens, while rays 3 and 4 converge at a farther point F_2. The same thing will happen if the point object were at a finite distance from the lens. That is, the focal length is different for rays at different distances from the principal axis. *This inability of a lens to bring to a focus a point object to a single point image is called spherical aberration.* Instead, the image is a small circle, CC' in Figure 30.1(a), called the *circle of least confusion.* A similar situation occurs in the case of a concave mirror, as is shown in Figure 30.1(b).

The way to reduce spherical aberration is to expose only a small portion of the lens at the center to the incident light. By using a lens of long focal length and small aperture, one can minimize the effect of this aberration.

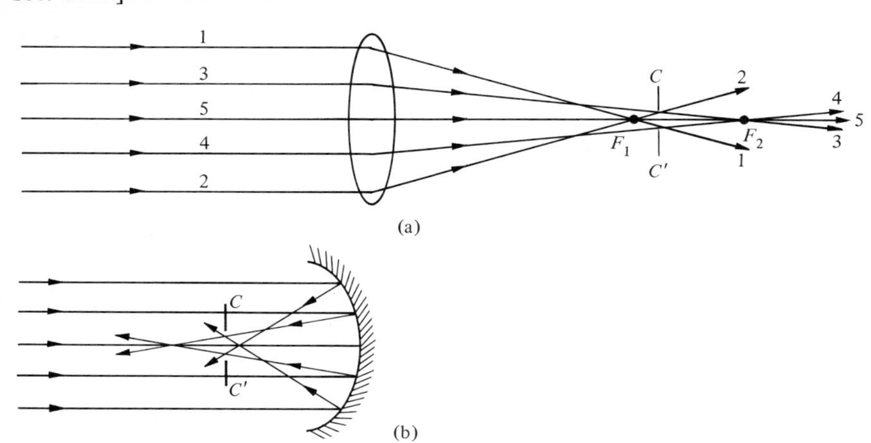

FIGURE 30.1. *Spherical aberration, showing the circle of least confusion CC′: (a) double convex lens and (b) a concave mirror.*

Astigmatism

Astigmatism or off-axis astigmatism is the result of imaging a point object that is located off-axis into a line image. Figure 30.2(a) shows the side view of the rays in the vertical plane starting from a point object O and converging to I_1 in the vertical plane. Similarly, Figure 30.2(b) shows the top view of the rays

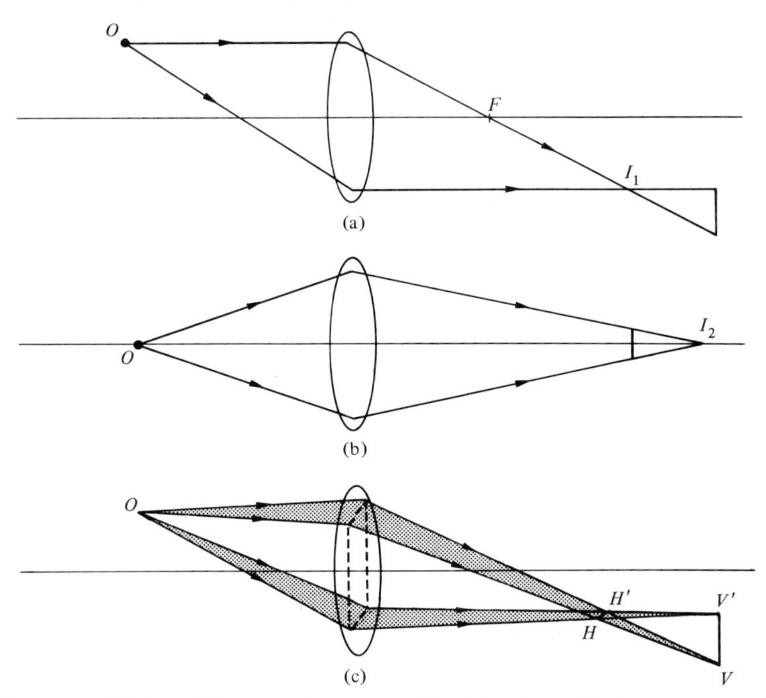

FIGURE 30.2. *Off-axis astigmatism: (a) side view, (b) top view, and (c) combined view.*

in the horizontal plane from point O and converging to I_2 in the horizontal plane. Thus, the image of O is the line VV' in the vertical plane and the line HH' in the horizontal plane, as shown in Figure 30.2(c). The lines HH' and VV' are called focal lines. The true image of O is a circle of least confusion between these two focal lines HH' and VV'.

In an ordinary single-lens system, the effect of astigmatism is minimized by trying to keep the object as close to the principal axis as possible. In more complex optical systems, a combination of different cylindrical lenses and different glass materials are used to reduce off-axis astigmatism.

Chromatic Aberration

White light consists of different colors or waves of different wavelength. The refractive index of the material of the lens is a function of the wavelength. Furthermore, the deviation or the amount of refraction of a ray depends upon the refractive index. Thus, when rays of white light parallel to the principal axis are incident on a simple lens, different colors are refracted by different amounts. Hence, different colors are brought to a focus at different distances from the lens, as shown in Figure 30.3(a). Since the refractive index of glass for violet light is larger than for red light, the violet rays are bent more than the red rays in a convex lens, as shown in Figure 30.3(a). That is, the focal length of a converging lens is smaller for violet light than for red. The chromatic aberration for the diverging lens is shown in Figure 30.3(b). As stated earlier, *this inability of a lens to bring to a focus in one plane all rays of different colors is chromatic aberration*. The net result is a diffused and colored image.

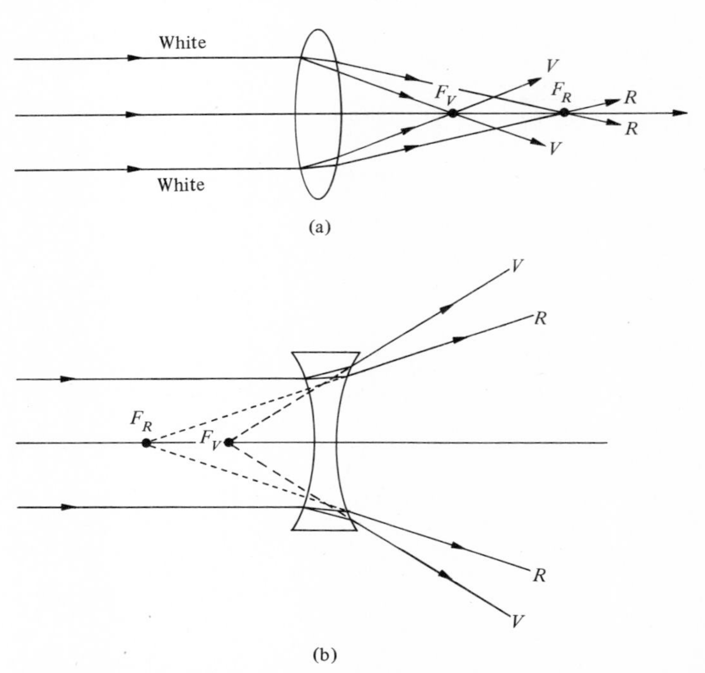

FIGURE 30.3. *Chromatic aberration of (a) a converging (double convex) lens and (b) a diverging (double concave) lens.*

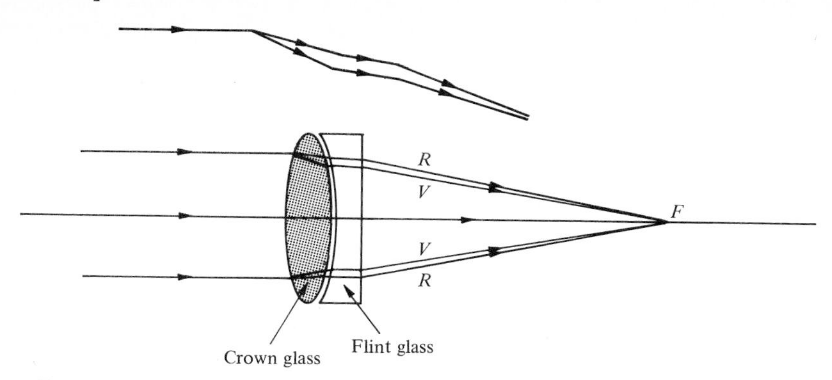

FIGURE 30.4. *Achromatic doublet which corrects for chromatic aberration.*

Since the converging and diverging lenses have opposite effects, as shown in Figures 30.3(a) and (b), it is possible to remove chromatic aberration by combining two such lenses. This is shown in Figure 30.4, where a converging lens made out of crown glass is combined with a diverging lens made out of flint glass. These two lenses produce equal and opposite dispersion, and are said to form an *achromatic doublet*. An achromatic doublet corrects only for two different wavelengths. In many situations, such as cameras, the lenses are corrected for blue and yellow, which automatically corrects for green as well, because the wavelength of green is in between that of yellow and blue.

In more complex optical systems, where much better achromatic corrections are needed, a combination of three or more lenses are used to produce zero net dispersion. Such systems of lenses producing zero net dispersion are called *achromatic lens systems*.

The defect *coma* is said to exist when the image of a point object off-axis results in a comet-like shape.

The defect *curvature of field* is said to exist when the image of a flat plane is not a flat plane but a curved surface.

The defect *distortion* is said to exist when the magnification for the off-axis points is different from those near or on the axis. For example, the image of a rectangular figure may take the form of a barrel.

30.2. The Photographic Camera

In theory a photographic camera is a simple optical instrument. In actual practice cameras may be very simple or technically very complex. Figure 30.5 shows an outline sketch of a simple camera. It essentially consists of a light-tight box, a converging lens (or a lens system) to form a real image of an object, and a light-sensitive plate or film at the back on which the image is received. The image distance q depends upon the object distance p. Thus, the distance q between the lens and the film is adjusted to obtain a sharp and well-focused image of the object. This adjustment is made by means of bellows, or one tube sliding over the other, as shown in Figures 30.5 and 30.6. Even in a simple-lens camera, the lens is corrected for chromatic aberration. An arrangement is made to take care of the off-axis astigmatism.

The film or the emulsion on the film is exposed to the light coming from the

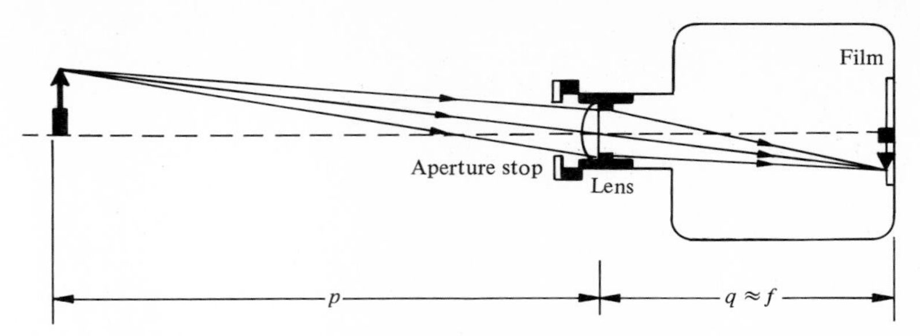

FIGURE 30.5. *Outline and ray diagram of a simple camera.*

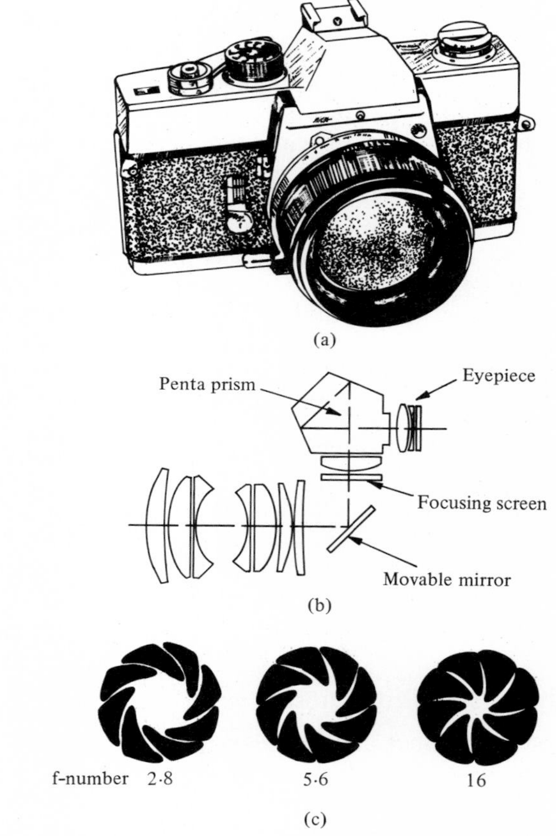

FIGURE 30.6. (*a*) *Sketch of a complex camera.* (*b*) *Lens system* (*c*) *Details of an adjustable diaphragm.*

object for a short interval of time. This light affects the silver halide grains in the emulsion. When this exposed film goes through the process of photographic developing, the affected silver halide grains are changed into black grains of metallic silver. The unaffected silver grains can be removed by treating the film chemically in a fixing bath. Thus, the *negative* is obtained.

More complex and sophisticated cameras, such as shown in Figure 30.6(a), are equipped with many more components. The lens system is usually complex, consisting of a combination of three or more converging and diverging lenses, such as one shown in Figure 30.6(b). Most of the cameras are also equipped with a shutter and an adjustable lens diaphragm. The purpose of the shutter is to allow light to enter the lens of the camera only for a precisely timed period, anywhere from (1/10) to (1/500) s. This has two functions. First, it allows the right amount of light to fall on the film. Second, it allows one to make reasonably sharp images of moving objects by using very short exposures. The purpose of the adjustable lens diaphragm shown in Figure 30.6(c) is not only to control the amount of light, but also to expose only the central portion of the lens. By covering the outer portions of the lens some of the aberrations are reduced.

Let us now talk in quantitative terms about the brightness of the image formed on the film. The amount of light received on the film is directly proportional to the area of the lens. If D is the diameter of the lens, the amount of light energy received is proportional to D^2. The image formed is usually very close to the focal point of the lens; that is, $q \approx f$. Thus, the magnification $m = q/p = f/p$. The area of the image formed on the film will be proportional to $m^2 = f^2/p^2 \propto f^2$. The illuminance I the amount of light per unit area at the image is

$$\text{illuminance} \propto \frac{\text{light energy received}}{\text{area of the image}}$$

or

$$I \propto \frac{D^2}{f^2} \propto \frac{1}{(f/D)^2} \tag{30.1}$$

It is a common practice to define the *f-number* equal to the ratio f/D. That is,

$$f\text{-number} = \frac{f}{D} \tag{30.2}$$

Hence,

$$I \propto \frac{1}{(f\text{-number})^2} \tag{30.3}$$

The f-number indicates the light-gathering power, hence is referred to as the *speed* of the lens. Thus, from Equation 30.3, the smaller the f-number, the higher the speed. Suppose that the lens has an f-number of 16. This means that the focal length f is 16 times the diameter D. Typical cameras are marked with f-numbers of $f/16$, $f/11$, $f/8$, $f/5.6$, $f/4$ and $f/2.8$. The f-number is controlled by adjusting the diaphragm (D corresponds to the portion of the lens exposed). In many expensive cameras the f-number may be as low as 1.6, which means that f is only 1.6 times the diameter of the lens. Such lenses require many corrections for different aberrations, hence increase the cost factor.

In many situations a regular lens is provided with an additional wide-angle lens (it has a shorter effective focal length; hence, I becomes larger) which enables it to focus a larger field of the object. Conversely, a telephoto lens increases the effective focal length, producing larger images of the objects at large distances (although I will be reduced).

Example 30.1 A lens of speed $f/2.8$ has a focal length of 10 cm. Calculate the diameter of the lens. For this speed the correct exposure time is (1/100) s. What will be the correct exposure time for a lens of speed $f/5.6$?

From Equation 30.2 f-number $= f/D$; that is, $2.8 = 10\ \text{cm}/D$:

$$D = \frac{10\ \text{cm}}{2.8} = 3.57\ \text{cm}$$

From Equation 30.3, $I \propto 1/(f\text{-number})^2$ and the total light energy received is It. That is, $I_1t_1 = I_2t_2$, or

$$\frac{t_1}{(f_1\text{-number})^2} = \frac{t_2}{(f_2\text{-number})^2}$$

$$t_2 = t_1\left(\frac{f_2\text{-number}}{f_1\text{-number}}\right)^2 = \left(\frac{1}{100}\ \text{s}\right)\left(\frac{5.6}{2.8}\right)^2 = \left(\frac{1}{100}\ \text{s}\right)(4) = \frac{1}{25}\ \text{s}$$

Exercise 30.1 A lens of speed $f/4$ has a focal length of 10 cm. What is the diameter of this lens? For this speed the correct exposure time is (1/100) s. What will be the correct exposure time for a lens of speed $f/8$? [*Ans.:* $D = 2.5$ cm, $t_2 = \frac{1}{25}$ s.]

30.3. The Eye and Defects of Vision

Eyes can be divided into two groups. One group, such as vertebrates, including human beings, possess eyes that use lens systems for forming images, while the other group, such as houseflies, possess eyes that make use of fiber optical bundles to convey the sensation of vision to the brain. As we shall see, the human eye is very similar in principle and working to a photographic camera described in the previous section—the eye uses a positive lens system and produces a real image on a light-sensitive screen.

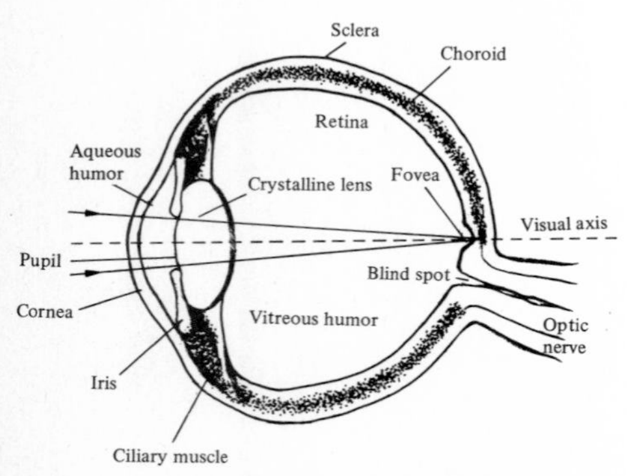

Figure 30.7. *The human eye.*

The Human Eye

As shown in Figure 30.7, the eye is almost spherical in shape, consisting of a jellylike mass enclosed in a hard shell, the *sclera*. The small projected portion in the front is called the *cornea*, which is transparent. The cornea is the first refracting element and produces most of the bending of light toward the optic or visual axis. The refractive index of the material is $n_c = 1.376$. The next element, which produces a bending, is the *crystalline lens*. It is a small convex lens with a structure of ~22,000 very fine layers. The lens material is inhomogeneous, with a refractive index of $n_{l_1} = 1.406$ at the center to $n_{l_2} = 1.386$ at the outer portions. This lens can change its shape—becoming thinner or thicker, changing its focal length. Thus, the crystalline lens provides the fine-focusing mechanism supplementing the focusing provided by the cornea.

The small chamber between the cornea and the lens is filled with a clear, waterlike fluid called *aqueous humor*, with $n_{ah} \approx 1.336$. Since $n_c \approx n_{ah}$, no bending of the light takes place in the aqueous humor chamber. Located in front of the lens in this chamber is a diaphragm, called the *iris*, which controls the amount of light that enters the lens through the portion called the *pupil*. The iris is responsible for giving the eye its characteristic color blue, brown, green, hazel, or gray. It is also provided with muscles, the *ciliary muscles*, which can dilate the pupil.

Behind the lens is another chamber, filled with a transparent jellylike substance called the *vitreous humor* ($n_{vh} = 1.337$). The inner surface of this chamber is covered with a layer of light-receptor cells called the *retina*. [Between the inner surface (the retina) and the outer surface (the sclera) there is another layer, which is dark and is called the *choroid*.] The retina takes the place of the light-sensitive film in a camera. A beam of light focused by the cornea and the lens is absorbed by the pinkish layer of the retina. The retina in the eye contains mostly two kinds of photoreceptor cells—*rods* (~125 million) and *cones* (~7 million). Figure 30.8 is an electron micrograph of the retina, showing the rods and cones. The rods are very sensitive and will respond to a very dim light but cannot distinguish between colors, and the images are not very sharp. Thus, it is similar to a high-speed, coarse-grain, black-and-white film. On the other hand, the cones are not very sensitive but can distinguish between colors. Thus, they are like a low-speed, fine-grain, color film.

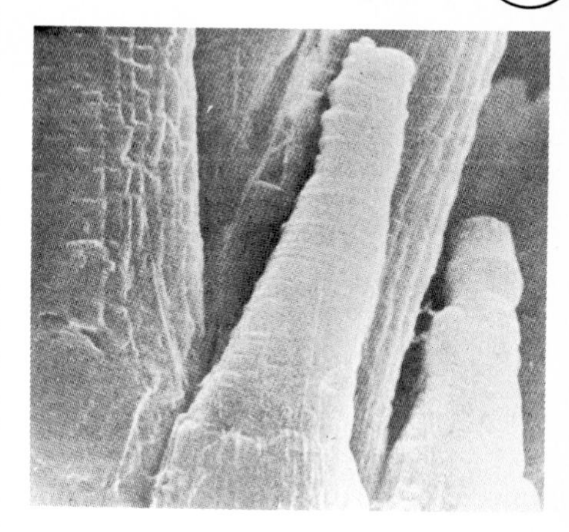

FIGURE 30.8. *Electron micrograph of the retinal area, showing two cones (in the foreground) and several rods (in the background).* [From E. R. Lewis, Y. Y. Zeevi, and F. S. Werblin, *Brain Res.*, **15**, 559 (1969).]

A large number of nerve fibers branching out from the *optic nerve* terminate in the photosensitive rods and cones discussed above. These rods and cones on the surface of the retina receive the optical image and transmit it through the optic nerve to the brain. The brain then analyzes the image and reports the identification of the object in front of the eye. There is a small depression where the visual axis crosses the retina. This depression is called the yellow spot or *macula*. At its center a region 0.25 mm in diameter is called the *fovea centralis*. In this region the cones are much thinner and more densely populated, and there are no rods. At the fovea, vision is much more acute than at any other portion of the retina. The eye muscles rotate the eyeball until the image is formed at the fovea. Thus, the fovea gives the sharpest image, while the rest of the retina reveals a general picture of the object.

The photosensitive rods and cones are absent at a point where the optic nerve enters the eye. No image is detected at this point; hence, this region is called a *blind spot*.

Accommodation

The lens of the eye is fixed at about 1.7 cm from the retina. When the object is at infinity, the image is formed at the retina, as shown in Figure 30.9(a). The muscles in this situation are completely relaxed. As the object moves closer to the eye, the distance p changes while q remains fixed (~1.7 cm); hence, f in the relation $1/p + 1/q = 1/f$ must also decrease for a fixed q. This is achieved by the ciliary muscles, which contract, changing the tension on the lens so that it bulges slightly under its own elastic forces, as shown in Figure 30.9(b). This leads to a decrease in the focal length. As the object is brought still closer, the ciliary muscles contract more and the lens surface achieves a still smaller focal length and larger power ($P = 1/f$). This fine adjustment by the lens of the eye (as the object moves from infinity closer to the eye) is called *accommodation*.

The closest point on which the eye can focus is called the *near point*. In a normal eye the distance of the near point depends upon age. It may be ~7 cm for a teenager, ~25 cm for an adult of about 35 years, ~100 cm for age 55,

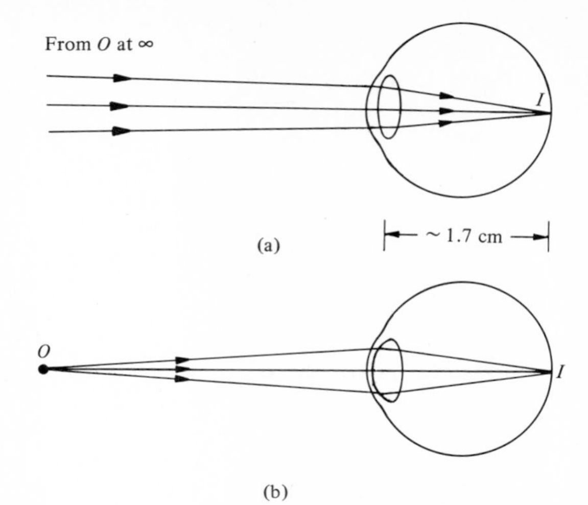

FIGURE 30.9. *Accommodation is achieved by causing a change in the lens configuration: (a) relaxed muscle and (b) contracted muscle to achieve accommodation.*

and more for older people. In our discussion we shall take 25 cm to be the near point. Of course, the *far point* is infinity.

As mentioned previously, it is the cornea and the fluid, aqueous humor, that play the major part in forming the image. In a typical eye this system has a power of +40 diopters ($f_c = 2.5$ cm). The lens has a variable power between +20 diopters ($f_{l_1} = 5$ cm) for a relaxed muscle condition and +24 diopters ($f_{l_2} = 4.17$ cm). Thus, the overall power of the eye is variable and is between +60 diopters ($f = 1.67$ cm) and +64 diopters ($f = 1.56$ cm). This variation allows the eye to see anywhere between the near point of 25 cm and infinity.

Defects of Vision

There are three major defects of vision: (1) myopia, (2) hyperopia, and (3) astigmatism.

MYOPIA—NEARSIGHTEDNESS. *Myopia* or *nearsightedness* means that the eye can see sharply nearby objects only (i.e., the image for only nearby objects is sharply in focus). In a myopic eye, the parallel rays come to focus in front of the retina, as shown in Figure 30.10(a). The far point, whose image is formed at the retina, is much closer than infinity, as shown in Figure 30.10(b). Thus, the objects between infinity and the far point appear blurred. An eye with this defect can see only objects between the near point (which is usually < 25 cm) and the far point, which is not infinity in this case.

The eye lens system has a short focal length or too large a power ($D = 1/f$)—the rays converge too fast. To increase the focal length or decrease the power, we must use a negative lens or diverging lens, such as a concave lens. This is demonstrated in Figure 30.10(c), where the rays coming from an object at infinity form an image on the retina with the help of the lens system of the eye. The image formed at O' is not real. The power of the negative lens to be used depends upon the location of the far point. Suppose that it is at a

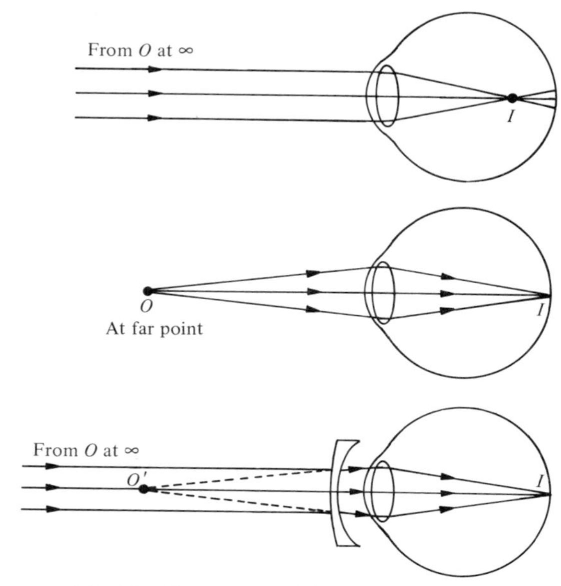

FIGURE 30.10. *Correction of the myopic (nearsighted) eye.*

distance d. Thus, if the object is at ∞, the image should be at $-d$, where $p = \infty$, $q = -d$, and

$$\boxed{\frac{1}{f} = \frac{1}{p} + \frac{1}{q} = \frac{1}{\infty} + \frac{1}{-d} = -\frac{1}{d}} \tag{30.4}$$

which gives the focal length of the diverging lens to be used. Thus, if d is in meters, the power of the lens to be used is $1/-d$: say d is 2 m, $f = -2$ m, and power $P = -\frac{1}{2}$ diopter.

HYPEROPIA—FARSIGHTEDNESS. A *hyperopic* (or *hypermetropic*) or *farsighted* eye can see objects that are far away. It cannot see objects that are close to the normal near point. For a farsighted eye, the near point may be farther away, say 1 m or more. In such cases the objects between 25 cm and 1 m cannot be seen clearly or focused on the retina.

As shown in Figure 30.11(a), the parallel rays coming from an object form an image far beyond the retina. This means that the lens has too long a focal length; hence, a small power—the rays do not converge fast enough. To make them converge on the retina, we must increase the power of the lens, that is, decrease the focal length. This is achieved by using a converging or positive lens such as a convex lens, as explained below [Figures 30.11(d) and (e)].

As shown in Figure 30.10(b), the object at the 25 cm point forms an image beyond the retina. Only when the object moves to point O in Figure 30.10(c) is the image formed at the retina. Suppose that a hyperopic eye has to see objects clearly which are at 25 cm, a normal reading distance. The convex lens placed in front of the eye must form its image at Q, as shown in Figure 30.11(d). Thus, $p = +25$ cm $= 0.25$ m, $q = -d$, the near point of the defective eye, and

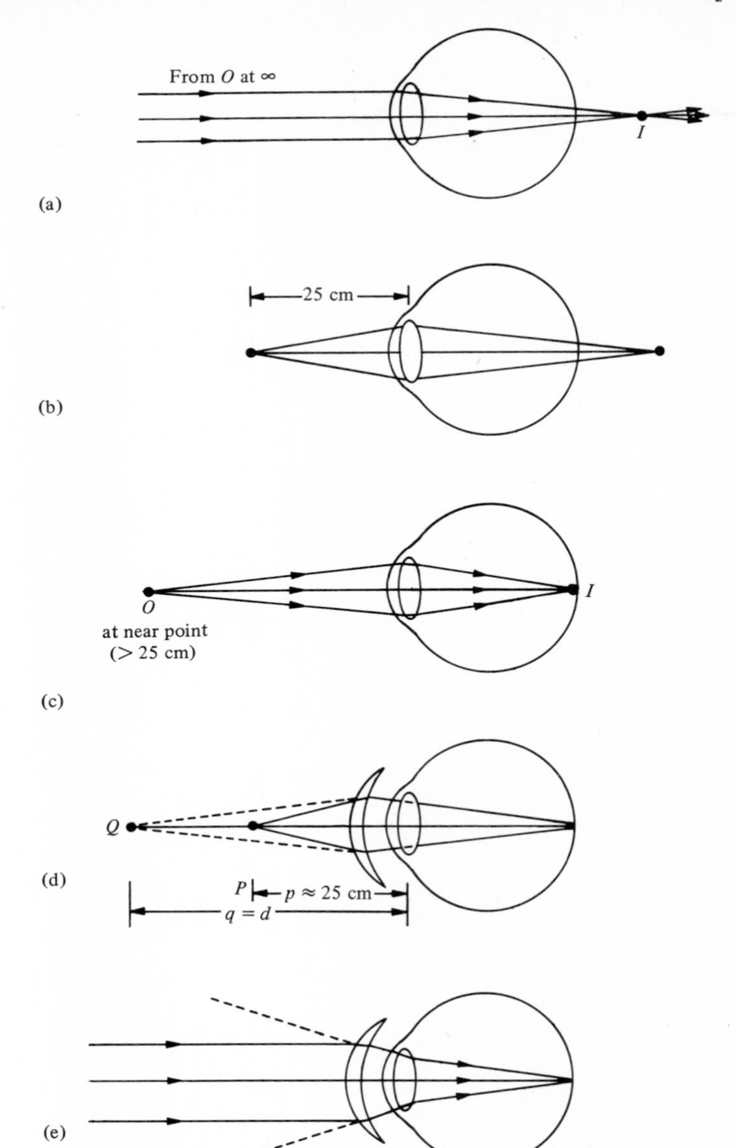

FIGURE 30.11. *Correction of the hypermetropic (farsighted) eye.*

$$\boxed{\frac{1}{f} = \frac{1}{p} + \frac{1}{q} = \frac{1}{0.25 \text{ m}} + \frac{1}{-d}} \tag{30.5}$$

The power of the lens may be calculated if d is known. Suppose that $d = 1.5$ m,

$$\frac{1}{f} = \frac{1}{0.25\text{ m}} + \frac{1}{-1.5\text{ m}} = \frac{1}{0.3\text{ m}}$$

That is, $f = 0.3$ m and $P = 3.33$ diopters.

ASTIGMATISM. An *astigmatic* eye cannot focus horizontal and vertical lines at the same time. Thus, the image formed is fuzzy in the vertical or horizontal direction, as shown in Figure 30.12. The reason for astigmatism is that the cornea is not a truly spherical surface; that is, the cornea is asymmetric. The curvature is different in different planes, containing the optic axis. This means that the focal length is not the same for different planes. Thus, the light coming from an object cannot form sharp images in different planes. Some of these astigmatism defects can be corrected by using cylindrical lenses.

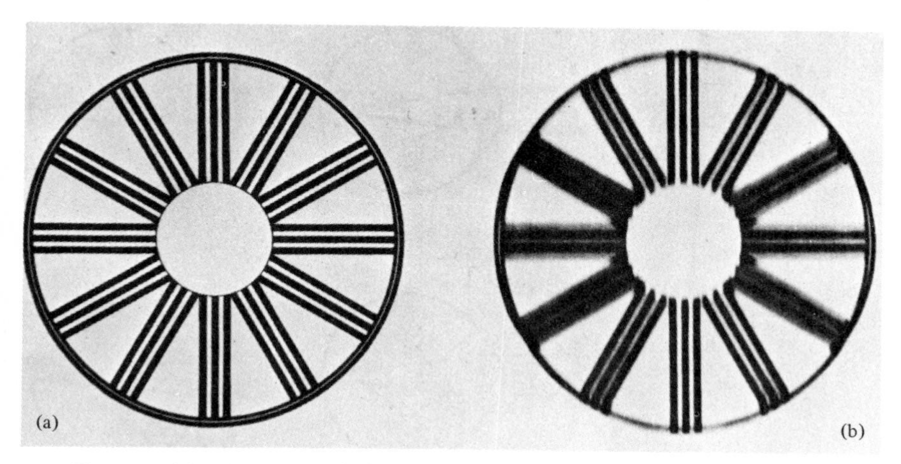

FIGURE 30.12. *Image of a wheel with spokes as viewed (a) by a normal eye and (b) by an astigmatic eye.* (Courtesy of Glenn A. Fry, Ohio State University.)

Diseases of the Eye

Of the several diseases of the eye we shall discuss only one of the most important ones—glaucoma. The aqueous humor produced by the ciliary body carries essential food to many parts of the eye, such as the lens and the cornea. The aqueous humor flows through the pupil into the interior chamber and leaves the eye through a microscopic channel called the *chamber angle* at the cornea. The rate of aqueous formation and its steady flow determines the intraocular pressure, which normally ranges from 12 to 22 mm Hg. The abnormal elevation of the pressure within the eyeball, the *intraocular pressure*, leads to a group of diseases of the eye called *glaucoma*. In these diseases the rate of production of the aqueous humor is normal, but the rate of elimination is reduced, owing to an obstruction. The blocking raises the pressure behind the iris and may lead to the swelling of the crystalline lens, which in turn may block the entrance of the chamber angle. (This leads to strong myopia.) This results in further increase in intraocular pressure and is called *primary glaucoma*.

Primary glaucoma may occur at any age and is associated with many definite symptoms: mild to severe pain in and behind the eye, occasional blurring of vision, and formation of color fringes around light sources. In many cases the treatment can be sought before irreversible visual damage takes place. But there is a chronic form of glaucoma which is painless and is not detected until it is too late. The elevated pressure reduces the blood supply to the nerve fibers at the retina. Eventually, the nerve fiber dies and this portion of the retina is no longer connected to the brain, causing blindness. If this defect is not detected in time and corrected, it spreads to other portions of the eye. Usually, chronic glaucoma does not start in both eyes at the same time.

The treatment of glaucoma (reducing the elevated pressure) may be medical—external in the form of drops or ointments, internal, or surgical. The purpose of surgery is to eliminate the "bottle-neck" effect either by repairing the channels or by creating new outlets for the aqueous humor.

Example 30.2 A man has a left eye that is myopic, with a far point of 2 m, while his right eye is hyperopic, with a near point of 75 cm. What will be the prescription for his eye glasses?

For the left eye: since the eye is myopic (or nearsighted) with a far point of 2 m, it can see clearly from 25 cm up to 2 m but cannot see objects beyond that. Thus, the lens must form images at 2 m for objects at infinity. That is, $p = \infty$, $q = -2$ m, and

$$\frac{1}{f} = \frac{1}{p} + \frac{1}{q} = \frac{1}{\infty} + \frac{1}{-2} = \frac{1}{-2} \qquad \text{or} \qquad f = -2 \text{ m}$$

$$\text{power} = \frac{1}{f} = \frac{1}{-2 \text{ m}} = -0.5 \text{ diopter}$$

The left eye needs a diverging lens (concave lens) of power -0.5 diopter.

For the right eye: since the eye is hyperopic (or farsighted) with a near point at 75 cm, it means that it cannot see clearly up to 75 cm. The comfortable distance for reading is 25 cm. Therefore, the objects at 25 cm should form clear images at 75 cm. Thus, $p = 25 \text{ cm} = 0.25$ m and $q = -75 \text{ cm} = -0.75$ m. Therefore,

$$\frac{1}{f} = \frac{1}{p} + \frac{1}{q} = \frac{1}{0.25} + \frac{1}{-0.75} = \frac{1}{0.375} \qquad \text{or} \qquad f = 0.375 \text{ m}$$

$$\text{power} = \frac{1}{0.375 \text{ m}} = 2.67 \text{ diopters}$$

Thus, the right eye needs a converging lens (convex lens) of power $+2.67$ diopters.

Exercise 30.2 In Example 30.2, if the far point of the left eye is at 1 m and the near point of the right eye is also 1 m, what will be the power of the lenses to be used in glasses to correct these defects? [*Ans.*: left eye: power = -1 diopter; right eye: power = $+3$ diopters.]

30.4. The Microscope

A microscope is used for producing at the retina of an eye an enlarged, distinct image of a small nearby object. This is achieved by means of a simple convex lens or a system of lenses, as we shall discuss below.

Angular Size

Suppose an object of size h is placed at A, the size of the image formed at the retina is OA' while the angle formed by the object at the lens of the eye is θ_1, as shown in Figure 30.13. If the object is now moved in closer, say to B or C, the angle θ_2 or θ_3 subtended at the eye becomes larger, and so does the size of the image OB' or OC' at the retina. From this we conclude that it is the angular size θ formed by the object at the eye that determines the size of the image.

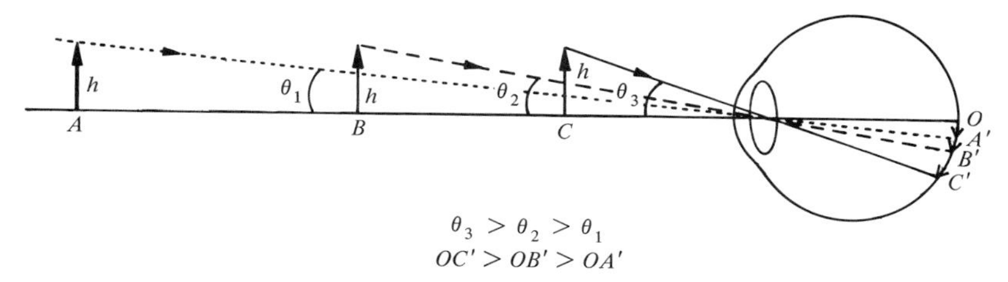

FIGURE 30.13. *The angular size of the object determines the size of the image formed at the retina.*

Of course, there is a limit on how close an object can be brought before the image formed on the retina becomes blurred. We know that if the object is placed at a distance of 25 cm, which is the distance of distinct vision in a normal eye, the image formed will be clear and distinct. As shown in Figure 30.14(a), the angular size in this case is θ and is given by

$$\theta = \frac{h}{25 \text{ cm}} \tag{30.6}$$

If the object is brought any closer, the image will be blurred, and moving it away will decrease the angle, hence the image size. But there is a way to increase the angular size without losing clarity.

The Simple Microscope

Suppose that we place the object at P close to the eye as shown in Figure 30.14(b) and interpose a small double convex lens between the eye and the object. The position P of the object is very close to F (the focal point of the convex lens) and such that the convex lens forms a virtual and erect image at Q, which is the point of distinct vision. Thus, as is clear from this figure, the angular size has been increased without sacrificing clarity. The lens used in this manner is called a simple microscope or a magnifier. (If the object is placed exactly at F, the image formed will be at a large distance, say infinity, which a normal eye can always see clearly.)

From Figure 30.14(b) we see that the linear size of the image is h', while its

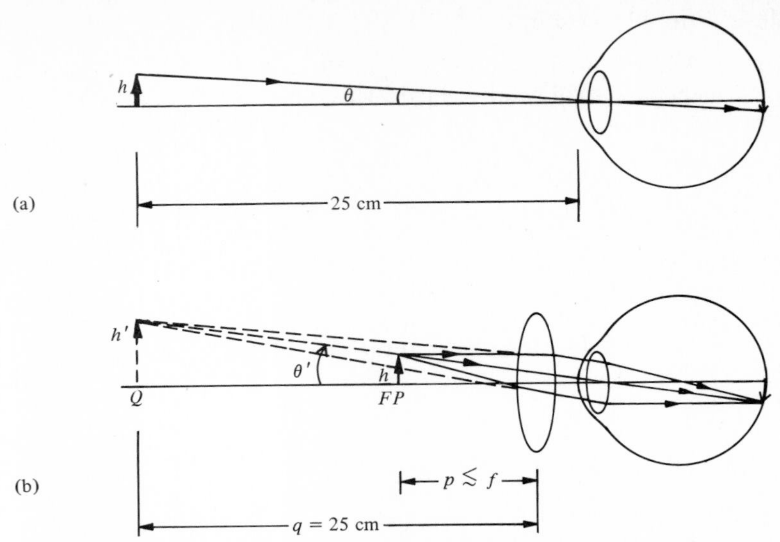

FIGURE 30.14. *(a) Angle subtended by an object when at a distance of distinct vision. (b) Ray diagram, showing the principle of a simple microscope.*

angular size θ' is given by

$$\theta' = \frac{h'}{25 \text{ cm}} \tag{30.7}$$

We define the *angular magnification* or the *magnifying power M* as the ratio of the angle θ' formed by the image to the angle formed by the object θ, both being at the distance of distinct vision, 25 cm. Thus,

$$M = \frac{\theta'}{\theta} = \frac{h'/25 \text{ cm}}{h/25 \text{ cm}} = \frac{h'}{h} \tag{30.8}$$

From the relation $1/p + 1/q = 1/f$, if $q = -25$ cm,

$$\frac{1}{p} - \frac{1}{25} = \frac{1}{f} \qquad \text{or} \qquad \frac{1}{p} = \frac{1}{f} + \frac{1}{25} = \frac{25 + f}{25f}$$

Also, from the relation for linear magnification,

$$m = \frac{h'}{h} = -\frac{q}{p} = -(-25)\left(\frac{25 + f}{25f}\right) = \frac{25 + f}{f} = 1 + \frac{25 \text{ cm}}{f} \tag{30.9}$$

Combining Equation 30.9 with Equation 30.8, we get

$$\boxed{M = 1 + \frac{25 \text{ cm}}{f}} \tag{30.10}$$

If the object is placed exactly at F instead of P, the image formed will be at infinity, and from Figure 30.14(b) we shall obtain $\theta' = h/f$ while $\theta = h/25$ cm.

Thus,

$$M = \frac{25 \text{ cm}}{f} \tag{30.11}$$

Using a magnifying lens we can get a usable angular magnification of up to 4, written 4×. A typical magnifying lens has a focal length of a few centimeters. For higher magnifying power the focal length must be reduced, which makes it hard to correct for aberrations of a simple lens. For higher M we must use a compound microscope, described below.

The Compound Microscope

The components and a ray diagram of a compound microscope are shown in Figure 30.15. It consists of an objective L_1, which has a very short focal length f_o (less than 1 cm), and an ocular or eyepiece L_2, which has a focal length f_e of a few centimeters (2 to 3 cm). An object to be viewed is placed at O, just beyond the focal length of the objective lens L_1. The image I_1 formed by the lens L_1 is real, inverted, and many centimeters (20 to 30 cm) from the lens. The position of the image I_1 is such that it is close to or at the focal length of the ocular lens L_2, which is a simple magnifier. The magnifier L_2 produces a magnified image of I_1 at I_2. Image I_2 is virtual and remains inverted.

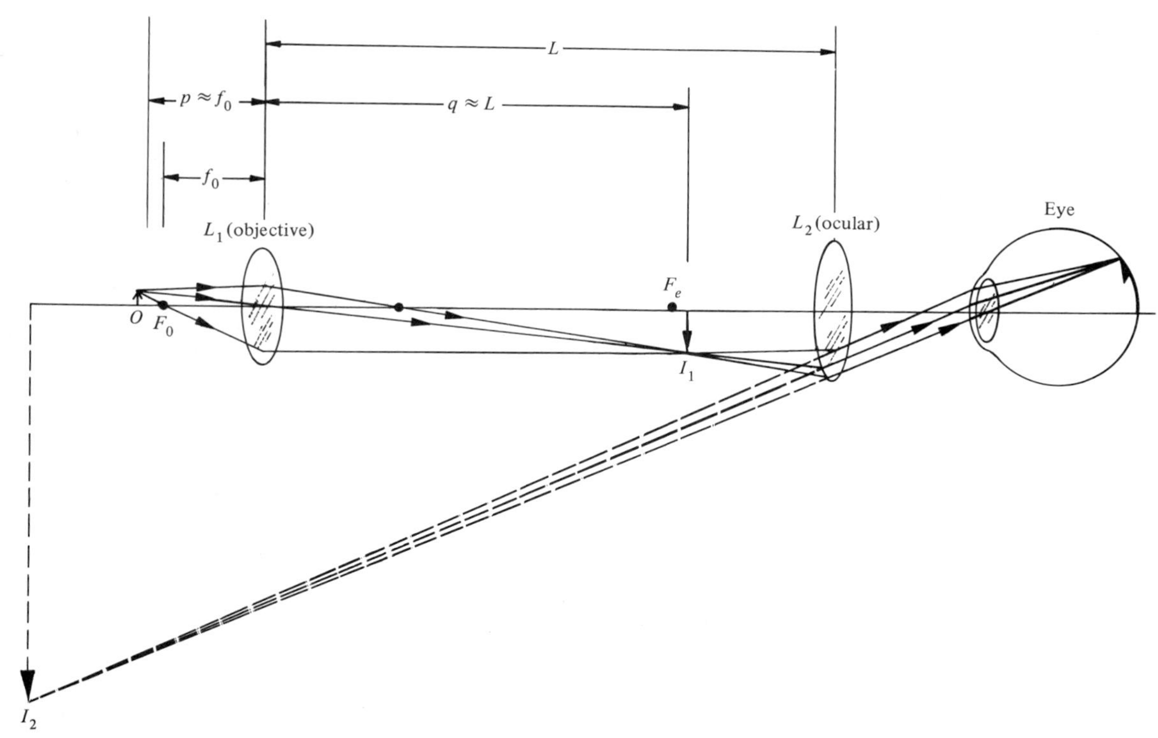

FIGURE 30.15. *Ray diagram of a typical compound microscope.*

The lateral magnification m_1 of the first image is

$$m_1 = -\frac{q}{p}$$

where $p \approx f_o$, while q is very close to the total length L of the microscope. Therefore (ignoring the minus sign),

$$m_1 \simeq \frac{L}{f_o} \tag{30.12}$$

The angular magnification produced by the ocular (a magnifying lens) as given by Equation 30.11 is

$$M_2 = \frac{25\text{ cm}}{f_e} \tag{30.13}$$

Thus, the overall magnification M is

$$\boxed{M = m_1 M_2 = \frac{L}{f_o}\frac{25\text{ cm}}{f_e}} \tag{30.14}$$

The smaller the values of f_o and f_e, the larger the value of M.

Example 30.3 An object 2 mm high is placed in front of a magnifying glass of focal length + 1.5 cm. Calculate (a) the angular size of the object when at a distance of distinct vision; (b) the maximum useful angular magnification; and (c) the angular magnification when the image is formed at infinity.

(a) When the object is at 25 cm and its linear size is 2 mm = 0.2 cm, its angular size (angle subtended at the unaided eye) is

$$\theta = \frac{0.2\text{ cm}}{25\text{ cm}} = 0.008\text{ rad}$$

(b) The magnification will be maximum when the image is at the distance of distinct vision, that is, $q = -25$ cm. Since $f = +1.5$ cm, the object distance p,

$$\frac{1}{p} = \frac{1}{f} - \frac{1}{q} = \frac{1}{1.5\text{ cm}} - \frac{1}{(-25\text{ cm})} = \frac{53}{75\text{ cm}} \quad \text{or} \quad p = \frac{75}{53}\text{ cm} = 1.42\text{ cm}$$

That is, the object must be placed at a distance of 1.42 cm in front of the lens. From the definition of linear magnification,

$$\frac{h'}{h} = -\frac{q}{p} = -\frac{-25\text{ cm}}{1.42\text{ cm}} = 17.6$$

$$h' = (17.6)h = (17.6)(2\text{ mm}) = 35.2\text{ mm}$$

Therefore, the angle θ' formed by the image on the eye will be

$$\theta' = \frac{h'}{25\text{ cm}} = \frac{3.52\text{ cm}}{25\text{ cm}} = 0.14\text{ rad}$$

while the angular magnification will be

$$M = \frac{\theta'}{\theta} = \frac{0.14}{0.008} = 17.5$$

We could have used the formula given by Equation 30.10; that is,

$$M = 1 + \frac{25 \text{ cm}}{f} = 1 + \frac{25 \text{ cm}}{1.5 \text{ cm}} = 17.7$$

which is in agreement with the result above.
(c) If the image is formed at infinity from Equation 30.11,

$$M = \frac{25 \text{ cm}}{f} = \frac{25 \text{ cm}}{1.5 \text{ cm}} = 16.7$$

EXERCISE 30.3 An object 2 mm long is placed in front of a lens of focal length +2 cm. Calculate (a) its maximum magnification; (b) its magnification when the image is formed at infinity; and (c) its magnification when the image is formed at 70 cm. [To solve (c), substitute $q = -70$ cm instead of -25 cm.] [*Ans.*: (a) $M = 13.5$; (b) $M = 12.5$; (c) $M = 1$.]

EXAMPLE 30.4 In a compound microscope the objective has a focal length of +1 cm, while the eyepiece has a focal length of +2 cm. An insect is placed at a distance of 1.1 cm in front of the objective. Calculate (a) the length of the tube of the microscope; and (b) the magnification.

(a) If $p = 1.1$ cm and $f_o = +1$ cm, q is given by

$$\frac{1}{q} = \frac{1}{f_o} - \frac{1}{p} = \frac{1}{1 \text{ cm}} - \frac{1}{1.1 \text{ cm}} = \frac{1}{1 \text{ cm}} - \frac{10}{11 \text{ cm}} = \frac{1}{11 \text{ cm}}, \quad q = 11 \text{ cm}$$

This image is formed at the focal point of the eyepiece. Thus, the total length L of the tube is

$$L = q + f_e = 11 \text{ cm} + 2 \text{ cm} = 13 \text{ cm}$$

(b) From Equation 30.4, the magnification is given by

$$M = m_1 M_2 = \frac{L}{f_o}\frac{25 \text{ cm}}{f_e} = \frac{13 \text{ cm}}{1 \text{ cm}}\frac{25 \text{ cm}}{2 \text{ cm}} = 162.5$$

For better accuracy, we should calculate M directly. Since $p = 1.1$ cm and $q = 11$ cm,

$$|m_1| = \frac{h'}{h} = \frac{q}{p} = \left(\frac{11 \text{ cm}}{1.1 \text{ cm}}\right) = 10$$

while

$$M_2 = \frac{25 \text{ cm}}{f_e} = \frac{25 \text{ cm}}{2 \text{ cm}} = 12.5$$

Therefore, $M = m_1 M_2 = 10 \times 12.5 = 125$, which differs somewhat from the calculated value above.

Exercise 30.4 Solve Example 30.4 for an object placed at a distance of 1.05 cm. What do you conclude by comparing the two results? [*Ans.*: $M = 250$, the nearer the object is to the focal point of the objective, the greater the magnification.]

30.5. The Telescope

The purpose of a telescope is to see a clear and distinct image of an object that is at a very large distance from the observer. Without the telescope, the angular size of the object at the eye is very small, hence, an objective lens of very long focal length is used to increase the angular size.

Astronomical Telescope

An outline sketch of an astronomical telescope is shown in Figure 30.16. Suppose that an object O is at a very large distance from the objective lens L_1. The rays incident on L_1 will be almost parallel and a real inverted image I_1 is formed at the focal point F_o of the lens L_1. This image I_1 is very close to and within the focal length of the eyepiece lens L_2. The lens L_2 then forms an enlarged, still inverted, image of I_1 at I_2. The image I_2 is formed somewhere between the near and far points of distinct vision of the normal eye. The total length L of the telescope tube is $L \simeq f_o + f_e$.

Let h' be the size of the image I_1. From the ray diagram, $AB = h' = CD$. The angle subtended by the object at the unaided eye is very close to angle θ (length L is very small compared to the object distance p). Thus,

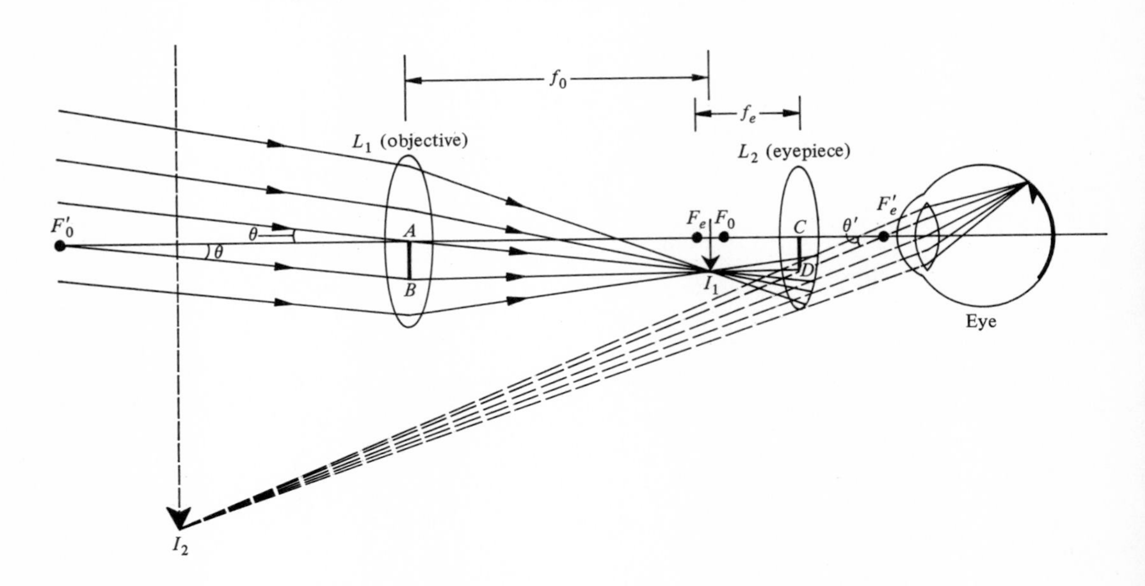

Figure 30.16. *The components and the ray diagram of an astronomical telescope.*

$$\theta = \frac{AB}{AF'_o} = \frac{h'}{f_o} \tag{30.15}$$

The angle θ' subtended by the image I_2 is given by

$$\theta' = \frac{CD}{CF'_e} = \frac{h'}{f_e} \tag{30.16}$$

Thus, the angular magnification M, using the previous equations, is

$$\boxed{M = \frac{\theta'}{\theta} = \frac{f_o}{f_e}} \tag{30.17}$$

The *angular magnification* of a telescope is the ratio of the focal length of the objective to that of the ocular. M increases with an increasing value of f_o and a decreasing value of f_e. The final image formed is inverted, but it does not make any difference in the case of astronomical objects.

The Galilean Telescope (Opera Glasses)

In viewing terrestrial objects it is desirable that the image be erect. This can be achieved by inserting a third convex lens between the objective and the eyepiece. Another alternative is to use the Galilean telescope, the arrangement of which is shown in Figure 30.17, in which the eyepiece is a diverging lens, that is, a concave lens. It is inserted before the position where the image is formed by the objective.

The total length L of the tube is approximately $f_o - f_e$ in this case, instead of $f_o + f_e$, even though the magnification is still $M = f_o/f_e$.

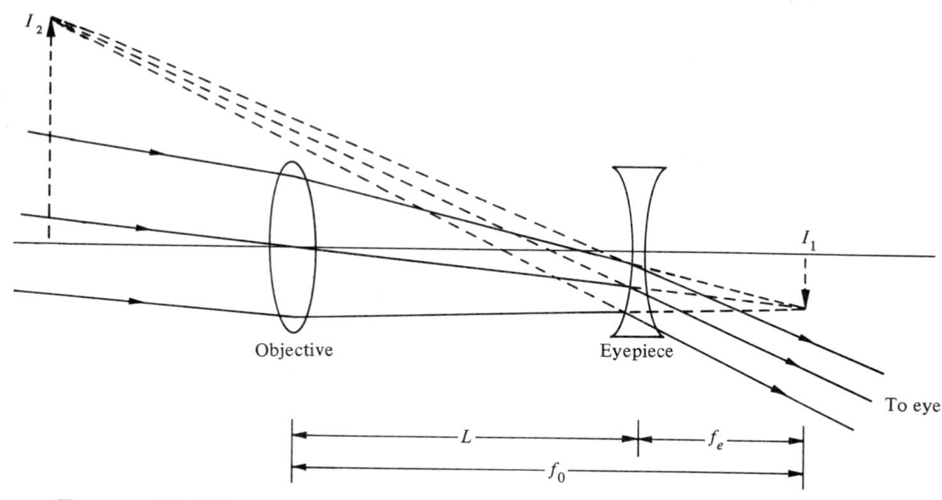

FIGURE 30.17. *Ray diagram of a Galilean telescope. Before the objective forms the image I_1, the eyepiece forms the final erect image I_2.*

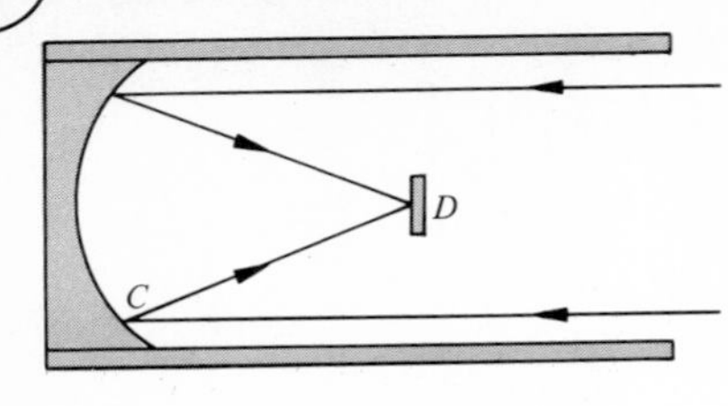

(a)

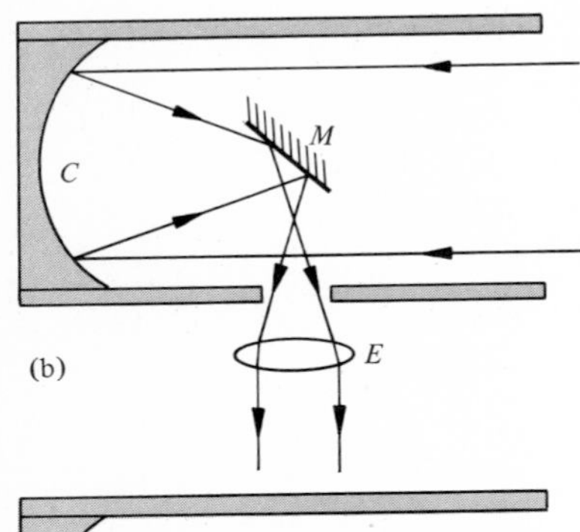

(b)

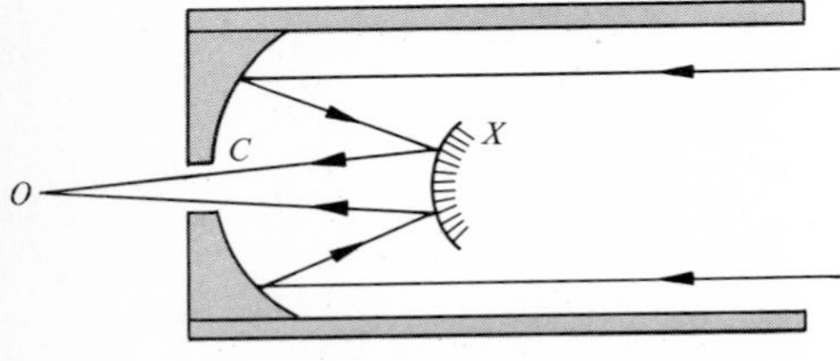

(c)

FIGURE 30.18. *Three different types of reflecting telescopes: (a) simple converging mirror, (b) Newtonian, and (c) Cassegrain.*

The Reflecting Telescope

In astronomical work it is necessary to use telescopes not only of very large focal lengths but also of very large diameters. Large diameters are necessary to gather light from very dim objects in outer space. There are many difficulties in the construction of large lenses for refracting telescopes of the type we have been talking about. The difficulties involve their large weight, smooth surface, changes due to variations in temperature, and correction for aberrations.

These problems have been overcome by using reflecting telescopes in which the objective lens is replaced by a concave mirror. (Also mirrors have fewer aberration problems than lenses.) Three different types of reflecting telescopes are shown in Figure 30.18. In Figure 30.18(a) the parallel rays of light after reflection from a concave mirror C are brought to a focus at a light-sensitive detector D. D may be a photographic film, spectrometer, or a human being making observations. The arrangements in Figure 30.18(b) is the Newtonian telescope, in which the rays after reflection from C but before converging to a point are reflected by the plane mirror M and are observed through the eyepiece E.

Figure 30.18(c) shows the arrangement of a Cassegrain telescope. A hole is made at the center of the concave mirror C in order to view the rays reflected from the convex mirror X.

Just to give you some idea of the size of the reflecting telescope, we mention two of them. The telescope built in 1921 at Mt. Wilson in California has a diameter of 100 in, while one at Mt. Palomar, also in California, built in 1951 has a diameter of 200 in. On the other hand, the largest refracting telescope in the world is at Yerkes Observatory, Williams Bay, Wisconsin; it has a diameter of 40 in.

Before closing this section we draw the components and ray diagrams of two more optical instruments: (a) the prism binoculars (Figure 30.19) and (b) the slide projector (Figure 30.20). The diagrams are self-explanatory. The binoculars are marked with two numbers separated by a multiplication sign, such as 7×35. The first number is the angular magnification, that is, 7, and the second number is the diameter of the lens in millimeters, 35 mm in this case. The magnification determines the power of the lens system and the diameter determines the light-collecting power.

EXAMPLE 30.5 In an astronomical telescope the power of the objective lens is +1.6 diopters, while that of the eyepiece is +50 diopters. Calculate (a) the length of the tube of the telescope; and (b) its angular magnification.

(a)
$$\text{Power of the objective} = \frac{1}{f_o} = +1.2 \text{ diopters}$$

$$f_o = \frac{1}{\text{power}} = \frac{1}{1.2 \text{ m}} = 0.83 \text{ m} = 83 \text{ cm}$$

$$\text{power of the eyepiece} = \frac{1}{f_e} = +50 \text{ diopters}$$

$$f_e = \frac{1}{\text{power}} = \frac{1}{50 \text{ m}} = 0.02 \text{ m} = 2 \text{ cm}$$

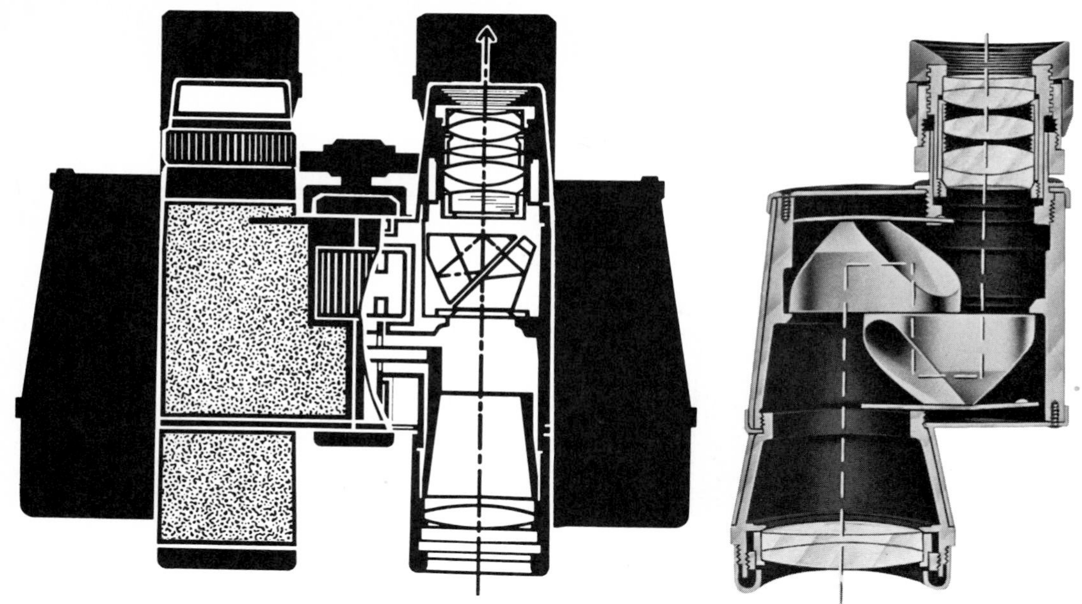

FIGURE 30.19. *Components and ray path of a prism binocular.* (Courtesy of Bushnell Optical Company, a division of Bausch & Lomb.)

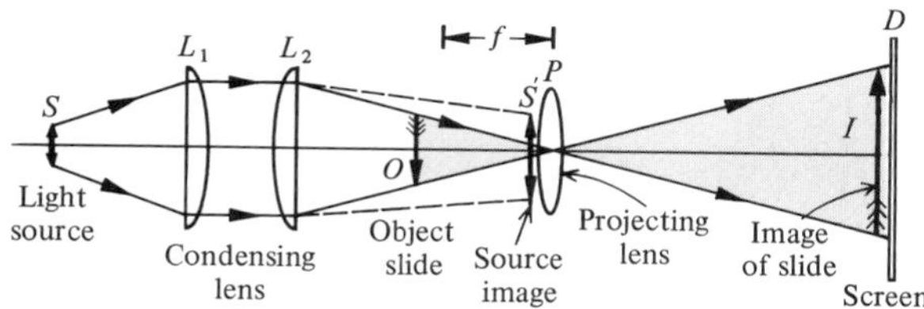

FIGURE 30.20. *Components and ray diagram of a slide projector.*

$L = f_o + f_e = (83 + 2)$ cm $= 85$ cm, which is the length of the tube.
(b) From Equation 30.17, the angular magnification is

$$M = \frac{f_o}{f_e} = \frac{83 \text{ cm}}{2 \text{ cm}} = 41.5$$

EXERCISE 30.5 To achieve an angular magnification of 50, what should be the focal length of the objective lens if the power of the eyepiece is 40 diopters? What is the length of the tube of the telescope? [*Ans.:* $f_o = 125$ cm, $L = 127.5$ cm.]

*30.6. Resolving Power

It will be proper to ask the following question at this stage. Is it possible to obtain magnification of any magnitude by using a proper set of large lenses? The answer is no. There are two limitations on the limit of performance of any optical instrument. First is the technical limitation. The bigger the lenses, the more care has to be taken in their construction, because even the slightest defects will show up quite magnified. Also, the lens aberrations become hard to remove. Even if we could remove all technical limitations, there is still a second more serious limitation—a more fundamental and basic limitation, which cannot be removed. This second limitation is concerned with the wave nature of light and is the topic of discussion in this section.

Suppose that light of wavelength λ is incident on a lens of diameter d or a mirror with a circular aperature of diameter d. The situation is similar to a single-slit diffraction, as discussed in Chapter 29. The light diffracted from this circular slit will show a diffraction pattern consisting of circular rings, as shown in Figure 30.21. According to the theory of diffraction, the diffraction angle $\Delta\theta$ in radians between the central maximum and the first minimum (the first dark

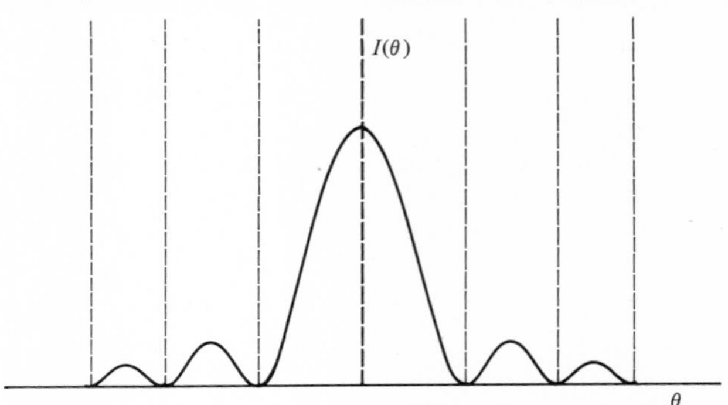

FIGURE 30.21. *The diffraction pattern of a circular aperture, showing Airy's disk.* (From M. Cagnet, M. Françon, and J. C. Thrierr, *Atlas of Optical Phenomena,* Springer-Verlag, Berlin, 1962.)

fringe) is given by

$$\sin \Delta\theta = 1.22 \frac{\lambda}{d} \tag{30.18a}$$

Since $\Delta\theta$ is small, we may also write $\sin \Delta\theta \approx \Delta\theta$,

$$\Delta\theta \approx 1.22 \frac{\lambda}{d} \tag{30.18b}$$

The central bright disk is called *Airy's disk.*

Suppose that we are looking at two bright objects side by side, say two stars through a telescope. Each star will produce its own diffraction pattern and Airy's disk. Let the angular separation between the two stars be $\Delta\phi$. If $\Delta\phi > \Delta\theta$, the two Airy's disks will be completely separated as shown in Figure 30.22(a)—the two stars are completely resolved. If $\Delta\phi = \Delta\theta$, the two

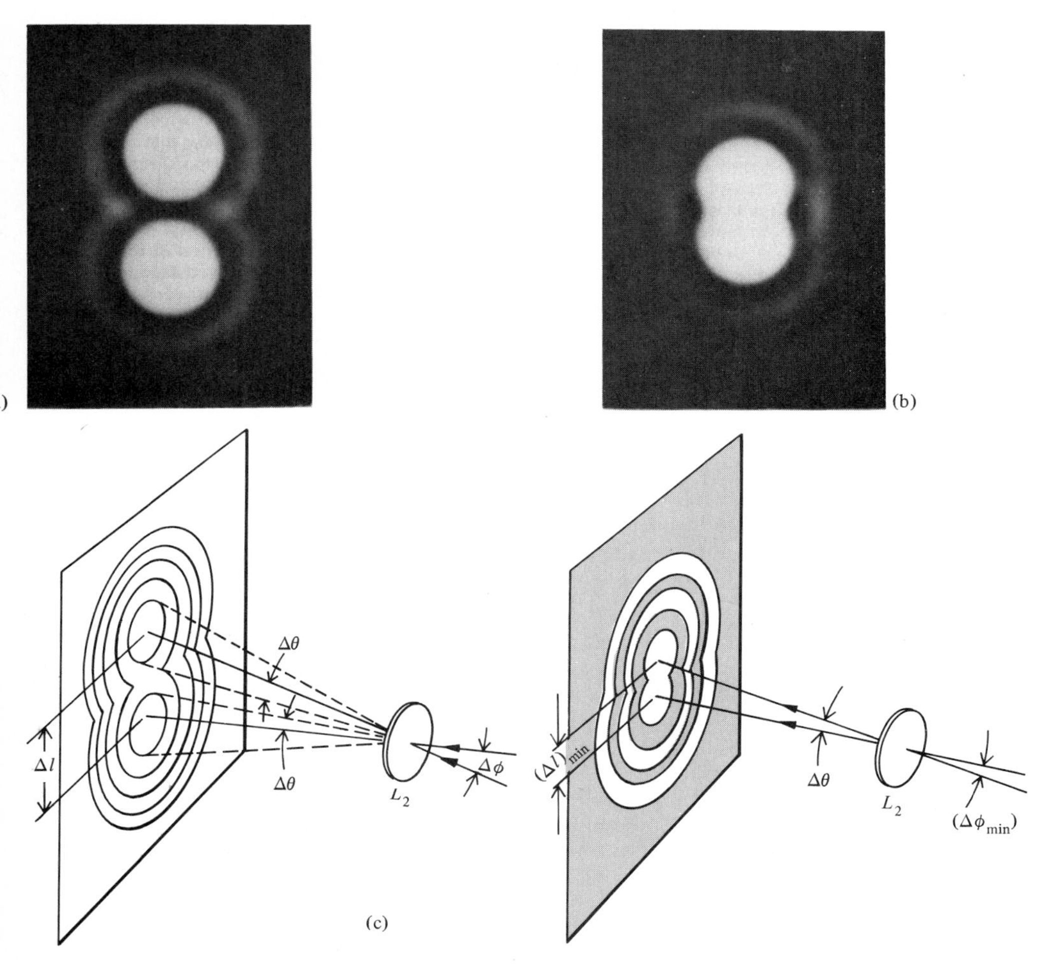

FIGURE 30.22. *Resolving two objects: (a)* $\Delta\phi > \Delta\theta$*, completely resolved; and (b)* $\Delta\phi = \Delta\theta$*, just resolved.* (From E. Hecht and A. Zajac, *Optics*, Addison-Wesley Publishing Company, Inc., Reading, Mass., 1974.)

Airy's disks are still separated; they are just resolved, as in Figure 30.22(b). On the other hand, if the two stars are still closer ($\Delta\phi < \Delta\theta$), the two Airy's disks may completely overlap, and in such situations the two stars cannot be resolved (the two appear as one).

According to *Rayleigh's criterion:* Two objects or stars are said to be just resolved if the center of one Airy's disk (central maximum) falls on the first minimum of the Airy's pattern of the other. Thus, the limit of angular resolution is given by

$$(\Delta\phi)_{\text{min}} = \Delta\theta = \frac{1.22\,\lambda}{d} \tag{30.19}$$

that is, only those stars or objects will be resolved whose angular separation $\Delta\phi$ is equal to or greater than $\Delta\theta$. If Δl is the distance between the images of the two objects, than $(\Delta\phi)_{\text{min}} = \Delta l/f$, where f is the focal length of the objective lens from Equation 30.19:

$$\boxed{(\Delta l)_{\text{min}} = 1.22\,\frac{f\lambda}{d}} \tag{30.20}$$

The stars or any other two objects cannot be resolved if their images are closer than Δl. Of course, Δl depends upon the wavelength λ. For visible light, λ is between 4000 and 7000 Å. f and d may be optimized, but the ultimate resolution depends upon λ. When the limit of resolution is reached, any further magnification does not reveal any more details of the image.

Finally, let us discuss the resolving power of the eye. The diameter of the iris is 0.5 cm and for green light $\lambda = 5000$ Å $= 5 \times 10^{-5}$ cm; from Equation 30.19,

$$\Delta\theta = \frac{1.22\,\lambda}{d} = \frac{(1.22)(5 \times 10^{-5}\text{ cm})}{0.5\text{ cm}} = 1.22 \times 10^{-4}\text{ rad}$$

An average eye is not capable of resolving objects with an angular separation of less than 5×10^{-4} rad. The reason for this is the size of the cones, which are closely packed at the fovea. The diameter of each cone is 2×10^{-6} m. For two points to be clearly resolved, their images formed at the retina must be such that the two excited cones are separated by at least one cone.

The resolution of the eye, 5×10^{-4} rad, leads to the following smallest size that a normal eye can resolve. Suppose that two point objects separated by y cm are placed 25 cm from the eye, as shown in Figure 30.23. Since θ is small,

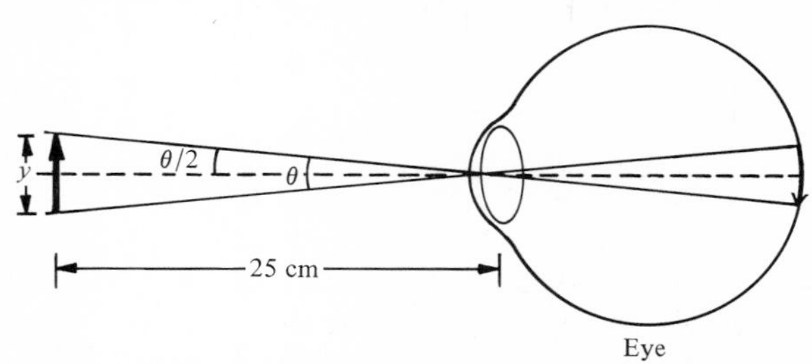

FIGURE 30.23. *Resolution of a normal eye.*

$$\tan\frac{\theta}{2} \approx \frac{\theta}{2} = \frac{y/2}{25\text{ cm}} \tag{30.21}$$

Since

$$\theta = 5 \times 10^{-4}\text{ rad}$$

$$y = (25\text{ cm})\theta \times (25\text{ cm})(5 \times 10^{-4}) = 0.12\text{ mm}$$

Thus, a normal eye can resolve objects that are separated by $\sim$0.1 mm and placed at the distance of distinct vision.

EXAMPLE 30.6 The objective of a microscope has a focal length of 1 cm and diameter of 0.5 cm. Using light of wavelength 6000 Å (yellow light), is it possible to resolve the two point objects if they are separated by a distance of 10^{-6} m?

From Equation 30.19, $\Delta\theta \approx 1.22\,\lambda/d$, where $\lambda = 6000\text{ Å} = 6000 \times 10^{-10}$ m and $d = 0.5\text{ cm} = 0.5 \times 10^{-2}$ m. Suppose that the two objects being viewed are a distance Δx apart. Since these objects are almost at the focal length of the objective, the angle subtended by Δx is

$$\Delta\phi \approx \frac{\Delta x}{f_o}$$

$\Delta\phi$ must be greater than $\Delta\theta$ for the two objects to be clearly resolved. That is,

$$\frac{\Delta x}{f_o} > 1.22\,\frac{\lambda}{d}$$

$$\Delta x > 1.22\,\frac{f_o\,\lambda}{d}$$

$$\Delta x > 1.22\,\frac{(1 \times 10^{-2}\text{ m})(6000 \times 10^{-10}\text{ m})}{0.5 \times 10^{-2}\text{ m}} = 1.46 \times 10^{-6}\text{ m}$$

The two objects will not be resolved, because 10^{-6} is smaller than 1.46×10^{-6} m.

EXERCISE 30.6 What will be your answer to Example 30.6 if the wavelength of light used is 4000 Å? [*Ans.:* $\Delta x > 9.76 \times 10^{-7}$ m, objects will be resolved.]

SUMMARY

The departures of an actual image from that of the ideal or perfect image are called *aberrations*. *Chromatic aberration* results from the inability of a lens to bring to a focus in one plane all rays of different colors. Those aberrations which still exist even if monochromatic light were used in image formation are called *monochromatic aberrations* and include spherical aberration, astigmatism, coma, curvature of field, and distortion. The inability of a lens to bring to a focus a point object to a single point image is called *spherical aberration*.

Astigmatism is the result of imaging a point object which is located off-axis into a line image.

In principle, a photographic camera consists of a light-tight box, converging lens, and light-sensitive plate or film. The f-number $= f/D$ and $I \propto 1/(f\text{-number})^2$ where I is the *illuminance*.

The most important components of a human *eye* are the sclera, cornea, crystalline lens, aqueous humor, pupil, ciliary muscles, vitreous humor, retina, charoid, rods, cones, optic nerve, macula, and fovea centralis. The fine adjustment that can be accomplished by the lens of the eye is called *accommodation*. The closest point on which the eye can focus is called the near point ($\sim$25 cm). The far point in a normal eye is at infinity. The three major defects of the eye are: (1) a *myopic* or nearsighted eye can see nearby objects only and is corrected by using a negative or diverging lens, $1/f = -1/d$; (2) a *hyperopic* or farsighted eye can see objects that are far away and is corrected by using a positive or converging lens, $1/f = 1/0.25\text{ m} - 1/d$; and (3) an *astigmatic* eye cannot focus horizontal and vertical lines at the same time and may be corrected by using cylindrical lenses. The abnormal elevation of the pressure within the eyeball leads to a group of diseases of the eye called *glaucoma*.

A *microscope* is used for producing at the retina of an eye, an enlarged, distinct image of a small nearby object. The angular size $\theta = h/25$ cm. The angular magnification or the magnifying power of a simple microscope is $M = \theta'/\theta = h'/h = 1 + 25\text{ cm}/f \simeq 25\text{ cm}/f$. The overall magnification of a compound microscope is $M = (L/f_o)(25\text{ cm}/f_e)$.

A *telescope* is used to see a clear and distinct image of an object which is at a very large distance from the observer. The angular magnification of an astronomical telescope is $M = f_o/f_e$. The *Galilean telescope* is used when it is desirable for the image to be erect. *Reflecting telescopes* are used frequently for viewing objects in outer space.

*According to *Rayleigh's criterion*, two objects or stars are said to be just resolved if the center of one Airy's disk (central maximum) falls on the first minimum of the Airy's disk of the other; $(\Delta\phi)_{\text{min}} = \Delta\theta = (1.22)(\lambda/d)$. If Δl is the distance between two objects $(\Delta l)_{\text{min}} = 1.22\, f\lambda/d$. The normal eye can resolve objects which are separated by $\sim$0.1 mm and placed at the distance of distinct vision.

QUESTIONS

1. Spherical aberrations can be reduced by using a lens of long focal length and small aperture. Explain how.
2. How do you experimentally distinguish that a lens has been corrected for distortion and curvature of field?
3. How will you infer that a camera lens has been corrected for various defects?
4. What is the effect of introducing a glass plate between the film and the lens of a camera?
5. How do you distinguish a telephoto lens from a wide-angle lens?
6. Does the lens of a human eye have the same aberrations as any other lens?
7. Are there any circumstances under which a human eye will produce a virtual image?

8. Why does the image formed on the retina become blurred when an object is brought closer than ~25 cm?
9. Why can't we see clearly under water? (*Hint:* Compare the refractive index of water with that of the cornea.)
10. If the eye lens had to be replaced, what power lens will have to be substituted? Will the eye still have accommodation?
11. Why do aged people usually need to use bifocal lenses?
12. Low-quality microscopes form images that are surrounded by colored rings. Explain.
13. Why can't a microscope be designed to magnify a water molecule so that it becomes visible?
14. Which of the following combinations may be used as a microscope or telescope: (a) $f_o = 1$ cm, $f_e = 2$ cm; (b) $f_o = 10$ cm, $f_e = 2$ cm; (c) $f_o = 0.2$ cm, $f_e = 1$ cm; and (d) $f_o = 20$ cm, $f_e = 4$ cm?
15. By interchanging the role of the two lenses, can a microscope be used as a telescope?
16. Why in cheap opera glasses are there colored fringes around the image?
17. What limits the ability of a telescope to image objects that are far out in space?

PROBLEMS

1. Which of the following lenses do you consider "fast"? (a) $f = 20$ cm, $D = 4$ cm; (b) $f = 20$ cm, $D = 5$ cm; (c) $f = 10$ cm, $D = 4$ cm; (d) $f = 10$ cm, $D = 5$ cm; (e) $f = 8$ cm, $D = 4$ cm; and (f) $f = 6$ cm, $D = 5$ cm.
2. An $f/6$ lens has a diameter of 2 cm. What is the focal length of the lens?
3. An $f/1.2$ has a focal length of 6 cm. What is the diameter of the lens?
4. A photograph is taken with a lens of speed $f/4$ and a time exposure of (1/100) s. If the same picture is to be repeated with an exposure of (1/50) s, what should be the speed of the lens?
5. A photograph is taken with a lens of speed $f/2.8$ and a time exposure of (1/200) s. If the same picture has to be repeated with a lens of speed $f/5.6$, what should be the time exposure?
6. In order to take a good photograph with a camera of lens speed $f/4$ and diameter of 4 cm, you need a time exposure of (1/100) s. If another lens has a speed of $f/8$ and a diameter of 2 cm, what should be the time exposure to take the photograph?
7. A bright point object at infinity is focused by a double convex lens of $f/9$ on a film. If the film is moved 2 cm away, calculate the diameter of the patch of light on the film.
8. A camera with a lens of $f/5.6$ is used to photograph a self-illuminated object at a distance of 4 m with a time exposure of (1/100) s. Should the time exposure be increased or decreased if the object is at a distance of 8 m? 2 m?
9. A myopic eye has a far point at 60 cm. What type and power of lens is needed to correct this eye? Draw a proper ray diagram.
10. A hyperopic eye has a near point at 120 cm. What type and power of lens is needed to correct this eye? Draw a proper ray diagram.

11. What will be the prescription of the lens if the near point of the left eye is 250 cm while the far point of the right eye is 40 cm?
12. A myopic eye has a far point at 50 cm. What is the power of the lens needed to see a clear image of distant objects? What will be the far point with a lens of +1.50 diopters?
13. A farsighted person can see objects clearly if they are at a distance greater than 1.20 m. What will be the power of the lens if he wishes to see objects that are placed a distance of 20 cm from his eye?
14. If the power of the lens of a left eye is −0.8 diopter and the right eye is +2.0 diopters, which eye is myopic and what is its far point? Which eye is hyperopic and what is its near point?
15. A woman with a myopic eye has a far point at 2 m. Where will be the far point if she uses a lens of −0.75 diopter?
16. A man with a hyperopic eye has a near point at 75 cm. Can he correct this defect by using a lens of +5.0 diopters?
17. Suppose that in a young adult the lens system of the eye can fluctuate between +56 diopters and +65 diopters. Calculate (a) the range of focal length and (b) the near point for the eye.
18. An old man can see objects that are as close as 50 cm and as far as 3 m. To correct his vision he must use bifocal glasses to see distant objects through the top half and to read from a distance of 25 cm from the lower half. Calculate the power of the lenses he must use.
19. Calculate the angular size of a quarter coin viewed from a distance of (a) 25 cm; (b) 1 m; and (c) 10 m. The diameter of a quarter is 2.4 cm.
20. Calculate the angular sizes of the sun and the moon as viewed by an observer on earth.
21. Compare the angular sizes of the following: (a) an actor 6 ft tall being viewed from a distance of 120 ft; and (b) a 6-in image on the TV screen viewed from 8 ft.
22. Calculate the maximum usable angular magnification produced by the following lenses: (a) focal length 5 mm; (b) focal length 1.2 cm; (c) a lens of power +2 diopters; (d) a lens of power +5 diopters; and (e) a lens of $f/9$ and diameter 2 cm.
23. If a magnifying glass produces an image of magnification 6, what is the power of the lens?
24. Construct an image of an object 1 cm tall placed 8 cm from a convex lens of focal length +10 cm. Measure the size of the image. Compare this with the calculated results.
25. An insect 4 mm tall is placed in front of a convex lens of focal length +1 cm. Calculate (a) the angular size of the insect when placed at +25 cm; (b) the maximum usable angular magnification; and (c) the magnification when the image is formed at infinity.
26. An object 2 mm tall is placed in front of a convex lens of focal length +1.5 cm. Calculate the magnification when the image is located (a) at the distance of distinct vision; (b) at +50 cm; and (c) at infinity.
27. In a compound microscope the objective has a focal length of 1 cm and the eyepiece has a focal length of 2 cm. The distance between the objective and the eyepiece is 21 cm. Calculate the angular magnification when the image is formed at infinity. In such situations, what is the position of the object?

28. A compound microscope has an objective of focal length +8 mm and forms an image at a distance of 20 cm. If the eyepiece has an angular magnification of 6, what is the total magnification?

29. Consider a microscope with objectives of focal length 16 mm, 8 mm, and 2 mm while the eyepiece has angular magnifications of 4× and 8×. If each objective forms an image at a distance of 20 cm, calculate the possible total magnification.

30. In a compound microscope $f_o = +2$ cm and $f_e = +2$ cm. If an object is placed at a distance of (a) 2.1 cm; and (b) 2.4 cm from the objective, calculate (1) the length of the microscope tube, and (2) the magnification in each case. Which of the two configurations of the microscope will you recommend and under what conditions?

31. A small microscope for examining gems has a total length of 6 cm, the objective has a focal length +0.6 cm, and the eyepiece a focal length 1.5 cm. Calculate (a) the position of the gem; (b) the linear magnification of the objective; (c) the angular magnification of the eyepiece; and (d) the total magnification.

32. Calculate the magnification of an astronomical telescope with an objective of focal length 200 cm and eyepiece with a focal length of 2 cm.

33. The largest refracting telescope at Yerkes Observatory has a focal length of 62 ft. Calculate the magnification if the eyepiece has a focal length of 1 in, $\frac{1}{2}$ in, or $\frac{1}{4}$ in.

34. What is the angular magnification of an astronomical telescope if the length of the telescope is 155 cm and the focal length of the eyepiece is 2 cm?

35. The distance between the objective and the eyepiece of an astronomical telescope is 1 m. If the magnification is 49, what are the focal lengths of the objective and the eyepiece?

36. The objective of the Lick Observatory telescope has a focal length of 58 ft. What should be the focal length of the eyepiece to obtain magnifications of 200, 400, and 600?

37. Consider an astronomical telescope with $f_o = 100$ cm, $f_e = 10$ cm. A building 50 m high is located 1 km away from the objective. Calculate (a) the size of the image formed by the objective; (b) the angular magnification produced by the eyepiece; and (c) the overall magnification.

38. Show that the angular magnification of a Galilean telescope (or opera glass) is given by $M = -f_o/f_e$.

39. The length of a certain Galilean telescope is 30 cm while the focal length of the eyepiece is 4 cm. Calculate the focal length of the objective and the total magnification.

40. Consider an opera glass with $f_o = +36$ cm and $f_e = -3$ cm. Calculate the distance between the lenses and the overall magnification. What is the size of the real image formed by the objective when an object of size 1 m is located at a distance of 30 m?

***41.** Is it possible to see finer details separated by a distance of 5×10^{-6} m using a microscope of $f_o = 8$ mm and diameter of 1 cm? The light used has a wavelength of 5500 Å.

***42.** What is the minimum angular distance between two stars so that they will look separated through a telescope having an objective lens of focal length 200 cm, diameter 30 cm, and an eyepiece of focal length 5 cm?

*43. The world's largest refracting telescope at Yerkes Observatory has $f_o = 62$ ft and a diameter of 40 in, while the eyepiece has a focal length of 3 in. What is the minimum angular distance between stars that can be resolved if the wavelength of light used is (a) 4000 Å; (b) 5000 Å; (c) 6000 Å; and (d) 7000 Å?

Part VII
Modern Physics

31 Quantum Physics

So far in this text we have treated matter such as electrons, protons, and neutrons as particles, while electromagnetic radiation, such as visible light, radio waves, x-rays, and gamma rays, have been treated as waves. We shall now show that a particle or wave characteristic is not an inherent property but depends upon the type of experimental observation. Thus, under proper circumstances particles can behave as waves while waves may behave as particles. That is, we are going to establish the dual nature*—particle and wave—for any system.*

31.1. Quantization of Radiation

AT PRESENT we know that all atoms are made up of multiples of electrons, protons, and neutrons. All the electrons have the same mass, so do all the protons and neutrons. Similarly, the charge comes in integer multiples of the charge of electrons (see Chapter 32); that is, $q = ne$, where $n = 1, 2, 3, \ldots$. We state this fact by saying that the *charge is quantized.* The question that we ask about radiation or electromagnetic radiation (such as visible light, radio waves, x-rays, and gamma rays, etc.) is this: Is it possible that there may be some characteristic of radiation which may show quantization? The answer is yes, and the energy of the electromagnetic waves shows the quantization property. Let us see how this came about.

In the seventeenth century two great scientists—Newton and Huygens—held opposite views about the nature of light. According to Newton's corpuscular theory, light consisted of tiny particles which are emitted by a source. According to Huygens' theory, light consists of waves. But it was Newton's theory that held ground until the beginning of the nineteenth century. The work of Young in the early nineteenth century demonstrated that diffraction and interference experiments could be explained only by means of a wave theory. The development of Maxwell's field equations for an electromagnetic field and their experimental verification by Hertz in 1887 proved beyond a doubt that electromagnetic energy (including light) flows continuously and hence consists of waves.

In the beginning of the twentieth century, experimental phenomena, such as (1) the blackbody radiation spectrum, (2) the photoelectric effect, (3) x-ray spectra, (4) Compton scattering, and (5) optical spectra, came into existence which could not be explained by the wave theory. As we shall see in this and the next chapter, if we assume that the flow of electromagnetic radiation energy is not continuous but is in the form of tiny discrete bundles of energy, called *photons,* these new experiments can be explained easily.

To start with, let us talk about blackbody radiation, which hypothesis together with quantum hypothesis was discussed in Section 5 of Chapter 15. In an effort to explain the observed radiation spectrum, Max Planck in 1901 introduced the following quantum hypothesis:

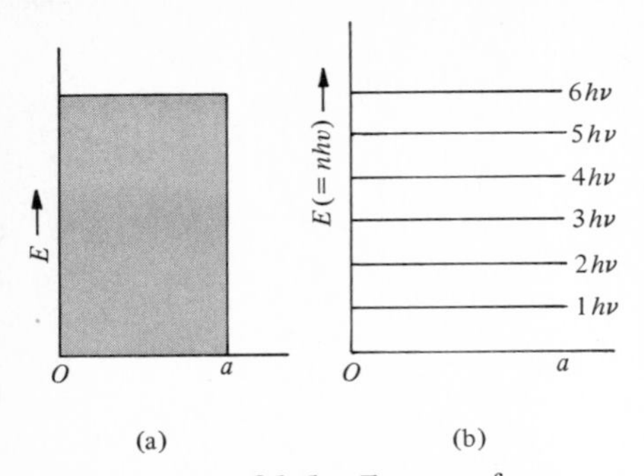

FIGURE 31.1. *Energy of a system according to (a) classical theory and (b) quantum theory.*

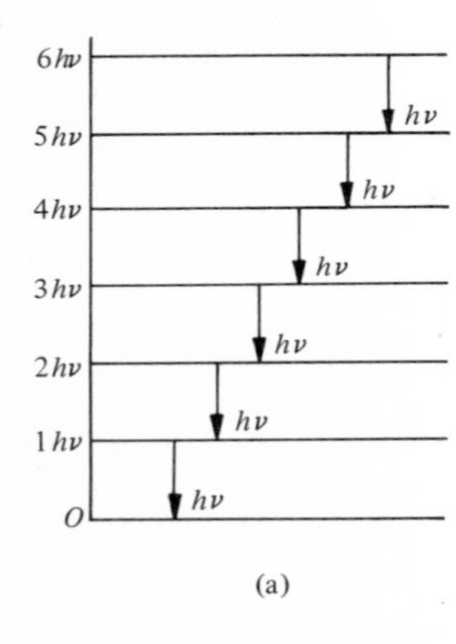

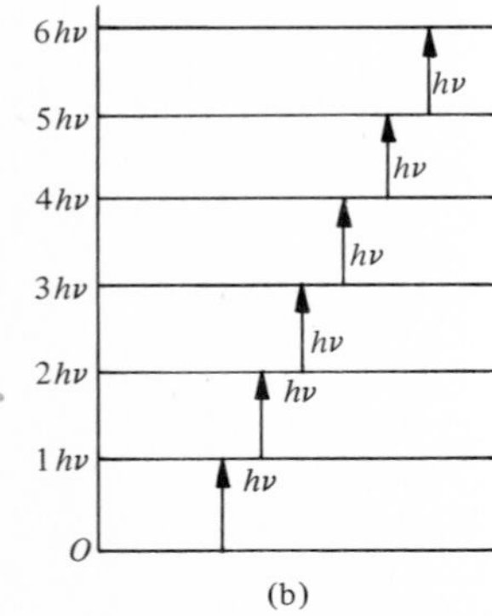

FIGURE 31.2. *(a) Emission of photons and (b) absorption of photons by a system.*

Planck's Hypothesis: ***Any physical system executing a simple harmonic motion in one dimension with frequency ν can have only those energies that satisfy the relation***

$$E = nh\nu \qquad \text{where } n = 1, 2, 3, \ldots \tag{31.1}$$

while h is a* universal constant called Planck's constant *and its value is h = 6.625 × 10^{-34} J-s.

According to Equation 31.1, the system can exist is only certain discrete energy states, called *quantum states,* and the integer n is called the *quantum number.* Also, according to Planck's hypothesis, the oscillating system can emit or absorb energy in bundles of the size $h\nu$, called *photons*. When the energy is emitted or absorbed in bundles of size $h\nu$, the *radiation is said to be quantized.* The theory that utilizes the ideas of Planck's hypothesis is called *quantum theory,* in contrast to the classical theory that utilizes wave concepts.

To demonstrate a distinction between classical theory and quantum theory as applied to the harmonic oscillator, we refer to Figures 31.1 and 31.2. In Figure 31.1 energy levels of an oscillator are shown as predicted by the classical theory as well as quantum theory. As is clear, in classical theory, the energy can have continuous values between 0 and E_{max}, while in quantum theory only discrete values (horizontal lines) are allowed. Since energy is emitted only when the system changes from one quantum state to another, the photon emitted will have energy $h\nu$. The vertical lines in Figure 31.2(a) show the energies of the emitted photons. Similarly, when a system absorbs energy, it changes its state; hence, the incident photon energy must be in bundles of the size $h\nu$ to accomplish this, as shown in Figure 31.2(b).

EXAMPLE 31.1 An FM station broadcasting at a frequency of 110 MHz has a power of 250 kW. Calculate the energy of each photon and the number of photons emitted per second.

Using Planck's hypothesis, the minimum photon energy for $n = 1$ in Equation 31.1 is

$$E = h\nu = (6.625 \times 10^{-34}\text{ J-s})(110 \times 10^{6}\text{ Hz})$$

$$= 7.29 \times 10^{-26}\text{ J} \times \frac{1\text{ eV}}{1.602 \times 10^{-19}\text{ J}}$$

$$= 4.55 \times 10^{-7}\text{ eV} \qquad \text{a very small energy}$$

The number of photons (or quanta of energy), N, emitted per second can be obtained by dividing power P by energy E of each photon. That is,

$$N = \frac{P}{E} = \frac{250\text{ kW}}{7.28 \times 10^{-26}\text{ J}} = \frac{250 \times 10^{3}\text{ J/s}}{7.28 \times 10^{-26}\text{ J}} = 3.43 \times 10^{30}/\text{s}$$

EXERCISE 31.1 A radio station is broadcasting at a power of 300 kW. If the energy of each photon emitted is 4.14 × 10^{-7} eV, calculate the broadcast frequency and the total number of photons emitted per second. [*Ans.:* 100 MHz, 4.52 × 10^{30}/s.]

31.2. Characteristics of Photons

As we have said, *photons are particles of light,* or, in general, particles of electromagnetic radiation. Photons are also called *quanta of the electromagnetic field.* Visible light, radio waves, x-rays, gamma rays, and so on, are all part of different regions of the large range of electromagnetic radiation shown in Figure 27.14. If the photon nature of radiation applies to any part of the electromagnetic spectrum, it should apply to all. Also, it means that all photons (like the photons of visible light) should be moving with a speed $v = c$ ($=3 \times 10^8$ m/s). Under the assumption that all electromagnetic radiation consists of particles, quanta of radiation, or photons, it should be possible to assign mass, energy, and momentum to such photons, as shown below.

From the special theory of relativity, if a particle of rest mass m_0 is moving with velocity v, its mass m is given by $m = m_0/\sqrt{1 - (v^2/c^2)}$. Since for a photon consisting of a packet of energy, $v = c$, the preceding relation yields the mass of the photon to be $m = m_0/\sqrt{1 - 1} = m_0/0 = \infty$. This infinite mass is not possible. On the other hand, this difficulty can be overcome if we assume that $m_0 = 0$, $m = 0/0$, which is an indeterminate quantity. The assumption that the rest mass of the photon is zero should not be of much concern, because photons are always found to be moving with the speed of light and never at rest. According to special relativity the photon of mass m has energy E given by

$$E = mc^2 \tag{31.2a}$$

while according to the photon hypothesis it is a bundle of energy given by

$$E = h\nu \tag{31.2b}$$

Combining these equations yields

$$m = \frac{E}{c^2} = \frac{h\nu}{c^2} \tag{31.3}$$

Also, from the relativistic relation $E^2 = p^2c^2 + m_0^2c^4$, if we substitute $m_0 = 0$ for a photon,

$$E = pc \tag{31.4}$$

Hence, the momentum of a photon is

$$p = \frac{E}{c} = \frac{h\nu}{c} = \frac{h}{\lambda} \tag{31.5}$$

or

$$p = \frac{E}{c} = \frac{mc^2}{c} = mc \tag{31.6}$$

In summary, the photon may be described by the following relations if we assume that $m_0 = 0$.

$$\boxed{E = h\nu = \frac{hc}{\lambda}, \qquad m = \frac{h\nu}{c^2}, \qquad p = \frac{h\nu}{c} = \frac{h}{\lambda}} \tag{31.7}$$

Very often we are given the wavelength of radiation in angstrom units, and we want to know the energy of photons in electron-volts. Thus, if we substitute the values of h and c in

$$E = \frac{hc}{\lambda} = \frac{(6.625 \times 10^{-34}\ \text{J-s})(3 \times 10^{8}\ \text{m/s})(10^{10}\ \text{Å/m})}{\lambda(1.6 \times 10^{-19}\ \text{J/eV})}$$

we get

$$\boxed{E = \frac{(1.24 \times 10^{4})\ \text{Å-eV}}{\lambda}} \tag{31.8}$$

where E is in electron volts and λ in angstrom units.

31.3. Waves Behaving as Particles

Of the five experiments mentioned in Section 31.1, photoelectric effect, x-ray spectra, and Compton scattering will be discussed here to show the quantum nature of electromagnetic radiation; that is, waves behave like particles.

Photoelectric Effect

When light or any other electromagnetic radiation is incident on a metallic surface, the electrons ejected from the surface are called *photo electrons* and the effect is called the *photoelectric effect*. To investigate different aspects of the photoelectric effect, the experimental arrangement shown in Figure 31.3 is used. A vacuum tube T contains a metallic plate A and a charge collecting plate C. The incident monochromatic light beam, after being collimated by the slit S and passing through the quartz window, is incident on the plate A. The electrons ejected from the surface A are attracted by the plate C when it is at a positive potential V with respect to A. The current i_p indicated by the meter G is due to the electrons collected by C.

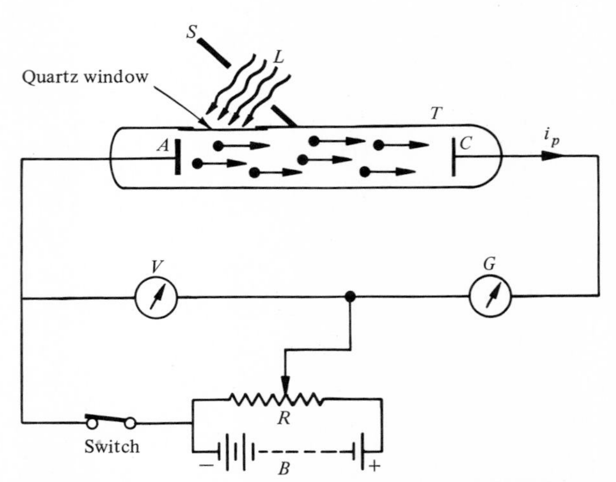

FIGURE 31.3. *Experimental arrangement used to investigate various characteristics of the photoelectric effect.*

By using a reversing switch, a negative potential may be applied to C which will repel the electrons emitted from A. The negative potential V_0, for which no electrons reach the plate C, that is, $i_p = 0$, is called the *stopping potential.* Under these circumstances the maximum kinetic energy $K_{max} = \frac{1}{2}mv^2_{max}$ must be equal to $e|V_0|$. That is,

$$|eV_0| = K_{max} = \tfrac{1}{2}mv^2_{max} \tag{31.9}$$

Let us now investigate the effect of radiation of different frequencies ν on the kinetic energy of the photo electrons from different metallic surfaces. Using the relations of Equation 31.9, K_{max} can be calculated by measuring the stopping voltages, V_0. Figure 31.4 shows plots of K_{max} versus ν for three different metallic surfaces. It is clear from this figure that the three plots are parallel straight lines with the same slopes a but different intercepts b. Since the equation of a straight line is given by $y = ax + b$, the plot of Figure 31.4 may be represented by

$$K_{max} = a\nu + b \tag{31.10}$$

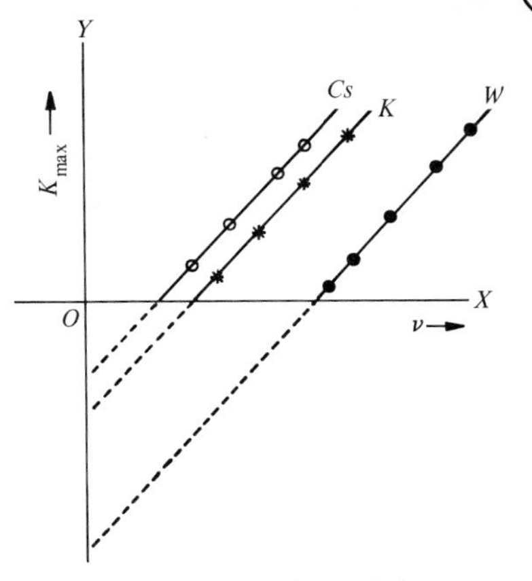

FIGURE 31.4. *Plots of the measured values of K_{max} (= eV_0) versus ν for various metals.*

This experimental result can be explained by adopting Planck's quantum hypothesis, as was done by Albert Einstein in 1905. According to him the electromagnetic radiation incident on a metallic surface consists of bundles or quanta of energy $h\nu$. Suppose that an electron in the metal surface absorbs this photon of energy $h\nu$. Before the electron can leave the surface, part of the incident energy must be used to overcome the binding of the electron in the metal. This energy is called the *work function* w and is different for different metals. The remaining energy $h\nu - w$ is carried away by the electron as its kinetic energy, K_{max}; that is

$$\boxed{K_{max} = h\nu - w} \tag{31.11}$$

Comparing this result with Equation 31.10 reveals that the constant slope a is equal to Planck's constant h and the intercept b is equal to the work function $-w$. The value of h obtained by this method agrees with the values obtained by other methods. According to Equation 31.11, for the electron to be just released from the surface, $K_{max} = 0$ and the photon must have a threshold energy $h\nu_0 = w$, that is, an energy equal to the work function (or binding energy of the electron) of the surface of the metal. According to classical theory, if the light shines long enough, no matter what the minimum frequency is, the electron should eventually be ejected. But this does not happen—classical theory fails and the quantum hypothesis is the answer.

EXAMPLE 31.2 Light of wavelength 5500 Å is incident on a metal surface. The stopping potential for the emitted electron is 0.42 V. Calculate (a) the maximum energy of the photoelectrons; (b) the work function; and (c) the threshold frequency.

(a) From Equation 31.9

$$K_{max} = e|V_0| = e(0.42 \text{ V}) = 0.42 \text{ eV}$$

(b) The work function from Equation 31.11 is

$$w = h\nu - K_{max} = \frac{hc}{\lambda} - K_{max}$$

where $K_{max} = 0.42$ eV and, from Equation 31.8, $hc/\lambda = (1.24 \times 10^4)$ Å-eV/λ. Therefore,

$$w = \frac{(1.24 \times 10^4)\text{ Å-eV}}{5500\text{ Å}} - 0.42\text{ eV} = 2.25\text{ eV} - 0.42\text{ eV} = 1.83\text{ eV}$$

(c) The threshold energy $h\nu_0$ is the minimum energy needed to eject an electron from the surface and must be equal to w. That is, $h\nu_0 = w$, or the threshold frequency is

$$\nu_0 = \frac{w}{h} = \frac{(1.83\text{ eV})(1.6 \times 10^{-19}\text{ J/eV})}{6.625 \times 10^{-34}\text{ J-s}} = 4.4 \times 10^{14}\text{ Hz}$$

Exercise 31.2 The maximum energy of the photoelectrons emitted from a metal surface of work function 1.78 eV is 0.36 eV. Calculate (a) the stopping potential; (b) the wavelength of the incident radiation; and (c) the threshold frequency. [*Ans.:* (a) $V_0 = 0.36$ V; (b) $\lambda = 5805$ Å; (c) $\nu_0 = 4.30 \times 10^{14}$ Hz.]

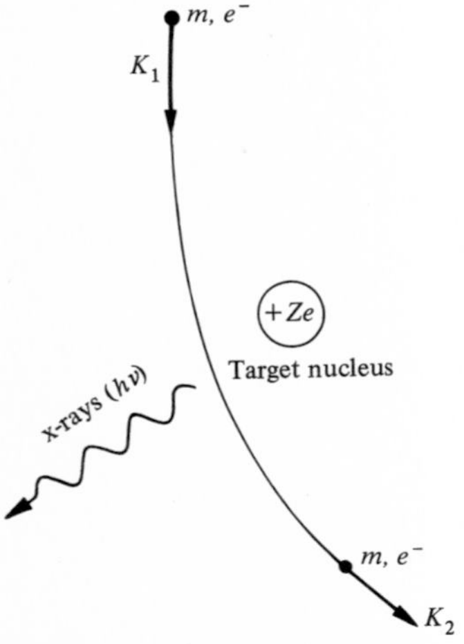

The Continuous X-Ray Spectrum

In 1895 Wilhelm Roentgen discovered that when fast-moving electrons hit a metallic target, a highly penetrating radiation of unknown nature was emitted. He named this radiation *x-rays*. Actually, x-rays are electromagnetic waves and, as is clear from Figure 27.14, their wavelengths range between 0.1 and 100 Å. As we shall explain below, since x-rays are emitted by slowing down charged particles, this process of producing radiation is also called *bremsstrahlung*, a German word meaning *slowing-down radiation*.

In Chapter 27 we explained how electromagnetic radiation is emitted from an accelerated charge. As shown in Figure 31.5, when a fast-moving electron approaches a target, it changes its path, owing to the attractive force between the electron and the positive charge of the nucleus of the atom. Thus, there will be an acceleration or deceleration of the electron, leading to the emission of electromagnetic waves. The energy loss of the electron appears in the form of x-rays.

Figure 31.5. *Production of x-rays as electrons decelerate in the field of the nucleus.*

Figure 31.6 shows an arrangement for producing x-rays. It consists of a high-vacuum tube called an *x-ray tube*. The electrons emitted from the indirectly heated cathode C (by filament F) are accelerated toward the anode T. If V is the potential difference between C and T, the kinetic energy K with which the electrons strike the target T is given by

$$K = eV \tag{31.12}$$

These electrons lose ~98 percent of their energy in collisions, hence producing heat. Only ~2 percent is utilized in producing x-rays. X-rays are penetrating radiation, hence are a health hazard. Thus, the x-ray tube is well shielded and a beam of x-rays is obtained by making a hole through the shielding.

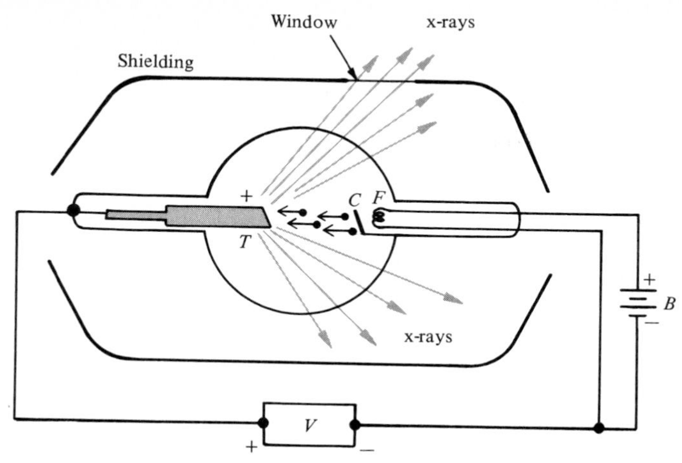

FIGURE 31.6. *Schematic of an x-ray tube. X-rays are produced by slowing down fast-moving electrons in the target T.*

The x-rays are emitted in all directions and with a continuous range of frequencies up to a maximum of ν_{max}. The following relation is found to be true for different targets:

$$\frac{\nu_{max}}{V} = \text{constant} \tag{31.13}$$

where V is the potential difference between the cathode and the anode. We can explain this relation with the help of quantum hypothesis, but not by the classical electromagnetic theory. Suppose that the incident electron loses all its kinetic energy $K = eV$ in a single collision, producing a photon of energy $h\nu_{max}$. Thus,

$$eV = h\nu_{max} \tag{13.14}$$

Since e and h are universal constants,

$$\boxed{\frac{\nu_{max}}{V} = \frac{e}{h} = \text{constant}} \tag{13.15}$$

which agrees with Equation 31.13. Using the relation $\lambda\nu = c = \lambda_{min}\nu_{max}$ in Equation 31.15, we get, after rearranging,

$$\lambda_{min} = \frac{hc}{eV} \tag{31.16}$$

Substituting for h, c, and e, we obtain a convenient and useful relation

$$\boxed{\lambda_{min} = \frac{1.24 \times 10^{-6}}{V} \text{ V-m}} \tag{31.17}$$

where V is in volts.

EXAMPLE 31.3 Calculate the minimum wavelength produced from a copper target in an x-ray tube that is operated at 10,000 V. Also, calculate the ratio of h/e.

Substituting $V = 10{,}000$ V in Equation 31.17,

$$\lambda_{min} = \frac{1.24 \times 10^{-6}\ \text{V-m}}{V} = \frac{1.24 \times 10^{-6}\ \text{V-m}}{10{,}000\ \text{V}} = 1.24 \times 10^{-10}\ \text{m}$$

Substituting this in Equation 31.16,

$$\frac{h}{e} = \frac{V\lambda_{min}}{c} = \frac{(10000\ \text{V})(1.24 \times 10^{-10}\ \text{m})}{3 \times 10^{8}\ \text{m/s}} = 4.13 \times 10^{-15}\ \text{V-s}$$

Knowing the value of e, h may be calculated.

EXERCISE 31.3 If the minimum wavelength of x-rays produced from a copper target is 1.31 Å, calculate the voltage at which the x-ray tube is operating. If $e = 1.602 \times 10^{-19}$ C, what is the value of h? [*Ans.:* $V = 9.47 \times 10^{3}$ V, $h = 6.62 \times 10^{-34}$ J-s.]

The Compton Effect

Suppose that x-rays are incident on a target containing free electrons. Such a situation is obtained by using a target such as carbon in which the energy needed to pull the electrons out is negligible as compared to x-ray energies of a few keV. The interaction between the incident x-rays and the electrons leads to the scattering of these x-rays. The phenomenon is called *Compton scattering* or the *Compton effect*. According to classical electromagnetic theory one would expect that the wavelength or frequency of the scattered x-rays are the same as that of the incident x-rays. But this is not true. Experimentally, we observe that scattered x-rays contain waves of two different wavelengths—λ, which is the same as the incident wavelength, called the *unmodified wavelength*, and a wavelength λ' (which is always $>\lambda$), called the *modified wavelength*. The value of λ' depends upon the angle at which the scattered x-rays are observed.

Figure 31.7 shows a schematic of an experimental arrangement for observing Compton scattering. A well-collimated beam of x-rays obtained from the x-ray tube S is incident on a target T. The measured wavelengths of the scattered x-rays at different angles are plotted as shown in Figure 31.8. The points are the experimental values while the continuous curves were calculated by Compton as explained below.

By using the quantum hypothesis, American physicist A. H. Compton in 1922 was able to explain the presence of modified wavelengths in x-ray scattering. (In 1927 a nobel prize was awarded to Compton for this work.) The collision between the incident x-ray and a free electron is shown schematically in Figure 31.9. Before the collision, as shown in Figure 31.9(a), the incident x-ray photon has energy $h\nu$, momentum $h\nu/c$, while the rest-mass energy of the electron is m_0c^2. As shown in Figure 31.9(b), after the collision the scattered photon has energy $h\nu'$ ($<h\nu$) and momentum $h\nu'/c$ ($<h\nu/c$). The remaining energy $h\nu - h\nu'$ appears as the kinetic energy K_e with momentum p_e of the

(Figure 31.9 is on p. 730.)

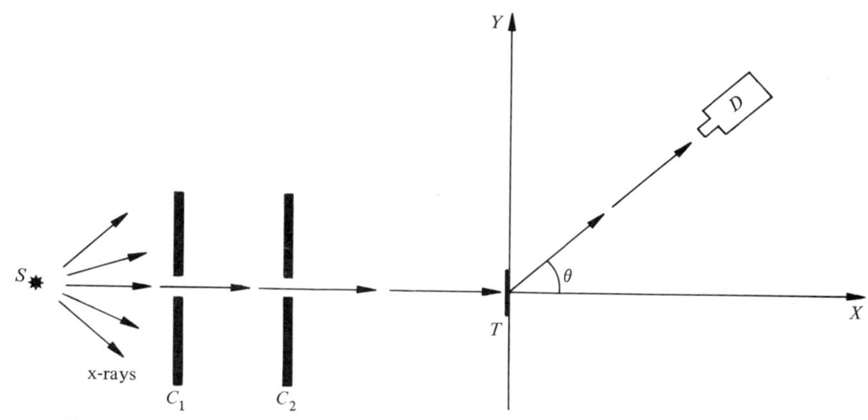

FIGURE 31.7. *Schematic sketch of an experimental arrangement for observing Compton scattering.*

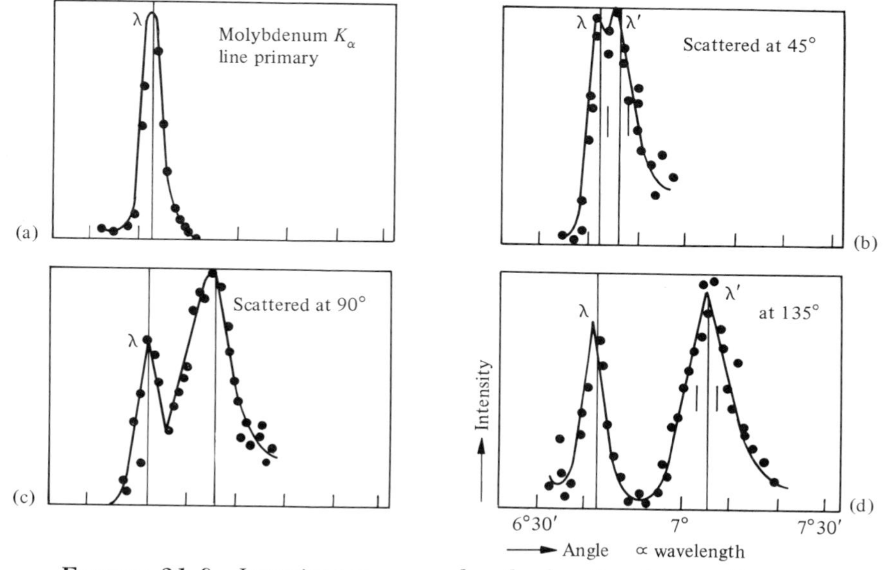

FIGURE 31.8. *Intensity versus wavelength of scattered x-rays (Compton scattering) at different angles. The modified wavelength λ' and unmodified wavelength λ are both shown.* [From A. H. Compton, *Phys. Rev.*, **22**, 411 (1923).]

recoil electron. If we apply the laws of conservation of momentum and energy to this collision as discussed in Chapter 7, we may write for momentum conservation

$$\frac{h\nu}{c} = \frac{h\nu'}{c}\cos\theta + p_e\cos\phi \tag{31.18}$$

$$0 = \frac{h\nu'}{c}\sin\theta - p_e\sin\phi \tag{31.19}$$

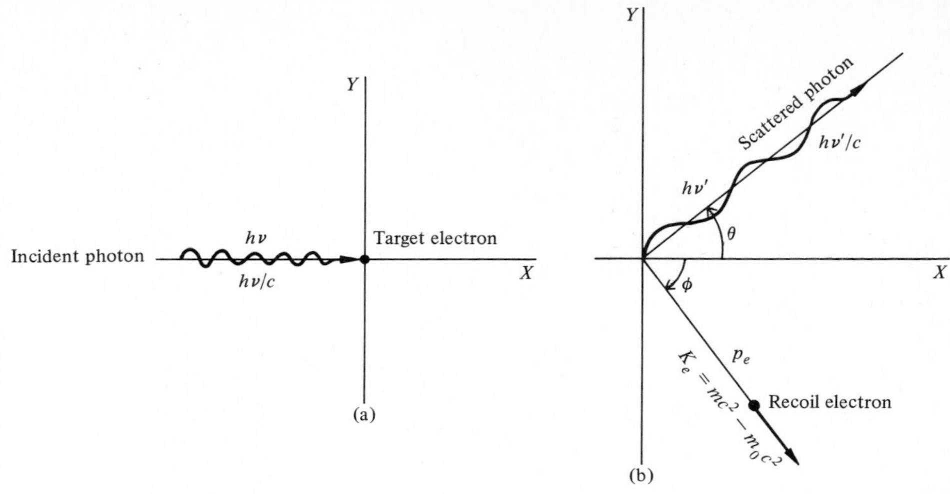

FIGURE 31.9. *The scattering of x-rays may be treated as a collision between an incident photon and a free electron at rest: (a) before collision and (b) after collision.*

and for energy conservation

$$h\nu = h\nu' + K_e \tag{31.20}$$

where

$$K_e = mc^2 - m_0c^2 = m_0c^2\left(\frac{1}{\sqrt{1 - v^2/c^2}} - 1\right) \tag{31.21}$$

and

$$p_e = mv = \frac{m_0 v}{\sqrt{1 - v^2/c^2}} \tag{31.22}$$

Equations 31.18, 31.19, and 31.20 contain four variables, θ, ϕ, $h\nu'$, and K_e. If we eliminate one, we can determine the relation between the other three. A detailed simplification gives the energy of the scattered photon to be

$$\boxed{h\nu' = \frac{h\nu}{1 + (h\nu/m_0c^2)(1 - \cos\theta)}} \tag{31.23}$$

where θ is the scattering angle of the scattered photon. By using the relation $\lambda\nu = \lambda'\nu' = c$ in the above equation, the change in wavelength $\Delta\lambda = \lambda' - \lambda$ is given by

$$\boxed{\Delta\lambda = \lambda' - \lambda = \frac{h}{m_0c}(1 - \cos\theta)} \tag{31.24}$$

Since $(1 - \cos\theta)$ is always $\geqslant 0$, $\Delta\lambda \geqslant 0$; that is, the modified wavelength λ' is always longer than the incident wavelength λ. It is important to note that

$\Delta\lambda = \lambda' - \lambda$ depends only on the rest mass m_0 and the scattering angle θ. Substituting the values of h, m_0, and c, where $h/m_0c = 0.02426$ Å is called the *Compton wavelength,* we obtain

$$\lambda' = \lambda + 0.02426(1 - \cos\theta) \text{ Å} \tag{31.25}$$

This agrees with the experimental values, as shown in Figure 31.8.

The presence of the unmodified wavelength can be explained if we consider Compton scattering between the incident photon and the whole atom. In this case m_0 must be replaced by M. In the case of the lightest atom (hydrogen), M is 1836 times m_0; hence, the Compton wavelength h/Mc is 0.000013 Å or smaller. Thus, $\Delta\lambda$ is negligible for any value of θ. This explains the presence of unmodified wavelengths together with modified wavelengths at all the angles as shown in Figure 31.9.

We have clearly demonstrated the validity of the quantum hypothesis. The quantum effects come into existence because of the finite value of h. For example, if h were almost equal to zero, that is, $h \to 0$, from Equation 31.24, λ' is almost equal to λ, $\lambda' \to \lambda$. Similarly, if m_0 were very large, that is, $m_0 \to \infty$, once again from Equation 31.24, λ' is almost equal to λ. This implies that in the macroscopic region in which h is very small compared to other quantities, that is, $h \to 0$ quantum theory reduces to classical theory. That is,

$$\lim_{h\to 0 \text{ or } m_0\to\infty} \text{quantum theory} \longrightarrow \text{classical theory}$$

But as far as the microscopic region is concerned, classical theory must be replaced by the quantum theory. Of course, this means that h must be reduced to zero in the macroscopic region; hence, the quantum theory must approach the classical theory.

31.4. de Broglie's Hypothesis

Let us briefly summarize the features of the theories of electromagnetic radiation. As we have shown in Chapter 29, *wave theory* can explain interference, diffraction, and polarization experiments. On the other hand, the quantum hypothesis or quantum theory as discussed in this chapter can successfully explain such experiments as blackbody radiation, photoelectric effect, x-ray spectra, and Compton scattering. Physicists accepted this dual nature—wave and particle (quantum)—of electromagnetic radiation until 1924.

Instead of solving this problem of the dual nature of electromagnetic radiation, French physicist Louis de Broglie in 1924 suggested in his doctoral dissertation that this dual nature be extended to the particles as well. That is, just like electromagnetic radiation, particles such as electrons, protons, and neutrons also have a dual nature—wave and particle. To unify these ideas, he suggested the following:

de Broglie's hypothesis:

1. ***The motion of a particle of momentum p is guided by a wave of wavelength λ given by***

$$\boxed{\lambda = \frac{h}{p}} \tag{31.26}$$

2. The square of the amplitude of a wave of wavelength λ at a given point in space is proportional to the probability of finding a particle of momentum p at that point where

$$\boxed{p = \frac{h}{\lambda}} \tag{31.27}$$

In both equations, p and λ are related through Planck's constant h.

The waves that guide the motion of the particles are called *de Broglie waves* or *matter waves.* We have already proved the dual nature of electromagnetic radiation. What remains to be proved is that, depending upon the circumstances, particles may behave like waves; that is, we must show that particles can also exhibit interference and diffraction phenomena. We do this in the next section.

EXAMPLE 31.4 Using the de Broglie hypothesis, calculate the de Broglie wavelengths associated with the following: (a) a stone of mass 50 g moving at a speed of 5 m/s; (b) a neutron moving with a speed of 3000 m/s; and (c) an electron with a kinetic energy of 100 eV.

(a) Substituting for $p = mv$ in Equation 31.26,

$$\lambda = \frac{h}{p} = \frac{h}{mv} = \frac{6.625 \times 10^{-34}\ \text{J-s}}{(50 \times 10^{-3}\ \text{kg})(5\ \text{m/s})} = 2.65 \times 10^{-33}\ \text{m}$$

too small a wavelength to be measured by any instrument.
(b) Once again substituting for $p = mv$ in Equation 31.26, where m is the mass of the neutron equal to 1.675×10^{-27} kg, we get

$$\lambda = \frac{h}{p} = \frac{h}{mv} = \frac{6.625 \times 10^{-34}\ \text{J-s}}{(1.675 \times 10^{-27}\ \text{kg})(3000\ \text{m/s})} = 1.32 \times 10^{-10}\ \text{m}$$

a wavelength that is measurable by x-ray spectroscopy.
(c) To use Equation 31.26, we must express momentum p in terms of the kinetic energy K (given to be 100 eV). Using the relation $K = \frac{1}{2}mv^2 = p^2/2m$ or $p = \sqrt{2mK}$, substituting this in Equation 31.26, and remembering that $1\ \text{eV} = 1.602 \times 10^{-19}$ J, we get

$$\lambda = \frac{h}{p} = \frac{h}{\sqrt{2mK}} = \frac{6.625 \times 10^{-34}\ \text{J-s}}{\sqrt{2(9.11 \times 10^{-31}\ \text{kg})(100\ \text{eV})(1.602 \times 10^{-19}\ \text{J/eV})}}$$
$$= 1.23 \times 10^{-10}\ \text{m} = 1.23\ \text{Å}$$

which is easily measurable.

EXERCISE 31.4 Calculate the de Broglie wavelength associated with (a) a football of mass 250 g moving with a speed of 20 m/s; and (b) an electron moving with a kinetic energy of 100 eV. [*Ans.:* (a) $\lambda = 1.33 \times 10^{-34}$ m; (b) $\lambda = 1.23$ Å.]

31.5. Particles Behaving as Waves

In order to establish the wave behavior of particles such as electrons, protons, neutrons, and helium atoms, we must show that beams of such particles will produce interference patterns. To be more specific, let us consider a beam of electrons that is obtained by accelerating electrons through a potential difference V. The electrons will acquire a kinetic energy $K = eV$. If m is the mass and v the velocity of the electrons, we may write

$$K = \tfrac{1}{2}mv^2 = eV \tag{31.28}$$

or

$$v = \sqrt{\frac{2eV}{m}} \tag{31.29}$$

Thus, the de Broglie wavelength of these electrons, remembering that $p = mv$, is

$$\lambda = \frac{h}{p} = \frac{h}{mv} = \frac{h}{\sqrt{2meV}} \tag{31.30}$$

Substituting the values of h, m, and e, we obtain

$$\boxed{\lambda = \sqrt{\frac{150\ \text{V}}{V}} \times 10^{-10}\ \text{m} = \sqrt{\frac{150\ \text{V}}{V}}\ \text{Å}} \tag{31.31}$$

where V must be expressed in volts. Thus, if $V = 150$ V, $\lambda = 10^{-10}$ m $= 1$ Å.

In 1927, C. J. Davisson and L. H. Germer, working at the Bell Telephone Laboratories, confirmed the existence of waves associated with electrons. In the same year, diffraction patterns, produced by a beam of fast-moving electrons incident on a thin foil, were obtained by G. P. Thomson (son of J. J. Thomson). Since then, many experiments have been performed with different particles. As an example, diffraction patterns of electrons incident on a metallic foil and that of x-rays of the same energies are shown in Figure 31.10. The similarity between the two reveals that electrons as well as x-rays are behaving as waves in such experiments.

31.6. Statistical Interpretation

We have been talking about matter waves, that is, the waves that guide the motion of the particles. We can calculate the wavelength of matter waves by using the de Broglie hypothesis, but we have not assigned any physical significance to these waves. When we talk about photons, we know that we are talking about electromagnetic waves associated with them. But what can we say about matter waves?

We know (see Chapter 17) that a typical wave motion along the X-axis is described by an equation of the form

$$y(x, t) = A \sin(kx - \omega t) \tag{31.32}$$

where $y(x, t)$ may be the displacement at x of water waves, waves on strings, or

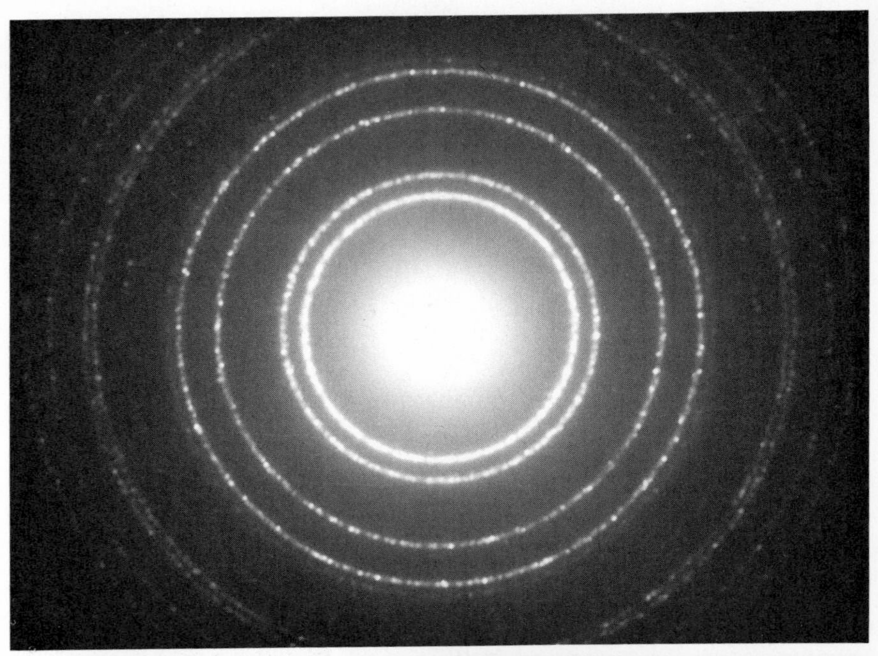

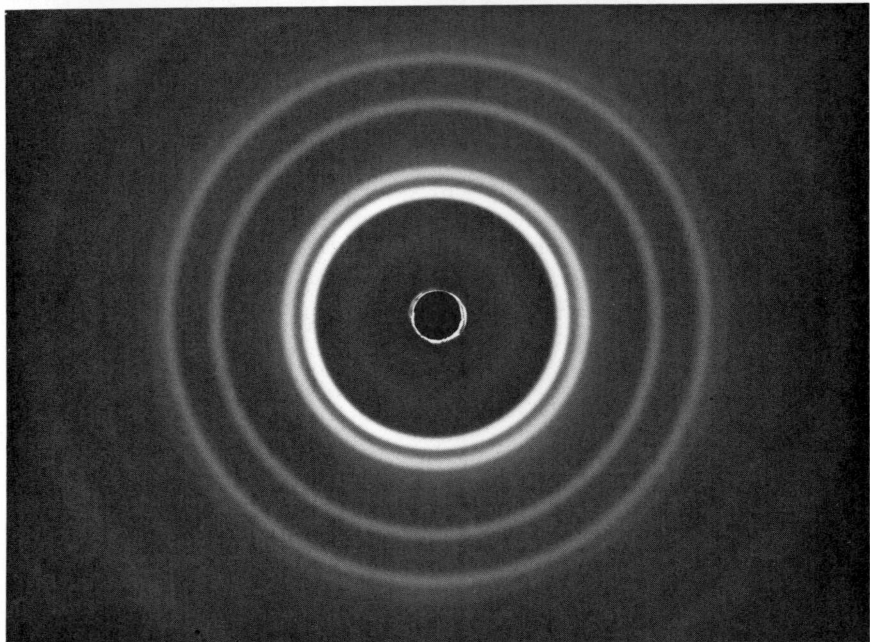

FIGURE 31.10. *Similarity between diffraction patterns produced by 600-eV electrons from an aluminum-foil target and x-rays of wavelength 0.71 Å, also from an aluminum-foil target, demonstrates the wave nature of electrons.* (Courtesy of Education Development Center, Inc., Newton, Mass.)

the electric field $E(x, t)$ of an electromagnetic wave. A is the amplitude, $k(= 2\pi/\lambda)$ is the propagation constant, and $\omega(= 2\pi\nu)$ is the angular frequency. In analogy to Equation 31.32, we may describe matter waves by an equation of the form

$$\Psi(x, t) = A \sin(kx - \omega t) \tag{31.33}$$

where $\Psi(x, t)$ is called the *wave function* of matter waves which guide the motion of the particle. Equation 31.33 is the simplest form of a wave function; in general, these wave functions have very complex forms. In 1926, Max Born, by analogy between $E(x, t)$ and $\Psi(x, t)$, gave the following statistical interpretation of the wave function $\Psi(x, y, z, t)$. *The product*

$$|\Psi(x, y, z, t)|^2 \, \Delta v \tag{31.34}$$

is a measure of the probability of observing a particle in a small volume element $\Delta v = \Delta x \Delta y \Delta z$ *at time* t.

For a stationary system, that is, a system that does not change with time t, the time-independent wave function is $\psi(x, y, z)$ and the probability of observing a particle in a volume element Δv at (x, y, z) is given by

$$|\psi(x, y, z)|^2 \, \Delta v \tag{31.35}$$

If we are dealing with a particle moving along the X-axis, its wave function according to Equation 31.33 is

$$\psi(x) = A \sin kx \tag{31.36}$$

and the probability of finding the particle between x and $x + \Delta x$ is

$$|\psi(x)|^2 \, \Delta x$$

Thus, the probability depends upon the magnitude of $\psi(x)$; the larger the wave function, the larger will be the probability.

Let us now demonstrate this statistical property (assigning probabilities) associated with material particles. Consider the results of a typical double-slit experiment. Let us replace the screen by a photographic plate and the source S by a very weak source of photons or electrons. Figures 31.11(a) to (d) show the patterns obtained on the photographic plates with increasing time exposures, (a) being the shortest and (d) being the longest. For short exposures, as is clear from Figure 31.11(a), the quantum nature of the phenomenon is quite clear. For very long exposure, the fluctuations cancel out the quantum effects and we obtain a typical double-slit pattern, as was shown in Figure 29.2(c). Whenever the intensity is large, $|\Psi|^2 \, \Delta v$ is large. Also, we may remark that only $|\Psi|^2$ has any physical meaning, not Ψ itself.

*31.7. Heisenberg's Uncertainty Principle

In the previous sections, we have been trying to combine the wave and particle characteristics of matter (material particles as well as radiation) into one by introducing the concept of wave function. As we shall see, this has led to many advantages in interpreting physical phenomena and mathematical interpretation. But these advantages are not without sacrifices. For example,

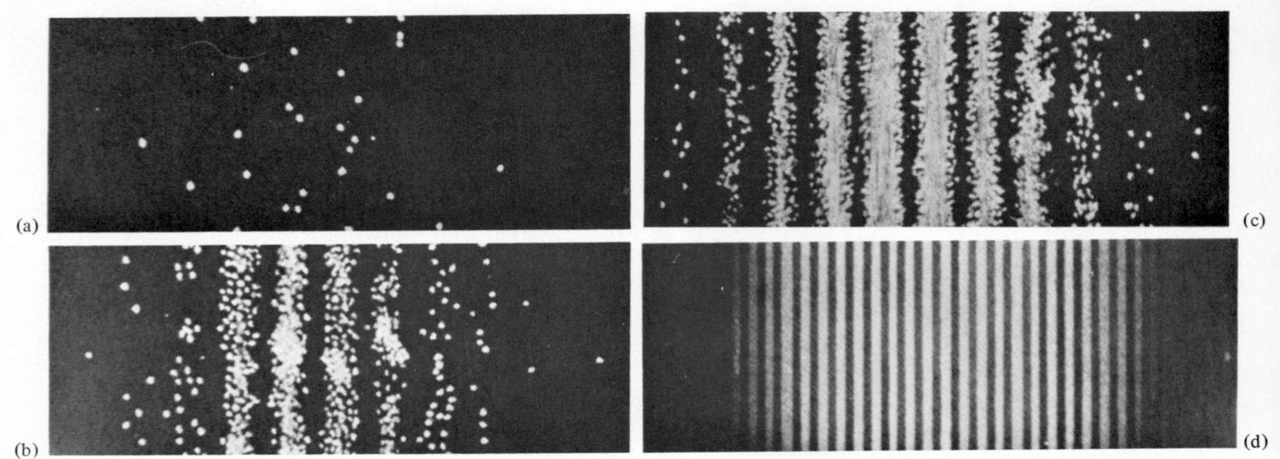

FIGURE 31.11. *Growth of a double-slit pattern on a photographic film when a beam of electrons or photons is incident on a double-slit system. The pattern obtained when the film is struck by (a) 28 electrons, (b) 1000 electrons, and (c) about 10,000 electrons. (The dots are much bigger than actual size.) Note that there are no dots in the region of interference minimia. (d) Double-slit pattern obtained when the film is exposed to millions of electrons or photons.* [Part (c) from Elisha R. Huggins, *Physics I*, copyright © 1968 by W. A. Benjamin, Inc., Menlo Park, Calif.]

the interpretation of $|\psi(x)|^2\,\Delta x$ as a probability of locating a particle between x and $x + \Delta x$ leads to uncertainties in the simultaneous measurement of its position x and linear momentum p, as explained below.

Consider a single slit of width d as shown in Figure 31.12. A beam of electrons starting from source S travels along the Y-axis and is diffracted by the slit. The uncertainty Δx in the position of an electron passing through the slit is equal to its width; that is,

$$\Delta x = d \tag{31.37}$$

Meanwhile the matter wave guiding the motion of the electron is disturbed while passing through the slit, hence produces a diffraction pattern on the screen instead of a single bright spot at the center. Any electron passing through the slit has the greatest chance of traveling within a cone of angle 2θ formed by the central maximum. Thus, if the initial momentum of the electron along the Y-axis is p, after passing through the slit, uncertainty Δp_x $(= p_x)$ in its momentum along the X-axis will be

$$\Delta p_x \approx p \sin\theta \tag{31.38}$$

But according to the theory of diffraction by a single rectangular slit, $d \sin\theta = \lambda$; that is, $\sin\theta = \lambda/d$. Substituting for $\sin\theta$ and $p = h/\lambda$ in Equation 31.38, we obtain

$$\Delta p_x \approx \frac{h}{\lambda}\frac{\lambda}{d} = \frac{h}{d}$$

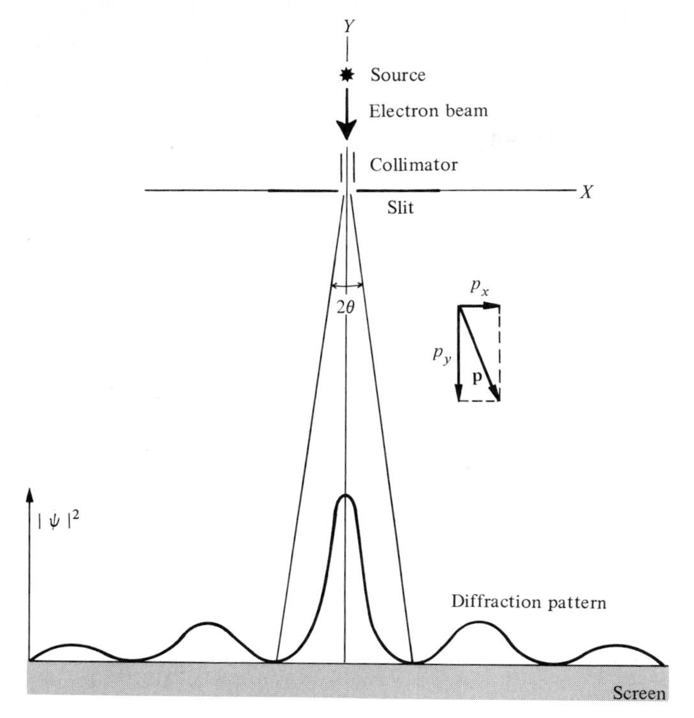

FIGURE 31.12. *Diffraction pattern of a beam of monoenergetic electrons incident on a single slit of width d.*

Combining this with Equation 31.37,

$$\Delta x \, \Delta p_x \simeq h \tag{31.39}$$

In more refined calculations h is replaced by $\hbar = h/2\pi$, and since h is the lowest limit, we may write Equation 31.39 in a general form as

$$\boxed{\Delta x \Delta p_x \geqslant \hbar} \tag{31.40}$$

Thus, if $\Delta x = 0$, $\Delta p_x = \infty$; that is, if we know the exact location of a particle, we will never know its momentum with any certainty. Similarly, if $\Delta p_x = 0$, $\Delta x = \infty$; that is, if the momentum of the particle is known precisely, its position is not known. Equation 31.40 is the *Heisenberg Uncertainty Principle* (given by Werner Heisenberg in 1927) and may be stated as follows:

> **Heisenberg's Uncertainty Principle:** ***It is impossible to know (or measure) precisely and simultaneously both the position and the momentum of a particle.***

Two other sets of quantities that cannot be measured precisely and simultaneously are

$$\Delta\theta \, \Delta L_\theta \geqslant \hbar \tag{31.41}$$

and

$$\Delta t \, \Delta E \geqslant \hbar \tag{31.42}$$

where $\Delta\theta$ and ΔL_θ are the uncertainties in the angular position and angular momentum of a particle, while Δt and ΔE are the time and energy uncertainties of a particle in a given state.

*31.8. Schrödinger's Wave Equation—Applications

The development and verification of the quantum hypothesis requires that the mathematical structures of classical or Newtonian mechanics described by Newton's laws and electromagnetic field theory described by Maxwell's equations must be replaced by a mathematical structure that is common to both. Wave mechanics or quantum mechanics, as proposed and developed by Schrödinger in 1926, fits the requirements.

Just as there is no theoretical derivation for Newton's laws, the same is true for the Schrödinger wave equation (SWE) as well. We shall not go into the development of the Schrödinger wave equation (SWE), but will describe some consequences of its application to certain situations.

Particle in a One-Dimensional Well

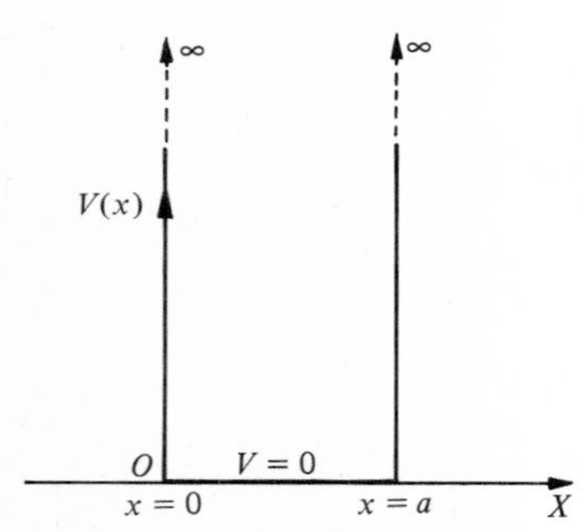

Figure 31.13. *One-dimensional infinite potential well. The motion of a particle inside the well is restricted between x = 0 and x = a by the infinite reflecting walls of the well.*

Let us consider a single particle whose motion is restricted by the reflecting walls of a one-dimensional infinite potential well (or barrier) as shown in Figure 31.13 and given by

$$V(x) = 0 \qquad \text{for} \qquad 0 < x < a$$
$$V(x) = \infty \qquad \text{for} \qquad x < 0 \text{ and } x > a$$

The total energy of the particle is all kinetic; hence, $E = K = \frac{1}{2}mv^2 = p^2/2m$ or $p = \sqrt{2mE}$. According to de Broglie's hypothesis, the wavelength associated with this particle is

$$\lambda = \frac{h}{p} = \frac{h}{\sqrt{2mE}} \tag{31.43}$$

while the wave function of this particle, from Equation 31.36, is

$$\psi(x) = A \sin kx \tag{31.44}$$

where $k = 2\pi/\lambda = 2\pi\sqrt{2mE}/h = \sqrt{2mE}/\hbar$. Substituting in Equation 31.44,

$$\psi(x) = A \sin \frac{\sqrt{2mE}}{\hbar} x \tag{31.45}$$

But $\psi(x)$ must be zero at the boundary (so that the particle does not go outside the well); that is, at $x = a$

$$\psi(a) = A \sin \frac{\sqrt{2mE}}{\hbar} a = 0$$

But A cannot be zero, because it results in a trivial solution. Thus,

$$\sin \frac{\sqrt{2mE}}{\hbar} a = 0$$

which is possible only if

$$\frac{\sqrt{2mE}}{\hbar} a = n\pi$$

or, after replacing E by E_n, we have

$$\boxed{E_n = \frac{n^2\pi^2\hbar^2}{2ma^2}} \quad \text{where } n = 1, 2, 3, \ldots \tag{31.46}$$

and n is called the *quantum number.*

Substituting for $n = 1, 2, 3, 4, \ldots$ gives the possible discrete energies that a particle may take when inside an infinite potential well, that is, $E_1 = \pi^2\hbar^2/2ma^2, E_2 = 4E_1, E_3 = 9E_1, \ldots$ for $n = 1, 2, 3, \ldots$, respectively, as shown in Figure 31.14. In contrast, classical theory yields a continuous range of energy values for the particle.

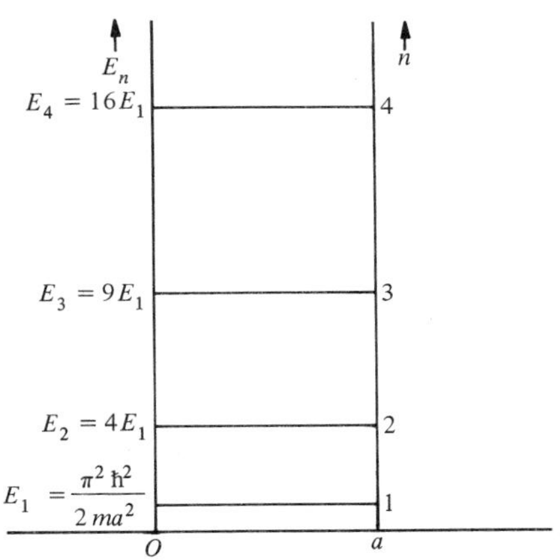

FIGURE 31.14. *Possible energy levels of a particle inside an infinite potential well according to quantum theory.*

Simple Harmonic Oscillator

According to classical theory, a particle executing simple harmonic motion obeys the rule $F = -kx$ and its potential energy is given by $V(x) = \frac{1}{2}kx^2$. When a particle is in such a parabolic potential well, as shown in Figure 31.15, it executes simple harmonic motion with any energy from zero to any positive value. But this is not true if we apply quantum mechanics. Only discrete energy values are possible. For a one-dimensional linear oscillator the allowed energy levels are given by

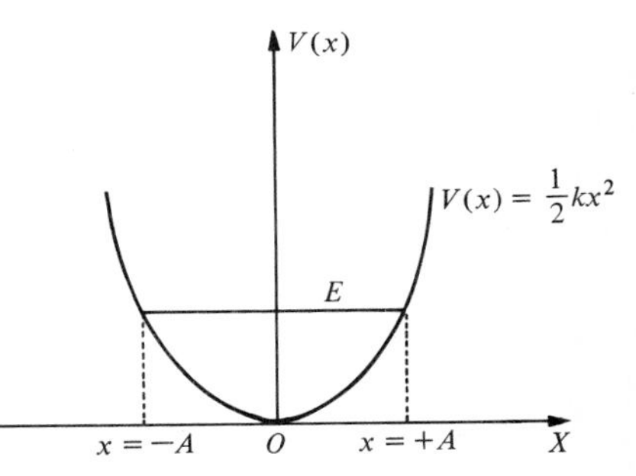

FIGURE 31.15. *Plot of parabolic potential $V(x) = \frac{1}{2}kx^2$ versus x for simple harmonic oscillator.*

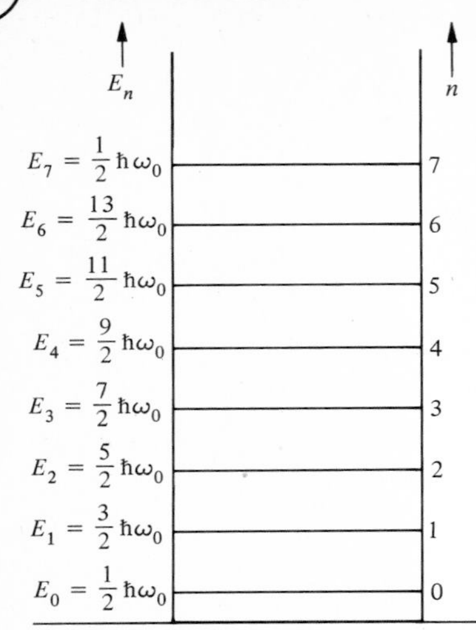

FIGURE 31.16. *Energy levels of a simple harmonic oscillator according to quantum mechanics.*

$$E_n = (n + \tfrac{1}{2})\hbar\omega \qquad \text{where } n = 0, 1, 2, 3, \ldots \tag{31.47}$$

and are schematically represented in Figure 31.16. Thus, the energy levels of an oscillator are $\frac{1}{2}\hbar\omega$, $\frac{3}{2}\hbar\omega$, $\frac{5}{2}\hbar\omega$, ...

An important observation may be made from Equation 31.47. If $n = 0$, $E_0 = \frac{1}{2}\hbar\omega$. That is, the lowest possible energy of a linear oscillator is not zero but $\frac{1}{2}\hbar\omega$. Also, for very large values of n, there is hardly any difference between E_{n+1} and E_n, and the quantum-mechanical results will agree with classical mechanics.

Finite Potential Barriers

Another situation of interest is shown in Figure 31.17, where a particle is incident on a rectangular barrier such that the energy of the particle $E < V_0$. According to classical theory, since $E < V_0$, the particle in region I can never penetrate the barrier (region II) and appear in region III. But according to quantum mechanics, the transmission is possible; that is, a small fraction of the particles incident from the left will cross the barrier and appear in region III. A similar situation occurs in optics also, as shown in Figure 31.18, and is demonstrated in the case of water waves in Figure 31.19.

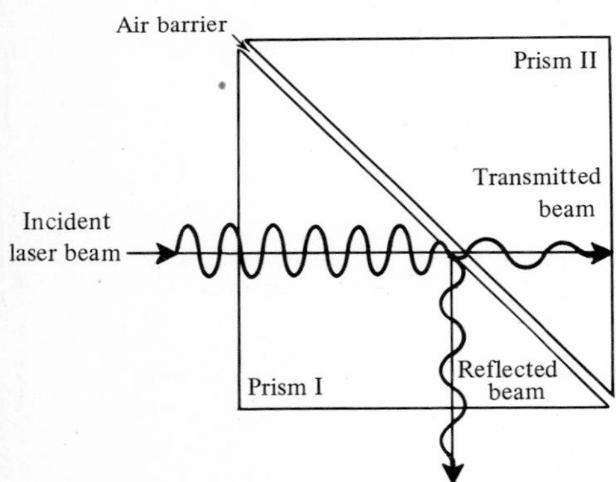

FIGURE 31.18. *Phenomenon of penetration of an optical barrier. Two 45° prisms I and II are placed as shown, with a thin layer of air as a barrier between the two. When a laser beam is incident on prism I, even though the angle of incidence is greater than the critical angle, not all the light is reflected. Part of the incident beam penetrates the air barrier and appears as a transmitted wave in prism II.*

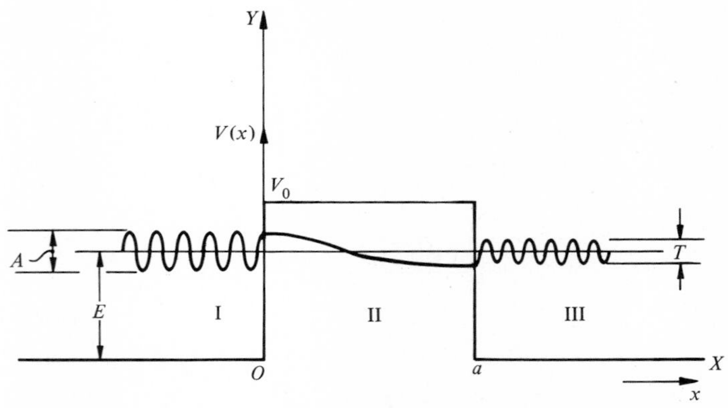

FIGURE 31.17. *Particle of energy E and amplitude A incident on a square-barrier potential of height V_0 ($E < V_0$) and width a. According to the classical theory, there should be no transmission through the barrier.*

SUMMARY

According to *Planck's hypothesis*, if a system executes a simple harmonic motion in one dimension with frequency ν, it can take only those energy values that are given by the relation $E = nh\nu$, $n = 1, 2, 3, \ldots$. Such states are called *quantum states* and n is called the *quantum number*. The system emits or absorbs energy in bundles of size $h\nu$, called a *photon*. Planck's hypothesis was applied in explaining the blackbody radiation spectrum, the photoelectric effect, x-ray spectra, Compton scattering, and optical line spectra.

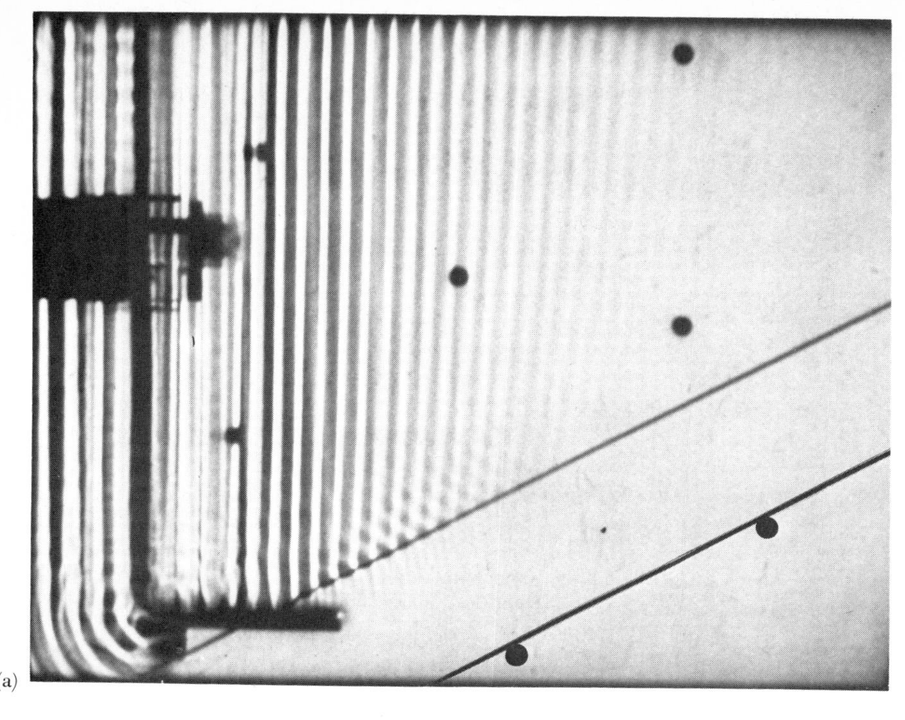

(a)

(b)

FIGURE 31.19. *Penetration of a barrier by water waves in a ripple tank. (a) Incident waves are totally reflected from a very wide gap of deep water. (b) Incident waves are partly transmitted when the deep water gap is very narrow.* (Courtesy Education Development Center, Inc., Newton, Mass.)

Photons are particles of electromagnetic radiation that have the characteristics $E = h\nu = hc/\lambda = 1.24 \times 10^4$ Å-eV$/\lambda$, $m = h\nu/c^2$, and $p = h\nu/c = h/\lambda$.

The ejection of electrons from a metal surface by the action of light (or electromagnetic radiation) is called the *photoelectric effect*. The potential for which no electrons reach the collector plate is called the *stopping potential* V_0; $K_{max} = \frac{1}{2}mv_{max}^2 = e|V_0|$. In general, $K_{max} = h\nu - w$, where w is the work function of the metal.

The process of radiation being produced by an accelerated charged particle is called *bremsstrahlung*—slowing-down radiation, $K = eV = h\nu_{max}$. The x-rays of minimum wavelength emitted are $\lambda_{min} = [1.24 \times 10^{-6}/V]$ V-m.

Compton scattering concerns the scattering of x-rays by free electrons. $h\nu' = h\nu/[1 + (h\nu/m_0c^2)(1 - \cos\theta)]$ and $\Delta\lambda = \lambda' - \lambda = (h/m_0c)(1 - \cos\theta)$, where λ and λ' are the unmodified and modified wavelengths and $h/m_0c =$ 0.02426 Å $=$ the Compton wavelength of the electron.

According to the *de Broglie hypothesis:* (1) if there is a particle of momentum p, its motion is associated with a wave of wavelength $\lambda = h/p$; and (2) if there is a wave of wavelength λ, the square of the amplitude of the wave at any point is proportional to the probability of observing a particle of momentum $p = h/\lambda$ at that point. The wavelength of a charged particle that has been accelerated through a potential difference V is $\lambda = \sqrt{150\text{ V}/V}$ Å.

The dual nature of particles is revealed in the microscopic region. The waves that guide the motion of particles are called *matter waves*. The wave function associated with a material particle is $\Psi(x, t) = A \sin(kx - \omega t)$. The wave function is a quantity such that the product $|\Psi(x, y, z, t)|^2\,\Delta v$ is the probability of observing a particle in a volume element $\Delta v = \Delta x\,\Delta y\,\Delta z$ at time t.

*According to the *Heisenberg uncertainty principle,* it is impossible to measure simultaneously and precisely both the position and the momentum of a particle, $\Delta x\,\Delta p_x \geqslant \hbar$. Also, $\Delta\theta\,\Delta L_\theta \geqslant \hbar$ and $\Delta t\,\Delta E \geqslant \hbar$.

* The *Schrödinger wave equation* plays the same role in quantum mechanics as Newton's laws play in mechanics and Maxwell's equations in electromagnetism. A particle in a well can have energies $E_n = n^2(\pi^2\hbar^2/2ma^2)$, where $n = 1, 2, 3, \ldots$. A simple harmonic oscillator can have energies $E_n = (n + \frac{1}{2})\hbar\omega$, where $n = 0, 1, 2, 3, \ldots$. According to classical mechanics, a particle with energy E cannot cross a barrier of height V_0 if $E < V_0$, but quantum mechanically there is some probability that it will cross.

QUESTIONS

1. What characteristics are common to both an electron and a photon?
2. Is it possible for two photons to have the same wavelengths but different energies?
3. Suppose that the work functions for all metals were the same. Could you still use the photoelectric effect to verify Planck's hypothesis?
4. Light of one wavelength shines on a surface. Why don't all the photoelectrons emitted have the same kinetic energy?
5. The threshold for electron ejection from surface A is higher than for surface B. Which has a higher work function?
6. Why is Compton scattering observed for x-rays and not for visible light waves?

7. Describe different methods (including the methods discussed in this chapter) for removing electrons from a metal surface.
8. Can x-rays be reflected, refracted, diffracted, and polarized just like any other waves? Explain.
9. Why do matter waves always manifest themselves in the microscopic region only? Explain.
10. Which has a matter wave of larger wavelength; (1) an electron of large energy or small energy; (2) an electron or a proton both having the same energy; or (3) an electron, proton, or neutron all having the same speed?
11. White light consists of discrete bundles of energy called photons. Why can't an eye detect this discontinuous nature of the beam?
12. How does the interference of photons differ from that of electrons?
*13. Is it possible to construct an instrument that could precisely measure the momentum and position of a particle, that is, $\Delta p = 0$, $\Delta x = 0$?
*14. Why is it not possible to detect the discrete nature of energy in a simple pendulum that executes simple harmonic motion?

PROBLEMS

1. What are the energies in eV of quanta of wavelengths $\lambda = 4000, 5500$, and 7000 Å?
2. Calculate the wavelengths of quanta of radiation of energies 0.1 eV, 1 eV, 10 eV, 100 eV, 1 keV, and 1 MeV.
3. For a normal eye to be sensitive to visible light it must receive at least 100 photons per second. Assuming that the wavelength of visible light is 5550 Å, how many watts of power are received by the eye when this light is barely visible?
4. Calculate the wavelength associated with photons of energies 1×10^{-19} J and 5×10^{-19} J. What are the energies of these photons in eV?
5. Calculate the mass, momentum, and wavelength of photons of energies (a) 1 eV; (b) 1 keV; and (c) 1 MeV. Calculate the corresponding frequencies.
6. Calculate the energy, momentum, and frequency of photons in the visible region of the electromagnetic spectrum.
7. For a certain surface the threshold of the photoelectrons is 5420 Å. Calculate the work function in electron volts.
8. The work function of a surface is 2.5 eV. What is the maximum wavelength of incident photons that will cause photoelectron emission?
9. When light of wavelength 3132 Å falls on a cesium surface, a photoelectron is emitted for which the stopping potential is 1.98 V. Calculate the maximum energy of the photoelectrons, the work function, and the threshold frequency.
10. The work function of tungsten is 4.53 eV. If ultraviolet light of wavelength 1500 Å is incident on this surface, does it cause photoelectron emission? If so, what is the maximum kinetic energy and velocity of the emitted electrons?
11. The work function of potassium is 2.20 eV. What should be the wavelength of the incident electromagnetic radiation so that the photoelectrons emitted from potassium will have a maximum kinetic energy of 4 eV?
12. What voltage should be applied to an x-ray tube to produce x-rays of wavelength 0.1 Å?

13. Electrons in an x-ray tube are accelerated through a potential difference of 3000 V. If these electrons were slowed down in a target, what will be the minimum wavelength of the x-rays produced?
14. What should be the operating voltage of an x-ray tube so that the emitted x-rays will have a maximum energy of 10 keV?
15. X-rays of wavelength 1 Å are scattered by a carbon target. Calculate (a) the wavelength of the scattered x-rays at 90°; (b) the energy of the incident and scattered photons; and (c) the maximum energy of the recoiling electron for this situation.
16. X-rays with energies of 200 keV are incident on a target and undergo Compton scattering. Calculate (a) the energy of the x-ray scattered at an angle of 60° to the incident direction; (b) the energy of the recoiling electron; and (c) the wavelengths of the incident and the scattered photons in this case.
17. Calculate the de Broglie wavelength associated with the following: (a) an electron with a kinetic energy of 1 eV; (b) an electron with a kinetic energy of 510 eV; (c) a neutron moving with a velocity of 2200 m/s; and (d) an automobile of 2000 kg moving with a velocity of 60 mi/h (96.54 km/h).
18. Calculate the de Broglie wavelength associated with (a) a proton; and (b) a neutron of kinetic energy 15 MeV.
19. What should be the kinetic energy of (a) a proton; and (b) an electron so that the de Broglie wavelength associated with either will be 5000 Å?
20. Through what potential difference should an electron be accelerated so that the de Broglie wavelength associated with it is 0.1 Å?
21. Calculate the de Broglie wavelength of (a) a hydrogen molecule; and (b) an oxygen atom, both at room temperature, that is, moving with thermal velocities ($v = 2200$ cm/s).
22. What should be the kinetic energy of an electron so that its wavelength is the same as that of x-rays produced in an x-ray tube operating at 60,000 V?

*23. A particle of mass 2 g moving with a speed of 5 m/s is located within 10^{-6} m. Calculate fractional uncertainty in its momentum, that is, $\Delta p/p$.

*24. Calculate Δp and $\Delta p/p$ for an electron moving with a velocity of 10^5 m/s. The uncertainty in its position is 10^{-6} m.

*25. Suppose that an electron absorbs energy and reemits within 10^{-8} s. What is the uncertainty in its energy before reemitting?

32

Atomic Structure and Spectra

The beginning of the present century saw the start of a new branch of physics—atomic structure and spectra. The discoveries of the electron, isotopes, quantization of charge, the nuclear model of the atom and the Bohr theory of the hydrogen atom had a profound effect on revealing the inner mysteries of the structure of atoms. We shall briefly outline some of those discoveries and see how, together with quantum mechanics, they led to the rules that predict atomic spectra and established the periodic table of elements.

32.1. Isotopes and Their Measurement

Discovery of Isotopes

As was pointed out in Chapter 2, the determination of the structure of the atom has been one of the basic problems. Dalton's atomic theory, suggested by the English chemist John Dalton in 1808, according to which the atom was a basic indivisible unit of all elements, proved to be incorrect. In 1886 Sir William Crookes suggested that certain elements have integral atomic masses, while those elements that have nonintegral atomic masses must be mixtures of two or more types of atoms. This was further supported when radioactivity was discovered by the end of the nineteenth century. Investigation of radioactive elements revealed that there were many elements that were chemically identical but had different masses. As an example, the element lead is found to have three different mass numbers, 206, 207, and 208. For elements that are chemically identical but have different masses, the chemist Frederick Soddy suggested the name *isotopes* (*iso* meaning equal and *topes* meaning place).

The existence of isotopes was experimentally established by J. J. Thomson in 1910. By using electric and magnetic deflections, the first element successfully investigated was neon. He found that neon was a mixture of two isotopes of atomic masses 10 u and 22 u, with abundances of 90 and 10 percent, respectively, resulting in the following equation:

$$\text{atomic mass of neon} = \frac{(20\ \text{u} \times 90) + (22\ \text{u} \times 10)}{90 + 10} = 20.20\ \text{u}$$

All subsequent measurements have confirmed that the atomic masses of different isotopes of different elements are very close to whole numbers. The formal definition of an isotope may be stated as follows:

Atoms that have the same atomic number Z but have different atomic masses are said to be isotopes of the element with atomic number Z.

Measurement of Isotopic Masses

After the original method of measuring isotopic masses developed by J. J. Thomson, many precise and improved instruments have been developed by

F. W. Aston, Arthur J. Dempster, Kenneth Bainbridge, Alfred Niers, and many others. The basic principle is to measure the value of q/M for positive ions by using the deflections produced by electric and magnetic fields. Knowing q, the charge on the ion, one can calculate M, the mass of the ion or the isotope. Aston's mass spectrograph has an accuracy of 1 part in 10,000; while some instruments have achieved an accuracy of 1 part in 100,000. We shall briefly describe one of them.

The Dempster Mass Spectrometer

Dempster's mass spectrometer for measuring the isotopic masses and their relative intensities (relative abundances) is shown in Figure 32.1. A beam of positive ions produced at F by bombarding salts with electrons (or by heating salt on a platinum strip) is allowed to enter an electric field between plates P and Q maintained at a potential difference of V. After being collimated by the slit S_1, the beam of positive ions enters a strong magnetic field B applied perpendicular to the plane of the paper. The beam, after being bent by this field into semicircular paths, is received by an electrometer E.

If q is the charge on the positive ions of mass M, and their velocity after passing through the electric field V is v, the kinetic energy $K = \frac{1}{2}Mv^2$ acquired by the ions is equal to qV. That is,

$$qV = \tfrac{1}{2}Mv^2 \tag{32.1}$$

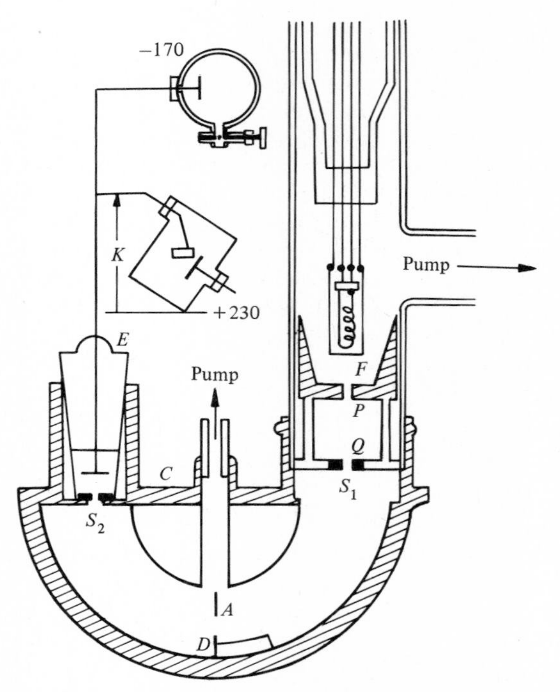

FIGURE 32.1. *Outline of an experimental arrangement of Dempster's mass spectrometer.* [From A. J. Dempster, *Phys. Rev.* **22**, 631 (1922).]

While in the magnetic field, the magnetic force qvB provides the necessary centripetal force Mv^2/R to make the ions move in a circle of radius R. Thus,

$$qvB = \frac{Mv^2}{R} \tag{32.2}$$

Eliminating v from the preceding two equations, we get

$$\boxed{\frac{q}{M} = \frac{2V}{B^2R^2}} \tag{32.3}$$

According to this relation, for fixed values of V and B, the value of R depends upon the value of q/M, as shown for three different cases in Figure 32.2. Usually, one keeps R fixed and the electric (or magnetic) field is changed to such a value that the beam is received at the slit S_2 and detected by the electrometer E. The current read by the electrometer is proportional to the number of positive ions reaching E. Each such value of V (or B) corresponds to a certain definite value of M. To measure unknown masses, the instrument is first calibrated using a standard mass.

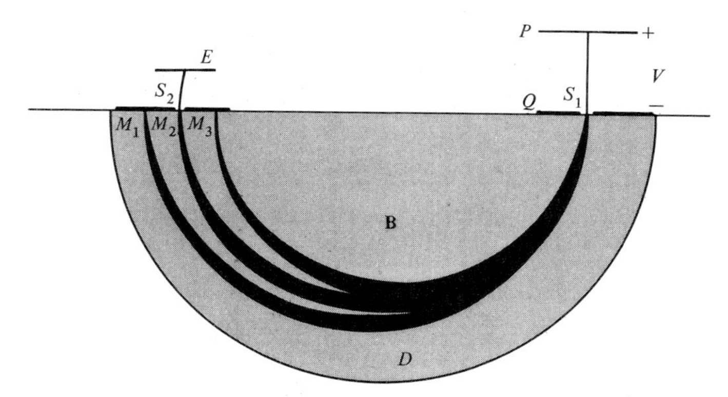

FIGURE 32.2. *Dispersion and focusing of a beam of ions of three isotopes of an element.*

A typical plot of current versus atomic weight (inversely proportional to the accelerating V or B fields) for two isotopes of potassium is shown in Figure 32.3. The relative areas under the two peaks are 18:1 and correspond to the two isotopes of atomic masses 39 and 41 units.

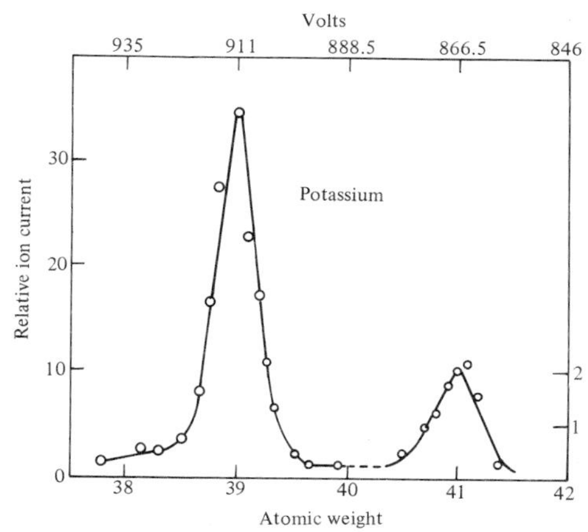

FIGURE 32.3. *Results obtained by Dempster mass spectrometer for two isotopes of potassium. The areas under the two peaks are proportional to the relative abundances of the isotopes. The small peak has been multiplied by a factor of 10.* [From A. J. Dempster, *Phys. Rev.* **22**, 631 (1922).]

EXAMPLE 32.1 A beam consisting of carbon atoms and protons is used in a Dempster mass spectrograph. In this analysis the charge on the carbon atom is the same as that on the proton. If the mass of carbon is assumed to be 12 u, what is the mass of the proton in terms of the mass of carbon and R_C and R_p?

From Equation 32.3 for carbon and a proton,

$$\frac{q}{M_C} = \frac{2V}{B^2}\frac{1}{R_C^2} \qquad \text{and} \qquad \frac{q}{M_p} = \frac{2V}{B^2}\frac{1}{R_p^2}$$

Dividing one by the other,

$$\frac{M_p}{M_C} = \frac{R_p^2}{R_C^2} \qquad \text{or} \qquad M_p = M_C\left(\frac{R_p}{R_C}\right)^2$$

Thus, knowing M_C and measuring R_C and R_p, M_p can be calculated.

EXERCISE 32.1 In Example 32.1, if $V = 4000$ V, $B = 0.5$ Wb/m^2, and $M_C = 12\text{ u} = 12 \times 1.67 \times 10^{-27}$ kg, calculate R_C. If $M_p = 1.67 \times 10^{-27}$ kg, what will be the value of R_p? [*Ans.*: 6.33 cm, 1.83 cm.]

32.2. The Charge of the Electron—Millikan Oil-Drop Method

In 1889, J. J. Thomson discovered the first subatomic particle—the electron. Hence, the electron was found to be one of the constituents of all matter. From his experiments he obtained the charge-to-mass ratio, q/m or e/m, for the electron. Measurement of e/m by Thomson revealed that the electrons coming from different materials are identical and that these have the same e/m value.

It is simplest to assume that all the electrons have the same mass m, and each electron carries the same charge e. Such an assumption implies that the charge is quantized and possibly that the smallest unit of charge is the charge of a single electron. According to this, the charge q on any other particle or body will be an integral multiple of this minimum charge e, that is,

$$q = ne \qquad \text{where } n = 1, 2, 3, \ldots \tag{32.4}$$

In order to calculate the mass m of the electron from the measured value of e/m, one must know e. In 1909, R. A. Millikan perfected a technique that resulted in precise measurement of the charge of the electron. We shall describe this technique shortly. The value of the unit electric charge found by Millikan is $e = 1.604 \times 10^{-19}$ C. The most recent and much more accurate measurements yield the following values for the charge e, mass m, and e/m for the electron:

$$\begin{aligned} e &= (1.602010 \pm 0.00007) \times 10^{-19}\text{ C} \\ m &= (9.107208 \pm 0.000\,246) \times 10^{-31}\text{ kg} \\ \frac{e}{m} &= 1.758796 \times 10^{11}\text{ C/kg} \end{aligned} \tag{32.5}$$

A schematic diagram of the Millikan oil-drop experiment is shown in Figure 32.4. Two parallel plates PP' are placed inside the container C. Enclosure in the container avoids disturbances due to air currents. The separation between the plates is d and the upper plate has a small hole H, as shown in the figure. A voltage V is applied to the plates; hence, the electric field E between the plates is $E = V/d$. An atomizer A is used for spraying oil drops into the container.

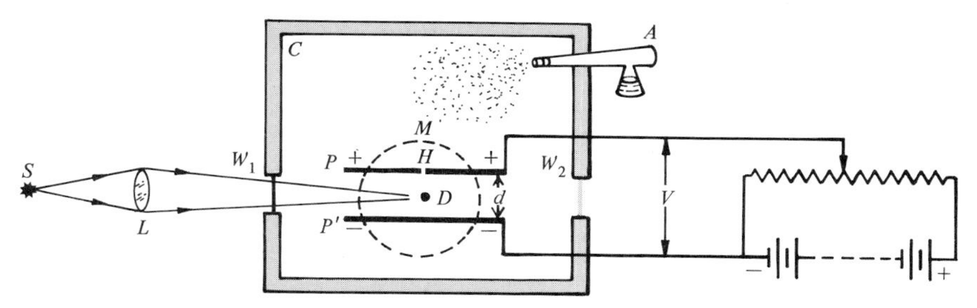

Figure 32.4. *Outline of an experimental arrangement of the Millikan oil-drop method for measuring electron charge.*

These oil drops are very small, with radii varying from 10^{-4} cm to about 6×10^{-4} cm and are actually in the form of mist. The drops produced by the atomizer are electrically charged; presumably the charge is produced as the drops rub against the walls of the atomizer. Some of these drops happen to pass through the hole in the upper plate and travel toward the lower plate. The light coming from the source S after passing through the lens L and the window W_1 illuminates the space between the plates. The motion of the drop is watched through a microscope M.

A given droplet between the two plates could be suspended in air if the gravitational force $F_g = mg$ acting on the drop was equal to the electrical force $F_E = qE$ as shown in Figure 32.5. Thus,

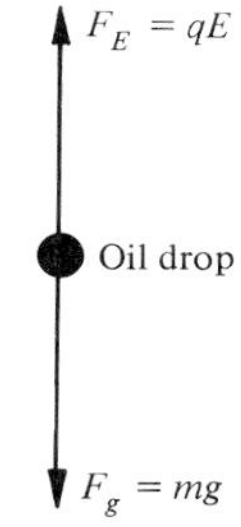

Figure 32.5. *Oil drop balanced by the gravitational force and the Coulomb force.*

$$\text{upward electrical force} = \text{downward gravitational force}$$

$$F_E = F_g \qquad \text{or} \qquad qE = mg$$

Since $E = V/d$, we may write

$$q\frac{V}{d} = mg$$

or

$$\boxed{q = \frac{mgd}{V}} \tag{32.6}$$

This gives the value of q in terms of m, g, d, and V. Millikan measured the charge on many drops and found that each charge was an integer multiple of a basic unit of charge e; that is, $q = ne$. The charge is quantized and comes in integer multiples of e. (Recently, there have been speculations about the existence of quarks, which are supposed to have one-third or two-thirds the charge of the electron. We shall not go into any details regarding these.)

32.3. Atomic Spectra

The branch of physics that deals with the investigation of the wavelengths and intensities of electromagnetic radiation emitted or absorbed by atoms is called *atomic spectra*. In general, there are three types of spectra: (1) continuous spectra, (2) band spectra, and (3) discrete (or line spectra). Blackbody

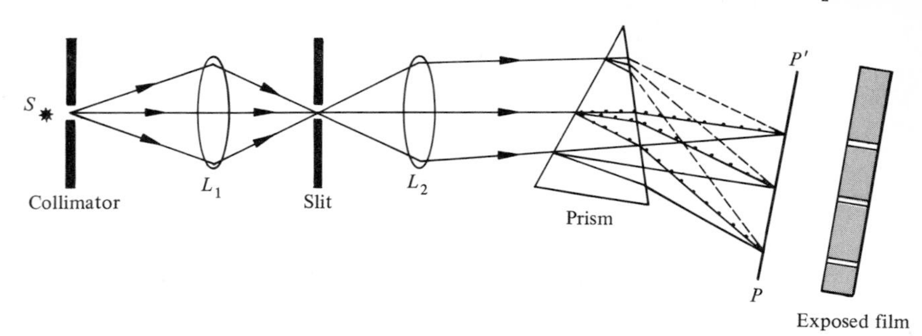

FIGURE 32.6. *Experimental arrangement for obtaining the line spectrum of an element using a prism for dispersion.*

radiation (see Chapter 15) is an example of (1); molecular spectra, to be investigated in Chapter 33, are an example of (2); and the atomic discrete spectra, we shall investigate in detail here, are an example of (3).

An experimental arrangement for observing atomic emission spectra is shown in Figure 32.6. The source S contains a small quantity of atoms in a gaseous state which are in a *normal* or *ground state*. By causing an electrical discharge, spark, or arcing, the atoms may be caused to absorb energy. (When using a solid as a source, the energy absorption process may be accomplished by heating the material.) The atom that has absorbed this energy is said to be in the *excited state*. The atoms stay in excited states no longer than 10^{-8} s, and then return to the ground state. During this transition from the excited state (higher-energy state) to the ground state (lowest-energy state), the extra energy is emitted in the form of electromagnetic radiation. This electromagnetic radiation is a characteristic of the atom, hence may be utilized to investigate atomic structure.

The electromagnetic radiation emitted by the source is analyzed by means of a diffraction grating or prism, as shown, which disperses different wavelengths by different amounts. After dispersion, the radiation is received on a photographic plate, where it leaves permanent impressions. The impressions on the plate are in the form of lines if the slit in front of the source S is a narrow rectangle. Thus, the spectrum is sometimes referred to as a *line spectrum*. If the setup on Figure 32.6 is used for wavelength measurements, it is called a *spectrometer;* and a *spectroscope* if used for only visual observation.

The fact that the spectrum of any element contains wavelengths that exhibit definite regularities was utilized in the second half of the nineteenth century in identifying different elements. These regularities were classified into certain groups called the *spectral series*. The first such series was identified by J. J. Balmer in 1885 in the spectrum of hydrogen. The series, called the *Balmer series,* is shown in Figure 32.7 and lies in the visible region of the electromagnetic spectrum.

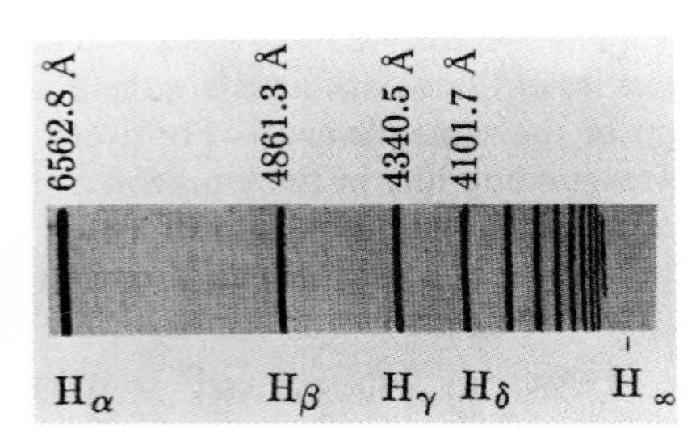

FIGURE 32.7. *Portion of the emission spectrum of hydrogen, corresponding to visible lines of the Balmer series.* (From G. Herzberg, *Atomic Spectra and Atomic Structure,* Dover Publications, New York, 1944.)

The results obtained by Balmer were expressed by J. R. Rydberg in 1896 in the following mathematical form:

$$\frac{1}{\lambda} = \bar{\nu} = R_H \left(\frac{1}{2^2} - \frac{1}{n^2} \right) \tag{32.7}$$

where $\bar{\nu}$ $(= 1/\lambda)$ is called the *wave number* and is equal to the reciprocal of the wavelength λ; $n = 3, 4, 5, 6, \ldots$; and R_H is the *Rydberg constant*. The most recent value of R_H is $109677.576 \pm 0.012 \text{ cm}^{-1}$. As we shall see, since then, many more series have been discovered and proved helpful in predicting the arrangement of the electrons in different atoms.

Before discussing theories to explain this empirical result, a word about absorption spectra is in order. The experimental arrangement for observing an absorption spectrum is shown in Figure 32.8. Source S emits radiation of all wavelengths. A beam of such radiation is made to pass through a tube containing, in the gaseous state, atoms of an element whose absorption spectrum is desired. After passing through the tube, the light when analyzed contains dark lines (absorption lines) against the background.

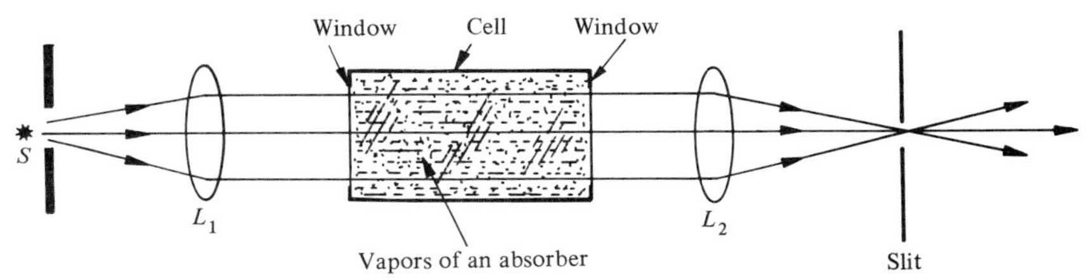

FIGURE 32.8. *Experimental arrangement for obtaining the absorption spectra of different elements. The element under investigation is in the form of vapors in the cell.*

32.4. Bohr's Theory of the Hydrogen Atom

In order to explain the empirical results obtained by Rydberg, Niels Bohr, in 1913, formulated a theory of the hydrogen atom utilizing classical physics and Planck's quantization hypothesis. This *semiclassical theory* is based on the following three postulates.

The Bohr Postulates:

Postulate I. ***An electron bound to the nucleus in an atom can move about the nucleus in certain circular orbits without radiating. These orbits are called the discrete stationary states of the atom.***

Postulate II. ***Only those stationary states are allowed for which orbital angular momentum of the electron is equal to an integral multiple of*** $\hbar$***. That is,***

$$\boxed{L = mvr = n\hbar} \tag{32.8}$$

where $n = 1, 2, 3, \ldots$ ***and n is called the* principal quantum number; *m and v are the mass and velocity of the electron; and*** $\hbar = h/2\pi$***, h being Planck's constant.***

Postulate III. ***Whenever an electron makes a transition, that is, jumps from an initial high-energy state*** E_i ***to a final lower-energy state*** E_f***, a photon of energy hν is emitted so that***

$$\boxed{h\nu = E_i - E_f} \tag{32.9}$$

where $\nu = c/\lambda$ ***is the frequency of the radiation emitted.***

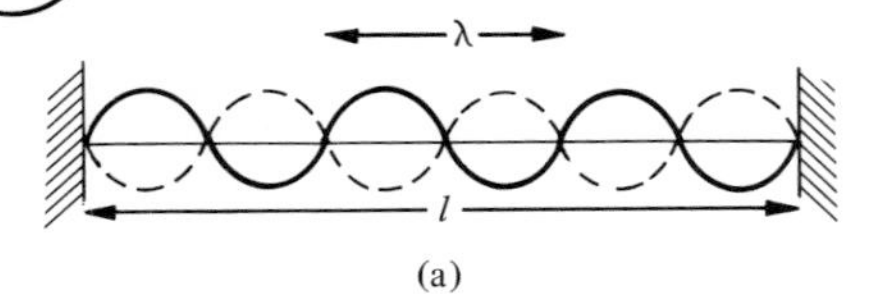

(a)

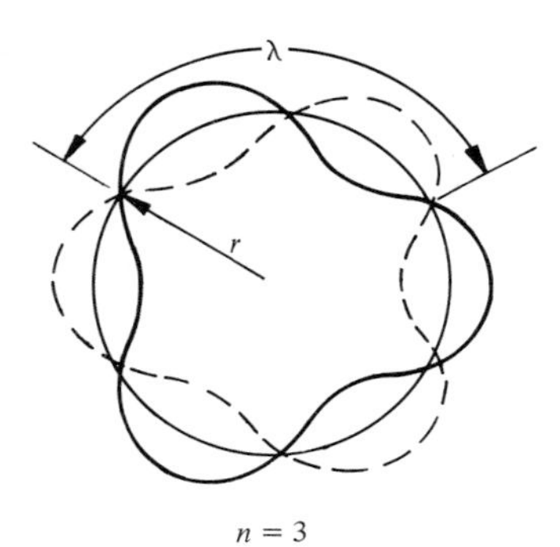

$n = 3$

(b)

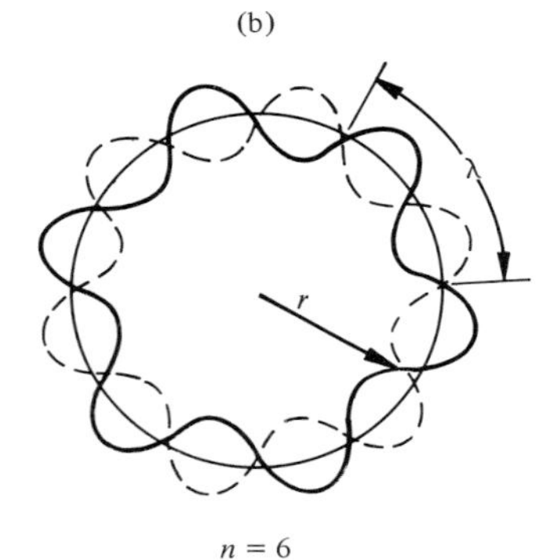

$n = 6$

(c)

FIGURE 32.9. *Formation of standing or stationary resonance vibrations (a) for $n = 3$ in a string and (b) and (c) on a circular loop for $n = 3$ and $n = 6$, respectively.*

At the time of the formulation of Bohr's theory, there was no justification for the first two postulates, while postulate III had some roots in Planck's hypothesis. Later, with the development of de Broglie's hypothesis, some justification could be seen in postulate II, as explained below.

Consider a string of length l as shown in Figure 32.9(a). If this is put into stationary vibrations, we must have $l = n\lambda$, where n is an integer. Suppose that the string is bent into a circle of radius r, as demonstrated for $n = 3$, and $n = 6$ in Figures 32.9(b) and (c), so that

$$l = 2\pi r = n\lambda \qquad \text{or} \qquad \lambda = 2\pi r/n \tag{32.10a}$$

From de Broglie's hypothesis, $\lambda = h/p = h/mv$. Therefore,

$$\frac{h}{mv} = \frac{2\pi r}{n} \qquad \text{or} \qquad mvr = \frac{nh}{2\pi} = n\hbar \tag{32.10b}$$

which is postulate II.

Energy Levels

Consider a hydrogen atom in which an electron moving with velocity v_n is in a stationary circular orbit of radius r_n. From postulate II, Equation 32.8,

$$v_n = \frac{n\hbar}{mr_n} \tag{32.11}$$

For this electron to stay in the circular orbit shown in Figure 32.10, the centripetal force $F_c = mv_n^2/r_n$ is provided by the Coulomb force $F_e = ke^2/r_n$. That is,

$$\frac{mv_n^2}{r_n} = \frac{ke^2}{r_n^2} \tag{32.12}$$

Eliminating v_n, we get (after substituting for v_n from Equation 32.11),

$$\boxed{r_n = \frac{n^2\hbar^2}{kme^2} = n^2 r_1} \tag{32.13}$$

where $r_1 = \hbar^2/kme^2$. Substituting this in Equation 32.11,

$$\boxed{v_n = \frac{1}{n}\frac{ke^2}{\hbar} = \frac{v_1}{n}} \tag{32.14}$$

where $n = 1, 2, 3, 4, \ldots$. For $n = 1$, $r_1 = \hbar^2/kme^2 = 0.528 \times 10^{-10}$ m $=$ 0.528 Å, which agrees with the experimentally measured values and is called the *first Bohr orbit radius of the hydrogen atom.* Thus, according to Bohr's theory, the radii and velocities of different stationary orbits of the electrons in the hydrogen atom are given by

$$r_n = r_1, 4r_1, 9r_1, 16r_1, \ldots \tag{32.15}$$

and

$$v_n = v_1, \frac{v_1}{2}, \frac{v_1}{3}, \frac{v_1}{4}, \ldots \tag{32.16}$$

Let us now calculate the total energy E_n of the electron in the nth Bohr orbit. E_n is the sum of the kinetic energy K and potential energy V. That is,

$$E_n = K + V = \frac{1}{2}mv_n^2 + \left(-\frac{ke^2}{r_n}\right) \tag{32.17}$$

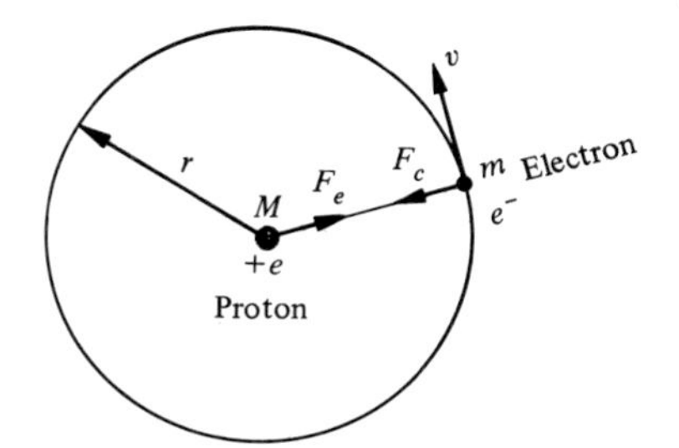

FIGURE 32.10. *For an electron moving in a circular orbit with the proton at the center, the centripetal force F_c is provided by the Coulomb force F_e; that is, $F_c = F_e$.*

Eliminating v_n and r_n by substituting their values derived above,

$$E_n = \frac{1}{2}m\left(\frac{k^2e^4}{n^2\hbar^2}\right) - ke^2\left(\frac{kme^2}{n^2\hbar^2}\right) = \frac{1}{n^2}\left(\frac{k^2e^4m}{\hbar^2}\right)\left(\frac{1}{2} - 1\right)$$

That is,

$$\boxed{E_n = -\frac{1}{n^2}\left(\frac{k^2e^4m}{2\hbar^2}\right) = -\frac{E_I}{n^2}} \tag{32.18a}$$

where

$$\boxed{E_I = \frac{k^2e^4m}{2\hbar^2} = 13.58 \text{ eV}} \tag{32.18b}$$

is the energy necessary to remove an electron from the first Bohr orbit. Thus, for $n = 1, 2, 3, 4, \ldots$, we get the allowed energy levels of a hydrogen atom to be

$$E_n = -E_I, -\frac{E_I}{4}, -\frac{E_I}{9}, -\frac{E_I}{16}, \ldots \tag{32.19}$$

The experimentally measured value of the binding energy of the electron in the hydrogen atom is -13.58 eV, in perfect agreement with the value predicted by the Bohr theory, Equation 32.18, for $n = 1$, $E_1 = -E_I$. The different energy levels corresponding to Equation 32.19 are shown in Figure 32.11.

Normally, the electron in the hydrogen atom is in the lowest energy state corresponding to $n = 1$, and this state is called the *ground state* or *normal state*. If the electron is any other state, but not in $n = 1$, it is said to be in the *excited state*. Suppose that an electron has to be transferred from a state $n = 1$ to an excited state n; the energy needed will be $E_n - E_1 = (E_n + E_I)$, which is called the *excitation energy*. If the electron has to be completely detached from the atom, $n = \infty$, $E_n = 0$, the energy needed to do this is equal to E_I and is called the *ionization energy*.

Hydrogen Emission Spectrum

The results derived above for the energy levels, Equation 32.18, together with Postulate III can be used to arrive at the hydrogen spectrum. Suppose that the electron in the hydrogen atom is in the initial excited state n_i with energy E_i and makes a transition to the final state n_f with energy E_f, where $E_f < E_i$. From Postulate III, the energy of the photon emitted is

$$h\nu = E_i - E_f \tag{32.20}$$

where

$$E_i = -\frac{E_I}{n_i^2} \quad \text{and} \quad E_f = -\frac{E_I}{n_f^2}$$

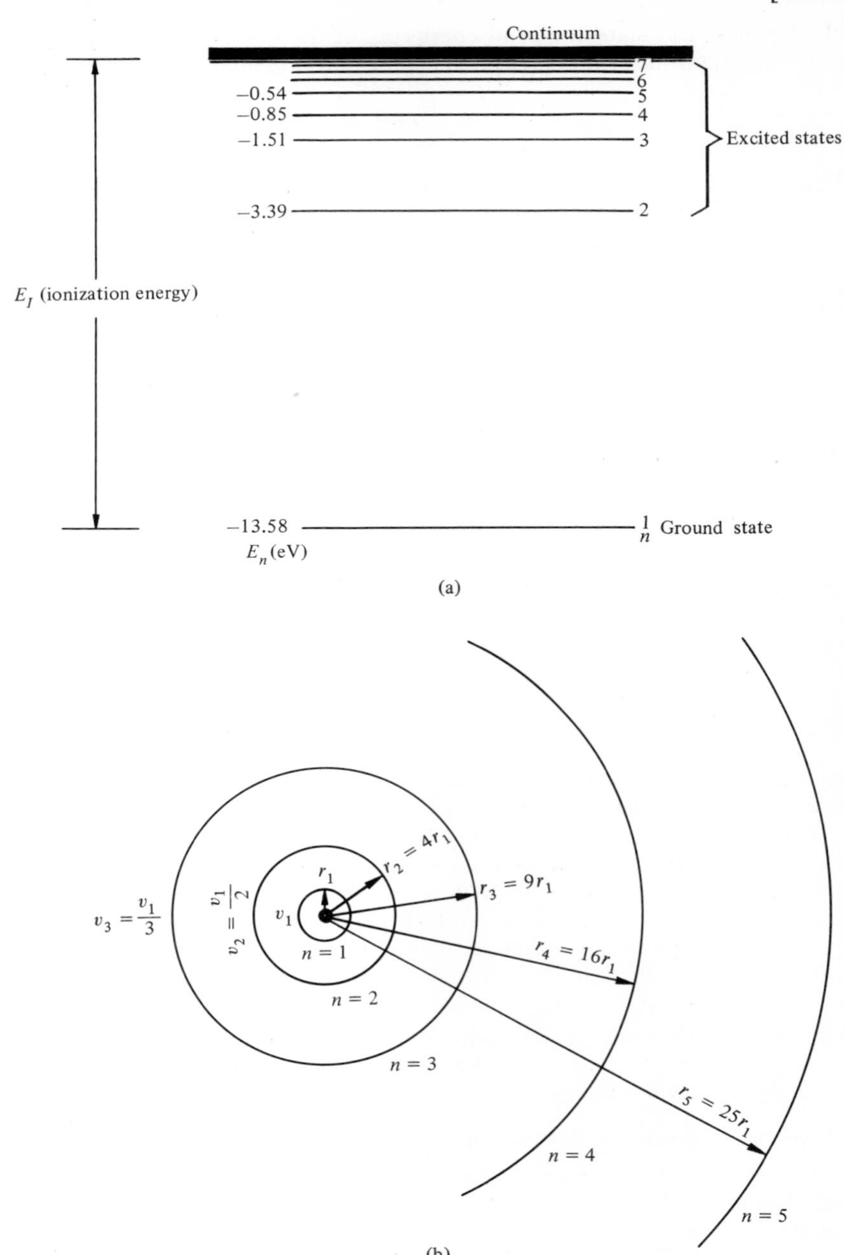

FIGURE 32.11. *(a) Ground state and possible excited energy states of the hydrogen atom corresponding to different values of the quantum number n. (b) Relative sizes of various circular orbits of the electron in a hydrogen atom.*

Therefore,

$$h\nu = E_I\left(\frac{1}{n_f^2} - \frac{1}{n_i^2}\right) \tag{32.21}$$

Substituting for $\nu = c/\lambda$ and the wave number $\bar{\nu} = 1/\lambda$, we obtain, after rearranging,

$$\bar{\nu} = \frac{1}{\lambda} = \frac{E_I}{hc}\left(\frac{1}{n_f^2} - \frac{1}{n_i^2}\right)$$

or

$$\boxed{\bar{\nu} = \frac{1}{\lambda} = R\left(\frac{1}{n_f^2} - \frac{1}{n_i^2}\right)} \tag{32.22}$$

where R is the Rydberg constant given by

$$R = \frac{E_I}{hc} = \frac{k^2e^4m}{4\pi\hbar^3c} = 1.0974 \times 10^{-3}\ \text{Å}^{-1} \tag{32.23}$$

which agrees very well with the latest measured value for the hydrogen atom,

$$R_{\text{expt}} = 109677.576 \pm 0.012\ \text{cm}^{-1} \tag{32.24}$$

Equation 32.22 reduces to the empirical result derived by Rydberg and given by Equation 32.7, provided that we substitute $n_f = 2$ and $n_i = 3, 4, 5, \ldots$. The transitions obtained, shown in Figure 32.12, are the Balmer series. A small sample of hydrogen usually contains billions of atoms. The resulting observed spectrum yields not only the Balmer series but many others, as shown in Figure 32.12. These series are usually named for the first observer. Not all the series are in the visible region. Thus, the Bohr theory has proved to be quite a success.

EXAMPLE 32.2 Calculate the energy, radius, and velocity of an electron in the $n = 3$ orbit of the hydrogen atom. What is the excitation energy for this state? What transitions will be observed when this electron returns to the ground state?

From Equation 32.18, the energy of the third excited state is

$$E_3 = -\frac{E_I}{n^2} = -\frac{13.58\ \text{eV}}{9} = -1.51\ \text{eV}$$

From Equations 32.13 and 32.14,

$$r_3 = (3)^2 r_1 = 9 \times 0.528\ \text{Å} = 4.752\ \text{Å}$$

$$v_3 = \frac{1}{n}\frac{ke^2}{\hbar} = \frac{1}{3}\frac{(9 \times 10^9\ \text{N-m/C}^2)(1.602 \times 10^{-19}\ \text{C})^2}{1.054 \times 10^{-34}\ \text{J-s}} = 7.305 \times 10^5\ \text{m/s}$$

The excitation energy E_x is given by

$$E_n - E_1 = E_n + E_I = E_3 + E_I = -1.51\ \text{eV} + 13.58\ \text{eV} = 12.07\ \text{eV}$$

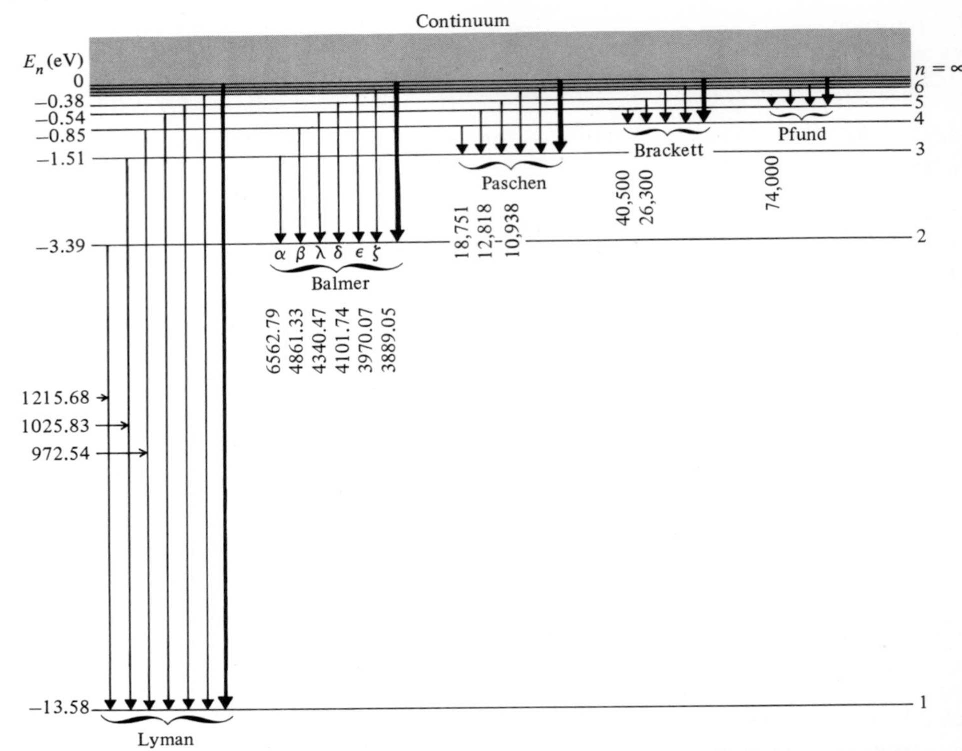

FIGURE 32.12. *Horizontal lines show the energy-level diagram of the hydrogen atom. Vertical lines show possible transitions corresponding to different series. Numbers with vertical lines are wavelengths of transitions, in angstroms. Boldface transitions indicate the series limits.*

From Figure 32.12 the transitions observed will be between the levels 3—2, 3—1, and 2—1, corresponding to wavelengths 6562.79, 1025.83, and 1215.68 Å, respectively.

EXERCISE 32.2 Repeat Example 32.2 for the fourth excited state of the hydrogen atom. [*Ans.*: $E_4 = -0.849$ eV, $r_4 = 8.45$ Å, $v_4 = 5.48 \times 10^5$ m/s, $E_x = 12.73$ eV, see Figure 32.12 for transitions.]

32.5. The Quantum Theory of the Hydrogen Atom

It is true that the Bohr theory of the hydrogen atom and its subsequent refinement by Sommerfeld succeeded in explaining many observed aspects of atomic spectra. But these theories have their shortcomings: First, as in any theory, one has to accept the basic postulates. Second, the theories cannot be extended to complex systems, nor can they explain many complex features.

Still a better way to handle the hydrogen problem is to make an assumption that the Schrödinger wave equation is applicable to this situation. Since the mathematics is too involved for this text, we simply summarize the results.

Let the proton of mass M in the hydrogen atom be located at the origin, and the electron, which has a mass m and charge e^-, be somewhere in space at (x, y, z). Since the proton is very heavy as compared to the electron, the proton may be supposed to be at rest. $\psi = \psi(x, y, z)$ is the wave function of the electron and depends upon three independent variables which are the three position coordinates x, y, and z of the electron. E is the total energy of the electron and V is its potential energy. The Schrödinger wave equation (S.W.E.) representing such a system can be split into three parts and each solved separately for three different wave functions. This results in three quantum conditions. Note that this is unlike the Bohr theory, where there is only one quantum condition. The resulting three quantum numbers are (1) the principal quantum number, n; (2) the angular momentum (or orbital momentum) quantum number, l; and (3) the magnetic quantum number, m_l.

Principal Quantum Number

The general solution of the first of the three parts of the S.W.E. gives the following expression for the energy levels of the hydrogen atom:

$$E_n = -\frac{1}{n^2}\left(\frac{mk^2e^4}{2\hbar^2}\right) = -\frac{E_I}{n^2} \qquad (32.25)$$

where n is the *principal quantum number*. The corresponding S.W.E. is satisfied only if

$$n = 1, 2, 3, 4, \ldots \qquad (32.26)$$

which is the result assumed by the Bohr theory.

Angular Momentum Quantum Number

The general solution of the second of the three parts of S.W.E. gives the expression for the magnitude of the angular momentum of the electron. The magnitude of the angular momentum L of the electron in an atom obtained from quantum mechanics is

$$L = \sqrt{l(l+1)}\hbar \qquad (32.27)$$

where l is called the *orbital quantum number* and its value depends upon the value of n. For a given value of n, l can have any of the following values:

$$l = 0, 1, 2, 3, \ldots, (n-1) \qquad (32.28)$$

Thus, if $n = 1$, $l = 0$; if $n = 2$, $l = 0$ or 1; if $n = 3$, $l = 0, 1, 2$; and so on. That is, for a given n, l can have n different values.

Also, unlike Bohr's theory, the magnitude of L is determined by l and not n. For a given quantum number n, there are n values of L given by Equations 32.27 and 32.28. Besides determining the energy values of the states, n also determines the maximum value of L; that is, $L_{max} = \sqrt{n(n-1)}\hbar$. The following notation is commonly used to denote different l-states of the electrons:

$$\begin{aligned} l &= 0, 1, 2, 3, 4, \ldots \\ \text{electron state notation} &= s, p, d, f, g, \ldots \end{aligned} \tag{32.29}$$

The corresponding atomic states (to be used later) are denoted by S, P, D, F, G, . . . , respectively. That is, the s state means $l = 0$, the p state means $l = 1$, the d state means $l = 2$; and so on. Similarly, 1s stands for an electron state with $n = 1$ and $l = 0$, 2s stands for an electron state with $n = 2$ and $l = 0$, 2p stands for a state with $n = 2$ and $l = 1$, 3d stands for $n = 3$ and $l = 2$, and so on. Knowing n and l, E_n and L can be calculated from Equations 32.25 and 32.27.

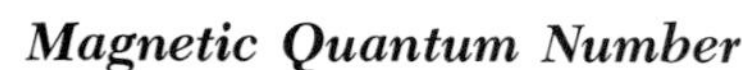

Magnetic Quantum Number

Knowing the energy and the angular momentum of an electron in an atom does not tell us anything about the direction of the angular momentum vector **L** in space. The answer to this question comes by solving the third part of the general S.W.E. Accordingly, the angular momentum **L** can take only those directions in space for which the resulting Z-component of **L** is given by the relation

$$\boxed{L_z = m_l \hbar} \tag{32.30}$$

where m_l is called the *magnetic quantum number*. For a given value of l, m_l can take the following $(2l + 1)$ values, that is,

$$\boxed{m_l = l, (l-1), (l-2), \ldots, 1, 0, -1, \ldots, -(l-1), -l} \tag{32.31}$$

Note that m_l determines the orientation of the angular momentum **L** in space, that is, the *space quantization of angular momentum*. The vector **L** rotating about the Z-axis makes an angle θ with it, as shown in Figure 32.13, and is given by

$$\cos\theta = \frac{L_z}{L} = \frac{m_l \hbar}{\sqrt{l(l+1)}\hbar} = \frac{m_l}{\sqrt{l(l+1)}} \tag{32.32}$$

Since m_l can take $(2l + 1)$ values, the vector **L** can have only $(2l + 1)$ orientations in space, the angle of orientation being given by Equation 32.32. Figure 32.14 illustrates this for different values of l.

Thus, any electron in an atom may be described by three quantum numbers, n, l, and m_l. These determine the energy, the angular momentum and the orientation of the angular momentum vector **L** of the electron in space. Furthermore, the plots of the square of the product of the three wave functions give us the probability of finding the electron in space. As an example, Figure 32.15 illustrates the probability of finding the 1s, 2s, and 2p electrons in space. The density of the points indicates the probability. Thus for the 1s electron, the probability is the highest for $r = a_0 = 0.53$ Å, which is the Bohr orbit radius. That is, the 1s electron is most likely to be found at $r = 0.53$ Å, with less likelihood of being found at other places. According to the Bohr theory, the 1s electron must be always at a radius of 0.53 Å.

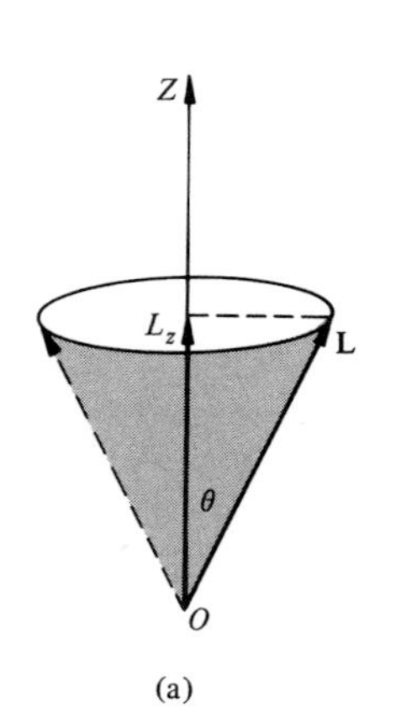

(a)

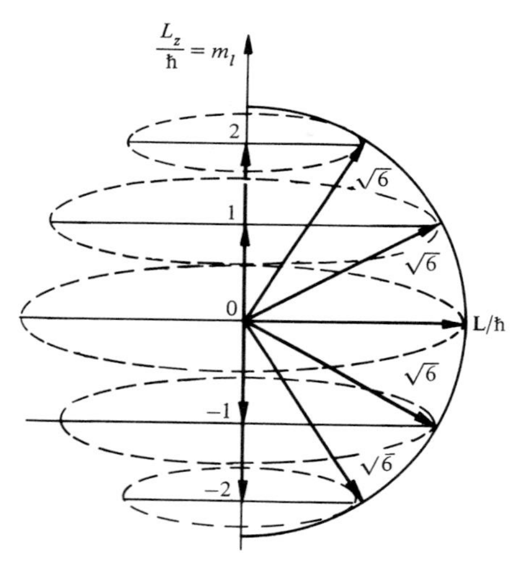

$L = \sqrt{l(l+1)}\hbar$

$L/\hbar = \sqrt{2(2+1)} = \sqrt{6}$

(b)

FIGURE 32.13. *(a) Rotation of the angular momentum vector **L** about the Z-axis and its relation to L_z. (b) According to quantum mechanics, only those values of θ are possible which satisfy the condition $\cos\theta = L_z/L$, where $L = \sqrt{l(l+1)}\,\hbar$ and $L_z = m\hbar$ as illustrated for $l = 2$.*

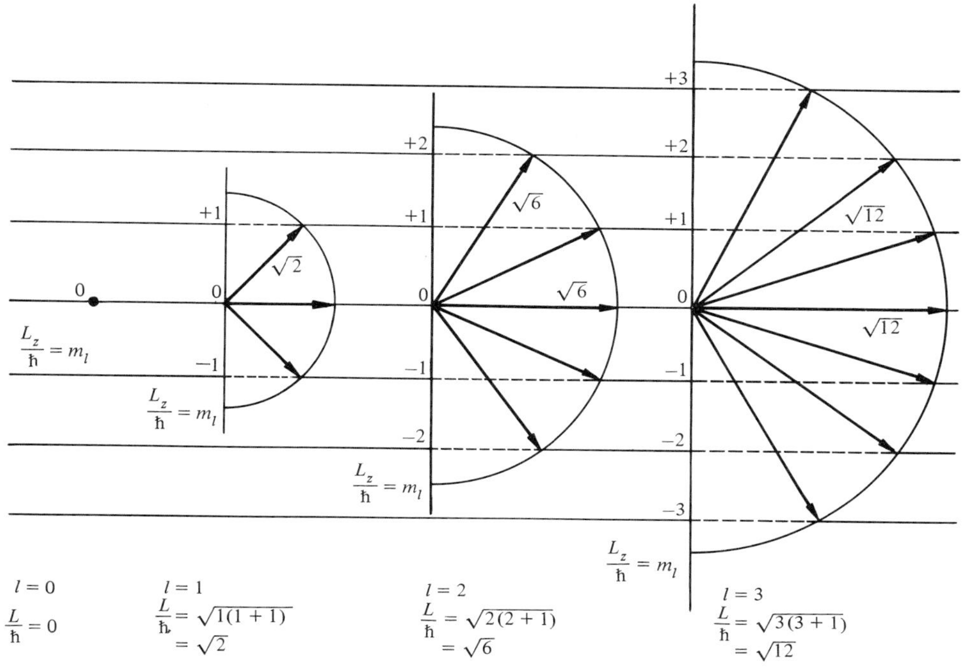

Figure 32.14. *Possible orientations of the angular momentum vectors for $l = 0, 1, 2,$ and 3.*

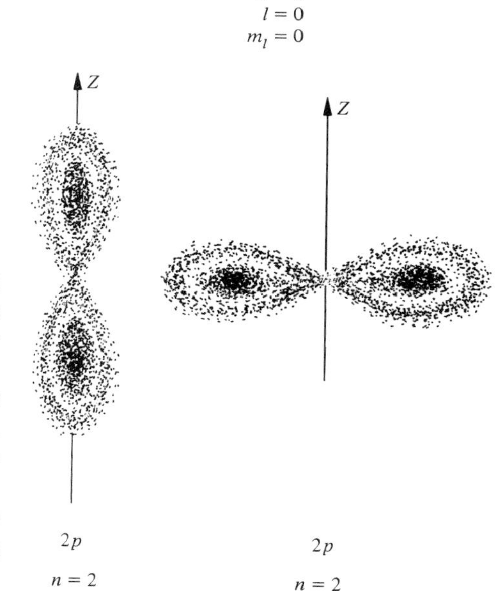

Figure 32.15. *Density of points indicates the probability of the electron in the hydrogen atom being in that quantum state.*

32.6. The Spinning Electron and the Vector Model of the Atom

Spin Quantum Numbers s and m_s

Further investigation of the experimently observed spectra of different elements revealed that it was not possible to explain all the transitions with the help of only three quantum numbers, n, l, and m_l. These new experimentally determined facts led Uhlenbech and Goudsmit, and independently Bichowsky and Urey, in 1925 to assign a new angular momentum to the electron. The electron in any state was assumed to be spinning about its own mechanical axis, as shown in Figure 32.16(a). Thus, the electron has an intrinsic angular momentum **S**, called the *spin angular momentum*.

To explain the experimentally observed facts, the spin angular momentum vector **S** was assigned the magnitude $S = \sqrt{s(s+1)}\hbar$, where s is called the *spin quantum number* and its value is $\frac{1}{2}$. Thus,

$$S = \sqrt{s(s+1)}\hbar = \sqrt{\tfrac{1}{2}(\tfrac{1}{2}+1)}\hbar = \frac{\sqrt{3}}{2}\hbar \tag{32.33}$$

The spin angular momentum vector **S**, just like the orbital angular momentum

(a)

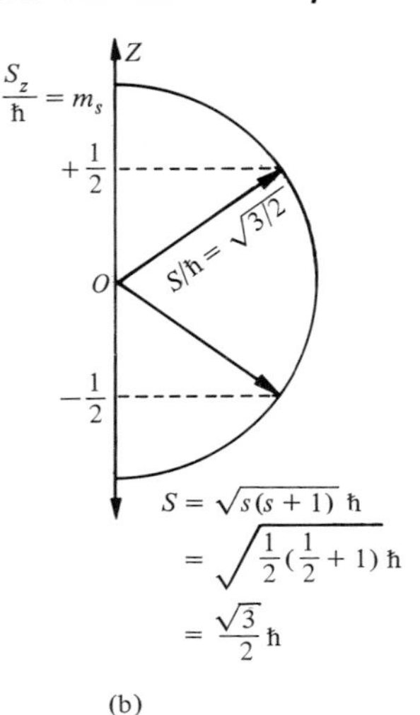

(b)

FIGURE 32.16. *(a) Electron in an atom spinning about its own axis with an intrinsic angular momentum* **S**. *(b) Space quantization of* **S** *according to the quantum-mechanical model.*

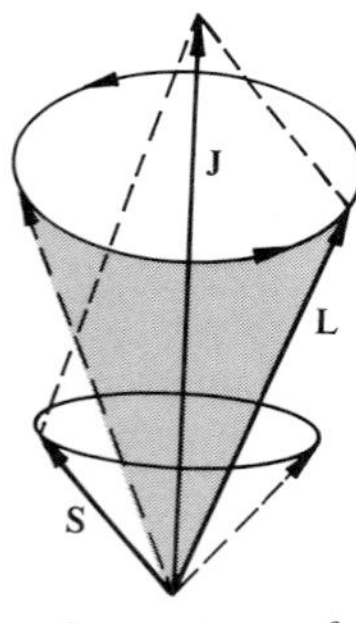

FIGURE 32.17. *Precession of angular momentum vectors* **L** *and* **S** *about their resultant total angular momentum vector* **J**.

L, is space-quantized. As shown in Figure 32.16(b), the Z-components of **S**, S_z, can take the following two values:

$$\boxed{S_z = m_s\hbar} \tag{32.34}$$

where

$$\boxed{m_s = +\tfrac{1}{2} \quad \text{or} \quad -\tfrac{1}{2}} \tag{32.35}$$

and m_s is called the *magnetic spin quantum number*.

The spin angular momentum is in addition to the orbital angular momentum **L** which the electron has because of its motion around the nucleus, and is similar to the motion of a planet (say Earth) spinning or rotating about its own axis. (Like the spinning motion of the electron, the spinning motion of the earth is in addition to its orbital motion around the sun.)

The Total Angular Momentum Quantum Number j

Thus, an electron in an atom has two angular momenta: the orbital angular momentum **L** and the spin angular momentum **S**. The vector addition of these two results in a new total angular momentum vector **J**, given by

$$\mathbf{J} = \mathbf{L} + \mathbf{S} \tag{32.36}$$

while its magnitude is

$$J = |\mathbf{J}| = \sqrt{j(j+1)}\hbar \tag{32.37}$$

where j is called the *total quantum number*. The possible values of j are

$$\boxed{\begin{aligned} j &= l + \tfrac{1}{2} \\ j &= l - \tfrac{1}{2} \end{aligned}} \tag{32.38}$$

where $j = l + \frac{1}{2}$ corresponds to the case when **S** adds parallel to **L**, and $j = l - \frac{1}{2}$ when **S** adds antiparallel to **L**. For example, if $l = 1$, $j = \frac{3}{2}$ or $\frac{1}{2}$; for $l = 2$, $j = \frac{5}{2}$ or $\frac{3}{2}$. However, for $l = 0$, $j = \frac{1}{2}$ only (because in the absence of any reference axis, $j = +\frac{1}{2}$ cannot be distinguished from $j = -\frac{1}{2}$).

Just like **L** and **S** the total angular momentum vector **J** also shows space quantization. As shown in Figure 32.17, the vectors **L** and **S** precess around **J**; while **J** will precess around some external arbitrary axis (not shown).

32.7. Pauli's Exclusion Principle and Electron Configuration

After introducing all the necessary quantum numbers, we are in a position to understand the arrangement of the electrons in the atoms. An atom may be considered a system containing indistinguishable, identical particles all having the same charge, mass, and spin. But the position of any electron in an atom can be specified by making use of a set of four quantum numbers n, l, m_l, and m_s (keep in mind that m_s can have two values, $+\frac{1}{2}$ and $-\frac{1}{2}$). That is, any quantum state in an atom is specified by these four quantum numbers.

The next question that arises is the following: Can we place more than one electron in a given quantum state? The answer is, no. The Austrian-born physicist Wolfgang Pauli (1900–1958) from experimental observations gave the following principle in 1925.

Pauli Exclusion Principle: ***No two electrons in any given atom can occupy the same quantum state; that is, no two electrons in an atom can have identical set of four quantum numbers*** (n, l, m_l, m_s).

Let us see how the electrons in an atom can be assigned quantum states. Suppose that $n = 1$; then $l = 0$, $m_l = 0$, and $m_s = \pm\frac{1}{2}$. Thus, corresponding to (n, l, m_l, m_s), there are two quantum states $(1, 0, 0, +\frac{1}{2})$ and $(1, 0, 0, -\frac{1}{2})$. That is, two electrons can be placed if $n = 1$. Similarly, if $n = 2$, $l = 0$ or 1; if $l = 0$, $m_l = 0$, $m_s = \pm\frac{1}{2}$; if $l = 1$, $m_l = +1, 0, -1$ and for each m_l, $m_s = \pm\frac{1}{2}$. Thus, for $n = 2$, there are eight distinct quantum states, as shown below:

$$\begin{aligned} &(2, 0, \ \ 0, +\tfrac{1}{2}), (2, 0, \ \ 0, -\tfrac{1}{2}) \\ &(2, 1, \ \ 1, +\tfrac{1}{2}), (2, 1, \ \ 1, -\tfrac{1}{2}) \\ &(2, 1, \ \ 0, +\tfrac{1}{2}), (2, 1, \ \ 0, -\tfrac{1}{2}) \\ &(2, 1, -1, +\tfrac{1}{2}), (2, 1, -1, -\tfrac{1}{2}) \end{aligned}$$

In general, as in Table 32.1, for any n there are N_t quantum states, given by

$$\boxed{N_t = 2n^2} \tag{32.39}$$

In a given atom the electrons that have the same principal quantum number n are said to be in the same *group* or *shell.* For a given n those electrons that have the same value of l are said to be in the same *subgroup, subshell,* or *sublevel.* Also according to x-ray notation, $n = 1$ is called the K-shell, $n = 2$ the L-shell, $n = 3$ the M-shell, $n = 4$ the N-shell, and so on. Also $l = 0, 1, 2, 3, 4, \ldots$ states are denoted by $S, P, D, F, G, \ldots$, respectively.

To assign quantum states to different electrons in an atom, start at the top of Table 32.1. Quantum numbers in any one horizontal line are assigned to one and only one electron. Thus, for hydrogen ($Z = 1$), which has a single electron in the ground state with quantum numbers $(1, 0, 0, \frac{1}{2})$. Thus, the ground-state configuration of hydrogen is $1s^1$ which means that $n = 1$, $l = 0$, and the superscript 1 means that there is only one electron in the $1s$-state. Similarly, for $_2$He ($Z = 2$) the two electrons are in states $(1, 0, 0, \frac{1}{2})$ and $(1, 0, 0, -\frac{1}{2})$; while the ground-state configuration of He is $1s^2$, meaning that $n = 1$, $l = 0$, and there are two electrons in the $1s$-states. As another example, sodium Na ($Z = 11$) has a ground-state configuration of $1s^2 2s^2 2p^6 3s^1$ and is obtained by reading across the eleventh line, that is, $(3, 0, 0, \frac{1}{2})$, with the last electron in the $3s$-state.

The filling of different energy levels occurs in such a way that the resulting configuration has minimum energy. For atoms with a large number of electrons, the calculations yield the following order in which the shells are filled:

$$1s^2 2s^2 2p^6 3s^2 3p^6 4s^2 3d^{10} 4p^6 5s^2 4d^{10} 5p^6 6s^2 4f^{14} 5d^{10} 6p^6 7s^2 6d^{10} \tag{32.40}$$

TABLE 32.1
Quantum States in Different Subshells and Shells

Electrons in an Atom	Quantum States of Electron n	l	m_l	m_s	Number of Electrons in Different l Subshells	Electron Configuration[a]
1	1	0	0	$+\frac{1}{2}$		$1s^1$
2	1	0	0	$-\frac{1}{2}$	2	$1s^2$
3	2	0	0	$+\frac{1}{2}$		$1s^22s^1$
4	2	0	0	$-\frac{1}{2}$	2	$1s^22s^2$
5	2	1	-1	$+\frac{1}{2}$		$1s^22s^22p^1$
6	2	1	-1	$-\frac{1}{2}$		$1s^22s^22p^2$
7	2	1	0	$+\frac{1}{2}$		$1s^22s^22p^3$
8	2	1	0	$-\frac{1}{2}$		$1s^22s^22p^4$
9	2	1	$+1$	$+\frac{1}{2}$		$1s^22s^22p^5$
10	2	1	$+1$	$-\frac{1}{2}$	6	$1s^22s^22p^6$
11	3	0	0	$+\frac{1}{2}$		$1s^22s^22p^63s^1$
12	3	0	0	$-\frac{1}{2}$	2	$1s^22s^22p^63s^2$
13	3	1	-1	$+\frac{1}{2}$		$1s^22s^22p^63s^23p^1$
14	3	1	-1	$-\frac{1}{2}$		$1s^22s^22p^63s^23p^2$
15	3	1	0	$+\frac{1}{2}$		$1s^22s^22p^63s^23p^3$
16	3	1	0	$-\frac{1}{2}$		$1s^22s^22p^63s^23p^4$
17	3	1	$+1$	$+\frac{1}{2}$		$1s^22s^22p^63s^23p^6$
18	3	1	$+1$	$-\frac{1}{2}$	6	$1s^22s^22p^63s^23p^6$

[a] For $Z > 18$, the electron configuration is to be according to Equation 32.40.

EXAMPLE 32.3 Write the electron configuration of Sc ($Z = 21$). What are the values of l, s, j, L, S, and J for the electron in the ground state?

From Equation 32.40, the electron configuration for $Z = 21$ is $1s^22s^22p^6$-$3s^23p^64s^23d^1$. The ground-state electron is $3d^1$; that is, $n = 3$, $l = 2$. Since $l = 2$ and $s = \frac{1}{2}$, $j = l + \frac{1}{2} = 2 + \frac{1}{2} = \frac{5}{2}$ or $j = l - \frac{1}{2} = 2 - \frac{1}{2} = \frac{3}{2}$. Using the relations $L = \sqrt{l(l+1)}\hbar$, $S = \sqrt{s(s+1)}\hbar$, and $J = \sqrt{j(j+1)}\hbar$,

$$l = 2: \quad L = \sqrt{2(2+1)}\hbar = \sqrt{6}\hbar$$

$$s = \tfrac{1}{2}: \quad S = \sqrt{\tfrac{1}{2}(\tfrac{1}{2}+1)}\hbar = \frac{\sqrt{3}}{2}\hbar$$

$$j = \tfrac{5}{2}: \quad J = \sqrt{\tfrac{5}{2}(\tfrac{5}{2}+1)}\hbar = \frac{\sqrt{35}}{2}\hbar$$

$$j = \tfrac{3}{2}: \quad J = \sqrt{\tfrac{3}{2}(\tfrac{3}{2}+1)}\hbar = \frac{\sqrt{15}}{2}\hbar$$

EXERCISE 32.3 Repeat Example 32.3 for an atom with $Z = 13$.

32.8. Spectra of One-Valence-Electron Atoms

Consider the atoms H (Z = 1), Li (Z = 3), Na (Z = 11), K (Z = 19), Rb (Z = 37), and Cs (Z = 55). The electron configuration of these atoms is such that in each case there is only one electron outside the closed shells. Such atoms are called *one-valence-electron atoms*. In this particular case, when the atoms are in the ground state, these valence electrons are in the S-states. The valence electrons play an important role for both chemists and physicists. For chemists, the valence electron determines the chemical properties. On the other hand, the optical spectra are characterized by this valence electron. Thus, the optical spectra of all the elements listed above are similar because in each case the optical electron, when in the ground state, is in the 1s state. To illustrate this point, let us take Na (Z = 11) as an example.

As is clear from Figure 32.18, the sodium atom may be pictured as consisting of three parts: (1) the nucleus, (2) the 10 electrons in states $1s^2 2s^2 2p^6$ forming closed shells, and (3) the optical electron in the ground state, 3s. The optical electron moves in the net field produced by the nucleus of charge +Ze and a core of electrons with charge −(Z − 1), where Z = 11. This core of electrons forms a shield between the nucleus and the optical electron. This shielding has the effect of making the energy levels of sodium dependent on l in addition to n. Thus, in the case of hydrogen, where there is no core electrons shielding the nucleus, 3S, 3P, and 3D have the same energies; 4S, 4P, 4D, and 4F have the same energies; and so on. This is not so in the case of sodium. All such levels as 3S, 3P, 3D, 4S, 4P, 4D, 4F, . . . have different energies. Thus, when the 3S optical electron in sodium is excited, it goes to energy levels 3P, 3D, 4S, 4P, ... Figure 32.19 shows the energy-level diagram of sodium together with that of hydrogen.

In sketching the energy-level diagram of sodium in Figure 32.19, we neglected the intrinsic spin of the electron. The single valence electron outside the closed shell can have two values of j: $j = l + \frac{1}{2}$ and $j = l - \frac{1}{2}$. Thus, each of the l-levels in Figure 32.19 is split into a doublet except for the S-level, corresponding to $l = 0$, which is still a singlet (when $l = 0$, $j = \frac{1}{2}$ and $j = -\frac{1}{2}$ cannot be distinguished unless there is a reference axis). Any particular state is denoted by the spectroscopic notation

$$\boxed{n^{2s+1}L_j}$$

where $s = \frac{1}{2}$; hence, $2s + 1 = 2(\frac{1}{2}) + 1 = 2$, denoting a doublet level. Here n stands for the principal quantum number; L for the orbital state, S, P, D, F, . . . ; and j is the total angular quantum number.

Once the electrons are in the excited state, in returning to the lower-energy states or ground state, emission of electromagnetic radiation takes place. Not all transitions are allowed. The selection rules that determine the transitions between different states are the following:

1. The value of n can change by any amount; there is no restriction.
2. The total quantum number j can change by zero or unity; that is,

$$\Delta j = j_i - j_f = 0, \pm 1$$

except that when $j_i = j_f = 0$, the transition is forbidden.

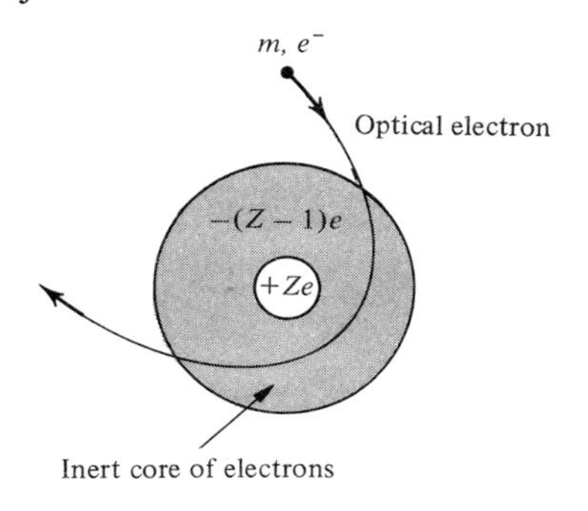

(a)

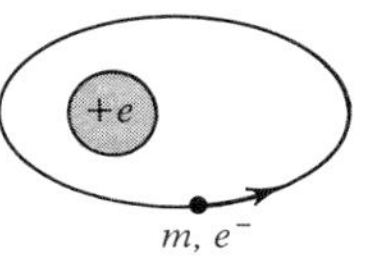

(b)

Figure 32.18. *(a) The structure of the sodium atom may be divided into three parts: the nucleus, the inert core of electrons shielding the nucleus, and the optical electron. (b) The hydrogen atom is similar to the sodium atom, except that it does not have a core of inert electrons. (Note that the more general orbit of an electron is elliptical and not circular.)*

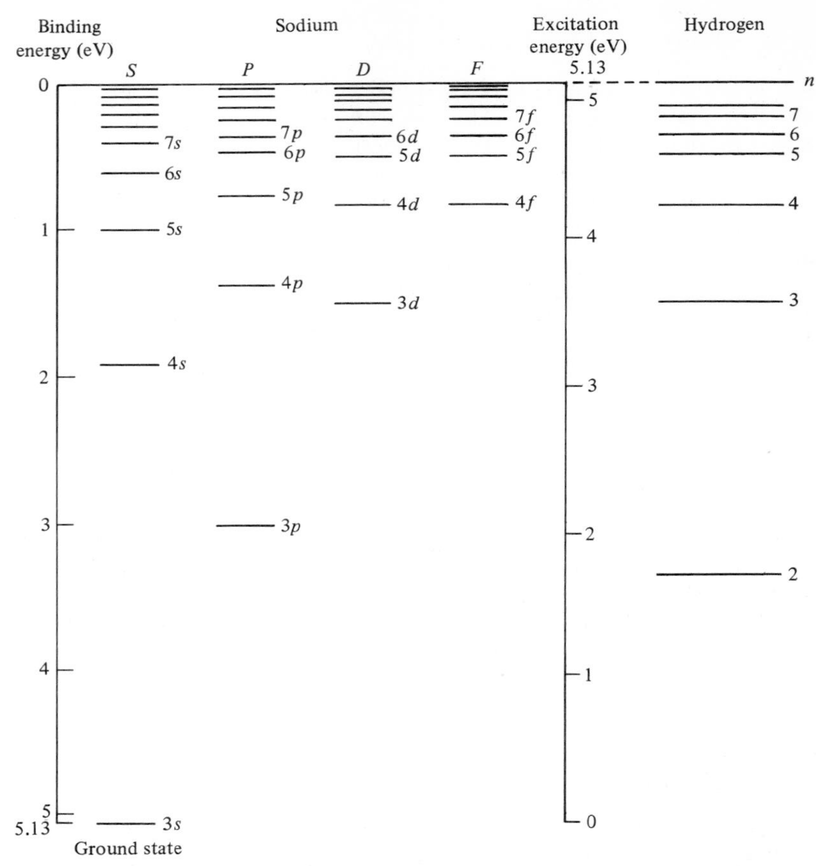

FIGURE 32.19. *Energy-level diagram of sodium. Short horizontal lines indicate the possible energy states to which the valence electron of sodium may be excited (the doublet nature of some levels is not shown). The energy-level diagram of hydrogen is shown for comparison.*

3. The emitted photon in any transition carries away one unit of angular momentum $= 1\hbar$. To conserve angular momentum, the orbital quantum number l must change by ± 1; that is,

$$\Delta l = l_i - l_f = \pm 1$$

Using these selection rules, the transitions in sodium are drawn as shown in Figure 32.20.

The spectra of other one-valence-electron atoms can be drawn by following the discussion above. One needs to know the net charge due to the nucleus and the shielding electrons, thereby giving the effective value of Z. Knowing the type of optical electron, we can calculate the energy level diagram. Also proper relativistic corrections are needed for fast moving electrons.

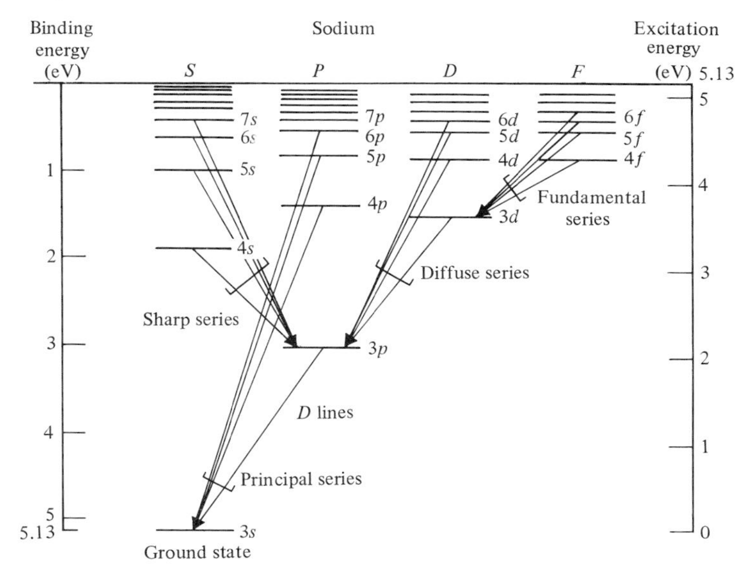

Figure 32.20. *The energy-level diagram of sodium (neglecting spin and relativistic corrections) is indicated by the short horizontal lines. The slanting lines are the possible transitions between different levels.*

32.9. Characteristic X-Ray Spectra and the Periodic Table

The study of the characteristic x-ray spectra has played a very important role in the study of atomic structure and the periodic table. Let us see how the characteristic x-rays are produced. We shall be using the x-ray notation in which $K, L, M, N, \ldots$ shells correspond to $n = 1, 2, 3, 4, \ldots,$ respectively. As an example, let us consider cadmium ($Z = 48$), whose energy-level diagram is shown in Figure 32.21. Suppose that a target made of cadmium or any other heavy atom is bombarded with fast-moving electrons. It is possible that in collision the electrons in the innermost shells, such as K or L, will be knocked out. Suppose that one of the electrons in the K-shell is removed, thereby producing a vacancy (or hole) in that shell. The electron from the L-shell jumps to occupy the hole in the K-shell, thereby emitting a photon of energy $h\nu_{K\alpha}$ called the K_α x-ray, given by

$$h\nu_{K_\alpha} = E_K - E_L \tag{32.41}$$

It is also possible that the electron from the M-shell or N-shell might also jump to occupy the hole in the K-shell. The photons emitted are the K_β and K_γ x-rays, with energies

$$\begin{aligned} h\nu_{K_\beta} &= E_K - E_M \\ h\nu_{K_\gamma} &= E_K - E_N \end{aligned} \tag{32.42}$$

and so on.

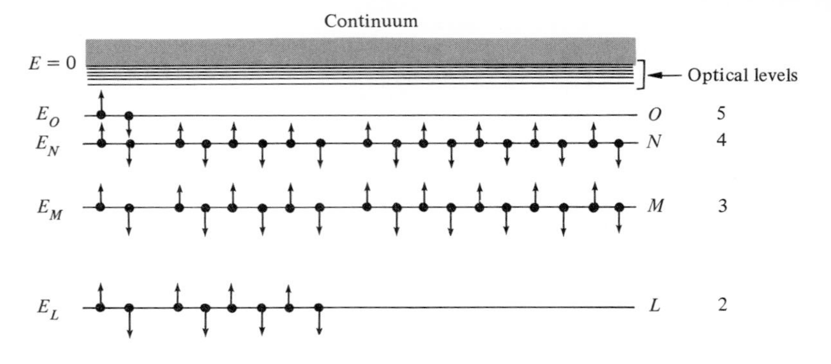

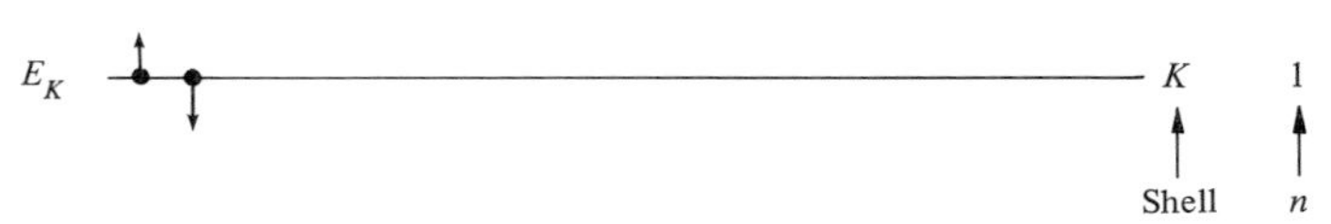

FIGURE 32.21. *Energy levels of cadmium (Cd) as used in an x-ray spectrum scheme. Note that the optical levels are near the continuum, while the other levels, K, L, M, N, and O are the x-ray levels.*

The photons emitted in such transitions are called *characteristic x-rays* because their energies depend upon the type of target material. The holes created in the L- and M-shells are occupied by transitions of electrons from higher states, creating more x-rays.

Characteristic x-rays are produced in the same type of experimental setup that was used to produce continuous x-rays (see Chapter 31). Thus, the characteristic x-rays appear as discrete lines superimposed on the continuous spectrum, as shown in Figure 32.22.

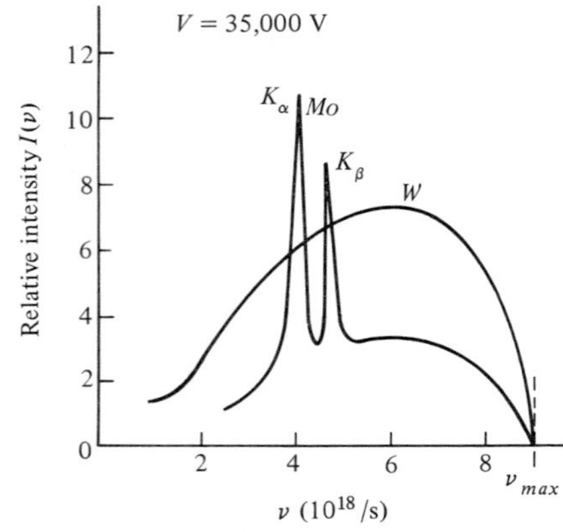

FIGURE 32.22. *X-ray spectra of tungsten (W) and molybdenum (Mo), showing the plots of relative intensity $I(\nu)$ versus ν.*

In 1913, H. G. B. Moseley investigated characteristic K x-ray radiation emitted from the elements Ca, Ti, V, Cr, Mn, Fe, Co, Ni, Cu, and Zn, which have Z = 20, 22, 23, 24, 25, 26, 27, 28, 29, and 30, respectively. He found that the wavelength of the K x-rays decreased smoothly with increasing Z. The experimentally measured values of $\sqrt{\nu_K}$ when plotted against Z resulted in a perfect straight line, but not when plotted against A. The relation $\sqrt{\nu_K} \propto Z$ is called *Moseley's law*. This had a significant effect on the periodic table.

The *periodic table* is an arrangement of elements in which the elements with similar physical and chemical properties occur at regular intervals. Table 32.2 is the form of the periodic table suggested by Dimitri Mendeleev (1834–1907) and Lothar Meyer (1830–1895). Originally, the elements were arranged in order of increasing mass number A. This resulted in some discrepancies. These discrepancies were removed when, in light of Moseley's law, the periodic table was rearranged in order of increasing Z.

Each box in Table 32.2 represents one element. Included in that box are the

Table 32.2

Periodic Table of the Elements[a]

Period ↓	Group →	I	II
1	$1s$	1 H 1.00 $1s^1$	
2	$2s$	3 Li 6.94 $2s^1$	4 Be 9.01 $2s^2$
3	$3s$	11 Na 22.99 $3s^1$	12 Mg 24.31 $3s^2$
4	$4s$	19 K 39.10 $4s^1$	20 Ca 40.08 $4s^2$
5	$5s$	37 Rb 85.47 $5s^1$	38 Sr 87.66 $5s^2$
6	$6s$	55 Cs 132.91 $6s^1$	56 Ba 137.34 $6s^2$
7	$7s$	87 Fr (223) $7s^1$	88 Ra 226.05 $7s^2$

$3d$	21 Sc 44.96 $3d^1$	22 Ti 47.90 $3d^2$	23 V 50.94 $3d^3$	24 Cr 52.00 $4s^13d^5$	25 Mn 54.9 $4s^23d^5$	26 Fe 55.85 $4s^23d^6$	27 Co 58.93 $4s^23d^7$	28 Ni 58.71 $4s^23d^8$	29 Cu 63.54 $4s3d^{10}$	30 Zn 65.37 $4s^23d^{10}$
$4d$	39 Y 88.91 $5s^24d^1$	40 Zr 91.22 $5s^14d^2$	41 Nb 92.91 $5s^14d^4$	42 Mo 95.94 $5s^14d^5$	43 Tc (99) $5s^24d^5$	44 Ru 101.1 $5s^14d^7$	45 Rh 102.91 $5s^14d^8$	46 Pd 106.4 $5s^04d^{10}$	47 Ag 107.87 $5s^14d^{10}$	48 Cd 112.40 $5s^24d^{10}$
$5d$	57–71 *	72 Hf 178.49 $6s^25d^2$	73 Ta 180.95 $6s^25d^3$	74 W 183.85 $6s^25d^4$	75 Re 186.2 $6s^25d^5$	76 Os 190.2 $6s^25d^6$	77 Ir 192.2 $6s^25d^7$	78 Pt 195.09 $6s^15d^9$	79 Au 197.0 $6s5d^{10}$	80 Hg 200.59 $6s^25d^{10}$
$6d$	89–103 †									

	III	IV	V	VI	VII	VIII
						2 He 4.00 $1s^2$
$2p$	5 B 10.81 $2p^1$	6 C 12.01 $2p^2$	7 N 14.01 $2p^3$	8 O 16.00 $2p^4$	9 F 19.00 $2p^5$	10 Ne 20.18 $2p^6$
$3p$	13 Al 26.98 $3p^1$	14 Si 28.09 $3p^2$	15 P 30.98 $3p^3$	16 S 32.07 $3p^4$	17 Cl 35.46 $3p^5$	18 Ar 39.94 $3p^6$
$4p$	31 Ga 69.72 $4p^1$	32 Ge 72.59 $4p^2$	33 As 74.92 $4p^3$	34 Se 78.96 $4p^4$	35 Br 79.91 $4p^5$	36 Kr 83.8 $4p^6$
$5p$	49 In 114.82 $5p^1$	50 Sn 118.69 $5p^2$	51 Sb 121.75 $5p^3$	52 Te 127.60 $5p^4$	53 I 126.90 $5p^5$	54 Xe 131.30 $5p^6$
$6p$	81 Tl 204.37 $6p^1$	82 Pb 207.19 $6p^2$	83 Bi 208.98 $6p^3$	84 Po (210) $6p^4$	85 At (210) $6p^5$	86 Rn 222 $6p^6$

$4f$	57 La 138.91 $6s^25d^1$	58 Ce 140.12 $5d^15f^2$	59 Pr 140.91 $5d^04f^3$	60 Nd 144.24 $5d^04f^4$	61 Pm (145)	62 Sm 150.35 $5d^04f^6$	63 Eu 152.0 $5d^04f^7$	64 Gd 157.25 $5d^14f^7$	65 Tb 158.92 $5d^14f^8$	66 Dy 162.50	67 Ho 164.92	68 Er 167.26	69 Tm 168.93 $5d^04f^{13}$	70 Yb 173.04 $5d^04f^{14}$	71 Lu 174.97 $5d^14f^{14}$
$5f$	89 Ac 227 $7s^26d^1$	90 Th 232.04 $6d^05f^2$	91 Pa 231	92 U 238.03 $5d^15f^3$	93 Np (237)	94 Pu (242)	95 Am (243) $6d^05f^7$	96 Cm (247)	97 Bk (249)	98 Cf (251)	99 Es (254)	100 Fm (253)	101 Md (256)	102 No (254)	103 Lr (257)

* Lanthanides (rare earths).
† Actinides.

[a] Each box contains information regarding atomic number, element symbol, atomic mass, and ground-state configuration respectively.

atomic number, symbol of the element, atomic weight, and the ground state of the outermost electron. These elements are divided into eight groups, I through VIII, each represented by a vertical column. Elements in a given group have similar physical and chemical properties. For example, the elements in group I (H, Li, Na, Rb, Cs, and Fr) are chemically very active. Except for H, these elements are alkali metals. On the other extreme, the elements of group VIII (He, Ne, Ar, Kr, Xe, and Rn) are chemically very inert. The horizontal lines are called *periods*. As should be clear by now, the properties of the elements in any given period change gradually from one element to the next.

The number of elements are different in different periods. The first period has two elements while the second and third period each contain eight elements, one element corresponding to each of the eight groups I through VIII. The periods fourth through seventh, in addition to the eight elements in each period, contain additional elements between groups II and III. In the fourth period there are 10 elements, starting with $_{21}$Sc and ending with $_{30}$Zn, which form the *first transition group*. These 10 elements are metals with similar characteristics. The fifth period contains a second set of 10 transition elements. The sixth period, in addition to a third set of transition elements, contains a set of 15 more elements, called *lanthanides* or *rare earths*. The rare earths are hard to distinguish from each other. Finally, the seventh period contains a group of 15 additional elements, called *actinides*. Actinides with $Z > 92$ (have been produced only artificially and are unstable) are called *transuranic elements*. The production and investigation of the properties of transuranic elements is an active field of research at this time.

SUMMARY

Atoms that have the same atomic number Z but different atomic masses are called *isotopes* of the same element of a given Z. The Dempster mass spectrometer used for measuring isotopic masses gives $q/M = 2V/B^2R^2$.

In the *Millikan oil-drop experiment*, an oil drop carrying charge is balanced between two parallel plates maintained at a potential difference V so that $q = mgd/V$ and $q = ne$.

The *Bohr theory* of hydrogen is based on three basic postulates and the levels are given by $E_n = -E_I/n^2 = -13.58\text{ eV}/n^2$, where $n = 1, 2, 3, \ldots$ is the *principal quantum number*. The frequencies of different transitions are given by $\bar{\nu} = 1/\lambda = R(1/n_f^2 - 1/n_i^2)$, where R is the *Rydberg constant*.

The *quantum theory of the hydrogen atom* gives the same energy levels for hydrogen as Bohr's, but in addition to n, it requires more quantum numbers and angular momenta. Thus, an electron with quantum number n can have angular momentum quantum number $l = 0, 1, 2, 3, \ldots, (n - 1)$ and angular momentum $L = \sqrt{l(l+1)}\hbar$. $l = 0, 1, 2, 3, \ldots$ electron states are denoted by $s, p, d, f, g, \ldots$, respectively. For a given l, an electron has a magnetic quantum number m_l such that $L_z = m_l\hbar$, where $m_l = l, (l-1), \ldots 1, 0, -1, \ldots, -l$. Also an electron has a spin quantum number $s = \frac{1}{2}$ such that $S = \sqrt{s(s+1)}\hbar = \frac{\sqrt{3}}{2}\hbar$ and $S_z = m_s\hbar$, where $m_s = \pm\frac{1}{2}$. Also, the vector addition of **L** and **S** gives a resultant called the *total angular momentum* $\mathbf{J} = \mathbf{L} + \mathbf{S}$, where $J = \sqrt{j(j+1)}\hbar$ and $j = l + \frac{1}{2}$ or $l - \frac{1}{2}$.

According to *Pauli's exclusion principle* no two electrons in an atom can have identical values for a set of four quantum number (n, l, m_l, m_s). For a given n the quantum states available are $N_t = 2n^2$.

The valence electron responsible for optical spectra is also called the *optical electron*. Any particular state is denoted by $n^{2s+1}L_j$. The selection rules for transition are $\Delta l = \pm 1$, $\Delta j = 0, \pm 1$, with the exception of $j_i = 0$ to $j_f = 0$.

Characteristic x-rays are emitted when electron transitions take place in the innermost shells. The relation $\sqrt{\nu_K} \propto Z$ is called *Moseley's law*. Elements with similar physical and chemical properties occur at regular intervals. The tabular arrangement of these elements is called the *periodic table*. The elements are arranged in order of increasing Z.

QUESTIONS

1. What effect did the discovery of isotopes have on understanding the structure of the atom?
2. Can we use the Dempster mass spectrometer for determination of the mass of light particles such as electrons and mesons?
3. In the Millikan oil-drop experiment, we measure $q = ne$. By measuring the charge on several different drops, how can one calculate the charge, e?
4. Why does the emission spectra of most elements, in general, exhibit more lines than the absorption spectrum?
5. How does the raising of the temperature of a source affect its optical spectra?
6. Does the gravitational force between an electron and a proton in a hydrogen atom have any effect on the spectrum of hydrogen? Explain.
7. What happens when a hydrogen atom absorbs a photon of energy greater than its binding energy (13.6 eV)? What happens to the extra energy?
8. If white light passes through a tube containing hydrogen (in its ground and first excited state), what wavelengths will be absorbed? What do you expect to see in the absorption spectrum?
9. What features of the quantum-mechanical model of a hydrogen atom are common to the Bohr model?
10. We have shown that an electron in any atom has a well-defined angular momentum, but its position at any time is not always well defined. Explain. (*Hint:* Use the Heisenberg uncertainty principle.)
11. When a group of elements have similar chemical properties, what does this indicate about their electronic structure?
12. Why in some elements are there electrons in the outer shells while the inner shells are not completely filled?

PROBLEMS

1. Oxygen has three stable isotopes. From the following information, calculate the atomic weight of oxygen:

Isotope	% Abundance	Isotopic Masses
$^{16}_{8}O$	99.76	15.99491 u
$^{17}_{8}O$	0.037	16.99914 u
$^{18}_{8}O$	0.203	17.99916 u

2. Sulfur has four stable isotopes. From the following information, calculate the atomic weight of sulfur.

Isotope	% Abundance	Isotopic Masses
$^{32}_{16}S$	95.0	31.97207 u
$^{33}_{16}S$	0.76	32.97146 u
$^{34}_{16}S$	4.22	33.96786 u
$^{36}_{16}S$	0.014	35.96709 u

3. In a Dempster mass spectrograph, calculate the distance between the point where the beam enters the magnetic field and the point where it strikes the photograph plate if $V = 2000$ V and $B = 0.3$ Wb/m^2.
(a) Singly and doubly ionized ^{7}Li atoms of mass 7.01601 u; and
(b) singly ionized ^{12}C atoms of mass 12.00000 u.

4. A beam of singly ionized boron atoms $^{10}B^+$ and $^{11}B^+$ are used in a Dempster mass spectrograph. Calculate the separation between these ions on a photographic plate if the potential difference is 2400 V and the magnetic field is 0.25 Wb/m^2.

5. An oil drop of mass 0.01 g between two parallel plates in a Millikan oil-drop experiment is balanced by an electric field of 1000 N/C. Calculate the magnitude of the charge on the oil drop and the direction of the electric field.

6. An oil drop carrying a charge of $+12e$ is balanced by an electric field of 1500 N/C. Calculate the mass of the drop.

7. Show that the force of the gravitational attraction between an electron and a proton in a hydrogen atom is negligible compared to the force of the electrostatic attraction between them. Assume the radius of the hydrogen atom to be 0.53 Å. (See Chapter 19).

8. Consider an electron in the hydrogen atom moving in the first Bohr orbit. Calculate the magnetic field produced by this electron at the center (see Chapter 24).

9. The range of transition in the Balmer series lies between 6563 and 3645 Å. Calculate this range in terms of energy units.

10. Suppose that the radii of the orbit of an electron in the hydrogen atom are 0.00000001, 0.000001, 0.0001, 0.1, and 1 m. Calculate the values of the quantum number n corresponding to these orbits. Which of these must be treated quantum mechanically?

11. Calculate the energies and the wavelengths corresponding to the following transitions of electrons from one energy state to another in the hydrogen atom: (a) $n_i = 2 \rightarrow n_f = 1$; (b) $n_i = 6 \rightarrow n_f = 5$; (c) $n_i = 11 \rightarrow n_f = 10$; (d) $n_i = 51 \rightarrow n_f = 50$; and (e) $n_i = 101 \rightarrow n_f = 100$. For what approximate value of n would you say that the discrete nature of the system is lost?

12. Calculate the ionization energy of the hydrogen atom from the following data. The wavelength of the Balmer series limit is 3645 Å and the wavelength of the first line in the Lyman series is 1215.7 Å (see Figure 32.12). How does this value compare with the one calculated from the wavelength of the series limit, $\lambda = 911.3$ Å of the Lyman series?

13. Suppose that an electron in the first excited state of the hydrogen atom stays in that state for 10^{-8} s. Using the uncertainty principle (Chapter 31), calculate the width (in eV) of this excited level. In this time interval, how many revolutions does this electron make before making a transition to the ground state?
14. Using Figure 32.13, calculate the angles between the angular momentum L and its components L_z along the Z-axis for $l = 1$ and $l = 2$.
15. What are the values of the orbital angular momentum for the f and g electrons according to the quantum-mechanical model?
16. What are the possible values of j for $l = 2$, that is, for a d electron in an atom? Also, calculate the possible values of the total angular momentum according to the quantum-mechanical model.
17. What are the possible values of j for $l = 3$, that is, for a g electron in an atom? Also, calculate the possible values of the total angular momentum according to the quantum-mechanical model.
18. Calculate different possible states for $n = 3$.
19. Calculate different possible states for $n = 4$. What are the angular momenta for different states?
20. What are the electronic configurations of the atoms of the following elements: Si (Z = 14), Ca (Z = 20), and Br (Z = 35)?
21. What are the ground-state configurations of electrons of atoms of the following elements: aluminum (Z = 13), sulfur (Z = 16), cobalt (Z = 27), and silver (Z = 47)?
22. Consider an atom with an electron configuration of $1s^2 2s^2 2p^1$. What are the possible values of l, s, and j and the corresponding values of the angular momenta L, S, and J, respectively?
23. Consider an atom with an electron configuration of $1s^2 2s^2 2p^6 3s^2 3p^6$-$4s^2 3d^1$. What are the possible values of l, s, and j and the corresponding values of L, S, and J, respectively?
24. Suppose that atoms could be formed with principal quantum numbers up to $n = 8$. How many elements could be formed?
25. With the help of Figure 32.19, identify the possible transitions which would take place if sodium atoms were left in the following excited states: (a) $4P$; (b) $4D$; (c) $4F$; and (d) $5P$.
26. The two most prominent lines in the spectrum of sodium are the D-lines, with wavelengths of 5890 and 5896 Å. Calculate the separation (in eV) between the two levels from which these lines originate.
27. The series limit of the principal series of a certain element is 43,486 cm^{-1}. Calculate the ionization energy for this element.
28. Before an atom emits an x-ray, it is at rest. To conserve momentum when it emits an x-ray, it must recoil. Given that the mass of the atom is M and the energy of the x-ray is $h\nu$, what is the recoil energy of the atom? (Use the conservation of momentum and the relation $E = \frac{1}{2}Mv^2 = p^2/2M$.)
29. The wavelength of the K_α x-ray from copper is 1.377 Å. What is the energy difference between the two levels from which this transition results?
30. The wavelength of the K_α x-ray from calcium is 3.35 Å. What is the energy difference between the two levels from which this transition results?

31. Using the following data, make a plot of $\sqrt{\nu}$ versus Z for the *K* x-rays emitted by these elements. From this plot, estimate *A* and *S* in the expression $\sqrt{\nu} = A(Z - S)$, where $A = \sqrt{C_n}$.

Element	Z	Å
Magnesium	12	9.87
Sulfur	16	5.36
Calcium	20	3.35
Chromium	24	2.29
Cobalt	27	1.79
Copper	29	1.59

33
Molecular and Solid-State Physics

Interaction between atoms when they are brought together leads to the formation of molecules and solids. In most instances, the interactions between nuclei may be neglected because they never come very close to each other. Actually, the formation of molecules and solids takes place through the interaction between different electrons in the atoms. Many scientists today are conducting complex experiments in hopes of finding the nature of such interactions that lead to different types of bonding. Such investigations perhaps can help mankind by solving many problems of biology, such as those involving cell division, heredity, and various diseases such as cancer, in addition to problems of physics, such as lasers, superconductivity, and many others.

33.1. Structure and Spectra of Molecules

A *diatomic molecule* is formed when two atoms combine to form a stable system while in *triatomic molecules* three atoms combine to form a stable system. In *polyatomic molecules* more than three atoms combine to form a stable system. The forces between the atoms (molecular forces) must be attractive to form stable bound molecules. For simplicity, we shall limit our discussion to diatomic molecules. Different molecules and solids can be classified into four groups according to the type of bonding. Of the four types of bonding: ionic, covalent, van der Walls, and metallic, the first two types are very common in molecules and will be briefly discussed. In some cases more than one type of bonding acts simultaneously.

Ionic Bonding

In this type of bonding, when two atoms are brought together, one or more electrons are transferred from one atom to the other, thereby forming positive and negative ions. These ions are attracted by electrostatic forces, forming stable molecules. Examples of such molecules are NaCl, KCl, and other salts. Figure 33.1 illustrates the formation of NaCl molecules. Such a molecule is

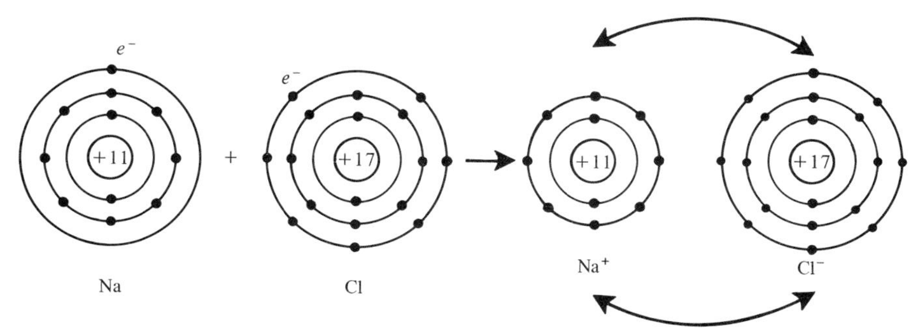

FIGURE 33.1. *Formation of sodium chloride molecule by ionic bonding.*

formed when sodium loses an electron and chlorine gains an electron, thereby forming Na^+ and Cl^- ions, respectively. These are then attracted to form stable molecules.

There are three different energies involved in the formation of the NaCl molecule.

1. *Energy needed to remove an electron from the sodium atom.* The sodium atom $_{11}Na$ consists of 11 electrons, with the eleventh electron in the 3s-state. The 3s-electron is very weakly bound and can be removed by adding energy of 5.1 eV;

$$Na + 5.1 \text{ eV} = Na^+ + e^- \tag{33.1}$$

Sodium is called *electropositive* because it easily loses an electron to form a positive ion.

2. *Energy given out when the Cl atom absorbs an electron.* The chlorine atom $_{17}Cl$ consists of 17 electrons and lacks one electron to complete the 3p-subshell to form a tightly bound system. Thus, a $_{17}Cl$ atom easily captures any available electron, thereby forming a Cl^- ion and releasing 3.8 eV of energy. That is,

$$Cl + e^- = Cl^- + 3.8 \text{ eV} \tag{33.2}$$

Thus, the Cl^- ion is electronegative and has an electron affinity of 3.8 eV.

3. *Energy due to the force of attraction between the ions.* Once the Na^+ and Cl^- ions have been formed, they are brought closer to form a molecule. The amount of energy emitted is equal to the Coulomb energy V' between $+e$ and $-e$ separated by a distance r:

$$V' = -\frac{1}{4\pi\epsilon_0}\left(\frac{e^2}{r}\right) \tag{33.3}$$

the plot of the total potential energy $V(r)$, which is the sum of the three terms above versus r, the distance between the two ions, is shown in Figure 33.2. To illustrate this point, let us consider a point at which $r = 2.4$ Å =

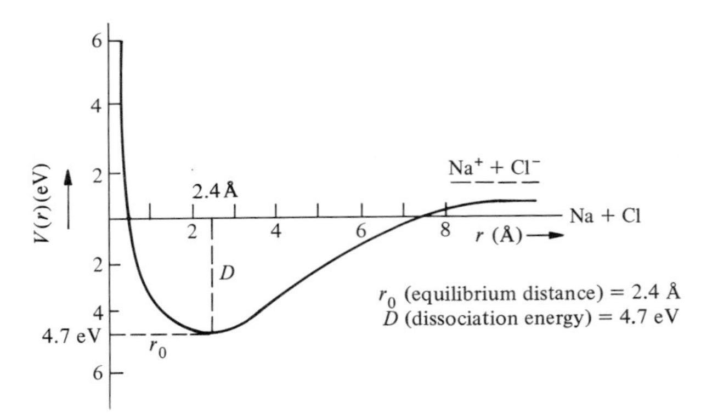

FIGURE 33.2. *Potential energy between Na^+ and Cl^- ions versus the distance between them. A stable NaCl molecule is formed when the ions are 2.4 Å apart.*

2.4×10^{-10} m. Equation 33.3 yields

$$V' = -9 \times 10^9 \frac{\text{N-m}^2}{\text{C}^2} \frac{(1.6 \times 10^{-19}\ \text{C})^2}{2.4 \times 10^{-10}\ \text{m}} \frac{1}{1.6 \times 10^{-19}\ \text{J/eV}} = -6\ \text{eV}$$

Thus, when a NaCl molecule is formed at a distance of 2.4 Å, 6 eV of energy is emitted because of the Coulomb force between the ions. Besides this, from Equation 33.2, 3.8 eV is emitted when Cl^- is formed, and from Equation 33.1, 5.1 eV is given to a sodium atom to form a Na^+ ion. That is, 6 eV + 3.8 eV − 5.1 eV = 4.7 eV energy is emitted when a NaCl molecule is formed at a distance of 2.4 Å. [Note that this amount of energy (4.7 eV) must be supplied to a NaCl molecule to break it back into two atoms.]

The plot of the total potential energy $V(r)$ versus distance r between the two ions is shown in Figure 33.2. As these ions are brought from infinity, decreasing $V(r)$ results in attractive force. At very short distances increasing $V(r)$ results in repulsive force between the ions. Between the two extremes there must be an *equilibrium distance* r_0, corresponding to which the energy is minimum and the molecular system is stable. This minimum energy corresponding to the equilibrium distance r_0 is called the *dissociation energy* and is equal to the energy needed to break the molecule into its components. Thus, in the case of a NaCl molecule as shown in Figure 33.2, the equilibrium distance is $r_0 = 2.4$ Å and the dissociation energy is 4.7 eV. It is worthwhile to note that in a NaCl molecule, the space surrounding a Na nucleus is positive whereas that surrounding Cl is negative. The existence of these negative and positive poles gives the ionic molecules the name *polar molecules*.

Covalent Bonding

In this completely different type of bonding, the attractive force between the two atoms of a diatomic molecule is due to the sharing of one or two pairs of electrons by both the atoms. These sharing electrons spend more time between the two atoms. Examples of such molecules are H_2 and N_2.

To have a better understanding, let us consider Figure 33.3. Two hydrogen atoms, each having one electron, will both share the two electrons when brought together. Such bondings can be understood only quantum mechanically (by solving the Schrödinger equation and applying Pauli's exclusion principle). The resulting attractive forces are called *exchange forces*.

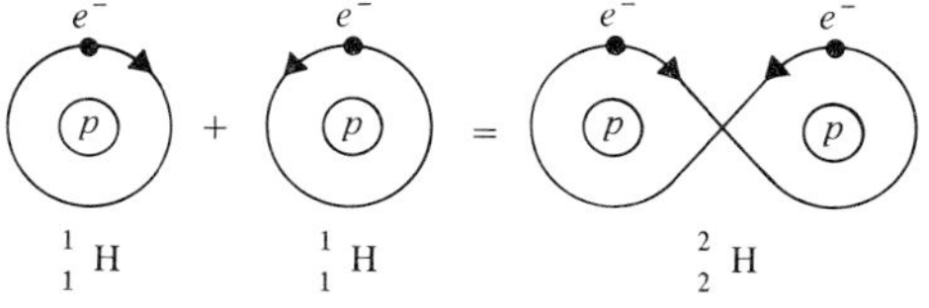

FIGURE 33.3. *When two hydrogen atoms are brought together to form a molecule, the two electrons are shared by both protons as shown.*

Without going into the details, we conclude that the plot of the potential energy $V(r)$ versus r for covalent diatomic molecules is similar to that for the ionic molecules. Hence, we may talk of an equilibrium distance r_0 and the dissociation energy. For example, for H_2 molecules, the equilibrium distance is 0.74 Å, while the dissociation energy is 4.476 eV (bonding energy = −4.476 eV).

Molecules such as H_2O are formed by another type of bonding, called *van der Waals' bonding*. These attractive forces of such bondings are very weak.

Molecular Spectra

Just as in the case of atomic spectra, the investigation of the emission and absorption spectra of molecules can yield information about the structure and energy levels of the molecules. It is simpler to separate the molecular energy levels into three different groups, resulting from three different types of motions.

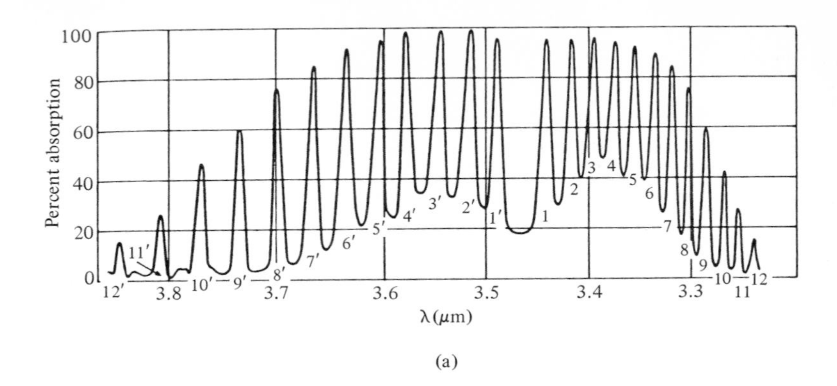

(a)

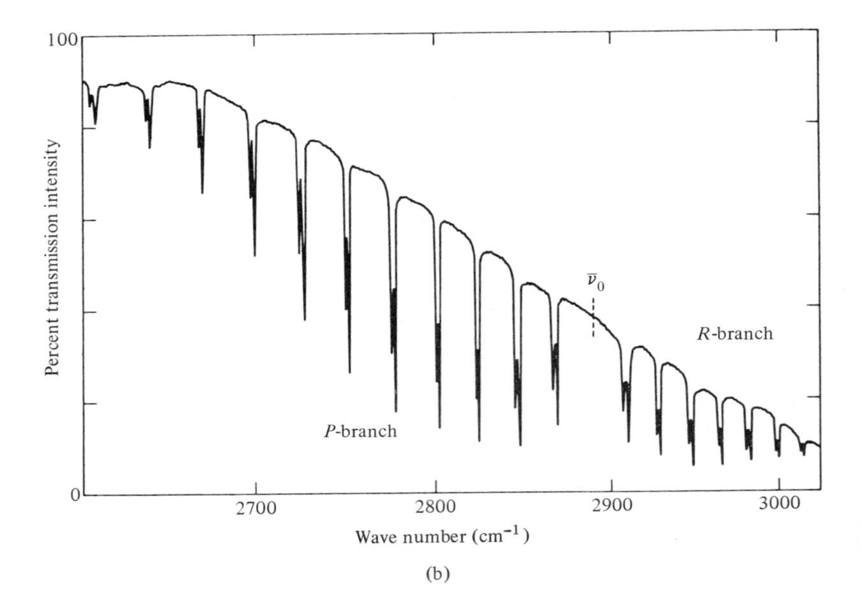

(b)

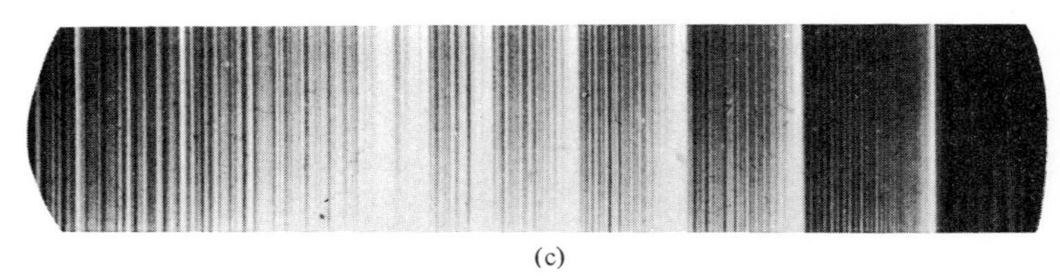

(c)

FIGURE 33.4. *(a) Pure rotational absorption spectrum of diatomic molecules of HCl in the gaseous phase. (b) Vibrational–rotational absorption spectrum of HCl molecules in the gaseous phase. The big dips are due to* $^1H^{35}Cl$ *molecules and the smaller dips to* $^1H^{37}Cl$ *molecules. (c) Photograph of a typical diatomic molecular spectrum (rotational, vibrational, and electronic), showing bands in a portion of the molecular spectrum of cyanogen (CN) in the ultraviolet region.* [Part (a) from G. Herzberg, *Molecular Spectra and Molecular Structure*, copyright © 1959 Litton Educational Publishing, Inc. Reprinted by permission of Van Nostrand Reinhold Company; part (b) from M. Alonso and E. J. Finn, *Fundamental University Physics*, Vol. III, Addison-Wesley Publishing Company, Inc., Reading, Mass., 1968.]

1. *Rotational energy levels* result from the rotational motion of the atoms of the molecules about their center of mass. These energy levels are in the range $\sim 10^{-4}$ eV. The emission and absorption spectra result in spectral lines separated by equal energy intervals, as shown in Figure 33.4(a).

2. *Vibrational energy levels* result from the vibrational motion of the molecules. The energy levels are equally spaced and are in the range $\sim 10^{-1}$ eV. Because of this high energy (as compared to 10^{-4} eV) when looking for vibrational spectra, the rotational spectra also appear side by side, as shown in Figure 33.4(b).

3. *Electronic energy levels* result from the excitation of electrons in atoms and are in the energy range ~ 1 to 10 eV. When the energy supplied is enough to cause excitation of the electronic energy levels of the molecules, the vibrational and rotational levels are excited. Thus, the resulting spectra consist of all three and appear as shown in Figure 33.4(c). Because of the nature of its appearance, such molecular spectra are called *band spectra.*

33.2. The Laser and the Maser

LASER stands for *light amplification by stimulated emission of radiation* and *MASER* stands for *microwave amplification by stimulated emission of radiation.* The idea of masers was introduced in 1954, those of lasers in 1958. We shall limit our discussion to lasers. The laser has not only made a great impact on the field of optics but to the other fields of physics, engineering, and medicine as well. As the name indicates, lasers are used for producing an intense, monochromatic, and unidirectional coherent beam of visible light. To understand the working of a laser, we must understand such terms as *stimulated emission, optical pumping,* and *population inversion.*

Consider a sample of free atoms some of which are in the ground state with energy E_1 and some in the excited state with energy E_2, as shown in Figure 33.5. If photons of energy $h\nu = E_2 - E_1$ are incident on this sample, the photons can interact with atoms in two different ways. In Figure 33.5(a) the incident photon is absorbed by an atom in the ground state E_1, thereby leaving the atom in the excited state E_2. This process is called *stimulated* or *induced absorption*. Once in the excited state, two things can happen to the atom. It may decay by *spontaneous emission*, as shown in Figure 33.5(b), in which the atom emits a photon of energy $h\nu = E_2 - E_1$ in any arbitrary direction. The other alternative for the atom in the excited state E_2 is to decay by *stimulated* or *induced emission*, as shown in Figure 33.5(c). In this case the incident photon of energy $h\nu = E_2 - E_1$ induces the atom to decay by emitting a photon that travels in the direction of the incident photon. For each incident photon we will have two photons going in the same direction. Thus, we have achieved two things: an amplified as well as an unidirectional coherent beam. From a practical point this is possible only if there is more stimulated or induced emission than spontaneous emission. This can be achieved as explained next.

Let us consider a simple case of a material whose atoms can reside in three different states, as shown in Figure 33.6: state E_1, which is the ground state; the excited state, E_3, in which the atoms can reside only for 10^{-8} s; and the metastable state, E_2, in which the atoms can reside for $\sim 10^{-3}$ s (much longer than 10^{-8} s). Also, let us assume that the incident photons of energy $h\nu'$

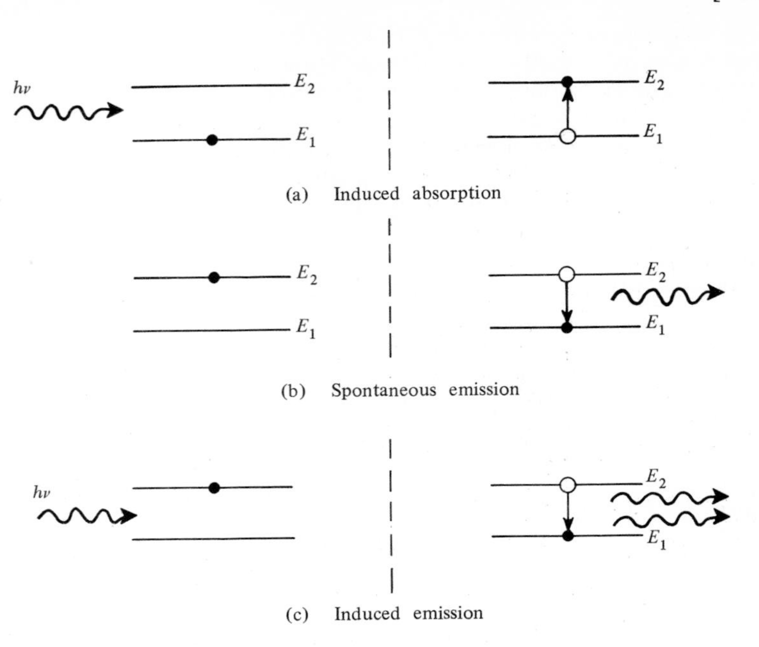

Figure 33.5. *Schematic representation of the interaction of photons with atoms in energy states E_1 and E_2. (a) Induced absorption, (b) spontaneous emission, and (c) stimulated or induced emission.*

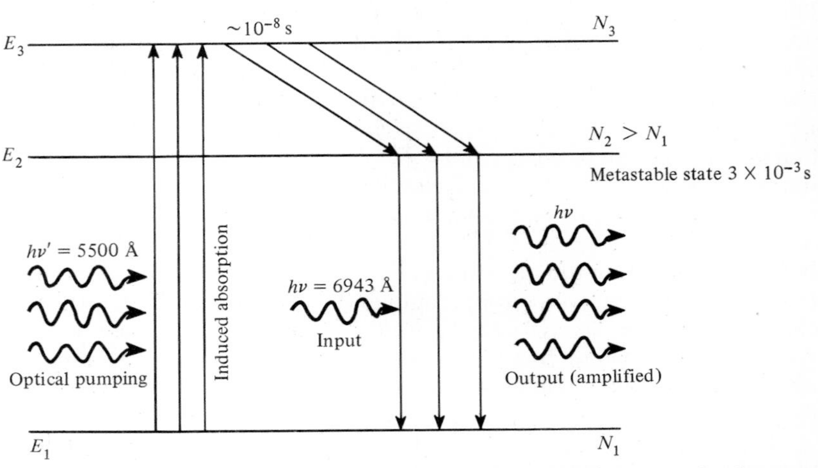

Figure 33.6. *Schematic representation of the working of a ruby laser, showing optical pumping, population inversion, and amplification.* (From Atam P. Arya, *Elementary Modern Physics*, Addison-Wesley Publishing Company, Inc., Reading, Mass., 1974.)

(= $E_3 - E_1$) raise the atoms from the ground state E_1 to the excited state E_3, but the excited atoms do not decay back to E_1. Thus, the only alternative for atoms in the excited state E_3 is to decay to state E_2. Since the lifetime of E_2 is much longer than E_3, the atoms reach state E_2 much faster than they leave state E_2. This eventually leads to the situation that the state E_2 contains more atoms than state E_1; that is, we have a *population inversion*. The process by which this is achieved (as shown in Figure 33.6) is called *optical pumping*.

Once the population inversion has been reached, the lasing action of a laser is simple to achieve. The atoms in metastable state E_2 are bombarded by photons of energy $h\nu = E_2 - E_1$, resulting in an induced emission, giving an intense, coherent beam in the direction of the incident photons.

Of the several available lasers, the two most common lasers are the ruby laser and the helium–neon laser. As an example, we first describe the ruby laser. It consists of 0.05 percent chromium (Cr) atoms in aluminum oxide. The energy-level diagram of these Cr ions (which produces the laser action) is shown in Figure 33.6. The optical pumping results when incident photons of wavelengths 5500 Å raise the Cr ions from state E_1 to E_3. Once in E_3 the ions decay to the metastable state E_2, producing population inversion. When the metastable state is exposed to ruby red light of wavelength 6943 Å, stimulated or induced emission results in a strong, coherent laser beam.

The physical construction of a ruby laser is as shown in Figure 33.7. An external light source is used to achieve optical pumping. On the other hand, a few spontaneous transitions of ruby red light from state E_2 are used to initiate further induced or stimulated emission from state E_2 itself. The ends of a cylindrical ruby laser are optically ground so that the beam is reflected many times before it emerges. A highly intense laser beam is thus obtained.

The other most commonly used laser is the Helium–Neon (He–Ne) laser. It consists of a discharge tube containing helium and neon in the ratio of 7 to 1 at a total pressure of about 1 torr (1 mm of Hg). Helium atoms are excited to metastable states by means of collisions with the electrons in the discharge tube. The excited He atoms collide with unexcited neon atoms, thereby transferring their energy and resulting in a population inversion in Ne atoms. The laser action takes place in these neon atoms with corresponding transitions of wavelengths 6328 Å, 11,523 Å, and 3.39 μm. (Beam is shown in Figure 29.18.)

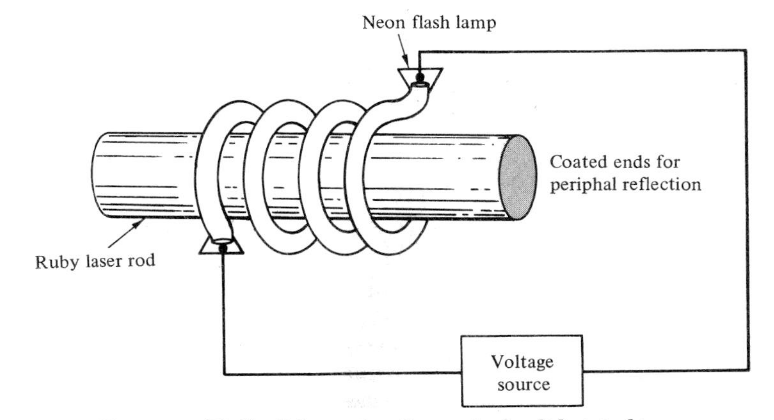

FIGURE 33.7. *Schematic of a typical solid-state laser.*

33.3. Holography

Truly three-dimensional images can be formed without the use of lenses by an altogether different and unusual method of producing an image called *holography*. The method is based on the principle of *wave-front reconstruction*. This principle was suggested by D. Gabor in 1948, but not much was done until the highly coherent light beam of the laser became available.

To make a hologram, a highly intense laser beam is divided into two parts by means of a mirror. (One may accomplish this also by using a half-silvered mirror, which works as a beam splitter.) One part of the beam illuminates the subject, as shown in Figure 33.8(a). The other beam, called the *reference beam*,

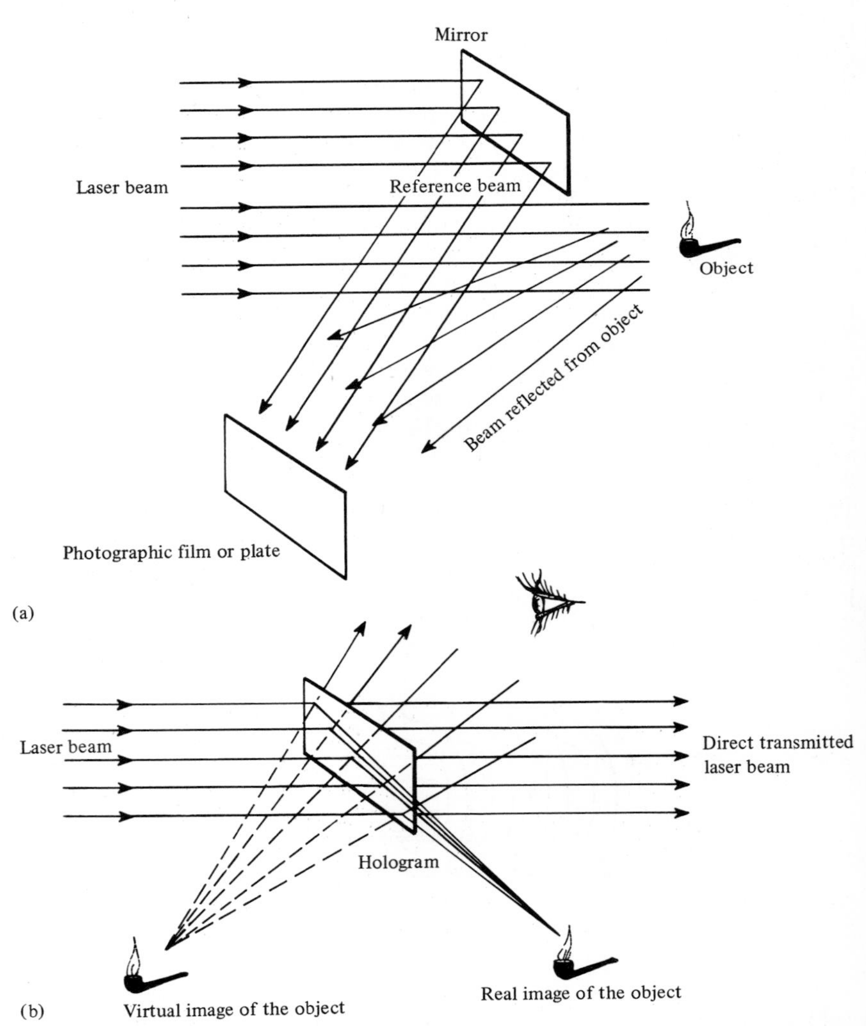

Figure 33.8. *(a) Experimental arrangement to produce a hologram. (b) Reconstruction of a real and virtual images by illuminating the hologram.*

after being reflected from the mirror is received on the photographic plates. Thus, the film is exposed simultaneously both to the reference beam as well as the beam reflected from the subject, as shown in Figure 33.8(a). These two beams are coherent (in phase) and highly directional. The resulting complicated interference pattern is formed in the film. This film, after developing, is a *hologram*. The density at any point of the hologram depends both on the amplitudes and phases of the interference waves that reach the point (unlike the ordinary interference pattern, which contains only the amplitude information). Thus, the film contains all the information needed to reproduce the wave field of the subject.

To view the subject, the image is reconstructed by illuminating the hologram with a single beam from a laser, as shown in Figure 33.8(b). Both a real and a virtual image are formed, as shown. When the viewer moves his eyes from side to side, the near parts of the subject seem to move with respect to the far parts. Thus, when viewing the hologram, one sees the image depth, hence a three-dimensional effect.

In holography, whether we use a positive or a negative transparency, the viewer always sees a positive image. This is because the eye is insensitive to a 180° phase change—the phase change by which the positive and the negative transparencies differ. It is also possible to make holograms for which the viewer can move in a 360° path around the image viewing all sides, true three-dimensional viewing.

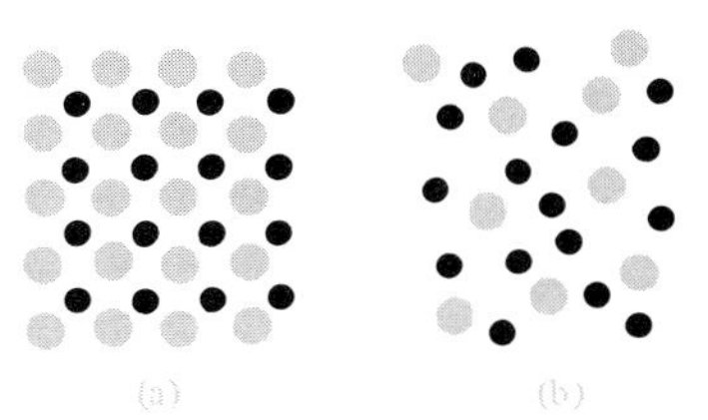

FIGURE 33.9. *(a) Crystalline solid and (b) amorphous solid.*

33.4. The Structure of Solids

Let us consider a large number of atoms or molecules. Bringing them together will result in the formation of a gas, liquid, or solid. If a gas is formed, the molecules are very far apart, the distances are not fixed, and the forces between them are very weak. On the other hand, when a solid is formed, the molecules are always at fixed distances, with forces stronger than in the case of gases. The formation of the liquid state is in between the other two states.

Solids formed may be in crystalline form or amorphous. In a *crystalline solid*, the atoms or the molecules are arranged in a regular three-dimensional pattern; in an *amorphous solid*, there is hardly any regularity, as shown in Figure 33.9. KI, NaCl, CsCl, and C are examples of crystalline solids, and plastic, glass, and wood are examples of amorphous solids; B_2O_3 (boron trioxide) is a solid that comes in both forms.

As in atoms, the *binding energy* or *cohesive energy* of a crystal is the amount of energy needed to separate it into its constituent atoms. Thus, the binding energy is a negative quantity and is proportional to the volume of the crystal. The binding or cohesive energy may be expressed in units of electron volts per atom or molecule or in units of kilocalories per mole (1 eV/atom = 23.52 kilocalories/mole).

Depending upon the strength of the bonding, the crystals have been classified into the following five categories: (1) ionic crystals, (2) covalent crystals, (3) hydrogen-bonded crystals, (4) molecular crystals, and (5) metallic crystals.

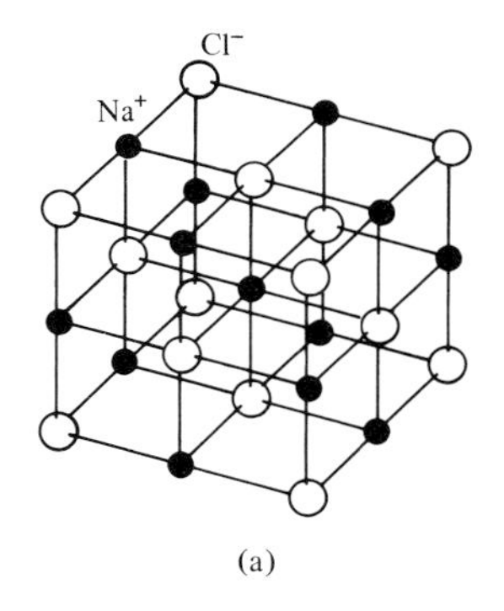

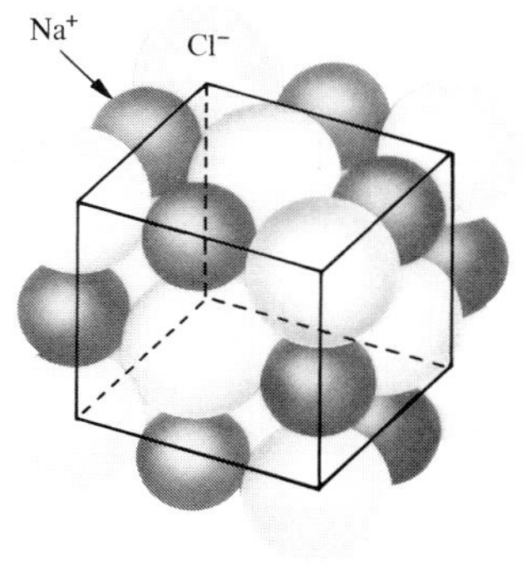

NaCl

FIGURE 33.10. *Structure of a face-centered-cubic crystal—NaCl: (a) geometrical arrangement and (b) scale model.*

Ionic Crystals

Figure 33.10 shows the structure of a typical ionic crystal such as NaCl. $CsCl_2$ is another example. The ionic bonding in crystals is the same as the ionic

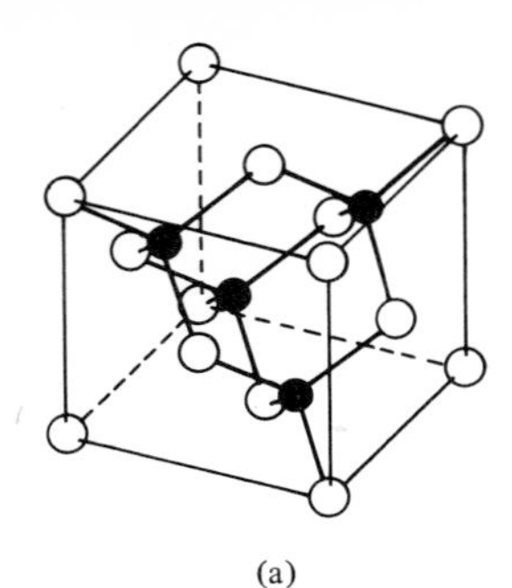

(a)

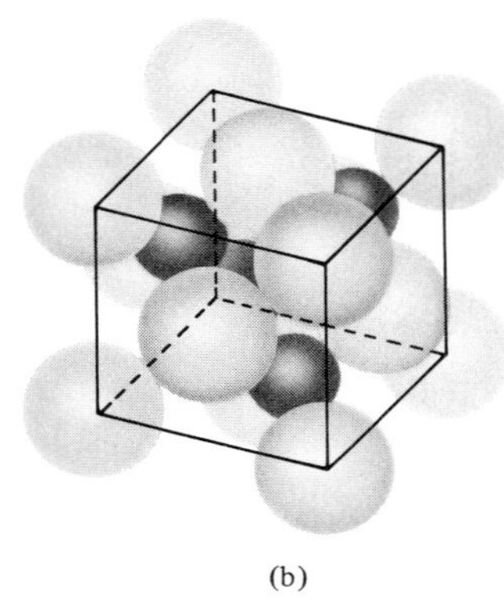

(b)

Diamond

FIGURE 33.11. *Covalent-bonded crystal with tetrahedral structure—diamond: (a) geometrical arrangement and (b) scale model.*

bondings in molecules discussed in Section 33.1. As before, the bonding is due to the force of attraction between oppositely charged ions.

The typical binding energies for ionic crystals are 5 to 10 eV/molecule. Owing to these ionic bonds, the ionic crystals are hard and brittle, with reasonably high melting points. Because ionic crystals lack free electrons, they are poor conductors of heat and electricity. The ionic molecules or crystals are easily soluble in water.

Covalent Crystals

Figure 33.11 shows the structure of a typical covalent crystal such as diamond. Silicon and germanium are two more examples. The covalent bonding in crystals is similar to molecular bonding such as in H_2; that is, the sharing electrons spend more time between the atoms.

Covalent bondings are very strong, with binding energies of 6 to 12 eV/atom. Diamond and Si, with binding energies of 7.4 eV/atom and 12.3 eV/atom, are the hardest things known. Covalent crystals are hard to break or deform, have very high melting points, and are insoluble. Lack of free electrons makes them poor conductors of heat and electricity. In most covalent crystals, the energy of the first excited state is more than the energy of the visible light—thus, most covalent crystals are transparent to visible light.

Hydrogen-bonded Crystals

Normally, hydrogen atoms are supposed to form covalent bonds. But hydrogen atoms, in combining with such electronegative atoms as oxygen, nitrogen, and fluorine, form *hydrogen bonds*. These are weak bonds with binding energies of $\sim$0.5 eV/molecule. Examples of hydrogen-bonded crystals are H_2O, NH_3, and HF. In most cases, chains of molecules are formed, such as the formation of ice shown in Figure 33.12. Other examples are HF·HF and HF·HF·HF.

Molecular Crystals

The outermost shells of the inert gases He, Ne, Ar, Kr, Xe, and Rn are completely filled with electrons; hence, covalent bonding cannot occur. These and other molecules, such as Cl_2, I_2, and CH_4, form *molecular crystals* when in solid forms. The bonds are very weak, with binding energies of $\sim$0.1 eV/molecule. The forces forming these crystals are called *van der Waals' forces*. Because of such weak bonding, these crystals are easily deformed. The lack of free electrons makes them poor conductors of heat and electricity. We may remind ourselves that such characteristics as friction, surface tension, and viscosity are explained by the van der Waals forces.

Metallic Crystals

When atoms with weakly bound outermost electrons are brought together, metallic crystals are formed. In the process of bonding, the kinetic energy increases at the cost of the potential energy. These weakly bound electrons become free to roam in the crystal. Thus, such crystals, because of the availability of free electrons, are good conductors of heat and electricity. Visible photons are easily absorbed by these electrons, thereby making the metals appear opaque.

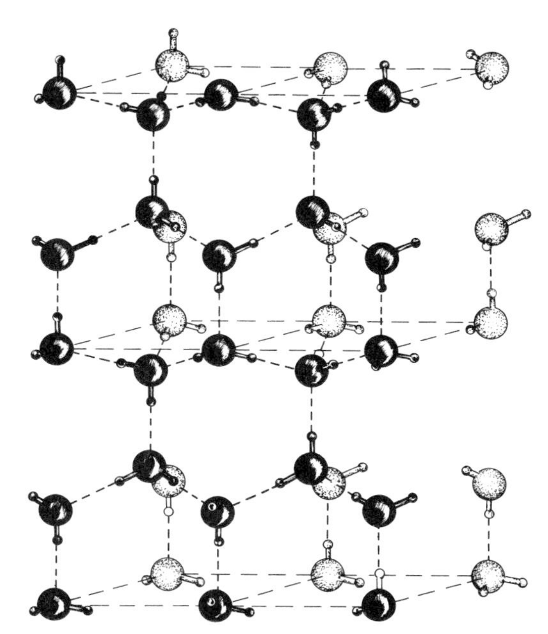

FIGURE 33.12. *Chain structure of water molecules in ice.* (From Linus Pauling, *The Nature of the Chemical Bond.* © 1939 and 1940, Third Edition © 1960 by Cornell University. Used by permission of Cornell University Press.)

The maximum energy of the free electrons in the crystal is called the *Fermi energy,* E_F.

33.5. The Band Theory of Solids

In Chapter 21, making use of the electrical conductivities of materials, we classified different materials into three different categories: conductors, insulators, and semiconductors. *Conductors* are good conductors of electricity, such as copper, silver, and other metals. *Insulators* are poor conductors of electricity, such as wood, diamond and quartz. *Semiconductors* are those which have electrical conductivities between the conductors and the insulators. Two typical examples are silicon (Si) and germanium (Ge). How this classification comes about can be understood according to the *band theory of solids,* explained below (also called the *tight-binding approximation theory*).

Suppose that an electron in each atom is tightly bound to its nucleus. Let Figure 33.13(a) represent the energy levels of such an atom. If a few such atoms are brought together, there will be discrete energy levels of slightly different energies corresponding to each state [as shown in Figure 33.13(b)]. In a solid there are a large number of atoms. In this case, the new energy levels are very, very close and lead to the formation of bands, as shown in Figure 33.13(c). The energy levels in all the bands are continuous, as shown for the $1s$-, $2s$-, $2p$-, and $3s$-bands. If there are N atoms, each l-band will contain $2(2l + 1)N$ states,

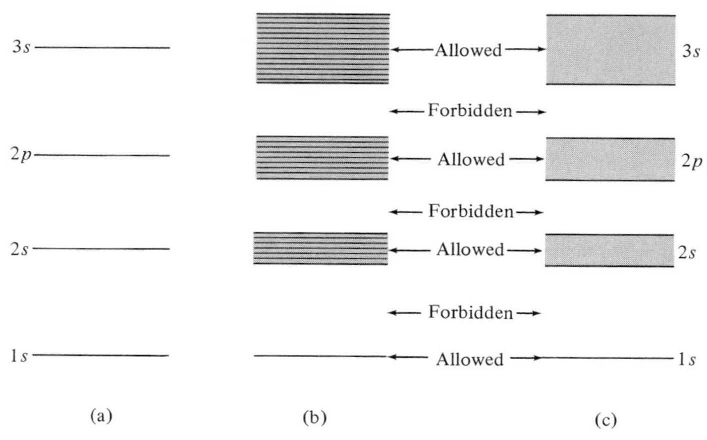

FIGURE 33.13. *(a) Energy levels of an isolated atom. (b) Solid with a small number of atoms; resulting in energy bands having discrete energy levels. (c) Solid containing a large number of atoms; resulting energy levels in the bands are almost continuous.*

where $(2l + 1)$ represents the number of m_l-states corresponding to a given l and 2 represents the two values of $m_s (= \pm\frac{1}{2})$ for a given s.

The band formation in different solids is different. In general, these may be classified into four categories. This classification helps in distinguishing among insulators, semiconductors, and conductors, as will be explained below. The electrons can reside or move about in these bands only, and these bands shown shaded are called *allowed bands*. The spacing or regions between the allowed bands are called *forbidden regions* or *gaps* because no electron can be placed there. The lower bands are usually filled. The uppermost band, which may be partially or completely filled, is called the *valence band*.

Insulators

If the band structure of a material is such that the forbidden region between the highest completely filled band and the lowest allowed empty band is very wide, ~3 to 7 eV, as in Figure 33.14(a), the material is an insulator. A few typical values of forbidden region widths for insulators are given in Table 33.1. Since the lowest band is completely filled with electrons, there cannot be any current. The only way to obtain current is to supply enough energy (by heating or applying an electric field) so that the electrons in the filled band will move up to the empty band and will be free to move. To produce these very high energies, high temperatures are needed. Thus, materials of this type are poor conductors of heat and electricity and are called *insulators*.

Semiconductors

If the band structure is such that the forbidden region between the highest completely filled band and the lowest empty band is very narrow, ~0.1 to 1 eV, as shown in Figure 33.14(b), the material is a *semiconductor*. A few examples are given in Table 33.1. As shown in Figure 33.15, the electrons from filled bands can be moved to the empty bands by supplying small amounts of energy. This can be accomplished by thermal excitations or by applying an

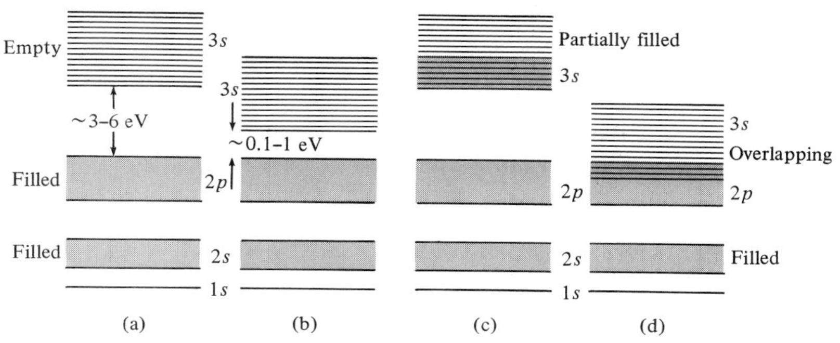

Figure 33.14. *Solids can be classified into three types, according to the way the energy bands are formed. (a) A solid is an insulator if the gap between the bands is very large. (b) A solid is a semiconductor if the gap between the bands is very small. (c) and (d) A solid is a conductor if the upper band is partially filled, as in (c), or if the filled and the empty bands overlap, as in (d).*

Table 33.1
Forbidden Energy Gaps at Room Temperature

Insulators	Gap (eV)	Semiconductors	Gap (eV)
Diamond	5.33	Silicon	1.14
Zinc sulfide	3.6	Germanium	0.67
Zinc oxide	3.2	Tellurium	0.33
Silver chloride	3.2	InSb	0.23
TiO_2	3	Tin (gray)	~0.1

electric field as shown in Figure 33.15(b). Once in the excited band, the electrons are free, hence will be responsible for conductivity.

The current in semiconductors is not only due to these electrons, but also to the *holes* (or vacancies) left behind at the site of the electrons when they leave the filled band. The electrons from nearby sites can be made to move to fill these holes—which is equivalent to saying that the hole has moved to a different energy state [Figure 33.15(c)]. The overall effect is that the holes behave like positive electrons and contribute to the electric current.

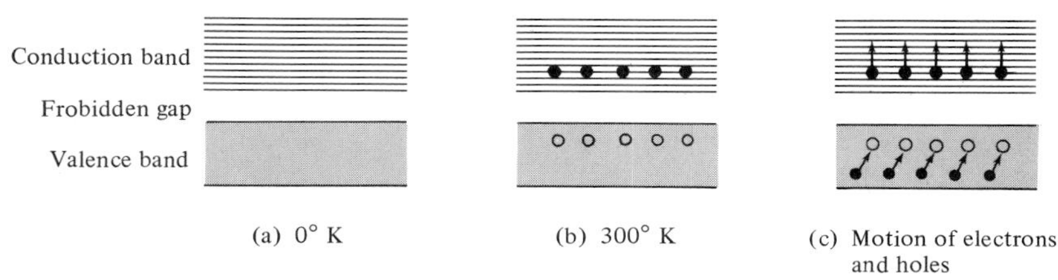

Figure 33.15. *Bands of electrons (shaded area) in a pure semiconductor. Filled circles denote thermally excited electrons and open circles denote holes.*

Conductors

If the band structure is such that the uppermost band (the valence band) is partially filled, as in Figure 33.14(c), or if completely filled it overlaps with the next allowed empty band, as shown in Figure 33.14(d), the material is a *conductor*. The corresponding bands are the *conduction bands*. In either case, there are unoccupied states available for the electrons to move, hence to produce current. All metals and some nonmetals exhibit such behavior and are good conductors of heat and electricity.

33.6. Semiconductors

Intrinsic and Impurity Semiconductors

Intrinsic semiconductors are those in which the transfer of electrons for the conduction and creation of holes can be accomplished by means of thermal excitation (by raising the temperature of the semiconductors), as shown in Figure 33.15. The electrons and the holes so created are called *intrinsic charge carriers*. Their contribution to the conductivity of the material is called *intrinsic conductivity*. This intrinsic conductivity of the semiconductor can be increased tremendously by adding certain types of impurities. Addition of impurities leads to the formation of impurity semiconductors of two different types—*n*-type and *p*-type as described below.

Let us consider a pure crystal of semiconductor silicon or germanium, each of which has four valence electrons. As an impurity, add a few atoms of phosphorus or arsenic, each of which has five valence electrons. The impurity atom replaces an atom of silicon or germanium at the lattice site. Four of the five valence electrons of impurity atom form covalent pair bonds with the four electrons of the neighboring atoms of the semiconductor, as shown in Figure 33.16(a). The fifth electron of the impurity atom is left alone and is very weakly bound, and it occupies a discrete energy level just below the conduction band, as shown in Figure 33.17(a). The separation between this new level

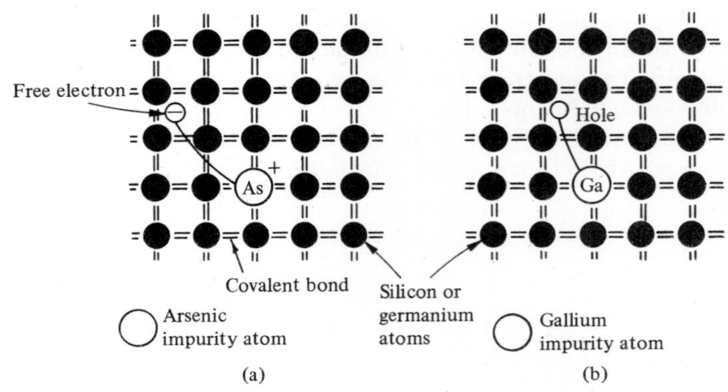

FIGURE 33.16. *(a) An n-type semiconductor; the impurity atom supplying the extra electron is called the donor atom. (b) A p-type semiconductor; holes are created by transferring electrons to the impurity atom, called an acceptor.* (From Atam P. Arya, *Elementary Modern Physics*, Addison-Wesley Publishing Company, Inc., Reading, Mass., 1974.)

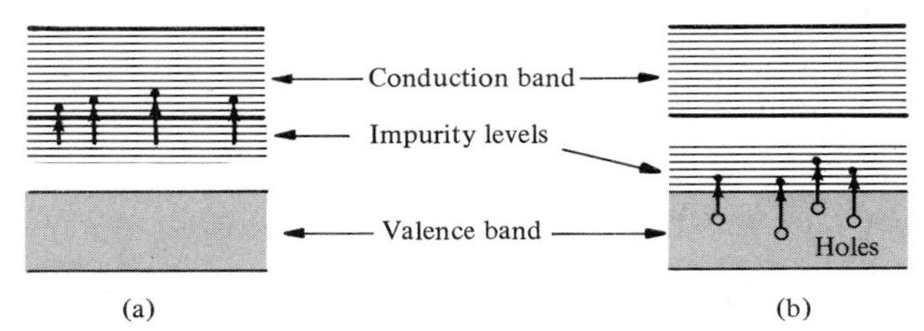

Figure 33.17. *Formation of impurity levels in (a) n-type materials and (b) p-type materials.*

and the conduction-band level may be only a few tenths of an electron volt. Thus, when the temperature is increased, the extra electrons jump into the conduction band, contributing to the electrical conductivity of the semiconductor. These electrons are *in addition to* the electron-hole pairs produced by thermal excitations of pure semiconductors. Thus, overall there are more electrons than holes to serve as charge carriers; hence, such materials are called *n-type semiconductors* or negative semiconductors.

Let us now add a few atoms of In, Al, or Ga (each with three valence electrons) as impurities into Ge or Si. Three valence electrons of the impurity atom form covalent pair bonds with three of the four electrons of neighboring atoms of the semiconductor. The fourth electron of the semiconductor cannot form a covalent pair bond. This lack of electron to form a pair is equivalent to a hole created at the cite of the impurity atom, as shown in Figure 33.16(b). It is possible that an electron from the nearest semiconductor atom may jump to fill the hole. Thus, the hole is transferred to the atom from which the electron came, as shown in Figure 33.17(b). Thereby, a negative charge is imposed on the impurity atom, and the resulting impurity levels are as shown. The holes behave like positive charge carriers. A crystal of this type always has an excess of holes or positive charge carriers and is called a *p-type semiconductor* or *positive semiconductor*.

The conductivity of an intrinsic semiconductor increases manyfold when only one impurity atom is added to 1 million intrinsic semiconductor atoms. The process by which impurities are diffused into a semiconductor is called *doping*.

Semiconductor Devices

The name "solid-state devices" is a misnomer for semiconductor devices. These devices have replaced electronic diodes and triodes, and are being used in rectifiers, amplifiers, photocells, and switching circuits, to name just some applications. We shall briefly describe a few of these.

Semiconductor Diode or *p-n* Junction Suppose that we place two crystals of Ge (or some other semiconductor), one a *p*-type and the other a *n*-type as shown in Figure 33.18. When these crystals are placed in contact, holes flow from left to right, and electrons from right to left. (In actual practice, one tiny crystal is treated in such a way that one side is a *p*-type and the other a *n*-type.) The flow of holes and electrons continues until there are positive and negative layers of charges on the right and the left sides of the junction. When equilibrium is reached, there will be a potential difference between the two layers. This is called a *contact potential*.

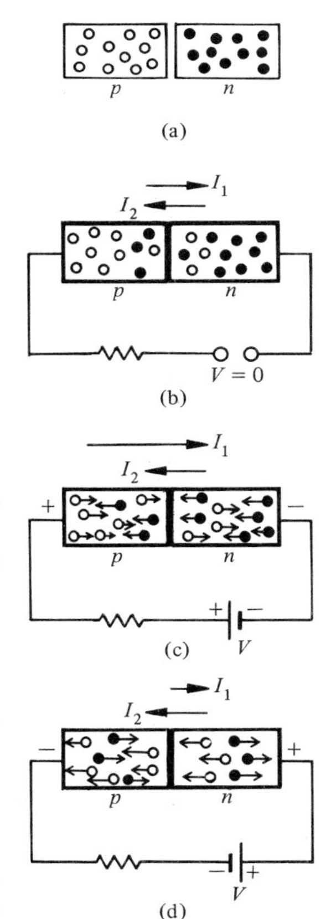

Figure 33.18. *Examples of p- and n-type materials (a) when not in contact, (b) when placed in contact, (c) when forward-biased, and (d) when reversed biased.*

By connecting the positive terminal of a voltage source to a *p*-type and the negative terminal to a *n*-type, as shown in Figure 33.18(c), a potential difference *V* is established between the two ends. The electrons in the *n*-type are pushed across the junction by the − terminal of the voltage source and are also attracted by the + terminal of the voltage source. On the other hand, the positive holes in the *p*-type are pushed in the opposite direction to that of the electrons, as shown in Figure 33.18(c). The dual motions of electrons and holes lead to a very large current, and the *p-n* junction or diode is said to be *forward-biased.*

Let us now consider the opposite case, shown in Figure 33.18(d). Here the negative terminal of the source is connected to the *p*-type and the positive terminal to the *n*-type. In this situation the electrons in the *p*-type should flow across the junction from the *p*-type to the *n*-type, but there are not many electrons in the *p*-type; hence, the current is very small. Similarly, the holes in the *n*-type should flow across the junction from the *n*-type to the *p*-type. Again, there are not many holes in the *p*-type; hence, the net current is very small and the diode is said to be *reverse-biased.*

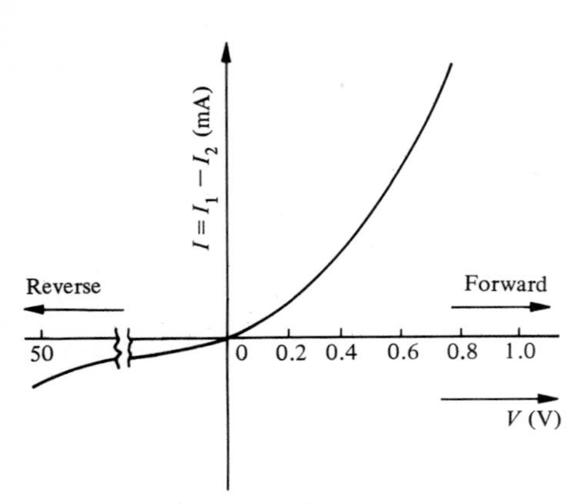

FIGURE 33.19. *Plot of resultant current I $(= I_1 - I_2)$ versus the applied voltage V across a p-n junction (V applied from p to n is forward-biased; from n to p is reversed-biased).*

The plot of the net current across the *p-n* junction versus the applied voltage *V* is shown in Figure 33.19 (both for forward-biased and reverse-biased). It is clear from this plot that there is more current in the direction from *p* to *n* than from *n* to *p*. This characteristic of the *p-n* junction is utilized in rectifiers, as shown in Figure 33.20 (compare this with the rectifier action of a vacuum tube shown in Figure 26.22).

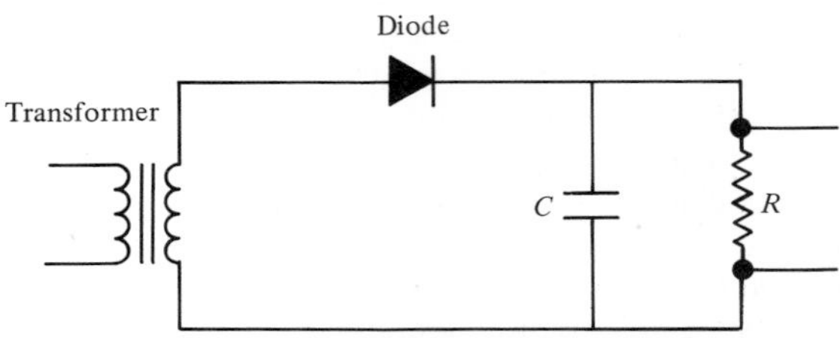

FIGURE 33.20. *A p-n junction when reversed-biased is used as a rectifier.*

JUNCTION TRANSISTOR. When a single piece of a semiconductor material consists of three parts with the central portion being *p*-type while either side is *n*-type, or if the central portion is *n*-type while either side is *p*-type as shown in Figure 33.21(a), we have a *n-p-n* or *p-n-p* junction transistor or simply a transistor. The central portion in either case is very thin, a few thousandths of a centimeter. The transistor plays the same role as a triode vacuum tube.

The *n-p-n* transistor shown in Figure 33.21(a) may be used in many electronic circuits in different ways. A typical use of a *n-p-n* transistor is as an amplifier, as shown in Figure 33.21(b). (It is equivalent to the triode amplifier shown in Figure 26.27). The central portion, the left portion, and the right portion play the role of base, emitter, and collector, respectively. The portion of the transistor to the left of the center is forward-biased and that on the right is reversed-biased. A circuit arrangement of this type is called *common-base operation.*

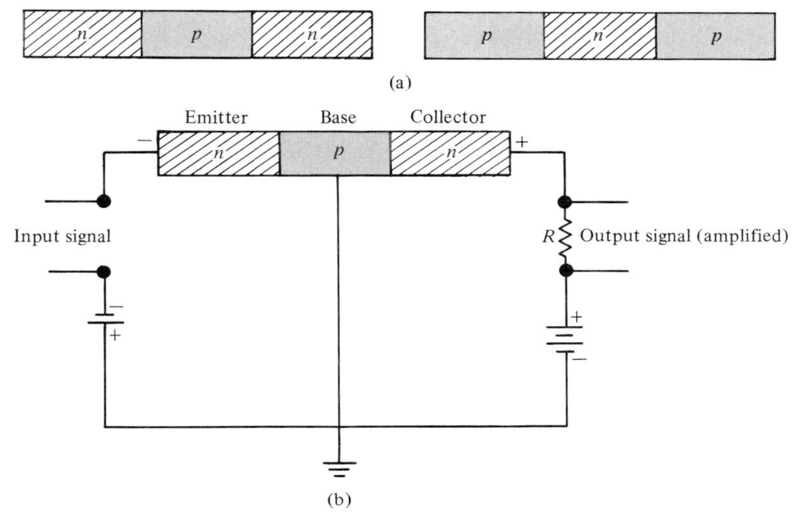

Figure 33.21. *Examples of (a) n-p-n and p-n-p transistors and (b) an n-p-n transistor being used as an amplifier.*

33.7. Superconductivity

So far we have been dealing with cases where the resistivity of a metal conductor decreases with decreasing temperature. In 1911, Kamerlingh Onnes measured the resistivity of mercury in the neighborhood of 4 K, the plot of which is shown in Figure 33.22. It is clear from this plot that the resistance of mercury drops very abruptly, to almost zero below 4.2 K. Since then, many metals and even many poor conductors such as lead are found to have abnormally low resistivity (or very high conductivity), similar to Figure 33.22. This property is called *superconductivity* and the material under such conditions is said to be in the *superconducting state.*

The temperature below which the material is in a superconducting state is called the *transition* or *critical temperature* T_c. Table 33.2 lists such temperatures for some materials.

Besides the property of superconductivity, another important characteristic

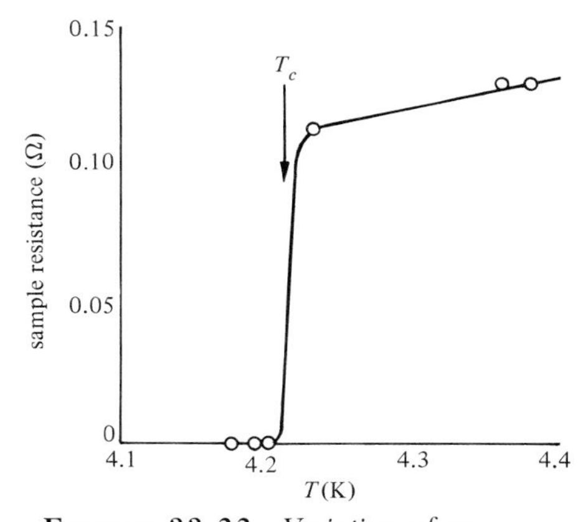

Figure 33.22. *Variation of resistance with temperature for mercury, showing the transition from the normal to the superconducting state at* T_c*K.* [From H. K. Onnes, *Leiden Commun.* **124C** (1911).]

Table 33.2
Some Superconducting Materials with Their Transition Temperatures

Element	T_c (K)	Compound	T_c (K)
Indium	0.14	$AnSb_2$	0.58
Titanium	0.40	$PdSb_2$	1.25
Aluminum	1.19	NiBi	4.25
Mercury	4.15	NbN	16.0
Lead	7.18	Nb_3Sn	18.05
Technetium	11.2	$Nb_3Al_{0.8}Ge_{0.2}$	20.05

of a superconducting state is the *Meissner effect* or *exclusion efflux*. Consider a superconductor and cool it until it is below the critical temperature. Now place it in an external magnetic field as shown in Figure 33.23. As is clear, the magnetic field cannot penetrate inside the conductor; that is, the magnetic flux is excluded from the conductor as shown in Figure 33.23(b). Now suppose that we switch off the field; the flux reduces to zero when $H = 0$, as shown in Figure 33.23(c).

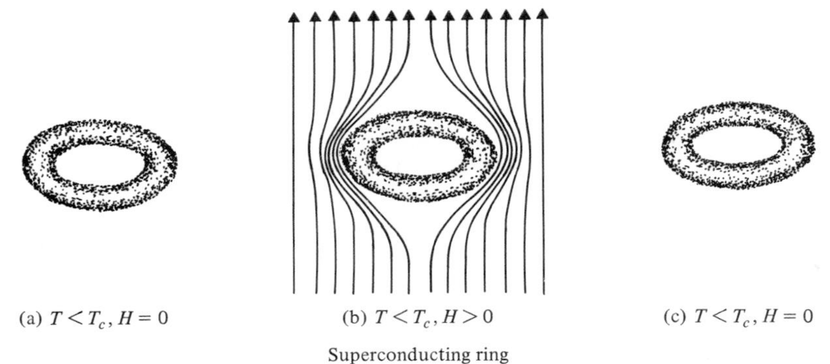

FIGURE 33.23. *(a) Superconductor at temperature $T < T_c$. (b) Superconductor at $T < T_c$ in an external magnetic field H. Note the exclusion of the flux from the interior of the ring. (c) Magnetic flux reduces to zero as H goes to zero.* (From Atam P. Arya, *Elementary Modern Physics*, Addison-Wesley Publishing Company, Inc., Reading, Mass., 1974.)

Let us consider one very interesting application of the exclusion of a magnetic flux [as shown in Figure 33.23(b)] to the case of a superconducting ring. To understand this, let us refer to Figure 33.24 and carefully follow a sequence. To start with, an ordinary conductor in Figure 33.24(a1) and a superconducting ring in Figure 33.24(b1) are brought to a low temperature $T > T_c$ and then placed in an external magnetic field H. The magnetic field lines can pass through the conductor as well as the superconductor, as shown. Now suppose that the temperature is lowered still further so that $T < T_c$, as shown in Figures 33.24(a2) and (b2). No change takes place in the case of an ordinary ring, but in the case of a superconducting ring the magnetic field lines cannot pass through the cross section of the ring, although there are still field lines passing through the hole, as shown.

Now suppose that we switch off the field. As shown in Figure 33.24(a3), for an ordinary conductor the magnetic field completely disappears. But this is not so for a superconducting ring, as shown in Figure 33.24(b3). Even though the flux outside the ring disappears, the flux inside the ring is trapped because it cannot penetrate through the superconductor. As the magnetic field outside the ring collapses, it induces a large current in the ring. It is this current that maintains the trapped flux, as in Figure 33.24(b3). Once the current is started in the ring and the battery is switched off (or the field is switched off), because of the extremely small resistivity, the current will continue indefinitely (without needing an emf source). Different experiments performed reveal that the current does not decrease appreciably over periods of years.

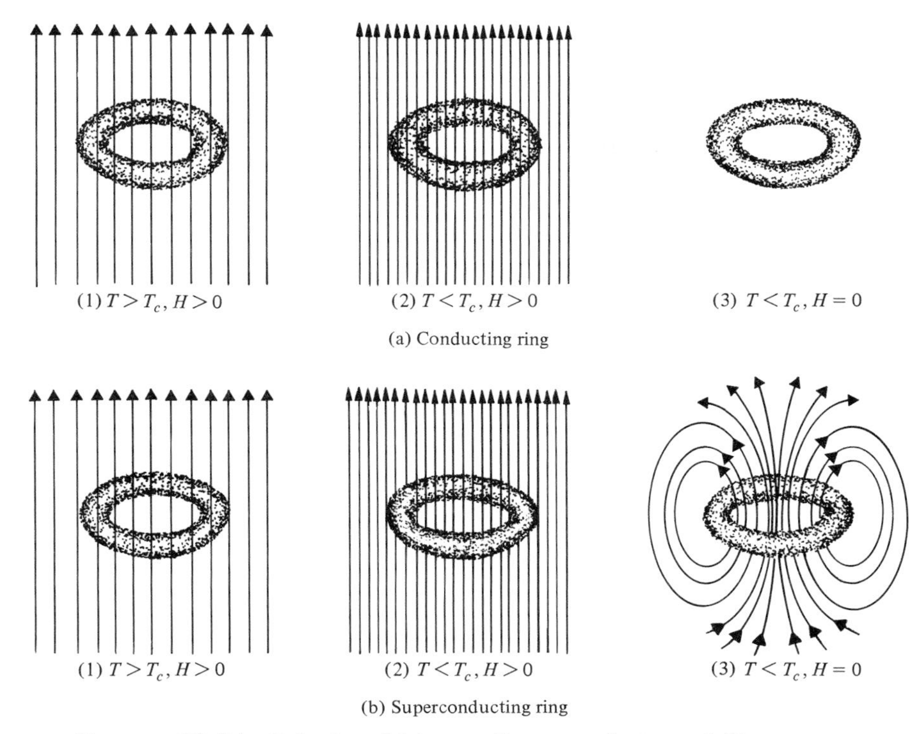

FIGURE 33.24. *Behavior of (a) an ordinary conductor and (b) a superconductor under different conditions. [The sequence of events (1) to (3) must be observed carefully.]* (From Atam P. Arya, *Elementary Modern Physics*, Addison-Wesley Publishing Company, Inc., Reading, Mass., 1974.)

At present, superconducting magnets capable of producing a field as high as 100 kG and maintaining currents over long periods are commercially available.

SUMMARY

There are four types of bonding by which atoms in a molecule or solid may be bound together. When the positive and negative ions in a molecule are held together by electrostatic force, it is *ionic bonding*. When one or two pairs of electrons are shared by the atoms in a molecule, *covalent bonding* results. The *equilibrium distance* r_0 between two atoms of a molecule is that distance at which the energy is minimum. This minimum energy is called the *dissociation energy*. The molecular spectra, depending upon the energy available, may be pure rotational, rotational-vibrational, or rotational-vibrational-electronic, leading to the formation of bands.

A *laser*—light amplification by stimulated emission of radiation—gives an intense coherent beam. *Population inversion* means an increased number of atoms in higher states, and the process for accomplishing this is called *optical pumping*. Using the principle of *wave-front reconstruction*, one can obtain a truly three-dimensional image by making a *hologram*.

The cohesive or *binding energy of a crystal* is a measure of the stability of the crystal. The crystals that compose solids may be classified into five categories: *ionic*, *covalent*, *hydrogen-bonded*, *molecular*, and *metallic*. The maximum energy that free electrons in a metal can have is called the *Fermi energy*.

The conductivity of semiconductors is between that of insulators and conductors. The tight-binding approximation leads to the formation of four types of band structures. In semiconductors the gaps between the filled and unfilled bands are between ~0.1 and 1 eV. For conductors either the band is not completely filled or the filled and unfilled bands overlap. In insulators there is a large gap between filled and empty bands.

Semiconductors may be classified into *intrinsic* and *impurity semiconductors*. In *n*-type semiconductors there are more electrons than holes, while in *p*-type there are more holes than electrons. A semiconductor diode is a *p-n* junction, and when reverse-biased can be used as a rectifier. A junction transistor *n-p-n* or *p-n-p* may be used as an amplifier.

In the neighborhood of 4 K, many metals exhibit abnormally high conductivity (or disappearance of resistivity). This property is called *superconductivity*, and the temperature below which it happens is the *critical temperature*. The exclusion of flux from the interior of a superconductor is the *Meissner effect*.

QUESTIONS

1. Which are stronger bonds—ionic or covalent—and why? Give an example other than one from the text.
2. What are conditions under which you can obtain a pure rotational spectrum of a molecule?
3. Unlike the line spectra of atoms, why do molecules exhibit band spectra?
4. What are the advantages of lasers over ordinary light sources?
5. How does a hologram differ from an ordinary film negative?
6. How do the binding energies of the crystalline solids differ from the molecular binding energies?
7. Give two examples of elements that form *p*-type materials.
8. Give two examples of elements that form *n*-type materials.
9. In the formation of *p*-type and *n*-type materials, are the energy levels raised or lowered?
10. Compare the relative numbers of conduction electrons and holes in (a) intrinsic; (b) *n*-type; and (c) *p*-type semiconductors.
11. What effect does the raising of the temperature have on a semiconductor?
12. Why are silicon and germanium frequently used in making semiconductor devices whereas tellurium is not?
13. The band structures of silicon and diamond are quite similar, but silicon has a metallic appearance whereas diamond is transparent. Explain this difference, keeping in mind that the energy gap between the valence and conduction band in silicon is 1.14 eV and in diamond is 5.33 eV.
14. A semiconductor has a negative coefficient of resistivity. Explain.
15. List some important uses of superconducting materials.
16. If the current is started in a ring-shaped superconductor, even when we switch off the field, the current continues to flow indefinitely inside the ring. Doesn't this violate the principle of the conservation of energy? Explain.

PROBLEMS

1. The dissociation energy of a NaCl molecule is 4.7 eV. Express this energy in units of kcal/mol.
2. For a HCl molecule the equilibrium distance is 1.27 Å and the ionization potential is 13.8 V. The electron affinity of chlorine is 3.8 eV. Calculate (a) the Coulomb energy between the two ions at the equilibrium distance; and (b) the dissociation energy.
3. The ionization energy of potassium is 4.34 eV, and the electron affinity of fluorine is 4.07 eV/electron. The equilibrium distance is 2.67 Å. Calculate (a) the Coulomb energy between the two ions at the equilibrium distance; and (b) the dissociation energy.
4. Repeat Problem 3 for the case when the distance between the ions is 53.3 Å.
5. In solids the potential energy at the equilibrium distance is $V_0(=E)$

$$V_0 = -\frac{\alpha}{4\pi\epsilon_0}\frac{e^2}{r_0}\left(1 - \frac{1}{n}\right)$$

where α is called the *Madelung constant* and n is a constant greater than 2. This is the lattice energy and may be given in terms of eV/ion pair or kcal/ion pair. Calculate this for the case of a NaCl crystal for which $\alpha = 1.7476$, $n = 9$, and $r_0 = 2.81$ Å.
6. Using the expression given in Problem 5, calculate the potential energy of calcium chloride given that the interatomic distance at equilibrium is 3.56 Å, $n = 11.5$, and $\alpha = 1.7476$.
7. Consider a crystal of size 1 cm^3. Treating this as an infinitely deep potential, the energy levels inside it are given according to (see Chapter 31) $E_n = n^2\hbar^2/8mL^2$. Show that the energy separation between adjacent levels in the crystal, that is, $\Delta E = E_{n+1} - E_n$, is of the order of 10^{-14} eV. (Assume that $n \approx 10$.)

34 Nuclear Science and Technology

According to the nuclear model of the atom, most of the mass of the atom is concentrated in a small spherical volume of radius 10^{-15} m. Unlike atomic, molecular, and solid-state physics, there are many fundamental questions that still remain unanswered in nuclear physics. For example: What is the exact form of the nuclear force that holds neutrons and protons together in the nucleus? Is there a single model of the nucleus which can explain all its characteristics? In spite of these difficulties, nuclear physics has made tremendous progress in the past 50 years.

In this chapter we shall briefly investigate some of the basic characteristics of nuclei and will see how these are applicable to such fields as nuclear energy and nuclear medicine.

34.1. Basic Properties

LET US first briefly summarize those basic properties of nuclei which are essential for understanding some of the topics we shall be discussing in this chapter.

Constituents and Binding Energy

Any nucleus consists of Z protons and N neutrons; that is, it consists of $A(= Z + N)$ *nucleons* (a generic name for neutrons and protons). A is called the *mass number* and Z is the *atomic number*. The neutrons are neutral while the charge of each proton is $+1e$; hence, the total charge on the nucleus is $+Ze$. The nucleus is surrounded by Z electrons to form a neutral atom.

The rest masses of a proton and neutron expressed in mass units and energy units are

$$m_p = 1.00727663 \text{ u} = 938.256 \text{ MeV}/c^2$$
$$m_n = 1.0086654 \text{ u} = 939.550 \text{ MeV}/c^2$$

where $1 \text{ u} = 1.66 \times 10^{-27}$ kg $= 931.478$ MeV$/c^2$, as defined in Chapter 8. When the nucleus is formed by bringing protons and neutrons together, the mass of the nucleus so formed is less than the sum of the masses of the constituent protons and neutrons. The energy equivalent of this mass difference is called the *binding energy* of the nucleus and may be calculated as follows.

Let us consider an atom denoted by ${}^A_Z X$ formed by bringing together Z protons, $(A - Z)$ neutrons, and Z electrons, but its mass $M(A, Z)$ is less than the sum of the masses of its constituents $[Zm_p + (A - Z)m_n + Zm_e]$. The energy equivalent ΔE of this mass difference ΔM is released when an atom is formed; that is,

$$\Delta E = \Delta M c^2 \tag{34.1}$$

We must supply energy equivalent to this amount if we want to break the atom into its constituents. This energy is called the *binding energy,* BE. Since BE $= -\Delta E$, we may write

$$\text{BE} = [Zm_p + (A - Z)m_n + Zm_e - M(A, Z)]c^2 \qquad (34.2)$$

The mass of the hydrogen atom, $m_H = 1.007825$ u, does not differ appreciably from the sum $m_p + m_e$; that is, the binding energy of the hydrogen atom is negligible. Thus, in Equation 34.2, after replacing $Z(m_p + m_e)$ by Zm_H, we may write the binding energy per nucleon as

$$\boxed{\frac{\text{BE}}{A} = [Zm_H + (A - Z)m_n - M(A, Z)]\frac{c^2}{A}} \qquad (34.3)$$

The plot of BE/A for different nuclei versus A is shown in Figure 34.1. We observe the following features from this plot.

1. All nuclei lie on a smooth curve except the nuclei ^{4}He, ^{12}C, and ^{16}O.
2. The maximum BE/A is around $A = 50$ and is ~8.8 MeV.
3. The average binding energy between $A \sim 20$ and $A \sim 160$ is almost constant and is equal to ~8.5 MeV/nucleon.
4. Up to $A = 20$, BE/A increases sharply with increasing A.
5. For $A > 140$, BE/A decreases slowly with increasing A with a value of ~7.2 MeV/nucleon around $A = 230$.

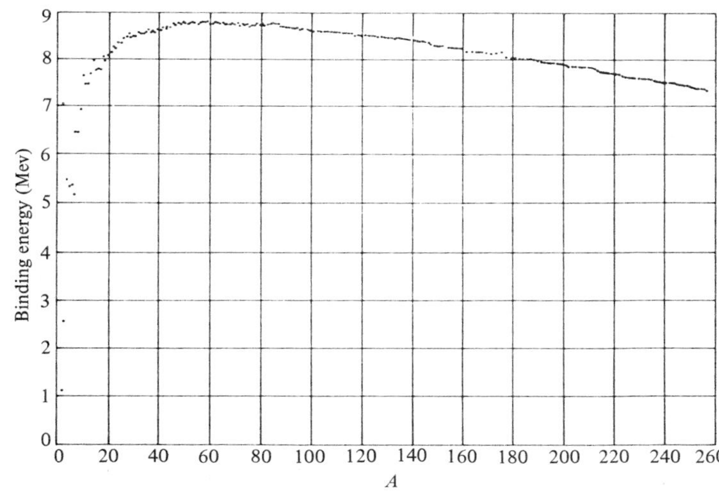

FIGURE 34.1. *Plot of binding energy per nucleon, BE/A, versus mass number A for different nuclei.* (From A. P. Arya, *Fundamentals of Nuclear Physics,* Allyn and Bacon, Inc., Boston, 1966.)

Two important factors that determine whether a given nucleus will be stable or unstable are (1) the neutron/proton ratio, and (2) the evenness and oddness of the number of neutrons and protons. Figure 34.2 represents a plot of N versus Z for all nuclei where solid squares represent stable nuclei, and all other unstable radioactive nuclei. The stable nuclei seem to be lying on a curve for

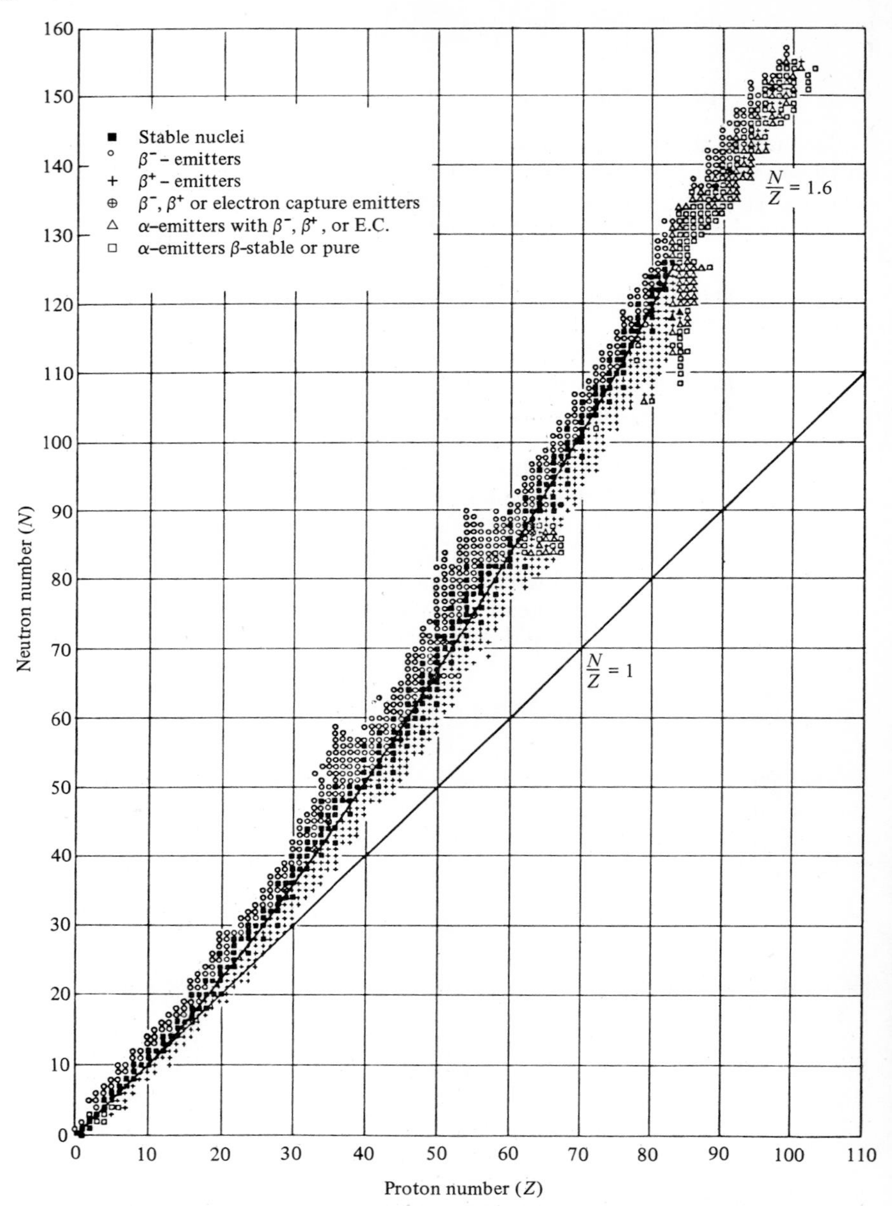

FIGURE 34.2. *Plot of N versus Z for all known nuclei. A curve through the stable nuclei shows that it starts with N/Z = 1 for nuclei of low mass numbers and reaches a value of N/Z = 1.6 for high mass numbers.* (From A. P. Arya, *Fundamentals of Nuclear Physics,* Allyn and Bacon, Inc., Boston, 1966.)

which $N/Z = 1$ for low-A nuclei and $N/Z = 1.6$ for heavy nuclei with $A > 230$. This is called the *stability curve* and indicates that heavy nuclei have many more neutrons than protons. On either side of this curve are located the radioactive isotopes.

All stable nuclei may be classified into four groups having an even or odd number of protons and neutrons, as shown in Table 34.1. It is clear that the most stable nuclei are even–even; only five nuclei are odd–odd, while the number of even–odd and odd–even stable nuclei are between the other two. This means that the *total binding energy depends not only on the ratio of the number of neutrons to the number of protons, but whether the number of neutrons and protons is odd or even.*

TABLE 34.1[a]
Classification of Stable Isotopes

A	Z	N	Number of Stable Nuclei
Even	Even	Even	156
Odd	Even	Odd	50
Odd	Odd	Even	48
Even	Odd	Odd	5
			259

[a] From A. P. Arya, *Fundamentals of Nuclear Physics*, Allyn and Bacon, Inc., 1966.

Size of the Nucleus

The size of any atom is almost constant, with a radius equal to $\sim 10^{-10}$ m. On the other hand, the size of the nucleus is not only much smaller than the size of the atom, but also depends upon the mass number A, that is, the number of nucleons in the nucleus. Assuming a spherical shape, the radius R of the nucleus of mass number A is

$$\boxed{R = r_0 A^{1/3} \simeq 1.35 \times 10^{-15} A^{1/3} \text{ m}} \tag{34.4}$$

Depending upon the type of experiment, the value of r_0 is found to be between 1.2×10^{-15} m and 1.48×10^{-15} m.

Angular Momentum of the Nucleus

Just as an electron in an atom has orbital angular momentum, spin angular momentum, and total angular momentum, any nucleon—neutron or proton—in the nucleus is assumed to have all three momenta. Once again in analogy with an electron, a neutron or proton has an intrinsic spin quantum number $= \frac{1}{2}$ with the corresponding spin angular momentum $= \sqrt{\frac{1}{2}(\frac{1}{2} + 1)}\hbar = \frac{\sqrt{3}}{2}\hbar$. The orbital quantum number of a nucleon is l with orbital angular momentum $= \sqrt{l(l + 1)}\hbar$, where $l = 0, 1, 2, 3, \ldots$. The vector addition of these two angular momenta yields the total angular momentum I of the nucleus, given by

$$I = \sqrt{i(i + 1)}\hbar \tag{34.5}$$

where $\hbar = h/2\pi$ and i is the *total quantum number*. Although misleading, i is usually referred to as the *nuclear spin*. The value of i can be an integer or half-integer. All experimental investigations yield

$$\begin{aligned} &\text{If } A \text{ is even:} \quad i = 0, 1, 2, 3, 4, \ldots \\ &\text{If } A \text{ is odd:} \quad i = \tfrac{1}{2}, \tfrac{3}{2}, \tfrac{5}{2}, \tfrac{7}{2}, \ldots \end{aligned} \tag{34.6}$$

For even–even nuclei (with both Z and N even), $i = 0$.

Nuclear Forces

From our brief knowledge of nuclear properties, it is possible to conclude that (1) nuclear forces must be very strong and attractive and (2) the range of the nuclear forces is short, as we shall explain below.

The fact that there are stable nuclei implies that the net forces between neutron–neutron, proton–proton, and neutron–proton pairs must be attractive. Furthermore, these attractive forces must be much stronger than the Coulomb forces, to overcome the repulsive forces between a large number of protons in heavy nuclei. This is also evident from the fact that BE/A for a nucleus is ~8 MeV/nucleons, as compared to a few keV for electron binding energies.

Also, we know that nuclear forces do not directly affect the structures of atoms and molecules. From this we must conclude that nuclear forces do not extend too far. Actually, they extend just a little beyond the nuclear surface, hence are short-range.

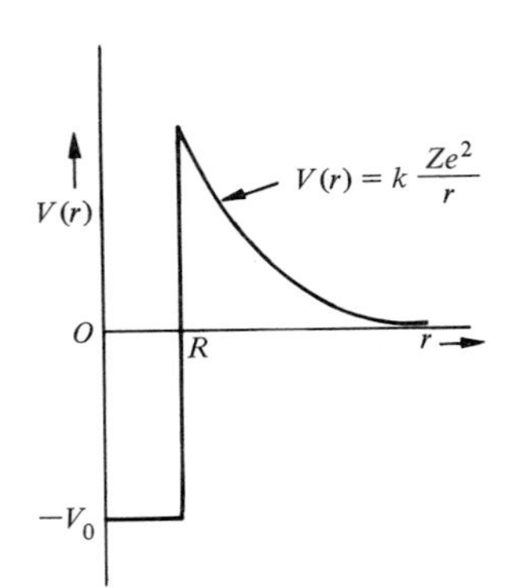

FIGURE 34.3. *Plot of potential V(r) versus r, the distance between the nucleus and a proton as the proton is brought from infinity toward the nucleus.*

Let us see how the nuclear potential (the nuclear force can be derived if the form of the nuclear potential is known) can be represented graphically. Suppose that a proton is brought from infinity toward a nucleus at 0. Let R be the radius of the nucleus. For $r > R$ the only force is the Coulomb repulsive force, according to which $V(r)$ varies as $1/r$. For $r < R$ the force is strongly attractive, hence $V(r)$ is negative. Over this short range of the nuclear force, we assume $V(r)$ to be constant, equal to $-V_0$. Thus, the plot of $V(r)$ versus r will be as shown in Figure 34.3. It may be pointed out that the nuclear potential form is approximate; the exact form is not known.

If we replace the proton by a neutron, since there will be no Coulomb force, the plot of $V(r)$ versus r will be as shown in Figure 34.4. Experiments yield a typical value of V_0 to be -42 MeV.

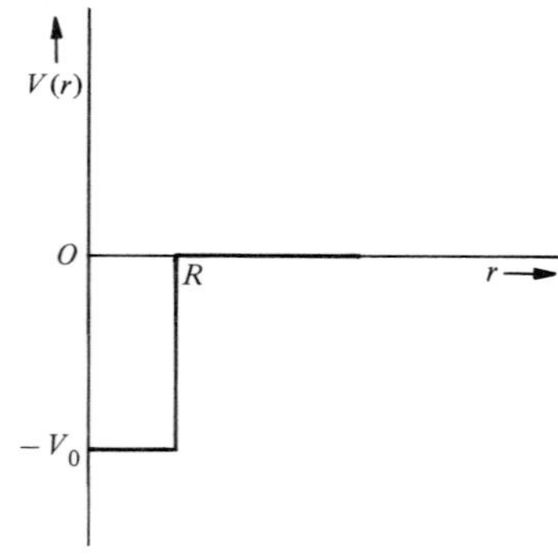

FIGURE 34.4. *Plot of potential V(r) versus r, the distance between the nucleus and a neutron as the neutron is brought from infinity.*

Nuclear Models

There is no single model of the nucleus that can explain all the nuclear characteristics. Some of the important models in use now are (1) the nuclear shell model, (2) the liquid-drop model, (3) the collective model, (4) the Fermi-gas model, (5) the compound-nucleus model, and (6) the optical model.

The nuclear shell model is analogous to the atomic shell model. For example, if we consider the protons, there are 2 protons in the first shell, 8 protons in the second shell, 18 in the third shell, and so on. Similarly, there are 2 neutrons in the first shell, 8 in the second shell, 18 in the third, and so on.

EXAMPLE 34.1 Calculate the amount of energy released when carbon 12 ($Z = 6$) is formed by bringing 6 protons, 6 neutrons, and 6 electrons together. The mass of ^{12}C is 12.000000 u.

From Equation 34.3,

$$\begin{aligned} BE &= [Zm_H + (A - Z)m_n - M(A, Z)]c^2 \\ &= [6(1.007825\ u) + (12 - 6)(1.008665\ u) - 12.000000\ u]c^2 \\ &= (6.046950\ u + 6.051990\ u - 12.000000\ u)c^2 = (0.121490\ u)c^2 \\ &= 0.121490 \times 931.478\ \frac{MeV}{c^2}\, c^2 = 113.04\ MeV \end{aligned}$$

$$\frac{BE}{A} = \frac{113.04\ MeV}{12} = 9.42\ MeV/nucleon$$

Exercise 34.1 Calculate the amount of energy released when ^{32}S ($Z = 16$) is formed. The mass of ^{32}S is 31.972074 u. [*Ans.*: 271.8 MeV.]

34.2 Radioactive Decay Law[a]

As was discussed in Chapter 2, the unstable nuclei that disintegrate by emitting radiation such as alpha, beta, or gamma rays are called *radioactive nuclei* and the process of such emission is called *radioactive decay*. We are interested in calculating the rate at which the radioactive nuclei disintegrate or decay. This can be done by using statistical methods.

Let us consider N radioactive nuclei. Each has a probability that it will decay in the next second. In time Δt, each nucleus will have a probability of $\lambda\,\Delta t$ to decay. Thus, out of N radioactive nuclei, the number ΔN that will decay in time between t and $t + \Delta t$ will be

$$\Delta N = -(\lambda\,\Delta t)N \tag{34.7}$$

The negative sign is required because N decreases as t increases. The probability constant λ is given the name *decay constant* or *disintegration constant*.

We defined *activity* A of a radioactive sample to be the number of disintegrations per second. That is, the decay rate from Equation 34.7 is given by

$$\boxed{A = \left|\frac{\Delta N}{\Delta t}\right| = \lambda N} \tag{34.8}$$

Thus, if we measure the activity $\Delta N/\Delta t$ of a sample, knowing N, we can calculate the decay constant, λ. The number of radioactive atoms present, N, changes with time, as explained below.

Suppose that at $t = 0$ we have N_0 radioactive atoms. As time passes, some of these decay. Let N be the remaining radioactive atoms at any time t. The plot of N versus t and the plot of activity $A = |\Delta N/\Delta t|$ versus t for any ensemble of radioactive nuclei are found to be of the form shown in Figure 34.5(a) as linear plots, and in Figure 34.5(b) as semilogarithmic plots. These plots suggest the following exponential relation among N_0, N, λ, and t:

$$\boxed{N = N_0 e^{-\lambda t}} \tag{34.9}$$

a. It is recommended that students should review Section 2.4

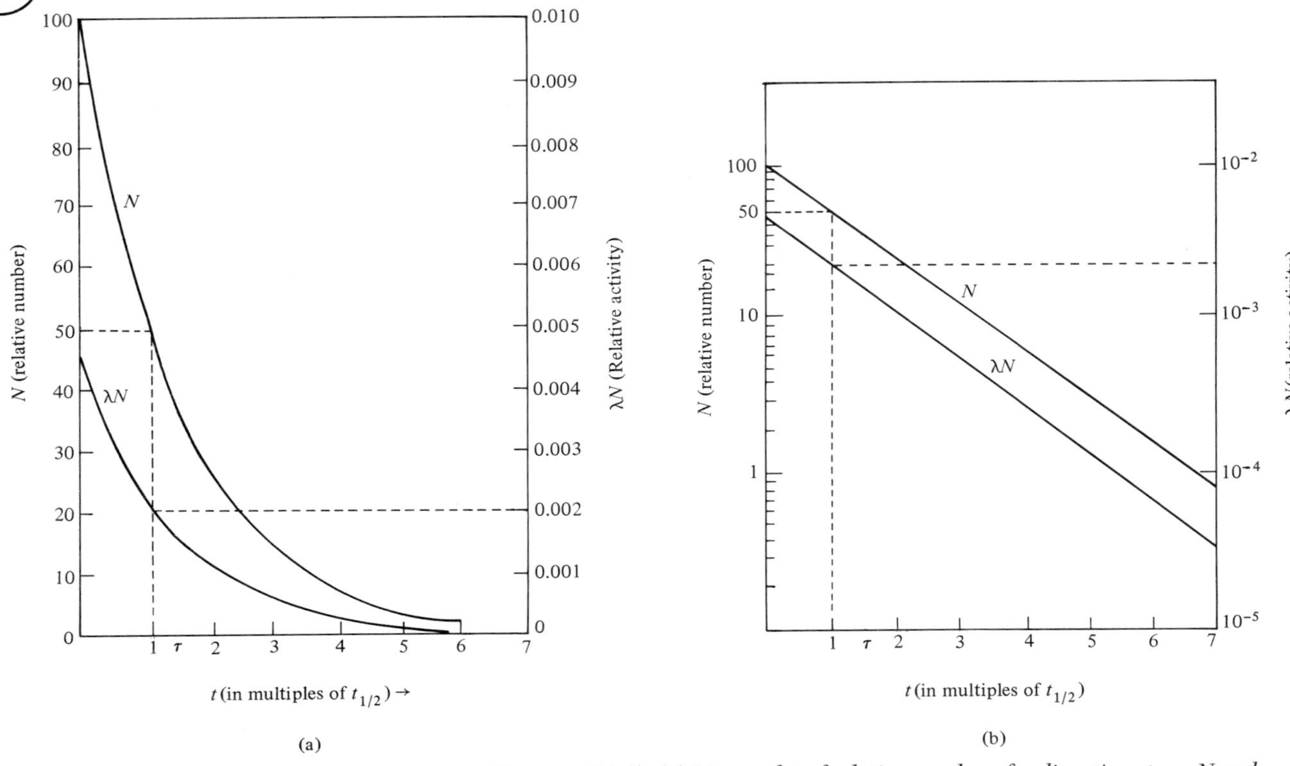

FIGURE 34.5. *(a) Linear plot of relative number of radioactive atoms N and activity λN versus time t for $^{105}_{44}Ru$. (b) Semilogarithmic plot of N and λN versus t.*

Combining Equations 34.8 and 34.9, we can calculate λ if N_0 and A are known.

According to the exponential form of the radioactive decay law, it will take an infinite time for a given sample to decay completely. That is, from Equation 34.9, $N = 0$ only if $t = \infty$. Furthermore, an individual radioactive atom may live for a time interval anywhere from zero to infinity. To overcome these difficulties, the concepts of half-life and mean life are used.

The *half-life* $t_{1/2}$ of any radioactive sample is that time interval in which the number of undecayed radioactive atoms reduces to one-half. Thus, if we have N_0 radioactive atoms at $t = 0$, then in $t = t_{1/2}$, N_0 will reduce to $N_0/2$. Thus, substituting $N = N_0/2$ for $t = t_{1/2}$ in Equation 34.9, we get

$$\frac{N_0}{2} = N_0 e^{-\lambda t_{1/2}} \quad \text{or} \quad \frac{1}{2} = e^{-\lambda t_{1/2}}$$

Taking logarithms on both sides,

$$\boxed{t_{1/2} = \frac{\ln 2}{\lambda} = \frac{0.693}{\lambda}} \tag{34.10}$$

Since the unit of λ is $1/s$, that of $t_{1/2}$ will be seconds. Note that in time $t_{1/2}$ the activity A_0 also reduces to $A_0/2$.

The *mean* or *average life* τ of a sample is defined as

$$\boxed{\tau = \frac{1}{\lambda}} \tag{34.11}$$

Example 34.2 illustrates these points.

The total number of radioactive atoms present in a given sample is not of much consequence; it is the disintegration or decay rate that matters. Thus, the units of radioactivity are in terms of the disintegration rate. A given sample is said to have an activity of 1 *curie* (Ci) if it has 3.7×10^{10} disintegrations per second. The smaller units are 1 *millicurie* (mCi) $= 10^{-3}$ Ci and 1 *microcurie* (μCi) $= 10^{-6}$ Ci. Another unit of activity is 1 *rutherford* (rd) $= 10^6$ disintegrations per second (dis/s).

EXAMPLE 34.2 One gram of the radioisotope gold 198, ${}^{198}_{79}\text{Au}$, decays with a half-life of 64.8 h by emitting electrons. Calculate (a) λ; (b) τ; (c) N_0; (d) A_0; and (e) N. Also make plots of N and A versus t.

(a) $t_{1/2} = 64.8 \text{ h} = 2.33 \times 10^5$ s. From Equation 34.10,

$$\lambda = \frac{0.693}{2.33 \times 10^5 \text{ s}} = 0.297 \times 10^{-5} \text{ s}^{-1}$$

(b) From Equation 34.11,

$$\tau = \frac{1}{\lambda} = \frac{1}{0.297 \times 10^{-5} \text{ s}^{-1}} = 3.37 \times 10^5 \text{ s} = 93.5 \text{ h}$$

(c) There are N_A (Avogadro's number) $= 6.02 \times 10^{23}$ atoms in 1 mole of gold. Therefore, 1 g of gold contains N_0 radioactive atoms at $t = 0$:

$$N_0 = \frac{N_A}{A} = \frac{6.02 \times 10^{23}}{198} = 3.13 \times 10^{21} \text{ atoms}$$

(d) The initial activity at $t = 0$, from Equation 24.8, is

$$A_0 = \lambda N_0 = 0.297 \times 10^{-5} \text{ s}^{-1} \times 3.13 \times 10^{21} \text{ atoms}$$

$$= \frac{0.93 \times 10^{16} \text{ dis/s}}{3.7 \times 10^{10} (\text{dis/s})/\text{Ci}} = 2.51 \times 10^5 \text{ Ci}$$

(e) For different values of t we can calculate N from Equation 34.9:

$$N = N_0 e^{-\lambda t} = (3.13 \times 10^{21}) e^{-(0.297 \times 10^{-5}/s) \cdot t}$$

while $A = \lambda N$. Plots for N and A versus t are shown on next page.

EXERCISE 34.2 Repeat Example 34.2 for 1 g of radioactive radium 226, which has a half-life of 1620 years. [*Ans.:* (a) $\lambda = 1.36 \times 10^{-11} \text{ s}^{-1}$;

(b) $\tau = 7.38 \times 10^{10}$ s = 2337 yr; (c) $N_0 = 2.66 \times 10^{21}$ atoms; (d) $A_0 = 3.62 \times 10^{10}$ disintegrations/s $\approx$ 1 Ci. The plots are as in Ex. 34.2.]

$${}^{198}_{79}\text{Au} \xrightarrow[t_{1/2} = 64.8 \text{ h}]{\beta^-} {}^{198}_{80}\text{Hg}$$

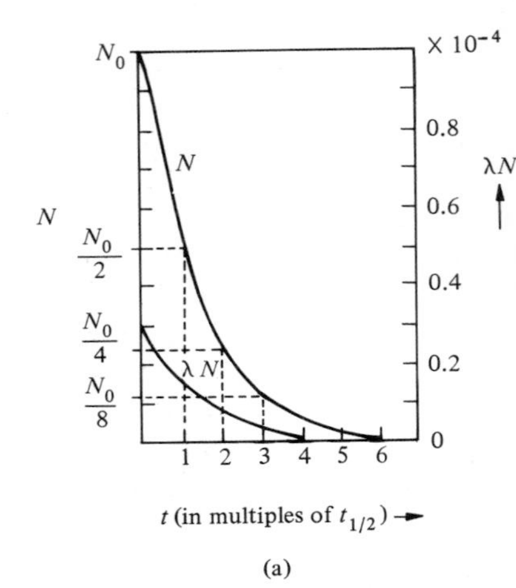

(a)

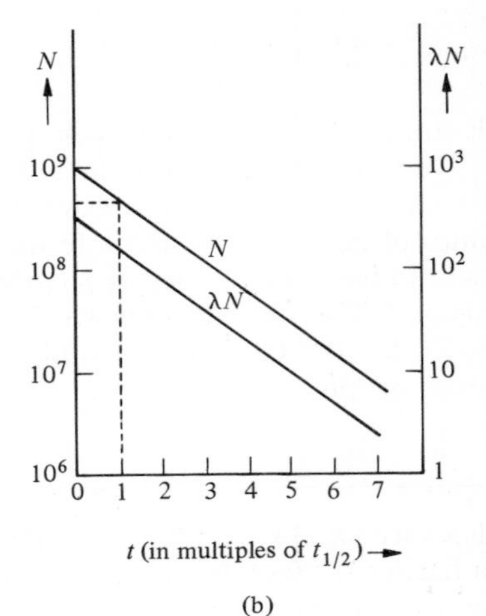

(b)

Ex. 34.2

Using the radioactive decay laws, W. F. Libby in 1952 devised a method of estimating the age of organic relics. This method is called *radioactive dating*. Cosmic rays coming from outer space and entering the earth's atmosphere consists mostly of high-energy protons and neutrons. The nitrogen 14 in the earth's atmosphere absorbs the neutron and changes into carbon 14, by the following reaction:

$${}^{14}_{7}\text{N} + {}^{1}_{0}n \longrightarrow {}^{14}_{6}\text{C} + {}^{1}_{1}p \tag{34.12}$$

where ${}^{1}_{1}p$ is the proton emitted in the reaction. The ^{14}C formed is radioactive and decays to ^{14}N by beta emission, with a half-life of 5730 years. The carbon dioxide in the air should contain traces of ^{14}C.

We know that living organisms consume carbon dioxide. When an organism dies, consumption stops. From here on, the amount of ^{14}C decreases, becoming half in 5730 years, but the amount of ^{12}C remains constant. The amount of ^{14}C can be determined by measuring the activity, $A = \lambda N$. Thus, knowing the relative amount of ^{12}C and ^{14}C, we can determine the age of the organism or relic.

Suppose that a living tree before it died had N_0 radioactive atoms. Its activity was $A_0 = \lambda N_0$. After the tree died, its activity decreased exponentially with time. That is, $A = |\Delta N/\Delta t| = \lambda N = \lambda N_0 e^{-\lambda t}$. The ratio of the two is

$$\frac{A}{A_0} = e^{-\lambda t} \tag{34.13}$$

where $\lambda = 0.693/t_{1/2} = 0.693/5730$ yr. Thus, knowing A and A_0, we can calculate t, the time the tree has been dead.

Example 34.3 Suppose that you come across a piece of wood that is supposed to be of great antiquity. How will you evaluate its age? The piece has a mass of 40 g and shows an activity of 280 disintegrations per minute. The living plant shows a ^{14}C activity of 12 disintegrations per minute per gram, while the half-life of ^{14}C is 5730 years.

Equation 34.13, $A/A_0 = e^{-\lambda t}$, may be written

$$t = \frac{\ln(A/A_0)}{\lambda}$$

where $A = 280$ disintegrations/40 g-min $= 7$ dis/g-min, $A_0 = 12$ dis/g-min, and $\lambda = 0.693/5730$ yr. Substituting in the equation above,

$$t = \frac{\ln(12/7)}{0.693/5730 \text{ yr}} = \frac{0.539 \times 5730}{0.693} \text{ yr} = 4457 \text{ yr}$$

That is, the wood was a part of a tree cut down 4457 years ago.

Exercise 34.3 What would be the activity of the piece of wood in Example 34.3 if it were 10,000 years old? [*Ans.*: 143 dis/min.]

34.3. Interaction of Radiation with Matter

Interaction of incident radiation, such as alpha, beta, or gamma rays or any other charged particles, with the atoms and molecules of the surroundings plays an important role in nuclear physics experimentation and theory. The energy loss by the incident radiation, its energy and intensity determination, and the identification of the type of incident radiation are made through the use of such interactions. If there is no medium, there is no energy loss, that is, radiation will travel indefinitely through a vacuum. The interaction of light charged particles is different from those of heavy particles, and these two are different from the manner in which gamma rays interact.

Interaction of Heavy Charged Particles with Matter

Examples of fast-moving heavy charged particles include α-particles, deuterons, and protons. While traveling through a medium, there is a Coulomb force (or interaction) between the incident charged particle and the electrons of the atoms of the medium; that is, there are collisions. In such collisions the incident charged particle loses part of its energy, which is used in the excitation and ionization of the atoms of the surrounding medium. The incident particle itself moves in a straight line undeviated by such collisions.

We define the *stopping power* as the amount of energy lost per unit length by the incident particle in a given medium. After losing all its energy, the particle comes almost to rest. We define the *average range* $\bar{R}$ of a charged particle as equal to the distance it travels from its point of origin to the point in the medium where it comes to a stop. The range of different charged particles, which can be calculated theoretically, is different in different materials and depends upon the energy. Figure 34.6(a) shows the range of protons and α-particles of different energies in aluminum.

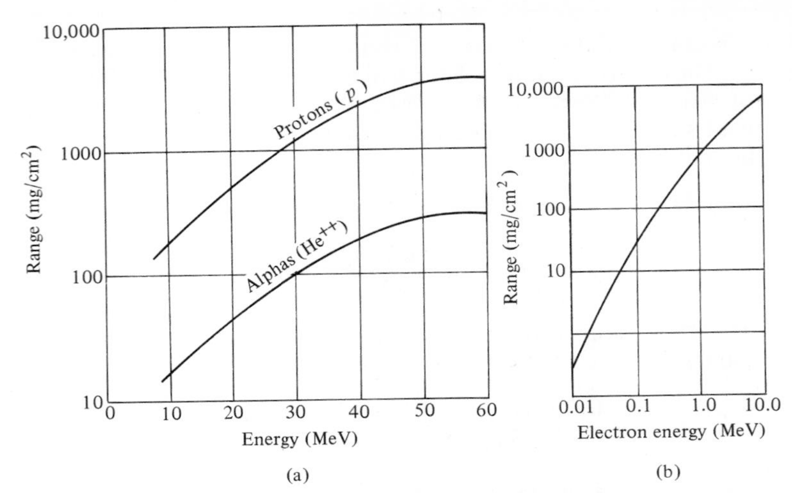

Figure 34.6. *(a) Range of protons and α-particles of different energies in aluminum. (b) Range of electrons in aluminum. To obtain the range in centimeter, divide the ordinate by density in units of mg/cm³.*

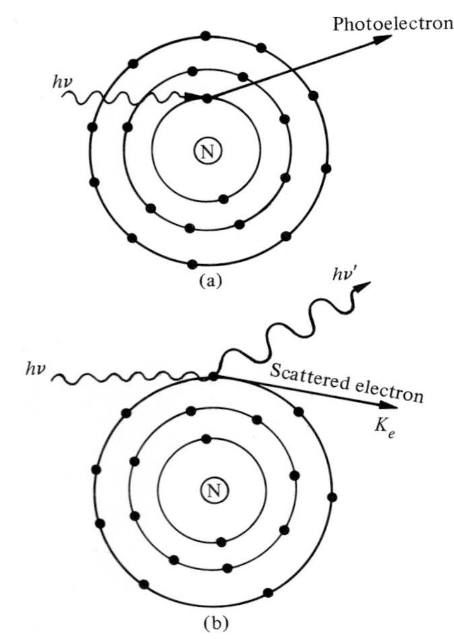

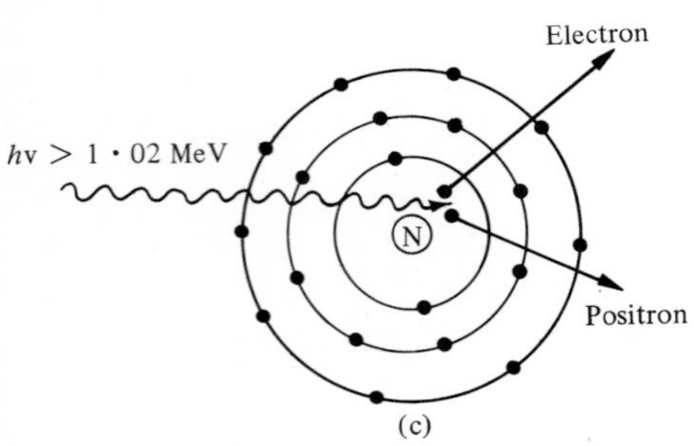

Figure 34.7. *The three most common processes by which γ-rays interact with matter and lose energy: (a) photoelectric effect, (b) Compton scattering, and (c) pair production.*

Interaction of Electrons with Matter

The methods by which light particles such as electrons interact with matter are much more complex. First, because of its small mass and high speeds, the collisions must be treated relativistically. Second, an electron can lose more than half its energy in a single collision. The problem is still further complicated by the fact that the electrons can lose energy by elastic collisions as well as by bremsstrahlung (x-rays) (Chapter 31). Even though the paths of the electrons in the medium are zigzag, they travel much longer distances in air as compared to heavy charged particles. Figure 34.6(b) shows the range of β-particles of different energies in aluminum.

Interaction of Gamma Rays with Matter

A γ-ray can lose most of its energy by interacting with matter in three different ways (and these are quite different from those of heavy charged particles and electrons). These processes, which are effective in different energy ranges, are: the photoelectric effect, Compton scattering, and pair production.

Photoelectric Effect. In this process the incident photon is absorbed by the atom. One of the electrons of the atom is ejected from the surface and is called a photoelectron (see Chapter 31), as shown in Figure 34.7(a). This process is most effective when the incident photon energies are between ~0.01 and ~0.5 MeV.

Compton Scattering. When a photon of energy $h\nu$ interacts with a free electron, the scattered photon has energy $h\nu'$ ($<h\nu$), and the kinetic energy of the scattered electron is $h\nu - h\nu'$, as shown in Figure 34.7(b) (see Chapter 31). This process is most effective when the incident photon energy is ~0.1 to ~10 MeV.

Pair Production. This process takes place only for an incident photon with energy $h\nu \geqslant 1.02$ MeV. When such a photon strikes a heavy nucleus, the photon disappears, and in its place an electron–positron pair is formed, as shown in Figure 34.7(c). The positron is a particle having the same mass as the electron but a positive charge of 1 unit. The conservation of energy requires that

$$h\nu \simeq 2m_0c^2 + K_+ + K_- \qquad (34.14)$$

where $h\nu$ is the incident photon energy, $2m_0c^2$ is the sum of the rest-mass energies of the positron and the electron, and K_+ and K_- are their respective kinetic energies. (We have neglected a small amount of recoil energy of the nucleus.) Figure 34.8 is a cloud-chamber photograph showing production of electron–positron pairs.

The existence of the positron was predicted theoretically by Paul A. M. Dirac in 1930 and discovered experimentally by Carl D. Anderson in 1932.

Unlike electrons, the positrons do not exist in free space. What happens to

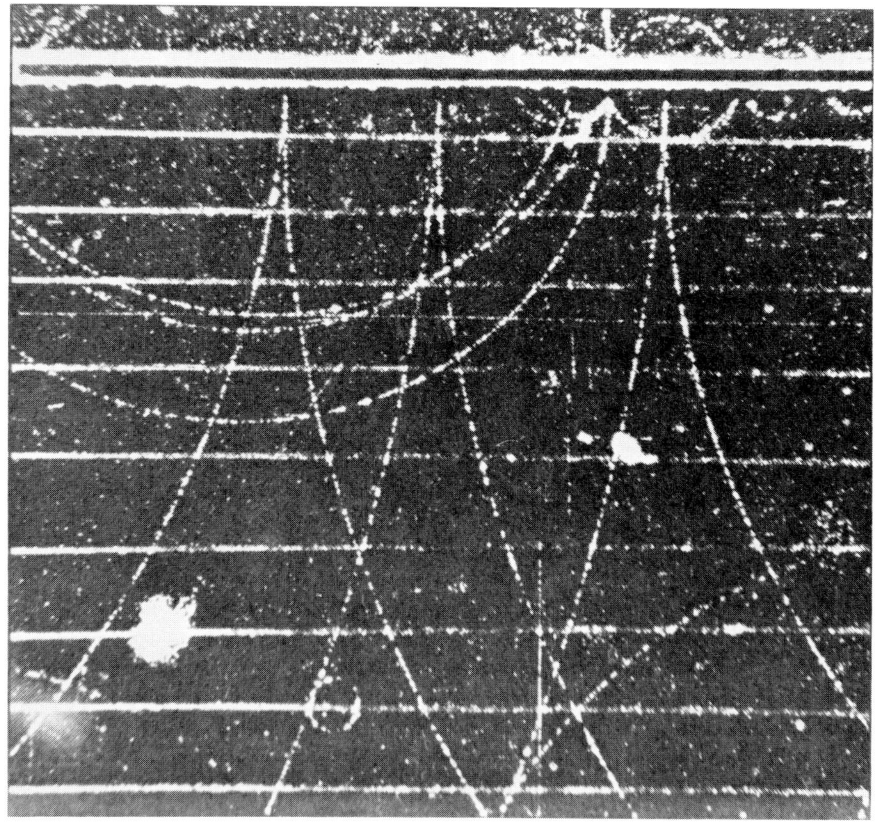

(b)

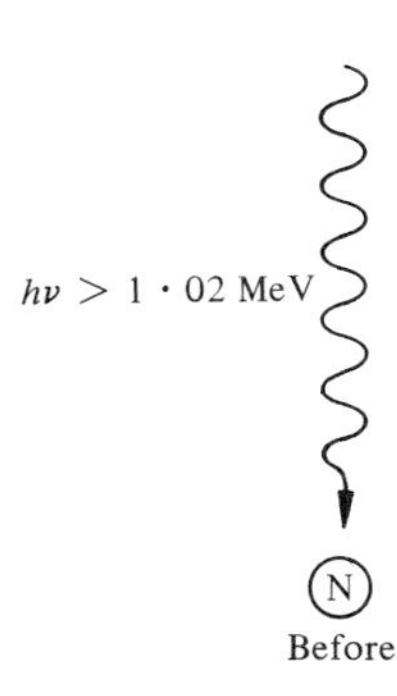

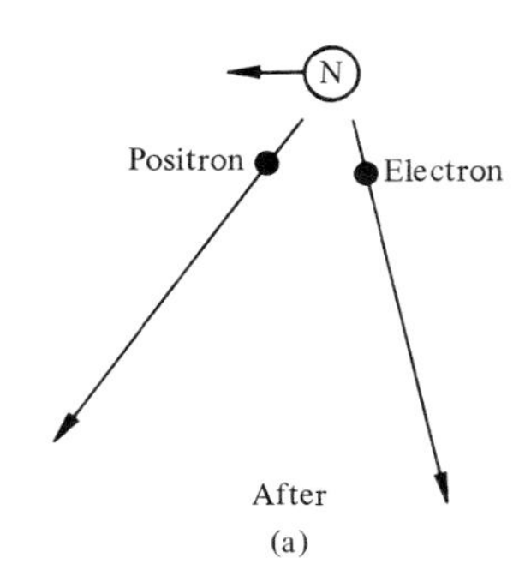

Figure 34.8. *(a) Schematic representation illustrating pair production. (b) Cloud-chamber photograph showing production of several electron–proton pairs when γ-rays enter from the top and interact with a lead sheet.* (Courtesy of Lawrence Berkeley Laboratory, University of California.)

the positrons created in electron–positrons pairs? The positrons are removed from circulation by a process called *pair annihilation*. A positron continues to keep on colliding with the surrounding atoms, losing its kinetic energy until it is almost at rest. The positron will now capture an electron to form a positronium atom. The positronium atom is unstable and within $\sim 10^{-10}$ s decays into two γ-rays of 0.51 MeV each. These statements have been firmly established experimentally.

If a beam of photons of intensity I_0 is incident on a material of thickness x, its intensity I after passing through the material is

$$\boxed{I = I_0 e^{-\mu x}} \tag{34.15}$$

where μ is called the *linear absorption coefficient* and depends upon the energy of the incident photons and the type of absorber. It can be calculated theoretically from the three processes above or determined experimentally.

Interaction of Neutrons with Matter

Neutrons are as heavy as protons but do not have any charge, hence do not interact very strongly with the atomic electrons. The most important process by which neutrons are slowed down is by elastic and inelastic scattering (see Chapter 7). When neutrons experience elastic collisions with protons, it slows them down considerably. For example, in a head-on elastic collision with a proton, a neutron may transfer all its energy to the proton. The slowed-down neutrons are easily captured by the nuclei of the matter through which they are passing.

34.4. Detectors of Nuclear Radiation

The purpose of radiation detectors is manyfold. They are used for measuring energy and intensity and to distinguish between different types of radiation. The working of most detectors is based on the interaction of radiation with matter, causing the ionization and excitation of the atoms. There are several types of detectors which are used in conjunction with complicated electronics. Without going into much detail we shall discuss some of these: (1) ionization chambers, (2) cloud chambers, (3) scintillation counters, and (4) solid-state detectors. Some of the other detectors, which we will not discuss here, are diffusion chambers, bubble chambers, spark chambers, Čerenkov detectors, and photographic emulsions. Most of these detectors are used for high-energy radiation.

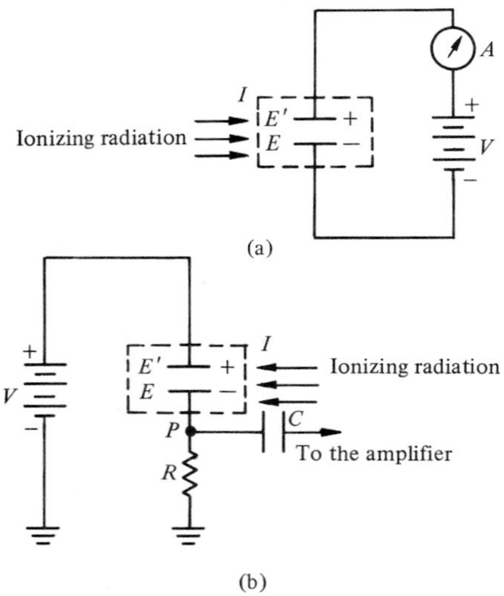

Ionization Chambers

Figure 34.9(a) shows one of the *simplest ionization chambers*. Two conducting plates E and E', are placed inside chamber I (filled with gas or air) at atmospheric pressure. The plates are maintained at a potential difference of 200 to 300 V by means of a voltage source V. The incident radiation produces positive and negative ions and are collected by the two plates, E and E'. The charge collected is measured by a current-measuring device as in Figure 34.9(a), or it may be converted into electrical pulses as shown in Figure 34.9(b).

FIGURE 34.9. *Ionization chamber with a meter A is arranged to detect and measure (a) a continuous beam of particles or x-rays and (b) individual particles.*

A modified form of the simple ionization chamber shown in Figure 34.10 is called a *proportional counter*. Physically, it consists of a cylinder and a wire along its axis which work as two electrodes, and a potential difference of 300 to 800 V is maintained between them. This detector can be used to distinguish between particles of different ionization powers.

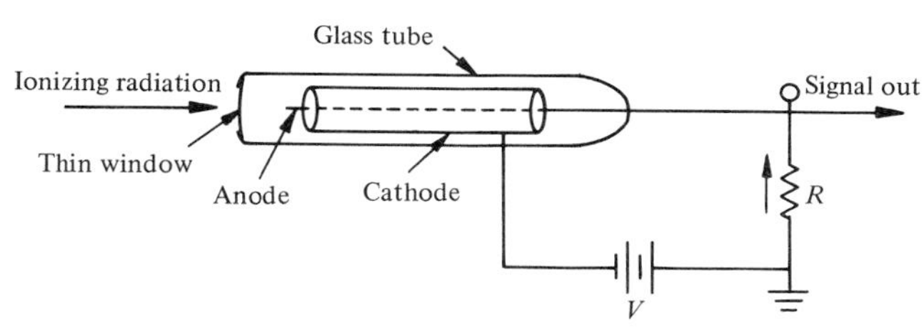

FIGURE 34.10. *Typical gas-filled detector, which may be operated as a proportional counter or as a Geiger counter. A thin mica window is used for detecting α- and β-rays.*

Still another modified form of a simple ionization chamber is a *Geiger–Müller counter* (GM counter) or simply a *Geiger counter*. The physical structure of a GM counter is the same as that of a proportional counter except that it operates at a very high potential, between 1000 and 1400 V. Also, the tube is filled with a mixture of 90 percent argon and 10 percent ethyl alcohol, Cl_2, or Br_2, maintained at a pressure of 10 cm. The Geiger counter will produce a pulse corresponding to each single particle that is incident on it, irrespective of energy.

Because of the simple construction and good efficiency for detection of low-energy γ-rays, x-rays,and β-rays, these detectors are commonly used in nuclear medical laboratories and hospitals, nuclear power plants, in uranium mines and by uranium prospectors, and in similar situations.

Cloud Chambers

The cloud chamber, invented by C. T. R. Wilcon in 1912, provides visible paths of charged particles. It is based on the principle of a supersaturated vapor condensing preferentially on charged particles.

Figure 34.11 shows the sketch of a cloud chamber. The chamber C is filled with dust-free air and saturated water vapor at room temperature. The piston is allowed to fall freely, resulting in the sudden expansion of the air/water vapor mixture; hence, the temperature falls below ambient room temperature. This results in the water vapor becoming supersaturated. A charged particle passing through the chamber at this moment produces ion pairs. The supersaturated vapors condense on those ions, producing a visible trail of droplets along the path of the charged particle. The chamber may be illuminated by lights, and a photograph of the track can be taken by a camera. After a photograph is taken the piston is pushed back to its original position and an electric field is applied to clear the chamber of all the ions. The chamber is then ready for recycling. In most cases arrangements are made to repeat the cycle automatically. Some photographs of tracks are shown in Figures 34.8 and 7.3.

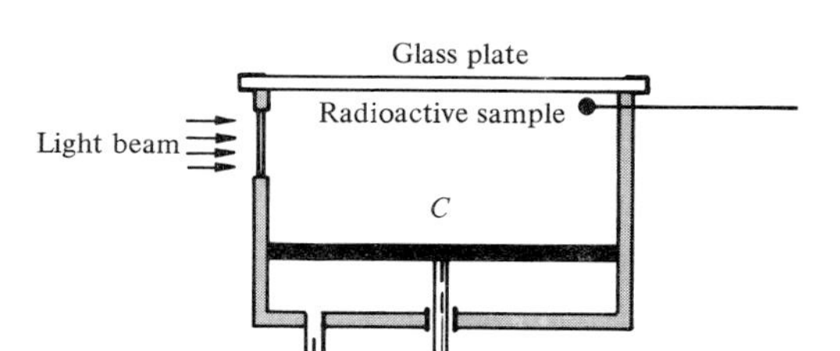

FIGURE 34.11. *Typical cloud chamber.* (From A. P. Arya, *Fundamentals of Nuclear Physics*, Allyn and Bacon, Inc., Boston, 1966.)

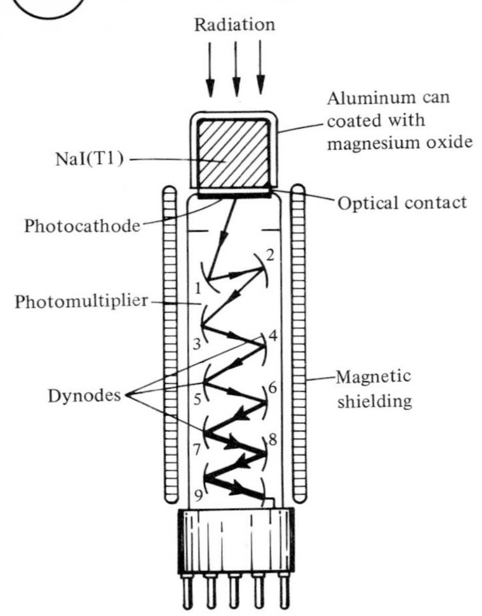

FIGURE 34.12. *Typical NaI(Tl) scintillation detector. The number of electrons ejected from the cathode is multiplied as they leave the successive dynodes, resulting in an electrical pulse at the anode.*

Scintillation Counters

Scintillators are substances which, when struck by a charged particle, x-ray, or γ-ray, produce a flash of light. Common examples are sodium iodide, cesium iodide, anthracene, and plastic scintillators. These light flashes can be converted into electrical pulses by means of an electronic system called a *scintillation counter,* shown in Figure 34.12.

Suppose that a γ-ray is incident on the NaI(Tl) (sodium iodide, thallium-activated) crystal shown in Figure 34.12. The electrons producing ionization caused by the γ-rays recombine with the atoms and the molecules of the crystal. In this recombination process, visible light in the wavelength region 3000 to 5500 Å is emitted. These visible photons falling on the photocathode of a photomultiplier produce photoelectrons. The photomultiplier operates on a very high voltage, anywhere between 700 and 2000 V. Such high voltages cause successive multiplication of the electrons while they travel from the cathode to the anode, resulting in a burst of electrons called an *electrical pulse.* The size of these pulses is proportional to the energy of the incident photon, and the number of these pulses are proportional to the number of incident particles.

One of the biggest advantages of NaI(Tl) detectors is that they are very efficient in counting. Also, large crystals of NaI(Tl) (2×2 in and 3×3 in are in common use) can be easily obtained. For the detection of charged particles such as α- and β-rays, plastic scintillators are commonly used.

Solid-State Detectors

Of the many semiconductor (see Chapter 33) devices recently made available, we shall discuss a *p-n* junction diode used for detection of charged particles. Incident particles enter the *n*-type side and stop in the depletion region, as shown in Figure 34.13. The electrons and the holes produced move toward the *n*-type (positive) layer and *p*-type (negative) layer, respectively. The arrival of those charges at the two layers produces a potential drop across the junction. The resulting electrical pulse, which is proportional to the energy of the incident particle, is detected by means of an electronic circuit.

Two modified forms are the lithium-drifted silicon Si(Li), in which lithium is drifted into a silicon *p-n* junction; and the lithium-drifted germanium Ge(Li), in which lithium is drifted into a germanium *p-n* junction. The lithium drifted into the junction creates a large depletion region. Usually, these detectors, especially Ge(Li), are kept at liquid nitrogen temperature (77 K) to prevent the lithium from drifting out.

A Si(Li) detector is generally used for β-rays, other charged particles, and

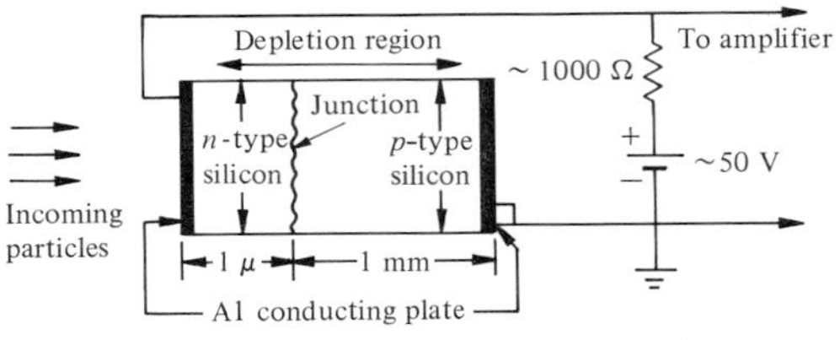

FIGURE 34.13. *Typical p-n junction detector, for detecting charged particles.* [From Friedland, Mayer, and Wiggins, *Nucleonics* **18**(2), 54 (1960).]

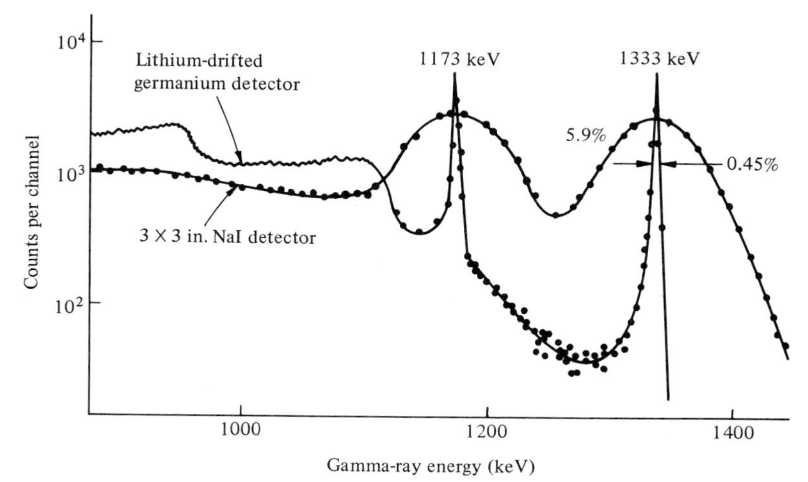

FIGURE 34.14. *Comparison of γ-ray spectra of ^{60}Co taken with a 3x3 in NaI(Tl) detector and a lithium-drifted germanium detector (at the temperature of liquid nitrogen).*

low-energy x-rays, while a Ge(Li) detector is used mostly for γ-rays. Figure 34.14 shows the gamma spectrum of radioactive cobalt 60. For comparison, the spectrum obtained by using a NaI(Tl) detector is also shown. The superiority of the semiconductor detector over the scintillation detector is quite clear (narrow peaks mean better resolution). One disadvantage of Ge(Li) is that it is only a few percent as efficient as a NaI(Tl) detector in detecting γ-rays.

34.5. Nuclear Fission and Power Reactors

Nuclear Fission

Fission is a process in which a heavy nucleus ($A > 230$) breaks into two or more fragments of comparable size, with the release of an enormous amount of energy. Nuclear fission was discovered by Otto Hahn and Fritz Strassmann in 1939. They bombarded uranium with slow neutrons and found that the resulting products were barium 139 and lanthanum 140. Further investigation revealed that it was uranium 235 which absorbed slow neutrons and underwent fission. In each such fission, the amount of energy released is ~200 MeV. Also, besides Ba and La, other elements produced in fission are different isotopes of the elements in the range $Z = 30$ (zinc) to $Z = 65$ (terbium).

^{235}U is one of three nuclei, the other two being ^{233}U and ^{239}Pu, which undergo fission by thermal neutrons (neutrons with energies ~0.025 eV). The above three and many other heavy nuclei can undergo fission by absorbing fast-moving neutrons, charged particles, and γ-rays. In the process of thermal fission, in addition to two fragments and an average of 2.5 neutrons, many γ-rays are emitted in each fission. Thus, fission of uranium 235 by slow neutrons may be represented as

$$^{235}\text{U} + {}^1_0n \rightarrow {}^{236}\text{U} \rightarrow \text{fission fragments} + \text{neutrons} + \text{radiation}$$

where fission fragments are in the range $A \simeq 70 - 160$ and $Z \simeq 30 - 65$ and include isotopes of barium, lanthanum, bromine, molybdenum, antimony, tellurium, cesium, iodine, krypton, and zenon. Most of the fission fragments are radioactive and decay by β^--emission and γ-rays.

It may be pointed out that natural uranium contains only 0.72 percent (1 part in 138) of the ^{235}U isotope. For use in nuclear power reactors as well as in nuclear weapons, natural uranium enriched in the isotope ^{235}U is employed.

Some idea of the amount of energy released in fission may be obtained from the following. According to the BE/A versus A curve shown in Figure 34.1, the binding energy of the heavy nucleus is less than the combined binding energies of the fission fragments. That is, the rest mass of the heavy nucleus is more than the rest masses of its fission fragments. The difference in mass appears as energy. For example, the binding energy of each nucleon in uranium is ~7.6 MeV, while the binding energy of the fission fragments ($A \sim 118$) is ~8.5 MeV/nucleon. Thus, the energy released in a single fission is

$$(2 \times 118 \times 8.5 \text{ MeV} - 236 \times 7.6 \text{ MeV}) \simeq 212 \text{ MeV}$$

Accurate calculation yields ~200 MeV/fission, which agrees well with the measured values. This energy appears in the form of kinetic energy of fission fragments, neutrons, β-rays, γ-rays, and neutrinos, as shown in Table 34.2.

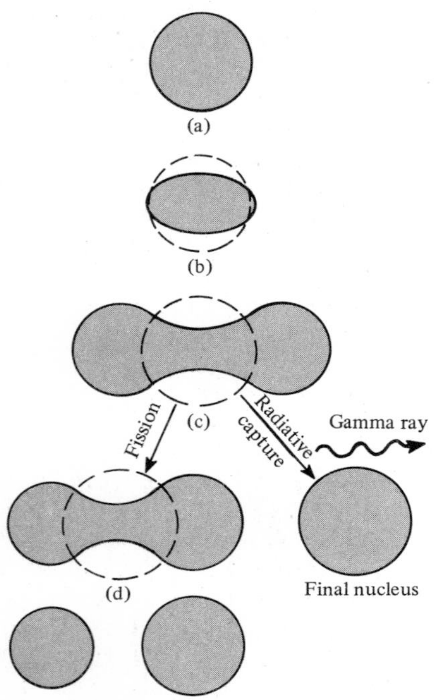

FIGURE 34.15. *A fissionable nucleus takes different shapes, depending on the degree of excitation resulting from the absorption of an incident particle. (a) Spherical nucleus before collision with incident particle. (b) After collision, nucleus undergoes deformation to an ellipsoid. (c) Nucleus still further deformed. (d) If the excitation is high, the nucleus will undergo fission. Otherwise, the particle is captured and a γ-ray is emitted.* (From Atam P. Arya, *Elementary Modern Physics*, Addison-Wesley Publishing Company, Inc., Reading, Mass., 1974.)

TABLE 34.2
Energy Release in Fission of ^{235}U by Thermal Neutrons

Kinetic Energy of:	MeV
Fission fragments	~165
Fission neutrons	~5
Gamma rays emitted in fission	~8
Gamma rays, β-rays, and neutrinos emitted in the decay of fission fragments	~22
Total energy per fission	~200

Theory of Fission

Although not complete, a somewhat satisfactory theory of fission which could explain certain aspects was developed by Niels Bohr and John Wheeler and is based on the liquid-drop model. According to them, the ^{235}U nucleus is assumed to be like a perfectly spherical liquid drop. Absorption of a neutron by this nucleus deforms its spherical shape. As the amount of energy absorbed increases, the deformation of the nucleus increases, as shown in Figure 34.15.

Suppose that the fissionable nucleus, after absorbing some energy, is in a state shown in Figure 34.15(c). At this stage two things can happen. The nucleus can get rid of its extra energy by emitting a γ-ray; hence, return to its original spherical shape. On the other hand, the deformation could increase and reach a shape shown in Figure 34.15(d), in which case it definitely undergoes fission.

Fission of other heavy nuclei by charged particles, γ-rays, or neutrons may be explained in a similar fashion. Sometimes the energy needed to cause fission may be as high as ~30 MeV.

Nuclear Power Reactors

The fundamental differences between nuclear fission and other nuclear reactions make the fission process useful in nuclear reactors that produce large amounts of power. Two outstanding features of fission are: (1) in each fission there is ~200 MeV of energy released, and (2) each fission produces on the average 2.5 neutrons.

To start with, let us assume that a small amount of uranium absorbs 100 neutrons. Of these, say 60 of them are lost or absorbed in nonfission processes, while the other 40 cause fission. Since in each fission there are 2.5 neutrons, there will be $40 \times 2.5 = 100$ neutrons again in the new generation. If this process continues, we say that a *self-sustained chain reaction* has been achieved. This is not always true; there may be more or less neutrons present than there were in the previous generation. We define the *reproductive factor* or *multiplication factor*, k, as the ratio

$$k = \frac{\text{number of neutrons in the present generation}}{\text{number of neutrons in the previous generation}} \tag{34.16}$$

As in the case discussed above, if $k = 1$, the number of neutrons remains the same in all generations—called a *critical system*. If $k > 1$, the number of neutrons in any generation is more than in the previous generation—called a *supercritical system*. On the other hand, if $k < 1$, the number of neutrons in a generation is less than in the previous generation—called a *subcritical system*.

In each nuclear fission a large amount of energy (~200 MeV) is produced. If $k > 1$, the number of neutrons increases in each succeeding generation and the energy produced also increases. This is the basic principle utilized in atomic bombs and other nuclear weapons. In the system if $k < 1$ (subcritical system), the number of neutrons decreases in each succeeding generation; hence, the energy or power also decreases. This eventually shuts down the system when the power level reaches zero. These types of systems are used in nuclear research. Finally, if $k = 1$, the system is critical; that is, the number of neutrons in each generation remains constant; hence, the power level of the system remains constant. This condition is the basic requirement in the operation of a nuclear power plant, as we shall explain shortly.

If we use natural uranium, which consists of the isotopes ^{235}U and ^{238}U in the ratio 1:138, it is not possible to achieve a supercritical or even a critical system. The reason being that more neutrons are absorbed in ^{238}U than in ^{235}U, or escape the system than are produced, leading to a shutdown. To increase the probability of neutron absorption in ^{235}U, uranium fuel is enriched in the isotope ^{235}U. For nuclear power plants enrichment of 3.5 to 7 percent is enough, whereas for uranium to be used in atomic bombs, much higher enrichment (~90 percent) of the ^{235}U isotope is needed. This enriching is an extremely costly process.

Let us now briefly discuss the working of a nuclear power plant. An outline sketch of a typical power plant is shown in Figure 34.16. The part that provides the power is called a *nuclear reactor*. The core of the reactor consists of a large number of fuel elements made out of uranium that has been enriched with ^{235}U. The details of the fuel elements surrounded by the moderator are shown in Figure 34.16(b). The moderator, which is usually water (sometimes heavy water), slows down the fast neutrons emitted in fission. Slow neutrons

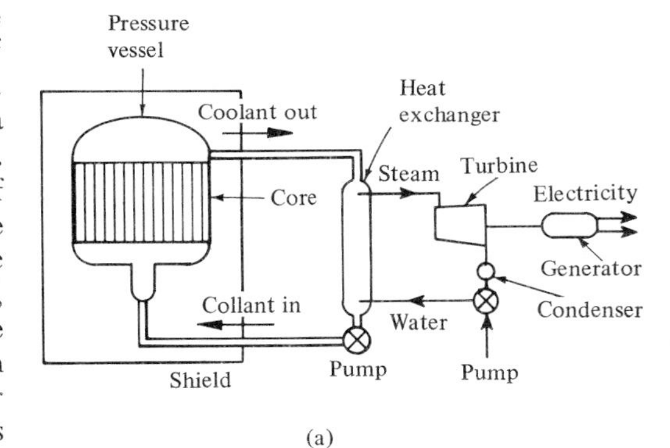

(a)

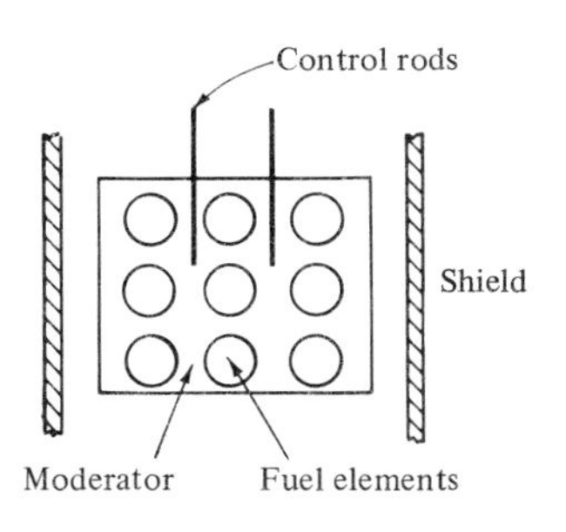

(b)

FIGURE 34.16. *(a) Typical nuclear power plant. (b) Arrangement of fuel elements in the core of a nuclear reactor.* (From A. P. Arya, *Fundamentals of Nuclear Physics*, Allyn and Bacon, Inc., Boston, 1966.)

have less probability of escaping the fuel system than do fast neutrons. To control the number of neutrons in the system, movable control rods made of cadmium (cadmium has a very high probability for neutron absorption) are immersed in the fuel elements, as shown. Initially an external source of neutrons starts the fission while the control rods are partially pulled out of the core. Under these conditions the system is supercritical, the number of neutrons increases, hence the power level increases. Once a desired power level has been reached, the control rods are pushed into the core until the system becomes critical and remains at that power level.

The fission produces heat in the reactor core. The coolant transfers this heat from the core to a heat exchanger, where steam is produced. This steam, produced at very high pressure, runs a turbine, and the electricity is obtained at the generator. The dead steam from the turbine condenses into water and is returned to the heat exchanger. To get some idea of the amount of power produced, let us consider the following example.

Example 34.4 How does the energy produced by ^{235}U compare with energy produced from coal?

Let us consider 1 g of ^{235}U, which contains $N_A/A = 6.02 \times 10^{23}/235 = 2.56 \times 10^{21}$ atoms. Since each fission produces 200 MeV of energy, the total energy available from 1 g of ^{235}U under fission is equal to

$$(200 \text{ MeV})(2.56 \times 10^{21})(1.6 \times 10^{-13} \text{ J/MeV}) = 8.2 \times 10^{10} \text{ J} = 8.2 \times 10^{10} \text{ W-s} \simeq 1 \text{ megawatt-day (1 MWD)}$$

It takes 3×10^6 g $(= 3 \times 10^3$ kg) of coal to produce 1 MWD of heat.

Exercise 34.4 How much energy is released in the fission of 1 g of ^{239}Pu? Assume the energy release per fission is 200 MeV? [*Ans.:* 8.06×10^{10} W-s.]

It may be pointed out that the uranium ore exists in quantities large enough to provide our power needs for many years. But the use of these fuels can be extended over a much longer period by using the *breeder concept*. According to this concept, as the nuclear reactor operates, ^{238}U absorbs fast neutrons and becomes ^{239}U, which decays into ^{239}Pu. ^{239}Pu itself is a fissionable fuel and can be used in nuclear reactors for power production.

34.6. Nuclear Fusion and Thermonuclear Energy

Fossil fuels such as oil and coal exist in limited quantities and will run out eventually. In the last two decades nuclear fission power plants have come into existence which can provide the power needs for many years and can be extended over much longer periods by utilizing the concept of breeder reactors using ^{239}Pu as fuel instead of ^{235}U. An alternative to all this power production is fusion or thermonuclear reactions, as we shall explain below. Fusion, if successful, can provide enough energy to meet our needs for an indefinite time.

Nuclear fusion is a process in which two very light nuclei (with $A \leqslant 8$) combine to form a heavy nucleus, with simultaneous release of a large amount of energy. (The process of nuclear fusion is opposite to that of fission and can be understood from the BE/A versus A curve in Figure 34.1.) Such nuclear fusion reactions are taking place in the sun and other stars, resulting in the emission of the energy we receive. The temperature in the interior of the stars is $\sim 10^7$ K, which leads to complete ionization of the atoms. Such a system will contain bare nuclei and electrons and is called *plasma*. These nuclei are moving with such high speeds that when they collide they can overcome Coulomb repulsion and combine together by attractive nuclear forces, with a release of energy. Such reactions, which are achieved by means of very high temperatures, are called *thermonuclear reactions*.

The following is a typical example of a thermonuclear reaction taking place in stars.

Proton–proton cycle:

$$\begin{aligned} {}^{1}\text{H} + {}^{1}\text{H} &\longrightarrow {}^{2}\text{H} + \beta^{+} + \nu + 0.42 \text{ MeV} \\ {}^{2}\text{H} + {}^{1}\text{H} &\longrightarrow {}^{3}\text{He} + \gamma + 5.49 \text{ MeV} \\ {}^{3}\text{He} + {}^{3}\text{He} &\longrightarrow {}^{4}\text{He} + 2{}^{1}\text{H} + 12.86 \text{ MeV} \\ &\text{Total energy released } 18.77 \text{ MeV} \end{aligned} \tag{34.17}$$

Another common cycle is that of carbon, in which a total energy of 24.68 MeV is released.

Two of the three isotopes of hydrogen—${}^{2}_{1}\text{H}$ (deuterium) and ${}^{3}_{1}\text{H}$ (tritium)—have been used to produce thermonuclear fusion in the laboratories by the following reactions:

$$\begin{aligned} {}^{2}_{1}\text{H} + {}^{2}_{1}\text{H} &\longrightarrow {}^{3}_{2}\text{He} + n + 3.2 \text{ MeV} \\ {}^{2}_{1}\text{H} + {}^{2}_{1}\text{H} &\longrightarrow {}^{3}_{1}\text{H} + {}^{1}_{1}\text{H} + 4.0 \text{ MeV} \\ {}^{2}_{1}\text{H} + {}^{3}_{1}\text{H} &\longrightarrow {}^{4}_{2}\text{He} + n + 17.6 \text{ MeV} \end{aligned} \tag{34.18}$$

Although deuterium is available in large quantities from the ocean, the problem involved is how to contain plasma at temperatures of $\sim 10^7$ K in order to maintain the thermonuclear reaction. No known material can maintain such a high temperature without vaporizing.

Alternatives that have been tried involve containing the plasma by means of magnetic fields. These techniques have been variously named *pinch effect*, *magnetic mirrors*, *astron*, and *stellarator*.

A recent development in the field of fusion is the use of the laser beam to initiate the thermonuclear reaction. These methods are still in the experimental stages but have shown some promise for future fusion reactors for power production.

34.7. Radiation and Applications—Medical and Industrial

Radiation absorbed by any part of the body will deposit energy and will cause damage to the tissue. Even sunlight, which contains ultraviolet rays and infrared rays, can cause damage. In the present discussion the term "radiation damage" will not include such radiation. By radiation damage we mean exposure to natural radioactivity, cosmic rays, medical and dental x-rays, and fallout from nuclear testing. The effects of nuclear radiation differ in one important aspect as far as the human body is concerned. Unless the human body is exposed to a large amount of radiation at one time, the ill effects of radiation are time delayed.

Units of Radiation Exposures

A quantity of radiation is measured by the amount of ionization produced by the passage of radiation through a medium. From the point of view of biological effects of radiation and use in medicine, it is more important and relevant to know the amount of energy that will be deposited when a body is exposed to radiation. The most commonly accepted units are the roentgen, rep, rad, and rem.

One *roentgen* is that amount of x- or γ-radiation that will result in absorption of 93 ergs of energy in 1 g of air. One *rep* (*roentgen equivalent physical*) is that amount of radiation that will result in absorption of 93 ergs of energy in 1 g of soft tissue. Since a human body contains denser materials than soft tissue, the rep unit is inadequate for measuring whole-body exposures to radiation. For this purpose a different unit, called a rad, is used. One *rad* (*radiation absorbed dose*) is equivalent to an energy absorption of 100 ergs/g in any medium from any type of radiation. In the MKS system, 1 rad indicates when an object of mass 1 kg absorbs 0.01 J of radiation energy.

$$\boxed{1 \text{ rad} = 0.01 \text{ J/kg}} \tag{34.19}$$

It is also known that different types of radiation have different effectiveness in causing radiation damage. That is, the biological effectiveness of the radiation depends not only on the number of rads to which a body is exposed, but also on the effectiveness of the type of radiation. For example, α-rays, which produce intense ionization, are much more harmful than γ-rays, which cause very low ionization. The *relative biological effectiveness* (*RBE*) has been measured for different types of radiation and is summarized in Table 34.3.

Thus when studying radiation damage to living cells, one must take into

TABLE 34.3
Relative Biological Effectiveness (RBE) Factors

Type of Radiation	RBE Factor
X-rays and γ-rays	1
Beta particles (β^-, β^+, CE)	1
Thermal neutrons	2–5
Fast neutrons	10
Protons[a]	8–10
Alpha particles[a]	10–20
Recoil ions	20

[a] Slower-moving protons or α-particles have higher RBE factors.

account the RBE of different radiations. For this purpose, the rad has been replaced by another unit, called the rem. The *rem* (*roentgen equivalent man*) is defined as the product of 1 rad and the RBE. That is,

$$\boxed{1 \text{ rem} = 1 \text{ rad} \times \text{RBE}} \tag{34.20}$$

For example, if a person receives a dose of 0.2 rad from a beam of protons, he has been exposed to 0.2 rad × 10 RBE = 2 rem.

Permissible Radiation Doses

The rem unit is used when recommending permissible amounts of radiation to human beings working in a radiation environment, to radiology workers, and to others. One important consideration in recommending any permissible dose is the length of time over which the body is exposed. For example, a dose of 400 rem in one short time exposure may be lethal, but the same dose spread over a period of 40 years may not show any significant damage. If an individual is exposed to a large single dose of radiation over a short interval of time, it is called an *acute exposure,* while a steady small dose of radiation over a long time is called a *chronic exposure.* It is found that on the average a typical individual in the United States receives a total dose of 180 mrem/yr (1 mrem = 10^{-3} rem) resulting from (1) ~100 mrem/yr from natural radioactivity and cosmic rays, (2) ~75 mrem/yr from dental and medical x-rays, and (3) ~5 mrem/yr from fallout from nuclear weapons testing.

It is true that any amount of radiation exposure is considered to be a health hazard. But there are situations where certain exposures cannot be avoided (for useful medical and industrial applications). Under such circumstances the exposure should be kept to a minimum. For this purpose, at present the *maximum permissible amount of radiation dose to which an individual may be exposed (without any ill effects) is set at 500 mrem per year.* It is assumed that such exposures are uniformly distributed over the whole year. It is also recommended that those persons under the age of 18 should have zero exposure. This is because at a young age, the body cells are growing and are very sensitive to radiation damage.

There are limits on the number of radioactive atoms that may be ingested from air and liquids. Permissible concentrations of an unknown mixture of radiotopes in air is 0.2×10^{-9} μCi/ml of β- and γ-emitters and 0.2×10^{-11} μCi/ml of α-emitters. In water it is set at less than 10^{-7} μCi/ml.

One may wonder at this stage what are the clinical symptoms of radiation sickness. For low long-term exposures (chronic exposures) there are basically no clinical symptoms, but in many cases cancer has been found. But high short-term exposures (acute exposures) do have clinical symptoms, as summarized in Table 34.4.

Biological Effects of Radiation Damage

The radiation effects on individuals can be divided into two categories: (1) *somatic effects,* those that concern particular individuals and are not propagated from generation to generation, and (2) *genetic effects,* which are propagated to offspring.

The damage to the human body from nuclear radiation is due to the

TABLE 34.4
Clinical Symptoms of Radiation Sickness

Time After Exposure	Lethal Dose (650 r)	Medium Lethal Dose (400 r)	Sublethal Dose (250–100 r)
First week	Nausea, vomiting within 2 h Diarrhea Inflammation of mouth and throat	Nausea, vomiting after 2 h	Possible nausea, vomiting
Second week	Fever Rapid loss in weight Death	Loss of hair Loss of appetite General discomfort	
Third week		Fever Severe reddening of mouth and throat	Loss of hair Loss of appetite General discomfort Sore throat Pallor Bleeding Diarrhea
Fourth week		Pallor Bleeding Diarrhea Rapid loss in weight Death 50% chance	Recovery likely

ionization or excitation of atoms in living cells by the photoelectric effect, bremsstrahlung, Compton effect, or recoiling atoms. The cells may be completely destroyed or altered by ionization. Radiation damage to the chromosomes in the reproductive organs of either of the sexes can cause genetic mutations. Radiation damage to the blood-producing cells in the spleen and bone marrow can increase the possibility of contracting leukemia. Some other delayed effects observed are bone cancer and eye cataracts. An acute overexposure to radiation destroys the infection–resistance mechanism, and thus may result in death. Besides causing cancer, radiation may shorten life expectancy when persons are exposed to large doses.

Besides external exposure, radiation damage can come from inhaling air containing radioisotopes and swallowing foods contaminated with radioisotopes. Different isotopes seek different positions in the body. For example, radioiodine 131 goes to the thyroid gland, strontium 89 and 90 to the bones, uranium 238 to the kidneys and lungs, and cadmium 109 to the liver. Thus, the damage to different portions will depend upon the types of isotopes swallowed or inhaled.

The genetic effects of radiation are hard to investigate because birth defects are also due to causes other than radiation. Some of the other damages due to radiation are listed in Table 34.4.

Medical and Industrial Uses of Radiation

So far we have been talking about the negative effects of radiation. It would be inappropriate not to reveal some of the many beneficial uses of nuclear radiation. These are spread throughout all aspects of life. It will be impossible to enumerate and describe all of them. We shall briefly describe only a few.

MEDICAL APPLICATIONS. There are indications that excessive doses of radiation cause cancer, but it is also used to *control* many types of cancer. Skin cancer is treated by x-rays or electron radiation. High-energy γ-rays from radioactive ^{60}Co are used for treating tumors and cancer—the treatment is referred to as cobalt therapy. Sometimes, pellets of ^{60}Co or some other radioactive source are placed by surgery in the body at the site of a malignant growth so that the radiation damage will destroy it.

A radioactive tracing technique is becoming a more common medical diagnostic tool. For example, radioactive tracing is used for locating brain damage and many other internal injuries to portions of the human body where a surgical diagnostic operation is not advisible and an x-ray photograph does not reveal the damage. The tracing technique is also used to study the flow of atoms through metabolic processes in plants and animals.

Another area where radioactive iodine, ^{131}I, is used is in the treatment of a hyperactive thyroid gland (hyperthyroidism). The hyperactive thyroid takes twice as much iodine as a normal one. Thus, the radioactive iodine reveals the presence of a hyperactive thyroid. Also, the current treatment for a cancerous thyroid is to continue the ingestion of radioiodine. The radiation destroys both normal and abnormal tissues, but the abnormal absorbs more; hence, more cancerous cells will be destroyed.

INDUSTRIAL APPLICATIONS. The industrial uses of radiation have not yet been fully exploited, but a few that are known will give us an idea of its scope.

The rate of chemical reactions can be increased and many complex compounds can be produced by irradiation. For example, a 3000-Ci source of ^{60}Co is used for the synthesis of ethyl bromide (C_2H_5Br). The sterilization of medical supplies can be accomplished more efficiently by exposing medical supplies to an electron beam.

When food for storage is exposed to radiation, such as γ-rays, the shelf life of food products is substantially increased, by killing bacteria and microorganisms more efficiently. These methods are still under investigation, and only a few products have been treated at present.

When cotton is treated by exposure to an electron beam, it acquires the characteristic of being permanent press. Also, irradiation for a few seconds can "cure paint"; that is, paint on metal or plastic will not peel or chip if treated with radiation. Before this method was used, paint needed a long heat treatment.

SUMMARY

An atomic nucleus consists of neutrons and protons. The energy equivalent of mass difference is called *binding energy*, that is BE/A =

$[Zm_H + (A - Z)\, m_n - M(A, Z)]c^2/A$. Even–even nuclei are more abundant. The radius of the nucleus is $R = 1.35 \times 10^{-15} A^{1/3}$ m. The angular momentum of the nucleus is $I = \sqrt{i(i+1)}\hbar$. Nuclear forces are attractive and short range.

According to the *radioactive decay law*, $N = N_0 e^{-\lambda t}$, activity $A = |\Delta N/\Delta t| = \lambda N$, $t_{1/2} = 0.693/\lambda$, and $\tau = 1/\lambda$. The units of radioactivity are 1 Ci $= 3.7 \times 10^{10}$ dis/s and 1 rd $= 10^6$ dis/s.

Stopping power is defined as the amount of energy lost per unit length by the incident particle in a given medium. The heavy charged particles lose energy by collisions, γ-rays by photoelectric effect, Compton scattering, and pair production.

Detectors in common use may be classified into ionization chambers, visual detectors, scintillation counters, semiconductor detectors, and high-energy detectors.

Fission is a process in which a heavy nucleus ($A > 230$) breaks into two or more fragments of comparable size, with the release of a large amount of energy. Fission of ^{235}U by slow neutrons yields ~200 MeV of energy per event. The liquid-drop model can be used to explain fission. For a critical reactor the reproduction or multiplication factor k is 1, for subcritical $k < 1$, and for supercritical $k > 1$.

Nuclear *fusion* is a process in which two very light nuclei ($A \leqslant 8$) combine to form a heavy nucleus, which releases a large amount of energy.

Units used in radiation exposures are roentgen, rep, rad, and rem. 1 rad $= 0.01$ J/kg and 1 rem $= 1$ rad $\times$ RBE. At present the maximum permissible dose to which an individual may be exposed is set at 500 mrem per year. The biological effects of radiation may be divided into somatic effects and genetic effects. There are several applications of radiation—including cancer treatment, radioactive tracing, and thyroid gland treatment. The industrial applications include increasing the rate of chemical reactions, curing paint, treating cotton to produce permanent press fibers, and in food storage.

QUESTIONS

1. How do you prove that nuclear forces are short range and attractive?
2. The sizes of all the atoms are almost the same, whereas the size of the nucleus depends on A. Can you think of a reason why this is so?
3. The atomic weight of most elements differs greatly from integer values. How do you account for this?
4. Can you devise an experiment to show that radioactivity is a nuclear process?
5. Do the mass number and the atomic number change when a nucleus decays by α-, β-, or γ-emission?
6. $^{232}_{90}$Th decays by emitting a series of α- and β-particles, eventually becoming stable $^{208}_{82}$Pb. How many (minimum) α- and β-particles are emitted?
7. Why are slow neutrons most commonly used (captured by nuclei) to produce artificial radioisotopes?
8. Why are fission products usually radioactive whereas fusion products are not?
9. Which is more dangerous: to be exposed to or to ingest radioactive material?

10. What can you say about the future of nuclear medicine and nuclear power?
11. When talking about nuclear radiation exposures, why don't we use the unit Ci instead of introducing r, rep, rad, and rem?

PROBLEMS

1. Calculate the amount of energy released when oxygen 16 is formed by bringing 8 protons, 8 neutrons, and 8 electrons together. Also, calculate the average binding energy per nucleon. The mass of oxygen 16 is 15.99492 u.
2. The mass of ^{107}Ag ($Z = 47$) is 106.90509 u. Calculate its average binding energy per nucleon.
3. Calculate the binding energy per nucleon for ^{28}Si ($Z = 14$), which has a mass of 27.97693 u.
4. The binding energy of the last neutron added to the nucleus (A, Z) is given by

$$BE(n) = [M(A - 1, Z) + m_n - M(A, Z)]c^2$$

   Using this relation, calculate the binding energy of the last added neutron in the following three isotopes of lead ($Z = 82$): ^{206}Pb, ^{207}Pb, and ^{208}Pb.
5. Calculate the binding energy of a neutron added to $^{235}_{92}U$ and $^{233}_{92}U$.
6. Calculate the binding energy of the last-added (a) neutron in $^{4}_{2}He$; and (b) proton in $^{16}_{8}O$ and $^{32}_{16}S$. Compare these values with the binding energy per nucleon.
7. What is the half-life of a radioactive sample if after 30 days only $\frac{1}{4}$ of the sample remains?
8. A radioactive sample has a half-life of 20 min. What fraction of radioactive sample remains after 2 h?
9. Calculate the activity of 1 mg of radium 226, which has a half-life of 1620 yr.
10. A sample of radioactive iodine 123 was observed to decay in the following manner:

Time Elapsed (h)	Counts/min	Time Elapsed (h)	Counts/min
0	99,968	25	26,448
5	77,105	30	20,789
10	58,860	35	15,567
15	44,932	40	12,246
20	34,646	45	9,163

   Make a plot of these data and calculate the following: (a) the half-life, decay constant, and mean life; (b) the activity of the sample after 26 h; and (c) the number of radioactive atoms after 26 h.
11. ^{144}Pm has a half-life of 365 days. How long will it take to reduce its activity from 1 mCi to 10 μCi?
12. The activity of a certain radioactive sample decreases by a factor of 8 in a time interval of 30 days. What are its half-life, mean life, and disintegration constant?

TABLE 34.5
Nuclear Masses

^{3}He	3.01603 u
^{4}He	4.00260 u
^{15}N	15.00011 u
^{16}O	15.99492 u
^{28}Si	27.97693 u
^{31}Si	30.97376 u
^{32}S	31.97207 u
^{107}Ag	106.90509 u
^{205}Pb	204.9747 u
^{206}Pb	205.9745 u
^{207}Pb	206.9759 u
^{208}Pb	207.9766 u
^{227}Th	227.0278 u
^{228}Th	228.0287 u
^{233}U	233.0395 u
^{234}U	234.0409 u
^{235}U	235.0439 u
^{236}U	236.0457 u
^{240}Pu	240.0540 u
^{242}Pu	242.0587 u
^{243}Pu	243.0606 u

13. What will be the mass of a 10-Ci sample of cobalt 60, given that its half-life is 5.26 y?

14. One gram of carbon from a living tree has an activity of 12 disintegrations/min. If 10 g of a wooden relic shows 20 disintegrations/min, calculate the age of the relic.

15. An α-particle is stopped in an ionization chamber, in which it produces 150,000 ion pairs. Each time the α-particle produces an ion pair, it loses 35 eV of energy. What is the kinetic energy of the particle? Calculate the amount of charge collected by each plate.

16. Calculate the binding energy of the thermal neutron added to the following nuclei: ^{227}Th, ^{233}U, ^{235}U, ^{239}Pu, and ^{242}Pu. Which of these will be classified as fissionable by thermal neutrons?

17. In the fission of ^{235}U, the mass ratio of the two fission fragments is 1.5. What is the ratio of the velocities of these two fragments? (Use the conservation-of-momentum principle.)

18. In each fission the energy released is 200 MeV. How much mass is converted into energy?

19. Energy released in a small atomic bomb explosion is approximately 10^{14} J. If an average energy of 200 MeV is released in each fission of ^{235}U, calculate the amount of ^{235}U that has undergone fission.

20. If only 10 percent of the energy released in fission is converted to electric power, calculate the number of fissions per hour needed to generate 20 MW of electric power. The energy released in each fission is 200 MeV.

21. A certain nuclear power plant converts 5 mg of mass into energy each day. Calculate the number of nuclei that undergo fission every day. Calculate the total power output.

22. Calculate the total energy liberated in (a) a proton–proton cycle; and (b) a carbon–nitrogen cycle when 1 g of material undergoes complete fusion.

23. How much energy is liberated when 1 g of hydrogen atoms is converted into helium by fusion? Compare this with the energy liberated in the fission of 1 g of ^{235}U.

24. The half-life of ^{131}I is 8 days. A thyroid patient has been given a dose of 100 μCi. How much radioactivity remains after (a) 8 days; (b) 40 days; and (c) 80 days?

Back Matter

Appendix A
SI Tables and Conversions

TABLE A.1
Conversion to and from the SI System
A. Conversion to SI

Name of Unit	Symbol	Multiply by	SI Unit	Symbol
inches	in	25.4	millimeters	mm
feet	ft	0.305	meters	m
yards	yd	0.914	meters	m
miles	mi	1.609	kilometers	km
square yards	yd^2	0.836	square meters	m^2
acres		0.405	hectares	ha
cubic yards	yd^3	0.765	cubic meters	m^3
quarts (liq)	qt	0.946	liters	l
ounces (avdp)	oz	28.350	grams	g
pounds	lb	0.454	kilograms	kg
Fahrenheit	°F	$\frac{5}{9}$ (after subtracting 32)	Celsius	°C

B. Conversion from SI

SI Unit	Symbol	Multiply by	To Find Unit	Symbol
millimeters	mm	0.039	inches	in
meters	m	3.281	feet	ft
meters	m	1.094	yards	yd
kilometers	km	0.621	miles	mi
square meters	m^2	1.196	square yards	yd^2
hectares	ha	2.471	acres	
cubic meters	m^3	1.308	cubic yards	yd^3
liters	l	1.057	quarts (liq)	qt
grams	g	0.035	ounces (avdp)	oz
kilograms	kg	2.205	pounds	lb
Celsius	°C	$\frac{9}{5}$ (then add 32)	Fahrenheit	°F

Table A.2
Multiples, Submultiples, and Prefixes (Applicable to All SI Units)

Multiples and Submultiples	Prefixes	Symbols
1 000 000 000 000 = 10^{12}	tera	T
1 000 000 000 = 10^{9}	giga	G
1 000 000 = 10^{6}	mega	M
1 000 = 10^{3}	kilo	k
100 = 10^{2}	hecto	h
10 = 10^{1}	deka	da
base unit: 1 = 10^{0}		
0.1 = 10^{-1}	deci	d
0.01 = 10^{-2}	centi	c
0.001 = 10^{-3}	milli	m
0.000 001 = 10^{-6}	micro	μ
0.000 000 001 = 10^{-9}	nano	n
0.000 000 000 001 = 10^{-12}	pico	p

Table A.3
Some Common Conversions

Length
- 1 m = 3.281 ft = 39.37 in; 1 cm = 0.3937 in
- 1 km = 1000 m = 0.6214 mi
- 1 ft = 30.48 cm; 1 in = 2.540 cm
- 1 yd = 0.9144 m
- 1 mi = 5280 ft = 1.609 km

Area
- 1 cm^2 = 0.155 m^2
- 1 m^2 = 10^4 cm^2 = 10.76 ft^2
- 1 in^2 = 6.452 cm^2
- 1 ft^2 = 929.0 cm^2 = 0.09290 m^2

Volume
- 1 m^3 = 1000 liters = 10^6 cm^3 = 1.308 yd^3 = 35.31 ft^3
- 1 liter = 1000 cm^3 = 0.001 m^3 = 61.03 in^3 = 0.0353 ft^3
- 1 ft^3 = 0.02832 m^3 = 7.481 gallons = 28.32 liters

Time
- 1 day = 86,400 s; 1 yr = 3.156×10^7 s

Velocity
- 1 m/s = 3.281 ft/s = 3.6 km/h
- 1 km/h = 0.2778 m/s = 0.6214 mi/h = 0.9113 ft/s
- 1 mi/h = 1.609 km/h = 0.447 m/s = 1.467 ft/s

Acceleration
- 1 m/s^2 = 100 cm/s^2 = 3.281 ft/s^2
- 1 ft/s^2 = 30.48 cm/s^2 = 0.3048 m/s^2

TABLE A.3 (Continued)
Some Common Conversions

Mass
- 1 kg = 1000 g = 0.0685 slug
- 1 slug = 14.59 kg = 32.17 lb mass
- 1 metric ton = 1000 kg

Density
- 1 g/cm^3 = 1,000 kg/m^3 = 1.940 slug/ft^3 = 62.43 lb-mass/ft^3
- 1 lb-mass/ft^3 = 0.0311 slug/ft^3 = 16.02 kg/m^3 = 0.01602 g/cm^3

Force
- 1 N = 10^5 dyn = 0.2248 lb
- 1 lb = 4.45 N = 4.45×10^5 dyn
- 1 ton = 2000 lb

Pressure
- 1 N/m^2 = 1.451×10^{-4} lb/in^2 = 0.209 lb/ft^2
- 1 lb/in^2 = 6.89×10^3 N/m^2 = 6.89×10^4 dyn/cm^2
- 1 atm = 76 cm Hg = 760 torr = 14.70 lb/in^2 = 1.013×10^5 N/m^2
  = 1.013 bar = 1.013×10^6 dyn/cm^2

Work and Energy
- 1 J = 10^7 ergs = 0.239 cal = 0.7376 ft-lb
- 1 ft-lb = 1.356 J
- 1 cal = 4.184 J = 3.086 ft-lb
- 1 Btu = 252 cal = 778 ft-lb = 1054 J
- 1 kilowatt-h (kWh) = 3.60×10^6 J
- 1 eV = 1.60×10^{-19} J

Power
- 1 W = 1 J/s = 0.738 ft-lb/s
- 1 hp = 0.746 kW = 550 ft-lb/s
- 1 Btu/h = 0.293 W

Specific Heat and Latent Heat
- 1 cal/g-C° = 4.184 J/g = 4184 J/kg-C°
- 1 cal/g = 4.184 J/g = 4184 J/kg = 1.80 Btu/kg-C°
- R = 8.314 J/mol-K = 1.99 cal/mol-K = 0.0821 (atm-l/mol-K)

Appendix B
Trigonometrical Relations

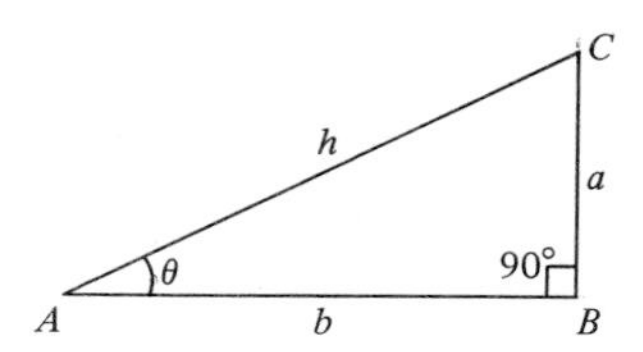

Consider a right-angle triangle ABC and angle θ for which $\sin\theta = \dfrac{a}{h}$

$\cos\theta = \dfrac{b}{h}$ $\quad\tan\theta = \dfrac{a}{b}$ $\quad\cot\theta = 1/\tan\theta = \dfrac{b}{a}$ $\quad\cos\theta = 1/\sin\theta = h/a$; $\sec\ \theta = 1/\cos\theta = h/b$. The values of these functions for different values of θ are listed in Table B.1.

The following relations hold good for any right-angle triangle:

$$\sin(90° - \theta) = \cos\theta$$
$$\cos(90° - \theta) = \sin\theta$$
$$\sin(180° - \theta) = \sin\theta$$
$$\cos(180° - \theta) = -\cos\theta$$
$$\sin(\theta_1 \pm \theta_2) = \sin\theta_1\cos\theta_2 \pm \cos\theta_1\sin\theta_2$$
$$\cos(\theta_1 \pm \theta_2) = \cos\theta_1\cos\theta_2 \mp \sin\theta_1\sin\theta_2$$
$$\cos^2\theta + \sin^2\theta = 1$$
$$\cos 2\theta = \cos^2\theta - \sin^2\theta = 2\cos^2\theta - 1 = 1 - 2\sin^2\theta$$
$$\sin 2\theta = 2\sin\theta\cos\theta$$

$$\sin\theta_1 \pm \sin\theta_2 = 2\sin\left(\frac{\theta_1 \pm \theta_2}{2}\right)\cos\left(\frac{\theta_1 \mp \theta_2}{2}\right)$$

$$\cos\theta_1 + \cos\theta_2 = 2\cos\left(\frac{\theta_1 + \theta_2}{2}\right)\cos\left(\frac{\theta_1 - \theta_2}{2}\right)$$

$$\cos\theta_1 - \cos\theta_2 = 2\sin\left(\frac{\theta_1 + \theta_2}{2}\right)\sin\left(\frac{\theta_1 - \theta_2}{2}\right)$$

If θ is expressed in radians,

$$\sin\theta = \theta - \frac{\theta^3}{3!} + \frac{\theta^5}{5!} - \frac{\theta^7}{7!} + \cdots$$

$$\cos\theta = 1 - \frac{\theta^2}{2!} + \frac{\theta^4}{4!} - \frac{\theta^6}{6!} + \cdots$$

Table B.1
Trigonometric Tables

Angle θ					Angle θ				
Degrees	Radians	sin θ	cos θ	tan θ	Degrees	Radians	sin θ	cos θ	tan θ
0	0.0000	0.0000	1.0000	0.0000					
1	0.0175	0.0175	0.9998	0.0175	46	0.8029	0.7193	0.6947	1.0355
2	0.0349	0.0349	0.9994	0.0349	47	0.8203	0.7314	0.6820	1.0724
3	0.0524	0.0523	0.9986	0.0524	48	0.8378	0.7431	0.6691	1.1106
4	0.0698	0.0698	0.9976	0.0699	49	0.8552	0.7547	0.6561	1.1504
5	0.0873	0.0872	0.9962	0.0875	50	0.8727	0.7660	0.6428	1.1918
6	0.1047	0.1045	0.9945	0.1051	51	0.8901	0.7771	0.6293	1.2349
7	0.1222	0.1219	0.9925	0.1228	52	0.9076	0.7880	0.6157	1.2799
8	0.1396	0.1392	0.9903	0.1405	53	0.9250	0.7986	0.6018	1.3270
9	0.1571	0.1564	0.9877	0.1584	54	0.9425	0.8090	0.5878	1.3764
10	0.1745	0.1736	0.9848	0.1763	55	0.9599	0.8192	0.5736	1.4281
11	0.1920	0.1908	0.9816	0.1944	56	0.9774	0.8290	0.5592	1.4826
12	0.2094	0.2079	0.9781	0.2126	57	0.9948	0.8387	0.5446	1.5399
13	0.2269	0.2250	0.9744	0.2309	58	1.0123	0.8480	0.5299	1.6003
14	0.2443	0.2419	0.9703	0.2493	59	1.0297	0.8572	0.5150	1.6643
15	0.2618	0.2588	0.9659	0.2679	60	1.0472	0.8660	0.5000	1.7321
16	0.2793	0.2756	0.9613	0.2867	61	1.0647	0.8746	0.4848	1.8040
17	0.2967	0.2924	0.9563	0.3057	62	1.0821	0.8829	0.4695	1.8807
18	0.3142	0.3090	0.9511	0.3249	63	1.0996	0.8910	0.4540	1.9626
19	0.3316	0.3256	0.9455	0.3443	64	1.1170	0.8988	0.4384	2.0503
20	0.3491	0.3420	0.9397	0.3640	65	1.1345	0.9063	0.4226	2.1445
21	0.3665	0.3584	0.9336	0.3839	66	1.1519	0.9135	0.4067	2.2460
22	0.3840	0.3746	0.9272	0.4040	67	1.1694	0.9205	0.3907	2.3559
23	0.4014	0.3907	0.9205	0.4245	68	1.1868	0.9272	0.3746	2.4751
24	0.4189	0.4067	0.9135	0.4452	69	1.2043	0.9336	0.3584	2.6051
25	0.4363	0.4226	0.9063	0.4663	70	1.2217	0.9397	0.3420	2.7475
26	0.4538	0.4384	0.8988	0.4877	71	1.2392	0.9455	0.3256	2.9042
27	0.4712	0.4540	0.8910	0.5095	72	1.2566	0.9511	0.3090	3.0777
28	0.4887	0.4695	0.8829	0.5317	73	1.2741	0.9563	0.2924	3.2709
29	0.5061	0.4848	0.8746	0.5543	74	1.2915	0.9613	0.2756	3.4874
30	0.5236	0.5000	0.8660	0.5774	75	1.3090	0.9659	0.2588	3.7321
31	0.5411	0.5150	0.8572	0.6009	76	1.3265	0.9703	0.2419	4.0108
32	0.5585	0.5299	0.8480	0.6249	77	1.3439	0.9744	0.2250	4.3315
33	0.5760	0.5446	0.8387	0.6494	78	1.3614	0.9781	0.2079	4.7046
34	0.5934	0.5592	0.8290	0.6745	79	1.3788	0.9816	0.1908	5.1446
35	0.6109	0.5736	0.8192	0.7002	80	1.3963	0.9848	0.1736	5.6713
36	0.6283	0.5878	0.8090	0.7265	81	1.4137	0.9877	0.1564	6.314
37	0.6458	0.6018	0.7986	0.7536	82	1.4312	0.9903	0.1392	7.115
38	0.6632	0.6157	0.7880	0.7813	83	1.4486	0.9925	0.1219	8.144
39	0.6807	0.6293	0.7771	0.8098	84	1.4661	0.9945	0.1045	9.514
40	0.6981	0.6428	0.7660	0.8391	85	1.4835	0.9962	0.0872	11.430
41	0.7156	0.6561	0.7547	0.8693	86	1.5010	0.9976	0.0698	14.301
42	0.7330	0.6691	0.7431	0.9004	87	1.5184	0.9986	0.0523	19.081
43	0.7505	0.6820	0.7314	0.9325	88	1.5359	0.9994	0.0349	28.636
44	0.7679	0.6947	0.7193	0.9657	89	1.5533	0.9998	0.0175	57.290
45	0.7854	0.7071	0.7071	1.0000	90	1.5708	1.0000	0.0000	∞

Appendix C
Commonly Used Mathematical Relations

Circle, circumference:	$2\pi r$
area:	πr^2
Sphere, surface area:	$4\pi r^2$
volume:	$\frac{4\pi}{3} r^3$

$\pi = 3.1416$

$e = 2.71828$

$e^{\circ} = 1$

$\ln e = 1; \qquad \text{if } y = e^x, \ x = \ln y$

$\ln 1 = 0$

$\ln xy = \ln x + \ln y$

$\ln (x/y) = \ln x - \ln y$

$\ln x^a = a \ln x$

$\ln x = \ln (10) \log x = 2.3026 \log x$

$e^x = 1 + x + \frac{x^2}{2!} + \frac{x^3}{3!} + \cdots$

$\ln (1 + x) = x - \frac{x^2}{2} + \frac{x^3}{3} - \frac{x^4}{4} + \cdots$

Quadratic Equation

Any quadratic equation can be put in the following form:

$$ax^2 + bx + c = 0$$

and the two roots are

$$x = \frac{-b \pm \sqrt{b^2 - 4ac}}{2a}$$

Binomial Expansion

If x is small as compared to 1,

$$(1 + x)^n = 1 + nx + \frac{n(n-1)}{2!} x^2 + \frac{n(n-1)(n-2)}{3!} x^3 + \cdots$$

$$(a + x)^n = a^n \left(1 + \frac{x}{a}\right)^n = a^n \left[1 + n\left(\frac{x}{a}\right) + \frac{n(n-1)}{2!}\left(\frac{x}{a}\right)^2 + \cdots\right]$$

Appendix D
Physical Constants

Gravitational constant	G	6.672×10^{-11} N-m^2/kg^2
Acceleration of gravity	g_E	9.80665 m/s^2
Earth, mass	M_E	5.98×10^{24} kg
radius	R_E	6.37×10^6 m
		3960 mi
distance to moon		3.844×10^8 m
		2.389×10^5 mi
distance to sun		1.496×10^{11} m
		9.30×10^7 mi
Moon	g_M	1.62 m/s^2
mass	M_M	7.35×10^{22} kg
radius	R_M	1.738×10^6 m
period		27.32 days
Sun mass		1.99×10^{30} kg
radius		6.96×10^8 m
Density of air (at STP)		1.293 kg/m^3
Molecular weight of air		28.97 kg/kmol
Avogadro's number	N_A	6.0220×10^{26} particles/kmol
Gas constant	R	8.314 J/mol-K
		8.206×10^{-2} l-atm/mol-K
Boltzmann's constant	$k = R/N_A$	1.3807×10^{-23} J/K
		8.617×10^{-5} eV/K
Standard temperature	T	273.15 K
Standard pressure	P	1.00 atm
Volume of ideal gas at STP	V	22.415 l/mol
Mechanical equivalent of heat	J	4.185×10^3 J/kcal
Heat of fusion of H_2O		79.7 cal/g
Heat of vaporization of H_2O		539.6 cal/g
Speed of sound in air at STP		331 m/s
Electron charge	e	1.60219×10^{-19} C
Coulomb constant	$k = 1/4\pi\epsilon_0$	8.98755×10^9 N-m^2/C^2
Permittivity of free space	ϵ_0	8.85419×10^{-12} C^2/N-m^2
Magnetic constant	$k = \mu_0/4\pi$	10^{-7} N/A^2(= Wb/A-m)
Permeability of free space	μ_0	$4\pi \times 10^{-7}$ N/A^2(= Wb/A-m)
Speed of light	c	2.997925×10^8 m/s
Planck's constant	h	6.6262×10^{-34} J-s
		4.1357×10^{-15} eV-s
	$\hbar = h/2\pi$	1.05459×10^{-34} J-s
		6.5822×10^{-16} eV-s

Unified mass	u	1.6606×10^{-27} kg
Mass of electron	m_e	5.486×10^{-4} u
		9.1095×10^{-31} kg
		511.0 keV/c^2
Mass of neutron	m_n	1.008665 u
		1.67482×10^{-27} kg
		939.550 MeV/c^2
Mass of proton	m_p	1.007277 u
		1.67265×10^{-27} kg
		938.28 MeV/c^2
Mass of hydrogen atom	m_{H}	1.007825 u
		1.67343×10^{-27} kg
		938.767 MeV/c^2
Energy equivalent of 1 u		931.5 MeV

Answers to Odd-Numbered Problems

Chapter 1

1. (a) 8854 m (b) 161 km (c) 5588 km (d) 169.4 m **3.** 744.2 m^2 **5.** 2.27 kg **7.** (a) 1000 kg/m^3 (b) 0.081 lb-mass/ft^3 **9.** 4.39×10^{17} s, 1×10^{-11} s **11.** (a) 1127–22540 m/s (b) 11.27–22.54 km/s (c) 40572–81144 km/h **13.** (a) L (b) M/L^3 (c) ML/T **15.** $R = 46$ km, makes an angle of 20° with the east. **17.** $V_x = 50$ units, $V_y = 86.6$ units **19.** $R = 64$ km, $\theta = 38.7°$ S of W **21.** $F_H = 48$ lb, $F_V = 36$ lb **23.** $F_R = 116.35$ lb, $\theta = 37.4°$ **25.** $F_R = 58.9$ lb, $\theta = 241.3°$ **27.** $\theta = 120.8°$

Chapter 2

1. 5.1×10^{34}; 1.1×10^8 kg **3.** 3.7×10^{19}, 6.2×10^{-8} kg, 62 kg/m^3 **5.** 1×10^{24} m^3, 6.20×10^7 m **7.** $R = (4.05, 5.40, 6.75, 8.10) \times 10^{-15}$ m $V = (7, 16, 32, 56) \times 10^{-44}$ m^3 **9.** 0.14204 u, 2.35847×10^{-28} kg, 2.35847×10^{-25} g **11.** $^{234}_{90}$Th, $^{236}_{92}$U, and $^{208}_{82}$Pb **13.** $^{32}_{16}$S, $^{137}_{56}$Ba, and $^{212}_{84}$Po **15.** $^{60}_{28}$Ni* and $^{60}_{28}$Ni **17.** Phosphorus-32 **19.** $^{13}_{6}$C, $^{23}_{13}$Al, $^{9}_{4}$Be, $^{15}_{7}$N **21.** $^{10}_{4}\text{Be} + ^{1}_{1}\text{H} \rightarrow ^{10}_{5}\text{B} + ^{1}_{0}n$

Chapter 3

1. (a) −0.6 m (b) −0.3 m/s (c) (See diagram in margin.) **3.** (a) 0.1 m/s (b) 0.05 m/s^2 **5.** $\bar{a} = 1.85$ m/s^2, $\bar{v} = 11.1$ m/s, $x = 1.332$ m **7.** −22 ft/s^2, 44 ft/s, 176 ft **9.** $v = 16$ m/s, $x = 6$ m, $t = 7.7$ s **11.** 1 m/s^2, 40 s **13.** (a) 24.5 s (b) 450 m (truck) 600 m (car) **15.** 3.19 s, −31.3 m/s **17.** 284 ft, 236 ft, 155 ft, 42.4 ft; −32.2 ft/s, −64.4 ft/s, −96.6 ft/s, −129 ft/s (a) −48.3 ft/s (b) −113 ft/s **19.** 2.55 s, −66.1 ft/s **21.** −11.2 m/s **23.** (a) 34.3 ft above ground (b) −48 ft/s, −16 ft/s (c) 1.5 s **25.** (a) 206 m/s, 13.8° (b) 2006 m (c) same as (a) **27.** (a) 8.84 m, 1.28 m (b) 10.2 m, 2.55 m (c) 8.84 m, 3.83 m **29.** (a) 2.47 s (b) 37.1 m (c) 28.5 m/s, −58.2° **31.** (a) 4.37 s (b) 75.6 m (c) 37.1 m **33.** (a) 0.928 s (b) 1.856 m (c) 5.78 m/s **35.** 32 m/s **37.** 335 km/h, 17.4° west of north **39.** 22.4 m/s, 26.6° with vertical

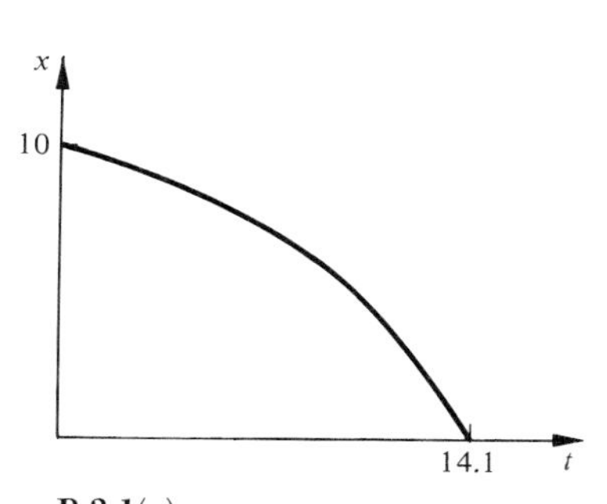

P.3.1(c)

Chapter 4

1. (a) 0.560 kg-m/s (b) 20 kg-m/s (c) 2.73×10^{-23} kg-m/s **3.** −4 kg-m/s, 4×10^3 N **5.** (a) 8.02×10^{-23} kg-m/s (b) -8.02×10^{-23} kg/s (c) -1.6×10^{-23} kg-m/s (d) 1.60×10^{-19} N **7.** 1500 lb **9.** −0.25 m/s^2, −12.5 N **11.** Yes, stops at 37.5 m **13.** -4.5×10^5 m/s^2, 900 N **15.** (a) 10 m/s^2 (b) 50 m/s (c) 125 m **17.** 0.165 N, 0.102 kg **19.** 267 N **21.** 1.64×10^{-22} m/s^2, which is negligible **23.** 213 lb **25.** 8200 N, 656 N **27.** 1.23 m/s^2 **29.** $f = 15680$ N in the direction opposite to motion. $F = 15680$ N in the direction of motion. **31.** (a) 53 N (b) applied and frictional force each 49 N (c) 0.816 m **33.** $a = -0.656$ m/s^2 downward, $N = 84.9$ N **35.** 0.839 **37.** $a = 3.27$ m/s^2, $T = 130.7$ N, $t = 1.11$ s

Chapter 5

1. 10 N-m, −8 N-m, 2 N-m **3.** (**a**) −2.0 N-m (**b**) 5.07 N-m (**c**) 8 N-m (**d**) 5.07 N-m (**e**) −2.0 N-m **5.** 5.00 N-m, −1.77 N-m, 3.23 N-m **7.** $T = 490\ \text{N}/\sin\theta$ **9.** $F = 226$ N, $T = 453$ N **11.** $R = 1960$ N, $T = 1697$ N **13.** 0.510, 0.306 **15.** 0.113 **17.** (**a**) 23.6° (**b**) 44.9 N (**c**) 19.6 N **19.** (**a**) 0, $+(a/2)N$, $-bF$ (**b**) 0, $-(a/2)N$, $-bF$ (**c**) $-bf$, $-(a/2)N$, 0 (**d**) $-bf$, $(a/2)N$, 0 **21.** 2450 N, 1666 N **23.** −2 N-m **25.** 78.1 N **27.** Yes, he can climb all the way. **29.** 16.9 ft **31.** (0.909 m, 0.727 m) **33.** $(R + r/3, 0)$ **35.** 1.23×10^{-10} m from hydrogen **37.** 11427 m **39.** (**a**) $(-1\ \text{m}, \frac{1}{6}\ \text{m})$ (**b**) 0.73 m/s^2

Chapter 6

1. 1000 J **3.** (**a**) 100 J (**b**) −12 J (**c**) 88 J **5.** 490 J, −490 J **7.** 9800 J **9.** 327 W, 980 N **11.** 8.95 hp, 6680 W, 6.68 kW **13.** 7.06 s **15.** 57.4 kW or 76.9 hp **17.** 29.4 kW, 19600 J **19.** (**a**) 4.55×10^{-17} J (**b**) 160 J (**c**) 300,000 J **21.** (**i**) 292 m (**ii**) 656 m **23.** 250 m **25.** Proof **27.** 14 m/s, −4900 N **29.** (**a**) 26.2 m/s (**b**) 25.8 m/s **31.** $v = \sqrt{2gR}$, 6.26 m/s **33.** (**a**) 1000 J (**b**) 188 J (**c**) 812 J (**d**) 472 J **35.** (**a**) 0.5 N, 5 N, 25 N (**b**) 0.0025 J, 0.25 J, 6.25 J **37.** 1.12×10^6 N/m **39.** 4.47 cm **41.** 87.6 m/s

Chapter 7

1. (**a**) 3.33×10^4 kg-m/s, 3.7×10^{-3} J (**b**) 1.80 kg-m/s, 27 J (**c**) 3.00 kg-m/s, 460 J (**d**) 9.1×10^{-25} kg-m/s, 4.55×10^{-19} J **3.** Proof **5.** (**a**) -1.76×10^6 m/s, 2.35×10^5 m/s (**b**) 0, 2×10^6 m/s (**c**) $\approx 2 \times 10^6$ m/s, $\approx 4 \times 10^6$ m/s **7.** 0.06 kg, 0.25 m/s **9.** (**a**) 0.92 MeV (**b**) 0.19 MeV **11.** $u_1 = 9.20 \times 10^3$ m/s, $u_2 = 1.13 \times 10^3$ m/s, $\phi = 42.6°$, $K_1 = 7.07 \times 10^{-20}$ J $= 0.442$ eV, $K_2 = 1.28 \times 10^{-20}$ J $= 0.08$ eV **13.** 44.7°, 1.96×10^{-17} J (122 eV) **15.** (**a**) 6.7 cm/s, 3.3 cm/s (**b**) 1.67 cm/s **17.** 80 s **19.** 1109 m/s **21.** −0.0067 N **23.** (**a**) 281 m/s (**b**) 393 J **25.** (**a**) 3.16 m/s (**b**) 158 m/s (**c**) 245 J **27.** (**a**) 2.43×10^5 m/s, 0.07 MeV (**b**) 4.07 MeV

Chapter 8

1. $S'(0, 0, 0, t, 0)$, $S(13.4t, d, 0, t', 13.4)$; $S'(1.34t', 0, 0, t, 1.34)$; $S(14.74t, d, 0, t', 14.34)$ **3.** Proof **5.** 0.714 h, 0.654 h; not possible to make round trip **7.** Proof **9.** 32.0 s, 31.2 m **11.** c **13.** $0.652c$, $-0.652c$ **15.** (**a**) 1.25×10^8 m/s (**b**) 2.98×10^8 m/s (**c**) 2.9998×10^8 m/s **17.** 2.99×10^8 m/s, $11.7\ m_0$ ($= 1.95 \times 10^{-26}$ kg), 5.83×10^{-18} kg-m/s **19.** 3.78 GeV, 4.71 GeV, 8.39×10^{-27} kg, 2.47×10^{-18} kg-m/s **21.** (**a**) 1.78×10^{-27} kg, 1 GeV/c (**b**) 1.78×10^{-27} kg, 347 MeV/c

Chapter 9

1. 15 rad/s **3.** (**a**) 4.71 rad/s, 270 deg/s (**b**) 8.17 rad/s, 468 deg/s **5.** $\omega = 4\pi$ rad/s, same for all points; $v = 6.28$ m/s, 3.14 m/s; $\theta = 240\pi$ rad $= 4.32 \times 10^4$ deg **7.** 4.5 rad/s^2, 144 rad, 2.25 m/s **9.** 1.125 rad/s^2 **11.** (**a**) 2.40 m/s (**b**) 1.20 m/s^2 (**c**) 9.60 m/s^2 (**d**) 9.67 m/s^2 **13.** 80 ft/s **15.** 0.23 **17.** 23.2° **19.** (**a**) 31.4 m/s (**b**) 986 m/s^2 (**c**) 493 N (**d**) 493 N **21.** 3.13 m/s, 7.00 m/s **23.** $I_0 = 8.1$ kg-m^2, $I_A = 12$ kg-m^2 **25.** $ML^2/3$, $L/2\sqrt{3}$, $L/\sqrt{3}$ **27.** 1000 N-m **29.** 3.47 rad/s^2, 17.6 rad/s **31.** 750 N-m, 1.125×10^6 J **33.** (**a**) 1.18×10^4 J (**b**) 37.7 s **35.** (**a**) T^{-1} (**b**) T^{-2} (**c**) ML^2T^{-2} (**d**) ML^2 (**e**) ML^2T^{-1} (**f**) ML^2T^{-2} **37.** (**a**) 1 kg-m/s^2 (**b**) 2 J **39.** 0.291 rev/s

Chapter 10

1. 0.5 N/cm, 12.5 N **3.** (a) 3 rad/s (b) 0.477 Hz (c) 2.10 s (d) 0.900 N/m (e) 0.25 m (f) $\pi/4$ rad **5.** (a) 400 N/m, 10 rad/s, 1.59 Hz, 0.628 s, 0.30 m (b) 120 N (c) 3 m/s, 30 m/s^2 (d) 2.83 m/s, -10 m/s^2 **7.** At $x = 5$ cm, $v = 1.16$ m/s, $a = -1.8$ m/s^2, $K = 1.35$ J, $U = 0.09$ J. At $x = 15$ cm, $v = 0.794$ m/s, $a = -5.4$ m/s^2, $K = 0.630$ J, $U = 0.81$ J $E = 1.44\text{ J} = K + U$. **9.** 1.41 Hz **11.** 1.02 **13.** (a) 0.175 s (b) 0.087 s **15.** (a) 0.262 s (b) 0.523 s **17.** 426 N/m **19.** (a) 0.518% (b) 2.076% (c) 4.720% **21.** $\sqrt{0.25\text{ g/l}}$ **23.** $\Delta T/T = -\Delta g/2g$ **25.** 0.581 kg-m^2 **27.** $I_2 = (T_2/T_1)^2 I_1$ **29.** $k = 1.22 \times 10^5$ N/m **31.** (a) 12×10^8 N/m (b) 5×10^{-2} (c) 3×10^8 N **33.** 2×10^9 N/m^2

Chapter 11

1. (a) 6.67×10^{-7} N (b) 3.55×10^{22} N (c) 2.03×10^{20} N **3.** (a) 1.86×10^{-38} N (b) 1.86×10^{-34} N (c) 1.86×10^{-30} N **5.** 3.34×10^{-9} N, 1.20° **7.** (a) 162 N (b) 0.028 N (c) 405 N **9.** $g = (7.56 \times 10^{-10}/a^2)$ m/s^2 along the negative X-axis **11.** $g_{(\text{Mercury})} = 3.42$ m/s^2 $g_{(\text{Jupiter})} = 24.7$ m/s^2 **13.** (See diagram in margin.) **15.** $T_M = 2.46T_E$ **17.** 2.948×10^4 m/s; 2.74×10^{33} J, -5.38×10^{33} J; 2.60×10^{33} J, -5.24×10^{33} J **19.** 2.00×10^{30} kg **21.** 19.5 km/s (from earth, 11.2 km/s) **23.** 3.08 km/s **25.** 3.98 h **27.** Earth (a) 11.1 km/s (b) 10.8 km/s (c) 10.4 km/s (d) 8.37 km/s Moon (a) 2.31 km/s (b) 2.09 km/s (c) 1.89 km/s (d) 1.21 km/s Mars (a) 5.20 km/s (b) 4.93 km/s (c) 4.64 km/s (d) 3.37 km/s **29.** $v_{e_1}/v_{e_2} = R_1/R_2$ (b) $v_{e_1}/v_{e_2} = \sqrt{M_1/M_2}$ **31.** 3.6×10^3 N **33.** 0.027

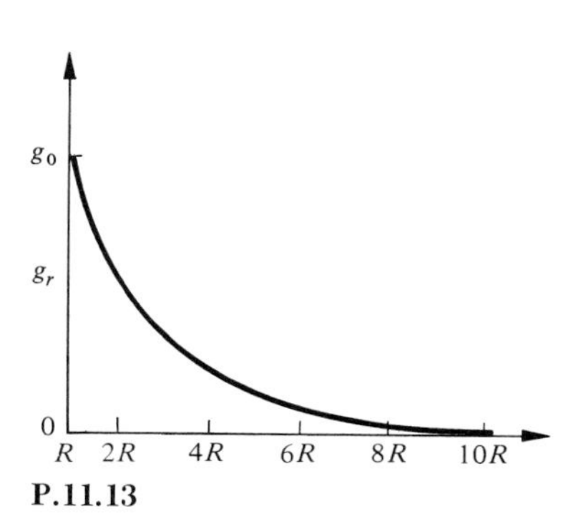

P.11.13

Chapter 12

1. (a) 1.07×10^5 N/m^2 (b) 1.19×10^5 N/m^2 **3.** 2.16 atm **5.** 20.6 m **7.** 2.08×10^4 N/m^2 **9.** 3.33×10^4 N/m^2, 1.34×10^5 N/m^2 **11.** 2.39×10^5 N/m^2 **13.** 2041 kg **15.** 1.0×10^{10} J **17.** $F_b = 392$ N, $V = 0.04$ m^3 **19.** 1.53×10^{-5} m^3, 1.03 N **21.** 229 kg/m^3 **23.** 0.08 or 8%, 0.125 m^3 **25.** (a) 4.56×10^4 N (b) 4.93×10^4 N **27.** 0.98 cm/s **29.** 8.85 m/s, 0.885 kg/s **31.** (a) 19.6 m/s (b) 0.082 slug/s **33.** 10 m/s, 1.45×10^4 N/m^2 **35.** 1.8×10^6 J **37.** 8000 N/m^2, 1.92×10^5 N **39.** 0.002 N, 2×10^{-5} J **41.** 16 **43.** (a) 14.8 cm (b) 3.70 cm (c) 1.48 cm (d) 0.37 cm; (a) 4.56×10^{-5} N (b) 0.182×10^{-5} N (c) 0.456×10^{-5} N (d) 0.0182×10^{-5} N **45.** Water level drops 2.15 cm **47.** 19.1°

Chapter 13

1. 107°C **3.** 48.9°C, -45.6°C **5.** -252.3°C, -422°F **7.** 0.866 atm **9.** 477 K = 204°C **11.** 338 K = 65°C **13.** 5 atm = 5.05×10^5 N/m^2 = 73.5 lb/in^2 **15.** 1.18 atm, 2730 K (= 2457°C) **17.** (a) 5 atm (b) 5.63 atm **19.** 16.1 ft^3 **21.** 67.5 m **23.** (a) 2.31×10^{-14} kg-m/s (b) 2.31×10^{-14} N **25.** (a) 7500 collisions/s (b) $\Delta p_x = 1.60 \times 10^{-22}$ kg-m/s, $\Delta p_y = 1.06 \times 10^{-22}$ kg-m/s, $\Delta p_z = 1.28 \times 10^{-22}$ kg-m/s (c) $F_x = 1.20 \times 10^{-18}$ N, $F_y = 5.32 \times 10^{-19}$ N, $F_z = 7.60 \times 10^{-19}$ N (d) $P_x = 1.20 \times 10^{-16}$ N/m^2, $P_y = 5.32 \times 10^{-17}$ N/m^2, $P_z = 7.66 \times 10^{-17}$ N/m^2 **27.** 513 m/s **29.** $\langle K \rangle = 6.13 \times 10^{-21}$ J = 0.0382 eV, $U = mgh = 5.21 \times 10^{-25}$ J, $U/\langle K \rangle = 8.50 \times 10^{-5} << 1$ **31.** $\frac{1}{4}$ **33.** 1.22×10^4 m/s, $v_{\text{rms}}/v_e = 6.29 \times 10^{-4}$, $\langle K \rangle = 1.24 \times 10^{-19}$ J = 0.775 eV **35.** 0.7056% of $^{235}UF_6$

Chapter 14

1. 1 kcal/kg-C°, 18 cal/mole-C°, 1 kcal/kmole-C°, 4186 J/kg-C°, 75348 J/kmole-C° **3.** 18.6×10^3 cal = 7.79×10^4 J **5.** 124.3 kcal = 5.203×10^5 J = 493 BTU **7.** 15.2 minute **9.** 1461 BTU/ft³, 2.15 m³ (=76.1 ft³) **11.** 0.966 kcal/kg-C° **13.** 0.202 kcal/kg-C° **15.** 0.187 C° **17.** 319°C **19.** (a) 1.67 J/s = 0.40 cal/s (b) 6.2×10^{-4} C°/s **21.** 3.41×10^5 J **23.** (a) 202 J (b) 509 J **25.** 496 cal **27.** $C_v = 3.71R$, $C_p = 4.78R$, polyatomic gas **29.** 4.77 kcal **31.** 45.2 g **33.** 29.3 g **35.** 14.5 kg **37.** (a) 169 J (=40.3 cal) (b) 2090 J (=499.2 cal)

Chapter 15

1. 6 mm, 9.6 mm **3.** 80.22 ft **5.** slow by 8.64 s/d **7.** 1.17×10^{-4} m² **9.** 86.7°C **11.** 100.024 cm³, 100.060 cm³, 100.120 cm³ **13.** 15×10^{-15}/C°, 2.2×10^{-5}/C° **15.** 8.14×10^7 N/m² **17.** 3.04×10^8 N/m² [=(1.92 + 1.12) × 10^8 N/m²] **19.** 0.1, 20.16 **21.** 0.0361 g/s **23.** 1431 cm² **25.** 3.9 cal/s **27.** 92.1 cal/s, it will freeze **29.** (a) 4.59×10^{-6} W/m² (b) 4.59×10^{-2} W/m² (c) 4.59×10^2 W/m² (d) 4.59×10^6 W/m² **31.** 54.7 W **33.** 1.63×10^{-4} m² **35.** (a) 228 W (b) 94.6 W (assuming $\epsilon = 1$)

Chapter 16

1. 43.7×10^3 J **3.** 285 J **5.** 0.314 **7.** 10.4 kJ, 2.48 kcal **9.** 7.253 kcal/K **11.** (a) 3.79 cal/K (b) 6.19 cal/K (c) 2.26 cal/K **13.** 0.024 kcal/kg-K **15.** 0.311 kcal/K **17.** (a) −1.34 kcal/K (b) 1.47 cal/K (c) 0.13 cal/K The process is not reversible. **19.** 20%, 400 K **21.** (a) 837 J (b) 36% (c) 10% **23.** Gasoline engine (η = 59%) **25.** (a) 64.4% (b) 33.84 kW (c) 59.2% **27.** 781 rev/min **29.** (a) 879 kcal (b) 79 kcal = 3.31×10^5 J **31.** 1.05 kcal, 5 kcal **33.** A better than B ($E_A = 7.11$, $E_B = 6.84$)

Chapter 17

1. 16 m, 0.017 s **3.** 935 m, 340 m from one mountain **5.** 1094 m **7.** 34000, 1.5 m **9.** (a) 0.03 m (b) 5 s (c) 0.2 **11.** (a) 0.01 m (b) 100 m (c) 1/100 s (d) 100 Hz (e) 10,000 m/s **13.** (a) 0.01 m (b) 20 m (c) 1/100 s (d) 100 Hz (e) 2000 m/s **15.** 44.3 m/s, 0.369 m **17.** 3.25 km/s **19.** 0.707 **21.** (a) 101 m/s (b) 143 m/s **23.** 1.43×10^5 N/m² **25.** (a) 32°C, 0.594 m/s-C° (b) 126°C, 0.532 m/s-C° (c) 822°C, 0.378 m/s-C° **27.** Plot is a straight line with a slope ~0.62 m/s-C° **29.** 0.0141 m **31.** 5×10^{-4} m

Chapter 18

1. 324 m/s; 108 Hz, 216 Hz, 324 Hz **3.** 2 m, 1 m, 0.67 m; 400 Hz, 800 Hz, 1200 Hz **5.** 41.8 N **7.** 605 N, 220 Hz, 330 Hz **9.** (a) 213 Hz, 1.60 m (b) 426 Hz, 0.53 m (c) 639 Hz, 0.53 m (d) same as (c) (e) 852 Hz, 0.40 m **11.** 144.5 N **13.** (a) $f_n = nf_1$, $f_1 = 286$ Hz $n = 1, 2, 3, 4, 5, \ldots$ (b) $f_{2n+1} = (2n + 1)f_1$, $f_1 = 143$ Hz, $n = 1, 2, 3, \ldots$ **15.** 708 Hz **17.** 2500 Hz, 7500 Hz, 12500 Hz; 1:3:5 **19.** 50 dB **21.** (a) 10^{-10} W/m² (b) 10^{-10} **23.** 27 dB **25.** 5000 W **27.** 12 beats/s **29.** 244 Hz or 236 Hz, (b) 236 Hz **31.** 515 Hz **33.** 5.18 Hz **35.** 7 Hz **37.** (a) 859 Hz (b) 741 Hz **39.** (a) 17.9 m/s (b) 17.0 m/s **41.** 68 m/s **43.** (a) 0.315 m, 0.365 m (b) 1079 Hz, 932 Hz (c) 1141 Hz, 878 Hz **45.** 30° **47.** 45.6°, 40.0 s

A N A

Fundamental

A N A N A N A

1st Overtone

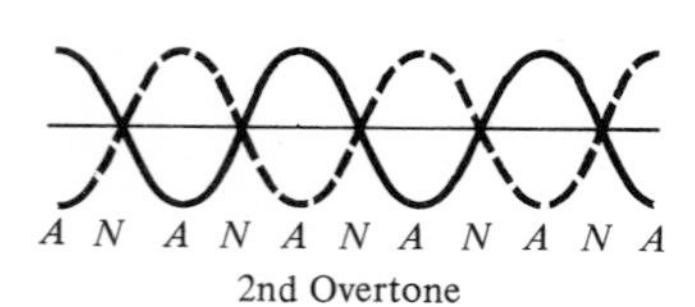

2nd Overtone

P.18.17

Chapter 19
1. (a) 6.25×10^{18} (b) 6.25×10^{12} **3.** 5.08 m **5.** -1.84×10^{-5} N **7.** 8.61×10^{-12} C **9.** 0.67 m **11.** (a) 0.13 N toward q_1 (attractive) (b) 0 **13.** 0.215 N, 45° **15.** 3.2×10^{-16} N, 3.51×10^{14} m/s^2; 3.2×10^{-16} N, 1.92×10^{11} m/s^2 **17.** (a) and (b) 1.25×10^3 N/C upward **19.** Proof **21.** (a) 1.6×10^{-15} N (b) 0.18×10^{16} m/s^2 (c) 0.85×10^7 m/s (d) 4.72×10^{-9} s (e) 3.2×10^{-17} J **23.** 5 m from the negative charge; 10 m from the positive **25.** 2.25×10^5 N/C toward + 100 μC charge **27.** (a) 0.358 N (b) 0 (c) 3.6×10^{-3} N **29.** 9.8×10^{-8} C, for positive charge field is upward; for negative charge field is downward

Chapter 20
1. 500 V **3.** 1.10×10^{-3} J, 6.88×10^{15} eV **5.** (a) 140 V (b) 60 V **7.** 50,000 V, 1 m, from A to B, A is at higher potential **9.** 4.5×10^4 V, 0.09 J **11.** 1.08×10^6 V **13.** 3.98×10^5 V, -0.398 J **15.** $8kq/a\sqrt{2}$, 0 **17.** (a) 10 MeV $= 1.6 \times 10^{-12}$ J, 4.38×10^7 m/s (b) 20 MeV $= 3.2 \times 10^{-12}$ J, 3.10×10^7 m/s **19.** (a) 20,000 eV (b) 3.2×10^{-15} J (c) 8.38×10^7 m/s (d) 8.14×10^7 m/s **21.** (a) 5.11×10^3 V, 4.24×10^7 m/s (b) -9.38×10^6 V, 4.24×10^7 m/s **23.** (a) 2.65×10^7 m/s, 6.19×10^5 m/s (b) 3.2×10^{-16} J $= 2$ keV for both (c) 4.997 cm from negative plate, essentially at the positive plate **25.** 0.01 C **27.** A square plate of 33.6 km on a side will be needed, which is impractical. **29.** (a) 28.3 pF (b) 141.5 pF **31.** (a) 6 μF, 90 μF **33.** 8.57 μF **35.** (a) 13.75 μF (b) $V_{10} = 24$ V, $V_5 = 18$ V, $V_{15} = 6$ V **37.** 0.250 μF, 0.125 J **39.** 10 μC, 5×10^{-5} J **41.** 8 J, 1.43 cal **43.** (a) 8.85×10^{-10} F (b) 4.43×10^{-7} C (c) 1.38 J/m^3 (d) 1.11×10^{-4} J (a) 1.77×10^{-10} F (b) 8.85×10^{-8} C (c) 0.276 J/m^3 (d) 2.22×10^{-5} J **45.** 0.4 F

Chapter 21
1. 20 V **5.** 18 min, 4.32×10^5 C **7.** 1.87 A, in the direction of the motion of positive charges **9.** 27.8 mA, 3.54×10^4 A/m^2, 7.38×10^{-6} m/s **11.** 8.38×10^{-2} Ω **13.** 200 Ω (same) **15.** 17.9 μΩ **17.** (a) 0.0228 Ω (b) 0.0713 Ω (c) 0.140 Ω (d) 0.624 Ω **19.** 1.56 A **21.** 29.5 Ω **23.** 42.9 Ω, 16.0 Ω, 7.56 Ω; (5.23×10^5, 1.34×10^5, 3.48×10^4)Ω/m **25.** 24 Ω **27.** 12.0 mV **29.** 0.0001/C, 10.07 Ω **31.** $R_{Cu}/R_W = 0.324$, 0.308 **33.** (a) 99.96 Ω (b) 99.76 Ω (c) 100.16 Ω (d) 101.96 Ω **35.** (a) 20 W (b) 4 V **37.** $6.75 **39.** 62.2 A, 3730 J **41.** 0.848 **43.** (a) 20 Ω (b) 6 A; between 605 W and 895 W **45.** 9.17 A, 4.76 times as long **47.** (a) 0.833 A (b) 144 Ω (c) 5.21×10^{18} electrons/s

Chapter 22
1. 120 mA **3.** 9.75 V **5.** (a) Positive terminals are connected with each other and the negative terminals likewise. (b) 10.76 Ω (c) 120 W (d) 5.6¢ **7.** (a) 36 Ω (b) 4 Ω **9.** 5 **11.** (a) 10 Ω (b) 7.5 Ω (c) 16.67 Ω (d) 25 Ω **13.** 10 Ω **15.** 9.84 Ω **17.** (a) 0.741 A, 7.41 V (b) 0.05 A, 2 V **19.** $I_{20} = 3$ A, $I_{60} = 1$ A **21.** (a) 5.94 V (b) 3.00 V (c) 8.97 V (d) 3.03 V **23.** (a) 432 Ω (b) 96 Ω **25.** 400 W **27.** 0.207 A, 0.269 A, 0.476 A **29.** 0.16 A, 1.42 A, 1.26 A **31.** $I_1 = -0.38$ A, $I_2 = 0.96$ A, and $I_3 = 0.58$ A **33.** 0.0001 Ω, 0.001 Ω, 0.01 Ω **35.** (a) 0.0200 Ω (b) 99950 Ω **37.** 2.2 Ω **39.** 1.79×10^{-8} m^2 **41.** 4.5 V **43.** 0.25, 7.35 μA, 1.47 μC **45.** (a) 0.0347 s (b) 0.0347 s

Chapter 23

1. 7.85×10^{-4} Wb, 0 **3.** 0.10 Wb (a) 0.0866 Wb (b) 0.0707 Wb (c) 0.05 Wb (d) 0 (e) −0.10 Wb **5.** Proof **7.** 2.4×10^{-15} N, 2.28×10^{15} m/s^2 **9.** 0, 6.4×10^{-12} N pointing out of paper, 6.4×10^{-12} N along the −Y-axis. **11.** (a) 1.28×10^{-13} N (east), 3.20×10^{-13} N (west), 1.92×10^{-13} N (west) (b) 1.28×10^{-13} N (east), 3.20×10^{-13} N (east), 4.48×10^{-13} N (east) **13.** 4×10^6 m/s, $E \perp B$, $v \perp$ both E and B. If E or B is zero, trajectory will deviate from straight line. **15.** 0.018 Wb/m^2 **17.** (a) 3.05×10^7 Hz (b) 6.13×10^7 m/s (c) 19.6 MeV **19.** 0.655 T, 3.29 MeV, 2.51×10^7 m/s **21.** $B = 5.69 \times 10^{-5}$ Wb/m^2, $\perp$ the plane of motion, 1.59 MHz, 6.29×10^{-7} s **23.** 1.2 N **25.** 0.0218 T toward the north **27.** 2×10^{-4} N-m, 2×10^{-3} A-m^2 **29.** circular loop **31.** 12°

Chapter 24

1. 4×10^{-6} Wb/m^2 **3.** (a) 0.04 mm (b) 0.4 mm (c) 4 mm **5.** At point P, $B = 3.33 \times 10^{-5}$ Wb/m^2 out of the page; at point Q, $B = 6.67 \times 10^{-5}$ Wb/m^2 out of the page. **7.** 1.5×10^{-5} Wb/m^2 into the page, 2.5×10^{-5} Wb/m^2 into the page. **9.** 2×10^{-8} Wb/m^2 **11.** 6.28×10^{-3} Wb/m^2 **13.** 7.96 A **15.** (a) 5×10^{-5} N/m, attractive (b) 5×10^{-5} N/m, repulsive **17.** (a) 2×10^{-5} Wb/m^2, 4×10^{-4} N/m (b) 4×10^{-5} Wb/m^2, 4×10^{-4} N/m The forces are attractive. **19.** (a) 1×10^{-5} Wb/m^2 (b) 3×10^{-5} Wb/m^2 **21.** 15.9 A **23.** (a) 0.012 Wb/m^2 (b) 0.00754 Wb/m^2 **25.** $I = 7.96 \times 10^{-3}$ A, field produced southward **27.** 1250 A **29.** (a) 0.000253 Wb (b) 0.0202 Wb **31.** (a) 0.125 A (b) 12.5 A **33.** 80,79

Chapter 25

1. 50.2 V **3.** 3.14 V **5.** 3.18×10^{-4} Wb/m^2 **7.** 10 V **9.** 6.28 V **11.** 5 V **13.** 15 mH **15.** 12 A/s **17.** 0.68 Wb **19.** 1.6 V (a) 1.6 Wb (b) 3.2 Wb **21.** 3 mV **23.** 8 H **25.** Proof **27.** $i = 1.2\text{ A}(1 - e^{-t/0.2\text{s}})$; 1.2 A **29.** $i = (0.12\text{ A})e^{-t/0.1\text{s}}$ **31.** Proof **33.** 1.41×10^{-15} F

Chapter 26

1. $v(t) = (148, 91.7, 0, -91.7, -148, 0)$V **3.** 155 V, 12.9 A, 9.17 A, 1.009 kW **5.** 3.87 A, 5.47 A, 58.1 V, 82.1 V; 3.87 A **7.** (a) 628 Ω (b) 2512 Ω (c) 10048 Ω **9.** (a) 3.18 kΩ (b) 796 Ω (c) 199 Ω **11.** 377 Ω, 2653 Ω **13.** 58.8 mA **15.** $X_C = 2653\ \Omega$, $Z = 2653\ \Omega$, $i_{\text{rms}} = 41.5$ mA, $V_R = 2.08$ V, $V_C = 110$ V, $\phi = -88.9°$ **17.** $X_L = 314\ \Omega$, $R_t = 52\ \Omega$, $Z = 318\ \Omega$; $i_{\text{rms}} = 0.692$ A, $V_L = 217$ V, $V_R = 34.6$ V **19.** (a) 2513 Ω, 15.9 Ω, $R = 100\ \Omega$, 2499 Ω (b) 88.0 mA (c) 221 V, 1.40 V, 8.80 V (d) 87.7° **21.** 0.130 H, 68.7°, 80 J, 0.36 **23.** 50.3 Hz **25.** 2.53×10^{16} F **27.** 60 V, 1/12 A **29.** 45 kW **31.** Step-down, $i_s = 2.4$ A, $i_p = 0.48$ A **33.** Step-down, 20 W **35.** 2

Chapter 27

1. Proof. **3.** Second equation will take the form $\Phi_B = kq_{\text{m}}$, where k is a constant and q_m is the strength of the magnetic monopole. **5.** 1.67×10^{-8} Wb/m^2 **7.** $E_z = (0.1\text{ V/m})\sin[(4\pi \times 10^7\text{ Hz})t - (0.419\ m^{-1})x]$ $B_y = -E_z/c$, $H_y = B_y/\mu_0$ **9.** 18.3 V/m **11.** 3.33×10^3 neutrons/cm^3 **13.** 3.98×10^3 J/m^3 **15.** 8.68×10^{-2} V/m, 2.89×10^{-10} Wb/m^2, 2.30×10^{-4} A/m, $(E, B, H) = (E_0, B_0, H_0)\sin[(\pi \times 10^7\text{ Hz})t - 10.105\text{ m}^{-1})x]$ **17.** 1.03×10^3 V/m, 3.43×10^{-6} Wb/m^2 **19.** $E_0 = 1.23 \times 10^3$ V/m **21.** 1.23 μV **23.** 388.18 rev/s

Chapter 28
1. 1 m **3.** 3 **5.** 1.97×10^8 m/s, 3875 Å **7.** 2000–3500 Å, 1.50×10^8 m/s **9.** 1.29×10^8 m/s **11.** 40.6° **13.** Proof **15.** 13.4°, 41.7°, 58.7° **17.** 1.33 m **19.** 1.30 **21.** 38°, 40.8° **23.** Proof **25.** 7.58° **27.** 0.501 m, 0.100 m **29.** $p = 30$ cm **31.** $R = 135$ cm, the image is virtual and erect **33.** (a) 45 cm (b) 15 cm **35.** $f = 40$ cm, $q = -40$ cm, concave mirror **37.** 0.267 m, 0.334 cm **39.** -120 cm, -0.833 diopters **41.** 0.333 m, 1.33 cm **43.** 20 cm **45.** (a) 120 cm (b) -120 cm (c) -24 cm (d) 24 cm **47.** Image is 2.83 cm behind the second lens, its size is 2.29 cm **49.** 120 cm from the mirror or 20 cm in front of the lens, the size is 24 cm.

Chapter 29
1. 0.0344°, 0.90 mm **3.** 6.25×10^{-7} m **5.** 0.455 mm **7.** (a) 0.88 mm (b) 1.76 mm **9.** 732 **11.** 0.606 mm **13.** 11.93 r, 6000 r **15.** 16320Å$/m$ where $m = 1, 2, 3, \ldots$ **17.** 7.75×10^{-8} m, blue and ultraviolet, and the film will appear yellowish-green **19.** 1.00×10^{-7} m **21.** 8.93×10^{-8} m **23.** 9.96×10^{-8} m **25.** 1.44 μm **27.** 20.3°, 44.0° **29.** 3 orders, 19.2°, 41.2°, 81.1° **31.** 2664 lines per cm **33.** Proof **35.** 687.3 Å, 0.44 cm **37.** $I_0/2$, $3I_0/8$ **39.** 67.5°, 58.9°, 53.7° **41.** 1.60 **43.** 2.24 g/cm^3

Chapter 30
1. (d), (e), and (f) **3.** 12 cm **5.** 1/50 s **7.** 0.22 cm **9.** -1.67 diopters **11.** Left $+3.60$ diopters; Right -2.50 diopters **13.** 4.17 diopters **15.** 1.33 m **17.** 1.79 cm and 1.54 cm; 16.4 cm **19.** (a) 5.50° (b) 1.38° (c) 0.138° **21.** (a) 2.86° (b) 3.58° **23.** 20 diopters **25.** (a) 0.016 radians = 9.2° (b) 26 (c) 25 **27.** 263, the object is at the focus of the objective lens. **29.** 46, 96, 396; 92, 192, 792 **31.** (a) 0.692 cm (b) 6.50 (c) 16.7 (d) 109 **33.** 744, 1488, 2976 **35.** 98 cm, 2 cm **37.** (a) 5.01 cm (b) 2.5 (c) 10 **39.** 34 cm, 8.5 **41.** $(\Delta 1)_{\text{min}} = 5.37 \times 10^{-7}$ m ($<5 \times 10^{-6}$ m), so it is possible **43.** (a) 0.0991 arc sec (b) 0.124 arc sec (c) 0.149 arc sec (d) 0.173 arc sec

Chapter 31
1. 3.10 eV, 2.25 eV, 1.77 eV **3.** 3.60×10^{-17} W **5.** (a) 1.78×10^{-36} kg, 5.33×10^{-28} kg-m/s, 12,400 Å, 2.41×10^{14} Hz (b) 1.78×10^{-33} kg, 5.33×10^{-25} kg-m/s, 12.4 Å, 2.41×10^{14} Hz (c) 1.78×10^{-30} kg, 5.33×10^{-22} kg-m/s, 0.0124 Å, 2.41×10^{20} Hz **7.** 2.29 eV **9.** 1.98 eV, 1.98 eV, 6263 Å **11.** 2000 Å **13.** 4.13 Å **15.** (a) 1.02426 Å (b) 12.4 keV, 12.2 keV (c) 200 eV **17.** (a) 12.3 Å (b) 0.544 Å (c) 1.80 Å (d) 1.24×10^{-38} m **19.** (a) 3.29×10^{-9} eV (b) 6.03×10^{-6} eV **21.** (a) 0.908 Å (b) 0.113 Å **23.** 1.06×10^{-26} **25.** 1.06×10^{-26} J

Chapter 32
1. 15.9937 u **3.** (a) 11.40 cm, 8.06 cm (b) 14.86 cm **5.** 9.80×10^{-8} C, downward **7.** $E_g/F_e = 4.40 \times 10^{-40}$ **9.** $(1.89 - 3.40)$ eV **11.** (a) 10.2 eV, 1216 Å (b) 0.166 eV, 74699 Å (c) 0.024 eV, 5.167×10^5 Å (d) 0.00021 eV, 5.90 mm (e) 0.00003 eV, 4.13 cm; Discrete nature is lost around $n = 20$. **13.** 6.56×10^{-8} eV, 5.21×10^7 rev **15.** $\sqrt{12}\,\hbar$, $\sqrt{20}\,\hbar$ **17.** 5/2, 7/2; $(\sqrt{35}/2)\hbar$, $(\sqrt{63}/2)\hbar$ **19.** (4, 0, 0) . . . (4, 3, -3) total of 16 states; $L = 0$, $\sqrt{2}\,\hbar$, $\sqrt{6}\,\hbar$, $\sqrt{12}\,\hbar$ **21.** $1s^22s^22p^63s^23p^1$, $1s^22s^22p^63s^23p^4$, $1s^22s^22p^63s^23p^64s^23d^7$, $1s^22s^22p^63s^23p^64s^23d^{10}4p^65s^14d^{10}$ **23.** $l = 2$, $s = \frac{1}{2}$,

$j = \frac{5}{2}$ or $\frac{3}{2}$; $L = (\sqrt{3}/2)\hbar$, $S = (\sqrt{3}/2)\hbar$, $J = (\sqrt{35}/2)\hbar$, $(\sqrt{15}/2)\hbar$
25. (**a**) $4P \rightarrow 3D \rightarrow 3P \rightarrow 3S$ or $4P \rightarrow 4S \rightarrow 3P \rightarrow 3S$ (**b**) $4D \rightarrow 4P$ and then as in part (**a**) (**c**) $4F \rightarrow 3D \rightarrow 3P \rightarrow 3S$ (**d**) $5P \rightarrow 5S \rightarrow 4P$ and then as in (**a**)
27. 5.39 eV **29.** 9.01 keV **31.** $A = 4.89 \times 10^7\ (\text{Hz})^{1/2}$, $S = 0.681$

Chapter 33
1. −82.5 kcal/mol **3.** (**a**) 5.39 eV (**b**) 5.12 eV
5. −7.96 eV/ion pair **7.** 7.92×10^{-14} eV

Chapter 34
1. 127.7 MeV, 7.98 MeV/nucleon **3.** 8.45 MeV/nucleon **5.** 6.43 MeV, 6.80 MeV **7.** 15 days **9.** 3.59×10^7 dis/s **11.** 2426 d
13. 8.84×10^{-3} g **15.** 5.25 MeV, 2.40×10^{-14} C **17.** $\frac{2}{3}$
19. 1.2 kg **21.** 1.28×10^{19} atoms/day, 4.74 kW **23.** 4.02×10^{24} MeV in fusion as compared to 5.12×10^{23} MeV in fission; $E_{(\text{fusion})} = 8E_{(\text{fission})}$

Index

A

E

G

H

I

Q

R

S

T

U

V